HANDBOOK OF
NANOPHYSICS

3

Handbook of Nanophysics

Handbook of Nanophysics: Principles and Methods

Handbook of Nanophysics: Clusters and Fullerenes

Handbook of Nanophysics: Nanoparticles and Quantum Dots

Handbook of Nanophysics: Nanotubes and Nanowires

Handbook of Nanophysics: Functional Nanomaterials

Handbook of Nanophysics: Nanoelectronics and Nanophotonics

Handbook of Nanophysics: Nanomedicine and Nanorobotics

HANDBOOK OF
NANOPHYSICS

Nanoparticles and Quantum Dots

Edited by

Klaus D. Sattler

CRC Press
Taylor & Francis Group
Boca Raton   London   New York

CRC Press is an imprint of the
Taylor & Francis Group, an **informa** business

CRC Press
Taylor & Francis Group
6000 Broken Sound Parkway NW, Suite 300
Boca Raton, FL 33487-2742

First issued in paperback 2020

© 2011 by Taylor and Francis Group, LLC
CRC Press is an imprint of Taylor & Francis Group, an Informa business

No claim to original U.S. Government works

ISBN-13: 978-0-367-57712-4 (pbk)
ISBN-13: 978-1-4200-7544-1 (hbk)

Library of Congress Cataloging-in-Publication Data

Handbook of nanophysics. Nanoparticles and quantum dots / editor, Klaus D. Sattler.
    p. cm.
  Includes bibliographical references and index.
  ISBN 978-1-4200-7544-1 (alk. paper)
  1. Semiconductor nanocrystals. 2. Quantum dots. I. Sattler, Klaus D. II. Title: Nanoparticles and quantum dots.

QC611.8.N33H36 2009
620'.5--dc22
          2009050386

Visit the Taylor & Francis Web site at
http://www.taylorandfrancis.com

and the CRC Press Web site at
http://www.crcpress.com

Contents

PART II Nanoparticle Properties

PART III Nanoparticles in Contact

PART IV Nanofluids

PART V Quantum Dots

Preface

The *Handbook of Nanophysics* is the first comprehensive reference to consider both fundamental and applied aspects of nanophysics. As a unique feature of this work, we requested contributions to be submitted in a tutorial style, which means that state-of-the-art scientific content is enriched with fundamental equations and illustrations in order to facilitate wider access to the material. In this way, the handbook should be of value to a broad readership, from scientifically interested general readers to students and professionals in materials science, solid-state physics, electrical engineering, mechanical engineering, computer science, chemistry, pharmaceutical science, biotechnology, molecular biology, biomedicine, metallurgy, and environmental engineering.

What Is Nanophysics?

Modern physical methods whose fundamentals are developed in physics laboratories have become critically important in nanoscience. Nanophysics brings together multiple disciplines, using theoretical and experimental methods to determine the physical properties of materials in the nanoscale size range (measured by millionths of a millimeter). Interesting properties include the structural, electronic, optical, and thermal behavior of nanomaterials; electrical and thermal conductivity; the forces between nanoscale objects; and the transition between classical and quantum behavior. Nanophysics has now become an independent branch of physics, simultaneously expanding into many new areas and playing a vital role in fields that were once the domain of engineering, chemical, or life sciences.

This handbook was initiated based on the idea that breakthroughs in nanotechnology require a firm grounding in the principles of nanophysics. It is intended to fulfill a dual purpose. On the one hand, it is designed to give an introduction to established fundamentals in the field of nanophysics. On the other hand, it leads the reader to the most significant recent developments in research. It provides a broad and in-depth coverage of the physics of nanoscale materials and applications. In each chapter, the aim is to offer a didactic treatment of the physics underlying the applications alongside detailed experimental results, rather than focusing on particular applications themselves.

The handbook also encourages communication across borders, aiming to connect scientists with disparate interests to begin interdisciplinary projects and incorporate the theory and methodology of other fields into their work. It is intended for readers from diverse backgrounds, from math and physics to chemistry, biology, and engineering.

The introduction to each chapter should be comprehensible to general readers. However, further reading may require familiarity with basic classical, atomic, and quantum physics. For students, there is no getting around the mathematical background necessary to learn nanophysics. You should know calculus, how to solve ordinary and partial differential equations, and have some exposure to matrices/linear algebra, complex variables, and vectors.

External Review

All chapters were extensively peer reviewed by senior scientists working in nanophysics and related areas of nanoscience. Specialists reviewed the scientific content and nonspecialists ensured that the contributions were at an appropriate technical level. For example, a physicist may have been asked to review a chapter on a biological application and a biochemist to review one on nanoelectronics.

Organization

The *Handbook of Nanophysics* consists of seven books. Chapters in the first four books (*Principles and Methods*, *Clusters and Fullerenes*, *Nanoparticles and Quantum Dots*, and *Nanotubes and Nanowires*) describe theory and methods as well as the fundamental physics of nanoscale materials and structures. Although some topics may appear somewhat specialized, they have been included given their potential to lead to better technologies. The last three books (*Functional Nanomaterials*, *Nanoelectronics and Nanophotonics*, and *Nanomedicine and Nanorobotics*) deal with the technological applications of nanophysics. The chapters are written by authors from various fields of nanoscience in order to encourage new ideas for future fundamental research.

After the first book, which covers the general principles of theory and measurements of nanoscale systems, the organization roughly follows the historical development of nanoscience. *Cluster* scientists pioneered the field in the 1980s, followed by extensive work on *fullerenes*, *nanoparticles*, and *quantum dots* in the 1990s. Research on *nanotubes* and *nanowires* intensified in subsequent years. After much basic research, the interest in applications such as the *functions of nanomaterials* has grown. Many bottom-up and top-down techniques for nanomaterial and nanostructure generation were developed and made possible the development of *nanoelectronics* and *nanophotonics*. In recent years, real applications for *nanomedicine* and *nanorobotics* have been discovered.

Acknowledgments

Many people have contributed to this book. I would like to thank the authors whose research results and ideas are presented here. I am indebted to them for many fruitful and stimulating discussions. I would also like to thank individuals and publishers who have allowed the reproduction of their figures. For their critical reading, suggestions, and constructive criticism, I thank the referees. Many people have shared their expertise and have commented on the manuscript at various stages. I consider myself very fortunate to have been supported by Luna Han, senior editor of the Taylor & Francis Group, in the setup and progress of this work. I am also grateful to Jessica Vakili, Jill Jurgensen, Joette Lynch, and Glenon Butler for their patience and skill with handling technical issues related to publication. Finally, I would like to thank the many unnamed editorial and production staff members of Taylor & Francis for their expert work.

Klaus D. Sattler
Honolulu, Hawaii

Editor

Klaus D. Sattler pursued his undergraduate and master's courses at the University of Karlsruhe in Germany. He received his PhD under the guidance of Professors G. Busch and H.C. Siegmann at the Swiss Federal Institute of Technology (ETH) in Zurich, where he was among the first to study spin-polarized photoelectron emission. In 1976, he began a group for atomic cluster research at the University of Konstanz in Germany, where he built the first source for atomic clusters and led his team to pioneering discoveries such as "magic numbers" and "Coulomb explosion." He was at the University of California, Berkeley, for three years as a Heisenberg Fellow, where he initiated the first studies of atomic clusters on surfaces with a scanning tunneling microscope.

Dr. Sattler accepted a position as professor of physics at the University of Hawaii, Honolulu, in 1988. There, he initiated a research group for nanophysics, which, using scanning probe microscopy, obtained the first atomic-scale images of carbon nanotubes directly confirming the graphene network. In 1994, his group produced the first carbon nanocones. He has also studied the formation of polycyclic aromatic hydrocarbons (PAHs) and nanoparticles in hydrocarbon flames in collaboration with ETH Zurich. Other research has involved the nanopatterning of nanoparticle films, charge density waves on rotated graphene sheets, band gap studies of quantum dots, and graphene foldings. His current work focuses on novel nanomaterials and solar photocatalysis with nanoparticles for the purification of water.

Among his many accomplishments, Dr. Sattler was awarded the prestigious Walter Schottky Prize from the German Physical Society in 1983. At the University of Hawaii, he teaches courses in general physics, solid-state physics, and quantum mechanics.

In his private time, he has worked as a musical director at an avant-garde theater in Zurich, composed music for theatrical plays, and conducted several critically acclaimed musicals. He has also studied the philosophy of Vedanta. He loves to play the piano (classical, rock, and jazz) and enjoys spending time at the ocean, and with his family.

Contributors

Javier Aizpurua
Centro de Física de Materiales
Spanish Scientific Research Council
Spanish Council for Scientific Research

and

Donostia International Physics Center
Donostia-San Sebastián, Spain

Leandro L. Araujo
Department of Electronic Materials
    Engineering
Research School of Physics
    and Engineering
The Australian National University
Canberra, Australian Capital Territory,
    Australia

Soumen Basu
Department of Chemistry
University of Alabama
Tuscaloosa, Alabama

Silvana Botti
Laboratoire des Solides Irradiés and ETSF
Ecole Polytechnique, CNRS, CEA-DSM
Palaiseau, France

and

Laboratoire de Physique de la Matière
    Condensée et Nanostructures
Université Claude Bernard Lyon I
    and CNRS
Villeurbanne, France

Garnett W. Bryant
Atomic Physics Division and the Joint
    Quantum Institute
National Institute of Standards
    and Technology
Gaithersburg, Maryland

Carlo Callegari
Sincrotrone Trieste
Basovizza, Trieste,
    Italy

Haisheng Chen
Institute of Particle Science
    and Engineering
University of Leeds
Leeds, United Kingdom

Ming-Shu Chen
State Key Laboratory of Physical
    Chemistry of Solid Surfaces
and
Department of Chemistry
College of Chemistry and Chemical
    Engineering
Xiamen University
Xiamen, China

Shun-Jen Cheng
Department of Electrophysics
National Chiao Tung University
Hsinchu, Taiwan

Silvano De Franceschi
Commissariat à l'Énergie Atomique
Grenoble, France

Al-Amin Dhirani
Lash Miller Chemical Laboratories
University of Toronto
Toronto, Ontario, Canada

Yulong Ding
Institute of Particle Science
    and Engineering
University of Leeds
Leeds, United Kingdom

Alain Dufresne
The International School of Paper, Print
    Media and Biomaterials
Grenoble Institute of Technology
St Martin d'Hères, France

Erik Dujardin
Centre d'Elaboration de Matériaux et
    d'Etudes Structurales
Centre National de la Recherche
    Scientifique
Toulouse, France

Eugène Duval
Laboratoire de Physico-Chimie des
    Matériaux Luminescents
Centre National de la Recherche
    Scientifique
Université Claude Bernard Lyon I
Villeurbanne, France

Jacob Eapen
Department of Nuclear Engineering
North Carolina State University
Raleigh, North Carolina

Gil de Aquino Farias
Departamento de Física
Universidade Federal do Ceará
Fortaleza, Brazil

Stéphane Fohanno
Faculté des Sciences
Université de Reims Champagne-Ardenne
Reims, France

Nicola Gaston
Industrial Research Ltd.
Lower Hutt, New Zealand

Christian Girard
Centre d'Elaboration de Matériaux et
    d'Etudes Structurales
Centre National de la Recherche
    Scientifique
Toulouse, France

Achim M. Goepferich
Department of Pharmaceutical
    Technology
University of Regensburg
Regensburg, Germany

Youssef Habibi
Department of Forest Biomaterials
North Carolina State University
Raleigh, North Carolina

Jan Petter Hansen
Department of Physics and
    Technology
University of Bergen
Bergen, Norway

Shaun C. Hendy
Industrial Research Ltd.
Lower Hutt, New Zealand

and

MacDiarmid Institute for Advanced
    Materials and Nanotechnology
School of Chemical and Physical
    Sciences
Victoria University of Wellington
Wellington, New Zealand

Anna F. E. Hezinger
Department of Pharmaceutical
    Technology
University of Regensburg
Regensburg, Germany

Vo Van Hoang
Department of Physics
Institute of Technology
National University of Ho Chi Minh
    City
Ho Chi Minh City, Vietnam

Andreas Hütten
Department of Physics
Bielefeld University
Bielefeld, Germany

Wolfgang Jäger
Department of Chemistry
University of Alberta
Edmonton, Alberta, Canada

Lucile Joly-Pottuz
Institut National des Sciences Appliquées
    de Lyon
University of Lyon
Villeurbanne, France

Sabre Kais
Department of Chemistry
Birck Nanotechnology Center
Purdue University
West Lafayette, Indiana

Nikola Kallay
Laboratory of Physical Chemistry
Department of Chemistry
University of Zagreb
Zagreb, Croatia

Dimitris Kechrakos
Department of Sciences
School of Pedagogical and
    Technological Education
Athens, Greece

Brian A. Korgel
Department of Chemical Engineering
Texas Materials Institute
Center for Nano- and Molecular Science
    and Technology
The University of Texas at Austin
Austin, Texas

Davor Kovačević
Laboratory of Physical Chemistry
Department of Chemistry
University of Zagreb
Zagreb, Croatia

Roman Krahne
Italian Institute of Technology
Genova, Italy

Kai Choong Leong
School of Mechanical and Aerospace
    Engineering
Nanyang Technological University
Singapore, Singapore

Xinlei L. Li
State Key Laboratory of
    Optoelectronic Materials
    and Technologies
Institute of Optoelectronic and
    Functional Composite Materials
School of Physics & Engineering
Zhongshan University
Guangzhou, China

Seth Lichter
Department of Mechanical Engineering
Northwestern University
Evanston, Illinois

Kongyong Liew
Key Laboratory of Catalysis and
    Materials Science of the State Ethnic
    Affairs Commission & Ministry
    of Education
College of Chemistry & Materials
    Science
South-Central University
    for Nationalities
Wuhan, China

and

Faculty of Industrial Science
    and Technology
University Malaysia Pahang
Kuantan, Malaysia

Eva Lindroth
Atomic Physics
Fysikum
Stockholm University
Stockholm, Sweden

Axel Lorke
Institute of Physics
Center for NanoIntegration Duisburg-
    Essen
University of Duisburg-Essen
Duisburg, Germany

Ingrid Mann
School of Science and Engineering
Kinki University
Higashi-Osaka, Japan

and

Belgian Institute for Space Aeronomy
Brussels, Belgium

Liberato Manna
Italian Institute of Technology
Genova, Italy

Alain Mermet
Laboratoire de Physico-Chimie des
    Matériaux Luminescents
Centre National de la Recherche
    Scientifique
Université Claude Bernard Lyon I
Villeurbanne, France

Thomas Michael
Institute of Physics
Martin-Luther-University
Halle, Germany

Seiji Mitani
National Institute for Materials Science
Tsukuba, Japan

Antaryami Mohanta
Department of Physics
Indian Institute of Technology
Kanpur, India

S. M. Sohel Murshed
Department of Mechanical, Materials
    and Aerospace Engineering
University of Central Florida
Orlando, Florida

Cong Tam Nguyen
Faculty of Engineering
Université de Moncton
Moncton, New Brunswick, Canada

Tarasankar Pal
Department of Chemistry
Indian Institute of Technology
Kharagpur, India

Rongjun Pan
Department of Information
    and Computing Science
Institute of Application of Nanoscience
    & Nanotechnology
Guangxi University of Technology
Liuzhou, China

Taras Plakhotnik
School of Mathematics and Physics
The University of Queensland
Brisbane, Queensland, Australia

Guillaume Polidori
Faculté des Sciences
Université de Reims Champagne-Ardenne
Reims, France

Tajana Preočanin
Laboratory of Physical Chemistry
Department of Chemistry
University of Zagreb
Zagreb, Croatia

Günter Reiss
Department of Physics
Bielefeld University
Bielefeld, Germany

Mark C. Ridgway
Department of Electronic Materials
    Engineering
Research School of Physics
    and Engineering
The Australian National University
Canberra, Australian Capital Territory,
    Australia

Alex Roxin
Center for Theoretical Neuroscience
Columbia University
New York, New York

Aaron E. Saunders
Department of Chemical and Biological
    Engineering
University of Colorado at Boulder
Boulder, Colorado

Lucien Saviot
Laboratoire Interdisciplinaire Carnot de
    Bourgogne
Centre National de la Recherche
    Scientifique
Université de Bourgogne
Dijon, France

Tom B. Sisan
Department of Physics and Astronomy
Northwestern University
Evanston, Illinois

Jeanlex Soares de Sousa
Departamento de Física
Universidade Federal do Ceará
Fortaleza, Brazil

Frank Stienkemeier
Institute of Physics
University of Freiburg
Freiburg, Germany

Koki Takanashi
Institute for Materials Research
Tohoku University
Sendai, Japan

Joerg K. Tessmar
Department of Pharmaceutical
    Technology
University of Regensburg
Regensburg, Germany

Raj K. Thareja
Department of Physics
Indian Institute of Technology
Kanpur, India

Steffen Trimper
Institute of Physics
Martin-Luther-University
Halle, Germany

Wilfred G. van der Wiel
NanoElectronics Group
MESA+ Institute for Nanotechnology
University of Twente
Enschede, the Netherlands

Derek Walton
Department of Physics and Astronomy
McMaster University
Hamilton, Ontario, Canada

Hefeng Wang
Department of Chemistry and Birck
    Nanotechnology Center
Purdue University
West Lafayette, Indiana

Liqiu Wang
Department of Mechanical Engineering
The University of Hong Kong
Hong Kong, China

Xiaohao Wei
Department of Mechanical Engineering
The University of Hong Kong
Hong Kong, China

Julia M. Wesselinowa
Department of Physics
University of Sofia
Sofia, Bulgaria

Hartmut Wiggers
Institute of Combustion and Gas Dynamics
Center for NanoIntegration Duisburg-
    Essen
University of Duisburg-Essen
Duisburg, Germany

Sanjeeva Witharana
Institute of Particle Science
    and Engineering
University of Leeds
Leeds, United Kingdom

Kay Yakushiji
National Institute of Advanced
    Industrial Science and Technology
Tsukuba, Japan

Chun Yang
School of Mechanical and Aerospace
    Engineering
Nanyang Technological University
Singapore, Singapore

Guowei W. Yang
State Key Laboratory of
    Optoelectronic Materials
    and Technologies
Institute of Optoelectronic and
    Functional Composite Materials
School of Physics & Engineering
Zhongshan University
Guangzhou, China

Vassilios Yannopapas
Department of Materials Science
University of Patras
Patras, Greece

Taeil Yi
Department of Mechanical Engineering
Northwestern University
Evanston, Illinois

Amir Zabet-Khosousi
Department of Chemistry
Lash Miller Chemical Laboratories
University of Toronto
Toronto, Ontario, Canada

I

Types of Nanoparticles

I

Types of
Nanoparticles

1

Amorphous Nanoparticles

Vo Van Hoang
National University of Ho Chi Minh City

1.1 Introduction

Nanoparticles have been extremely interesting objects in modern materials science and nanophysics over the past decades due to their enormous technological importance. Although for various substances there is a possibility to change the nanoparticles into either a crystalline or an amorphous state by using reasonable synthesis methods, much attention has been paid to the former rather than the latter (Günter 2004). There is no comprehensive work related to amorphous nanoparticles and this motivates us to write this chapter on the *Handbook of Nanophysics*. It is well known that crystalline nanoparticles have a well-defined crystal structure with a large fraction of their atoms located on the surface, including a structural disorder in the vicinity of the surface when compared to that of a perfect crystal, which provide them with unique properties that are different from their crystalline bulk counterparts (Changsheng et al. 1999). In contrast, amorphous nanoparticles have a disordered structure, which may be divided into two parts, i.e., the core with structural characteristics close to that of the corresponding amorphous bulk-counterparts and a surface exhibiting a more porous structure due to the presence of large amounts of structural defects (Hoang and Khanh 2009). Due to their disordered structure, amorphous nanoparticles can have more advanced applications than a crystal structure with well-defined properties. Indeed, it was found that catalytic amorphous Fe_2O_3 nanoparticles are more active than the nanocrystalline polymorphs of the same diameter thanks to the "dangling bonds" and a higher surface-bulk ratio (Srivastava et al. 2002). Due to surface effects, the structure and the properties of amorphous nanoparticles are also different from those of their corresponding amorphous bulk-counterparts. Therefore, amorphous nanoparticles have attracted a great interest and have been under intensive investigation in the recent years (Libor et al. 2007, Wu et al. 2007). Much attention has been paid to the synthesis and the characterization of amorphous nanoparticles; therefore, important methods for the synthesis of amorphous nanoparticles have been listed in a subsequent section of the chapter. On the other hand, in order to get structural information about amorphous nanoparticles, one can use several diffraction techniques. However, more detailed information of the microstructure of amorphous nanoparticles at the atomic level can be provided by a computer simulation. Therefore, we also discuss the results obtained by a computer simulation of amorphous nanoparticles. Moreover, the physicochemical properties of amorphous nanoparticles have been under intensive investigation by both experiments and computer simulations (Hoang 2007a, Libor et al. 2007, Wu et al. 2007, Hoang and Odagaki 2008, Hoang and Khanh 2009). In particular, amorphous nanoparticles can have advanced catalytic properties compared with traditional crystalline catalysts or good magnetic materials, etc., leading to their potential applications in various areas of technology (Srivastava et al. 2002, Libor et al. 2007, Wu et al. 2007). Therefore, applications of amorphous nanoparticles have also been given considerable attention in the chapter.

1.2 Synthesis and Characterization

1.2.1 Methods of Synthesis

There are various methods of synthesis of amorphous nanoparticles used in practice, and selected methods have been presented in Table 1.1. Our aim here is not to review the methods

TABLE 1.1   Selected Methods of Synthesis of Amorphous Nanoparticles

Synthesis Methods	Substances	References
Hydrolysis followed by condensation	Fe_2O_3, GeO_2	Kan et al. (1996) and Tracy et al. (2007)
Thermally induced solid-state decomposition	Fe_2O_3, Ni–B	Zboril et al. (2004a,b) and Zhong et al. (2008)
Sol–gel method	TiO_2	Gonzalez (1998)
Precipitation	Fe_2O_3	Subrt et al. (1998)
Microemulsion technique followed by precipitation and heating precipitation	Fe_2O_3	Ayyub et al. (1988)
Microwave pyrolysis	Fe_2O_3	Palchik et al. (2000)
Sonochemical synthesis	Fe_2O_3, Fe_3O_4, Ni, Ag	Cao et al. (1997), Abu and Gedanken (2005), Koltypin et al. (1996), and Suwen et al. (2001)
Microwave irradiation	Fe_2O_3	Liao et al. (2000)
Chemical reduction	(Fe,Co,Ni)–(B,P), MoS_2, Fe–Ni–B, Fe–Cr–B, $(Fe_xNd_{1-x})_{0.6}B_{0.4}$	Wu et al. (2007), Lianxia et al. (2008), Zysler et al. (2001), Fiorani et al. (1995), and Tortarolo et al. (2004)
Electroless deposition	Ni–W–P	Jianhua et al. (2008)
Gas phase condensation	SnO_2	Jimenez et al. (1999)
Laser ablation condensation	Al_2O_3, Co	Chiennan et al. (2007) and Changsheng et al. (1999)
Heavy ion irradiation	YCo_2	Ghidini et al. (1995)

of synthesis of amorphous nanoparticles; hence, we present only some substances for each method. Note that, we focus attention only on the methods of the synthesis of nanopowders of amorphous nanoparticles without the presence of matrices or other supported materials. It was found that the size, the shape, and the size distribution of amorphous nanopowders depend on the method of synthesis used in practice (Libor et al. 2007). It seems that chemical reduction has often been used for the synthesis of amorphous nanoparticles of alloys rather than for other substances (Table 1.1). Moreover, syntheses based on ultrasound or microwave irradiation have also often been used for the preparation of amorphous nanoparticles in addition to the precipitation methods, and much attention has been paid to sonochemical synthesis in the recent years.

1.2.2 Characterization

The amorphous state of nanoparticulate samples can be defined by using different techniques, including scanning (SEM) and transmission (TEM) electron microscopy, differential scanning calorimetry (DSC), thermogravimetric analysis (TGA), x-ray diffraction (XRD), and magnetic measurements. Particularly, and in order to characterize magnetic amorphous nanoparticles such as Fe_2O_3 ones, Mössbauer spectroscopy and various magnetic measurements, a long scale of other experimental techniques yielding important information on chemical purity, local structure, size, morphology, or stability have been used (Libor et al. 2007). However, two methods which have been widely used in order to characterize the amorphous nature of nanoparticulate samples are XRD and selected area electron diffraction (SAED) as part of TEM analysis. The absence of Bragg peaks in the XRD pattern is an identification of the amorphous nature of a nanoparticulate sample, which is different from that of nanocrystalline polymorphs, i.e., for the latter, broadened diffraction peaks usually appear (Figure 1.1).

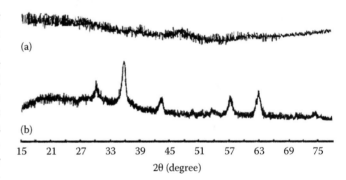

FIGURE 1.1   XRD pattern of amorphous Fe_2O_3 nanoparticles (a) and nanocrystalline γ-Fe_2O_3 (b). (From Prozorov, R. et al., *Phys. Rev. B*, 59, 6956, 1999. With permission.)

However, the application of XRD for the detection of the amorphous phase is limited if the samples contain the crystalline matrix or ultrasmall nanocrystalline polymorphs. Further evidence of the existence of the amorphous phase is provided by the SAED pattern. The broad, diffusive ring suggests a typical amorphous structure of nanoparticulate samples (Figure 1.2). Note that the indication of an amorphicity given by the SAED pattern is usually related to a very small number of particles involved in such an analysis and it is its limitation. Therefore, in order to detect the amorphous nature of nanoparticulate samples, additional indirect approaches emerging from the monitoring of thermal and magnetic behaviors, for example, are applicable. However, the obtained data are strongly affected by the sample character and the measurement conditions (Libor et al. 2007).

The morphology and the size of amorphous nanoparticles have been determined by SEM and TEM. In particular, the SEM of $Fe_{82}P_{11}B_7$ amorphous nanoparticles produced by chemical reduction shows that the sample consists of nearly spherical particles with a diameter ranging from 150 to 350 nm (Jianyi et al.

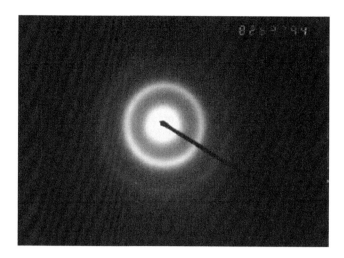

FIGURE 1.2 The SAED image of amorphous Ni–B alloy nanoparticles. (From Zhong, G.Q. et al., *J. Alloy Compd.*, 465, L1, 2008. With permission.)

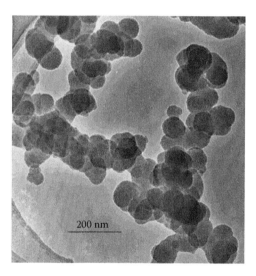

FIGURE 1.3 TEM of amorphous B nanoparticles synthesized by an arc decomposing diborane, showing an average diameter of 75 nm and a narrow size distribution. (From Si, P.Z. et al., *J. Mater. Sci.*, 38, 689, 2003. With permission.)

1992). Similarly, the TEM of amorphous B nanoparticles synthesized by the arc decomposing diborane shows that nanopowder consists of nearly spherical particles with an average diameter of 75 nm and a narrow size distribution (Figure 1.3). The narrow size distribution and the ideal spherical shape should be attributed to the high temperature of the arc (Si et al. 2003).

1.3 Structural Properties

1.3.1 Experiments

The interplay between the structure of amorphous nanoparticles and their physicochemical properties is of great interest. While the structure of crystalline nanoparticles is well defined, our knowledge of the structure of amorphous nanoparticles is still limited. However, it is evident that they have a short-range structure like that observed for the corresponding amorphous bulk counterparts.

One can use traditional experimental techniques such as TEM, XRD, x-ray absorption spectroscopy, infrared spectroscopy, Mössbauer spectroscopy, etc., in order to study the structure of amorphous nanoparticles. In particular, valuable information about the short-range structure and the magnetic behavior of amorphous magnetic nanoparticles such as Fe_2O_3, Fe_3O_4, etc., can be obtained via Mössbauer spectroscopy (Libor et al. 2007). Generally, a room temperature Mössbauer spectrum of amorphous Fe_2O_3 nanoparticles reveals a broadened doublet that was thought to be related to the nonequivalent surface and bulk Fe atoms of the system. Further, the ratio of the spectral lines corresponding to the surface and the bulk Fe atoms should strongly relate to the particle size. However, the published data are not consistent with this relation (Libor et al. 2007). In addition, TEM, XRD, x-ray absorption spectroscopy, and infrared spectroscopy have been used for the structural characterization of partially amorphous SnO_2 nanoparticles, i.e., it was found that the original powder was partially amorphous and was formed by very fine particles ($d \sim 8$–10 nm) linked in a fractal-like structure (Jimenez et al. 1999).

In a structural analysis of disordered materials including liquid and amorphous nanoparticles, the radial distribution function (RDF), $g(r)$, is no doubt of the chosen value. It yields the central information about the short-range order and serves as a key test for different structures. For simplicity, we discuss about $g(r)$ for monatomic fluids. One can measure the structure factor $S(k)$ by the elastic scattering of x-rays or neutrons, and then $g(r)$ can be obtained via the following relation:

$$g(r) = 1 + \frac{1}{2\pi^2 N/V} \int_0^\infty [S(k) - 1] \frac{\sin kr}{kr} k^2 \mathrm{d}k \qquad (1.1)$$

Here, we have N atoms in volume V and k is the wave-vector. A schematic explanation of $g(r)$ of a monatomic fluid can be seen in Figure 1.4.

The radial distribution function, $g(r)$, can be interpreted as the (not normalized) conditional probability to find another particle a distance r away from the origin, given that there is a particle at the origin. Now, we discuss the physical interpretation of the information that can be gotten from $g(r)$. At a small-enough distance, r, the function $g(r)$ is essentially zero since atoms cannot strongly overlap their electronic shells. Based on Figure 1.4, one can define the first coordination shell by the atoms between $r = 0$ and the first minimum at R_1 between the peaks of the first and the second maximum in $g(r)$. An average coordination number Z of an atom in the system can be defined by

$$Z = \int_0^{R_1} g(r) 4\pi r^2 \mathrm{d}r \qquad (1.2)$$

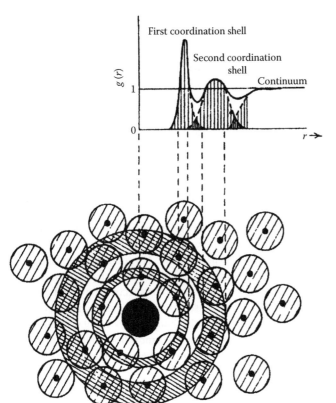

FIGURE 1.4 Schematic explanation of $g(r)$ of a monatomic fluid. The atom at the origin is highlighted by a black sphere. The dashed regions between the concentric circles indicate which atoms contribute to the first and second coordination number shells, respectively. (From Ziman, J.M., *Models of Disorder. The Theoretical Physics of Homogeneously Disordered Systems*, Cambridge University Press, Cambridge, U.K., 1979. With permission.)

However, more detailed information on the local structure of disordered materials such as the interatomic distance, the coordination number and the bond-angle distributions, etc., can be provided by a computer simulation. Note that one can directly calculate $g(r)$ via the coordinates of all atoms in the models obtained by a computer simulation, i.e.,

$$g(r) = \frac{dn}{N/V \, 4\pi r^2 dr} \qquad (1.3)$$

Here, dn is the number of atoms belonging to the spherical shell formed by two spheres with the radii of r and $r + dr$ away from the central atom. The function has been averaged over all atoms in the system. Similarly, the partial RDF (PRDF), $g_{ij}(r)$, in a binary system can be interpreted, sitting on one atom of species i, as the conditional probability of finding one atom of the species j in a spherical shell between r and $r + dr$.

In order to get detailed information about the microstructure of amorphous nanoparticles, a combination of experiment and computer simulation is needed. That is, a detailed atomic structure of amorphous TiO_2 nanoparticles has been studied via synchrotron wide-angle x-ray scattering (WAXS) where the atomic

RDF derived from the Fourier transform of the WAXS data was used for the reverse Monte-Carlo (RMC) simulations of the atomic structure of the samples (Hengshong et al. 2008). The atomic structure of 2 nm amorphous TiO_2 nanoparticles has been studied in detail via analysis of PRDFs, bond-length distribution, coordination number, and bond-angle distributions. In addition, the structural characteristics of the core and the surface shell of nanoparticles have been also analyzed. It was found that 2 nm amorphous TiO_2 nanoparticles consist of a highly distorted surface shell and a small strained anatase-like crystalline core. The reduction in the coordination number of Ti atoms in amorphous TiO_2 nanoparticles compared with that observed in the corresponding amorphous bulk indicates the surface effects in the former. On the other hand, the shortening of the Ti–O bond in amorphous TiO_2 nanoparticles was suggested to be related to the distorted surface shell in the nanoparticulate samples. Unfortunately, no more similar work related to the atomic structure of amorphous nanoparticles has been found in literature yet, and our understanding of their microstructure is still limited.

1.3.2 Computer Simulations

Nanoparticles are interesting objects for computer simulations due to their small size, and detailed simulations of amorphous nanoparticles have been done (Hoang 2007a,b, Hoang and Odagaki 2008, Hoang and Khanh 2009). Thanks to the results obtained by the computer simulations, our understanding of the atomic structure of liquid and amorphous nanoparticles has been substantially improved. The detailed size (and temperature) dependence of the atomic structure and the various thermodynamic properties of amorphous nanoparticles of different substances have been studied. In particular, the structural properties of amorphous nanoparticles have often been studied in spherical models of different sizes ranging from 2 to 5 nm. Models have been obtained by cooling from the melt via classical MD simulation with the pair interatomic potentials. The structural properties of amorphous nanoparticles have been analyzed in detail through PRDFs, interatomic distances, coordination number, and bond-angle distributions or radial density profile $\rho(R)$. (That is, the dependence of particle density $\rho(R)$ on the distance R from the center of the nanoparticle. This quantity is determined as follows: we find the number of atoms belonging to the spherical shell with the thickness $2dR$ formed by two spheres with the radii of $R - dR$ and $R + dR$. Then we calculate the quantity $\rho(R)$). It was found that the peaks in PRDFs of amorphous nanoparticles are broader than those for the bulk, indicating that the structure of nanoparticles is more heterogeneous than that for the bulk due to the contribution of the surface structure of the former (see, for example, Figure 1.5). Moreover, the structural characteristics of amorphous nanoparticles are size dependent, and the mean coordination number increases toward the value of the bulk if the particle size increases due to the reduction of the surface-to-volume ratio (Figure 1.6). Note that for spherical models of nanoparticles, the non-periodic boundary conditions were

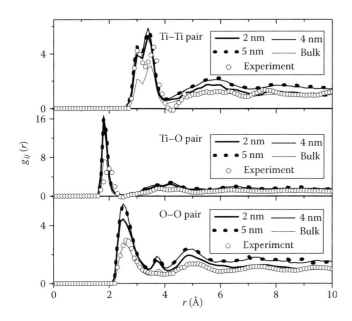

FIGURE 1.5 PRDFs of amorphous TiO_2 nanoparticles of three different sizes obtained at 350 K compared with the experimental data for the bulk. (From Hoang, V.V. et al., *Eur. Phys. J. D*, 44, 515, 2007. With permission.)

shell. The structure of the former is relatively size-independent and close to that of the corresponding bulk while the structure of the latter is strongly size dependent and more porous compared with that of the bulk or of the core of nanoparticles. This means that the surface plays a key role in the size dependence of the structure of amorphous nanoparticles. It was found that the surface shell of amorphous nanoparticles contains large amounts of structural defects that might be the origin of a variety of surface phenomena of amorphous nanoparticles, including catalysis, adsorption, optical properties, and so forth (Hoang 2007a,b). Similar results have been obtained for amorphous nanoparticles of different substances such as TiO_2, SiO_2, GeO_2, Fe_2O_3, or monatomic simple nanoparticles (Hoang and Khanh 2009 and references therein). Note that there is no common rule for the determination of the surface shell of amorphous nanoparticles. From the structural point of view, it can be considered that atoms belong to the surface if they do not have full coordination for all atomic pairs in principle. In contrast, atoms belong to the core if they can have full coordination for all atomic pairs in principle, like that located in the bulk. Therefore, one can assume that the outermost spherical shell of the thickness equaling the largest radius of coordination spheres used in the system is a surface shell and the remaining part is a core of nanoparticles (Figure 1.7). In addition, it was found that stoichiometries in the surface shell and in the core of amorphous nanoparticles of binary substances are quite different. It can lead to the formation of additional defects in amorphous nanoparticles.

used. In contrast, models obtained in a cube under periodic boundary condition were considered as the corresponding bulk counterparts.

Moreover, calculations also show that amorphous nanoparticles consist of two distinct parts: the core and the surface

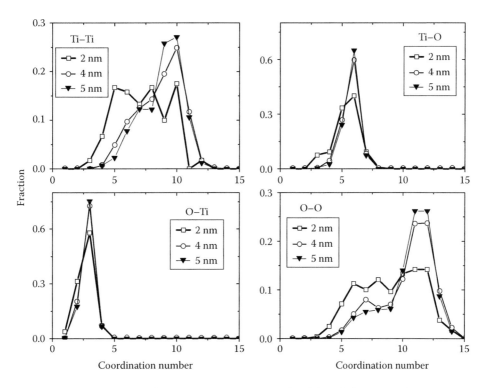

FIGURE 1.6 Coordination number distributions of amorphous TiO_2 nanoparticles of three different sizes obtained at 350 K. (From Hoang, V.V. et al., *Eur. Phys. J. D*, 44, 515, 2007. With permission.)

FIGURE 1.7 Schematic illustration of surface and core of amorphous nanoparticles. (The black sphere is a core of nanoparticles; the outermost white spherical shell with a thickness equal to the largest radius of coordination spheres used in simulation is a surface.)

1.4 Physicochemical Properties

1.4.1 Catalytic Properties

Amorphous nanoparticles of various substances exhibit superior catalytic behaviors. In particular, amorphous nanoparticles consisting of transition metals (M) and metalloid elements (B, P) can be potential alternatives to Raney nickel or noble metals in catalytic hydrogenation. Since Raney nickel shows many serious disadvantages (i.e., short lifetime and environment pollution), amorphous nanoparticles of M–(B, P) alloys owning effective catalytic behavior are inexpensive and environmentally benign (Wu et al. 2007). Besides the specific amorphous structure, the surface of nanoparticles can play an important role in their catalytic performance. It was widely accepted that the promoting effect of alloying B or P is attributed to the modification of the catalyst's structural characteristics resulting in the short-range order and the long-range disorder of the structure, the homogeneous dispersion of the active sites, and the high-concentration coordinately unsaturated sites (Wu et al. 2007). Indeed, in accordance with our research related to the amorphous nanoparticles mentioned above, the structural defects including unsaturated sites are mainly concentrated in the surface shell of amorphous nanoparticles and might play a key role in their catalytic performance. On the other hand, we also found the existence of the dangling bonds at the surface of amorphous nanoparticles (i.e., due to the breaking bonds at the surface). This also might enhance the catalysis of amorphous nanoparticles. Note that the structure of the surface of amorphous nanoparticles is size dependent. This means that the catalytic behavior of amorphous nanoparticles might also be size dependent. Indeed, it was found that amorphous Fe_2O_3 nanoparticles are more active in catalysis than the nanocrystalline polymorphs of the same diameter thanks to the dangling bonds and to the higher surface-to-volume ratio of the former (Libor et al. 2007). A similar situation can be suggested for the catalysis of amorphous nanoparticles of M–(B, P) alloys. It is essential to note that among the MB and MP alloy nanoparticulate catalysts, NiB and NiP amorphous nanoparticulate catalysts were studied

and reported most extensively and thoroughly (Wu et al. 2007). In addition, the modification of NiB alloy with other transition metals (Cu, Co, Fe, Mo, etc.) or P element can promote the catalytic activity and selectivity. Amorphous nanoparticles of metal–metalloid alloys of noble metals also show excellent catalytic behavior (Wu et al. 2007).

1.4.2 Optical Properties

As the size of condensed matter is reduced to nanoscale levels, the electron and phonon states are influenced due to confinement. It is true for both the crystalline and the amorphous phases of nanoparticles. Indeed, changes in the phonon spectra of amorphous Si nanoparticles during crystallization were found, and it is size dependent (Sirenko et al. 2000). Furthermore, the photoluminescence (PL) properties of ultrasmall amorphous Si nanoparticles with sizes smaller than 2 nm have been studied. It was indicated that the surface structure has a large influence on the PL properties of amorphous Si nanoparticles with a size smaller than 2 nm (Xie et al. 2007). The continuously tunable emission in a range from 400 to 460 nm and the stability of luminescence are new features of such small amorphous Si nanoparticles. These results can be expected to have applications in nanodevices and biomaterials. Photoluminescence has also been found for amorphous SiO_2 nanoparticles of different sizes of 7 and 15 nm compared with those of the bulk counterparts (Yuri et al. 2002). Three PL bands that peaked in the red (~1.9 eV), green (~2.35 eV), and blue (~2.85 eV) spectral ranges were found for the 15 nm nanoparticles. Similar red and green PL bands were observed for 7 nm nanoparticles, whereas the blue band peaked at ~3.25 eV (Yuri et al. 2002). The red and green PL bands for the bulk peaked at almost the same spectral positions as those for SiO_2 nanoparticles. This indicates the similarity of light-emitter types in both the bulk and nanoparticles (Yuri et al. 2002). The strong red photoluminescence of amorphous SiO_2 nanoparticles has been attributed to the defects at their inner surfaces and it was pointed out that the intrinsic-point defects are the origin of optical band-gap narrowing in fumed silica nanoparticles. This indicates the important role of structural defects contained in the surface shell in the structure and the properties of amorphous nanoparticles in general.

1.4.3 Thermodynamic Properties

Thermodynamic properties of liquid and amorphous nanoparticles are of great interest. However, the information obtained by experimental studies in this direction is still limited. That is, DSC curves have been observed for amorphous nanoparticles of various substances in order to detect the existence of amorphous phase in the samples. DSC curves of 20 nm amorphous Co nanoparticles in oxygen and in Ar ambient conditions have been found and discussed (Changsheng et al. 1999), i.e., when heating amorphous Co nanoparticles in O_2 ambient conditions there is a sharp exothermic reaction in the temperature range from 207.2°C to 297.2°C with a peak at 260°C and

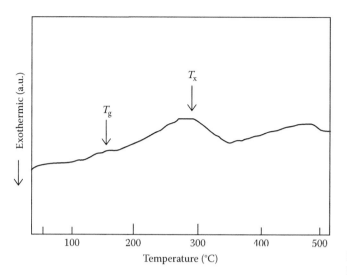

FIGURE 1.8 DSC curves of amorphous Co nanoparticles in Ar ambient conditions. (From Changsheng, X. et al., *NanoStruct. Mater.*, 11, 1061, 1999. With permission.)

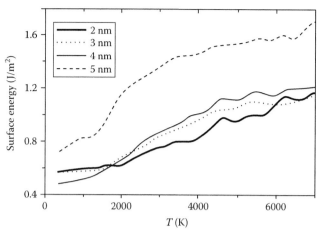

FIGURE 1.9 Temperature dependence of the surface energy of simulated liquid and amorphous TiO_2 nanoparticles. (From Hoang, V.V., *Nanotechnology*, 19, 105706, 2007b. With permission.)

an exothermic enthalpy of 100.08 kJ/mol. In contrast, when the Ar ambient condition was used, Co nanoparticles transformed from the amorphous solid into a supercooled liquid state at about 167°C and kept the supercooled liquid states from 167°C to 277°C followed by crystallization at 277°C with the exothermic heat of around 23.2 kJ/mol (Figure 1.8), which is larger than that of the fusion of the bulk Co. On the other hand, the size dependence of a glass-transition temperature (T_g, i.e., the temperature at which the transition from a supercooled liquid into a glassy state occurs) in nanoscaled systems, including in liquid nanoparticles is also of great interest. While the glass-transition temperature is typically lower in a confined geometry, experiments have also found cases where T_g decreases (Alcoutlabi and McKenna 2005). The finite size effects on T_g cannot be interpreted as readily as that on the melting temperature T_m because of the lack of a consensus on the nature of the glass transition in general.

Comprehensive work related to the thermodynamic properties of liquid and amorphous nanoparticles have been done by computer simulation. Indeed, the temperature dependence of potential energy, surface energy or the diffusion constant of liquid and amorphous SiO_2, TiO_2, Fe_2O_3 or simple monatomic nanoparticles have been found by MD simulation and discussed in detail (Hoang 2007a,b, Hoang and Odagaki 2008, Hoang and Khanh 2009). It was found that the surface energy of liquid and amorphous SiO_2 nanoparticles almost monotonously decreases with decreasing temperature. A similar tendency has been found for the surface energy of TiO_2 and monatomic simple nanoparticles, whereas at room temperature, the surface energy for TiO_2 nanoparticles is around 0.50–0.70 J/m² depending on the nanoparticle size, which is very close to that experimentally obtained for crystalline TiO_2 nanoparticles (Figure 1.9). Furthermore, it was found that T_g increases with a decrease in the size of Fe_2O_3, TiO_2,

and monatomic simple nanoparticles (Hoang 2007a,b, Hoang and Odagaki 2008, Hoang and Khanh 2009) (Figure 1.10). In contrast, for simulated SiO_2 nanoparticles, T_g decreases with decreasing nanoparticle size (Hoang 2007a). This exhibits a complex size dependence of T_g of simulated liquid nanoparticles like the situation faced in practice. Note that T_g was found via the intersection of the low- and high-temperature extrapolation of the potential energy of nanoparticles.

On other hand, the diffusion constant (D) of atoms in liquid nanoparticles is also of great interest and it can be determined via a mean-squared displacement (MSD) of atoms, which is given by

$$\langle r^2(t) \rangle = \frac{1}{N} \sum_{i=1}^{N} \left[r_i(t) - r_i(0) \right]^2 \tag{1.4}$$

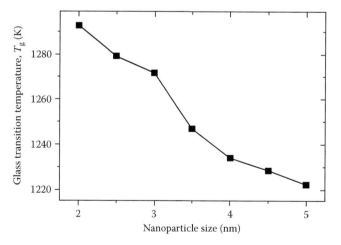

FIGURE 1.10 Size dependence of the T_g of simulated liquid TiO_2 nanoparticles. (From Hoang, V.V., *Nanotechnology*, 19, 105706, 2007b. With permission.)

The diffusion constant can be determined via the Einstein relation

$$D = \lim_{t \to \infty} \frac{\langle r^2(t) \rangle}{6t} \qquad (1.5)$$

Here

N is the atomic number

$r(t)$ is the position of the atom at a time t

$r(0)$ is the position at the time origin

The temperature dependence of the diffusion constant (D) of species in liquid TiO_2 nanoparticles was found in that it follows an Arrhenius law at relatively low temperatures and it deviates from an Arrhenius one at higher temperatures (Figure 1.11). The form of an Arrhenius law is given below:

$$D = D_0 e^{-\frac{E}{k_B T}} \qquad (1.6)$$

where

E is an activation energy

k_B is a Boltzmann constant

In addition, it was found by MD simulation that close to the surface of liquid SiO_2 nanoclusters, the diffusion constant is somewhat larger than that in the bulk, and that with decreasing temperature, the relative difference grows (Roder et al. 2001). Surface dynamics in liquid and amorphous nanoparticles can play an important role in various thermodynamic properties of nanoparticles, and it is worth carrying out a study in this direction.

1.4.4 Magnetic Properties

Over the past decades, amorphous nanoparticles of magnetic substances (mainly nanoparticles of 3d metal oxides, pure 3d metals, and their alloys) have been under intensive investigation due to their specific magnetic behavior, which is markedly different from that exhibited by the corresponding bulk counterparts. The size, the morphology, the local structure of amorphous nanoparticles, the surface effects, together with interparticle interactions, are the key factors influencing the macroscopic magnetic properties of nanosized systems such as magnetization, magnetic susceptibility, coercive field, and magnetic transition temperature (Libor et al. 2007). It was found that such nanoparticles exhibit new phenomena such as superparamagnetism, high field irreversibility, high saturation field, or shifted hysteresis loops after field cooling. Among experimental techniques, Mössbauer spectroscopy, magnetization, and magnetic susceptibility measurements are the most popular tools for studying magnetic properties of amorphous nanoparticles. In particular, the thermal evolution of the shape of the Mössbauer spectrum of amorphous Fe_2O_3 nanopowders was explained by the strong interaction between superparamagnetic particles with a significant shift of the magnetic regime from inhomogeneous blocking to the glass collective state as in spin glass (Figure 1.12). The fast temperature variation of the spectral area of superparamagnetic fraction means that the transition to superparamagnetism retains the memory of the collective state. Experimental results of amorphous Fe_2O_3 nanoparticles confirm the model. In addition, the Mössbauer spectra of amorphous Fe_2O_3 nanoparticles

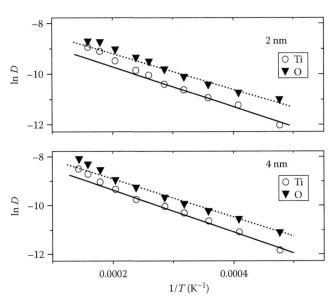

FIGURE 1.11 $1/T$ dependence of the logarithm of the diffusion constant of atomic species in simulated liquid TiO_2 nanoparticles. The straight lines just serve as a guide for the eyes. (From Hoang, V.V., *Nanotechnology*, 19, 105706, 2007b. With permission.)

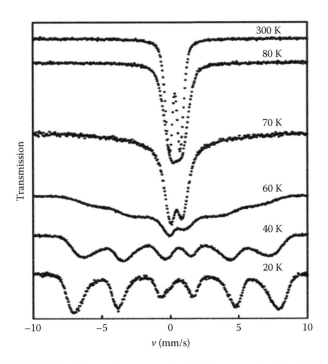

FIGURE 1.12   Temperature-dependent Mössbauer spectra (20–300 K) of the amorphous nanopowders prepared from Prussian blue. (From Zboril, R. et al., *Cryst. Growth Des.* 1, 1317, 2004a. With permission.)

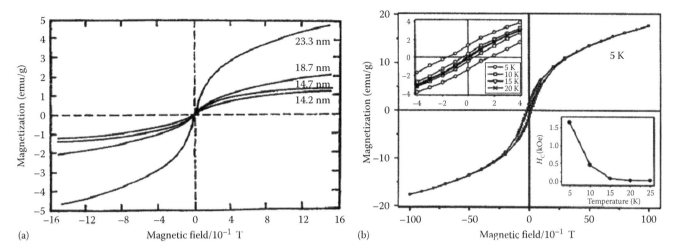

FIGURE 1.13 (a) Room temperature magnetization curves of amorphous Fe_2O_3 nanoparticles with different sizes. (From Cao, X. et al., *J. Mater. Res.*, 12, 402, 1997. With permission.) (b) Hysteresis loop for amorphous Fe_2O_3 nanopowder recorded at 5 K. The left inset shows the low-field region of the hysteresis loop measured at different temperatures. The right inset shows the temperature dependence of the coercive field H_C. (From Mukadam, M.D. et al., *J. Magn. Magn. Mater.*, 269, 317, 2004. With permission.)

in the external field applied parallel to the γ-ray direction have been obtained. Independently of the different degrees of interparticle interaction, the spectra display negligibly changes when compared with those recorded in a zero-applied field at the same temperature. In particular, the intensities of lines 2 and 5 remain almost unchanged (Libor et al. 2007). It becomes one of the principal markers in the identification of an amorphous phase of Fe_2O_3 nanoparticles and distinguishes them from their nanocrystalline counterparts. The unchanged-line intensities in the zero-field and the in-field spectra can be interpreted through a spin-glass-like behavior (Zboril et al. 2004a).

The magnetic susceptibility (χ) and the magnetization (M) of amorphous nanoparticles are also of great interest. It was found that the magnetic susceptibility and the magnetization curves of amorphous Fe_2O_3 nanoparticles significantly depend on the synthetic route, i.e., on the particle size distribution and the degree of interparticle interaction. Susceptibility versus temperature (χ vs. T) for amorphous Fe_2O_3 nanoparticles shows a maximum at about 50 K, which corresponds to the magnetic-transition temperature of the system. Above a magnetic-ordering temperature, temperature dependence of reciprocal susceptibility ($1/\chi$ vs. T) fulfills the Curie–Weiss law and it indicates the paramagnetic or superparamagnetic behavior of amorphous Fe_2O_3 nanoparticles (Libor et al. 2007).

Magnetization measurements for amorphous nanoparticles as a function of applied magnetic field or temperature are also under much attention (i.e., it indicates hysteretic/non-hysteretic behavior, saturation vs. nonsaturation or the value of the coercive field H_C, saturation magnetization, and remanent magnetization). At room temperature, magnetization curves observed for amorphous Fe_2O_3 nanoparticles are not hysteretic and do not saturate even at a high applied field (Figure 1.13a). Such behavior is expected in the unblocked regime of superparamagnetic particles, when the magnetic moments of particles can align in

its various easy directions during measurement time. For superparamagnetic materials in a magnetic field (H), one can use the following simple relation employing Boltzmann statistics:

$$M = M_S \left(\coth \frac{\mu H}{k_B T} - \frac{k_B T}{\mu H} \right) \tag{1.7}$$

Here, the expression in parentheses represents the Langevin function. M is the total magnetic moment of particles per unit volume, μ is the magnetic moment of a single nanoparticle, M_S is the saturation magnetization, and k_B is the Boltzmann constant. As a result, the saturation of magnetization at a define temperature is reached at a higher magnetic field for smaller particles. Indeed, such particle-size-dependent magnetic behavior was supported by the experimental data for amorphous Fe_2O_3 nanoparticles (Cao et al. 1997). The decrease of magnetization with decreasing nanoparticle size was explained in terms of a non-collispin arrangement at or the surface of nanoparticles. Furthermore, below blocking-temperature magnetization cannot relax in the time window of the measurement and a hysteretic behavior occurs (Figure 1.13b). The non-saturation behavior of magnetization at a low temperature and at a high field indicates the random orientation of spins in systems like the spin-glass systems with competing exchange interactions below the spin-freezing temperature.

1.5 Applications

Due to the specific, unique, isotropic disordered structure, the high concentration of coordinatively unsaturated sites (i.e., structural defects), the dangling bonds at the surfaces, and the high surface-bulk ratio, which can lead to catalytic activity and are selected superior to their nanocrystalline counterparts, amorphous nanoparticles can have potential applications in technology. In particular, amorphous-alloy nanoparticles have gained increasing attention as novel catalytic materials since 1980; the catalytic properties of metal–metalloid amorphous nanoparticles have especially been under intensive testing for applications in practice. Among these materials, amorphous NiB nanoparticles have been mostly investigated and used. On the other hand, amorphous NiB nanoparticles have also been used in enantioselective hydrogenation or reduction of ketones and hydrodesulfurization (Wu et al. 2007). Recently, it was found that amorphous nanoparticles of metal–metalloid alloys can be also used in the partial oxidation of methane to syngas, in ethanol dehydrogenation, in the synthesis of hydrogen peroxide from carbon monoxide, water, and oxygen, in the catalytic growth of carbon nanofibers or boron nitride tubes, in desulfurization, in the reduction of alkyl halides or nitro compounds, and in the coupling reaction of alkenes and hydrogen fuel cells. In addition, the amorphous nanoparticles of MB, MP alloys have been widely used as hydrogen-storage materials and anticorrosive materials (Wu et al. 2007). Excellent catalytic properties have also been found for amorphous nanoparticles of other materials such as Fe_2O_3, TiO_2, SiO_2, Ni, etc. (Koltypin et al. 1996, Hoang 2007a,b, Libor et al. 2007).

Amorphous nanoparticles of semiconductors such as Si, SiO_2, GeO_2, TiO_2, etc. have been under intensive investigation for the past decades due to their enormous technological importance for advanced quantum-confined electronic and optoelectronic devices (Sirenko et al. 2000, Yuri et al. 2002, Hoang 2007a,b, Xie et al. 2007). In particular, amorphous SiO_2 nanoparticles have been used for gene delivery, as a carrier for indomethacin in solid-state dispersion, and for drug-release control (Hoang 2007a). On the other hand, due to their specific photoluminescence ability, ultrasmall amorphous Si nanoparticles can be expected to have applications in nanodevices and biomaterials (Xie et al. 2007).

As noted above, amorphous nanoparticles of magnetic substances are also of great interest due to their specific magnetic behavior, which can lead to potential applications as novel magnetic materials. In particular, amorphous Fe_2O_3 nanoparticles have been presented as an advanced material applicable in various fields of modern nanotechnology including the manufacture of magnetic storage media and magnetic fluids. Generally, amorphous nanoparticles of metal oxides have great industrial potential in solar energy transformation, electronics, electrochemistry, catalysis, in optical and humidity sensors, or in sorption and purification processes. (Libor et al. 2007).

Additional advanced applications of amorphous nanoparticles are related to the formation of nanocomposites, i.e., materials containing amorphous nanoparticles dispersed in some matrix. It was shown that this is the most frequent form of the stabilization of an amorphous metal-oxide phase (Libor et al. 2007). That is, nanocomposites of amorphous ferric-oxide nanoparticles with an SiO_2 matrix are good candidates for use in the field of magneto-optical sensors and magnetic devices due to their attractive properties, including soft magnetic behavior, low density and electric resistivity (Casas et al. 2001). Further, the modification of the Si surface by amorphous Fe_2O_3 nanoparticles was used for the synthesis of magnetic nanocomposites, exhibiting several unique properties. Indeed, the incorporation of amorphous Fe_2O_3 nanoparticles onto a high-quality Si wafer, followed by annealing the composite, leads to the multiple functionality (magnetic, metallic, semiconducting, insulating, and optical properties) of materials (Prabhakaran and Shafi 2001).

1.6 Conclusion

An overview of the various aspects of amorphous nanoparticles including synthesis, characterization, structure, important chemico-physical properties and selected popular applications has been given. Note that although amorphous nanoparticles can be obtained in practice from a wide range of substances (pure elements or compounds), in the present chapter, we focused attention mainly on the most important classes of amorphous nanoparticles of the following substances: metals and alloys, oxides, and semiconductors. It is clearly seen that due to the disordered structure and the high surface-to-volume ratio, in addition to the large amount of structural defects in the surface shell, amorphous nanoparticles have unique physicochemical properties different from those of their crystalline counterparts, and this leads to their advanced applications in technology rather than the use of a crystal structure with well-defined properties.

Acknowledgment

This work was supported by the Foundation for Science and Technology of the University of Ho Chi Minh City (Vietnam) under Grant of Q2008-18-1.

References

Abu, M.R., Gedanken, A. 2005. Sonochemical synthesis of stable hydrosol of Fe_3O_4 nanoparticles. *J. Colloid Interface Sci.* 284: 489–494.

Alcoutlabi, M., McKenna, G.B. 2005. Effects of confinement on material behaviour at the nanometre size scale. *J. Phys.: Condens. Matter* 17: R461–R524.

Ayyub, P., Nultani, M., Barma, M., Palkar, V.R., Vijayaraghavan, R. 1988. Size-induced structural phase transitions and hyperfine properties of microcrystalline Fe_2O_3. *J. Phys. C: Solid State Phys.* 21: 2229–2245.

Cao, X., Prozorov, R., Koltypin, Yu., Kataby, G., Felner, I., Gedanken, A. 1997. Synthesis of pure amorphous Fe_2O_3. *J. Mater. Res.* 12: 402–406.

Casas, L., Roig, A., Rodriguez, E., Molins, E., Tejada, J., Sort, J. 2001. Silica aerogel-iron oxide nanocomposites: Structural and magnetic properties. *J. Non-Cryst. Solids* 285: 37–43.

Changsheng, X., Junhui, H., Run, W., Hui, X. 1999. Structure transition comparison between the amorphous nanosize particles and coarse-grained polycrystalline of cobalt. *NanoStruct. Mater.* 11: 1061–1066.

Chiennan, P., Pouyan, S., Shuei-Yuan, C. 2007. Condensation, crystallization and coalescence of amorphous Al_2O_3 nanoparticles. *J. Cryst. Growth* 299: 393–398.

Fiorani, D., Romero, H., Suber, L. et al. 1995. Synthesis and characterization of amorphous $Fe_{80-x}Cr_xB_{20}$ nanoparticles. *Mater. Sci. Eng. A* 204: 165–168.

Ghidini, M., Nozieres, J.P., Givord, D., Gervais, B. 1995. Magnetic processes in amorphous YCo_2 nanoparticles obtained by heavy ion irradiation. *J. Magn. Magn. Mater.* 140–144: 483–484.

Gonzalez, R.J. 1998. Raman, infrared, X-ray, and EELS studies of nanophase titania. PhD thesis, Virginia Polytechnic Institute State University, Middleburg, VA.

Günter, S., ed. 2004. *Nanoparticles from Theory to Application.* Weinheim, Germany: Wiley-VCH Verlag GmbH & Co. KGaA.

Hengshong, Z., Bin, C., Jillian, F.B. 2008. Atomic structure of nanometer-sized amorphous TiO_2. *Phys. Rev. B* 78: 214106–214117.

Hoang, V.V. 2007a. Molecular dynamics simulation of amorphous SiO_2 nanoparticles. *J. Phys. Chem. B* 111: 12649–12656.

Hoang, V.V. 2007b. The glass transition and thermodynamics of liquid and amorphous TiO_2 nanoparticles. *Nanotechnology* 19: 105706–105711.

Hoang, V.V., Khanh, B.T.H.L. 2009. Static and thermodynamic properties of liquid and amorphous Fe_2O_3 nanoparticles. *J. Phys.: Condens. Matter* 21: 075103–075111.

Hoang, V.V., Odagaki, T. 2008. Molecular dynamics simulation of monatomic amorphous nanoparticles. *Phys. Rev. B* 77: 125434–125444.

Hoang, V.V., Zung, H., Trong, N.H.B. 2007. Structural properties of amorphous TiO_2 nanoparticles. *Eur. Phys. J. D* 44: 515–524.

Jianhua, Z., Jiangping, H., Tao, W., Xui, C. 2008. A hybrid approach of template synthesis and electroless depositing for Ni-W-P nanoparticles. *J. Solid State Electrochem.*: DOI 10.1007/s10008-008-0677-1.

Jianyi, S., Zheng, H., Yuanfu, H., Yi, C. 1992. Investigation of amorphous $Fe_{82}P_{11}B_7$ ultrafine particles produced by chemical reduction. *J. Phys.: Condens. Matter* 4: 6381–6388. 3710–3716.

Jimenez, V.M., Caballero, A., Fernandez, A., Espinos, J.P., Ocana, M., Gonzalez-Elipe, A.R. 1999. Structural characterization of partially amorphous SnO_2 nanoparticles by factor analysis of XAF and FT-IR spectra. *Solid State Ionics* 116: 117–127.

Kan, S.H., Yu, S., Peng, X.G. et al. 1996. Formation process of nanometer-sized cubic ferric oxide single crystals. *J. Colloid Interface Sci.* 178: 673–680.

Koltypin, Y., Katabi, G., Cao, X., Prozorov, R., Gedanken, A. 1996. Sonochemical preparation of amorphous nickel. *J. Non-Cryst. Solids* 201: 159–162.

Lianxia, C., Haibin, Y., Wuyou, F. et al. 2008. Simple synthesis of MoS_2 inorganic fullerene-like nanomaterials from MoS_2 amorphous nanoparticles. *Mater. Res. Bull.* 43: 2427–2433.

Liao, X., Zhu, J., Zhong, W., Chen, H.Y. 2000. Synthesis of amorphous Fe_2O_3 nanoparticles by microwave irradiation. *Mater. Lett.* 50: 341–346.

Libor, M., Radek, Z., Aharon, G. 2007. Amorphous iron (III) oxide—A review. *J. Phys. Chem. B* 111: 4003–4018.

Mukadam, M.D., Yusuf, S.M., Sharma, P., Kulshreshtha, S.K. 2004. Magnetic behavior of field induced spin-clusters in amorphous Fe_2O_3. *J. Magn. Magn. Mater.* 269: 317–326.

Palchik, O., Felner, I., Kataby, G., Gedanken, A. 2000. Amorphous iron oxide prepared by microwave heating. *J. Mater. Res.* 15: 2176–2181.

Prabhakaran, K., Shafi, K.V.P.M. 2001. Nanoparticle-induced light emission from multi-functionalized silicon. *Adv. Mater.* 13: 1859–1862.

Prozorov, R., Yeshurun, Y., Prozorov, T., Gedanken, A. 1999. Magnetic irreversibility and relaxation in assembly of ferromagnetic nanoparticles. *Phys. Rev. B* 59, 6956–6965.

Roder, A., Kob, W., Binder, K. 2001. Structure and dynamics of amorphous silica surfaces. *J. Chem. Phys.* 114: 7602–7614.

Si, P.Z., Zhang, M., You, C.Y. et al. 2003. Amorphous boron nanoparticles and BN encapsulating boron nano-peanuts prepared by arc-decomposing diborane and nitriding. *J. Mater. Sci.* 38: 689–692.

Sirenko, A.A., Fox, J.R., Akimov, I.A., Xi, X.X., Ruvimov, S., Liliental-Weber, Z. 2000. In situ Raman scattering studies of the amorphous and crystalline Si nanoparticles. *Solid State Commun.* 113: 553–558.

Srivastava, D.N., Perkas, N., Gedanken, A., Felner, I. 2002. Sonochemical synthesis of mesoporous iron oxide and accounts of its magnetic and catalytic properties. *J. Phys. Chem. B* 106: 1878–1883.

Subrt, J., Bohacek, J., Stengl, V., Grygar, T., Bezdicka, P. 1998. Uniform particles with a large surface area formed by hydrolysis $Fe_2(SO_4)_3$ with urea. *Mater. Res. Bull.* 34: 905–914.

Suwen, L., Weiping, H., Siguang, C., Sigalit, A., Aharon, G. 2001. Synthesis of X-ray amorphous silver nanoparticles by the pulse sonoelectronchemical method. *J. Non-Cryst. Solids* 283: 231–236.

Tortarolo, M., Zysler, R.D., Troiani, H., Romero, H. 2004. Magnetic order in amorphous $(Fe_xNd_{1-x})_{0.6}B_{0.4}$ nanoparticles. *Physica B* 354: 117–120.

Tracy, M.D., Mark, A.S., Michael, T. 2007. Germania nanoparticles and nanocrystals at room temperature in water. *Langmuir* 23: 12469–12472.

Wu, Z., Li, W., Zhang, M., Tao, K. 2007. Advances in chemical synthesis and application of metal-metalloid amorphous alloy nanoparticulate catalysts. *Front. Chem. Eng. China* 1: 87–95.

Xie, Y., Wu, X.L., Qiu, T., Chu, P.K., Siu, G.G. 2007. Luminescence properties of ultrasmall amorphous Si nanoparticles with sizes smaller than 2 nm. *J. Cryst. Growth* 304: 476–480.

Yuri, D.G., Sheng-Hsien, L., Yit-Tsong, C. 2002. Time-resolved photoluminescence study of silica nanoparticles as compared to bulk type-III fused silica. *Phys. Rev. B* 66: 035404–035413.

Zboril, R., Machala, L., Mashlan, M., Sharma, V. 2004a. Iron(III) oxide nanoparticles in the thermally induced oxidative decomposition of Prussian blue, $Fe_4[Fe(CN)_6]_3$. *Cryst. Growth Des.* 4: 1317–1325.

Zboril, R., Machala, L., Mashlan, M., Tucek, J., Muller, R., Schneeweiss, O. 2004b. Magnetism of amorphous Fe_2O_3 nanopowders synthesized by solid-state reactions. *Phys. Stat. Sol. C* 1: 3710–3716.

Zhong, G.Q., Zhou, H.L., Jia, Y.Q. 2008. Preparation of amorphous Ni-B alloys nanoparticles by room temperature solid-solid reaction. *J. Alloy Compd.* 465: L1–L3.

Ziman, J.M. 1979. *Models of Disorder. The Theoretical Physics of Homogeneously Disordered Systems.* Cambridge, U.K.: Cambridge University Press.

Zysler, R.D., Ramos, C.A., Romero, H., Ortega, A. 2001. Chemical synthesis and characterization of amorphous Fe-Ni-B magnetic nanoparticles. *J. Mater. Sci.* 36: 2291–2294.

2

Magnetic Nanoparticles

Günter Reiss
Bielefeld University

Andreas Hütten
Bielefeld University

2.1 Introduction

Since the beginning of this century, science and engineering has seen a rapid increase in interest for materials at the nanoscale. Nanomaterials have attracted a strong interest because of their physical, electronic, and magnetic properties, which is a result of their small size, and where both surface effects become dominant and quantum size effects occur.

Within the field of nanomaterials under worldwide research is the subset of magnetic nanomaterials.

Depending on their size and the subsequent change in their magnetic property, magnetic nanoparticles are used in different applications [Reiss 2005]. Since the relaxation time of magnetic nanoparticles can be changed by varying the size of the nanoparticles or by using different kinds of materials, magnetic nanoparticles have been (and will be in the future) a very useful tool in different kinds of applications from biomedical to data storage systems. An example of a high-resolution transmission electron microscope image of a magnetic nanoparticle with the composition $Fe_{47}Co_{53}$ is shown in Figure 2.1 [Hütten 2005].

As can be seen from this highly resolved image, these small particles with—in this case about 10 nm in diameter—are highly crystalline; the outer rim, however, is usually not very well resolved even in high-resolution imaging. Nevertheless, Figure 2.1 demonstrates that it is possible nowadays to resolve the internal structure of the nanoparticles down to the atomic level. These possibilities of characterization, which are not discussed in this chapter, contributed considerably to the rapid development of research and applications dealing with magnetic nanoparticles.

This chapter discusses the synthesis and the basic physical properties of magnetic nanoparticles [Billas 1994], describes some of the methods used to characterize magnetic particles, and discusses current research for the application of these particles.

2.2 Historical Background

Although, generally, nanoparticles are considered an invention of modern science, they actually have a very long history. Specifically, nanoparticles were used by artisans as far back as in the ninth century in Mesopotamia for generating a glittering effect on the surface of pots.

Even these days, pottery from the Middle Ages and the Renaissance often retain a distinct gold- or copper-colored metallic glitter. This so-called luster is caused by a metallic film that was applied to the transparent surface of a glazing. The luster can still be visible if the film has resisted atmospheric oxidation and other weathering.

Studies of magnetic nanostructures started at the beginning of the twentieth century, in which amorphous or nanocrystalline materials were investigated. The preparation and characterization of particulate magnetic nanoparticles started in the 1970s and encompassed a broad range of synthetic and investigative techniques involving tools from and the knowledge of chemistry, physics, and engineering. There are two basic approaches to nanoparticle preparation and assembly: "bottom up" and "top down." The bottom-up approach takes molecules or a cluster of molecules (nanoparticles) and assembles them up into a pretailored architecture. This approach relies on the energetics of the assembly process to guide it. Typical examples are templated film growth, self- and directed-assembly of colloidal particles, and spinodal wetting/dewetting. The top-down approach, on the other hand, relies on micromachining materials to the desired sizes and patterns, and is generally subtractive in nature. Typical examples of the top-down approach include photolithography,

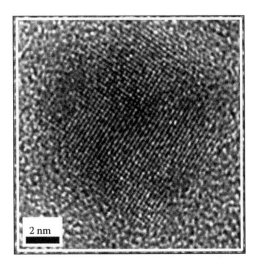

FIGURE 2.1 A high-resolution transmission electron microscope image of an $Fe_{47}Co_{53}$ nanoparticle supported by an electron transparent carbon foil. The lattice planes of the crystalline structure are clearly resolved.

mechanical machining/polishing, laser beam and electron beam processing, and electrochemical removal. Generally, the preparation and use of magnetic nanoparticles is as rapidly evolving as the wide field of nanotechnology [Roco 1999]. In this chapter, we concentrate mainly on particles produced by the bottom-up approach because this enables preparation of large amounts.

2.3 State of the Art

2.3.1 Preparation

In contrast with many nonmagnetic particles such as Au, magnetic particles are usually either metallic and very sensitive to oxidation or consist of oxides with—in many cases—a complex distribution of phases within the particles.

This progress in the preparation techniques was driven by various emerging applications of nanoparticles in, for example, surface protection, data storage, and biotechnology. This chapter concentrates on magnetic particles and thus shows application examples for data storage and biotechnological applications.

Many routes to the preparation of magnetic nanoparticles have been worked out successfully. In this chapter, we discuss gas phase preparation [Masala 2004] and organometallic routes [Coperet 2005].

2.3.1.1 Gas Phase Preparation

Magnetic nanoparticles can be prepared by an inert gas condensation process [Gleiter 1989, Kruis 1998], in which a supersaturated vapor of the material is created either by metal evaporation or by sputtering a metal target by Ar^+ ions at energies of some hundred electron volts and at pressures of roughly 1 mbar. Within this supersaturated metal vapor, particle nuclei are formed by homogeneous nucleation. The nanoparticles then grow by successive aggregation and Ostwald ripening.

For this type of preparation, modified UHV sputtering or evaporation systems are used (Figure 2.2).

The particles are grown in a nucleation and aggregation ultrahigh vacuum chamber. From there, they are ejected into high vacuum (10^{-6} to 10^{-4} mbar) by differential pumping between two apertures, which provides a free particle beam. A quadrupole mass spectrometer can be used to fractionate the particle beam with respect to size. Prior to deposition onto substrates, the nanoparticles beam can be subjected to optical heating in a light furnace.

With this method, highly monodisperse metallic particles can be produced, which do not suffer from oxidation due to vacuum conditions. The amount of particles that can be obtained with this method, however, is usually small. Upscaling to larger yields is on the way.

2.3.1.2 Coprecipitation

Coprecipitation [Kim 2001] is a facile and convenient way to synthesize, for example, oxides of magnetic 3d transition metals (TMs) (e.g., Fe_3O_4 or γ-Fe_2O_3) from aqueous salt solutions by the addition of a base under inert atmosphere at room temperature or at elevated temperature [Sun 2006]. The size, shape, and composition of magnetic nanoparticles depend largely on the type of salts used (e.g., chlorides, sulfates, and nitrates), the Fe^{2+}/Fe^{3+} ratio, the reaction temperature, the pH value, and the ionic strength of the media.

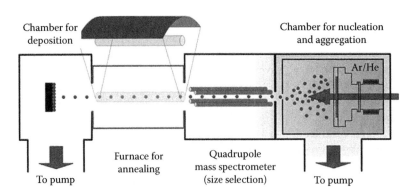

FIGURE 2.2 Sketch of a four-chamber ultrahigh vacuum apparatus used for the gas phase preparation of nanoparticles.

As an example, the preparation of elongated particles is based on the growth of goethite (α-FeOOH) from an iron sulfate solution using NaOH. The growth process is controlled by the addition of Al^{3+} added as a salt at various stages of the preparation process. The purpose of Al is to reduce and control the growth rate. Other additives such as yttrium are included in the final stages of the precipitation of the goethite to produce a complex particle that consists of a core composed mainly of α-FeOOH with a surface coating of (Fe, Al, Y)OOH.

Generally, Co is coprecipitated into the goethite throughout the process to produce a core material having a high saturation magnetization M_S. This leads to the production of a complex core with an appropriate surface layer. These particles are then dehydrated by heating to transform the core to α-$(Co,Fe)_2O_3$ (hematite) with Al and other additives in a complex oxide on the surface. The hematite is then reduced by heating in a hydrogen atmosphere to produce an α-CoFe core with controlled reoxidation, leading to a complex FeAlY oxide surface layer. This serves not only to protect the particles from unintentional further oxidation, but also to aid their dispersability. The resulting slurry can then be coated onto a base film providing an ultrasmooth coating with the particles aligned by the application of a magnetic field during the drying process [Chadwick 2008].

2.3.1.3 Thermal Decomposition

Monodisperse magnetic nanocrystals with small size can be synthesized through the thermal decomposition of organometallic compounds in high-boiling organic solvents containing stabilizing surfactants, which can also be used for determining the shape of the particles [Puntes 2002, Dumestre 2004].

The preparation method synthesizing nanoparticles from 3d TMs, with sizes basically ranging from 4 to 8 nm, by thermolysis of metal carbonyls precursors given in [Puntes 2001] can be summarized as follows: 0.1–0.3 g of trioctylphosphine oxide and 0.2 mL of oleic acid are dissolved in 12 mL of 1,2-dichlorobenzene. The solution subsequently is heated to reflux.

Separately, 0.45–0.5 g of dicobaltoctacarbonyl, $Co_2(CO)_8$, is dissolved in 3–6 mL of 1,2-dichlorobenzene. During vigorous stirring, the second solution is then injected into the refluxing bath. After a reaction time of 30 min, the mixture is cooled to room temperature.

In order to produce magnetic nanocrystals with mean particle diameters larger than 8 nm, the preparation method has to be changed so as to perform successive precursor addition after the rapid initial injection. For example, to double the diameter, one-eighth of the total precursor is rapidly injected at once and the remaining precursor material then is consecutively added. This recipe is based on the combination of, first, the adaptation of the production of silica particles [van Blaaderen 1992] and, second, obeying the predictions of LaMer's model [Murray 2000], which is based on the temporal evolution of the concentration of monomers. Monomers, in terms of the model, are the initial building blocks for particles (e.g., Co atoms for Co particles).

To double the particle size, precursor is successively added after initial injection. This further precursor addition has to be so slow so that the nucleation threshold will not be reached and hence no new nuclei can be formed. Figure 2.3 shows the results of these procedures for Co particles.

In Figure 2.3a, the TEM image shows particles that have been prepared by a single injection of Co precursors; the resulting diameter of the particles is typically in the range between 3 and 5 nm (4.2 nm in this case). By a careful reinjection of precursor molecules, an increase in the mean particle diameter up to 10 nm can be obtained.

To obtain alloyed nanoparticles, at least two precursors have to be used. For Fe–Co alloys, $Co_2(CO)_8$ and ironpentacarbonyl, $Fe(CO)_5$, can be injected in the same way as for the single-material particles. Starting from different mixtures of $Co_2(CO)_8$ and $Fe(CO)_5$, alloyed nanoparticles with compositions from $Fe_{90}Co_{10}$ to $Fe_{10}Co_{90}$ in incremental steps of 10 atom% could be experimentally realized besides pure Fe or Co nanoparticles. An example for an Fe–Co alloyed particle was already shown in Figure 2.1.

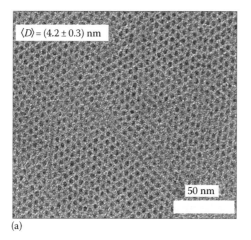
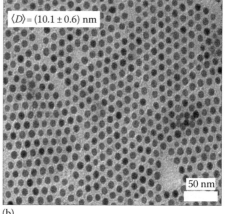

(a) (b)

FIGURE 2.3 (a) TEM image of Co nanoparticles prepared by a single injection of Co precursors; the mean particle diameter is 4.2 nm. (b) Co particles produced by multiple injections of precursor molecules (TEM image). An increase of the mean particle to 10 nm is obtained.

2.3.1.4 Microemulsion

Surfactants, with hydrophilic and hydrophobic parts, dissolved in organic solvents form spheroidal aggregates called reverse micelles, which can serve as microreactors [Pileni 1989, 1993]. Water is essential to form large surfactant aggregates, although they can be formed both in the presence and in the absence of water. It is then readily solubilized in the polar inner core, forming a so-called "water pool," characterized by water–surfactant molar ratio. Aggregates containing a small amount of water are usually called reverse micelles, whereas microemulsions correspond to droplets containing a large amount of water molecules. Usually, metal salts dissolved in water are added to a mixture of nonpolar liquids consisting of the oil phase and possible cosurfactants. Then the molar ratio of water to surfactant determines the resulting size of the micelles.

Within theses micelles, metallic nanoparticles are formed by adding an agent that reduces the metallic salts. Using the microemulsion technique, metallic cobalt, cobalt/platinum alloys, and gold-coated cobalt/platinum nanoparticles have been synthesized in reverse micelles of cetyltrimethlyammonium bromide, using 1-butanol as the cosurfactant and octane as the oil phase [Petit 1998]. The metal and alloy nanoparticles are then formed within the reverse micelle by the reduction of metallic salts using sodium borohydride as a reducing agent.

2.3.2 Basic Properties

The physical properties of nanoparticles strongly depend on their size [Bansmann 2005]. For ultrasmall particles with only a few to around 1000 atoms, the small size gives rise to a quantum mechanical splitting of the electronic states, which determines the properties. Therefore, such free clusters and nanoparticles are not just small pieces of material with physical properties nearly identical to the bulk. Their electronic, optical, and magnetic properties are clearly size-dependent with a nonlinear behavior between the two general limits given by the atomic and the bulk-like behavior.

2.3.2.1 Magnetic Nanoparticles with Only a Few Atoms

In magnetic nanoclusters, magnetism therefore develops as a material is built from individual atoms to the solid state [Shi 1996]. The most widely used model to describe the delocalized electrons in metallic clusters is that of a free-electron gas, known as the jellium model [Kohn 2003]. The positive charge is regarded as being smeared out over the entire volume of the cluster, while the valence electrons are free to move within this homogeneously distributed, positively charged background. The calculated potential for the electrons in a spherical jellium approximation typically looks like the example in Figure 2.4a.

Here, the inner part of the effective potential resembles the bottom of a wine bottle. The electronic energy levels are grouped together to form shells. If we look at the jellium model's predictions for lead clusters with 4 valence electrons, one would expect preferred clusters with 2, 5, and 10 atoms, because then the states shown in Figure 2.4a would be filled. The corresponding probability for finding clusters with specific numbers of atoms is shown in Figure 2.4b. In the case of lead, magic numbers were observed at 7 and 10 atoms. Pb has 4 valence electrons so 10 atoms corresponds to filling the 2p shell. Note, however, that 7 Pb atoms with

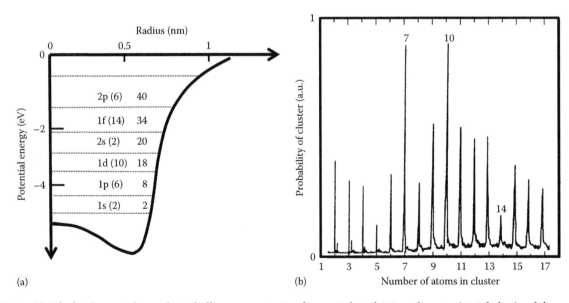

(a)   (b)

FIGURE 2.4 (a) Calculated potential in a spherical jellium approximation for a particle with 1.5 nm diameter (straight line) and the energy levels for the electrons (dotted lines) with abbreviations, number of electrons in the level, and total number of electrons if the level is filled. (b) The probability for finding lead clusters within a particle beam as a function of the number of atoms in the cluster. Distinct maxima for clusters with 2, 7, 10, and 14 atoms can be observed.

28 electrons do not match the jellium model. This can be attributed to a nonspherical electron distribution, that is, a preference for a particular structure (a pentagonal bipyramid).

Consequently, also the magnetic properties of clusters are very sensitive to the details of the electronic correlations and to temperature [Lau 2002, Stahl 2003]. In isolated atoms, almost all elements show a nonvanishing magnetic moment given by Hund's rules, while in the solid state, only a few of them (some TMs of the Fe group, the lanthanides, and actinides) preserve a nonvanishing magnetization. Finite clusters constitute a new state of matter with its own fascinating characteristics [Gruner 2006].

The magnetism of TM clusters represents one of the fundamental challenges, since atomic and bulk behaviors are intrinsically different. Atomic magnetism is due to electrons that occupy localized orbitals, while in TM solids, the electrons responsible for magnetism are itinerant, conducting d-electrons. Consequently, the magnetic properties of nanoparticles are very sensitive to size, composition, and local atomic environment, thus showing a wide variety of intriguing phenomena.

First calculations on the electronic structure and magnetic properties of small iron and nickel clusters started already in the 1980s. Lee [Lee 1985] proposed a narrowing of the d-bands with decreasing number of atoms in Fe clusters and an enhanced spin polarization in Fe clusters compared with the bulk. Shortly later, these calculations were extended, including different geometric structures in the size range from 2 atoms per cluster up to about 50 atoms. Very small Fe clusters with less than 10 atoms showed magnetic moments of about $3\mu_B$ with decreasing values for larger clusters. The first measurements on magnetic phenomena of 3d metal clusters in a molecular beam were performed in the beginning of the 1990s by the groups of [Billas 1994] and [Bloomfield 1993].

For free mass selected clusters, it is possible to observe how the magnetic properties change from the monomer to the bulk. These include a significant increase in the magnetic moment per atom relative to the bulk in 3d TMs, the appearance of magnetism in paramagnetic metals [Bloomfield 1993], and ferrimagnetism in antiferromagnetic materials. Lowered magnetic moments per atom in ferromagnetic rare-earths have been ascribed to canted atomic moments and both lowered and increased Curie temperatures have been observed. Particularly, large changes are observed in very small clusters; [Knickelbein 2002], for example, observed a magnetic moment per atom very close to the atomic limit of $6\mu_B$ per atom in 12-atom clusters that reduces to a value close to the bulk limit of $2.2\mu_B$ per atom on addition of a single atom to produce a 13-atom cluster.

In Figure 2.5, examples for measured magnetic moments per atom in Fe clusters are shown in the dependence of the number of atoms in the cluster [Shi 1996].

A significant increase of the moment as compared with the Fe bulk value of $2.2\mu_B$ can be found especially in very small clusters. Additionally, the shape of the clusters, which is also determined by the number of atoms, will have an influence on their magnetic properties [Pellarin 1994].

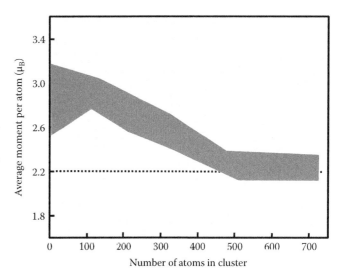

FIGURE 2.5 Magnetic moments per atom in Fe clusters in dependence of the number of atoms in the cluster. The gray-shaded area corresponds to the reported values. The dotted line marks the bulk magnetic moment of $2.2\mu_B$.

2.3.2.2 Nanoparticles with Many Atoms

In contrast with the very small clusters, nanoparticles with more than around 1000 atoms usually do not exhibit quantum effects due to their small size. Nevertheless, still considerable deviations from the bulk properties can be found in these particles up to radii of around 1 µm. The simplest reason for this is a change in the outer shell of the particles due to, for example, oxidation or chemical bonds to organic shell molecules.

As an example, the magnetization curve for Co particles synthesized by thermal decomposition of $Co_2(CO)_8$ with oleic acid as organic shell [Hütten 2004] is shown in Figure 2.6.

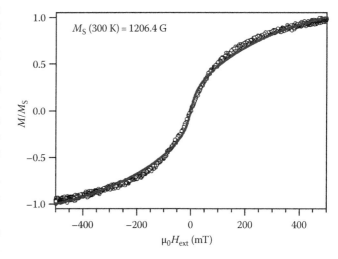

FIGURE 2.6 Magnetization curve for Co particles synthesized by thermal decomposition of $Co_2(CO)_8$ with oleic acid as organic shell (o); the gray solid line is a fit with a Langevin function.

TABLE 2.1 Upper Diameters for the Crossover between Superparamagnetic and Ferromagnetic (Thermally Blocked) State for Different Ferromagnetic Materials at Room Temperature

Material	Critical Diameter (nm)
fcc-Co	15.8
hcp-Co	7.8
Fe_3O_4	26.2
Fe_2O_3	34.9
FeCo	23.6

Here, the diameter of the particles amounts to about 3.3 nm. The ratio between the measured saturation magnetization and the bulk value is about 0.85 at room temperature. Thus, 15% of the Co atoms nominally do not contribute to the magnetic moment of the particles, which is related with the formation of an oxidic shell around the metal core of the particles.

When considering the magnetic behavior of such particles, they can be divided into particles that are superparamagnetic [Mørup 2007] and ferromagnetic. For superparamagnetic particles, the magnetic anisotropy energy given by $E_A = KVM_S$ (K, effective anisotropy constant; V, particle volume; and M_S, saturation magnetization) is of the order of the thermal energy $E_T = k_B T$ (k_B, Boltzmann constant and T, temperature). Thus, the magnetic relaxation time related with thermal excitations are shorter compared with the typical timescale of the measurement. Thermally blocked particles appear to be ferromagnetic because they have magnetic relaxation times that are larger when compared with a typical timescale of measurement being used to study the particle system.

Because the volume enters the relation of the two energies, there is a threshold diameter for the superparamagnetic state, which depends on the temperature. Typical values for these critical diameters are given in Table 2.1.

If nanoparticles with varying sizes are demobilized in a solid matrix, the thermally blocked nanoparticles will exhibit both remanence and coercivity while the superparamagnetic particles will not show any remanence and coercivity.

As an additional feature of the magnetic states of small particles, the so-called single-domain limit is important to understand their properties. The spatial distribution of the magnetization is influenced by several energy contributions, which can be summarized in an anisotropy energy E_{an}, an exchange energy E_{ex}, a magnetostatic energy E_{dp}, and the Zeman energy E_{ze} in an external magnetic field. Because these energies depend on the size and the shape of the particles in a different way, complex domain patterns can result for differently shaped particles. As an example, we show in Figure 2.7a,b the simulated magnetization pattern of two ferromagnetic squares of Permalloy ($Ni_{20}Fe_{80}$), one with 300 nm and the other one with 30 nm edge length, respectively.

The simulation was close to the remanent state (external field $H = 5$ Oe) by solving the Landau–Lifshitz–Gilbert equation describing the time-dependent behavior of the magnetization [OOMMF 1999]:

$$\frac{d\vec{M}}{dt} = -\gamma\mu_0 \vec{M} \times \vec{H} + \frac{\alpha}{M_S}\left(\vec{M} \times \frac{d\vec{M}}{dt}\right)$$

where
  $\vec{M}$ is the magnetization
  γ is the gyromagnetic ratio
  μ_0 is the Bohr magneton
  M_S is the saturation magnetization
  α is the damping parameter

As can be seen in this figure, the magnetization of the larger square splits up into four regions, where the magnetization at

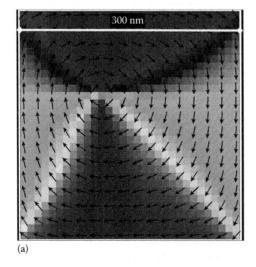 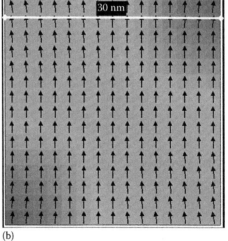

(a)   (b)

FIGURE 2.7 (a) Simulations of the magnetization pattern [OOMMF 1999] of a ferromagnetic square of Permalloy ($Ni_{20}Fe_{80}$) with 300 nm edge length at a magnetic field of 5 Oe pointing from bottom to top. The gray code is related with the directions of the magnetization which is indicated additionally by arrows. (b) The same as in Figure 2.7a for a smaller square with 30 nm edge length.

the rim of the structure tries to follow the outer contour of the ferromagnet. This is caused by the magnetostatic energy, which favors a magnetization state with low stray field outside the ferromagnet. In the middle of the pattern, a vortex is formed, where the magnetization points perpendicular to the surface.

In contrast to this domain splitting, the 30 nm large square has a rather homogeneous magnetization with only small bending at the rims of the structure. Thus, this size is below the so-called single-domain limit, where it is energetically favorable to avoid magnetic domains. The reason for this is that the domain wall energy caused by the anisotropy and the exchange energy in this case would be larger than the magnetostatic energy of the large stray field of the single-domain state. Because the term defining the domain wall energy E_{DW} varies with the anisotropy energy and the exchange energy roughly as $E_{DW} \propto \sqrt{E_{ex}E_{an}}$ (note, that the width of the wall is approximately $w = 2\sqrt{E_{ex}/E_{an}}$) and the magnetostatic energy is a complex function of the saturation magnetization and the shape of the ferromagnetic sample, the size limit for single-domain behavior also varies strongly for different materials and can be as large as some hundreds of nanometers for highly anisotropic materials such as FePt in the L_{10} structure. For low anisotropic materials such as Permalloy, however, the single-domain state is reached only if the size is smaller than about 50 nm.

Usually, the term "magnetic nanoparticles" relates to the case of either very small clusters or particles, which are in the single-domain state. Larger particles, however, can be at the border between the single-domain and multidomain states. Figure 2.8 shows an overview of the different possible characteristic.

The maximum coercivity occurs when the particles are as large as possible but still in the single-domain state. After the transition into multidomains where the magnetization in the nanoparticles splits up into several magnetic domains, coercivity decreases with particle sizes. There the magnetization process is dominated by domain wall motion at correspondingly low magnetic fields.

Figure 2.8 also determines the possible applications of magnetic nanoparticles: While several purposes like data storage need a very high stability of the magnetization, others like ferrofluids

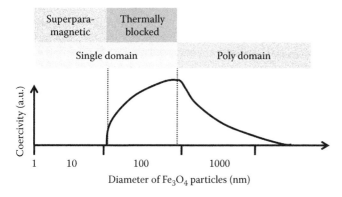

FIGURE 2.8 The magnetic characteristic of magnetic nanoparticles as a function of their size for magnetite (Fe_3O_4). The lower diagram shows the typical variation of the particle's coercivity with the diameter. The upper panels display the different regimes of magnetic behavior of the particles.

prefer superparamagnetic particles with nonhysteretic magnetization curves, as shown already in Figure 2.6. In Section 2.3.3, some of the most important fields of application are discussed.

2.3.3 Applications

Magnetic nanoparticles are envisioned for a wide variety of uses. In the medical realm, scientists hope to use single nanoparticles to deliver anticancer drugs or radionuclide atoms to a targeted area of the body, or to enhance the contrast in magnetic resonance imaging. The particles also could assist in the development of advanced data storage, and further down the road, in spintronic devices. In general, the higher the nanoparticle's magnetic moment—the measure of a material's magnetic strength—the more valuable it is for these applications.

2.3.3.1 Data Storage

Continuing increases in the areal density of hard disk drives and tapes will be limited by thermal instability of the thin film medium. Patterned media, in which data are stored in an array of single-domain magnetic particles, have been suggested as a means to overcome this limitation and to enable disk recording densities of up to 150 Gb/cm² (1 Tb/in.²) to be achieved. However, the implementation of patterned media requires fabrication of sub-50 nm features over large areas and the design of recording systems that differ substantially from those used in conventional hard drives.

The magnetic nanoparticles for this type of application have to fulfill several requirements. The most important are the stability of the stored information for at least 10 years and its read- and writability [Weller 1999].

While the first requirement can be met by choosing particles such as FePt with a very high anisotropy [Rellinghaus 2006], the writability deteriorates if the energy barrier for changing the magnetization direction becomes too large. In today's media, the signal-to-noise ratio needed for high-density recording is thus achieved by statistically averaging over a large number of weakly interacting magnetic grains per bit.

Critical to the application of any material as a data storage medium is the switching field distribution that is controlled by the particle size distribution, the magnetic anisotropy (K), the magnetization reversal mechanism of the particles, and particularly, the alignment of the particle easy axes on the tape [Chadwick 2008] or the disk.

During the development of magnetic particles for data storage, elongated shapes were introduced in order to induce a high anisotropy. In the early 1980s, the particles had axial ratios of approximately 10 or greater with H_C in the range of 1200–1500 Oe. Today, the most advanced materials have an axial ratio between 4 and 6 and H_C of almost 3000 Oe when aligned on tape [Ross 2001].

A relatively new development is the use of single particles in "patterned media": A patterned recording medium, shown schematically in Figure 2.9, consists of a regular array of magnetic elements, each of which has uniaxial magnetic anisotropy.

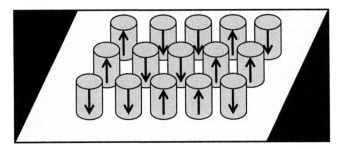

FIGURE 2.9 Sketch of a particulate media for magnetic data storage consisting of a regular array of ferromagnetic elements with uniaxial anisotropy. Each element can store one bit.

The easy axis can be parallel or perpendicular to the substrate. Each element stores one bit, depending on its magnetization state.

For a patterned media system to be viable, it must offer significantly higher data densities than can be achieved in a conventional hard drive. Hard disk drives using longitudinal media will be able to reach 15 Gb/cm² or higher, so patterned media designs must be capable of reaching densities of 30 Gb/cm² and beyond. This implies a periodicity in Figure 2.9 of 50 nm or smaller, for instance, 25 nm elements with 25 nm separation. A 50 nm period corresponds to a density of 40 Gparticles/cm².

2.3.3.2 Ferrofluidic Applications

One of the most important fields of already realized application of magnetic nanoparticles relies on their ability to change the viscosity of so-called ferrofluids [Odenbach 2002]. There, magnetic particles are suspended in either water or organic solvents. If a spatially inhomogeneous magnetic field H acts on this ferrofluid, the particles experience a force F_M. Usually, superparamagnetic particles are used, and thus the force is given by $\vec{F}_M = -\vec{\nabla}(\vec{M} \cdot \vec{H})$, where $\vec{M} = \chi \vec{H}$ is the particle's magnetic moment and χ is the nearly constant initial susceptibility.

Figure 2.10 displays the viscosity $\eta'(H)$ that changes during the application of a magnetic field by several tens of percentage.

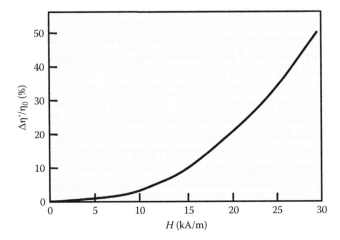

FIGURE 2.10 Magnetic field dependence of the viscosity $\eta'(H)$ of a ferrofluid with magnetite particles normalized to its value η_0 at zero field.

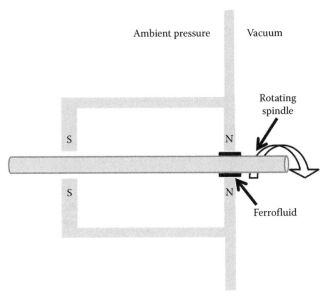

FIGURE 2.11 Sketch of a vacuum sealing of a rotational feedthrough realized by the use of a ferrofluid as sealing agent. The ferrofluid separates the vacuum from the ambient atmosphere and is held in place by a magnetic field gradient.

Although simple theoretical calculations show that the change of the viscosity due to the interaction of the particle's magnetization with an external field gradient should be only of the order of 1%, up to 100% changes are measured at a magnetic field of the order of 100 kA/m. The explanation of this effect is still not complete because the magnetization of the particles is usually not highly anisotropic, so that it is not fixed with respect to the shape of the particles. This, however, would be a prerequisite for the increase of the viscosity. Several authors [Odenbach 2002] assume the formation of chains of particles caused by their dipolar interaction. These chains could largely hinder the fluid flow and thus enhance the viscosity.

The applications of such ferrofluids are mainly in the field of suspensions [Raj 1980], sealings, and heat sinks. The possibility to keep the ferrofluid in place by just applying a magnetic field enables, for example, sealings for rotational feedthroughs that are able to withstand a pressure difference of up to 1 bar (see Figure 2.11). Other related applications are in loudspeakers to remove the heat created by the oscillating coil, especially in high end systems.

2.3.3.3 Biotechnological and Medical Applications

The availability of the wide range of materials, sizes, and shapes led to speculation from the 1960s onward that magnetic nanoparticles may have applications in biology and medicine [Hütten 2004, Reiss 2005]. The possibility that a particle can be manipulated by a magnetic field gradient as already discussed in the foregoing section leads to the additional vision of targeting specific locations in, for example, a microfluidic system [Hung 2007] or even in living bodies. Within the years from 2000 to 2009, a rapidly increasing number of publications and conferences were dedicated to these fascinating possibilities.

2.3.3.3.1 Hyperthermia and Drug Delivery

If magnetic particles are irradiated with an oscillating magnetic field, they absorb energy from the electromagnetic wave and heat up. The temperature enhancement that occurs in a magnetic nanoparticle system under this influence has found applications in, for example, hardening of adhesives, thermosensitive polymers, as well as in biomedicine.

In the latter case, hyperthermia as therapeutical part of a cancer therapy [Hergt 2006], drug targeting via thermosensitive magnetic nanoparticles, and the application of catheters, which are magnetically controllable, are important.

In all cases, the temperature enhancement needed for a special application should be obtained with the smallest possible amount of magnetic nanoparticles. Therefore, their specific loss power measured in watts per gram of magnetic material must be as high as possible. This is particularly important for applications where the target concentration is very low as in the case of antibody targeting of tumors.

The absorption of energy from the electromagnetic wave is due to several processes like hysteretic losses, relaxation processes, and viscous losses. In general, the absorption increases both with the frequency f and with the amplitude H_0 of the applied oscillating magnetic field. In Figure 2.12, the heat absorption per cycle is shown as a function of the field amplitude for dextran-coated magnetite particles in an aqueous fluid for different particle core diameters.

In metallic nanoparticles, values in the range of 700 W/g have been found, i.e., larger than all data reported above for magnetic iron oxides with exception of reports for the bacterial magnetosomes. One can expect future values beyond 1 kW/g. For application of such metallic nanoparticles in hyperthermia, however, the problem of stable aqueous suspensions of metallic particles will have to be solved. One example for a possible route to achieve solubility in water is sketched in Figure 2.13.

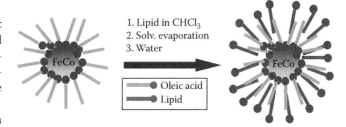

FIGURE 2.13 Principle of obtaining aqueous solutions of magnetic nanoparticles with originally nonpolar organic shell: the addition of lipids with polar end groups enhances the stability in water. The polar ends are pointing outward toward the solvent.

The addition of lipids with polar end groups could largely enhance the ability of the particles with this outer shell to be stable in water and thus enable in vivo applications like hyperthermia.

Drug delivery and effective application of drugs within the body is another heavily investigated application for magnetic nanoparticles. Researchers are particularly investigating ferromagnetic nanoparticles with respect to the treatment of various cancers by directly delivering the drugs to the region of the carcinoma. The goal is to have medicines functionally bound to magnetic nanoparticles and utilize a magnetic gradient to guide the nanoparticle to the affected region. Once at the affected region, the heating of the particles by the electromagnetic wave is used for either cracking the bond between particle and drug or for an activation and completion of a chemical reaction between the applied medicine and the adversely affected area. Current research is attempting to develop magnetic nanoparticles with large magnetic moments and resistance to physical breakdown within the body. In Figure 2.14, we show as an example the magnetophoretic mobility of metallic nanoparticles as a function of the particle size for different materials. Here, the

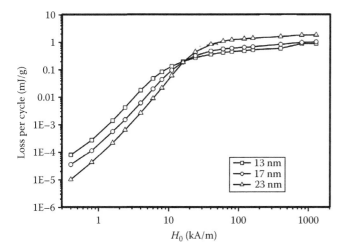

FIGURE 2.12 Energy loss absorbed by dextran-coated magnetite nanoparticles of different diameter per cycle of the magnetic field as a function of the field amplitude.

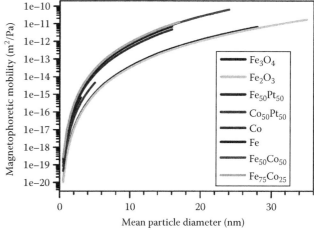

FIGURE 2.14 The calculated magnetophoretic mobility of different superparamagnetic nanoparticles at room temperature in *o*-dichlorobenzene.

magnetophoretic mobility reflects the capability of the particles to follow a magnetic field gradient.

Clearly, high moment materials such as Fe–Co alloys have the highest mobility due to the scaling of the force with the volume of the particles. Thus, they have a large potential for being successful within application needing a manipulation of magnetic nanoparticles.

2.3.3.3.2 Biochips Using Immunoassays

The last example of applications is also related with linking biological molecules to magnetic nanoparticles. The advantages of using magnetic beads as labels in bioassays in vitro have been well established. Magnetic particles—called magnetic beads for this type of applications—are not affected by reagent chemistry or photobleaching and are therefore stable over time. In addition, the magnetic background in a biomolecular sample is usually insignificant, and magnetic beads can be manipulated remotely by magnetism. This is in contrast to the manipulation of particles by, for example, laser tweezers, where the dielectric response to an electromagnetic wave is used. In this case, the background is much larger because almost all materials exhibit a dielectric polarizability.

For these applications, the attachment of selected molecules to the magnetic particle is necessary. To achieve selectivity of the link between particle and biomolecule, large efforts have been made in the recent years on the functionalization of the particle's outer shell [Li 2008]. The principle of functionalization [Berry 2003] is shown in Figure 2.15.

The particles produced by, for example, thermal precursor decomposition have an outer organic shell consisting of a surfactant like oleic acid. Next, a molecule binding to the surfactant is introduced that carries an end group capable of specifically binding other molecules. Frequently used molecules are biotin, which binds with high specificity to avidin. Then, a biomolecule functionalized with the corresponding binding partner will be specifically linked with the magnetic particle and thus can be manipulated as well as detected.

This detection of magnetic nanoparticles, however, needs additional devices such as giant magnetoresistance or tunneling magnetoresistance sensors [Baselt 1998, Reiss2 2005]. Such sensors are already used in, for example, read heads of hard

disk drives and therefore readily available. Figure 2.16a shows as an example a tunneling magnetoresistance sensor covered with magnetic particles (0.8 μm diameter); Figure 2.16b displays the change of the resistance of such sensors as a function of an applied magnetic field for different amounts of coverage of the sensor's surfaces by magnetic particles.

These results demonstrate that it is nowadays possible to not only manipulate and functionalize magnetic particles for biomedical purposes but also to detect them quantitatively by appropriate magnetic sensor devices. Therefore, also the presence of biomolecules linked specifically to these particles can be evaluated.

Research and development concentrates in the moment to fully integrate sensing and manipulation devices into microfluidic environments, opening the way to, for example, handheld diagnostic labs on a chip for detection of antibodies, DNA fragments, or other biomolecules.

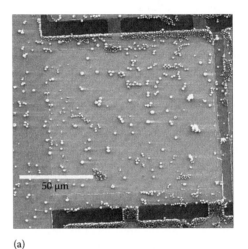

(a)

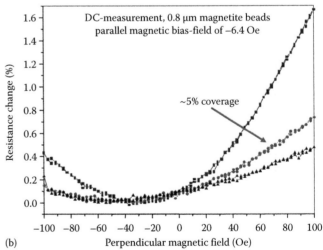

(b)

FIGURE 2.16 (a) An SEM image of the surface of a tunneling magnetoresistance sensor covered with 0.8 μm diameter magnetic particles. (b) The resistance change of a TMR sensor as a function of an applied magnetic field for differently dense surface coverage of the sensor by magnetic particles.

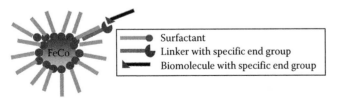

FIGURE 2.15 Principle of biofunctionalization of magnetic nanoparticles: the particles coated with an organic surfactant are functionalized by a molecule binding to the surfactant. The outer end group of this linker molecule supplies specific binding to corresponding end groups attached to biomolecules.

2.4 Critical Discussion

Although the use of magnetic nanoparticles in data storage and for in vitro procedures offers bright perspectives, their level of toxicity is still not completely evaluated. In 2004, tests found extensive brain damage to fish exposed to fullerenes for a period of just 48 h at a relatively moderate dose of 0.5 ppm.

Earlier studies in 2002 indicated nanoparticles accumulated in the bodies of lab animals, and still other studies showed that nanoparticles travel freely through soil and could be absorbed by animals living there. This is a potential link up of the food chain to humans and presents one of the possible dangers of nanotechnology.

Other nanoparticles have also been shown to have adverse effects. Cadmium selenide nanoparticles, also called quantum dots, can cause cadmium poisoning in humans.

Complicating the dangers of nanotechnology, size, and shape of nanoparticles could also affect the level of toxicity, preempting the ease of uniform categories even when considering a single element. In general, experts report that smaller particles are more bioactive and toxic. Their ability to interact with other living systems increases because they can easily cross the skin, lung, and, in some cases, the blood–brain barrier. Once inside the body, there may be further biochemical reactions like the creation of free radicals that damage cells.

There is no doubt that nanoparticles have interesting and useful properties; applications for in vivo investigations or treatments, however, still need the level of long-term toxicity of magnetic nanoparticles to be investigated carefully. Also releasing nanoparticles in the environment must be considered to be unsafe in the moment.

Nevertheless, the large benefits of, for example, ultradense data storage or a cancer therapy restricted to the area of the tumor justify both the careful use of magnetic nanoparticles as well as the intensive efforts for increased safety in their further commercialization.

2.5 Summary

Magnetic nanoparticles can be now synthesized using a variety of methods with atomic precision for gas phase separation of very small particles and standard deviations of the radius of only a few percentage for larger particles produced by, for example, thermal precursor decomposition. Mainly the chemical methods also allow for a cost-effective high-volume production that is necessary to realize the applications of the particles in various fields.

While the principles to understand the physical properties of such small magnetic particles are well developed, the interpretation of, for example, the magnetism in nanometer-sized objects of only a few atoms is not yet completed. This is due to their role as an object being intermediate between atomic and bulk like properties. Nevertheless, strong and nonmonotonous variations of the magnetic moment per atom due to quantum size effects are observed when particles of different sizes are investigated.

For larger particles, the border between the ferromagnetic (thermally blocked) and the superparamagnetic state, where thermal fluctuations of the particle's magnetic moment are faster than the observation time, is crossed. Typically, the particles used for applications are close to this border. In both cases, however, the unique ability to manipulate the particles by magnetic field gradients very selectively within fluidic systems provides outstanding possibilities for applications.

2.6 Future Perspectives

Depending on the physical properties of the magnetic nanoparticles, a wide variety of applications is either already realized or being developed. While ferrofluids are commercialized in sealing systems or as heat-conducting media in high-end multimedia devices, other applications still need intensive research and development for improving the particle properties.

For data storage in particles, the stability of the magnetic moment given by the shape and the crystalline magnetic anisotropy needs to be developed in order to obtain ultimate data storage density. Here, systems consisting of 3d ferromagnets and nonmagnetic 3d TMs such as FePt offer perspectives for data retention of 10 years. Prerequisite for a large anisotropy in such particles is the degree of ordering of the constituents in the crystal structure (L_{10} in this case), which is size dependent [Miyazaki 2005] and can be improved by, for example, annealing procedures.

For biophysical purposes, reasonable surface protections accompanied with functionalization of the organic shell of the particles are necessary. Moreover, the magnetic cores should have a magnetic moment as large as possible, because the force acting on the particles from a magnetic gradient field scales with this property. Therefore, $Fe_{50}Co_{50}$ nanoparticles [Hütten 2005] are superior to all other systems known from the magnetophoretic mobility point of view. Here, however, the anisotropy should be small in order to avoid agglomeration of the particles in fluids. Again, the degree of crystalline order is a key to obtain this desired property. Another issue for the future perspectives of magnetic nanoparticles is the realization of an optical control by making the particles fluorescent [Corr 2008].

In general, research and development on magnetic nanoparticles has created applications that are already in use. The further development will concentrate on the preparation of particles with specifically tailored properties such as high or low anisotropy or high magnetic moment to fulfill the requirements of applications in data storage and biotechnology. Because both fields offer a huge market volume, possible threats created by magnetic nanoparticles to the health of people handling preparation and use of these systems urgently need to be investigated. Similar to the tailoring of the properties specific for different applications, however, it should be possible to create magnetic cores and coatings of, for example, Au- [Babincova 2000, Kouassi 2006, Zelenáková 2008] or carbon-based outer shells that are harmless to the environment.

Acknowledgments

The authors are indebted to D. Sudfeld, H. Brückl, J. Schotter, I. Ennen, A. Weddemann, and P. Jutzi for many fruitful discussions and contributions. The work was in part supported by the Deutsche Forschungsgemeinschaft (SFB 613) and the German Ministry for Research and Education (BMBF).

References

[Babincova 2000] Babincova M., Leszczynska D., Sourivong P., Babinec P. 2000. Selective treatment of neoplastic cells using ferritin-mediated electromagnetic hyperthermia, *Med. Hypotheses* 54: 177–179.

[Bansmann 2005] Bansmann J., Baker S.H., Binns C. et al. 2005. Magnetic and structural properties of isolated and assembled clusters, *Surf. Sci. Rep.* 56: 189–275.

[Baselt 1998] Baselt D.R., Lee G.U., Natesan M. et al. 1998. A biosensor based on magnetoresistance technology, *Biosens. Bioelectron.* 13: 731–739.

[Berry 2003] Berry C.C., Curtis A.S.G. 2003. Functionalisation of magnetic nanoparticles for applications in biomedicine, *J. Phys. D: Appl. Phys.* 36: R198–R206.

[Billas 1994] Billas I.M.L., Châtelain A., de Heer W.A. 1994. Magnetism from the atom to the bulk in iron, cobalt, and nickel clusters, *Science* 265: 1682.

[Bloomfield 1993] Cox A.J., Louderback J.G., Bloomfield L.A. 1993. Experimental observation of magnetism in rhodium clusters, *Phys. Rev. Lett.* 71: 923–926.

[Chadwick 2008] Chadwick S.J.F., Virden A.E., Haehnel V. et al. 2008. Development of metal particle (MP) technology for flexible recording media, *J. Phys. D: Appl. Phys.* 41: 134018–134026.

[Coperet 2005] Coperet C., Chaudret B. 2005. *Surface and Interfacial Organometallic Chemistry and Catalysis*, Springer, Berlin, Germany.

[Corr 2008] Corr S.A., Rakovich Y.P., Gun'ko Y.K. 2008. Multifunctional magnetic-fluorescent nanocomposites for biomedical applications, *Nanoscale Res. Lett.* 3(3): 87–104.

[Dumestre 2004] Dumestre F., Chaudret B., Amiens C. et al. 2004. Superlattices of iron nanocubes synthesized from Fe[N(SiMe$_3$)$_2$], *Science* 303: 821.

[Gleiter 1989] Gleiter H. 1989. Nanocrystalline materials, *Prog. Mater. Sci.* 33: 223–315.

[Gruner 2006] Gruner M.E., Rollmann G., Sahoo S. et al. 2006. Magnetism of close packed Fe$_{147}$ clusters, *Phase Transit.* 79: 701.

[Hergt 2006] Hergt R., Dutz S., Müller R., Zeisberger M. 2006. Magnetic particle hyperthermia: Nanoparticles magnetism and materials development for cancer therapy, *J. Phys.: Condens. Matter* 18: S2919–S2934.

[Hung 2007] Hung L.-H., Lee A.P. 2007. Microfluidic devices for the synthesis of nanoparticles and biomaterials, *J. Med. Biol. Eng.* 27(1): 1–6.

[Hütten 2004] Hütten A., Sudfeld D., Ennen I. et al. 2004. New magnetic nanoparticles for biotechnology, *J. Biotechnol.* 112: 47.

[Hütten 2005] Hütten A., Sudfeld D., Ennen I. et al. 2005. Ferromagnetic FeCo nanoparticles for biotechnology, *J. Magn. Magn. Mater.* 293: 93.

[Kim 2001] Kim D.K., Zhang Y., Voit W. et al. 2001. Synthesis and characterization of surfactant-coated superparamagnetic monodispersed iron oxide nanoparticles, *J. Magn. Magn. Mater.* 225: 30–36.

[Knickelbein 2002] Knickelbein M.B. 2002. Adsorbate-induced enhancement of the magnetic moments or iron clusters, *Chem. Phys. Lett.* 353: 221–225.

[Kohn 2003] Kohn W. 2003. *Electronic Structure of Matter— Wave Functions and Density Functional*, Nobel Lectures, Chemistry 1996–2000, I. Grenthe, (Ed.), p. 213. World Scientific, Singapore.

[Kouassi 2006] Kouassi G.K., Irudayaraj J. 2006. Magnetic and gold-coated magnetic nanoparticles as a DNA sensor, *Anal. Chem.* 78: 3234–3241.

[Kruis 1998] Kruis F.E., Fissan H., Peledt A. 1998. Synthesis of nanoparticles in the gas phase for electronic, optical and magnetic applications: A review, *J. Aerosol Sci.* 29: 5-65-6, 511–535.

[Lau 2002] Lau J.T., Föhlisch A., Martins M. et al. 2002. Spin and orbital magnetic moments of deposited small iron clusters studied by x-ray magnetic circular dichroism spectroscopy, *New J. Phys.* 4: 98.1–98.12.

[Lee 1985] Lee K., Callaway J., Kwong K. et al. 1985. Electronic structure of small clusters of nickel and iron, *Phys. Rev. B* 31: 1796–1803.

[Li 2008] Li Z., Tan B., Allix M. et al. 2008. Direct coprecipitation route to monodisperse dual-functionalized magnetic iron oxide nanocrystals without size selection, *Small* 4(2): 231–239.

[Masala 2004] Masala O., Seshadri R. 2004. Synthesis routes for large volumes of nanoparticles, *Annu. Rev. Mater. Res.* 34: 41–81.

[Miyazaki 2005] Miyazaki T., Kitakami O., Okamoto S. et al. 2005. Size effect on the ordering of L$_{10}$ FePt nanoparticles, *Phys. Rev. B* 72: 144419.

[Mørup 2007] Mørup S., Hansen M.F. 2007. Superparamagnetic particles, *Handbook of Magnetism and Advanced Magnetic Materials*, John Wiley & Sons, Chichester, U.K.

[Murray 2000] Murray C.B., Kagan C.R., Bawendi M.G. 2000. Synthesis and characterization of monodisperse nanocrystals and close-packed nanocrystal assemblies, *Annu. Rev. Mater. Sci.* 30: 545–610.

[Odenbach 2002] Odenbach S. 2002. *Ferrofluids: Magnetically Controllable Fluids and Their Applications*, Springer, Berlin, Germany.

[OOMMF 1999] Donahue M.J., Porter D.G. 1999. OOMMF User's Guide, Version 1.0, Interagency Report NISTIR 6376, National Institute of Standards and Technology, Gaithersburg, MD.

[Pellarin 1994] Pellarin M., Baguenard B., Vialle J.L. et al. 1994. Evidence for icosahedral atomic shell structure in nickel and cobalt clusters—Comparison with iron clusters, *Chem. Phys. Lett.* 217: 349.

[Petit 1998] Petit C., Taleb A., Pileni M.P. 1998. Self-organization of magnetic nanosized cobalt particles, *Advanced Materials*, 10, 259–261.

[Pileni 1989] Pileni M.P. 1989. *Structure and Reactivity in Reverse Micelles*, Elsevier, Amsterdam, the Netherlands.

[Pileni 1993] Pileni M.P. 1993. Reverse micelles as microreactors, *J. Phys. Chem.* 97(27): 6961–6973.

[Puntes 2001] Puntes V.F., Krishnan K.M., Alivisatos P. 2001. Synthesis, self-assembly, and magnetic behavior of a two-dimensional superlattice of single-crystal epsilon-Co nanoparticles, *Appl. Phys. Lett.* 78: 2187–2189.

[Puntes 2002] Puntes V.F., Zanchet D., Erdonmez C.K. et al. 2002. Synthesis of hcp-Co nanodisks, *J. Am. Chem. Soc.* 124: 12874–12880.

[Raj 1980] Raj K., Moskowitz R. 1980. A review of damping applications of ferrofluids, *IEEE Trans. Magn.* 16(2): 358–363.

[Reiss 2005] Reiss G., Hütten A. 2005. Magnetic nanoparticles: Applications beyond data storage, *Nat. Mater. News Views* 4: 725–726.

[Reiss2 2005] Reiss G., Brückl H., Hütten A. et al. 2005. Magnetoresistive sensors and magnetic nanoparticles for biotechnology, *J. Mater. Res.* 20: 3294.

[Rellinghaus 2006] Rellinghaus B., Mohn E., Schultz L., Gemming T., Acet M., Kowalik A., Kock B.F. 2006. On the L10 ordering kinetics in Fe-Pt nanoparticles, *IEEE Trans. Magn.* 42(10): 3048–3050.

[Roco 1999] Roco M.C., Williams R.S., Alivisatos P. 1999. Nanotechnology research directions: IWGN workshop report. *Vision for Nanotechnology R&D in the Next Decade*, National Science and Technology Council (U.S.). Committee on Technology, Interagency Working Group on Nanoscience, Engineering, and Technology. Springer, Berlin, Germany.

[Ross 2001] Ross C.A. 2001. Patterned magnetic recording media, *Annu. Rev. Mater. Res.* 31: 203–235.

[Shi 1996] Shi J., Gider S., Babcock K. et al. 1996. Magnetic clusters in molecular beams, metals and semiconductors, *Science* 271: 937–941.

[Stahl 2003] Stahl B., Ellrich J., Theissmann R. et al. 2003. Electronic properties of 4-nm FePt particles, *Phys. Rev. B* 67: 014422.

[Sun 2006] Sun S. 2006. Recent advances in chemical synthesis, self-assembly, and applications of FePt nanoparticles, *Adv. Mater.* 18: 393.

[van Blaaderen 1992] van Blaaderen A., Vrij A. 1992. Synthesis and characterization of colloidal dispersions of fluorescent, monodisperse silica spheres, *Langmuir* 8: 2921–2931.

[Weller 1999] Weller D., Moser A. 1999. Thermal effect limits in ultrahigh-density magnetic recording, *IEEE Trans. Magn.* 35: 4423.

[Zelenáková 2008] Zelenáková A., Zeleniák V., Degmová J. et al. 2008. The iron-gold magnetic nanoparticles: Preparation, characterization and magnetic properties, *Rev. Adv. Mater. Sci.* 18: 501–504.

3

Ferroelectric Nanoparticles

Julia M. Wesselinowa
University of Sofia

Thomas Michael
Martin-Luther-University

Steffen Trimper
Martin-Luther-University

3.1 Introduction

From their discovery, ferroelectrics were more of academic interest, of little application and theoretical relevance. The recognition of the relationship between lattice dynamics and ferroelectricity as well as the modeling of ferroelectric phase transitions has intensified the investigations of ferroelectrics. The focus changed further, when thin-film ferroelectrics were developed and applied in different devices in 1980s. Since that time, there has been a renewed effort in the fabrication, application, and theoretical understanding of ferroelectric materials scaled down up to nanometers. This chapter reviews the physical behavior of such ferroelectric nanoparticles.

3.1.1 Ferroelectric Properties

The main properties of ferroelectrics in bulk material (Blinc and Zeks 1974, Lines and Glass 2004, Strukov and Levanyuk 1998) are summarized in this section. The appearance of multistable degenerated states with spontaneous macroscopic polarization $P = \sigma_s$ below a critical temperature T_c, which can be switched by an electric field, is the general feature of ferroelectricity. The system is paraelectric above the phase transition temperature. The system can undergo a first- or a second-order phase transition. In the first case, the polarization, as the order parameter of the system, exhibits a discontinuous change from the paraelectric to the ferroelectric phase. A second-order transition is characterized by a continuous change of the polarization. Most ferroelectric materials reveal a first-order transition near to a second-order one which is characterized by a small jump in the polarization

as well as a drastic increase of the corresponding dielectric susceptibility $\varepsilon(T)$. The transition is often masked by intrinsic fields, depolarization effects, and defects. This chapter is not focused on the behavior in the immediate vicinity of the phase transition. The discussion of the critical fluctuations, relevant near to a second-order transition, is beyond the scope of this chapter. The access to the polarized states by the application and variation of an external electric field E is a further important feature of ferroelectrics. In particular, the intrinsic polarization is reversible through the application of a field E. Ferroelectrics are polar substances of either solid (crystalline or polymeric) or liquid crystals. The coercive field denotes the critical electric field, which switches the polarization. The electric displacement as a function of the applied field E reveals a hysteresis curve. The occurrence of the spontaneous polarization is related to lattice distortions in case of ferroelectric materials with a crystal structure. Hence, ferroelectric transitions belong to the wide class of structural phase transitions. These are usually divided into two subclasses: displacive and order–disorder ones (Strukov and Levanyuk 1998). Clearly, the labels refer to limiting cases, but the division is still convenient. This classification is based more on a microscopic picture than on the previous macroscopic characterization. The order parameter dynamics of the displacive ferroelectrics are assigned to a phonon-dominated process. This is related to the shift of some atoms or atomic groups within an elementary cell of the corresponding material. A ferroelectric prototype is barium titanate (BTO) with the chemical formula $BaTiO_3$. Ions are mutually shifted below the phase transition temperature. As a result, the centers of the positive and negative charges are separated and give rise to electric dipole moments. Its average is related to the macroscopic polarization. Whereas the order parameter dynamics of displacive ferroelectrics are phonon-like, the order–disorder ferroelectrics exhibit a relaxation dynamics. The prototype of that class is hydrogen-bonded potassium dihydrogen phosphate (KDP) KH_2PO_4. The protons can adopt two positions within a double-well potential, which is created by the other ions. Protons are distributed uniformly above the phase transition temperature. On the other hand, the protons favor a certain well in the low-temperature phase. The averaged number of protons within that well is assumed to be a measure of the spontaneous polarization. Generalizing the model, one can think of whole groups of atoms or molecules that offer a flip-dynamics between two or more equilibrium positions. Lattice distortions should be included in a more refined approach within this class. Both limiting cases are characterized by the occurrence of a soft-mode behavior (Blinc and Zeks 1974, Lines and Glass 2004, Strukov and Levanyuk 1998) from a microscopical point of view. A low-lying elementary excitation energy $\omega(\vec{q},T)$ exists and depends on the wave vector $\vec{q}$ and the temperature T. This mode becomes soft at a special wave vector $\vec{q}_c$ when the temperature is approaching to the critical one:

$$\lim_{T \to T_c} \omega(\vec{q}_c,T) = 0. \qquad (3.1)$$

The critical wave vector for the ferrodistorsive (including ferroelectric) phase transitions is located in the center of the Brillouin zone $\vec{q}_c = 0$. Antiferrodistorsive systems exhibit a critical wave vector at the boundary of the Brillouin zone $\vec{q}_c = \pi/a$ (see Section 3.4 for further details). Most ferroelectric families are not oxides, though these are studied mostly because of their robustness and practical applications. The key principle to the operation of devices, such as nonvolatile ferroelectric random access memories (FRAMs) (Evans and Womack 1988), is the response of the ferroelectric materials to an electric field.

3.1.2 Ferroelectric Nanomaterial

Since ferroelectrics in lower dimensions promise a drastic increase of the storage density of RAM, nanoscale ferroelectrics have attracted extensive attention. The anticipated benefit depends on whether the phase transition and the polarized low-temperature state (or multistate) still exist when the system is scaled down up to less than 100 nm. The challenge in low-dimensional finite ferroelectric structures concerns the synthesis, the experimental characterization of their size-dependent properties, and the theoretical description. Nanostructures are observed in a wide variety of realizations such as nanoparticles, nanorods, nanowires, nanocubes, and nanotubes. Generally, the size of nanoscale material is assumed to be less than 100 nm. A notable number of review articles are addressed to ferroelectric nanostructures (see, e.g., Ahn et al. 2004, Hu et al. 1999, Patzke et al. 2002, Rao and Nath 2003, Scott 2006, Xia et al. 2003). Ferroelectric nanoparticles of different shapes (spherical, nonspherical, cylindrical, and ellipsoidal) and their nanocomposites are actively studied in modern physics and material science. Size effects and

the possible disappearance of ferroelectricity at a critical particle volume have initiated the growing scientific interest and are applicable in many fields of nanotechnology (Spaldin 2004). The challenge of developing nanoscaled devices for a diversity of applications is inseparably linked with the ability to synthesize and characterize these nanostructures in order to exploit their optical, electronic, thermal, and mechanical properties. Comparatively, very less effort has been spent on the fabrication of technologically important ternary perovskite transition metal nanostructures (see, e.g., Urban et al. 2003).

Perovskite structures, including $BaTiO_3$, $SrTiO_3$, $BaZrO_3$, and $SrZrO_3$, and their complexes, such as $Ba_xSr_{1-x}TiO_3$, $Ca_xSr_{1-x}TiO_3$, and $BaTi_xZr_{1-x}O_3$, are noteworthy for their advantageous dielectric, piezoelectric, electrostrictive, pyroelectric, and electro-optic properties. Corresponding applications in the electronics industry are electromechanical devices, pyroelectric detectors, imaging devices, optical memories, modulators, deflectors, transducers, actuators, capacitors, dynamic RAM, field effect transistors, logic circuitry, and high-k dielectric constant materials. Such properties and applications for perovskite oxides are described in literature (e.g., in Dawber et al. 2005, Hill 2000, Millis 1998, Scott 2008). Advanced applications for high-k dielectric and ferroelectric materials in the electronic industry necessitate the understanding of the underlying physics in a reduced dimensionality up to the nanoscale.

Lead zirconate titanate (PZT) has been extensively used in electronic devices such as nonvolatile FRAMs and as promising candidate for sensors, transducers, and capacitors (Ramesh et al. 2001, Schafer et al. 1997) due to its ferroelectric properties. The crucial dependence of the properties of ferroelectric materials on the particle size is one of the main problems using ferroelectric nanoparticles in the development of nanometer-sized electronic devices, as mentioned above. In view of this, the fabrication of PZT nanoparticles in a free-standing form is fundamental in order to determine the finite size effect on their ferroelectric properties. One of the most important dielectric materials is BTO. It is the basic substance for electronic devices like MLCC (multilayer ceramic capacitor). In terms of the miniaturization of devices, the downsizing of MLCC has been developed and upgraded permanently. As a result, the thickness of the BTO layers in MLCC is expected to become thinner up to a value below 0.5 μm. A further downsizing from a few hundred to a few tens of nanometers is required to reach a higher performance. Consequently, the particle size of the corresponding BTO raw materials will decrease to about a few tens of nanometers. However, the continual scaling down of ferroelectric fine particles is confronted with the reduction of ferroelectricity with decreasing particle size. The final disappearance of ferroelectricity below a certain critical size is known as the "size effect" (Fridkin 2006). This phenomenon found in materials such as BTO, $SrTiO_3$ (STO), and PZT is of high interest in industry as well as in basic research. However, the estimation of the critical size is not unambiguous. The critical size of BTO nanoparticles has been reported in a wide range between 10 and 110 nm. The spreading is originated to the different measurement techniques (Ishikawa and Uemori 1999, Uchino et al. 1989).

The critical size of 10–20 nm is observed by Ohno et al. (2004) and Wada et al. (2005a).

The reduction of the sintering temperature of BTO in another aspect would enable the substitution of expensive nobel metal electrodes by cheaper ones. Both these production requirements—the size effects and the sintering temperature—emphasize the establishment of novel low-temperature synthetic approaches. BTO has good dielectric and ferroelectric properties and is widely used in thermistors, MLCC, and electro-optic devices. Recent developments in microelectronic and communication technology involve the miniaturization of MLCC. A further miniaturization and advanced high dielectric constant ceramic particles of better quality require a uniform size (Venigalla 2001). High permittivities in combination with miniaturization can be obtained by controlling the microstructure. It is determined in a decisive manner by homogeneity, composition, surface area, and particle size of the primary powder material. The manufacture of reliable MLCC requires high-purity, agglomerate-free, highly crystalline, and superfine ceramic (Wilson 1995). The bulk properties of BTO ceramics have been widely investigated. The strong dependence of the electrical properties of nanoscale particles on the grain size and crystalline structure raised a renewed interest in BTO more recently. Tetragonal BTO is used in ferroelectrics, and cubic BTO is applied in capacitors. A better understanding of the nanostructure of BTO ultrafine particles in both phases is of interest as well as the correlation of properties with the particle size. Perovskite oxides, including BTO and STO, exhibit typical nonlinear optical coefficients and large dielectric constants, as reported by Song et al. (1996). These effects depend on the ratios of metallic elementals, the impurity concentration, the microstructure, and finite size effects. Therefore, a considerable effort to control synthesis of crystalline materials and thin films of these ferroelectric oxides was pursued (see Wang et al. 2001, Wills et al. 1992, Zhang et al. 1994, Zhao et al. 1997).

There is a permanent need for relatively simple and cost-effective manufacturing processes of perovskite nanostructures. In view of the drawbacks mentioned with the prior applied methods, the shape of the nanostructure has to be controlled in a reproducible manner.

3.2 Preparation of Ferroelectric Nanoparticles

The progress in studying and applying modern ferroelectrics is closely related to the preparation of such materials. Hence, one observes an increasing interest in preparing nanosized particles of metals, oxides, sulfides, etc. using microemulsions. Here, the precipitation of nanomaterial is carried out in aqueous cores dispersed in an apolar solvent and stabilized by surfactant or cosurfactant molecules, respectively. The extension of the reaction chamber may be controlled by a different amount of water in the aqueous cores. These cores are about 5–10 nm in size. The obtained material is homogeneous as the desired stoichiometry is maintained. Besides the adjustment of the particle size, the morphology of the produced nanoparticles is also controlled by a proper choice of the composition of the microemulsion system. Oxide powders, such as BTO, offer problems due to chemical inhomogeneities and varying reactivities if they are produced by high-temperature solid-state reactions. In addition, there exist a wide range of grain sizes, typically in between 0.5 and 3.0 μm. Otherwise, the control of the size, the shape, and the ability of agglomeration is limited. Thus, alternative routes are necessary for the synthesis of nanomaterials. They are based on novel low-temperature processes that provide high-purity ultrafine powders with a definite morphology and size of the particles. Various low-temperature routes involving organometallic precursors like alkoxides, acetates, oxalates, nitrates, and citrates of Ba and Ti have been used to obtain well-defined BTO. This section summarizes some experimental techniques to fabricate ferroelectric nanoparticles with a desired spectrum of properties.

3.2.1 Sol–Gel Method

The preferred procedure for the preparation of ferroelectric nanoparticles is the sol–gel method, which is based on low-temperature processes by using chemical precursors (Hench and West 1990). This method yields fine nanoparticles that exhibit high chemical reactivity, as well as a better purity, homogeneity, and physical properties as those manufactured by conventional high-temperature processes. The sol–gel method is a cost-effective and convenient route to prepare mono- and multicomponent glasses and ceramics, which would not be available by conventional methods. Reasons are the usage of homogeneous liquid solutions and the ability to form gels at room temperature. The term sol–gel goes back to the late eighteenth century. The sol–gel method provides a great variability of compositions, mostly oxides, in various forms, including powders, fibers, coatings, thin films, monoliths, composites, and porous membranes. Organic/inorganic hybrids can be likewise composed, in which a gel (usually silica) is impregnated with polymers or organic dyes to provide materials with specific properties. One of the most attractive features of the sol–gel process is the fabrication of composites that cannot be created with conventional methods. Another benefit of the methods is the maintenance of the final product with a fixed mixing level of the solution, often on the molecular scale.

Nanoparticles composed of BTO have been prepared by the sol–gel method (Kobayashi et al. 2004, Ohno et al. 2004, Viswanath and Ramasamy 1997). They were synthesized by the hydrolysis of complex alkoxide precursors that were prepared in a reflux of metallic barium and tetraethylorthotitanate in solvent. The hydrolysis was performed by the addition of water/ethanol solution to the precursor solution. The particle size, measured by transmission electron microscopy, became smaller as the reflux time was increased. This process is accomplished by a further sharpening of the size distribution. As water concentration and benzene content in the hydrolysis were increased, the particle size was enhanced with the crystallite size.

Tetragonal freestanding PZT nanoparticles consisting of titanium and zirconium alkoxides and lead acetate by using triethanolamine (or 2-methoxyethanol) as a polymerizing agent (Faheem and Joya 2007, Faheem and Shoaib 2006, Fernández-Osorio et al. 2007) were also obtained by the sol–gel technique. The metal ions may interact chemically with triethanolamine in the precursor solution and gel under refluxing conditions. Drying and aging treatments lead to the development of a precursor-polymeric gel network. A single-phase perovskite structure was formed at 470°C.

Nanocrystalline La-doped PZT materials, obtained by the sol–gel method as powders, exhibit some features that offer the increasing facilities of their application in electronic and optoelectronic devices such as segment displays, light shutters, coherent modulators, color filters, linear gate arrays, and image storages (Haertling 1999, Plonska et al. 2003). As a result, such materials have been widely investigated.

3.2.2 Two-Step Thermal Decomposition Method

BTO fine particles were prepared by using the two-step thermal decomposition method of barium titanyl oxalate (Hoshina et al. 2006, Takizawa et al. 2007, Wada et al. 2003, 2005a). At the second step within this method, the intermediate compound $Ba_2Ti_2O_5CO_3$ was decomposed into BTO and CO_2 under various degrees of vacuum pressure. As a result, the particle size of the prepared BTO nanoparticles is diminished under reduced pressure. Moreover, the dielectric constant of these BTO nanoparticles was measured by applying the powder dielectric measurement method using the slurry. The dielectric constant of BTO particles increases with decreasing pressure for the same particle size. Notice that the control of mesoscopic and nanoscopic nanoparticles by vacuum pressure is decisive for the dielectric properties of BTO nanoparticles. As reported by Wada et al. (2005a), the BTO nanoparticles prepared by the two-step thermal-decomposition method were free of defects and impurities.

3.2.3 Laser Ablative Technology

Another important method is the laser ablative technology (Seol et al. 2002) that is aimed to prepare monodisperse PZT nanoparticles of sizes in between 4 and 20 nm in diameter. Laser ablation of PZT ceramic targets in oxygen ambiance produces amorphous and irregularly shaped PZT nanoparticles. A subsequent online thermal treatment, performed on the PZT nanoparticles and dispersed in the gas phase, allows to fabricate compaction and crystallization of the nanoparticles without an additional particle growth. The amorphous nanoparticles began to crystallize above 600°C, and revealed a perovskite structure at 900°C. The crystallized nanoparticles can be classified with regards to its size by a differential mobility analyzer in order to get monodisperse, highly pure, and single-crystalline PZT nanoparticles. BTO nanoparticles were also prepared by laser ablation of a

Ba–Ti–O ceramic target using a differential mobility analyzer (Fujita et al. 2006). Using a complex method of producing ferroelectric metal oxide crystalline particles, Seol et al. (2002) and Fujita et al. (2006) have proposed an apparatus including a particle-producing device, a heat treatment device, and a particle-collecting device. The particle-producing device allows to fabricate nanoparticles of a ferroelectric metal oxide from a particle source placed in a vessel by a laser ablation method. The particle source is irradiated with a laser beam. Hereby, the nanoparticles are dispersed in an oxygen atmosphere (gas phase). The nanoparticles produced in the vessel and dispersed in a carrier gas are supplied through a connecting pipe into a vessel included in the heat treatment device. In this device, the nanoparticles are subjected to a heat treatment. The material dispersed in the oxygen gas atmosphere is heated up to predetermined temperatures within a fixed time, whereas the nanoparticles together with the carrier gas flow through the vessel. The heat-treated nanoparticles are supplied together with the carrier gas through a connecting pipe into a vessel included in the particle-collecting device. Here, the nanoparticles are concentrated on a plate by a collector.

3.2.4 Other Methods

This section mentions other relevant methods. Ishikawa et al. (1988) and Tsunekawa et al. (2000) have reported on the influence of size effect on the ferroelectric phase transition in $PbTiO_3$ (PTO) and BTO ultrafine particles, respectively. The samples were synthesized by an alkoxide method.

A wet chemical synthesis technique is applied by Qi et al. (2005) in order to find large-scaled barium strontium titanate $Ba_{1-x}Sr_xTiO_3$ (BST) nanoparticles near room temperature and under ambient pressure.

Well-ordered large-area arrays of ferroelectric La-substituted $Bi_2Ti_3O_{12}$ (BLT) nanostructures were prepared by pulsed-laser deposition using gold nanotube membranes as shadow masks by Lee et al. (2005b).

Another method for the preparation of BTO nanoparticles is the hydrothermal technique (Zhu et al. 2005). By applying this method, BTO nanoparticles were synthesized by combustion spray pyrolysis using a 1:1 molar ratio of oxidizer and fuel (Lee et al. 2004). To prepare the solution of precursors consisting of $Ba(NO_3)_2$, $TiO(NO_3)_2$, CH_6N_4O, and NH_4NO_3 with the molar ratio of 1:1:4:2.75, the substances were mixed in distilled water with 10% ethyl alcohol. A 0.01 M solution was ultrasonically sprayed into a quartz tube heated at 800°C. The concentration of droplets was decreased and large particles were removed by passing the droplets through a metal screen filter. The synthesized particles were well crystallized to tetragonal BTO. Nanosized BTO particles were prepared by citric acid-assisted spray pyrolysis by Lee et al. (2005a). Great differences were found in the structure and the morphology of BTO particles during the calcination, when the spray solution was controlled by an organic additive citric acid. Ferroelectric lead bismuth tantalate $(PbBi_2Ta_2O_9)$ nanoparticles were successfully synthesized using a colloid-emulsion process (Lu and Saha 2001). Monophasic $PbBi_2Ta_2O_9$ was obtained through calcining the precursor powder at 750°C for 2 h. The precursor powders are soft agglomerates with primary nanosized particles. A novel approach to prepare nanopowders of BTO by a solution reaction was established by Peng and Chen (2003). A solution including a titanate group was formed by using metatitanate, hydrogen peroxide, and ammonia as the reactants. By controlling the reaction conditions, one was able to get dispersed and uniform nanopowders of BTO from the solution. A series of titanates nanopowders such as nickel titanate, calcium titanate, and lead titanate can be prepared using this approach. A transparent and stable monodispersed suspension of nanocrystalline BTO was prepared by dispersing a piece of BTO gel into a mixed solvent of 2-methoxyethanol and acetylacetone (Li et al. 2004). The results of high-resolution transmission electron microscopy and size analyzer confirmed BTO nanoparticles in the suspension with an average size of 10 nm and a narrow size distribution. BTO nanoparticles have been synthesized though a chemical route using polyvinyl alcohol by Jana et al. (2004, 2005). As a result, tetragonal BTO ultrafine particles less than 50 nm in diameter could be produced by the gas evaporation method by Kodama et al. (2005). The conventional gas evaporation of a powder of mixed TiO_2 and $BaCO_3$ and the gas flush evaporation of BTO powders have been performed.

3.3 Experimental Results

The great progress in preparation methods of ferroelectric thin films and nanoparticles is accompanied with the ongoing miniaturization of devices based on these materials. Hence, the study of the size dependence of ferroelectric properties including the possible disappearance of ferroelectricity at a finite critical volume attracts a high scientific interest. First investigations on small particles date back to 1950s (Anliker et al. 1952, Jaccard et al. 1953, Kaenzig 1950). Nowadays, ferroelectric nanoparticles of different shapes are actively studied in nanophysics and nanotechnology. In this section, the main experimental results for different quantities as polarization, coercive field, hysteresis, dielectric constant, and others are reviewed. Significant differences are observed in comparison to bulk materials. Furthermore, surface and doping effects on phase transitions are more pronounced for nanoparticles. Especially, the physical behavior is strongly influenced by defect configurations. This yields desired properties and improvements for upcoming practical applications.

3.3.1 Polarization and Curie Temperature

The effects of the particle size on the physical properties of ferroelectric materials, especially of nanocomposites used in a variety of electronic devices, have been extensively investigated in experiments by x-ray diffraction (Chattopadhyay et al. 1995, Hoshina et al. 2006, Yu et al. 2003b), Raman scattering (Ishikawa et al. 1988), electron paramagnetic resonance, as well as nuclear magnetic resonance measurements (Erdem et al.

2005). The properties of ferroelectric materials vary considerably from substances to substances. Hence, a broad spectrum of investigations exists for different materials. The effect of the particle size on the crystal structure of BTO was studied in Frey et al. (1998), whereas Ishikawa et al. (1988) investigated the effect of the particle size on the Curie temperature in PTO nanoparticles. The size dependence of the dielectric properties of PZT was reported by Huang et al. (2001). Yu et al. (2003a) observed the shift of the ferroelectric phase transition in $SrBi_2Ta_2O_9$ (SBT) nanoparticles. Ohno et al. (2007) elucidated the size effects in lead zirconate titanate $Pb(Zr_{0.4}Ti_{0.6})O_3$ (PZT40) nanoparticles by x-ray diffraction.

The critical temperature T_c can decrease or increase when the particle size d is reduced. For example, Colla et al. (1997, 1999) obtained an anomalous large transition temperature in KDP nanoparticles that increases further with decreasing d (Figure 3.1). Embedded into the main opal pores, the transition temperature can be determined rather exactly as the maximum of the real part of the dielectric permittivity $\varepsilon'(T)$ versus the temperature. The shift of T_c is about 8 K compared with the single crystal. However, such a large shift in T_c could not be observed in other low-dimensional ferroelectric systems.

Generally, a lowering of the dimension is accompanied by an increase of fluctuations (Landau et al. 1980). Consequently, the phase transition temperature is expected to decrease monotonically for a smaller characteristic size of the nanoparticle. Such an effect was observed, for example, in nanoparticles of BTO (Ohno et al. 2004, Schlag and Eicke 1994, Schlag et al. 1995), PTO (Figure 3.2) (Chattopadhyay et al. 1995, Ishikawa et al. 1988), $LiTaO_3$ (Satapathy et al. 2007), BST (Zhang et al. 1999), and SBT (Yu et al. 2003a).

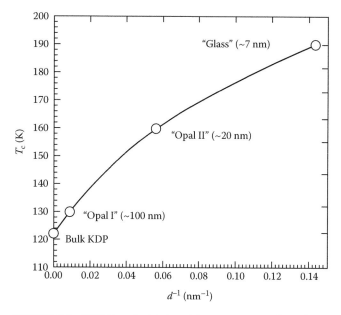

FIGURE 3.1 KDP ferroelectric transition temperature dependency on the particle size. (From Colla, E. et al., *Solid State Commun.*, 103(2), 127, 1997. With permission.)

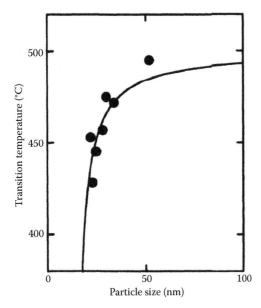

FIGURE 3.2 The transition temperature, at which the Raman line disappears, versus the particle size; observed values are denoted by full circles, the solid curve is obtained by an empirical expression $T_c = 500 - 588.5/(D - 12.6)(°C)$, where D is the particle diameter in nanometers. (From Ishikawa, K. et al., *Phys. Rev. B*, 37(10), 5852, 1988. With permission.)

Another behavior of the Curie point was obtained for $Bi_4Ti_3O_{12}$ nanoparticles (Jiang et al. 1998). The transition temperature decreases with increasing grain size when the grain size exceeds 25 nm (see Figure 3.3). In case the grain size is below 25 nm, T_c decreases instead of increasing with further decreasing grain size, that is, a maximum occurs in the grain-size dependence of

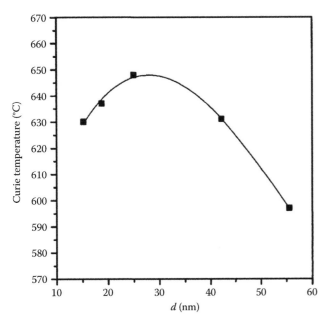

FIGURE 3.3 Dependence of the Curie temperature (T_c) on the grain size (d) for nanocrystalline $Bi_4Ti_3O_{12}$. (From Jiang, A. et al., *J. Appl. Phys.*, 83(9), 4878, 1998. With permission.)

$T_c(d)$. The results can be understood by considering the special crystal structure of bismuth titanate (BiTO). Meng et al. (1996) have exploited high-temperature Raman spectra of nanocrystalline $Bi_4Ti_3O_{12}$. The observed enhancement of the phase transition temperature for smaller grains is associated with the effect of charge transfer in the Bi–O–Ti system.

Owing to the rapid development of a nanostructure-based technology, the determination of the critical size in ferroelectric material is an essential problem that became crucial for applied research (Fridkin 2006). The critical size is defined as the maximal thickness of a film or the maximal size of a crystal, in which ferroelectricity as a collective effect disappears. Referring to this fact, particles with a size smaller as the critical one do not offer a ferroelectric hysteresis loop or a peak in the dielectric constant. The critical size is no universal quantity and varies for different substances. For example, SBT nanoparticles exhibit a critical size of 2.6 nm (Yu et al. 2003a), below which ferroelectricity disappears. Otherwise, in PZT40 nanoparticles, the critical size is about 35 nm (Ohno et al. 2007). Many other studies were addressed to reveal the existence of a critical particle size and the change of the macroscopic properties like the shift of T_c as function of the size (Anliker et al. 1954, Chattopadhyay et al. 1995, Du et al. 2004, Jaccard et al. 1953, Jana et al. 2005, Nagarajan et al. 2004, Wada et al. 2005b, Wang and Smith 1995, Wang et al. 1994a, Yu et al. 2003a, Zhong et al. 1994b). These experiments were performed by applying x-ray and electron diffraction, specific heat measurements, and Raman scattering on particles of various size (see Zhong et al. 1994b).

The grain-size decrease is connected to a reduction of the tetragonal axial ratio c/a. Moreover, the ferroelectric polarization decreases also for BTO nanocrystals as observed by Uchino et al. (1989) and Yashima et al. (2005).

The ferroelectric phase vanishes at a critical size of 48 nm (Zhang et al. 2001). Below 100 nm in PTO, the tetragonality c/a shows a strong dependence on the grain size. As the grain size is scaled down to a critical size of 7.0 nm, the ratio c/a is rapidly decreased to 1 and the ferroelectric tetragonal phase is transformed into a paraelectric cubic phase. The relationship between the tetragonality c/a and the grain size d in PTO or $PbZrO_3$ nanoparticles is in good agreement with an empirical formula given by Chattopadhyay et al. (1995, 1997)

$$\frac{c}{a} \approx 1 - \exp^{-\alpha d} \quad \alpha \simeq 1. \tag{3.2}$$

A similar relationship between the orthorhombic distortion a/b and the grain size d is found for BiTO and PTO nanoparticles (Jiang and Bursill 1999, Zhu et al. 2008)

$$\frac{a}{b} = \left(\frac{a}{b}\right)_\infty - \left[\left(\frac{a}{b}\right)_\infty - 1\right]\exp\left[-C(d - d_c)\right], \tag{3.3}$$

where

$(a/b)_\infty$ is the orthorhombic distortion of the single crystal
C is a constant
d_c is the critical grain size

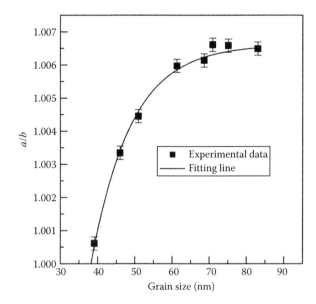

FIGURE 3.4 Orthorhombic distortion of BTO nanocrystals versus the grain size. Solid square stands for the experimental data and solid line is to Equation 3.2. (From Zhu, K. et al., *Solid State Commun.*, 145(9–10), 456, 2008. With permission.)

For $d = d_c$, it results $a/b = 1$, and the orthorhombic phase is transformed into a tetragonal phase. A critical grain size for the disappearance of ferroelectricity in BiTO is found to be $d_c = 38$ nm (Zhu et al. 2008) (compare Figure 3.4).

The multiple ion occupation of A and/or B sites in ABO_3 compounds is expected to offer a change of the Curie temperature and other physical quantities. This kind of substitution affects immediately the lattice parameters, the tetragonal distortion c/a, as well as the polarization and T_c. A direct evidence of A-site-deficient SBT and the enhancement of the ferroelectric quantities as T_c is discussed by Noguchi et al. (2001).

Otherwise, the substitution of La in PZT nanopowders and thin films lead to a marked decrease of T_c (Iijima et al. 1986, Plonska et al. 2003, Tyunina et al. 1998). The Curie temperature is lowered for higher Ba or Sr concentration in lead lanthanum zirconate titanate (PLZT) ceramics (Ramam and Lopez 2008, Ramam and Miguel 2006). The addition of Pt particles to a PZT matrix reduces the critical temperature (Duan et al. 2000). The occurrence of vacancies, dislocations, and defects in nanoparticles has a strong influence on the static and dynamic properties, for a study of the dielectric properties of Fe-ion-doped BTO nanoparticles (see Wang et al. 2000).

Otherwise, the macroscopic behavior is directly triggered by the microscopic quantities such as the elementary excitations and their damping. Thus, an evidence for the occurrence of a soft-mode behavior, see Equation 3.1, has been given in Wada et al. (2005b) and Zhong et al. (1994b). Using x-ray or Raman-scattering methods for PTO fine particles, a soft-mode behavior was detected, designated as $E(1TO)$. The mode is shifted toward a low-frequency region with decreasing temperature. The damping associated with the excitations in SBT (Wang et al. 1996), BTO (Wang et al. 1997), or SBT (Yu et al. 2003a) nanoparticles of various size increases

with decreasing particle sizes. Furthermore, the lattice vibration of smaller particles becomes softer compared with that of larger ones. This is a consequence of the reduction of the soft-mode frequency at an arbitrary temperature for smaller particles. Supplementary, this result implies a lowering of T_c with the shrinking of the particles. For a more theoretical consideration, see also Section 3.4.

3.3.2 Hysteresis

One important application as nonvolatile memories is known as FRAMs. The device is composed of ferroelectric capacitor materials. The processing issues involved in the high-density integration process are highly dependent on the ferroelectric and electrode-barrier materials. Hence, the selection of materials is a decisive factor in determining the performance of the device (Dawber et al. 2005). In view of fundamental ferroelectric properties, there are two potential ferroelectric materials for FRAM applications, namely, PZT and SBT (Evans and Womack 1988). They possess a high remanent polarization σ_r; low coercive electric field E_c, which characterizes the polarization reversal; and low dielectric loss. The use of ferroelectric thin films and small particles in high-density nonvolatile RAMs is based on the ability of ferroelectrics being positioned in two opposite polarization states by an external electric field (Auciello et al. 1998). An important question is whether this property, well established in bulk material, still exists in reduced dimensions. Therefore, it is of great interest to study the size dependence of E_c for small ferroelectric particles. This coercive electric field usually increases significantly with decreasing film thickness or particle size (Jeong et al. 2006, Nagarajan et al. 2004, Pertsev et al. 2003, Ren et al. 1996) (Figures 3.5 and 3.6).

The strength of the coercive field is related to the ease of domain nucleation and domain wall motion, whereas the permittivity is

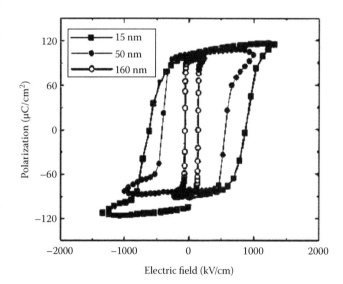

FIGURE 3.5 Ferroelectric measurements as a function of film thickness. Hysteresis loops for 15, 50, and 160 nm thick PZT films. The loops are sharp and well saturated down to 15 nm with $2P_r \propto 150°C$ μC/cm². (From Nagarajan, V. et al., *Appl. Phys. Lett.*, 84(25), 5225, 2004. With permission.)

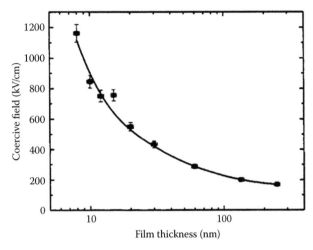

FIGURE 3.6 Coercive field of PZT 52/48 epitaxial films measured at 20 kHz and plotted versus the film thickness. (From Pertsev, N. et al., *Appl. Phys. Lett.*, 83(16), 3356, 2003. With permission.)

coupled to the density of domain walls and their mobility at low fields. A diversity of different explanations has been proposed in the past for this size effect. The surface pinning of domain walls plays an important role. Internal electric fields influence the domain nucleation in depleted films. Ferroelectric hysteresis at room temperature is measured in single-crystalline, monodisperse PZT nanoparticles of 9 nm in diameter by Seol et al. (2004).

The coercive field of a ferroelectric particle is stress sensitive. The increase of the internal compressive stress for thinner PZT films leads to the increase of the coercive field and the breakdown electrical strength (Lebedev and Akedo 2002). Moreover, the tensile stress gives rise to a decrease of σ_r and E_c. Chu et al. (2004) have reported the dislocation-induced polarization instability of (001)-oriented PZT nanoislands. These were grown on compressive perovskite substrates with an average height of ≈9 nm. Misfit strain is identified as one possible extrinsic origin for the polarization instability. Misfit dislocations in epitaxial PZT nanostructures involve strain fields. A negative vertical shift of the piezoelectric hysteresis loop of ferroelectric nanostructures has been described and discussed in terms of imprint due to interfacial effects (Alexe et al. 2001, Hesse and Alexe 2005, Ma and Hesse 2004).

In order to obtain high remanent polarization and low coercive field, there are many experiments with doped ferroelectric thin films and small particles. The materials are modified by adding oxide group softeners, hardeners, and stabilizers. Softeners (donors) reduce the coercive field strength and the elastic modulus and increase the permittivity, the dielectric constant, and the mechanical losses. Doping of hardeners (acceptors) gives higher conductivity, reduces the dielectric constant, and increases the mechanical quality factor (Desu and Payne 1990). The increase of Ba concentration in PLZT ceramics done by Ramam and Miguel (2006) and the substitution of La in PZT nanopowders and thin films lead to a marked decrease in σ_r and E_c (Iijima et al. 1986, Plonska et al. 2003, Tyunina et al. 1998). An enhancement of the dielectric constant and lower E_c were observed by the addition of PT particles to a PZT matrix (Duan et al. 2000).

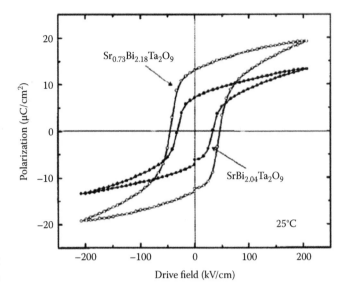

FIGURE 3.7 Polarization hysteresis loops measured at 25°C using dense ceramics. (From Noguchi, Y. et al., *Phys. Rev. B*, 63(21), 214102, 2001. With permission.)

Multiple ion occupation of A and/or B sites in ABO_3 compounds can affect the lattice parameters and the tetragonal distortion (c/a). As a consequence, a change of the hysteresis is expected, too. Direct evidence of A-site-deficient SBT and its enhanced ferroelectric characteristics is given by Noguchi et al. (2001) (see Figure 3.7).

3.3.3 Dielectric Constant

Many measurements have shown that the dielectric constant of ferroelectrics depends strongly on the grain size (Hoshina et al. 2007). The Curie temperature decreases and the Curie peak becomes lower and broader and eventually disappears with decreasing grain sizes in BST ceramics (Zhang et al. 1999) (see Figure 3.8).

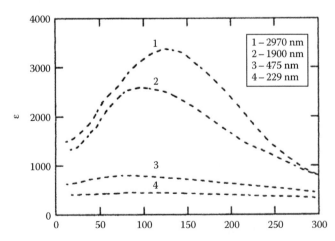

FIGURE 3.8 The temperature dependence of the dielectric constant of the $Ba_xSr_{1-x}TiO_3$ with different mean sizes. (From Zhang, L. et al., *J. Phys. D Appl. Phys.*, 32(5), 546, 1999. With permission.)

The dielectric properties, lattice constants, and microstructure of BTO ceramics with grain sizes of 0.3–100 μm have been reported by Arlt et al. (1985). The permittivity shows a pronounced maximum at grain size of 0.8–1 μm at temperatures below the Curie point. At grain sizes smaller than 0.7 μm, the permittivity decreases strongly. The crystal lattice changes gradually from a tetragonal to pseudocubic one. Similar dielectric measurements of BTO nanoparticles (Kim et al. 2005) show a broad peak below 100°C, which is possible due to the ferroelectric phase transition. The maximum of the dielectric constant at a temperatures T_m is lowered by 70 K (BTO) and by 130 K (SBT) (Higashijima et al. 1999, Kohiki et al. 2000). A lowering of T_m from the paraelectric–ferroelectric transition temperature T_c occurs compared with bulk material. The nanocrystals seem to reveal a single domain structure, and the system is in a superparaelectric state. However, there has been no report on a frequency dependence of T_m as an indication of the superparaelectric state for nanominiaturized ferroelectrics.

Low-power nonvolatile memory devices and low-field optical switching devices of Pb-free ferroelectrics (Ashkin et al. 1966, Miller and Nordland 1970, Tangonan et al. 1977) are desired. A promising candidate is LiTaO₃ (Gopalan and Gupta 1996), due to the high stability of the ferroelectric phase. The nanocrystals exhibit a high T_c and a large spontaneous polarization. The lowered T_m depends on the frequency. For nanosized LaTiO₃ ferroelectrics with insignificant cooperative effects between the particles, see Kohiki et al. (2003). The diameter is about ≈20 Å. The maximum temperature T_m in the real part of the dielectric function is apparently lower than the paraelectric–ferroelectric transition temperature of bulk LiTaO₃ for a fixed frequency of applied field. The maximum temperature of the imaginary part rose with increasing frequency. Since the bulk LiTaO₃-material shows no relaxor behavior, such superparaelectric behavior is obviously a consequence of the miniaturization of LiTaO₃ crystals and an insignificant cooperative interaction between the nanoparticles.

Regarding the size dependence of the ferroelectric transition in an ensemble of PTO nanoparticles produced by coprecipitation, see Chattopadhyay et al. (1995). Several methods like dielectric measurements, variable temperature x-ray diffraction, and differential scanning calorimetry are used to monitor the phase transition. The transition temperature T_c decreases gradually with a decrease in the size from 80 to 30 nm. The transition becomes increasingly diffusive (see Figure 3.9). The peak in the dielectric constant and in the heat capacity disappears below that size. Nevertheless, the ferroelectric ordering is probably persistent up to about 7 nm.

Three peaks are found in the curves of the dielectric response as a function of temperature in nanocrystalline Bi₄Ti₃O₁₂ (Jiang et al. 1998). The first peak is shifted to higher temperatures with decreasing grain size. The second peak decreases gradually in its intensity and finally disappears with increasing grain size. The last one corresponds to the ferroelectric transition temperature. It increases at first with decreasing grain size from 56 to 25 nm. With further decreasing grain size, the peak shifts to lower temperatures. It seems that the mechanism is correlated

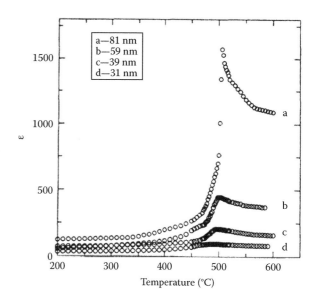

FIGURE 3.9 Temperature dependence of the dielectric function $\varepsilon(T)$ for PbTiO₃ samples with different average size (all measured at 1 MHz). (From Chattopadhyay, S. et al., *Phys. Rev. B*, 52(18), 13177, 1995. With permission.)

with competing effects of the released internal stresses and the clamped domain walls due to the diffusion of oxygen vacancies.

The ferroelectric properties can be efficiently controlled by doping with different elements. It is possible to tailor the parameters such as the maximum dielectric constant ε_m, transition temperature T_c, and dε/dT by a suitable doping. Doping of either A-site ions or Ti ions modifies T_c and the nature of the ferroelectric–paraelectric transition in BTO. A-site doping with cation can cause a decrease as well as an increase in T_c. A significant broadening of the transition is observed. The TiO₆ octahedra are distributed with B-site doping resulting configuration in the system (Hennings et al. 1982, Langhammer et al. 2000). A specific doping with 3d transition elements in BTO stabilizes a different structural configuration in the system (Langhammer et al. 2000). In addition, the incorporation of transition metal impurities in BTO is important for the use of cheaper metal electrode in multilayer BTO ceramic capacitors. Some doped BTO ceramics are sensitive to the grain size. The permittivity peaks and the transition temperature of Ba(Zr$_x$Ti$_{1-x}$)O₃ (BZT) ceramics are greatly suppressed with the decrease of grain size (Hennings 1987, Hennings and Schreinemacher 1994, Tang et al. 2004).

The temperature of the dielectric constant maximum T_m increases. The corresponding ε_m value decreases with increasing frequency. It is suggested that the BZT ceramics with fine-grain size show a transition from a normal ferroelectric to "relaxorlike" ferroelectric. However, the grain size reported is situated in the micrometer range. The dielectric behavior of Fe- and Ni-ion-doped BTO nanoparticles has been discussed by Jana et al. (2005) and Kundu et al. (2008), respectively. The dielectric permittivity in doped specimens is enhanced by an order of magnitude compared with undoped BTO ceramics. A reduction of the dielectric permittivity with decreasing grain size occurs due to crystal distortion assisted by the surface atoms. Moreover, a significant broadening of the phase transition and a shift of T_m to lower temperatures had been observed.

Complex perovskite-type ferroelectrics with disordered cation arrangements, in general, reveal a very diffused or smeared phase transition. A modified empirical expression including the diffuseness of the ferroelectric phase transition was proposed by Uchino and Nomura (1982)

$$\frac{1}{\varepsilon} - \frac{1}{\varepsilon_m} = \frac{(T - T_m)^\gamma}{C_1}. \qquad (3.4)$$

Here, γ and C_1 are assumed to be constant. The parameter γ yields information on the character of the phase transition: for $\gamma = 1$, a conventional Curie–Weiss law is obtained, whereas $\gamma = 2$ describes a complete diffuse transition. Experimentally, Tang et al. (2004) obtained for $\gamma = 1.82$, 1.78, and 1.64 in BZT ceramics with grain sizes of 2, 15, and 60 μm, respectively. In the fine-grained sample, the fitted value of γ decreases from 1.89 to 1.81 in case the frequency is increased from 100 Hz to 100 kHz. Such a behavior implies clearly that the fine-grained BZT ceramics exhibit features of diffuse phase transition and relaxor-like ferroelectric behavior.

3.3.4 Spectroscopic Observation of Excitations

As already pointed out in the introduction, the macroscopic properties of nanoparticles as well as bulk materials are governed by their elementary excitations. The phase transition in displacive ferroelectrics, like BTO, results from an instability of one of the normal vibrational modes of the lattice (Cochran 1959). The nuclei move in a slightly anharmonic potential. In this approach, the frequency of the relevant soft phonon decreases on approaching the critical temperature. The restoring force for the mode displacements tends to zero until the phonon has condensed out at the stability limit. The static atomic displacements on going from the paraelectric to the ferroelectric phase thus represent the frozen-in mode displacements of the unstable phonon. The order parameter of such a transition is the static component of the eigenvector of the unstable phonon. As the ferroelectric state is characterized by a macroscopic spontaneous polarization, the soft phonon must be both polar and of long wavelength ($q \to 0$). The potential field in order–disorder ferroelectrics like KDP is strongly anharmonic. The permanent electric dipoles are moving between at least two equilibrium positions. The soft collective excitations are rather unstable pseudospin waves than phonons (Blinc 1960). All ferroelectric materials exhibit both kind of behavior. The ratio is material dependent. This section summarizes the results of spectroscopic studies, as Raman and infrared spectroscopy. The basis for such investigations consists of microscopic properties of excitation modes, discussed also in Section 3.4. The lowest mode offers the so-called soft-mode behavior. The dispersion of the elementary excitation $\omega(\vec{q}; T)$ for a fixed wave vector $q = q_c$ tends to zero for $T \to T_c$ (see Equation 3.1). This property is a characteristic for bulk material. In nanomaterial, one observes a similar behavior. However, because of the lacking translational

invariance, the frequency is size dependent and reveals no dispersion. Recent spectroscopic observations are given for important nanosized ferroelectric materials, subsequently.

Low-frequency Raman spectroscopy was performed on BiTO nanocrystals as function of the grain size (32–83 nm) (Zhu et al. 2008). Four Raman modes were found in the frequency range in between 15 and 75 cm^{-1} for the 83 nm sample. This is in agreement with the BTO single crystal. The intensities of several modes with higher frequency decrease with decreasing grain size. Hence, the ferroelectricity weakens below 69 nm and the ferroelectric phase is transformed into the paraelectric phase below a size of 38 nm. The soft-mode frequency ω^2 ($q = 0$, T) for BiTO crystals is proportional to $T - T_c$. At $T = T_c$, the mode has zero energy indicating that a reordering of the microscopic constituents is quite easily possible (Kojima et al. 1994, Kojima and Shimada 1996).

Furthermore, the phase transition temperature T_c decreases with decreasing grain size d and is proportional to ($T_{c\infty} - 1/d$) for ultrafine ferroelectric particles (Jiang and Bursill 1999, Uchino et al. 1989, Zhong et al. 1994b). Here, $T_{c\infty}$ is the temperature of the phase transition for single crystals. So $\omega^2(\vec{q}, T)$ for BTO nanocrystals is proportional to ($T_{c\infty} - T - 1/d$). Since the low-frequency mode of BTO nanocrystals was measured at room temperature, the relationship between ω^2 and d can be expressed as

$$\omega^2 = \omega_0^2 \left(1 - \frac{d_0}{d}\right), \qquad (3.5)$$

where

ω_0 is the soft-mode frequency of the single crystal with $d \to \infty$
d_0 is the grain size at soft-mode transition point ($\omega = 0$)

For BTO, $\omega = 0$ for $d = d_0 = 23$ nm was obtained by Zhu et al. (2008). The critical value is slightly smaller than the before predicted one of 38 nm. This fact suggests that the size-driven phase transition is still of first order, same to that of the temperature-driven phase transition for the BTO nanocrystals.

The dielectric properties of BTO are dominated by a displacive behavior. Especially the elementary excitations are due to phonon excitations, in which the low-lying phonon mode reveals a soft-mode behavior. The direct observation of soft modes in BTO is difficult. One promising method is the measurement of Raman scattering spectra that are obtained for BTO nanoparticles by several authors (Huang et al. 2007, Wada et al. 2005a, Zhu et al. 2008) (see Figure 3.10).

BTO powders with various crystallite sizes were studied thoroughly. A tetragonal phase was detected for ultrafine powders with an average crystallite size above 30 nm. The lifetime of phonons assigned to the tetragonal phase decreases with decreasing crystallite size below a critical size of about 100 nm (Shiratori et al. 2007a,b).

A discontinuous change of the damping factor occurs at a certain temperature within the Raman spectra. This is nearly consistent with the cell volume expansion temperature from the x-ray diffraction measurement (Hoshina et al. 2006).

Another method to analyze the phonon behavior of BTO nanoparticles are far-infrared reflection measurements. A high

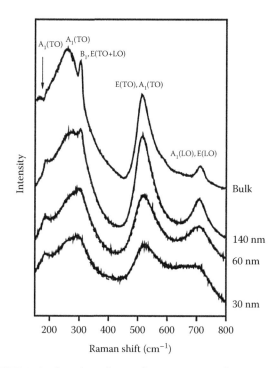

FIGURE 3.10 Size dependence of Raman spectra for $BaTiO_3$ bulk (>1 μm) and nanoparticles of diameter 140, 60, and 30 nm, respectively. (From Huang, T.-C. et al., *J. Phys.: Condens. Matter*, 19(47), 476212, 2007. With permission.)

dielectric constant is obtained for dense colloidal crystals of the particles (Hoshina et al. 2007). This is originated from the softening of the TO mode. Moreover, the result in the temperature dependence of far-infrared reflection suggested that the BTO particles with 58 nm can have a very broad phase transition.

Combined Raman spectroscopy and thermal analysis on SBT nanoparticles indicates the existence of a new intermediate ferroelectric phase within a sequence of the phase transitions, designated as ferroelectric–ferroelectric–paraelectric ones (Ke et al. 2007). Two anomalies were observed in the temperature dependence of the specific heat. Moreover, the size effect was addressed to inner compressive stress in nanoparticles for this special transition behavior. The results show that the SBT nanoparticles keep the ferroelectricity until the particle size is decreased to 4.2 nm.

Raman spectra for PZT40-nanoparticles of various sizes, studied by Ohno et al. (2007), yield a decrease of the soft mode around a size of 35 nm. The authors have suggested the existence of a critical size for the PZT40 particles. The temperature dependence of Raman spectra has revealed clearly that the Curie temperature will be shifted toward lower temperatures owing to size effects. The intrinsic dielectric constant for PZT40 nanoparticles calculated by the Lyddane–Sachs–Teller relation increased with decreasing particle size. These results show again that Raman scattering is a powerful tool to investigate ferroelectric materials, and especially ferroelectric nanoparticles.

A further application of Raman spectroscopy for ultrafine PTO particles is due by Ishikawa et al. (1988). A soft mode has been detected, denoted as the $E(1TO)$ mode, that shifts toward the low-frequency region with decreasing temperature. The line shapes become broad as the temperature approaches T_c. The lattice vibration of smaller particles is softer than that of the larger ones because the soft-mode frequency at an arbitrary temperature decreases as the size decreases. The damping factor increases near T_c. Smaller particles have larger damping factors (compare Figure 3.11a and b). Recent investigations provide an emergence of the orthorhombic phase at room temperature when the PTO

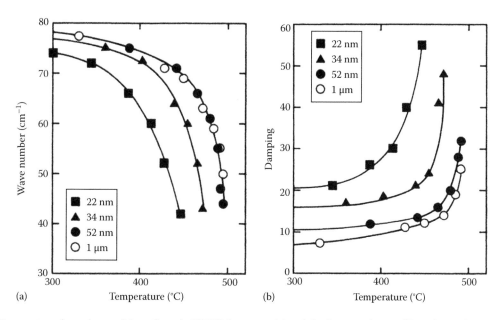

FIGURE 3.11 Temperature dependence of the soft-mode $E(1TO)$ frequency (a) and the damping factors (b) in $PbTiO_3$ fine particles of different size. (From Ishikawa, K. et al., *Phys. Rev. B*, 37(10), 5852, 1988. With permission.)

particle size is reduced to 11 nm (Deng and Zhang 2005). The doping effects of structural transformation, ferroelectricity, and soft-mode character in Ba-doped PTO nanoparticles were examined by x-ray diffraction and Raman spectroscopy by Lee et al. (2008). With increasing Ba concentration, tetragonality c/a reduces, transition temperature T_c decreases, and the $E(1TO)$ soft mode softens. A critical Ba doping concentration of $x = 0.4$ was found.

3.4 Theoretical Approach

The broad variety of experimental activities in the field of ferroelectric nanoparticles is accomplished by numerous theoretical studies that cover the topic on different levels and scales. First-principles methods (Ghosez and Junquera 2006) are based on the determination of the quantum mechanical ground state, where the energy of the lowest state is obtained by minimization of the total energy with respect to the associated electronic and nuclear coordinates. Among the different ab initio methods, the density functional theory (DFT) has become a reference. Other variational methods are introduced by Morozovska (2006). The quantum mechanics-based methods, overviewed by Ghosez and Junquera (2006), yield the electronic polarization as the central quantity of ferroelectric materials. The method is limited to a reduced number of atoms. Furthermore, the method does not include finite temperature effects relevant near to the phase transition. To overcome these problems, one should start from a microscopic many-body Hamiltonian including the interaction between the constituents (Kleemann et al. 1999a,b, 2000, Michael et al. 2007, Prosandeev et al. 1999, Wang et al. 1998a,b, 2000, Zhang et al. 2000, Zhong et al. 1999). The application of quantum statistical methods as Green's function technique allows to calculate the main characteristics of ferroelectrics like polarization, dielectric functions, the hysteresis, the susceptibility, and other relevant quantities. The problem confronted with is that the underlying Hamiltonian includes unknown coupling parameters that have to be determined by fitting experimental results or alternatively by ab initio calculations. The behavior of the material in the vicinity of the paraelectric–ferroelectric phase transition may be studied by the Landau expansion (Khare and Sa 2008) as an adequate tool. Generally, this thermodynamic approach can be exceeded by the inclusion of fluctuations that play an important role on the mesoscopic length scale.

Because the approaches mentioned earlier are tested successfully for bulk material, there is the hope to carry the methods for thin films and nanoparticles, too. So first-principles techniques play a decisive role in finding out the different dielectric properties of small ferroelectric particles and thin films. To that aim, ab initio methods have been improved permanently since the 1990s, even for analyzing ferroelectric properties. Nowadays, the most prominent method is the DFT, which is based on the Kohn–Sham energy functional (Kohn 1999). The application of DFT to ferroelectric oxide nanostructures is in the focus of the review article by Ghosez and Junquera (2006). To overcome, at least to some extent, the limitations due to the

small number of particles and to include finite temperature effects, an effective Hamiltonian was proposed by Rabe and Joannopoulos (1987).

A more generalized version of the method with regards to ferroelectricity has been offered by Zhong et al. (1994a). The parameters involved in that expansion are calculated via a linear-response theory and the total energy within DFT. Other approaches are shell-model calculations (Tinte et al. 1999) or a phenomenological model to simulate PZT structures by chemical rules from the DFT (Grinberg et al. 2002). Recently, the ground-state polarization of BTO nanosized films and cells is studied using an atomic-level simulation approach, in which the parameters are obtained by first-principles calculations (Stachiotti 2004).

Whereas the first principle studies are mainly focused on a microscopic understanding of the composition and the structure of ferroelectric nanomaterial, the many-body models and their quantum or classical statistical analyses are aimed at the understanding of macroscopic properties like the temperature-dependent polarization, the phase transition temperature and its shift due to finite size effects, and the existence of a critical particle size. Mostly, the characterization of ferroelectric properties including nanoparticles on a macroscopic or mesoscopic level is based on the application of the Landau theory, also known as Landau–Devonshire expansion in the field of ferroelectricity. On a more microscopic level, one uses lattice dynamic models for ferroelectrics of displacive type or the Ising model in a transverse field for the order–disorder type ferroelectrics.

3.4.1 Landau Theory

The Landau theory is an excellent method to understand the phase transition properties of bulk materials. In the last years, this thermodynamic approach has been extended to study the surface and size effects of thin films or nanostructures composed of ferroelectric substances. Concerning the analytical access to the description of ferroelectric nanoparticles, the Landau-type phenomenological theories are still a powerful technique (Akdogan and Safari 2002, Baudry 2006, Huang et al. 2001, Ishikawa and Uemori 1999, Jiang and Bursill 1999, Li et al. 1996, Wang and Smith 1995, Wang et al. 1994a,b, 1996, Zhong et al. 1994b).

In order to apply the Landau theory to a finite-size and inhomogeneous ferroelectric, the total free energy is given by the density of the free energy (Charnaya et al. 2001, Wang and Smith 1995, Wang et al. 1994b). If the ferroelectric exhibits a second-order phase transition, the total free energy can be written as:

$$F = \int dV \left(\frac{1}{2} A(T - T_{c\infty})P^2 + \frac{1}{4} BP^4 + \frac{1}{2} D(\nabla P)^2 - E_{ext}P \right)$$
$$+ \int \frac{D}{2\delta} P^2 dS, \tag{3.6}$$

where

    P is the polarization as the one-component order parameter

    $T_{c\infty}$ the Curie temperature of the bulk crystal and A as well as B, D, and δ are material parameters

The volume and surface integrals give the free energy of the interior and surface, respectively. Compared with the free energy expression for an infinite and homogeneous ferroelectric, the gradient term and the surface term were added. The quantity δ is the extrapolation length describing the difference between the surface and the bulk. The coefficient B is positive, and D is connected with the correlation length ξ, $D = \xi^2 |A(T - T_{c\infty})|$. E_{ext} is an external electric field that couples linearly to the polarization.

The spontaneous polarization is obtained by minimizing the free energy. Furthermore, the system has to be subjected to a boundary condition. It results:

$$D\nabla^2 P = A(T - T_{c\infty})P + BP^3 - E_{\text{ext}},$$

$$\frac{\partial P}{\partial n} + \frac{P}{\delta} = 0, \tag{3.7}$$

where n is the unit length along the normal direction of the surface. The susceptibility is defined as

$$\chi = \frac{1}{\varepsilon_0}\frac{\partial P}{\partial E_{\text{ext}}}, \tag{3.8}$$

where one is often interested in the zero-field susceptibility at $E_{\text{ext}} = 0$, so the susceptibility obeys the differential equation

$$D\nabla^2\chi = (A + 3BP^2)\chi - \frac{1}{\varepsilon_0}, \tag{3.9}$$

with the corresponding boundary condition

$$\frac{\partial\chi}{\partial n} + \frac{\chi}{\delta} = 0. \tag{3.10}$$

In the framework of linear response theory, the polarization P in Equation 3.9 is the spontaneous polarization, which can be obtained from Equation 3.7 for zero external field. In case of ferroelectric films, Equation 3.7 have been simplified and solved by Tilley and Zeks (1984). For a ferroelectric nanoparticle of arbitrary shape, two simplifications can be made to solve the basic Equation 3.7. At first, the particles are assumed to be spherical with the diameter as $d = 2r$. Second, the polarization is directed into a single direction and their magnitude depends only on the radius r. Then, Equation 3.7 can be formulated in spherical coordinates

$$D\left(\frac{d^2P}{dr^2} + \frac{2}{r}\frac{dP}{dr}\right) = A(T - T_{c\infty})P + BP^3,$$

$$\frac{dP}{dr} + \frac{P}{\delta} = 0. \tag{3.11}$$

If the ferroelectric material undergoes a first-order phase transition, characterized by $B < 0$ in Equation 3.6, one has to include a higher order term into the Landau expansion given by Equation 3.6 with the result

$$F = \int dV\left(\frac{1}{2}A(T - T_{0\infty})P^2 + \frac{1}{4}BP^4 + \frac{1}{6}CP^6 + \frac{1}{2}D(\nabla P)^2 - E_{\text{ext}}P\right)$$

$$+ \int \frac{D}{2\delta}P^2 dS. \tag{3.12}$$

Here, the coefficient C in front of the sixth-order term has to be positive to stabilize the ferroelectric state. The bulk Curie–Weiss temperature $T_{0\infty}$ is lower than the temperature $T_{c\infty}$ introduced in Equation 3.6. Similarly to the second-order phase transitions, one can consider the spherical symmetric case by assuming that the magnitude of the polarization depends only on the radius $P(r)$. The corresponding Euler–Lagrange equation together with the boundary condition reads

$$D\left(\frac{d^2P}{dr^2} + \frac{2}{r}\frac{dP}{dr}\right) = A(T - T_{c\infty})P + BP^3 + CP^5,$$

$$\frac{dP}{dr}\frac{P}{\delta} = 0. \tag{3.13}$$

A significant modification occurs to the extrapolation length. The quantity δ measures the strength of the surface effect. It depends not only on the different interaction constants at the surface and in the bulk but also on the coordination number at the surface. With regard to the microscopic models in Section 3.4.2, let us relate the parameters of the Landau expansion to microscopical quantities. There, the microscopic theory is formulated on a lattice, such as a simple cubic lattice with a lattice constant a_0. The interaction between the constituents at the surface is denoted as J_s, whereas J characterizes the interaction within the bulk material (for details see the forthcoming section). Then the parameter δ in Equation 3.6 can be expressed as

$$\frac{1}{\delta} = \frac{5J - 4J_s}{a_0 J}. \tag{3.14}$$

In ferroelectric films, the coordinate number on the surface is always four in case of a simple cubic lattice. If the interaction parameters J_s and J are kept constant, then δ is thickness independent. However, for spherical and cylindrical nanoparticles, even if J_s and J are size independent, the parameter δ will depend on the size because the smaller the coordination number at the surface, the smaller the diameter d. The averaged surface coordinate number for a sphere reads

$$n_{\text{av}} = 4\left(1 - \frac{a_0}{d}\right). \tag{3.15}$$

Combining Equations 3.14 and 3.15, one obtains

$$\frac{1}{\delta_s} = \frac{5J - n_{av}J_s}{a_0 J} = \frac{5}{d} + \frac{1}{\delta_f}\left(1 - \frac{a_0}{d}\right), \qquad (3.16)$$

where δ_f denotes the extrapolation length at infinite size (f means film). It can be seen that even if $\delta_f < 0$, δ_s becomes positive if

$$d < 5|\delta_f| + a_0. \qquad (3.17)$$

The size dependence of δ leads to some interesting features for nanomaterial, which is different from ferroelectric films. As $d \gg a_0$ in most cases, the above expression can be simplified to

$$\frac{1}{\delta_s} = \frac{5}{d} + \frac{1}{\delta_f}. \qquad (3.18)$$

Similarly, the extrapolation length for a cylindrical nanoparticle is given by

$$\frac{1}{\delta_c} = \frac{5}{2d} + \frac{1}{\delta_f}. \qquad (3.19)$$

To get the spatial distribution of the polarization, Equation 3.11 in case of a second-order transition, or Equation 3.13 for a first-order transition, respectively, should be solved numerically. When $\delta > 0$, the polarization at the surface is reduced compared with that one in the bulk. In case $\delta < 0$, the polarization at the surface is enhanced. Let us point out that for $\delta < 0$, the polarization can even exist above the bulk Curie temperature. The situation is comparable to ferroelectric thin films (Tilley and Zeks 1984). In that case, the ferroelectricity is enhanced if the surface polarization is stronger. It is reduced when the surface ferroelectricity is weaker, and as a consequence, there exits a size-driven phase transition. However, for spherical particles (Rychetsky and Hudak 1997, Wang et al. 1994a, Zhong et al. 1994b) and the cylindrical ones (Wang and Smith 1995) (see Figure 3.12a and b), Landau theory predicts that the ferroelectricity is always suppressed at small sizes and a size-driven phase transition always exists. An experimental evidence supporting this assertion is not yet available in a convincing manner. As a result of the degradation of ferroelectric properties, the phase transition temperature in spherical nanoparticles is significantly lower than the bulk one. Using a more refined technique for a microscopical model, one can get both an enhancement and a reduction. The Landau expansion can be generalized by including other degrees of freedom, in particular elastic degrees of freedom (Morozovska 2006). Very often the ferroelectric phase transition is coupled to lattice distortions that can be taken into account on the mesoscopic level by expanding the free energy, also in terms of stress components. Such a coupling between the order parameter P and the elastic degrees of freedom can lead to a noticeable enhancement of the ferroelectric properties.

In nanocylinders (Yadlovker and Berger 2005) and in nanorods (Morozovska 2006), such an enhancement could be demonstrated. Since the depolarization field value depends on the shape of a particle, the enhancement of the polarization can be expected the smaller the depolarization field is. Morozovska (2006) has investigated the size effects and the influence of the depolarization field on the phase diagrams of cylindrical ferroelectric nanoparticles (Figure 3.13). The corresponding equations were solved by a direct variational method. It was shown that the transition temperature could be higher for nanorods and nanowires than that of the bulk material. An opposite behavior was observed for nanodiscs.

The achieved results explain the observed enhancement in the Rochelle salt nanorods (Yadlovker and Berger 2005) of a radius $\approx 30\,nm$, the piezoelectric properties conservation in lead–zirconate–titanate nanorods (Mishina et al. 2002) with a radius of about $5–10\,nm$. Moreover, the results are in a good agreement with the first principles calculations in BTO nanowires (Geneste et al. 2006). The authors observed that the possible reason of the enhancement of the polar properties in confined ferroelectric nanowires and nanorods is the occurrence of an effective surface pressure coupled to the polarization via the electrostrictive interaction and the decrease of the depolarization field observed in prolate cylindrical particles. The predicted effects could be

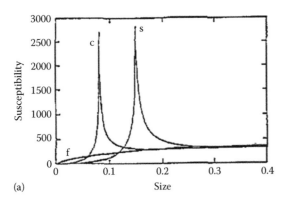

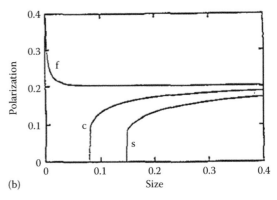

FIGURE 3.12 Size dependence of the relative susceptibility (a) and of the spontaneous polarization (b) at the surface for $\delta_f = -43\,nm$. The meaning of symbols in the figure are f—film geometry, c—cylinder geometry, and s—sphere geometry. (From Wang, C.L. and Smith, S.R.P., *J. Phys.: Condens. Matter*, 7(36), 7163, 1995. With permission.)

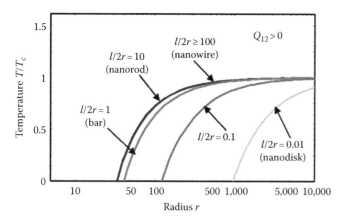

FIGURE 3.13 Transition temperature size dependence for different ratios $1/2r = 100$, 10, 1, 0.1, and 0.01. (From Morozovska, A.N., *Phys. Rev. B*, 73(21), 214106, 2006. With permission.)

very useful for the elaboration of modern nanocomposites with perfect polar properties.

The effect of bond contraction in the surface layers is intensively studied. Compressive stress is induced on the inner part of a grain and results in a size effect for ferroelectric materials in the nanometer size range, that is, when the surface–volume ratio becomes very large. Huang et al. (2001) have investigated the grain-size effect induced by the surface bond contraction based on the Landau–Devonshire phenomenological theory. The elastic Gibbs free energy is expressed as a Taylor series in powers of the order parameters and the stress. Useful results on intrinsic properties of nanosized $Pb(ZrTi)O_3$ have been obtained. It was found that, due to the surface-bond contraction, the phase stability is affected by the grain size and the size-dependent properties show differences in different phases. Recently, the effect of long-range elastic interactions on the toroidal moment of polarization in a two-dimensional ferroelectric particle was investigated by using a phase field model by Wang and Zhang (2006). The phase field simulations exhibit vortex patterns with purely toroidal moments of polarization and negligible macroscopic polarization when the spontaneous strains are low and the simulated ferroelectric size is small.

3.4.2 Microscopic Models

A more refined method in describing both ferroelectric bulk material as well as thin films and nanoparticles is based on a many-body Hamiltonian. This section is devoted to a quantum statistical modeling of the collective behavior of ferroelectric systems. The approach covers the entire regime from the phase transition between the paraelectric and the ferroelectric phase up to the low-temperature properties. The starting point is an appropriate Hamilton operator that includes the relevant degrees of freedom. On the basis of this Hamiltonian, the elementary excitation and their damping are calculated. These collective phenomena determine the macroscopic behavior of the system such as the order parameter, the susceptibility, the dielectric function, and other quantities. Physically, the search for a microscopical approach

can be traced back to the observation made by Cochran (1959). The phase transition in ferroelectrics arises from an instability of a low-lying frequency mode. In ferroelectrics of displacive type, such an unstable mode is realized by one normal lattice vibration mode. In order–disorder ferroelectrics, the soft mode is given by the pseudospin excitation, which is discussed here. The underlying model is an Ising model in a transverse field abbreviated as TIM. This model is a promising candidate to figure out ferroelectric properties from a microscopic point of view and to apply all the well-known quantum statistical techniques elaborated in detail for magnetic systems. In the same manner as for magnets for which the excitation energy of the spin waves is considered, the macroscopic ferroelectric properties can be derived from the corresponding modes such as phonon-like modes in displacive type ferroelectrics or pseudospin wave modes in order–disorder ferroelectrics. The phase transition in displacive ferroelectrics is related to the rearrangement of a few atoms in the unit cell, in which the position of the other ones remain unchanged. The relevant unit moves in a slightly anharmonic potential. The main process in order–disorder ferroelectrics consists of the reordering of polar groups. The simplest realization is given by the rearrangement of the protons in strongly anharmonic double-well potentials of hydrogen-bonded ferroelectrics such as KDP. Hence, let us discuss the origin of the Hamiltonian and the results achieved with this quantum statistical approach. Originally, the TIM had been proposed by Blinc and de Gennes for the description of ferroelectrics of KDP type (Blinc and Zeks 1974).

In this hydrogen-bonded ferroelectrics, the transverse field represents the proton tunneling between the two equilibrium positions of the protons within the O–H–O bonds. The approximative applicability of the TIM to displacive type ferroelectrics such as $BaTiO_3$ (BTO) had been demonstrated by Pirc and Blinc (2004) and Cao and Li (2003). The idea behind this application is, following the rules of the order–disorder model, that the paraelectric phase in BTO is associated with the position of the Ti ions. Instead of occupying the body center positions as in an ideal cubic perovskite structure, the Ti ions are randomly displaced along the cube diagonals that cause the appearance of the disordered phase. In the case of a small tunneling field compared with the interaction constant, one may use the TIM as a model for order–disorder ferroelectrics without tunneling motion. Such a situation is encountered in $NaNO_2$ and triglycine sulfate. Therefore, the TIM seems to be a rather universal model that can be used, at least, approximatively for a broad class of ferroelectric material.

The simple idea behind the TIM assumes the existence of polar groups with two alignments, such as protons in one minimum of a double-well potential. This alignment is described by the z-component of a spin variable S^z. The mapping of the relevant mechanism onto a virtual spin operator is one of the key ideas for this model. No real spins are considered. Both eigenvalues of $S^z = \pm 1/2$ represent the two allowed positions. In so far, the spin components play the role of "dipolar" coordinates. The entire system is arranged on a lattice, so the two possible orientations of the microscopic dipole S_i^z are used as the dynamical variable. The interaction between the wells situated at different positions is assumed to be realized by the Ising model. However, as pointed out by Blinc and Zeks (1974), one should take into account a tunneling between the two positions signalized by the eigenvalues of the S^z. Taking into account the ability for tunneling, the resulting Hamiltonian of the TIM reads

$$H = -\frac{1}{2}\sum_{ij} J_{ij} S_i^z S_j^z - \sum_i \Omega_i S_i^x - \mu E \sum_i S_i^z. \qquad (3.20)$$

The components of a spin-$\frac{1}{2}$ operator S_i^z and S_i^x at a certain lattice site i interact via the interaction parameter $J_{ij} \equiv J(r_i - r_j)$ and are influenced by the tunneling term Ω_i. These energies have to be included from experimental results or ab initio calculations. It is important to note that the interaction strength depends on the distance between the pseudospins. Consequently, the interaction strength is determined by the lattice parameters, the lattice symmetry, and the number of nearest neighbors. The sum is performed over all lattice points of the infinite extended bulk material. An external electric field E couples linearly to dipole moment. This Hamiltonian had been successfully adopted for bulk material (Kuehnel et al. 1977, Wesselinowa 1990, 1994, Wesselinowa and Apostolov 1997, Wesselinowa et al. 1994). Recently, the applicability of the model was extended to thin films (Wesselinowa 2001, 2002a–d, 2005a,b, Wesselinowa and Dimitrov 2007, Wesselinowa and Kovachev 2007, Wesselinowa et al. 2006, Wesselinowa and Trimper 2001, 2002, 2003, 2004a,b, Wesselinowa et al. 2005). The Hamiltonian in Equation 3.20 describes systems undergoing a second-order phase transition. Taking into account four-spin interactions, it can be applied to first-order phase transitions (Wesselinowa 2002d, Wesselinowa and Marinov 1992), which are not considered here. Because of the surface and size effects in nanoparticles, the interaction parameter between nearest neighbors are different for bulk and surface constituents. Likewise, the tunneling frequency Ω_i is different for bulk and surface atoms. The interaction between the pseudospins (this name is used to stress that there is no real spin related to S_i^z) between groups at the surface shell is denoted as $J_{ij} = J_s$, whereas the bulk interaction strength is J_b. In the same manner, Ω_b and Ω_s represent transverse fields in the bulk and surface shell, respectively.

The Hamiltonian is likewise the starting point to include further degrees of freedom as impurities and doping. Modern tools of statistical mechanics as two time temperature Green's functions (Economou 2006) give access to both static and dynamic properties of condensed matter on the nanoscale. This covers macroscopic as well as microscopic quantities. This Green's function contains all the information about the system. It is defined by

$$G_{lm}(t) = \ll S_l^+(t); S_m^-(0) \gg \equiv i\Theta(t-t')\langle [S_l^+(t)S_m^-(0) - S_m^-(0)S_l^+(t)]\rangle. \qquad (3.21)$$

Since the lack of translational invariance in nanomaterial, the Green's function has to be investigated in the real space.

The Heavyside function $\Theta(t)$ defines its retarded nature. The average is defined in the conventional way as

$$\langle S^z \rangle = \frac{Tr S^z \exp(-\beta H)}{Tr \exp(-\beta H)}. \qquad (3.22)$$

The ordered phase of the system described by Equation 3.20 is characterized by $\langle S^x \rangle \neq 0$ and $\langle S^z \rangle \neq 0$ (compare Blinc and Zeks 1974). Therefore, it is appropriate to introduce a new coordinate system by rotating the original one by an angle θ in the x–z plane (Kuehnel et al. 1977). This rotation angle is determined by the requirement $\langle S^{x'} \rangle = 0$ in the new coordinate system. Instead of S^x, S^y, and S^z, a new set including Pauli operators S_l^+, S_m^-, and $S^{z'}$ is used in the rotated system.

Now, let us consider a spherical particle characterized by fixing the origin at a certain pseudospin in the center of the particle. Rest of them within the particle are ordered in shells, which are numbered by $n = 0, 1, ..., N$. Here, $n = 0$ denotes the central pseudospin and $n = N$ represents the surface of the system (see Figure 3.14) (Michael et al. 2007).

After the Fourier transformation, the equation of motion of the Green's function in random phase approximation (RPA) reads

$$\omega G_{lm} = 2\langle S_l^z \rangle \delta_{lm} + \left[2\Omega_l \sin\theta_l + \mu E \cos\theta_l + \sum_j J_{lj} \cos\theta_l \cos\theta_j \langle S_j^z \rangle \right.$$
$$+ \frac{1}{2}\sum_j J_{lj} \sin\theta_l \sin\theta_j (\langle S_l^+ S_j^- \rangle + \langle S_l^- S_j^- \rangle) \Big] G_{lm}$$
$$- \frac{1}{2}\sum_j J_{lj}[\sin\theta_l \sin\theta_j \langle S_l^z \rangle + 2\cos\theta_j \cos\theta_l \langle S_l^+ S_j^- \rangle]G_{jm}. \qquad (3.23)$$

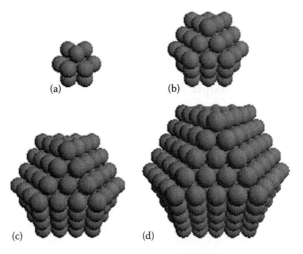

(a)   (b)

(c)   (d)

FIGURE 3.14 Ferroelectric nanoparticles of different size composed of shells. Each sphere represents a pseudospin situated in the center, where (a) consists of one central spin plus $N = 1$ shell, (b) $N = 2$, (c) $N = 3$, and (d) $N = 4$.

The poles of the Green's function give the transverse excitation energies. Within the applied RPA, the transverse spin-wave energy is found as

$$\omega_n = 2\Omega_n \sin\theta_n + \frac{1}{N'}\sum_j J_{nj}\cos\theta_n\cos\theta_j\langle S_j^z\rangle + \mu E\cos\theta_n,$$

(3.24)

where N' is the number of sites in any of the shells. In the same manner (see Tserkovnikov 1971), the damping of the spin-wave is given by

$$\gamma_n = \frac{\pi}{4}\sum_j J_{nj}^2(\cos\theta_n\cos\theta_j - 0.5\sin\theta_n\sin\theta_j)^2$$
$$\times \bar{n}_j(1-\bar{n}_j)\delta(\omega_n - \omega_j + \omega_j - \omega_n),$$

(3.25)

where $\bar{n}_n = \langle S_n^- S_n^+\rangle$ is the correlation function. It is calculated via the spectral theorem and using the excitation energy in the RPA (Equation 3.24).

To complete the soft-mode energy ω_n of the nth shell, one needs the rotation angle θ_n, which follows from the condition $\langle S^x\rangle = 0$. The angle is determined by the equation

$$-\Omega_n\cos\theta_n + \frac{1}{4}\sigma_n J_n\cos\theta_n\sin\theta_n + \frac{1}{2}\mu E\sin\theta_n = 0.$$

(3.26)

Using the standard procedure for Green's function, we get the relative polarization of the nth shell as

$$\sigma_n = \langle S_n^z\rangle = \frac{1}{2}\tanh\frac{\omega_n}{2T}.$$

(3.27)

The following investigations of ferroelectric nanoparticles are based on these analytical expressions. The required interaction parameters for the nonsurface and nondoped cases were chosen due to former calculations for BTO systems (Wesselinowa 2001). The interaction strength reads $J_b = 150\,K$; the tunneling integral is $\Omega_b = 10\,K$.

This part is addressed to the theoretical description of ferroelectric nanoparticles of various sizes without an electric field. The influence of the surface, size effects, and the occurrence of distortions (e.g., via doping) of the particles is discussed.

The existence of a surface in nonbulk system changes all physical quantities. The number of nearest neighbors at the surface differs from that in the inner part. Hence, the appearing strain/stress of especially ferroelectric nanoparticles results in a change of the interaction constant at the surface J_s. These surface effects influence the temperature-dependent polarization of spherical particles composed of shells. The variation of the coupling at the surface changes the polarization accordingly. A lowered surface interaction strength $J_s < J_b$ leads to a reduced polarization σ for almost the whole temperature range. σ vanishes continuously at a lower critical temperature T_c. Hence, the phase transition is a pronounced second-order one. The opposite case $J_s > J_b$ yields a larger dipole moment and an enhanced phase transition temperature T_c. This reflects the observation that both the bulk and the surface coupling contribute to the ordering of the pseudospins.

The shell-resolved polarization σ_n is given in Figure 3.15a and b. The particle (see Figure 3.14d) is composed of eight shells ($N = 8$). The index n denotes the considered shell of the particle, for example, $n = 8$ represents the surface shell. The reduction of the local polarization σ_n depending on the position within the particle is clearly visible. The behavior is contrary for weaker or stronger surface couplings, respectively. The smaller the strength J_s compared with the bulk value, the faster is the decrease of

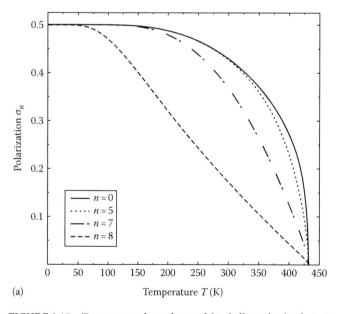

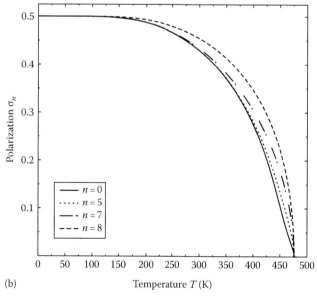

(a) Temperature T (K)

(b) Temperature T (K)

FIGURE 3.15 Temperature dependence of the shell-resolved polarization σ_n for a particle with eight shells. The surface energy is $J_s = 50\,K$ (a) and $J_s = 325\,K$ (b). The nonsurface interaction $J_b = 150\,K$ is fixed.

the polarization in the outer shells. A higher surface interaction provides smaller values in the inner shells. This reflects the importance of the inclusion of surface effects. The case $J_s < J_b$ (see Figure 3.15a) could explain the decrease of the polarization and the phase transition temperature in small particles of BTO (Ohno et al. 2006, Schlag and Eicke 1994) and PTO (Chattopadhyay et al. 1995, Zhong et al. 1993). The second case $J_s > J_b$ (compare Figure 3.15b) is responsible for the increase of the polarization and T_c in small KDP particles (Colla et al. 1997) and KNO_3 thin films (Scott et al. 1987). An ab initio study of the polarization as a function of temperature is also given in Tenne et al. (2006).

There is a long-standing debate on how physical properties like the polarization or the critical temperature are affected by the size of the system, especially in the nanometer scale. The dependence of the polarization on the size within the microscopic model will be considered now. The size is controlled by the number of shells N. Obviously, the polarization is enhanced with the increasing particle size (see Figure 3.16a). Summarizing all the data, the phase transition temperature versus the number of shells is shown in Figure 3.16b. The ferroelectric particles exhibit a fast increase of T_c with an ascending number of shells. In the limit of very large numbers N, the critical temperature approaches nearly to the constant bulk value. The result is in qualitative agreement with the experimental data of small particles composed of BTO (Ohno et al. 2006) and PTO (Chattopadhyay et al. 1995, Zhong et al. 1993). However, the chosen set of parameters does not lead to an indication for a pronounced critical size effect.

Apart from macroscopic quantities, the method yields microscopic features of the nanoparticles as the energy of the elementary excitations (compare Equation 3.24) and its damping (see Equation 3.25). In Figure 3.17a, the temperature dependence of the excitation energy is plotted for a different number of shells when the relation $J_s < J_b$ is fulfilled.

A lowering of the excitation energy is observed for increasing temperatures. The larger the particles, the higher the energies. The nanoparticle shows a typical soft-mode behavior as already observed in the bulk material. Apparently, the excitation energy is shifted to smaller values in comparison to the bulk material, when the number of shells decreases. The result implies a lowering of the force constant in the small particle, which was observed for PTO particles (Fu et al. 2000, Ishikawa et al. 1988, Zhong et al. 1993). Consequently, this leads to the decrease of the phase transition temperature between the tetragonal and the cubic phase.

Because of the higher order interactions between the constituents and/or the scattering at defects or due to the inclusion of phonon degrees of freedom, the elementary excitation can be damped. Such a damping (Equation 3.25) could be manifested in a finite lifetime of the excitations. The temperature dependence of the damping is plotted in Figure 3.17b. When the particle size is lowered, the damping increases. At low temperatures, the excitations are underdamped, the damping is extremely small, accordingly. In approaching the critical temperature, the damping increases strongly but remains finite (see Figure 3.17b). This behavior is in contrast to the behavior of bulk material, for example, PTO (Burns and Scott 1970), where the linewidth of the soft mode diverges at the ferroelectric-to-paraelectric transition. The softmode becomes overdamped close to the phase transition. Such a behavior is in agreement with experimental data for PTO (Fu et al. 2000, Ishikawa et al. 1988), BTO (Wada et al. 2005b), and SBT (Yu et al. 2003a) particles. The enhanced damping in small nanoparticles offers an explanation of the broadened peak observed in the dielectric constant of PTO particles (Chattopadhyay et al. 1995) and (Ba,Sr) TiO_3 thin films (Parker et al. 2002, Tenne et al. 2001). A broadened dielectric anomaly leads also to a smearing out of the critical regime. The insert

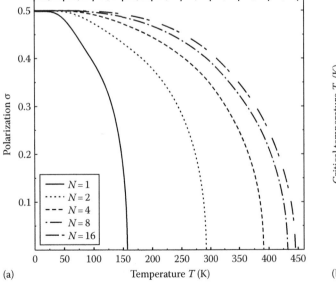

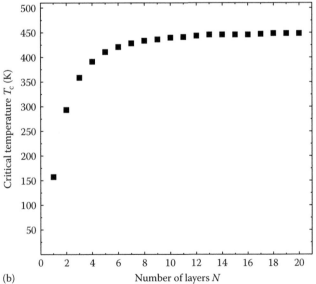

FIGURE 3.16 Temperature dependence of the polarization (a) and the critical temperature (b) depending on the number of shells N. The interactions strengths $J_b = 150\,\text{K}$, $J_s = 50\,\text{K}$ are fixed.

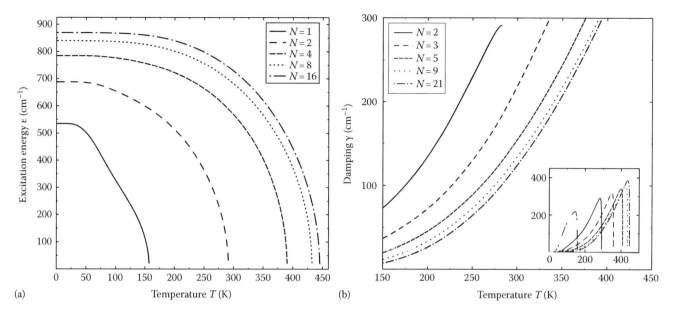

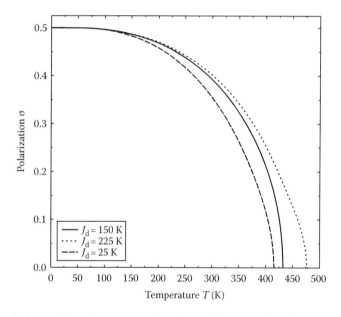

FIGURE 3.17 Temperature dependence of the excitation energy (a) and the related damping (b) for a different number of shells N with $J_b = 150\,\text{K}$, $J_s = 50\,\text{K}$.

shows the overall development of the damping. Very close to the critical point a sudden decrease was observed, which is only plotted in the insert for the sake of completeness. Fluctuation effects, predominantly occurring in the vicinity of the phase transition, are slightly suppressed through the selected approximation. The results near the critical temperature should be considered as an extrapolation.

Experiments show a clear influence of impurities or defects on physical properties. The simplest way to incorporate defect configurations into the model is to assume a variation of the interaction strength J. Microscopically, the substitution of defects into the material leads to a change of the coupling parameter. The defect coupling between neighbors J_d is altered and in general is different from the surface value J_s as well as the bulk one J_b. Physically, this variation of the coupling parameter is originated by the appearance of local stress and by the substitution of ions with different radii in comparison to the host material, consequently, different distances between them (smaller radii corresponds to a larger distance) as well as by localized vacancies. The polarization, excitation energy, as well as its damping should depend on the defect concentration. Furthermore, the defect can be situated at different shells within the nanoparticle (Michael et al. 2008).

The influence to the polarization for a field-free particle with eight shells in the absence of an electric field can be seen in Figure 3.18. The first two shells are defect. The temperature dependence deviates from the defect-free case. A smaller interaction in defect shells results in a lowering of the polarization and the critical temperature (dashed curve). The polarization as well as critical temperature are enlarged for impurities with a larger radius compared with the constituent ions (dotted curve). This is equivalent to an increased interaction energy, compared with the unperturbed case (solid curve).

FIGURE 3.18 Temperature dependence of the averaged polarization σ for a ferroelectric nanoparticle with $J_b = 150\,\text{K}$, $J_s = 50\,\text{K}$. From the total number of $N = 8$ shells, the first two shells are defect shells: $J_d = J_b$ (solid curve); $J_d = 225\,\text{K}$ (dotted curve); $J_d = 25\,\text{K}$ (dashed curve). (From Michael, T. et al., *Ferroelectrics*, 363, 110, 2008. With permission.)

The temperature regime of the energy of the elementary excitations ε for different numbers of defect shells n_d (Michael et al. 2007) results in graphs equivalent to Figure 3.17a. All up to the n_dth shell are defect ones. The bulk coupling is stronger than the defect and the surface coupling, that is, $J_b > J_s > J_d$. The excitation energy depends on both the number of defects n_d and the corresponding coupling J_d. An enhanced defect concentration reduces the energy of the excitations in the present choice of parameters. The corresponding behavior of the damping of excitations is shown in Figure 3.17b. An experimental evidence of the lowering of the soft-mode frequency for La-doped nanocrystalline PTO was given in Meng et al. (1994). Similar results are found for Er- and La-substituted PTO thin films (Yakovlev et al. 2006). The Raman peak width is broadened in comparison to undoped specimen. This is in accordance with a larger damping of the excited modes. The results reveal that different mechanisms such as surfaces, stress, and defects contribute additively to the damping coefficient. Insofar, the damping is always enhanced in comparison to the bulk and materials without defects.

The dependence of the averaged polarization σ of spherical nanoparticles on the number of defect shells n_d at a fixed temperature is shown in Figure 3.19a.

The Curie temperature of the nanoparticle depends likewise on the number of shells, which is depicted in Figure 3.19b. The polarization and the phase transition temperature show a dependency on the growth direction of the defects. Two different strength of the coupling are considered. A defect coupling smaller than the bulk and surface couplings (squares) leads to a decreasing of the polarization. The same behavior is observed for the Curie temperature with increasing number of defect shells. Higher defect strengths (diamonds) enlarge both physical quantities. A secondary effect occurs by an approach of making the nanoparticle a defect. The full squares and diamonds correspond to the case, in which the sequence of defects starts at the center. Subsequently, the next shells are assumed to be defect configurations. The procedure is performed until the surface shell is reached and becomes itself defect, too. The opposite realization is drawn as open symbols. Here, the configuration of the surface shell is a defect. Then, subsequently, the other shells become defect until the center is reached. The two different realizations are denoted as up- and downprocess, respectively.

Both approaches in common are the increase or decrease of polarization as well as T_c with the growing number of defect shells for the particular interaction strength. For the downprocess, the slope is stronger than for the upprocess. Both responses to the doping were experimentally observed. For Sr-deficient and Bi-excess SBT, the Bi substitution with A-site vacancies is responsible for the higher Curie temperature and polarization (Noguchi et al. 2001). This is governed by the bonding characteristics with oxide ions. The influence of the orbital hybridization on T_c is very large, and Bi substitution results in a higher transition temperature. A decrease in the Curie temperature and polarization was found in PLZT for the increase of the Ba (Ramam and Miguel 2006) and La concentration (Plonska et al. 2003). The Curie point shifts to lower temperatures in BZT5 nanoparticles (Ohno et al. 2006). This effect in ABO_3 structures is addressed to induced A-site vacancies, which weaken the coupling between neighboring BO_6 octahedral (Kim and Jang 2000).

The inclusion of an electric field and the theoretical observation of the associated hysteresis loops are discussed in the following text. Let us consider the hysteresis loop for different surface configurations represented by the interaction constant J_s at a fixed temperature $T = 300\,K$ and fixed N (Michael et al. 2006). The results for a particle with $N = 8$ shells are shown in Figure 3.20a.

The coercive field E_c and the remanent polarization σ_r are sensitive to variations of the interaction parameter at the surface. If the coupling at the surface is smaller as that in the bulk (dashed line), both quantities are reduced in comparison to the case for $J_s = J_b$ (solid line). In other words, the coercive field is lowered

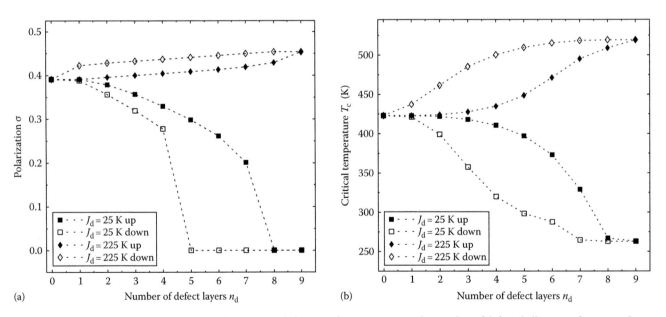

FIGURE 3.19 Dependence of the averaged polarization and the critical temperature on the number of defect shells $n = n_d$ for a particle size $N = 8$. The interaction strength reads $J_b = 150\,K$, $J_s = 50\,K$, and two different J_d values: $25\,K$ (squares) and $225\,K$ (diamonds) were chosen. The full symbols denote the up-process; the open symbols the down-process, see the text. (From Michael, T. et al., *Ferroelectrics*, 363, 110, 2008. With permission.)

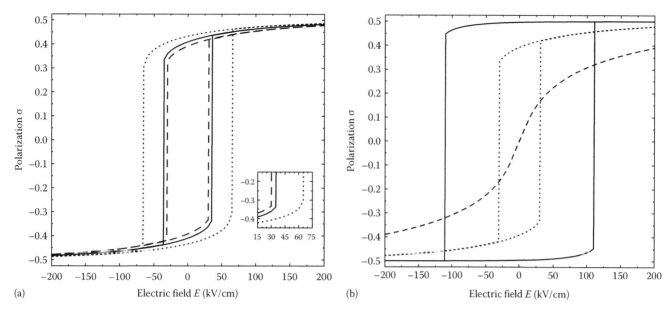

FIGURE 3.20 (a) Influence of the surface coupling strength J_s on the hysteresis at fixed temperature $T = 300\,\text{K}$ for $J_s = 150\,\text{K}$ (solid curve), $350\,\text{K}$ (dotted curve), and $50\,\text{K}$ (dashed curve); the inset offers the low field behavior. (b) Temperature dependence of the hysteresis with $J_s = 50\,\text{K}$: $T = 100\,\text{K}$ (solid curve), $300\,\text{K}$ (dotted curve), and $500\,\text{K}$ (dashed curve). The particle size and the nonsurface interaction are specified as $N = 8$ and $J_b = 150\,\text{K}$, respectively.

when the critical temperature of the system is decreased. This was observed in small BTO (Schlag and Eicke 1994) and PTO particles (Chattopadhyay et al. 1995). In the opposite case (dotted line), both the coercive field and the remanent polarization increase. This is in agreement with observations made in small KDP particles (Colla et al. 1997), in which the polarization and the critical temperature increase compared with the bulk material.

The temperature dependence of the hysteresis loop for eight layers is shown in Figure 3.20b. With increasing temperature, the hysteresis loop is more compact and lower, the coercive field decreases, and for $T \geq T_c$, the hysteresis loop vanishes (dashed line). Apart from the hysteresis loops obtained by the microscopic model, see results based on a thermodynamic approach (Baudry 1999, Baudry and Tournier 2001, 2005).

There are several experimental indications for a significant influence of doping effects on the hysteresis loop. The behavior of the polarization depending on an external electric field is influenced by the presence of defects. The coercive field and the remanent polarization of the ferroelectric particle are reduced or enhanced due to the different interaction strength within the defect shell. This results in different hysteresis loops comparable to Figure 3.20a. The variation of the interaction strength J can also be interpreted as the appearance of local stress, originated by the inclusion of different kinds of defects. The case $J_d > J_b$ (dotted curve) corresponds to a compressive stress, leading to an enhancement of E_c, which has been observed in thin PZT films (Duan et al. 2000). It is also in accordance with the experimental results observed through the substitution of doping ions, such as Bi in SBT (Liu et al. 2005) or by increasing the Ba contents in PLZT ceramics (Das et al. 2003). Referring to the case of smaller defect coupling, that is, tensile stress, the coercive field and the

remanent polarization are reduced (dashed curve). This may explain the experimentally observed decrease of the coercive field and the remanent polarization in small ferroelectric particles by the substitution of doping ions. This is realized by substituting La in PTO (Noguchi et al. 2002) and PZT (Kim and Jang 2000, Sakai et al. 2003) nanopowders. A single isolated defect layer offers only a weak influence on the hysteresis curve. Because of that, the first five layers of the nanoparticle are defect ones (compare also Figure 3.14). Here, the number of ferroelectric constituents is large enough to give a significant contribution to the polarization and, consequently, to the hysteresis loop. Apparently, one observes a change of the shape of the hysteresis loop due to defects.

Obviously E_c should depend on the number of the inner defect shells, that is, on the concentration of the defects. The result is shown in Figure 3.21 for a particle with eight shells. Notice that, for instance, $n_d = 5$ means that all shells until the fifth layers are defect layers. The squares in Figure 3.21 demonstrate that the coercive field strength E_c decreases with increasing number of defect shells. For the defect coupling, we assume $J_d = 25\,\text{K}$, that is, $J_d < J_b$. The result is in reasonable accordance to the experimental data reported in Kim and Jang (2000), Noguchi et al. (2002), and Sakai et al. (2003). A similar result is also obtained for the remanent polarization P_r. An increase of the La content in PTO and PZT ceramics decreases the coercive field E_c. The opposite behavior is offered as the diamonds. With increasing number of defect shells, the coercive field E_c (respectively P_r) increases. The open squares and diamonds represent the filling of the particles with defect shells beginning from the surface shell (downprocess), whereas the full symbols stands for the upprocess. One observes that the increase or decrease of E_c is more pronounced and stronger for the downprocess. This finding is in a quite good

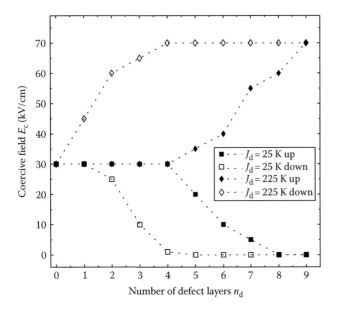

FIGURE 3.21 Dependence of the coercive field E_c on the number of defect shells n_d for $N = 8$, $J_b = 150\,K$, $J_s = 50\,K$, and different J_d values: 25 K (squares) and 225 K (diamonds). The full symbols denote the up-process, and the open symbols the down-process. (From Michael, T. et al., *Ferroelectrics*, 363, 110, 2008. With permission.)

agreement with the experimental data offered in Das et al. (2003) and Liu et al. (2005).

Let us point out to a promising way to study properties of ferroelectric nanoparticles using an alternative approach that had been used very successfully in magnetic nanoparticles (Tserkovnyak et al. 2005). Motivated by the progress of a multi-scale approach in such magnetic materials, the dynamics of the Ising model in a transverse field as a basic model for ferroelectric order–disorder phase transition is reformulated in terms of a mesoscopic model and inherent microscopic parameters (Trimper et al. 2007). To that aim, we have determined the effective field $h(\vec{x}, t)$ of the Ising model in a transverse field and have shown that the propagating part obeys the equation

$$\frac{\partial \vec{S}(\vec{x},t)}{\partial t} = \vec{h}(\vec{x},t) \times \vec{S}(\vec{x},t). \qquad (3.28)$$

The effective field is expressed in a continuous approximation as

$$\vec{h}(\vec{x},t) = (\Omega, 0, J\kappa S^z(\vec{x},t)), \quad \text{with } \kappa = a^2 \nabla^2 + z \qquad (3.29)$$

where

J is the coupling strength between the z nearest neighbors
Ω is the transverse field
a is the lattice spacing in a simple cubic lattice

From Equation 3.28, it follows the excitation energy $\varepsilon_l(\vec{q})$:

$$\varepsilon_l(\vec{q}) = Jz\sqrt{m_z^2 + m_x^2 \frac{a^2 \vec{q}^2}{z}}. \qquad (3.30)$$

The dispersion relation (Equation 3.30) reveals the typical soft-mode behavior

$$\lim_{T \to T_c} \varepsilon(\vec{q} = 0) = 0$$

in accordance to the microscopic behavior. In a scaling form, the dispersion reads

$$\varepsilon_l(\vec{q}, \xi_c) = \xi_c^{-1} f_l(q\xi_c),$$

where ξ_c is the correlation length, $f_l(x) \propto \sqrt{1+x^2}$ is the scaling function, and the critical exponents fulfill $\nu = \beta$. As pointed out in the microscopic approach, the dispersion relation can be damped. In that case, Equation 3.28 has to be modified resulting in

$$\frac{\partial \vec{S}(\vec{x},t)}{\partial t} = \vec{h}(\vec{x},t) \times \vec{S}(\vec{x},t) + \vec{D}(\vec{S}), \qquad (3.31)$$

where the damping part $\vec{D}$ is obtained in Trimper et al. (2007). We get

$$\frac{\partial \vec{S}}{\partial t} = \vec{h} \times \vec{S} - \frac{1}{\tau_1} \vec{h} - \frac{1}{\tau_2} \vec{S} \times (\vec{S} \times \vec{h}). \qquad (3.32)$$

Two damping terms arise with the prefactors τ_1 and τ_2. The set of Equation 3.32 is studied in detail. The results for the excitation energy and its damping are depicted in Figure 3.22. Following the line for magnetic nanomaterial, we plan to apply the approach to ferroelectric nanoparticles.

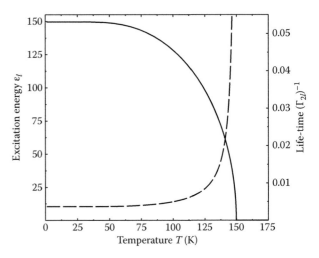

FIGURE 3.22 Excitation energy $\varepsilon\ (\vec{q} = 0)$ (solid curve) and the life-time $(\Gamma_{2l})^{-1}$ (dashed curve) at $\vec{q} = 0$ as function of the temperature, $\Omega = 10\,K$, $J = 25\,K$.

3.5 Conclusions

In this chapter, we have offered a special insight into a very vital field of current research, namely, the study of ferroelectric nanomaterials. Our intention was to present both experimental results as well as a more refined theoretical description. Obviously, such a review covers only selected aspects of an extended field of interest. For this reason, the considerations focus on properties of ferroelectric nanoparticles that should have a significant impact on future research. Thus, the manuscript is concentrated on such properties of ferroelectric nanoparticles that are embedded in the well-established concepts of solid state physics. This includes the description of the collective properties of a many-body system in terms of elementary excitations of quasi-particles as well as a statistical modeling of macroscopic quantities as polarization, susceptibility, hysteresis, and so on. Like in bulk materials, the theoretical approach may be roughly divided in three levels, a macroscopic, a mesoscopic, and a microscopic one. However, in the physics of nanoparticles, the analysis is often relied on a multiscale approach, in which the macroscopic properties are understood by analyzing the underlying microscopic interactions. Because this approach is considered relevant and reasonable in case of bulk material, we have adopted a similar concept for nanoparticles, too. The special problems, related to low-dimensional systems, are described in detail in the chapter. Aside from the theoretical modeling, the basic experimental findings as well as some important methods in preparing nanomaterial has been summarized.

Acknowledgments

We acknowledge support by the Martin-Luther University Halle, the International Max Planck Research School for Science and Technology of Nanostructures in Halle, and the DFG-SFB 418.

References

Ahn, C., Rabe, K., and Triscone, J. 2004. Ferroelectricity at the nanoscale: Local polarization in oxide thin films and heterostructures. *Science*, 303(5657):488–491.

Akdogan, E. K. and Safari, A. 2002. Phenomenological theory of size effects on the cubic-tetragonal phase transition in $BaTiO_3$ nanocrystals. *Jpn. J. Appl. Phys.*, 41(Part 1, No. 11B):7170–7175.

Alexe, M., Harnagea, C., Hesse, D., and Gosele, U. 2001. Polarization imprint and size effects in mesoscopic ferroelectric structures. *Appl. Phys. Lett.*, 79(2):242–244.

Anliker, M., Kanzig, W., and Peter, M. 1952. Das verhalten sehr kleiner ferroelektrischer teilchen. *Helv. Phys. Acta*, 25(5):474–475.

Anliker, M., Brugger, H., and Kanzig, W. 1954. Das verhalten von kolloidalen seignetteelektrika 3. bariumtitanat $BaTiO_3$. *Helv. Phys. Acta*, 27(2):99–124.

Arlt, G., Hennings, D., and Dewith, G. 1985. Dielectric-properties of fine-grained barium-titanate ceramics. *J. Appl. Phys.*, 58(4):1619–1625.

Ashkin, A., Boyd, G., Dziedzic, J., Smith, R., Ballman, A., Levinstein, J. J., and Nassau, K. 1966. Optically-induced refractive index inhomogeneities in $LiNbO_3$ and $LiTaO_3$. *Appl. Phys. Lett.*, 9(1):72.

Auciello, O., Scott, J. F., and Ramesh, R. 1998. The physics of ferroelectric memories. *Phys. Today*, 51(7):22–27.

Baudry, L. 1999. Theoretical investigation of the influence of space charges on ferroelectric properties of $PbZrTiO_3$ thin film capacitor. *J. Appl. Phys.*, 86(2):1096–1105.

Baudry, L. 2006. Surface polarization enhancement and switching properties of small ferroelectric particles. *Ferroelectrics*, 333(1):27–39.

Baudry, L. and Tournier, J. 2001. Lattice model for ferroelectric thin film materials including surface effects: Investigation on the "depolarizing" field properties. *J. Appl. Phys.*, 90(3):1442–1454.

Baudry, L. and Tournier, J. 2005. Model for ferroelectric semiconductors thin films accounting for the space varying permittivity. *J. Appl. Phys.*, 97(2):024104.

Blinc, R. 1960. On the isotopic effects in the ferroelectric behaviour of crystals with short hydrogen bonds. *J. Phys. Chem. Solids*, 13(3–4):204–211.

Blinc, R. and Zeks, B. 1974. *Soft Modes in Ferroelectrics and Antiferroelectrics*. North-Holland, Amsterdam, the Netherlands.

Burns, G. and Scott, B. 1970. Raman studies of underdamped soft modes in $PbTiO_3$. *Phys. Rev. Lett.*, 25(3):167–170.

Cao, H. and Li, Z. 2003. Thermodynamic properties of temperature graded ferroelectric film. *J. Phys.: Condens. Matter*, 15(36):6301–6310.

Charnaya, E. V., Pogorelova, O., and Tien, C. 2001. Phenomenological model for the antiferro-electric phase transition in thin films and small particles. *Physica B*, 305(2):97–104.

Chattopadhyay, S., Ayyub, P., Palkar, V. R., and Multani, M. 1995. Size-induced diffuse phase-transition in the nanocrystalline ferroelectric $PbTiO_3$. *Phys. Rev. B*, 52(18):13177–13183.

Chattopadhyay, S., Ayyub, P., Palkar, V. R., Gurjar, A., Wankar, R., and Multani, M. 1997. Finite-size effects in antiferroelectric $PbZrO_3$ nanoparticles. *J. Phys.: Condens. Matter*, 9(38):8135–8145.

Chu, M.-W., Szafraniak, I., Scholz, R., Harnagea, C., Hesse, D., Alexe, M., and Gösele, U. 2004. Impact of misfit dislocations on the polarization instability of epitaxial nanostructured ferroelectric perovskites. *Nat. Mater.*, 3(2):87–90.

Cochran, W. 1959. Crystal stability and the theory of ferroelectricity. *Phys. Rev. Lett.*, 3(9):412–414.

Colla, E., Fokin, A., and Kumzerov, Y. 1997. Ferroelectrics properties of nanosize KDP particles. *Solid State Commun.*, 103(2):127–130.

Colla, E., Fokin, A., Koroleva, E., Kumzerov, Y., Vakhrushev, S., and Savenko, B. 1999. Ferroelectric phase transitions in materials embedded in porous media. *Nanostruct. Mater.*, 12(5–8):963–966.

Das, R. R., Bhattacharya, P., Perez, W., and Katiyar, R. 2003. Influence of Ca on structural and ferroelectric properties of laser ablated $SrBi_2Ta_2O_9$ thin films. *Jpn. J. Appl. Phys. 1*, 42(1):162–165.

Dawber, M., Rabe, K. M., and Scott, J. F. 2005. Physics of thin-film ferroelectric oxides. *Rev. Mod. Phys.*, 77(4):1083–1130.

Deng, Y. and Zhang, M. 2005. Orthorhombic-to-tetragonal phase transition in 6.8-nm $PbTiO_3$ nanoparticles. *Int. J. Mod. Phys. B*, 19(15–17):2669–2675.

Desu, S. and Payne, D. 1990. Interfacial segregation in perovskites. 1. Theory. *J. Am. Ceram. Soc.*, 73(11):3391–3397.

Du, Y., Chen, G., and Zhang, M. 2004. Grain size effects in $Bi_4Ti_3O_{12}$ nanocrystals investigated by Raman spectroscopy. *Solid State Commun.*, 132(3–4):175–179.

Duan, N., ten Elshof, J., Verweij, H., Greuel, G., and Dannapple, O. 2000. Enhancement of dielectric and ferroelectric properties by addition of Pt particles to a lead zirconate titanate matrix. *Appl. Phys. Lett.*, 77(20):3263–3265.

Economou, E. N. 2006. *Green's Functions in Quantum Physics*. Springer Series in Solid-State Science. Springer, Berlin, Germany.

Erdem, E., Boettcher, R., Glaesel, H., Hartmann, E., Klotzsche, G., and Michel, D. 2005. Size effects in $BaTiO_3$ nanopowders studied by EPR and NMR. *Ferroelectrics*, 316:43–49.

Evans, J. and Womack, R. 1988. An experimental 512-bit nonvolatile memory with ferroelectric storage cell. *IEEE J. Solid-State Circ.*, 23(5):1171–1175.

Faheem, Y. and Joya, K. 2007. Phase transformation and freestanding nanoparticles formation in lead zirconate titanate derived by sol-gel. *Appl. Phys. Lett.*, 91(6):063115-1–063115-3.

Faheem, Y. and Shoaib, M. 2006. Sol-gel processing and characterization of phase-pure lead zirconate titanate nanopowders. *J. Am. Ceram. Soc.*, 89(6):2034–2037.

Fernández-Osorio, A. L., Vázquez-Olmos, A., Mata-Zamora, E., and Saniger, J. M. 2007. Preparation of free-standing $Pb(Zr_{0.529}Ti_{0.48})O_3$ nanoparticles by sol-gel method. *J. Sol-Gel Sci. Technol.*, 42(2):145–149.

Frey, M. H., Xu, Z., Han, P., and Payne, D. A. 1998. The role of interfaces on an apparent grain size effect on the dielectric properties for ferroelectric barium titanate ceramics. *Ferroelectrics*, 206:337–353.

Fridkin, V. M. 2006. Critical size in ferroelectric nanostructures. *Phys. Usp.*, 49(2):193–202.

Fu, D., Suzuki, H., and Ishikawa, K. 2000. Size-induced phase transition in $PbTiO_3$ nanocrystals: Raman scattering study. *Phys. Rev. B*, 62(5):3125–3129.

Fujita, J., Suzuki, K., Wada, N., Sakabe, Y., Takeuchi, K., and Ohki, Y. 2006. Dielectric properties of $BaTiO_3$ thin films prepared by laser ablation. *Jpn. J. Appl. Phys. 1*, 45(10A):7806–7812.

Geneste, G., Bousquet, E., Junquera, J., and Ghosez, P. 2006. Finite-size effects in $BaTiO_3$ nanowires. *Appl. Phys. Lett.*, 88(11):112906.

Ghosez, P. and Junquera, J. 2006. First-principles modeling of ferroelectric oxides nanostructures. In *Handbook of Theoretical and Computational Nanotechnology*, Chap. 134, pp. 1–149, M. Rieth and W. Schommers (eds.). American Scientific Publisher, Stevenson Ranch, CA.

Gopalan, V. and Gupta, M. 1996. Observation of internal field in $LiTaO_3$ single crystals: Its origin and time-temperature dependence. *Appl. Phys. Lett.*, 68(7):888–890.

Grinberg, I., Cooper, V., and Rappe, A. 2002. Relationship between local structure and phase transitions of a disordered solid solution. *Nature*, 419(6910):909–911.

Haertling, G. 1999. Ferroelectric ceramics: History and technology. *J. Am. Ceram. Soc.*, 82(4):797–818.

Hench, L. L. and West, J. K. 1990. The sol-gel process. *Chem. Rev.*, 90(1):33–72.

Hennings, D. 1987. Barium titanate based ceramic materials for dielectric use. *Int. J. High Technol. Ceram.*, 3(2):91–111.

Hennings, D. and Schreinemacher, H. 1994. High-permittivity dielectric ceramics with high endurance. *J. Eur. Ceram. Soc.*, 13(1):81–88.

Hennings, D., Schnell, A., and Simon, G. 1982. Diffuse ferroelectric phase transitions in $Ba(Ti_{1-y}Zr_y)O_3$ ceramics. *J. Am. Ceram. Soc.*, 65(11):539–544.

Hesse, D. and Alexe, M. 2005. Interfaces in nanosize perovskite titanate ferroelectrics—Microstructure and impact on selected properties. *Z. Metallkd.*, 96(5):448–451.

Higashijima, H., Kohiki, S., Takada, S., Shimizu, A., and Yamada, K. 1999. Optical and dielectric properties of quantum-confined $SrBi_2Ta_2O_9$ mesocrystals. *Appl. Phys. Lett.*, 75(20):3189–3191.

Hill, N. A. 2000. Why are there so few magnetic ferroelectrics? *J. Phys. Chem. B*, 104(29):6694–6709.

Hoshina, T., Kakemoto, H., Tsurumi, T., Wada, S., and Yashima, M. 2006. Size and temperature induced phase transition behaviors of barium titanate nanoparticles. *J. Appl. Phys.*, 99(5):054311.

Hoshina, T., Yasuno, H., Kakemoto, H., Tsurumi, T., and Wada, S. 2007. Particle size and temperature dependence of THz-region dielectric properties for $BaTiO_3$ nanoparticles. *Ferroelectrics*, 353:55–62.

Hu, J., Odom, T., and Lieber, C. 1999. Chemistry and physics in one dimension: Synthesis and properties of nanowires and nanotubes. *Acc. Chem. Res.*, 32(5):435–445.

Huang, H., Sun, C., Zhang, T., and Hing, P. 2001. Grain-size effect on ferroelectric $Pb(Zr_{1-x}Ti_x)O_3$ solid solutions induced by surface bond contraction. *Phys. Rev. B*, 63(18):184112.

Huang, T.-C., Wang, M.-T., Sheu, H.-S., and Hsieh, W.-F. 2007. Size-dependent lattice dynamics of barium titanate nanoparticles. *J. Phys.: Condens. Matter*, 19(47):476212.

Iijima, K., Takayama, R., Tomita, Y., and Ueda, I. 1986. Epitaxial-growth and the crystallographic, dielectric, and pyroelectric properties of lanthanum-modified lead titanate thin-films. *J. Appl. Phys.*, 60(8):2914–2919.

Ishikawa, K. and Uemori, T. 1999. Surface relaxation in ferroelectric perovskites. *Phys. Rev. B*, 60(17):11841–11845.

Ishikawa, K., Yoshikawa, K., and Okada, N. 1988. Size effect on the ferroelectric phase transition in $PbTiO_3$ particles. *Phys. Rev. B*, 37(10):5852–5855.

Jaccard, C., Kanzig, W., and Peter, M. 1953. Das verhalten von kolloidalen seignetteelektrika. 1. Kaliumphosphat KH_2PO_4. *Helv. Phys. Acta*, 26(5):521.

Jana, A., Mandal, T., Ram, S., and Kundu, T. 2004. Synthesis of $BaTiO_3$ nanoparticles through a novel chemical route with polymer precursor. *Ind. J. Phys. Pt-A*, 78A(1):97–99.

Jana, A., Kundu, T. K., Pradhan, S. K., and Chakravorty, D. 2005. Dielectric behavior of Fe-iondoped $BaTiO_3$ nanoparticles. *J. Appl. Phys.*, 97(4):044311.

Jeong, S.-J., Ha, M.-S., and Koh, J.-H. 2006. Shape effect on dielectric and piezoelectric properties in multilayer actuator. *Ferroelectrics*, 332:83–87.

Jiang, B. and Bursill, L. A. 1999. Phenomenological theory of size effects in ultrafine ferroelectric particles of lead titanate. *Phys. Rev. B*, 60(14):9978–9982.

Jiang, A., Li, G., and Zhang, L. 1998. Dielectric study in nanocrystalline $Bi_4Ti_3O_{12}$ prepared by chemical coprecipitation. *J. Appl. Phys.*, 83(9):4878–4883.

Kaenzig, W. 1950. Atomic positions and vibrations in the ferroelectric $BaTiO_3$ lattice. *Phys. Rev.*, 80(1):94–95.

Ke, H., Jia, D., Wang, W., and Zhou, Y. 2007. Ferroelectric phase transition investigated by thermal analysis and Raman scattering in $SrBi_2Ta_2O_9$ nanoparticles. *Nanosci. Technol.*, Pts. 1 and 2, 121–123:843–846.

Khare, P. and Sa, D. 2008. Landau theory of ferroelectric transition in long cylindrical nanoparticles. *Eur. Phys. J. B*, 63(2):205–209.

Kim, T. and Jang, H. 2000. B-site vacancy as the origin of spontaneous normal-to-relaxor ferro-electric transitions in la-modified $PbTiO_3$. *Appl. Phys. Lett.*, 77(23):3824–3826.

Kim, C., Park, J., Moon, B., Seo, H., Choi, Y., Yeo, K., Chung, S., Son, S., and Kim, J. 2005. Synthesis and nanodomain patterns of $BaTiO_3$ nanoparticles. *J. Korean Phys. Soc.*, 46(1):308–310.

Kleemann, W., Dec, J., Kahabka, D., Lehnen, P., and Wang, Y. G. 1999a. Phase transitions and precursor phenomena in doped quantum paraelectrics. *Ferroelectrics*, 235:33–46.

Kleemann, W., Wang, Y. G., Lehnen, P., and Dec, J. 1999b. Phase transitions in doped quantum paraelectrics. *Ferroelectrics*, 229:39–44.

Kleemann, W., Dec, J., Wang, Y. G., Lehnen, P., and Prosandeev, S. 2000. Phase transitions and relaxer properties of doped quantum paraelectrics. *J. Phys. Chem. Solids*, 61(2):167–176.

Kobayashi, Y., Nishikata, A., Tanase, T., and Konno, M. 2004. Size effect on crystal structures of barium titanate nanoparticles prepared by a sol-gel method. *J. Sol-Gel Sci. Technol.*, 29(1):49–55.

Kodama, S., Kido, O., Suzuki, H., Saito, Y., and Kaito, C. 2005. Characterization of nanoscale $BaTiO_3$ ultrafine particles prepared by gas evaporation method. *J. Cryst. Growth*, 282(1–2):60–65.

Kohiki, S., Takada, S., Shimizu, A., Yamada, K., Higashijima, H., and Mitome, M. 2000. Quantum-confinement effects on the optical and dielectric properties for mesocrystals of $BaTiO_3$ and $SrBi_2Ta_2O_9$. *J. Appl. Phys.*, 87(1):474–478.

Kohiki, S., Nogami, S., Kawakami, S., Takada, S., Shimooka, H., Deguchi, H., Mitome, M., and Oku, M. 2003. Large frequency dependence of lowered maximum dielectric constant temperature of $LiTaO_3$ nanocrystals dispersed in mesoporous silicate. *Appl. Phys. Lett.*, 82(23):4134–4136.

Kohn, W. 1999. Nobel lecture: Electronic structure of matter-wave functions and density functionals. *Rev. Mod. Phys.*, 71(5):1253–1266.

Kojima, S. and Shimada, S. 1996. Soft mode spectroscopy of bismuth titanate single crystals. *Physica B*, 220:617–619.

Kojima, S., Imaizumi, R., Hamazaki, S., and Takashige, M. 1994. Raman-scattering study of bismuth layer-structure ferroelectrics. *Jpn. J. Appl. Phys. 1*, 33(9B):5559–5564.

Kuehnel, A., Wendt, S., and Wesselinowa, J. M. 1977. Dynamic behavior of Ising-model in a transverse field. *Phys. Status Solidi B*, 84(2):653–664.

Kundu, T. K., Jana, A., and Barik, P. 2008. Doped barium titanate nanoparticles. *B Mater. Sci.*, 31:501–505.

Landau, L. D., Lifshitz, E. M., and Pitaevski, L. P. 1980. *Statistical Physics, Part 2*. Pergamon, Oxford, U.K.

Langhammer, H., Muller, T., Felgner, K., and Abicht, H. 2000. Crystal structure and related properties of manganese-doped barium titanate ceramics. *J. Am. Ceram. Soc.*, 83(3):605–611.

Lebedev, M. and Akedo, J. 2002. What thickness of the piezoelectric layer with high breakdown voltage is required for the microactuator? *Jpn. J. Appl. Phys. 1*, 41(5B):3344–3347.

Lee, S., Son, T., Yun, J., Kwon, H., Messing, G., and Jun, B. 2004. Preparation of $BaTiO_3$ nanoparticles by combustion spray pyrolysis. *Mater. Lett.*, 58(22–23):2932–2936.

Lee, K., Kang, Y., Jung, K., and Kim, J. 2005a. Preparation of nanosized $BaTiO_3$ particle by citric acid-assisted spray pyrolysis. *J. Alloy Compd.*, 395(1–2):280–285.

Lee, S. K., Hesse, D., Alexe, M., Lee, W., Nielsch, K., and Gösele, U. 2005b. Growth and characterization of epitaxial ferroelectric lanthanum-substituted bismuth titanate nanostructures with three different orientations. *J. Appl. Phys.*, 98(12):124302.

Lee, C. T., Zhang, M. S., and Yin, Z. 2008. Doping effects of structural transformation and soft mode in $Ba_xPb_{1x}TiO_3$ nanoparticles. *J. Mater. Sci.*, 43(8):2675–2679.

Li, S., Eastman, J., Li, Z., Foster, C., Newnham, R., and Cross, L. 1996. Size effects in nanostructured ferroelectrics. *Phys. Lett. A*, 212(6):341–346.

Li, J., Wu, Y., Tanaka, H., Yamamoto, T., and Kuwabara, M. 2004. Preparation of a monodispersed suspension of barium titanate nanoparticles and electrophoretic deposition of thin films. *J. Am. Ceram. Soc.*, 87(8):1578–1581.

Lines, M. E. and Glass, A. 2004. *Principles and Applications of Ferroelectrics and Related Materials*. Clarendon Press, Oxford, U.K.

Liu, J., Zhang, S., Zeng, H., Yang, C., and Yuan, Y. 2005. Coercive field dependence of the grain size of ferroelectric films. *Phys. Rev. B*, 72(17):172101.

Lu, C. and Saha, S. 2001. Fabrication of ferroelectric lead bismuth tantalate nanoparticles by colloid emulsion route. *Br. Ceram. Trans.*, 100(3):120–123.

Ma, W. and Hesse, D. 2004. Polarization imprint in ordered arrays of epitaxial ferroelectric nanostructures. *Appl. Phys. Lett.*, 84(15):2871–2873.

Meng, J., Zou, G., Li, J., Cui, Q., Wang, X., Wang, Z., and Zhao, M. 1994. Investigations of the phase-transition in nanocrystalline $Pb_{1-x}La_xTiO_3$ system. *Solid State Commun.*, 90(10):643–645.

Meng, J., Huang, Y., and Zou, G. 1996. Temperature dependence of the Raman active modes in nanocrystalline $Bi_4Ti_3O_{12}$. *Solid State Commun.*, 97(10):887–890.

Michael, T., Trimper, S., and Wesselinowa, J. M. 2006. Size and doping effects on the coercive field of ferroelectric nanoparticles: A microscopic model. *Phys. Rev. B*, 74(21):214113.

Michael, T., Trimper, S., and Wesselinowa, J. M. 2007. Size effects on static and dynamic properties of ferroelectric nanoparticles. *Phys. Rev. B*, 76(9):094107.

Michael, T., Trimper, S., and Wesselinowa, J. M. 2008. Impact of defects on the properties of ferroelectric nanoparticles. *Ferroelectrics*, 363:110–119.

Miller, R. and Nordland, W. 1970. Absolute signs of second-harmonic generation coefficients of piezoelectric crystals. *Phys. Rev. B*, 2(12):4896–4902.

Millis, A. 1998. Lattice effects in magnetoresistive manganese perovskites. *Nature*, 392(6672):147–150.

Mishina, E., Morozov, A., Sigov, A., Sherstyuk, N., Aktsipetrov, O., Lemanov, V., and Rasing, T. 2002. A study of the structural phase transition in strontium titanate single crystal by coherent and incoherent second optical harmonic generation. *J. Exp. Theor. Phys.*, 94(3):552–567.

Morozovska, A. N. 2006. Ferroelectricity enhancement in confined nanorods: Direct variational method. *Phys. Rev. B*, 73(21):214106–214118.

Nagarajan, V., Prasertchoung, S., Zhao, T., Zheng, H., Ouyang, J., Ramesh, R., Tian, W. et al. 2004. Size effects in ultrathin epitaxial ferroelectric heterostructures. *Appl. Phys. Lett.*, 84(25):5225.

Noguchi, Y., Miyayama, M., and Kudo, T. 2001. Direct evidence of A-site-deficient strontium bismuth tantalate and its enhanced ferroelectric properties. *Phys. Rev. B*, 63(21):214102.

Noguchi, Y., Miyayama, M., Oikawa, K., Kamiyama, T., Osada, M., and Kakihana, M. 2002. Defect engineering for control of polarization properties in $SrBi_2Ta_2O_9$. *Jpn. J. Appl. Phys. 1*, 41(11B):7062–7075.

Ohno, T., Suzuki, D., Suzuki, H., and Ida, T. 2004. Size effect for barium titanate nano-particles. *J. Soc. Powder Technol.*, 42:85–91.

Ohno, T., Suzuki, D., Ishikawa, K., Horiuchi, M., Matsuda, T., and Suzuki, H. 2006. Size effect for $Ba(Zr_xTi_{1x})O_3$ ($x = 0.05$) nano-particles. *Ferroelectrics*, 337(1):25–32.

Ohno, T., Suzuki, D., Ishikawa, K., and Suzuki, H. 2007. Size effect for lead zirconate titanate nano-particles with PZT(40/60) composition. *Adv. Powder Technol.*, 18(5):579–589.

Parker, C., Maria, J., and Kingon, A. 2002. Temperature and thickness dependent permittivity of $(Ba,Sr)TiO_3$ thin films. *Appl. Phys. Lett.*, 81(2):340–342.

Patzke, G. R., Krumeich, F., and Nesper, R. 2002. Oxidic nanotubes and nanorods—Anisotropic modules for a future nanotechnology. *Angew. Chem. Int. Ed.*, 41(14):2446–2461.

Peng, Z. and Chen, Y. 2003. Preparation of $BaTiO_3$ nanoparticles in aqueous solutions. *Microelectron. Eng.*, 66(1–4):102–106.

Pertsev, N., Contreras, J., Kukhar, V., Hermanns, B., Kohlstedt, H., and Waser, R. 2003. Coercive field of ultrathin $Pb(Zr_{0.52}Ti_{0.48})O_3$ epitaxial films. *Appl. Phys. Lett.*, 83(16):3356–3358.

Pirc, R. and Blinc, R. 2004. Off-center Ti model of barium titanate. *Phys. Rev. B*, 70(13):134107.

Plonska, M., Czekaj, D., and Surowiak, Z. 2003. Application of the sol-gel method to the synthesis of ferroelectric nanopowders $(Pb_{1-x}La_x)(Zr_{0.65}Ti_{0.35})_{1-0.25x}O_3$, $0.06 \leq x \leq 0.1$. *Mater. Sci. Poland*, 21(4):431–437.

Prosandeev, S., Kleemann, W., Westwanski, B., and Dec, J. 1999. Quantum paraelectricity in the mean-field approximation. *Phys. Rev. B*, 60(21):14489–14491.

Qi, J., Wang, Y., Chen, W., Li, L., and Chan, H. 2005. Direct large-scale synthesis of perovskite barium strontium titanate nano-particles from solutions. *J. Solid State Chem.*, 178(1):279–284.

Rabe, K. M. and Joannopoulos, J. 1987. Ab initio determination of a structural phase transition temperature. *Phys. Rev. Lett.*, 59(5):570–573.

Ramam, K. and Lopez, M. 2008. Microstructure, dielectric and electromechanical properties of PLSZFT nanoceramics for piezoelectric applications. *J. Mater. Sci. Mater. El.*, 19(11):1140–1145.

Ramam, K. and Miguel, V. 2006. Microstructure, dielectric and ferroelectric characterization of Ba doped PLZT ceramics. *Eur. Phys. J. Appl. Phys.*, 35(1):43–47.

Ramesh, R., Aggarwal, S., and Auciello, O. 2001. Science and technology of ferroelectric films and heterostructures for non-volatile ferroelectric memories. *Mater. Sci. Eng. R*, 32(6):191–236.

Rao, C. and Nath, M. 2003. Inorganic nanotubes. *Dalton T*, 1(1):1–24.

Ren, S., Lu, C., Liu, J., Shen, H., and Wang, Y. 1996. Size-related ferroelectric-domain-structure transition in a polycrystalline $PbTiO_3$ thin film. *Phys. Rev. B*, 54(20):R14337–R14340.

Rychetsky, I. and Hudak, O. 1997. The ferroelectric phase transition in small spherical particles. *J. Phys.: Condens. Matter*, 9(23):4955–4965.

Sakai, T., Watanabe, T., Funakubo, H., Saito, K., and Osada, M. 2003. Effect of La substitution on electrical properties of highly oriented $Bi_4Ti_3O_{12}$ films prepared by metalorganic chemical vapor deposition. *Jpn. J. Appl. Phys. 1*, 42(1):166–169.

Satapathy, S., Gupta, P. K., Srivastava, H., Srivastava, A. K., Wadhawan, V. K., Varma, K. B. R., and Sathe, V. G. 2007. Effect of capping ligands on the synthesis and on the physical properties of the nanoparticles of LiTaO₃. *J. Cryst. Growth*, 307(1):185–191.

Schafer, J., Sigmund, W., Roy, S., and Aldinger, F. 1997. Low temperature synthesis of ultrafine Pb(Zr,Ti)O₃ powder by sol-gel combustion. *J. Mater. Res.*, 12(10):2518–2521.

Schlag, S. and Eicke, H.-F. 1994. Size driven phase-transition in nanocrystalline BaTiO₃. *Solid State Commun.*, 91(11):883–887.

Schlag, S., Eicke, H.-F., and Stern, W. 1995. Size driven phase transition and thermodynamic properties of nanocrystalline BaTiO₃. *Ferroelectrics*, 173:351–369.

Scott, J. F. 2006. Nanoferroelectrics: Statics and dynamics. *J. Phys.: Condens. Matter*, 18(17):R361–R386.

Scott, J. F. 2008. Ferroelectric nanostructures for device applications. In: *Handbook of Advanced Dielectric Piezoelectric and Ferroelectric Materials: Synthesis, Properties and Applications*, vol. 5, Cambridge University Press, Cambridge, U.K.

Scott, J. F., Zhang, M., Godfrey, R., Araujo, C., and McMillan, L. 1987. Raman-spectroscopy of submicron KNO₃ films. *Phys. Rev. B*, 35(8):4044–4051.

Seol, K., Tomita, S., Takeuchi, K., Miyagawa, T., Katagiri, T., and Ohki, Y. 2002. Gas-phase production of monodisperse lead zirconate titanate nanoparticles. *Appl. Phys. Lett.*, 81(10):1893–1895.

Seol, K., Takeuchi, K., and Ohki, Y. 2004. Ferroelectricity of single-crystalline, monodisperse lead zirconate titanate nanoparticles of 9nm in diameter. *Appl. Phys. Lett.*, 85(12):2325–2327.

Shiratori, Y., Pithan, C., Dornseiffer, J., and Waser, R. 2007a. Raman scattering studies on nanocrystalline BaTiO₃ part I—Isolated particles and aggregates. *J. Raman Spectrosc.*, 38(10):1288–1299.

Shiratori, Y., Pithan, C., Dornseiffer, J., and Waser, R. 2007b. Raman scattering studies on nanocrystalline BaTiO₃ part II—Consolidated polycrystalline ceramics. *J. Raman Spectrosc.*, 38(10):1300–1306.

Song, T., Kim, J., and Kwun, S.-I. 1996. Size effects on the quantum paraelectric SrTiO₃ nanocrystals. *Solid State Commun.*, 97(2):143–147.

Spaldin, N. 2004. Fundamental size limits in ferroelectricity. *Science*, 304(5677):1606–1607.

Stachiotti, M. G. 2004. Ferroelectricity in BaTiO₃ nanoscopic structures. *Appl. Phys. Lett.*, 84(2):251.

Strukov, B. and Levanyuk, A. 1998. *Ferroelectric Phenomena in Crystals*. Springer-Verlag, Berlin, Germany.

Takizawa, K., Hoshina, T., Kakemoto, H., Tsurumi, T., Kuroiwa, Y., and Wada, S. 2007. Control of mesoscopic particle structure in barium titanate nanoparticles and their dielectric properties. *Key Eng. Mater.*, 350:59–62.

Tang, X., Wang, J., Wang, X., and Chan, H. 2004. Effects of grain size on the dielectric properties and tunabilities of sol-gel derived Ba(Zr₀.₂Ti₀.₈)O₃ ceramics. *Solid State Commun.*, 131(3–4):163–168.

Tangonan, G., Barnoski, M., Lotspeich, J., and Lee, A. 1977. High optical power capabilities of Ti-diffused LiTaO₃ waveguide modulator structures. *Appl. Phys. Lett.*, 30(5):238–239.

Tenne, D. A., Clark, A., James, A. R., Chen, K., and Xi, X. X. 2001. Soft phonon modes in Ba₀.₅Sr₀.₅TiO₃ thin films studied by Raman spectroscopy. *Appl. Phys. Lett.*, 79(23):3836–3838.

Tenne, D. A., Bruchhausen, A., Lanzillotti-Kimura, N. D., Fainstein, A., Katiyar, R. S., Cantarero, A., Soukiassian, A. et al. 2006. Probing nanoscale ferroelectricity by ultraviolet Raman spectroscopy. *Science*, 313(5793):1614–1616.

Tilley, D. R. and Zeks, B. 1984. Landau theory of phase-transitions in thick-films. *Solid State Commun.*, 49(8):823–827.

Tinte, S., Stachiotti, M. G., Sepliarsky, M., Migoni, R., and Rodriguez, C. 1999. Atomistic modelling of BaTiO₃ based on first-principles calculations. *J. Phys.: Condens. Matter*, 11(48):9679–9690.

Trimper, S., Michael, T., and Wesselinowa, J. M. 2007. Ferroelectric soft modes and Gilbert damping. *Phys. Rev. B*, 76(9):094108.

Tserkovnikov, Y. A. 1971. Decoupling of chains of equations for two-time green's functions. *Theor. Mater. Phys.*, 7(2):250.

Tserkovnyak, Y., Brataas, A., Bauer, G., and Halperin, B. 2005. Nonlocal magnetization dynamics in ferromagnetic heterostructures. *Rev. Mod. Phys.*, 77(4):1375–1421.

Tsunekawa, S., Ishikawa, K., Li, Z., Kawazoe, Y., and Kasuya, A. 2000. Origin of anomalous lattice expansion in oxide nanoparticles. *Phys. Rev. Lett.*, 85(16):3440–3443.

Tyunina, M., Levoska, J., Sternberg, A., and Leppavuori, S. 1998. Relaxor behavior of pulsed laser deposited ferroelectric (Pb₁₋ₓLaₓ)(Zr₀.₆₅Ti₀.₃₅)O₃ films. *J. Appl. Phys.*, 84(12):6800–6810.

Uchino, K. and Nomura, S. 1982. Critical exponents of the dielectric-constants in diffused-phase-transition crystals. *Ferroelectr. Lett.*, 44(3):55–61.

Uchino, K., Sadanaga, E., and Hirose, T. 1989. Dependence of the crystal structure on particle size in barium titanate. *J. Am. Ceram. Soc.*, 72(8):1555–1558.

Urban, J. J., Spanier, J. E., Ouyang, L., Yun, W. S., and Park, H. 2003. Single-crystalline barium titanate nanowires. *Adv. Mater.*, 15(5):423–426.

Venigalla, S. 2001. Advanced materials and powders digest—Barium titanate. *Am. Ceram. Soc. Bull.*, 6:63–64.

Viswanath, R. and Ramasamy, S. 1997. Preparation and ferroelectric phase transition studies of nanocrystalline BaTiO₃. *Nanostruct. Mater.*, 8(2):155–162.

Wada, S., Narahara, M., Hoshina, T., Kakemoto, H., and Tsurumi, T. 2003. Preparation of nm-sized BaTiO₃ particles using a new 2-step thermal decomposition of barium titanyl oxalate. *J. Mater. Sci.*, 38(12):2655–2660.

Wada, S., Hoshina, T., Yasuno, H., Nam, S., Kakemoto, H., Tsurumi, T., and Yashima, M. 2005a. Size dependence of dielectric properties for nm-sized barium titanate crystallites and its origin. *J. Korean Phys. Soc.*, 46(1):303–307.

Wada, S., Hoshina, T., Yasuno, H., Ohishi, M., Kakemoto, H., Tsurumi, T., and Yashima, M. 2005b. Dielectric properties of nm-sized barium titanate fine particles and their size dependence. *Adv. Electron. Ceram. Mater.*, 26:89.

Wang, C. L. and Smith, S. R. P. 1995. Landau theory of the size-driven phase transition in ferroelectrics. *J. Phys.: Condens. Matter*, 7(36):7163–7171.

Wang, J. and Zhang, T. 2006. Effect of long-range elastic interactions on the toroidal moment of polarization in a ferroelectric nanoparticle. *Appl. Phys. Lett.*, 88(18):182904.

Wang, Y. G., Zhong, W.-L., and Zhang, P. 1994a. Size driven phase transition in ferroelectric particles. *Solid State Commun.*, 90(5):329–332.

Wang, Y. G., Zhong, W.-L., and Zhang, P. 1994b. Size effects on the curie temperature of ferroelectric particles. *Solid State Commun.*, 92(6):519–523.

Wang, C. L., Smith, S. R. P., and Tilley, D. R. 1996. Size effect on the dielectric susceptibility of ferroelectrics. *Ferroelectrics*, 186:33–36.

Wang, Y. G., Zhong, W.-L., and Zhang, P. 1997. Ferroelectric films described by transverse Ising model with long-range interactions. *Solid State Commun.*, 101(11):807–810.

Wang, Y. G., Kleemann, W., Dec, J., and Zhong, W.-L. 1998a. Dielectric properties of doped quantum paraelectrics. *Europhys. Lett.*, 42(2):173–178.

Wang, Y. G., Kleemann, W., Zhong, W.-L., and Zhang, L. 1998b. Impurity-induced phase transition in quantum paraelectrics. *Phys. Rev. B*, 57(21):13343–13346.

Wang, C. L., Xin, Y., Wang, X. S., and Zhong, W.-L. 2000. Size effects of ferroelectric particles described by the transverse Ising model. *Phys. Rev. B*, 62(17):11423–11427.

Wang, X., Zhang, Z., and Zhou, S. 2001. Preparation of nanocrystalline SrTiO$_3$ powder in sol-gel process. *Mater. Sci. Eng. B Solid*, 86(1):29–33.

Wesselinowa, J. M. 1990. The dielectric function and reflectivity in KDP-type ferroelectrics. *Phys. Status Solidi B*, 160(2):697–704.

Wesselinowa, J. M. 1994. Phase-transitions of PbHPO$_4$-type and PbDPO$_4$-type ferroelectrics investigated with a greens-function technique. *Phys. Rev. B*, 49(5):3098–3103.

Wesselinowa, J. M. 2001. On the theory of thin ferroelectric films. *Phys. Status Solidi B*, 223(3):737–743.

Wesselinowa, J. M. 2002a. Dielectric susceptibility of ferroelectric thin films. *Solid State Commun.*, 121(9–10):489–492.

Wesselinowa, J. M. 2002b. Dynamical properties of thin ferroelectric films described by the transverse Ising model. *Phys. Status Solidi B*, 231(1):187–191.

Wesselinowa, J. M. 2002c. Electric field dependence of phase transitions in ferroelectric thin films. *Phys. Status Solidi B*, 229(3):1329–1333.

Wesselinowa, J. M. 2002d. Properties of ferroelectric thin films with a first-order phase transitions. *Solid State Commun.*, 121(2–3):89–92.

Wesselinowa, J. M. 2005a. Dielectric function of antiferroelectric thin films. *Phys. Status Solidi B*, 242(7):1528–1536.

Wesselinowa, J. M. 2005b. Effects of spin-phonon interaction on the dynamical properties of thin ferroelectric films. *J. Phys.: Condens. Matter*, 17(19):3001–3014.

Wesselinowa, J. M. and Apostolov, A. 1997. On the origin of the central peak in hydrogen-bonded ferroelectrics. *Solid State Commun.*, 101(5):343–346.

Wesselinowa, J. M. and Dimitrov, A. B. 2007. Influence of substrates on the statical and dynamical properties of ferroelectric thin films. *Phys. Status Solidi B*, 244(6):2242–2253.

Wesselinowa, J. M. and Kovachev, S. 2007. Hardening and softening of soft phonon modes in ferroelectric thin films. *Phys. Rev. B*, 75(4):045411.

Wesselinowa, J. M. and Marinov, M. 1992. On the theory of 1st-order phase-transition in order-disorder ferroelectrics. *Int. J. Mod. Phys. B*, 6(8):1181–1192.

Wesselinowa, J. M. and Trimper, S. 2001. Critical behaviour of the transverse Ising model with modified surface exchange. *Int. J. Mod. Phys. B*, 15(4):379–384.

Wesselinowa, J. M. and Trimper, S. 2002. Critical behaviour of ferroelectric thin films. *Int. J. Mod. Phys. B*, 16(3):473–480.

Wesselinowa, J. M. and Trimper, S. 2003. Layer polarizations and dielectric susceptibilities of antiferroelectric thin films. *Mod. Phys. Lett. B*, 17(25):1343–1347.

Wesselinowa, J. M. and Trimper, S. 2004a. Central peak in the excitation spectra of thin ferroelectric films. *Phys. Rev. B*, 69(2):024105.

Wesselinowa, J. M. and Trimper, S. 2004b. Thickness dependence of the dielectric function of ferroelectric thin films. *Phys. Status Solidi B*, 241(5):1141–1148.

Wesselinowa, J. M., Apostolov, A., and Filipova, A. 1994. Anharmonic effects in potassium-dihydrogen-phosphate-type ferroelectrics. *Phys. Rev. B*, 50(9):5899–5904.

Wesselinowa, J. M., Trimper, S., and Zabrocki, K. 2005. Impact of layer defects in ferroelectric thin films. *J. Phys.: Condens. Matter*, 17(29):4687–4699.

Wesselinowa, J. M., Michael, T., Trimper, S., and Zabrocki, K. 2006. Influence of layer defects on the damping in ferroelectric thin films. *Phys. Lett. A*, 348(3–6):397–404.

Wills, L., Wessels, B., Richeson, D., and Marks, T. 1992. Epitaxial growth of BaTiO$_3$ thin films by organometallic chemical vapor deposition. *Appl. Phys. Lett.*, 60(1):41–43.

Wilson, J. M. 1995. Barium-titanate. *Am. Ceram. Soc. Bull.*, 74(6):106–110.

Xia, Y., Yang, P., Sun, Y., Wu, Y., Mayers, B., Gates, B., Yin, Y., Kim, F., and Yan, H. 2003. One-dimensional nanostructures: Synthesis, characterization, and applications. *Adv. Mater.*, 15(5):353–389.

Yadlovker, D. and Berger, S. 2005. Uniform orientation and size of ferroelectric domains. *Phys. Rev. B*, 71(18):184112.

Yakovlev, S., Solterbeck, C.-H., Skou, E., and Es-Souni, M. 2006. Structural and dielectric properties of Er substituted sol-gel fabricated PbTiO$_3$ thin films. *Appl. Phys. A*, 82(4):727–731.

Yashima, M., Hoshina, T., Ishimura, D., Kobayashi, S., Nakamura, W., Tsurumi, T., and Wada, S. 2005. Size effect on the crystal structure of barium titanate nanoparticles. *J. Appl. Phys.*, 98(1):014313.

Yu, T., Shen, Z., Toh, W., Xue, J., and Wang, J. 2003a. Size effect on the ferroelectric phase transition in $SrBi_2Ta_2O_9$ nanoparticles. *J. Appl. Phys.*, 94(1):618–620.

Yu, T., Shen, Z., Xue, J., and Wang, J. 2003b. Effects of mechanical activation on the formation of $PbTiO_3$ from amorphous Pb-Ti-O precursor. *J. Appl. Phys.*, 93(6):3470–3474.

Zhang, J., Beetz, C., and Krupanidhi, S. 1994. Photoenhanced chemical-vapor deposition of $BaTiO_3$. *Appl. Phys. Lett.*, 65(19):2410–2412.

Zhang, L., Zhong, W., Wang, C., Zhang, P., and Wang, Y. G. 1999. Finite-size effects in ferroelectric solid solution $Ba_xSr_{1-x}TiO_3$. *J. Phys. D Appl. Phys.*, 32(5):546–551.

Zhang, L., Zhong, W.-L., and Kleemann, W. 2000. A study of the quantum effect in $BaTiO_3$. *Phys. Lett. A*, 276:162–166.

Zhang, M., Yin, Z., Chen, Q., Zhang, W., and Chen, W. 2001. Study of structural and photoluminescent properties in barium titanate nanocrystals synthesized by hydrothermal process. *Solid State Commun.*, 119(12):659–663.

Zhao, J., Fuflyigin, V., Wang, F., Norris, P., Bouthilette, L., and Woods, C. 1997. Epitaxial electro-optical $BaTiO_3$ films by single-source metal—Organic chemical vapour deposition. *J. Mater. Chem.*, 7(6):933–936.

Zhong, W.-L., Jiang, B., Zhang, P., Ma, J. M., Cheng, H., Yang, Z., and Li, L. 1993. Phase transition in $PbTiO_3$ ultra-fine particles of different sizes. *J. Phys.: Condens. Matter*, 5(16):2619–2624.

Zhong, W., Vanderbilt, D., and Rabe, K. M. 1994a. Phase transitions in $BaTiO_3$ from first principles. *Phys. Rev. Lett.*, 73(13):1861–1864.

Zhong, W.-L., Wang, Y. G., Zhang, P., and Qu, B. 1994b. Phenomenological study of the size effect on phase transitions in ferroelectric particles. *Phys. Rev. B*, 50:698.

Zhong, W., Wang, C. L., Zhang, P., and Wang, Y. G. 1999. Ferroelectric particles described by the transverse Ising model. *Ferroelectrics*, 229(1–4):11–19.

Zhu, X., Zhu, J., Zhou, S., Liu, Z., Ming, N. B., and Hesse, D. 2005. Microstructural characterization of $BaTiO_3$ ceramic nanoparticles synthesized by the hydrothermal technique. *Solid State Phenom.*, 106:41–46.

Zhu, K., Zhang, M., Deng, Y., Zhou, J., and Yin, Z. 2008. Finite-size effects of lattice structure and soft mode in bismuth titanate nanocrystals. *Solid State Commun.*, 145(9–10):456–460.

4

Helium Nanodroplets

Carlo Callegari
Sincrotrone Trieste

Wolfgang Jäger
University of Alberta

Frank Stienkemeier
University of Freiburg

4.1 Introduction

These days, the prefix nano- (in words such as "nanotechnology") evokes at first the idea of machines and foremost the idea of objects that do the same thing as their macroscopic counterparts, only they do it better, cheaper, and faster (in science fiction, usually with unexpected catastrophic consequences).

In such applications, size reduction is the goal. The accompanying change of properties, and the shift of balance between mechanical and electrostatic forces are well-recognized consequences, which may be desirable or not but appear at first sight to be of lesser importance than the function of the device.

This perspective changes once one recognizes that the very change of properties just mentioned does affect, profoundly, not only the function of an object but also the way it is assembled. It is not by chance that living cells more closely resemble a fuel cell than an internal combustion engine; it is not by chance that cellular structures are "self-assembled" through clever use of chemical forces.

We all easily accept that a molecule has different properties than its constituent atoms and, more in general, that a molecule cannot be further subdivided without losing its identity. The application of the same statement to bulk matter, say a gold crystal, is ill defined: one cannot exactly say to what extent the crystal should be fractionated before it becomes something else or vice versa when a small set of atoms begins to show collective properties. Yet we came to recognize that nanoaggregates have properties that are neither those of the single atom nor those of the bulk material. This recognition has resulted in a different branch of nanoscience, which uses nanoparticles as building blocks for the construction of meta-materials. This, in fact, has been done,

without a clue about nanoscience, for thousands of years now, for example, in the coloring of glass and ceramics.

At a yet more abstract level, nanoscience studies not so much what properties can be tailored onto a nanoparticle but rather why these properties come about in the first place (Jortner, 1992). There is no fixed recipe to define, let alone to predict, at which size a certain property deviates from "bulklike"; clearly two important parameters are the surface-to-volume ratio (surface atoms have about half as many nearest neighbors as do those in the interior, resulting in altered lattice parameters, dangling bonds, and reconstruction phenomena) and the onset of space quantization. For very small aggregates, both surface effects and space quantization can result in "magic numbers" associated with the completion of a shell. In the first case, the magic numbers reflect geometric constraints. The second (Knight et al., 1984) most often reflects occupation constraints for electrons, the same that are responsible for the regular structure of the periodic table of the elements (de Heer, 1993; Johnston, 2002).

In short, the investigation of nanoparticles is luckily not just "stamp collecting," that is, the organization of a vast amount of information based on some useful but arbitrary scheme: particles can instead be classified based on some underlying fundamental physical principle.

Many methods are available to produce and study particulate matter. Looking at a very familiar phenomenon, smoke, we can easily appreciate one of the key ingredients: the production of a strongly-out-of-equilibrium distribution of constituents (atoms/molecules) that can interact with each other and aggregate. The aggregation process must be strongly competitive with the supply, so that growth comes to a sharp stop and the formation of bulk matter is avoided; it is also important that the size of aggregates is not too broadly dispersed. We will meet these very concepts again when we discuss the formation of helium nanodroplets. The reader will rightfully suspect that the number densities involved are by necessity much smaller than those of solid matter.

Having recognized the rare and transient nature of nanoparticles, we come at a fork: we can choose to either collect and immobilize them onto a suitable substrate or investigate them for the brief time available before they turn into something else. Both cases are considered in the books by Haberland (1994a,b), which remain an excellent reference about clusters altogether (see also Castleman and Bowen, 1996). In the first case, there is a clear advantage in terms of the time available to the experimenter, which may be virtually infinite; the price to be paid is the strong interaction with the substrate, necessary to immobilize the particles that would otherwise diffuse and coalesce. This approach is naturally not suitable for our main theme: helium nanodroplets. It should, however, be mentioned that a close analogon, the investigation of liquid helium in specialized porous materials, has been up to the present moment a subject of great interest (Adams et al., 1984; Beamish et al., 1983; Kim and Chan, 2004b).

In the second case, transient sources, the number density, and the lifetime of the aggregates are in general the limiting factors of an experiment. Although very serious, these factors have not prevented the very successful use of transient aggregation sources for the investigation of the most disparate materials. Among the various implementation of such sources, a special place is occupied by those collectively referred to as molecular beams (also encompassing atomic beams), which we consider from here on. In the molecular-beams community, aggregates are traditionally referred to as clusters, only in recent years prefixed by "nano";

those made of helium are more often called (nano)droplets to reflect their unique liquid nature.

A helium droplet machine (Figure 4.1) is the direct descendant of the venerable molecular beam machine; the reader interested in the many common aspects is referred to the excellent books by Scoles (1988, 1992), Pauly (2000a,b), and Campargue (2001). Molecular beams have a long tradition, dating back almost 100 years, when Dunoyer (1911a,b) used the straight propagation of sodium vapor in vacuum as an explicit demonstration of its atomic nature. Note that in those early experiments, clustering was neither anticipated nor would it have been desired; typical densities in the source were low enough that one had a collision-free effusive source. Cluster sources, instead, rely on a high density of the gas, so that an expansion into vacuum can be obtained. Because the forward speed almost invariably exceeds the local speed of sound, this is referred to as a supersonic expansion. The term can be misleading: the forward velocity does not change much during the expansion (see Section 4.2.1); it is the speed of sound that drops dramatically. As mentioned previously, supersaturation, not vacuum, is the requisite for aggregation; vacuum is however necessary for the subsequent collision-free propagation of the clusters. As a rule of thumb, consider that a typical value of the mean free path for a gas at 1 Torr is 0.1 mm. Supersaturation means that, somewhere along the expansion, the local values of temperature and pressure lie below the dew point; stated differently, the balance between temperature (gauged against binding energy) and density (collisions) must favor condensation over evaporation. Following an accepted standard, we say "temperature" often meaning its energy equivalent; the two are related by the Boltzmann constant k_B.

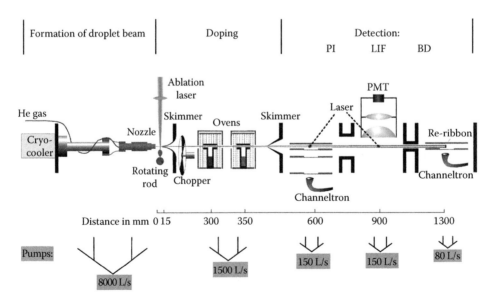

FIGURE 4.1 Typical He droplet machine. From left to right one can recognize in the source chamber the cryocooler, the source (nozzle), a laser ablation setup for doping with refractory materials, and the skimmer admitting the center portion of the beam into the doping chamber. There one sees the chopper (for differential measurements, usually in combination with a gated counter or a lock-in amplifier) and the doping ovens (for gaseous species these may consist of a simple metal box connected to a reservoir, and are usually called pickup cells). In the detection chamber, one may have any of the detectors mentioned in the main text. Shown here are a channeltron combined with a laser (photoionization); a photomultiplier combined with a laser (laser-induced fluorescence), and a channeltron combined with an ionizing surface (beam depletion).

Clusters are characterized in the first place by the forces binding them. Ionic clusters (e.g., NaCl) and covalent clusters (notably, fullerenes) have been extensively studied; despite the general difficulty to vaporize their precursor materials, they are quite accessible with the proper source, which is also suitable for refractory metals and semiconductors (Dietz et al., 1981; Kroto et al., 1985; Martin, 1983; Milani and deHeer, 1990); the greater binding strength of these clusters compensates for the lower precursor densities attainable. Hydrogen-bond clusters have also been studied, notably water, pure and mixed with other molecules, because of its importance in chemistry, biology, and atmospheric science (CR100–11; Keutsch and Saykally, 2001; Zwier, 1996).

The easiest to generate and characterize are however clusters of nonrefractory metals and van der Waals clusters. The guinea pigs in the first class are alkali, alkaline earths, silver, and gold: the smallest of these clusters are nonmetallic clusters, and the onset of metallic behavior is of great interest. Similarly, the occurrence of closed electronic shells, which is directly related to such aspects as stability, reactivity, and catalytic properties, is of interest. Geometric factors are also important and become predominant for large clusters.

The guinea pigs in the second class are rare gases. Being composed of closed-shell atoms, these are the prototype systems where geometric factors dominate. Rare gases at high pressure (i.e., density) are a common staple of almost every laboratory, thus their clusters are easy to obtain in a supersonic expansion. The heavier ones, argon and up, bind strongly enough that clusters can be obtained already from a room temperature expansion. For He and Ne, cooling of the expansion source to cryogenic temperatures is necessary.

All clusters can be described, to a different level of accuracy, in terms of pairwise interactions between their constituents; van der Waals clusters are the best benchmark of this approximation (Xie et al., 1989). To the extent that all pair potentials $v(r)$, with r the distance between the two interacting partners, are described by the same functional form—parametrized by the interaction radius σ and energy ε, typically the 6–12 Lennard–Jones potential $v(r) = 4\varepsilon[(r/\sigma)^{12} - (r/\sigma)^6]$—all properties of the cluster can be obtained from scaling laws containing those parameters. In thermodynamics, this is known as the law of corresponding states (Hill, 1986) and is of fundamental importance. Let us immediately note that helium is a special case because of strong quantum effects.

In relation to clusters, scaling laws have two very important applications: first they predict that under similarly scaled expansion conditions (source temperature T_0, pressure p_0, and diameter d) the same cluster size distribution should result. Second, they predict a scaling of the temperature of a cluster with the depth of the pair potential (Gspann, 1982; Klots, 1987). We will return later to the physics behind a cluster's temperature; for now it suffices to say that the latter is very roughly equal to ε/k_B (and to a fraction of that for helium, because of quantum effects). For argon clusters, this means a temperature of ~40 K. We shall see how such low temperatures make clusters technically interesting. Helium occupies a special place because

of several interrelated properties (Wilks and Betts, 1987): let us mention here that the strength of a van der Waals potential is determined by the polarizability of the interacting partners, which increases with the number of electrons (Hirschfelder et al., 1954). Helium has thus the weakest pair potential of all the rare gases [$\varepsilon/k_B = 10.995 \pm 0.005$ K; Anderson, 2001, 2004], while at the same time quantum effects are the largest because of its small nuclear mass: the zero point energy of a dimer is so large that the potential between two ^{4}He atoms barely supports one bound state, that between two ^{3}He no bound state at all. For this reason, bulk helium is liquid down to 0 K, becomes superfluid below $\approx$2 K (^{4}He), and its clusters are the coldest of all (0.38 K for ^{4}He and 0.15 K for ^{3}He). Note that ^{3}He clusters are energetically stable only above a minimum size estimated between 20 and 40 atoms (Guardiola and Navarro, 2000, and references therein).

Not surprisingly, He droplets are model systems to learn about the microscopic mechanisms of superfluidity, and increasingly gain popularity as "nanocryostats" to cool other species. In fact, historically, foreign atoms and molecules (referred to as dopants) have initially been introduced, first in Ar clusters (Gough et al., 1985) and later in He droplets (Goyal et al., 1992), as a handle to make the droplet spectroscopically active. Only later the enormous potential of nanodroplets as nano-cryo-laboratories, and the potential of spectroscopy as a diagnostic tool of the complexes formed, became a common notion (Lehmann and Scoles, 2000) and started to be exploited to a significant extent. Spectroscopy remains the best probe of He droplets; the absence of permanent electric and magnetic dipole moments, as well as the stiff electronic structure, rules out almost every conceivable spectroscopy of pure helium. The atoms do have of course intense electric-dipole-allowed electronic transitions, whose energies lie, however, between the 2S–2P transition: 20 eV, and the ionization limit: 24 eV (Ralchenko et al., 2009), corresponding to photon wavelengths of 50–60 nm. These are not available in the laboratory and require a synchrotron source; synchrotron spectroscopy of pure helium droplets is a well-established method (Joppien et al., 1993a,b; Karnbach et al., 1993; Kim et al., 2006; Möller et al., 1999; Peterka et al., 2003, 2006, 2007; von Haeften et al., 1997, 2001, 2002, 2005a), nicely complementing neutron scattering as a probe of the structure of pure bulk He, but will not be discussed further here.

Before we concentrate on the spectroscopy of doped helium droplets, we briefly review the concepts and methods associated with their production and doping, and their basic properties. A typical He droplet machine is shown in Figure 4.1.

4.2 Methods

4.2.1 Production

As said, helium droplets are produced by the condensation of supersaturated gas in a supersonic expansion. The expansion into vacuum is usually considered isentropic; it is thus accompanied by substantial cooling. The formation process is conceptually well understood, and sophisticated models based on the kinetic theory of gases have been developed (Knuth, 1997). Three different regimes are possible depending on the helium state prior to the expansion (already liquid or still gas) and in the latter case on whether the expansion isentrope crosses the liquid-gas line from the liquid side (supercritical expansion) or from the gas side (subcritical expansion). The most important quantities characterizing each regime are the probability distribution for the number of He atoms N in a droplet, and the associated average droplet size $\langle N \rangle$ (following a somewhat established pattern we indicate with n a small number of He atoms, <100, that can still be discriminated in spectroscopic experiments, and with N a large number whose value only makes sense in coarse studies of size-dependent experimental quantities, and in theoretical calculations). Droplet sizes become progressively smaller going from $\langle N \rangle > 10^8$ in the first regime to $\langle N \rangle < 10^5$ in the third one. The latter is the most important one for several reasons: the requirements on the source temperature T_0 are less stringent (10–20 K, sometimes up to 35 K); droplets sized between 10^3 and 10^4 atoms each are computationally tractable, present interesting finite-size effects, yet are large enough to efficiently pick up dopants and accommodate them without evaporating away in the process; finally for a given gas flux to be pumped away, smaller droplets means that a larger number of them is available. Very large droplets are useful to aggregate hundreds of atoms in them (Section 4.4.1) whereas smaller droplets with a countable number of He atoms, <100, are the best to study the evolution of superfluidity (Section 4.3.2); these are produced in a pulsed aggregation source (at pressures that may exceed 100 atm and temperatures as low as 77 K), which is the only one where the dopant molecules, at dilutions below 1%, are co-expanded with the helium and act as condensation seeds ("bottom up" approach). In all other cases, the nonhelium fraction of a seeded expansion would condense on its way to the nozzle; the pickup technique (Gough et al., 1985; Scheidemann et al., 1990) is used instead ("top down" approach, Section 4.2.3).

Experimentally, while the strongly out-of-equilibrium formation phase (restricted to the first few hundred micrometers/few hundred nanoseconds after formation) has remained rather inaccessible, the final outcome of the expansion has been well characterized (Buchenau et al., 1990; Harms and Toennies, 1998, 1999; Harms et al., 2001, 1997b; Knuth and Henne, 1999). Besides allowing fine-tuning of the theory, experiments have revealed that helium droplets are not monodispersed: their number size follows a log-normal distribution with standard deviation comparable to the mean value $\langle N \rangle$ in a subcritical expansion. In an expansion from the liquid (Knuth and Henne, 1999), the distribution is an exponential. The sensitivity to initial conditions is high, notably to source temperature; in an experiment the latter is used as main control parameter, and is actively stabilized within 0.1 K by resistive heating.

After the formation phase ceases, the droplets have cooled to their limit temperature (more below) and have formed a jet with a typical velocity of a few hundred meters per second ($\approx \sqrt{(5/2)k_{\mathrm{B}}T_0}$; for more accurate formulae see Buchenau et al., 1990) and a velocity spread typically less than 1% [technically one refers to the reciprocal quantity, termed "speed ratio," which

thus reaches values up to a few hundred (Scoles, 1988)]. Just like any molecular beam, a droplet beam thus offers to the experimenter a much appreciated collisionless environment.

Only the center portion of the jet is admitted from the first vacuum chamber (source chamber) into the doping/detector chamber, through a high-precision, electroformed metal skimmer (the source–skimmer distance is typically ≈ 1 cm). This is done in order to avoid interference from the large amount of background gas, which would quickly destroy the collisionless jet; one can easily appreciate the significance of the operation by considering that the droplet source operates at the maximum possible flux (i.e., the source chamber pressure is as high as the pumps can tolerate), whereas the pressure in the doping/detector chamber is $<10^{-7}$ Torr so that the probability of collision of a He droplet with background gas be much less than 1, and the droplets remain clean and undisturbed. Compared to a conventional molecular beam machine, a He droplet machine requires some form of cryocooling, high-pressure He gas (up to 100 atm; correspondingly small nozzle diameters, $d = 5$–10 μm, are used), and larger pumps (several thousand liters per second; collection and recycling of the gas must be built in if the rare ^{3}He isotope is used). The source is cooled either by a closed-cycle refrigerator—typically capable of >1 W at ~10 K, with newer models reaching down to ≈ 4 K—or with a liquid-helium-cooled continuous-flow cryostat. The former have a higher initial cost and may cause vibration problems (minor, in our experience), but are easier and cheaper to operate.

The operation of the source in continuous mode remains the easiest and most popular, first because most experiments are carried out with continuous lasers, second because pulsed sources mean moving parts and repeated making–breaking of a seal, which are troublesome at high pressure and low temperature. Pulsed sources, however, ideally match pulsed lasers and/or time-resolved experiments: the increase of signal-to-noise ratio may reach 100-fold (Slipchenko et al., 2002). Commercially available pulsed valves have been modified for use as a droplet beam source (Slipchenko et al., 2002); the heat load contributed by the electrical power used to operate the valve needs to be carefully considered in the design, but is not a serious obstacle. Pulsed sources are the standard in the "bottom up" approach, where on the one hand cooling of the source is less extreme [~200 K (Tang et al., 2002)], and where the detection of the free-induction-decay calls for pulsed operation. Further technical details on sources can be found in Callegari et al. (2001) and Xu and Jäger (1997).

4.2.2 Properties

One of the basic questions for clusters other than helium is whether they are solid or liquid. While their temperature often lies well below the bulk solidification temperature, the finiteness of the system and the nucleation rate (particularly as affected by the presence of barriers) may act against solidification (Knuth et al., 1995; Levi and Mazzarello, 2001; Maris et al., 1983). Let us also note that surface atoms are more mobile; thus it is possible to have a solid-like core and a liquid-like surface (Levi and Mazzarello, 2001). This is not a philosophical speculation: it is relevant today in fields ranging from atmospheric science to the search for superfluid hydrogen. Helium droplets, however, are certainly liquid, as even bulk helium near 0 K requires pressures >25 bar to solidify [let us mention that the search for super-solid helium is presenting our colleagues with many surprises (Galli and Reatto, 2008; Kim and Chan, 2004a,b)]. This peculiar property of helium can be seen as a macroscopic manifestation of its large zero-point-energy that easily overcomes the weak localization due to the He–He van der Waals attraction. Like all free rare-gas clusters in vacuum, He droplets cool by evaporative cooling (absorption and emission of blackbody radiation is vanishingly small), which is a self-limiting process; in principle arbitrarily small temperatures would be possible, in practice the rate of change slows down exponentially, so all experiments in the world, looking at the same timescale, measure the same temperature T_d (Section 4.3.3). The latter is reached less than 1 μs after formation: 0.38 K for ^{4}He, 0.15 K for ^{3}He (the difference reflects the lighter mass giving a higher zero-point energy) (Brink and Stringari, 1990; Gspann, 1982; Guirao et al., 1991). These values are well below the superfluid transition temperature for the bosonic ^{4}He (bulk value ≈2 K) and well above for the fermionic ^{3}He, where fermion pairing has to occur first (bulk value ≈1 mK). Thus, although the temperature is not an experimental parameter under experimenter's control, one has nevertheless two chemically identical systems one of which is superfluid, the other not. Practically, ^{3}He is so expensive that only few chosen comparative experiments have been performed with it.

A free He droplet in vacuum is a sphere of liquid held together by the weak van der Waals attraction between its atoms; its surface is very diffuse (10%–90% width: 6–8 Å) (Harms et al., 1998; Toennies and Vilesov, 2004); its density below the surface is uniform and close to the bulk liquid value (^{4}He: 0.0218 Å^{-3}; ^{3}He: 0.0163 Å^{-3}), by which one estimates for a given number size N a droplet radius $R/\text{Å} = 2.22N^{1/3}$ for ^{4}He and $2.44N^{1/3}$ for ^{3}He. This also defines the cross section for pickup of dopant atoms or molecules (Section 4.2.3), which is taken equal to the geometric cross section (Harms et al., 1998, 2001). Except for some experiments looking for "transparency" of He droplets to He atoms (Harms and Toennies, 1998, 1999), the sticking probability is assumed to be unity. The binding energy of one ^{4}He atom in the liquid (i.e., the energy expenditure to evaporate it) amounts to 7 K, less even than the 11 K well depth of the He–He pair potential, yet considerably larger than the 1 mK binding energy of a dimer (Anderson, 2001). These extreme deviations from a classical behavior are a direct manifestation of the large zero-point energy. The binding energy per atom is an experimentally relevant parameter, as it determines the cooling capacity available for bringing dopants from room temperature or above, to 0.38 K.

4.2.3 Doping

A simple calculation (Lewerenz et al., 1995) based on geometric cross section and Poisson statistics shows that a column density of some ~10^{-4} Torr cm maximizes the probability that one dopant be picked up by the droplet. Thus small gas cells, possibly heated up to 1500 K (but in any case to much smaller temperatures than a conventional effusive cell) positioned just after the skimmer have sufficed for loading the droplets with the most diverse materials (Küpper and Merritt, 2007). Laser ablation has been used for more refractory materials (Claas et al., 2003). Virtually all species solvate inside the droplet, because their van der Waals interaction with helium easily overcomes that between He atoms. Alkali atoms and their complexes (Section 4.4.4) are an exception, because of their diffuse valence electron: they reside on the surface; alkaline earth metal atoms are deeply buried into the surface but not fully solvated; complexes of an alkali–metal atom and a closed shell molecule do form a "buoy" as one would intuitively expect (Douberly and Miller, 2007). All sorts of van der Waals complexes are easily assembled from constituents forced onto the same droplet, upon sequential pick up from two separately controlled cells. While the details of the assembly process are not known, the timescale for complex formation should reasonably be determined by the motion across a droplet (nanoseconds at typical thermal velocities). The collision energy, solvation energy, and, when applicable, the binding energy of complexes, all are disposed of into the droplet, whose temperature is quickly restored to 0.38 K by resumed evaporative cooling. One often observes that van der Waals complexes are formed in metastable geometries (Choi et al., 2006), and infers that their formation occurs along the lowest-potential-energy path as a consequence of the very efficient cooling (Section 4.4.1). The formation of aggregates of alkali–metal atoms (dimers, trimers, so far) is barrierless irrespective of the spin state, so based on thermodynamics alone one would expect only the stablest low-spin configuration (singlet, doublet, respectively) to be formed, and no metastable high-spin configuration (triplet, quartet, respectively). Kinetics does however dominate and high-spin ones are observed in much greater abundance (Section 4.4.4).

4.2.4 Detection

The direct detection of pure helium droplets is quite feasible. Photoexcitation (Joppien et al., 1993a), photoionization (Fröchtenicht et al., 1996), electron-bombardment ionization followed by mass spectrometry of He$_n^+$ ions (Buchenau et al., 1990, and references therein), and the bolometric detection of the kinetic energy carried by the beam (Goyal et al., 1992) have been successfully used. Nonresonant scattering of visible light, easily done on large H$_2$ clusters (Reho et al., 1997), is conceivably also applicable to He droplets. Photoionization (Fröchtenicht et al., 1996), electron-bombardment ionization (Scheidemann et al., 1990), and surface ionization (Callegari et al., 1998) of the dopant are applicable to doped droplets in addition to the above methods. All of the above methods can be combined with a source of resonant excitation. This is typically a laser (in which case laser-induced fluorescence may also be an option) but can also be a microwave source. Resonant spectroscopy of doped droplets has provided the large part of what we experimentally know about helium droplets, and encompasses many diverse methods and results, which we present later in more detail. For now let us note that all nonfluorescence methods rely on some excitation-induced change of the droplets' overall flux, thus go back to the detection of pure droplets. Because the energy of one microwave photon is insufficient to cause any change of said flux, microwave experiments set themselves apart: detection relies either on many hundred cycles of absorption–relaxation per droplet or on direct detection of the free-induction decay. Several review articles (Choi et al., 2006; Küpper and Merritt, 2007; Makarov, 2004; Stienkemeier and Lehmann, 2006; Tiggesbäumker and Stienkemeier, 2007; Toennies and Vilesov, 2004), and sometimes an entire journal issue (JCP115-22; JPCA111-31; JPCA111-49), have been published on the spectroscopy of doped helium droplets, here we want to follow a thread based on the properties and applications of He droplets. Because of the strong association between a particular spectroscopic method and the droplet properties that it can measure,

the end result is practically the same; we should mention the importance of nonspectroscopic methods (mass spectrometry above all) and of the theory supporting experimental measurements. In short, mass spectrometry in combination with kinetic theory of gases and with density-functional calculations tells us about droplet formation, size distribution, shape and mechanical properties of a droplet. Rotational and ro-vibrational spectroscopies, again in combination with density functional and quantum Monte Carlo calculations, tell us about the response of superfluid helium to small displacements around the equilibrium position of the first few helium shells surrounding the dopant. The difference with spectra in ^{3}He is striking (Figure 4.2). Because rotational energies are close to $k_B T_d$, rotationally resolved spectra deliver the temperature of the droplet [via the line intensities, that is, the level populations: indeed these spectra were the first (Hartmann et al., 1995), and for a long time the only, available "thermometer," now complemented by spin-polarization measurements, Section 4.4.5]. Rotationally

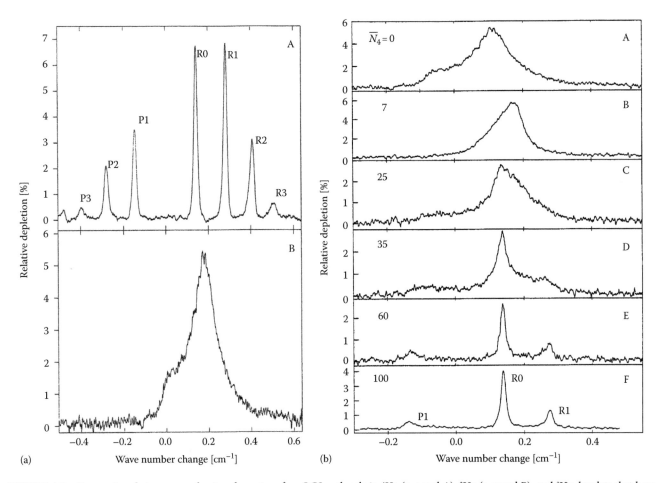

FIGURE 4.2 Comparison between ro-vibrational spectra of an OCS molecule in ^{4}He (a, panel A), ^{3}He (a, panel B), and ^{3}He droplets that have picked up 0, 7, 25, 35, 60, and 100 ^{4}He atoms (b, panels A through F). The well-resolved P and R branches in ^{4}He droplets indicate that rotational coherence is preserved. The line spacing (2*B*, see Section 4.3.1) is ~1/3 of that of the free molecule, indicating an ~3 times larger moment of inertia. Lack of a Q branch, as expected for a linear molecule, indicates that the symmetry of the rotor is not affected by the helium. The structure collapses into a single peak in ^{3}He droplets, indicating rotational diffusion, and is recovered if ≈60 ^{4}He atoms are picked up by a ^{3}He droplet and act as a buffer layer. These results provide consistent evidence of the microscopic superfluidity of ^{4}He droplets. Note the different intensity patterns in panels A (a) and F (b), consistent with a temperature of 0.38 and 0.15 K, respectively. (From Grebenev, S. et al., *Science*, 279, 2083, 1998. With permission.)

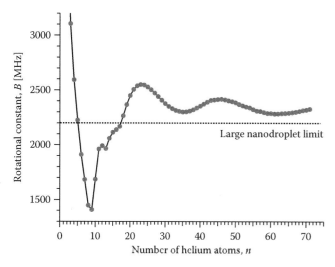

FIGURE 4.3 The evolution of the rotational constant B of He_n–OCS clusters with the number of helium atoms, n. The turnaround in B at $n = 9$ indicates decoupling of helium density from the rotational motion of the OCS molecule and marks the onset of microscopic superfluidity. The oscillatory behavior at larger n may be a signature of a helium solvation shell that builds up around the OCS molecule. There is at least one further maximum before the B-value approaches the limiting helium droplet value.

resolved spectra also deliver the moments of inertia I of the molecule, which are the sum of those of the bare molecule plus the contribution of the coherent motion of the helium. For very small droplets ($n < 100$), regular oscillations of I as a function of n directly reflect the closure of solvation shells and the onset of superfluidity (Figure 4.3). Vibrational spectroscopy is the most flexible method to characterize van der Waals complexes formed in He droplets. Electronic (visible) spectroscopy tell us about the response of the helium to large impulsive displacements brought about by the change of the dopants' electronic wavefunction; because these changes can be greatly varied by choice of the dopant species and transition, one may observe sharp zero-phonon lines ($\Delta v/v \sim 10^{-4}$), typically in organic molecules, as well as broad multiphonon bands ($\Delta v/v \sim 10^{-2}$), typically in atoms. Incidentally, the few species known to be bound to the surface of a droplet have only been investigated so far by electronic spectroscopy. As a detection method, electronic spectroscopy has allowed the detection of electron spin resonance (ESR) transitions, which in turn directly tell us about small deformations of the dopants' wavefunction in the ground state. The availability of nanosecond, picosecond, and femtosecond lasers makes electronic spectroscopy (including photoelectron spectroscopy) the best suited to time-resolved studies of the dynamics in He droplets: because displacements and the corresponding velocities are large, dynamics will often not dramatically change as a consequence of superfluidity (or lack thereof); let us note however that in the bulk the existence of threshold values of related quantities, such as the Landau velocity, are one of the most interesting aspects of superfluidity (see, e.g., Wilks and Betts, 1987).

4.3 Superfluidity

Superfluidity in He droplets was historically first demonstrated for large ones: $N = 1{,}000$–$10{,}000$. *A posteriori* superfluidity of such large droplets seems an obvious fact; the experiments measuring it (Grebenev et al., 1998; Hartmann et al., 1996), their findings, and even a precise definition of what microscopic superfluidity is, were however far from trivial. Strictly speaking, superfluidity is a macroscopic property associated to a phase transition with long-range order, only observable in extended systems. The smearing of sharp phase transitions is a concept very familiar to cluster scientists. Besides, macroscopic experiments measure properties (viscosity, critical flow velocity, thermal conductivity, etc.) that are difficult to define and/or measure at the atomic scale. Let us also be reminded that the temperature of a droplet is not an experimentally tunable parameter, so the unfolding of a measured quantity across the critical temperature cannot be followed. Besides, the dopant is presumably most sensitive to the properties of the first few layers surrounding it, which are those most perturbed by the dopant–helium interaction. The quantum-mechanical indistinguishability of the bosonic He atoms is a prerequisite of superfluidity, and localization due to the strong attractive He–dopant interaction works against it.

Borrowing concepts from matrix spectroscopy (Rebane, 1970; Sild and Haller, 1988), such as zero-phonon lines and single-phonon excitations, the first experiment used the electronic excitation of a molecule, glyoxal (HOCCOH), to look at the energy gap in the elementary excitation spectrum of an ^{4}He droplet (Hartmann et al., 1996). This gap is considered a signature of superfluidity, and is predicted to occur already at small droplet sizes (Rama Krishna and Whaley, 1990a,b); let us note however that it becomes an ill-defined quantity when the droplet is so small that its excitation modes become discrete. Experimentally, it has been later observed with spin-singlet Na_2 molecules (Higgins et al., 1998), and with a variety of organic molecules (Section 4.4.6); in the latter case multiple zero-phonon lines are often observed, in general denoting the existence of several conformers. Nonclassical inertial properties are a hallmark of superfluidity: we mentioned that early experiment attempted to investigate the transparency of droplets to colliding He atoms (Harms and Toennies, 1998, 1999). The moment of inertia of the rotating fluid is of special value to the experimentalists and theorists alike: it turned out to be easily accessible experimentally (Hartmann et al., 1995), and being an extensive quantity, it remains well defined also at the microscopic scale. A macroscopic amount of superfluid helium cannot be set into rotation, because of the prescription of irrotational flow (this remains true until the rotational velocity exceeds the limit for the formation of vortices; considerations of energy costs and rapid decay suggest that vortices will be an unlikely, albeit very interesting, observation in He nanodroplets). The first spectroscopic experiment assessed the free rotation of molecules (specifically, SF_6) in ^{4}He (Hartmann et al., 1995); there being no means to control the droplets' temperature, superfluidity could only be inferred. The experiment was later repeated with another molecule (OCS) in a machine suitable for the production of ³He droplets (Grebenev et al., 1998). As we said, ³He droplets are not superfluid at their 0.15 K limit temperature, because ³He is a fermion; the mass difference between the two isotopes brings about, through the different zero-point energy, interesting thermodynamic differences (two of which we mentioned: the limit temperature, and the fact that small ³He droplets cannot be bound). Isotopically purified ³He still contains small amounts of ⁴He, which in a supersonic expansion act as condensation seeds; the resulting droplets are ⁴He-enriched relative to the original mixture (note however that a more practical way of tuning the amount of ⁴He is by subsequent pickup in a gas cell). Remarkably, in a droplet the two isotopes phase-separate into a core of the stronger-bound ⁴He and an outer layer of ³He (Barranco et al., 1997; Navarro et al., 2004), the former solvating the dopant, the latter setting the temperature of the whole droplet. It was found that free dopant rotation is a prerogative of ⁴He (Figure 4.2), and that approximately two solvation layers of ⁴He in ³He suffice to recover it. Theory had predicted that inertial manifestations of superfluidity in pure droplets would be observable at a comparable number of atoms (Rama Krishna and Whaley, 1990a,b). These experiments probe at once the minimum number of ⁴He atoms necessary to "protect" the rotating molecule from the mechanical coupling to the outer ³He as well as the minimum number of ⁴He necessary to observe deviations from classical inertia. The effect of the surrounding ³He layer on the spectroscopic properties of the ⁴He coated OCS molecule is unclear, however. Meanwhile, one of us (WJ, Section 4.3.2) has succeeded to aggregate a countable number of ⁴He atoms $n = 1$–10^2 around a dopant molecule in a seeded expansion (McKellar et al., 2007, 2006; Tang et al., 2002; Topic et al., 2006; Xu and Jäger, 2001, 2003; Xu et al., 2003, 2006). These complexes show a classical inertia until enough He atoms are present to form a structure that closes onto itself and encompasses the dopant (a ring, or a full solvation shell); for larger values of n a marked decoupling of the molecule and the helium is observed, as reflected in a smaller moment of inertia. The decoupling is not monotonic with n, and its local maxima can be associated with the completion of a ring or shell.

In the large droplet limit, the ro-vibrational spectra of all molecules have a number of common features, notably (a) a gas-phase-like appearance of the spectra, that is, the observation of rotational fine structure, which is accepted to be a manifestation of microscopic superfluidity; (b) increased linewidths of the observed molecular transitions compared to the corresponding gas-phase values (250 MHz to 2 GHz compared to a few tens of kHz); and (c) an increased moment of inertia, that is, a decreased rotational constant, of the dopant molecule, as if it drags some helium density around with it.

Experiments and theory on small and large droplets combine to give a well-defined general picture. Most of the fundamental interesting questions are however still open, such as: How many helium atoms are required for the onset of microscopic superfluidity, and what observable could be used as an indicator? Through which channels does the excitation energy flow from the dopant molecule to the helium surrounding? Which mechanisms are responsible for the increased linewidths? What determines the degree of renormalization of the rotational constant? In the following sections (Sections 4.3.1 through 4.3.3), we discuss what systematic experiments are being performed to address these questions.

4.3.1 Rotation Hamiltonian

Because the rotation of molecules plays such an important role in the study of helium droplets, we briefly summarize here the minimum formalism used to interpret the spectra. It is advantageous to deal with high-symmetry molecules, and indeed most of the molecules investigated in He droplets are symmetric tops, linear molecules, or more rarely spherical tops. The mode being excited is characterized by the vibrational quantum number v and rotational quantum numbers J, K (corresponding to the total angular momentum and to its projection along the high-symmetry molecular axis, respectively); prime and double prime superscripts (e.g., v', J''), when present, explicitly indicate the upper and lower states, respectively. Asymmetric tops are more complicated to treat, and so are the cases where other quantum numbers appear accounting for a nonzero orbital or spin angular momentum; we also ignore anharmonicities, centrifugal distortions, and cross-terms, although they do appear in detailed models of some spectra featuring sufficiently sharp lines even in He droplets; for all these important refinements we refer the interested reader to specialized monographs and textbooks (Brown and Carrington, 2003; Herzberg, 1989–1991; Hougen, 2001; Lefebvre-Brion and Field, 1986). Let us note that the formalism has been developed to interpret the spectra of gas-phase molecules, which due to their extremely high resolution do require highly refined Hamiltonians; the full formalism has to be applied to doped clusters containing a countable number of He atoms, whose spectra have to be interpreted as those of an extremely floppy, usually asymmetric, molecule. Just as in solids one goes from the discrete-level structure of the constituent atoms to the band structure of the bulk, here as the number of He atoms increases one goes from discrete molecular levels to a band structure; rotation of the dopant within the helium can however be seen as a localized excitation carrying most of the oscillator strength. One recovers a spectrum with different molecular constants, but the same symmetry as the Hamiltonian of the bare molecule. This is a remarkable observation, although a posteriori one to be expected: it indicates that the molecule imposes its symmetry onto the helium, in contrast to most other matrices where the symmetry of the trapping site enters the interpretation of the spectra, usually lowering the symmetry of the dopant. The very fact that rotational resolution is possible implies rotational coherence, that is, the lifetime of a rotational state is longer than the rotational period. A large helium droplet does degrade spectral resolution, thus relaxes the requirements on the level of detail of the Hamiltonian at the price of washing out most of the information extractable from a spectrum.

It is known that any rigid body has three mutually perpendicular axes of rotation (principal axes of inertia) along which the tensor of inertia is diagonal; the axes are named *a*, *b*, *c* such that for the moments of inertia there holds: $I_a \leq I_b \leq I_c$. These are extensive quantities (i.e., the moment of inertia of a composite system is the sum of those of its parts) and are measured in amu Å^2 (amu = atomic mass unit). The associated rotational constants *A*, *B*, *C* are the reciprocal of $I_a \leq I_b \leq I_c$, and have the units of frequency or wavenumber via the conversion constants 505 379 MHz amu Å^2 $\equiv$ 16.857629 cm^{-1} amu Å^2; the constants *A*, *B*, *C* are directly related to the separation of rotational lines in a spectrum, as we shall see. For a symmetric top molecule, one of the principal axes coincides with the high-symmetry axis and the choice of the other two within the plane perpendicular to the high-symmetry axis is arbitrary. There holds either $A \geq B = C$ (prolate, i.e., "cigar shaped," *a* is the high-symmetry axis) or $A = B \geq C$ (oblate, i.e., "pancake shaped," *c* is the high-symmetry axis). A linear molecule can be seen as the special case $A \to \infty$, $K = 0$, a spherical top as the case $A = B = C$; in the latter case choice of the principal axes is fully arbitrary, although it is practical to refer them to high-symmetry axes of the molecule.

The ro-vibrational energy levels are given by

$$\frac{E_{JK}}{h} = \nu_0 \nu + BJ(J+1) + (A-B)K^2 \quad \text{[prolate]} \tag{4.1}$$

$$\frac{E_{JK}}{h} = \nu_0 \nu + BJ(J+1) + (C-B)K^2 \quad \text{[oblate]} \tag{4.2}$$

where ν_0 is the vibrational frequency of the mode. Selection rules depend on the relative orientation of the high-symmetry axis and transition dipole moment. One has

$$\Delta J = 0, \pm 1 \quad \Delta K = 0 \quad \text{for } K \neq 0 \tag{4.3}$$

$$\Delta J = \pm 1 \quad \Delta K = 0 \quad \text{for } K = 0 \tag{4.4}$$

for a parallel band, and

$$\Delta J = 0, \pm 1 \quad \Delta K = \pm 1 \tag{4.5}$$

for a perpendicular band; the latter give more congested spectra and have been more rarely considered in He droplets. Transitions with $\Delta J = -1$, 0, +1 are termed P, Q, R branch, respectively. For a symmetric-top parallel-band, one has a comb of lines spaced by $2B$, with all the lines of the Q branch coinciding at the position $\nu_0(\nu' - \nu'')$ (band center) in the simplifying assumptions we made above. Note that based on Equation 4.4 the Q branch is missing in a linear molecule. Also note that typically many rotational states are thermally populated, and that the populations of rotational states determine the intensities of the rotational lines, which can be fitted to extract the rotational temperature.

While there are many important practical differences between rotational and ro-vibrational parallel-band spectra (spectral domain: microwave versus infrared; integrated intensity: dependent on $|\mu_e|^2$ vs $|d\mu_e/dq|^2$ with μ_e the molecule's electric dipole moment and q the normal-mode coordinate), the shape of the R branch (more precisely the relative positions and intensities of the lines within the branch) is the same in our approximation (note that a purely rotational excitation can only be of the R-type).

4.3.2 Small Droplets

The systematic study, both experimental and theoretical, of smaller He$_n$-molecule clusters with increasing number, *n*, of helium atoms allows the cluster properties to be determined with "atomic resolution." The experimental approach is the generation of smaller clusters using a pulsed supersonic molecular expansion and their spectroscopic characterization. The sizes of clusters produced in this manner can be controlled to a certain degree by the variation of sample pressure and nozzle temperature, with higher pressure and lower temperature favoring the production of larger clusters. A dopant highly diluted in He (concentration <0.1%) under extreme conditions (pressures of more than 100 atm, temperatures as low as 77 K) can generate He$_n$-molecule clusters containing more than 100 helium atoms (McKellar et al., 2007, 2006).

Initial microwave and infrared experiments of He$_n$-OCS clusters with *n* from 1 to 8 (Tang and McKellar, 2003; Tang et al., 2002; Xu and Jäger, 2003) revealed that the cluster rotational constants drop below the nanodroplet value for *n* = 6, 7, 8. This observation was important since it implied that the rotational constant (moment of inertia) needed to *increase* (*decrease*) at some point with the addition of further helium atoms to reach the helium droplet value. This turnaround in rotational constant would mark the onset of microscopic superfluidity. The turnaround was later observed for He$_n$-OCS at *n* = 9 (McKellar et al., 2006, 2007; see Figure 4.3). The increase in rotational constant implies that the helium begins to decouple from the rotational motion of the dopant molecule. Path integral quantum Monte Carlo simulations show that this decoupling coincides with a full coating of the entire OCS molecule with helium density (Blinov et al., 2004; Moroni et al., 2003).

Figure 4.4 shows the helium density for selected He$_n$-OCS clusters. At *n* = 10, the entire molecule is coated, allowing, in Feynman path integral language, for long-range exchange paths between helium atoms. As seen in Figure 4.3, the cluster rotational constants overshoot the nanodroplet value for *n* > 17 and then show broad oscillations, which appear to slowly converge to the limiting nanodroplet value. The oscillations could indicate the appearance of a helium solvation shell; however, thus far there exist no theoretical simulations for confirmation. More recently, He$_n$-CO clusters have been studied using microwave and millimeter wave spectroscopy and the turnaround in rotational constant was found at *n* = 3 (Surin et al., 2008). This implies that a coating of the carbon monoxide molecule with four helium atoms is sufficient to induce superfluidity in the helium layer. At *n* = 6, the moment of inertia is smaller than that of He$_1$-CO, implying that the equivalent of less than one helium atom is rotating with the CO molecule in He$_6$-CO.

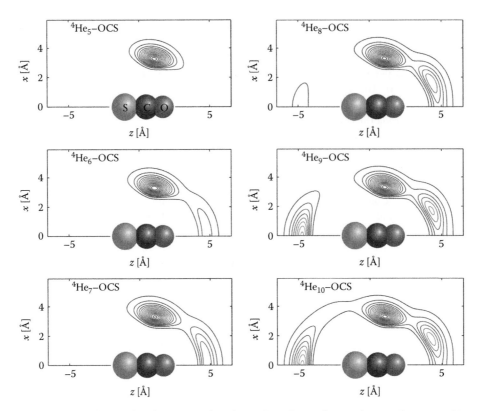

FIGURE 4.4   Contour plots of helium density distributions in selected He_n–OCS clusters from path integral quantum Monte Carlo calculations. For He_5–OCS, the cut through a helium doughnut ring around the equator of the OCS molecule is clearly visible. For $n = 6$ and 7, helium density builds up at the oxygen end. For $n = 8$ and 9, helium density accumulates at the sulfur pole, and for $n = 10$ the whole OCS molecule is coated with helium density. These density data are in excellent agreement with the experimental information from isotopic studies, which show that helium atoms 6 and 7 move to the oxygen end and helium atom 8 to the sulfur atom. The full coating at $n = 10$ allows for long-range exchanges of helium atoms and the helium becomes superfluid. This is also where experimentally the turnaround in the B rotational constant was found. For regions not enclosed by contour plots, the helium density is insignificant. For example, for He_5–OCS the values of the density spilled toward the oxygen and sulfur ends are at least six orders of magnitude smaller than the density in the donut domain.

4.3.3 Large Droplets

One of the messages from Section 4.3.2 is that superfluidity builds up at sizes between 10 and 100 He atoms. What more do we then learn from larger droplets?

First some practical arguments: large droplets are easier to make, dope, and detect. Also, acquiring the spectra of a class of similar molecules often does not require any modification of the experimental setup (including the laser), but rather the willingness to invest time and effort in the measurement; finally, irrespective of the purpose of the experiment, rotational constants are always one of its outcomes. In the end, there is simply a larger amount of ro-vibrational spectra, approximately 50 different molecules at the time of writing, that have been measured for large droplets, and this number is bound to increase.

From the physical point of view, by looking at a set of molecules in large droplets one looks no longer at the completion of solvation shells, but rather at how small changes of the overall solvation structure, brought about by the set of slightly different helium–dopant interactions, does affect superfluidity. This information is contained in the rotational constants and in their variation with factors such as chemical substitution, isotopic

substitution, rotational (J) and vibrational (v) quantum number. Most of these changes can be efficiently parametrized with further terms in the rotational Hamiltonian. Isotopic substitution, in particular, by changing the speed of rotation of otherwise equal molecules, neatly highlights dynamics factors in the rotor-He coupling: rotational spectra of HCN and DCN show that the lighter rotor is more decoupled from the helium (Conjusteau et al., 2000); this effect has been neatly captured in Quantum Monte Carlo calculations where the moment of inertia of the bare molecule can be arbitrarily tuned (Lee et al., 1999). Just like in conventional molecular spectroscopy, with a wise choice of molecular parameters one captures most of the physics of the problem and condenses it into these few highly informative numbers (Callegari et al., 2000a, 2001; Grebenev et al., 2000a,b; Harms et al., 1997a; Hartmann et al., 1999; Lehmann, 2001; von Haeften et al., 2005b). The molecular parameters in large droplets lend themselves to be interpreted with models that treat the helium as a continuum fluid ("superfluid hydrodynamics") either with numerically calculated He densities (Callegari et al., 1999; Lehmann and Callegari, 2002) or with simplified analytical wavefunctions (Lehmann, 2001) and that are very suitable for computationally inexpensive, often quantitative predictions. It always remains very desirable to compare the experimental results to "exact" Quantum Monte Carlo calculations, when available.

In large droplets, the rotational motion of the dopant and the motion of its center-of-mass relative to that of the helium become clearly distinct, albeit not fully decoupled. It is then in principle possible to study the microscopic flow of helium around a moving object, as well as the confinement effects brought about by the boundaries of the droplet; both of these are of great fundamental interest, and at present poorly understood (Lehmann, 1999). This information is contained in the shape of the rotational line, which in some favorable cases is split into two or more peaks (Nauta and Miller, 1999b) reflecting, for example, different orientations of the dopant. Spectral lines are also broadened by the finite life of ro-vibrational states: when this is the dominant broadening mechanism (unfortunately, seldom the case) one can thus extract the lifetime of a state (Nauta and Miller, 2001; Slipchenko and Vilesov, 2005; von Haeften et al., 2006). This is a very interesting quantity, which has been found to span several orders of magnitude (picoseconds to milliseconds, see Figure 4.11 in Choi et al., 2006), in any case always long enough to preserve rotational resolution. Since measurements in ^{3}He show rotational diffusion instead of rotational coherence, there is no question that the long ro-vibrational lifetime is a direct consequence of superfluidity. More interesting is the question of how relaxation in ^{4}He is accelerated when specific relaxation channels become energetically accessible, or slowed when either selection rules or poor coupling closes some relaxation channels.

Mostly in relation to electronic transitions of the dopant, large droplets also have the most favorable length scale to study finite-size effects in the excitation of collective modes of the helium (phonons).

4.4 Applications

4.4.1 Helium Droplets as Nanocryostat

The cooling capabilities of He droplets have been exploited to cool down large molecules, with the main goal to either simplify their spectra (Section 4.4.6), or to assemble and stabilize exotic aggregates. The Miller group, in particular, has provided some beautiful examples of the latter (Figure 4.5). They were able to demonstrate, using infrared spectroscopy, that hydrogen cyanide molecules self-assemble in helium droplets to form linear chains (Nauta and Miller, 1999a). This is in contrast to the situation in free-jet expansions, where hydrogen cyanide molecules form folded aggregations, for example, a cyclic structure for the trimer. The rationale for this behavior is that the dipole–dipole interactions between a hydrogen cyanide molecule and an existing chain orients them in a "head-to-tail" fashion, already at distances of about 3 nm. Upon aggregation, the condensation energy is dissipated into the helium droplet and the molecular assembly is trapped in a linear configuration, which corresponds to a local minimum on the interaction potential energy surface.

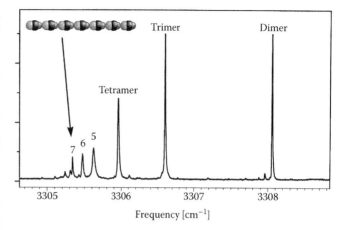

FIGURE 4.5 Infrared pendular spectra of HCN linear chains assembled in He nanodroplets. The number of molecules in the chain is determined by the shift from the monomer peak (at 3311.20 cm^{-1}, not shown), which is well known from gas-phase data. The regular progression shows that the linear chain structure remains the norm also for a high number of monomer units, unlike in the gas phase where cyclic structures are favored. (From Nauta, K. and Miller, R.E., *Science*, 283, 1895, 1999a. With permission.)

The low temperature of the helium bath prevents the system from isomerizing into more stable folded structures.

Nauta and Miller (2000) have investigated in a similar fashion the aggregation of water molecules in helium droplets. They found that the water aggregates with up to six water molecules have cyclic structures in the helium environment. For the hexamer, the cyclic structure corresponds to a higher energy isomer. In gas-phase experiments, a cage structure was found for the hexamer in the gas phase, which corresponds to the global energy minimum. The cyclic structures are apparently formed through some sort of insertion mechanism despite the low-temperature helium environment. The observation of the cyclic water hexamer is significant, as it is the smallest ice-like cluster and probably also a structural motif of liquid water. It is the unique properties of the helium matrix that makes it possible to study higher energy isomers of molecular assemblies, which are usually not accessible in gas-phase experiments.

The opportunities offered by this rare combination of low temperature, high mobility, and high spectroscopic resolution (essential for diagnostics) have been extensively exploited by the Miller group. Through the complexation of HCN with a small number of Mg atoms, they indirectly measure the onset of metalization in Mg clusters (Nauta et al., 2001). The vibrational dynamics of molecules adsorbed at metal surfaces and metal clusters are of significant interest, for example, for the field of catalysis. The magnesium atoms were produced using an oven at ~300°C and then captured by the helium droplets. A second pickup cell was used to capture hydrogen cyanide as an adsorbate molecule. Nauta and Miller were able to identify the spectra of HCN–Mg$_n$ clusters with up to four magnesium atoms. In analyzing the resulting spectroscopic parameters, they found strong evidence for the presence of nonadditive many-body interactions in the metal clusters. For example, the redshift of the vibrational band origin of the C–H stretch shows an unusual nonmonotonic dependence on the cluster size, while there is a smooth behavior in rare gas clusters with HCN in helium droplets. Further evidence for nonadditive behavior was found in the structural data, which could be extracted from the determined rotational constants. For example, the N–Mg distance contracts by 0.3 Å in going from $HCN–Mg_2$ to $HCN–Mg_3$, indicating that these systems cannot be described by pairwise additive interactions alone. Nauta and Miller caution that these strong nonadditive effects are likely not indicative of the onset of metallic behavior, which is only expected to occur at larger cluster sizes with about 18 magnesium atoms.

Even more powerful is the combination with a pyrolysis radical source (Küpper and Merritt, 2007). Complexes of many atoms with either HF or HCN (which are not only interesting workhorse molecules in the reaction studied, but also act as the infrared tag) have been characterized: Cl, Br, I, Al, Ga, In, Ge, Na, K, Rb, Cs, Mg, Ca, Sr, Zn, Cu, and Au. It is useful that the IR-active molecule is light, so that the complex thus formed still exhibits a rotationally resolved spectrum, and more structural information can be gained. Further complexes have been observed with molecular radicals: NO, CH_3, C_2H_5, and C_3H_5 (Küpper and Merritt, 2007).

4.4.2 Helium Droplets as Chemical Nanoreactor

The capability of introducing different chemical species into the helium droplets opens up the possibility to let a chemical reaction occur at the low temperature of the nanodroplet. Vilesov and coworkers (Lugovoj et al., 2000) studied the highly exothermic, chemiluminescent reaction $Ba + N_2O \rightarrow BaO + N_2$, by introducing first Ba atoms and then N_2O molecules into the helium droplets, using two pickup cells. The BaO molecule is produced in an electronically excited state, and the resulting chemiluminescent emission was monitored from 400 to 900 nm, in the range corresponding to the $A^1\Sigma^+ \rightarrow X^1\Sigma^+$ electronic transition. Two main spectroscopic signatures were observed: a broad feature in the region from 400 to 600 nm and clearly resolved vibrational structure from 600 to 900 nm. The interpretation is that the broad feature results from "hot" BaO molecules, which have left the helium droplet before their emission life time of 360 ns and show essentially a gas-phase spectrum. The resolved vibrational structure results from BaO molecules that have recoiled into the interior of the helium droplet and thermalized with the helium bath at 0.38 K. Only few vibrational and rotational levels remain significantly populated, leading to the observed clearly resolved vibrational structure. This scenario suggests that the reaction occurs at the surface of the helium droplet. This is consistent with the finding that Ba atoms reside at the surface of the helium droplets, partly embedded in a "dimple," similar to the case of alkali atoms (See Section 4.4.4) (Stienkemeier et al., 1999). Vilesov and coworkers (Lugovoj et al., 2000) carried out further experiments, where they introduced about 15 xenon atoms into the helium droplets, prior

to the pickup of Ba and N_2O. In the observed spectrum, only the "cold" sharp vibrational transitions remain and the "hot" broad feature has disappeared. In this case, the xenon atoms reside in the center of the helium droplet and their attractive interactions with the Ba atoms also pulls these into the droplet. As a result, the reaction takes place within the droplet, and essentially all produced BaO emits within the droplet, after thermalization.

The reaction of alkalis (Na, K, Rb, Cs) with water clusters embedded in helium nanodroplets has been studied using femtosecond photo-ionization as well as electron impact ionization. Unlike Na and K, Rb and Cs were found to completely react with water in spite of the ultracold helium droplet environment (Müller et al., 2009a). Several reaction intermediates have been identified in the mass spectra, which are apparently stabilized in the cold helium environment.

The Drabbels research group has studied photodissociation reactions of CH_3I and CF_3I embedded in helium droplets (Braun and Drabbels, 2007a,b,c). These experiments are described in some more detail in Section 4.4.7.

4.4.3 Microwave Spectroscopy of Doped Helium Droplets

Much of the spectroscopic work on smaller molecular systems embedded in helium droplets to date has been done in the infrared range, where typically ro-vibrational transitions are probed. Studies of pure rotational transitions, which often fall into the microwave or millimeter wave ranges, can help to separate the effects of vibrations and rotations on, for example, relaxation dynamics and line-broadening mechanisms.

A sensitive spectroscopic detection method is based on the evaporation of helium atoms upon resonant excitation and subsequent relaxation of the dopant molecule within the helium droplet. The loss of helium atoms can be monitored using a liquid helium cooled bolometer, which measures essentially the kinetic energy of the helium droplet beam, or a mass spectrometer, whose signal is sensitive to the change in ionization cross section that accompanies the change in droplet size. This beam depletion technique works well in the infrared range, where, for example, one photon at 2000 cm⁻¹ is sufficient to evaporate about 400 helium atoms, assuming a binding energy of 5 cm⁻¹ for each helium atom, thus causing a large fractional change in kinetic energy or size. The situation is different in the microwave range. At 8 GHz, for example, 18 photons are needed to evaporate only one helium atom! Microwave spectroscopy on doped helium droplets requires thus the repeated excitation and relaxation of the same droplet on a sufficiently fast timescale.

It was not clear at all if the rotational relaxation rate would be fast enough before the first such study was done (Callegari et al., 2000b; Reinhard et al., 1999) on the rotational spectra of cyanoacetylene, H–C≡C–C≡N, in the range from 10.5 to 14 GHz. In this study, a microwave amplifier providing up to 3.8 W of output power was used. In these experiments, the microwave power was amplitude modulated, and the signal was detected using lock-in techniques. Under nonsaturated conditions, the measured line widths were found to be of similar width as those of corresponding ro-vibrational, infrared transitions. This implies that vibrational relaxation and dephasing are not the dominant line-broadening mechanisms for dopant molecules in helium droplets. The authors found, from microwave power dependence studies, estimates for the upper and lower rotational relaxation times of 20 and 2 ns, respectively. Microwave–microwave double resonance experiments in the same study provided evidence that the line width is dominated by "dynamic" inhomogeneous broadening. The rotational dopant states split in the droplet into substates, which could be caused, for example, through coupling between molecule rotation and translation within the helium droplet. In this sense, the sublevel structure corresponds to particle-in-a-spherical-box states. The inhomogeneous broadening is dynamic in the sense that the rotational relaxation rate is comparable to, or slower than, the substate relaxation rate.

Further evidence for the existence of such sublevel structures comes from recent microwave experiments on ammonia, NH_3, embedded in helium droplets (Lehnig et al., 2007). The umbrella inversion motion of ammonia leads to a tunneling splitting of rotational levels, which is at 23.69 GHz for the $J, K = 1,1$ state in the gas phase. For this study, a Fabry-Pérot microwave resonator was implemented into the helium droplet instrument. The setup is such that the droplet beam enters and exits the resonator through holes near the centers of the mirrors and traverses the resonator coaxially. A microwave amplifier can deliver up to 57 W, and power levels up to 2.8 kW can be achieved in the resonator. The observed ammonia transition has a peculiar line shape, consisting of a broad feature with a width of ~1.5 GHz, and a sharp peak on top, only 15 MHz wide (see Figure 4.6). This is by far the narrowest spectral feature observed in doped helium droplets this far. A similar line shape was also found in the corresponding transition of the $^{15}NH_3$ isotopologue, thus confirming its molecular origin.

This line shape is interpreted in terms of a series of transitions between the sublevels of the two ammonia inversion states, similar to the P—, Q—, and R— branches of a vibrational band. The sublevel structures were modeled using a particle-in-a-box Hamiltonian, and the structures were assumed to be identical for both inversion states. This assumption is justified by the similarity of the probability densities of the two lowest inversion states of ammonia and the fact that the rotational wavefunction state is the same in both states. The observed transition was simulated by populating the levels according to a Boltzmann distribution at 0.38 K and by allowing all possible transitions. Transitions between sublevel states with the same quantum number fall on top of each other ("Q-branch"), since the substructures are identical, and form the sharp peak. The lines were convoluted with a Lorentzian line width of 30 MHz. The amazing agreement between simulation and experiment in Figure 4.5 is clear evidence for the splitting of molecular energy levels into sublevel structures in the helium droplet environment. The lack of such distinctive line shape in microwave spectra of other molecules in helium droplets is

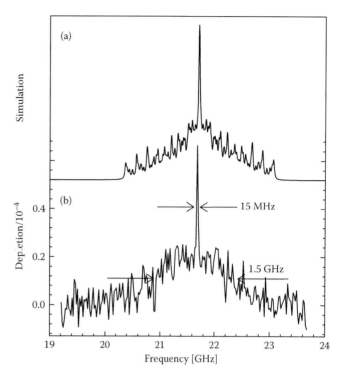

FIGURE 4.6 Shown are the (a) simulated and (b) experimental spectra of a tunneling inversion transition of ammonia embedded in helium droplets. The simulated spectrum was obtained by assuming sublevel structures that correspond to particle-in-a-box states. No selection rules were imposed, and the lines were convoluted with a Lorentzian lineshape with a width of 30 MHz. With the particle-in-a-box parameter $f = 8$ MHz and an effective mass of ammonia solvated with 30 helium atoms, a box length of 67 Å is obtained. This value is in good accord with the diameter of 62 Å of helium droplets with a mean size $\langle N \rangle = 2700$.

a consequence of the change in rotational quantum number in the observed transitions. The different rotational wavefunctions lead to different coupling with the center of mass motion and thus different sublevel structures for the two states involved in the transition.

Very recently, the research group of one of us (WJ) has succeeded in measuring the rotational spectra of carbonyl sulfide (OCS) embedded in helium droplets (Lehnig et al., 2009). Four transitions, involving rotational quantum numbers J from 0 to 4, were measured. The line widths were found to increase with increasing J-value, which is indicative of a distribution of the effective rotational constant, B. The line shapes are also reminiscent of the log-normal distribution of droplet sizes. However, a droplet size dependence of the rotational constant can be excluded as the sole reason for the increase in line width with J, since the droplet size was found to have only a small effect on the width. At present, it is unclear how the energy level substructure found for the case of ammonia can be reconciled with these findings. In particular, the mechanism responsible for the distribution of the effective B values is unknown. It currently appears that there are several mechanisms that affect line shapes and line widths in the spectra of molecules embedded in helium droplets.

4.4.4 Atoms

Atoms do not have internal degrees of freedom in need of cooling. As related to He droplets, they are mostly interesting as a probe of the helium itself, or as building blocks of aggregates under cold-controlled conditions. Several metal atoms have been studied in bulk liquid and solid helium (Tiggesbäumker and Stienkemeier, 2007), with some effort related to the difficulty of injecting the atoms into the helium in significant amounts. Interestingly, even bare electrons can be injected, in fact easily, into liquid helium. The high energy (≈20 eV) of the first unoccupied electronic level of a He atom cause the conduction band of the bulk to be also high (≈1 eV) (Rosenblit and Jortner, 2006; Woolf and Rayfield, 1965). The stablest state of an electron in He is thus not the delocalized one (Springett et al., 1968); rather the electron sits in a bubble whose radius, 17.2 Å (Poitrenaud and Williams, 1972, 1974), minimizes the sum of the localization energy of the electron and the surface energy of the bubble. This is an excellent experimental realization of a particle in a spherical box. Its excitation spectrum happens to lie in the near-infrared/visible; it has been characterized both experimentally and theoretically (Fowler and Dexter, 1968; Grimes and Adams, 1990, 1992; Jortner et al., 1965; Northby and Sanders, 1967, see also Section 4.4.5).

The study of electrons attached to He droplets has been summarized by Northby (2001); notably, photoelectron detachment spectra have been assigned to electronic transitions of the bubble. Let us note that in droplets the electron can be either delocalized over the droplet surface or localized in a bubble state. Both states are at best metastable and are thought to require large droplets to be reasonably long lived; the former is energetically favorable, but reckoned to be short lived already in weak electric fields (Northby and Kim, 1994). The barrier for an electron bubble to escape through the droplet surface is believed to be high enough to make the bubble state the one to account for all experimental observations, although the details have not been fully clarified.

It is easy to rationalize that, as compared to a lone electron, an atom must occupy a tighter bubble, created in the helium by the same repulsive forces, this time between the outer electrons of the metal atom and those of the helium. With the exception of alkali atoms (more below) the interaction of an atom with He easily overcomes that of the helium being displaced: in other words, there is a net gain of solvation energy. The investigation of atoms in He droplets is primarily the study of the valence-electron(s) bubble. In this picture, the energy levels of the valence electron(s) are essentially those of the bare atom, with the helium bubble as a perturbation.

Given the tight confinement, it is easy to accept that the perturbation broadens and shifts the electronic transitions to higher energies, typically by a few percent (Figure 4.7). The main transitions of Li (Bünermann et al., 2007; Callegari et al., 1998; Stienkemeier et al., 1996), Na (Bünermann et al., 2007; Callegari et al., 1998; Mayol et al., 2005; Stienkemeier et al., 1996, 2004), K (Bünermann et al., 2007; Callegari et al., 1998; Stienkemeier et al., 1996), Rb (Auböck et al., 2008a; Brühl et al., 2001; Bünermann

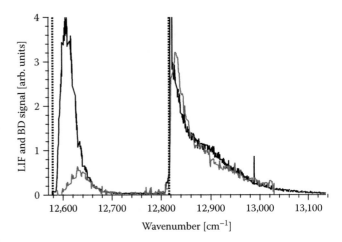

FIGURE 4.7 Laser-induced fluorescence (black) and beam depletion (gray) spectra of Rb atoms on the surface of He droplets. Dashed vertical lines show the positions of the spin–orbit split doublet (so-called D lines) for the gas-phase atom. The negligible shift and the broadening of ~100 cm⁻¹ are typical of the surface-bound alkali atoms. The dissimilarity of the two spectra shows that at excitation energies near the lower D line, atoms do not desorb from the droplet. This observation is peculiar to Rb, and has been exploited in combination with a circularly polarized laser to accomplish optical pumping on He nanodroplets (see Auböck et al., 2008a).

et al., 2007), Cs (Bünermann et al., 2004, 2007), Mg (Diederich et al., 2001; Przystawik et al., 2008; Reho et al., 2000a), Ca (Stienkemeier et al., 1997, 2000), Ba (Stienkemeier et al., 1999, 2000), Sr (Stienkemeier et al., 1997, 2000), Ag (Bartelt et al., 1996; Diederich et al., 2002; Federmann et al., 1999a,b; Przystawik et al., 2008), Al (Reho et al., 2000b), Eu (Bartelt et al., 1996, 1997), and In (Bartelt et al., 1996) have been measured in He droplets and all follow this general pattern. Alkali metal atoms are an exception because of their diffuse valence electron: they reside on the surface of the droplet where the attractive van der Waals forces suffice to keep them weakly bound to the droplet; Mg is an intermediate case and can be considered as "buried" near the surface, rather than fully solvated. The detailed helium distribution around a dopant is easily calculated from the He–dopant pair potential with density functional codes (Barranco et al., 2006, and references therein). Reliable empirical formulae, also based on the pair potential, to guess the location of a dopant have been proposed by Ancilotto et al. (1995) and Perera and Amar (1990).

Regardless of the location of the dopant, the shift and broadening of electronic transitions can be described simply but effectively with the introduction of a single, physically meaningful, effective coordinate, whose choice is dictated by the symmetry of the problem. For a solvated atom, this is the radius R of the solvation bubble. For a surface atom, it is its distance from the surface (Bünermann et al., 2007; Stienkemeier et al., 1996), more conveniently measured from the center of the droplet. In the latter case, the electronic states can be thought of as those of a pseudodiatomic van der Waals molecule in which the whole droplet plays the role of a giant rare-gas atom; by extension, one speaks of an "internuclear" axis, which is often the appropriate quantization axis *z* for the problem. Atomic states are still a convenient label for the excitation, complemented by the appropriate labels for nonrotating diatomic molecules.

Fully solvated atoms are an interesting probe of the surrounding helium: the choice of the atom can be used to "tune" the strength of the interaction upon excitation. Often the excited state is orbitally degenerate in the bare atom (e.g., in the p ← s excitation of Ag), and it is interesting to study how degeneracy may be lifted by dynamic deformations of the bubble (Dupont-Roc, 1995; Kinoshita et al., 1995), and how the latter compete with the spin–orbit interaction, when present.

Highly excited atoms (Rydberg atoms) are interesting because they probe the interaction of a quasi-free electron with the helium. Clearly for extremely high quantum numbers, the electron orbital is almost exclusively located *outside* the droplet. This system has since long intrigued theorists and many interesting properties have been predicted (Ancilotto et al., 2007; Golov and Sekatskii, 1993); there is experimental evidence of its realization (Loginov, 2008).

The assembly of many atoms onto the same droplet can be used to look for the onset of metallic behavior, typically with alkaline earth atoms where the valence band originates from their closed outer s-shell. Apparently He clusters are better than gas-phase experiments in that unwanted compounds of highly reactive elements (such as Mg, easily forming MgO) are not observed (see Tiggesbäumker and Stienkemeier, 2007, and references therein).

Another interesting observation relates to the well-known fact that the attractive atom–helium interaction implies an increased helium density in the first solvation shell. The resulting structure is informally known as a snowball, the word being borrowed from the description of positive ions in bulk He, for which the interaction is strong enough that the first solvation shell is solid beyond reasonable doubt. This layer may result in a high enough energy barrier that the atoms do not coalesce but form instead a metastable superstructure. This is a well-known occurrence in the bulk (Gordon et al., 1989a, 1993); in droplets its observation is presumed for Mg (Przystawik et al., 2008).

Alkali atoms were the first ever investigated on helium droplets (Stienkemeier et al., 1995a,b), and remain the true workhorse of atomic spectroscopy in He droplets. This is related to a number of favorable properties. The high vapor pressure at temperatures easily attainable with resistive heating; the simple one-valence-electron structure, which for the theorist means well-isolated energy levels and hydrogen-like electronic wavefunctions, for the experimentalist a strong excitation transition (*n*P ← *n*S, with *n* the electronic-ground-state principal quantum number; *n* = 2, 3, 4, 5, 6 for Li, Na, K, Rb, Cs, respectively), and correspondingly strong fluorescence emission, conveniently located in the visible portion of the electromagnetic spectrum and well covered by high-resolution tunable lasers. The larger members of the family, K, Rb, Cs, are all within the reach of the powerful and versatile $Ti:Al_2O_3$ laser, thus experiments based on broadband tunability, power, or pulsed operation down to

the femtosecond range, are easily feasible. As said, their equilibrium position is at the surface of a droplet, and their spectrum is minimally perturbed as compared to the free atoms [Figure 4.7; as a measure of this, consider that the spin-orbit splitting (Ralchenko et al., 2009) remains resolvable down to the second smallest value in the series Na, $17.196\,cm^{-1}$ or 0.1% of the transition energy; only for Li, $0.34\,cm^{-1}$ or 0.002%, it is unresolved]. The limited loss of resolution means that these spectra can be combined with detailed statistical models, and effects such as the deformation of the He surface caused by the dopant atom, or the motion of the dopant in the surface potential can be accurately unraveled (Bünermann et al., 2007).

Alkali-atom-doped droplets have also been the first systems where the dynamics after excitation has been time-resolved. The surface location plays an important role in determining the outcome of these experiments. The lifetime of the excited state is essentially the same as for free atoms, thus in the tens of nanoseconds. Within several hundred picoseconds at most, however, excited atoms are ejected from the droplet, either "naked" or after having formed an exciplex with one (very seldom more than one) helium atom. The emission frequency is a strong function of the state of the atom so one can use it to select the different "reaction channels," and to a certain extent to follow the time-evolution of the helium surface. This has been more sensitively done with the vibrational frequency of K_2 molecules as measured in fs pump-probe experiments (Section 4.4.7).

The first experiments were based on time-correlated photon counting, and afforded to look at times down to ~100 ps. While this is not fast compared to surface rearrangement and desorption, it turned out to be well suited to observe excimer formation. The latter process depends sensitively on the height of barriers in the reaction path; these exist because of the role of spin–orbit coupling in shaping the alkali–helium excimer potential compounded with the need to extract a He atom out of the droplet surface. By tuning the excitation energy one can thus tune the excimer formation time, and effectively study a simple photoassociation reaction at very low temperatures.

By tuning the pressure in the pickup cell, one can maximize the probability that one, or two, or more, atoms be picked up by each droplet. At the temperature of the droplet (0.38 K) even weak van der Waals forces between these atoms are sufficient to make their complexes thermodynamically favorable: so far all experimental observations confirm that complex formation is indeed the norm in multiply-doped helium droplets. Open shell atoms have a nonzero electron spin (1/2, for alkalis), and since there is no reason to believe that the droplet may cause an orientation effect, one can expect that complexes will be formed in all possible spin multiplicities, their relative abundance determined by simple spin statistics. If there is the possibility of interconversion, for example, through repeated breaking and forming of the complex, then the final abundance is determined by thermodynamic equilibrium. This occurs normally in a dense vapor of alkali atoms at high temperature: because singlet dimers are covalently bound (binding energy of several thousands cm^{-1}) they greatly outnumber triplet dimers (van der Waals bound, few hundreds cm^{-1}); for the same reason trimers are observed in the doublet spin multiplicity, but never in the quartet one. In He droplets, it is experimentally observed that interconversion does not occur: temperatures are too low for repeated breaking and forming; in addition, no magnetic interaction exists to mix states of different spin multiplicity. It thus appears that in He all spin multiplicities should be observed, for example, triplet dimers in a 3:1 proportion to singlet dimers, but this is not the case. One needs to consider that the energy of formation of the dimer may partly work toward its direct detachment from the droplet, with the rest deposited in the droplet and ultimately lost by the evaporation of helium atoms (there are no experiments quantifying this, but based on bulk values one assumes the evaporation of one helium atom every about 5 cm^{-1} of deposited energy). Both processes work to decrease the number of doped droplets available for spectroscopy. Thus the opposite situation as in the gas phase is experimentally observed: on He droplets triplet dimers greatly outnumber singlet ones and trimers are observed in the quartet spin multiplicity, but never in the singlet one (Auböck et al., 2008b; Higgins et al., 1996a,b, 2000; Nagl et al., 2008a,b; Reho et al., 2001).

The spectra of these systems (except the quartet trimers) are well known in the gas phase, where they have been measured with the greatest accuracy. In He droplets, they are severely broadened, but their vibrational structure generally remains visible, and has been used to learn about the fine details of the interaction with the helium [e.g., by the presence or absence of a zero-phonon line and a phonon gap (Higgins et al., 1998)], of the desorption dynamics (through the time-dependence of the vibrational frequency, Section 4.4.7), and, in trimers, to learn about Jahn–Teller distortions (Auböck et al., 2008b; Higgins et al., 1996b, 2000; Reho et al., 2001).

We said that there is no interconversion between different spin multiplicities. This is true in the lowest electronic state. In excited states, both triplet dimers and quartet trimers undergo spin flip processes (clearly identified by the fact that the photon energy of the emitted fluorescence is higher than that of the exciting photon, a very basic example of conversion of chemical energy); in addition, trimers dissociate into a dimer and an atom, with many output channels whose branching ratios depend strongly on small changes of the energy of the exciting photon. These processes are interesting as prototype of very simple photoinitiated chemistry proceeding from well-defined initial states (Higgins et al., 1996a, 1998).

The high-spin structure of these molecules (and indeed already the single spin of the atom) lends themselves to magnetic studies. Let us note right away that at typical ESR frequencies (~10 GHz) the energy separation between Zeeman states is comparable to $k_B T$ at 0.38 K, so in a moderately strong magnetic field (a few tenths of a Tesla) a substantial spin polarization must exist, provided that spin relaxation is fast enough. All this will be considered in Section 4.4.5.

Larger aggregates of alkali atoms formed on He droplets have been investigated by mass spectroscopy (see Tiggesbäumker and Stienkemeier, 2007). Potassium cluster ions at masses as large

as 70 atoms have been reported, showing that very large clusters can indeed be assembled. Not much could be said about the electronic structure of these clusters, nor about related aspects (the spin state; the location on the He droplet: surface or solvated). Interestingly, the ion abundances show magic numbers (e.g., Na_9^+, Na_{21}^+), which correspond to electron shell filling at two electrons per shell, only possible with low spin states; no investigation has been made as to whether this also was the spin state of the neutral parent cluster, or a spin-flip occurred. Magic numbers and the associated shell closure are used to infer the onset of electron delocalization (i.e., metallic behavior) in Mg clusters at approximately 20 atoms (Diederich et al., 2005). Metallic behavior in Mg clusters, when interacting with acetylenic molecules, has been studied in the Miller group (Dong and Miller, 2004; Moore and Miller, 2004; Nauta et al., 2001; Stiles et al., 2004), exploiting the structural information provided by ro-vibrational spectra and, once more, the assembly capabilities of helium droplets.

Ionized single atoms, as seen in mass spectra, exhibit a surrounding helium "snowball," which should be particularly stable for closed geometric shells. A number of theoretical techniques have been applied to studying the solvation of positive ions in He droplets (Coccia et al., 2007; Galli et al., 2001; Marinetti et al., 2007; Nakayama and Yamashita, 2000; Rossi et al., 2004). Mainly alkali and alkaline earth ions have been addressed so far since reliable Me$^+$–He potentials are available for these species. Using variational Monte Carlo simulations, it has been found that all alkali and alkaline-earth cations form snowball structures featuring shells of He atoms with high average density. In addition to a modulated radial density profile around the impurity ions, snowballs are characterized by angular correlations in the first He shell as well as a high degree of radial localization of He atoms. This solid-like order is compatible in some cases with icosahedron packing.

Associated magic numbers have been experimentally confirmed as steps in mass spectra of ionized alkali-doped helium nanodroplets (Müller et al., 2009b). A general observation in mass spectrometric investigations of coinage metal clusters formed within He droplets is that "naked" clusters are detected, with no accompanying helium atoms attached, in stark contrast with what expected from the large electrostrictive force at play (see "snowball" above). It is also observed that the structure found in the mass spectra is to a large degree independent of the ionization method. All this must correlate to fundamental properties of the helium droplets that certainly warrant further investigation.

4.4.5 Magnetic Studies

Magnetic studies in He nanodroplets merge matrix spectroscopy with a most venerable field: molecular beam magnetic resonance (MBMR) spectroscopy.

The use of inert matrices for magnetic studies has a long tradition, especially for complexes that needed the stabilizing action of the matrix (Weltner et al., 1995): spin-resonance spectroscopy was often used to identify unusual compounds stabilized in the matrix [e.g., alkali clusters (Lindsay et al., 1976; Thompson et al., 1983), or Mn clusters (Baumann et al., 1983)]. Magnetic methods are invaluable in support of other types of spectroscopy, such as infrared and visible, where the perturbation induced by the matrix, especially in relation to multiple types of trapping sites, may lead to ambiguities in the interpretation of the spectra. When individual Zeeman states cannot be resolved in optical spectra, circular dichroism can provide accurate information on dopant–matrix interaction, based on general symmetry arguments (Piepho and Schatz, 1983); spin-resonance measurements directly provide the multiplicity of the target species, and by their ability to discriminate inequivalent spins, considerable information on its symmetry (Weltner et al., 1995). Like all rare gas matrices, helium is also closed shell, thus magnetically inert (more precisely, very weakly diamagnetic, as all substances are when no stronger effects are present). The common isotope ^{4}He also has zero nuclear spin, so it is truly nonmagnetic, whereas ^{3}He has nuclear spin 1/2: albeit weak, the resulting magnetic interaction is significant at short distances (interatomic collisions) and has been successfully used in spin-exchange schemes (Bouchiat et al., 1960; Grover, 1978; Middleton et al., 1995). All magnetic studies in nanodroplets are so far limited to ^{4}He, so in the following we restrict discussion to this isotope; there is no doubt however that experiments in ^{3}He droplets will be extremely interesting; even more, experiments in mixed droplets because of the known surface segregation of the lighter ^{3}He isotope.

Already in the 1970s, Reichert and collaborators performed ESR measurements of electrons injected in bulk He with standard ESR methods, finding long relaxation times and little shifts relative to the free electron (Reichert and Dahm, 1974; Reichert and Jarosik, 1983; Reichert et al., 1979; Zimmermann et al., 1977). E. B. Gordon and collaborators used spin resonance, among other methods, to study atoms and molecules injected in bulk He. They studied in particular impurity-helium solids: highly porous structures formed by condensing a jet of impurity-helium gas mixture into liquid helium (Gordon et al., 1982, 1985, 1989b). Optically detected methods were applied by Kanorsky, Weis, and collaborators (Arndt et al., 1993, 1995; Kanorsky et al., 1996, 1998; Lang et al., 1995, 1999; Nettels et al., 2003a,b; Ulzega et al., 2007; Weis et al., 1995), by Yabuzaki and collaborators (Kinoshita et al., 1994; Takahashi et al., 1995b; Yabuzaki et al., 1995), and are reviewed in Kanorsky and Weis (1998); Moroshkin et al. (2006, 2008). Shimoda and collaborators look at the spin polarization of atoms injected in liquid helium, with alkalis as a test system and short-lived radioactive isotopes as the main goal (Furukawa et al., 2006; Takahashi et al., 1995a, 1996).

MBMR "was the first extremely high-resolution spectroscopic technique developed. Many fundamental nuclear, atomic, and molecular properties were first observed in these experiments. Molecular beam magnetic resonance experiments still provide the definitive information on the electronic structure of small nonpolar molecules" (Yokozeki and Muenter, 1980). With these achievements in mind, one of us (CC) has proposed and implemented magnetic methods to counteract the well-known loss of spectral resolution in He nanodroplets (as compared to gas-phase spectra), to study spin relaxation (or lack thereof), and to use resonance shifts as a probe of the minute changes of electronic structure of the dopant brought about by the helium and, in perspective, by complexation with another dopant.

Relatively simple magnetic circular dichroism (MCD) experiments prove that electron spin relaxation is slow for alkali-atom dopants (Auböck et al., 2008a; Nagl et al., 2007); so much in fact, as compared to the transit time of the droplets in the magnetic field, that only a *lower* limit (>2 ms) can be given for the relaxation time. This is not unexpected, based on the long relaxation times observed in the bulk, and on the more general observation that no obvious coupling mechanism exists between the alkali spin and the helium nanodroplet, here acting as a thermal bath. Given this long relaxation time, the preparation of a spin-polarized ensemble of doped droplets, by selective photodissociation of one spin state, is trivial. By clever choice of atom (Rb so far, but Cs should be even better, due to the larger spin–orbit constant) and photon energy, one can even close the desorption channel of the optical excitation (see Figure 4.7) and optical pumping between the Zeeman levels of the electronic ground state becomes possible (Auböck et al., 2008a).

Fast spin thermalization is in contrast observed for alkali dimers (Auböck et al., 2007; Nagl et al., 2007) where coupling mechanisms must exist; these have yet to be identified but the most reasonable link between spin and thermal bath is the rotation of the molecule; for dimers (Auböck et al., 2007; Nagl et al., 2007) as well as trimers (Auböck et al., 2008b) one can assume complete thermalization of the spin, thus one can give an *upper* limit of ~40 μs for the relaxation time. An interesting remark is that these spectra provide a thermometer of a different type than the traditional rotational spectrum of a solvated molecule. Upon more accurate measurements, the consistency, or lack thereof, of the two methods can be used to test possible biases, as well as differences between the interior and surface temperature, whose eventuality has been suggested by Lehmann (2003, 2004). Another by-product of MCD spectra is the strength of the interaction between the dopant and the helium in the excited state, in the form of a "crystal field" splitting (Auböck et al., 2007). For the more complex spectra of the trimers, where proper assignment of an electronic band must account for three perturbations, all of comparable strength (spin–orbit coupling, Jahn–Teller distortion, and the above-mentioned "crystal field"), MCD spectra are invaluable (Auböck et al., 2008b).

The above knowledge is more than sufficient to cover the prerequisite steps toward optically detected magnetic resonance. We have just succeeded to detect the ESR spectrum of K and Rb atoms in a magnetic field of ≈3.4 kG, at microwave frequencies of ≈9.4 GHz (Koch et al., 2009a,b, 2010). We observe sharp, probably instrument-limited, lines ($\Delta\nu/\nu \approx 10^{-5}$). At the present accuracy, the *g* factor is not significantly affected. The hyperfine splitting constant instead, is larger than that of the free atom by a small but clearly measurable amount; this clearly reflects an increased

Fermi-contact term due to "compression" by the droplet of the alkali valence electron wavefunction. Rabi oscillations are also detected, attesting to the long coherence of the spin in helium.

4.4.6 Spectroscopy of Organic Molecules and Nanostructures

Larger organic molecules and their complexes have been isolated in the cold helium droplet environment. So far, the focus of most of the studies lies in the electronic properties. The spectroscopic work with helium droplets in the visible range has been reviewed some time ago (Stienkemeier and Vilesov, 2001). Studies even include the UV or XUV photon energy range (Peterka et al., 2007; von Haeften et al., 2001). In terms of probing larger molecules and complexes, several results exploring different directions have been published: biomolecules (Dong and Miller, 2002), metal clusters (Tiggesbäumker and Stienkemeier, 2007; Diederich et al., 2002), and heterogeneous structures (Nauta et al., 2001). Organic molecules include among others tetracene (Hartmann et al., 2001), perylene (Carçabal et al., 2004), pentacene (Lehnig and Slenczka, 2005), and phthalocyanine (Lehnig and Slenczka, 2005). This line of work has also been extended to high-resolution fluorescence emission spectroscopy (Lehnig and Slenczka, 2003, 2005).

The idea is to characterize and also to synthesize organic structures having peculiar properties in a bottom-up approach. The experiments in one of our groups (FS) target on complexes that are characterized by correlations of the constituents that lead to collective or excitonic configurations. In particular, crystalline aggregates have been studied that are of practical interest because of their semiconducting or opto-electronic properties (Wewer and Stienkemeier, 2003, 2005). Prominent representatives (Figure 4.8) are oligoacenes, perylene derivatives (e.g., PTCDA, PDI), or thiophene derivatives (α-quarterthiophene, α-sexithiophene).

The uniqueness of doing spectroscopic studies in helium droplets can be summarized as follows:

1. The weak perturbation by the helium environment leads to solvent shifts of the order of $10\,cm^{-1}$ and broadenings $\lesssim 1\,cm^{-1}$. Hence vibrationally resolved vibronic spectra of larger molecules and their complexes can be recorded. For some molecules, such as tetracene, even rotational contours visibly determine the lineshape of vibronic bands. In this way, detailed information about the geometric and electronic structure can be obtained.

2. Even at high spectral resolution, a high number of populated states, and corresponding hot bands, still hinders the assignment and the interpretation of spectra of larger organic molecules. At room temperature, these molecules and their aggregates have a large number of soft modes that are populated and reduce the value of experimental measurements. Gas-phase studies and cooling in supersonic jets have partially overcome this issue but have only been successful when applying elaborate double-resonance techniques in combination with detailed

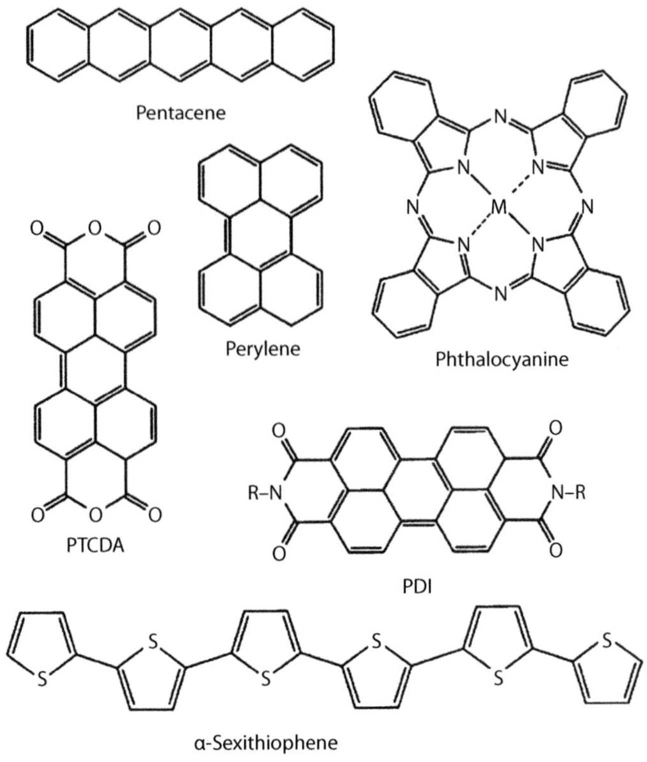

FIGURE 4.8 Representative organic molecules whose properties evolve toward those of a semiconductor when aggregated in complexes nanostructures or films.

theoretical works (Chin et al., 2002; Hunig et al., 2003). At the sub-Kelvin temperature of helium droplets, molecules are virtually frozen in the vibrational ground state. This simplifies assignment and also sets well-defined conditions for exciting and probing the electronic structure with quantum-state selectivity. In order to compare the broadening and shifting of spectra, Figure 4.9 shows the absorption of PTCDA molecules measured in different environments. Only the helium droplet spectrum nicely resolves the full vibronic progression of the $S_0 \rightarrow S_1$ transition. Attaching PTCDA to molecular hydrogen or argon clusters (cf. Figure 4.9) already induces significant broadening and shifting; the main vibrational modes, however, are still visible. The spectrum in an organic solvent (DMSO) at room temperature appears as if only a progression of one mode (often called "effective") is present. Figure 4.10 clearly demonstrates that this effective mode is a convolution of the many individual vibrational modes: The spectrum in DMSO can nicely be reproduced just by shifting ($1600\,cm^{-1}$) and broadening the high-resolution spectrum obtained by helium nanodroplet isolation spectroscopy. The top spectrum in Figure 4.9 shows the absorption of a PTCDA film on mica. Here, in addition to molecular absorption, excitonic transitions contribute to determine the electronic spectrum of the aggregated molecules. Since such excitonic transition can also be measured when PTCDA molecules are aggregated into

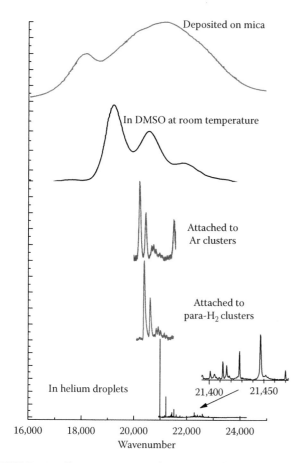

FIGURE 4.9 Absorption spectra of 3,4,9,10-perylenetetracarboxylic-dianhydride (PTCDA) recorded in different environments. The bottom spectrum shows a laser-induced fluorescence absorption spectrum in helium nanodroplets, followed by spectra recorded with doped large molecular hydrogen and argon clusters, respectively. The top two spectra are the absorption of PTCDA molecules in DMSO (Bulovic et al., 1996) and the absorption of a PTCDA film on mica (Proehl et al., 2005).

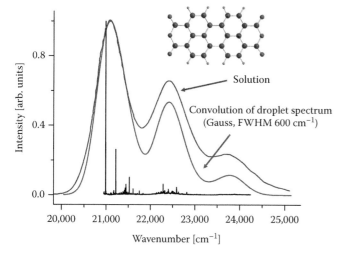

FIGURE 4.10 Excitation spectrum of PTCDA embedded in helium nanodroplets (bottom trace) and its broadened spectrum by convolution with a 600 cm^{-1} Gauss function. For comparison, a spectrum of PTCDA molecules in DMSO (Bulovic et al., 1996) is shown, shifted by +1800 cm^{-1}.

complexes in helium droplets, the different contributions of electronic transitions can easily be disentangled.

3. Having the molecules and molecular structures attached to a helium droplet beam has several advantages as far as detection methods are concerned. First, one deals with a continuously regenerating target; hence photobleaching or other degrading mechanisms are not of importance. Furthermore, special detection schemes can be utilized in order to obtain excitation properties. Several beam depletion methods are at hand, monitoring energy deposition, the destruction of droplets or dopants, or desorption mechanisms. On the other hand, photo ionization or electron impact ionization can be utilized for efficient ion detection. These techniques, combined with mass selection (e.g., quadrupole fields or time-of-flight measurements), can be used to obtain mass-specific properties. However, in comparison with methods using bulk material one should keep in mind that the droplet beam is very dilute and techniques requiring high-density targets, like monitoring the direct absorption of light, usually cannot be applied.

4. The versatility of doping helium droplets allows in particular for forming heterogeneous nanostructures. Atoms and molecules having very different properties like refractory metals, complex molecules, radicals, or even ions can be loaded in a specific order. In this way, for example, specific donor–acceptor systems or core shell complexes can be studied.

In general, vibronic bands of organic molecules embedded in helium nanodroplets are characterized by the interaction with the helium matrix, that is, the lines are composed of a narrow zero phonon line (ZPL) and a phonon wing (PW). Because of their different saturation behavior, PWs only become prominent at higher laser power, in particular when using pulsed lasers. In many cases, the ZPLs are split into different components, indicating discrete and long-lived states of the solvation structure of the surrounding helium matrix. Since vibrational modes of localized helium atoms are not expected to exist in superfluid helium, the experiments confirm the existence of a solid-like (snowball) solvation shell (Lehnig and Slenczka, 2004, 2005). Depending on the molecule, different helium layer configurations have been assigned and one was able to derive relaxation probabilities.

4.4.7 Dynamics in Helium Droplets

Both the superfluid properties of helium droplets and the potential to study even complex structures in well-defined states at millikelvin temperatures aroused much attention to understand dynamical processes in these systems. Time-dependent experiments in connection with a diversity of theoretical approaches have unraveled many puzzles, in particular as far as the energy and angular momentum dissipation and cooling is concerned. A review article has been devoted in particular to this topic (Stienkemeier and Lehmann, 2006). Many experimental results come from femtosecond real-time studies where one observes dynamical processes in the range from tens of femtoseconds to the nanosecond range. In brief, one triggers the system via an excitation with a femtosecond laser pulse and then probes the evolution of the system with a delayed second femtosecond pulse [pump-probe technique (Zewail, 1994)]. As an example, one may look at the vibrational motion of dimer molecules. In Figure 4.11, a pump-probe signal of rubidium dimers is plotted. The corresponding wave packet motion takes place in the first excited triplet state of Rb_2. High-precision measurements of this kind can be performed even for the weakly bound triplet dimers. In general, wave packet oscillation in helium droplets doped with alkali dimers have been observed for pump-probe delay times extending more than a nanosecond. Extracting vibrational frequencies by Fourier analysis of the spectra in the frequency domain leads to an unprecedented precision. As an example, Figure 4.12 plots the fast Fourier transform (FFT) spectra comparing two different isotopes of Rb dimers. Vibrational spacings can be determined in absolute numbers within one hundredth of a wave number (Mudrich et al., 2009a). These measurements allow a detailed determination of interaction potentials and are stringent tests of these as provided by up-to-date *ab initio* potentials.

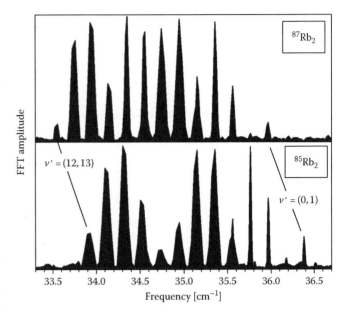

FIGURE 4.12 Fourier transformation of a wave packet motion in the first excited triplet state $b^3 \Sigma_g^+$ of Rb dimers formed on helium nanodroplets for two different isotopes. Spectra recorded at different photon energies have been overlapped, which is responsible for the varying envelope intensity. Each peak represents the frequency difference between consecutive vibrational states, as indicated by the pair of vibrational quantum numbers V'. (From Mudrich, M. et al., *Phys. Rev. A* 80, 042512, 2009.)

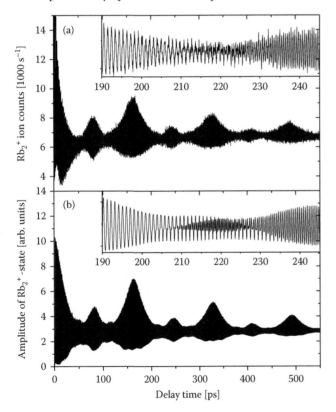

FIGURE 4.11 Pump-probe spectra of rubidium dimers formed on helium nanodroplets (Mudrich et al., 2009a). The measured oscillation (a) represents the vibrational motion of an induced wave packet. The different maxima correspond to the so-called revivals and fractional revivals coming from the dephasing and rephasing of the contributions of the coherently excited vibrational states and (b) compares the spectrum with the outcome of quantum calculation, propagating a corresponding wave packet in time. (From Schlesinger, M. and Strunz, W., unpublished results, 2009.)

Properties of the dynamics of the helium environment can be studied when looking at perturbations in the wave packet motion. In this way, one has observed desorption times of surface-bound molecules. By employing femtosecond pump-probe techniques, also fragmentation dynamics of metal clusters attached to helium droplets and corresponding energy dissipation mechanisms have been observed (Claas et al., 2009). By shining high-intensity lasers on, for example, Mg clusters attached to helium droplets, not only the decomposition but also the charging of the fragments has been investigated (Döppner et al., 2001).

An experimental strategy to directly investigate the translational dynamics of neutral species embedded in helium nanodroplets has been pursued by creating fragments from a photo-dissociation process with well-defined velocity distributions inside a helium nanodroplet (Braun, 2004; Braun and Drabbels, 2004). The comparison of the fragments' initial and final (after having left the droplet) velocity distribution provides detailed insight into the translational dynamics and the interaction with the helium environment. The photo-dissociation of CH_3I and CF_3I has been probed inside helium droplets. Based on the observed speed distributions and anisotropy parameters, it is concluded that the CF_3 fragments escape via a direct mechanism, only partially transferring their excess kinetic energy to the droplet. So far these experimental approaches have only probed superfluid ^4He droplets. A direct comparison to nonsuperfluid ^3He droplets is planned and gives hope to provide more insight into the quantum properties of the bosonic versus fermionic nanoclusters.

4.5 Summary and Outlook

Helium nanodroplets constitute a fascinating medium with extraordinary properties, which include a tunable size range, a low temperature of 0.38 K (0.15 K), the property of superfluidity (^{4}He isotope only), and an extremely weak perturbation of embedded atomic or molecular systems. Their ability to readily pick up one or more atoms or molecules, together with their transparency in most of the frequency range of interest, make them an ideal matrix for spectroscopic studies. The resulting spectra are gas-phase-like, with only slightly modified spectroscopic parameters and increased linewidths. These spectra give thus information about the solvated molecule, but also about the properties of the helium droplets themselves. Spectroscopic studies have been carried out in frequency ranges from the microwave to the vacuum ultraviolet (VUV), and femtosecond pump-probe experiments have provided insight into the dynamical properties of doped droplets. Molecular excited degrees of freedom are thermalized quickly within the droplets; this allows the targeted assembly and investigation of nanostructures within the droplets. Many of the specific droplet properties and droplet–dopant interactions that lead to the differences to the corresponding gas-phase spectra remain elusive. Additional experiments, together with theoretical modeling, will be needed to further our understanding about these intriguing nanosized entities. In the future, we can anticipate the refining of existing experimental techniques for the investigations of more complex molecular systems, and the development of new ones. Other promising applications of helium nanodroplets may include, for example, the assembly, transport, and surface deposition of engineered nanoclusters, as demonstrated by Vilesov and coworkers (Mozhayskiy et al., 2007).

Acknowledgments

We would like to acknowledge the cooperation of our colleagues and coworkers in many exciting and successful experiments.

CC thanks Olivier Allard, Gerald Auböck, Wolfgang Ernst, Andreas Hauser, Markus Koch, Johannes Lanzersdorfer, Johann Nagl, and Alexandra Pifrader as well as Francesco Ancilotto, Marcel Drabbels, Kevin Lehmann, and John Muenter.

WJ thanks Yunjie Xu, Bob McKellar, Pierre-Nicholas Roy, Nicholas Blinov, Wendy Topic, and James Song.

FS thanks Oliver Bünermann, Matthieu Dvorak, Philipp Heister, and Marcel Mudrich.

A special thank-you goes to Giacinto Scoles for continued inspiration, and for bringing us to helium droplets.

References

Adams, E. D., K. Uhlig, Y.-H. Tang, and G. E. Haas, 1984. Solidification and superfluidity of ^{4}He in confined geometries. *Phys. Rev. Lett.* 52: 2249–2252.

Ancilotto, F., P. B. Lerner, and M. W. Cole, 1995. Physics of solvation. *J. Low Temp. Phys.* 101: 1123–1146.

Ancilotto, F., M. Pi, R. Mayol, M. Barranco, and K. Lehmann, 2007. Squeezing a helium nanodroplet with a Rydberg electron. *J. Phys. Chem. A* 111: 12695–12701.

Anderson, J. B. 2001. An exact quantum Monte Carlo calculation of the helium–helium inter-molecular potential. II. *J. Chem. Phys.* 115: 4546–4548.

Anderson, J. B. 2004. Comment on "an exact quantum Monte Carlo calculation of the helium–helium intermolecular potential" [*J. Chem. Phys.* 115, 4546 (2001)]. *J. Chem. Phys.* 120: 9886–9887.

Arndt, M., S. I. Kanorsky, A. Weis, and T. W. Hänsch, 1993. Can paramagnetic atoms in superfluid-helium be used to search for permanent electric-dipole moments. *Phys. Lett. A* 174: 298–303.

Arndt, M., S. I. Kanorsky, A. Weis, and T. W. Hänsch, 1995. Long electronic spin relaxation-times of Cs atoms in solid ^{4}He. *Phys. Rev. Lett.* 74: 1359–1362.

Auböck, G., J. Nagl, C. Callegari, and W. E. Ernst, 2007. Triplet state excitation of alkali molecules on helium droplets: Experiments and theory. *J. Phys. Chem. A* 111: 7404–7410.

Auböck, G., J. Nagl, C. Callegari, and W. E. Ernst, 2008a. Electron spin pumping of Rb atoms on He nanodroplets via nondestructive optical excitation. *Phys. Rev. Lett.* 101: 035301.

Auböck, G., J. Nagl, C. Callegari, and W. E. Ernst, 2008b. Observation of relativistic $E \otimes e$ vibronic coupling in Rb_3 and K_3 quartet states on helium droplets. *J. Chem. Phys.* 129: 114501.

Barranco, M., M. Pi, S. M. Gatica, E. S. Hernandez, and J. Navarro, 1997. Structure and energetics of mixed ^{4}He-^{3}He drops. *Phys. Rev. B* 56: 8997–9003.

Barranco, M., R. Guardiola, S. Hernandez, R. Mayol, J. Navarro, and M. Pi, 2006. Helium nanodroplets: An overview. *J. Low Temp. Phys.* 142: 1–81.

Bartelt, A., J. D. Close, F. Federmann, N. Quaas, and J. P. Toennies, 1996. Cold metal clusters: Helium droplets as a nanoscale cryostat. *Phys. Rev. Lett.* 77: 3525–3528.

Bartelt, A., J. D. Close, F. Federmann, K. Hoffmann, N. Quaas, and J. P. Toennies, 1997. The UV-absorption of europium atoms embedded in helium nanodroplets. *Z. Phys. D* 39: 1–2.

Baumann, C. A., R. J. van Zee, S. V. Bhat, and W. Weltner, 1983. ESR of Mn_2 and Mn_5 molecules in rare-gas matrices. *J. Chem. Phys.* 78: 190–199.

Beamish, J. R., A. Hikata, L. Tell, and C. Elbaum, 1983. Solidification and superfluidity of ^{4}He in porous Vycor glass. *Phys. Rev. Lett.* 50: 425–428.

Blinov, N., X. Song, and P.-N. Roy, 2004. Path integral Monte Carlo approach for weakly bound van der Waals complexes with rotations: Algorithm and benchmark calculations. *J. Chem. Phys.* 120: 5916–5931.

Bouchiat, M. A., T. R. Carver, and C. M. Varnum, 1960. Nuclear polarization in He^3 gas induced by optical pumping and dipolar exchange. *Phys. Rev. Lett.* 5: 373–375.

Braun, A. 2004. Photodissociation studies of CH_3I and CF_3I in fluid ^{4}Helium nanodroplets. PhD thesis, EPFL, Lausanne, Switzerland.

Braun, A. and M. Drabbels, 2004. Imaging the translational dynamics of CF_3 in liquid helium droplets. *Phys. Rev. Lett.* 93: 253401.

Braun, A. and M. Drabbels, 2007a. Photodissociation of alkyl iodides in helium nanodroplets. I. Kinetic energy transfer. *J. Chem. Phys.* 127: 114303.

Braun, A. and M. Drabbels, 2007b. Photodissociation of alkyl iodides in helium nanodroplets. II. Solvation dynamics. *J. Chem. Phys.* 127: 114304.

Braun, A. and M. Drabbels, 2007c. Photodissociation of alkyl iodides in helium nanodroplets. III. Recombination. *J. Chem. Phys.* 127: 114305.

Brink, D. and S. Stringari, 1990. Density of states and evaporation of helium clusters. *Z. Phys. D* 15: 257–263.

Brown, J. M. and A. Carrington, 2003. *Rotational Spectroscopy of Diatomic Molecules*. Cambridge Molecular Science Series. Cambridge, NY: Cambridge University Press.

Brühl, F. R., R. A. Trasca, and W. E. Ernst, 2001. Rb-He exciplex formation on helium nanodroplets. *J. Chem. Phys.* 115: 10220–10224.

Buchenau, H., E. L. Knuth, J. Northby, J. P. Toennies, and C. Winkler, 1990. Mass spectra and time-of-flight distributions of helium cluster beams. *J. Chem. Phys.* 92: 6875–6889.

Bulovic, V., P. E. Burrows, S. R. Forrest, J. A. Cronin, and M. E. Thompson, 1996. Study of localized and extended excitons in 3,4,9,10-perylenetetracarboxylic dianhydride (PTCDA) I. Spectroscopic properties of thin films and solutions. *Chem. Phys.* 210: 1–12.

Bünermann, O., M. Mudrich, M. Weidemüller, and F. Stienkemeier, 2004. Spectroscopy of Cs attached to helium nanodroplets. *J. Chem. Phys.* 121: 8880–8886.

Bünermann, O., G. Droppelmann, A. Hernando, R. Mayol, and F. Stienkemeier, 2007. Unraveling the absorption spectra of alkali metal atoms attached to helium nanodroplets. *J. Phys. Chem. A* 111: 12684–12694.

Callegari, C., J. Higgins, F. Stienkemeier, and G. Scoles, 1998. Beam depletion spectroscopy of alkali atoms (Li, Na, K) attached to highly quantum clusters. *J. Phys. Chem. A* 102: 95–101.

Callegari, C., A. Conjusteau, I. Reinhard, K. K. Lehmann, G. Scoles, and F. Dalfovo, 1999. Superfluid hydrodynamic model for the enhanced moments of inertia of molecules in liquid ^{4}He. *Phys. Rev. Lett.* 83: 5058–5061. [Erratum: 84, 1848 (2000)].

Callegari, C., A. Conjusteau, I. Reinhard, K. K. Lehmann, and G. Scoles, 2000a. First overtone helium nanodroplet isolation spectroscopy of molecules bearing the acetylenic CH chromophore. *J. Chem. Phys.* 113: 10535–10550.

Callegari, C., I. Reinhard, K. K. Lehmann, G. Scoles, K. Nauta, and R. E. Miller, 2000b. Finite size effects and rotational relaxation in superfluid helium nanodroplets: Microwave-infrared double-resonance spectroscopy of cyanoacetylene. *J. Chem. Phys.* 113: 4636–4646.

Callegari, C., K. K. Lehmann, R. Schmied, and G. Scoles, 2001. Helium nanodroplet isolation rovibrational spectroscopy: Methods and recent results. *J. Chem. Phys.* 115: 10090–10110.

Campargue, R., ed. 2001. *Atomic and Molecular Beams: The State of the Art 2000*. Berlin, Germany: Springer.

Carçabal, P., R. Schmied, K. K. Lehmann, and G. Scoles, 2004. Helium nanodroplet isolation spectroscopy of perylene and its complexes with oxygen. *J. Chem. Phys.* 120: 6792–6793.

Castleman, A. and K. Bowen, 1996. Clusters: Structure, energetics, and dynamics of intermediate states of matter. *J. Phys. Chem.* 100: 12911–12944.

Chin, W., M. Mons, I. Dimicoli, F. Piuzzi, B. Tardivel, and M. Elhanine, 2002. Tautomer contribution's to the near UV spectrum of guanine: Towards a refined picture for the spectroscopy of purine molecules. *Eur. Phys. J. D* 20: 347–355.

Choi, M. Y., G. E. Douberly, T. Falconer et al. 2006. Infrared spectroscopy of helium nanodroplets: Novel methods for physics and chemistry. *Int. Rev. Phys. Chem.* 25: 15–75.

Claas, P., S.-O. Mende, and F. Stienkemeier, 2003. Characterization of laser ablation as a means for doping helium nanodroplets. *Rev. Sci. Instrum.* 74: 4071–4076.

Claas, P., C. P. Schulz, and F. Stienkemeier, 2009. Fragmentation dynamics of potassium clusters attached to helium droplets. Unpublished results.

Coccia, E., E. Bodo, F. Marinetti et al. 2007. Bosonic helium droplets with cationic impurities: Onset of electrostriction and snowball effects from quantum calculations. *J. Chem. Phys.* 126: 124319.

Conjusteau, A., C. Callegari, I. Reinhard, K. K. Lehmann, and G. Scoles, 2000. Microwave spectra of HCN and DCN in ^{4}He nanodroplets: A test of adiabatic following. *J. Chem. Phys.* 113: 4840–4843.

CR100–11 2000. Thematic issue on van der Waals molecules. *Chem. Rev.* 100 (11).

de Heer, W. A. 1993. The physics of simple metal clusters: Experimental aspects and simple models. *Rev. Mod. Phys.* 65: 611–676.

Diederich, T., T. Döppner, J. Braune, J. Tiggesbäumker, and K. H. Meiwes-Broer, 2001. Electron delocalization in magnesium clusters grown in supercold helium droplets. *Phys. Rev. Lett.* 86: 4807–4810.

Diederich, T., J. Tiggesbäumker, and K. H. Meiwes-Broer, 2002. Spectroscopy on rare gas-doped silver clusters in helium droplets. *J. Chem. Phys.* 116: 3263–3269.

Diederich, T., T. Döppner, T. Fennel, J. Tiggesbäumker, and K. H. Meiwes-Broer, 2005. Shell structure of magnesium and other divalent metal clusters. *Phys. Rev. A* 72: 023203.

Dietz, T. G., M. A. Duncan, D. E. Powers, and R. E. Smalley, 1981. Laser production of supersonic metal cluster beams. *J. Chem. Phys.* 74: 6511–6512.

Dong, F. and R. E. Miller, 2002. Vibrational transition moment angles in isolated biomolecules: A structural tool. *Science* 298: 1227–1230.

Dong, F. and R. E. Miller, 2004. Laser spectroscopy of cyanoacetylene-Mg_n complexes in helium nanodroplets: Multiple isomers. *J. Phys. Chem. A* 108: 2181–2191.

Döppner, T., T. Diederich, J. Tiggesbäumker, and K. H. Meiwes-Broer, 2001. Femtosecond ionization of magnesium clusters grown in ultracold helium droplets. *Eur. Phys. J. D* 16: 13–16.

Douberly, G. and R. E. Miller, 2007. Rotational dynamics of HCN–M (M = Na, K, Rb, Cs) van der Waals complexes formed on the surface of helium nanodroplets. *J. Phys. Chem. A* 111: 7292–7302.

Dunoyer, L. 1911a. Sur la réalisation d'un rayonnement matériel d'origine purement thermique. Cinétique expérimentale. *Le Radium* 8: 142–147.

Dunoyer, L. 1911b. Sur la théorie cinétique des gaz et la réalisation d'un rayonnement materiel d'origine thermique. *Compt. Rend.* 152: 592–595.

Dupont-Roc, J. 1995. Excited p-states of alkali atoms in liquid-helium. *Z. Phys. B* 98: 383–386.

Federmann, F., K. Hoffmann, N. Quaas, and J. D. Close, 1999a. Rydberg states of silver: Excitation dynamics of doped helium droplets. *Phys. Rev. Lett.* 83: 2548–2551.

Federmann, F., K. Hoffmann, N. Quaas, and J. P. Toennies, 1999b. Spectroscopy of extremely cold silver clusters in helium droplets. *Eur. Phys. J. D* 9: 11–14.

Fowler, W. B. and D. L. Dexter, 1968. Electronic bubble states in liquid helium. *Phys. Rev.* 176: 337–343.

Fröchtenicht, R., U. Henne, J. P. Toennies, A. Ding, M. Fieber-Erdmann, and T. Drewello, 1996. The photoionization of large pure and doped helium droplets. *J. Chem. Phys.* 104: 2548–2556.

Furukawa, T., Y. Matsuo, A. Hatakeyama et al. 2006. Measurement of a long electronic spin relaxation time of cesium atoms in superfluid helium. *Phys. Rev. Lett.* 96: 095301.

Galli, D. E. and L. Reatto, 2008. Solid ^{4}He and the supersolid phase: From theoretical speculation to the discovery of a new state of matter?—A review of the past and present status of research. *J. Phys. Soc. Jpn.* 77: 111010.

Galli, D. E., M. Buzzacchi, and L. Reatto, 2001. Pure and alkali-ion-doped droplets of ^{4}He. *J. Chem. Phys.* 115: 10239–10247.

Golov, A. and S. Sekatskii, 1993. A new-type of excimer atom: Electron + ionized helium cluster. *Z. Phys. D* 27: 349–355.

Gordon, E. B., A. A. Pel'menev, O. F. Pugachev, and V. V. Khmelenko, 1982. ESR studies of atoms trapped in superfluid helium. I. Technique. ESR spectra of nitrogen atoms. *Sov. J. Low Temp. Phys.* 8: 299–302.

Gordon, E. B., A. A. Pel'menev, O. F. Pugachev, and V. V. Khmelenko, 1985. EPR study of atoms trapped in superfluid helium. II. Spectra of hydrogen and deuterium atoms. *Sov. J. Low Temp. Phys.* 11: 307–311.

Gordon, E. B., V. V. Khmelenko, A. A. Pelmenev, E. A. Popov, and O. F. Pugachev, 1989a. Impurity-helium van der Waals crystals. *Chem. Phys. Lett.* 155: 301–304.

Gordon, E. B., A. A. Pel'menev, E. A. Popov, O. F. Pugachev, and V. V. Khmelenko, 1989b. On the existence of impurity-helium van der Waals crystals. *Sov. J. Low Temp. Phys.* 15: 48–49.

Gordon, E. B., V. V. Khmelenko, A. A. Pelmenev, E. A. Popov, O. F. Pugachev, and A. F. Shestakov, 1993. Metastable impurity-helium solid phase. Experimental and theoretical evidence. *Chem. Phys.* 170: 411–426.

Gough, T. E., M. Mengel, P. A. Rowntree, and G. Scoles, 1985. Infrared spectroscopy at the surface of clusters: SF_6 on Ar. *J. Chem. Phys.* 83: 4958–4961.

Goyal, S., D. L. Schutt, and G. Scoles, 1992. Vibrational spectroscopy of sulfur-hexafluoride attached to helium clusters. *Phys. Rev. Lett.* 69: 933–936.

Grebenev, S., J. P. Toennies, and A. F. Vilesov, 1998. Superfluidity within a small helium-4 cluster: The microscopic Andronikashvili experiment. *Science* 279: 2083–2086.

Grebenev, S., M. Hartmann, M. Havenith, B. Sartakov, J. P. Toennies, and A. F. Vilesov, 2000a. The rotational spectrum of single OCS molecules in liquid ^{4}He droplets. *J. Chem. Phys.* 112: 4485–4495.

Grebenev, S., M. Hartmann, A. Lindinger et al. 2000b. Spectroscopy of molecules in helium droplets. *Phys. B* 280: 65–72.

Grimes, C. C. and G. Adams, 1990. Infrared-spectrum of the electron bubble in liquid-helium. *Phys. Rev. B* 41: 6366–6371.

Grimes, C. C. and G. Adams, 1992. Infrared-absorption spectrum of the electron bubble in liquid helium. *Phys. Rev. B* 45: 2305–2310.

Grover, B. C. 1978. Noble-gas NMR detection through noble-gas-rubidium hyperfine contact interaction. *Phys. Rev. Lett.* 40: 391–392.

Gspann, J. 1982. Electronic and atomic impacts on large clusters. In *Physics of Electronic and Atomic Collisions*, ed. S. Datz, pp. 79–96. Amsterdam, the Netherlands: North-Holland.

Guardiola, R. and J. Navarro, 2000. Variational study of ^{3}He droplets. *Phys. Rev. Lett.* 84: 1144–1147.

Guirao, A., M. Pi, and M. Barranco, 1991. Finite size effects in the evaporation rate of ^{3}He clusters. *Z. Phys. D* 21: 185–188.

Haberland, H., ed. 1994a. *Clusters of Atoms and Molecules I. Theory, Experiment, and Clusters of Atoms*, Volume 52 of *Springer Series in Chemical Physics*. Berlin, Germany: Springer.

Haberland, H., ed. 1994b. *Clusters of Atoms and Molecules II. Solvation and Chemistry of Free Clusters, and Embedded, Supported, and Compressed Clusters*, Volume 56 of *Springer Series in Chemical Physics*. Berlin, Germany: Springer.

Harms, J. and J. P. Toennies, 1998. Observation of anomalously low momentum transfer in the low energy scattering of large ^{4}He droplets from ^{4}He and ^{3}He atoms. *J. Low Temp. Phys.* 113: 501–508.

Harms, J. and J. P. Toennies, 1999. Experimental evidence for the transmission of ^{3}He atoms through superfluid ^{4}He droplets. *Phys. Rev. Lett.* 83: 344–347.

Harms, J., M. Hartmann, S. Sartakov, J. P. Toennies, and A. Vilesov, 1997a. Rotational structure of the IR spectra of single SF_6 molecules in liquid ^{4}He and ^{3}He droplets. *J. Mol. Spectrosc.* 185: 204–206.

Harms, J., J. P. Toennies, and E. L. Knuth, 1997b. Droplets formed in helium free-jet expansions from states near the critical point. *J. Chem. Phys.* 106: 3348–3357.

Harms, J., J. P. Toennies, and F. Dalfovo, 1998. Density of superfluid helium droplets. *Phys. Rev. B* 58: 3341–3350.

Harms, J., J. P. Toennies, M. Barranco, and M. Pi, 2001. Experimental and theoretical study of the radial density distributions of large ^{3}He droplets. *Phys. Rev. B* 63: 184513.

Hartmann, M., R. E. Miller, J. P. Toennies, and A. Vilesov, 1995. Rotationally resolved spectroscopy of SF_6 in liquid helium clusters. A molecular probe of cluster temperature. *Phys. Rev. Lett.* 75: 1566–1569.

Hartmann, M., F. Mielke, J. P. Toennies, A. F. Vilesov, and G. Benedek, 1996. Direct spectroscopic observation of elementary excitations in superfluid He droplets. *Phys. Rev. Lett.* 76: 4560–4563.

Hartmann, M., N. Pörtner, B. Sartakov, J. P. Toennies, and A. F. Vilesov, 1999. High resolution infrared spectroscopy of single SF_6 molecules in helium droplets. I. Size effects in ^{4}He droplets. *J. Chem. Phys.* 110: 5109–5123.

Hartmann, M., A. Lindinger, J. P. Toennies, and A. F. Vilesov, 2001. Hole-burning studies of the splitting in the ground and excited vibronic states of tetracene in helium droplets. *J. Phys. Chem. A* 105: 6369–6377.

Herzberg, G. 1989–1991. *Molecular Spectra and Molecular Structure.* Malabar, FL: R. E. Krieger Publishing Company. Reprint with corrections. Originally published: 2nd edn. New York: Van Nostrand, 1945. 1. Spectra of diatomic molecules—2. Infrared and Raman spectra of polyatomic molecules—3. Electronic spectra and electronic structure of polyatomic molecules.

Higgins, J., C. Callegari, J. Reho et al. 1996a. Photoinduced chemical dynamics of high-spin alkali trimers. *Science* 273: 629–631.

Higgins, J., W. E. Ernst, C. Callegari et al. 1996b. Spin polarized alkali clusters: Observation of quartet state of the sodium trimer. *Phys. Rev. Lett.* 77: 4532–4535.

Higgins, J., C. Callegari, J. Reho et al. 1998. Helium cluster isolation spectroscopy of alkali dimers in the triplet manifold. *J. Phys. Chem. A* 102: 4952–4965.

Higgins, J., T. Hollebeek, J. Reho et al. 2000. On the importance of exchange effects in the three-body interactions: The lowest quartet state of Na_3. *J. Chem. Phys.* 112: 5751–5761.

Hill, T. L. 1986. *An Introduction to Statistical Thermodynamics.* New York: Dover Publications, Inc.

Hirschfelder, J. O., C. F. Curtiss, and R. B. Bird, 1954. *Molecular Theory of Gases and Liquids.* New York: Wiley.

Hougen, J. 2001. The calculation of rotational energy levels and rotational line intensities in diatomic molecules (version 1.1). Online. Originally published as *The Calculation of Rotational Energy Levels and Rotational Line Intensities in Diatomic Molecules,* J.T. Hougen, NBS Monograph 115 (June 1970), URL http://physics.nist.gov/DiatomicCalculations

Hünig, I., K. Seefeld, and K. Kleinermanns 2003. REMPI and UV-UV double resonance spectroscopy of tryptophan ethylester and the dipeptides tryptophan-serine, glycine-tryptophan and proline-tryptophan. *Chem. Phys. Lett.* 369: 173–179.

JCP115–22 2001. Special Topic: Helium nanodroplets: A novel medium for chemistry and physics. *J. Chem. Phys.* 115 (22).

Johnston, R. L. 2002. *Atomic and Molecular Clusters.* Masters Series in Physics and Astronomy. London, U.K.: Taylor & Francis.

Joppien, M., R. Karnbach, and T. Möller, 1993a. Electronic excitations in liquid-helium: The evolution from small clusters to large droplets. *Phys. Rev. Lett.* 71: 2654–2657.

Joppien, M., R. Muller, and T. Möller, 1993b. Excitation and decay processes in helium clusters studied by fluorescence spectroscopy. *Z. Phys. D* 26: 175–177.

Jortner, J. 1992. Cluster size effects. *Z. Phys. D* 24: 247–275.

Jortner, J., N. R. Kestner, S. A. Rice, and M. H. Cohen, 1965. Study of the properties of an excess electron in liquid helium. I. The nature of the electron–helium interactions. *J. Chem. Phys.* 43: 2614–2625.

JPCA111–31, 2007. Roger E. Miller memorial issue. *J. Phys. Chem. A,* 111 (31).

JPCA111–49, 2007. Giacinto Scoles festschrift. *J. Phys. Chem. A,* 111 (49).

Kanorsky, S. I. and A. Weis, 1998. Optical and magneto-optical spectroscopy of point defects in condensed helium. *Adv. Atom. Mol. Opt. Phys.* 38: 87–120.

Kanorsky, S. I., S. Lang, S. Lucke, S. B. Ross, T. W. Hänsch, and A. Weis, 1996. Millihertz magnetic resonance spectroscopy of Cs atoms in body-centered-cubic ^{4}He. *Phys. Rev. A* 54: R1010–R1013.

Kanorsky, S., S. Lang, T. Eichler, K. Winkler, and A. Weis, 1998. Quadrupolar deformations of atomic bubbles in solid ^{4}He. *Phys. Rev. Lett.* 81: 401–404.

Karnbach, R., M. Joppien, J. Stapelfeldt, J. Wörmer, and T. Möller, 1993. CLULU: An experimental setup for luminescence measurements on van der Waals clusters with synchrotron radiation. *Rev. Sci. Instrum.* 64: 2838–2849.

Keutsch, F. N. and R. J. Saykally, 2001. Water clusters: Untangling the mysteries of the liquid, one molecule at a time. *Proc. Natl. Acad. Sci. USA* 98: 10533–10540.

Kim, E. and M. H. W. Chan, 2004a. Observation of superflow in solid helium. *Science* 305: 1941–1944.

Kim, E. and M. H. W. Chan, 2004b. Probable observation of a supersolid helium phase. *Nature* 427: 225–227.

Kim, J. H., D. S. Peterka, C. C. Wang, and D. M. Neumark, 2006. Photoionization of helium nanodroplets doped with rare gas atoms. *J. Chem. Phys.* 124: 214301.

Kinoshita, T., Y. Takahashi, and T. Yabuzaki, 1994. Optical pumping and optical detection of the magnetic resonance of alkalimetal atoms in superfluid helium. *Phys. Rev. B* 49: 3648–3651.

Kinoshita, T., K. Fukuda, Y. Takahashi, and T. Yabuzaki, 1995. Optical-properties of impurity atoms in pressurized superfluid helium. *Z. Phys. B* 98: 387–390.

Klots, C. E. 1987. Temperatures of evaporating clusters. *Nature* 327: 222–223.

Knight, W. D., K. Clemenger, W. A. de Heer, W. A. Saunders, M. Y. Chou, and M. L. Cohen, 1984. Electronic shell structure and abundances of sodium clusters. *Phys. Rev. Lett.* 52: 2141–2143.

Knuth, E. L. 1997. Size correlations for condensation clusters produced in free-jet expansions. *J. Chem. Phys.* 107: 9125–9132.

Knuth, E. L. and U. Henne, 1999. Average size and size distribution of large droplets produced in a free-jet expansion of a liquid. *J. Chem. Phys.* 110: 2664–2668.

Knuth, E. L., F. Schünemann, and J. P. Toennies, 1995. Supercooling of H_2 clusters produced in free-jet expansions from supercritical states. *J. Chem. Phys.* 102: 6258–6271.

Koch, M., G. Auböck, C. Callegari, and W. E. Ernst, 2009a. Coherent spin manipulation and ESR on superfluid helium nanodroplets. *Phys. Rev. Lett.* 103: 035302.

Koch, M., J. Lanzersdorfer, C. Callegari, J. S. Muenter, and W. E. Ernst, 2009b. Molecular beam magnetic resonance in doped helium nanodroplets. A setup for optically-detected ESR/NMR in the presence of unresolved Zeeman splittings. *J. Phys. Chem. A* 113: 13347–13356.

Koch, M., C. Callegari, and W. E. Ernst, 2010. Alkali-metal electron spin density shift induced by a helium nanodroplet. *Mol. Phys.*, in press.

Kroto, H. W., J. R. Heath, S. C. O'Brien, R. F. Curl, and R. E. Smalley, 1985. C_{60}: Buckminsterfullerene. *Nature* 318: 162–163.

Küpper, J. and J. M. Merritt, 2007. Spectroscopy of free radicals and radical containing entrance-channel complexes in superfluid helium nanodroplets. *Int. Rev. Phys. Chem.* 26: 249–287.

Lang, S., S. I. Kanorsky, M. Arndt, S. B. Ross, T. W. Hänsch, and A. Weis, 1995. The hyperfine-structure of Cs atoms in the bcc phase of solid ^{4}He. *Europhys. Lett.* 30: 233–237.

Lang, S., S. Kanorsky, T. Eichler, R. Müller-Siebert, T. W. Hänsch, and A. Weis, 1999. Optical pumping of Cs atoms in solid ^{4}He. *Phys. Rev. A* 60: 3867–3877.

Lee, E., D. Farrelly, and K. Whaley, 1999. Rotational level structure of SF_6-doped ^4He_N clusters. *Phys. Rev. Lett.* 83: 3812–3815.

Lefebvre-Brion, H. and R. W. Field, 1986. *Perturbations in the Spectra of Diatomic Molecules*. Orlando, FL: Academic Press.

Lehmann, K. K. 1999. Potential of a neutral impurity in a large ^{4}He cluster. *Mol. Phys.* 97: 645–666.

Lehmann, K. K. 2001. Rotation in liquid ^{4}He: Lessons from a highly simplified model. *J. Chem. Phys.* 114: 4643–4648.

Lehmann, K. K. 2003. Microcanonical thermodynamic properties of helium nanodroplets. *J. Chem. Phys.* 119: 3336–3342.

Lehmann, K. K. 2004. Bias in the temperature of helium nanodroplets measured by an embedded rotor. *J. Chem. Phys.* 120: 513–515.

Lehmann, K. K. and C. Callegari, 2002. Quantum hydrodynamic model for the enhanced moments of inertia of molecules in helium nanodroplets: Application to SF_6. *J. Chem. Phys.* 117: 1595–1603.

Lehmann, K. K. and G. Scoles, 2000. Nanomatrices are cool. *Science* 287: 2429–2430.

Lehnig, R. and A. Slenczka, 2003. Emission spectra of free base phthalocyanine in superfluid helium droplets. *J. Chem. Phys.* 118: 8256–8260.

Lehnig, R. and A. Slenczka, 2004. Microsolvation of phthalocyanines in superfluid helium droplets. *Chem. Phys. Chem.* 5: 1014–1019.

Lehnig, R. and A. Slenczka, 2005. Spectroscopic investigation of the solvation of organic molecules in superfluid helium droplets. *J. Chem. Phys.* 122: 244317.

Lehnig, R., N. V. Blinov, and W. Jäger, 2007. Evidence for an energy level substructure of molecular states in helium droplets. *J. Chem. Phys.* 127: 241101.

Lehnig, R., P. L. Raston, and W. Jäger, 2009. Rotational spectroscopy of single carbonyl sulfide molecules embedded in superfluid helium nanodroplets. *Faraday Discussion* 142: 297–309.

Levi, A. C. and R. Mazzarello, 2001. Solidification of hydrogen clusters. *J. Low Temp. Phys.* 122: 75–97.

Lewerenz, M., B. Schilling, and J. P. Toennies, 1995. Successive capture and coagulation of atoms and molecules to small clusters in large liquid helium clusters. *J. Chem. Phys.* 102: 8191–8207.

Lindsay, D. M., D. R. Herschbach, and A. L. Kwiram, 1976. E.S.R. spectra of matrix isolated alkali atom clusters. *Mol. Phys.* 32: 1199–1213.

Loginov, E. 2008. Photoexcitation and photoionization dynamics of doped liquid helium-4 nanodroplets. PhD thesis, EPFL, Lausanne, Switzerland. URL library.epfl.ch/theses/?nr=4207

Lugovoj, E., J. P. Toennies, and A. Vilesov, 2000. Manipulating and enhancing chemical reactions in helium droplets. *J. Chem. Phys.* 112: 8217–8220.

Makarov, G. N. 2004. Spectroscopy of single molecules and clusters inside helium nanodroplets. Microscopic manifestation of ^{4}He superfluidity. *Phys. Usp.* 47: 217–247.

Marinetti, F., E. Coccia, E. Bodo et al. 2007. Bosonic helium clusters doped by alkali metal cations: Interaction forces and analysis of their most stable structures. *Theor. Chem. Acc.* 118: 53–65.

Maris, H. J., G. M. Seidel, and T. E. Huber, 1983. Supercooling of liquid H_2 and the possible production of superfluid H_2. *J. Low Temp. Phys.* 51: 471–487.

Martin, T. P. 1983. Alkali halide clusters and microcrystals. *Phys. Rep.* 95: 167–199.

Mayol, R., F. Ancilotto, M. Barranco, O. Bünermann, M. Pi, and F. Stienkemeier, 2005. Alkali atoms attached to ^{3}He nanodroplets. *J. Low Temp. Phys.* 138: 229–234.

McKellar, A. R. W., Y. J. Xu, and W. Jäger, 2006. Spectroscopic exploration of atomic scale superfluidity in doped helium nanoclusters. *Phys. Rev. Lett.* 97: 183401.

McKellar, A., Y. Xu, and W. Jäger, 2007. Spectroscopic studies of OCS-doped ^{4}He clusters with 9–72 helium atoms: Observation of broad oscillations in the rotational moment of inertia. *J. Phys. Chem. A* 111: 7329–7337.

Middleton, H., R. D. Black, B. Saam et al. 1995. MR imaging with hyperpolarized ^{3}He gas. *Magn. Reson. Med.* 33: 271–275.

Milani, P. and W. A. deHeer, 1990. Improved pulsed laser vaporization source for production of intense beams of neutral and ionized clusters. *Rev. Sci. Instrum.* 61: 1835–1838.

Möller, T., K. von Haeften, T. Laarman, and R. von Pietrowski, 1999. Photochemistry in rare gas clusters. *Eur. Phys. J. D* 9: 5–9.

Moore, D. T. and R. E. Miller, 2004. Structure of the acetylene-magnesium binary complex from infrared laser spectroscopy in helium nanodroplets. *J. Phys. Chem. A* 108: 9908–9915.

Moroni, S., A. Sarsa, S. Fantoni, K. E. Schmidt, and S. Baroni, 2003. Structure, rotational dynamics, and superfluidity of small OCS-doped He clusters. *Phys. Rev. Lett.* 90: 143401.

Moroshkin, P., A. Hofer, S. Ulzega, and A. Weis, 2006. Spectroscopy of atomic and molecular defects in solid ^{4}He using optical, microwave, radio frequency, and static magnetic and electric fields (Review). *Low Temp. Phys.* 32: 981–998.

Moroshkin, P., A. Hofer, and A. Weis, 2008. Atomic and molecular defects in solid ^{4}He. *Phys. Rep.* 469: 1–57.

Mozhayskiy, V., M. N. Slipchenko, V. K. Adamchuk, and A. F. Vilesov, 2007. Use of helium nanodroplets for assembly, transport, and surface deposition of large molecular and atomic clusters. *J. Chem. Phys.* 127: 094701.

Mudrich, M., Ph. Heister, Th. Hippler, Ch. Giese, O. Dulieu, and F. Stienkemeier, 2009. High-resolution spectroscopy of triplet states of Rb$_2$ by femtosecond pump-probe photoionization of doped helium nanodroplets, *Phys. Rev. A* 80: 042512.

Müller, S., S. Krapf, Th. Koslowski, M. Mudrich, and F. Steinkemeier, 2009a. Cold reactions of alkali-metal and water clusters inside helium nanodroplets, *Phys. Rev. Lett.* 102: 183401.

Müller, S., M. Mudrich, and F. Steinkemeier, 2009b. Alkali-helium snowball complexes formed on helium nanodroplets, *J. Chem. Phys.* 131: 044319.

Nagl, J., G. Auböck, C. Callegari, and W. E. Ernst, 2007. Magnetic dichroism of potassium atoms on the surface of helium nanodroplets. *Phys. Rev. Lett.* 98: 075301.

Nagl, J., G. Auböck, A. W. Hauser, O. Allard, C. Callegari, and W. E. Ernst, 2008a. Heteronuclear and homonuclear high-spin alkali trimers on helium nanodroplets. *Phys. Rev. Lett.* 100: 063001.

Nagl, J., G. Auböck, A. W. Hauser, O. Allard, C. Callegari, and W. E. Ernst, 2008b. High-spin alkali trimers on helium nanodroplets: Spectral separation and analysis. *J. Chem. Phys.* 128: 154320.

Nakayama, A. and K. Yamashita, 2000. Theoretical study on the structure of Na$^+$-doped helium clusters: Path integral Monte Carlo calculations. *J. Chem. Phys.* 112: 10966–10975.

Nauta, K. and R. E. Miller, 1999a. Nonequilibrium self-assembly of long chains of polar molecules in superfluid helium. *Science* 283: 1895–1897.

Nauta, K. and R. E. Miller, 1999b. Stark spectroscopy of polar molecules solvated in liquid helium droplets. *Phys. Rev. Lett.* 82: 4480–4483.

Nauta, K. and R. E. Miller, 2000. Formation of cyclic water hexamer in liquid helium: The smallest piece of ice. *Science* 287: 293–295.

Nauta, K. and R. E. Miller, 2001. Rotational and vibrational dynamics of methane in helium nanodroplets. *Chem. Phys. Lett.* 350: 225–232.

Nauta, K., D. T. Moore, P. L. Stiles, and R. E. Miller, 2001. Probing the structure of metal cluster-adsorbate systems with high-resolution infrared spectroscopy. *Science* 292: 481–484.

Navarro, J., A. Poves, M. Barranco, and M. Pi, 2004. Shell structure in mixed ^{3}He-^{4}He droplets. *Phys. Rev. A* 69: 23202.

Nettels, D., R. Müller-Siebert, S. Ulzega, and A. Weis, 2003a. Multi-photon processes in the Zeeman structure of atomic Cs trapped in solid helium. *Appl. Phys. B* 77: 563–570.

Nettels, D., R. Müller-Siebert, and A. Weis, 2003b. Relaxation mechanisms of multi-quantum coherences in the Zeeman structure of atomic Cs trapped in solid He. *Appl. Phys. B* 77: 753–764.

Northby, J. A. 2001. Experimental studies of helium droplets. *J. Chem. Phys.* 115: 10065–10077.

Northby, J. A. and C. Kim, 1994. Lifetime of an electron on the liquid helium surface in the presence of an electric field. *Phys. B* 194: 1229–1230.

Northby, J. A. and T. M. Sanders, 1967. Photoejection of electrons from bubble states in liquid helium. *Phys. Rev. Lett.* 18: 1184–1186.

Pauly, H. 2000a. *Atom, Molecule, and Cluster Beams I*, Volume 28 of *Springer Series on Atomic, Optical, and Plasma Physics*. Berlin, Germany: Springer.

Pauly, H. 2000b. *Atom, Molecule, and Cluster Beams II*, Volume 32 of *Springer Series on Atomic, Optical, and Plasma Physics*. Berlin, Germany: Springer.

Perera, L. and F. G. Amar, 1990. Spectral shifts and structural classes in microsolutions of rare gas clusters containing a molecular chromophore. *J. Chem. Phys.* 93: 4884–4897.

Peterka, D. S., A. Lindinger, L. Poisson, M. Ahmed, and D. M. Neumark, 2003. Photoelectron imaging of helium droplets. *Phys. Rev. Lett.* 91: 043401.

Peterka, D., J. Kim, C. Wang, and D. Neumark, 2006. Photoionization and photofragmentation of SF$_6$ in helium nanodroplets. *J. Phys. Chem. B* 110: 19945–19955.

Peterka, D., J. Kim, C. Wang, L. Poisson, and D. Neumark, 2007. Photoionization dynamics in pure helium droplets. *J. Phys. Chem. A* 111: 7449–7459.

Piepho, S. B. and P. N. Schatz, 1983. *Group Theory in Spectroscopy: with Applications to Magnetic Circular Dichroism*. Wiley-Interscience monographs in chemical physics. New York: Wiley.

Poitrenaud, J. and F. I. B. Williams, 1972. Precise measurement of effective mass of positive and negative charge carriers in liquid helium II. *Phys. Rev. Lett.* 29: 1230–1232.

Poitrenaud, J. and F. I. B. Williams, 1974. Erratum: Precise measurement of effective mass of positive and negative charge carriers in liquid helium II. *Phys. Rev. Lett.* 32: 1213.

Proehl, H., R. Nitsche, T. Dienel, K. Leo, and T. Fritz, 2005. *In situ* differential reflectance spectroscopy of thin crystalline films of PTCDA on different substrates. *Phys. Rev. B* 71: 165207.

Przystawik, A., S. Göde, T. Döppner, J. Tiggesbäumker, and K.-H. Meiwes-Broer, 2008. Light-induced collapse of metastable magnesium complexes formed in helium nanodroplets. *Phys. Rev. A* 78: 021202.

Ralchenko, Y., A. Kramida, J. Reader, and NIST ASD Team, 2009. NIST atomic spectra database, v. 3.1.5. online. URL http://physics.nist.gov/asd3

Rama Krishna, M. V. and K. B. Whaley, 1990a. Collective excitations of helium clusters. *Phys. Rev. Lett.* 64: 1126–1129.

Rama Krishna, M. V. and K. B. Whaley, 1990b. Microscopic studies of collective spectra of quantum liquid clusters. *J. Chem. Phys.* 93: 746–759.

Rebane, K. K. 1970. *Impurity Spectra of Solids; Elementary Theory of Vibrational Structure*. New York: Plenum Press.

Reho, J., C. Callegari, J. Higgins, W. E. Ernst, K. K. Lehmann, and G. Scoles, 1997. Spin–orbit effects in the formation of the Na–He excimer on the surface of He clusters. *Faraday Discuss. Chem. Soc.* 108: 161–174.

Reho, J., U. Merker, M. R. Radcliff, K. K. Lehmann, and G. Scoles, 2000a. Spectroscopy of Mg atoms solvated in helium nanodroplets. *J. Chem. Phys.* 112: 8409–8416.

Reho, J. H., U. Merker, M. R. Radcliff, K. K. Lehmann, and G. Scoles, 2000b. Spectroscopy and dynamics of Al atoms solvated in superfluid helium nanodroplets. *J. Phys. Chem. A* 104: 3620–3626.

Reho, J. H., J. Higgins, M. Nooijen, K. K. Lehmann, G. Scoles, and M. Gutowski, 2001. Photoinduced nonadiabatic dynamics in quartet Na_3 and K_3 formed using helium nanodroplet isolation. *J. Chem. Phys.* 115: 10265–10274.

Reichert, J. F. and A. J. Dahm, 1974. Observation of electron spin resonance of negative ions in liquid helium. *Phys. Rev. Lett.* 32: 271–274.

Reichert, J. F. and N. C. Jarosik, 1983. Magnetic-resonance studies of negative ions in liquid ^3He- ^4He mixtures. *Phys. Rev. B* 27: 2710–2721.

Reichert, J. F., N. Jarosik, R. Herrick, and J. Andersen, 1979. Observation of electron spin resonance of negative ions in liquid ^3He. *Phys. Rev. Lett.* 42: 1359–1361.

Reinhard, I., C. Callegari, A. Conjusteau, K. K. Lehmann, and G. Scoles, 1999. Single and double resonance microwave spectroscopy in superfluid ^4He clusters. *Phys. Rev. Lett.* 82: 5036–5039.

Rosenblit, M. and J. Jortner, 2006. Electron bubbles in helium clusters. I. Structure and energetics. *J. Chem. Phys.* 124: 194505.

Rossi, M., M. Verona, D. E. Galli, and L. Reatto, 2004. Alkali and alkali-earth ions in ^4He systems. *Phys. Rev. B* 69: 212510.

Scheidemann, A., B. Schilling, J. P. Toennies, and J. A. Northby, 1990. Capture of foreign atoms by helium clusters. *Phys. B* 165: 135–136.

Schlesinger, M. and W. Strunz, 2009. Unpublished results.

Scoles, G., ed. 1988. *Atomic and Molecular Beam Methods*, Volume 1. New York: Oxford University Press.

Scoles, G., ed. 1992. *Atomic and Molecular Beam Methods*, Volume 2. New York: Oxford University Press.

Sild, O. and K. Haller, eds. 1988. *Zero-Phonon Lines and Spectral Hole Burning in Spectroscopy and Photochemistry*. Berlin, Germany: Springer.

Slipchenko, M. N. and A. F. Vilesov, 2005. Spectra of NH_3 in He droplets in the 3 μm range. *Chem. Phys. Lett.* 412: 176–183.

Slipchenko, M. N., S. Kuma, T. Momose, and A. F. Vilesov, 2002. Intense pulsed helium droplet beams. *Rev. Sci. Instrum.* 73: 3600–3605.

Springett, B. E., J. Jortner, and M. H. Cohen, 1968. Stability criterion for the localization of an excess electron in a nonpolar fluid. *J. Chem. Phys.* 48: 2720–2731.

Stienkemeier, F. and K. K. Lehmann, 2006. Spectroscopy and dynamics in helium nanodroplets. *J. Phys. B* 39: R127–R166.

Stienkemeier, F. and A. F. Vilesov, 2001. Electronic spectroscopy in He droplets. *J. Chem. Phys.* 115: 10119–10137.

Stienkemeier, F., J. Higgins, W. E. Ernst, and G. Scoles, 1995a. Laser spectroscopy of alkali-doped helium clusters. *Phys. Rev. Lett.* 74: 3592–3595.

Stienkemeier, F., J. Higgins, W. E. Ernst, and G. Scoles, 1995b. Spectroscopy of alkali atoms and molecules attached to liquid He clusters. *Z. Phys. B* 98: 413–416.

Stienkemeier, F., J. Higgins, C. Callegari, S. I. Kanorsky, W. E. Ernst, and G. Scoles, 1996. Spectroscopy of alkali atoms (Li, Na, K) attached to large helium clusters. *Z. Phys. D* 38: 253–263.

Stienkemeier, F., F. Meier, and H. O. Lutz, 1997. Alkaline earth metals (Ca, Sr) attached to liquid helium droplets: Inside or out? *J. Chem. Phys.* 107: 10816–10818.

Stienkemeier, F., F. Meier, and H. O. Lutz, 1999. Spectroscopy of barium attached to superfluid helium clusters. *Eur. Phys. J. D* 9: 313–315.

Stienkemeier, F., M. Wewer, F. Meier, and H. Lutz, 2000. Langmuir-Taylor surface ionization of alkali (Li, Na, K) and alkaline earth (Ca, Sr, Ba) atoms attached to helium droplets. *Rev. Sci. Instrum.* 71: 3480–3484.

Stienkemeier, F., O. Bünermann, R. Mayol, F. Ancilotto, M. Barranco, and M. Pi, 2004. Surface location of sodium atoms attached to ^3He nanodroplets. *Phys. Rev. B* 70: 214509.

Stiles, P. L., D. T. Moore, and R. E. Miller, 2004. Structures of HCN-Mg_n ($n = 2 - 6$) complexes from rotationally resolved vibrational spectroscopy and ab initio theory. *J. Chem. Phys.* 121: 3130–3142.

Surin, L. A., A. V. Potapov, B. S. Dumesh et al. 2008. Rotational study of carbon monoxide solvated with helium atoms. *Phys. Rev. Lett.* 101: 233401.

Takahashi, N., T. Shimoda, Y. Fujita, T. Itahashi, and H. Miyatake, 1995a. Snowballs of radioactive ions-nuclear spin polarization of core ions. *Z. Phys. B* 98: 347–351.

Takahashi, Y., K. Fukuda, T. Kinoshita, and T. Yabuzaki, 1995b. Sublevel spectroscopy of alkali atoms in superfluid-helium. *Z. Phys. B* 98: 391–393.

Takahashi, N., T. Shimoda, H. Miyatake et al. 1996. Freezing-out of nuclear polarization in radioactive core ions of microclusters, "snowballs" in superfluid helium. *Hyperfine Interact.* 97–8: 469–477.

Tang, J. and A. R. W. McKellar, 2003. High-resolution infrared spectra of carbonyl sulfide solvated with helium atoms. *J. Chem. Phys.* 119: 5467–5477.

Tang, J., Y. J. Xu, A. R. W. McKellar, and W. Jäger, 2002. Quantum solvation of carbonyl sulfide with helium atoms. *Science* 297: 2030–2033.

Thompson, G. A., F. Tischler, and D. M. Lindsay, 1983. Matrix ESR spectra of polyatomic alkali metal clusters. *J. Chem. Phys.* 78: 5946–5953.

Tiggesbäumker, J. and F. Stienkemeier, 2007. Formation and properties of metal clusters isolated in helium droplets. *Phys. Chem. Chem. Phys.* 9: 4748–4770.

Toennies, J. P. and A. F. Vilesov, 2004. Superfluid helium droplets: A uniquely cold nanomatrix for molecules and molecular complexes. *Angew. Chem. Int. Ed.* 43: 2622–2648.

Topic, W., W. Jäger, N. Blinov, P. N. Roy, M. Botti, and S. Moroni, 2006. Rotational spectrum of cyanoacetylene solvated with helium atoms. *J. Chem. Phys.* 125: 144310.

Ulzega, S., A. Hofer, P. Moroshkin, R. Müller-Siebert, D. Nettels, and A. Weis, 2007. Measurement of the forbidden electric tensor polarizability of Cs atoms trapped in solid ^{4}He. *Phys. Rev. A* 75: 042505.

von Haeften, K., A. R. B. de Castro, M. Joppien, L. Moussavizadeh, R. von Pietrowski, and T. Möller, 1997. Discrete visible luminescence of helium atoms and molecules desorbing from helium clusters: The role of electronic, vibrational, and rotational energy transfer. *Phys. Rev. Lett.* 78: 4371–4374.

von Haeften, K., T. Laarmann, H. Wabnitz, and T. Möller, 2001. Observation of atomiclike electronic excitations in pure ^{3}He and ^{4}He clusters studied by fluorescence excitation spectroscopy. *Phys. Rev. Lett.* 87: 153403.

von Haeften, K., T. Laarmann, H. Wabnitz, and T. Möller, 2002. Bubble formation and decay in ^{3}He and ^{4}He clusters. *Phys. Rev. Lett.* 88: 233401.

von Haeften, K., T. Laarmann, H. Wabnitz, and T. Möller, 2005a. The electronically excited states of helium clusters: An unusual example for the presence of Rydberg states in condensed matter. *J. Phys. B* 38: S373–S386.

von Haeften, K., A. Metzelthin, S. Rudolph, V. Staemmler, and M. Havenith, 2005b. High-resolution spectroscopy of NO in helium droplets: A prototype for open shell molecular interactions in a quantum solvent. *Phys. Rev. Lett.* 95: 215301.

von Haeften, K., S. Rudolph, I. Simanovski, M. Havenith, R. E. Zillich, and K. B. Whaley, 2006. Probing phonon-rotation coupling in helium nanodroplets: Infrared spectroscopy of CO and its isotopomers. *Phys. Rev. B* 73: 054502.

Weis, A., S. Kanorsky, M. Arndt, and T. W. Hänsch, 1995. Spin physics in solid helium: experimental results and applications. *Z. Phys. B* 98: 359–362.

Weltner, W. Jr., R. J. van Zee, and S. Li, 1995. Magnetic molecules in matrices. *J. Phys. Chem.* 99: 6277–6285.

Wewer, M. and F. Stienkemeier, 2003. Molecular versus excitonic transitions in PTCDA dimers and oligomers studied by helium nanodroplet isolation spectroscopy. *Phys. Rev. B* 67: 125201.

Wewer, M. and F. Stienkemeier, 2005. Laser-induced fluorescence spectroscopy of N, N′-dimethyl 3,4,9,10-perylene tetracarboxylic diimide monomers and oligomers attached to helium nanodroplets. *Phys. Chem. Chem. Phys.* 7: 1171–1175.

Wilks, J. and D. S. Betts, 1987. *An Introduction to Liquid Helium*, 2nd edn. Oxford, NY: Oxford University Press.

Woolf, M. A. and G. W. Rayfield, 1965. Energy of negative ions in liquid helium by photoelectric injection. *Phys. Rev. Lett.* 15: 235–237.

Xie, J., J. A. Northby, D. L. Freeman, and J. D. Doll, 1989. Theoretical studies of the energetics and structures of atomic clusters. *J. Chem. Phys.* 91: 612–619.

Xu, Y. and W. Jäger, 1997. Evidence for heavy atom large amplitude motions in RG-cyclopropane van der Waals complexes (RG = Ne, Ar, Kr) from rotation-tunneling spectroscopy. *J. Chem. Phys.* 106: 7968–7980.

Xu, Y. and W. Jäger, 2001. Fourier transform microwave spectroscopic investigation of a very weakly bound ternary complex: OCS-He$_2$. *Chem. Phys. Lett.* 350: 417–422.

Xu, Y. and W. Jäger, 2003. Rotational spectroscopic investigation of carbonyl sulfide solvated with helium atoms. *J. Chem. Phys.* 119: 5457–5466.

Xu, Y., W. Jäger, J. Tang, and A. R. W. McKellar, 2003. Spectroscopic studies of quantum solvation in ^{4}He$_N$-N$_2$O clusters. *Phys. Rev. Lett.* 91: 163401.

Xu, Y., N. Blinov, W. Jäger, and P.-N. Roy, 2006. Recurrences in rotational dynamics and experimental measurement of superfluidity in doped helium clusters. *J. Chem. Phys.* 124: 081101.

Yabuzaki, T., T. Kinoshita, K. Fukuda, and Y. Takahashi, 1995. Laser spectroscopy and optical pumping of alkali atoms in superfluid liquid helium. *Z. Phys. B* 98: 367–369.

Yokozeki, A. and J. S. Muenter, 1980. Laser fluorescence state selected and detected molecular beam magnetic resonance in I$_2$. *J. Chem. Phys.* 72: 3796–3804.

Zewail, A. H., ed. 1994. *Femtochemistry, Ultrafast Dynamics of the Chemical Bond*, Volume 3 of *World Scientific Series in 20th Century Chemistry*. Singapore: World Scientific.

Zimmermann, P. H., J. F. Reichert, and A. J. Dahm, 1977. Study of the electron spin resonance of negative ions field emitted into liquid helium. *Phys. Rev. B* 15: 2630–2650.

Zwier, T. S. 1996. The spectroscopy of solvation in hydrogen-bonded aromatic clusters. *Annu. Rev. Phys. Chem.* 47: 205–241.

5

Silicon Nanocrystals

Hartmut Wiggers
University of Duisburg-Essen

Axel Lorke
University of Duisburg-Essen

5.1 Introduction

Silicon is probably one of the most investigated materials worldwide and is the second most common element on earth. The semiconductor industry has relied upon silicon for decades and maximized its technical performance through increasingly sophisticated techniques, while decreasing the required functional size according to Moore's law.

Apart from semiconductor applications based on traditional silicon technology, there has been a continuous interest in nanocrystalline silicon since the 1990s as a result of a report of Canham on luminescing nanocrystalline silicon (Canham 1990). This strong luminescence was somewhat surprising as bulk silicon is a very inefficient emitter because of the indirect nature of its band gap. However, by reducing its size to below 10 nm, the situation changes dramatically. This is due to the fact that below 10 nm the confinement of the electrons and holes becomes more and more important, resulting in quantum mechanical effects. This confinement affects the optical and electronic properties of nanosized silicon and opens the way to new (opto)-electronic devices.

The influence of quantum confinement on the properties of semiconductor nanostructures has been intensively investigated over the last 20 years. In direct band-gap semiconductors, spectroscopic studies have revealed an increase of the band gap with decreasing size and a discrete character of the electronic states. Advances in the synthesis and characterization of quantum dots, made from III–V and II–VI semiconductors such as GaAs and CdS, have made them perfect model systems for investigating size-dependent confinement effects. However, applications based on these tunable properties are held back by concerns regarding the toxicity and the unknown environmental impact of these heavy metal based materials. These are important concerns that bring the silicon nanoparticles into play. In spite of its inferior physical properties, numerous scientists are searching for possibilities to use silicon as a basic material for optoelectronic and photovoltaic devices by utilizing the quantum confined properties of nanostructured silicon.

5.2 Synthesis of Silicon Nanocrystals

5.2.1 Porous Silicon

Silicon nanocrystals can be prepared in different ways: Via top-down as well as bottom-up routes, via wet-chemical steps or vacuum technologies, from bulk material or from liquid or gaseous precursors. In connection with the findings of Canham, the most widely used top-down method is the formation of porous silicon by means of a wet chemical route. It has attracted much interest due to the simplicity of the preparation procedure. Porous silicon can be prepared by anodic etching or stain etching of single-crystalline silicon wafers in hydrofluoric acid as shown in Figure 5.1.

Porous silicon from these etching procedures initially consists of hydrogen-terminated silicon nanowires or small silicon nanocrystals in the size regime of a few nm, which are interconnected via very small point contacts. Silicon nanocrystals can be synthesized through a subsequent oxidation step, which results in single crystalline silicon nanoparticles, embedded in a separating and electrically isolating SiO_2 matrix. To remove these small crystals from the supporting wafer, a simple etching step with HF is required. The formation of silicon nanocrystals from porous silicon is depicted schematically in Figure 5.2.

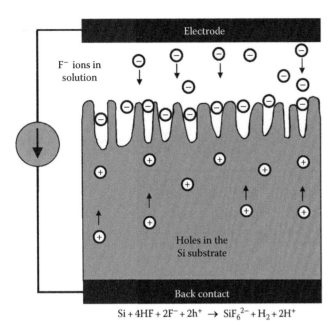

$$Si + 4HF + 2F^- + 2h^+ \rightarrow SiF_6^{2-} + H_2 + 2H^+$$

FIGURE 5.1 Anodic etching of silicon wafers for the formation of porous silicon.

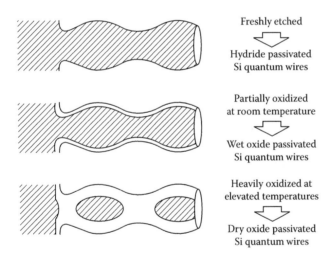

FIGURE 5.2 Schematic representation of the formation of silicon nanocrystals from freshly etched porous silicon. (From Hamilton, B., *Semicond. Sci. Technol.*, 10, 1187, 1995. With permission.)

In case of anodic etching, the size of the nanocrystals can be adjusted to be between 3 and 10 nm by the pH of the hydrofluoric acid and the applied current. Porous films up to a few hundred nm in thickness are accessible.

A modified and very simple method to prepare porous silicon is a stain etch procedure instead of the anodic etching. The etching solution consists of hydrofluoric acid, nitric acid, and water. After a fast initial etching step, the etching rate decreases dramatically because of diffusion limitations. Hence, stain etching is recommended for the formation of a thin porous layer. Usually, the particle size distribution of silicon nanocrystals within the porous silicon is very broad. More details concerning the preparation of porous silicon can be found in Cullis et al. (1997).

5.2.2 Thin Layer Formation of Silicon Nanocrystals

A number of processes resulting in crystalline silicon nanoparticles employ the formation and subsequent annealing of either SiO or an amorphous silicon layer, usually embedded in a silica matrix. These methods were developed with respect to compatibility with traditional semiconductor technology and very-large-scale integration (VLSI). A thin layer formation can be realized by ion implantation of silicon into a SiO_2 matrix (Shimizu-Iwayama et al. 1998), Chemical Vapor Deposition (CVD) of substoichiometric silicon oxide films, molecular-beam epitaxy (MBE) of silicon combined with controlled oxidation (Lockwood et al. 1996), reactive evaporation of SiO, and by sputtering techniques. Nanocrystal formation from silicon-rich layers, produced by one of the methods mentioned above, requires a thermal treatment that generally involves two steps: (1) the diffusion and the nucleation of the silicon phase and (2) the subsequent growth of the initially formed crystals by diffusion. The investigation of nanocrystal growth in silicon-rich SiO_2 deposited by CVD was described by Nesbit (1985) and indicated a diffusion controlled growth given by

$$D(T) = D_0 e^{E_A/kT} \tag{5.1}$$

with $E_A = 1.9$ eV and $D_0 = 1.2 \times 10^{-9}$ cm²/s. The minimum crystal size that was observed from the thin layer formation of silicon nanocrystals is about 2.5 nm in diameter. It seems, that a minimum excess of silicon in the silicon rich layer is required to start the initial nanocrystal formation. A further development of thin layer formation is the controlled formation of 3D nanocrystal stacks via a superlattice approach. The use of Si/SiO₂ superlattices was first introduced by Lu et al. (1995) to very precisely grow nanometer-thick amorphous silicon layers in between sheets of SiO_2. The size of the resulting silicon nanocrystals after the annealing step is controlled by the thickness of the silicon layer. With this approach, stacks of hundreds of layers are made possible. A reactive evaporation-based method developed by Zacharias et al. utilizes the thermal decomposition of thin SiO layers prepared between layers of SiO_2 (Zacharias et al. 2002). A high-temperature annealing step of the initially amorphous SiO_x films results in a phase separation described by

$$SiO_x \rightarrow \frac{x}{2} SiO_2 + \left(1 - \frac{x}{2}\right) Si \tag{5.2}$$

and in the formation of silicon nanocrystals embedded in a separating SiO_2 matrix. The nanocrystal sizes can be controlled independently using a SiO layer thickness equal to or slightly below the desired crystal size. As an example, Figure 5.3 shows a transmission electron microscope (TEM) image of an as-prepared as well as an annealed SiO/SiO₂ superlattice.

One main advantage of the substrate-supported thin layer methods is the possibility to produce silicon nanocrystals with

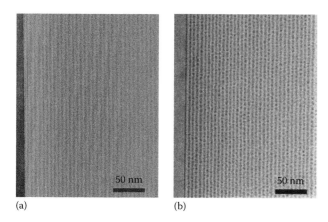

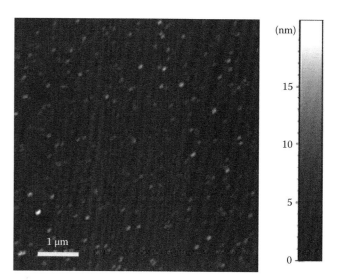

FIGURE 5.3 (a) TEM image of an as-prepared sample with 3 nm thick SiO layers. (b) TEM image of a sample with 3 nm thick SiO layers after annealing at 1100°C under N$_2$ atmosphere. (From Heitmann, J. et al., *J. Non-Cryst. Solids*, 299, 1075, 2002. With permission.)

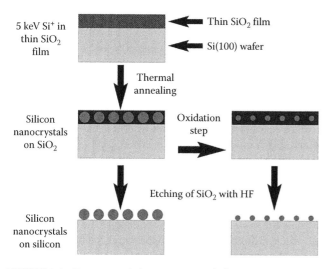

FIGURE 5.4 Formation of silicon nanocrystals from ion implantation.

FIGURE 5.5 6 × 6 μm non-contact atomic force microscope image of an etched sample (see Figure 5.4) with d_{Si} = 3.2 nm. (From Biteen, J.S., Plasmon-enhanced silicon nanocrystal luminescence for opto-electronic applications, PhD thesis, California Institute of Technology, Pasadena, CA, 2006.)

tunable size and narrow size distribution. For this reason, these methods play an important role regarding the investigation of size-dependent optical and electronic properties of silicon nanocrystals.

A similar quality of silicon nanoparticles is available from silicon ion implantation in thin silicon dioxide films (see Figure 5.4). The concentration of silicon in the resulting silicon-rich SiO$_2$ can be controlled by the dosage of implanted silicon atoms. As can be seen from Figure 5.5, the HF etching step does not wash away the particles but keeps them sticking on the surface due to van der Waals forces.

5.2.3 Gas Phase Synthesis

Whereas the previously described manufacturing techniques are substrate-based, gas phase formation of silicon nanocrystals does not require any support. Silicon particles can be produced in aerosol processes based on homogeneous reactions in the gas phase, which produce a supersaturated silicon vapor. Mainly two precursor materials are used for silicon particle gas phase formation, trichlorosilane (SiHCl$_3$, TCS) and monosilane (SiH$_4$). Trichlorosilane is broadly used in the Siemens process for the formation of polycrystalline silicon for the semiconductor and photovoltaic industry. In order to obtain elemental silicon, a temperature of about 1150°C is required, resulting in the formation of silicon and hydrogen chloride.

$$SiHCl_3(g) + H_2(g) \rightarrow Si\,(solid) + 3HCl(g) \qquad (5.3)$$

Due to the heat of reaction of 222 kJ/mol, this precursor material requires much more energy than the pyrolysis of monosilane ($_fH^0 = -34$ KJ/mol), which can be thermally decomposed to silicon and hydrogen according to

$$SiH_4(g) \rightarrow Si\,(solid) + 2H_2(g) \qquad (5.4)$$

The decomposition of monosilane starts already at 400°C, producing amorphous silicon. At about 700°C, the reaction product begins to crystallize during the formation process, leading to crystalline silicon particles. Due to the lower process temperature and the avoidance of corrosive hydrogen chloride, gas phase formation of silicon nanoparticles is mostly carried out using monosilane. Depending on the process parameters, particles are formed by nucleation, surface reactions, coagulation, and/or coalescence (see Figure 5.6).

Gas phase synthesis routes open the possibility of an easy scale-up of the process for higher production rates. Furthermore, they allow for online measurement techniques for in-situ determination of particle sizes. There are various routes for gas phase synthesis of silicon nanoparticles: thermal decomposition in a hot wall reactor (Onischuk et al. 1997), laser decomposition of

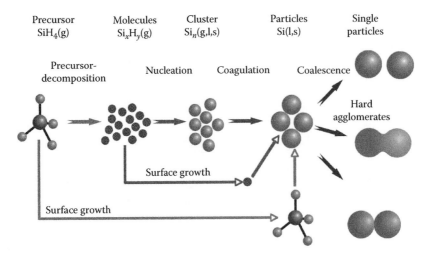

FIGURE 5.6 Schematic representation of silicon nanoparticle formation from the gas phase.

silane (Ledoux et al. 2000), nonequilibrium plasmas (Mangolini et al. 2005), as well as thermal plasmas (Rao et al. 1998, Giesen et al. 2005). Moreover, compared to other techniques, doping is easier during gas phase synthesis due to the fact that gaseous dopant precursor such as phosphine (PH_3) or diborane (B_2H_6) can be added to the gas mixture at any ratio required. The specific requirements needed for doping of silicon nanoparticles will be discussed later.

Compared to other gas phase techniques, plasma synthesis of silicon nanoparticles is able to produce large quantities of non-agglomerated, spherical particles. This is due to the fact that particles, suspended in a plasma, are negatively charged because of the high mobility of electrons relative to that of the ions. Therefore, if particles approach each other, they will experience interparticle Coulomb forces, preventing them from agglomeration.

5.2.4 Other Techniques

A chemical route to silicon nanocrystals was developed by Kauzlarich and colleagues, using the reaction of $SiCl_4$ with Mg_2Si in ethylene glycol dimethyl ether (Bley and Kauzlarich 1996). To stabilize the as-prepared particles and to prevent them from agglomeration and growth, usually stabilizing ligands (e.g., alkyl chains) are required. One crucial disadvantage of the liquid phase formation is the purity of the formation process. Even the highest purity of the used chemicals contains more than sufficient foreign ions for an uncontrolled doping of the nanoparticles.

Laser ablation is another technique for the formation of silicon nanoparticles by collecting the material ejected from a laser heated substrate (Riabinina et al. 2007). The nanoparticle size is controlled by the ambient gas pressure, laser pulse energy density, and the distance from the laser beam. Laser ablation usually produces nanoparticles with a broad particle size distribution, and particles with a specific size must be selected from the produced particle ensemble.

5.3 Quantum Size Effects

When the size of an object is reduced below a characteristic length scale, its physical properties can become drastically different from the corresponding bulk material and may even be tunable by appropriately choosing the size and shape. Examples for such characteristic length scales are the ballistic mean free path, which determines whether charge transport is governed by random scatterers or by reflection from the sample boundaries, and the phase coherence length, which gives an upper limit for the observation of interference phenomena. On the smallest length scales (typically well below 100 nm and often as low as a few nm), all physical properties are dominated by the so-called quantum size effects. The electronic levels, which exhibit a continuous spectrum for bulk materials, become discrete when the de Broglie wavelength is of the order of the size of the nanostructures. Correspondingly, the density of states disintegrates into a discrete set of sharp peaks (see Figure 5.7).

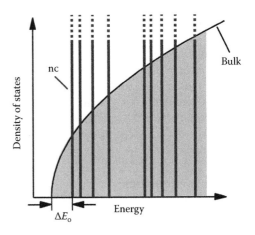

FIGURE 5.7 Sketch of the density of states for bulk and nanocrystalline (nc) material. The shift of the lowest energy state is indicated by ΔE_0.

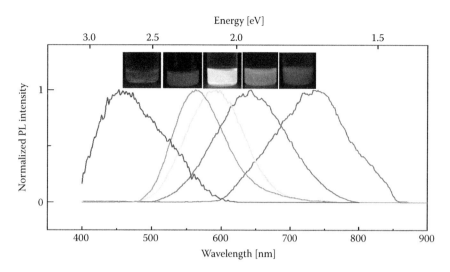

Energy [eV]

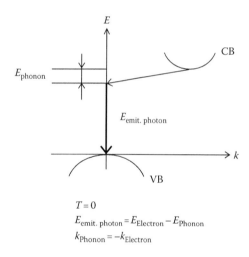

FIGURE 5.8 **(See color insert following page 9-8.)** Normalized PL emission spectra and the corresponding red ($\lambda = 735$ nm), orange ($\lambda = 641$ nm), yellow ($\lambda = 592$ nm), green ($\lambda = 563$ nm), and blue ($\lambda = 456$ nm) emission color from etched Si-NPs. (From Gupta, A. et al., *Adv. Funct. Mater.*, 19(5), 696, 2009. With permission.)

Since almost all physical properties (electronic, optical, magnetic, thermal) are affected by the density of states, quantum size effects reveal themselves in many nanomaterial characteristics. Another important change that occurs as the size of the material is reduced is the shift of the lowest energy state, ΔE_o. In semiconductor nanoparticles, for example, this shift (upward for electrons and downward for holes) leads to an increase of the band gap energy as the particle diameter is reduced. This makes it possible to tune the light emission of Si nanoparticles, as discussed below (see Figure 5.8). In the most simple picture, the shift of the lowest energy state can be estimated by treating the nanoparticle as an infinite quantum well. This leads to ΔE_o proportional to d^{-2}, where d is the particle diameter. More detailed calculations of the optical properties of Si nanoparticles, using linear combination of atomic orbitals (LCAO) theory, confirmed the qualitative behavior $\Delta E_o \sim d^{-n}$, however, with an exponent $n \neq 2$ (Delerue et al. 1993). To a good approximation, the optical gap, which determines the photoluminescence energy, follows the semi-empirical formula

$$E_g \text{ (eV)}(d) = E_{g0} \text{ (eV)} + \frac{3.73}{d \text{ (nm)}^{1.39}} \qquad (5.5)$$

where E_{g0} is the energy gap of bulk Si.

Another consequence of quantum confinement is a strongly increased radiative recombination efficiency in nanocrystalline silicon. Bulk silicon is an indirect semiconductor and as such exhibits only very weak luminescence because of the mismatch between the electron and the hole momentum k (see also Figure 5.9, below). In quantum confined systems, the energy shift is associated with a shift in momentum, which increases the overlap between electron and hole wave functions in k-space. For particles with a gap above 2.3 eV, emission becomes quite efficient, with a radiative recombination time of the order of 0.01 ms (Delerue et al. 1993, Meier et al. 2007). This leads to a

FIGURE 5.9 Schematic band structure of bulk silicon and the possible optical emission. (From Kovalev, D. et al., *Phys. Status Solidi B Basic Res.*, 215, 871, 1999. With permission.)

photoluminescence intensity, which is 3–4 orders of magnitude larger than that of bulk silicon and makes nanoparticles very attractive for optical applications, as discussed in the following.

5.4 Light Emission from Silicon Nanocrystals

Since the finding of Canham et al., there has been an enormous interest in the optical properties, especially in the generation of light from silicon nanocrystals. This is mostly due to two important facts: The prospect of replacing III–V devices with silicon-based light-emitting devices has been the driving force of silicon luminescence research from the beginning. As silicon is the dominant material in microelectronics, it would be highly advantageous to also use silicon as a key material for light emitting devices since this would make it possible to integrate both logic and optoelectronic circuits using just a single, cheap, and nontoxic material. Secondly, the desired color of the light emitted from such silicon-based devices could be tuned by making use of quantum size effects.

5.4.1 Photoluminescence

The photoluminescence, depending on the size of the silicon crystallites and on the morphology of the crystallite ensemble, can be continuously tuned over a very wide spectral range from the silicon band-gap at 1.12 eV to the blue region as shown in Figure 5.8 (Gupta et al. 2009), see also Pi et al. (2008).

These spectra are inhomogeneously broadened with a typical full width at half maximum (FWHM) of up to a few hundred meV. No distinct emission features that allow a determination of the nature of the luminescence are observed. This has stimulated a long-standing discussion about the mechanism for light emission, and the high-efficiency PL from Si nanostructures is still under debate. Canham has argued that the visible luminescence results from quantum confinement effects, leading to energy increases to levels well above the bulk energy gap E_g. The wavelength of the emitted light is thus a direct consequence of the reduced particle size. This conclusion has been supported by many authors, using experimental data as well as theoretical calculations. However, the topic remains controversial, and there have been other suggestions concerning the origin of the visible luminescence. For example, "surface-state" models ascribe it to the recombination of carriers trapped at surface sites. One is that the nanocrystals are surrounded by amorphous surface layer of Si, and the visible luminescence can be understood by the removal of k selection rules and the nature of the density-of-states of both the valence and the conduction bands, which correspond to the amorphous state (Matsumoto et al. 1992, Vasquez et al. 1992). Other explanations propose that the formation of a hydride species or the formation of siloxene derivatives causes

the strong luminescence of silicon nanocrystals (Brandt et al. 1992, Prokes et al. 1992).

Here, we discuss the evidence that supports the "quantum-confinement" models that explain the luminescence by recombination across the fundamental nanostructure band gap. The band structure of bulk silicon is schematically shown in Figure 5.9.

The top of the valence band (VB) is located at the center of the Brillouin zone, while the bottom of the conduction band is at about 3/4 of the Brillouin zone boundary. Since photons only carry negligible momentum, this mismatch in k makes a direct optical transition between the bottom of the conduction band and the top of the valence band impossible. Optical transitions are allowed only if they are accompanied by the emission or absorption of phonons to conserve the crystal momentum. The relevant phonon modes that assist this momentum transfer include transverse optical phonons (TO) with energy $E_{TO} \approx 56$ meV, longitudinal optical phonons (LO) with $E_{LO} \approx 53.5$ meV, and transverse acoustic phonons (TA) with $E_{TA} \approx 18.7$ meV. The largest contribution to the PL in bulk silicon is due to TO phonon-assisted recombination.

The increase in the radiative rate of silicon nanocrystals is attributed to a confinement-induced relaxation of momentum conservation, which opens an additional radiative decay channel via zero-phonon, pseudodirect transitions. Such a pseudodirect, no-phonon emission has been observed from single dot luminescence spectroscopy (Sychugov et al. 2005), see the left graph in Figure 5.10. In this case, the PL spectrum of a single silicon dot exhibits a sharp signal, and its FWHM is slightly bigger than k_BT, suggesting that some scattering processes contribute to the signal.

In the right graph of Figure 5.10, a satellite peak at lower energy is observed separated by about 60 meV from the main line. This value is very close to the TO phonon energy for bulk silicon (56 meV), and it is assumed that the spectrum of the single particle shown in Figure 5.10 can be ascribed to no-phonon (main signal) and phonon-assisted (satellite peak) luminescence. At room temperature, these single crystals exhibit a quite broad emission line with a FWHM up to $\Delta E = 150$ meV.

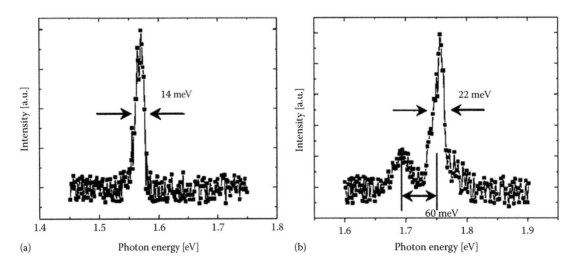

FIGURE 5.10 PL spectra of two different single silicon nanocrystals at $T = 80$ K. FWHM of Lorentzian fits are shown. (From Sychugov, I. et al., *Phys. Rev. Lett.*, 94, 4, 2005. With permission.)

As mentioned above, spatial confinement shifts the energy of the electronic states to higher values in a similar way in both direct and indirect band-gap semiconducting nanocrystals.

From Equation 5.5, it follows that the band-gap energy has a strong dependence on the particle size, shifting the emission spectra to the blue with decreasing particle size. When measuring the photoluminescence of silicon nanocrystal ensembles, as shown in Figure 5.8, not only the (homogeneous) broadening of the single particle emission but also the inhomogeneous broadening of the particle ensemble has to be taken into account. Meier et al. developed a model that describes the photoluminescence line width of such a particle ensemble, exhibiting a lognormal size distribution with a geometric standard deviation σ. Additionally, taking into account the increasing oscillator strength of smaller particles, the model was able to very well account for experimental results (Meier et al. 2007).

From Figure 5.11, it is obvious that both the homogeneous and the inhomogeneous broadening influence the PL spectra and that the particle size distribution has the dominant influence on the FWHM of the measured PL spectra. Nevertheless, only the combination of both, the energy distribution of each particle and

the particle size distribution, lead to such comparatively broad emission spectra with distinct emission features missing.

5.4.2 Electroluminescence

While photoluminescence (i.e., light emission under optical excitation) is a versatile tool to study the underlying physical mechanisms of radiative recombination in Si nanoparticles, electroluminescence (i.e., light emission from electrical excitation) is of much more relevance for optoelectronic applications.

However, only a few groups have reported on the electroluminescence characteristics of nanocrystalline (nc)-silicon. This is mainly because electroluminescence (EL) requires the formation of excitons via the injection of electrons and holes from contacting electrodes. Particularly in granular media such as nc-Si, electrical carrier injection is more difficult to achieve than optical carrier generation. Additionally, the emitted light from the active layer may be absorbed in the conducting layer, which is indispensable for the carrier injection in the Si-based light-emitting diode (LED). Therefore, the electrical injection of carriers and the efficient extraction of emitted light are main issues toward the fabrication of Si-based visible LEDs.

Green et al. have shown that the extraction efficiency of light even from bulk silicon can be enhanced by texturizing a silicon surface (Green et al. 2001), resulting in a power conversion efficiency of up to 1%. The respective electroluminescence spectra of these devices are typical of band-to-band recombination in silicon. Electroluminescence of nanocrystalline silicon was first reported by Koshida and Koyama, see Figure 5.12 (Koshida and Koyama 1992). Their device was based on porous silicon, contacted using a semitransparent gold layer. While this first prototype had an external quantum efficiency (EQE) of only 10^{-5}%, some five orders of magnitude lower than the likely PL efficiency of the same layer, worldwide progress in improving the EL efficiency has yielded devices with more than 1% EQE by means of thin porous silicon–indium–tin–oxide (Si-ITO) junctions (Gelloz and Koshida 2000).

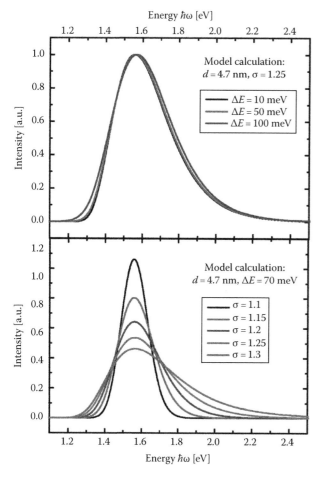

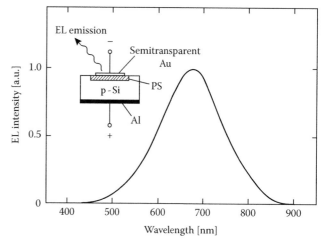

FIGURE 5.11   (See color insert following page 9-8.) Comparison between the influence of the homogeneous broadening ΔE and that of the inhomogeneous broadening (described by the geometrical standard deviation σ on the ensemble) on the width of the PL spectra. (From Meier, C. et al., *J. Appl. Phys.*, 101, 8, 2007. With permission.)

FIGURE 5.12   Schematic diagram of one of the first porous Si LEDs (inset) and its visible EL spectrum. (From Cullis, A.G. et al., *J. Appl. Phys.*, 82, 909, 1997. With permission.)

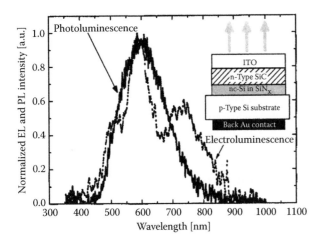

FIGURE 5.13 Comparison between the photoluminescence and the electroluminescence of the nc-Si device shown in the inset. (From Cho, K.S. et al., *Appl. Phys. Lett.*, 86, 071909, 2005. With permission.)

It is obvious that the realization of electroluminescent devices with light emission in the visible is not limited to porous silicon. Thin films of silicon nanocrystals as produced by CVD, ion implantation, pulsed laser deposition, etc., can also be used as an active medium for nc-Si LEDs. Light-emitting diodes with a very high EQE of 1.6% were produced from silicon nanocrystals, embedded in a silicon nitride matrix formed by plasma-enhanced chemical

vapor deposition (Cho et al. 2005). As can be seen from Figure 5.13, the electroluminescence characteristic closely follows that of the photoluminescence of the Si nanoparticles, which shows that the electroluminescence from the device mainly originates from electron–hole pair recombination in the nc-Si. Furthermore, it was shown that the injection of the charge carriers can be described by Fowler–Nordheim tunneling. When the formation of contacting electrodes to the active layer of the electroluminescent devices is performed very carefully, it is observed that the PL-spectrum and the EL-spectrum reveal nearly the same optical spectrum.

As mentioned above, a critical challenge for electroluminescent silicon nanocrystal devices is to provide for an efficient electrical carrier injection. All devices discussed so far are driven by DC voltages in the range of a few volts. A different concept for inducing electroluminescence has been developed by Walters et al. (2005). The authors developed a scheme for electrically pumping dense silicon nanocrystal arrays by a field-effect electroluminescence mechanism. Both, electrons and holes are injected from the same semiconductor channel across a tunneling barrier. In contrast to simultaneous carrier injection in conventional pn-junction light-emitting-diode structures, the carriers are sequentially injected using an alternating voltage. The observed light emission is strongly correlated with the injection of carriers into nanocrystals that have been previously loaded with charge of the opposite sign. Figure 5.14 shows a schematic of the working principle of this device.

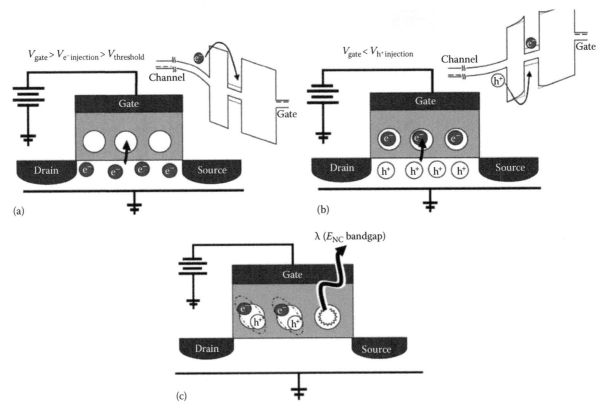

FIGURE 5.14 Schematic of the field-effect electroluminescence mechanism in a silicon nanocrystal floating-gate transistor structure. The inset band diagrams depict the relevant tunneling processes. The array of silicon nanocrystals embedded in the gate oxide of the transistor can be sequentially charged with electrons (a) and holes (b) to induce excitons that can radiatively recombine (c). (From Walters, R.J. et al., *Nat. Mater.*, 4, 143, 2005. With permission.)

Just like the DC-electroluminescence discussed above, the AC-driven electroluminescence increases dramatically with increasing drive voltage. This can again be ascribed to Fowler–Nordheim tunneling, which is exponentially dependent on the electric field inside the tunneling barrier. Here, the tunneling barrier is given by the oxide between the channel and the silicon nanocrystals (see Figure 5.14), and the field is directly proportional to the driving gate voltage.

According to the present state of research, it is now commonly agreed that not only do quantum confined excitons play an important role in the radiative emission of silicon nanocrystals but localized states at the silicon surface, for instance, at the Si/SiO$_2$ interface, also have to be taken into account. These paramagnetic defects are caused by missing bonds between silicon and its environment (the so-called dangling bonds). One of the best known defects in oxidized nc-Si is the Si dangling bond at the interface between nc-Si and the surrounding SiO$_2$ (P$_b$ center). At room temperature, the P$_b$ center acts as a nonradiative recombination center, thereby reducing the band-edge luminescence. Therefore, by decreasing the density of the P$_b$ centers, further improvement in the luminescence efficiency of oxidized nc-Si is expected.

A lot of effort has been made to characterize the electronic nature of the Si/SiO$_2$ interface, and it has been shown in multiple publications that electron spin resonance (ESR) measurements are a valid tool to characterize the amount and nature of dangling bonds at the surface of silicon nanoparticles. A recent paper that discusses the origin of photoluminescence from Si nanocrystals has shown that photoluminescence can be maximized by complete passivation of the Si-nc surface with hydrogen, while the density of paramagnetic defects in such passivated crystals originating from P$_b$ center is negligible (Godefroo et al. 2008). Unfortunately, heating or irradiating such samples reintroduces some defects, and as a consequence, the luminescence diminishes. To compensate such defects, one simple idea is to introduce additional charge carriers, which form lone pairs of electrons with the silicon dangling bond independent of any excitation to avoid the formation of any P$_b$ center. Such an appropriate dopant for silicon is phosphorous.

5.5 Electrical Properties of Silicon Nanocrystals

5.5.1 Doping of Silicon Nanoparticles

Fujii et al. reported an interesting experimental connection between luminescence, dangling bonds, and doping (Fujii et al. 2000). The effect of P doping has been studied in oxide films containing oxide-passivated Si nanocrystals in phosphosilicate glass (PSG). As-prepared, co-sputtered films of silicon and PSG were annealed in nitrogen to form Si-nanocrystals with a diameter of about 3.5 nm. The concentration of phosphorus within the surrounding glass matrix was adjusted from 0 to 1.7 mol %. The samples show an emission near 1.4 eV, which was attributed to the recombination of free electron–hole pairs in nc-Si (band-edge PL). The luminescence increases, and the ESR dangling bond signal decreases, as the phosphorous concentration C_p increases (see Figure 5.15). For the samples with C_p smaller

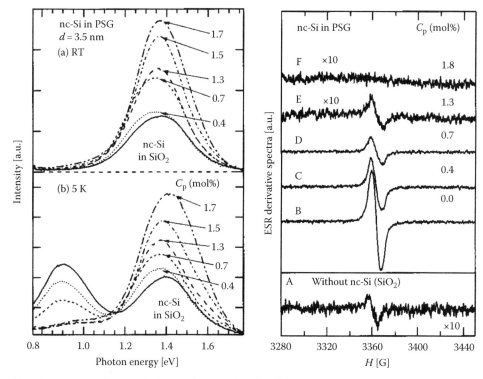

FIGURE 5.15 Left: Photoluminescence from nc-Si dispersed in PSG thin films (a) at room temperature and (b) at 5 K. The phosphorus concentration (C_p) is changed from 0 to 1.7 mol %. Right: ESR derivative spectra of a pure SiO$_2$ film (A) and SiO$_2^-$ (B) and PSG-films (C-F) containing nc-Si. For the samples containing nc-Si, C_p is changed from 0 to 1.8 mol %. (From Fujii, M. et al., *J. Appl. Phys.*, 87, 1855, 2000. With permission.)

than 1.3 mol %, a broad peak appears at around 0.9 eV in addition to the 1.4 eV peak. The 0.9 eV peak is generally assigned to the recombination of electron–hole pairs via P$_b$ centers.

A very similar behavior was also found for phosphorous-doped silicon nanoparticles, synthesized from the gas phase by thermal decomposition of a gaseous precursor mixture in a microwave plasma reactor (Stegner et al. 2007). A nominal doping level between 0 and 1.5×10^{20} cm^{-3} was adjusted by choosing the respective ratio of the precursors, silane and phosphine. At room temperature, the ESR signal from the Si dangling-bonds (Si-dbs) decreases when the P doping level is increased, indicating a charge transfer from donors to Si-dbs, see Figure 5.16. This compensation effect is also quantified by varying the density of Si-dbs via temperature programmed desorption (TPD) of H from Si–H bonds on the surface of Si nanocrystals. When heating the samples up to 550°C for a few minutes, the intensity of the Si-dbs ESR increases by a factor of five due to thermal desorption of the silicon-terminating hydrogen.

Although the structural location of P is not clear, the electrical conductivity of undoped and P-doped Si-NCs was studied using films composed of densely packed Si-nc. As can be seen from Figure 5.17, a pronounced doping effect on the electrical conductivity of such films with a strong increase in conductivity by several orders of magnitude is observed (Pereira et al. 2007, Stegner et al. 2008).

Additionally, electrically detected magnetic resonance (EDMR) studies have demonstrated the direct participation of P donor and Si-dangling bond states in the electronic transport through Si-nc networks: P donors and Si-dbs contribute to conductivity via spin-dependent hopping. This leads to the conclusion that doping with phosphorus results in a compensation of defects, which significantly increases both photoluminescence

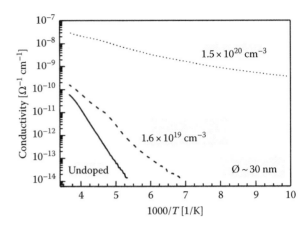

FIGURE 5.17 Arrhenius plot of conductivity vs temperature for films of Si-NCs (diameter ≈30 nm) doped at different levels. (From Stegner, A.R. et al., *Phys. Rev. Lett.*, 100, 4, 2008. With permission.)

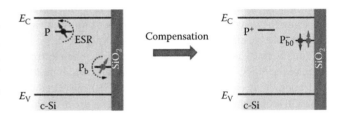

FIGURE 5.18 Schematic representation of the charge compensation of a P$_b$-center located at the Si/SiO$_2$ interface by electron transfer from a phosphorus atom.

and electrical conductivity of silicon nanocrystals. The corresponding mechanism is an electron transfer from a P atom dopant to the surface dangling bond (P$_b$ center), creating a lone pair, which does not show an ESR signal or trap the (optically excited) electrons (see Figure 5.18).

5.6 Future Perspective

During the last two decades, the main focus for the application of silicon nanoparticles was on the optical properties of the material, e.g., for optoelectronics, electroluminescence devices, and silicon LEDs (Fiory and Ravindra 2003). Nevertheless, there are a number of different and highly interesting fields of application that turn out to become more and more important. In the following, we will dwell on just a few of them.

One of these fields is the so-called bulk heterojunction hybrid solar cells. These solar cells use blends of inorganic nanocrystals with semiconducting polymers as a photovoltaic layer (Huynh et al. 2002). The basis of the bulk heterojunction concept is very similar to that used in pure organic solar cells. Electron-hole pairs created upon photoexcitation are separated into free charge carriers at interfaces. In the heterojunction solar cells, this interface is located between an organic and an inorganic semiconducting material. Electrons will move to the material with the higher

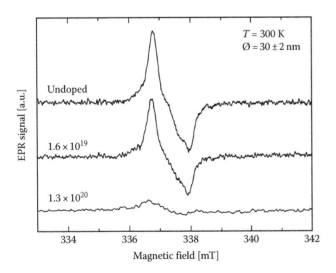

FIGURE 5.16 Room temperature ESR spectra of P-doped Si nanoparticles with a mean particle diameter of 30 nm and different nominal doping levels. The ESR intensity was normalized to the sample mass. (From Stegner, A.R. et al., *Phys. B: Condens. Matter*, 401, 541, 2007. With permission.)

electron affinity, and the hole to the material with the lower ionization potential, which also acts as the electron donor. So far, heterojunction solar cells have been demonstrated with various, semiconducting polymer blends containing CdSe, CuInS$_2$, CdS, or PbS nanocrystals, and first attempts combining silicon thin-films and regio-regular poly(3-hexylthiophene) (P3HT) have been made (Alet et al. 2006) based on the charge separation between P3HT as an organic electron donor and silicon as an inorganic electron acceptor. It is expected that in the future, hybrid inorganic/organic solar cells will gain a remarkable market share for several reasons:

- Inorganic semiconductor materials can have higher absorption coefficients (especially in the near-infrared), charge carrier mobility, and photoconductivity than many organic semiconductor materials.
- In comparison with organic semiconductors, the n- or p-type doping level of nanocrystalline materials can easily be varied by the synthesis route.
- Making use of quantum size effects, band-gap tuning of the nanoparticles can be used for the realization of complex device architectures, such as tandem solar cells, with stacks of multiple active layer.
- A substantial interfacial area for charge separation is provided by the nanocrystals due to their high surface to volume ratio.
- Cost-effective production processes are accessible by use of inexpensive printing technologies.

A second field with a promising application potential is lithium ion batteries with anodes containing silicon nanocrystals. Up to now, these anodes mainly consist of different carbon species such as graphite, soot, and some stabilizing binder. The maximum uptake for lithium in graphite corresponds to a charge density of 372 mAh/g, whereas the maximum uptake of lithium in silicon is 4.4 times the molar content of silicon, resulting in a storage capacity of 4200 mAh/g. Unfortunately, silicon containing electrodes degrade during alloying/dealloying with an abrupt increase in internal resistance that is caused by a breakdown of the conductive network. This results from the volume expansion and the contraction of the Si particles during the alloying of up to 400% and a subsequent amorphization of the crystalline silicon (Ryu et al. 2004). Chan et al. have shown that it is possible to overcome this limitation by using silicon nanowires with a few ten nm in diameter (Chan et al. 2008). A facile strain relaxation, which is only possible in the nanometer regime, allows the silicon nanowires to increase in diameter and length without breaking and enables for the synthesis of high capacity anode materials. Nevertheless, this method is limited to thin-film devices due to the fact that the silicon nanowires are electrically contacted at one end and their length is in the range of a few micrometers. Strain relaxation is also known for nanometer-sized silicon particles, and it seems to be possible to chemically and electrically bond them to a conductive matrix (Hochgatterer et al. 2008). Using specific connectors between silicon nanoparticles and the matrix, excellent long-term cycling behavior of a

Si-graphite-composite is achieved, and a considerable amount of silicon remains electrochemically active. This is possible only if the properties of the silicon nanoparticles are maintained and a stable connection between the strongly swelling Si particles and the graphite matrix, which prevents the electrode from disintegration, is established.

A third field for future applications of silicon nanocrystals is thermoelectric devices. The effectiveness of a thermoelectric material is linked to the dimensionless thermoelectric figure of merit *ZT*, defined as

$$ZT = \frac{S^2 \cdot \sigma}{\kappa_T} \cdot T \qquad (5.6)$$

where
   S is the Seebeck coefficient
   σ is the electrical conductivity
   κ_T is the total thermal conductivity
   T is the absolute temperature

The quantities *S*, σ, and κ_T for conventional, three-dimensional crystalline systems are interrelated in such a way that it is very difficult to control these variables independently to increase *ZT*. This is because an increase in *S* usually results in a decrease in σ, and a decrease in σ produces a decrease in the electronic contribution to κ_T. However, if the dimensionality of the material is reduced, the size becomes available as a new variable to control the material properties. When the relevant length scale becomes small enough to give rise to quantum-confinement effects, the density of electronic states is dramatically altered, as described above, making it possible to influence the interrelation between *S*, σ, and κ_T and optimize these parameters with respect to maximum thermoelectric efficiency.

Since already the SiGe alloy is a good thermoelectric material, it can be expected that tailored Si-Ge nanocomposite materials have even further improved thermoelectric properties. Silicon and SiGe powders and particles have already been demonstrated to lead to respectable results in sintered structures. Preliminary results indicate that a random assemblage of silicon and germanium nanoparticles in heterogeneous nanoscale composites has a lower thermal conductivity than a silicon-germanium alloy of the same silicon-to-germanium ratio (Dresselhaus et al. 2007). Silicon-germanium composites and alloys combine several desirable properties for thermoelectric applications: The raw material is relatively cheap and available in industrial quantities. At high temperatures, they have a competitive figure of merit. Promising Si-Ge nanocomposites produced by ball milling have been fabricated and studied for thermoelectric applications (Dresselhaus et al. 2007). The compatibility of standardized silicon technology implies the possibility of high integration for thin films. Last but not least, silicon and germanium are theoretically and experimentally perfectly characterized as both bulk and nanoscale material so that a reliable data base for modeling is available.

Acknowledgments

The authors would like to thank Christof Schulz, Cedrik Meier, Stephan Lüttjohann, Anoop Gupta, Ingo Plümel, Matthias Offer, and Andreas Gondorf for the productive and rewarding joint research on silicon nanoparticles within the Collaborative Research Centre "Nanoparticles from the gas phase" and the Research Training Group "Nanotronics" and Martin Stutzmann, Martin Brandt, Dmitry Kovalev, and André Ebbers for fruitful collaboration and discussions. Financial support by the Deutsche Forschungsgemeinschaft is gratefully acknowledged.

References

Alet, P. J., S. Palacin, P. R. I. Cabarrocas, B. Kalache, M. Firon, and R. de Bettignies (2006) Hybrid solar cells based on thin-film silicon and P3HT. *European Physical Journal—Applied Physics*, 36, 231–234.

Biteen, J. S. (2006) Plasmon-enhanced silicon nanocrystal luminescence for optoelectronic applications. PhD thesis, California Institute of Technology, Pasadena, CA.

Bley, R. A. and S. M. Kauzlarich (1996) A low-temperature solution phase route for the synthesis of silicon nanoclusters. *Journal of the American Chemical Society*, 118, 12461–12462.

Brandt, M. S., H. D. Fuchs, M. Stutzmann, J. Weber, and M. Cardona (1992) The origin of visible luminescence from porous silicon: A new interpretation. *Solid State Communications*, 81, 307–312.

Canham, L. T. (1990) Silicon quantum wire array fabrication by electrochemical and chemical dissolution of wafers. *Applied Physics Letters*, 57, 1046–1048.

Chan, C. K., H. L. Peng, G. Liu, K. McIlwrath, X. F. Zhang, R. A. Huggins, and Y. Cui (2008) High-performance lithium battery anodes using silicon nanowires. *Nature Nanotechnology*, 3, 31–35.

Cho, K. S., N. M. Park, T. Y. Kim, K. H. Kim, G. Y. Sung, and J. H. Shin (2005) High efficiency visible electroluminescence from silicon nanocrystals embedded in silicon nitride using a transparent doping layer. *Applied Physics Letters*, 86, 071909.

Cullis, A. G., L. T. Canham, and P. D. J. Calcott (1997) The structural and luminescence properties of porous silicon. *Journal of Applied Physics*, 82, 909–965.

Delerue, C., G. Allan, and M. Lannoo (1993) Theoretical aspects of the luminescence of porous silicon. *Physical Review B*, 48, 11024–11036.

Dresselhaus, M. S., G. Chen, M. Y. Tang, R. Yang, H. Lee, D. Wang, Z. Ren, J.-P. Fleurial, and P. Gogna (2007) New directions for low-dimensional thermoelectric materials. *Advanced Materials*, 19, 1043–1053.

Fiory, A. T. and N. M. Ravindra (2003) Light emission from silicon: Some perspectives and applications. *Journal of Electronic Materials*, 32, 1043–1051.

Fujii, M., A. Mimura, S. Hayashi, K. Yamamoto, C. Urakawa, and H. Ohta (2000) Improvement in photoluminescence efficiency of SiO$_2$ films containing Si nanocrystals by P doping: An electron spin resonance study. *Journal of Applied Physics*, 87, 1855–1857.

Gelloz, B. and N. Koshida (2000) Electroluminescence with high and stable quantum efficiency and low threshold voltage from anodically oxidized thin porous silicon diode. *Journal of Applied Physics*, 88, 4319–4324.

Giesen, B., H. Wiggers, A. Kowalik, and P. Roth (2005) Formation of Si-nanoparticles in a microwave reactor: Comparison between experiments and modelling. *Journal of Nanoparticle Research*, 7, 29–41.

Godefroo, S., M. Hayne, M. Jivanescu, A. Stesmans, M. Zacharias, O. I. Lebedev, G. Van Tendeloo, and V. V. Moshchalkov (2008) Classification and control of the origin of photoluminescence from Si nanocrystals. *Nature Nanotechnology*, 3, 174–178.

Green, M. A., J. H. Zhao, A. H. Wang, P. J. Reece, and M. Gal (2001) Efficient silicon light-emitting diodes. *Nature*, 412, 805–808.

Gupta, A., M. T. Swihart, and H. Wiggers (2009) Luminescent colloidal dispersion of silicon quantum dots from microwave plasma synthesis: Exploring the photoluminescence behavior across the visible spectrum. *Advanced Functional Materials*, 19(5), 696–703.

Hamilton, B. (1995) Porous silicon. *Semiconductor Science and Technology*, 10, 1187–1207.

Heitmann, J., R. Scholz, M. Schmidt, and M. Zacharias (2002) Size controlled nc-Si synthesis by SiO/SiO$_2$ superlattices. *Journal of Non-Crystalline Solids*, 299, 1075–1078.

Hochgatterer, N. S., M. R. Schweiger, S. Koller, P. R. Raimann, T. Wohrle, C. Wurm, and M. Winter (2008) Silicon/graphite composite electrodes for high-capacity anodes: Influence of binder chemistry on cycling stability. *Electrochemical and Solid State Letters*, 11, A76–A80.

Huynh, W. U., J. J. Dittmer, and A. P. Alivisatos (2002) Hybrid nanorod-polymer solar cells. *Science*, 295, 2425–2427.

Koshida, N. and H. Koyama (1992) Visible electroluminescence from porous silicon. *Applied Physics Letters*, 60, 347–349.

Kovalev, D., H. Heckler, G. Polisski, and F. Koch (1999) Optical properties of Si nanocrystals. *Physica Status Solidi B—Basic Research*, 215, 871–932.

Ledoux, G., O. Guillois, D. Porterat, C. Reynaud, F. Huisken, B. Kohn, and V. Paillard (2000) Photoluminescence properties of silicon nanocrystals as a function of their size. *Physical Review B*, 62, 15942–15951.

Lockwood, D. J., Z. H. Lu, and J. M. Baribeau (1996) Quantum confined luminescence in Si/SiO$_2$ superlattices. *Physical Review Letters*, 76, 539–541.

Lu, Z. H., D. J. Lockwood, and J. M. Baribeau (1995) Quantum confinement and light-emission in SiO$_2$/Si superlattices. *Nature*, 378, 258–260.

Mangolini, L., E. Thimsen, and U. Kortshagen (2005) High-yield plasma synthesis of luminescent silicon nanocrystals. *Nano Letters*, 5, 655–659.

Matsumoto, T., M. Daimon, T. Futagi, and H. Mimura (1992) Picosecond luminescence decay in porous silicon. *Japanese Journal of Applied Physics Part 2—Letters*, 31, L619–L621.

Meier, C., A. Gondorf, S. Luttjohann, A. Lorke, and H. Wiggers (2007) Silicon nanoparticles: Absorption, emission, and the nature of the electronic bandgap. *Journal of Applied Physics*, 101, 8.

Nesbit, L. A. (1985) Annealing characteristics of Si-rich SiO_2-films. *Applied Physics Letters*, 46, 38–40.

Onischuk, A. A., V. P. Strunin, M. A. Ushakova, and V. N. Panfilov (1997) On the pathways of aerosol formation by thermal decomposition of silane. *Journal of Aerosol Science*, 28, 207–222.

Pereira, R. N., A. R. Stegner, K. Klein, R. Lechner, R. Dietinueller, H. Wiggers, M. S. Brandt, and M. Stutzmann (2007) Electronic transport through Si nanocrystal films: Spin-dependent conductivity studies. *Physica B: Condensed Matter*, 401, 527–530.

Pi, X. D., R. W. Liptak, J. D. Nowak, N. Pwells, C. B. Carter, S. A. Campbell, and U. Kortshagen (2008) Air-stable full-visible-spectrum emission from silicon nanocrystals synthesized by an all-gas-phase plasma approach. *Nanotechnology*, 19, 5.

Prokes, S. M., O. J. Glembocki, V. M. Bermudez, R. Kaplan, L. E. Friedersdorf, and P. C. Searson (1992) SiH_x excitation: An alternate mechanism for porous Si photoluminescence. *Physical Review B*, 45, 13788–13791.

Rao, N. P., N. Tymiak, J. Blum, A. Neuman, H. J. Lee, S. L. Girshick, P. H. McMurry, and J. Heberlein (1998) Hypersonic plasma particle deposition of nanostructured silicon and silicon carbide. *Journal of Aerosol Science*, 29, 707–720.

Riabinina, D., C. Durand, F. Rosei, and M. Chaker (2007) Luminescent silicon nanostructures synthesized by laser ablation. *Physica Status Solidi A—Applications and Materials Science*, 204, 1623–1638.

Ryu, J. H., J. W. Kim, Y. E. Sung, and S. M. Oh (2004) Failure modes of silicon powder negative electrode in lithium secondary batteries. *Electrochemical and Solid State Letters*, 7, A306–A309.

Shimizu-Iwayama, T., N. Kurumado, D. E. Hole, and P. D. Townsend (1998) Optical properties of silicon nano-clusters fabricated by ion implantation. *Journal of Applied Physics*, 83, 6018–6022.

Stegner, A. R., R. N. Pereira, K. Klein, H. Wiggers, M. S. Brandt, and M. Stutzmann (2007) Phosphorus doping of Si nanocrystals: Interface defects and charge compensation. *Physica B: Condensed Matter*, 401, 541–545.

Stegner, A. R., R. N. Pereira, K. Klein, R. Lechner, R. Dietmueller, M. S. Brandt, M. Stutzmann, and H. Wiggers (2008) Electronic transport in phosphorus-doped silicon nano-crystal networks. *Physical Review Letters*, 100, 4.

Sychugov, I., R. Juhasz, J. Valenta, and J. Linnros (2005) Narrow luminescence linewidth of a silicon quantum dot. *Physical Review Letters*, 94, 4.

Vasquez, R. P., R. W. Fathauer, T. George, A. Ksendzov, and T. L. Lin (1992) Electronic structure of light-emitting porous Si. *Applied Physics Letters*, 60, 1004–1006.

Walters, R. J., G. I. Bourianoff, and H. A. Atwater (2005) Field-effect electroluminescence in silicon nanocrystals. *Nature Materials*, 4, 143–146.

Zacharias, M., J. Heitmann, R. Scholz, U. Kahler, M. Schmidt, and J. Blasing (2002) Size-controlled highly luminescent silicon nanocrystals: A SiO/SiO_2 superlattice approach. *Applied Physics Letters*, 80, 661–663.

6

ZnO Nanoparticles

Raj K. Thareja
Indian Institute of Technology Kanpur

Antaryami Mohanta
Indian Institute of Technology Kanpur

6.1 Introduction

Nanomaterials have been getting increasing attention due to their potential applications in many different fields such as coatings, catalysts, sensors, magnetic data storage, solar energy devices, ferrofluids, cell labeling, special drug delivery systems, etc. (Byrappa et al. 2008). Nanomaterials can be classified into a group intermediate between molecules and bulk materials with dimensions of the order of 10^{-9} m (nm), and which can have physical and chemical properties different from that of molecules and bulk materials even if the ingredients are the same. After the observation of the size-quantization effect in semiconductors (Rossetti et al. 1983, Byrappa et al. 2008), efforts are on to study the size-dependent properties of materials. The semiconductor materials exhibit the same physical properties irrespective of their size above a particular value called the threshold value for that material. Below this threshold, the band gap of the semiconductor materials increases with a decrease in their size, for example, in ZnO quantum-particle thin films, the band gap increases with a decrease of the particle size, and the enhancement of the band gap is significant when the particle size is smaller than 3 nm (Wong and Searson 1999). The decrease in size enhances the surface area relative to the volume. This results in an increase in surface atoms, which has a strong influence on the electronic and magnetic properties of the materials. The potential interest of studying nanostructured materials is

therefore due essentially to the ease of tunability of the physical properties by varying the particle size and shape. Recently, II–VI semiconductor nanoparticles have been extensively studied for their applications in displays, high-density storage devices, photovoltaics, biological labels, etc. (Sarigiannis et al. 2000, Amekura et al. 2006). One of the major efforts is on optimizing the emission properties of the wide-band-gap II–VI semiconductor materials due to the increasing demand for high-brightness light sources operating in the ultraviolet (UV) region. Among the II–VI wide-band-gap semiconductor materials, ZnO is one of the most promising candidates for the UV emitter applications due to its wide-band-gap of ~3.37 eV (at 300 K) and a high exciton binding energy of 60 meV. A chief competitor for ZnO is GaN, a wide-band-gap (~3.4 eV, 300 K) semiconductor material (III–V group) with similar optoelectronic applications as those of ZnO. GaN is widely used for green, blue, UV, and white light-emitting devices (Özgür et al. 2005). Although some optoelectronic devices (laser diode and light-emitting diodes) using GaN have already been reported (Nakamura et al. 1997), ZnO has several fundamental advantages over GaN, such as a higher free excitonic binding energy (60 meV) when compared to GaN (21–25 meV); the possibility of wet chemical processing; and more resistance to radiation damage (Look 2001). The impetus to ZnO has been due to band-gap engineering for the fabrication of efficient ZnO-based emitters such as quantum well laser diodes and light-emitting diodes (Fukuda 1998). The solid solution of

ZnO with MgO can produce a wide-band-gap semiconductor alloy (Mg, Zn) O for application in quantum well–related devices (Narayan et al. 2002, 2003, Thareja et al. 2005). The active layer of ZnO in the quantum well–related devices (laser diode or light-emitting diodes) due to its large excitonic binding energy (~60 meV) promises an efficient excitonic emission at room temperature. A laser emission from ZnO-based structures at room temperature and beyond has been reported (Thareja and Mitra 2000, Mitra and Thareja 2001, Özgür et al. 2005). The carrier and photon confinement in a small region are essential ingredients to achieve lasers and light-emitting diodes with low-threshold current densities. Therefore, a clear understanding of nanoscale semiconductor materials is imperative to achieve efficient optoelectronic devices. The chapter is organized as follows. Section 6.2 describes the crystal structure of ZnO. Band structure is discussed in Section 6.3. An overview of bulk semiconductors is presented in Section 6.4. Quantum well, quantum wire, and quantum dot are discussed in Sections 6.5 through 6.7, respectively. A brief mathematical note on nanoparticles is presented in Section 6.8. An overview of synthesis of ZnO nanoparticles by various research groups are briefly summarized in Section 6.9. The structural properties of ZnO nanoparticles are given in Section 6.10. A detailed discussion on optical properties including the concept of free excitons and polaritons, bound exciton complexes, donor–acceptor-pairs, and the photoluminescence process and Raman spectroscopy is presented in Section 6.11. A brief summary of some applications of ZnO is given in Section 6.12.

6.2 Crystal Structure

ZnO is a II–VI semiconducting material that exists in three forms: (1) hexagonal wurtzite (B4), (2) cubic zinc blende (B3), and (3) cubic rocksalt (B1). The Wurtzite structure is the most stable phase of ZnO in ambient conditions unlike other II–VI semiconductors that exist both in hexagonal wurtzite and the cubic zinc-blende structure, for example, ZnS (Klingshirn 2007). The zinc-blende structure is achieved by growing ZnO on a cubic substrate. However, the wurtzite form of ZnO can be converted to the rocksalt (NaCl) structure at relatively high pressures, and a reverse transition from cubic rocksalt (B1) to the hexagonal wurtzite (B4) structure occurs on removing the external pressure (Cai and Chen 2007). The most stable hexagonal wurtzite structure of ZnO is shown in Figure 6.1. The interpenetration of two hexagonal-closed-packed (hcp) lattices consisting of one type of atoms results in this hexagonal wurtzite structure. The lattice parameters a and b of this structure lie in the x–y plane and have equal length, and c is parallel to the z-axis. The values of the lattice parameters at room temperature are $a = b \approx 0.3249$ nm and $c \approx 0.5206$ nm. The ratio c/a (≈ 1.602) deviates slightly from the ideal value $c/a = \sqrt{(8/3)} = 1.633$ (Klingshirn 2007). The zinc blende and wurtzite-type structures are covalently bonded with sp^3 hybridization. However, the group IV element semiconductors such as carbon, silicon, and germanium have essentially covalent binding, and I–VII insulators, for example NaCl, have almost ionic binding. An intermixture of the ionic binding to

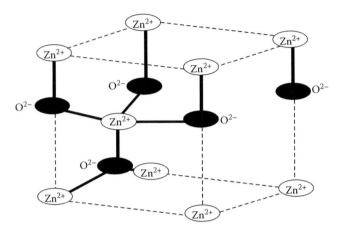

FIGURE 6.1 Hexagonal wurtzite structure of ZnO.

the covalent binding is noticed while moving from group IV element semiconductors to I–VII insulators through III–V and II–VI compound semiconductors (Klingshirn 2007). Thus, ZnO belonging to the II–VI group has the ionicity that lies at the border line between covalent and ionic semiconductors. However, the zinc blende or the wurtzite structures of ZnO lead to its classification as covalently bonded material (Schröer et al. 1993).

6.3 Band Structure

The band theory of solids describes the electronic states in crystals. Since atoms in solids are closely packed, the overlap of outer orbitals of the atoms results in the splitting of each atomic energy level of the constituent atoms. This results in a band of closely spaced discrete levels. In case of covalently bonded solid-like semiconductors, the uppermost energy levels of individual constituent atoms broaden into bands of levels. This can be realized by considering a single covalent bond of two atoms. When two atoms are brought sufficiently close to each other, the outer valence electron of one atom can arrange itself into a low-energy level (bonding) or into a high-energy level (antibonding). This means that each level of isolated atoms now splits into levels due to the two possible arrangements of electrons around the two atoms. In solids, there are large numbers of atoms coupled together that result in the formation of bands of closely spaced discrete energy levels. Several approaches have been used to describe the band structure of solids. The Kronig–Penney model is the one which approximates the periodic nature of potential by a square wave potential. A simpler and more appropriate approach is the coupled-mode approach (Feynman et al. 1964, Coldren and Corzine 1995). This deals with the general solution of the Schrödinger wave equation. Let us first consider the coupling between two similar atoms. The Schrödinger wave equation is therefore

$$H\psi = i\hbar \left(\frac{\partial \psi}{\partial t} \right) \qquad (6.1)$$

where

$H\left(=-\dfrac{\hbar^2}{2m}\nabla^2+V(\vec{r})\right)$ is the Hamiltonian

ψ represents the state of the coupled system which can be expressed as the linear combination of the orthonormal wave functions $\{\psi_1, \psi_2\}$ of the isolated atoms, i.e.,

$$\psi(\vec{r},t)=a_1(t)\psi_1(\vec{r})+a_2(t)\psi_2(\vec{r}) \qquad (6.2)$$

where $a_1(t)$ and $a_2(t)$ are the time-dependent coefficients and have the form:

$$a_m(t)=b_m\exp\left(-\dfrac{iEt}{\hbar}\right); \quad m=1, 2. \qquad (6.3)$$

b_m is a time-independent constant.

Substituting Equation 6.2 in Equation 6.1, we get

$$a_1(t)H\psi_1(\vec{r})+a_2(t)H\psi_2(\vec{r})=i\hbar\left(\dfrac{\partial a_1(t)}{\partial t}\right)\psi_1(\vec{r})+i\hbar\left(\dfrac{\partial a_2(t)}{\partial t}\right)\psi_2(\vec{r})$$

$$(6.4)$$

Multiply $\psi_1^*(\vec{r})$ on both sides of Equation 6.4 and integrate over space to get

$$a_1(t)\int\psi_1^*(\vec{r})H\psi_1(\vec{r})\mathrm{d}\tau+a_2(t)\int\psi_1^*(\vec{r})H\psi_2(\vec{r})\mathrm{d}\tau$$

$$=i\hbar\left(\dfrac{\partial a_1(t)}{\partial t}\right)\int\psi_1^*(\vec{r})\psi_1(\vec{r})\mathrm{d}\tau+i\hbar\left(\dfrac{\partial a_2(t)}{\partial t}\right)\int\psi_1^*(\vec{r})\psi_2(\vec{r})\mathrm{d}\tau$$

or

$$H_{11}a_1(t)+H_{12}a_2(t)=i\hbar\left(\dfrac{\mathrm{d}a_1(t)}{\mathrm{d}t}\right) \qquad (6.5)$$

where

$$\int\psi_m^*(\vec{r})H\psi_n(\vec{r})\mathrm{d}\tau=H_{mn}\delta_{mn}$$

$$\int\psi_m^*(\vec{r})\psi_n(\vec{r})\mathrm{d}\tau=\delta_{mn}; \quad\text{with }\delta_{mn}=0 \quad\text{for }m\neq n \quad\text{and}$$

$$\delta_{mn}=1 \quad\text{for }m=n$$

Similarly, we can obtain

$$H_{21}a_1(t)+H_{22}a_2(t)=i\hbar\left(\dfrac{\mathrm{d}a_2(t)}{\mathrm{d}t}\right) \qquad (6.6)$$

Suppose the energy of the state is $H_{11}=H_{22}=E_o$ and the coupling energy is $H_{21}=H_{12}=\delta E$; then we can write Equations 6.5 and 6.6 as

$$i\hbar\left(\dfrac{\mathrm{d}a_1(t)}{\mathrm{d}t}\right)=E_o a_1(t)+\delta E a_2(t) \qquad (6.7)$$

$$i\hbar\left(\dfrac{\mathrm{d}a_2(t)}{\mathrm{d}t}\right)=E_o a_2(t)+\delta E a_1(t) \qquad (6.8)$$

Using Equation 6.3 in Equations 6.7 and 6.8, we can get

$$E=E_o\pm\delta E \qquad (6.9)$$

This clearly indicates that the atomic energy level E_o of an isolated atom now splits into two discrete levels on either side of E_o by an amount δE. In a crystal, large numbers of atoms are coupled together to form energy bands of closely spaced discrete levels. In order to understand this behavior, let us consider the case of the one-dimensional chain of atoms with a separation of a under the assumption of nearest neighbors interaction only. Therefore, by taking $H_{mm}=E_1$ and $H_{m\,m\pm1}=\delta E$, we can write

$$i\hbar\left(\dfrac{\mathrm{d}a_m}{\mathrm{d}t}\right)=\delta E a_{m-1}+E_1 a_m+\delta E a_{m+1} \qquad (6.10)$$

From Equations 6.3 and 6.10; we have

$$E=E_1+\delta E\left[\dfrac{b_{m-1}+b_{m+1}}{b_m}\right] \qquad (6.11)$$

Since b_m corresponds to the mth lattice site, i.e., x_m, and $b_{m\pm1}$ corresponds to the $(m\pm1)$th lattice sites, i.e., $x_m\pm a$, we can consider $b_m=A\exp(ik_x x_m)$ and $b_{m\pm1}=A\exp[ik_x(x_m\pm a)]$.

Hence, Equation 6.11 gives

$$E=E_1+2\delta E\cos k_x a \qquad (6.12)$$

Equation 6.12 indicates that the allowed energy values lie within a band of energies between $E=E_1\pm2\delta E$ as shown in Figure 6.2. A similar treatment is to be done to estimate the allowed energy band when atoms of higher energy levels E_2 are bonded to form a crystal. The allowed energy values in this case can be obtained from the following expression:

$$E'=E_2+2\delta E'\cos k_x a \text{ (for indirect band-gap semiconductors)}$$

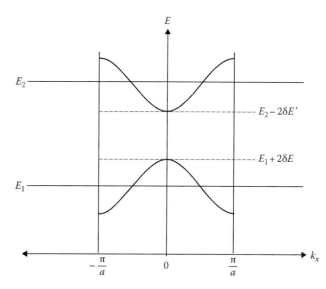

FIGURE 6.2 E–k diagram for a one-dimensional crystal.

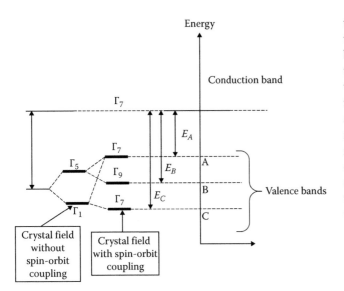

FIGURE 6.3 Valence band ordering in ZnO.

and

$$E' = E_2 - 2\delta E' \cos k_x a \text{ (for direct band-gap semiconductors)}$$

where $\delta E'$ is the coupling energy. Figure 6.2 shows the E–k diagram for a direct band-gap semiconductor crystal which illustrates the formation of bands. It is obvious that the E–k extrema (i.e., the conduction-band minimum and the valence-band maximum) can be approximated by parabolas. The width of the band is dependent on the coupling strength, and increases with an increase of the coupling energy.

ZnO is a compound semiconductor belonging to the II–VI group. The electronic configuration of Zn is $1s^2 2s^2 2p^6 3s^2 3p^6 3d^{10} 4s^2$ and that of O is $1s^2 2s^2 2p^4$. The valence orbital of Zn is 4s and that of O is 2p. The lowest conduction band and the uppermost valence band are formed due to the antibonding level of Zn (Zn4s) and the bonding level of O (O2p), respectively. Therefore, the conduction band of ZnO is predominantly s-like and the valence band, p-like. The Zn3d orbital strongly interacts with the O2p orbital that causes variation in the band gap due to p-d interaction (Schröer et al. 1993). The valence band of ZnO due to the occupied O2p orbital is split into three bands due to the influence of the crystal field and the spin-orbit coupling (Mang et al. 1995). The only influence of the crystal field without spin-orbit coupling is to split the valence band into two bands, Γ_5 and Γ_1. The combined influence of the crystal field and the spin-orbit coupling gives rise to three twofold degenerate valence bands, named A(Γ_7), B(Γ_9), and C(Γ_7) from the top to the bottom and is illustrated in Figure 6.3.

6.4 Bulk Semiconductor

In semiconductors, the valence band is completely filled at the absolute zero temperature leaving the conduction band empty. However, as the temperature increases, electrons from

the valence band are thermalized and get excited through the conduction band leaving holes at the top of the valence band. Therefore, it is worthwhile to look at the available electron states in the conduction band and the hole states in the valence band. Let us denote the density of states as $D(E)$, the total number of available states per unit energy range at E. In semiconductors, electrons of low energies with an effective mass m^* are free to move where the E–k diagram is approximated by parabolas. Within this picture, electrons in semiconductors can be treated as free electrons with plane wave solutions confined in a three-dimensional potential box. The time-independent Schrödinger wave equation for a free particle of energy E is given by

$$-\frac{\hbar^2}{2m^*}\left(\frac{\partial^2}{\partial x^2} + \frac{\partial^2}{\partial y^2} + \frac{\partial^2}{\partial z^2}\right)\psi(x,y,z) = E\psi(x,y,z) \quad (6.13)$$

The solution of this wave equation is of the type

$$\psi(x,y,z) = A e^{i(k_x x + k_y y + k_z z)} \quad (6.14)$$

with

$$|\vec{K}| = k = \sqrt{(k_x^2 + k_y^2 + k_z^2)} = \sqrt{\frac{2m^* E}{\hbar^2}} \quad (6.15)$$

and A is an arbitrary constant. The wave-function represented by Equation 6.14 satisfies the following periodic boundary conditions in x, y, z with period L,

$$\left.\begin{array}{l}\psi(x+L,y,z) = \psi(x,y,z) \\ \psi(x,y+L,z) = \psi(x,y,z) \\ \psi(x,y,z+L) = \psi(x,y,z)\end{array}\right\} \quad (6.16)$$

Using these boundary conditions, we can obtain the allowed values of $\vec{K}$ as

$$(k_x, k_y, k_z) = 0, \pm\frac{2\pi}{L}, \pm\frac{4\pi}{L}, \pm\frac{6\pi}{L}, \cdots, \pm\frac{2n\pi}{L} \quad (6.17)$$

That is, there is one allowed wave vector $\vec{K}$ in each volume element $(2\pi/L)^3$ of a three-dimensional k-space. The number of states between k and $k + dk$ is given by

$$D(k)\,dk = 2\frac{\left((4\pi/3)(k+dk)^3 - (4\pi/3)k^3\right)}{(2\pi/L)^3} \quad (6.18)$$

The factor 2 in Equation 6.18 represents two states for each k value due to the two possible spin orientations of the electron.

Equation 6.18 can be rewritten after neglecting the terms containing a higher order in dk, as

$$D(k)\,dk = \left(\frac{L^3}{\pi^2}\right) k^2 \, dk \qquad (6.19)$$

The number of states between E and $E + dE$ are

$$D(E)\,dE = \frac{V}{2\pi^2}\left(\frac{2m^*}{\hbar^2}\right)^{3/2} E^{1/2}\, dE \quad \text{for } E \geq 0, \qquad (6.20)$$

where $V\ (=L^3)$ is the volume. The minimum energy of the electron is the energy at the bottom of the conduction band, i.e., E_c. If $\rho_c(E)$ is the density of the states of electrons in the conduction band per unit volume, then

$$\rho_c(E) = \frac{1}{2\pi^2}\left(\frac{2m_e^*}{\hbar^2}\right)^{3/2} (E - E_c)^{1/2} \quad \text{for } E \geq E_c \qquad (6.21)$$

Using Equation 6.15, the energy momentum relation for the conduction band can be written as

$$E = E_c + \frac{\hbar^2 k_x^2}{2m_e^*} + \frac{\hbar^2 k_y^2}{2m_e^*} + \frac{\hbar^2 k_z^2}{2m_e^*} \qquad (6.22)$$

Similarly, the maximum energy of a hole is the energy at the top of the valence band, i.e., E_v. If $\rho_v(E)$ is the density of the states of holes in the valence band per unit volume, then

$$\rho_v(E) = \frac{1}{2\pi^2}\left(\frac{2m_h^*}{\hbar^2}\right)^{3/2} (E_v - E)^{1/2} \quad \text{for } E \leq E_v \qquad (6.23)$$

and the energy-momentum relation for the valence band can be written as

$$E = E_v - \left(\frac{\hbar^2 k_x^2}{2m_h^*} + \frac{\hbar^2 k_y^2}{2m_h^*} + \frac{\hbar^2 k_z^2}{2m_h^*}\right) \qquad (6.24)$$

where m_e^* and m_h^* are the effective masses of the electrons and the holes, respectively, $m_e^* = 0.24m_o$ and $m_h^* = 0.45m_o$ for ZnO (Wong et al. 1998).

6.5 Quantum Well

A quantum well is a structure of double heterojunction in which a thin layer of a semiconductor material of thickness comparable to or smaller than the de Broglie wavelength is sandwiched between two semiconductor materials of a wider band gap than that of the thin layer (Saleh and Teich 1991). An example for a quantum well structure can be a thin layer of ZnO surrounded

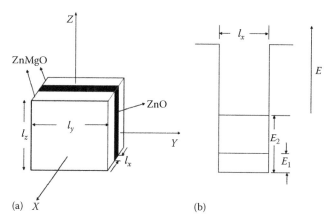

FIGURE 6.4 (a) A typical geometry of the quantum well structure, (b) Energy band diagram in a quantum well.

by two ZnMgO semiconductor alloys. In a quantum well, heterojunction is used for carrier confinement due to discontinuities, and the geometry of a thin layer of lower band-gap material sandwiched between two wide-band-gap semiconductor materials is responsible for photon confinement due to wave guiding. A typical geometry of a quantum well structure is shown in Figure 6.4a and the rectangular potential well formed due to the sandwiching a thin layer in a quantum well is shown in Figure 6.4b. The sufficiently deep rectangular potential wells in the conduction band and the valence band can be approximated as a one-dimensional infinitely deep potential well in which the particles of mass m^* (m_e^* for electrons in conduction band and m_h^* for holes in valence band) are free to move. Therefore, the free-particle Schrödinger wave equation in a one-dimensional infinitely deep potential well will have the form

$$-\frac{\hbar^2}{2m^*}\left(\frac{d^2\psi(x)}{dx^2}\right) = E\psi(x) \qquad (6.25)$$

The general solution of the above equation can be of the form

$$\psi = A\sin k_x x + B\cos k_x x \qquad (6.26)$$

with

$$k_x = \left(\frac{2m^* E}{\hbar^2}\right)^{1/2} \qquad (6.27)$$

This wave function must vanish at the boundary of the one-dimensional infinitely deep potential well, i.e.,

1. $\Psi\,(x = 0) = 0$.
2. $\Psi\,(x = l_x) = 0$.

On application of the first boundary condition, we get $B = 0$ and hence Equation 6.26 becomes

$$\psi = A\sin k_x x \qquad (6.28)$$

Using second boundary condition, we get

$$\sin k_x l_x = 0 \quad \text{or} \quad k_x l_x = n_x \pi, \quad n_x = 1, 2, 3, \ldots \quad (6.29)$$

From Equations 6.27 and 6.29, we have

$$E_{n_x} = \frac{n_x^2 \pi^2 \hbar^2}{2m^* l_x^2}, \quad n_x = 1, 2, 3, \ldots \quad (6.30)$$

In case of an infinitely deep ZnO quantum well of width $l_x = 10\,\text{nm}$, the allowed energy levels of electrons of effective mass $m_e^* = 0.24m_o$ are 15, 60, 225, 240 ... meV. The separation between the energy levels increases if the width of the well decreases. From Figure 6.4a, it is obvious that the movement of carriers (electrons and holes) gets restricted along x-axis within a distance of l_x; whereas carriers can move freely along y- and z-axis over a larger distance of l_y and l_z ($l_y, l_z \gg l_x$), respectively. The energy-momentum relation in conduction band for bulk semiconductor is given by Equation 6.22. For a quantum well, $l_x \ll l_y, l_z$; k_x takes discrete values in accordance with Equations 6.27 and 6.30. However, k_y and k_z have the values similar to that of bulk semiconductors. Therefore, the energy-momentum relation for electrons in conduction band of a quantum well becomes

$$E = E_c + E_{n_x} + \frac{\hbar^2 k_y^2}{2m_e^*} + \frac{\hbar^2 k_z^2}{2m_e^*}; \quad n_x = 1, 2, 3, \ldots \quad (6.31)$$

Similarly, the energy-momentum relation for holes in valence band of a quantum well is

$$E = E_v - \left[E_{n_x} + \frac{\hbar^2 k_y^2}{2m_h^*} + \frac{\hbar^2 k_z^2}{2m_h^*} \right]; \quad n_x = 1, 2, 3, \ldots \quad (6.32)$$

From Equations 6.31 and 6.32, it can be concluded that a quantum well can be treated as a two-dimensional bulk semiconductor where bottom of the conduction band is $E_c + E_{n_x}$, and the top of the valence band is $E_v - E_{n_x}$ for each $n_x = 1, 2, 3, \ldots$.

In two-dimensional bulk semiconductors,

$$k = \sqrt{(k_y^2 + k_z^2)} = \sqrt{\frac{2m^* E}{\hbar^2}}; \quad (6.33)$$

and the allowed values of k_y and k_z in bulk semiconductor are

$$(k_y, k_z) = 0, \pm \frac{2\pi}{L}, \pm \frac{4\pi}{L}, \ldots \quad (6.34)$$

where
$L = l_y$ for k_y
$L = l_z$ for k_z

There is one allowed two-dimensional wave vector $\vec{K}$ in surface element $(2\pi)^2/l_y l_z$ of two-dimensional k-space. Thus, the number of states between k and $k + \mathrm{d}k$ is given by

$$D(k)\mathrm{d}k = 2 \frac{\pi(k + \mathrm{d}k)^2 - \pi k^2}{\left(\frac{(2\pi)^2}{l_y l_z} \right)} \quad (6.35)$$

Neglecting the terms containing higher order in $\mathrm{d}k$, Equation 6.35 can be rewritten as

$$D(k)\mathrm{d}k = \left(\frac{l_y l_z}{\pi} \right) k \, \mathrm{d}k \quad (6.36)$$

The number of states between E and $E + \mathrm{d}E$ can be obtained using Equation 6.33 as

$$D(E)\mathrm{d}E = \left(\frac{m^* l_y l_z}{\pi \hbar^2} \right) \mathrm{d}E \quad (6.37)$$

In a quantum well structure, for each value of k_x, i.e., for each value of quantum number n_x, the density of states per unit area is $(m^*/\pi\hbar^2)$ and the density of states per unit volume is $(m^*/l_x \pi\hbar^2)$. If $\rho_c(E)$ is the density of states of electrons in the conduction band per unit volume, and $\rho_v(E)$ is the density of states of holes in the valence band per unit volume; we have

$$\rho_c(E) = \begin{cases} \left(\dfrac{m_e^*}{l_x \pi \hbar^2} \right); & E > E_c + E_{n_x} \\ 0; & E < E_c + E_{n_x} \end{cases} \quad (6.38)$$

and

$$\rho_v(E) = \begin{cases} \left(\dfrac{m_h^*}{l_x \pi \hbar^2} \right); & E < E_v - E_{n_x} \\ 0; & E > E_v - E_{n_x} \end{cases} \quad (6.39)$$

Equations 6.38 and 6.39 imply that the density of electrons in the conduction band and that of holes in the valence band per unit volume are constant for each quantum number n_x provided $E > E_c + E_{n_x}$ and $E < E_v - E_{n_x}$, respectively. The density of states profile in a quantum well is shown in Figure 6.5 which shows a stairway distribution.

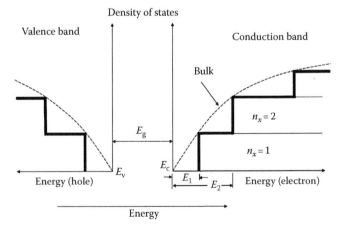

FIGURE 6.5 Density of states in a quantum well.

6.6 Quantum Wire

A quantum wire is a thin wire-like structure of a semiconductor material of diameter comparable to or smaller than the de Broglie wavelength which is surrounded by a wider band-gap semiconductor material. The wire behaves as a two-dimensional potential well for carriers (electrons in the conduction band and holes in the valence band) along the x- and the y-axis. A typical geometry of a quantum wire structure is shown in Figure 6.6. In a quantum wire, electrons and holes are confined along the x- and the y-axis within a distance of l_x and l_y as shown in Figure 6.6; whereas they extend over large distances of l_z along the z-axis in the plane of the confining layer. Therefore, it can be treated in a manner similar to as if electrons and holes are confined along the x and the y-axis, and along the z-axis they behave as if they are in the bulk semiconductor. The energy-momentum relation for a quantum wire can thus be obtained by following the procedure as that of a quantum well structure. Following Equations 6.31 and 6.32, we can write the energy-momentum relation for electrons in the conduction band in a quantum wire as

$$E = E_c + E_{n_x} + E_{n_y} + \frac{\hbar^2 k^2}{2m_e^*} \tag{6.40}$$

and the energy-momentum relation for holes in the valence band as

$$E = E_v - \left[E_{n_x} + E_{n_y} + \frac{\hbar^2 k^2}{2m_h^*} \right] \tag{6.41}$$

where

$$E_{n_x} = \left(\frac{n_x^2 \pi^2 \hbar^2}{2m^* l_x^2} \right)$$

$$E_{n_y} = \left(\frac{n_y^2 \pi^2 \hbar^2}{2m^* l_y^2} \right); \quad n_x, n_y = 1, 2, 3 \dots \tag{6.42}$$

$m^* = m_e^*$ (for electrons) and $m^* = m_h^*$ (for holes)

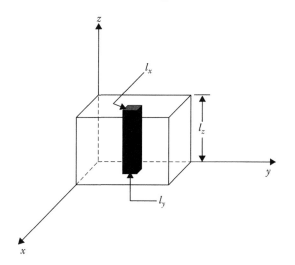

FIGURE 6.6 A typical geometry of a quantum wire. Electrons and holes are confined along the x- and the y-axes.

and $k = k_z$ is a wave-vector component along the z-direction (along the axis of the wire).

Equations 6.40 and 6.41 indicate that a quantum wire can be treated as a one-dimensional bulk semiconductor where the bottom of the conduction band is $E_c + E_{n_x} + E_{n_y}$ and the top of the valence band is $E_v - \left[E_{n_x} + E_{n_y} \right]$ for each pair of quantum numbers $(n_x, n_y) = 1, 2, 3 \dots$

In a one-dimensional bulk semiconductor,

$$k = \sqrt{k_z^2} = \sqrt{\frac{2m^* E}{\hbar^2}} \tag{6.43}$$

and the allowed values of k_z have been obtained in the bulk semiconductor as

$$k_z = 0, \pm \frac{2\pi}{l_z}, \pm \frac{4\pi}{l_z} \dots$$

There is one allowed one-dimensional wave vector $\vec{K}$ in the linear element of the one-dimensional k-space. Thus, the number of states between k and $k + dk$ is given by

$$D(k)dk = \left(\frac{l_z}{\pi} \right) dk \tag{6.44}$$

The number of states between E and $E + dE$ can be obtained using Equation 6.43 as

$$D(E)dE = \left(\frac{l_z}{\sqrt{2}\pi\hbar} \right) \left(\frac{m^*}{E} \right)^{1/2} dE \tag{6.45}$$

In a quantum wire structure, for each pair of (k_x, k_y), i.e., for each pair of quantum numbers (n_x, n_y), an energy sub-band is associated with a density of states of $\left(1/\sqrt{2}\pi\hbar \right)\left(m^*/E \right)^{1/2}$ per unit length of the wire and of $(1/l_x l_y)(1/\sqrt{2}\pi\hbar)(m^*/E)^{1/2}$ per unit volume of the wire. If $\rho_c(E)$ is the density of the states of electrons in the conduction band and $\rho_v(E)$ is the density of the states of holes in the valence band per unit volume, then we have

$$\rho_c(E) = \begin{cases} \left[\left(\frac{1}{l_x l_y} \right)\left(\frac{1}{\sqrt{2}\pi\hbar} \right)\left(\frac{m_e^*}{E - E_c - E_{n_x} - E_{n_y}} \right) \right]^{1/2}; & E > E_c + E_{n_x} + E_{n_y} \\ 0; & E < E_c + E_{n_x} + E_{n_y} \end{cases} \tag{6.46}$$

and

$$\rho_v(E) = \begin{cases} \left[\left(\frac{1}{l_x l_y} \right)\left(\frac{1}{\sqrt{2}\pi\hbar} \right)\left(\frac{m_h^*}{E_v - E_{n_x} - E_{n_y} - E} \right) \right]^{1/2}; & E < E_v - [E_{n_x} + E_{n_y}] \\ 0; & E > E_v - [E_{n_x} + E_{n_y}] \end{cases} \tag{6.47}$$

The density of the state distribution is shown in Figure 6.7.

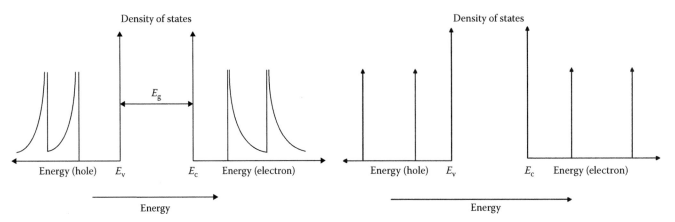

FIGURE 6.7 Density of states in a quantum wire.

FIGURE 6.9 Density of states for a quantum dot.

6.7 Quantum Dot

In Sections 6.5 and 6.6, we discussed about the quantum well and the quantum wire. In order to have a clear insight of ZnO nanoparticles, a discussion on the zero-dimensional quantum-dot structure is inevitable. For a semiconducting material, a quantum dot structure is a small box with sides comparable to or smaller than the de Broglie wavelength which is surrounded by a wider band-gap semiconductor material. This box behaves as a three-dimensional potential well for carriers (electrons in the conduction band and the holes in the valence band). A typical geometry of a quantum dot structure is shown in Figure 6.8. In a quantum dot, carriers are narrowly confined in all three directions along each side of the box l_x, l_y, and l_z along the x-, the y-, and the z-axis, respectively. Therefore, the energy is quantized along all three directions and can be written for electrons in the conduction band as

$$E = E_c + E_{n_x} + E_{n_y} + E_{n_z} \qquad (6.48)$$

and for holes in the valence band as

$$E = E_v - [E_{n_x} + E_{n_y} + E_{n_z}] \qquad (6.49)$$

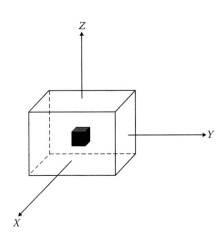

FIGURE 6.8 A typical geometry of a quantum dot.

where

$$E_{n_x} = \frac{n_x^2 \pi^2 \hbar^2}{2m * l_x^2}$$

$$E_{n_y} = \frac{n_y^2 \pi^2 \hbar^2}{2m * l_y^2}$$

$$E_{n_z} = \frac{n_z^2 \pi^2 \hbar^2}{2m * l_z^2} \quad \text{with } n_x, n_y, n_z = 1, 2, 3, \dots$$

$m*$ is the mass of the carriers (the electrons in the conduction band and the holes in the valence band).

The energy levels are discrete and well separated. The density of the states is therefore represented by delta functions as shown in Figure 6.9. As the carrier (the electrons in the conduction band and the holes in the valence band) motion is restricted, the conduction band and the valence band split into sub bands which become narrower with the increasing restriction in more dimensions. Finally, the density of states will be represented by the delta functions where the carrier motion is restricted in all three directions, the case of the quantum dot.

6.8 Nanoparticles

Nanoparticles are usually defined according to their size. Particles with size more than 1 nm and less than or comparable to 100 nm are classified as nanoparticles. Bulk materials have fixed physical properties irrespective of their size. However, nanoparticles may or may not have the same physical properties as that of bulk materials. Quantum dots are referred to as nanoparticles in the case of semiconductors which have quantum-confinement property. Nanoclusters are also nanoparticles whose size lies between 1 and 10 nm with a narrow size distribution which always show the effect of the quantum confinement. In general, semiconductors have a nonzero (small) band gap. Quantum dots and nanoclusters may have a band gap larger than that of the bulk that increases with decreasing size. The absorption peak corresponding to the threshold for the absorption of light in the quantum dot is blue-shifted on decreasing size (Rama Krishna and Friesner 1991, Thareja and Shukla 2007). Similarly, the photoluminescence peak position of nanoclusters also shows a blue-shift with respect to that of bulk materials (Mohanta et al. 2008). According to the effective mass approximation (Wong et al. 1998), the band gap of nanoparticles showing a quantum confinement effect is related to the band gap of bulk material as

$$E_{\text{nano}} = E_{\text{g}} + \frac{\pi^2 \hbar^2}{2R^2} \left(\frac{1}{m_{\text{e}}^*} + \frac{1}{m_{\text{h}}^*} \right) \qquad (6.50)$$

where

E_{g} is the band gap of bulk material
R is the radius of the nanoparticles showing the quantum confinement effect
m_{e}^* and m_{h}^* are the effective masses of the electrons and the holes, respectively

In semiconductors, the optical spectra may have photon energies less than that of the band gap due to the excitonic recombination. An exciton is an electron–hole pair bounded by Coulombic attraction. If we consider the case of free excitons (Mott-Wannier excitons), then the electron and the hole attract each other via the Coulomb potential;

$$V(r) = \frac{-e^2}{\varepsilon r} \qquad (6.51)$$

where

r is the distance between the electron–hole pair
ε is the dielectric constant

An exciton can be treated as hydrogen-like and therefore the energies of the exciton states can be written as (Mang et al. 1995)

$$E_{\text{ex}}(n) = E_{\text{g}} - \frac{E_{\text{e–b}}}{n^2} \qquad (6.52)$$

where

$E_{\text{ex}}(n)$ is the exciton energy
$n = 1, 2, 3, \ldots$ is the exciton principal-quantum number
E_{g} is the band-gap energy
$E_{\text{e–b}}$ is the exciton-binding energy

The Hamiltonian of an exciton confined to a nanoparticle of radius R can be written as (Kayanuma 1988)

$$H = \frac{p_{\text{e}}^2}{2m_{\text{e}}^*} + \frac{p_{\text{h}}^2}{2m_{\text{h}}^*} - \frac{e^2}{k|\vec{r}_{\text{e}} - \vec{r}_{\text{h}}|} \qquad (6.53)$$

where $\vec{r}_i$, $\vec{p}_i$, and m_i^* are the coordinate, the momentum, and the effective mass of the electron ($i = e$) and the hole ($i = h$), respectively. Kayanuma (1988) and Brus (1984) derived the following expression for the exciton energy

$$E_{\text{ex}} = E_{\text{g}} + \frac{\hbar^2 \pi^2}{2R^2} \left(\frac{1}{m_{\text{e}}^*} + \frac{1}{m_{\text{h}}^*} \right) - \frac{1.786e^2}{\varepsilon R} - 0.248 \frac{\mu e^4}{2\hbar^2 \varepsilon^2} \qquad (6.54)$$

Using Equation 6.50, we can write

$$E_{\text{ex}} = E_{\text{nano}} - \frac{1.786e^2}{\varepsilon R} - 0.248 \frac{\mu e^4}{2\hbar^2 \varepsilon^2} \qquad (6.55)$$

where

E_{nano} is the energy band gap of nanoparticles
μ is the reduced effective mass

$$\mu = \frac{1}{\dfrac{1}{m_{\text{e}}^*} + \dfrac{1}{m_{\text{h}}^*}}.$$

Equation 6.55 gives a relation between the exciton energy and the band gap of quantum size nanoparticles. It is obvious that the exciton energy is dependent on the radius of nanoparticles, i.e., the size of the nanoparticles, and decreases with an increase of size. The exciton energy obtained from Equation 6.55 for spherical nanoparticles agrees with the experimental results; and deviates in case of nonspherical nanoparticles (Rama Krishna and Friesner 1991).

6.9 Synthesis of ZnO Nanoparticles

A significant progress has been made on the growth and synthesis of ZnO nanoparticles following various techniques (Koch et al. 1985, Mohanta et al. 2008). There are mainly two approaches that have been used for the synthesis of nanomaterials; the bottom-up approach, and the top-down approach. The bottom-up approach is a chemical synthesis method which involves the controlled arrangement of small building blocks (atomic and molecular species) to form larger structures. The structures thus obtained have an authentic size distribution and are normally reproducible. However, the top-down approach is a physical synthesis method in which bulk materials of micrometer size are graved to achieve nanometer-size particles through mechanical milling. The most popular methods of the top-down approach are ball milling and ion-beam milling. Through these methods it is simple and easy to fabricate nanomaterials; however, the synthesized nanomaterials have a nonuniform shape and size, and are usually not reproducible. ZnO nanoparticles have been synthesized for a wide range of applications (Koch et al. 1985, Mohanta et al. 2008). In the following, we summarize the preparation route of ZnO nanomaterials. Koch et al. (1985) prepared extremely small (<2.5 nm) ZnO particles by precipitation in alcoholic solution. Spanhel et al. (1987) reported an increase in the growth rate of the colloidal ZnO particles on addition of water. The colloids so obtained contain 0.002 M colloidal ZnO and about 0.01 M excess OH⁻ ions. The hydroxyl group is present due to the starting material NaOH. Mahamuni et al. (Reetz and Helbig 1994, Mahamuni et al. 1999) synthesized stable hydroxyl (OH) group-free zinc oxide quantum dots at room temperature following an electrochemical route. Guo et al. (2000) prepared highly monodisperse stable ZnO nanoparticles using poly(vinyl pyrrolidone) [PVP] as the capping agent. The synthesis is due to the reaction of zinc acetate [(CH$_3$COO)$_2$ Zn·2H$_2$O] with sodium hydroxide [NaOH] in the presence of poly(vinyl pyrrolidone) with a molar ratio of Zn^{2+}/PVP = 5:3 where 2-propanol is the reaction media. The size of the ZnO quantum dots are related to the Zn^{2+}/PVP ratio (Yang et al. 2001). The mean size of the ZnO quantum dots obtained using high-resolution transmission microscopy were 2.8 ± 0.4, 2.6 ± 0.3, 3.6 ± 0.5, and 4 ± 0.5 nm corresponding to the Zn^{2+}/PVP ratio of 5:5, 5:3, 5:1, and 5:0 (no PVP), respectively. The surface modification of ZnO quantum dots by polyvinyl pyrrolidone (PVP) yielded the spherical shape of dots; however the ZnO quantum dots without a PVP modification were ellipsoid in shape.

Stable ZnO nanoparticles (~diameter 4–5 nm) have also been synthesized (Spanhel and Anderson 1991, Hoyer and Weller 1994, Yang et al. 2005) by the direct solvent evaporation of an alcohol solution containing potassium hydroxide (KOH) and zinc acetate. Bahnemann et al. (1987) prepared transparent colloidal suspensions of small zinc oxide particles in water, 2-propanol and acetonitrile. Wong and coworkers (1998) aged the colloids at two different temperatures ~35°C and 65°C and obtained two different-sized particles of diameters 3.8 ± 0.2 and 4.2 ± 0.2 nm, respectively.

ZnO nanoparticles have been synthesized by using pulsed-laser ablation besides chemical synthesis. Ou et al. (2008) obtained ZnO nanoparticles by ablating Zn target (with purity of 99.999%) in oxygen atmosphere at a pressure of 100 kPa using a pulsed Nd:YAG laser. The nanoparticles with a size distribution between 10 and 100 nm with an average size of ~60 nm were synthesized. Pulsed laser ablation of ZnO in liquid has also been used for producing ZnO nanoparticles. Zeng et al. (2006) prepared ZnO–Zn nanocomposites by the laser ablation of a zinc-metal target in pure water and in the aqueous solution of sodium dodecyl sulfate (SDS). The nanoparticles so obtained were nearly spherical with an average diameter of 20 nm. A high-resolution transmission-electron microscopy (HRTEM) study showed that the nanoparticles so obtained have core-shell structure with zinc core and a ZnO shell. The thickness of the ZnO shell is strongly dependent on the laser power and is found to decrease with a decreasing ablating laser power. Thareja and Shukla (2007) prepared ZnO nanoparticles in various liquid media; deionized water (DIW), isopropanol, and acetone by pulsed laser ablation. Spherical nanoparticles of 14–20 nm were obtained in case of water and isopropanol. However, two types of particles, (1) spherical nanoparticles of size around 100 nm, and (2) a platelet-like structure of 1 μm in diameter and 40 nm in width, were obtained in acetone. He et al. (2007) have fabricated ZnO

nanoparticles in surfactant-free aqueous solutions by a pulsed-laser ablation (PLA) of zinc and studied the dependence of particle size on pH. The chemicals with different pH [HCL (pH 5.36), NaOH (pH 11.98), and 10 mM NaCl (pH 7.15)] were used to study the pH-dependent particle size. The average particle size and the standard deviation of the particle size produced in HCL and NaOH solutions were smaller (15 ± 6 nm for HCL, pH 5.36; 20 ± 8 nm for NaOH, pH 11.98) than those prepared in deionized water (23 ± 11 nm, pH 7.51). However, the average particle size and the standard deviation of the particle size produced in 10 mM NaCl solution (pH 7.15) was 26 ± 15 nm lager than that of in DIW (23 ± 11 nm, pH 7.51). The formation of ZnO nanoparticles is also observed in the gas phase in the ZnO plasma. Mohanta et al. (2008) observed the photoluminescence from ZnO nanoparticles formed in the ZnO plasma in air. These ZnO nanoparticles are formed due to the cooling of highly energetic ZnO plasma species by colliding with the molecules of the ambient (air in this case).

The techniques to obtain ZnO nanoparticles discussed above are based on the bottom-up approach. The simplest top-down technique for producing nanoparticles is ball milling (Damonte et al. 2004, Giri et al. 2007). ZnO nanoparticles of size in the range 7–35 nm from commercial ZnO powder of particle size >300 nm have been reported (Giri et al. 2007). Conventionally, ZnO powder is milled in a mechanical milling machine at say 300 rpm in a stainless vial under atmospheric pressure and temperature. Homogeneity in size with particle size distribution of 50–110 nm is achieved after 1 h of milling of commercial ZnO powder of size ≅ 500 nm (Damonte et al. 2004). However, particles become indistinguishable with an increase in the milling time due to a kind of accretion between them.

6.10 Structural Properties of ZnO Nanoparticles

The structural properties of ZnO are determined by the x-ray diffraction technique. The x-ray diffraction spectrum of bulk ZnO shows several diffraction peaks corresponding to the (100), (002), (101), (102), (110), (103), (200), (112), (201), (004), and (202) planes, as shown in Figure 6.10. The lattice parameters are obtained from the peak position of the x-ray diffraction spectra. For wurtzite ZnO, the lattice constant a mostly ranges from 3.2475 to 3.2501 Å and c from 5.2042 to 5.2075 Å (Özgür et al. 2005). The particle size (t) can be estimated from the diffraction spectrum using the Debye-Scherrer formula:

$$t = \frac{0.9\lambda}{\beta\cos\theta};$$

where

λ is the wavelength of the x-ray used
β is the full width at half maximum (FWHM) of the diffraction peaks
θ is the Bragg diffraction angle

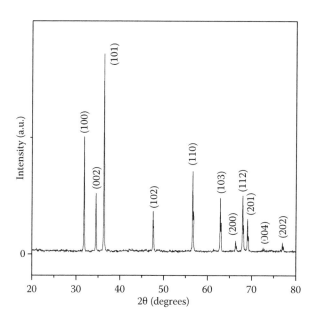

FIGURE 6.10　X-ray diffraction spectrum of bulk ZnO.

The FWHM of the diffraction peaks in the x-ray diffraction spectrum increases with a decrease in the particle size. Therefore, the diffraction profiles are observed to be broader in the case of nanoparticles than that of bulk ZnO. Zhou et al. (2002) observed broader diffraction profiles of ZnO-quantum dots in comparison to that of bulk wurtzite ZnO. The x-ray diffraction spectrum of ZnO nanoparticles obtained from the ball milling technique shows a broadening of the diffraction profiles with a slight up-shift of the XRD peak positions with respect to the commercial bulk ZnO powder (size > 300 nm). This is attributed to the decrease in particle size and the possibility of strain due to the ball-milling process (Damonte et al. 2004, Giri et al. 2007).

6.11　Optical Properties of ZnO

In optical excitations, an electron–hole pair is created in a semiconductor material by the absorption of a photon that recombines emitting a photon. The advantage of this technique is that it can be used to excite high resistivity materials where electroluminescence would be inefficient or impractical. This is also useful for materials where contact or junction technology is not adequately developed. The technique is used to characterize the semiconductor materials prior to the fabrication of any optoelectronic devices. The optical transitions in semiconductors are connected with both extrinsic and intrinsic effects. Intrinsic effects involve the optical transition between electrons in the conduction band and the holes in the valence band including the recombination of electron–hole pairs bounded by the Coulomb interaction. The interaction of the electron and the hole via the attractive Coulomb potential forms a series of hydrogen or positronium-like states below the band gap. These are called free-excitons (Wannier excitons) and are characterized by the fact that the average distance between the electron and the hole, i.e., the exciton Bohr radius is larger than the lattice constant.

On the other hand, the extrinsic effects in optical transitions are related to dopants or defects that create discrete electronic states in the band gap that have strong influences in the absorption and the luminescence spectra. Excitons can be bound to these dopants or defects to form bound exciton complexes (BEC).

6.11.1　Free Excitons and Polaritons

A free exciton is an electron–hole pair i.e., a pair of opposite charges bounded by the Coulomb potential (Pankove 1975). This indicates that the electron–hole pair system (exciton) is similar to the hydrogen-like atom. In hydrogen-like atoms, the reduced mass of the nucleus and the electron is equal to the mass of the electron as the mass of the nucleus is larger in comparison to that of the electron. However, the reduced effective mass of the hole and the electron in the case of the free exciton is not equal to the effective mass of the electron, and is less than the effective masses of the hole and the electron. This is because the effective masses of the electron and the hole are comparable, for example, in ZnO, $m_e^* = 0.24m_o$, and $m_h^* = 0.45m_o$, where m_e^* and m_h^* are the effective masses of the electron and the hole, and m_o is the mass of the electron. The free exciton is a mobile pair and can move through out the crystal. Moreover, excitonic complexes similar to positronium-like molecules can be formed by combining two free holes and two free electrons. Such a complex has a lower energy than two free excitons.

Polariton is another complex which has strong influences on the optical properties of semiconductors. A polariton is a complex that results from the interaction between an exciton and a photon. The dispersion curve of a photon is a straight line, whereas that of a free exciton is a parabola. The coupling between these two results in the dispersion curve of the coupled state of the exciton and the photon and is known as the exciton polariton. The lower part of the dispersion curve of the lower-polariton branch (LPB) behaves as that of photons and the upper part of the dispersion curve of the lower-polariton branch behaves as that of excitons. The finite transverse-longitudinal splitting Δ_{LT} indicates the presence of a longitudinal eigenmode (Klingshirn 2007). The upper-polariton branch (UPB) bends and follows the photon-like dispersion curve.

6.11.2　Bound Exciton Complexes

The free excitons and polaritons have been discussed in Section 6.11.1. However, there is a finite possibility where a free hole can combine with an electron of a neutral donor to form a positively charged excitonic ion. The electron remains bound to the donor and travels around the donor. The hole which is combined with the electron also travels about the donor. These complexes are called bound exciton complexes. On the other hand, an electron can get bound to a neutral acceptor and is called a neutral acceptor bound exciton complex. Furthermore, an exciton can get bound to an ionized donor to form an ionized donor bound exciton. The abbreviations often used for ionized bound excitons, neutral donor bound excitons and neutral acceptor bound excitons are D+X, D°X, and A°X, respectively. The bound exciton does not have the freedom to translate throughout the crystal. The electron and the hole remain in the same unit cell. The bound excitonic transitions are observed in the absorption and the luminescence bands at low temperatures. In bulk ZnO, the bound excitonic transitions cover a wide range from 3.348 to 3.374 eV (Özgür et al. 2005). In case of good quality samples, the line width of bound excitons is less than 1 meV. The two-electron satellite (TES) transition is an important characteristic of the neutral donor-bound exciton transition which appears in the spectral region of 3.32–3.34 eV (Özgür et al. 2005). In this transition, the donor remains in an excited state after a radiative recombination of an exciton bound to a neutral donor. This results in a smaller transition energy than that of the donor bound exciton energy of an amount equal to the energy difference of the ground and first excited states of the donor. At low temperatures, the luminescence spectra are dominated by the transition of bound excitonic complexes and the two-electron satellite transitions. However, with increasing temperature, the bound excitons get thermalized and disappear.

6.11.3 Donor–Acceptor Pairs

A donor and an acceptor can interact with each other by the Coulomb potential and form a pair of donor and acceptor called the donor–acceptor-pair (DAP) that remains stationary in the crystal. The binding energies of the donor and the acceptor decrease due to the coulomb interaction between a donor and an acceptor. As the distance between the donor and the acceptor decreases, the Coulomb attraction increases. The binding energy becomes zero for a fully ionized state. It corresponds to the impurity level (donor and acceptor) at the band edge. The separation of the energy levels of the donor and the acceptor can be represented by the following equation (Pankove 1975):

$$E_{DAP} = E_g - E_D - E_A + \frac{e^2}{\varepsilon r};\qquad (6.56)$$

where

E_g is the energy band-gap
E_D is the ionization energy of the isolated donor
E_A is the ionization energy of the isolated acceptor
r is the distance of separation between the donor and the acceptor

The last term on the right-hand side of Equation 6.56 is a measure of the shift of the donor and the acceptor levels due to the Coulomb attraction. The DAP transition is observed in the optical spectra at low temperatures. As the temperature increases, the intensity of the DAP peak in the optical spectra decreases and finally disappears at high temperatures.

The donor–acceptor pair is of two types: type-I donor–acceptor pair and type-II donor–acceptor pair; they are distinguished by the manner of occupation of the impurities in the lattice sites (Pankove 1975). In the case of the type-I donor–acceptor pair,

the donor and the acceptor occupy the same sublattice, for example, iodine (I) and nitrogen (N) occupy O sites in ZnO to form the donor and the acceptor, respectively. On the other hand, if the donor and the acceptor occupy the opposite lattice sites, then they form a type-II donor–acceptor pair, for example, copper (Cu) on the Zn site and fluorine (F) on the O site form the donor and the acceptor, respectively.

6.11.4 Photoluminescence

Photoluminescence (PL) is a process through which a system gets excited to a higher-energy level by absorbing a photon, and then spontaneously emits a photon and decays to a lower-energy level. The energy and the momentum remain conserved in this process. Photoluminescence spectroscopy is usually used to characterize surfaces, interfaces, impurity levels, and is also used to identify alloy disorder and surface roughness (Gfroerer 2000). This is a nondestructive technique since the sample is excited optically and no electrical contacts and junctions are involved. The technique is simple and requires no or very little control of the environment. The structure of the electronic energy levels of photo-excited materials can be obtained by analyzing the transition energies from the PL spectrum. An estimation of the relative rates of radiative and non-radiative recombination can be made from the PL intensity. The variation of the PL peak position and the intensity with temperature and applied voltage can be used to make further characterizations of the electronic states and the bands of the material. The PL process strongly depends on the nature of the optical excitation. If the wavelength of the incident light is such that the absorption is weak at the surface, the light penetrates deeper into the material and the PL is predominately through bulk recombination. Time-resolved photoluminescence (TRPL) using pulsed optical excitation is useful to characterize the rapid processes in semiconductors. The PL signal so obtained is used to determine recombination rates. The photoluminescence spectroscopy technique is limited to the radiative transitions. However, it is difficult to characterize poor-luminescent indirect-band-gap semiconductor materials. ZnO is a II–VI direct band-gap semiconductor material and photoluminescence spectroscopy is widely used to characterize the various forms of ZnO. The photoluminescence peak of quantum size ZnO nanoparticles gets blue-shifted with respect to that of bulk ZnO showing the quantum confinement effect. Figure 6.11 shows the room temperature steady-state PL spectra of ZnO-thin films of varying particle size (Wong and Searson 1999). It is obvious that both band-to-band emission and the visible emission band show a blue shift with a decrease in particle size due to an increase in the band gap (Figure 6.11). The visible emission band due to the recombination between the oxygen vacancies, which act as deep-level electron donor states, and the valence band in bulk ZnO has been reported. Both the UV and the visible green emission band in case of ZnO quantum dots capped with polyvinyl pyrrolidone (PVP) molecules has also been reported. The UV emission band is commonly attributed to band-to-band or near-band-edge emissions which has an excitonic origin and

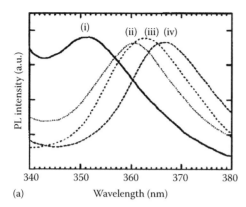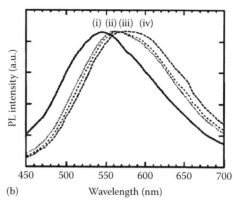

FIGURE 6.11 Steady state photoluminescence spectra, (a) band-to-band transition and (b) visible emission of ZnO thin films of quantum-sized particles of radii (i) 20.6 Å, (ii) 23.6 Å, (iii) 24.6 Å, and (iv) 26.8 Å. (From Wong, E.M. and Searson, P.C., *Appl. Phys. Lett.*, 74, 2939, 1999. With permission.)

the green visible emission band is due to the surface states associated with oxygen vacancies (Yang et al. 2001). The violet PL band at 425 nm from the ZnO shell layer of Zn/ZnO core-shell nanoparticles prepared by laser ablation in liquid media has also been observed (Zeng et al. 2006). The PL peak intensity of the violet emission band at 425 nm increases with a decrease of shell thickness, however, the PL peak position remains unchanged. The violet emission band is different in nature from th UV and the green emission band and arises due to an electron–hole recombination between the localization defect level of interstitial zinc and the valence band (Zeng et al. 2006).

At low temperatures, the transitions from various luminescence centers (impurities, excitons, etc.) are distinguishingly observed as the line width of the transition lines become narrower with a decrease in temperature. As the temperature increases, the line width broadens according to the relation (Klingshirn 2007)

$$\Gamma(T) = \Gamma(0) + a_1 T + \frac{b_1}{\left[\exp\left(\dfrac{\hbar\varpi_{LO}}{k_B T}\right) - 1\right]};$$

where

  $\Gamma(0)$ is the temperature-independent contribution to the linewidth
  $a_1 T$ is the acoustic phonon contribution which varies linearly with temperature and the last term on the right-hand side is due to the scattering of LO phonons

At low temperatures ($k_B T \ll \hbar\varpi_{LO}$), the second scattering term is negligible and, therefore, the line width increases linearly with the temperature due to scattering by acoustic phonons because of negligible LO phonon population. At high temperatures ($b_1 \gg a_1$), the first scattering term is negligible and the scattering is dominated by LO-phonons. In bulk ZnO, the most dominant emission peaks in the low temperature PL spectra are the neutral donor-bound excitons due to the presence of unintentional impurities and/or defects. The acceptor-bound excitons are also

observed in the low temperature PL spectra of bulk ZnO. Teke et al. (2004) observed many sharp lines of donor and acceptor bound excitons in the spectral range of 3.348–3.374 eV. The binding energies of the donor-bound excitons that are obtained from the PL spectra, ranges from 10 to 20 meV. As the temperature increases, the bound excitonic peaks and their phonon replicas disappear from the PL spectra due to the thermal quenching process. Figure 6.12 shows the PL profile at 6 K of ZnO nanowires where the quantum-confinement effect is not observed due to the large diameters of the wires (Mohanta and Thareja 2008a). It contains bound excitons (BX), free excitons ($FX_A^{n=1}$, $FX_A^{n=2}$), phonon replicas ($FX_A^{n=1} - mLO$, $m = 1,2,3$) of free exciton ($FX_A^{n=1}$), and donor–acceptor pairs. At 6 K, the bound exciton emission peak dominates. As the temperature increases, the intensity of the bound exciton decreases and finally disappears at high temperatures. However, the intensity of the first-phonon replica ($FX_A^{n=1} - 1LO$) of the free exciton increases with an increase of temperature and dominates at 125 K and beyond, as shown in Figure 6.13. At room temperature,

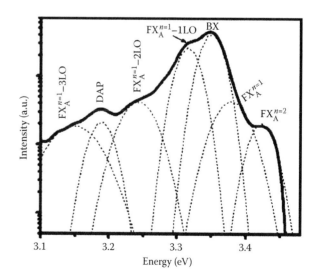

FIGURE 6.12 PL profile of ZnO nanowires at 6 K. The dotted lines are Gaussian fitting to the emission peaks. (From Mohanta, A. et al., *J. Appl. Phys.*, 104, 044906, 2008. With permission.)

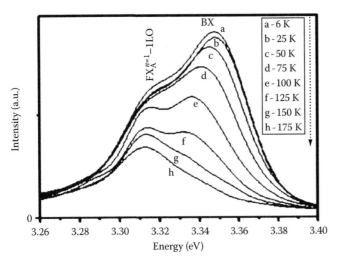

FIGURE 6.13 Evolution of a bound exciton (BX) and a first phonon replica ($FX_A^{n=1} - 1LO$) of a free exciton. (From Mohanta, A. et al., *J. Appl. Phys.*, 104, 044906, 2008. With permission.)

the LO-phonon replicas of free exciton transition dominates with a first-LO-phonon replica of the free exciton at the maximum (Shan et al. 2005, Mohanta and Thareja 2008a). This shows that the LO-phonon-exciton coupling becomes more efficient as the temperature rises. The LO-phonon energy in ZnO is 71–73 meV, therefore the LO-phonon replicas occur at a spectroscopic energy separation of 71–73 meV at a low temperature (10 K) (Teke et al. 2004). Shan et al. (2005) observed the energy separation of 63 meV between the free exciton and its first LO-phonon replica at room temperature which is explained by following the relation for the emission lines involving phonon and exciton emission:

$$E_m = E_0 - m\hbar\varpi_{LO} + \Delta E$$

where
   E_m corresponds to the spectroscopic-energy positions
   E_0 is the exciton energy
   $\hbar\varpi_{LO}$ is the phonon energy
   ΔE is the kinetic energy of the free excitons due to the temperature of the sample

Mohanta and Thareja (2008a) observed a reduced spectroscopic-energy separation of 47 meV between the free exciton and its first LO-phonon replica at room temperature due to the effect of the localized heating of the sample by the Nd:YAG laser pulse that can be understood from the following relation:

$$E_m = E_0 - m\hbar\varpi_{LO} + \Delta E + E_L$$

where E_L is the additional energy due to the localized heating of the sample by the laser pulse.

Fonoberov et al. (2006) undertook a photoluminescence study of ZnO quantum dots (~4 nm in diameter) both at low and room temperatures. Figure 6.14 shows the PL spectra of ZnO quantum dots (~4 nm in diameter) at various temperatures (8.5–150 K). It contains donor-bound excitons (D, X), acceptor-bound excitons (A, X), and a LO-phonon replica of acceptor-bound excitons (A, X) that are assigned according to their spectral positions. At low temperatures, the acceptor bound exciton emission dominates in the PL spectra of ZnO quantum dots. The longitudinal optical-phonon energy is observed to be 72 meV that is well in agreement with the reported value (Teke et al. 2004). The arrow shown in Figure 6.14 indicates the position of the confined exciton energy (3.462 eV) for ZnO quantum dots with the diameter of 4.4 nm. The peak energy position of quantum dots for a 4 nm quantum dot lies outside the range of energies shown in Figure 6.14. The PL peak energies of (D, X)

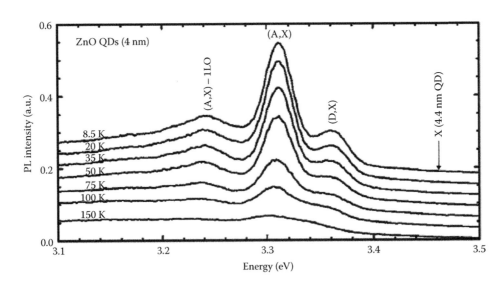

FIGURE 6.14 PL spectra of ZnO quantum dots (4 nm) at temperatures from 8.5 nm to 150 K. (From Fonoberov, V.A. et al., *Phys. Rev. B*, 73, 165317-1, 2006. With permission.)

and (A, X) decrease with an increase in temperature according to Varshni's Law (Varshni 1967):

$$E(T) = E(0) - \frac{\alpha T^2}{\beta + T}$$

where

$E(0)$ is the energy at temperature $T = 0\,\text{K}$

α and β are the Varshni's thermal coefficients

The peak position of (D, X) in 4 nm ZnO quantum dots is blue-shifted by 5 meV from that in bulk ZnO due to the quantum-confinement of donor-bound excitons. However, acceptor-bound-exciton energies in 4 nm ZnO quantum dots decrease from the bulk value of about 10 meV at temperatures up to 70 K. This cannot be explained by the quantum-confinement model. This could be possible due to (1) lowering of the impurity potential near the quantum dot surface (Fonoberov and Balandin 2004c), (2) additional binding at low temperatures similar to that in a charged donor–acceptor pair (Look et al. 2002, Fonoberov and Balandin 2004c, Xiu et al. 2005). As the temperature increases, the intensity of donor-bound-exciton decreases and finally disappears at a high temperature. However, the acceptor-bound-exciton emission peak remains dominated up to room temperature. The blue-shift of the UV PL peak of ZnO quantum dots (4 nm) from that of bulk ZnO due to the quantum confinement effect is insignificant as the quantum confinement of acceptor-bound excitons in ZnO quantum dots does not induce the significant blue-shift of the UV emission peak of the acceptor-bound exciton because acceptors are the deep impurities for ZnO (Look et al. 2002, Fonoberov and Balandin 2004c, Xiu et al. 2005).

Zeng et al. (2007) studied the temperature-dependent violet-blue photoluminescence of Zn/ZnO core/shell nanoparticles. The temperature-dependent behavior of this violet-blue emission band observed in Zn/ZnO core/shell nanoparticles is quite different from that of the UV emission band and the green visible emission band commonly observed in various ZnO nanostructures (Wong and Searson 1999, Yang et al. 2001). The temperature dependence of the PL peak energy does not follow Varshni's law; it shows a red- blue shift with an increasing temperature. Zeng et al. (2007) explained this abnormal red-blue shift behavior with temperature following the localization model proposed by Li and co-workers (Li et al. 2005) that is represented by the following equations:

$$E(T) = E_0 - \frac{\alpha T^2}{\beta + T} - x k_B T,$$

$$x e^x = \left(\frac{\tau_r}{\tau_{tr}}\right)\left[\left(\frac{\sigma}{k_B T}\right)^2 - x\right] e^{\frac{(E_0 - E_a)}{k_B T}}$$

where

E_0 is the average value of the localized state levels

E_a represents a special energy level below which the localized states are occupied by the excitons at 0 K similar to the Fermi level in the Fermi–Dirac distribution function

σ is the standard deviation of the energy distribution width for the localized electronic state

k_B is the Boltzmann constant

τ_{tr} and τ_r are the carrier transfer time and the carrier recombination time, respectively

$x(T)$ is the temperature-dependent dimensionless coefficient

This violet-blue emission band originates from the electron–hole recombination between the localization defect level of the interstitial zinc and the valence band, and the red-blue shift behavior with an increasing temperature is a result of the competition between the electron localization effect at the zinc interstitial level and the temperature-induced band-gap shrinkage (Varshni 1967). There are very few reports on photoluminescence from the gas phase ZnO nanoparticles (Ozerov et al. 2005, Mohanta et al. 2008). Mohanta et al. (2008) observed the photoluminescence from ZnO nanoclusters in air by passing a fourth harmonic (266 nm) of an Nd:YAG laser referred to as the probe pulse through ZnO plasma created by the third harmonic (355 nm) of an Nd:YAG laser perpendicular to its expansion axis at various distances and at various time delays with respect to the ablating pulse (355 nm). The laser-ablated plasma consists of ions, electrons, and neutrals. These highly energetic plasma species expand in an ambient medium and collide with the molecules of the ambient (air) that results in a slowing down of the species inducing a rapid cooling of the plasma subsequently resulting in the formation of ZnO nanoclusters suspended in the vapor phase. Figure 6.15a shows the emission spectrum containing Zn I transition lines (Striganov 1968) at 330 nm (4s4d ³D–4s4p ³P), 334 nm (4s4d ³D–4s4p ³P), 468 nm (4s5s ³S–4s4p ³P), and 472 nm (4s5s ³S–4s4p ³P) at a 1 μs delay with respect to the ablating pulse (355 nm) without a passage of the probe pulse through the plasma. When the probe pulse (266 nm) is passed through the ZnO plasma at a 1 μs delay with respect to the ablating pulse (355 nm), a weak band is observed along with the Zn I transition lines as shown in Figure 6.15b. With an increase in the delay (>1 μs) of the probe pulse (266 nm) with respect to the ablating pulse (355 nm), the intensity of the band increases and the intensity of the Zn I transition lines decreases. At a delay of 5 μs, the Zn I transition lines are suppressed leaving only an emission band peaked at 3.229 eV as shown in Figure 6.16. This band with its typical asymmetric shape falls in the spectral region of the PL band of ZnO (Acquaviva et al. 2007, Mohanta and Thareja 2008b) and is attributed to the near band-edge excitonic recombination in ZnO clusters. These clusters are formed by cooling due to collisions of plasma species with the molecules of the ambient (Ozerov et al. 2005, Mohanta et al. 2008). The PL peak position is blue-shifted by 42 meV with respect to the PL peak position of the bulk ZnO (3.187 eV) and demonstrates the quantum-confinement effect. The FWHM of the PL profiles of

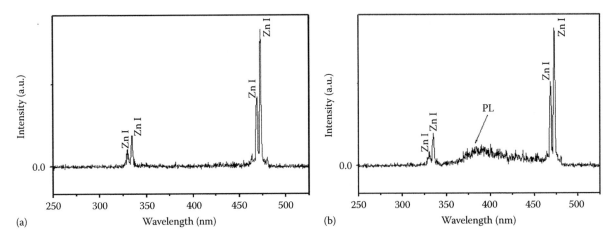

FIGURE 6.15 (a) Emission spectrum of Zn I lines at 330 nm (4s4d ^{3}D–4s4p ^{3}P), 334 nm (4s4d^{3}D–4s4p ^{3}P), 468 nm (4s5s ^{3}S–4s4p ^{3}P), and 472 nm (4s5s ^{3}S–4s4p ^{3}P) at a 1 μs delay with respect to the ablating pulse (355 nm) without a passage of the probe pulse (266 nm). (b) PL spectrum of ZnO nanoclusters along with Zn I transition lines at a 1 μs delay with respect to the ablating pulse (355 nm) with a passage of the probe pulse (266 nm).

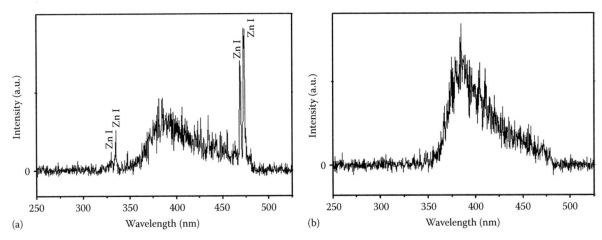

FIGURE 6.16 (a) PL spectrum of ZnO nanoclusters along with Zn I transition lines when a probe pulse is passed at a 1.5 μs delay with respect to the ablating pulse. (b) PL profile peaked at 3.229 eV when a probe pulse is passed at a 5 μs delay with respect to the ablating pulse.

the gas phase ZnO nanoparticles is larger than that of bulk ZnO. The PL peak position shows a red-shift with an increasing ablating intensity at a fixed probe intensity that is attributed to the temperature-induced band-gap shrinkage that arises due to an increase of the electron temperature with an increase in ablating intensity.

Laser emission from ZnO has also been observed (Thareja and Mitra 2000, Mitra and Thareja 2001, Mitra et al. 2001, Burin et al. 2002, Cao 2003). Figure 6.17 shows the evolution of the emission intensity with an increasing excitation intensity for a ZnO film of 1.5 μm thickness. There is a sharp rise in the output intensity above the threshold intensity of ~2.4 MW/cm². As the excitation intensity increases, the FWHM of the emission spectra decreases and above the threshold intensity, the emission spectra becomes 10 times or even more narrow than that below threshold. The emission spectrum becomes narrower due to preferential amplification at frequencies close to the maximum of the gain spectrum. Due to the local variation of the particle density and the spatial distribution in the film, there exist small regions of higher disorder and strong scattering and of lower disorder and

weaker scattering. Light can be confined in these regions forming closed-loop feedback paths through multiple scattering and interference (Wiersma 2000). Laser oscillations occur once the optical gain in a cavity exceeds the losses of a cavity. The various peaks observed in the emission spectrum (Figure 6.17a) are the cavity-resonant frequencies. The threshold excitation intensity is observed to depend on the excitation area. The lasing threshold-excitation intensity decreases with an increase of the excitation area and below a critical limit the laser oscillations are stopped. Laser emission in this case is observed in all directions, unlike the case of the conventional laser which has a well-defined cavity, and is hence referred to as the random laser.

6.11.5 Raman Spectroscopy

Raman spectroscopy has been a useful nondestructive spectroscopic technique to study the vibrational properties of ZnO nanostructures (Alim et al. 2005). The Raman scattering process involves the interaction of photons with the optical

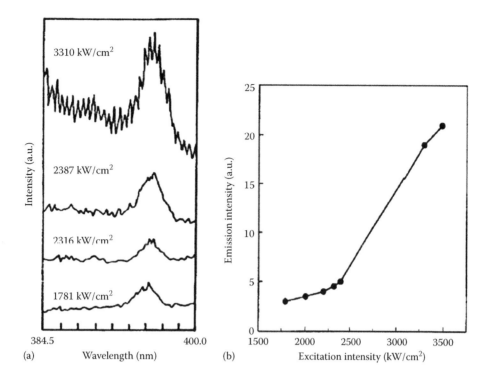

FIGURE 6.17   (a) Emission spectra from an optically pumped ZnO film of 1.5 µm thickness. (b) Variation of the peak intensity with an excitation intensity. (From Mitra, A. and Thareja, R.K., *J. Appl. Phys.*, 89, 2025, 2001. With permission.)

modes of the lattice vibration. ZnO nanoparticles have been characterized by both resonant and nonresonant scattering processes (Zhou et al. 2002, Wang et al. 2003). The longitudinal optical (LO) and transverse optical (TO) phonon frequencies are split into two frequencies with symmetries A_1 and E_1 due to the wurtzite crystal structure of ZnO (Alim et al. 2005). Besides these two longitudinal optical (LO) and transverse optical (TO) phonon modes, two additional nonpolar Raman-active phonon modes with symmetry E_2 exist in ZnO where the vibration of the Zn sublattice corresponds to the low frequency E_2 mode and the oxygen atoms are involved with the high frequency E_2 mode (Alim et al. 2005). However, in the case of ZnO nanoparticles, the Raman spectra show a shift from the phonon frequencies of the bulk. The origin of this shift is still under debate. Three main mechanisms have been suggested for the peak shift of phonon frequencies. They are spatial confinement within the boundaries of the nanocrystals, due defects that are responsible for the phonon localization, and the localized heating by the laser. Rajalakshmi et al. (2000) used the first mechanism (optical-phonon confinement) to explain the phonon frequency shifts in ZnO nanostructures. However, Fonoberov and Balandin (2004a,b) had theoretically shown that the mechanism related to optical phonon confinement cannot be applicable for ionic ZnO quantum dots of sizes larger than 4 nm (Alim et al. 2005). In order to have a clear understanding of the above concept of the phonon frequency shift in ZnO nanostructures, Alim et al. studied both resonant and nonresonant Raman spectroscopy of ZnO quantum dots with a diameter of 20 nm and bulk ZnO. They concluded that the first two mechanisms cause only a few cm^{-1} shifts of phonon frequencies and the third mechanism, laser-induced heating, causes a peak shift as large as tens of cm^{-1}.

6.12 Applications of ZnO

ZnO is a potential candidate of futuristic optoelectronic devices (laser diode and light-emitting diode) in the UV range due to its wide direct band-gap (~3.37 eV) (Özgür et al. 2005). Due to the high sensitivity of the surface conductivity of ZnO to various gases, it can be used for gas sensors (Comini et al. 2002). ZnO nanostructures can also be used as field emitters due to the strong enhancement of the electric field (Wan et al. 2003). Besides these, it is useful for liquid crystal displays (Oh et al. 2006), solar cells (Caputo et al. 1997), and transparent thin film transistors (Hoffmann et al. 2003). ZnO, due to its strong influences on the vulcanization process (Brown 1957, 1976) has been used as an additive to rubber for the fabrication of tires of cars. ZnO has been mixed in concrete in order to achieve a high resistance of concrete against water (Brown 1957). It is also used as sunscreen lotion to block UV radiations, talcum powder that absorbs moisture, and as varistors (Chen et al. 1997).

Acknowledgments

This work is partly supported by the Department of Science and Technology, New Delhi. The authors thank Drs. M. K. Harbola and Monica Katiyar for their critical review of this chapter.

References

Acquaviva, S., D'Anna, E., and De Giorgi, M. L. 2007. Atomic and molecular emissions of the laser-induced plasma during zinc and zinc oxide target ablation. *J. Appl. Phys.* 102: 073109-1–073109-7.

Alim, K. A., Fonoberov, V. A., and Balandin, A. A. 2005. Origin of the optical phonon frequency shifts in ZnO quantum dots. *Appl. Phys. Lett.* 86: 053103-1–053103-3.

Amekura, H., Plaksin, O. A., Umeda, N., Takeda, Y., Kishimoto, N., and Buchal, Ch. 2006. A short review and present status of ZnO nanoparticles formation by ion implantation combined with thermal oxidation. *MRS Proceeding* 908E: OO8.1.1–OO8.1.6.

Bahnemann, D. W., Kormann, C., and Hoffmann, M. R. 1987. Preparation and characterization of quantum size zinc oxide: A detailed spectroscopic study. *J. Phys. Chem.* 91: 3789–3798.

Brown, H. E. 1957. *Zinc Oxide Rediscovered.* New York: The New Jersey Zinc Company.

Brown, H. E. 1976. *Zinc Oxide, Properties and Application.* New York: The New Jersey Zinc Company.

Brus, L. E. 1984. Electron-electron and electron–hole interactions in small semiconductor crystallites: The size dependence of the lowest excited electronic state. *J. Chem. Phys.* 80: 4403–4409.

Burin, A. L., Ratner, M. A., Cao, H., and Chang, S. H. 2002. Random laser in one dimension. *Phys. Rev. Lett.* 88: 093904-1–093904-4.

Byrappa, K., Ohara, S., and Adschiri, T. 2008. Nanoparticles synthesis using supercritical fluid technology—Towards biomedical applications. *Adv. Drug Deliv. Rev.* 60: 299–327.

Cai, J. and Chen, N. 2007. First-principles study of the wurtzite-to-rocksalt phase transition in zinc oxide. *J. Phys.: Condens. Matter* 19: 266207–266218.

Cao, H. 2003. Lasing in random media. *Waves Random Media* 13: R1–R39.

Caputo, D., Forghieri, U., and Palma, F. 1997. Low-temperature admittance measurement in thin film amorphous silicon structures. *J. Appl. Phys.* 82: 733–741.

Chen, C. S., Kuo, C. T., Wu, T. B., and Lin, I. N. 1997. Microstructures and electrical properties of V_2O_5-based multi-component ZnO varistors prepared by microwave sintering process. *Jpn. J. Appl. Phys.* 36: 1169–1175.

Coldren, L. A. and Corzine, S. W. 1995. *Diode Lasers and Photonic Integrated Circuits.* New York: John Wiley & Sons, Inc.

Comini, E., Fagila, G., Sberveglieri, G., Pan, Z., and Wang, Z. L. 2002. Stable and highly sensitive gas sensors based on semiconducting oxide nanobelts. *Appl. Phys. Lett.* 81: 1869–1871.

Damonte, L. C., Mendoza Zélis, L. A., Soucase, B. M., and Hernández Fenollosa, M. A. 2004. Nanoparticles of ZnO obtained by mechanical milling. *Powder Technol.* 148: 15–19.

Feynman, R. P., Leighton, R. B., and Sands, M. L. 1964. *The Feynman Lectures on Physics.* New York: Addison-Wesley.

Fonoberov, V. A. and Balandin, A. A. 2004a. Interface and confined optical phonons in wurtzite nanocrystals. *Phys. Rev. B* 70: 233205-1–233205-4.

Fonoberov, V. A. and Balandin, A. A. 2004b. Interface and confined polar optical phonons in spherical ZnO quantum dots with wurtzite crystal structure. *Phys. Stat. Sol. (c)* 1: 2650–2653.

Fonoberov, V. A. and Balandin, A. A. 2004c. Origin of ultraviolet photoluminescence in ZnO quantum dots: Confined excitons versus surface-bound impurity exciton complexes. *Appl. Phys. Lett.* 85: 5971–5973.

Fonoberov, V. A., Alim, K. A., and Balandin, A. A. 2006. Photoluminescence investigation of the carrier recombination processes in ZnO quantum dots and nanocrystals. *Phys. Rev. B* 73: 165317-1–165317-9.

Fukuda, M. 1998. *Optical Semiconductor Devices.* New York: Wiley & Sons.

Gfroerer, T. H. 2000. Photoluminescence in analysis of surfaces and interfaces, *Encyclopedia of Analytical Chemistry*, ed. R. A. Meyers, pp. 9209–9231. New York: John Wiley & Sons Ltd.

Giri, P. K., Bhattacharyya, S., Singh, D. K., Kesavamoorthy, R., Panigrahi, B. K., and Nair, K. G. M. 2007. Correlation between microstructure and optical properties of ZnO nanoparticles synthesized by ball milling. *J. Appl. Phys.* 102: 093515-1–093515-8.

Guo, L., Yang, S., Yang, C. et al. 2000. Highly monodisperse polymer-capped ZnO nanoparticles: Preparation and optical properties. *Appl. Phys. Lett.* 76, 2901–2903.

He, C., Sasaki, T., Usui, H., Shimizu, Y., and Koshizaki, N. 2007. Fabrication of ZnO nanoparticles by pulsed laser ablation in aqueous media and pH-dependent particle size: An approach to study the mechanism of enhanced green photoluminescence. *J. Photochem. Photobiol. A: Chem.* 191: 66–73.

Hoffmann, R. L., Norris, B. J., and Wager, J. F. 2003. ZnO-based transparent thin-film transistors. *Appl. Phys. Lett.* 82: 733–735.

Hoyer, P. and Weller, H. 1994. Size-dependent redox potentials of quantized zinc oxide measured with an optically transparent thin layer electrode. *Chem. Phys. Lett.* 221: 379–384.

Kayanuma, Y. 1988. Quantum-size effects of interacting electrons and holes in semiconductor microcrystals with spherical shape. *Phys. Rev. B* 38: 9797–9805.

Klingshirn, C. 2007. ZnO: From basics towards applications. *Phys. Stat. Sol. (b)* 244: 3027–3073.

Koch, U., Fojtik, A., Weller, H., and Henglein, A. 1985. Photochemistry of semiconductor colloids. Preparation of extremely small ZnO particles, fluorescence phenomena and size quantization effects. *Chem. Phys. Lett.* 122: 507–510.

Li, Q., Xu, S. J., Xie, M. H., and Tong, S. Y. 2005. Origin of the S-shaped temperature dependence of luminescent peaks from semiconductors. *J. Phys.: Condens. Matter* 17: 4853–4858.

Look, D. C. 2001. Recent advances in ZnO materials and devices. *Mater. Sci. Eng. B* 80: 383–387.

Look, D. C., Reynolds, D. C., Litton, C. W., Jones, R. L., Eason, D. B., and Cantwell, G. 2002. Characterization of homoepitaxial p-type ZnO grown by molecular beam epitaxy. *Appl. Phys. Lett.* 81: 1830–1832.

Mahamuni, S., Borgohain, K., Bendre, B. S., Leppert, V. J., and Risbud, S. H. 1999. Spectroscopic and structural characterization of electrochemically grown ZnO quantum dots. *J. Appl. Phys.* 85: 2861–2865.

Mang, A., Reimann, K., and Rübenacke, St. 1995. Band gaps, crystal-field splitting, spin-orbit coupling, and exciton binding energies in ZnO under hydrostatic pressure, *Solid State Commun.* 94: 251–254.

Mitra, A. and Thareja, R. K. 2001. Photoluminescence and ultraviolet laser emission from nanocrystalline ZnO thin films. *J. Appl. Phys.* 89: 2025–2028.

Mitra, A., Thareja, R. K., Ganesan, V., Gupta, A., Sahoo, P. K., and Kulkarni, V. N. 2001. Synthesis and characterization of ZnO thin films for UV laser. *Appl. Surf. Sci.* 174: 232–239.

Mohanta, A. and Thareja, R. K. 2008a. Photoluminescence study of ZnO nanowires grown by thermal evaporation on pulsed laser deposited ZnO buffer layer. *J. Appl. Phys.* 104: 044906-1–044906-6.

Mohanta, A. and Thareja, R. K. 2008b. Photoluminescence study of ZnCdO alloy. *J. Appl. Phys.* 103: 024901-1–024901-5.

Mohanta, A., Singh, V., and Thareja, R. K. 2008. Photoluminescence from ZnO nanoparticles in vapor phase. *J. Appl. Phys.* 104: 064903-1–064903-6.

Nakamura, S., Pearton, S., and Fasol, G. 1997. *The Blue Laser Diode.* New York: Springer.

Narayan, J., Sharma, A. K., and Muth, J. F. 2002. U.S. Patent No. 6,423,983, B1.

Narayan, J., Sharma, A. K., and Muth, J. F. 2003. U.S. Patent No. 6,518,077: licensed by Kopin Corp.

Oh, B. Y., Jeong, M. C., Moon, T. H., Lee, W., Myoung, J. M., Hwang, J. Y., and Seo, D. S. 2006. Transparent conductive Al-doped ZnO films for liquid crystal displays. *J. Appl. Phys.* 99: 124505.

Ou, Q., Shinji, K., Ogino, A., and Nagatsu, M. 2008. Enhanced photoluminescence of nitrogen-doped ZnO nanoparticles fabricated by Nd: YAG laser ablation. *J. Phys. D: Appl. Phys.* 41: 205104-1–205104-5.

Ozerov, I., Bulgakov, A. V., Nelson, D. K., Castell, R., and Marine, W. 2005. Production of gas phase zinc oxide nanoclusters by pulsed laser ablation. *Appl. Surf. Sci.* 247: 1–7.

Özgür, Ü., Alivov, Ya. I., Liu, C. et al. 2005. A comprehensive review of ZnO materials and devices. *J. Appl. Phys.* 98: 041301-1–041301-103.

Pankove, J. I. 1975. *Optical Processes in Semiconductor.* Englewood Cliffs, NJ: Prentice-Hall, Inc.

Rajalakshmi, M., Arora, A. K., Bendre, B. S., and Mahamuni, S. 2000. Optical phonon confinement in zinc oxide nanoparticles. *J. Appl. Phys.* 87: 2445–2448.

Rama Krishna, M. V. and Friesner, R. A. 1991. Quantum confinement effects in semiconductor clusters. *J. Chem. Phys.* 95: 8309–8322.

Ramakrishna, G. and Ghosh, H. N. 2003. Effect of particles size on the reactivity of quantum size ZnO nanoparticles and charge transfer dynamics with adsorbed catechols, *Langmuir* 19: 3006–3012.

Reetz, M. T. and Helbig, W. 1994. Size-selective synthesis of nanostructured transition metal clusters. *J. Am. Chem. Soc.* 116: 7401–7402.

Rossetti, R., Nakahara, S., and Brus, L. E. 1983. Quantum size effects in the redox potentials, resonance Raman spectra, and electronic spectra of CdS crystallites in aqueous solution. *J. Chem. Phys.* 79: 1086–1088.

Saleh, B. E. A. and Teich, M. C. 1991. *Fundamental of Photonics.* New York: John Wiley & Sons, Inc.

Sarigiannis, D., Peck, J. D., Mountziaris, T. J., Kioseoglou, G., and Petrou, A. 2000. Vapor phase synthesis of II-VI semiconductor nanoparticles in a counter flow jet reactor. *MRS Proceeding* 616: 41–46.

Schröer, P., Krüger, P., and Pollmann, J. 1993. First-principles calculation of the electronic structure of the wurtzite semiconductors ZnO and ZnS. *Phys. Rev. B* 47: 6971–6980.

Shan, W., Walukiewicz, W., Ager III, J. W. et al. 2005. Nature of room-temperature photoluminescence in ZnO. *Appl. Phys. Lett.* 86: 191911-1–191911-3.

Spanhel, L. and Anderson, M. A. 1991. Semiconductor clusters in the Sol-Gel process: quantized aggregation, gelation, and crystal growth in concentrated ZnO colloids. *J. Am. Chem. Soc.* 113: 2826–2833.

Spanhel, L., Weller, H., and Henglein, A. 1987. Photochemistry of semiconductor colloids. 22. Electron Injection from Illuminated CdS into Attached TiO_2 and ZnO Particles. *J. Am. Chem. Soc.* 109: 6632–6635.

Striganov, A. R. 1968. *Tables of Spectral Lines of Neutral and Ionized Atoms.* Moscow, Russia: Commission on spectroscopy of the Academy of sciences of the USSR.

Teke, A., Özgür, Ü., Doğan, S. et al. 2004. Excitonic fine structure and recombination dynamics in single-crystalline ZnO. *Phys. Rev. B* 70: 195207-1–195207-10.

Thareja, R. K. and Mitra, A. 2000. Random laser action in ZnO. *Appl. Phys. B* 71: 181–184.

Thareja, R. K. and Shukla, S. 2007. Synthesis and characterization of zinc oxide nanoparticles by laser ablation of zinc in liquid. *Appl. Surf. Sci.* 253: 8889–8895.

Thareja, R. K., Saxena, H., and Narayanan, V. 2005. Laser ablated ZnO for thin films of ZnO and $Mg_xZn_{(1-x)}$ O. *J. Appl. Phys.* 98: 034908–034917.

Varshni, Y. P. 1967. Temperature dependence of the energy gap in semiconductors. *Physica* 34: 149–154.

Wan, Q., Yu, K., Wang, T. H., and Lin, C. L. 2003. Low-field electron emission from tetrapod-like ZnO nanostructures synthesized by rapid evaporation. *Appl. Phys. Lett.* 83: 2253–2255.

Wang, Z., Zhang, H., Zhang, L., Yuan, J., Yan, S., and Wang, C. 2003. Low temperature synthesis of ZnO nanoparticles by solid-state pyrolytic reaction. *Nanotechnology* 14: 11–15.

Wiersma, D. 2000. Laser Physics: The smallest random laser. *Nature* 406: 132–133.

Wong, E. M. and Searson, P. C. 1999. ZnO quantum particle thin films fabricated by electrophoretic deposition. *Appl. Phys. Lett.* 74: 2939–2941.

Wong, E. M., Bonevich, J. E., and Searson, P. C. 1998. Growth kinetics of nanocrystalline ZnO particles from colloidal suspensions. *J. Phys. Chem. B* 102: 7770–7775.

Xiu, F. X., Yang, Z., Mandalapu, L. J., Zhao, D. T., Liu, J. L., and Beyermann, W. P. 2005. High-mobility Sb-doped p-type ZnO by molecular-beam epitaxy. *Appl. Phys. Lett.* 87: 152101-1–152101-3.

Yang, C. L., Wang, J. N., Ge, W. K. et al. 2001. Enhanced ultraviolet emission and optical properties in polyvinyl pyrrolidone surface modified ZnO quantum dots. *J. Appl. Phys.* 90: 4489–4493.

Yang, R. D., Tripathy, S., Li, Y., and Sue, H. J. 2005. Photoluminescence and micro-Raman scattering in ZnO nanoparticles: The influence of acetate adsorption. *Chem. Phys. Lett.* 411: 150–154.

Zeng, H., Cai, W., Hu, J., Duan, G., Liu P., and Li, Y. 2006. Violet photoluminescence from shell layer of Zn/ZnO core-shell nanoparticles induced by laser ablation. *Appl. Phys. Lett.* 88: 171910-1–171910-3.

Zeng, H., Li, Z., Cai, W., and Liu, P. 2007. Strong localization effect in temperature dependence of violet-blue emission from ZnO nanoshells. *J. Appl. Phys.* 102: 104307-1–104307-4.

Zhou, H., Alves, H., Hofmann, D. M., Kriegseis, W. et al. 2002. Behind the weak excitonic emission of ZnO quantum dots: ZnO/Zn(OH)$_2$ core-shell structure. *Appl. Phys. Lett.* 80: 210–212.

Tetrapod-Shaped Semiconductor Nanocrystals

Roman Krahne
Italian Institute of Technology

Liberato Manna
Italian Institute of Technology

7.1 Introduction

Nanoscience promises innovative solutions in a large variety of sectors, ranging from cost-effective optoelectronic devices to energy generation to highly performing materials and interfaces. One of the most studied building blocks of nanoscience are colloidal inorganic nanocrystals, since their properties and interparticle interactions can be controlled on a high level by tailoring their size, composition, and surface functionalization. Indeed, semiconductor, metal, and magnetic nanocrystals have been already applied in biological and biomedical research (i.e., fluorescent of magnetic tagging, hyperthermia, and biosensing), electro-optical devices such as light-emitting diodes and lasers, photovoltaic cells, catalysis and gas sensing. This trend has been possible via breakthrough advances in the wet-chemical syntheses and assembly of robust and easily processable nanocrystals of a wide range of materials, sizes, and shapes. Also, the design of architectures of such nanocrystals constructed by self-assembly has been investigated, as assemblies represent new materials on which chemical and physical interactions among nanocrystals can be investigated.

Several branched nanocrystals have been also reported by many groups, and one peculiar shape occurring in several inorganic nanocrystals is the tetrapod, which basically consists of a nanocrystal in which four arms are joined together at a central region and protrude from it at roughly tetrahedral angles (see Figure 7.1). This shape has been observed in many semiconductor nanocrystals of the II–VI group, like ZnO (Kitano et al.,

1991; Fujii et al., 1993; Takeuchi et al., 1994; Nishio et al., 1997; Iwanaga et al., 1998; Dai et al., 2003; Yan et al., 2003; Chen et al., 2004; Wang et al., 2005; Yu et al., 2005), ZnSe (Hu et al., 2005), ZnS (Zhu et al., 2003), CdS (Jun et al., 2001; Chen et al., 2002; Shen and Lee, 2005), CdSe (Manna et al., 2000; Peng and Peng, 2001), CdTe (Bunge et al., 2003; Manna et al., 2003; Yu et al., 2003; Zhang and Yu, 2006), and $CdSe_xTe_{1-x}$ alloys (Li et al., 2006). Recently, II–VI semiconductor tetrapods have been fabricated, in which the central "core" region and the arms were made of two different types of II–VI semiconductors, such as ZnTe/CdSe (Xie et al., 2006), ZnTe/CdS (Xie et al., 2006; Carbone et al., 2007), ZnSe/CdS (Carbone et al., 2007), and CdSe/CdS (Talapin et al., 2007a) tetrapods (here the first compound denotes the material of the core and the second that of the arm). The tetrapod shape has been observed additionally in other materials [such as Au (Chen et al., 2003), iron oxide (Cozzoli et al., 2006), CoO (Zhang et al., 2007), PbSe (Na et al., 2008), and others].

Tetrapod-shaped nanocrystals have attracted considerable interest in the last years due to their optical/electronic (Pang et al., 2005; Peng et al., 2005; Wang, 2005; Malkmus et al., 2006; Tarì et al., 2006; Al Salman et al., 2007; Nobile et al., 2007) and mechanical properties (Fang et al., 2007), chemical reactivity (Liu and Alivisatos, 2004; Mokari et al., 2004), and hence for their potential applications in fields such as photovoltaics (Sun et al., 2003; Zhou et al., 2006; Gur et al., 2007; Zhong et al., 2007), single nanoparticle transistors (Cui et al., 2005), electromechanical devices (Fang et al., 2007), and recently also in scanning probe microscopy (Nobile et al., 2008). This chapter reviews various

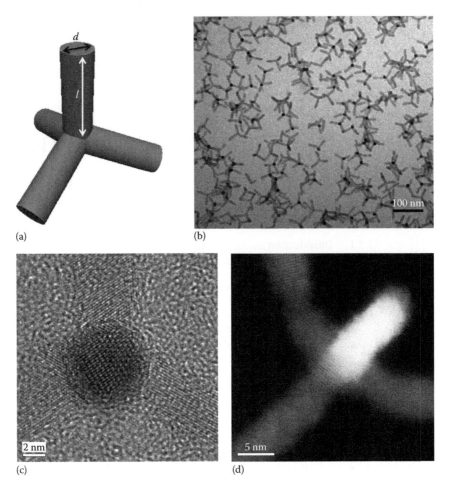

FIGURE 7.1 (a) A model of a tetrapod in which the arms are built as cylinders. (b) A low-resolution transmission electron microscopy image of several tetrapod-shaped nanocrystals having cadmium telluride (CdTe) arms and deposited on a thin amorphous carbon film. (Adapted from Fiore, A. et al., *J. Am. Chem. Soc.*, 131(6), 2274, 2009. With permission.) (c) A "phase contrast" (Williams and Carter, 2004) high-resolution TEM image of a single CdTe tetrapod taken by having the electron beam aligned with the tetrapod arm that is pointing upward. The other three arms (i.e., those touching the substrate), can be also seen. (Adapted from Fiore, A. et al., *J. Am. Chem. Soc.*, 131(6), 2274, 2009. With permission.) (d) A "high angle angular dark field" (Williams and Carter, 2004) TEM image of a tilted CdTe tetrapod taken in scanning mode (STEM). Here the bright regions come from the heavy atoms that belong to the nanocrystal, while the dark regions are areas where no heavy atoms are present (here one the carbon atoms of the supporting carbon film are present). In the image, the brighter regions are those of the arm pointing upward. On the top side of this image, one can also see the tip of an arm that belongs to another tetrapod.

aspects connected with tetrapod-shaped nanocrystals based on semiconductors and synthesized by chemical approaches in the liquid phase. In practice, we will focus on the II–VI class of semiconductors, as most syntheses, studies, and applications so far have been limited to these materials, and also because for these materials, the quantum confinement resulting from the tetrapod shape has an impact on their physical properties.

The organization of the chapter is as follows: we will first explain some basic concepts of crystal structures and defects, which will be useful for a discussion of the structural models that rationalize the tetrapod shape in semiconductors. We will then give a brief overview of the synthesis routes to these nanomaterials. In connection to this, we will try to explain the main mechanisms according to which the growth of tetrapods takes place. We will then discuss the various properties of tetrapods (optical, electron transport, and mechanical) and how these have

been exploited so far in various potential applications. Assembly of tetrapods will be also reviewed briefly. We will close the chapter with an outlook on these materials.

7.2 Structural Models and Synthetic Approaches

7.2.1 A Few Useful Crystallographic Concepts

7.2.1.1 Hexagonal Close-Packed and Face Cantered Cubic (fcc) Lattices

In order to understand the structural models of tetrapods, some basic concepts of crystal structures and of planar defects need to be introduced. This will help us to understand better also the various properties of tetrapods. Of primary relevance for our discussion is a detailed description of the wurtzite and sphalerite crystal structures, in which tetrapods of II–VI semiconductors form. A way of understanding the similarities and the differences between the wurtzite and the sphalerite structures is by looking at their lattices as if they were built by close-packed arrangements of hard spheres. Let us first focus therefore on describing how such arrangements can be realized.

Lattices based on close-packed arrangements of spheres can be built up according to the reasoning that follows and which is described graphically in Figure 7.2. First of all, a single close-packed layer of spheres (which we shall call "A") can be realized by placing each sphere in contact with six others and so on. This layer may serve either as the basal (001) plane of the *hexagonal close packed* (*hcp*) structure or as the (111) plane of the *face centered cubic* (*fcc*) structure. A second layer "B" of spheres can be only assembled in a close-packed configuration by placing each sphere of this layer in contact with three spheres of the bottom "A" layer, such that each sphere of the "B" layer actually sits right on the top of a hole created by three underneath touching spheres of the "A" layer. The third layer "C" may be added in two ways. We will obtain the *fcc* structure if the spheres of the third layer are added such that in projection they are sitting over the holes of the first layer that are not occupied by the spheres of the "B" layer. Overall, when such sequence is repeated, this will

correspond therefore to an "ABCABC" stacking of planes (see Figure 7.2a through d, here, and in all the figures that follow, the spheres are actually not in contact with each other in order to make the drawings easier to understand). We will obtain instead the *hcp* structure when the spheres in the third layer are placed in projection directly over the centers of the spheres of the first "A" layer. In this case, the "A" and "C" layers will be equivalent, and when this arrangement will be repeated, it will correspond to an "ABAB" stacking of planes. Lattices based on such two possible types of arrangements, and their associated unit cells, are displayed in Figure 7.2e through h. Here a close-packed "ABCABC" type of lattice could be, for example, that of metallic gold, whereas a "ABAB" type of lattice could be that of metallic cobalt.

7.2.1.2 Sphalerite and Wurtzite Crystal Structures

Both the cubic sphalerite and the hexagonal wurtzite structures can be understood as each being composed of two interpenetrating sublattices, one made of anions and the other made of cations, respectively (see Figure 7.3). In the sphalerite structure, in each sublattice, there is a "ABCABC" stacking sequence of atoms along any of its 111 directions. In the wurtzite structure, on the other hand, the stacking sequence for both cation and anion

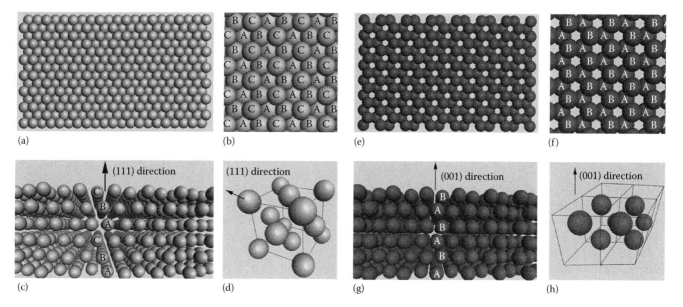

(a) (b) (e) (f)

(c) (d) (g) (h)

FIGURE 7.2 Two possible ways or realizing a close-packed assembly of hard spheres. They are shown in this figure and are related to the *fcc* (panels a through d) and *hcp* (panels e through h) crystal structures, respectively. In panel (a), three layers of close-packed hard spheres are seen from the top. Panel (b) represents a magnified view of the same structure, with each sphere labeled according to the layer to which it belongs. The spheres of the bottom layer are indicated by A. On top of this layer, a second layer of close-packed spheres is deposited (these are labeled as B). Once this second layer is in place, a third layer is deposited. There are two choices for placing these spheres: either as shown in panel (b) or alternatively on sites such that in projection they hide the "A" spheres. In the first case this arrangement would lead to a sequence of stacking of planes of "ABC" type. If repeated ("ABCABC..."), this would represent the *fcc* crystal structure. Such sequence of planes can also be seen from a "side view" in panel (c). The crystallographic direction of the *fcc* crystal structure along which this stacking of planes is realized is the (111) direction. In panel (c), the unit cell of the *fcc* structure is reported and the (111) direction is highlighted. If, on the other hand, the sequence of planes is such that the third layer of spheres is exactly projected on the "A" layer of spheres, such as indicated in panels (e) and (f), and this sequence is repeated (ABAB...), the *hcp* crystal structure is obtained. Such sequence of planes can be seen also from a "side view" in panel (g). The crystallographic direction of the *hcp* crystal structure along which this stacking of planes is realized is the (001) direction. In panel (c), four adjacent unit cells of the *hcp* structure are displayed and the (001) direction is highlighted.

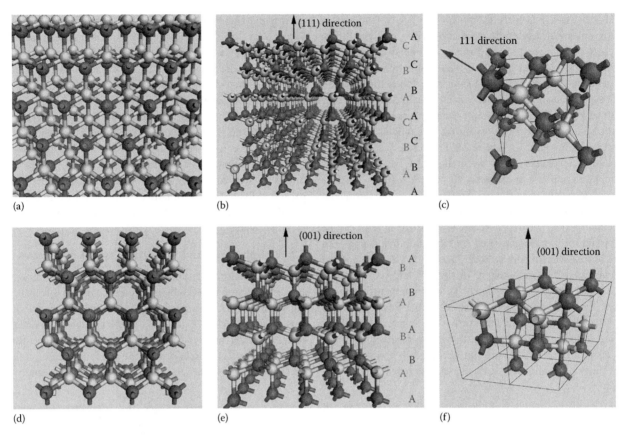

FIGURE 7.3 The crystalline structures of wurtzite and sphalerite are binary, that is, they are composed of two interpenetrating sublattices. Each sublattice is made of one type of atom (either Cd or Se) and is assembled either in an ABC or an AB stacking sequence. The relative arrangement of the two sublattices is shown in panels (a through c) for an ABC stacking sequence (which describe the sphalerite structure) and in panels (d through f) for an AB stacking sequence (which describe the wurtzite structure). In both cases, the overall structure can be built by placing each atom of the second sublattice just on top of an "A" site of the first sublattice, so that it would form one bond with the underlying "A" atom of the first sublattice and three bonds with three nearest neighboring atoms in the layer above (again belonging to the first sublattice). Notice that in the wurtzite structure, the two opposite directions along the AB stacking sequence (the *c* axis of the structure) are not equivalent. Hence, there is no plane of symmetry perpendicular to the *c* axis in this structure.

sublattices is "ABAB" along the 001 direction. Because of the lower symmetry of the wurtzite structure with respect to sphalerite, there will be many facets of the wurtzite crystals that will be crystallographically different from each other. Additionally, in the wurtzite structure, the 001 axis (i.e., the *c axis*) has a three-fold rotational symmetry, but there is no plane of symmetry perpendicular to it. This axis is therefore polar and one can define a direction of polarity along this axis. The lack of inversion symmetry along such axis has interesting implications on the growth of wurtzite nanocrystals, as growth rates along the 001 direction and the $00\bar{1}$ direction can be significantly different, as we will discuss later in detail (Shiang et al., 1995). On the basis of the same reasoning, also in sphalerite crystals, all four (111) axes are polar (the sphalerite phase too does not have a center of symmetry).

There are close similarities between the wurtzite and the sphalerite structures. With respect to any atom of the lattice, the nearest neighboring atoms have exactly the same arrangement in both structures (in both structures, there is tetrahedral coordination), whereas differences in the relative positions of

atoms arise only when comparing second neighboring atoms. One can find similarities and differences between the wurtzite and the sphalerite structures also by looking at the arrangements of atoms and bonds at the various crystal facets. This is better shown in Figure 7.4, in which also the four-index Miller–Bravais notation for hexagonal systems is introduced for indexing the various wurtzite facets (see bottom part of panel e), instead of the more conventional Miller notation (the reader can find the explanation of this notation in the caption of Figure 7.4) (Hurlbut et al., 1998; Williams and Carter, 2004). Henceforth, we will use such four index notations whenever dealing with wurtzite crystals. In a sphalerite crystal, four of the eight (111) facets are equivalent to the (0001) facet of the wurtzite structure, while the remaining four (111) facets are equivalent to the $(000\bar{1})$ facet of wurtzite, both in terms of atomic arrangements at the surface and of dangling bonds, as can be seen in Figure 7.4a through c. Therefore, both in the wurtzite and sphalerite structures, there is the possibility that between these two groups of four equivalent facets, differences in chemical reactivity and growth rates arise under suitable conditions.

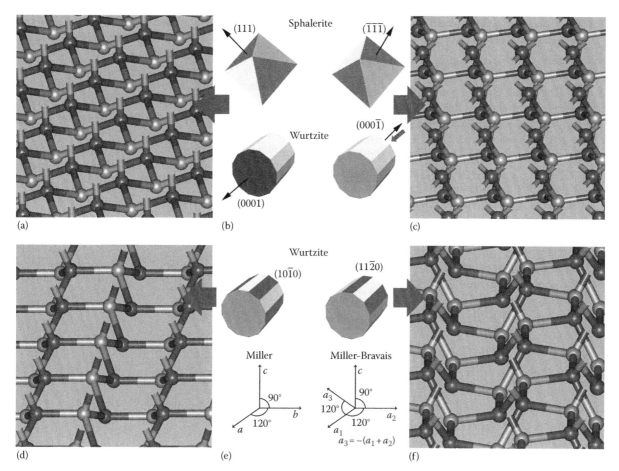

FIGURE 7.4 Similarities and differences in the arrangement of surface atoms between wurtzite and sphalerite crystals. Panel (b) shows models of an octahedral-shaped sphalerite crystal, terminated by the eight (111) facets, and of a prism-shaped wurtzite crystal, terminated by the prismatic (10$\bar{1}$0) and (11$\bar{2}$0) types of facets and by the basal (0001) and (000$\bar{1}$) facets. Four of the eight (111) facets of sphalerite are identical to the (0001) facet of sphalerite. These facets are all painted in dark gray in the models of the left side of panel (b). The corresponding arrangement of atoms on the surface is shown in panel (a). If we assume that the cations here are those colored in dark gray, then these types of facets expose alternating layers of cations (each carrying one dangling bond) and anions (each carrying three dangling bonds). On these, cations and anions are never present together, and therefore the facets, as a consequence of the dangling bonds, have a net residual charge (either positive or negative). These facets are therefore "polar." The other four facets of the sphalerite structure are, on the other hand, equivalent to the (000$\bar{1}$) facet of the wurtzite structure, see right side of panel (b) and also panel (c) for the arrangement of atoms and dangling bonds on the facets. These types of facets expose alternating layers of cations (each carrying three dangling bond this time) and anions (each carrying one dangling bonds this time). Therefore also, these facets are polar, and we can now see how different they are from the previous group of facets. Panels (d) and (f) show, on the other hand, the arrangement of atoms and dangling bonds for two types of facets that are present only in the wurtzite structure. These are the (10$\bar{1}$0) and the (11$\bar{2}$0) facet [panels (d) and (f), respectively], and they are also shown on the prism-shaped wurtzite crystal as the gray facets [panel (e)]. Two interesting observations can be made on these types of facets. First of all, they are nonpolar, as in each alternating layer of atoms that can be exposed on these facets there are both cations and anions, and in equal numbers. Second, cations and anions form six-member rings of atoms that are arranged in "boat" conformations. These "boat" conformations are not found in the sphalerite structure. In such conformations, the distances between certain cation–anion couples [for instance, those at the opposite sides of each "boat" in panel (d)] are shorter than those found on "chair" conformations, such as those shown in panel (a) for atoms arranged in six-member rings. The "chair" conformation is the only type of conformation seen in sphalerite crystals, whereas wurtzite has both chairs and boats. In the case of more ionic types of lattices (as in most II–VI semiconductors), the wurtzite structure is more stable than the sphalerite structure as the "boat" arrangements in the lattice bring third neighbor cations and anions a little bit closer to each other, thus contributing to lower the lattice energy. In more covalent lattices, on the other hand (like for most of the III–V semiconductors), atoms tend to stay as far away as possible from their second and third neighbors. Therefore, subject to constraints of bond lengths and of tetrahedral coordination, they prefer the sphalerite structure. The lower part of panel (e) briefly shows for an hexagonal structure (such as the wurtzite) the difference between the Miller notation, which is based on the three conventional a, b, and c crystallographic axes, and the more extensively used Miller–Bravais notation, which is based on four axes: a_1, a_2, a_3, and c. In this latter notation, the first three axes are all laying on one plane, and therefore are not linearly independent. Indeed $a_3 = -(a_1 + a_2)$. The c axis is the same in both notations. Therefore, the third index in the four-index notation is equal to the inverted sum of the first two indexes. As an example, the (110) facet in the conventional notation would be the (11$\bar{2}$0) facet in the four-index notation, or the (1$\bar{1}$3) facet in the conventional notation would be the (1$\bar{1}$03) facet in the four index notation.

7.2.1.3 Deviation from Real Wurtzite Structure and Intrinsic Dipole Moment

This section shortly discusses a simple model describing the emergence of an intrinsic electric dipole moment in wurtzite crystals, which is relevant for the discussion of the optical and electronic properties of rod and tetrapod-shaped nanocrystals, and which can be easily explained by structural considerations. The wurtzite crystal structure that we have described in Section 7.2.1.2 is an idealized structure, in the sense that actual "wurtzite" crystals form in a phase that differs slightly from this ideal structure. Let us discuss in more detail this concept. Figure 7.5a shows the "ideal" wurtzite cell and how all the atoms are arranged in a perfect tetrahedral coordination, in which all bonds are exactly of the same length ℓ and all bond angles are $\theta = 109.47°$. In this case, it is possible to show by simple geometric considerations that the lattice constants a and c are related to the bond length ℓ by the expressions $a = \sqrt{8/3}\,\ell$ and $c = 8/3\ell$, so that the ratio of the two lattice constants is $c/a = \sqrt{8/3}$. In terms of the parameter $u = 3/8$, these expressions can be written as

$$a = \frac{\ell}{\sqrt{u}}, \quad c = \frac{\ell}{u}, \quad \frac{c}{a} = \frac{1}{\sqrt{u}} = 1.633 \tag{7.1}$$

In a real solid crystallizing in the wurtzite structure, however, the parameter u is never exactly equal to 3/8 and therefore the c/a ratio is not equal to 1.633, but slightly smaller or larger than this value (this can be estimated experimentally with a high degree of accuracy from x-ray powder diffraction data). In other words, a real unit cell will be either a little bit squeezed or a little bit pulled along the c direction, as a consequence of deviation of the bonding geometry of atoms from the perfect tetrahedral coordination. This deformation leads to the emergence of an electric dipole moment per unit cell and which is oriented along the c axis. Let us see why this occurs. The geometric explanation is depicted in Figure 7.5, in which we suppose that we are dealing with CdS.

Because of the difference in electronegativity between the two types of atoms in wurtzite crystals, a net charge is localized on each atom [the "Born effective charge" (Pasquarello and Car, 1997)] and each bond has an associated small electric dipole aligned along its the axial direction. For each "CdS$_4$" molecule of the lattice, for perfect tetrahedral coordination, there will be four equivalent dipoles departing from the Cd atom (indicated with red arrows in the figure), each pointing toward an S atom. Considering e^* as the module of the Born effective charge on each atom, the magnitude of each dipole will be equal to $e^*\ell/4$ and the vector sum of all these dipoles will be zero. However, if the cell deviates from ideality, the sum of these dipoles will not be zero any more as the various bond lengths and angles will be different from each other. We follow here the description given by Nann and Schneider (2004). The projections of the dipoles along the c axis can be described in terms of the parameter u and of the bond lengths ℓ_1 and ℓ_2 (as from Figure 7.5b) as

$$(\mu_1)_{//C} = \frac{e^*}{4}\ell_1 = \frac{e^*}{4}uc, \quad (\mu_2)_{//C} = \frac{e^*}{4}(\ell_2)_{//C} = \frac{e^*}{4}\left(\frac{1}{2} - u\right)c, \tag{7.2}$$

Now, the total dipole along the c axis will be equal to

$$(\mu_{TOT})_{//C} = \frac{e^*}{4}\left[u - 3\left(\frac{1}{2} - u\right)\right]c \tag{7.3}$$

Indeed, if $u = 3/8$, the above sum is zero, otherwise it could be either negative or positive, and the dipole will point either in the positive or in the negative direction along the c axis. This net dipole will depend clearly on the effective Born charge and on the degree of distortion of the cell [examples of μ_{TOT} are 0.071, 0.139, and 0.345 Debye for CdSe, CdS, and ZnO, respectively (Nann and Schneider, 2004)]. Since a net dipole moment is associated

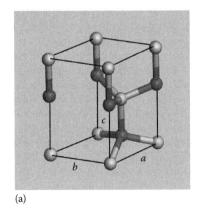

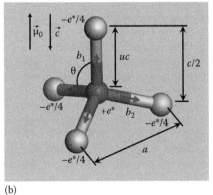

(a)   (b)

FIGURE 7.5 (a) The idealized wurtzite cell in which all bond lengths are the same (as well as bond angles). (b) In a real wurtzite structure, each atom does not have a perfect tetrahedral coordination. This figure highlights also all the parameters needed to estimate the dipole moment arising from such distortion from the ideal structure.

with a single unit cell, the overall dipole moment in a bulk crystal or in a nanocrystal will scale according to the volume of the crystal. This has been confirmed by several reports. Other studies do not show evidence of this volume dependence and report that even in cubic nanocrystals (i.e., sphalerite ZnSe nanocrystals), there is a net dipole moment, which clearly cannot be explained by the above model. As said before, the sphalerite structure does not have a unique axis of symmetry, and any dipole developing along a given polar direction would be canceled by symmetry by other dipoles developing along the other polar directions. Indeed, the total dipole, especially in nanocrystals, will depend on many other factors. The presence of random surface charges (which are independent of crystal structure!), for example, can create dipoles that are much bigger that this intrinsic lattice related dipole (Nann and Schneider, 2004). Additionally, solvents, shape effects, and the presence of surfactants can introduce effects such as screening, so that the estimate of the total dipole cannot be straightforward, unless of course one can measure it experimentally, as has been done for some nanocrystals, like the rod-shaped CdSe wurtzite nanocrystals (Li and Alivisatos, 2003a).

7.2.1.4 Wurtzite–Sphalerite Dimorphism

It is relatively easy to understand that, in several cases, the energy difference between the wurtzite and the sphalerite structures is small (Yeh et al., 1992). In the case of CdS and CdSe, this is of the order of ~1 meV/atom (Yeh et al., 1992). In general, the relative stability of the two phases depends on the specific semiconductor (the cubic phase being the more stable phase in the more covalent semiconductors), but additionally in nanocrystals, it can also depend on the conditions under which they are grown (Jun et al., 2001). CdS, CdSe, and CdTe are *dimorphous* compounds because they can exist both in the wurtzite and in the sphalerite structures.

If we recall the previous reasoning on the different sequences of stacking, we see now how we can actually build a mixed wurtzite–sphalerite crystal. This is obviously realized if the stacking sequence is of the ABC type for a certain number of layers, thus creating a sphalerite domain, and AB for a certain number of other layers, thus creating a wurtzite domain. An example is shown in Figure 7.6a. Multiple wurtzite–sphalerite

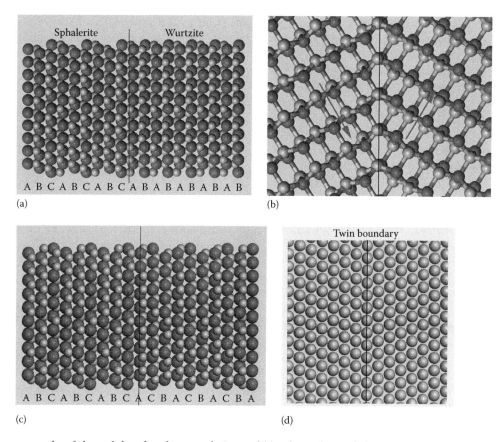

(a)

(b)

(c)

(d)

FIGURE 7.6 Some examples of planar defects found in crystals. In panel (a), a dimorphous sphalerite–wurtzite crystal is shown. Here the stacking sequence of planes changes from "ABCABC…" to "ABAB…." The significance of "stacking fault" is therefore quite clear. In panel (b), a twin plane joining two wurtzite domains is shown. This particular twin boundary is along the 112 plane (or the 11$\bar{2}$0 plane in Miller–Bravais notation). Here the arrows in the two domains indicate the polarity. Panel (c) shows a "rotation twin" in a sphalerite crystal. Here the stacking sequence at some point is inverted from "ABCABC…" to "CBACBA…." This twin can be meant as built by "cutting" a crystal along a (111) plane, by rotating by 180° one of the two domains along the (111) crystallographic direction, and by joining the two domains again. In panel (d), the same type of twin boundary is shown, but for an *fcc* crystal (for example metallic gold).

domains can be realized by continuing this construction and therefore by switching from the ABC to the AB sequence and back at wish. This clearly can be done without actually implying any periodicity in the spatial extension or in the repetition of both types of domains. If, on the other hand, there is a periodicity in the alternation of such domains, then the crystal is said to exhibit *polytypism* (which is therefore a particular form of polymorphism), since it is made by an ordered mixture of sphalerite and wurtzite stacking of planes (Lawaetz, 1972). A typical polytypic material is SiC (Bechstedt et al., 1997).

When a change in the stacking sequence of planes takes place, a planar defect is said to be formed, which can be considered as the boundary between two different crystal structures, and this is called a *stacking fault*. The formation of a stacking fault in the present case indeed does not require the breaking, stretching, or bending of chemical bonds. The energy of formation of a stacking fault is therefore relatively small in many polymorphic materials, and in such cases, this can be related to the small difference in the total energies of formation of the two structures.

7.2.1.5 Twinning in Sphalerite and Wurtzite Crystals

Twinning is another type of crystal defect, of which many subclasses exist. In one possible case of a twinned crystal, a plane separates two crystal domains that can be considered as the mirror image of each other with respect to the twin plane (Hurlbut et al., 1998) (this would be a reflection twin). A reflection twin forming in an *fcc* crystal along the (111) direction is shown in Figure 7.6d. Here in practice the sequence of planes is inverted at the twin boundary. The twin boundary here acts therefore as a mirror plane for the two twinned domains. In Figure 7.6c, a similar type of twin boundary (i.e., an inversion in the stacking sequence) is shown for a sphalerite crystal. In the example, the exact sequence is ABCABCABCĀ CBACBACBA, where Ā indicates the layer crossed by the twin boundary. This type of twin is actually called "rotation twin," since each domain can be thought of rotated by 180° with respect to the other domain along an axis perpendicular to the twin plane.

Twins form during crystal growth. A twin boundary can occur as a result of a kinetic control in the growth of a crystal (i.e., it can be triggered by some sort of erroneous attachment of atoms to a growing facet), or perhaps because in the overall energy balance of the crystal, this still represents a favorable event, or by a combination of these and yet other effects (Vere et al., 1983; Randle, 1997; Hurlbut et al., 1998; Dai et al., 2001; Elechiguerra et al., 2006; Yang et al., 2006). In general, the generation of a twin boundary, being this a planar defect, requires a certain amount of energy (Hurlbut et al., 1998). Close to the twin boundary, there might be considerable stretching and bending of atomic bonds, or even the occurrence of some broken bonds, as the geometry of atomic bonding there could deviate considerably from the low-energy case of a perfect crystal. In those materials for which the energy of formation of twins if somehow low (e.g., metals like gold, silver, and platinum), twin boundaries are frequently encountered (Dai et al., 2001; Elechiguerra et al., 2006; Xiong et al., 2007; Tao et al., 2008). For these materials,

even multiple twinned nanocrystals are observed, and such multidomain nanocrystals are frequently formed in very peculiar shapes, such as regular decahedra, elongated prisms with pentagonal cross section, icosahedra, and other types of branched geometries (Burt et al., 2005; Elechiguerra et al., 2006; Lim et al., 2007; Maksimuk et al., 2007; Xiong et al., 2007). The different orientations of the various twins with respect to each other follow precise crystallographic rules, depending on the type of twin. Also in the former case of a rotation twin in sphalerite, its energy of formation is quite low as again it does not involved breaking or distortion of bonds.

In wurtzite crystals, an important type of twin defect (which will be of relevance for the discussion that will follow on tetrapods) is shown in Figure 7.6b. In this case, a boundary is formed by joining two wurtzite domains, each cuts along a (11$\bar{2}$2) facet. This is actually a particularly complex type of twin boundary, since for each couple of domains sharing a twin plane, there is a head-to-tail arrangement of the crystal polarities of the two domains (see arrows in Figure 7.6b). Also, the twin plane does not actually represent a plane of symmetry for the two domains that are joined by it, as it was for the (111) twin boundary in an *fcc* crystal discussed above. This particular type of boundary has higher energy of formation than that of a stacking fault, but still not very high because it does not involve the breaking of bonds and it requires little lattice distortion. Typical energies of formation for such boundary are 40 mJ/m² for ZnO, 51 mJ/m² for InN, 109 mJ/m² for AlN, 107 mJ/m² for GaN (Yan et al., 2005), and 70 mJ/m² for CdTe (Carbone et al., 2006).

7.2.2 Structural Models of II–VI Semiconductor Tetrapods

7.2.2.1 Polymorph Model

After the former introductory section, we are now in a position to better understand the structural features of tetrapods and the models proposed for their structure and formation. In general, there are two models that are invoked to explain the growth and the structure of tetrapods of II–VI semiconductors. The most credited and simplest explanation for the formation of these nanocrystals (both in solution phase and in gas phase approaches) is the so-called "polymorphic modification," according to which they nucleate in the cubic sphalerite phase, after which at some point, the size evolution continues in the hexagonal wurtzite phase (Manna et al., 2000, 2003; Peng and Peng, 2001; Yu et al., 2003; Gong et al., 2006; Ding et al., 2007). Because of the intrinsic similarities between the sphalerite and the wurtzite structures, as discussed above, the growth of wurtzite domains that takes place along four of the eight 111 crystallographic directions of a sphalerite nucleus does not generate strain at each sphalerite core–wurtzite arm interface. This is because along these directions, there is a perfect match in lattice parameters between the two structures, and the only relevant structural difference among them is a change in the stacking sequence of atomic planes (as discussed in Section 7.2.1.4). This is the most simple and popular model, and the

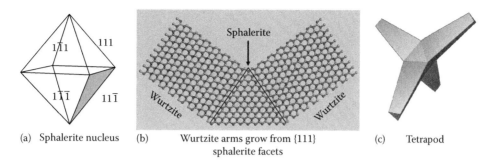

(a) Sphalerite nucleus   (b)   Wurtzite arms grow from {111}   (c)   Tetrapod
sphalerite facets

FIGURE 7.7 In the polymorphic model of the tetrapod shape observed in many nanocrystals, as sketched in (c), the central core region is supposed to have a cubic sphalerite phase (a). This "nucleus" has four equivalent {111} facets and four equivalent {$\bar{1}11$} facets. In one of these two sets, the facets are identical to the (0001) facet of the wurtzite structure, whereas in the other set, the facets are identical to the (000$\bar{1}$) facet of the wurtzite structure. We recall that in the wurtzite structure, there can be a relatively large difference in the growth rate between the (0001) and the (000$\bar{1}$) facet. Therefore, also in the cubic nucleus, one set would enclose all the "faster growing" facets, whereas the other set would enclose all the "slower growing" facets. A tetrapod shape is formed by generation of stacking faults on the four fast growing facets, after which the growth on these facets continues in the hexagonal phase, leading to the development of four arms. The continuation of growth of wurtzite arms on top of the {111} sphalerite facets is sketched in (b).

one that has been supported the most by electron microscopy observation of various nanocrystals (especially those grown in the solution phase) and was also confirmed indirectly by successful growth of uniform tetrapods starting from cubic sphalerite nanocrystals as seeds (see later in the following sections for more details). Also other branched shapes such as dipods and tripods or even multibranched nanostructures have been interpreted as resulting from such phase change occurring at some point during growth (Jun et al., 2001; Manna et al., 2003) (Figure 7.7).

7.2.2.2 Multiple Twin Model

Another popular model that rationalizes the tetrapod shape [for instance, in ZnO and ZnSe (Iwanaga et al., 1993, 1998; Takeuchi et al., 1994; Nishio et al., 1997; Dai et al., 2003; Hu et al., 2005)] is based instead on a twinning mechanism and proposes that the initial nucleus is formed by eight wurtzite domains connected to each other through (11$\bar{2}$2) twin boundaries of the type discussed in the above paragraphs. Ideally, the multiple twin nucleus that is formed is then terminated by four (0001) and four (000$\bar{1}$) wurtzite facets. The growth rate between these two groups of facets can be remarkably different (Manna et al., 2000; Kudera et al., 2005); hence, four out of the eight domains that constitute the nucleus are "fast growing" and the remaining four are "slow growing." Therefore, the initial nucleus evolves to a tetrapod (Figure 7.8c). This more elaborate model has been supported by the statistical analysis of the interleg angles in ZnO tetrapods (Iwanaga et al., 1998) (which agree with the angles that are generated by complete relaxation of the octahedral nucleus, as shown in Figure 7.8g). It has been confirmed in part also by transmission electron microscopy (Dai et al., 2003), and has been observed recently in CdTe nanocrystals (Carbone et al., 2006). In particular, in ZnO micro/nanocrystals, the interleg angles have been found to deviate indeed from the perfect tetrahedral geometry, and this can be explained as a consequence of cracking of the octa-twin nucleus due to the release of internal strain (Iwanaga et al., 1998) (see Figure 7.8g).

The multiple twin model explains also the formation of nanostructures with a smaller number of branches like dipods or tripods (see Figure 7.8d and e), if the initial nucleus is composed of a smaller number of twins. Such structures, however, can be also rationalized by the polymorph model, if one assumes that only two or three facets of the initial sphalerite nucleus evolve into arms. On the other hand, we should also point out that more complex branched shapes than the tetrapod have been observed both in ZnO micro/nanocrystals (Nishio et al., 1997) and in several cadmium chalcogenides nanocrystals (Carbone et al., 2006), which cannot really be explained by invoking the polymorph model. Examples of such structures are some of the nanocrystals of Figure 7.25c, which present more than four branches. These structures, indeed, can be explained by considering that other types of wurtzite twins can be formed in addition to the (11$\bar{2}$2) type. Nishio et al. have made a detailed account of the various types of twin defects occurring in wurtzite structures, in the specific case of ZnO multipods grown from the gas phase (Nishio et al., 1997).

7.2.3 Synthetic Approaches to II–VI Semiconductor Tetrapods

7.2.3.1 Synthesis of Colloidal Nanoparticles

In order to understand how tetrapod-shaped nanocrystals are synthesized in solution, we will give here a short description of the synthesis of colloidal nanoparticles carried out at high temperatures in organic surfactants, a technique that has been exploited widely up to now, especially for the II–VI class of semiconductors (Donega et al., 2005). In this synthesis scheme, inorganic or organometallic precursors are injected in a mixture of surfactants that are heated in a flask at a temperature that is sufficiently high to cause thermal decomposition of the precursors and hence to induce homogeneous nucleation of nanoparticles. In Figure 7.9a, a typical batch type laboratory-scale setup for the synthesis is shown, from which one can see that the synthesis is carried out under inert atmosphere. For the synthesis

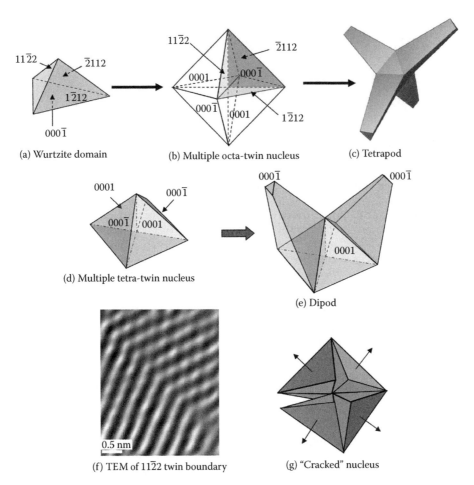

FIGURE 7.8 (a) A pyramid-shaped wurtzite crystal. (b) A multiple octa-twin nucleus formed by connecting eight of such pyramid-shaped crystals. (c) Continuation of growth from this nucleus leads to a tetrapod. (d) A multiple tetra-twin nucleus would evolve, on the other hand, into a dipod (e). (f) A TEM image of $(11\bar{2}2)$ twin boundary observed in CdTe nanocrystals. (Adapted from Carbone, L. et al., *J. Am. Chem. Soc.*, 128(3), 748, 2006. With permission.) All the wurtzite domains in the multiple twin nuclei have to sustain a considerable strain in order to have all their boundaries matched. This strain can be released by formation of cracks along the twin boundaries. (Adapted from Hu, J.Q. et al., *Small*, 1(1), 95, 2005.) When this "relaxed" nucleus, as shown in (g), evolves to a tetrapod shape, the angles between the arms are not those of a perfect tetrahedron. This deviation from tetrahedral angles has been observed experimentally. (From Iwanaga, H. et al., *J. Cryst. Growth*, 183(1–2), 190, 1998.)

of II–VI semiconductor nanocrystals, precursors are generally introduced in the reaction bath either as organometallic precursors or as inorganic precursors, like metal salts or even metal oxides (Dushkin et al., 2000; Qu et al., 2001; Donega et al., 2005). The latter are usually mixed with the surfactants and heated up with them, such that they eventually decompose and form metal complexes with the surfactants. Organometallic precursors, on the other hand, are usually diluted further in liquid surfactants (phosphines, amines, or carboxylic acids), and often are swiftly injected in the reaction flask. The result of decomposition of precursors leads therefore to the formation of new reactive species, often referred to as "the monomers," directly in the reaction environment. Because of the high temperature and a sudden rise in the concentration of the monomers in solution, the nucleation of nanocrystals takes place, followed by nanocrystal growth. During growth, unreacted monomers will diffuse from the bulk of the solution to the surface of nanocrystals,

eventually reacting at their surface and therefore contributing to nanocrystal growth.

The presence of surfactants is crucial for the controlled growth of colloidal nanoparticles. Surfactants, which are often bulky molecules formed by one or more hydrocarbon chains and by a polar head group, are introduced in the reaction environment for several reasons. One of them, as just stated above, is to form new chemical species following decomposition of the precursors. These new species, being made of chemical elements bound to one or more surfactant molecules, are in general also bulky and have therefore a slow diffusion coefficient, and they often have a limited chemical reactivity. This makes their overall attitude to induce nucleation and growth of nanocrystals more controllable by playing with parameters such as temperature and concentration. One additional and perhaps even more important role of surfactants is their dynamic binding to the surface of the growing nanocrystals. During growth,

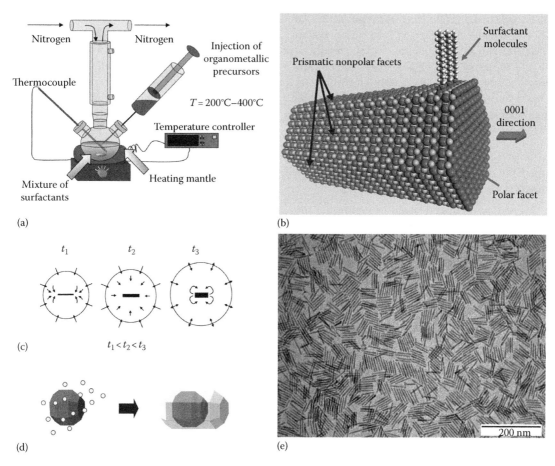

FIGURE 7.9 **(See color insert following page 9-8.)** (a) A sketch of a typical setup for the synthesis of colloidal nanoparticles. In a typical one-pot synthesis, precursors are injected in a flask containing hot coordinating solvents. The choice of coordinating solvents is dictated by several reasons, such as the conditions of growth, the precursor reactivity, and the desired nanoparticle shape and size. In order to avoid reaction with oxygen, the synthesis is carried out under inert atmosphere (such as nitrogen or argon). The growth temperature is monitored by a controller (via a thermocouple) that feedbacks a heating mantle. For the synthesis of II–VI semiconductor nanocrystals, in general precursors are introduced in the reaction bath either as organometallic precursors [i.e., $Cd(CH_3)_2$, $Zn(C_2H_5)_2$, S:TOP, Se:TOP, Te:TOP, where TOP stands for trioctylyphosphine, $S(Si(CH_3)_3)_2$] or as inorganic precursors (metal salts or even metal oxides, such as $Cd(CH_3COO)_2$, $Cd(NO_3)$, CdO), (Dushkin et al., 2000; Qu et al., 2001; Donega et al., 2005). (b) Model of a wurtzite CdTe nanorod in which three of the prismatic nonpolar facets and the 0001 polar facet are shown. Some surfactant molecules (one example is octadecylphosphonic acids, of which three molecules are shown in this model), under specific conditions, bind selectively to the nonpolar facets, depressing growth of these facets (Manna et al., 2005; Rempel et al., 2005; Barnard et al., 2007). (c) Different stages of anisotropic growth of rod-shaped nanoparticles. In each stage, a "rod" is shown enclosed in its surrounding diffusion layer. (d) A cartoon sketching the concept of seeded growth of nanorods. (e) A low-resolution TEM images of wurtzite CdS nanorods "seeded" with spherical CdSe nanocrystal seeds. Here also the phase of the nanocrystal seeds was wurtzite.

surfactant molecules (which are often present in large amounts in the reaction environment) continuously adsorb and desorb from the surface of nanocrystals, allowing them to grow, or even to be dismantled, in a controlled way. Obviously, the choice of a surfactant that binds too strongly to the surface of nanocrystals will prevent their growth, whereas on the other hand, a weakly binding surfactant would cause fast uncontrolled growth and even interparticle aggregation. Clearly, also the temperature has a strong influence on the growth, by modulating the adsorption/deadsorption rate of the surfactants on/from the nanocrystal surface as well as the monomer diffusion rate.

Surfactants also guarantee the stability of nanocrystals, as they bind to the nanocrystal surface atoms via their polar head, whereas their hydrocarbon tail(s), protruding outward, effectively make(s) the overall nanocrystals behave as an hydrophobic object for the external environment (see Figure 7.9b). The surfactants therefore allow not only for the stabilization of the nanocrystals in the reaction mixture but also for their solubility in a wide range of nonpolar or moderately polar organic solvents (after nanocrystals are isolated from the reaction bath and purified). Surfactants that are typically used in nanocrystal synthesis are alkyl-amines, phosphines, phosphine-oxide, phosphonic acids, thiols, and carboxylic acids, all containing from moderately long to significantly long alkyl chains (up to C_{16}–C_{20}).

In a practical experiment, nanocrystal growth is sustained until the nanocrystals reach the desired size/shape, after which the reaction is quenched by removing the heating mantle. For semiconductor nanoparticles (and in some cases, also for metals), particle size and size distribution during growth can be monitored almost in real time by absorption and/or emission spectra on aliquots taken from the solution (Peng et al., 1998). Particle shape is more difficult to be monitored in such a way, although some indications about nanoparticle shape can be obtained by inspecting both absorption and emission spectra (Hu et al., 2001).

7.2.3.2 Shape Control of Semiconductor Nanocrystals: Nanorods

Although in this section we discuss about nanorods, the concepts that we highlight are of crucial importance for understanding the growth of tetrapods.

Surfactant binding to the surface of nanocrystals is driven by the minimization of interfacial energy between the inorganic nanocrystal and the solution phase. We have additionally learned in Section 7.2 of this chapter that different facets in a crystal can have varying arrangements of atoms. Therefore, such facets can often exhibit different chemical affinity for adsorbate molecules, which in the present case are the surfactants. Minimization of interfacial energy by surfactants is therefore facet dependent. Facets to which surfactants stick stronger will be on average more covered by surfactants during synthesis (hence will be more stable), and therefore their growth rate will be slower with respect to facets to which surfactants will bind less strongly. Since slower growing, hence more stable facets, will tend to develop a larger surface area, the overall "shape" of the nanocrystal, or better its habit, will be dictated with the relative stabilities of its various facets [from the Wulff's theorem (Markov, 2003) that relates the overall crystal habit with relative facet stability].

If a nanocrystal forms in a crystallographic phase that does not have unique crystallographic directions, such as those belonging to cubic space groups (like the sphalerite phase), its final shape might range from a roughly spherical one, to a shape that could be for instance a truncated octahedron, or a truncated cube but in general it will not show a preferential growth direction, so that it will not have a prismatic habit (Jun et al., 2006). More interesting is the case in which a nanocrystal forms in a phase that does have a unique axis of symmetry (such as the hexagonal wurtzite or the tetragonal anatase phase), since in this case, a careful choice of surfactants might lead to anisotropic growth (Manna et al., 2000; Peng et al., 2000; Jun et al., 2006). In the recent years, many groups have indeed reported the synthesis of rod-shaped wurtzite nanocrystals of several II–VI semiconductors, promoted by surfactants such as alkyl-phosphonic acids, alkyl-carboxylic acids, and alkyl-amines, which appear to depress the growth rate of the prismatic nonpolar facets of the wurtzite structure (see Figure 7.9b). These reports have been supported recently also by computational studies (Mann et al., 2005; Rempel et al., 2005; Barnard et al., 2007).

It is important to point out that not only thermodynamic but also kinetic factors are important in the growth of nanocrystals. Several studies so far have shed light on the various parameters

that are responsible for the size and shape evolution in nanocrystals, ranging from isotropic (i.e., spheres, cubes) to anisotropic (i.e., rods, wires, branched nanostructures) (Manna et al., 2000; Peng and Peng, 2001; Lee et al., 2003). These concepts can be easily explained by considering that, during nanocrystal growth, the concentration of monomers close the surface of nanocrystals is lower than in the bulk of the solution, and therefore a net concentric diffusion field forms around each nanocrystal, sustained by a gradient in monomer concentration between the solution bulk and the surface of nanocrystals. This allows identifying an "ideal" spherical shell around each nanocrystal, the so-called diffusion layer, where the concentration of monomers drops steadily from that of the solution bulk value to that at the surface of the nanocrystal, as shown in Figure 7.9c (Reiss, 1951; Sugimoto, 1987; Park et al., 2007). The most reactive, hence fastest growing sites of a nanocrystal, such as the fast growing direction in a rod-shaped wurtzite nanocrystal, will likely find themselves in a region of higher concentration of monomers than the rest of the nanocrystal surface, since in the presence of a high concentration of monomers, the spatial extent of the diffusion layer will be relatively small (see cartoon at time t_1 of Figure 7.9c) (Peng and Peng, 2001). This will cause the most reactive sites of nanocrystals to grow much faster than other regions of the nanocrystals (Xu and Xue, 2007). Additionally, faster consumption of monomers near these reactive regions should intensify monomer diffusion toward these regions, thus promoting their growth further.

At lower concentration of monomers, on the other hand, there will be a lower flux of monomers to the growing nanocrystals, the diffusion layer will become more extended in space, and the differences between the growth rates among the various facets will be less significant, that is, the growth of nanoparticles will be more under thermodynamic control (see cartoon at time t_2 of Figure 7.9c) (Peng and Peng, 2001). Finally, at very low concentrations of monomers, the situation will be reversed. Atoms will start detaching from the most unstable facets and will feed other facets. Over time, the overall habit of the crystals will actually evolve toward the shape that minimizes the overall surface energy under the new environmental conditions. For rod-shaped nanocrystal, this will mean that their aspect ratio will start decreasing (see cartoon at time t_3 of Figure 7.9c) (Peng and Peng, 2001).

There is one major critical issue of all the syntheses of anisotropic nanocrystals, in addition to the above-mentioned care that must be taken of working under kinetic control to achieve large aspect ratio nanorods. Most of these syntheses are indeed very fast, and shape evolution takes place in a few seconds. Any overlap of the nucleation stage with the growth stage (i.e., while some rods have already formed and are therefore continuing to grow, new rods nucleate) inevitably leads to a final sample with broad distributions of rod lengths and diameters. One way of getting around this problem is by the so-called "seeded-growth" approach, in which preformed, nearly monodisperse nanocrystal seeds are coinjected with the precursors in the reaction flask (see bottom sketch of Figure 7.9d). Seeded growth of shape-controlled colloidal nanocrystals is a well-established procedure, especially for metals (Jana et al., 2001a,b; Nikoobakht and El-Sayed, 2003; Habas et al., 2007). This approach has been reported so far by a few groups (including ours) to prepare II–VI semiconductor nanorods with narrow distribution of rod diameters and lengths, such as CdSe/CdS core/shell heterostructures (Carbone et al., 2007; Talapin et al., 2007a) (see bottom sketch of Figure 7.9e). Here also the phase of the nanocrystal seeds was wurtzite. This method has been extended also to tetrapods, as we will see in Section 7.2.3.3. The major advantage of the method is indeed that it overcomes the nucleation stage, with all its associated problems of overlap of nucleation with growth that inevitably lead to broad distributions of sizes and shapes. Indeed, as the homogeneous nucleation is bypassed by the presence of the seeds, all nanocrystals undergo almost identical growth conditions since their formation, and therefore, they maintain a narrow distribution of lengths and diameters during their evolution. Furthermore, the material of the seed and that of the rod that will encase this seed can be clearly different, and this yields nanorod structures (and, as we see in the next section, also tetrapod structures) with more tunable properties than those traditionally formed of a single material.

7.2.3.3 Shape Control of Semiconductor Nanocrystals: Tetrapods

We are now in the position to understand the growth of colloidal tetrapods. This combines the concepts of anisotropic growth as described in the previous section with the possibility of a growth regime that allows to switch from one crystal phase to another phase. If we stick to the polymorph model of a tetrapod, then this shape, as anticipated in Section 7.3, arises from the fact that under certain conditions (appropriate temperature ranges during injection and during size evolution, concentration of chemical precursors, and mixtures of surfactants), nanocrystals actually nucleate in the cubic sphalerite phase, and at a certain point, they continue growing in the hexagonal wurtzite phase (Manna et al., 2000, 2003; Peng and Peng, 2001; Yu et al., 2003; Gong et al., 2006; Ding et al., 2007), and consequently, start developing four arms. These arms grow in rod shapes because the synthesis conditions favor anisotropic growth of the wurtzite domains. The reasons for this switch and why it occurs so frequently in various types of materials are not fully understood at present.

Before proceeding further we need to remind the reader that, as already pointed out in the introduction, the tetrapod shape is not unique of II–VI semiconductors, and has been observed indeed in other types of materials. Clearly, the mechanism of the tetrapod shape evolution in those materials cannot be based on the wurtzite–sphalerite polymorphic modification. This is easily understood first because such materials do not crystallize in neither of these two phases, but also because in many of them [like iron oxide (Cozzoli et al., 2006), copper oxide (Xu and Xue, 2007), or lead selenide (Na et al., 2008)], tetrapods are, on the other hand, single crystals, that is, there is no difference in crystal structure not in crystallographic orientation between the central region and the arms. In all these cases, the tetrapod shape can be explained as arising from the fastest growth rate of reactive

corners present on the initially formed crystals, since they can protrude out in regions of higher monomer concentration within the monomer diffusion layer that surrounds each nanocrystal. As in the previously discussed case of nanorods, such shape evolution can be therefore interpreted according to the so-called Mullins–Sekerka instability (Mullins and Sekerka, 1964).

It is also true that even for certain II–VI semiconductors, like CdTe in order to grow tetrapod-shaped nanocrystals, one does not need to rely strictly on the wurtzite–sphalerite dimorphic model. A recent report by Cho et al. (2008) has indicated that for CdTe it is possible to synthesize tetrapods entirely in the sphalerite phase, when using a mixture of alkyl amines, phosphonic acids, and alkyl phosphines. In that report, the authors showed that when using tellurium atoms coordinated with tributylphosphine in the synthesis, the tetrapod arms had the usual wurtzite phase. When using the bulkier trioctylphosphine (TOP), the arms were entirely in the sphalerite phase. The explanation given by the authors was that the bulkier Te–TOP precursor reduced the growth rate of the tetrapods such that their size evolution was more under thermodynamic control than when using the smaller, hence more reactive, Te–TOP complex. Thermodynamic control ensured the formation of the sphalerite phase, which is indeed more stable than the wurtzite phase in CdTe. A critical point of this type of interpretation is, however, that strong thermodynamic control would lead to spherical shapes rather than tetrapods. Indeed, as in the previously discussed cases of single crystalline tetrapods of various materials, the growth of single-crystal sphalerite CdTe tetrapods can be explained by the Mullins–Sekerka instability (Mullins and Sekerka, 1964), whereas the predominance of sphalerite phase might be due to a somewhat more stabilizing role of the specific mixture of surfactants for the sphalerite phase rather than for the wurtzite phase.

Several reports have clearly appeared on the liquid–phase synthesis tetrapod-shaped nanocrystals of II–VI semiconductors in the last years (Bunge et al., 2003; Manna et al., 2003; Yu et al., 2003; Carbone et al., 2006; Li et al., 2006; Zhang and Yu, 2006; Asokan et al., 2007; Cho et al., 2008) (not all of them are mentioned here). They differ from each other for the type of materials synthesized (which, however, were mainly Cd-chalcogenides) and for the synthesis conditions (mainly the types of surfactants employed). From all these reports, it emerges that the fabrication of such nanoparticle shapes in high yields in the liquid phase is difficult due to the inherent mechanism of their formation. Many syntheses yield, indeed, mixtures of rods, dipods, tripods, tetrapods, and even hyperbranched nanoparticles, and the reason is that one cannot strictly identify reaction conditions that promote nucleation entirely in the cubic sphalerite phase and growth entirely in the hexagonal wurtzite phase (if one wants to stick to the polymorph modification model). If, for example, nucleation of both wurtzite and sphalerite nuclei takes place, the final samples are contaminated with rods. In addition to this, often concerted growth of arms out of a nucleus does not take place, and therefore even in samples rich in tetrapods, there is a considerable distribution of arm lengths. This should represent an issue when tetrapods are used in practical applications such

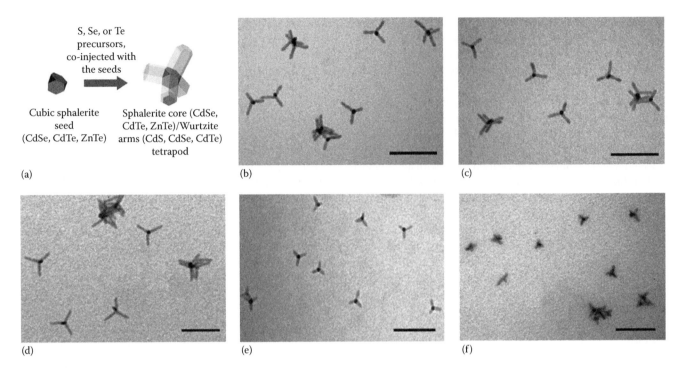

FIGURE 7.10   (a) Sketch highlighting the seeded growth approach to tetrapod-shaped nanocrystals, based on a combination of different materials for the core and for the arms. (b–f) TEM images of tetrapod-shaped nanocrystals prepared by the seeded growth approach. The images are referred to tetrapods with CdTe cores (i.e., seeds) and CdTe arms (b), CdSe cores and CdTe arms (c), ZnTe cores and CdTe arms (d), ZnTe cores and CdS arms (e), ZnTe cores and CdSe arms (f). All scale bars are 100 nm long. (Adapted from Fiore, A. et al., *J. Am. Chem. Soc.*, 131(6), 2274, 2009. With permission.)

as photovoltaic devices, single nanocrystal transistors, atomic force microscopy (AFM)-functionalized tips, and others, as are discussed in the following sections. More recent methods to improve the yield of tetrapods have included the seeded growth starting from noble metal nanoparticles (Yong et al., 2006) and the coinjection of Se or Te precursors in the case of CdS nanocrystals to enhance the probability of formation of sphalerite nuclei at the early stages of tetrapod formation (Hsu and Lu, 2008).

Fortunately, also in this case, the "seeded growth" approach has contributed to improve the yield of tetrapods. Seeded growth has been exploited to grow ZnTe/CdSe (Xie et al., 2006), ZnTe/CdS (Xie et al., 2006; Carbone et al., 2007), ZnSe/CdS (Carbone et al., 2007), and CdSe/CdS (Talapin et al., 2007a) tetrapods (here the first compound denotes the material of the seed, which then forms the central core of the tetrapod, the second that of the material that forms the arms of the tetrapod). In such cases, preformed nuclei in the sphalerite phase are coinjected together with the precursors needed to grow the "arms" of the tetrapods in a hot mixture of surfactants that promotes wurtzite growth (see Figure 7.10) (Xie et al., 2006; Carbone et al., 2007; Talapin et al., 2007a). A more controlled and "concerted" growth of wurtzite arms on top of such seeds is usually observed (especially by employing large seeds), and this favors the formation of arms with more uniform lengths per each tetrapod.

Our group has recently reported a more general approach to synthesize tetrapod-shaped colloidal nanocrystals made of various combinations of group II–VI semiconductors, using preformed seeds in the sphalerite structure, onto which mainly hexagonal

wurtzite arms were formed, by coinjection of the seeds and chemical precursors into a hot mixture of surfactants (Fiore et al., 2009). For the core region of the tetrapod, hence the seed, we could chose among CdSe, ZnTe, and CdTe, as nanocrystals of these materials could be prepared in the sphalerite phase and furthermore they gave good yields in terms of tetrapods when used as seeds, whereas the best materials for arm growth were CdS and CdTe (See Figure 7.10). In addition to tetrapods, many branched heterostructured nanocrystals have been prepared and studied so far by several groups. These works aimed mainly at exploring the optical and electronic properties of such nanocrystals. Finally, we need to mention that seeded growth to form branched nanostructures is not limited at all to semiconductors. This approach has been exploited even to grow star-shaped Au nanocrystals, starting from multiple-twinned Au nanoparticles as seeds (Nehl et al., 2006).

7.3 Physical Properties of Tetrapods

7.3.1 Introduction

For what concerns the basic understanding of the optical properties of tetrapods, we can adopt a much simplified picture in which a tetrapod can be regarded as four cylinders that are connected at tetrahedral angles at a central branch point. This section focuses on the optical and electrical properties of the tetrapods, and an interesting question will be in what respect the properties of tetrapods differ from those of four isolated rods. The optical and electronic properties are governed by the electronic

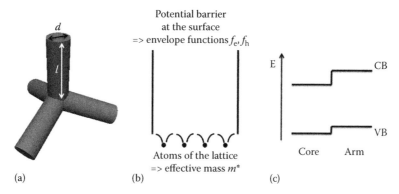

FIGURE 7.11 (a) An illustration of a tetrapod in which the length *l* and the diameter *d* as the dominant parameters for the confinement are highlighted. (b) Scheme illustrating the effective mass and the envelope function approximation. (c) The energy bands related to the sphalerite (core) and wurtzite (arm) crystal structures in CdTe or CdSe form a type II band offset at the interface (Fiore et al., 2009).

Band Alignment at Heterojunction Interfaces

Morе generally, in a heterojunction separating two different types of intrinsic semiconductors, there are two possible configurations of the band offsets of the two components. In one configuration, the band edges of the first material are both localized inside the band gap of the second material (an arrangement that is called of type I). In that case, electron–hole pairs would stay confined in the first semiconductor material. Another possible configuration is the one in which only one of the band edges of the first material is localized in the gap of the second material (an arrangement that is called of type II). In that case, electron–hole pairs that are generated in either semiconductor are separated at the hetero-junction. In materials like CdTe, a type II heterojunction is actually realized between a region with sphalerite structure and a region with wurtzite structure (i.e., in the present case, between the core of the tetrapod and its arms). The band offset in this case is very small, of the order of few tens of milli electron volts (Madelung et al., 1982). A more striking case, as we shall see later in this chapter, in when indeed the core region of the tetrapods has a different chemical composition of the arms (for instance, the core is made of CdSe and the arms are made of CdTe). This configuration leads to new interesting optical properties of tetrapods, such as the possibility or radiative recombination from oppositely charged carriers that are separately localized in core (electrons) and the arms (holes), because of the strongly staggered, type-II arrangement of the band edges. The energy of the light emitted from the recombination of these carriers can be in the infrared region.

structure of the nanocrystals. Although the exact calculation of the energy-level structure of tetrapods is very complicated, we can obtain useful information from some basic approximations. The energy-level structure of small nanocrystals will differ from that of the corresponding bulk material by quantum effects resulting from the finite size. To evaluate the impact of the size, the Bohr radius gives a convenient length scale. The Bohr radius of a particle is defined as $a_B = \varepsilon\,(m/m^\star)a_0$ (Ashcroft and Mermin, 1976). Here ε is the dielectric constant of the medium (i.e., the nanocrystal material), m and a_0 are the electron mass and Bohr radius, respectively, and $m^\star$ is the effective mass of the particle. We note that in this section, we refer with the term "particle" to electrons, holes, and other "quasiparticles" as the excitons. If the size-related parameters are in the range or smaller than the Bohr radius of the particle of interest (e.g., an exciton or an electron), we can expect a significant impact of the confinement on the energy-level structure related to that particle. The simple sketch of a tetrapod in Figure 7.11 shows that the dominant parameters for the tetrapod shape are the diameter and the length of the arms. For high aspect ratio of the arms, we would expect the diameter to be the dominant parameter for the confinement effects. The impact of the crystal lattice on the energy-level structure of the particle can be considered in the effective mass approximation, in which the particle can then be treated as moving freely within the nanocrystal lattice with this effective mass.* One can picture this as if the nanocrystal lattice exerts a drag on the particle. In addition to the influence of the crystal lattice, we have to consider the confinement resulting from the

---

* This concept can be applied if the nanocrystal dimensions are much larger than the crystal lattice constant, which is the case for the tetrapods under discussion.

TABLE 7.1 Parameters for CdTe, CdSe, and ZnO That Are the Most Common Tetrapod Materials

	Exciton Bohr Radius (nm)	Electron Effective Mass (× e)	Hole Effective Mass hh/lh	Band Gap (eV)
CdTe	7.5	0.096	0.81/0.12	1.475
CdSe	4.9	0.13	0.45	1.84
ZnO	1.0	0.19	1.21	3.37

Source: Landolt-Boernstein, *Group III Condensed Matter*, Vol. 41C, Springer–Verlag GmbH, Germany, 1998.

finite size of the tetrapods. The simplest confinement potential is given for a particle that can move freely within the nanocrystal and encounters infinitely high barriers at the nanocrystal surface (in one dimension such a potential is referred to as "particle in a box") (Eisberg and Resnick, 1985). To model the shape of the tetrapods, we have to consider two types of geometries: the core that can be approximated by a sphere, and the arms that can be modeled as cylinders with diameter d and length l. Because of the different crystal structure in the core and the arms, we have to implement a small offset in potential in between these two regions. Assuming a sphalerite core and wurtzite arms, we find a type II potential offset at the arm–core interface, as sketched in Figure 7.11c (Yeh et al., 1992; Klimov et al., 1999).

We can combine the finite size effects and the influence of the crystal lattice by replacing the free particle mass with its effective mass in the solutions that were obtained for the confinement potential. This theoretical approach that treats the particles as moving freely (leading to parabolic bands in k-space) within the confinement boundaries of the nanocrystal is called envelope function approximation (Bastard, 1991).

Typical sizes of CdTe and CdSe tetrapods are 5–15/15–150 nm for the arm diameter/length. Comparison of the values with the exciton Bohr radius in CdTe and CdSe materials shows that the dominant confinement effects should originate from the arm diameter (Table 7.1).

7.3.2 Optical Spectroscopy on Colloidal Nanocrystals

The contribution of the electronic levels to the optical absorption can be obtained by calculating the optical transition probabilities from the ground state $|0\rangle$ to the various electron–hole pair states. This transition probability can be written as

$$P = |\langle \Psi_e | \vec{e} * \hat{p} | \Psi_h \rangle|^2 \qquad (7.4)$$

where

Ψ$_e$ and Ψ$_h$ are the wave functions of the electrons and holes, respectively
$\vec{e}$ is the polarization vector of the incident light
$\hat{p}$ is the momentum operator

In the envelope function approximation, the wave functions can be described as products of the Bloch functions of the crystal

lattice and the envelope functions describing the confinement potential. The momentum operator acts only on Bloch functions and therefore P can be stated as

$$P = |\langle u_c | \vec{e} * \hat{p} | u_e \rangle|^2 |\langle f_e | f_h \rangle|^2 \qquad (7.5)$$

with

u_c and u_e Bloch functions of the (bulk) crystal lattice
f_e and f_h the envelope functions related to the electron and hole confinement (Klimov, 2003)

The second part of Equation 7.5 contains the selection rules for the optical transitions.

The incident light generates bound electron–hole pairs that are called excitons (see Figure 7.12a). The peaks in the absorption spectrum correspond to the transitions that are optically allowed by the selection rules (Figure 7.12b). The excited exciton states have very short life times in nanocrystals with high symmetry (Efros et al., 1996). Therefore, the photogenerated carriers relax into the exciton ground state which, due its low transition probability, has a much longer life time. Consequently, the radiative emission signal, for example, in spherical nanocrystals, is dominated by the exciton ground state (Figure 7.12c).

7.3.2.1 The Stokes Shift

For colloidal semiconductor nanocrystals, the optical emission peak occurs at slightly lower energy than the lowest energy peak observed in absorption experiments, an effect that is referred to as the Stokes shift (Efros et al., 1996). The origin of the Stokes shift lies in the complex electronic structure of the excitons in semiconductor nanocrystals and the respective transition probabilities in between the levels. For example, in spherical wurtzite CdSe nanocrystals, the degeneracy of the band edge exciton level is lifted by the deviations from the spherical shape, the anisotropy of the crystal lattice, and the exchange interaction. In this case, it was found that an angular momentum quantum number of "2" can be assigned to the lowest level of these degenerate states, which does not allow an optical excitation of this state in the electric dipole approximation (Norris et al., 1996). Consequently, this state is called the dark exciton state. However, the photogenerated electron–hole pairs can relax into the dark exciton state and then recombine with the assistance of optical phonons. The low efficiency of this recombination process leads to a long life time of the emitting state. For tetrapod-shaped nanocrystals, we would expect that a significant contribution to the Stokes shift will originate from the shape anisotropy, that is, the tetrapod shape and the resulting distribution of the electron–hole wave functions.

7.3.2.2 Steady-State Absorption and Emission Experiments on Tetrapods

This section discusses absorption and photoluminescence experiments of tetrapods. From Section 7.3.2.1, we know that optical absorption experiments provide us information about the allowed excitonic transitions of the material. The most straightforward

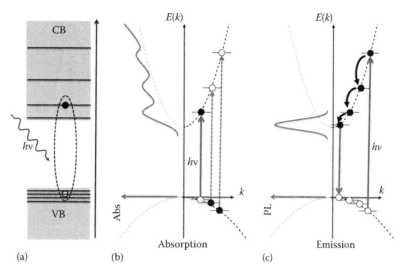

FIGURE 7.12 Schematic illustrations of optical probing processes. (a) An electron–hole pair (exciton) is created by incident photons and bound by the Coulomb interaction. (b–c) In the effective mass approximation, the electronic bands can be regarded as parabolic, and the confinement due to the nanocrystal shape leads to discrete energy levels. Absorption experiments probe the allowed transitions according to Equation 7.5, photoluminescence probes the emitting transitions. In nanocrystals, the photogenerated carriers relax very quickly (on the order of few picoseconds) into the lowest energy state as indicated by the black arrows (the arrows indicating the relaxation of the holes are not shown).

optical absorption and emission experiments on colloidal nanocrystals are performed in solution at room temperature, typically using commercial fluorescence spectrophotometers, in which a quartz cuvette containing the nanocrystals dissolved in solution can be comfortably inserted in the optical path. This type of experiment probes a large fraction of all the nanoparticles present in the solution. We therefore expect broadening of the signal due to the size distribution of the nanocrystals, and have to keep in mind that signals can also originate from undesired contaminants that are present in the solution. We note that the emission intensity of CdTe or CdSe tetrapods is much smaller than the emission of comparable rods or spherical nanocrystals of the same material (the quantum yield of tetrapods is around 1%).

Absorption spectra of CdTe tetrapods recorded in solution are displayed in Figure 7.13 (Manna et al., 2003b). In the absorption spectra, we can identify peaks that can be correlated to the exciton level structure, and in particular, the lowest energy peak which corresponds to the band edge exciton can clearly be identified in all spectra. We find that the observed band gap of the tetrapods is much larger than the band gap of the CdTe bulk material [which is 1.5 eV with 830 nm at room temperature (Madelung et al., 1982)], which is due to the confinement effects. Figure 7.13 shows that the absorption spectra, and in particular the band edge exciton energy, depend mostly on the arm diameter of the tetrapods and that the arm length has little influence, as we would have expected from our confinement estimate based on the exciton Bohr radius. Higher energy peaks are more difficult to resolve in tetrapods, for example, in nanospheres, due to the more complex geometry of the tetrapods that leads to a correspondingly complex exciton level structure.

Figure 7.14 shows the absorption and emission spectra of CdTe tetrapod samples with different size recorded in solution at room temperature. The emission signal is observed at lower energies than the absorption peaks due to the previously discussed Stokes shift. We find that the Stokes shift increases with decreasing tetrapod size like it was observed in spherical

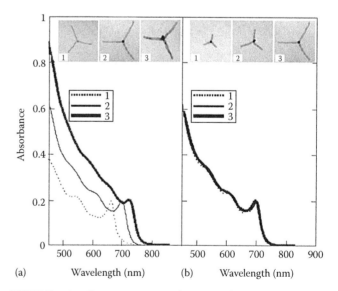

FIGURE 7.13 Absorption spectra of various solutions, each containing tetrapods of given average arm diameter and lengths. The spectra were recorded at room temperature. The insets show transmission electron microscopy images of representative isolated tetrapods taken from each sample. (a) Spectra of tetrapods with comparable arm lengths and differing diameters, (b) spectra of tetrapods with comparable arm diameters and differing lengths. We find that the absorption spectra depend strongly on the arm diameter and that the influence of the arm length is negligible. (Reprinted from Manna L. et al., *Nature Mater.*, 2(6), 382, 2003. With permission.)

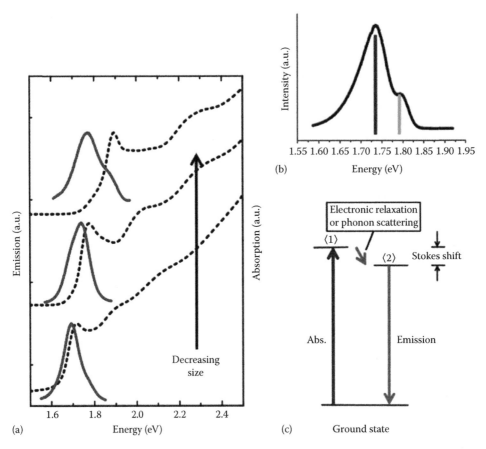

FIGURE 7.14 (a) Emission (solid line) and absorption (dotted line) spectra of CdTe tetrapods with different dimensions, tetrapod size is decreasing from bottom to top. (b) Emission spectrum of CdTe tetrapods at $T = 4$ K. (c) Schematic illustration of the excitation and recombination processes and the origin of the Stokes shift. (Adapted from Krahne, R. et al., *J. Nanoelectron. Optoelectron.*, 1(1), 104, 2006b. With permission.)

nanocrystals. However, in tetrapods, the Stokes shift is larger than in spherical nanocrystals. For comparison, in spherical nanocrystals with 5 nm diameter, the Stokes shift is 50 meV, whereas for tetrapods with arm diameter of 4.7 nm, the Stokes shift is 100 meV. One reason for the large Stokes shift of tetrapods could be the large anisotropy of the nanocrystal shape. However, part of the energy difference can also be attributed to the comparatively broader size distribution found in tetrapod samples with respect to spherical nanocrystals. In inhomogeneous samples, the emission peak energy is dominated by the larger nanocrystals present in the solution, which leads to a red shift of the luminescence. This effect is called the nonresonant Stokes shift and refers to the energy difference between the full luminescence peak of the nanocrystal solution and the lowest absorption peak as it is the case in Figure 7.14a. The resonant Stokes shift, on the other hand, can be measured by fluorescence line narrowing experiments, in which only the largest nanocrystals in the solution are selectively excited (Efros et al., 1996). This eliminates essentially the size distribution effects, and therefore the resonant Stokes shift reveals more accurately the energy difference between the dark and the bright exciton states.

A closer inspection of the emission spectra of the tetrapod samples reveals a double peak structure (Tarì et al., 2005). A detailed analysis of the emission spectra shows that a decrease in arm width

leads to an increase both in energy spacing between the two peaks and of the intensity of the high energy peak. Photoluminescence experiments at cryogenic temperatures* resolve even more clearly the double peak structure of the emission signal as can be seen in Figure 7.14b. This double peak in the emission, which can be related to spatial distribution of the electron and hole wave functions, appears to be a very peculiar property of the tetrapods that is not observed neither in dots nor in rods. In tetrapods, the central branch point invokes a specific symmetry in the exciton ground and excited states that leads to diverse recombination dynamics. In order to understand this, we have to look at the electronic structure of tetrapods in more detail.

Figure 7.15a shows the band structure superimposed on a TEM image of a tetrapod. The different crystal structures at the branch point, sphalerite in the core and wurtzite in the arms, lead to a type II stacking of the energy bands at the arm–core interface that enhances the confinement of the electrons in the core. The results of two theoretical models that calculate the spatial distribution of the electron and hole density for the first and second exciton states in the tetrapods are shown in Figure 7.15b

---

* For measurements at cryogenic temperature, the tetrapods was casted from solution onto a substrate (silicon or silicon oxide) and the solvent was allowed to evaporate.

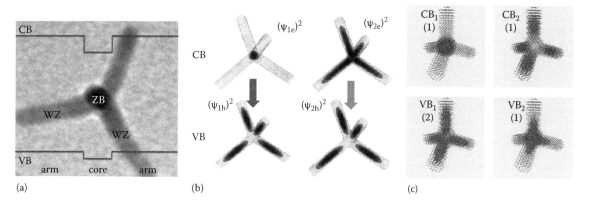

FIGURE 7.15 (a) A sketch of the band offset between the tetrapod arms and core superimposed on a TEM image of a tetrapod. (b) Calculated charge density distribution of the first (ψ_1) and second (ψ_2) exciton states in the envelope function approximation for tetrapods that correspond in size to the sample from which the emission displayed in Figure 7.14b was recorded. (Adapted from Tarì, D. et al., *Appl. Phys. Lett.*, 87(22), 224101, 2005. With permission.) (c) Calculated wave function charge densities in the atomistic approach with a semiempirical pseudopotential method. CB and VB indicate the conduction and valence band, respectively; the subscript refers to the first and second exciton state and the numbers in the brackets give the degeneracy of the state. (Adapted from Li, J.B. and Wang, L.W., *Nano Lett.*, 3(10), 1357, 2003. With permission.)

and c. The theoretical distributions displayed in Figure 7.15b are obtained by the calculation of the electronic structure in the envelope approximation (Tarì et al., 2005) and by considering a band offset at the core as illustrated in Figure 7.15a. This method allows for the modeling of tetrapod sizes that are comparable to the experiments. We find that the electrons are localized in the core for the first exciton state, whereas they are delocalized over the arms and core for the second exciton state. The hole wave functions are distributed in the arms for both the first and second excited state. As a result, the electron wave functions of the first and second exciton state have only a small overlap that significantly supresses intraband relaxation and promotes the direct radiative recombination of the second exciton state. The transitions that correspond to the two peaks observed in the emission spectrum of Figure 7.14b are indicated by the blue and green arrows in Figure 7.15b. We see that the wave function localization resulting from the tetrapod shape is the origin for

the appearance of the double peak structure in emission. Figure 7.15c shows the carrier densities obtained by Li and Wang who use an atomistic model of the tetrapods and calculate the electronic states in a semiempirical pseudopotential method (Li and Wang, 2003). There is good agreement between the results of the two models, in particular for the localization of the electrons in the first and second exciton state. The atomistic approach gives also the higher exciton states and their degeneracy.

Heterostructured tetrapods as described in Section 7.4.3 provide additional parameters to tailor the optical emission properties and the electron and hole wave function distributions (Talapin et al., 2007a,b; Fiore et al., 2009). Different material combinations can lead to different band structure stackings as illustrated in Figure 7.16c. For CdSe/CdS tetrapods, the core has a smaller bandgap than the arms that leads to a significantly enhanced emission efficiency of the tetrapods (Talapin et al., 2007a,b) (Figure 7.16b). In CdSe/CdTe tetrapods, the band

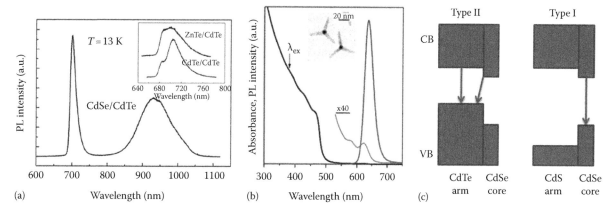

FIGURE 7.16 Optical spectra of core/shell tetrapods: (a) Main plot: PL spectra of CdSe/CdTe tetrapods as recorded at a temperature T of 13 K. Inset: PL emission from ZnTe/CdTe and CdTe/CdTe tetrapods, also recorded at $T = 13$ K. (Adapted from Fiore, A. et al., *J. Am. Chem. Soc.*, 131(6), 2274, 2009. With permission.) (b) Emission (gray) and absorption (black) spectra of CdSe/CdS tetrapods. (Adapted from Talapin, D.V. et al., *Nano Lett.*, 7(5), 1213, 2007a. With permission.) (c) Sketches of the conduction and valence band level alignments for CdSe/CdTe and CdSe/CdS tetrapods. The red arrow illustrates the type II recombination process.

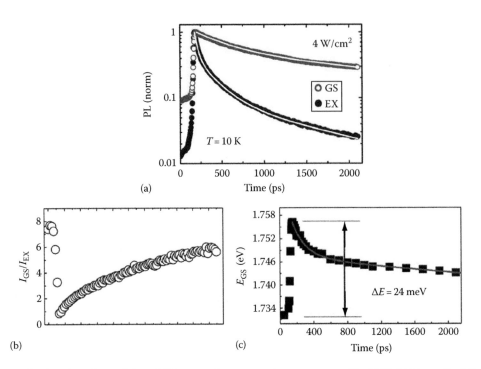

FIGURE 7.17 Time-resolved PL data obtained from CdTe tetrapods at cryogenic temperatures ($T = 10\,\text{K}$). (a) Normalized time traces of the PL signal of the first (GS) and second (EX) exciton peak. (b) Intensity ratio of the first and second exciton peak, and (c) dynamical energy shift of the first exciton peak. (Adapted from Morello, G. et al., *Appl. Phys. Lett.*, 92(19), art. no. 191905, 2008. With permission.)

alignment forms a type II interface that leads to emission in the visible through direct carrier recombination in the arms and to type II emission in the infrared via recombination of the holes localized in the CdTe arms and the electrons localized in the CdSe core (see Figure 7.16a) (Fiore et al., 2009).

Mauser et al. (2008) reported polarized emission from CdSe/CdS core/shell tetrapods which they explained by asymmetries in the tetrapod shape, which should lead to localization of the electrons in the arm with the largest width.

7.3.2.3 Time-Resolved Exciton Dynamics in Tetrapods

In time-resolved photoluminescence experiments, the nanocrystals are excited by a pulsed laser source and the emission is recorded with respect to a delay time relative to the excitation pulse. Therefore, time-resolved photoluminescence experiments can give more insight into the relaxation dynamics of the exciton states in tetrapods. Figure 7.17a shows the normalized time-resolved PL traces of the first (GS) and second (EX) exciton states of a CdTe tetrapod sample (Morello et al., 2008) recorded at cryogenic temperature. The steady-state emission of this sample was similar to that displayed in Figure 7.14b. The comparable rise times related to the two exciton states show that they have independent excitation channels. Then, the second exciton state decays much faster than the first exciton state, and both decay traces have to be fitted with multiple exponentials curves (of the type $\sum_{i=1}^{n} A_i \exp(-(t - t_0)/\tau_i)$, where A_i and τ_i are the weight and decay time of the ith decay mechanism, whereas t_0 denotes the point in time where the PL has reached its maximum),

and, consequently, multiple time constants contribute to the relaxation process. Best fits can be obtained with bi- and triexponential functions. This decay with three time constants is due to Auger-like recombination processes* (tens of picoseconds), to the intrinsic emission of the two states (hundreds of picoseconds), and to the emission from defect states (few nanoseconds). Figure 7.17b shows an interesting correlation in time between the two exciton peaks. The intensity of the second state increases rapidly in the first 140 ps, which is accompanied by a blue shift in energy of the first exciton peak (see Figure 7.17c). In the following, the intensity of the second state decreases, and, at the same time, the blue shift of the ground state is reduced. This dynamical blue shift of the ground state indicates the screening of the internal polarization field present in the tetrapods by the photogenerated carriers in the second exciton state. In general, internal electric fields lead to a red shift of the optical emission. In tetrapods, these internal electric fields are due to the wurtzite lattice structure of the arms that induces a dipole moment, and to the spatial separation of electrons and holes due to the type II band offset at the core region.

While time-resolved PL measurements elucidate the recombination dynamics, time-resolved absorption experiments can give information about the dynamics related to the population of the exciton states. Transient absorption spectra can be obtained by a

---

* In the Auger recombination process in colloidal nanocrystals, the photogenerated electron–hole pairs scatter on third particles, either phonons or other excitons. As a consequence, one of the carriers can get trapped, for example, at the surface, leading to a separation of the electron–hole pair and a nonradiative relaxation (Klimov and McBranch, 1997).

pump and probe technique as described in (Malkmus et al., 2006). Here the sample was excited by a laser pulse at an energy high above the band gap, for example at 480 nm as illustrated in Figure 7.18c, and then the time-resolved absorption was obtained by a second broad band probe pulse with a specific delay time. The plotted signal is the difference between the probe pulse and a reference pulse that was recorded before the pump pulse. Negative transient absorption (ΔA) occurs for states that were filled by the pump pulse, that is, these states are photobleached. Photoinduced absorption occurs if the energy-level structure or the selection rules for the optical transitions have been modified by the pump pulse excitation, that is, by the photoinduced population of energy levels.

The transient absorption spectra in Figure 7.18a and b reveal different relaxation dynamics for different energy ranges. Higher energy states, for example at 600 nm, decay much faster than lower energy states near the band gap (680 nm for the dots, 700 nm for the tetrapods). These fast relaxation processes are generally attributed to intraband transitions. For a more detailed review on time-resolved absorption experiments on colloidal nanocrystals, the reader can refer to Klimov (2000). The comparison of spectra of dots and tetrapods in the low-energy range near the band gap reveals the specific features of the relaxation dynamics in tetrapods. The dot spectra in Figure 7.18a show maximum bleaching already at very short time (0.2 ps) after the pump pulse followed by a rapid decay of the bleach signal due to the very fast carrier

relaxation dynamics present in the dots. For tetrapods, the maximum in the bleach occurs much later, at 2 ps after the pump pulse, and is followed by a comparatively slow decay of the bleach signal. Multiexponential fitting to the spectra yields a biexponential decay for the dots with time constants of 1 and 25 ps, and four time constants for the tetrapod signal, 0.8, 1.4, 3.6, and 32 ps as depicted in Figure 7.18d. From these decay-associated spectra, we see that the bleaching occurs only after 3.6 ps at the low-energy states, which reflects the time that the high-energy carriers need to relax into these states. The parallel and perpendicular polarized transient absorption spectra in Figure 7.18d show that the three faster decay components show a polarization anisotropy, whereas the slowest component is completely isotropic. This indicates that the faster components are localized in the tetrapod arms (which are anisotropic) and that the slowest transition can be related to the isotropic tetrapod core. Taking into account that the faster transitions occur at higher energy levels and the slow component at energies near the band gap, we can conclude from the transient absorption experiments that the higher excitonic states are localized in the arms and the lowest energy state is localized at the core. This experimental result is in good agreement with the theoretical data presented in Figure 7.15 from Li and Wang (2003) and Tarì et al. (2005).

Also on tetrapod heterostructures, time-resolved absorption experiments have been reported. Peng et al. (2005) succeeded

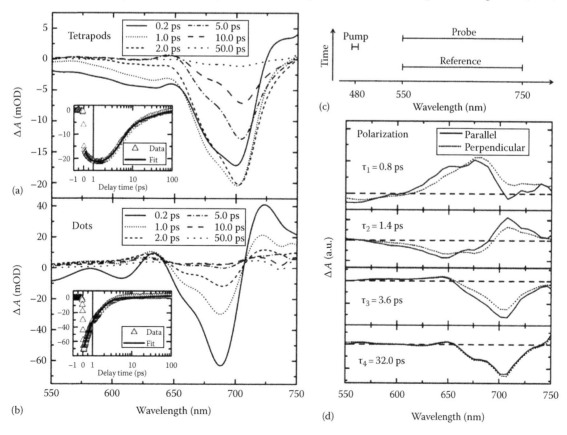

FIGURE 7.18 (a and b) Transient absorption spectra of CdTe tetrapods and dots. The insets show the decay behavior near the band gap (680 nm for dots, 700 nm for tetrapods). (c) Illustration of the pump and probe energy ranges. (d) Parallel and perpendicular polarized decay associated spectra of tetrapods for different decay time constants. (Adapted from Malkmus, S. et al., *J. Phys. Chem. B*, 110(35), 17334, 2006. With permission.)

to grow core/shell CdTe/CdSe tetrapods and branched CdTe nanostructures from a CdSe rod. In the latter case, the CdSe rod becomes embedded in one, thereby prolonged, arm of the tetrapod. At the type II band structure interface between the CdSe and CdTe, efficient charge separation of the photogenerated carriers occurs, where the electrons are collected in the CdSe and the holes in the CdTe. Consequently, emission from these heteronanostructures was not observed. Time-resolved pump and probe absorption experiments in the visible and infrared spectrum on such tetrapod heteronanostructures enabled to study the recombination dynamics of electron and holes separately, revealing faster relaxation times for the holes. Charge separation effects in the recombination dynamics have also been reported by Fiore et al. (2009) for core/shell tetrapods with CdTe arms and ZnTe, CdSe, and CdTe as core materials (see Figure 7.16c).

7.3.2.4 Tetrapods as Active Material for Photovoltaic Applications

The branched shape and the low optical emission intensity of the tetrapods make them promising candidates for the active material layer in thin film photovoltaic applications. The branched shape can increase the absorption cross section and, at the same time, the arms can provide percolation pathways to harvest the photogenerated charges more effectively at the electrodes. An ideal device design of a photovoltaic cell based on tetrapods is sketched in Figure 7.19a. Some pioneering works on tetrapod-based solar cells have already been reported (Sun et al., 2005; Gur et al., 2006; Zhou et al., 2006). In such devices, the anode consists of a transparent indium-tin-oxide layer that was evaporated on a glass substrate and coated with a thin layer of PEDOT. Then the active layer consisting of the polymer and tetrapods was deposited. For this, either the tetrapods were casted from a solvent solution onto the surface (the solvent is allowed to evaporate (Gur et al., 2006)), followed by polymer deposition by spin coating, or the tetrapods and the polymer were mixed prior to deposition at a certain ratio and are then spin coated onto the device (Sun et al., 2005; Zhou et al., 2006). In the last step, a layer of aluminum was evaporated that functions as the cathode.

The highest energy conversion efficiency of 2.8% was reported by the group of Neil Greenham (Sun et al., 2005) that used CdSe tetrapods with 50 nm long arms embedded in poly(p-phenylen-vinylene) derivative OC1C10-PPV matrix with thickness of about 150 nm. Zhou et al. (2006) investigated CdSe$_x$Te$_{1-x}$ ternary compound tetrapods for solar cell devices and found that CdSe is so far the most favorable material for obtaining high-power conversion efficiency.

7.3.3 Optical Phonons in Tetrapods

This section discusses some crystal lattice vibration modes (phonons) of tetrapod-shaped nanocrystals. Vibration modes of ionic materials can be classified into acoustic and optical phonon modes. Acoustic phonons correspond to sound waves, here the atoms (e.g., Cd and Se) oscillate in parallel phase (Kittel, 1996). For optical phonons, the anions and cations oscillate against each other, creating a time varying electric dipole moment and therefore these modes can be excited directly by light. In a three-dimensional lattice, the atoms can furthermore oscillate along the propagation direction of the phonon wave (longitudinal) and perpendicular to this direction (transversal). Standard abbreviations for the respective phonon modes are longitudinal-acoustical (LA), transversal-acoustical (TA), longitudinal-optical (LO), and transversal-optical (TO). Typical dispersion relations of these phonon modes are depicted in Figure 7.20a.

So far, acoustic phonons have not been observed in tetrapod-shaped nanocrystals, neither are the authors aware of any theoretical work in this respect. Optical phonon modes in nanocrystal can be detected by Raman (Trallero-Giner et al., 1998)- and Fluorescence–Line–Narrowing (FLN) (Nirmal et al., 1994) spectroscopy. For nanocrystals surrounded by a dielectric medium, also surface-optical phonon modes can be induced.

Raman spectroscopy is sensitive to the inelastic scattering processes of the photogenerated excitons. The inelastic scattering process consists of the creation or annihilation of quasiparticles, for example, the emission and absorption of phonons. Resonant Raman scattering on collective excitations, like phonons, can be described in three steps as shown in Figure 7.20c: (1) the incident

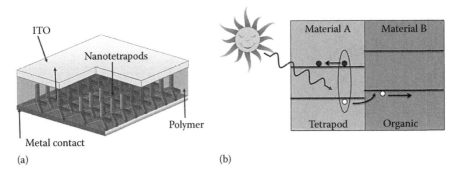

(a)   (b)

FIGURE 7.19 Hybrid nanocrystal/polymer composites can be interesting candidates for future photovoltaic devices. (a) Illustration of an organic/inorganic device structure for a photovoltaic cell based on a tetrapod array in the active layer. (b) The type II band alignment of the two materials (organic—inorganic) leads to a spatial separation of the photogenerated carriers, which is illustrated by a photogenerated electron–hole pair in the tetrapod.

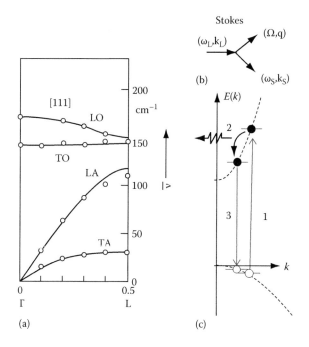

FIGURE 7.20 (a) Phonon dispersion relation in CdTe bulk material (Landolt-Boernstein, 1998). (b) Raman scattering process creating a phonon. (c) Schematic illustration of the three-step process in which photogenerated excitons scatter at crystal lattice vibrations.

light generates an exciton, (2) the exciton scatters and emits a phonon, and (3) the radiative recombination of the exciton takes place.

Figure 7.21a shows a typical Raman spectrum revealing signals of the LO, TO, and SO phonon modes. The properties of the optical phonons in tetrapods can be described using a nanowire model, that is, there is no specific signature of the branch point on these vibration modes. Resonant Raman experiments (see Figure 7.21b and c) show that the phonon excitations are in resonance with the higher exciton levels in which the carriers are distributed in the arms of the tetrapods. The dominant excitation in the Raman spectrum is the LO phonon mode, which in tetrapods is found at slightly lower energies than in the bulk material. This behavior can be intuitively understood by the decreasing dispersion of the LO and TO phonon energy in k-space near the Brillouin center (see Figure 7.20a) (Ashcroft and Mermin, 1976). The finite size of the nanocrystals allows for a transfer in momentum to the phonon excitation by the relation $q = 2\pi/a$, where a is the confinement in the direction of interest (diameter or length for rod-shaped nanocrystals). Also, the confinement leads to a broadening of the phonon peak that originates from variations in confinement length. For anisotropic nanocrystals with large aspect ratio (for example nanorods or nanowires), the phonon component along the wire can be regarded as bulk-like and the perpendicular component as the confined mode. For polar nanocrystal lattices like the wurtzite crystal structure, also long-range dipolar interactions can alter the LO and TO phonon energies. Mahan et al. (2003) found a

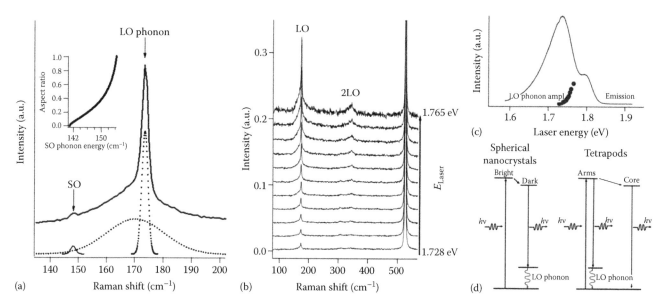

FIGURE 7.21 (a) Resonant Raman spectrum of CdTe tetrapods at cryogenic temperature (solid) and Lorentz fits (dotted) to the data. The LO dominant peak at 173 cm^{-1} originates from the LO phonon, and the small shoulder at 148 cm^{-1} is the fundamental SO phonon mode. The broad signal centered at 170 cm^{-1} could arise from confined TO phonons. The inset shows the dependence of the SO phonons on the inverse aspect ratio of the tetrapod arms. (Adapted from Krahne, R. et al., *Nano Lett.*, 6(3), 478, 2006a. With permission.) (b) Resonant Raman spectra of CdTe tetrapods for different laser excitation energies where the resonant enhancement of the LO phonon intensity is clearly visible (the peak at 514 cm^{-1} is the phonon of the Si substrate). (c) Plot of the LO phonon intensity and the photoluminescence recorded under comparable experimental conditions. We find that the LO phonon excitation is in resonance with the second, high energy exciton peak for which the carriers are localized in the tetrapod arms. (Adapted from Krahne, R. et al., *Nano Lett.*, 6(3), 478, 2006a. With permission.) (d) Schematic illustration of the different scattering processes in dots and tetrapods.

splitting of the LO and TO modes into parallel and perpendicular components that depends on the aspect ratio of the nanowire, which originates from the oscillating dipoles in the vibration of a polar lattice. In this model, the shape effect leads to an increasing blue shift of the perpendicular LO or TO phonon component with increasing aspect ratio of the nanowire. This behavior could be reflected in the Raman spectra of large aspect ratio tetrapods (12/80 nm for arm diameter/length) from Krahne et al. (2006b) displayed in Figure 7.21a. Here we find a sharp phonon line at 173.5 cm^{-1} with a broad underlying signal centered at 170 cm^{-1}. These values agree very well with the predictions of Mahan et al. (2003) for the parallel component of the LO and the perpendicular component of the TO phonon in CdTe nanowires, and the difference in peak widths could reflect the bulk like and the confined character of the phonon modes, respectively. If we record the LO phonon resonance at cryogenic temperature and plot it together with the corresponding emission spectrum (Figure 7.21b), we find that the phonon excitations are not in resonance with the lowest emitting exciton state (the dark exciton), but with the high-energy emission peak. Therefore, the phonons are in resonance with excitonic transitions, for which the carriers are localized in the arms of the tetrapods. The arms of the tetrapods make up for the largest portion of the tetrapod crystal, and consequently, it is not surprising that the crystal lattice vibrations are in resonance with excitations in the arms. This resonant behavior has strong impact on the spectra obtained in FLN experiments that for spherical nanocrystals can be used to detect the LO phonon excitations (Efros et al., 1996). In FLN, the nanocrystal sample is excited at the red edge of the lowest absorption peak, and phonon replica of the band edge emission of spherical nanocrystals can be detected. For tetrapod-shaped nanocrystals, these phonon replica of the band edge emission do not appear in the FLN spectrum because the phonons are not resonant with the lowest exciton energy transition that originates from the tetrapod core. The corresponding scattering processes are illustrated in Figure 7.21d.

The mode observed at the low-energy side of the LO phonon can be attributed to SO phonons. In rod- and tetrapod-shaped nanocrystals, the SO phonon energy depends on their aspect ratio (Gupta et al., 2003; Krahne et al., 2006a,b). The SO phonon energy can be calculated in a nanowire model as

$$\omega_{SO}^2 = \omega_{TO}^2 + \frac{\varpi_p^2}{\varepsilon_\infty + \varepsilon_m f(x)}; \quad x = q \cdot r \quad (7.6)$$

where

$\varpi_p^2 = \varepsilon_\infty(\omega_{LO}^2 - \omega_{TO}^2)$ is the screened ion-plasma frequency
$f(x) = I_0(x)K_1(x)/I_1(x)K_0(x)$ (with I and K Bessel functions)
ε_∞ and ε_m are the bulk CdTe high frequency and surrounding medium dielectric constants, respectively

The aspect ratio dependence of the SO phonons is plotted in the inset of Figure 7.21a, which shows that the energy decreases with increasing aspect ratio and that for aspect ratio equal to one,

the SO phonon energy for spherical nanocrystals is recovered (Ruppin and Englman, 1970; Chamberlain et al., 1995).

7.3.4 Electrical Properties of Tetrapods

The single electron transistor (SET) represents an ideal system for the investigation and exploitation of quantum effects in the electrical conduction of nanostructures, such as charging energies, electronic level spacing, and coupling of the electrical and mechanical properties. In a SET device, a conductive island is coupled via tunnel junctions to source and drain electrodes and capacitively to a third gate electrode (Grabert and Devoret, 1992). Figure 7.22a shows a schematic representation of a SET. In a small conductive island, discrete energy levels arise due to the finite charging energy that is needed to add another electron to the island. This charging energy can be written as $E_C = q^2/2C$, with q the charge and C the capacitance of the island, and this effect is referred to as Coulomb blockade. For semiconductor nanostructures, the Coulomb blockade is superimposed on their more complex electronic level structure, which was discussed in the optical spectroscopy section (Steiner et al., 2004).

In a theoretical work, Wang (2005) showed that surface effects, that is, the type of molecules that passivate the tetrapod surface, should have a significant impact on the electronic structure of tetrapods. In his atomistic pseudopotential method, he

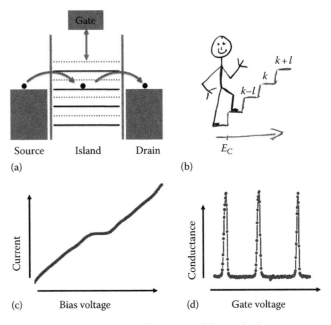

FIGURE 7.22 (a) Schematic illustration of the single electron transistor (SET) action. Source and drain electrodes are coupled to a conductive island via tunnel junctions. The electric potential of the island can be shifted via an external voltage applied to the gate electrode. (b) The finite size of the island results in discrete energy steps in order to charge it with an additional electron (or hole). (c) Typical experimental source-drain IV of a SET. (d) The conduction peaks arise from single electron tunneling when the Fermi levels of source and drain electrode align with an electronic level of the island.

integrated a term representing the surface polarization potential and found that the band gap and the charging energies depend strongly on the surface polarization potential.

The branched shape of tetrapods makes them interesting candidates for active elements in electronic and optoelectronic applications. On the one hand, the different arms can be exploited for a multiterminal device geometry; on the other hand, the branch point and the small diameter of the arms and the core should lead to novel quantum phenomena in the electrical conduction properties. Moreover, tetrapods deposited by drop casting on a substrate surface have the appealing property to self-align with three arms touching the surface and the fourth arm pointing vertically upward. Planar lithography techniques for the electrode fabrication, for example electron beam lithography (Sze, 1982), allow straightforward contact fabrication to the three base arms, as shown in the insets of Figure 7.23. Current voltage (*I–V*) measurements at cryogenic temperatures on CdTe tetrapod- and rod-shaped nanocrystals show Coulomb blockade, that is, a zero current plateau that corresponds in magnitude to the Coulomb

blockade energy (Cui et al., 2005). At first glance, it is surprising that the observed zero current plateau does not correspond to the magnitude of the band gap of the semiconductor nanocrystal. A generally accepted explanation is that the difference in the work functions between CdTe and the metal electrodes, typically Au or Pd, leads to a pinning of the Fermi energy within the valence band of the CdTe nanocrystals, in which the level density is too high to be experimentally resolved. Cui et al. (2005) found that in a certain number of their three terminal contacted tetrapod devices, the zero current plateau related to one of the arms was significantly higher than that of the other two arms. The origin of the difference in conductivity could be crystal defects, or increased mechanical strain related to this arm. This configuration allowed to exploit the high-resistance arm as a gate electrode as shown in Figure 7.22c. Here an AC voltage, with an amplitude inferior to the zero current region to avoid leakage, was applied to the third arm, and the effect on the source-drain conductivity at fixed bias for different values of the planar back gate was recorded. The high efficiency of the AC modulation suggests that the gating mechanism is effective in the tetrapod arm, i.e., that the gate voltage drops somewhere near the arm–core interface.

Another peculiar property of the branched shape of the tetrapods is that the conduction can be dominated by the electronic level structure of the core and the arms separately, or by the electronic structure of the tetrapod as a whole. These two cases are illustrated in Figure 7.24a. In the first case, the arm–core–arm pathway can be regarded as three conductive islands connected in series. To pass current, the energy levels of the three islands have to align within the thermal and source-drain bias energy window, such that the carriers can tunnel (or "hop") from one island to the other.

In Figure 7.24b, we see that for the lowest source-drain voltage bias of 1 mV, no conduction occurs, that is, the thermal energy alone is not sufficient. Increasing the bias leads to largely spaced conduction peaks that reflect the energy-level structure of the core, where the separation of the energy levels is larger due to the small size and thus increased confinement effects. At high bias, subsets of conduction, peaks appear that can be related to the more dense energy-level structure of the arms. Here the tetrapod acts as three conductive islands in series. In Figure 7.24c, even at the lowest source-drain bias, all the conduction peaks are already present. This points to the delocalization of the conduction charges over the entire tetrapod volume, leading to a dense level structure. In this case, the tetrapod acts as one conductive island that is connected to the source-drain electrodes. Electrostatic calculations that deduct the charging energies from the size of the tetrapod or its arms and core, respectively, confirm the above-described conduction mechanisms (Grabert and Devoret, 1992).

7.3.5 Mechanical Properties of Tetrapods

The mechanical properties of tetrapods can be studied by AFM. Here the main questions of interest are as follows: (1) Is a tetrapod after deposition on a substrate distorted due to adhesion

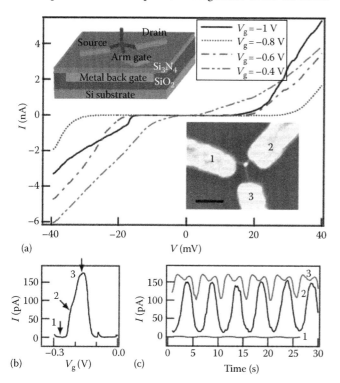

FIGURE 7.23 Single electron transistor based on a CdTe tetrapod: (a) The upper inset shows a schematic illustration of the device structure, in which the three tetrapod base arms are contacted by planar electrodes and an additional planar back gate is implemented in the substrate structure. The main panel displays two-terminal *I–V* curves for different voltages applied to the planar back gate, demonstrating transistor action. The lower inset shows an SEM image of a contacted tetrapod. (b) Source-drain conduction versus planar back gate voltage at fixed source-drain bias. The peak corresponds to electronic level alignment as discussed in Figure 7.22. (c) Source-drain current for the indicated back gate values in (b) and fixed source-drain voltage when a sinusoidal voltage modulation is applied to the third (gate) arm. (Adapted from Cui, Y. et al., *Nano Lett.*, 5(7), 1519, 2005. With permission.)

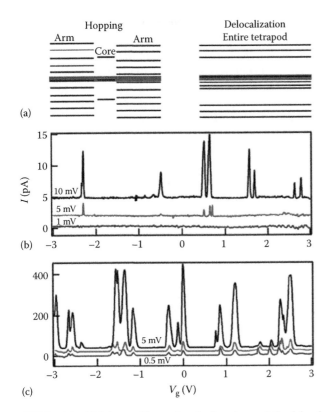

(a)

(b)

(c)

FIGURE 7.24 (a) Schemes illustrating the hopping and the delocalization models for the conduction process inside a tetrapod. The blue stripes indicate the voltage range of the thermal energy window. (b and c) Plots of current versus planar back gate voltage for different values of source-drain bias (1, 5, and 10 mV in (b) and 0.5, 1, and 5 mV in (c). (Adapted from Cui, Y. et al., *Nano Lett.*, 5(7), 1519, 2005. With permission.)

forces? (2) How hard can one push onto a tetrapod before it breaks, and how does the tetrapod respond to pressures below this threshold? (3) How do the electrical and optical properties depend on the exerted pressure?

Figure 7.25 shows different microscopy images of tetrapods from the same synthesis that confirm that typically tetrapods casted onto a surface align with three arms touching the surface,

and the fourth arm pointing vertically upward (appearing as a bright/dark spot in top-down AFM/TEM images, respectively). The tetrapod size of this sample obtained from TEM image analysis yields 10 and 100 nm for the arm diameter and length, respectively. By tapping mode AFM measurements, we obtained a height of 120 nm of, for example, the tetrapod in the upper center in Figure 7.25a (the instabilities in the feedback signal are most likely due to bending deformations of the tetrapod arm during the measurement). For an undistorted tetrapod, the arms should branch out from the core at tetrahedral angles of 109.5°, which for 100 nm long arms leads to a core–substrate distance of 33 nm. Thus, the height of an undistorted tetrapod should be around 138 nm. The measured height of tetrapods by AFM can be considerably lower due to two possibilities: the base arms are closer to the substrate surface due to attractive forces, and/or the vertical tetrapod arm had been broken during the AFM scan, as it was surely the case for the tetrapod with the lower contrast at the bottom left of Figure 7.25a. The SEM image recorded from a tilted angle in Figure 7.25c shows several tetrapods with undamaged vertical arms, in which the core–substrate distance is much smaller than the expected 33 nm, and therefore confirms the distortion of the tetrapod base arms.

Other than being imaged by AFM, tetrapods can also be used as probes in scanning probe microscopy (Nobile et al., 2008). In the simplest configuration, a single tetrapod is positioned on a previously flattened AFM tip with one arm pointing vertically downward, as shown in Figure 7.26a and b. In this geometry, the high aspect ratio of the tetrapod arms can be exploited for enhanced resolution in AFM topography imaging, as demonstrated in Figure 7.26c, in which a tetrapod deposited on an SiO$_2$ surface was imaged with a tetrapod-functionalized tip.

Fang et al. (2007) used tapping mode AFM and the force-volume technique to study the mechanical properties of CdTe tetrapods that had 8 and 130 nm arm diameter and length, respectively. By taking tapping mode AFM images of several tetrapods with different load forces, the regimes for elastic (below 90 nN) and inelastic (130 nN and above) deformation could be identified. Then force–volume maps were recorded for different

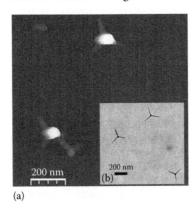

(a)

(b)

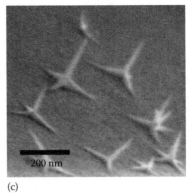

(c)

FIGURE 7.25 Images of CdTe tetrapods casted on different substrates. (a) AFM image of tetrapods drop casted on an Si substrate surface. (b) TEM image of the tetrapods on a carbon-coated TEM grid. (c) Tilted view SEM image of tetrapods on a gold-coated surface. (Adapted from Nobile, C. et al., *Small*, 4(12), 2123, 2008. With permission.)

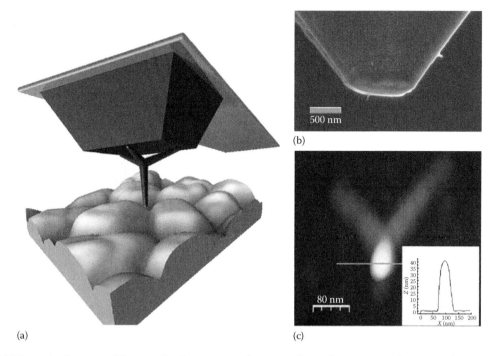

(a)

(b)

(c)

FIGURE 7.26 (a) Schematic of a tetrapod functionalized scanning probe tip together with a 3D image of Au grains that was recorded with such a tip. (b) SEM image of a single CdTe tetrapod positioned on a probe tip. (c) AFM topography image of a CdTe tetrapod that was imaged with the tetrapod-functionalized probe tip shown in (b). (Adapted from Nobile, C. et al., *Small*, 4(12), 2123, 2008. With permission.)

thresholds of load force. In the force–volume technique, a force–distance curve is recorded at every pixel of the map in x-y space. In order to measure the properties of a single tetrapod, the pixel density was chosen such that at least one curve was recorded on the top of the vertical arm of the tetrapod. A topography image and three force–distance curves up to different threshold limits are shown in Figure 7.27. Here the separation plotted on the x-axis is already corrected for the bending of the cantilever and therefore resembles the actual compression of the tetrapod.

Fang et al. (2007) found that a load of 130 nN leads to irreversible, plastic deformation of the tetrapod, that is, the fracture or breaking of one or more arms. In this case, the compression plus the arm diameter correspond to the initial height, meaning that the core is fully pushed onto the substrate surface. For the elastic regime, the spring constant for the tetrapod deformation can be obtained by dividing the load by the compression distance: 50 nN/4 nm = 12.5 N/m and 90 nN/9 nm = 10 N/m. To understand the nature of the deformation, the authors modeled two kinds of responses of the tetrapod to the applied force: freely sliding contact points of the base arms with the surface, as shown in Figure 7.28a (bottom section), and fixed contact points that lead to buckling of the arms, as shown in Figure 7.28a (top section). The simulation used the valence–force–field method containing nearest neighbor bond stretching, bond angle bending, and bond length/bond angle terms fitted to the experimental bulk elastic constants. This atomistic model considered a tetrapod with the same aspect ratio as the experimental tetrapods, but with a size reduced by a factor of 3.

The assumption of fixed contact points for the base arms of the tetrapods seems to agree better, both quantitatively and

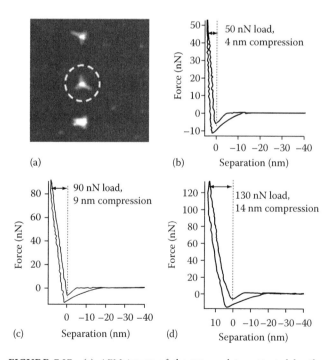

(a)

(b) 50 nN load, 4 nm compression

(c) 90 nN load, 9 nm compression

(d) 130 nN load, 14 nm compression

FIGURE 7.27 (a) AFM image of the tetrapod investigated by the force–volume technique. (b–d) Force–volume curves recorded on top of the vertical tetrapod arm with different load thresholds. The initial maximum height measured in (a) was 21 nm. (Reprinted from Fang, L. et al., *J. Chem. Phys.*, 127(18), 184704, 2007. With permission.)

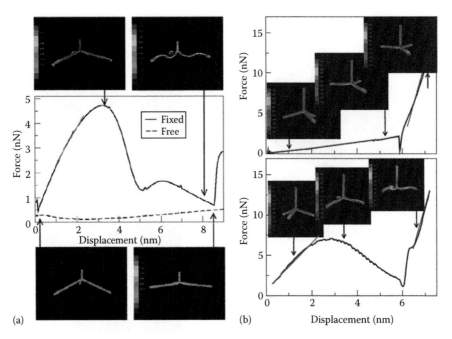

FIGURE 7.28 Simulation of the elastic deformation of CdTe (a) and CdSe (b) tetrapods caused by a force exerted on the vertical arm. Panel (a): The parameters used for the CdTe tetrapods correspond to the sample studied by the force–volume technique shown in Figure 7.27, where *a a* spring constant of 10–12.5 N/m was experimentally obtained. Full line shows the force curve related to buckling, and the dotted line corresponds to sliding of the tetrapod arms. (Adapted from Fang, L. et al., *J. Chem. Phys.*, 127(18), 184704, 2007. With permission.) Panel (b): Model for CdSe tetrapods with arm diameter/length of 2.6/21 nm and core diameter of 3.3 nm. (Adapted from Schrier, J. et al., *J. Nanosci. Nanotechnol.*, 8(4), 1994, 2008. With permission of American Scientific Publishers.)

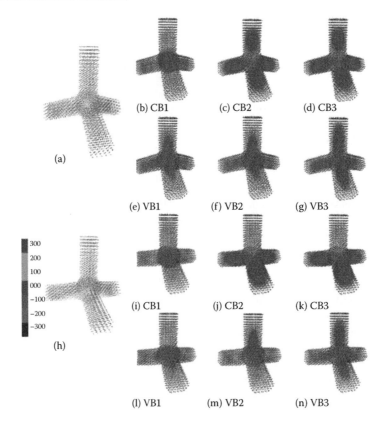

FIGURE 7.29 Electron and hole wave function states obtained from atomistic calculations as described for Figure 7.15c for unstrained (a–g) and strained (h–n) CdSe tetrapods. For the strained tetrapods, an applied force of 6.2 nN was taken that results in completely flattened base arms against the surface. (Adapted from Schrier, J. et al., *J. Nanosci. Nanotechnol.*, 8(4), 1994, 2008. With permission of American Scientific Publishers.)

qualitatively, with the experimental results. On the one hand, it gives a higher spring constant for the regime of small forces, which is closer to the experimental value than the result for free sliding arms. On the other hand, it yields a decrease in spring constant for large displacement that occurs also in the experiment. A similar calculation for CdSe tetrapods, displayed in the right panel of Figure 7.28, considers undamaged tetrapods and spans a larger force range. The linear regime, approximated with the red line, reflects the elastic deformation of the base arms, whereas for displacements larger than 6 nm, the compression of the vertical arm is dominant, leading to a much higher spring constant.

The elastic deformation of the tetrapods modifies also their electronic structure, and consequently their electrical and optical spectra. An atomistic calculation that combines the methods to calculate the electronic structure with the method to simulate the deformation shows how the wave function distribution of the electronic states is affected by the applied force. In Figure 7.29, we see that especially the electron wave functions are sensitive to the deformation, which leads to a stronger localization in the three base arms. The effect of the strain on the optical transitions is a red shift and the lifting of the degeneracy of the previously doubly degenerated levels.

Conductive AFM measurements with a TiN-coated tip on the tetrapods described in Figure 7.27 that were immobilized on a precleaned Au surface show a zero current region that extends up to 2.2 V at minimum pressure (Fang et al., 2007) (see Figure 7.30). At first, we notice that the zero current region is orders of magnitude larger than the one reported in Cui et al. (2005) shown in Figure 7.23. This is most likely due to the use of a low work function material like TiN as one electrode that leads to a different pinning of the Fermi level, that is, within the CdTe tetrapod band gap. We note that scanning tunneling spectroscopy studies, which use tungsten tips, on CdSe nanorods also report zero current plateaus that correspond to the nanocrystal band gap (Millo et al., 2004).

Figure 7.30 shows a significant decrease of the zero current region with increasing applied load force. However, to deduct the energy gap of the tetrapod from source-drain current measurements is problematic. On the one hand, the leverage of the applied voltage across the device is unknown. On the other hand, from our discussion related to Figure 7.22, we know that a gate potential has significant impact on the source-drain *I–V*. In the case of the conductive AFM measurement, the value of the gate potential is arbitrarily defined by charges present in the vicinity of the tetrapod, and therefore it is also unknown.

7.4 Assembly of Tetrapods

7.4.1 Some Self-Assembly Concepts for Spherical and Rod-Shaped Nanocrystals

If one considers colloidal nanocrystals as building blocks, the main focus of research for what concerns their assembly has been so far directed at their organization in ordered superstructures, either promoted by self-assembly processes or by deliberately driving nanoparticle organization by means of external perturbations. Examples include the preparation of long-range ordered superlattices of nearly monodisperse spherical nanocrystals, obtained on slow evaporation of the solvent from concentrated colloidal solutions (Redl et al., 2003; Shevchenko et al., 2006; Chen et al., 2007). More elaborate examples in this direction include the self-assembly of combinations of spherical nanocrystals of different sizes and materials in binary or ternary superlattices (Redl et al., 2003; Shevchenko et al., 2006; Chen et al., 2007). These structures are interesting from the fundamental point of view as they mimic the organization of atoms into crystals, and because it should be possible to extract useful collective properties arising from ordered superstructure organization, so that they can be implemented in practical materials and devices.

Self-assembly of shape-controlled nanocrystals, such as nanorods, is even more demanding than for spherical nanocrystals, because it requires both positional and orientational ordering of individual nanorods. As a result, nanorod assemblies with long-range order have proven to be more difficult to fabricate by means of slow solvent evaporation methods alone (Li and Alivisatos, 2003b), though, liquid crystalline-like self-assembly in both smectic and nematic phase-like superstructures was reported for rod-shaped CdSe nanocrystals (Li et al., 2002, Li and Alivisatos, 2003b), and self-organization of CdSe nanorods into

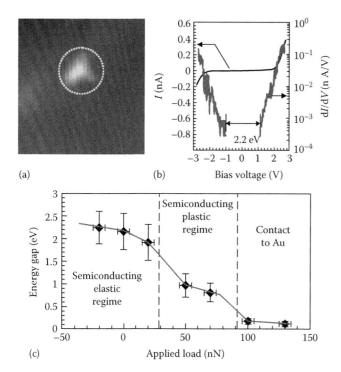

FIGURE 7.30 Conductive AFM measurements on CdTe tetrapods immobilized on a precleaned Au surface. (a) Topography image, (b) *I–V* and logarithmic plot of the differential conductance that reveals the zero current plateau related to the energy gap, (c) measured and calculated energy gap versus applied load force. (Reprinted from Fang, L. et al., *J. Chem. Phys.*, 127(18), 184704, 2007. With permission.)

3D superlattices was observed by destabilization of the solvent evaporation from the corresponding colloidal solution on slow diffusion of a nonsolvent (Talapin et al., 2004) Additionally, colloidal nanorods have been aligned in both vertically (Ryan et al., 2006; Ahmed and Ryan, 2007; Carbone et al., 2007) and laterally ordered arrays (Talapin et al., 2004; Sun and Sirringhaus, 2006) using a wide variety of techniques, which exploited inter-rod van der Waals or magnetic forces, interactions with applied electric fields, or via substrate templating effects (Talapin et al., 2007b; Wetz et al., 2007; Querner et al., 2008) and via depletion forces (Barnov et al., 2010).

7.4.2 Approaches for the Controlled Assembly of Tetrapods

Even less studied is the assembly of branched nanostructures such as the tetrapod-shaped nanocrystals that we have studied in this chapter, mainly because it is much harder to realize superstructures with such nanocrystals and clearly one has to specify here what is really meant by assembly of complex-shaped nanocrystals. One cannot really fabricate a 3D ordered superlattice of tetrapods as is done with spherical nanocrystals, that is, based on close-packed organization of the building blocks. Therefore, by assembly here one means mainly a "somehow" controlled organization of tetrapods on a substrate. So far, only minor efforts have been undertaken in this direction, and indeed were limited to the controlled deposition of tetrapods on substrates (Cui et al., 2004; Fang et al., 2007). We have already seen that tetrapods self-align when deposited on a planar surface, with three arms touching the surface and the fourth pointing vertically upward. The degree of order can be enhanced by specifically patterned substrate surfaces. As an example, Cui et al. (2004) fabricated nanoscale trenches in a polymer film on Au-coated Si substrates, after which they immersed the patterned substrates vertically into a solvent solution containing CdTe tetrapods, and found that the capillary forces during solvent evaporation lead to oriented assemblies of the tetrapods inside the trenches (see Figure 7.31a and b). Another example of assembly of tetrapods can be obtained via electrostatic trapping (Nobile et al., 2008). By this method, the tetrapods are forced toward the region of strongest electric field, for example, onto the extremity of a metallized AFM tip (as it has been already shown in Figure 7.26), or in between electrode pairs. This approach can be used to position single tetrapods in between electrodes with gaps of few tens of nanometers.

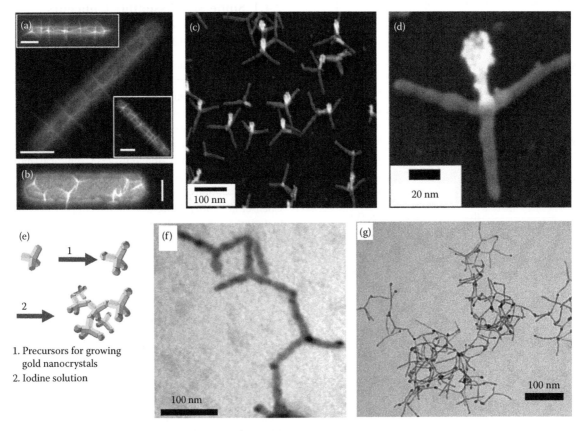

1. Precursors for growing
   gold nanocrystals
2. Iodine solution

FIGURE 7.31 (a and b) SEM images of tetrapod assemblies in nanotrenches. The scale bars here are all 200 nm long. (Reprinted from Cui, Y. et al., *Nano Lett.*, 4(6), 1093, 2004. With permission.) (c–d) SEM images of tetrapods deposited on a substrate and selectively decorated with Au nanoparticles at the tip of their vertically standing arms. (Adapted from Liu, H.T. and Alivisatos, A.P., *Nano Lett.*, 4(12), 2397, 2004. With permission.) (e) A sketch of the "nanosoldering" approach to connect gold-tipped tetrapods into network structures. (f and g) TEM images of two network structures obtained by connecting ZnTe(core)/CdTe(arm) tetrapods via Au tips. (Reprinted from Figuerola, A. et al., *Adv. Mater.*, 21, 550, 2009. With permission.)

Assembly of tetrapods in 3D structures has been proposed recently by our group. II–VI semiconductor tetrapods were organized into network structures using gold domains as linkers, which resulted in an end-to-end connection in between the arms of different tetrapods (Figuerola et al., 2009). This approach exploits the shape anisotropy of nanocrystals to grow small metallic Au nanoparticles on selected locations of their surface, basically at their tips [an approach that was reported for the first time by Banin and coworkers (Mokari et al., 2004)]. Small amounts of molecular iodine are used to destabilize the Au domains grown on the arm tips and to induce the coalescence of Au domains belonging to different nanocrystals, thus forming larger Au particles, each of them bridging two or more tetrapods through their tips (see Figure 7.31e through g). This strategy introduces an inorganic and robust junction between nanocrystals and hence avoids the use of molecular organic spacers for the assembly (Salant et al., 2006). It works also in connecting nanorods in chain-like structures (see Figure 7.31f).

Site-selective decoration of one of the tetrapod tips with Au nanoparticles was also achieved by spin coating a polymer onto a substrates covered with tetrapods, such that the tetrapods were partially protected (Liu and Alivisatos, 2004). The Au nanoparticles were attached to the uncovered tips of the vertical arms via dithiol linkers. The authors also demonstrated that it was possible to break of the uncovered, gold-decorated vertical arms and by this way, they obtained CdTe rods with Au particles on only one end (see Figure 7.31c and d).

7.5 Conclusions and Outlook

Research on tetrapod-shaped colloidal nanocrystals is being boosted by more and more refined synthesis approaches to such type of nanoparticles. In the last few years, it has been possible to synthesize tetrapods with considerably narrow distributions of arm lengths and diameters, and even more interesting, to fabricate tetrapods whose central region is of different chemical composition than that of the arms. Additional advances in synthesis and functionalization have been the selective growth of metal domains at tetrapod tips, or the attachment of metal domains selectively only on one arm. All these high-quality samples have certainly paved the way to several interesting experiments aimed at assessing their structure and their physical properties, as discussed in this chapter. Perhaps, one interesting development will come from an advanced synthesis of tetrapods in which each of the arms will be of a different material. This will introduce both novel functionalities (electrons and hole could be localized in different arms, or one arm could act as an effective gate in a single-tetrapod device), but also chirality in nanocrystals.

For what concerns the assembly of tetrapods, important directions toward which research will likely orient will be (a) the controlled deposition of single layer of tetrapods on a substrate, for thin film photovoltaic applications; (b) the controlled anchorage of tetrapods on substrates, for applications, in field emitters, and

single nanocrystal transistors. Another direction could be the realization of complex 3D networks of branched nanocrystals joined to each other via their tips, which would lead to open framework superstructures. Applications of these assemblies could be in various areas. Nanocomposites realized by this approach would enlarge the toolkits of materials available to scientists and engineers in addition to the more traditional mesoporous materials like zeolites or the sol-gel-derived porous monoliths, with interesting applications in lightweight, high-performing materials, catalysis, or even in tissue engineering, and as such structures could be additionally envisaged to act as scaffolds.

References

Ahmed, S. and K. M. Ryan. 2007. Self-assembly of vertically aligned nanorod supercrystals using highly oriented pyrolytic graphite. *Nano Letters* 7(8): 2480–2485.

Al Salman, A., A. Tortschanoff, M. B. Mohamed et al. 2007. Temperature effects on the spectral properties of colloidal CdSe nanodots, nanorods, and tetrapods. *Applied Physics Letters* 90(9): 093104/1–093104/3.

Ashcroft, N. W. and N. D. Mermin. 1976. *Solid State Physics*. Orlando, FL: Saunders College Publishers.

Asokan, S., K. M. Krueger, V. L. Colvin, and M. S. Wong. 2007. Shape-controlled synthesis of CdSe tetrapods using cationic surfactant ligands. *Small* 3(7): 1164–1169.

Baranov, D., Fiore, A., Van Huis et al. 2010. Assembly of colloidal semiconductor nanorods in solution by depletion attraction. *Nano Letters* 10(2): 743–749.

Barnard, A. S. and H. Xu. 2007. First principles and thermodynamic modeling of CdS surfaces and nanorods. *Journal of Physical Chemistry C* 111(49): 18112–18117.

Bastard, G. 1991. *Wave Mechanics Applied to Semiconductor Heterostructures*. New York: John Wiley & Sons.

Bechstedt, F., P. Kackell, A. Zywietz et al. 1997. Polytypism and properties of silicon carbide. *Physica Status Solidi B—Basic Research* 202(1): 35–62.

Bunge, S. D., K. M. Krueger, T. J. Boyle et al. 2003. Growth and morphology of cadmium chalcogenides: The synthesis of nanorods, tetrapods, and spheres from CdO and $Cd(O_2CCH_3)(2)$. *Journal of Materials Chemistry* 13(7): 1705–1709.

Burt, J. L., J. L. Elechiguerra, J. Reyes-Gasga, J. M. Montejano-Carrizales, and M. Jose-Yacaman. 2005. Beyond Archimedean solids: Star polyhedral gold nanocrystals. *Journal of Crystal Growth* 285(4): 681–691.

Carbone, L., S. Kudera, E. Carlino et al. 2006. Multiple wurtzite twinning in CdTe nanocrystals induced by methylphosphonic acid. *Journal of the American Chemical Society* 128(3): 748–755.

Carbone, L., C. Nobile, M. De Giorgi et al. 2007. Synthesis and micrometer-scale assembly of colloidal CdSe/CdS nanorods prepared by a seeded growth approach. *Nano Letters* 7(10): 2942–2950.

Chamberlain, M. P., C. Tralleroginer, and M. Cardona. 1995. Theory of one-phonon Raman-scattering in semiconductor microcrystallites. *Physical Review B* 51(3): 1680–1693.

Chen, M., Y. Xie, J. Lu et al. 2002. Synthesis of rod-, twinrod-, and tetrapod-shaped CdS nanocrystals using a highly oriented solvothermal recrystallization technique. *Journal of Materials Chemistry* 12(3): 748–753.

Chen, S. H., Z. L. Wang, J. Ballato, S. H. Foulger, and D. L. Carroll. 2003. Monopod, bipod, tripod, and tetrapod gold nanocrystals. *Journal of the American Chemical Society* 125(52): 16186–16187.

Chen, Z., Z. W. Shan, M. S. Cao, L. Lu, and S. X. Mao. 2004. Zinc oxide nanotetrapods. *Nanotechnology* 15(3): 365–369.

Chen, Z. Y., J. Moore, G. Radtke, H. Sirringhaus, and S. O'Brien. 2007. Binary nanoparticle superlattices in the semiconductor-semiconductor system: CdTe and CdSe. *Journal of the American Chemical Society* 129(50): 15702–15709.

Cho, J. W., H. S. Kim, Y. J. Kim et al. 2008. Phase-tuned tetrapod-shaped CdTe nanocrystals by ligand effect. *Chemistry of Materials* 20(17): 5600–5609.

Cozzoli, P. D., E. Snoeck, M. A. Garcia et al. 2006. Colloidal synthesis and characterization of tetrapod-shaped magnetic nanocrystals. *Nano Letters* 6(9): 1966–1972.

Cui, Y., M. T. Bjork, J. A. Liddle et al. 2004. Integration of colloidal nanocrystals into lithographically patterned devices. *Nano Letters* 4(6): 1093–1098.

Cui, Y., U. Banin, M. T. Bjork, and A. P. Alivisatos. 2005. Electrical transport through a single nanoscale semiconductor branch point. *Nano Letters* 5(7): 1519–1523.

Dai, Z. R., S. H. Sun, and Z. L. Wang. 2001. Phase transformation, coalescence, and twinning of monodisperse FePt nanocrystals. *Nano Letters* 1(8): 443–447.

Dai, Y., Y. Zhang, and Z. L. Wang. 2003. The octa-twin tetraleg ZnO nanostructures. *Solid State Communications* 126(11): 629–633.

Ding, Y., Z. L. Wang, T. J. Sun, and J. S. Qiu. 2007. Zinc-blende ZnO and its role in nucleating wurtzite tetrapods and twinned nanowires. *Applied Physics Letters* 90(15): 153510.

Donega, C. D., P. Liljeroth, and D. Vanmaekelbergh. 2005. Physicochemical evaluation of the hot-injection method, a synthesis route for monodisperse nanocrystals. *Small* 1(12): 1152–1162.

Dushkin, C. D., S. Saita, K. Yoshie, and Y. Yamaguchi. 2000. The kinetics of growth of semiconductor nanocrystals in a hot amphiphile matrix. *Advances in Colloid and Interface Science* 88(1–2): 37–78.

Efros, A. L., M. Rosen, M. Kuno et al. 1996. Band-edge exciton in quantum dots of semiconductors with a degenerate valence band: Dark and bright exciton states. *Physical Review B* 54(7): 4843–4856.

Eisberg, R. and R. Resnick. 1985. *Quantum Physics of Atoms, Molecules, Solids, Nuclei, and Particles*. New York: John Wiley & Sons.

Elechiguerra, J. L., J. Reyes-Gasga, and M. J. Yacaman. 2006. The role of twinning in shape evolution of anisotropic noble metal nanostructures. *Journal of Materials Chemistry* 16(40): 3906–3919.

Fang, L., J. Y. Park, Y. Cui et al. 2007. Mechanical and electrical properties of CdTe tetrapods studied by atomic force microscopy. *The Journal of Chemical Physics* 127(18): 184704.

Figuerola, A., I. R. Franchini, A. Fiore et al. 2009. End-to-end assembly of shape-controlled nanocrystals via a nanowelding approach mediated by gold domains. *Advanced Materials* 21: 550–554.

Fiore, A., R. Mastria, M. Lupo et al. 2009. Tetrapod-shaped colloidal nanocrystals of II-VI semiconductors prepared by seeded growth. *Journal of the American Chemical Society* 131(6): 2274–2282.

Fujii, M., H. Iwanaga, M. Ichihara, and S. Takeuchi. 1993. Structure of Tetrapod-like ZnO crystals. *Journal of Crystal Growth* 128(1–4): 1095–1098.

Gong, J. F., S. G. Yang, H. B. Huang et al. 2006. Experimental evidence of an octahedron nucleus in ZnS tetrapods. *Small* 2(6): 732–735.

Grabert, H. and M. H. Devoret. 1992. *Single Charge Tunneling—Coulomb Blockade Phenomena in Nanostructures*. New York: Kluwer Academic/Plenum Publisher.

Gupta, R., Q. Xiong, G. D. Mahan, and P. C. Eklund. 2003. Surface optical phonons in gallium phosphide nanowires. *Nano Letters* 3(12): 1745–1750.

Gur, I., N. A. Fromer, and A. P. Alivisatos. 2006. Controlled assembly of hybrid bulk-heterojunction solar cells by sequential deposition. *Journal of Physical Chemistry B* 110(50): 25543–25546.

Gur, I., N. A. Fromer, C. P. Chen, A. G. Kanaras, and A. P. Alivisatos. 2007. Hybrid solar cells with prescribed nanoscale morphologies based on hyperbranched semiconductor nanocrystals. *Nano Letters* 7(2): 409–414.

Habas, S. E., H. Lee, V. Radmilovic, G. A. Somorjai, and P. Yang. 2007. Shaping binary metal nanocrystals through epitaxial seeded growth. *Nature Materials* 6(9): 692–697.

Harrison, W. A. 1989. *Electronic Structure and the Properties of Solids*. New York: Dover Publications.

Hsu, Y. S. and S. Y. Lu. 2008. Dopant-induced formation of branched CdS nanocrystals. *Small* 4(7): 951–955.

Hu, J. T., L. S. Li, W. D. Yang et al. 2001. Linearly polarized emission from colloidal semiconductor quantum rods. *Science* 292(5524): 2060–2063.

Hu, J. Q., Y. S. Bando, and D. Golberg. 2005. Sn-catalyzed thermal evaporation synthesis of Tetrapod-branched ZnSe nanorod architectures. *Small* 1(1): 95–99.

Hurlbut, C. S., C. Klein, and J. D. Dana. 1998. *Manual of Mineralogy*, 21st edn. New York: John Wiley & Sons.

Iwanaga, H., M. Fujii, and S. Takeuchi. 1993. Growth-model of Tetrapod zinc-oxide particles. *Journal of Crystal Growth* 134(3–4): 275–280.

Iwanaga, H., M. Fujii, and S. Takeuchi. 1998. Inter-leg angles in tetrapod ZnO particles. *Journal of Crystal Growth* 183(1–2): 190–195.

Jana, N. R., L. Gearheart, and C. J. Murphy. 2001a. Seed-mediated growth approach for shape-controlled synthesis of spheroidal and rod-like gold nanoparticles using a surfactant template. *Advanced Materials* 13(18): 1389–1393.

Jana, N. R., L. Gearheart, and C. J. Murphy. 2001b. Wet chemical synthesis of high aspect ratio cylindrical gold nanorods. *Journal of Physical Chemistry B* 105(19): 4065–4067.

Jun, Y. W., S. M. Lee, N. J. Kang, and J. Cheon. 2001. Controlled synthesis of multi-armed CdS nanorod architectures using monosurfactant system. *Journal of the American Chemical Society* 123(21): 5150–5151.

Jun, Y. W., J. S. Choi, and J. Cheon. 2006. Shape control of semiconductor and metal oxide nanocrystals through nonhydrolytic colloidal routes. *Angewandte Chemie-International Edition* 45(21): 3414–3439.

Kitano, M., T. Hamabe, S. Maeda, and T. Okabe. 1991. Growth of large Tetrapod-like ZnO crystals. 2. Morphological considerations on growth-mechanism. *Journal of Crystal Growth* 108(1–2): 277–284.

Kittel, C. 1996. *Introduction to Solid State Physics*, 7th edn. New York: John Wiley & Sons.

Klimov, V. I. 2000. Optical nonlinearities and ultrafast carrier dynamics in semiconductor nanocrystals. *Journal of Physical Chemistry B* 104(26): 6112–6123.

Klimov, V. I. 2003. *Semiconductor and Metal Nanocrystals: Synthesis and Electronic and Optical Properties*. New York: CRC Press.

Klimov, V. I. and D. W. McBranch. 1997. Auger-process-induced charge separation in semiconductor nanocrystals. *Physical Review B* 55(19): 13173–13179.

Klimov, V. I., D. W. McBranch, C. A. Leatherdale, and M. G. Bawendi. 1999. Electron and hole relaxation pathways in semiconductor quantum dots. *Physical Review B* 60(19): 13740–13749.

Krahne, R., G. Chilla, C. Schuller et al. 2006a. Confinement effects on optical phonons in polar tetrapod nanocrystals detected by resonant inelastic light scattering. *Nano Letters* 6(3): 478–482.

Krahne, R., G. Chilla, C. Schuller et al. 2006b. Shape dependence of the scattering processes of optical phonons in colloidal nanocrystals detected by Raman Spectroscopy. *Journal of Nanoelectronics and Optoelectronics* 1(1): 104–107.

Kudera, S., L. Carbone, M. F. Casula et al. 2005. Selective growth of PbSe on one or both tips of colloidal semiconductor nanorods. *Nano Letters* 5(3): 445–449.

Landolt-Boernstein 1998. *Group III Condensed Matter*, Vol. 41C. Springer-Verlag GmbH, Germany.

Lawaetz, P. 1972. Stability of the wurtzite structure. *Physical Review B-Condensed Matter* 5(10): 4039–4045.

Lee, S. M., S. N. Cho, and J. Cheon. 2003. Anisotropic shape control of colloidal inorganic nanocrystals. *Advanced Materials* 15(5): 441–444.

Li, L. S. and A. P. Alivisatos. 2003a. Origin and scaling of the permanent dipole moment in CdSe nanorods. *Physical Review Letters* 90(9): 097402–097405.

Li, L. S. and A. P. Alivisatos. 2003b. Semiconductor nanorod liquid crystals and their assembly on a substrate. *Advanced Materials* (*Weinheim, Germany*) 15(5): 408–411.

Li, J. B. and L. W. Wang. 2003. Shape effects on electronic states of nanocrystals. *Nano Letters* 3(10): 1357–1363.

Li, L. S., J. Walda, L. Manna, and A. P. Alivisatos. 2002. Semiconductor nanorod liquid crystals. *Nano Letters* 2(6): 557–560.

Li, Y. C., H. Z. Zhong, R. Li et al. 2006. High-yield fabrication and electrochemical characterization of tetrapodal CdSe, CdTe, and CdSe$_x$Te$_{1-x}$ nanocrystals. *Advanced Functional Materials* 16(13): 1705–1716.

Lim, B., Y. J. Xiong, and Y. N. Xia. 2007. A water-based synthesis of octahedral, decahedral, and icosahedral Pd nanocrystals. *Angewandte Chemie—International Edition* 46(48): 9279–9282.

Liu, H. T. and A. P. Alivisatos. 2004. Preparation of asymmetric nanostructures through site selective modification of tetrapods. *Nano Letters* 4(12): 2397–2401.

Madelung, O., M. Schulz, and H. Weiss. 1982. *Landolt-Börnstein, New Series, Group III*, Vol. 17b. Berlin, Germany: Springer-Verlag.

Mahan, G. D., R. Gupta, Q. Xiong, C. K. Adu, and P. C. Eklund. 2003. Optical phonons in polar semiconductor nanowires. *Physical Review B* 68(7): art. no. 73402.

Maksimuk, S., X. Teng, and H. Yang. 2007. Roles of twin defects in the formation of platinum multipod nanocrystals. *Journal of Physical Chemistry C* 111(39): 14312–14319.

Malkmus, S., S. Kudera, L. Manna, W. J. Parak, and M. Braun. 2006. Electron-hole dynamics in CdTe tetrapods. *Journal of Physical Chemistry B* 110(35): 17334–17338.

Manna, L., E. C. Scher, and A. P. Alivisatos. 2000. Synthesis of soluble and processable rod-, arrow-, teardrop-, and tetrapod-shaped CdSe nanocrystals. *Journal of the American Chemical Society* 122(51): 12700–12706.

Manna, L., D. J. Milliron, A. Meisel, E. C. Scher, and A. P. Alivisatos. 2003. Controlled growth of tetrapod-branched inorganic nanocrystals. *Nature Materials* 2(6): 382–385.

Manna, L., L. W. Wang, R. Cingolani and A. P. Alivisatos. 2005. First-principles modeling of unpassivated and surfactant-passivated bulk facets of wurtzite CdSe: A model system for studying the anisotropic growth of CdSe nanocrystals. *Journal of Physical Chemistry B* 109(13): 6183–6192.

Markov, I. V. 2003. *Crystal Growth for Beginners: Fundamentals of Nucleation, Crystal Growth, and Epitaxy*. Singapore: World Scientific.

Mauser, C., T. Limmer, E. Da Como et al. 2008. Anisotropic optical emission of single CdSe/CdS tetrapod heterostructures: Evidence for a wavefunction symmetry breaking. *Physical Review B* 77(15): 153303.

Millo, O., D. Katz, D. Steiner et al. 2004. Charging and quantum size effects in tunnelling and optical spectroscopy of CdSe nanorods. *Nanotechnology* 15(1): R1–R6.

Mokari, T., E. Rothenberg, I. Popov, R. Costi, and U. Banin. 2004. Selective growth of metal tips onto semiconductor quantum rods and tetrapods. *Science* 304(5678): 1787–1790.

Morello, G., D. Tari, L. Carbone et al. 2008. Radiative recombination dynamics in tetrapod-shaped CdTe nanocrystals: Evidence for a photoinduced screening of the internal electric field. *Applied Physics Letters* 92(19): art. no. 191905.

Mullins, W. W. and R. F. Sekerka. 1964. The stability of a planar interface during solidification of a dilute binary alloy. *Journal of Applied Physics* 35: 444.

Na, Y. J., H. S. Kim, and J. Park. 2008. Morphology-controlled Lead Selenide nanocrystals and their in situ growth on carbon nanotubes. *Journal of Physical Chemistry C* 112(30): 11218–11226.

Nann, T. and J. Schneider. 2004. Origin of permanent electric dipole moments in wurtzite nanocrystals. *Chemical Physics Letters* 384(1–3): 150–152.

Nehl, C. L., H. Liao, and J. H. Hafner. 2006. Synthesis and optical properties of star-shaped gold nanoparticles. *Nano Letters* 6: 683–686.

Nikoobakht, B. and M. A. El-Sayed. 2003. Preparation and growth mechanism of gold nanorods (NRs) using seed-mediated growth method. *Chemistry of Materials* 15(10): 1957–1962.

Nirmal, M., C. B. Murray, and M. G. Bawendi. 1994. Fluorescence-line narrowing in CdSe quantum dots—Surface localization of the photogenerated exciton. *Physical Review B—Condensed Matter* 50(4): 2293–2300.

Nishio, K., T. Isshiki, M. Kitano, and M. Shiojiri. 1997. Structure and growth mechanism of tetrapod-like ZnO particles. *Philosophical Magazine A-Physics of Condensed Matter Structure Defects and Mechanical Properties* 76(4): 889–904.

Nobile, C., S. Kudera, A. Fiore et al. 2007. Confinement effects on optical phonons in spherical, rod-, and tetrapod-shaped nanocrystals detected by Raman spectroscopy. *Physica Status Solidi A—Applications and Materials Science* 204(2): 483–486.

Nobile, C., P. D. Ashby, P. J. Schuck et al. 2008. Probe tips functionalized with colloidal nanocrystal Tetrapods for high-resolution atomic force microscopy imaging. *Small* 4(12): 2123–2126.

Norris, D. J., A. L. Efros, M. Rosen, and M. G. Bawendi. 1996. Size dependence of exciton fine structure in CdSe quantum dots. *Physical Review B—Condensed Matter* 53(24): 16347–16354.

Pang, Q., L. J. Zhao, Y. Cai et al. 2005. CdSe nano-tetrapods: Controllable synthesis, structure analysis, and electronic and optical properties. *Chemistry of Materials* 17(21): 5263–5267.

Park, J., J. Joo, S. G. Kwon, Y. Jang, and T. Hyeon. 2007. Synthesis of monodisperse spherical nanocrystals. *Angewandte Chemie-International Edition* 46(25): 4630–4660.

Pasquarello, A. and R. Car. 1997. Dynamical charge tensors and infrared spectrum of Amorphous SiO_2. *Physical Review Letters* 79: 1766.

Peng, Z. A. and X. G. Peng. 2001. Mechanisms of the shape evolution of CdSe nanocrystals. *Journal of the American Chemical Society* 123(7): 1389–1395.

Peng, X. G., J. Wickham, and A. P. Alivisatos. 1998. Kinetics of II-VI and III-V colloidal semiconductor nanocrystal growth: "Focusing" of size distributions. *Journal of the American Chemical Society* 120(21): 5343–5344.

Peng, X. G., L. Manna, W. D. Yang et al. 2000. Shape control of CdSe nanocrystals. *Nature* 404(6773): 59–61.

Peng, P., D. J. Milliron, S. M. Hughes et al. 2005. Femtosecond spectroscom of carrier relaxation dynamics in type IICdSe/CdTe tetrapod heteronanostructures. *Nano Letters* 5(9): 1809–1813.

Qu, L. H., Z. A. Peng, and X. G. Peng. 2001. Alternative routes toward high quality CdSe nanocrystals. *Nano Letters* 1(6): 333–337.

Querner, C., M. D. Fischbein, P. A. Heiney, and M. Drndic. 2008. Millimeter-scale assembly of CdSe nanorods into smectic superstructures by solvent drying kinetics. *Advanced Materials* 20(12): 2308–2314.

Randle, V. 1997. *The Role of the Coincidence Site Lattice in Grain Boundary Engineering*. Cambridge, U.K.: Woodhead Publishing Limited.

Redl, F. X., K. S. Cho, C. B. Murray, and S. O'Brien. 2003. Three-dimensional binary superlattices of magnetic nanocrystals and semiconductor quantum dots. *Nature* 423(6943): 968–971.

Reiss, H. J. 1951. The growth of uniform colloidal dispersions. *Journal of Chemical Physics* 19(4): 482–487.

Rempel, J. Y., B. L. Trout, M. G. Bawendi, and K. F. Jensen. 2005. Properties of the CdSe(0001), (0001), and (1120) single crystal surfaces: Relaxation, reconstruction, and adatom and admolecule adsorption. *Journal of Physical Chemistry B* 109(41): 19320–19328.

Ruppin, R. and R. Englman. 1970. Optical phonons of small crystals. *Reports on Progress in Physics* 33: 149–196.

Ryan, K. M., A. Mastroianni, K. A. Stancil, H. T. Liu, and A. P. Alivisatos. 2006. Electric-field-assisted assembly of perpendicularly oriented nanorod superlattices. *Nano Letters* 6(7): 1479–1482.

Salant, A., E. Amitay-Sadovsky, and U. Banin. 2006. Directed self-assembly of gold-tipped CdSe nanorods. *Journal of the American Chemical Society* 128(31): 10006–10007.

Schrier, J., B. Lee, and L. W. Wang. 2008. Mechanical and electronic-structure properties of compressed CdSe tetrapod nanocrystals. *Journal of Nanoscience and Nanotechnology* 8(4): 1994–1998.

Shen, G. Z. and C. J. Lee. 2005. CdS multipod-based structures through a thermal evaporation process. *Crystal Growth & Design* 5(3): 1085–1089.

Shevchenko, E. V., D. V. Talapin, N. A. Kotov, S. O'Brien, and C. B. Murray. 2006. Structural diversity in binary nanoparticle superlattices. *Nature* 439(7072): 55–59.

Shiang, J. J., A. V. Kadavanich, R. K. Grubbs, and A. P. Alivisatos. 1995. Symmetry of annealed wurtzite CdSe nanocrystals—Assignment to the C-3v point group. *Journal of Physical Chemistry* 99(48): 17417–17422.

Steiner, D., D. Katz, O. Millo et al. 2004. Zero-dimensional and quasi one-dimensional effects in semiconductor nanorods. *Nano Letters* 4(6): 1073–1077.

Sugimoto, T. 1987. Preparation of monodispersed colloidal particles. *Advances in Colloid and Interface Science* 28(1): 65–108.

Sun, B. Q. and H. Sirringhaus. 2006. Surface tension and fluid flow driven self-assembly of ordered ZnO nanorod films for high-performance field effect transistors. *Journal of the American Chemical Society* 128(50): 16231–16237.

Sun, B. Q., E. Marx, and N. C. Greenham. 2003. Photovoltaic devices using blends of branched CdSe nanoparticles and conjugated polymers. *Nano Letters* 3(7): 961–963.

Sun, B. Q., H. J. Snaith, A. S. Dhoot, S. Westenhoff, and N. C. Greenham. 2005. Vertically segregated hybrid blends for photovoltaic devices with improved efficiency. *Journal of Applied Physics* 97(1): 014914.

Sze, S. M. 1982. *Semiconductor Devices: Physics and Technology.* New York: John Wiley & Sons.

Takeuchi, S., H. Iwanaga, and M. Fujii. 1994. Octahedral multiple-twin model of Tetrapod ZnO crystals. *Philosophical Magazine A-Physics of Condensed Matter Structure Defects and Mechanical Properties* 69(6): 1125–1129.

Talapin, D. V., E. V. Shevchenko, C. B. Murray et al. 2004. CdSe and CdSe/CdS nanorod solids. *Journal of the American Chemical Society* 126(40): 12984–12988.

Talapin, D. V., J. H. Nelson, E. V. Shevchenko et al. 2007a. Seeded growth of highly luminescent CdSe/CdS nanoheterostructures with rod and tetrapod morphologies. *Nano Letters* 7(10): 2951–2959.

Talapin, D. V., E. V. Shevchenko, C. B. Murray, A. V. Titov, and P. Kral. 2007b. Dipole-dipole interactions in nanoparticle superlattices. *Nano Letters* 7(5): 1213–1219.

Tao, A. R., S. Habas, and P. Yang. 2008. Shape control of colloidal metal nanocrystals. *Small* 4(3): 310–325.

Tarì, D., M. De Giorgi, F. Della Sala et al. 2005. Optical properties of tetrapod-shaped CdTe nanocrystals. *Applied Physics Letters* 87(22): 224101.

Tarì, D., M. De Giorgi, P. P. Pompa et al. 2006. Exciton transitions in tetrapod-shaped CdTe nanocrystals investigated by photomodulated transmittance spectroscopy. *Applied Physics Letters* 89(9): art. no. 094104.

Trallero-Giner, C., A. Debernardi, M. Cardona, E. Menendez-Proupin, and A. I. Ekimov. 1998. Optical vibrons in CdSe dots and dispersion relation of the bulk material. *Physical Review B* 57(8): 4664–4669.

Vere, A. V., S. Cole, and D. J. Williams. 1983. The origins of twinning in CdTe. *Journal of Electronic Materials* 12(3): 551–561.

Wang, L. W. 2005. Charging effects in a CdSe nanotetrapod. *Journal of Physical Chemistry B* 109(49): 23330–23335.

Wang, F. Z., Z. Z. Ye, D. W. Ma, L. P. Zhu, and F. Zhuge. 2005. Rapid synthesis and photoluminescence of novel ZnO nanotetrapods. *Journal of Crystal Growth* 274(3–4): 447–452.

Wetz, F., K. Soulantica, M. Respaud, A. Falqui, and B. Chaudret. 2007. Synthesis and magnetic properties of Co nanorod superlattices. *Materials Science & Engineering C-Biomimetic and Supramolecular Systems* 27(5–8): 1162–1166.

Williams, D. B. and C. B. Carter. 2004. *Transmission Electron Microscopy: A Textbook for Materials Science.* Berlin, Germany: Springer.

Xie, R. G., U. Kolb, and T. Basché. 2006. Design and synthesis of colloidal nanocrystal heterostructures with tetrapod morphology. *Small* 2(12): 1454–1457.

Xiong, Y. J., B. Wiley, and Y. N. Xia. 2007. Nanocrystals with unconventional shapes—A class of promising catalysts. *Angewandte Chemie—International Edition* 46(38): 7157–7159.

Xu, J. S. and D. F. Xue. 2007. Five branching growth patterns in the cubic crystal system: A direct observation of cuprous oxide microcrystals. *Acta Materialia* 55(7): 2397–2406.

Yan, H. Q., R. R. He, J. Pham, and P. D. Yang. 2003. Morphogenesis of one-dimensional ZnO nano- and microcrystals. *Advanced Materials* 15(5): 402–405.

Yan, Y. F., M. M. Al-Jassim, M. F. Chisholm et al. 2005. [1$\bar{1}$00]/ (1102) twin boundaries in wurtzite ZnO and group-III-nitrides. *Physical Review B* 71(4): art. no. 041309.

Yang, Y. M., X. L. Wu, L. W. Yang, and F. Kong. 2006. Twinning defects in spherical GeSi alloy nanocrystals. *Journal of Crystal Growth* 291(2): 358–362.

Yeh, C. Y., Z. W. Lu, S. Froyen, and A. Zunger. 1992. Zinc-blende-wurtzite polytypism in semiconductors. *Physical Review B* 46(16): 10086–10097.

Yong, K. T., Y. Sahoo, M. T. Swihart, and P. N. Prasad. 2006. Growth of CdSe quantum rods and multipods seeded by noble-metal nanoparticles. *Advanced Materials* 18(15): 1978–1982.

Yu, W. W., Y. A. Wang, and X. G. Peng. 2003. Formation and stability of size-, shape-, and structure-controlled CdTe nanocrystals: Ligand effects on monomers and nanocrystals. *Chemistry of Materials* 15(22): 4300–4308.

Yu, K., Y. S. Zhang, R. L. Xu et al. 2005. Efficient field emission from tetrapod-like zinc oxide nanoneedles. *Materials Letters* 59(14–15): 1866–1870.

Zhang, J. Y. and W. W. Yu. 2006. Formation of CdTe nanostructures with dot, rod, and tetrapod shapes. *Applied Physics Letters* 89(12): art. no. 123108.

Zhang, Y. L., X. H. Zhong, J. Zhu, and X. Song. 2007. Alcoholysis route to monodisperse CoO nanotetrapods with tunable size. *Nanotechnology* 18(19): 195605.

Zhong, H. Z., Y. Zhou, Y. Yang, C. H. Yang, and Y. F. Li. 2007. Synthesis of type II CdTe-CdSe nanocrystal heterostructured multiple-branched rods and their photovoltaic applications. *Journal of Physical Chemistry C* 111(17): 6538–6543.

Zhou, Y., Y. C. Li, H. Z. Zhong et al. 2006. Hybrid nanocrystal/polymer solar cells based on tetrapod-shaped CdSe$_x$Te$_{1-x}$ nanocrystals. *Nanotechnology* 17(16): 4041–4047.

Zhu, Y. C., Y. Bando, D. F. Xue, and D. Golberg. 2003. Nanocable-aligned ZnS tetrapod nanocrystals. *Journal of the American Chemical Society* 125(52): 16196–16197.

8

Fullerene-Like CdSe Nanoparticles

Silvana Botti

*Ecole Polytechnique,
CNRS, CEA-DSM*

and

*Université Claude Bernard
Lyon I, CNRS*

8.1 Introduction

Cadmium selenide (CdSe) is a binary compound made of cadmium and selenium that crystallizes in the hexagonal closed-packed wurtzite structure. Its optical band gap measures 1.85 eV at low temperature (Dai et al. 2007b). Current research on CdSe has focused mostly on nanoparticles, that is, small portions cut out from bulk CdSe, with diameters between 1 and 100 nm. The interest in these nanosized systems can be understood by their special properties, significantly different from the properties of the parent bulk compound, that open the possibility of novel technological applications. Furthermore, the very small size of these nanoparticles makes them particularly suited for miniaturization purposes. In fact, while the miniaturization of conventional silicon-based electronics is approaching fundamental performance limits, researchers are actively working to find new nanosized materials that are able to overcome these limits.

All nanoparticles exhibit a fundamental property known as "quantum confinement" (Bawendi et al. 1990), due to the modification of the energy states of electrons confined in a very small volume. Quantum confinement is dependent on the confinement volume, that is, on the size of the nanoparticle. This means that the electronic properties of CdSe nanoparticles can be tailored by controlling their size. As a consequence, CdSe nanoparticles have size-tunable absorption and luminescence spectra. This characteristic makes them particularly attractive to be employed in optical devices, such as in light-emitting diodes that have to cover a large part of the visible spectrum (Coe et al. 2002, Bowers et al. 2005). Along the same lines, CdSe nanoparticles have already proved to be excellent components for a variety of applications, such as in optically pumped lasers (Tessler et al. 2002), photovoltaic cells (Greenham et al. 1996, Klimov 2003), telecommunications (Harrison et al. 2000), and biomedicine as chemical markers (Bruchez et al. 1998, Michalet et al. 2005).

The common requirement that makes all these different applications of CdSe nanoparticles possible is the high proficiency achieved in the control of a remarkably narrow size distribution [even lower than 5% (Murray et al. 1993)] during the synthesis process. In fact, it is the size distribution that determines the sharpness of the optical peaks. A further advantage of CdSe nanocrystals is the degree of efficiency attained in their synthesis, the high quality of the resulting samples, and the fact that the optical gap is in the visible range. In most common experimental setups, CdSe nanoparticles are formed by kinetically controlled precipitation and are terminated with capping organic ligands, like the trioctyl phosphine oxide (TOPO) molecule, which provide stabilization of the otherwise reactive dangling orbitals at the surface (Murray et al. 1993). High-quality colloidal CdSe nanoparticles have been routinely synthesized for more than a decade: their sizes range from 1 nm to hundreds of nanometers and their core displays the same symmetry as wurtzite.

The electronic states of any nanoobject are also sensitive to the overall cluster shape, and more specifically to the deformations due to surface reconstruction, to the presence of defects, and to the symmetry properties of the arrangement of atoms in the core (Peng et al. 2000). These geometrical details are, of course, more critical when the cluster is very small, that is, when the surface/volume ratio is the largest. In particular, defects and dangling bonds are essentially localized at the surface. Moreover, for practical uses, further requirements, such as a high chemical stability of the nanostructure and an enhanced photoluminescence intensity, are of utmost importance. Unfortunately, these characteristics are inhibited by the presence of defects. As a consequence, often the quantum yields for very small CdSe nanoparticles in solution turn out to be less than 1% (Bruchez et al. 1998, Chan and Nie 1998). The reason is that these colloidal nanoparticles contain a large number of defects, especially at the surface, where radiationless recombination of the charge carriers can occur. Therefore, controlling the quality of the growth of small clusters, especially the formation of dangling bonds at their surface, is essential for any kind of application.

In this context, the recent synthesis and probable identification of the very small, and highly stable, $(CdSe)_{33}$ and $(CdSe)_{34}$ nanoparticles grown in a solution of toluene (Kasuya et al. 2004, 2005) came as a breakthrough. The experimental absorption spectra of these nanoparticles at low temperature exhibit sharp peaks, similar to the ones that characterize TOPO-capped clusters of the same size (Murray et al. 1993). However, the surfactant molecules employed in the synthesis process are, in this case, removed by laser vaporization. Furthermore, an x-ray analysis indicates that the coordination number of Se is between 3 (the coordination of a fullerene) and 4 (the coordination of the bulk crystal). In view of this, and in absence of direct structural data, the nonpassivated compound nanoparticles were predicted to have a core-cage structure, composed by a puckered fullerene-like $(CdSe)_{28}$ cage accommodating a $(CdSe)_n$ ($n = 5,6$) wurtzite unit inside (see Figure 8.1). Further ab initio calculations of structural and optical properties validated this interpretation (Kasuya et al. 2004, Botti and Marques 2007).

These very small fullerene-like systems, in the size range of 1–2 nm, are particularly interesting, as they have an increased probability to take the form of magic-sized nanocrystals, leading to ultrastable single-sized ensembles, which are in principle characterized by very sharp absorption peaks. The concept of magic size has been well known for several years in the field of metal clusters, but it is less common for semiconductor nanoparticles.

Furthermore, the recent discovery of CdSe and other fullerene-like semiconducting cluster has renewed the interest for the so-called "cluster-assembled materials." In fact, cluster-assembled materials form one of the most promising frontiers in the design of nanodevices. They are composed by three-dimensional arrays of ultrastable size-selected nanoparticles, organized in a similar way as atoms are organized to form a crystal. Cluster-assembled materials ideally combine the properties of the single nanoobject with novel collective behaviors arising from the periodic arrangement of the solid. Of course, the interaction between clusters cannot be too strong, in order not to destroy the discrete nature of the optical transitions. This means that the surface of the cluster has to be well saturated, with no dangling bonds. Unfortunately, up to now, most attempts to design cluster-assembled matter have led to metastable materials, which can be stabilized only by a dielectric matrix that prevents the individual clusters from reacting with their neighbors. Only few cluster materials are known at present; the most famous are made of carbon fullerenes (C_{60} and C_{70}). However, the recently synthesized CdSe fullerenes are very small clusters (1.5 nm of diameter), are extremely stable, and can be produced in macroscopic quantities: all these characteristics point to the possibility of using them to produce new cluster-assembled materials.

8.2 Synthesis and Spectroscopic Characterization

Numerous approaches (Katari et al. 1994, Hines and Guyot-Sionnest 1996, Chen et al. 1997, Dabbousi et al. 1997, Peng et al. 1997, Mikulec and Bawendi 2000, Peng et al. 2000, Peng and Peng 2001, Talapin et al. 2001, Gaponik et al. 2002a,b, Reiss et al. 2002, Yu et al. 2003a,b, Zhang et al. 2003, Zhong et al. 2004, Dai et al. 2006, Pradhan et al. 2006) have been developed to synthesize highly crystalline and monodisperse II–VI semiconductor nanocrystals, following the path opened by Murray et al. (1993). However, these approaches are mostly suitable to produce regular-sized nanocrystals (>2 nm) but cannot be commonly employed to synthesize magic-sized small clusters (1–2 nm). In particular, in the magic-sized regime, a large percentage of the atoms are at the surface, which makes the control of dangling bonds much more important.

Very small CdSe nanocrystals have been synthesized by the overlayering method (Soloviev et al. 2000), the etching preparation starting from larger nanocrystals (Landes et al. 2001), and the reverse-micelle approach (Kasuya et al. 2004). Peculiar optical properties were obtained by magic-sized nanoparticles grown by hot injection (Bowers et al. 2005): these ultrasmall clusters exhibit broadband emission (420–710 nm) throughout most of the visible light spectrum, while not suffering from self-absorption. This property makes them ideal materials to produce white-light light-emitting diodes. In general, it is assumed that these clusters are saturated with ligands, even if there is no direct information about the reconstruction at the surface.

However, ligand-free fullerene-like core-cage particles were for the first time produced by Kasuya et al. (2004, 2005) only in 2004. Since then, other groups tested new reproducible and controllable methods to grow magic-sized small CdSe clusters. The exact control of the size of the nanocrystal and the sharpness of

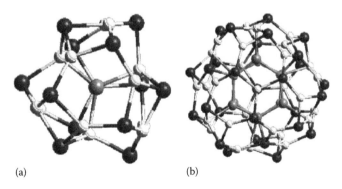

(a)   (b)

FIGURE 8.1 Structures of the $(CdSe)_n$ corecage nanoparticles calculated to be most stable by Kasuya et al. (2004), viewed down a threefold symmetry axis. (a) $(CdSe)_{13}$ has four-membered and 10 six-membered rings on the cage of 12 Se (dark gray) and 13 Cd (white) ions with a Se (light gray) ion inside. (b) $(CdSe)_{34}$ has a truncated-octahedral morphology formed by a $(CdSe)_{28}$ cage (Se, dark gray; Cd, white) with 6 four-membered and 8×3 six-membered rings. A $(CdSe)_6$ cluster (Se and Cd, light gray) encapsulated inside this cage provides additional network and stability. (Adapted from Kasuya, A. et al., *Nat. Mater.*, 3, 99, 2004.)

the optical peaks are both essential for any practical application. Of course, the stability in time of the clusters is also an important parameter to consider.

Kudera et al. (2007) reported a method for controlling the sequential growth of CdSe clusters in solution that yields only magic-sized nanocrystals of progressively larger sizes. The resulting nanoobjects are characterized by sharp optical absorption spectra with peaks at well-defined energies, in agreement with the ones reported by Kasuya et al. (2004). Also the cluster sizes, estimated by x-ray diffraction analysis, are compatible with the findings of Kasuya et al. (2004). Further, transmission electron microscopy analysis revealed that all clusters are roughly spherical and that they are not aggregated. The mechanism of growth is determined by the competition between the attachment and the detachment of single atoms at the surface. Once a cluster has grown to a magic size, its structure is so stable that no atom can detach from it. Therefore, it can only grow further, but it cannot shrink. This growth mechanism is compatible with the creation of cage-like structures, even if there is no direct proof of the fact that fullerene-like clusters are actually produced in this experiment. Unfortunately, these clusters have rather weak luminescence properties. Kudera et al. (2007) also proved that the optical properties of their clusters could be improved by passivating their surfaces with a ZnS shell.

Dai et al. (2007a) reported an injection approach for the synthesis of nanocrystals with long existence period, using cheap cadmium oleate as the source of cadmium. The resulting CdSe clusters are saturated by ligands. They exhibit strong and fixed absorption features and a narrow red-shifted emission. Higher injections/growth temperatures favor a white light emission, but also transform the magic-sized nanocrystals into regular-sized ones. This same approach was also used by the same authors to synthesize CdTe clusters.

On the other hand, Ouyang et al. (2008) used a noninjection one-pot synthetic approach to achieve colloidal CdSe ensembles consisting of single-sized nanocrystals exhibiting bright band-gap photoluminescence emission. Their systematic study suggests that the growth of large CdSe clusters is favored by long ligands at high growth temperature, whereas the growth of small CdSe magic-sized clusters is favored by the same authors ligands at low growth temperature.

Finally, Kucur et al. (2008) reported an efficient top-down synthesis in an amine-rich solution of small stable CdSe nanocrystals. They are produced by the decomposition of initial nanocrystals within several days. The most stable clusters were characterized by spectroscopic methods, and the comparison of absorption and photoluminescence spectra with previous studies suggests a predominant cage-like structure. The analysis of the absorption peaks revealed a preferred synthesis of $(CdSe)_{33,34}$ clusters. The emission decay rate of these clusters is comparable with that of organic dyes.

Despite the important contributions coming from all these recent studies, the preparation and understanding of highly luminescent, thermodynamically stable, small-sized CdSe clusters is still at the beginning. We are optimistic, however, that the next few years will bring new optimized techniques for the production of these clusters that will open the way for the development of the exciting and innovative applications that have already been foreseen.

8.3 Ab Initio Calculations

From the theoretical side, it is desirable to obtain from reliable calculations all possible complementary information on the atomic arrangement and surface deformation of CdSe clusters, in order to understand and complement experimental evidences. In fact, experimental measurements alone are usually not able to provide conclusive results concerning the surface reconstruction and the role of passivating ligands. Moreover, theoretical calculations can give a deeper insight on how surface reconstructions produce modifications of the electronic states, and consequently of the optical properties at the basis of all technological applications.

For ligand-terminated small- and regular-sized CdSe clusters, transmission electron microscopy data (Murray et al. 1993, Shiang et al. 1995), molecular dynamics simulations or first-principles techniques without self-consistency (Rabani 2001, Sarkar and Springborg 2003), and self-consistent ab initio structural relaxations (Puzder et al. 2004, Botti and Marques 2007) agree on predicting an atomic arrangement of the inner Cd and Se atoms analogous to the one in the wurtzite CdSe crystal. The extent to which the cluster surface retains the crystal geometry is more controversial as the surface cannot be easily resolved experimentally. Generally, if the surface is properly passivated, the reconstruction is assumed to be small and limited to the outermost layer (and eventually the layer just beneath it), which is in agreement with molecular dynamics simulations (Rabani 2001). However, Puzder et al. (2004) predicted for clusters with diameters up to 1.5 nm a strong surface reconstruction, remarkably similar in vacuum and in the presence of passivating ligands.

The core-cage structures proposed by Kasuya et al. (2004) are significantly different from all bulk-derived arrangements previously studied. These geometries were found to be particularly stable by first-principles total energy calculations (Kasuya et al. 2004, Botti and Marques 2007). Furthermore, calculations of optical spectra (Botti and Marques 2007) have offered a definitive proof for the identification of the observed nanoparticles with the fullerene-like structures, through the comparison between measured (Kasuya et al. 2004) and simulated spectra. In fact, as the electronic states (and, as a consequence, absorption or emission peaks) are strongly modified by changes of size and shape, optical spectroscopy can thus be a powerful tool (especially if it can be combined with other spectroscopic techniques) to probe the atomic arrangement of synthesized nanoparticles.

Below, we will discuss how the well-known density functional theory (DFT) (Hohenberg and Kohn 1964) has been applied to access information concerning the structural and electronic properties of CdSe fullerenes. Moreover, we will see how the comparison between theoretical and experimental results provides a deeper insight into the properties of complex nanostructured materials.

We chose to restrict our discussion to DFT, as it is the most popular and versatile method available in condensed-matter physics, computational physics, and computational chemistry. Compared with empirical or semiempirical approaches, DFT has a total absence of parameters fitted to experimental data. This characteristic is essential to guarantee predictive power to any theory. Furthermore, within first principles (i.e., parameter-free) approaches, DFT is relatively light from a computational perspective. In fact, in contrast with traditional methods in electronic structure theory, in particular Hartree–Fock theory and its descendants, DFT is not aiming at finding a good approximation for the complicated many-electron wavefunction: the electronic density becomes the key quantity at the heart of the theory. Whereas the many-body wavefunction is dependent on $3N$ variables (without considering spin), three spatial variables for each of the N electrons, the density is only a function of three variables and is a simpler quantity to deal with, both conceptually and practically. For practical purposes, DFT is usually implemented in the Kohn–Sham scheme (Kohn and Sham 1965), which makes use of a noninteracting system yielding the same density as the original problem. For a review on the basics of DFT, we suggest the reader to look at the rich literature on the subject (Parr and Yang 1989, Dreizler and Gross 1995, Fiolhais et al. 2003).

8.3.1 Structures of Energetically Stable CdSe Nanoparticles

The atomic positions of CdSe nanoparticles can be routinely obtained by geometry optimization using any quantum chemistry or solid state physics code. The starting point of any structural optimization procedure is to consider a series of candidate structures with different geometries and sizes. Here we consider $(CdSe)_n$ aggregates with sizes ranging up to about 1.5 nm. To build these atomic arrangements, it is possible to start from three different kinds of ideal geometries: (1) bulk fragments cut into the infinite wurtzite crystal, (2) octahedral fullerene-like cages made of four- and six-membered rings, and (3) the core-cage structures of Kasuya et al. (2004), composed of puckered CdSe fullerene-type cages that include $(CdSe)_n$ wurtzite units of adequate size to form a three-dimensional network. Following Botti and Marques (2007), we can assume that the Cd–Se distance before structural relaxation is the distance in the CdSe wurtzite crystal, calculated within DFT (Soler et al. 2002) in the same approximations used for the nanoparticles: its value (0.257 nm) compares well with the experimental value (0.263 nm).

In the following text, we will analyze as an illustration the structural calculations of Botti and Marques (2007), comparing them with the analogous DFT calculations for wurtzite-like clusters of Puzder et al. (2004) and for core-cage clusters of Kasuya et al. (2004). Botti and Marques (2007) used an implementation of DFT (Soler et al. 2002) within the local density approximation

(LDA) (Perdew and Zunger 1981) for the exchange and correlation potential and norm-conserving pseudopotentials (Hamann 1989, Troullier and Martins 1991). Puzder et al. (2004) used a similar technique, but with another implementation of DFT (Gygi, F. 1999). Finally, Kasuya et al. (2004) performed DFT calculations (Kresse and Furthmuller 1996) using ultrasoft pseudopotentials (Vanderbilt 1990) and the generalized gradient approximation (Perdew et al. 1996) for the exchange-correlation potential.

Atomic arrangements after optimization using DFT are depicted in Figure 8.2 (see Botti and Marques 2007). All clusters suffer contraction on geometry minimization. For example, $(CdSe)_{33,34}$ clusters experience a size reduction of about 1%–1.5%. The theoretical results are in agreement with the x-ray analysis of Kasuya et al. (2004). However, as the relaxation affects mainly the outermost atoms, the overall effect is more pronounced in smaller structures, in which the average Cd–Se distance decreases up to 4%. This contraction does not conserve the overall shape, as Cd atoms are pulled inside the cluster and Se atoms are puckered out. As a consequence, Cd–Cd average distances can be reduced by 30%, whereas Se–Se distances remain essentially unvaried. This is clearly visible in Figure 8.3, in which the relaxed distance of Cd (circles) and Se (diamonds) atoms from the center of the cluster is plotted for $(CdSe)_{33,34}$ clusters as a function of their distance before relaxation. If the atoms remained in their initial position, all data points would fall on the straight line $y = x$. The fact that most Cd atoms lie below the line, while most Se atoms are above it, shows that in our simulation, Cd atoms prefer to move inward and Se atoms outward. That puckering happens independently of the cluster size (Kasuya et al. 2004, Puzder et al. 2004, Botti and Marques 2007).

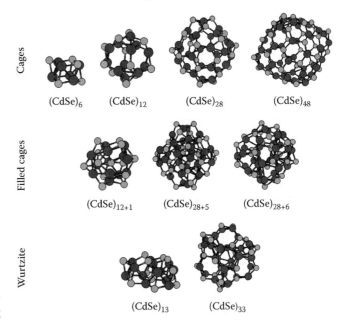

FIGURE 8.2 Examples of relaxed cages, relaxed filled cages, and relaxed wurtzite structures of $(CdSe)_n$ with a diameter smaller than 2 nm. Cd atoms are in dark gray and Se atoms are in light gray. (Adapted from Botti, S. and Marques, M.A.L., *Phys. Rev. B*, 75, 035311, 2007.)

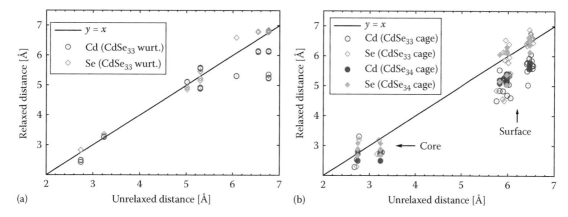

FIGURE 8.3 Distance of Cd atoms (circles) and Se atoms (diamonds) from the center of the cluster after geometry optimization, as a function of their distance before optimization. An atom that lies on the straight line $y = x$ did not change its position. In panel (a), results of the analysis for $(CdSe)_{33,34}$ core-cage clusters and in panel (b), for the $(CdSe)_{33}$ wurtzite cluster. (Adapted from Botti, S. and Marques, M.A.L., *Phys. Rev. B*, 75, 035311, 2007.)

All wurtzite fragments get significantly distorted on relaxation and break their original symmetry. However, the strong modification of bond lengths and angles concerns essentially the surface layer (Puzder et al. 2004, Botti and Marques 2007). In particular, we can see in Figure 8.3a that the wurtzite-type $(CdSe)_{33}$ is already large enough to conserve a bulk-like crystalline core. In fact, the spread of the points from the straight line is pronounced only for the external shell of atoms. The calculated overall contraction of the cluster is consistent with experimental data (Zhang et al. 2002). Also the empty cages $[(CdSe)_{12}, (CdSe)_{28},$ and $(CdSe)_{48}]$ get puckered, but conserve their overall shape. Their binding energies are smaller by about 0.05 eV per CdSe unit with respect to the binding energies of the corresponding filled cages (see Figure 8.4a), showing the importance of preserving the three-dimensional sp^3 Cd–Se network.

Models based only on the wurtzite structure of bulk CdSe fail to predict the existence of stable "magic clusters" with well-defined sizes and number of atoms. In contrast, the core-cage structures proposed by Kasuya et al. can appear only for well-defined sizes and number of atoms, as fullerene cages can be built only for 12, 16, 28, 48, 76, etc. atoms and only some of these cages can be filled conveniently with wurtzite-coordinated CdSe units. To optimize the core-cage structures $[(CdSe)_{12+1=13},$ $(CdSe)_{28+5=33},$ and $(CdSe)_{28+6=34}]$ Botti and Marques (2007) created different starting arrangements assuming different orientations for the encapsulated $CdSe_n = 1,5,6$ units. In the relaxed assemblies, the distributions of bond lengths and angles result very similar despite of the distinct initial configurations. The fact that the surfaces of core-cage clusters do not show neither strong reconstruction nor deleterious dangling bonds, in contrast with surfaces of wurtzite-like cluster not cured by passivation, explains why fullerene-like CdSe clusters are particularly nonreactive and prevent them from merging together to form larger clusters. This is crucial to have promising building blocks for three-dimensional cluster solids.

Figure 8.4b shows the DFT Kohn–Sham gap between the highest occupied and lowest unoccupied molecular orbitals (HOMO–LUMO) for a series of clusters of different types:

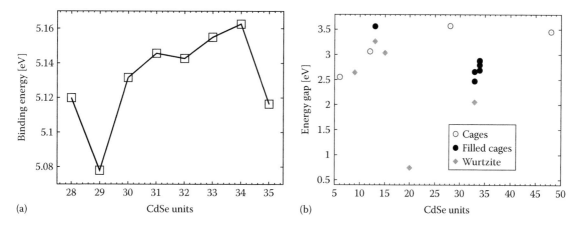

FIGURE 8.4 (a) Calculated binding energies per CdSe unit as a function of the number of CdSe units. The binding energies are calculated per CdSe molecule of $(CdSe)_n$ composed of a cage-like $(CdSe)_{28}$ with $(CdSe)_m$ inside ($n = 28 + m$, $m = 0, 1, \ldots, 7$). (Data from Kasuya, A. et al., *Nat. Mater.*, 3, 99, 2004. With permission.) (b) HOMO–LUMO gaps as a function of the number of CdSe units. The empty (filled) circles refer to cage (core-cage) clusters, whereas the diamonds refer to wurtzite-based structures. (Adapted from Botti, S. and Marques, M.A.L., *Phys. Rev. B*, 75, 035311, 2007.)

wurtzite, cages, and filled cages. Both empty and filled cages exhibit much larger HOMO–LUMO gaps than their wurtzite counterparts, indicating therefore that there are no dangling bonds at their surface. In Figure 8.4a, we show the results from Kasuya et al. (2004) for the binding energy of the filled cages. The two most stable structures are clearly $(CdSe)_{33}$ and $(CdSe)_{34}$. It is curious that the first is significantly more deformed under optimization than $(CdSe)_{34}$, but it turns out to have a very similar binding energy. The filled cage structure made of 13 units gives as well a relative minimum in the total energy per pair (Botti and Marques 2007). In the case of $(CdSe)_{13}$ and $(CdSe)_{33}$, it is possible to compare the total energies of the different three-dimensional isomers (Botti and Marques 2007): the core-cage nanoparticles have a slightly higher binding energy per CdSe unit [0.15 eV for $(CdSe)_{13}$ and 0.05 eV for $(CdSe)_{33}$]. However, we should not forget that the energy differences we are discussing here are all very tiny, sometimes of the same order of magnitude as the accuracy of the calculations. That fact confirms how difficult it can be to extract structural information from a single number (the total energy) and leads to the conclusion that the simple analysis of total energy differences cannot be considered conclusive to demonstrate the existence of fullerene-like CdSe clusters.

8.3.2 Optical Absorption Spectra

From the relaxed geometries, it is possible to obtain the optical spectra at zero temperature using time-dependent density functional theory (TDDFT) (Runge and Gross 1984, Gross and Kohn 1985). TDDFT is an exact reformulation of time-dependent quantum mechanics, in which the fundamental variable is no longer the many-body wavefunction but the time-dependent density. It can be viewed as an extension of DFT to the time-dependent domain to describe what happens when a time-dependent perturbation is applied. For a review on the subject of TDDFT, we suggest the reader to have a look at the rich literature on the subject (Marques and Gross 2004, Marques et al. 2006, Botti et al. 2007).

For the calculation of the photoabsorption cross section, Botti and Marques (2007) employed a real-time TDDFT approach (Marques et al. 2003, Castro et al. 2006), based on the explicit propagation of the time-dependent Kohn–Sham equations. In this approach, one first excites the system from its ground state by applying a delta electric field $E_0\delta(t)\mathbf{e}_m$. The unit vector $\mathbf{e}_m$ determines the polarization direction of the field and E_0 its magnitude, which must be small if one is interested in linear response. The reaction of the noninteracting Kohn–Sham system to this sudden perturbation can be readily computed: each ground state Kohn–Sham orbital $\varphi_i^{GS}(\mathbf{r})$ is instantaneously phase-shifted: $\varphi_i(\mathbf{r}, t = 0^+) = e^{iE_0\mathbf{e}_m \cdot \mathbf{r}}\varphi_i^{GS}(\mathbf{r})$. The Kohn–Sham equations are then propagated forward in real time, and the time-dependent density $n(\mathbf{r}, t)$ can then be computed. The induced dipole moment variation is an explicit functional of the density:

$$\delta \mathbf{D}^m(t) = \delta\langle\hat{\mathbf{R}}\rangle(t) = \int d^3 r[n(\mathbf{r}, t) - n(\mathbf{r}, t = 0)]\mathbf{r}. \quad (8.1)$$

The superindex m reminds that the perturbation has been applied along the mth Cartesian direction. The components of the dynamical dipole polarizability tensor $\alpha(\omega)$ are directly related to the Fourier transform of the induced dipole moment function:

$$\alpha_{mn}(\omega) = \frac{\delta D_n^m(\omega)}{E_0}. \quad (8.2)$$

The spatially averaged absorption cross section is trivially obtained from the imaginary part of the dynamical polarizability:

$$\sigma(\omega) = \frac{4\pi\omega}{c}\Im[\alpha(\omega)], \quad (8.3)$$

where α is the spatial average, or trace, of the tensor

$$\alpha(\omega) = \frac{1}{3}\mathrm{Tr}[\underline{\alpha}(\omega)]. \quad (8.4)$$

Here we will discuss the results for the excitation energies and the optical spectra of Botti and Marques (2007), obtained using TDDFT within the adiabatic local density approximation (ALDA) (Gross and Kohn 1985). These are the only calculations on CdSe clusters available in literature that go beyond the simple application of Fermi's golden rule, that is, the sum of independent single-particle transitions from occupied to empty states (in this case, Kohn–Sham one-particle states). It is well known that the simpler approach of taking the differences of eigenvalues between Kohn–Sham orbitals gives peaks at lower frequencies in disagreement with the experimental spectra (Castro et al. 2002). On the other hand, TDDFT within the ALDA typically reproduces the low energy peaks of the optical spectra with an average accuracy below 0.2 eV. The accuracy in reproducing transitions of intermediate energy is known to be somewhat deteriorated, due to the wrong asymptotic behavior of the LDA exchange-correlation potential. For this reason, we focus the analysis of the spectra on the lowest energy peaks.

Figure 8.5 displays the photoabsorption spectra of the empty cages of different diameters, as calculated by Botti and Marques (2007). It is clear from the figure that the absorption threshold is systematically blue-shifted with respect to the bulk optical gap ($\simeq 1.8$ eV). This blue shift is due to the well-known quantum confinement effects, so it is not surprising that the shift increases with decreasing cluster size. We can compare the absorption threshold with the Kohn–Sham HOMO–LUMO gap shown in the right panel of Figure 8.4: the Kohn–Sham gap is systematically smaller than the TDDFT absorption threshold. This is a common observation as the Kohn–Sham transition energies are usually at lower frequencies than the experimental peaks.

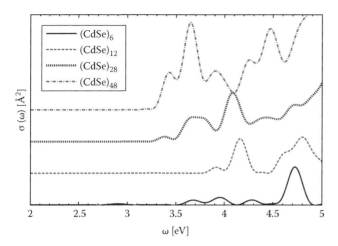

FIGURE 8.5 Calculated photoabsorption cross section σ(ω) of the empty cages $(CdSe)_6$, $(CdSe)_{12}$, $(CdSe)_{28}$, and $(CdSe)_{48}$. The spectra were shifted vertically for visualization purposes. (Adapted from Botti, S. and Marques, M.A.L., *Phys. Rev. B*, 75, 035311, 2007.)

We note that the TDDFT optical gaps include both electron–electron and electron–hole corrections to the Kohn–Sham gap at the level of the ALDA.

We should keep in mind that the opening of the gap due to confinement can be counterbalanced by a closing of the gap due to surface reconstruction. This leads to a nontrivial dependence of the absorption gap as a function of the cluster size. This effect is already present at the Kohn–Sham level (see Figure 8.4a) and it persists in TDDFT spectra. In fact, the calculated absorption curves are strongly dependent not only on the cluster size but also on the details of its atomic arrangement. This is evident if we compare the optical response of the different isomers of $(CdSe)_{13}$ in Figure 8.6 and of $(CdSe)_{33}$ in Figure 8.7 (Botti and Marques 2007).

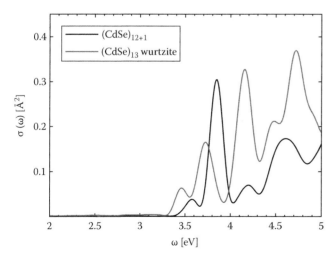

FIGURE 8.6 Calculated photoabsorption cross section σ(ω) of the isomers of $(CdSe)_{13}$. (Adapted from Botti, S. and Marques, M.A.L., *Phys. Rev. B*, 75, 035311, 2007.)

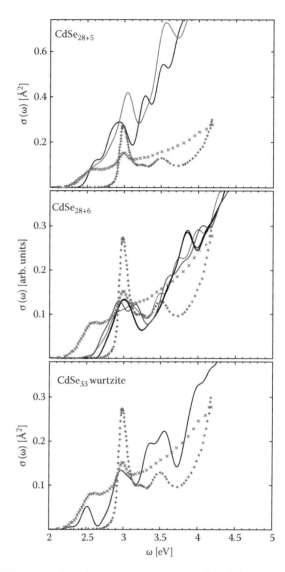

FIGURE 8.7 Photoabsorption cross section σ(ω) of the isomers of $(CdSe)_{33,34}$. The experimental data (Kasuya et al. 2004) in arbitrary units (dots: sample I at 45°C and crosses: sample II at 80°C) are compared with calculated spectra from Botti and Marques (2007). The different solid curves correspond to distinct relaxed geometries obtained starting from different filled cages.

The absorption threshold is lower in wurtzite-type clusters since the HOMO–LUMO gap is reduced due to the presence of defect states in the gap as a consequence of the strong surface deformation. For a similar reason, the larger surface deformation of the core-cage $(CdSe)_{33}$ aggregate in comparison with the more stable $(CdSe)_{34}$ structure explains why the first starts absorbing at lower energies than the second. Finally, we note that the similar curves of different tones of gray in Figure 8.7 correspond to distinct core-cage geometries obtained in various optimization simulations. We conclude that the dependence of the relevant peak positions and shapes on the different atomic arrangements is not negligible, but the peak positions and oscillator strengths are sufficiently defined for the purpose to distinguish different geometries by comparing photoabsorption spectra.

A comparison between calculated (Botti and Marques 2007) and measured spectra (Kasuya et al. 2004) is possible for nanoparticles made of 33 and 34 CdSe units (see Figure 8.7). The dots refer to room temperature absorption data for mass-selected nanoparticles prepared in toluene at 45°C (sample I), whereas the crosses correspond to analogous data for the solution prepared at 80°C (sample II). Both samples are characterized by strong absorption at 3 eV. For sample II, the experimental data show the appearance of a broad peak extending to lower energies. This peak turns out to move to even lower energies when the temperature and the time in the synthesis process increase. In a simple quantum confinement picture, these findings suggest that larger particles, possibly reconstructed bulk fragments, are formed when the temperature increases. Moreover, the sharp peak at about 3 eV, which is always present, was hypothesized to be the signature of the highly resistant fullerene-like clusters.

The calculated spectra (Botti and Marques 2007) shown in Figure 8.7 prove the presence of fullerene-like core-cage structures. The theoretical optical response of all model core-cage $(CdSe)_{34}$ clusters is indeed characterized by a well-defined absorption peak at 3 eV. Also the core-cage $(CdSe)_{33}$ cluster and the $(CdSe)_{33}$ reconstructed bulk fragment can contribute to this peak. However, they cannot be present in sample I, as that would be signaled by the appearance of a broader peak at lower energy, which is absent in the experimental spectrum. On the other hand, a peak at about 2.5 eV, connected to the peak at 3 eV by a region of increasing absorption, is present in the spectrum for sample II. Our calculations show that the $(CdSe)_{33}$ wurtzite fragment is responsible for the peak at 2.5 eV, whereas the broad absorption region between 2.5 and 3 eV can be explained by the presence of $(CdSe)_{33}$ core-cage structures. This is in disagreement with the intuition of Kasuya et al. (2004) that bulk fragments of about 2.0 nm gave rise to the broad absorption below 3 eV.

In summary, by comparing our theoretical spectra with measurements, Botti and Marques (2007) could confirm the existence of the stable core-cage fullerene-like structures hypothesized in the seminal work of Kasuya et al. (2004).

8.4 Conclusions

The use of CdSe fullerene-like nanoparticles for technological applications in the field of cluster-assembled materials is a promising challenge for materials science. For this purpose, there is much work in progress to optimize the production procedures of magic-size small CdSe clusters. Concerning the characterization and the understanding of electronic excitations in these novel nanostructured materials, the combination of experimental and theoretical spectroscopic techniques has proved to be essential to extract reliable and conclusive information on their structural and optical properties.

Acknowledgments

I thank Miguel Marques for the critical reading of the manuscript. I acknowledge financial support from the EC Network of Excellence NANOQUANTA (NMP4-CT-2004-500198) and the French ANR (JC05_46741 and NT05-3_43900).

References

Bawendi, M., Steigerwald, M., and Brus, L., 1990. The quantum-mechanics of larger semiconductor clusters (quantum dots), *Annual Review of Physical Chemistry* 41: 477–496.

Botti, S. and Marques, M. A. L., 2007. Identification of fullerene-like CdSe nanoparticles from optical spectroscopy calculations, *Physical Review B* 75: 035311.

Botti, S., Schindlmayr, A., Del Sole, R., and Reining, L., 2007. Time-dependent density-functional theory for extended systems, *Reports on Progress in Physics* 70: 357–407.

Bowers, M., McBride, J., and Rosenthal, S., 2005. White-light emission from magic-sized cadmium selenide nanocrystals, *Journal of the American Chemical Society* 127: 15378–15379.

Bruchez, M., Moronne, M., Gin, P., Weiss, S., and Alivisatos, A., 1998. Semiconductor nanocrystals as fluorescent biological labels, *Science* 281: 2013–2016.

Castro, A., Marques, M., Alonso, J., Bertsch, G., Yabana, K., and Rubio, A., 2002. Can optical spectroscopy directly elucidate the ground state of C-20? *Journal of Chemical Physics* 116: 1930–1933.

Castro, A., Appel, H., Oliveira, M. et al., 2006. Octopus: A tool for the application of time-dependent density functional theory, *Physica Status Solidi B—Basic Solid State Physics* 243: 2465–2488.

Chan, W. and Nie, S., 1998. Quantum dot bioconjugates for ultrasensitive nonisotopic detection, *Science* 281: 2016–2018.

Chen, W., Wang, Z., Lin, Z., and Lin, L., 1997. Absorption and luminescence of the surface states in ZnS nanoparticles, *Journal of Applied Physics* 82: 3111–3115.

Coe, S., Woo, W., Bawendi, M., and Bulovic, V., 2002. Electroluminescence from single mono-layers of nanocrystals in molecular organic devices, *Nature* 420: 800–803.

Dabbousi, B., RodriguezViejo, J., Mikulec, F. et al., 1997. (CdSe) ZnS core-shell quantum dots: Synthesis and characterization of a size series of highly luminescent nanocrystallites, *Journal of Physical Chemistry B* 101: 9463–9475.

Dai, Q., Li, D., Chen, H. et al., 2006. Colloidal CdSe nanocrystals synthesized in noncoordinating solvents with the addition of a secondary ligand: Exceptional growth kinetics, *Journal of Physical Chemistry B* 110: 16508–16513.

Dai, Q., Li, D., Chang, J. et al., 2007a. Facile synthesis of magic-sized CdSe and CdTe nanocrystals with tunable existence periods, *Nanotechnology* 18:405603.

Dai, Q., Song, Y., Li, D. et al., 2007b. Temperature dependence of band gap in CdSe nanocrystals, *Chemical Physics Letters* 439: 65–68.

Dreizler, R. and Gross, E. K. U., 1995. *Density Functional Theory.* New York: Plenum Press.

Fiolhais, C., Marques, M. A. L., and Nogueira, F., eds., 2003. *A Primer in Density Functional Theory*, Vol. 602 of *Lecture Notes in Physics.* Berlin, Germany: Springer.

Gaponik, N., Talapin, D., Rogach, A., Eychmuller, A., and Weller, H., 2002a. Efficient phase transfer of luminescent thiol-capped nanocrystals: From water to nonpolar organic solvents, *Nano Letters* 2: 803–806.

Gaponik, N., Talapin, D., Rogach, A. et al., 2002b. Thiol-capping of CdTe nanocrystals: An alternative to organometallic synthetic routes, *Journal of Physical Chemistry B* 106: 7177–7185.

Greenham, N., Peng, X., and Alivisatos, A., 1996. Charge separation and transport in conjugated-polymer/semiconductor-nanocrystal composites studied by photoluminescence quenching and photoconductivity, *Physical Review B* 54: 17628–17637.

Gross, E. and Kohn, W., 1985. Local density-functional theory of frequency-dependent linear response, *Physical Review Letters* 55: 2850–2852.

Gygi, F., 1999. GP code version 1.16.0 (F. Gygy, LLNL 1999–2004).

Hamann, D., 1989. Generalized norm-conserving pseudopotentials, *Physical Review B* 40: 2980–2987.

Harrison, M., Kershaw, S., Burt, M. et al., 2000. Colloidal nanocrystals for telecommunications. Complete coverage of the low-loss fiber windows by mercury telluride quantum dots, *Pure and Applied Chemistry* 72: 295–307, *First IUPAC Workshop on Advanced Material (WAM1)*, Hong Kong, Peoples Republic of China, July 14–18, 1999.

Hines, M. and Guyot-Sionnest, P., 1996. Synthesis and characterization of strongly luminescing ZnS-Capped CdSe nanocrystals, *Journal of Physical Chemistry* 100: 468–471.

Hohenberg, P. and Kohn, W., 1964. Inhomogeneous electron gas, *Physical Review B* 136: B864–B871.

Kasuya, A., Sivamohan, R., Barnakov, Y. et al., 2004. Ultra-stable nanoparticles of CdSe revealed from mass spectrometry, *Nature Materials* 3: 99–102.

Kasuya, A., Noda, Y., Dmitruk, I. et al., 2005. Stoichiometric and ultra-stable nanoparticles of II-VI compound semiconductors, *European Physical Journal D* 34: 39–41, *12th International Symposium on Small Particles and Inorganic Clusters*, Nanjing, Peoples Republic of China, September 06–10, 2004.

Katari, J., Colvin, V., and Alivisatos, A., 1994. X-ray photoelectron-spectroscopy of CdSe nanocrystals with applications to studies of the nanocrystal surface, *Journal of Physical Chemistry* 98: 4109–4117.

Klimov, V., 2003. Nanocrystal quantum dots, *Los Alamos Science* 28: 214.

Kohn, W. and Sham, L., 1965. Self-consistent equations including exchange and correlation effects, *Physical Review* 140: 1133–1138.

Kresse, G. and Furthmuller, J., 1996. Efficient iterative schemes for ab initio total-energy calculations using a plane-wave basis set, *Physical Review B* 54: 11169–11186.

Kucur, E., Ziegler, J., and Nann, T., 2008. Synthesis and spectroscopic characterization of fluorescent blue-emitting ultrastable CdSe clusters, *Small* 4: 883–887.

Kudera, S., Zanella, M., Giannini, C. et al., 2007. Sequential growth of magic-size CdSe nanocrystals, *Advanced Materials* 19: 548.

Landes, C., Braun, M., Burda, C., and El-Sayed, M., 2001. Observation of large changes in the band gap absorption energy of small CdSe nanoparticles induced by the adsorption of a strong hole acceptor, *Nano Letters* 1: 667–670.

Marques, M. and Gross, E., 2004. Time-dependent density functional theory, *Annual Review of Physical Chemistry* 55: 427–455.

Marques, M., Castro, A., Bertsch, G., and Rubio, A., 2003. Octopus: A first-principles tool for excited electron-ion dynamics, *Computer Physics Communications* 151: 60–78.

Marques, M. A. L., Ullrich C., Nogueira F., Rubio, A., Burke, K., and Gross, E. K. U., eds., 2006. *Time-Dependent Density Functional Theory*, Vol. 706 of *Lecture Notes in Physics.* Berlin, Germany: Springer.

Michalet, X., Pinaud, F., Bentolila, L. et al., 2005. Quantum dots for live cells, in vivo imaging, and diagnostics, *Science* 307: 538–544.

Mikulec, F. and Bawendi, M., 2000. Synthesis and characterization of strongly fluorescent CdTe nanocrystal colloids, in Komarneni, S. and Parker, J. C., and Hahn, H., eds., *Nanophase and Nanocomposite Materials III*, Vol. 581 of *Materials Research Society Symposium Proceedings*, Boston, MA, 139–144.

Murray, C., Norris, D., and Bawendi, M., 1993. Synthesis and characterization of nearly monodisperse CdE (E = S, Se, Te) semiconductor nanocrystallites, *Journal of the American Chemical Society* 115: 8706–8715.

Ouyang, J., Zaman, M. B., Yan, F. J. et al., 2008. Multiple families of magic-sized CdSe nanocrystals with strong bandgap photoluminescence via noninjection one-pot syntheses, *Journal of Physical Chemistry C* 112: 13805–13811.

Parr, R. G. and Yang, W., 1989. *Density-Functional Theory of Atoms and Molecules.* New York: Oxford University Press.

Peng, Z. and Peng, X., 2001. Formation of high-quality CdTe, CdSe, and CdS nanocrystals using CdO as precursor, *Journal of the American Chemical Society* 123: 183–184.

Peng, X., Schlamp, M., Kadavanich, A., and Alivisatos, A., 1997. Epitaxial growth of highly luminescent CdSe/CdS core/shell nanocrystals with photostability and electronic accessibility, *Journal of the American Chemical Society* 119: 7019–7029.

Peng, X., Manna, L., Yang, W. et al., 2000. Shape control of CdSe nanocrystals, *Nature* 404: 59–61.

Perdew, J. and Zunger, A., 1981. Self-interaction correction to density functional approximations for many-electron systems, *Physical Review B* 23: 5048–5079.

Perdew, J., Burke, K., and Ernzerhof, M., 1996. Generalized gradient approximation made simple, *Physical Review Letters* 77: 3865–3868.

Pradhan, N., Xu, H., and Peng, X., 2006. Colloidal CdSe quantum wires by oriented attachment, *Nano Letters* 6: 720–724.

Puzder, A., Williamson, A., Gygi, F., and Galli, G., 2004. Self-healing of CdSe nanocrystals: First-principles calculations, *Physical Review Letters* 92: 217401.1–217401.4.

Rabani, E., 2001. Structure and electrostatic properties of passivated CdSe nanocrystals, *Journal of Chemical Physics* 115: 1493–1497.

Reiss, P., Bleuse, J., and Pron, A., 2002. Highly luminescent CdSe/ZnSe core/shell nanocrystals of low size dispersion, *Nano Letters* 2: 781–784.

Runge, E. and Gross, E., 1984. Density-functional theory for time-dependent systems, *Physical Review Letters* 52: 997–1000.

Sarkar, P. and Springborg, M., 2003. Density-functional study of size-dependent properties of CdmSen clusters, *Physical Review B* 68: 235409.1–235409.7.

Shiang, J., Kadavanich, A., Grubbs, R., and Alivisatos, A., 1995. Symmetry of annealed wurtzite CdSe nanocrystals: Assignment to the C-3v point group, *Journal of Physical Chemistry* 99: 17417–17422.

Soler, J., Artacho, E., Gale, J. et al., 2002. The SIESTA method for ab initio order-N materials simulation, *Journal of Physics-Condensed Matter* 14: 2745–2779.

Soloviev, V., Eichhofer, A., Fenske, D., and Banin, U., 2000. Molecular limit of a bulk semi-conductor: Size dependence of the "band gap" in CdSe cluster molecules, *Journal of the American Chemical Society* 122: 2673–2674.

Talapin, D., Haubold, S., Rogach, A., Kornowski, A., Haase, M., and Weller, H., 2001. A novel organometallic synthesis of highly luminescent CdTe nanocrystals, *Journal of Physical Chemistry B* 105: 2260–2263.

Tessler, N., Medvedev, V., Kazes, M., Kan, S., and Banin, U., 2002. Efficient near-infrared polymer nanocrystal light-emitting diodes, *Science* 295: 1506–1508.

Troullier, N. and Martins, J., 1991. Efficient pseudopotentials for plane-wave calculations, *Physical Review B* 43: 1993–2006.

Vanderbilt, D., 1990. Soft self-consistent pseudopotentials in a generalized eigenvalue formalism, *Physical Review B* 41: 7892–7895.

Yu, W., Qu, L., Guo, W., and Peng, X., 2003a. Experimental determination of the extinction coefficient of CdTe, CdSe, and CdS nanocrystals, *Chemistry of Materials* 15: 2854–2860.

Yu, W., Wang, Y., and Peng, X., 2003b. Formation and stability of size-, shape-, and structure-controlled CdTe nanocrystals: Ligand effects on monomers and nanocrystals, *Chemistry of Materials* 15: 4300–4308.

Zhang, J., Wang, X., Xiao, M., Qu, L., and Peng, X., 2002. Lattice contraction in free-standing CdSe nanocrystals, *Applied Physics Letters* 81: 2076–2078.

Zhang, H., Cui, Z., Wang, Y. et al., 2003. From water-soluble CdTe nanocrystals to fluorescent nanocrystal-polymer transparent composites using polymerizable surfactants, *Advanced Materials* 15: 777.

Zhong, X., Zhang, Z., Liu, S., Han, M., and Knoll, W., 2004. Embryonic nuclei-induced alloying process for the reproducible synthesis of blue-emitting $Zn_xCd_{1-x}Se$ nanocrystals with long-time thermal stability in size distribution and emission wavelength, *Journal of Physical Chemistry B* 108: 15552–15559.

9
Magnetic Ion–Doped Semiconductor Nanocrystals

Shun-Jen Cheng
National Chiao Tung University

9.1 Introduction

Semiconductor quantum dots (QDs) are manufactured nanostructures with strong three-dimensional (3D) spatial confinement on length scales that are comparable to or even smaller than the effective Bohr radius, which is typically of the order of nanometers [Alivisatos 1996, Banin and Millo 2003]. The size effects of nanostructures result in strong quantization of electronic structures and material and physical properties that differ significantly from those of bulk systems. Because of their novel properties, semiconductor QDs have been extensively adopted as promising nanomaterials for various applications from optoelectronics to biotechnology [Bruchez et al. 1998, Klimov et al. 2000].

While most dot-based applications exploit the electrical and/or optical properties of dots, in the emerging fields of magnetoelectronics and spintronics, the fabrication of magnetic nanodevices made of *magnetic* QDs that exhibit both semiconductor and magnetic properties are highly desirable. Some spin devices that are based on magnetic QDs have been suggested for efficiently detecting or manipulating individual spins in spin-related applications [Recher et al. 2000, Efros et al. 2001, Fernandez-Rossier and Aguado 2007].

Magnetic semiconductors can be realized by incorporating magnetic ions (typically Mn^{2+}) into semiconductor compounds [Furdyna 1988, Dietl 2002]. The technology for fabricating bulk II–VI and III–V magnetic semiconductors has been developed [Ohno et al. 2000, Chiba et al. 2003, Jungwirth et al. 2006] and the material and physical properties have also been extensively investigated for decades. Nevertheless, making semiconductor QDs magnetic by incorporating magnetic dopants into the host materials of dots remains challenging. In 1994, Bhargava et al. became the first to report the successful doping of magnetic Mn ions by organometallic reactions in semiconductor (ZnS) nanocrystals [Bhargava et al. 1994]. II–VI and III–V self-assembled QDs doped with controlled numbers of magnetic ions Mn^{2+} have been recently fabricated [Maksimov et al. 2000, Dorozhkin et al. 2003, Besombes et al. 2004, 2005, Gurung et al. 2004, Gould et al. 2006, Leger et al. 2006, Mariette et al. 2006, Wojnar et al. 2007]. Magnetic ion dopants have been demonstrated to be able to be incorporated into a variety of colloidal semiconductor nanocrystal materials, including ZnO [Radovanovic et al. 2002, Schwartz et al. 2003, Norberg et al. 2004], ZnS [Bol and Meijerink 1998, Radovanovic and Gamelin 2001, Sarkar et al. 2007], ZnSe [Suyver et al. 2000, Norris et al. 2001, Norman et al. 2003, Erwin et al. 2005, Lad et al. 2007], CdS [Feltin et al. 1999], CdSe [Archer et al. 2007, Mikulec et al. 2000, Jian et al. 2003, Erwin et al. 2005], and PbSe colloidal nanocrystals [Ji et al. 2003]. Rich physical phenomena, such as giant Zeeman splitting [Hoffman et al. 2000, Norberg and Gamelin 2006], magnetic polarons [Maksimov et al. 2000, Dorozhkin et al. 2003, Wojnar et al. 2007, Cheng 2008], zero-field magnetization [Gurung et al. 2004, Gould et al. 2006, Sarkar et al. 2007], and rich fine structures of exciton–Mn complexes [Besombes et al. 2004, 2005, Leger et al. 2006, Mariette et al. 2006] have been observed in those magnetic nanostructures. The underlying physics of most of the physical phenomena are attributable to the intriguing spin interactions between magnetic ions and quantum-confined carriers.

Chemically synthesized colloidal nanocrystals (NCs) have certain advantages over self-assembled QDs in the engineering of quantum confinement owing to their controllability of size and shape. The diameters of nanocrystals can be controlled over a wide range, typically from 1 to 10 nm, using delicate fabrication processes [Brus 1991, Alivisatos 1996, Katz et al. 2002, Banin and Millo 2003]. Significant effects of size and shape on the electronic and optical properties of nanocrystals and nanorods have been identified using optical and resonant tunneling spectroscopies [Nirmal et al. 1996, Norris et al. 1996, Klein et al. 1997, Banin et al. 1999, Hu et al. 2001, Katz et al. 2002]. The engineered quantum confinement has also a pronounced effect on magnetic properties of NCs. Even without any paramagnetic dopants, quantum size effects have been observed to enhance substantially both paramagnetism and spontaneous magnetization in various nonmagnetic NCs [Neeleshwar et al. 2005, Madhu et al. 2008, Seehra et al. 2008]. The controllability of quantum confinement allows for the engineering of electronic structures and particle interactions, further influencing the spin interactions that dominate the magnetic properties of magnetically doped QDs [Fernandez-Rossier and Brey 2004, Abolfath et al. 2007, Cheng 2008].

Moreover, many physical properties of QDs are sensitive to the number of electrons or valence holes resident in the dots, which is electrically or optically tunable by using the techniques of bias control [Reimann and Manninen 2002], photochemistry [Liu et al. 2007], or photoexcitation [Leger et al. 2006, Cheng and Hawrylak 2008]. In spite of the complexity of the particle–particle interactions, the electronic properties of interacting electrons in many semiconductor QDs simply follow Hund's rules [Reimann and Manninen 2002], which enable the spin and orbital properties of the few electron ground states (GSs) to be determined, even without the need for complex many-body calculations.

The validity of Hund's rules for magnetic ion–doped QDs, however, remains an open question [Cheng 2005]. Doping magnetic ions into QDs, accompanied by spin interactions with electrons, can affect the spin and orbital properties of the few electron states of QDs. Magnetically doped QDs provide a playground for fundamental studies of the particle–particle interactions and the relevant underlying principles in QDs. On the other hand, spin interactions between magnetic ions and carriers can give rise to the magnetic ordering of magnetic ions in magnetic semiconductors [Furdyna 1988]. In III–V DMSs, magnetic ion dopants, typically Mn^{2+} with spin 5/2, act as acceptors, not only providing the sp–d spin interaction with itinerant carriers but also adding an attractive potential to them [Dietl 2002]. The spin interactions between carriers and localized magnetic dopants in III–V DMSs can be further enhanced as holes are bound by Mn^{2+} acceptors due to the high local density at the Mn site, forming magnetic polarons (MPs) [Bhatt et al. 2002]. Fascinating magnetic properties, such as the high T_c ferromagnetism of III–V DMSs, especially in the insulating regime or in the regime near the metal-insulating transition, are related to the formation of bound magnetic polarons. Unlike in III–V DMSs, such bound

magnetic polarons, however, are not necessarily formed stably in II–VI DMSs because divalent Mn ions are isoelectronic in II–VI materials. However, recent experimental and theoretical studies suggest that the magnetic properties of II–VI DMSs can be optimized by reducing the dimensionality of the DMS material, such as in QDs, with the stable formation of magnetic polarons improved by quantum confinement [Fernandez-Rossier and Brey 2004].

The electronic structure of nonmagnetic colloidal nanocrystals has been extensively studied theoretically using various approaches, from pseudopotential [Rama et al. 1992, Tomasulo and Ramakrishna 1996, Wang and Zunger 1996, Fu and Zunger 1997], tight-binding [Lippens and Lannoo 1990, Albe et al. 1998, Hill et al. 1999, Perez-Conde and Bhattacharjee 2001, Viswanatha et al. 2005], and $k \cdot p$ methods [Richard et al. 1996, Fu et al. 1998, Efros and Rosen 2000] to effective mass theory [Hu et al. 1990, Bhattacharjee and Benoit a la Guillaume 1997]. Sophisticated microscopic approaches, such as pseudopotential or tight-binding methods allow for the consideration of the electronic structure of nanostructures down to the atomistic scale. They however also create more complications in the analysis of physics. The macroscopic $k \cdot p$ and effective mass theories generally provide simpler descriptions of the electronic structures of QDs, which is usually in qualitative agreement with those given by the microscopic approaches [Lippens and Lannoo 1990, Albe et al. 1998]. In the framework of the macroscopic theories, the electrical, optical, and magnetic properties of magnetically doped semiconductor QDs have been theoretically studied by using the local mean field theory [Chang et al. 2004, Fernandez-Rossier and Brey 2004, Govorov 2004, Govorov and Kalameitsev 2005, Abolfath et al. 2007] and the configuration interaction (CI) method combined with exact diagonalization (ED) techniques [Bhattacharjee and Perez-Conde 2003, Cheng 2005, Climente et al. 2005, Qu and Hawrylak 2005, 2006, Cheng 2008, Nguyen and Peeters 2008, Qu and Vasilopoulos 2006].

This chapter presents a theoretical description of, and the fundamental theory that governs, the electronic and magnetic properties of Mn^{2+}-doped II–VI NCs. Section 9.2 introduces the electronic structure and magnetic properties of nonmagnetic NCs, developed in the framework of effective mass approximation and the theory of atomic magnetism. Section 9.3 discusses models of substitutional divalent Mn impurities in II–VI semiconductors (SCs) and the relevant spin interactions. In Section 9.4, an analysis of NCs that contain a single electron coupled to many Mn ions is conducted to illustrate carrier-mediated magnetism in QDs using a simplified model. Section 9.5 presents the generalized Hamiltonian for magnetic NCs containing arbitrary number of charged carriers and magnetic ions, and introduces two numerical approaches, beyond the simple model, for calculating the electronic structure and the magnetic properties of magnetically doped NCs. The configuration interaction method combined with exact diagonalization techniques allows for accurate calculations of the energy spectra and the magnetic properties of magnetic NCs doped with small number of Mn ions. It contrasts with local mean field theory, which is often employed for magnetic NCs doped with numerous magnetic ions. Section 9.6 draws conclusions.

9.2 Electronic Structure and Magnetic Properties of Nonmagnetic Nanocrystals

9.2.1 Electronic Structure

The electronic structure of nonmagnetic (Mn-free) NCs is described first. The Schrödinger equation for a single electron confined in an NC is

$$h_0\phi(\vec{r}) = \epsilon\phi(\vec{r}),\qquad(9.1)$$

where

$$h_0 = \frac{|\vec{p}|^2}{2m^\star} + V_0(\vec{r})\qquad(9.2)$$

is the single electron Hamiltonian that consists of the kinetic energy term and the confining potential of NC

$\epsilon(\phi(\vec{r}))$ is the eigen energy (wave function) of the electron
$\vec{r}(\vec{p})$ denotes the coordinate position (linear momentum) of the electron
$m^\star$ is the effective mass for electron

Taking the hard wall spherical model [Hu et al. 1990, Bhattacharjee and Benoit a la Guillaume 1997] in which the effective confining potential V_0 for a sphere-like NC of radius a is modeled by

$$V_0(\vec{r}) = \begin{cases} 0 & r \le a \\ \infty & r > a \end{cases},\qquad(9.3)$$

the eigen energy and the wave function of the eigen states for Equations 9.1 through 9.3 are explicitly given by

$$\epsilon_{nlm_l m_s} = \frac{\hbar^2\alpha_{nl}^2}{2m^\star a^2},\qquad(9.4)$$

and

$$\phi_{nlm_l m_s}(\vec{r}) = \langle\vec{r}\,|\,nlm_l m_s\rangle = \sqrt{\frac{2}{a^3}}\,\frac{J_l\left(\frac{\alpha_{nl}}{a}r\right)}{J_{l+1}(\alpha_{nl})}Y_{lm_l}(\theta,\phi),\qquad(9.5)$$

respectively, where
$J_l(r)$ is the spherical Bessel function
α_{nl} is the nth zero of J_l
$Y_{lm}(\theta,\phi)$ is the spherical harmonic function

Figure 9.1 schematically depicts the model. In the central force problem, the eigen states for Equation 9.1 are labeled using the

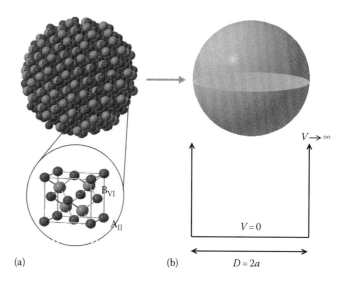

FIGURE 9.1 Schematics of (a) the zinc blende atomistic structure and (b) the continuous hard wall model of a spherical II–VI semiconductor nanocrystal (NC).

following set of quantum numbers: the principal quantum number n, the angular momentum $l = L/\hbar$, the z-component of orbital angular momentum $m_l = -l, -l + 1, \ldots, l - 1, l$, and the z-component of electron spin $m_s = \pm 1/2$.

Equation 9.4 shows that the eigen energies of symmetric NCs are a function of n and l only. For some set of n and l, the orbitals with different $-l \le m_l \le l$ and $m_s = \pm 1/2$ form a $2 \times (2l + 1)$ degenerate electronic shell. Because of the characteristic shell structure, QDs are referred to as artificial atoms. Figure 9.2 plots the energy levels and the corresponding charge densities of the two lowest electronic shells of a spherical NC in the $E - m_l$ plot. Table 9.1 presents expressions for the eigen energies and the wave functions of the low-lying states.

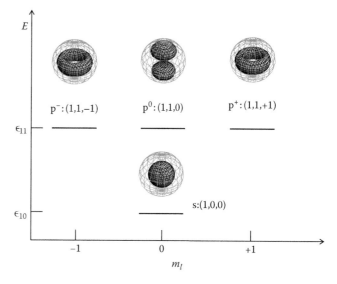

FIGURE 9.2 Schematic of the energy diagram (ϵ_{nl} vs. m_l) of a spherical NC at zero magnetic field and the charge density distributions of the s- and p-orbitals.

TABLE 9.1 Expressions for the Eigen Energies and the Wave Functions of the Low Lying Single Particle States of a Spherical NC with Radius *a* within Hard Wall Spherical Model

(n, l, m_l)	ϵ_{nl}	$\phi_{n,l,m_l}(r, \theta, \phi)$	Degeneracy
(1,0,0)	$\dfrac{\hbar^2 \alpha_{10}^2}{2m^* a^2}$	$\phi_{100}(\vec{r}) = \dfrac{1}{2}\sqrt{\dfrac{1}{\pi}} \cdot R_{10}(r)$	1
(1,1,−1)	$\dfrac{\hbar^2 \alpha_{11}^2}{2m^* a^2}$	$\phi_{11-1}(\vec{r}) = \sqrt{\dfrac{3}{8\pi}} \cdot R_{11}(r)\sin(\theta)\exp(-i\phi)$	3
(1,1,0)	$\dfrac{\hbar^2 \alpha_{11}^2}{2m^* a^2}$	$\phi_{110}(\vec{r}) = \dfrac{1}{2}\sqrt{\dfrac{3}{\pi}} \cdot R_{11}(r)\cdot\cos(\theta)$	3
(1,1,1)	$\dfrac{\hbar^2 \alpha_{11}^2}{2m^* a^2}$	$\phi_{111}(\vec{r}) = -\sqrt{\dfrac{3}{8\pi}} \cdot R_{11}(r)\sin(\theta)\exp(i\phi)$	3
(1,2,−2)	$\dfrac{\hbar^2 \alpha_{12}^2}{2m^* a^2}$	$\phi_{12-2}(\vec{r}) = \sqrt{\dfrac{15}{32\pi}} \cdot R_{12}(r)\sin^2\theta\exp(-2i\phi)$	5
(1,2,−1)	$\dfrac{\hbar^2 \alpha_{12}^2}{2m^* a^2}$	$\phi_{12-1}(\vec{r}) = \sqrt{\dfrac{15}{8\pi}} \cdot R_{12}(r)\sin\theta\cos\theta\exp(-i\phi)$	5
(1,2,0)	$\dfrac{\hbar^2 \alpha_{12}^2}{2m^* a^2}$	$\phi_{120}(\vec{r}) = \dfrac{1}{4}\sqrt{\dfrac{5}{\pi}} \cdot R_{12}(r)(3\cos^2\theta - 1)$	5
(1,2,1)	$\dfrac{\hbar^2 \alpha_{12}^2}{2m^* a^2}$	$\phi_{121}(\vec{r}) = -\sqrt{\dfrac{15}{8\pi}} \cdot R_{12}(r)\sin\theta\cos\theta\exp(i\phi)$	5
(1,2,2)	$\dfrac{\hbar^2 \alpha_{12}^2}{2m^* a^2}$	$\phi_{122}(\vec{r}) = \sqrt{\dfrac{15}{32\pi}} \cdot R_{12}(r)\sin^2\theta\exp(2i\phi)$	5

Note: The radial function R_{nl} is defined as $R_{nl} = \sqrt{2/a^3}\left[J_l((\alpha_{nl}/a)r)/J_{l+1}(\alpha_{nl})\right]$ in terms of Bessel function J_l and the zeros (α_{nl}) of the Bessel function ($\alpha_{10} = \pi$, $\alpha_{11} = 4.493$, and $\alpha_{12} = 5.764$).

Rescaling Equation 9.4 by the effective Rydberg energy Ry^*, the single electron spectrum is reformulated as

$$\frac{\epsilon_{nlm}}{Ry^*} = \frac{\hbar^2}{2m^*}\left(\frac{1}{a_B^*}\right)^2\left(\frac{a_B^*}{a}\right)^2 \alpha_{nl}^2/Ry^* = \left(\frac{a_B^*}{a}\right)^2 \alpha_{nl}^2, \quad (9.6)$$

where a_B^* is the effective Bohr radius. For most semiconductors, a typical value of a_B^* is a few nm and that of Ry^* is a few tens of meV. Equation 9.6 indicates that the energetic quantization of an NC with a radius comparable to the effective Bohr radius $a \sim a_B^*$ is of the order of $\alpha_{nl}^2 Ry^* \sim 10^1 Ry^*$, one order of magnitude greater than that of the effective Rydberg. Notably, the typical energy separation between adjacent electronic shells, ∼ hundreds of meV, is one order of magnitude larger than the strength of the Coulomb interactions between carriers and two orders of magnitude larger than the strengths of the spin interactions between carriers and magnetic ions. Figure 9.3a shows the optical absorption spectra for CdSe colloidal nanocrystals, in which the energetic separation between absorption peaks is approximately 400–600 meV. Table 9.2 summarizes some relevant energy scales for nonmagnetic and magnetic NCs. Restated, NCs are so strongly quantized that the particles usually have difficulty being transferred between different electronic shells via Coulomb interactions and/or spin interactions with magnetic ions. The weak inter-shell couplings (i.e., correlation interactions) ensure that the few single particle states given by Equation 9.5 are a good basis for the expansion of the undetermined eigen states of a few interacting electrons in nonmagnetic and/or magnetic NCs.

However, the interactions between particles on the same shell play are crucial to the spin and orbital arrangement of the few particle states. Hund's rules state that spin electrons on an electronic shell should be arranged to maximize the total spin *S* (the first rule). Then, once the total spin *S* has been determined, the arrangement of electrons on the orbitals of the shell should be determined to maximize have the total angular momentum *L* whenever possible [Blundell 2001]. Figure 9.4 plots the total spin *S* and total angular momentum *L* as functions of the number of electrons, as determined by Hund's rules.

9.2.2 Magnetic Properties

Next, the magnetic response of (nonmagnetic) NCs to external applied magnetic fields is considered. The Hamiltonian for a

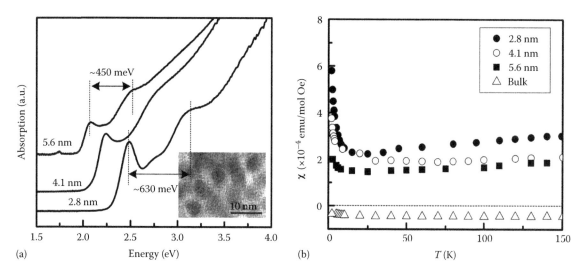

FIGURE 9.3 (a) Measured optical absorption spectra for CdSe colloidal nanocrystal quantum dots of diameter d = 2.8, 4.1, and 5.6 nm. Inset: The high-resolution transmission electron microscopy (HRTEM) image of d = 5.6 nm CdSe NCs. (b) Measured magnetic susceptibility χ as a function of temperature for the bulk and d = 2.8, 4.1, and 5.6 nm CdSe NCs. (Courtesy of Prof. Yang Yuan Chen, Institute of Physics, Academia Sinica, Taiwan.)

TABLE 9.2 Relevant Energy Scales of Mn-Doped NCs

Physical Quantities	Energy Scale (meV)
Single electron energy quantization	$>10^2$
Direct Coulomb interaction	10^2
Exchange Coulomb interaction	10^1
Electron–Mn interaction	10^0
Mn–Mn interaction (nearest neighbor)	10^0
Orbital Zeeman energy (B = 1 T)	10^{-1}–10^0
Spin Zeeman energy (B = 1 T)	10^{-2}–10^{-1}

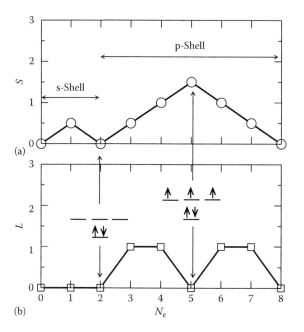

FIGURE 9.4 (a) The total spin S and (b) the total orbital angular momentum L of interacting electrons in a spherical NC vs. the total number of electrons N_e according to Hund's rules.

single electron confined in an NC in an external magnetic field $\vec{B} = (0, 0, B)$ is

$$h_B = \frac{(\vec{p} + e\vec{A})^2}{2m^\star} + V_0(\vec{r}) - g_s\mu_B s_z B, \qquad (9.7)$$

where

$\vec{A}$ is the vector potential due to magnetic field

s_z is the electron spin projection operator, the last term is the spin Zeeman energy in terms of the g-factor of electron g_s

The Bohr magneton is defined as $\mu_B \equiv |e|\hbar/2m_0$
Taking the vector potential $\vec{A} = B/2 \, (-y, x, 0)$ in symmetric gauge, the Hamiltonian is rewritten as

$$h_B = h_0 - g_s\mu_B B\hat{s}_z - g_l\mu_B^\star B\hat{l}_z + \frac{e^2 B^2 (x^2 + y^2)}{8m^\star}$$

$$= h_0 - g_s m^\star \frac{\hbar\omega_c}{2}\hat{s}_z - g_l \frac{\hbar\omega_c}{2}\hat{l}_z + \frac{\hbar\omega_c}{8}\left(\frac{r}{l_B}\right)^2, \qquad (9.8)$$

where

$l_z = L_z/\hbar = (1/i)\left[(\partial/\partial y)x - (\partial/\partial x)y\right](s_z)$ is defined as the z-projection operator for orbital (spin) angular momentum

$g_l = -1(g_s)$ is the g-factor for the orbital magnetic moment of electron (spin angular momentum of electron in CdSe)

$\omega_c = |e|B/m^\star$ the cyclotron frequency of the electron

$\mu_B^\star = |e|\hbar/2m_0 m^\star$ is the effective Bohr magneton

The second (third) B-linear term on the right hand side of Equation 9.8 is referred to as the spin (orbital) Zeeman term due to the coupling between the spin (orbital) magnetic moment of the electron and the magnetic field. Both terms make paramagnetic contributions (Curie paramagnetism) to the magnetic response of NC. In contrast, the last B-quadratic term contributes diamagnetism. For a small dot in a weak magnetic field (with long magnetic length $l_B \equiv \sqrt{\hbar/(eB)} \gg a$), the diamagnetism term is negligible and the Hamiltonian of Equation 9.8 can be approximated as

$$h_B \approx h_0 - \vec{\mu} \cdot \vec{B}, \tag{9.9}$$

where the magnetic moment operator $\vec{\mu}$ is defined as

$$\vec{\mu} = g_l \mu_B^* \vec{l} + g_s \mu_B \vec{s}. \tag{9.10}$$

In the approximation, the magneto-energy spectrum of an electron in an NC is expressed as $\langle h_B \rangle \approx \epsilon_{nlm_l m_s} + (g_l \mu_B^* m_l + g_s \mu_B m_s)B$ with $m_l = -l, -l+1, \ldots, l-1, l$ and $m_s = \pm 1/2$. Figure 9.5 schematically depicts the energy spectra of an NC in magnetic fields. The magnetization of a single electron in an NC subject to a magnetic field and thermal fluctuations is defined by the averaged magnetic moment, and, according to Equation 9.9, expressed as

$$M \equiv \langle \mu_z \rangle = g_l \mu_B^* \langle m_l \rangle + g_s \mu_B \langle m_s \rangle. \tag{9.11}$$

Classical statistics yields the averaged magnetic moment from the orbital motion of the electron [Blundell 2001]

$$M_l = g_l \mu_B^* \langle m_l \rangle = \frac{\displaystyle\sum_{m_l=-l}^{+l} g_l \mu_B^* m_l \exp(-g_l \mu_B^* m_l B/kT)}{\displaystyle\sum_{m_l} \exp(-g_l \mu_B^* m_l B/kT)}, \tag{9.12}$$

and that from electron spin

$$M_s = g_s \mu_B \langle m_s \rangle = \frac{\displaystyle\sum_{m_s=-1/2}^{+1/2} g_s \mu_B m_s \exp(-g_s \mu_B m_s B/kT)}{\displaystyle\sum_{m_s} \exp(-g_s \mu_B m_s B/kT)} \tag{9.13}$$

For Equations 9.12 and 9.13, compact analytical expressions are available [Blundell 2001]. The following presents the derivations. First, the average of the orbital angular momentum projection, Equation 9.12, is rewritten as

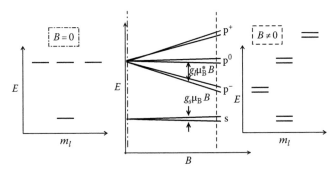

FIGURE 9.5 Schematic diagram of the energy spectrum (E vs. B) of a single electron in an NC in magnetic fields B (center). The schematic E vs. m_l diagram for zero magnetic field (left) and finite magnetic field (right).

$$\langle m_l \rangle = \frac{\displaystyle\sum_{m_l=-l}^{l} m_l \exp(m_l x)}{\displaystyle\sum_{m_l=-l}^{l} \exp(m_l x)} \tag{9.14}$$

where $x \equiv g_l \mu_B B/kT$ is defined as a dimensionless variable. Equation 9.14 can be rewritten as

$$\langle m_l \rangle = \frac{1}{Z_l} \frac{\partial Z_l}{\partial x} = \frac{kT}{g_l \mu_B^*} \frac{\partial \ln Z_l}{\partial B}. \tag{9.15}$$

where $Z_l \equiv \displaystyle\sum_{m_l=-l}^{l} \exp(m_l g_l \mu_B B/kT)$ is defined as the partition function. The partition function can be formulated as

$$Z = \sum_{m_l=-l}^{l} e^{m_l x}$$

$$= e^{-lx}(1 + e^x + e^{2x} + \cdots + e^{2lx})$$

$$= e^{-lx} \frac{1 - \exp[(2l+1)x]}{1 - \exp(x)}$$

$$= \frac{\sinh[(2l+1)x/2]}{\sinh(x/2)} \tag{9.16}$$

Substituting Equation 9.16 into Equation 9.15, the averaged magnetization is explicitly expressed as

$$M_l = kT \frac{\partial \ln Z}{\partial B} = M_l^{sat} B_l \left(\frac{g_l \mu_B^* lB}{kT} \right), \tag{9.17}$$

where

$M_l^{sat} = g_l \mu_B^* l$ is the saturation magnetization
B_l is the Brillouin function, which is defined as

$$B_l(y) = \frac{2l+1}{2l} \coth\left(\frac{2l+1}{2l} y \right) - \frac{1}{2l} \coth \frac{y}{2l}. \tag{9.18}$$

Figure 9.6 plots the Brillouin functions $B_l(y)$ for $l = 1, 2$, and 3. Equation 9.17 describes the magnetization of an electron moving in an orbital with l in a quantum confinement system as a function of an external magnetic field and temperature.

In the limit of high field ($y \equiv \mu_B^* B/kT \gg 1$), the magnetization approaches the maximum value $M_l^{sat} = g_l \mu_B^* l$. In the low-field regime ($y \ll 1$), where $B_l(y) \approx \frac{l+1}{3l} y + O(y^3) + \cdots$, the magnetization is approximated as

$$M_l \approx \frac{(g_l \mu_B^*)^2 l(l+1)B}{3kT}, \tag{9.19}$$

and the magnetic susceptibility in a low field approximated as

$$\chi_{l,0} \equiv \left. \frac{\partial M_l}{\partial B} \right|_{\mu_B B/kT \to 0} \approx \frac{(g_l \mu_B^*)^2 l(l+1)}{3kT} = \frac{(\mu_l^{eff})^2}{3kT}, \tag{9.20}$$

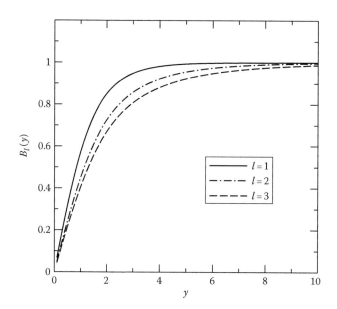

FIGURE 9.6 The Brillouin functions $B_l(y)$ for $l = 1, 2,$ and 3.

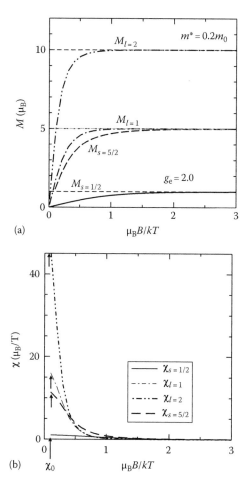

FIGURE 9.7 Magnetization of paramagnets with orbital angular momenta $l = 1, 2,...$ or spin angular momenta $s = 1/2, 5/2$ as a function of $\mu_B B/kT$ following the Brillouin function (a). (b) Shows the corresponding magnetic susceptibility χ as a function of $\mu_B B/kT$.

where $\mu_l^{eff} \equiv g_l \mu_B^* \sqrt{l(l+1)}$ is defined as the effective moment, which characterizes the magnetism of the system. Equations 9.19 is Curie's law describing the linear behavior of magnetization as a function of B in the low-field regime.

Likewise, the averaged magnetic moment can be derived from electron spin as

$$M_s = M_s^{sat} B_s \left(\frac{g_s \mu_B s B}{kT} \right), \qquad (9.21)$$

where $M_s^{sat} = g_s \mu_B s = g_s \mu_B/2$, and the corresponding low-field magnetic susceptibility is

$$\chi_{s,0} \approx \frac{(\mu_s^{eff})^2}{3kT}, \qquad (9.22)$$

where the effective moment from electron spin is $\mu_s^{eff} = |g_s| \mu_B \sqrt{s(s+1)} = (\sqrt{3}/2)|g_s| \mu_B$. Figure 9.7 shows the magnetization M and the corresponding magnetic susceptibility χ of paramagnets according to Equations 9.17 and 9.21. According to Equations 9.22 and 9.22, the low-field magnetic susceptibility χ_0 of a paramagnet with well-defined angular momentum is inversely proportional to temperature T (see Figure 9.8).

Notably, in most semiconductor materials, conduction electrons have effective masses that are lighter than that of a free electron, typically $m^* \sim 10^{-1} m_0 \ll m_0$, leading to $\mu_l^{eff} \gg \mu_s^{eff}$. Thus, the orbital moments of electrons dominate the total magnetization of a QD as long as the electrons on the bound orbitals of the QD have a finite orbital angular momentum [Cheng 2005].

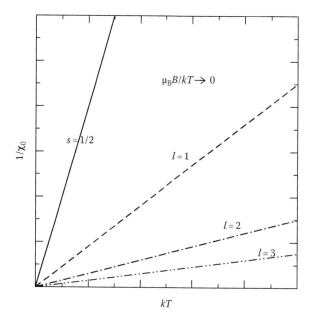

FIGURE 9.8 According to Curie's law, low-field magnetic susceptibility χ_0 is inversely proportional to temperature T (see text). Thus, a straight line is obtained in the $1/\chi_0$ vs. T plot.

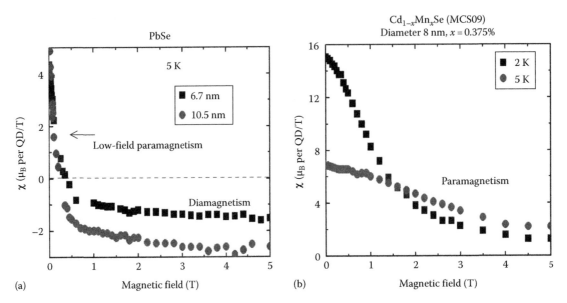

FIGURE 9.9 Measured magnetic susceptibilities χ as a function of applied magnetic fields for an ensemble of (a) nonmagnetic PbSe NCs (b) diluted magnetic semiconductor $Cd_{1-x}Mn_xSe$ NCs. The nonmagnetic PbSe QDs exhibit significant paramagnetism ($\chi > 0$) at low magnetic fields ($B < 0.5\,T$) due to the quantum size effects. By contrast, the PbSe dots, behaving like PbSe bulk systems, exhibit diamagnetism ($\chi < 0$) at high fields $B > 0.5\,T$. With magnetic ion Mn^{2+} dopants, the magnetic CdMnSe QDs exhibit paramagnetism over a wide range of applied magnetic fields. (Courtesy of Prof. Wen-Bin Jian, Department of Electrophysics, National Chiao Tung University, Hsinchu, Taiwan.)

Experimentally, pronounced paramagnetism due to the quantum size effects has been observed for some nonmagnetic QDs [Neeleshwar et al. 2005, Madhu et al. 2008, Seehra et al. 2008]. Figure 9.3b shows the positive susceptibilities $\chi > 0$ (paramagnetism) as a function of temperature for CdSe NCs, which further increase as the size of the NCs decreases. By contrast, CdSe bulk lacking quantum confinement exhibits diamagnetism $\chi < 0$. Figure 9.9a shows the measured magnetic susceptibilities χ as a function of magnetic field for an ensemble of nonmagnetic PbSe NCs. The size effects of QD lead to pronounced paramagnetism at low magnetic fields ($B < 0.5\,T$).

Figure 9.10 plots the calculated magnetic susceptibility as a function of applied magnetic field and the low-field susceptibility as a function of electron number N_e of nonmagnetic CdSe NCs with a radius of 5 nm, determined numerically using exact diagonalization techniques (see Section 9.5.1). In Figure 9.10b, pronounced low-field paramagnetism is observed for $N_e = 3$ because of the finite total orbital angular momentum ($|L| = 1$). Notably, the low-field magnetic susceptibility (χ_0) of NC follows a similar N_e dependence to that of total angular momentum of NC given by Hund's rules because of the dominance of the contribution of the orbital moments to χ_0 (inset of Figure 9.10) [Cheng 2005].

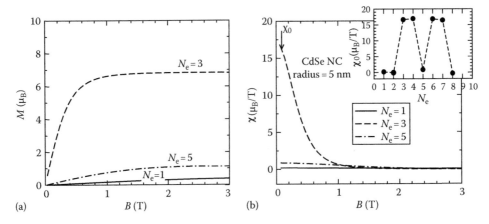

FIGURE 9.10 Calculated magnetizations (a) and magnetic susceptibilities (b) vs. applied magnetic fields of nonmagnetic CdSe NCs charged with electron number $N_e = 1, 3, 5$. Inset: the low-field magnetic susceptibilities χ_0 as a function of N_e. As seen, the χ_0 vs. N_e follows the similar relationship of L vs. N_e according to Hund's rules. In the calculation, we take the following material parameters for the CdSe NCs: the effective mass of electron $m^* = 0.15\,m_0$, the dielectric constant $\kappa = 8.9$, and the g-factor of electron $g_e = 1.2$.

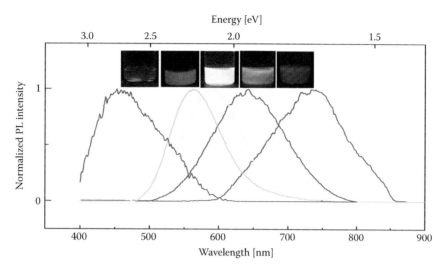

FIGURE 5.8 Normalized PL emission spectra and the corresponding red ($\lambda = 735$ nm), orange ($\lambda = 641$ nm), yellow ($\lambda = 592$ nm), green ($\lambda = 563$ nm), and blue ($\lambda = 456$ nm) emission color from etched Si-NPs. (From Gupta, A. et al., *Adv. Funct. Mater.*, 19(5), 696, 2009. With permission.)

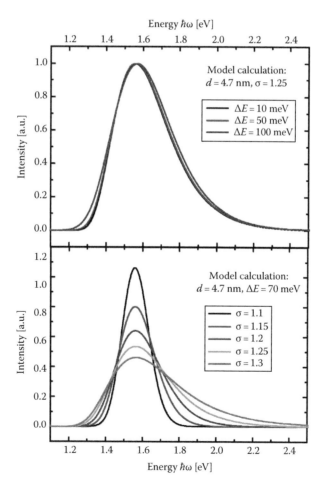

FIGURE 5.11 Comparison between the influence of the homogeneous broadening ΔE and that of the inhomogeneous broadening (described by the geometrical standard deviation σ on the ensemble) on the width of the PL spectra. (From Meier, C. et al., *J. Appl. Phys.*, 101, 8, 2007. With permission.)

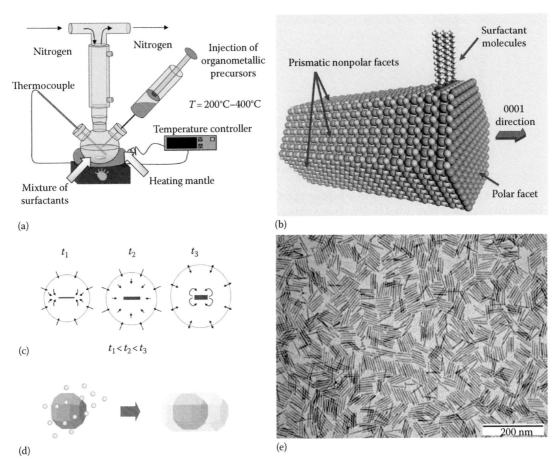

(a)

(b)

(c)

$t_1 < t_2 < t_3$

(d)

(e)

200 nm

FIGURE 7.9 (a) A sketch of a typical setup for the synthesis of colloidal nanoparticles. In a typical one-pot synthesis, precursors are injected in a flask containing hot coordinating solvents. The choice of coordinating solvents is dictated by several reasons, such as the conditions of growth, the precursor reactivity, and the desired nanoparticle shape and size. In order to avoid reaction with oxygen, the synthesis is carried out under inert atmosphere (such as nitrogen or argon). The growth temperature is monitored by a controller (via a thermocouple) that feedbacks a heating mantle. For the synthesis of II–VI semiconductor nanocrystals, in general precursors are introduced in the reaction bath either as organometallic precursors [i.e., $Cd(CH_3)_2$, $Zn(C_2H_5)_2$, S:TOP, Se:TOP, Te:TOP, where TOP stands for trioctylphosphine, $S(Si(CH_3)_3)_2$] or as inorganic precursors (metal salts or even metal oxides, such as $Cd(CH_3COO)_2$, $Cd(NO_3)$, CdO), (Dushkin et al., 2000; Qu et al., 2001; Donega et al., 2005). (b) Model of a wurtzite CdTe nanorod in which three of the prismatic nonpolar facets and the 0001 polar facet are shown. Some surfactant molecules (one example is octadecylphosphonic acids, of which three molecules are shown in this model), under specific conditions, bind selectively to the non-polar facets, depressing growth of these facets (Manna et al., 2005; Rempel et al., 2005; Barnard et al., 2007). (c) Different stages of anisotropic growth of rod-shaped nanoparticles. In each stage, a "rod" is shown enclosed in its surrounding diffusion layer. (d) A cartoon sketching the concept of seeded growth of nanorods. (e) A low-resolution TEM images of wurtzite CdS nanorods "seeded" with spherical CdSe nanocrystal seeds. Here also the phase of the nanocrystal seeds was wurtzite.

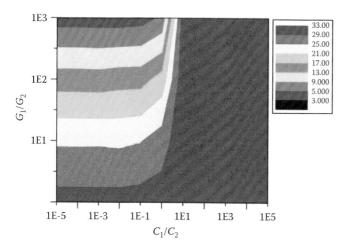

FIGURE 13.11 Contour plot of the TMR maximum as a function of the ratios of C_1/C_2 and G_1/G_2 in F/N-nanoparticle/F junctions. (Adapted from Wang, H. et al., *Phys. Stat. Sol. (b)*, 244, 4443, 2007.)

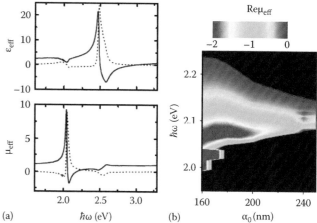

FIGURE 14.11 (a) Real (solid lines) and imaginary (dashed lines) part of the effective permittivity, ϵ_{eff}, and permeability, μ_{eff}, of a hexagonal array, with lattice constant $a = 200\,nm$, of silver–silica–silver nanosandwiches, with $S = 50\,nm$, $h_1 = h_3 = 20\,nm$, and $h_2 = 40\,nm$ (a, S, h_1, h_2, h_3 are defined in Figure 14.10), on a quartz substrate. (b) A map of the negative effective permeability of different hexagonal arrays of the nanoparticles described earlier. (Reprinted from Tserkezis, C. et al., *Phys. Rev. B*, 78, 165114-1, 2008. With permission.)

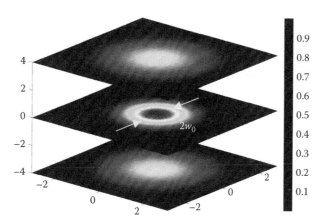

FIGURE 15.1 Cross sections in *x-y* plane of a Gaussian beam propagating in vertical *z*-direction. The vertical bar shows coding of the amplitude of the electric field in the wave.

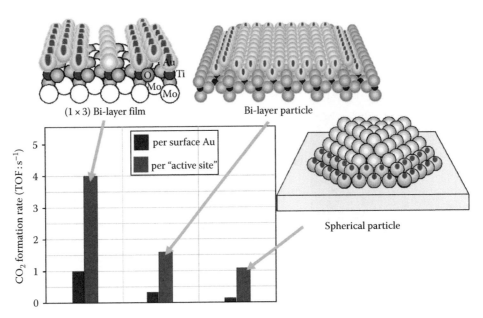

FIGURE 17.12 Comparison of catalytic activities for CO oxidation on the Mo(112)–(1 × 3)–(Au, TiO_x), Au/TiO_2(110), and Au supported on high-surface-area TiO_2 with a mean particle size of ~3 nm. The corresponding structural models were shown with red and blue marks to indicate the active sites. (From Valden, M. et al., *Science*, 281, 1647, 1998; Chen, M.S. and Goodman, D.W., *Science*, 306, 252, 2004. With permission.)

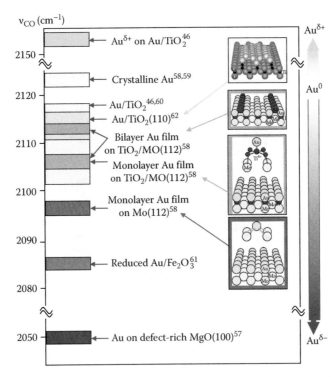

FIGURE 17.13 Comparison of the stretching frequencies for CO adsorption on various supported Au catalysts. The indicated reference number in the figure was originated in Ref. [31]. (From Chen, M.S. and Goodman, D.W., *Acc. Chem. Res.*, 39, 739, 2006. With permission.)

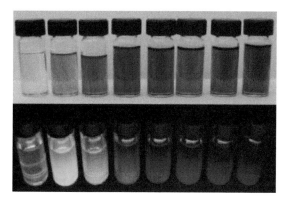

FIGURE 23.2 The Lycurgus cup seen in reflected light, green (a) and transmitted light, red (b). (Courtesy of the Trustees of the British Museum. With permission.)

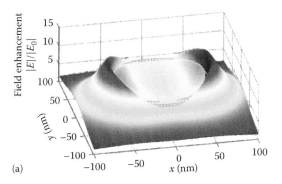

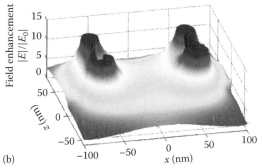

FIGURE 24.10 Field enhancement in a gold nanoring for two cross sections through the center of the ring: top view (a) and side view (b).

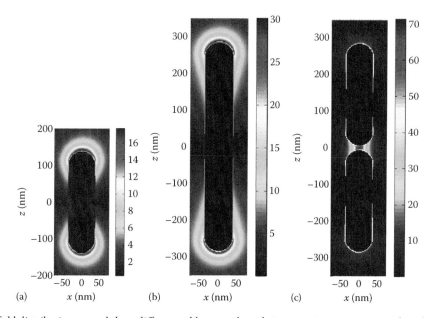

FIGURE 24.11 Near-field distribution around three different gold nanorods at their respective resonance wavelengths. (a) A single nanorod of total length 280 and width 80 nm at a wavelength of $\lambda = 940$ nm. (b) A single gold nanorod with a total length of 570 and width 80 nm at a wavelength of $\lambda = 1695$ nm. (c) A pair of nanorods each of total length of 280 nm, longitudinally aligned with a 10 nm gap. The incident field is polarized along the rod axes.

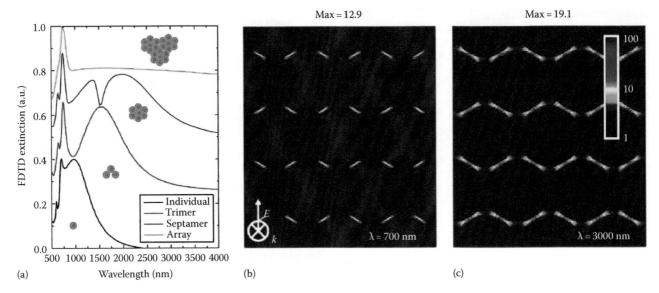

(a)   (b)   (c)

FIGURE 24.17 (a) Extinction cross section of an individual gold nanoshell (black), a trimer (red), a septamer (blue), and an infinite hexagonal array (green curve) of gold nanoshells. (b) Near-field distribution for the infinite array at a wavelength of $\lambda = 700\,\text{nm}$ and (c) at a wavelength of $\lambda = 3000\,\text{nm}$. The gold nanoshells have an inner radius of 150 nm and an outer radius of 172 nm. The separation between nanoshells is 8 nm. Similar local field enhancements are achieved between particles both for visible and infrared wavelengths. (Adapted from Le, F. et al., *ACS Nano*, 2, 707, 2008. With permission.)

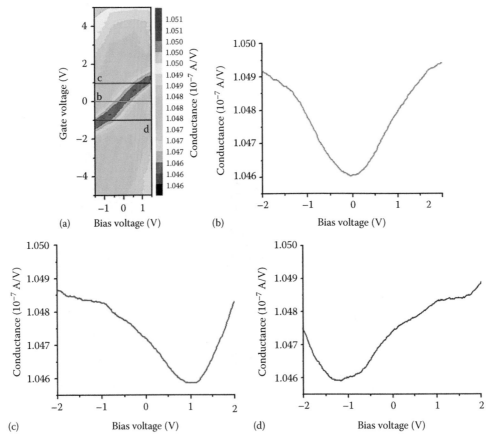

FIGURE 25.13 (a) Differential conductance map as a function of V_b and V_g at 77 K. The map is obtained using a four-layer film of butanedithiol-linked Au NPs. (b–d) Differential conductance versus V_b at various V_gs, (b) $V_g = 0$ V, (c) $V_g = +1$ V, and (d) $V_g = -1$ V. (Reprinted from Suganuma, Y. et al., *Nanotechnology*, 16, 1196, 2005. With permission.)

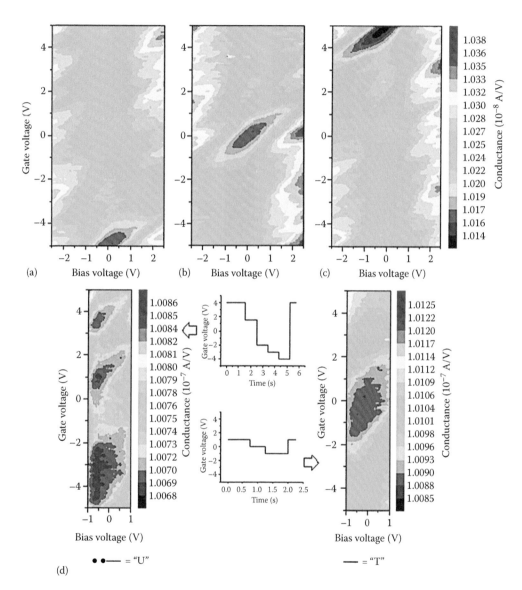

FIGURE 25.14 Differential conductance maps of a four-layer film of butanedithiol-linked Au NPs as a function of V_b and V_g. The maps are obtained at 77 K after applying various gate voltages to the film as the film was slowly cooled. (a–c) Constant V_gs are applied to the film during cooling: (a) $V_g = -5\,V$, (b) $V_g = 0$, and (c) $V_g = +5\,V$. (d) Cyclic V_g (shown in the center) are applied to the film during cooling. The stored information in the conductance maps reading from $V_g = +5\,V$ toward $-5\,V$ resembles "••–" (left) and "–" (right), which correspond to "U" and "T", respectively, in Morse code. (Reprinted from Suganuma, Y. et al., *Nanotechnology*, 16, 1196, 2005. With permission.)

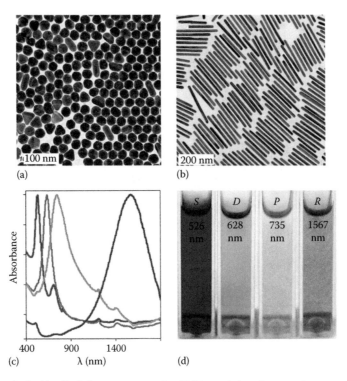

(a)

(b)

(c) λ (nm)

(d)

FIGURE 27.9 Morphological control of gold colloid plasmonic properties. (a) Transmission electron microscopy (TEM) image of gold nanospheres. (b) TEM image of high aspect ratio gold nanorods. (c) Normalized extinction spectra of solutions of isolated spheres (S, red), nanodisks (d, blue), platelets (P, turquoise), and nanorods (R, brown). The photograph shows the corresponding solutions in deuterated water. (Adapted from Khanal, B.P. and Zubarev, E.R., *J. Am. Chem. Soc.*, 130, 12634, 2008. With permission.)

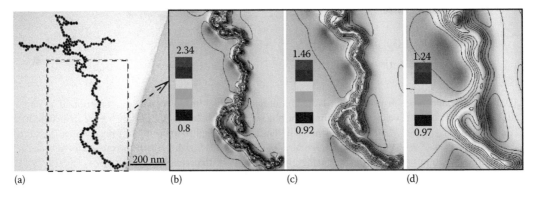

(a)   (b)   (c)   (d)

FIGURE 27.21 (a) TEM image showing a self-assembled Au nanoparticle chain network deposited on a substrate. (b), (c), and (d) sequence of three optical near-field intensity maps computed in three consecutive planes parallel to the sample. The plane–sample distances are 20, 30, and 50 nm respectively. (From Girard, C. et al., *New J. Phys.*, 10, 105016, 2008. With permission.)

Although spin is regarded as a minor contributor to the magnetic moment of nonmagnetic QDs, it is essential in magnetic ion–doped semiconductors. While the orbital moments of magnetic QDs are likely to be quenched by the scatterings of carriers with magnetic ion impurities [Cheng 2005], the magnetic ordering of magnetic ion dopants is induced by the mediation of the spin interactions between carriers and magnetic impurities. Such a carrier-mediated magnetism is the underlying mechanism of spontaneous magnetization of many magnetic semiconductors and could be further improved by quantum confinement of QD [Fernandez-Rossier and Brey 2004].

9.3 Divalent Magnetic Impurities in II–VI Semiconductors

An isolated Mn^{2+} ion with a half-filled d-shell has spin $M = 5/2$. According to Equation 9.21, it exhibits magnetization that is described by the Brillouin function, $M_{Mn} = (5/2)g_{Mn}\mu_B B_{5/2}$ $(5g_{Mn}\mu_B B/2kT)$. The total magnetization of a magnetic semiconductor with N_{Mn} Mn impurities however is not simply the sum of the magnetic moments provided by each Mn ion, $N \cdot M_{Mn}$, because relevant spin interactions that involve Mn ions occur among Mn ions [Cheng 2008].

In Mn-doped II–VI semiconductors, Mn^{2+} ions substitute divalent cations. Figure 9.11 schematically depicts the atomistic structure of a zinc blende semiconductor NC doped with magnetic Mn^{2+} ions. Since Mn ions are isoelectronic in II–VI compounds, they neither introduce nor bind charged carriers. Thus, a magnetic ion can be characterized by its spin alone, and its electrostatic potential negligible [Furdyna 1988, Dietl 2002]. The effective spin interactions between Mn ions are known to be antiferromagnetic (AF) and short ranged [Furdyna 1988, Larson et al. 1988, Shen et al. 1995]. The AF Mn–Mn interactions result from the mediation of superexchange interaction, an indirect exchange interaction mediated by anions [Furdyna 1988]. A widely adopted model for the effective Mn–Mn interaction is the Heisenberg-like Hamiltonian

$$h_{MM} = -J_{MM}(R_{IJ})\vec{M}_I \cdot \vec{M}_J \tag{9.23}$$

with an AF coupling constant $J_{MM} = J_{MM}^{(0)} \exp\{-\lambda[(R_{IJ}/a_0)-1]\} < 0$, decaying rapidly with increasing Mn–Mn distance increases, where $J_{MM}^{(0)} < 0$ (typically $J_{MM}^{(0)} \sim 10^{-1} - 10^0$ meV) is the strength

of the nearest-neighbor (NN) Mn–Mn interaction, $R_{IJ} \equiv |\vec{R}_I - \vec{R}_J|$ is the distance between magnetic ions, a_0 is the lattice constant of the NC material, and $\lambda \sim 5$ [Qu and Hawrylak 2005].

By contrast, the spin interaction between a conduction electron and Mn ions is ferromagnetic (FM) [Bhattacharjee 1992, Mizokawa and Fujimori 1997]. The contact e–Mn interaction is described by

$$h_{eM} = -J_{eM}^{(0)} \vec{s} \cdot \vec{M}_I \delta(\vec{r} - \vec{R}_I), \tag{9.24}$$

with the FM coupling constant $J_{eM}^{(0)} > 0$ ($\sim 10^1$ meV $\cdot$ nm^3). The FM interaction causes the spins of the conduction electrons and the Mn ions to align in the same direction, and magnetic ordering of Mn ion spins is induced by the mediation of the e–Mn spin interaction. The competition between both interactions plays an essential role in the magnetism of magnetically doped semiconductors. Besides the spin effects, magnetic ions act as impurities causing the backscattering of carriers and reducing the total orbital moment of QD. As an illustration, we examine the following matrix element

$$\left\langle M_z' \middle| \left\langle n'l'm_l'm_s' \middle| H_{eM} \middle| nlm_l m_s \right\rangle \middle| M_z \right\rangle$$
$$= -J_{eM}^{(0)} \phi_{n'l'm_l'}^\star(\vec{R})\phi_{nlm_l}(\vec{R})$$
$$\left\langle M_z' \middle| \left\langle m_s' \middle| s_z M_z + \frac{1}{2}\left(s_+ M_- + s_- M_+\right) \middle| m_s \right\rangle \middle| M_z \right\rangle, \tag{9.25}$$

where $s_+ \equiv s_x + is_y$ ($s_- \equiv s_x - is_y$) and $M_+ \equiv M_x + iM_y$ ($M_- \equiv M_x - iM_y$) are defined as the raising (lowering) operators of spin and orbital angular momentum, respectively. The first term in Equation 9.25 describes the z-components of electron spins as an effective field that acts on Mn spins M_z, and the last two terms involving operators $M_\pm$ are responsible for electron spin flips, which is compensated by Mn spin flips. The minus sign in the equation indicates that a carrier gains energy from the spin-exchange interaction if its spin is aligned with those of Mn ions. As revealed by Equation 9.25, the effective strength of a spin interaction between an Mn ion and a quantum-confined electron, given by

$$J_{ii'}^{eM}(\vec{R}) \equiv J_{eM}^{(0)} \phi_i^\star(\vec{R})\phi_{i'}(\vec{R}) \propto a^{-3}, \tag{9.26}$$

is determined by the local carrier density at the positions of Mn ions. The effective strength of the e–Mn spin interaction in a QD increases as the size of the QD decreases. Accordingly, the e–Mn interaction can transfer an electron between different orbitals as long as the wave functions of the orbitals overlap at the site of the Mn ion. For instance, $\langle -m_l | H_{eM} | m_l \rangle \neq 0$ for two orbitals with opposite angular momenta. Restated, an Mn^{2+} ion as an impurity in NCs could cause backscattering of particles and reverse the direction of motion of a particle with some finite angular momentum. This effect quenches the orbital angular

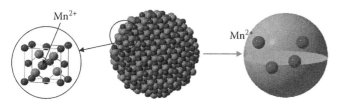

FIGURE 9.11 Schematic of the (zinc-blende) atomistic structure of a semiconductor NC doped with magnetic Mn^{2+} ions.

momentum. Such *orbital quenching* suppresses the magnetism and, as shown in Ref. [Cheng 2005], also leads to magnetic anisotropy of charged NCs in the low-field regime.

9.4 Carrier-Mediated Magnetism in Magnetic Nanocrystals

Consider a simple illustrative example of Mn-doped NCs charged with a single electron. The Hamiltonian for such quantum-confined, single-electron-many-Mn magnetic polarons (at zero field) is

$$H = h_0(\vec{r}, \vec{p}) - \sum_I J_{eM}^{(0)} \vec{s} \cdot \vec{M}_I \delta(\vec{r} - \vec{R}_I) - \frac{1}{2} \sum_{I \neq J} J_{MM}(R_{IJ}) \vec{M}_I \cdot \vec{M}_J. \tag{9.27}$$

Since the typical energy quantization of NCs (of the order of 10^2 meV) is two orders of magnitude larger than those of the e–Mn and Mn–Mn interactions (~10^0 meV), an electron in magnetic NCs is nearly frozen in the lowest orbital. Neglecting higher shell scatterings of electron yields an effective spin Hamiltonian,

$$H_{eff} = \langle \phi_{100} | H | \phi_{100} \rangle = \epsilon_{100} - \sum_I J_c(\vec{R}_I) \vec{s} \cdot \vec{M}_I$$
$$- \frac{1}{2} \sum_{I \neq J} J_{MM}(R_{IJ}) \vec{M}_I \cdot \vec{M}_J, \tag{9.28}$$

where $J_c(\vec{R}) = J_{eM}^{(0)} |\phi_{100}(\vec{R})|^2 = J_{eM}^{(0)} (\pi/2a^3)[\text{sinc}(\pi R/a)]^2$. For further analysis, constant e–Mn and Mn–Mn interactions are assumed [Gould et al. 2006] and the effective Hamiltonian is written as

$$H'_{eff} = -J_c \vec{s} \cdot \vec{M} + \frac{J_M}{2} \sum_{I \neq J} \vec{M}_I \cdot \vec{M}_J, \tag{9.29}$$

where

$\vec{M} = \sum_I \vec{M}_I$ is the total spin of the Mn's

$J_c \geq 0$ ($J_M \geq 0$) is the effective e–Mn (Mn–Mn) interaction constant

the energy offset ϵ_{100} is omitted for brevity

Since Equation 9.29 commutes with the total angular momentum $\vec{J} \equiv \vec{M} + \vec{s}$, J (and M) can be chosen as the quantum numbers to label the magnetic polaron states as $|J, M\rangle$ with $J = M \pm 1/2$ [Gould et al. 2006]. The analytical solutions of the eigen energies of the states $|J = M \pm 1/2, M\rangle$ are given by

$$E\left(M \pm \frac{1}{2}, M\right) = \mp \frac{J_c}{2} M + \frac{J_M}{2} \left[M(M+1) - 35 \frac{N_{Mn}}{4} \right]$$

where $M = 0, 1, \ldots, 5N_{Mn}/2$ ($M = 1/2, 3/2, \ldots, 5N_{Mn}/2$) for even (odd) N_{Mn}. The total Mn spin of the ground states,

$$M_{GS} = \text{integer part}\left[|J_c/(2J_M)| + c\right] - c, \tag{9.30}$$

can be derived, with the upper limit $M_{GS} \leq 5N_{Mn}/2$, where $c = 0(1/2)$ for an even (odd) number of Mn's [Cheng 2008]. Notable is that the total spin of the Mn's coupled to a quantum-confined electron is determined by the ratio of the effective e–Mn and Mn–Mn coupling constants, J_c/J_M. Therefore, the net magnetization of magnetic NC is determined by the competition between e–Mn and Mn–Mn interactions. The former can be tuned by controlling the NC sizes, while the latter depends on Mn concentration and distribution. Accordingly, we have the condition $|J_c/J_M| \geq 5N_{Mn}$ for the formation of ferromagnetic magnetic polarons (with maximum total Mn spin), and that $|J_c/J_M| < 2$ under which e–Mn complexes in NCs retain antiferromagnetism (vanishing Mn spin, $M = 0$).

9.5 Numerical Approaches

The Hamiltonian for a magnetic nanocrystal that contains an arbitrary number of electrons and Mn impurities in magnetic fields is

$$H = \sum_i h_B(\vec{r}_i, \vec{p}_i) + \frac{1}{2} \sum_{i \neq j} \frac{e^2}{4\pi\epsilon_0 \kappa |\vec{r}_i - \vec{r}_j|}$$
$$- \sum_{i,I} J_{eM}^{(0)} \vec{s}_i \cdot \vec{M}_I \delta(\vec{r}_i - \vec{R}_I) - \frac{1}{2} \sum_{I \neq J} J_{MM}(R_{IJ}) \vec{M}_I \cdot \vec{M}_J$$
$$- g_{Mn} u_B B \cdot \sum_I M_I^z - g_e u_B B \cdot \sum_i s_i^z, \tag{9.31}$$

where

the first term is the total kinetic energy of electrons

the second term the Coulomb interactions between electrons

the third (fourth) term describes the e–Mn (Mn–Mn) interactions

the last two terms are the spin Zeeman energy of the Mn's and the electrons, respectively

subscript $i(I)$ denotes the ith electron (the Ith Mn)

$\vec{M}_I(M_I^z)$ denotes the spin (z-component) of the Ith Mn ion ($M = 5/2$)

$g_M = -2.0(g_e)$ is the g-factor of Mn (electron)

κ is the dielectric constant of the host material

The other spin-related terms such as those of spin orbital coupling and the hyperfine interaction are neglected in Equation 9.31 because their strengths are much weaker than those of the Mn-related spin interactions [Cerletti et al. 2005]. Finding an exact solution to Equation 9.31 is nontrivial since the number of e–Mn configurations rapidly increases with the number of Mn's. Two theoretical approaches, exact diagonalization and local mean field theory, will be introduced to solve the problem.

9.5.1 Exact Diagonalization

For straightforward implementation of the configuration inter-action (CI) method [Cheng 2005, 2008, Qu and Hawrylak 2005], the Hamiltonian Equation 9.31 is usually transferred into the second quantized Hamiltonian

$$H = \sum_{n\sigma} \epsilon_n c_{n\sigma}^+ c_{n\sigma} + \frac{1}{2} \sum_{nmkl} \sum_{\sigma\sigma'} V_{nmkl}^{ee} c_{n\sigma}^+ c_{m\sigma'}^+ c_{k\sigma'} c_{l\sigma}$$

$$- \sum_{I} \sum_{n,n'} \frac{J_{nn'}^{eM}(\vec{R}_I)}{2} \Big[(c_{n'\uparrow}^+ c_{n\uparrow} - c_{n'\downarrow}^+ c_{n\downarrow}) M_I^z + c_{n'\downarrow}^+ c_{n\uparrow} M_I^+ + c_{n'\uparrow}^+ c_{n\downarrow} M_I^- \Big]$$

$$\frac{1}{2} \sum_{I\neq J} J_{MM}(R_{IJ}) \vec{M}_I \cdot \vec{M}_J - g_{Mn} u_B B \sum_I M_I^z$$

$$- \frac{1}{2} g_e u_B B \sum_n (c_{n\uparrow}^+ c_{n\uparrow} - c_{n\downarrow}^+ c_{n\downarrow}), \tag{9.32}$$

where

$c_{n\sigma}^+ (c_{n\sigma})$ is the creation (annihilation) operator for an electron on orbital $|n\sigma\rangle$

n is an orbital index

$\sigma = \uparrow/\downarrow$ denotes electron spin $s_z = +\frac{1}{2}/-\frac{1}{2}$

$V_{nmkl}^{ee} \equiv \iint d^3 r_1 d^3 r_2 \phi_n^*(\vec{r}_1) \phi_m^*(\vec{r}_2)(e^2/4\pi\kappa |\vec{r}_1 - \vec{r}_2|) \phi_k(\vec{r}_2) \phi_l(\vec{r}_1)$ is defined as a Coulomb matrix element

In the implementation of the CI method, a number N_s of lowest energy single electron states is initially selected and all possible e–Mn configurations $c_{n_1,\sigma_1}^+ c_{n_2,\sigma_2}^+ \ldots c_{n_{N_e}\sigma_{N_e}}^+ |vac\rangle \otimes |M_1^z, M_2^z, \ldots, M_{N_{Mn}}^z\rangle$. In the basis of

N_c chosen e–Mn configuration, the $N_c \times N_c$ Hamiltonian matrix is generated and then directly diagonalized to find the eigen states and energy spectrum $\{E_i\}$ of the e–Mn complex. The main numerical difficulty arises from the fact that the total number of e–Mn configurations $N_c \propto 6^{N_{Mn}}$ rapidly increases with the number of Mn ions. The convergence of the results is tested by increasing the number and choice of single electron orbitals. Advanced eigen solvers, such as LANCZOS and ARPACK, are usually employed to find the low-lying eigenstates and eigen energies of large matrices with high accuracy. The magnetization of magnetic NCs at temperature T is numerically calculated using the definition of magnetization [Blundell 2001]

$$\mathbf{M} = -\left(\frac{\partial F}{\partial B}\right)_{T,V} = \frac{1}{\beta}\left(\frac{\partial \ln Z}{\partial B}\right)_T, \tag{9.33}$$

where

$\beta \equiv 1/(k_B T)$, $Z = \sum_i \exp(-E_i\beta)$ is the canonical ensemble equilibrium partition function

E_i is the ith eigen energy

$F \equiv -\ln Z/\beta$ is the Helmholtz free energy [Blundell 2001]

The magnetic susceptibility defined as the partial derivative of magnetization with respect to magnetic field $\chi \equiv \partial \mathbf{M}/\partial B$ is obtained using standard three-point numerical derivation.

Figure 9.12a shows the low-lying energy spectrum (relative to the ground state energy), numerically calculated by exact diago-nalization, of singly charged NCs doped with two long-range interacting Mn^{2+} ions positioned at $\vec{R}_1 = (X_1, Y_1, Z_1) = (a/2, 0, 0)$ and $\vec{R}_2 = (-a/2, 0, 0)$, respectively. With the long spatial separation between Mn's ($a \gg a_0$), the AF Mn–Mn interaction $J_{MM} \to 0$ and the predominant FM e–Mn interactions give rise to the magnetic

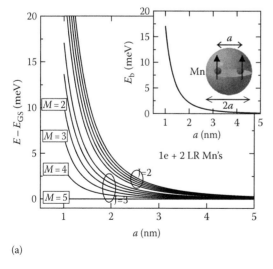

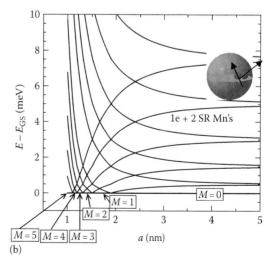

(a)   (b)

FIGURE 9.12 The energy spectra relative to the ground state (GS) energies of singly charged NCs of radius a doped with (a) two long-range (LR) interacting Mn ions positioned at $\vec{R}_1 = (X_1, Y_1, Z_1) = (a/2, 0, 0)$ and $\vec{R}_2 = (-a/2, 0, 0)$, and (b) two NN short-range (SR) interacting Mn ions at $\vec{R}_1 \sim \vec{R}_2 \sim (a/2, 0, 0)$. The results are calculated by using exact diagonalization. The GSs of the NCs containing the long-ranged Mn's are stable in the ferromagnetic phases. By contrast, the ground states of the NCs with short-ranged Mn's undergo a series of magnetic phase transitions, from antiferromagnetism (AF) to ferromagnetism (FM) as the NC sizes decrease.

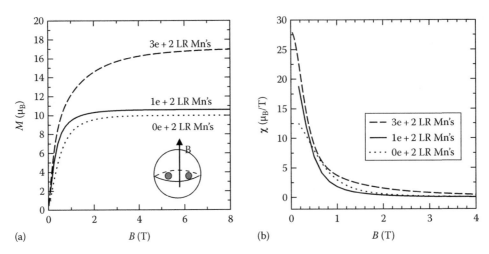

FIGURE 9.13 Exact diagonalization results of (a) the magnetizations M and (b) the magnetic susceptibilities χ vs. applied magnetic fields B of magnetic CdSe NCs charged with $N_e = 0, 1, 3$ electrons and doped with two long-ranged Mn ions located at $\vec{R}_1 = (a/2, 0, 0)$ and $\vec{R}_2 = (-a/2, 0, 0)$, respectively.

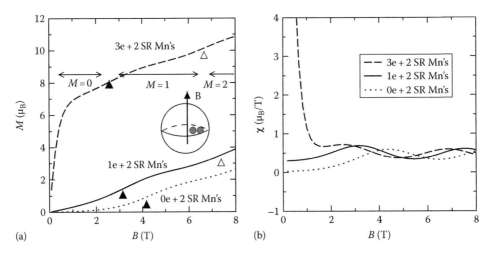

FIGURE 9.14 Exact diagonalization results of (a) the magnetizations M and (b) the magnetic susceptibilities χ vs. applied magnetic fields B of magnetic CdSe NCs charged with $N_e = 0, 1, 3$ electrons and doped with two short-ranged Mn ions located at $\vec{R}_1 \sim \vec{R}_2 = (a/2, 0, 0)$.

ordering of Mn spins, leading to the FM ground states with a maximum total spin $M_{GS} = 5N_{Mn}/2 = 5$. Figure 9.12b shows the calculated relative energy spectrum of singly charged NCs that contain two NN Mn's at $\vec{R}_1 \sim \vec{R}_2 \sim (a/2, 0, 0)$. By contrast, the GSs of the NCs with the short-ranged Mn cluster undergo a series of magnetic state transitions, from antiferromagnetism ($M = 0$) to ferromagnetism ($M = 5$), as the NC sizes decrease because the strength of the FM e–Mn interaction increases as the NC sizes decrease (see Equation 9.26), eventually overwhelming the strong AF Mn–Mn interactions.

Figure 9.13 (Figure 9.14) shows the ED results of the magnetizations M and magnetic susceptibilities χ as a function of applied magnetic fields B for charged and uncharged NCs with the two long-ranged (short-ranged) Mn ions, with reference to Figure 9.12a (Figure 9.12b). The magnetization and susceptibility of the uncharged NC doped with the two long-ranged Mn ions are like those of a $M = 5$ paramagnet, according to

Curie's law, due to the FM GSs (Figures 9.8 and 9.13). In general, the magnetizations and susceptibilities of the Mn-doped NCs charged with more electrons are increased by the additional magnetic contribution from the electron spin and orbital moments.

By contrast, the magnetizations and magnetic susceptibilities of the NCs with short-ranged Mn's exhibit behaviors that differ markedly from those given by Curie's law (Figure 9.14). The positive magnetic susceptibilities (paramagnetism), rather than monotonically decaying like those of the paramagnet of NCs with long-ranged Mn's, lasts over a wide range of magnetic field and oscillate as magnetic field increases. The oscillation results from the series of the transitions of the magnetic ground states of the magnetic NCs due to the strong AF interactions between the short-ranged Mn's in the dots. Experimentally, significant paramagnetism in an ensemble of CdMnSe NC was observed in high applied magnetic fields $B > 5$ T (Figure 9.5).

9.5.2 Mean Field Theory Approximation

It is difficult to perform exact diagonalization studies of NCs with many Mn's because of the very large number of e–Mn configurations that are required for numerical convergence. Instead, *local* mean field theory (LMFT) is often used to study the magnetic semiconductor QDs that contain many Mn ions [Fernandez-Rossier and Brey 2004, Abolfath et al. 2007]. Furthermore, the local mean field theory has been combined with density functional theory to consider many-body physics in charged magnetic QDs [Abolfath et al. 2007]. For simplicity of illustration, cases that involve a single electron are considered here and the LMFT presented below is formulated for singly charged magnetic NCs only.

In the spirit of mean field theory, the e–Mn and Mn–Mn interactions in Equation 9.31 are replaced by an Ising-like coupling between electron spin and a local effective field $h_{sd}(\vec{r})$ provided by the magnetization of Mn's. In the theory, the effective Hamiltonian of a singly charged NC with many Mn's is given by

$$H_\sigma^{MF} = h_0 - s_z h_{sd}(\vec{r}),\tag{9.34}$$

where

h_0 is the single electron Hamiltonian of an Mn-free NC
the z-component of electron spin is $s_z = \frac{1}{2} / -\frac{1}{2}$ for $\sigma = \uparrow/\downarrow$
the local field h_{sd} experienced by the spin electron is given by

$$h_{sd}(\vec{r}) = J_{sd} n_{Mn} \langle M_z(\vec{r}) \rangle,\tag{9.35}$$

where n_{Mn} denotes the density of Mn ions. The averaged local magnetization of Mn's is modeled by

$$\langle M_z(\vec{r}) \rangle = M B_M \left(\frac{M b(\vec{r})}{kT} \right),\tag{9.36}$$

where B_M is the Brillouin function and

$$b(\vec{r}) = \frac{J_{sd}}{2}(n_\uparrow - n_\downarrow) - J_{eff}^{AF} \langle M_z(\vec{r}) \rangle\tag{9.37}$$

is the local mean field that is experienced by the Mn, $n_\sigma = \sum_i f(E_{i\sigma}; T) |\psi_\sigma^{MF}(\vec{r})|^2$ is the mean density of electrons with spin σ subject to thermal fluctuations, and J_{eff}^{AF} is the effective field due to the AF interaction with neighbor Mn's, where $f(E_{i\sigma}; T) = \exp(-E_{i\sigma}/kT)/\sum_{j\sigma'} \exp(-E_{j\sigma'}/kT)$ is the occupancy probability of state $|i, \sigma\rangle$ and ψ_σ^{MF} is the single electron wave function, which satisfies the Schrödinger equation

$$H_\sigma^{MF}\psi_\sigma^{MF} = E_\sigma^{MF}\psi_\sigma^{MF}.\tag{9.38}$$

In principle, the coupled Equations 9.35 through 9.38 must be solved self-consistently, and then the local field $b(\vec{r})$ and the averaged local magnetization per Mn ion $\langle M_z(\vec{r}) \rangle$ can be determined. The magnetism of an Mn-doped NC is characterized by the averaged Mn magnetization $\langle M \rangle_{MF} \equiv 1/\Omega_{QD} \int \langle M_z(\vec{r}) \rangle d^3$, where Ω_{QD} is the volume of NC [Fernandez-Rossier and Brey 2004, Abolfath et al. 2007].

9.6 Summary

This chapter presented theoretical descriptions of the electronic structure and the magnetic properties of Mn^{2+}-doped II–VI semiconductor nanocrystals. The introduction briefly reviewed recent experimental findings and the current state of fundamental research into the magnetism of nonmagnetic and magnetic nanocrystals. A generalized theory for the magnetism in nanocrystal QDs with an arbitrary number of interacting charged carriers and magnetic dopants was developed, based on the theory of atomic magnetism within the framework of effective mass approximation. To highlight the underlying physics, an analysis of the magnetism of singly charged magnetic nanocrystals was performed using a simplified constant interaction model. The analysis provides explicit expressions for the magnetization of magnetic QDs as a function of size and Mn density. Two numerical approaches, exact diagonalization technique and local mean field theory, were described at the end of this chapter.

The developed theory was successfully applied to interpret some recent observations of the magnetic responses of nonmagnetic and magnetic semiconductor QDs. The orbital moments were shown to dominate the quantum-size induced paramagnetism observed in nonmagnetic nanocrystal ensembles. By contrast, spin interactions are essential in the magnetism of magnetic Mn-doped nanocrystals. The magnetic behavior of a magnetically doped quantum dot is determined by the competition between the ferromagnetic e–Mn spin interactions and the antiferromagnetic Mn–Mn interactions. The strength of the former is determined by the size of nanocrystals while that of the latter ones is related to the density and the spatial distribution of Mn ions.

Exact diagonalization studies reveal the signatures of the magnetic ground states and the corresponding dominant spin interactions of nanocrystals doped with few magnetic ions. They show the controllability of magnetizations in magnetic ion nanocrystals by the engineering of magnetic ion dopants and nanocrystal size.

Much room exists for further improvement and extension of theoretical research in the rapidly emerging field of this work. For example, microscopic methods for magnetic nanocrystals with a typical size of a few nm, only one order of magnitude larger than the size of a unit crystalline cell, are needed. The empirical tight-binding theory may be a suitable method for exploring more atomistic effects in magnetic semiconductor nanocrystals. Besides, since the number of magnetic ions in a dot is quite small (typically ~10^0–10^1), the discreteness of Mn spatial distribution, which is actually disregarded in widely used mean field theory, should substantially affect the magnetic behavior of magnetic nanocrystals. The validity of the mean field theory for few Mn-doped semiconductor nanostructures is also worthy of further study.

Appendix 9.A: List of Symbols

Table 9.A.1 lists the symbols frequently used throughout this chapter.

TABLE 9.A.1 List of Symbols

h	Hamiltonian for a single particle
H	Hamiltonian for interacting many particles
m^*	Effective mass of electron
a	Radius of a spherical nanocrystal
ϕ	Single particle wave function
$\vec{s}$	Spin angular momentum
$\vec{l}$	Orbital angular momentum
n	Principal quantum number
l	Orbital angular momentum quantum number
m	Magnetic quantum number
g_s	g-Factor for electron spin angular momentum
g_l	g-Factor for electron orbital angular momentum
B	Magnetic field
M	Magnetization
χ	Magnetic susceptibility
k	Boltzmann constant
T	Temperature
Z	Partition function
J_{eM}	Electron–Mn interaction
J_{MM}	Mn–Mn interaction
Ry^*	Effective Rydberg
a_B^*	Effective Bohr radius
μ_B^*	Effective Bohr magneton

Acknowledgments

The author would like to thank the National Science Council of Taiwan, the National Center of Theoretical Sciences in Hsinchu, and the National Center for High-Performance Computing of Taiwan for their support. Wen-Bin Jian (National Chiao Tung University, Taiwan), Yang Yuan Chen (Academia Sinica, Taiwan), Yung Liou (Academia Sinica), Pawel Hawrylak (National Research Council of Canada), and Fanyao Qu (National Research Council of Canada) are appreciated for their valuable discussions, as well as Shu-Kai Lu for collecting relevant literature.

References

[Abolfath et al. 2007] R. M. Abolfath, P. Hawrylak, and I. Zutić, Tailoring magnetism in quantum dots, *Phys. Rev. Lett.* **98**, 207203 (2007).

[Albe et al. 1998] V. Albe, C. Jouanin, and D. Bertho, Confinement and shape effects on the optical spectra of small CdSe nanocrystals, *Phys. Rev. B* **58**, 4713 (1998).

[Alivisatos 1996] A. P. Alivisatos, Semiconductor clusters, nanocrystals, and quantum dots, *Science* **271**, 933 (1996).

[Archer et al. 2007] P. I. Archer, S. A. Santangelo, and D. M. Gamelin, Direct observation of sp-d exchange interactions in colloidal Mn^{2+}- and Co^{2+}-doped CdSe quantum dots, *Nano Lett.* **7**, 1037 (2007).

[Banin and Millo 2003] U. Banin and O. Millo, Tunneling and optical spectroscopy of semiconductors nanocrystals, *Annu. Rev. Phys. Chem.* **54**, 465 (2003).

[Banin et al. 1999] U. Banin, Y. Cao, D. Katz, and O. Millo, Identification of atomic-like electronic states in indium arsenide nanocrystal quantum dots, *Nature* **400**, 542 (1999).

[Besombes et al. 2004] L. Besombes, Y. Leger, L. Maingault, D. Ferrand, H. Mariette, and J. Cibert, Probing the spin state of a single magnetic ion in an individual quantum dot, *Phys. Rev. Lett.* **93**, 207403 (2004).

[Besombes et al. 2005] L. Besombes, Y. Leger, L. Maingault, D. Ferrand, H. Mariette, and J. Cibert, Carrier-induced spin splitting of an individual magnetic atom embedded in a quantum dot, *Phys. Rev. B* **71**, 161307(R) (2005).

[Bhargava et al. 1994] R. N. Bhargava, D. Gallagher, X. Hong, and A. Nurmikko, Optical properties of manganese-doped nanocrystals of ZnS, *Phys. Rev. Lett.* **72**, 416 (1994).

[Bhatt et al. 2002] R. N. Bhatt, M. Berciu, M. P. Kennett, and X. Wan, Diluted magnetic semiconductors in the low carrier density regime, *J. Supercond.: INM* **15**, 71 (2002).

[Bhattacharjee 1992] A. K. Bhattacharjee, Interaction between band electrons and transition metal ions in diluted magnetic semiconductors, *Phys. Rev. B* **46**, 5266 (1992).

[Bhattacharjee and Benoit a la Guillaume 1997] A. K. Bhattacharjee and C. Benoit a la Guillaume, Exciton magnetic polaron in semimagnetic semiconductor nanocrystals, *Phys. Rev. B* **55**, 10613 (1997).

[Bhattacharjee and Perez-Conde 2003] A. K. Bhattacharjee and J. Perez-Conde, Optical properties of paramagnetic ion-doped semiconductor nanocrystals, *Phys. Rev. B* **68**, 045303 (2003).

[Blundell 2001] *Magnetism in Condensed Matter*, S. Blundell, Oxford University Press, New York (2001).

[Bol and Meijerink 1998] A. A. Bol and A. Meijerink, Long-lived Mn^{2+} emission in nanocrystalline $ZnS:Mn^{2+}$, *Phys. Rev. B* **58**, R15997 (1998).

[Bruchez et al. 1998] M. Bruchez, Jr., M. Moronne, P. Gin, S. Weiss, and A. P. Alivisatos, Semiconductor nanocrystals as fluorescent biological labels, *Science* **281**, 2013 (1998).

[Brus 1991] L. Brus, Quantum crystallites and nonlinear optics, *Appl. Phys. A* **53**, 465 (1991).

[Cerletti et al. 2005] V. Cerletti, W. A. Coish, O. Gywat, and D. Loss, Recipes for spin-based quantum computing, *Nanotechnology* **16**, R27 (2005), and references therein.

[Chang et al. 2004] K. Chang, S. S. Li, J. B. Xia, and F. M. Peeters, Electron and hole states in diluted magnetic semiconductor quantum dots, *Phys. Rev. B* **69**, 235203 (2004).

[Cheng 2005] S. J. Cheng, Magnetic response of magnetic ion-doped nanocrystals: Effects of single Mn^{2+} impurity, *Phys. Rev. B* **72**, 235332 (2005).

[Cheng 2008] S. J. Cheng, Theory of magnetism in diluted magnetic semiconductor nanocrystals, *Phys. Rev. B* **77**, 115310 (2008).

[Cheng and Hawrylak 2008] S. J. Cheng and P. Hawrylak, Controlling magnetism of semi-magnetic quantum dots with odd-even exciton numbers, *Europhys. Lett.* **81**, 37005 (2008).

[Chiba et al. 2003] D. Chiba, M. Yamanouchi, F. Matsukura, and H. Ohno, Electrical manipulation of magnetization reversal in a ferromagnetic semiconductor, *Science* **301**, 943 (2003).

[Climente et al. 2005] J. I. Climente, M. Korkusinski, P. Hawrylak, and J. Planelles, Voltage control of the magnetic properties of charged semiconductor quantum dots containing magnetic ions, *Phys. Rev. B* **71**, 125321 (2005).

[Dietl 2002] T. Dietl, Ferromagnetic semiconductors, *Semicond. Sci. Technol.* **17**, 377 (2002).

[Dorozhkin et al. 2003] P. S. Dorozhkin, A. V. Chernenko, V. D. Kulakovskii, A. S. Brichkin, A. A. Maksimov, H. Schoemig, G. Bacher, A. Forchel, S. Lee, M. Dobrowolska, and J. K. Furdyna, Longitudinal and transverse fluctuations of magnetization of the excitonic magnetic polaron in a semimagnetic single quantum dot, *Phys. Rev. B* **68**, 195313 (2003).

[Efros and Rosen 2000] Al. L. Efros and M. Rosen, The electronic structure of semiconductor nanocrystals, *Annu. Rev. Mater. Sci.* **30**, 475 (2000).

[Efros et al. 2001] Al. L. Efros, E. I. Rashba, and M. Rosen, Paramagnetic ion-doped nanocrystal as a voltage-controlled spin filter, *Phys. Rev. Lett.* **87**, 206601 (2001).

[Erwin et al. 2005] S. C. Erwin, L. Zu, M. I. Haftel, A. L. Efros, T. A. Kennedy, and D. J. Norris, Doping semiconductor nanocrystals, *Nature* **436**, 91 (2005).

[Feltin et al. 1999] N. Feltin, L. Levy, D. Ingert, and M. P. Pileni, Magnetic properties of 4-nm $Cd_{1-y}Mn_yS$ nanoparticles differing by their compositions, y, *J. Phys. Chem. B* **10**, 4 (1999).

[Fernandez-Rossier and Brey 2004] J. Fernandez-Rossier and L. Brey, Ferromagnetism mediated by few electrons in a semimagnetic quantum dot, *Phys. Rev. Lett.* **93**, 117201 (2004).

[Fernandez-Rossier and Aguado 2007] J. Fernandez-Rossier and R. Aguado, Single-electron transport in electrically tunable nanomagnets, *Phys. Rev. Lett.* **98**, 106805 (2007).

[Fu and Zunger 1997] H. Fu and A. Zunger, Local-density-derived semiempirical nonlocal pseudopotentials for InP with applications to large quantum dots, *Phys. Rev. B* **55**, 1642 (1997).

[Fu et al. 1998] H. Fu, L.-W. Wang, and A. Zunger, Applicability of the $k \cdot p$ method to the electronic structure of quantum dots, *Phys. Rev. B* **57**, 9971 (1998).

[Furdyna 1988] J. K. Furdyna, Diluted magnetic semiconductors, *J. Appl. Phys.* **64**, R29 (1988).

[Gould et al. 2006] C. Gould, A. Slobodskyy, D. Supp, T. Slobodskyy, P. Grabs, P. Hawrylak, F. Qu, G. Schmidt, and L. W. Molenkamp, Remanent zero field spin splitting of self-assembled quantum dots in a paramagnetic host, *Phys. Rev. Lett.* **97**, 017202 (2006).

[Govorov 2004] A. O. Govorov, Optical probing of the spin state of a single magnetic impurity in a self-assembled quantum dot, *Phys. Rev. B* **70**, 035321 (2004).

[Govorov 2005] A. O. Govorov and A. V. Kalameitsev, Optical properties of a semiconductor quantum dot with a single magnetic impurity: Photoinduced spin orientation, *Phys. Rev. B* **71**, 035338 (2005).

[Gurung et al. 2004] T. Gurung, S. Mackowski, H. E. Jackson, L. M. Smith, W. Heiss, J. Kossut, and G. Karczewski, Optical studies of zero-field magnetization of CdMnTe quantum dots: Influence of average size and composition of quantum dots, *J. Appl. Phys.* **96**, 7407 (2004).

[Hill et al. 1999] N. A. Hill, S. Pokrant, and A. J. Hill, Optical properties of Si-Ge semiconductor nano-onions, *J. Phys. Chem. B* **103**, 3156–3161 (1999).

[Hoffman et al. 2000] D. M. Hoffman, B. K. Meyer, A. I. Ekimov, I. A. Merkulov, Al. L. Efros, M. Rosen, G. Couino, T. Gacoin, and J. P. Boilot, Giant internal magnetic fields in Mn doped nanocrystal quantum dots, *Solid State Commun.* **114**, 547 (2000).

[Hu et al. 1990] Y. Z. Hu, M. Lindberg, and S. W. Koch, Theory of optically excited intrinsic semiconductor quantum dots, *Phys. Rev. B* **42**, 1713 (1990).

[Hu et al. 2001] J. Hu, L. Li, W. Yang, L. Manna, L. Wang, and A. P. Alivisatos, Linearly polarized emission from colloidal semiconductor quantum rods, *Science* **292**, 2060 (2001).

[Ji et al. 2003] T. Ji, W. B. Jian, and J. Fang, The first synthesis of $Pb_{1-x}Mn_xSe$ nanocrystals, *J. Am. Chem. Soc.* **12**, 8448 (2003).

[Jian et al. 2003] W. B. Jian, J. Fang, T. Ji, and J. He, Quantum-size-effect-enhanced dynamic magnetic interactions among doped spins in $Cd_{1-x}Mn_xSe$ nanocrystals, *Appl. Phys. Lett.* **83**, 16 (2003).

[Jungwirth et al. 2006] T. Jungwirth, J. Sinova, J. Masek, J. Kucera, and A. H. MacDonald, Theory of ferromagnetic (III,Mn)V semiconductors, *Rev. Mod. Phys.* **78**, 809 (2006).

[Katz et al. 2002] D. Katz, T. Wizansky, O. Millo, E. Rothenberg, T. Mokari, and U. Banin, Size-dependent tunneling and optical spectroscopy of CdSe quantum rods, *Phys. Rev. Lett.* **89**, 086801 (2002).

[Klein et al. 1997] D. L. Klein, R. Roth, A. K. L. Lim, A. Paul Alivisatos, and P. L. McEuen, A single-electron transistor made from a cadmium selenide nanocrystal, *Nature* **389**, 699 (1997).

[Klimov et al. 2000] V. I. Klimov, A. A. Mikhailovsky, S. Xu, A. Malko, J. A. Hollingsworth, C. A. Leatherdale, H.-J. Eisler, and M. G. Bawendi, Optical gain and stimulated emission in nanocrystal quantum dots, *Science* **290**, 314 (2000).

[Lad et al. 2007] A. D. Lad, Ch. Rajesh, M. Khan, N. Ali, I. K. Gopalakrishnan, S. K. Kulshreshtha, and S. Mahamuni, Magnetic behavior of manganese-doped ZnSe quantum dots, *J. Appl. Phys.* **101**, 103906 (2007).

[Larson et al. 1988] B. E. Larson, K. C. Hass, H. Ehrenreich, and A. E. Carlsson, Theory of exchange interactions and chemical trends in diluted magnetic semiconductors, *Phys. Rev. B* **37**, 4137 (1988).

[Leger et al. 2006] Y. Leger, L. Besombes, J. Fernández-Rossier, L. Maingault, and H. Mariette, Electrical control of a single Mn atom in a quantum dot, *Phys. Rev. Lett.* **97**, 107401 (2006).

[Lippens and Lannoo 1990] P. E. Lippens and M. Lannoo, Comparison between calculated and experimental values of the lowest excited electronic state of small CdSe crystallites, *Phys. Rev. B* **41**, 6079 (1990).

[Liu et al. 2007] W. K. Liu, K. M. Whitaker, A. L. Smith, K. R. Kittilsved, B. H. Robinson, and D. R. Gamelin, Room-temperature electron spin dynamics in free-standing ZnO quantum dots, *Phys. Rev. Lett.* **98**, 186804 (2007).

[Madhu et al. 2008] C. Madhu, A. Sundaresan, and C. N. R. Rao, Room-temperature ferromagnetism in undoped GaN and CdS semiconductor nanoparticles, *Phys. Rev. B* **77**, 201306(R) (2008).

[Maksimov et al. 2000] A. A. Maksimov, G. Bacher, A. McDonald, V. D. Kulakovskii, A. Forchel, C. R. Becker, G. Landwehr, and L. W. Molenkamp, Magnetic polarons in a single diluted magnetic semiconductor quantum dot, *Phys. Rev. B* **62**, R7767 (2000).

[Mariette et al. 2006] H. Mariette, L. Besombes, C. Bougerol, D. Ferrand, Y. Leger, L. Maingault, and J. Cibert, Control of single spins in individual magnetic quantum dots, *Phys. Stat. Sol. (b)* **243**, 2709 (2006).

[Mikulec et al. 2000] F. V. Mikulec, M. Kuno, M. Bennati, D. A. Hall, R. G. Griffin, and M. G. Bawendi, Organometallic synthesis and spectroscopic characterization of manganese-doped CdSe nanocrystals, *J. Am. Chem. Soc.* **122**, 2532 (2000).

[Mizokawa and Fujimori 1997] T. Mizokawa and A. Fujimori, p-d exchange interaction for 3d transition-metal impurities in II-VI semiconductors, *Phys. Rev. B* **56**, 6669 (1997).

[Neeleshwar et al. 2005] S. Neeleshwar, C. L. Chen, C. B. Tsai, Y. Y. Chen, C. C. Chen, S. G. Shyu, and M. S. Seehra, Size-dependent properties of CdSe quantum dots, *Phys. Rev. B* **71**, 201307(R) (2005).

[Nirmal et al. 1996] N. Nirmal, B. O. Dabbousi, M. G. Bawendi, J. J. Macklin, J. K. Trautman, T. D. Harris, and L. E. Brus, Fluorescence intermittency in single cadmium selenide nanocrystals, *Nature* **383**, 802 (1996).

[Nguyen and Peeters 2008] N. T. T. Nguyen and F. Peeters, Magnetic field dependence of the many-electron states in a magnetic quantum dot: The ferromagnetic-antiferromagnetic transition, *Phys. Rev. B* **78**, 045321 (2008).

[Norberg and Gamelin 2006] N. S. Norberg and D. R. Gamelin, Giant Zeeman effects in colloidal diluted magnetic semiconductor quantum dots with homogeneous dopant speciation, *J. Appl. Phys.* **99**, 08M104 (2006).

[Norberg et al. 2004] N. S. Norberg, K. R. Kittilsved, J. E. Amonette, R. K. Kukkadapu, D. A. Schwartz, and D. R. Gamelin, Synthesis of colloidal Mn²⁺: ZnO quantum dots and high-T_C ferromagnetic nanocrystalline thin films, *J. Am. Chem. Soc.* **126**, 9387 (2004).

[Norman et al. 2003] T. J. Norman, Jr., D. Magana, T. Wilson, C. Burns, J. Z. Zhang, D. Cao, and F. Bridges, Optical and surface structural properties of Mn²⁺-doped ZnSe nanoparticles, *J. Phys. Chem. B* **107**, 6309 (2003).

[Norris et al. 1996] S. A. Empedocles, D. J. Norris, and M. G. Bawendi, Photoluminescence spectroscopy of single CdSe nanocrystallite quantum dots, *Phys. Rev. Lett.* **77**, 3873 (1996).

[Norris et al. 2001] D. J. Norris, N. Yao, F. T. Charnock, and T. A. Kennedy, High-quality manganese-doped ZnSe nanocrystals, *Nano Lett.* **1**, 3 (2001).

[Ohno et al. 2000] H. Ohno, D. Chiba, F. Matsukura, T. Omiya, E. Abe, T. Dietl, Y. Ohno, and K. Ohtani, Electric-field control of ferromagnetism, *Nature (London)* **408**, 944 (2000).

[Perez-Conde and Bhattacharjee 2001] J. Perez-Conde and A. K. Bhattacharjee, Exciton states and optical properties of CdSe nanocrystals, *Phys. Rev. B* **63**, 245318 (2001).

[Qu and Hawrylak 2005] F. Qu and P. Hawrylak, Magnetic exchange interactions in quantum dots containing electrons and magnetic ions, *Phys. Rev. Lett.* **95**, 217206 (2005).

[Qu and Hawrylak 2006] F. Qu and P. Hawrylak, Theory of electron mediated Mn-Mn interactions in quantum dots, *Phys. Rev. Lett.* **96**, 157201 (2006).

[Qu and Vasilopoulos 2006] F. Qu and P. Vasilopoulos, Influence of exchange interaction on spin-dependent transport through a single quantum dot doped with a magnetic ion, *Phys. Rev. B* **74**, 245308 (2006).

[Radovanovic and Gamelin 2001] P. V. Radovanovic and D. R. Gamelin, Electronic absorption spectroscopy of Cobalt ions in diluted magnetic semiconductor quantum dots: Demonstration of an isocrystalline core/shell synthetic method, *J. Am. Chem. Soc.* **123**, 12207 (2001).

[Radovanovic et al. 2002] P. V. Radovanovic, N. S. Norberg, K. E. McNally, and D. R. Gamelin, Colloidal transition-metal-doped ZnO quantum dots, *J. Am. Chem. Soc.* **124**, 15192 (2002).

[Rama et al. 1992] M. V. Rama Krishna and R. A. Friesner, Prediction of anomalous red-shift in semiconductor clusters, *J. Chem. Phys.* **96**, 873 (1992).

[Recher et al. 2000] P. Recher, E. V. Sukhorukov, and D. Loss, Quantum dot as spin filter and spin memory, *Phys. Rev. Lett.* **85**, 1962 (2000).

[Reimann and Manninen 2002] S. M. Reimann and M. Manninen, Electronic structure of quantum dots, *Rev. Mod. Phys.* **74**, 1283 (2002).

[Richard et al. 1996] T. Richard, P. Lefebvre, H. Mathieu, and J. Allegre, Effects of finite spin-orbit splitting on optical properties of spherical semiconductor quantum dots, *Phys. Rev. B* **53**, 7287 (1996).

[Sarkar et al. 2007] I. Sarkar, M. K. Sanyal, S. Kar, S. Biswas, S. Banerjee, S. Chaudhuri, S. Takeyama, H. Mino, and F. Komori, Ferromagnetism in zinc sulfide nanocrystals: Dependence on manganese concentration, *Phys. Rev. B* **75**, 224409 (2007).

[Schwartz et al. 2003] D. A. Schwartz, N. S. Norberg, Q. P. Nguyen, J. M. Parker, and D. R. Gamelin. Magnetic quantum dots: Synthesis, spectroscopy, and magnetism of Co- and Ni-doped ZnO nanocrystals, *J. Am. Chem. Soc.* **125**, 13205 (2003).

[Seehra et al. 2008] M. S. Seehra, P. Dutta, S. Neeleshwar, Y. Y. Chen, C. L. Chen, S. W. Chou, C. C. Chen, C. L. Dong, and C. L. Chang, Size-controlled Ex-nihilo ferromagnetism in Capped CdSe quantum dots, *Adv. Mater.* **9999**, 1–5 (2008).

[Shen et al. 1995] Q. Shen, H. Luo, and J. K. Furdyna, Spatial dependence of exchange interaction in Heisenberg Antiferromagnet $Zn_{1-x}Mn_xTe$, *Phys. Rev. Lett.* **75**, 2590 (1995).

[Suyver et al. 2000] J. F. Suyver, S. F. Wuister, J. J. Kelly, and A. Meijerink, Luminescence of nanocrystalline ZnSe:Mn^{2+}, *Phys. Chem. Chem. Phys.* **2**, 5445 (2000).

[Tomasulo amd Ramakrishna 1996] A. Tomasulo and M. V. Ramakrishna, Quantum confinement effects in semiconductor clusters. II, *J. Chem. Phys.* **105**, 3612 (1996).

[Viswanatha et al. 2005] R. Viswanatha, S. Sapra, T. Saha-Dasgupta, and D. D. Sarma, Electronic structure of and quantum size effect in III-V and II-VI semiconducting nanocrystals using a realistic tight binding approach, *Phys. Rev. B* **72**, 045333 (2005).

[Wang and Zunger 1996] L.-W. Wang and A. Zunger, Pseudopotential calculations of nanoscale CdSe quantum dots, *Phys. Rev. B* **53**, 9579 (1996).

[Wojnar et al. 2007] P. Wojnar, J. Suffczyñski, K. Kowalik, A. Golnik, G. Karczewski, and J. Kossut, Microluminescence from $Cd_{1-x}Mn_xTe$ magnetic quantum dots containing only a few Mn ions, *Phys. Rev. B* **75**, 155301 (2007).

10

Nanocrystals from Natural Polysaccharides

Youssef Habibi
North Carolina State University

Alain Dufresne
Grenoble Institute of Technology

10.1 Introduction

The term "nanotechnology" was introduced by Eric Drexler in mid-1980s to describe the manufacturing of machines and tools on the molecular scale. Over the years, this term has been adapted more accurately to characterize "nanoscale technology," which is related to processes involving products in the range of 0.1–100 nm. Nanotechnology or nanoscience has become one of the most important and exciting fields in physics, chemistry, engineering, and biology. It changes, and will continue to change, the nature of almost every manufactured object by offering not only better products but also new ways of processing. This new industrial revolution is best described by the quote of the Nobel Laureate, Richard Smalley: "Just wait, the next century is going to be incredible; these little nanothings will revolutionize our industries and our lives."

The term "nanocomposites" is used to refer to multiphase materials in which at least one of the constituent phases has one dimension in the nanoscale size range, and, therefore, are related to the large field of nanotechnology. This topic has attracted a great interest because of its intellectual appeal of creating and utilizing building blocks on the nanometer scale. Furthermore, the technical innovations permit to design and create new nanocomposites and structures with unprecedented flexibility, improvements in their physical properties, and significant industrial impact.

Some nanofilled polymer composites such as carbon black and fumed silica-filled polymers have been used for more than a century. A variety of clays such as montmorillonite and organoclays have been used to obtain unusual nanocomposites. Nowadays, exploring these new nanofillers is one of the many challenges for the nanocomposites community in order to develop new nanocomposite materials with specific properties.

Mother nature has been, and is still, a wonderful source of inspiration for the creation of new materials and products. The cell wall, shellfish exoskeleton, or cuticles represent biological nanocomposites that can be mimicked. In these nanocomposites, polysaccharide nanocrystals (mostly from cellulose and chitin) play the role of nanofillers and are located on segments along the elementary fibrils, which are embedded in a matrix of other biopolymers. Polysaccharides are probably the most promising sources for the production of nanoparticles as huge quantities of these nanoparticles are potentially available, often as waste products from agriculture. These abundant renewable raw materials are increasingly used in nonfood applications and they can also be used for the preparation of crystalline nanoparticles with different geometrical characteristics providing a wide range of potential nanoparticle properties. Moreover, polysaccharide surfaces provide the potential for significant surface modification using well-established carbohydrate chemistry, which allows tailoring the surface functionality of the nanoparticles. In this field, the scientific and technological challenges that need to be tackled and overcome are tremendous.

10.2 Brief Background on Polysaccharide Structures

10.2.1 Cellulose

Cellulose, discovered and isolated by Anselme Payen in 1838 (Payen, 1838), is often said to be the most abundant polymer on earth. It is certainly one of the most important structural elements in plants that helps to maintain their structure and is also

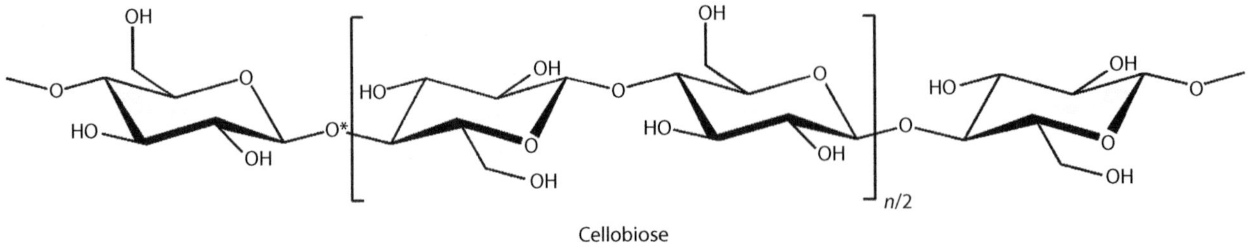

Cellobiose

FIGURE 10.1 Molecular structure of cellulose (n = DP).

important to other living species such as bacteria, fungi, algae, amoebas, and even animals. It is a ubiquitous structural polymer that confers its mechanical properties to higher plant cells. Several reviews have been published on cellulose research, structure, and applications (Sarko, 1987; Chanzy, 1990; O'Sullivan, 1997; French et al., 2004).

Cellulose is a high-molecular-weight homopolysaccharide composed of β-1,4-anhydro-D-glucopyranose units that do not lie exactly in plane with the structure, rather they assume a chair conformation, with successive glucose residues rotated though an angle of 180° about the molecular axis with hydroxyl groups in an equatorial position.

This repeated segment is frequently taken to be the cellobiose dimer (Figure 10.1). In nature, cellulose chains have a degree of polymerization (DP) of approximately 10,000 glucopyranose units in wood cellulose and 15,000 in native cotton cellulose (Sjöström, 1981). One of the most specific characteristics of cellulose is that each of its monomer contains three hydroxyl groups. These hydroxyl groups and their hydrogen bonding ability play a major role in directing crystalline packing and in governing important physical properties of these highly cohesive materials.

In the plant cell walls, cellulose fiber biosynthesis results from the combined action of biopolymerization spinning and crystallization. All these events are orchestrated by specific enzymatic terminal complexes (TC) that act as biological spinnerets, resulting in the linear association of cellulose chains to form cellulose microfibrils. Depending on the origin, the microfibril diameters range from about 2 to 20 nm with lengths that can reach several tens of microns. The cellulose obtained from nature is referred to as cellulose I or native cellulose. In cellulose I, the chains within the unit cell are in a parallel conformation (Woodcock and Sarko, 1980). Crystalline cellulose I is not the most stable form of cellulose; special treatments of native cellulose results in other forms of cellulose, namely, cellulose II, III, and IV (Marchessault and Sundararajan, 1983), which also allow for the possibility of conversion from one form to another (O'Sullivan, 1997).

10.2.2 Starch

Starch is a natural polysaccharide produced by many plants and utilized as storage for nutrients. It is the major carbohydrate reserve in plant tubers and seed endosperm where it is found as

granules (Buléon et al., 1998). By far the largest source of starch is corn (maize) with other commonly used sources being wheat, potato, tapioca, rice, and peas. Native starch occurs in the form of discrete and partially crystalline microscopic granules, and chemically, starches are composed of a number of glucose molecules linked together with α-D-(1 → 4) and/or α-D-(1 → 6) linkages. Starch is a combination of two main structural components called amylose and amylopectin (Figure 10.2).

The relative content of amylose and amylopectin varies between species and between cultivars of the same species. Waxy starches are mainly composed of amylopectin and contain only 0%–8% of amylose, whereas standard starches are made of around 75% amylopectin and 25% amylose.

Amylose molecules consist of single, mostly unbranched chains with 500–20,000 D-glucose units α-(1–4) linked dependent on the source (a very few α-1 → 6 branches and linked phosphate groups may be found (Hoover, 2001)). Amylose can form an extended shape (hydrodynamic radius 7–22 nm (Parker and Ring, 2001)) but generally tends to form a rather stiff left-handed single helix or an even stiffer parallel left-handed double helical junction zones. Amylopectin (colored by elemental iodine) is a larger molecule and differs from amylose in that branching occurs, with an α-1,6 linkage every 24–30 glucose monomer units.

X-ray diffraction analysis shows that starch is a semicrystalline polymer (Katz, 1934) and that native starches can be classified into three groups depending on their diffraction pattern type: A, B, and C. A-type is characteristic of cereal starches (wheat and maize starch), B-type is typical of tuber and amylose-rich cereal starches, and C-type is characteristic of leguminous starches and corresponds to a mixture of A and B crystalline types. V-type, from German *Verkieiterung* (gelatinization), is observed during the formation of complexes between amylose and a complexing molecule (iodine, alcohols, cyclohexane, fatty acids, and others). Water is an important component of the crystalline organization of starch. The appearance of x-ray diffraction pattern of starch depends on the water content of granules during the measurement. The more hydrated the starch, the thinner the diffraction pattern rings are up to a given limit. The determination of starch crystallinity is difficult because of both the influence of water content and the absence of a 100% crystalline standard. It ranges between 15% and 45% depending on the botanical origin of starch (Zobel, 1988). The crystalline to amorphous transition occurs at 60°C–70°C in water and this

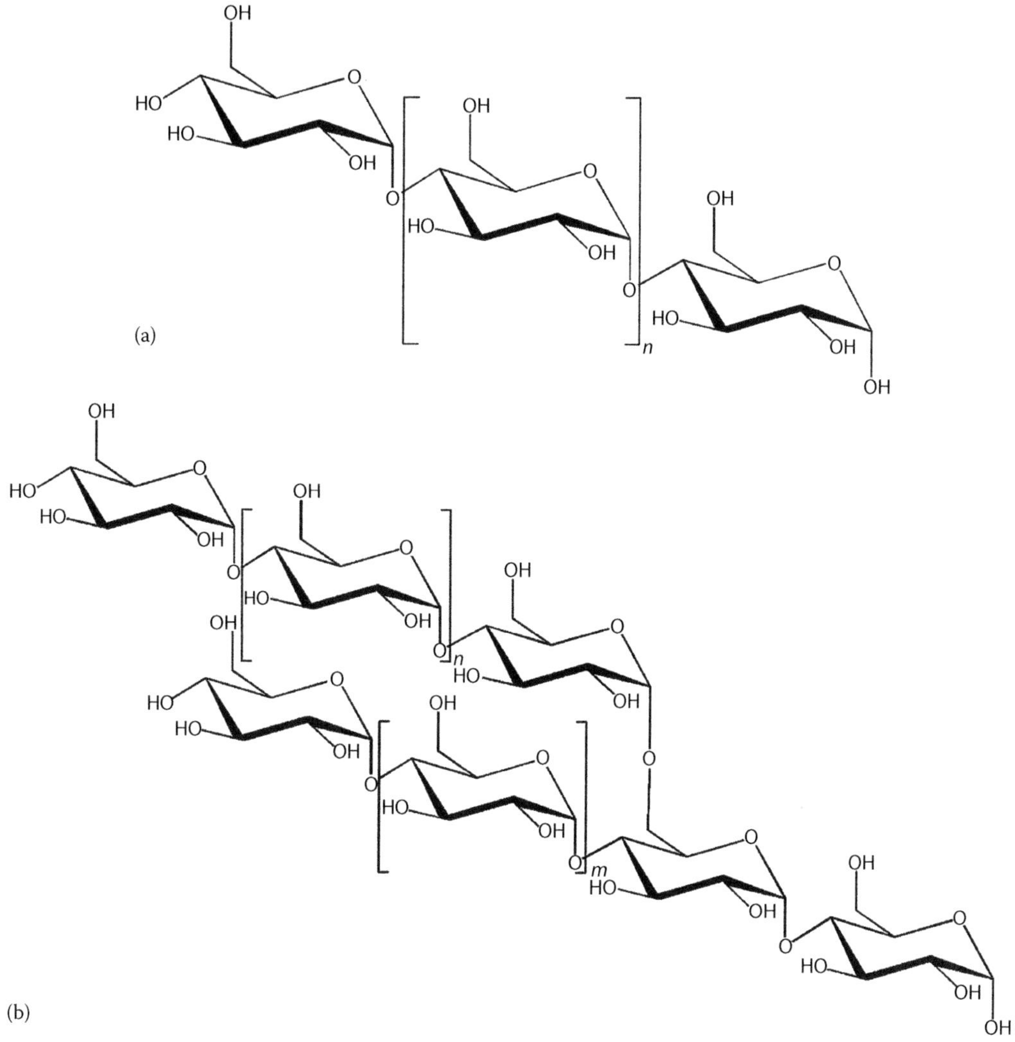

FIGURE 10.2 Chemical structures of (a) amylose and (b) amylopectine.

process is called gelatinization. In the amorphous state, hydrolysis is faster and this is why cooking starch-containing foods makes them easier to digest.

Under the optical microscope, starch granules show a distinctive Maltese cross effect (also known as "extinction cross" and birefringence) under polarized light. The semicrystalline nature of starch granules can be also visualized from transmission electron microscopic (TEM) observation of a hydrolyzed granule. The starch granule is composed of alternating hard crystalline and soft semicrystalline shells that results in a display of the so-called onion-like structure with more or less concentric growth rings between 120 and 400 nm thick (Yamaguchi et al., 1979). A model in which lamellae are organized into spherical structures termed "blocklets" has been proposed by Gallant et al. (1997). The blocklets range in diameter from around 20 to 500 nm depending on starch type (botanical source) and location in the granule. The crystalline lamellae around 9–10 nm thick are made of parallel arrays of double helices from the amylopectin linear side chains (Tang et al., 2006).

10.2.3 Chitin

Chitin is one of the main components in the cell walls of fungi, the exoskeleton of shellfish, insects and other arthropods, and in some other animals. It was first identified in 1884 and is considered as the second most important natural polymer in the world. Zooplankton cuticles (in particular small shrimps called krill) are the most important source of chitin. However, shellfish canning industry waste (shrimp or crab shells) in which the chitin content ranges between 8% and 33% constitutes the main source of this biopolymer.

Chitin is a polysaccharide composed of N-acetyl-D-glucose-2-amine units (Figure 10.3). These are linked together in β-1,4 fashion, similar to the glucose units in cellulose. Because of their similarities, chitin may be thought of as cellulose, with one hydroxyl group on each monomer replaced by an acetylamino group. This substitution allows for increased hydrogen bonding between adjacent polymer chains, giving the material an increased strength.

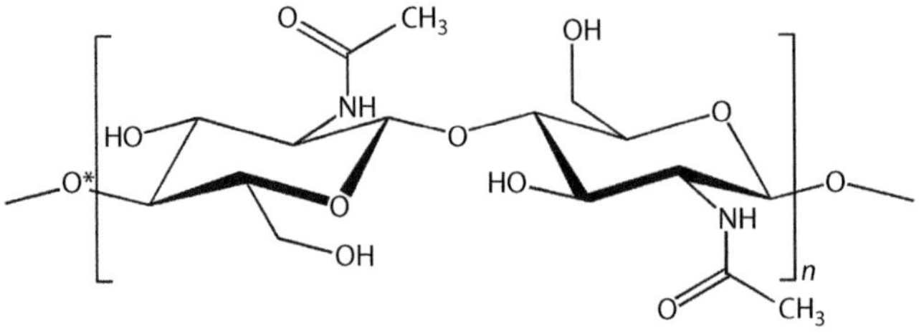

FIGURE 10.3 Chemical structure of chitin.

Native chitin is highly crystalline, and depending on its origin, occurs in three forms identified as α-, β- and γ-chitin, which can be differentiated by infrared and solid-state NMR spectroscopy together with x-ray diffraction (Salmon and Hudson, 1997). In both α and β forms, the chitin chains are organized as sheets in which they are tightly held by a number of intrasheet hydrogen bonds. In α-chitin, all chains are arranged in an antiparallel fashion whereas the β-form consists of a parallel arrangement; from detailed analysis, it seems that the γ-chitin is just a variant of the α-form (Atkins, 1985). α-Chitin is the most abundant and most stable form since it constitutes arthropod cuticles and mushroom cellular walls. It occurs in fungal and yeast cell walls, krill, lobster and crab tendons and shells, shrimp shells, and insect cuticles. In addition to the native chitin, the α-form systematically results from recrystallization from solution (Persson et al., 1992; Helbert and Sugiyama, 1998), in vitro biosynthesis (Bartnicki-Garcia et al., 1994), or enzymatic polymerization (Sakamoto et al., 2000). The rarer β-chitin is found in association with proteins in squid pens (Rudall and Kenchington, 1973), tubes synthesized by pogonophoran and vestimetiferan worms (Blackwell et al., 1965; Gaill et al., 1992), aphrodite chaetae (Lotmar and Picken, 1950), and lorica built by some seaweeds or protozoa (Herth et al., 1977).

Chitin has been known to form microfibrillar arrangements embedded in a protein matrix and these microfibrils have diameters ranging from 2.5 to 2.8 nm (Revol and Marchessault, 1993). Crustacean cuticles possess chitin microfibrils with diameters as large as 25 nm (Brine and Austin, 1975; Mussarelli, 1977). Although it has never been specifically measured, the stiffness of chitin nanocrystals is at least 150 GPa, based on the observation that cellulose is about 130 GPa and the extra bonding in the chitin crystallite causes further stiffening (Vincent and Wegst, 2004).

10.3 Nanocrystals from Natural Polysaccharides

10.3.1 Acid Hydrolysis of Polysaccharides

Stable aqueous suspensions of polysaccharide nanocrystals can be prepared by acid hydrolysis of the biomass. Throughout the chapter, different descriptors of the resulting colloidal suspended particles will be used, including whiskers, monocrystals, and nanocrystals. The designation "whiskers" is used to designate elongated rodlike nanoparticles. These crystallites have also

often been referred in literature as microcrystals or microcrystallites, despite their nanoscale dimensions. Most of the studies reported in the literature refer to cellulose nanocrystals. Recent reviews reported the properties and application in nanocomposite field of cellulosic whiskers (Azizi Samir et al., 2005; Dufresne, 2008).

The procedure for the preparation of such colloidal aqueous suspensions is described in detail in the literature for cellulose and chitin. The biomass is generally first submitted to a chemical treatment with alkaline solutions and a bleaching agent in order to purify cellulose or chitin by removing other constituents. The pure material is then disintegrated in water, and the resulting suspension is submitted to a hydrolysis treatment with acid. The amorphous regions of cellulose or chitin act as structural defects and are responsible of the transverse cleavage of the microfibrils into short monocrystals under acid hydrolysis (Battista et al., 1956). Under controlled conditions, this transformation consists of the disruption of amorphous regions surrounding and embedded within cellulose or chitin microfibrils while leaving the microcrystalline segments intact. The resulting suspension is subsequently diluted with water and washed by successive centrifugations. Dialysis against distilled water is then performed to remove free acid in the dispersion.

This general procedure is adapted depending on the nature of the substrate. The geometrical characteristics of the nanocrystals depend on the origin of the substrate and acid hydrolysis process conditions such as time, temperature, and purity of materials. Dong et al. (1998) studied the effect of preparation conditions (time, temperature, ultrasound treatment) on the resulting cellulose nanocrystal structure from sulfuric acid hydrolysis of cotton fiber. They reported a decrease in nanocrystals length and an increase in their surface charge with prolonged hydrolysis time. The concentration of the acid was also found to affect the morphology of whiskers prepared from sugar-beet pulp as reported by Azizi Samir et al. (2004b). Reaction time and acid-to-pulp ratio on nanocrystals obtained by sulfuric acid hydrolysis of black spruce acid sulfite pulp were also investigated by Beck-Candanedo et al. (2005). They reported that longer hydrolysis times produced shorter and less polydisperse nanoparticles. Optimized conditions have been stated by Bondenson et al. using MCC, derived from Norway spruce (*Picea abies*), as starting material and the processing parameters have been optimized by using a response surface methodology. The authors show that with an acid concentration of 63.5 wt%, it is possible to obtain cellulose nanocrystals with a length ranging between 200 and 400 nm and a width less than 10 nm in approximately 2 h with a yield of 30 wt% (Bondeson et al., 2006). Similar results have been reported by Elazzouzi-Harfaoui et al. (2008).

Aqueous suspensions of starch nanocrystals can be prepared according to the "lintnerization" procedure described in the literature (Robin et al., 1974; Battista, 1975). Acid hydrolysis is a chemical treatment largely used in industry to prepare glucose syrups from starch. Classically, the acid hydrolysis of starch is performed in aqueous medium with hydrochloric acid (Lintner, 1886) or sulfuric acid (Nageli, 1874) at 35°C. Residues from hydrolysis are called "lintners" and "nägeli" or amylodextrin, respectively.

The degradation of native starch granules by acid hydrolysis depends on many parameters. It includes the botanical origin of starch, namely crystalline type, granule morphology (shape, size, surface state), and relative proportion of amylose and amylopectin. It also depends on the acid hydrolysis conditions, namely acid type, acid concentration, starch concentration, temperature, duration of hydrolysis, and stirring. The degradation of starch from different origins by hydrochloric acid has been studied in detail by Robin et al. (1975). The kinetics of lintnerization involves two main steps. For lower times ($t < 8$–15 days), the hydrolysis kinetics is fast and corresponds to the hydrolysis of amorphous domains. For higher times ($t > 8$–15 days), the hydrolysis kinetics is slow and corresponds to the hydrolysis of crystalline domains. The critical time corresponding to fast/slow hydrolysis conditions depends on the botanical origin of starch (Singh and Ray, 2000; Jayakody and Hoover, 2002). It has been also reported that hydrolysis is faster when using hydrochloric acid rather than sulfuric acid (Muhr et al., 1984). Higher temperature favors the hydrolysis reaction but it is restricted to the gelatinization temperature of starch in the acid medium. Gelatinization corresponds to an irreversible swelling and solubilization phenomenon when native granules are heated above 60°C in excess water. As for temperature, the acid concentration favors the hydrolysis kinetics; above a given acid concentration, granule gelatinization occurs, around 2.5–3 N hydrochloric acid (Robin, 1976).

The main drawbacks for the use of such hydrolysis residues in composite applications are the duration (40 days of treatment) and the yield (0.5 wt%) of the hydrochloric acid hydrolysis step (Battista, 1975). Response surface methodology was used by Angellier et al. (2004) to investigate the effect of five selected factors on the selective sulfuric acid hydrolysis of waxy maize starch granules in order to optimize the preparation of aqueous suspensions of starch nanocrystals. These predictors were temperature, acid concentration, starch concentration, hydrolysis duration, and stirring speed. The preparation of aqueous suspensions of starch nanocrystals was achieved after 5 days of 3.16 M H_2SO_4 hydrolysis at 40°C, 100 rpm, and with a starch concentration of 14.69 wt% with a yield of 15.7 wt%.

10.3.2 Morphology of Polysaccharide Nanocrystals

Cellulose whiskers can be prepared from different cellulosic sources as shown in the TEM images in Figure 10.4. The constitutive nanocrystals occur as elongated rodlike particles or whiskers. Each rod can be considered as a cellulosic crystal with no apparent defect. The precise physical dimensions of the crystallites depend on several factors, including the source of the cellulose, the hydrolysis conditions, and the ionic strength. Moreover, complications in size heterogeneity are inevitable owing to the diffusion-controlled nature of the acid hydrolysis. The typical geometrical characteristics for crystallites derived from different species and reported in the literature are collected in Table 10.1. The length is generally on

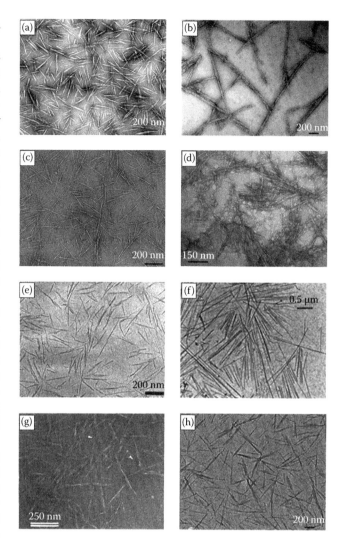

FIGURE 10.4 Transmission electron micrographs from dilute suspension of cellulose nanocrystals from: (a) ramie (Reproduced from Habibi, Y. et al., *J. Mater. Chem.*, 18, 5002, 2008b. With permission.), (b) bacterial (Reproduced from Grunnert, M. and Winter, W.T., *J. Polym. Environ.*, 10, 27, 2002. With permission.), (c) sisal (Reproduced from Garcia de Rodriguez, N. et al., *Cellulose*, 13, 261, 2006. With permission.), (d) microcrystalline cellulose (Reprinted from Kvien, I. et al., *Biomacromolecules*, 6, 3160, 2005. With permission.), (e) sugar beet pulp (Reprinted from Azizi Samir, M.A.S. et al., *Macromolecules*, 37, 4313, 2004b. With permission.), (f) tunicin (Reprinted from Angles, M.N. and Dufresne, A., *Macromolecules*, 33, 8344, 2000. With permission.), (g) wheat straw (Reproduced from Helbert, W. et al., *Polym. Compos.*, 17, 604, 1996. With permission.), and (h) cotton. (Reprinted from Fleming, K. et al., *J. Am. Chem. Soc.*, 122, 5224, 2000. With permission.)

the order of a few hundred nanometers and the width is on the order of a few nanometers. The aspect ratio of these whiskers is defined as the ratio of the length to the width. The high axial ratio of the rods is important for the determination of anisotropic phase formation and reinforcing properties.

The precise shapes and dimensions of cellulose whiskers have been generally accessed from TEM observations. Revol (1982) reported that the cross-section of cellulose crystallites in

TABLE 10.1   Geometrical Characteristics of Polysaccharide Nanocrystals from Various Sources: Length (L) and Cross Section (D) of Rodlike Particles Obtained from Acid Hydrolysis of Cellulose or Chitin

Nature	Source	L (nm)	D (nm)	References
Cellulose	Algal (Valonia)	>1,000	10–20	Revol (1982) and Hanley et al. (1992)
	Bacterial	100–several 1,000	5–10 × 30–50	Tokoh et al. (1998), Grunnert and Winter (2002), and Roman and Winter (2004)
	Cladophora	—	20 × 20	Kim et al. (2000)
	Cotton	100–300	5–10	Fengel and Wegener (1983), Dong et al. (1998), Ebeling et al. (1999), Araki et al. (2000), and Podsiadlo et al. (2005)
	Cottonseed linter	170–490	40–60	Lu et al. (2005)
	MCC	150–300	3–7	Kvien et al. (2005)
	Ramie	200–300	10–15	Habibi et al. (2007, 2008b)
	Sisal	100–500	3–5	Garcia de Rodriguez et al. (2006)
	Sugar beet pulp	210	5	Azizi Samir et al. (2004b)
	Tunicin	100–several 1,000	10–20	Favier et al. (1995a,b)
	Wheat straw	150–300	5	Helbert et al. (1996)
	Wood	100–300	3–5	Fengel and Wegener (1983), Araki et al. (1998, 1999), and Beck-Candanedo et al. (2005)
Chitin	Crab shell	80–600	8–50	Nair and Dufresne (2003b), Nge et al. (2003), and Lu et al. (2004)
	Riftia tubes	500–10,000	18	Morin and Dufresne (2002)
	Shrimp	50–300	5–70	Sriupayo et al. (2005a,b)
	Squid pen	150–800	10	Paillet and Dufresne (2001)

Valonia ventricosa was almost square, with an average side of 18 nm. Scattering techniques such as the small-angle scattering investigation of aqueous suspensions of cellulosic whiskers was also used. From this technique, tunicin whiskers were found to have a rectangular 88 × 182 Å² cross-sectional shape (Terech et al., 1999). This result is in good agreement with previous crystallographic data (Belton et al., 1989; Sugiyama et al., 1991). The investigation of the dynamic properties of cotton and tunicin whisker suspensions was performed using polarized and depolarized dynamic light scattering (de Souza Lima et al., 2003). From the determination of their translational and rotational diffusion coefficients, lengths and cross-section diameters of 255 and 15 nm for cotton, and 1160 and 16 nm for tunicin, were reported. In situ small angle neutron scattering (SANS) measurements of the magnetic and shear alignment of cellulose whiskers aqueous suspensions (Orts et al., 1998) support the hypothesis that cellulose nanocrystals are twisted rods, perhaps due to strain in their crystalline microstructure (Revol et al., 1993).

Chitin whiskers also occur as rodlike nanoparticles. Figure 10.5 shows TEM micrographs obtained from dilute suspensions of chitin fragments from different origins. The typical geometrical characteristics for crystallites derived from different species were previously reported in Table 10.1. The dimensions of chitin whiskers extracted from squid pen (Paillet and Dufresne, 2001) and crab shell (Nair and Dufresne, 2003b) were found to be close to those reported for cotton whiskers. For Riftia tubes, the average length of nanocrystals was around 2.2 μm and the aspect ratio was 120 (Morin and Dufresne, 2002). Riftia tubes are secreted by a vestimetiferan worm called Riftia and were collected at a depth of 2500 m on the East Pacific ridge.

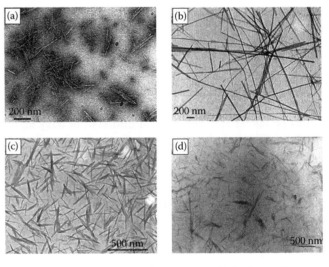

FIGURE 10.5   Transmission electron micrographs from dilute suspension of chitin nanocrystals from (a) squid pen (Reprinted from Paillet, M. and Dufresne, A., *Macromolecules*, 34, 6527, 2001. With permission.), (b) Riftia tubes (Reprinted from Morin, A. and Dufresne, A., *Macromolecules*, 35, 2190, 2002. With permission.), (c) crab shell (Reprinted from Nair, K.G. and Dufresne, A., *Biomacromolecules*, 4, 657, 2003b. With permission.), and (d) shrimps. (Reproduced from Sriupayo, J. et al., *Polymer*, 46, 5637, 2005a. With permission.)

Starch can also be used as a source for the production of polysaccharide nanocrystals. Experiments were performed using potato pulp (Dufresne et al., 1996; Dufresne and Cavaille, 1998), smooth yellow pea (Dubief et al., 1999), and waxy maize (Putaux et al., 2003; Angellier et al., 2004; Putaux, 2005; Kristo and Biliaderis, 2007),

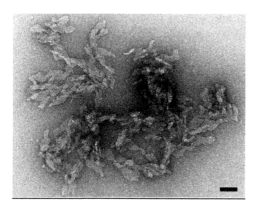

FIGURE 10.6 Transmission electron micrographs from dilute suspension of nanocrystals from waxy maize starch (scale bar 50 nm). (Reprinted from Angellier, H. et al., *Biomacromolecules*, 5, 1545, 2004. With permission.)

i.e., almost pure amylopectin, as the starch source. For the two former sources, hydrochloric acid was used whereas for the latter, sulfuric acid was used except in the study of Kristo (Kristo and Biliaderis, 2007). Compared to cellulose or chitin, the morphology of constitutive nanocrystals obtained from starch is completely different. Figure 10.6 shows TEM micrographs obtained from dilute suspensions of waxy maize starch nanocrystals. They consist of 5–7 nm thick platelet-like particles with a length ranging from 20 to 40 nm and a width in the range 15–30 nm and marked 60°–65° acute angles were observed. The detailed investigation on the structure of these platelet-like nanoparticles was reported (Putaux et al., 2003; Putaux, 2005; Kristo and Biliaderis, 2007). TEM observations show that during acid hydrolysis, branching points are first hydrolyzed in amorphous domains, in which starch nanocrystals lie parallel to the incident electron beam. When the acid hydrolysis is progressing, the amorphous regions between crystalline lamellae become completely hydrolyzed and nanocrystals are seen lying flat on the carbon film. Such nanocrystals are generally observed in the form of aggregates having an average size around 4.4 μm, as measured by laser granulometry (Angellier et al., 2005c).

10.3.3 Stability of Aqueous Suspensions

The stability of resulting suspensions depends on the dimensions of the dispersed particles, their size polydispersity, and surface charge. The use of sulfuric acid for polysaccharide nanocrystals preparation leads to more stable aqueous suspension than those prepared using hydrochloric acid (Araki et al., 1998; Angellier et al., 2005c). Indeed, the H_2SO_4-prepared nanoparticles present a negatively charged surface while the HCl-prepared nanoparticles are not charged. A comparison between the effects of the two acids was performed with waxy maize starch (Angellier et al., 2005c). It was found that the use of sulfuric acid rather than hydrochloric acid allows for reducing the possibility of agglomeration of starch nanoparticles and limits their flocculation in aqueous medium. Small angle light scattering (SALS) experiments were performed on 3.4 wt% H_2SO_4-prepared starch nanocrystal aqueous

suspensions in order to evaluate the kinetic of sedimentation of the nanoparticles (Angellier et al., 2005b). It was shown that there was no sedimentation of the nanocrystals for a period of at least 12 h. However, the intensity of scattered light slightly increased, revealing that starch nanocrystals tend to aggregate in aqueous medium but not sufficiently to induce a sedimentation phenomenon.

During acid hydrolysis of most clean polysaccharide sources via sulfuric acid, acidic sulfate ester groups are likely formed on the nanoparticle surface. This creates electric double layer repulsion between the nanoparticles in suspension, which plays an important role in their interaction with a polymer matrix and with each other. The density of charges on the polysaccharide nanocrystals surface depends on the hydrolysis conditions and can be determined by elemental analysis or conductimetric titration to accurately determine the sulfur content. The sulfate group content increases with acid concentration, acid-to-polysaccharide ratio, and hydrolysis time.

Based on the density and size of the cellulose crystallites, Araki et al. (1998, 1999) estimated for a nanocrystal with dimensions of $7 \times 7 \times 115$ nm^3 that the charge density is 0.155e·nm^{-2}, where e is the elementary charge. With the following conditions (cellulose concentration of 10 wt% in 60% sulfuric acid at 46°C for 75 min), the charge coverage was estimated at 0.2 negative ester groups per nm (Revol et al., 1992). Other typical values of the sulfur content of cellulose microcrystals prepared by sulfuric acid hydrolysis were reported (Marchessault et al., 1961; Revol et al., 1994). It was shown that even at low levels, the sulfate groups caused a significant decrease in degradation temperature and an increase in char fraction, confirming that the sulfate groups act as flame retardants (Roman and Winter, 2004). For high thermostability in the crystals, low acid concentrations, small acid-to-cellulose ratios, and short hydrolysis times should be used.

Another way to achieve charged whiskers consists in the oxidation of the hydroxyl groups on the whiskers surface (Araki et al., 2001; Habibi et al., 2006) or the postsulfation of HCl-prepared MCC (Araki et al., 1999).

10.4 Polysaccharide Nanocrystal–Reinforced Polymer Nanocomposites

10.4.1 Processing

Because of the hydrophilic nature of polysaccharide nanocrystals, a high level of dispersion of the nanocrystals within the host matrix is obtained when nanocomposites are processed in aqueous medium. This is indispensable for homogenous composites processing and therefore restricts the choice of the matrix to hydrosoluble polymers, or aqueous polymer suspensions, i.e., latexes. The possibility of dispersing polysaccharide nanocrystals in nonaqueous media is an alternative and it presents other possibilities for nanocomposites processing (Capadona et al., 2007; van den Berg et al., 2007). The dispersion of polysaccharide nanocrystals in nonpolar media can be obtained by chemically modifying their surface.

Nanocomposite materials are generally obtained by the casting technique. Twin extrusion has been also reported (Mathew et al., 2006). Recently, Habibi et al. (Habibi and Dufresne, 2008; Habibi et al., 2008) developed an interesting way to process polysaccharide nanocrystal–reinforced nanocomposites. This process consists of transforming polysaccharide nanocrystals into a co-continuous material through long-chain surface grafting before nanocomposite processing. This surface chemical modification, via polymer chain grafting, can be carried out utilizing either grafting onto or grafting from approaches (Thielemans et al., 2006; Habibi and Dufresne, 2008; Habibi et al., 2008). These chains act as long "plasticizing" tails and create a co-continuous phase between the nanocrystal and the matrix. The processing methods, such as hot pressing, extrusion, injection molding, or thermoforming, can be used to process nanocomposites from these co-continuous materials.

10.4.2 Microstructure

In addition to visual examination, different techniques have been used to control the microstructure of polysaccharide nanocrystal–reinforced nanocomposites and to access the dispersion of the nanocrystals within the host polymeric matrix. This allows for conclusions about the homogeneity of the composite, presence of voids, dispersion level of the nanoparticles within the continuous matrix, presence of aggregates, sedimentation, and possible orientation of rodlike particles.

Polarized optical microscopy was used to observe and follow the growth of polyoxyethylene (POE) spherulites in tunicin whisker-reinforced films (Azizi Samir et al., 2004c). It was observed that the spherulites exhibited a less birefringent character in the presence of tunicin whiskers, most probably due to a weakly organized structure. It was suggested that the cellulosic filler most probably interfered with the spherulite growth and that during growth the whiskers are ejected and then occluded in interspherulitic regions. The high viscosity of the filled medium most probably restricts this phenomenon and limits the size of the spherulites.

TEM and scanning electron microscopy (SEM) observations can also be performed to investigate the microstructure and dispersion quality of the nanoparticles in the nanocomposite films. SANS and small angle x-ray scattering (SAXS) have been used to conclude about the organization of tunicin whiskers in plasticized polyvinylchloride (PVC) without aggregates (Chazeau et al., 1999). Atomic force microscopy (AFM) imaging has also been recently used to investigate the microstructure of cellulose nanocrystal–reinforced polymer nanocomposites (Kvien et al., 2005).

10.4.3 Mechanical Properties

Nanoscale dimensions and impressive mechanical properties make polysaccharide nanocrystals, particularly when they occur as high aspect ratio rodlike nanoparticles, ideal candidates to improve the mechanical properties of host material. This lies in the fact that their axial Young's modulus is potentially stronger

than steel and similar to Kevlar. For cellulose nanocrystals, the theoretical value of Young's modulus high crystalline cellulose was estimated to be 167.5 GPa (Tashiro and Kobayashi, 1991). Recently, Raman spectroscopy has been used to measure the elastic modulus of native cellulose crystals from tunicin, resulting in a value of 143 GPa (Sturcova et al., 2005).

In recent years, a great interest has focused on investigating the use of polysaccharide nanocrystals, especially cellulose whiskers, as a reinforcing phase in a polymeric matrix, evaluating the mechanical properties of the resulting composites and elucidating the origin of the mechanical reinforcing effect. The dynamic mechanical analysis (DMA) is a powerful tool to investigate the linear mechanical behavior of materials in a broad temperature/frequency range, and it is strongly sensitive to the morphology of heterogeneous systems. Nonlinear mechanical properties are generally accessed through classical tensile or compressive tests (Chazeau et al., 2000). Nanoindentation was also reported to be a suitable method for mechanical characterization of cellulose based nanocomposites (Zimmermann et al., 2005).

The first demonstration of the reinforcing effect of cellulose whiskers in a nanocomposite was reported by Favier et al. (1995a,b). The authors observed a substantial improvement in the storage modulus after adding tunicin whiskers, even at low content, into the host poly(S-co-BuA) matrix, using DMA in the shear mode. This increase was especially significant above the glass–rubber transition temperature of the thermoplastic matrix because of its poor mechanical properties in this temperature range. Figure 10.7 shows the isochronal evolution of the logarithm of the relative storage shear modulus ($\log G'_T/G'_{200}$, where

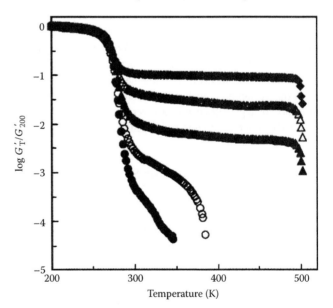

FIGURE 10.7 Logarithm of the normalized storage shear modulus ($\log G'_T/G'_{200}$, where G'_{200} corresponds to the experimental value measured at 200 K) vs. temperature at 1 Hz for tunicin whiskers reinforced poly(S-co-BuA) nanocomposite films obtained by water evaporation and filled with 0 wt% (●), 1 wt% (○), 3 wt% (▲), 6 wt% (△), and 14 wt% (◆) of cellulose whiskers. (Reprinted from Azizi Samir, M. A. S. et al., *Biomacromolecules*, 6, 612, 2005. With permission.)

G'_{200} corresponds to the experimental value measured at 200 K) at 1 Hz as a function of temperature for such composites prepared by water evaporation.

The Halpin-Kardos model, which is the classical model usually used for randomly dispersed short fiber-reinforced composites (Halpin and Kardos, 1972), failed to describe the unusual reinforcing effect of tunicin whiskers in poly(S-*co*-BuA) (Favier et al., 1995a,b). Using this model, the cellulose whiskers seemed to act as fibers much longer than expected from geometrical observation. The outstanding properties observed for these systems were ascribed to a mechanical percolation phenomenon.

Percolation for the statistical-geometry model was first introduced in 1957 by Hammersley (1957). It is a statistical theory that can be applied to any system involving a great number of species that are likely to be connected. The aim of the statistical theory is to forecast the behavior of a noncompletely connected set of objects. By varying the number of connections, this approach allows for a description of the transition from a local to an infinite "communication" state. The percolation threshold is defined as the critical volume fraction separating these two states. Various parameters, such as particle interactions (Balberg and Binenbaum, 1983), orientation (Balberg et al., 1984), or aspect ratio (de Gennes, 1976) can modify the value of the percolation threshold. The use of this approach to describe and predict the mechanical behavior of cellulosic whisker based nanocomposites suggests the formation of a rigid network of whiskers, which should be responsible for the unusual reinforcing effect observed at high temperatures. The modeling consists of three important steps:

1. First, the calculation of the percolation threshold (v_{Rc}) should be carried out. The volume fraction of cellulose nanoparticles required to achieve geometrical percolation can be calculated using a statistical percolation theory for cylindrical shape particles according to their aspect ratio and the effective skeleton of whiskers (Favier et al., 1997b). The latter corresponds to the infinite length of a branch of nanoparticles connecting the sample ends. Favier et al. (1997b) used computer simulation and showed that about 0.75 vol% tunicin whiskers (assuming L/d = 100) are needed to get a 3-D geometrical percolation. The authors calculated the effective skeleton by eliminating the finite length branches. The following relation was found between the percolation threshold and the aspect ratio of rodlike particles:

$$v_{Rc} = \frac{0.7}{L/d} \tag{10.1}$$

For wheat straw, cellulose whisker reinforced poly (S-*co*-BuA) the v_{Rc} value was found to be around 2 vol% (Dufresne et al., 1997). This value is about half (4.4 vol%) of the value observed for NR reinforced with chitin whiskers obtained from crab shell, presenting an aspect ratio close to 16 (Nair and Dufresne, 2003b).

In the case of starch nanocrystals, the critical volume fraction at percolation is difficult to determine due to the ill-defined geometry of the percolating species, but was reported around 6.7 vol% (i.e., 10 wt%) for waxy maize starch nanocrystal-reinforced natural rubber (Angellier et al., 2005b). This value is smaller than the one reported for poly(S-*co*-BuA) filled with potato starch nanocrystals (around 20 vol%) (Dufresne et al., 1998). This difference may be due to a higher surface area of the waxy maize starch nanocrystals and the particular morphology of starch nanocrystals that aggregate by forming a "lace net."

2. The second step is the estimation of the modulus of the percolating filler network. The modulus is different from that of the individual nanoparticles, and depends on the origin of the polysaccharide, preparation procedure of the nanocrystals, and the nature and strength of interparticle interactions. This modulus can be assumed to be that of a paper sheet for which the hydrogen bonding forces provide the basis of its stiffness. For tunicin (Favier et al., 1995b) and wheat straw cellulose whiskers (Helbert et al., 1996), the tensile modulus was around 15 and 6 GPa, respectively. The tensile modulus of chitin whiskers was found to be around 0.5 and 2 GPa for squid pen (Paillet and Dufresne, 2001) and *Riftia* tubes (Morin and Dufresne, 2002), respectively.

3. The description of the composite requires the use of a model involving three different phases: the matrix, the filler percolating network, and the nonpercolating filler phase. The simplest model consists of two parallel phases, namely the effective whisker skeleton and the rest of the sample. In their study of the mechanical behavior of poly(methyl methacrylate) and poly(S-*co*-BuA) blends, Ouali et al. (1991) extended the classical phenomenological series-parallel model of Takayanagi et al. (1964) and proposed a model in which the percolating filler network is set in parallel with a series part composed of the matrix and the nonpercolating filler phase (Figure 10.8).

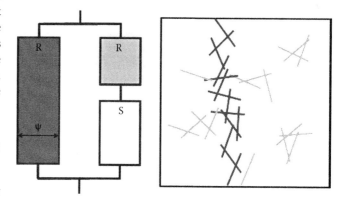

FIGURE 10.8 Schematic representation of the series-parallel model: R and S refer to the rigid (cellulosic filler) and soft (polymeric matrix) phases, respectively, and ψ is the volume fraction of the percolating rigid phase. Dark grey and clear grey rods correspond to percolating and unpercolating nanoparticles, respectively.

In this approach, the elastic tensile modulus E_c of the composite is given by the following equation:

$$E_c = \frac{(1 - 2\psi + \psi v_R)E_S E_R + (1 - v_R)\psi E_R^2}{(1 - v_R)E_R + (v_R - \psi)E_S} \tag{10.2}$$

The subscripts S and R refer to the soft and rigid phase, respectively. The adjustable parameter, ψ, involved in the Takayanagi et al. model corresponds in the Ouali et al. prediction to the volume fraction of the percolating rigid phase. With b being the critical percolation exponent, ψ can be written as

$$\psi = 0 \qquad\qquad \text{for } v_R < v_{Rc}$$
$$\psi = v_R \left(\frac{v_R - v_{Rc}}{1 - v_{Rc}} \right)^b \qquad \text{for } v_R > v_{Rc} \tag{10.3}$$

where $b = 0.4$ (Stauffer and Aharony, 1992) for a 3-D network.

At high temperatures when the polymeric matrix could be assumed to have zero stiffness ($E_S \sim 0$), the calculated stiffness of the composite is simply the result of the percolating fillers network and the volume fraction of percolating filler phase:

$$E_c = \psi E_R \tag{10.4}$$

In the former study of Favier et al. dealing with tunicin whisker–reinforced poly(S-*co*-BuA), a good agreement between experimental and predicted data was reported when using the series-parallel model of Takayanagi modified to include a percolation approach. It was then suspected that the stiffness of the material was due to infinite aggregates of cellulose whiskers. Above the percolation threshold, the cellulosic nanoparticles can connect and form a 3-D continuous pathway through the nanocomposite film. The formation of this cellulose network was expected to result from strong interactions between whiskers, like hydrogen bonds (Favier et al., 1997a). This phenomenon is similar to the high mechanical properties observed for a paper sheet, which result from the hydrogen-bonding forces that hold the percolating network of fibers. This mechanical percolation effect allows for the explanation of both the high reinforcing effect and the thermal stabilization of the composite modulus for casted films. A unified description of the moduli of nanocomposites containing elongated filler particles over a range of volume fractions spanning the filler percolation threshold has been recently provided (Chatterjee, 2006).

The existence of such 3-D percolating nanoparticles network was evidenced by performing successive tensile tests on crab shells chitin whiskers (Nair and Dufresne, 2003b) and waxy maize starch nanocrystal (Angellier et al., 2006b) reinforced natural rubber.

Any factor that affects the formation of the percolating nanocrystals network, or interferes with it, changes the mechanical performances of the composite. Three main parameters were reported to affect the mechanical properties of such materials, viz. the morphology and dimensions of the nanoparticles, the

processing method, and the microstructure of the matrix and matrix–filler interactions. The effect of these parameters on the mechanical performances of nanocomposites reinforced by polysaccharide nanocrystals are reported and discussed below.

10.4.3.1 Morphology and Dimensions of the Nanoparticles

Cellulose and chitin nanocrystals occur as rodlike nanoparticles contrarily to starch nanocrystals that consist of nanometer scale aggregated platelet-like particles. For rodlike particles, the geometrical aspect ratio is an important factor since it determines the percolation threshold value according to Equation 10.1. This factor is linked to the source of cellulose or chitin and whisker preparation conditions. Fillers with high aspect ratios give the best reinforcing effect because a lower content is needed to achieve percolation.

The flexibility and tangling possibility of the nanofibers play an important role. This was exemplified by Azizi Samir et al. (2004b). In this study, the authors reported the mechanical properties of poly(S-*co*-BuA) reinforced with cellulose rodlike nanoparticles extracted from cellulose microfibrils from sugar beet with different hydrolysis conditions. These cellulose microfibrils, almost 5 nm in width and practically infinite in length, were submitted to a hydrolysis treatment using different sulfuric acid concentrations. As the acid concentration increased, the length of the nanoparticles decreased. DMA experiments performed on poly(S-*co*-BuA) reinforced with these nanoparticles did not show significant differences by varying their length. However, from nonlinear mechanical tensile tests, it was observed that as the length decreased, both the modulus and the strength of the composite decreased, whereas the elongation at break increased. This result showed strong influence of entanglements on the mechanical behavior of the nanocomposites.

10.4.3.2 Processing Method

The processing method governs the possible formation of a continuous nanocrystal network and the final properties of the nanocomposite material. Slow processes such as casting/evaporation were reported to give the highest mechanical performance materials compared to freeze-drying/molding and freeze-drying/extruding/molding techniques. This effect was observed for tunicin whisker–reinforced poly(S-*co*-BuA) (Favier et al., 1995b), Riftia tubes chitin whisker reinforced polycaprolactone (Morin and Dufresne, 2002), and crab shells chitin whisker–reinforced natural rubber (Nair and Dufresne, 2003a). It was related to the probable orientation of these rodlike nanoparticles during film processing due to shear stresses induced by freeze-drying/molding or freeze-drying/extruding/molding techniques.

During slow water evaporation, because of Brownian motions in the suspension or solution (whose viscosity remains low until the end of the process when the latex particle or polymer concentration becomes very high), the rearrangement of the nanoparticles is possible. They have adequate time to interact and connect to form a percolating network, which is the basis of their reinforcing effect. The resulting structure (after the coalescence of latex particles or and/or interdiffusion of polymeric chains) is completely relaxed and direct contacts between the nanocrystals are then created. Conversely, during the freeze-drying/hot-pressing process, the nanoparticle arrangement in the suspension is first frozen, and then, during the hot-pressing stage, the particle rearrangements are strongly limited due to the polymer melt viscosity. Thus, in this case, contacts are made through a certain amount of polymer matrix. However, although the freeze-drying/hot-pressing process limits the possibility of creation of hydrogen bonds, it is expected that for high polysaccharide nanoparticle content some bonds may evenly be created.

Hajji et al. (1996) studied the tensile behavior of poly(S-*co*-BuA)/tunicin whisker composites prepared by different methods. The authors classified processing methods in ascending order of their reinforcement efficiency (both tensile modulus and strength): extrusion < hot pressing < evaporation. This evolution was associated to probable fracture and/or orientation of whiskers during processing.

10.4.3.3 Microstructure of the Matrix and Matrix–Filler Interactions

The microstructure of the matrix and the resulting competition between matrix–filler and filler–filler interactions also affect the mechanical behavior of the polysaccharide nanocrystal-reinforced nanocomposites. Classical composite science tends to privilege the former as a fundamental condition for optimal performance. In polysaccharide nanocrystal–based nanocomposites, the opposite trend is generally observed when the materials are processed via casting/evaporation method. The higher the affinity between the polysaccharide filler and the host matrix is the lower the mechanical properties are. This unusual behavior is ascribed to the originality of the reinforcing phenomenon of polysaccharide nanocrystals resulting from the formation of a percolating network thanks to hydrogen bonding forces.

Strong interactions between cellulose nanocrystals prepared from cottonseed linters and the glycerol plasticized starch matrix were reported to play a key role in reinforcing properties (Lu et al., 2005). In nonpercolating systems, for instance, for materials processed from freeze-dried cellulose nanocrystals, strong matrix–filler interactions enhance the reinforcing effect of the filler. This observation was reported using EVA matrices with different vinyl acetate contents and then different polarities (Chauve et al., 2005). The improvement of matrix–filler interactions by using cellulose whiskers coated with a surfactant was shown to play a major role on the nonlinear mechanical properties, especially on the elongation at break (Ljungberg et al., 2005).

Grunnert and Winter found a higher reinforcing effect for unmodified cellulose whiskers than for trimethylsilylated whiskers (Grunnert and Winter, 2002). Apart from the fact that 18% of the weight of the silylated crystals was due to the silyl groups,

they attributed this difference to restricted filler–filler interactions. Similar results and loss of mechanical properties were reported for natural rubber–based nanocomposites reinforced with both unmodified and surface chemically modified chitin whiskers (Nair et al., 2003c) and starch nanocrystals (Angellier et al., 2005a). When cellulose nanocrystals grafted with high molecular weight PCL were used as filler in PCL matrix, the final nanocomposite shows a lower modulus but significantly higher strain at break compared to the one filled with unmodified nanocrystals (Habibi and Dufresne, 2008). This unusual behavior clearly reflects the restricted filler–filler interactions that drop the modulus and the high filler–matrix compatibilization resulting from the formation of a percolating network held by chain entanglements and possible co-crystallization between the grafted chains and the matrix. A strong interaction between the filler and the matrix is the origin of the higher strain at break. In a similar system, Habibi et al. demonstrated a significant improvement in terms of Young's modulus and storage modulus when short chains of PCL were grafted to the cellulose nanocrystals but with high grafting density (Habibi et al., 2008). The PCL chains were long enough to behave as compatibilizer between the filler and the matrix but do not restrict the filler–filler interactions and consequently the formation of the percolating network between the cellulose whiskers.

The transcrystallization phenomenon reported for semicrystalline poly(hydroxyoctoanoate) PHO on cellulose whiskers resulted in a disastrous decrease of the mechanical properties (especially above the melting temperature of the matrix) when compared to that obtained for fully amorphous PHO (Dufresne et al., 1999). In these systems, the filler–matrix interactions and the distance away from the surface at which the molecular mobility of the amorphous PHO phase is restricted were quantified using a physical model predicting the mechanical loss angle (Dufresne, 2000). The determination of the ratio of experimental and predicted magnitude of the main relaxation process allows for the removal of the filler reinforcement effect and for keeping only the interfacial effect, and was used to calculate the thickness of the interphase. It was shown that when using semicrystalline PHO as the matrix, the molecular mobility of amorphous PHO chains was only slightly affected by the presence of tunicin whiskers, owing to a possible transcrystallization phenomenon, leading to the coating of the nanoparticles with the crystalline PHO phase. The thickness of the transcrystalline layer, around 2.7 nm, was found to be independent of the cellulose whiskers content. In contrast, when using an amorphous PHO as the matrix, the flexibility of polymeric chains in the surface layer was lowered by the conformational restrictions imposed by cellulose surface. This results in a broader interphase and in a broadening of the main relaxation process of the matrix.

Similar transcrystallization was reported for plasticized starch–reinforced with cellulose whiskers (Angles and Dufresne, 2001). This strong loss of performance demonstrates the event of outstanding importance of the filler–filler interactions to ensure the mechanical stiffness and thermal stability of these composites.

10.4.4 Thermal Properties

The characterization of the thermal properties of materials is important to determine the temperature range of processing and use. The main thermal characteristics of polymeric systems are the glass–rubber transition, melting point, and thermal stability.

10.4.4.1 Glass Transition Temperature T_g

In most studies, no modification of glass transition temperature (T_g) values has been reported when increasing the amount of whiskers, regardless the nature of the polymeric matrix. This result appears to be surprising because of the high specific area of these nanoparticles that is around $170\,m^2 \cdot g^{-1}$ for tunicin whiskers (Dufresne, 2000).

In glycerol-plasticized starch-based composites, peculiar effects of tunicin whiskers on the T_g of the starch-rich fraction were reported depending on moisture conditions (Angles and Dufresne, 2000). For low loading level (up to 3.2 wt%), a classical plasticization effect of water was reported. However, an antiplasticization phenomenon was observed for higher whisker content (6.2 wt% and up). These observations were discussed according to the possible interactions between hydroxyl groups on the cellulosic surface and starch, the selective partitioning of glycerol and water in the bulk starch matrix or at the whisker surface, and the restriction of amorphous starch chain mobility in the vicinity of the starch crystallite–coated filler surface. For glycerol-plasticized starch-reinforced with cellulose nanocrystals prepared from cottonseed linter (Lu et al., 2005), an increase of T_g with filler content was reported and attributed to cellulose–starch interactions. For tunicin whiskers/sorbitol-plasticized starch (Mathew and Dufresne, 2002), the values of T_g were found to increase slightly up to about 15 wt% whiskers and to decrease for higher whiskers loading. The crystallization of amylopectin chains upon whisker addition and migration of sorbitol molecules to the amorphous domains were proposed to explain the observed modifications.

For waxy maize starch nanocrystal–reinforced natural rubber, a decrease in the onset glass transition temperature with the increase of the nanoparticles content was reported (Angellier et al., 2005b). Using a glycerol plasticized starch matrix, it was reported that a temperature increase of the main relaxation process was associated with the glass–rubber transition of amylopectin-rich domains with the increasing of the starch nanocrystals content (Angellier et al., 2006a). The reduction in the molecular mobility of matrix amylopectin chains for filled materials was explained by the establishment of hydrogen bonding forces between both components. A similar observation was reported for polyvinyl acetate (PVA) (Garcia de Rodriguez et al., 2006; Roohani et al., 2008) and carboxymethyl cellulose (CMC) (Choi and Simonsen, 2006) reinforced with cellulose whiskers. For waxy maize starch nanocrystal–reinforced glycerol plasticized starch the increase of T_g led to a considerable slowing down of the retrogradation of the matrix (Angellier et al., 2006a).

This is a very interesting result since retrogradation and crystallization of thermoplastic starch during aging is one of the main drawbacks of this material and lead to an undesired change in thermomechanical properties.

10.4.4.2 Melting Temperature (T_m) and Crystallinity

In semicrystalline polymeric matrix–based nanocomposites, the melting temperature (T_m) and heat of fusion (ΔH_m) of the thermoplastic matrix can be determined from DSC measurements. X-ray diffraction can also be used to elucidate the eventual modifications on the crystalline structure of the matrix after the addition of polysaccharide nanocrystals.

Melting temperature (T_m) values were reported to be nearly independent on the filler content in plasticized starch (Angles and Dufresne, 2000; Mathew and Dufresne, 2002) and in POE-based materials (Azizi Samir et al., 2004a,c,d) filled with tunicin whiskers. The same observation was reported for polycaprolactone reinforced with Riftia tubes chitin whiskers (Morin and Dufresne, 2002) and cellulose acetate butyrate (CAB) reinforced with native bacterial cellulose whiskers (Grunnert and Winter, 2002). However, for the latter system, T_m values were found to increase when the amount of trimethylsilylated whiskers increased. Similar observations were reported in the case of polycaprolactone reinforced with polycaprolactone grafted cellulose nanocrystals (Habibi and Dufresne, 2008; Habibi et al., 2008). This difference is related to the stronger filler–matrix interaction in the case of chemically modified whiskers.

A significant increase in crystallinity of sorbitol plasticized starch (Mathew and Dufresne, 2002) was reported when increasing cellulose whiskers content. This phenomenon was ascribed to an anchoring effect of the cellulosic filler, probably acting as a nucleating agent. For POE-based composites, the degree of crystallinity of the matrix was found to be roughly constant up to 10 wt% tunicin whiskers (Azizi Samir et al., 2004a,c,d) and to decrease for higher loading level (Azizi Samir et al., 2004c). Incorporation of shrimp shells chitin whiskers did not have any effect on the crystallinity of PVA (Sriupayo et al., 2005a) and chitosan (Sriupayo et al., 2005b).

It seems that the nucleating effect of cellulosic nanocrystals is mainly governed by surface chemical considerations. Indeed, both untreated and surfactant-coated whiskers were also reported to be very good nucleating agents for isotactic polypropylene (iPP). The unmodified whiskers have the largest nucleating effect (Ljungberg et al., 2006). On the contrary, whiskers grafted with maleated polypropylene did not modify the crystallization of iPP. It was shown from both x-ray diffraction and DSC analyses that the crystallization behavior of films containing unmodified and surfactant-modified whiskers displayed two crystalline forms (α and β), whereas the neat matrix and the nanocomposite reinforced with nanocrystals grafted with maleated polypropylene only crystallized in the α form. It was suspected that the more hydrophilic the whisker surface was, the more it appeared to favor the appearance of the β phase. It was observed that native bacterial fillers impede the crystallization of the CAB matrix whereas silylated ones help to nucleate the crystallization (Grunnert and Winter, 2002). A decrease of the degree of crystallinity of polycaprolactone was reported when adding Riftia tubes chitin whiskers (Morin and Dufresne, 2002). It was suggested that during crystallization, the rodlike nanoparticles are most probably first ejected and then occluded in intercrystalline domains, hindering the crystallization of the polymer. In iPP reinforced with tunicin whiskers, a mechanical coupling between the polypropylene crystallites and filler–filler interactions was reported (Ljungberg et al., 2006).

For tunicin whisker–filled semicrystalline matrices such as poly(β-hydroxyoctanoate) (PHO) (Dufresne et al., 1999) and glycerol-plasticized starch (Angles and Dufresne, 2000) a transcrystallization phenomenon was reported. It consists on a preferential crystallization of the amorphous polymeric matrix chains during cooling at the surface of nanoparticles. For glycerol-plasticized starch-based systems, the formation of the transcrystalline zone around the whiskers was assumed to be due to the accumulation of plasticizer in the cellulose–amylopectin interfacial zones improving the ability of amylopectin chains to crystallize. These specific crystallization conditions were evidenced at high moisture content and high whiskers content by DSC and wide angle x-ray scattering (WAXS). It was displayed through a shoulder on the low-temperature side of the melting endotherm and the observation of a new peak in the x-ray diffraction pattern. This transcrystalline zone could originate from a glycerol-starch V structure. In addition, the inherent restricted mobility of amylopectin chains was put forward to explain the lower water uptake of cellulose–starch composites for increasing filler content.

10.4.4.3 Thermal Stability

Thermogravimetric analysis (TGA) experiments were performed to determine the water content of tunicin whiskers/plasticized starch nanocomposites (Angles and Dufresne, 2000) and investigate the thermal stability of tunicin whiskers/POE nanocomposites (Azizi Samir et al., 2004a,c). No significant influence of the cellulosic filler on the degradation temperature of the POE matrix was reported. Shrimp shell chitin whiskers did not much affect the thermal stability of chitosan (Sriupayo et al., 2005b) but were found to improve it when using a PVA matrix (Sriupayo et al., 2005a). Cotton cellulose nanocrystal content appeared to have an effect on the thermal behavior of CMC plasticized with glycerin (Choi and Simonsen, 2006) suggesting a close association between the filler and the matrix. The thermal degradation of unfilled CMC was observed from its melting point (270°C), and had a very narrow temperature range of degradation. Cellulose nanocrystals were found to degrade at a lower temperature (230°C) than CMC, but showed a very broad degradation temperature range. However, the degradation temperature of cellulose Whisker–reinforced CMC composites was observed between these two limits.

10.5 Conclusions

Polysaccharide nanocrystals are building blocks biosynthesized to provide structural properties to living organisms. They can be isolated from biomass through acid hydrolysis with concentrated mineral acids under strictly controlled conditions of time and temperature. Acid action results in an overall decrease of amorphous material by removing polysaccharide material closely bonded to the crystallite surface and breaks down the amorphous regions. A leveling-off degree of polymerization is achieved corresponding to the residual highly crystalline regions of the original material, i.e., cellulose or chitin fiber, or starch granule. Dilution of the acid and dispersion of the individual crystalline nanoparticles complete the process and yield an aqueous suspension of polysaccharide nanoparticles. These nanoparticles occur as rodlike nanocrystals that can display chiral nematic properties depending on the mineral acid chosen for the hydrolysis in the case of cellulose- or chitin-based materials, or platelet-like nanoparticles when using starch granules as the raw material.

Polysaccharide nanocrystals are inherently low-cost and renewable materials, which are available from a variety of natural sources. These nanosized particles are self-assembling into well-defined architectures with a wide range of aspect ratios, e.g., ~200 nm long and 5 nm in lateral dimension and up to several microns long and 18 nm in lateral dimension for cellulose and chitin. They display very interesting thermomechanical properties, e.g., strength, modulus and dimensional stability, thermal stability, and heat distortion temperature, in addition to their permeability to gases and water, surface appearance, and optical clarity in comparison to conventionally fillers. They are an attractive nanomaterial for multitude of potential applications in a diverse range of fields. Indeed, nanotechnology has applications across most economic sectors and allows the development of new enabling science with broad commercial potential. Possible and suggested areas of application include optically variable films and ink-iridescent pigments for security papers. Polysaccharide nanocrystal–reinforced polymer nanocomposites display outstanding mechanical properties and can be used to process high-modulus thin films. Nowadays, nanocomposite polymer electrolyte-reinforced with cellulosic nanoparticles are successfully prepared. There are many other appealing expectations regarding their potential. The growing literature studying polysaccharide nanocrystals, mainly from cellulose, is a clear indication of this evolution. Practical applications of such fillers and transition into industrial technology require a favorable ratio between the expected performances of the composite material and its cost. To exploit their potential, research and development investments must be made in science and engineering that will fully determine the properties and characteristics of polysaccharides at the nanoscale, develop the technologies to manipulate self-assembly and multifunctionality, and develop these new technologies to the point where industry can produce advanced and cost-competitive polysaccharide nanoscale products. There are still significant scientific and technological challenges to take up.

References

Angellier, H., L. Choisnard, S. Molina-Boisseau, P. Ozil, and A. Dufresne. 2004. Optimization of the preparation of aqueous suspensions of waxy maize starch nanocrystals using a response surface methodology. *Biomacromolecules* 5: 1545–1551.

Angellier, H., S. Molina-Boisseau, and A. Dufresne. 2005a. Mechanical properties of waxy maize starch nano-crystal reinforced natural rubber. *Macromolecules* 38: 9161–9170.

Angellier, H., S. Molina-Boisseau, L. Lebrun, and A. Dufresne. 2005b. Processing and structural properties of waxy maize starch nanocrystals reinforced natural rubber. *Macromolecules* 38: 3783–3792.

Angellier, H., J.-L. Putaux, S. Molina-Boisseau, D. Dupeyre, and A. Dufresne. 2005c. Starch nanocrystal fillers in an acrylic polymer matrix. *Macromol. Symp.* 221: 95–104.

Angellier, H., S. Molina-Boisseau, P. Dole, and A. Dufresne. 2006a. Thermoplastic starch-waxy maize starch nanocrystals nanocomposites. *Biomacromolecules* 7: 531–539.

Angellier, H., S. Molina-Boisseau, and A. Dufresne. 2006b. Waxy maize starch nanocrystals as filler in natural rubber. *Macromol. Symp.* 233: 132–136.

Angles, M. N. and A. Dufresne. 2000. Plasticized starch/tunicin whiskers nanocomposites. 1. Structural analysis. *Macromolecules* 33: 8344–8353.

Angles, M. N. and A. Dufresne. 2001. Plasticized starch/tunicin whiskers nanocomposite materials. 2. Mechanical behavior. *Macromolecules* 34: 2921–2931.

Araki, J., M. Wada, S. Kuga, and T. Okano. 1998. Flow properties of microcrystalline cellulose suspension prepared by acid treatment of native cellulose. *Colloids Surf. A* 142: 75–82.

Araki, J., M. Wada, S. Kuga, and T. Okano. 1999. Influence of surface charge on viscosity behavior of cellulose microcrystal suspension. *J. Wood Sci.* 45: 258–261.

Araki, J., M. Wada, S. Kuga, and T. Okano. 2000. Birefringent glassy phase of a cellulose microcrystal suspension. *Langmuir* 16: 2413–2415.

Araki, J., M. Wada, and S. Kuga. 2001. Steric stabilization of a cellulose microcrystal suspension by poly(ethylene glycol) grafting. *Langmuir* 17: 21–27.

Atkins, E. 1985. Conformations in polysaccharides and complex carbohydrates. *J. Biosci.* 8: 375–387.

Azizi Samir, M. A. S., F. Alloin, W. Gorecki, J.-Y. Sanchez, and A. Dufresne. 2004a. Nanocomposite polymer electrolytes based on poly(oxyethylene) and cellulose nanocrystals. *J. Phys. Chem. B* 108: 10845–10852.

Azizi Samir, M. A. S., F. Alloin, M. Paillet, and A. Dufresne. 2004b. Tangling effect in fibrillated cellulose reinforced nanocomposites. *Macromolecules* 37: 4313–4316.

Azizi Samir, M. A. S., F. Alloin, J.-Y. Sanchez, and A. Dufresne. 2004c. Cellulose nanocrystals reinforced poly(oxyethylene). *Polymer* 45: 4149–4157.

Azizi Samir, M. A. S., F. Alloin, J.-Y. Sanchez, N. El Kissi, and A. Dufresne. 2004d. Preparation of cellulose whiskers reinforced nanocomposites from an organic medium suspension. *Macromolecules* 37: 1386–1393.

Azizi Samir, M. A. S., F. Alloin, and A. Dufresne. 2005. Review of recent research into cellulosic whiskers, their properties and their application in nanocomposite field. *Biomacromolecules* 6: 612–626.

Balberg, I. and N. Binenbaum. 1983. Computer study of the percolation threshold in a two-dimensional anisotropic system of conducting sticks. *Phys. Rev. B* 28: 3799–3812.

Balberg, I., N. Binenbaum, and N. Wagner. 1984. Percolation thresholds in the three-dimensional sticks system. *Phys. Rev. Lett.* 52: 1465–1468.

Bartnicki-Garcia, S., J. Persson, and H. Chanzy. 1994. An electron microscope and electron diffraction study of the effect of calcofluor and congo red on the biosynthesis of chitin in vitro. *Arch. Biochem. Biophys.* 310: 6–15.

Battista, O. A. 1975. *Microcrystal Polymer Science.* New York: McGraw-Hill.

Battista, O. A., S. Coppick, J. A. Howsmon, F. F. Morehead, and W. A. Sisson. 1956. Level-off degree of polymerization. Relation to polyphase structure of cellulose fibers. *Ind. Eng. Chem.* 48: 333–335.

Beck-Candanedo, S., M. Roman, and D. G. Gray. 2005. Effect of reaction conditions on the properties and behavior of wood cellulose nanocrystal suspensions. *Biomacromolecules* 6: 1048–1054.

Belton, P. S., S. F. Tanner, N. Cartier, and H. Chanzy. 1989. High-resolution solid-state carbon-13 nuclear magnetic resonance spectroscopy of tunicin, an animal cellulose. *Macromolecules* 22: 1615–1617.

Blackwell, J., K. D. Parker, and K. M. Rudall. 1965. Chitin in pogonophore tubes. *J. Mar. Biol.* 45: 659–661.

Bondeson, D., A. Mathew, and K. Oksman. 2006. Optimization of the isolation of nanocrystals from microcrystalline cellulose by acid hydrolysis. *Cellulose* 13: 171–180.

Brine, C. J. and P. R. Austin. 1975. Renatured chitin fibrils, films and filaments. In *ACS Symposium Series: Marine Chemistry in the Coastal Environment*, ed. T. D. Church, pp. 505–518. Washington, DC: American Chemical Society.

Buléon, A., P. Colonna, V. Planchot, and S. Ball. 1998. Starch granules: Structure and biosynthesis. *Int. J. Biol. Macromol.* 23: 85–112.

Capadona, J. R., O. van den Berg, L. A. Capadona et al. 2007. A versatile approach for the processing of polymer nanocomposites with self-assembled nanofibre templates. *Nat. Nanotechnol.* 2: 765–769.

Chanzy, H. 1990. Aspects of cellulose structure. In *Cellulose Sources and Exploitation: Industrial Utilization, Biotechnology, and Physico-Chemical Properties*, eds. J. F. Kennedy, G. O. Phillips, and P. A. Williams. Chichester, U.K.: Ellis Horwood.

Chatterjee, A. P. 2006. A model for the elastic moduli of three-dimensional fiber networks and nanocomposites. *J. Appl. Phys.* 100: 054302/1–054302/8.

Chauve, G., L. Heux, R. Arouini, and K. Mazeau. 2005. Cellulose poly(ethylene-co-vinyl acetate) nanocomposites studied by molecular modeling and mechanical spectroscopy. *Biomacromolecules* 6: 2025–2031.

Chazeau, L., J. Y. Cavaille, and P. Terech. 1999. Mechanical behaviour above Tg of a plasticised PVC reinforced with cellulose whiskers: A SANS structural study. *Polymer* 40: 5333–5344.

Chazeau, L., J. Y. Cavaille, and J. Perez. 2000. Plasticized PVC reinforced with cellulose whiskers. II. Plastic behavior. *J. Polym. Sci., Part B: Polym. Phys.* 38: 383–392.

Choi, Y. and J. Simonsen. 2006. Cellulose nanocrystal-filled carboxymethyl cellulose nanocomposites. *J. Nanosci. Nanotechnol.* 6: 633–639.

de Gennes, P. G. 1976. On a relation between percolation theory and the elasticity of gels. *J. Phys. Lett.* 37: L1–L2.

de Souza Lima, M. M., J. T. Wong, M. Paillet, R. Borsali, and R. Pecora. 2003. Translational and rotational dynamics of rodlike cellulose whiskers. *Langmuir* 19: 24–29.

Dong, X. M., J. F. Revol, and D. G. Gray. 1998. Effect of microcrystallite preparation conditions on the formation of colloid crystals of cellulose. *Cellulose* 5: 19–32.

Dubief, D., E. Samain, and A. Dufresne. 1999. Polysaccharide microcrystals reinforced amorphous poly(beta-hydroxyoctanoate) nanocomposite materials. *Macromolecules* 32: 5765–5771.

Dufresne, A. 2000. Dynamic mechanical analysis of the interphase in bacterial polyester/cellulose whiskers natural composites. *Compos. Interfaces* 7: 53–67.

Dufresne, A. 2008. Polysaccharide nano crystal reinforced nanocomposites. *Can. J. Chem.* 86: 484–494.

Dufresne, A. and J.-Y. Cavaille. 1998. Clustering and percolation effects in microcrystalline starch-reinforced thermoplastic. *J. Polym. Sci., Part B: Polym. Phys.* 36: 2211–2224.

Dufresne, A., J.-Y. Cavaille, and W. Helbert. 1996. New nanocomposite materials: Microcrystalline starch reinforced thermoplastic. *Macromolecules* 29: 7624–7626.

Dufresne, A., J. Y. Cavaille, and W. Helbert. 1997. Thermoplastic nanocomposites filled with wheat straw cellulose whiskers. Part II: Effect of processing and modeling. *Polym. Compos.* 18: 199.

Dufresne, A., M. B. Kellerhals, and B. Witholt. 1999. Transcrystallization in Mcl-PHAs/cellulose whiskers composites. *Macromolecules* 32: 7396–7401.

Ebeling, T., M. Paillet, R. Borsali et al. 1999. Shear-induced orientation phenomena in suspensions of cellulose microcrystals, revealed by small angle x-ray scattering. *Langmuir* 15: 6123–6126.

Elazzouzi-Hafraoui, S., Y. Nishiyama, J.-L. Putaux, L. Heux, F. Dubreuil, and C. Rochas. 2008. The shape and size distribution of crystalline nanoparticles prepared by acid hydrolysis of native cellulose. *Biomacromolecules* 9: 57–65.

Favier, V., G. R. Canova, J. Y. Cavaille, H. Chanzy, A. Dufresne, and C. Gauthier. 1995a. Nanocomposite materials from latex and cellulose whiskers. *Polym. Adv. Technol.* 6: 351–355.

Favier, V., H. Chanzy, and J. Y. Cavaille. 1995b. Polymer nanocomposites reinforced by cellulose whiskers. *Macromolecules* 28: 6365–6367.

Favier, V., G. R. Canova, C. Shrivastavas, and J. Y. Cavaillé. 1997a. Mechanical percolation in cellulose whisker nanocomposites. *Polym. Eng. Sci.* 37: 1732–1739.

Favier, V., R. Dendievel, G. Canova, J. Y. Cavaille, and P. Gilormini. 1997b. Simulation and modeling of three-dimensional percolating structures: Case of a latex matrix reinforced by a network of cellulose fibers. *Acta Mater.* 45: 1557–1565.

Fengel, D. and G. Wegener. 1983. *Wood, Chemistry, Ultrastructure, Reactions.* New York: Walter de Gruyter.

Fleming, K., D. Gray, S. Prasannan, and S. Matthews. 2000. Cellulose crystallites: A new and robust liquid crystalline medium for the measurement of residual dipolar couplings. *J. Am. Chem. Soc.* 122: 5224–5225.

French, A. D., N. R. Bertoniere, R. M. Brown et al. 2004. Cellulose. In *Kirk-Othmer Concise Encyclopedia of Chemical Technology* (5th edn.), ed. A. Seidel, Vol. 5, pp. 360–394. New York: John Wiley & Sons, Inc.

Gaill, F., J. Persson, J. Sugiyama, R. Vuong, and H. Chanzy. 1992. The chitin system in the tubes of deep sea hydrothermal vent worms. *J. Struct. Biol.* 109: 116–128.

Gallant, D. J., B. Bouchet, and P. M. Baldwin. 1997. Microscopy of starch: Evidence of a new level of granule organization. *Carbohydr. Polym.* 32: 177–191.

Garcia de Rodriguez, N. L., W. Thielemans, and A. Dufresne. 2006. Sisal cellulose whiskers reinforced polyvinyl acetate nanocomposites. *Cellulose* 13: 261–270.

Grunnert, M. and W. T. Winter. 2002. Nanocomposites of cellulose acetate butyrate reinforced with cellulose nanocrystals. *J. Polym. Environ.* 10: 27–30.

Habibi, Y. and A. Dufresne. 2008. Highly filled bionanocomposites from functionalized polysaccharide nanocrystals. *Biomacromolecules* 9: 1974–1980.

Habibi, Y., H. Chanzy, and M. R. Vignon. 2006. TEMPO-mediated surface oxidation of cellulose whiskers. *Cellulose* 13: 679–687.

Habibi, Y., L. Foulon, V. Aguié-Béghin, M. Molinari, and R. Douillard. 2007. Langmuir–Blodgett films of cellulose nanocrystals: Preparation and characterization. *J. Colloid Interface Sci.* 316: 388–397.

Habibi, Y., A.-L. Goffin, N. Schiltz, E. Duquesne, P. Dubois, and A. Dufresne. 2008. Bionanocomposites based on poly(ε-caprolactone)-grafted cellulose nanocrystals by ring opening polymerization. *J. Mater. Chem.* 18: 5002–5010.

Hajji, P., J. Y. Cavaille, V. Favier, C. Gauthier, and G. Vigier. 1996. Tensile behavior of nanocomposites from latex and cellulose whiskers. *Polym. Compos.* 17: 612–619.

Halpin, J. C. and J. L. Kardos. 1972. Moduli of crystalline polymers derived from composite theory. *J. Appl. Phys.* 43: 2235.

Hammersley, J. M. 1957. Percolation processes II: The connective constant. *Proc. Cambridge Philos. Soc.* 53: 642–645.

Hanley, S. J., J. Giasson, J. F. Revol, and D. G. Gray. 1992. Atomic force microscopy of cellulose microfibrils—Comparison with transmission electron-microscopy. *Polymer* 33: 4639–4642.

Helbert, W. and J. Sugiyama. 1998. High-resolution electron microscopy on cellulose II and a-chitin single crystals. *Cellulose* 5: 113–122.

Helbert, W., J. Y. Cavaille, and A. Dufresne. 1996. Thermoplastic nanocomposites filled with wheat straw cellulose whiskers. Part I: Processing and mechanical behavior. *Polym. Compos.* 17: 604–611.

Herth, W., A. Kuppel, and E. Schnepf. 1977. Chitinous fibrils in the lorica of the flagellate chrysophyte *Poteriochromonas stipitata* (syn *Ochromonas malhamensis*). *J. Cell. Biol.* 73: 311–321.

Hoover, R. 2001. Composition, molecular structure, and physicochemical properties of tuber and root starches: A review. *Carbohydr. Polym.* 45: 253–267.

Jayakody, L. and R. Hoover. 2002. The effect of lintnerization on cereal starch granules. *Food Res. Int.* 35: 665–680.

Katz, J. R. 1934. X-ray investigation of gelatinization and retrogradation of starch in its importance for bread research. *Bakers Wkly.* 81: 34–37.

Kim, U. J., S. Kuga, M. Wada, T. Okano, and T. Kondo. 2000. Periodate oxidation of crystalline cellulose. *Biomacromolecules* 1: 488–492.

Kristo, E. and C. G. Biliaderis. 2007. Physical properties of starch nanocrystal-reinforced pullulan films. *Carbohydr. Polym.* 68: 146–158.

Kvien, I., B. S. Tanem, and K. Oksman. 2005. Characterization of cellulose whiskers and their nanocomposites by atomic force and electron microscopy. *Biomacromolecules* 6: 3160–3165.

Lintner, C. J. 1886. Diastase. *J. Prak. Chem.* 34: 378–94.

Ljungberg, N., C. Bonini, F. Bortolussi, C. Boisson, L. Heux, and J. Y. Cavaillé. 2005. New nanocomposite materials reinforced with cellulose whiskers in atactic polypropylene: Effect of surface and dispersion characteristics. *Biomacromolecules* 6: 2732–2739.

Ljungberg, N., J.-Y. Cavaillé, and L. Heux. 2006. Nanocomposites of isotactic polypropylene reinforced with rod-like cellulose whiskers. *Polymer* 47 6285–6292.

Lotmar, W. and L. E. R. Picken. 1950. A new crystallographic modification of chitin and its distribution. *Experientia* 6: 58–59.

Lu, Y., L. Weng, and L. Zhang. 2004. Morphology and properties of soy protein isolate thermoplastics reinforced with chitin whiskers. *Biomacromolecules* 5: 1046–1051.

Lu, Y., L. Weng, and X. Cao. 2005. Biocomposites of plasticized starch reinforced with cellulose crystallites from cottonseed linter. *Macromol. Biosci.* 5: 1101–1107.

Marchessault, R. H. and P. R. Sundararajan. 1983. Cellulose. In *The Polysaccharides*, ed. G. O. Aspinall. New York: Academic Press.

Marchessault, R. H., F. F. Morehead, and M. J. Koch. 1961. Hydrodynamic properties of neutral suspensions of cellulose crystallites as related to size and shape. *J. Colloid Sci.* 16: 327–344.

Mathew, A. P. and A. Dufresne. 2002. Morphological investigation of nanocomposites from sorbitol plasticized starch and tunicin whiskers. *Biomacromolecules* 3: 609–617.

Mathew, A. P., A. Chakraborty, K. Oksman, and M. Sain. 2006. The structure and mechanical properties of cellulose nanocomposites prepared by twin screw extrusion. In *ACS Symposium Series: Cellulose Nanocomposites: Processing, Characterization and Properties*, ed. K. Oksman and M. Sain, pp. 114–131. Washington, DC: American Chemical Society.

Morin, A. and A. Dufresne. 2002. Nanocomposites of chitin whiskers from Riftia tubes and poly(caprolactone). *Macromolecules* 35: 2190–2199.

Muhr, A. H., J. M. V. Blanshard, and D. R. Bates. 1984. The effect of lintnerization on wheat and potato starch granules. *Carbohydr. Polym.* 4: 399–425.

Mussarelli, R. A. A. 1977. *Chitin*. New York: Pergamon Press.

Nageli, C. W. 1874. Beitage zur naheren kenntniss der starke group. *Annalen der chemie* 173: 218–227.

Nair, K. G. and A. Dufresne. 2003a. Crab shell chitin whisker reinforced natural rubber nanocomposites. 2. Mechanical behavior. *Biomacromolecules* 4: 666–674.

Nair, K. G. and A. Dufresne. 2003b. Crab shell chitin whisker reinforced natural rubber nanocomposites. 1. Processing and swelling behavior. *Biomacromolecules* 4: 657–665.

Nair, K. G., A. Dufresne, A. Gandini, and M. N. Belgacem. 2003c. Crab shell chitin whiskers reinforced natural rubber nanocomposites. 3. Effect of chemical modification of chitin whiskers. *Biomacromolecules* 4: 1835–1842.

Nge, T. T., N. Hori, A. Takemura, H. Ono, and T. Kimura. 2003. Phase behavior of liquid crystalline chitin/acrylic acid liquid mixture. *Langmuir* 19: 1390–1395.

Orts, W. J., L. Godbout, R. H. Marchessault, and J. F. Revol. 1998. Enhanced ordering of liquid crystalline suspensions of cellulose microfibrils: A small-angle neutron scattering study. *Macromolecules* 31: 5717–5725.

O'Sullivan, A. C. 1997. Cellulose: The structure slowly unravels. *Cellulose* 4: 173–207.

Ouali, N., J.-Y. Cavaillé, and J. Perez. 1991. Elastic, viscoelastic and plastic behavior of multiphase polymer blends. *Plast. Rubber Compos. Process. Appl.* 16: 55.

Paillet, M. and A. Dufresne. 2001. Chitin whisker reinforced thermoplastic nanocomposites. *Macromolecules* 34: 6527–6530.

Parker, R. and S. G. Ring. 2001. Aspects of the physical chemistry of starch. *J. Cereal Sci.* 34: 1–17.

Payen, A. 1838. Mémoire sur la composition du tissu propre des plantes et du ligneux. *CR Hebd. Seances Acad. Sci.* 7: 1052–1056.

Persson, J. E., A. Domard, and H. Chanzy. 1992. Single crystals of a-chitin. *Int. J. Biol. Macromol.* 14: 221–224.

Podsiadlo, P., S.-Y. Choi, B. Shim, J. Lee, M. Cuddihy, and N. A. Kotov. 2005. Molecularly engineered nanocomposites: Layer-by-layer assembly of cellulose nanocrystals. *Biomacromolecules* 6: 2914–2918.

Putaux, J.-L. 2005. Morphology and structure of crystalline polysaccharides: Some recent studies. *Macromol. Symp.* 229: 66–71.

Putaux, J.-L., S. Molina-Boisseau, T. Momaur, and A. Dufresne. 2003. Platelet nanocrystals resulting from the disruption of waxy maize starch granules by acid hydrolysis. *Biomacromolecules* 4: 1198–1202.

Revol, J. F. 1982. On the cross-sectional shape of cellulose crystallites in *Valonia ventricosa. Carbohydr. Polym.* 2: 123–134.

Revol, J. F. and R. H. Marchessault. 1993. In vitro chiral nematic ordering of chitin crystallites. *Int. J. Biol. Macromol.* 15: 329–335.

Revol, J. F., H. Bradford, J. Giasson, R. H. Marchessault, and D. G. Gray. 1992. Helicoidal self-ordering of cellulose microfibrils in aqueous suspension. *Int. J. Biol. Macromol.* 14: 170–172.

Revol, J. F., L. Godbout, X. M. Dong, D. G. Gray, H. Chanzy, and G. Maret. 1994. Chiral nematic suspensions of cellulose crystallites; phase separation and magnetic field orientation. *Liq. Cryst.* 16: 127–134.

Robin, J. P. (1976). *Comportement du grain d'amidon à l'hydrolyse acide ménagée.* Paris, France: Universite Pierre et marie Curie.

Robin, J. P., C. Mercier, R. Charbonniere, and A. Guilbot. 1974. Lintnerized starches. Gel filtration and enzymic studies of insoluble residues from prolonged acid treatment of potato starch. *Cereal Chem.* 51: 389–406.

Robin, J. P., C. Mercier, F. Duprat, R. Charbonniere, and A. Guilbot. 1975. Lintnerized starches. Chromatographic and enzymic studies of insoluble residues from hydrochloric acid hydrolysis of cereal starches, particularly waxy maize [starch]. *Staerke* 27: 36–45.

Roman, M. and W. T. Winter. 2004. Effect of sulfate groups from sulfuric acid hydrolysis on the thermal degradation behavior of bacterial cellulose. *Biomacromolecules* 5: 1671–1677.

Roohani, M., Y. Habibi, N. M. Belgacem, G. Ebrahim, A. N. Karimi, and A. Dufresne. 2008. Cellulose whiskers reinforced polyvinyl alcohol copolymers nanocomposites. *Eur. Polym. J.* 44: 2489–2498.

Rudall, K. M. and W. Kenchington. 1973. The chitin system. *Biol. Rev.* 48: 597–633.

Sakamoto, J., J. Sugiyama, S. Kimura et al. 2000. Artificial chitin spherulites composed of single crystalline ribbons of a-chitin via enzymatic polymerization. *Macromolecules* 33: 4155–4160.

Salmon, S. and S. M. Hudson. 1997. Crystal morphology, biosynthesis, and physical assembly of cellulose, chitin, and chitosan. *Polym. Rev.* 37: 199–276.

Sarko, A. 1987. Cellulose-how much do we know about its structure? In *Wood and Cellulosics: Industrial Utilisation, Biotechnology, Structure, and Properties*, ed. J. F. Kennedy. Chichester, U.K.: Ellis Horwood.

Singh, D. K. and A. R. Ray. 2000. Biomedical applications of chitin, chitosan, and their derivatives. *J. Macromol. Sci., Rev. Macromol. Chem. Phys.* C40: 69–83.

Sjöström, E. 1981. *Wood Chemistry: Fundamentals and Applications.* New York: Academic Press.

Sriupayo, J., P. Supaphol, J. Blackwell, and R. Rujiravanit. 2005a. Preparation and characterization of alpha -chitin whisker-reinforced poly(vinyl alcohol) nanocomposite films with or without heat treatment. *Polymer* 46: 5637–5644.

Sriupayo, J., P. Supaphol, J. Blackwell, and R. Rujiravanit. 2005b. Preparation and characterization of α-chitin whisker-reinforced chitosan nanocomposite films with or without heat treatment. *Carbohydr. Polym.* 62: 130–136.

Stauffer, D. and A. Aharony. 1992. *Introduction to Percolation Theory.* London, U.K.: Taylor & Francis.

Sturcova, A., G. R. Davies, and S. J. Eichhorn. 2005. Elastic modulus and stress-transfer properties of tunicate cellulose whiskers. *Biomacromolecules* 6: 1055–1061.

Sugiyama, J., R. Vuong, and H. Chanzy. 1991. Electron diffraction study on the two crystalline phases occurring in native cellulose from an algal cell wall. *Macromolecules* 24: 4168–4175.

Takayanagi, M., S. Uemura, and S. Minami. 1964. Application of equivalent model method to dynamic rheo-optical properties of a crystalline polymer. *J. Polym. Sci.* 5: 113–122.

Tang, H., T. Mitsunaga, and Y. Kawamura. 2006. Molecular arrangement in blocklets and starch granule architecture. *Carbohydr. Polym.* 63: 555–560.

Tashiro, K. and M. Kobayashi. 1991. Theoretical evaluation of three-dimensional elastic constants of native and regenerated celluloses: Role of hydrogen bonds. *Polymer* 32: 1516–1526.

Terech, P., L. Chazeau, and J. Y. Cavaille. 1999. A small-angle scattering study of cellulose whiskers in aqueous suspensions. *Macromolecules* 32: 1872–1875.

Thielemans, W., M. N. Belgacem, and A. Dufresne. 2006. Starch nanocrystals with large chain surface modifications. *Langmuir* 22: 4804–4810.

Tokoh, C., K. Takabe, M. Fujita, and H. Saiki. 1998. Cellulose synthesized by acetobacter xylinum in the presence of acetyl glucomannan. *Cellulose* 5: 249–261.

van den Berg, O., J. R. Capadona, and C. Weder. 2007. Preparation of homogeneous dispersions of tunicate cellulose whiskers in organic solvents. *Biomacromolecules* 8: 1353–1357.

Vincent, J. F. V. and U. G. K. Wegst. 2004. Design and mechanical properties of insect cuticle. *Arthropod Struct. Dev.* 33: 187–199.

Woodcock, C. and A. Sarko. 1980. Packing analysis of carbohydrates and polysaccharides. 11. Molecular and crystal structure of native ramie cellulose. *Macromolecules* 13: 1183–1187.

Yamaguchi, M., K. Kainuma, and D. French. 1979. Electron microscopic observations of waxy maize starch. *J. Ultrastruct. Res.* 69: 249–261.

Zimmermann, T., E. Pöhler, and P. Schwaller. 2005. Mechanical and morphological properties of cellulose fibril reinforced nanocomposites. *Adv. Eng. Mater.* 7: 1156–1161.

Zobel, H. F. 1988. Starch crystal transformations and their industrial importance. *Starch/Staerke* 40: 1–7.

II

Nanoparticle Properties

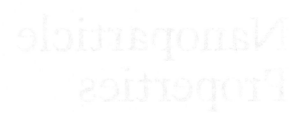

II

Nanoparticle Properties

11

Acoustic Vibrations in Nanoparticles

Lucien Saviot
Université de Bourgogne

Alain Mermet
Université Claude Bernard Lyon I

Eugène Duval
Université Claude Bernard Lyon I

11.1 Introduction (Broad Overview)

The purpose of this chapter is to present current experimental observations of vibrations of nanoparticles and theoretical models to describe them. Only so-called acoustic vibrations will be considered, i.e., those that are more strongly affected by reducing the size of a solid to nanometric dimensions. A good knowledge of these vibrations is required to describe various properties of nanoparticles such as their specific heat but also their optical properties where the coupling between electrons and vibrations can play a significant role. For example, vibrations are an important player in the dephasing mechanism of charged carriers, which significantly affects the performance of optoelectronic devices. The same vibrations can be used as a way to characterize nanometer-scale objects. This is the nanoscale equivalent of hitting an object in everyday life and listening to the sound it makes in order to figure out what it is made of. For example, it is very easy to recognize whether a bottle is full or empty in this way, or even to know what a wall is made of depending on whether it sounds hollow or not. The very same result could be obtained for a metallic core–shell nanoparticle where the vibrations were found to "sound hollow" when "hitting" the shell leading to an original characterization of the interface between the core and the shell (Portalès et al. 2002). Therefore, these vibrations can be used to characterize nanoparticles but their knowledge is also required to design efficient devices.

Raman scattering, a spectroscopic technique whereby light is inelastically scattered by atomic vibrations, has been the main tool to study such vibrations over the years. Recently, it has started to reveal its full potential with experiments on very high quality samples. This is due to the fact that vibrations are very sensitive to the exact microscopic structure. As the detection of light inelastically scattered by a single nanoparticle is not yet possible for very small nanoparticles, more monodisperse systems are needed. Having systems for which the shape and size of the nanoparticles are controlled as well as their crystallinity and environment enables the observation of experimental features that would otherwise be hidden by inhomogeneous broadening.

11.2 Background (History and Definitions)

11.2.1 Available Experimental Techniques

In order to investigate acoustic vibrations of nanoparticles, it is natural to turn to vibrational spectroscopies. Indeed, the first experimental technique used to observe the acoustic vibrations of nanoparticles was inelastic light scattering (Raman scattering) (Weitz et al. 1980b; Duval et al. 1986). Since then, experimental evidences have been obtained using a variety of different techniques. Other usual vibrational spectroscopies such as infrared absorption (Murray et al. 2006; Liu et al. 2008) or inelastic neutron scattering (Saviot et al. 2008) have had limited success. On the other hand, other optical techniques such as photoluminescence, hole-burning (Zhao and Masumoto 1999; Palinginis et al. 2003), and femtosecond pump-probe experiments (Del Fatti et al. 1999; Bragas et al. 2006) have provided very valuable results. This is due to the possibility of having resonant excitations with optical techniques. Tuning the incident photon energy to match an electronic transition in the nanoparticle (plasmon, exciton, or electron–hole pair) results in an enhanced Raman intensity from the nanoparticle and a reduced signal from the environment. This is a very important aspect for matrix-embedded nanoparticles, for example, because the volume concentration of the nanoparticles is usually very small. It also provides a valuable way to study the electronic states of nanoparticles through their coupling with acoustic vibrations.

Most of the techniques mentioned above are commonly used to investigate optical phonons. Because the frequencies of acoustic phonons are much lower than for optical phonons, some existing experimental setups are not suitable to observe acoustic vibrations confined in nanoparticles. This is the case of Raman spectrometers using a notch filter to remove the elastically scattered photons as it prevents the detection for wavenumbers below approximately ±100 cm⁻¹ (depending on the filter), where the nanoparticles' vibrations typically show up. On the opposite, interferometric setups such as the ones used for Brillouin spectroscopy are powerful tools to study acoustic vibrations of nanoparticles as they were designed for this frequency range. It should be noted that Brillouin and Raman spectroscopy differ only in the experimental setup design. They both measure inelastically scattered photons but generally vary in the way scattered photons are analyzed experimentally. Brillouin measurements use interferometry while Raman measurements use dispersive optics.

While a major part of the published results rely on low-frequency Raman measurements, time-resolved optical investigations through femtosecond pump-probe experiments have played a significant part during the last 10 years for metallic nanoparticles. The selective heating of the nanoparticle after absorption of photons from the intense pump laser beam is faster than the oscillation period of the nanoparticles. As a result, to accommodate this out-of-equilibrium situation, the nanoparticle starts oscillating in order to increase its volume. These volume changes are monitored using a less intense probe laser beam, which allows the determination of the optical absorption changes. Unlike inelastic light scattering, it is easier to study larger nanoparticles because their oscillation periods are larger. It is also possible to study colloids (Martini and Hartland 1998), which is not the case with Raman spectroscopy due to the intense inelastic scattering by liquids in the low-frequency region.

11.2.2 Models

In order to interpret different experimental data, models to describe the vibrations are needed. In this part, we first detail the most used model which provides analytic displacements corresponding to the different vibration modes. More advanced models that are needed for more accurate predictions are then briefly introduced.

11.2.2.1 Lamb's Model

As first unveiled by experiment (Duval et al. 1986), the vibrational modes of nanoparticles are well described at first order by the eigenmodes of free elastic nanospheres. Horace Lamb was the first to mathematically describe the eigenmodes of a free homogeneous elastic sphere, regardless of its size (Lamb 1882). In fact, back in 1882, he illustrated his theoretical developments with the vibration modes of a centimeter steel ball and those of the Earth. This model is based on the simplest assumptions: the particle is described as a continuous sphere in the frame of the elasticity theory and it is made of an isotropic material whose longitudinal and transverse sound speeds are v_L and v_T, respectively. Within the continuous elastic medium approximation, these speeds do not

depend on the propagation direction because of isotropy and in this case the elastic wave equation is given by Equation 11.1 where $\vec{u}(\vec{r}, t)$ is the displacement at time t of the point located at $\vec{r}$.

$$v_L^2 \vec{\nabla} \cdot (\vec{\nabla} \cdot \vec{u}) - v_T^2 \vec{\nabla} \times (\vec{\nabla} \times \vec{u}) = \ddot{\vec{u}} \qquad (11.1)$$

A complete derivation of the solutions of this system can be found elsewhere (see, for example, Eringen and Suhubi 1975) and yields the following result:

$$\vec{u} = \vec{\nabla}\phi + \vec{\nabla} \times (\psi \vec{r}) + \vec{\nabla} \times \vec{\nabla} \times (\zeta \vec{r}) \qquad (11.2)$$

where

$$\begin{cases} \phi(r,t) = A & j_\ell(qr) Y_\ell^m(\theta,\varphi) \exp(-i\omega t) \\ \psi(r,t) = B & j_\ell(Qr) Y_\ell^m(\theta,\varphi) \exp(-i\omega t) \\ \zeta(r,t) = C & j_\ell(Qr) Y_\ell^m(\theta,\varphi) \exp(-i\omega t) \end{cases} \qquad (11.3)$$

j_ℓ are the spherical Bessel functions
Y_ℓ^m the spherical harmonics
ℓ is an integer, $-\ell \le m \le \ell$
$Q = \omega/v_T$
$q = \omega/v_L$

For spherically symmetric systems, the solutions can be separated into spheroidal eigenmodes ($B = 0$) and torsional eigenmodes ($A = 0$ and $C = 0$). Both these types of modes will be labeled with the integer ℓ. Boundary conditions are needed in order to finish the resolution and actually calculate the eigenfrequencies.

In our simple case, we assume that the surface of the nanoparticle is free, i.e., that there is no force applied at the surface of the sphere of radius R. This condition results in a linear system with three equations and three unknowns (A, B, and C) to be solved. Nonzero solutions for this system exist only if the determinant is zero. The eigenfrequencies can then be obtained by searching the roots of the following equations where the unknown is either qR or QR and using the relation $(qR)v_L = (QR)v_T$ when both are present.

$$\frac{\tan qR}{qR} = \frac{1}{1 - (v_\ell^2/4v_t^2)q^2R^2} \qquad (11.4)$$

$$-\frac{Q^2R^2}{2}\left(2l^2 - l - 1 - \frac{Q^2R^2}{2}\right) j_l(qR) j_l(QR)$$

$$+ (l^3 + 2l^2 - Q^2R^2)qRj_{l+1}(qR) j_l(QR)$$

$$+ \left(l^3 + l^2 - 2l - \frac{Q^2R^2}{2}\right) QRj_l(qR) j_{l+1}(qR) \qquad (11.5)$$

$$+ (2 - l^2 - l)qRQRj_{l+1}(qR) j_{l+1}(qR) = 0$$

For the spheroidal eigenmodes, the solutions are the roots of Equations 11.4 and 11.5 for $\ell = 0$ and $\ell > 0$, respectively. For $\ell > 0$, these modes have both a radial and a non-radial displacement and the displacements are due to two different contributions ($A \neq 0$ and $C \neq 0$). It is worth noting that only the $\ell = 0$ modes induce a volume change of the nanoparticle during its oscillation (breathing mode).

For the torsional modes, the roots of Equation 11.6 have to be considered for $\ell \neq 0$ (there exists no torsional mode with $\ell = 0$). These modes have a non-radial displacement only and the volume of the sphere does not change during the oscillation.

$$(\ell - 1)j_l(QR) - QRj_{l+1}(QR) = 0 \qquad (11.6)$$

Within the frame of continuum elasticity, the *eigenfrequencies are proportional to the inverse of the dimension of the particle* even for nonspherical particles and anisotropic continuous media. This rule is analogous to the well-known one governing the frequencies of vibrations of a one-dimensional string. For a homogeneous and continuous sphere, it is convenient to write the eigenfrequencies as $v = S\, v/D$ where S is obtained by solving the previous equations, D is the diameter of the sphere, and v is the transverse sound speed (or the longitudinal one for the spheroidal $\ell = 0$ modes). As an example, we give two typical relations that may be used *in a first approximation* for the frequencies of the $\ell = 0$ and the $\ell = 2$ fundamental modes, in the case of free nanoparticles:

$$v_{\ell=0}^{n=1} = 0.9\,\frac{v_L}{D} \qquad v_{\ell=2}^{n=1} = 0.84\,\frac{v_T}{D} \qquad (11.7)$$

11.2.2.2 Group Theory and Raman Selection Rules

The sole use of frequency as identification criterion for the interpretation of nanoparticles vibration modes might be error-prone since several vibrations can exist in a given narrow frequency range. The knowledge of the symmetry of the modes is therefore required in order to derive the selection rules that restrict the number of modes observable by a given technique. For example, to interpret Raman spectra, point group theory can be used to classify the vibrations of objects much smaller than the wavelength of light into Raman active and inactive ones.

The symmetry of the particle results in a number of irreducible representations, which characterize the eigenmodes of vibrations. Only some of these representations correspond to Raman active modes. The procedure to follow is well known and commonly applied to molecules. However, its application to spherical isotropic nanoparticles is a bit special since such particles are invariant under every rotation whose axis goes through the center of the particle and no molecule having such a high symmetry can exist. The symmetry group and the associated irreducible representations and selection rules have been derived by Eugène Duval (1992) for nanoparticles whose diameter is small compared to the wavelength of the incident photons. This condition typically applies to *Raman* scattering, as it is conventionally conceived, i.e., as a molecular spectroscopy. Recently, these selection rules have been extended to nanoparticles with much larger dimensions (Montagna 2008) which case is more relevant of *Brillouin* scattering; in the following we will essentially consider selection rules that pertain to small nanoparticles, i.e., to Raman scattering.

The point group associated to an isotropic spherical nanoparticle is the group of the proper and improper rotations ($O(3)$). The irreducible representations are noted as $D_g^{(\ell)}$ and $D_u^{(\ell)}$. The irreducible representation corresponding to a spheroidal vibration is $D_g^{(\ell)}$ for even ℓ and $D_u^{(\ell)}$ for odd ℓ. The irreducible representation corresponding to a torsional vibration is $D_g^{(\ell)}$ for odd ℓ and $D_u^{(\ell)}$ for even ℓ. For a nanoparticle whose diameter is small compared to the wavelength of light, the Raman active vibrations have the $D_g^{(0)}$ or $D_g^{(2)}$ irreducible representation due to the symmetry of the polarizability tensor. Therefore, *Raman active modes are spheroidal modes with either $\ell = 0$ or $\ell = 2$*. Polarization rules enable to distinguish between these two families as only the scattering by $\ell = 0$ modes is polarized (i.e., not observable when the polarization of the incident photons and that of the detector are perpendicular). Because these $\ell = 0$ modes have a *longitudinal* character (radial motions only), this polarization rule is equivalent to the polarized scattering from longitudinal acoustic modes in classical Brillouin scattering.

Some vibrations are represented in Figure 11.1. They show that the value of ℓ is closely related to the number of extrema for $m = 0$. The vibrations for $\ell = 1$ are more complex since they correspond to harmonics of the rotations (torsional) and translations (spheroidal) which have zero frequency for a free sphere. This is the reason why the displacement close to the center of the sphere and the displacement at the surface are out of phase in these cases.

Because the observation of the spheroidal modes with $\ell = 0$ or $\ell = 2$, respectively, called "breathing mode" (or radial mode) and "quadrupolar mode," has been reported for a variety of experimental techniques, the corresponding displacements of the fundamental modes is detailed here and is represented in Figure 11.1. The $\ell = 0$ modes are the simplest ones as their associated displacement is purely radial. In the fundamental mode of this type of oscillation, all constituting points of the sphere

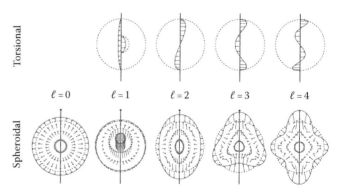

FIGURE 11.1 Displacements for the fundamental vibrations with $\ell \leq 4$ and $m = 0$. The z axis is shown by a long vertical arrow. The equilibrium surface of the sphere is represented by a dashed circle. For the spheroidal vibrations, the deformed shape of the sphere is represented by a continuous line and the vector field represents the displacements of the points in the $x = 0$ plane. For the torsional vibrations, the displacement of the meridian at $y = 0$ is shown. For the torsional $\ell = 1$ plot, the displacements of an inner meridian are also shown. For all vibrations, the displacements for other points of the sphere not shown in this figure are obtained by rotation around the z axis.

simultaneously move in and out from the center. The surface of the sphere moves as if the particle was alternately inflating and deflating. The external shape is always a sphere with an oscillating radius. Regarding the $\ell = 2$ modes, each of them consists of five degenerate vibrations ($m = \pm 2, \pm 1, 0$), i.e., five different displacements which occur at the same frequency. These five displacements correspond to a stretching along one or two direction(s) and a shrinking along one or two perpendicular direction(s). For example, the $\ell = 2$, $m = 0$ mode corresponds to a stretching along one direction (z) with a simultaneous shrinking in the perpendicular plane (x, y) over half a period of vibration. Over the second half, the sphere shrinks along z and expands in the (x, y) plane. It is followed by a shrinking along z and a stretching in the (x, y) plane. The amplitude along z is twice that in the (x, y) plane.

11.2.2.3 Illustrations

Typical low-frequency Raman scattering spectra obtained for anatase TiO_2 nanopowders prepared by continuous hydrothermal synthesis (Pighini et al. 2007) are displayed in Figure 11.2. Such spectra can be recorded in a few minutes using a Raman setup with a microscope and a multichannel detector. The intense elastic scattering at vanishing Raman shift is not shown because it cannot generally be recorded without damaging the detector. As explained before, notch filters and similar devices used to suppress this elastic part cannot be used as they also remove the low-frequency part of the Raman spectrum.

The Raman shifts are commonly expressed as wavenumbers in units of cm^{-1}. One cm^{-1} corresponds to a frequency of 30 GHz. The spectra clearly demonstrate the following points:

- Raman peaks exist in the low-frequency range. Such peaks do not exist for the bulk material and can therefore be safely attributed to confined acoustic vibrations.
- The frequency of the peaks shifts toward larger frequencies when the average diameter decreases.

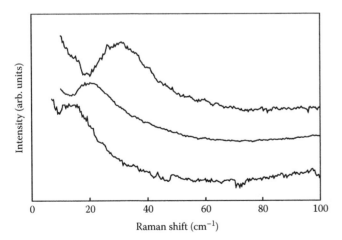

FIGURE 11.2 Low-frequency Raman scattering spectra of anatase TiO_2 nanopowders. The average diameter of the nanoparticles as determined from the broadening of the x-ray diffraction peaks is 3.4, 5.0, and 5.7 nm ± 1 nm from top to bottom.

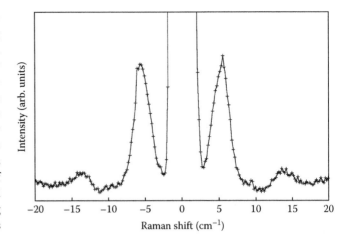

FIGURE 11.3 Low-frequency Raman scattering spectra of gold nanoparticles embedded in a glass. The Stokes and anti-Stokes parts (positive and negative Raman shifts, respectively) are shown and correspond to inelastic light scattering involving the annihilation and creation of a vibration, respectively.

Because these Raman peaks can be observed even when the polarizations of the incident and detected photons are crossed, they are attributed to spheroidal vibrations with $\ell = 2$ in agreement with the Raman selection rules. Using the elastic parameters for bulk anatase TiO_2 from Iuga et al. (2007), the wavenumber of the fundamental spheroidal $\ell = 2$ modes is approximately $\omega(cm^{-1}) \simeq 110/d(nm)$. The average diameters calculated using the different spectra and this formula are in remarkable agreement with the ones determined by x-ray diffraction and high-resolution transmission electron microscopy.

As another example of the detection of nanoparticles' vibration modes, Figure 11.3 displays the low-frequency Raman spectrum recorded from a ruby shade bulky glass containing Au nanoparticles, using a quintuple monochromator (single channel detection). As discussed later (see Section 11.3.2.1), low-frequency Raman scattering by noble metal nanoparticles is very intense due to the resonant visible excitation. The typical spectrum of Figure 11.3 shows essentially two peaks. The lower frequency one, which is also the most intense one due to efficient coupling with plasmonic excitations, is assigned to the fundamental of the spheroidal $\ell = 2$ mode (quadrupolar mode). The higher frequency and much less intense peak arises from the fundamental of the spheroidal $\ell = 0$ mode (breathing mode). This identification follows from the observation that when performing the Raman experiment with crossed light polarizations, the higher frequency peak vanishes. The continuous rise of intensity observed at the increasing frequency ends of the spectrum comes from the Raman scattering of the embedding medium, i.e., the glass. The large intensity of the quadrupolar mode is a definite asset in the characterization of metallic nanoparticles when buried in an embedding medium. From its frequency position, one derives that the Au nanoparticles have an average diameter of 6.8 nm. It is worth noting that the frequency ratio of the two peaks slightly differs from the free sphere predictions (Equation 11.7) while it conforms to that expected taking into account the effect the matrix (see Section 11.3.1.3) (Stephanidis et al. 2007b).

The short preparation and acquisition times as well as the reliability of the determination of size by low-frequency Raman scattering make it a powerful characterization technique. As will be demonstrated in the following, more than just the size can be investigated by this mean.

11.2.2.4 Numerical Methods for Systems Lacking Spherical Symmetry

While the model proposed by Lamb has been successfully used to interpret a variety of experimental data, recent works have focused on systems for which the isotropic or spherical assumptions are not valid. For instance, this can be the case of nanocrystals whose atomic crystalline structures imply different sound speeds along different crystallographic directions. To solve such cases, one has to use a numerical method. It goes beyond the scope of this chapter to examine in details the various available options so only one such method will be used. It was originally conceived to predict the frequencies in resonant ultrasound experiments (Visscher et al. 1991). This approach still relies on continuum elasticity. The shape and elastic tensor for a given object are defined. Then the displacements of the eigenmodes are expanded on a $x^i y^j z^k$ basis. For free boundary conditions, the eigensolution problem is turned into a real generalized symmetric-definite eigenproblem through the use of Hamilton's principle. Such a problem is efficiently solved on any modern computer.

The models presented before all rely on the elasticity theory for continuous media. Of course, using continuum elasticity is bound to failure for small enough systems made of a very small number of atoms. In that case, atomistic models are required. However, such calculations are more complex to handle. It is therefore interesting to know the smallest system for which a continuous descriptions is still accurate enough so that atomistic calculations are not required. A simple way to answer this question is to compare the wavelength of the acoustic waves involved in the continuous description and to compare it to the lattice parameter of the material the nanoparticle is made of. Continuous models are expected to be reliable when the wavelength is much larger than the interatomic distance. Depending on the definition which is chosen for "much larger," this provides for example a minimum radius for spherical nanoparticles. A different picture is obtained when focusing on the number of surface atoms. Indeed, the continuous description works well for bulk atoms only. Therefore another limit is obtained by considering that the number of bulk atoms should be much larger then the number of surface atoms. It should be noted that this limit depends on the material the nanoparticle is made of and the environment of the nanoparticle. A less arbitrary value can be obtained with atomistic calculations. Such results for silicon and germanium nanoparticles (Cheng et al. 2005a,b; Combe et al. 2007; Ramirez et al. 2008) indicate that the lowest eigenfrequencies are in good agreement for atomistic and continuous models down to diameters as small as 3–4 nm. Moreover, projecting the discrete displacements onto the continuous one also reveals a good agreement for the wavefunctions obtained with both approaches.

Bearing in mind the good agreement between atomistic and continuous medium approaches, the following developments will only refer to the continuum approach as it offers the advantage of being easily applicable to larger systems.

11.3 Presentation of State of the Art

11.3.1 Narrow Particle Distributions (Ideally Single Particle Measurements)

11.3.1.1 Influence of Shape

Although the study of nanoparticle vibration modes has long focused on spherical shapes or at least on *assumed* spherical shapes, this case remains an ideal one. In real life, the shape of the nanoparticles produced with either bottom-up or top-down approaches is hardly ever a perfect sphere. While it is safe to forget about minor deviations from this shape, some preparation techniques provide samples with controlled non-spherical shapes. The simplest deviation from the sphere consists in changing the size in just one direction, i.e., having a spheroid. Only spheroids made of an elastically isotropic material are considered here.

11.3.1.1.1 Numerical Modeling

As for spherical systems, two complementary points of view will be considered: the calculation of the eigenfrequencies and point group symmetry. Regarding symmetry, the point group associated to a spheroid is $D_{\infty h}$ (same group as the dihydrogen molecule as both are invariant under the same symmetry operations) and the Raman active vibrations correspond to the A_{1g}, E_{1g}, and E_{2g} irreducible representations. To illustrate the effect of a spheroidal transformation, we will focus on the case of the $\ell = 2$ mode of a silver nanosphere, which happens to be by far the most easily detected one through Raman scattering (see Section 11.3.2.1).

For a silver sphere, the main Raman peak comes from the spheroidal mode with $\ell = 2$. It is therefore interesting to follow this mode as the shape is changed. Group theory considerations allow to predict how the different vibration eigenmodes of a sphere transform under a spheroidal deformation. The spheroidal $\ell = 2$ mode degeneracy is lifted into the three Raman active modes: A_{1g}, E_{1g}, and E_{2g} whose degeneracy is 1, 2, and 2, respectively. Except for the spheroidal modes with $\ell = 0$ which transform into A_{1g} modes, all the modes degeneracy is partly lifted. For spheroids with R_z/R significantly different from 1 (R being the sphere radius and R_z the semi-axis of the spheroid along the deformed direction), modes having the same irreducible representation can mix and it is therefore more difficult to relate them with a unique sphere eigenmode.

Figure 11.4 presents the variation of all the lowest Raman active vibrations as a function of the dimension of the silver spheroid. These frequencies were calculated using the method presented in Section 11.2.2.4 and the irreducible representations corresponding to each mode were computed from the displacements. As explained before, we are mainly interested

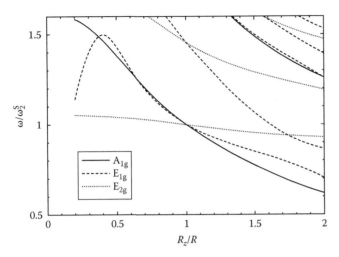

FIGURE 11.4 Reduced frequency of Raman active vibration eigenmodes of isotropic silver spheroids as a function of their aspect ratio. R is the radius of the reference sphere and the length of the degenerate semi-axis of the spheroid. R_z is the length of the nondegenerate semi-axis. ω_2^S is the frequency of the spheroidal $\ell = 2$ vibration of the sphere of radius R.

in the changes in the spheroidal mode with $\ell = 2$ of the sphere, i.e., in the branches going through the point at $R_z/R = 1$ and $\omega/\omega_2^S = 1$. The different points predicted by group theory can be observed:

- The three branches going through the $R_z = R$ and $\omega = \omega_2^S$ point have the symmetry predicted for the spheroidal mode with $\ell = 2$ after its degeneracy is lifted by the spheroidal deformation.
- Branches having the same irreducible representation can mix as can be seen from the anti-crossing pattern between the two lowest E_{1g} modes close to $R_z/R = 1.7$.

The branches that do not go through the point at $R_z = R$ and $\omega = \omega_2^S$ originates from other vibrational modes of the sphere and their Raman scattering cross section is expected to be very small.

11.3.1.1.2 Frequency Approximation in the Case of Small Eccentricities

To get a better understanding of what the different branches correspond to in terms of particle shape oscillatory deformations, we will focus on spheroids which are almost spheres, i.e., which have small eccentricities.

Looking at the displacements corresponding to these modes for small deviations from the sphere, it is possible to understand the frequency variations. The lowest A_{1g} mode corresponds to a stretching along the z direction accompanied by a shrinking in the xy plane. Therefore it can be seen as a vibration confined along the z direction and its frequency varies roughly as $1/R_z$. The E_{2g} vibrations correspond to a stretching in the xy plane without changes along the z axis and therefore their frequencies hardly changes with R_z. The E_{1g} vibrations correspond to a stretching in the xz and yz plane without changes along the z axis and therefore their frequencies vary with a slower eccentricity dependence

than the previous mode. These rough approximations are in agreement with the dependence observed in Figure 11.4.

Using perturbation theory (Mariotto et al. 1988), it is possible to obtain more accurate expressions for the frequencies of these three branches. The exact variations for $|R_z - R| \ll R$ are $\Omega(1 + 4\beta/21)$, $\Omega(1 - 2\beta/21)$, and $\Omega(1 - 4\beta/21)$ for the E_{1g}, E_{2g}, and A_{1g} modes, respectively, where $\beta = 2(R_z - R)/(R_z + R)$ and Ω is the frequency of the spheroidal mode with $\ell = 2$ for a spherical particle having the same volume. Note that the length of the degenerate semi-axis is constant in Figure 11.4 and therefore the volume varies linearly with R_z. Using $\Omega = \omega_2^S \sqrt[3]{R/R_z}$ results in expressions which are in very good agreement with Figure 11.4 close to $R_z = R$.

11.3.1.2 Influence of the Inner Structure of the Nanoparticles

11.3.1.2.1 Core–Shell Systems

Semiconductor nanoparticles are sometimes called quantum dots to emphasize the possibility of tailoring their optical properties only by changing their size. A complementary way to achieve this result is to have a more complex structure, for example using core–shell system where a spherical nanoparticle consists of a spherical core made of one material and a spherical shell made of a different material. The vibrations of such systems have already been investigated experimentally and thus require a specific theoretical modeling, as for simple spheres. Extending the model by Lamb (see Section 11.2.2.1) to all core–shell and even onion-like systems made of several shells is straightforward (Portalès et al. 2002). It basically comes down to using both Bessel and Hankel spherical functions inside the shells (Hankel functions diverge at the origin and are not used for the core) and writing additional boundary conditions at both inner and outer interfaces. Calculation details can be found elsewhere (Portalès et al. 2002) and we will focus on the underlying physics here. To do so, we will consider a particularly interesting core–shell configuration where the elastic properties of the materials in the core and in the shell are significantly different. This is the case for instance of 5 nm radius silica core covered with a 1 nm thick gold shell. Once again, we focus on the Raman active $\ell = 2$ mode of the spheroidal vibrations. Figure 11.5 presents the frequencies of these vibrations together with the ones of a Au shell having the same dimension (i.e., silica core removed) and those of a plain silica sphere. Because gold's density is much greater than that of silica, these frequencies were calculated for a silica sphere whose surface cannot move and a gold shell whose surfaces are free to move. This figure demonstrates that the vibrations of this core–shell system are similar to the vibrations of the core and the shell alone. The main result is that the frequency spectrum becomes denser in a core–shell system. Typically, the eigenfrequencies of both isolated components (core and shell separately) are found back in the core–shell system yet with a different value. This difference can be mild to significant as the frequencies of both systems become close. In that case, there can be strong mixing of both modes and approximating the core–shell system as a separated core and shell is not appropriate anymore.

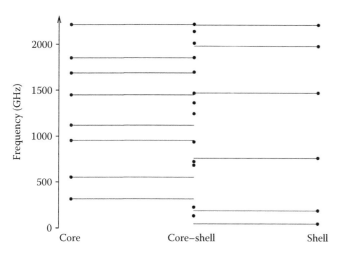

FIGURE 11.5 Frequencies of the first few spheroidal $\ell = 2$ eigenmodes of a core–shell nanoparticle made of a 5 nm radius silica core and a 1 nm thick gold shell. The frequencies of the eigenmodes of the core with a blocked surface and the ones of the shell with both surfaces free are also plotted for comparison.

11.3.1.2.2 Influence of Crystallinity

Besides the spherical shape and homogeneous approximations discussed above, another restrictive approximation is the isotropy of the nanoparticles. It is well known that crystals and therefore also *nano*crystals are anisotropic elastic materials because the strain resulting from a given stress depends on the direction of that stress. For example, in a material having a cubic crystalline structure, stresses applied along the [100] and [110] direction result in different strains. This results in different sound velocities along different crystallographic directions so that the use of orientationally averaged sound velocities becomes too crude of an approximation. In other words, it becomes important to take into account all the parameters of the elastic tensor C_{ij}. This refinement of nanoparticles' vibrations as they become single nanocrystals was recently brought about due to recent advances in the possibility of controlling the crystallinity of synthesized nanoparticles (Hartland 2007; Tang and Ouyang 2007; Portalès et al. 2008).

11.3.1.2.2.1 Elastic Anisotropy
The numerical approach presented before (see Section 11.2.2.4) can be used to take into account elastic anisotropy. As before, point group theory provides a valuable point of view to understand how the well-known isotropic solutions transform. The elastic tensor symmetry is closely related to the crystalline structure of the material. In the following, we will focus on the case of cubic materials, namely silver and gold.

Let's consider a single-domain spherical nanoparticle made of gold. Single domain means that the lattice is perfect inside the gold sphere, i.e., no crystal twinning exists. In that case, the elastic tensor is identical everywhere inside the sphere which is of course the simplest case to take into account. The point group associated with the elastic tensor of a cubic structure is O_h (same symmetry operations as the cube or the SF_6 molecule). Because we consider a particle whose shape is a sphere, there is no further lowering

TABLE 11.1 Irreducible Representations of the Isotropic Eigenvibrations of a Sphere Whose Degeneracy Is Lifted by a Cubic Elastic Anisotropy

$O(3)$	O_h
Spheroidal, $\ell = 0$	A_{1g}
Spheroidal, $\ell = 1$	T_{1u}
Spheroidal, $\ell = 2$	$E_g + T_{2g}$
Spheroidal, $\ell = 3$	$A_{2u} + T_{1u} + T_{2u}$
…	
Torsional, $\ell = 1$	T_{1g}
Torsional, $\ell = 2$	$E_u + T_{2u}$
Torsional, $\ell = 3$	$A_{2g} + T_{1g} + T_{2g}$
…	

Note: Only the A_{1g}, E_g, and T_{2g} modes are Raman active.

of the symmetry for a sphere made of a cubic material. For the O_h point group symmetry, the Raman active vibrations have the A_{1g}, E_g, and T_{2g} irreducible representations and as for the case of the transformation of spheres into spheroids (see Section 11.3.1.1), one can derive how the degeneracy of the isotropic case is lifted taking into account the crystallinity of the sphere. Table 11.1 summarizes these results. It shows several important things:

- The degeneracy of the spheroidal mode with $\ell = 2$ is lifted into E_g and T_{2g} modes which are both Raman active. As this $\ell = 2$ mode is the main low-frequency peak observed in Raman scattering from silver or gold nanoparticles, and as such it is often an essential tool to derive the sizes of the nanoparticles, it is important to bear in mind that this mode can be split into two components when one deals with single-domain crystalline nanoparticles.
- Modes which are not Raman active in the isotropic approximation can transform into Raman active ones. This is the case of the T_{2g} branch of the isotropic torsional mode with $\ell = 3$ for example. Such branches can, in principle, modify the Raman spectrum by providing additional peaks or by mixing with other Raman active modes whose Raman scattering cross section is significant. (As before for spheroidal nanoparticle shapes, only modes having the same irreducible representation can mix.)

In order to quantify the difference we can expect when taking into account elastic anisotropy, we need to evaluate the frequencies of the different modes associated with the different irreducible representations. Figure 11.6 presents the evolution of the frequencies of some Raman active modes for a gold nanoparticle with radius 5 nm as its inner atomic structure evolves from an orientationally averaged crystalline structure (as a model of polycrystalline or multiply twinned nanoparticles, see Section 11.3.1.2.2.2) to a single crystalline face-centered cubic spherical one. To model this evolution the elastic tensor is varied linearly from the isotropic case to the anisotropic one using $C_{ij} = (1 - x)C_{ij}^I + xC_{ij}^A$ where x varies from 0 to 1, C_{ij}^I is the isotropic elastic tensor obtained by 3D averaging of the sound velocities,

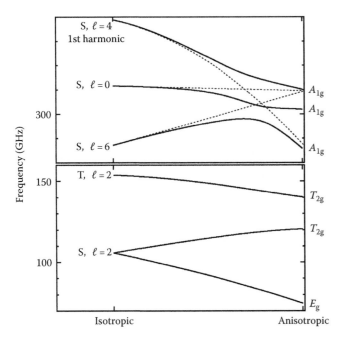

FIGURE 11.6 Evolution of the frequency of Raman active modes with anisotropy in the frequency range of the spheroidal mode with $\ell = 2$ (bottom) and $\ell = 0$ (top). The dotted lines are guides for the eyes. S and T refer to spheroidal and torsional modes, respectively. See text for details.

and C_{ij}^{A} is the elastic tensor of anisotropic gold. The lowest part of the figure shows the splitting of the spheroidal mode with $\ell = 2$ into the E_g (lowest frequencies) and T_{2g} branches. The ratio of the frequencies of the T_{2g} and E_g for anisotropic gold is close to 1.6. This ratio is quite big and clearly demonstrates that the isotropic approximation is not reliable in this particular case. Experimental results on this system are in very good agreement with these calculations (Portalès et al. 2008).

In order to understand the underlying physics leading to this significant splitting, it is instructive to inspect the deformations associated with the T_{2g} and E_g vibrations issued from the fundamental spheroidal mode with $\ell = 2$. One T_{2g} vibration corresponds to a stretching along the [110] direction accompanied by an out of phase stretching along [$\bar{1}$10] and no motion along [001]. Since the degeneracy of T_{2g} is three, the two other vibrations are obtained by changing the [110] direction by [101] and [011] and rotating the other two directions accordingly. The E_g vibrations involve displacements along the [100] directions only. As a result, we can adopt a simplified point of view and consider that the T_{2g} and E_g vibrations originate from the confinement of transverse acoustic waves propagating in the [100] and [110], respectively. A closer inspection reveals that amongst the two transverse acoustic waves propagating along a [110] direction, the displacement of the E_g vibrations corresponds to the wave propagating with sound speed $\sqrt{(C_{11} - C_{12})/2/\rho}$ where ρ is the mass density of gold. For T_{2g}, the associated sound velocity is $\sqrt{C_{44}/\rho}$ (degenerate). Because in the spherical isotropic case the frequency of this mode varies as $\nu = S v_T / R$, the frequencies of the E_g mode and of the T_{2g} mode

are derived by replacing the isotropic transverse sound velocity with the appropriate respective values. Then we assume that the frequency ratio of the T_{2g} and E_g modes equals the ratio of the respective sound velocities. For gold, this sound velocity ratio is 1.68, which is a good approximation of the exact ratio (1.6) derived from the numerical approach. While this calculation is of course less accurate, it clearly shows how the dependence of the sound velocities on the propagation direction (the elastic anisotropy) is responsible for the splitting of the spheroidal mode with $\ell = 2$.

The upper branch of the bottom plot in Figure 11.6 is the T_{2g} branch starting as the isotropic torsional mode with $\ell = 3$. The converging directions of the two T_{2g} branches with the anisotropy increase means that the two respective modes progressively mix. Such mixing is more clearly evidenced in the upper part of the figure, which shows the different A_{1g} branches in the spheroidal $\ell = 0$ frequency range. The three branches start from the spheroidal modes with $\ell = 6, 0$, and 4 from bottom to top, respectively (note that only the isotropic $\ell = 0$ breathing mode is Raman active). These three branches tend to converge with increasing anisotropy so that the mixing effect has modified the filiation of the modes at the anisotropic end. Indeed, amongst the three A_{1g} anisotropic modes, the one which is the closest to a breathing mode is not in the continuation of the $\ell = 0$ branch but rather in the continuation of the $\ell = 4$ branch so that, in a way, the mixing made the breathing mode "jump" from one branch to another. Such patterns are more common for modes at higher frequencies because the number of modes (or branches) increases with frequency. Dotted lines in the upper part of Figure 11.6 are a guide for the eyes to show how the three branches could behave without any branch repelling effect.

11.3.1.2.2.2 Applying the Isotropic and Anistropic Cases   Even if elastic anisotropy can have a dramatic influence on the vibrations of nanoparticles, it is worth using an isotropic approximation in many cases.

- Some materials have a very small anisotropy, which can safely be neglected. One measure for anisotropy is the Zener anisotropy factor $A = 2C_{44}/(C_{11} - C_{12})$ for a cubic material. This factor is 1 for an isotropic material and the more it differs from 1, the more anisotropic the material is. (For gold, $A = 2.92$.)
- The effects of anisotropy can be reduced for non-single-domain nanoparticles. For such objects, the elastic tensor orientation changes inside the nanoparticle. If the twinnings are randomly located inside the different nanoparticles, a simple approximation consists in using a single position independent tensor obtained by averaging the real tensor over the different crystal orientations. The validity of this approximation should increase with the number of twinnings. While a single twinned nanoparticle has a well-defined symmetry and well-defined vibrations, an ensemble of twinned nanoparticle is expected to behave on average as an isotropic nanoparticle. Even if to the best

of our knowledge no such calculations have been reported until now, we can expect in such a situation that the isotropic approximation is a good one. However, an inhomogeneous broadening of the frequencies is expected due to the random distribution of twinnings. It should be noted that multiply twinned particles (MTP) are quite common for some materials, for example, silver. The reasons detailed above explain why the isotropic approximation can be used successfully for silver nanoparticles despite the significant elastic anisotropy of silver.

- For non-monodisperse systems, in particular, when the shapes of the nanoparticles are broadly distributed, the relevance of taking into account elastic anisotropy is highly questionable.

Several different methods exist (Norris 2006) to approximate an isotropic material out of a known anisotropic one. For example, because the sound speeds appear explicitly in the resolution of the motion equations, it is natural to take their average considering all the propagation directions (Saviot et al. 2004b). Differences due to the different ways of averaging over all the propagation directions exist and lead to different eigenfrequencies. In general, the larger the elastic anisotropy, the more the eigenfrequencies calculated with different averaging methods differ. It is important to understand that directional averaging is just a convenient approximation because an exact form for the eigenfunctions exists only in the isotropic case. Using an isotropic approximation for a strongly anisotropic material can never provide accurate results.

11.3.1.3 Influence of Surrounding Medium

The previous sections focused on isolated nanoparticles where the surrounding medium was neglected. As useful as this assumption can be for a rough estimate of the eigenfrequencies knowing the size of a nanoparticle (and vice versa), it hardly ever sticks to reality. Indeed, many applications deal with nanoparticles embedded in a matrix: quantum dots in a liquid, nanocrystals in vitroceramics, noble metal nanoparticles in colored glasses. The case of glasses colored with metallic nanoparticles is a widespread example. Because the glass is naturally transparent, the optical properties of the sample in the visible range are mainly those of the semiconductor or metallic particles. The electronic band gap of the glass matrix is larger than the one of semiconductor nanoparticles or than the surface plasmon frequency of a silver or gold nanoparticle. Therefore, the electronic excitations are confined inside the nanoparticles. However, this confinement does not exist for acoustic vibrations as some acoustic phonons inside the matrix have the same frequencies as the eigenvibration of a nanoparticle. This is similar to the mixing of modes in the case of core–shell nanoparticles where a given vibration can have a significant displacement both in the core and the shell. It is therefore necessary to be able to model the vibrations of the entire particle + matrix system in order to know whether it is safe to ignore the matrix or not.

The main difference between an isolated nanoparticle and a nanoparticle embedded inside an infinite matrix from the point of view of the eigenmodes of vibration is that the first system has a finite size while the second one has an infinite size. Therefore, there is a discrete set of vibrations in the first case (as was shown before) and a continuum in the second case. Indeed, the presence of a nanoparticle inside an infinite matrix hardly changes the vibrational properties of the matrix. However, neglecting the presence of the nanoparticle is not possible because the resonant nature of most optical measurements dramatically enhance the interaction of light with the nanoparticle.

11.3.1.3.1 Complex Frequency Model

The goal of this approach is to focus on the vibrations of the embedded nanoparticle without trying to provide an exact description of the entire particle + matrix system. To do so, it is convenient to consider that the true eigenmodes of an isolated nanoparticle transform into pseudomodes for a matrix-embedded nanoparticle. These pseudomodes manifest as resonances in the scattering of an incoming acoustic wave inside the matrix by the nanoparticle (Dubrovskiy and Morochnik 1981). The full set of pseudomode frequencies for an embedded sphere with elastic isotropy are obtained by locating the roots of equations that are very similar to those of the isotropic sphere problem (Saviot and Murray 2004a). The main modifications are

- The consideration of the usual mechanical boundary conditions to take into account the particle–matrix interface. Rigorously, this implies an *a priori* knowledge of the quality of the contact between the surface of the nanoparticle and its environment. Although this aspect has never really been thoroughly investigated experimentally, we will see that assuming a full contact between the nanoparticle and its environment is more satisfactory than assuming the nanoparticle as free.
- The use of the Hankel functions of the first kind instead of Bessel functions inside the matrix: This corresponds to an outgoing wave inside the matrix which represents the energy flowing away from the nanoparticle as a result of the presence of the matrix.
- The use of complex frequency ω in the exp $(i\omega t)$ terms, e.g., $\omega = \omega_0 + i\gamma$ where γ expresses the damping of the vibration.

The solution of the problem tells us that pseudomodes can be labeled as spheroidal and torsional exactly as in the isolated sphere case. The real part of the frequencies corresponds to the position of a given resonance and the imaginary part to its frequency width (homogeneous width). For an isolated sphere, this width is zero. For an embedded sphere, this width corresponds to the inverse of the lifetime of the pseudomode. In other words, in cases where there is not much reflection of the acoustic wave at the particle–matrix interface, energy flows easily inside the matrix and the lifetime of the corresponding pseudomode is small, i.e., the damping is large (weak acoustic impedance mismatch). For a larger reflection coefficient, or equivalently stronger impedance mismatch, the lifetime increases and the damping of the pseudomode decreases.

It is interesting to note that in addition to the introduction of a vibrational damping, taking into account the existence of an embedding medium also generates new modes, in comparison with the case of a free nanoparticle. Indeed, the translation and the rotation of an isolated particle have a vanishing frequency because there is no restoring force. This is not true for a matrix-embedded nanosphere. The translations and rotations transform into spheroidal pseudomodes with $\ell = 1$ ("rattling mode") and torsional pseudomodes with $\ell = 1$ ("libration mode") respectively. The frequencies of these pseudomodes are not zero. Additional modes with larger imaginary parts are also solutions of the same set of equations. However, at least for systems where the elastic properties of the nanoparticle and the matrix are significantly different, these additional solutions can be shown to be due to the matrix only and have therefore been labeled "matrix pseudomodes" (Murray and Saviot 2004).

Although quite useful for evaluating the changes of the vibration eigenfrequencies, both in terms of frequency broadening and absolute positions, the main drawback of the complex frequency model (CFM) approach is that it doesn't provide usable wavefunctions. Indeed, it is straightforward to check that simply introducing the complex frequency results in diverging displacements. While it is a very valuable tool to predict the frequencies of the pseudomodes, it is not suitable for more detailed calculations involving the displacement such as calculations of electron-vibration coupling which is a major ingredient of Raman scattering and other optical techniques.

11.3.1.3.2 Core–Shell Model

To avoid the shortcomings of the CFM approach, another model has been proposed (Murray and Saviot 2004) to describe as exactly as possible the particle + matrix system. As a result, eigenmodes are real modes and not pseudomodes. In order to do so, we consider a finite core–shell system having the spherical symmetry (including elastic isotropy), the idea behind this scheme being to approach the situation of a matrix-embedded nanoparticle by asymptotically growing the thickness of the shell. The consideration of a core–shell system allows to write exact expressions for the displacements as in the case of core–shell nanoparticles. We do not consider an infinite matrix but a spherical one (radius R_m) having a spherical nanoparticle at its center (radius R_p). Of course, because we are interested in an infinite matrix, we will choose R_m so that R_m is much larger than R_p.

Looking at the frequency spectrum of eigenmodes in that case is not very interesting. All the eigenfrequencies of the system vary roughly as $1/R_m$ and therefore decrease toward zero as R_m is increased. This is due to the natural trend of a frequency spectrum that transforms from a discrete one to a continuous one as the size of the system tends to infinity. However amongst all these modes, the ones we are really interested in are those for which the displacement inside the nanoparticle is significant as only such modes can couple efficiently with electronic states in the case of resonant Raman scattering for example.

A simple way to distinguish between the vibrations of the matrix that hardly penetrate inside the nanoparticle and those who resonantly excite it is to calculate the mean-square displacement inside the nanoparticle only. This quantity is a measure of the amplitude of the vibration inside the nanoparticle even for complex deformations. As such the electron-vibration interaction increases with the mean-square displacement. To do that, the eigenmodes of the system are orthonormalized using Equation 11.8 where the integral is over the volume of the particle + matrix system, $\vec{u}_i(\vec{R})$ is the displacement for the eigenmode i of the point whose equilibrium position is $\vec{R}, \rho(\vec{R})$ is the mass density, and M_{p+m} is the mass of the core–shell particle + matrix system.

$$\int \rho(\vec{R}) \vec{u}_i(\vec{R}) \cdot \vec{u}_j(\vec{R}) \, d^3\vec{R} = \delta_{ij} M_{p+m} \tag{11.8}$$

Then the mean-square displacement is calculated according to Equation 11.9.

$$\langle u_i^2 \rangle_p = \frac{1}{(4/3)\pi R_p^3} \int\limits_{R < R_p} \left\| \vec{u}_i(\vec{R}) \right\|^2 d^3\vec{R} \tag{11.9}$$

In order to show how this quantity (or similar ones) provide valuable tools to interpret experimental results and how it compares with the previous CFM approach, we consider in the following a simple example.

11.3.1.3.3 Example Calculation

Let's consider the case of a gold nanoparticle (radius 5 nm) inside a silica matrix, modeled as a silica shell of radius 500 nm. Figure 11.7 presents the variation of the mean-square displacement with frequency calculated using core–shell model (CSM) for the spheroidal $\ell = 0$ and $\ell = 2$ Raman active modes. CFM pseudomode frequencies for this system are indicated with arrows: for each pseudomode, a vertical arrow indicates the frequency while the double headed horizontal arrow indicates the corresponding damping. "Matrix pseudomodes" (i.e., CFM pseudomodes with large damping) have been omitted from this figure. This example clearly demonstrates that

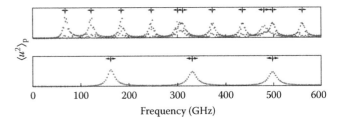

FIGURE 11.7 Mean-square displacement inside an isotropic gold nanosphere (radius 5 nm) located at the center of a silica shell (external radius 500 nm) for the spheroidal modes with $\ell = 0$ (bottom) and $\ell = 2$ (top) obtained using CSM. The vertical arrows at the top of each graph represent the position and width of the pseudomodes for the same system calculated using CFM.

- The mean-square displacement calculated using CSM reaches local maximum values at the frequencies of the pseudomodes predicted from CFM.
- The homogeneous frequency widths predicted by both models are in very good agreement.
- While all the points for the spheroidal $\ell = 0$ modes lie on a single curve, those for the $\ell = 2$ mode are mainly distributed over two different curves. This comes from the fact that the spheroidal modes with $\ell \neq 0$ are a linear combination of terms which correspond to longitudinal and transverse waves for an infinite system (see Equation 11.2 for $C = 0$).

As a conclusion on the influence of the surrounding medium on the vibration modes of nanoparticles, CFM calculations are enough if one only needs to evaluate the frequencies and damping of the pseudomodes. CSM is required when the wavefunctions are needed. It should be noted that even for isotropic materials, it is not easy to figure out a simple rule to know in advance how much a given vibration is affected by the matrix. In particular, predictions based on acoustic impedances calculated for simple cases (plane waves instead of spherical waves for example) can lead to very inaccurate estimations (Saviot and Murray 2004b).

After having given an overview of how theory can handle the eigenmodes of nanoparticles' vibrations, we turn to the description of some important, or at least widespread, experimental cases of nanoparticle vibration spectroscopy.

11.3.2 Resonant Raman Scattering

11.3.2.1 Noble Metal Nanoparticles

The vibration modes of metallic nanoparticles are easily detectable by Raman scattering as the incident light is in resonance with the excitation of the free electrons of the nanoparticle (Weitz et al. 1980a; Mariotto et al. 1988; Palpant et al. 1993). Under the application of an electric field like that of visible light the free electrons of metal collectively oscillate like a plasma. Like for vibrations, the electronic plasma oscillations in a spherical metallic nanoparticle reflect spherical symmetry, and are characterized by the quantum number ℓ. Plasma oscillations are quantized in terms of "plasmons." For most noble metal nanoparticles (whether embedded in a liquid or in a transparent solid matrix), the plasmons corresponding to $\ell = 1$ (so-called dipolar plasmon or surface plasmon resonance) have frequencies lying in the visible spectrum. The excitation of the dipolar plasmon is observed by the presence of a very strong light absorption band. The maximum of the absorption band is around $\lambda \simeq 420\,nm$ for silver, 530 nm for gold, and 570 nm for copper in oxide matrices (Kriebig and Volmer 1995; Cottancin et al. 2006). Typically, glass samples containing silver nanoparticles appear yellow, red with gold, and dark red with copper. The spectral widths of these absorption bands lie between 50 and 100 nm. It is noted that the frequency of this absorption maximum is very weakly dependent on the nanoparticle size (Muskens et al. 2006). As for vibration again, the dipolar plasmon degeneracy is lifted by an

ellipsoidal distortion so that in the case of spheroids, the dipolar plasmon is split into a nondegenerate component whose electronic dipolar moment is oriented along the rotation symmetry axis, and into a double-degenerate one with a dipolar moment perpendicular to the symmetry axis (Berger 1993; Muskens et al. 2006). A prolate distortion shifts the nondegenerate plasmon to low frequency, while the double-degenerate one slightly shifts to high frequency. (These trends are opposite for an oblate distortion.) Due to the different orientations of the electronic dipolar moments of the two components, they can be selectively excited by controlling the polarization of the incident light: when the polarization of the electric field is parallel (perpendicular) to the symmetry axis, one favors the response of the lower (higher) energy component.

Although the dipolar plasmon is the essential feature that is responsible for the absorption of visible light from noble metal nanoparticles, one should not omit an additional component which shows up as an increasing background at the high-frequency end of the spectrum, rather than a peak in the absorption spectra. This component is due to interband transitions from the $(n-1)d$ levels to the ns ones; in the case of silver, from the $4d^{10}$ levels to the $5s^1$ free-electron levels. The plasmon absorption band is well separated from the interband absorption in silver. They are much less separated and closer in gold, and strongly overlap in copper (Cottancin et al. 2006).

The dipolar plasmon corresponds to a strong *electronic dipolar moment* with which the incident excitation light strongly interacts. Therefore, the selection rules for resonant Raman scattering by metallic nanoparticle vibrations are the same as for nonresonant Raman scattering (Duval 1992; Goupalov et al. 2006): the radial $\ell = 0$ and the quadrupolar $\ell = 2$ modes are resonant Raman active. However, the nature of the plasmon–phonon coupling mechanism for radial vibrations is different from that for quadrupolar ones (Bachelier and Mlayah 2004). The coupling with the quadrupolar vibrations comes from the strong dependence of the plasmon frequency on the particle shape (as noted above concerning the ellipsoidal distortion): as the vibrational quadrupolar modes modulate the nanoparticle shape, they modulate the electronic dipolar plasmon frequency, and therefore both types of excitation strongly couple with each other. The origin of the coupling with the radial vibrations is more complex. In fact, the plasmon frequency is dependent on the interband-transition contribution to the metal dielectric susceptibility (Del Fatti et al. 2000). Because the radial vibrations modulate the nanoparticle volume, they also modulate the dielectric susceptibility and, then, become coupled to the plasmon.

Thanks to the strong plasmon–phonon coupling it is possible to observe the Raman scattering from nanoparticle vibrations from samples with metal molar content less than 0.01%. In fact, the nanoparticle vibrations can be observed by Raman spectroscopy, when a color, even very weak, due to nanoparticle absorption is visible to the naked eye. The stronger coupling of the plasmon excitations with the quadrupolar modes than with the radial ones explains why their Raman signal is typically 10 times larger. On the other hand, because the coupling of the radial modes with the plasmon is effective through the interband dielectric susceptibility, it is expected that the radial modes are more Raman active when the incident light, that excites the plasmon, excites also the electrons in the *ns* free-electron band. As a result, the interband transition is more strongly excited when the excitation light is shifted to the blue, and therefore it is expected that the Raman scattering cross section of the radial modes is enhanced in that case. This was experimentally verified for silver nanoparticles where the radial modes contributions were found to become more and more Raman visible when the incident light is shifted to the blue (Portales et al. 2001).

As noted above, a prolate ellipsoidal distortion of the nanoparticle splits the dipolar plasmon into a nondegenerate blue-shifted component polarized along the rotation symmetry axis, and into a somewhat less red-shifted double-degenerate one. The same scenario applies to the vibrational excitations so that the quadrupolar mode is also split into three components, among which an ellipsoidal vibration polarized along the symmetry axis and shifted to low frequency in the case of a prolate distortion (irreducible representation A_{1g}—see Section 11.3.1.1). Due to their common polarization this nondegenerate mode is strongly coupled to the nondegenerate plasmon. Therefore, when the Raman experiment is performed in such conditions that the laser light polarization is parallel to the symmetry axis of the prolate distorted nanoparticle and the wavelength of light is shifted to the red, the nondegenerate split component of the quadrupolar mode should be more easily observed. It was confirmed by experiment for silver nanoparticles in amorphous silica: as expected, the more the incident wavelength is shifted to the red, the more distorted are the excited nanoparticles and the lower is the frequency of the Raman observed nondegenerate mode (Portales 2001). One should however keep in mind that in the most frequent cases of nanoparticles grown in glasses, the orientation of the ellipsoidal nanoparticles is random, so that the nondegenerate vibration mode is observed by Raman scattering whatever is the incident light polarization (Weitz et al. 1980a; Portales et al. 2001). One notable exception was obtained with self-assembled nanocolumns produced by pulsed laser deposition. In these works, the samples consist of elongated spheroids oriented along the same direction and enabled the careful study of the Raman intensity as a function of the angle between the spheroid nondegenerate axis and the polarization of incident light (Margueritat et al. 2006, 2007). Finally, an alternative use of the plasmon resonance for low-frequency Raman scattering experiments was demonstrated for a nanocomposite. The alternative deposition of silver and cobalt enabled the preparation of spherical cobalt nanoparticles at a controllable distance of silver spheroids. While the low-frequency Raman intensity of cobalt nanoparticles alone is too small to be detected with a visible excitation, it was demonstrated that using the silver plasmon resonance enables the observation of the vibrations of cobalt nanoparticles through surface-enhanced Raman scattering (SERS) (Margueritat et al. 2008).

The enhanced low-frequency Raman scattering from metallic nanoparticles has offered the possibility of performing many detailed studies of acoustic vibrations in nanoparticles, like the effects of shape anisotropy (Portales 2001; Margueritat et al. 2006, 2007) or of nanocrystalline anisotropy (Stephanidis et al. 2007b; Portalès et al. 2008).

In fact, as we now see in the following section, another type of nanoparticles has significantly contributed to a better understanding of nanoparticles' vibrations thanks to a strong light-vibration coupling: semiconductor nanoparticles.

11.3.2.2 Semiconductor Nanoparticles

The effect of resonance on the Raman scattering from acoustic vibrations was also observed in semiconductor nanoparticles, in particular in CdS, CdSe, and $CdS_{1-x}Se_x$ nanoparticles (Saviot et al. 1996, 1998; Ivanda et al. 2003). While with metallic nanoparticles resonance occurs with plasmonic excitations, in the case of semiconductor nanoparticles the resonance is related to the electron–hole excitation. As before, it was experimentally and theoretically shown that radial ($\ell = 0$) and quadrupolar ($\ell = 2$) vibrations are resonant Raman active (Gupalov and Merkulov 1999a,b; Goupalov et al. 2006). In fact, the effect of resonance also occurs for optical modes (Chamberlain et al. 1995).

The threshold of absorption corresponding to electron–hole pair excitation is obviously dependent on the semiconductor energy gap. Light absorption by the electron–hole excitations in semiconductor nanoparticles is strongly dependent on the size of the nanoparticles due to the effect of confinement on the energy gap. For nanoparticles, the energy gap increases with the decrease of the nanoparticle size. For example, for CdSe nanoparticles embedded in a borosilicate glass, the wavelength of the light that excites the electron–hole lowest energy level increases from 480 to 620 nm when the nanoparticle radius increases from 1.8 to 3.8 nm (Saviot 1995). Therefore, the effect of resonance with the electron–hole excitation, results in a size-selective observation of vibrations by resonant Raman scattering, as clearly demonstrated for CdS-doped glasses (Saviot et al. 1998).

11.3.3 Application to Other Systems

As mentioned in the beginning of this chapter, nanoparticles' vibration modes are well described by the mathematical treatment of H. Lamb that was initially performed on a free sphere (Lamb 1882) with no particular restriction on its size as long as the medium could be considered as homogeneous and isotropic. As a matter of fact, this treatment has proved to apply equally well to nanospheres (like quantum dots) and macrospheres (like planets), at least at first order.

As far as the field of condensed matter is concerned, we have already considered the specific cases of noble metal and semiconductor nanoparticles as they have been the focus of extensive experimental and theoretical investigations due to their enhanced Raman light scattering cross sections. Comparable size systems like inorganic nanocrystals (Duval et al. 1986; Pighini et al. 2007) do also occupy a substantial field in low-frequency vibrational spectroscopy. For all these objects, the typical sizes range between a few nanometers and less than about 20 nm. Over the past decade, comparatively much larger objects have also been investigated. They are essentially submicrometer hard spheres (few hundreds of nanometers) made of polymer (PMMA, polystyrene) or of silica. The combination of large particle sizes with hard condensed matter cohesion leads to very low eigenmode frequencies that compare to those of the sound modes probed through Brillouin light scattering. Accordingly, vibration modes from submicrometer spheres are experimentally probed with Brillouin scattering setups like a multipass Fabry–Perot interferometer (Lindsay et al. 1981).

Large nanosphere systems investigated by Brillouin scattering can be divided into two categories: colloidal spheres and opals. While colloidal systems refer to spheres embedded in a liquid, opals usually refer to dry assemblies. Apart from a size difference, these systems notably differ from smaller nanoparticles through their high particle volume fractions and high monodispersity. Among these two cases, only that of dry opals which consist of highly ordered and compact arrangements of submicrometer beads can resolve their eigenmodes (Kuok et al. 2003; Cheng et al. 2005c). In comparison with Raman scattering from small nanoparticles (~10 nm and lower), Brillouin scattering from dry opals allows to detect many more eigenmodes (up to 13 components (Cheng et al. 2005c)) than just the first modes of the $\ell = 0$ and $\ell = 2$ spheroidal vibrations. As shown by M. Montagna (2008), inelastic light scattering from spheres whose sizes nearly compare with the wavelength of the exciting light allows more modes to be active (including torsional ones), thereby lifting the rather restrictive Raman selection rules that apply to smaller nanoparticles. Another remarkable feature with submicrometer spheres is that it becomes possible to investigate single isolated ones using micro-Brillouin spectroscopy (Li et al. 2006).

Hard sphere colloidal suspensions, or "wet opals," are comparatively more complex systems than dry opals since the embedding liquid generates a coupling between the nanosphere modes and a lower acoustic impedance mismatch. However, they are more flexible systems since they offer the possibility of studying these two effects by either changing the nature of the infiltrating liquid or the volume concentration of the particles (Liu et al. 1990; Penciu et al. 2002, 2003; Cheng et al. 2006). Unlike dry opals, the Brillouin spectra of hard sphere colloidal suspensions do not directly display the inelastic components associated with selected eigenmodes of the spheres; instead, they typically show two components that arise from the hybridization of propagating acoustic modes with the localized eigenmodes of the spheres. This hybridization occurs as the sound mode momentum Q becomes comparable to the inverse size of the spheres ($\sim \pi/d$), leading to the opening of a phononic band gap in the phonon dispersion curve of the liquid at frequencies close to those of the sphere eigenmodes. Cheng et al. (2006) have illustrated how the phononic gap can be tuned by changing either the size of the spheres or the nature of the embedding liquid.

Interestingly, a similar hybridization scheme between nanosphere modes and propagating plane acoustic waves was used to interpret the phonon spectra of systems whose local structures are suspected to feature a nanotexturation, although less defined than that of opals or colloidal assemblies. This is the case of icosahedral quasicrystals for which the atomic arrangement is often described in terms of small clusters of icosahedral symmetry that serve as building blocks of the quasicrystalline structure. As small as they are (i.e., ~1 nm), it turns out that the eigenmodes of such clusters well account for the optical modes detected through inelastic neutron scattering from these materials (Duval et al. 2005b). The smallness of the quasicrystal clusters certainly sets here a limit to the application of Lamb's theory in condensed matter physics. It is worth underlining at this stage that the hybridization of nanosphere eigenmodes with sound modes is all the more strong as the mechanical frontier between the nanospheres and their environment is less defined. For instance, in the case of quasicrystals, the icosahedral clusters are not independent clusters like in opals but rather clusters that are separated from each other by defect regions ("glue regions") made of atoms that share chemical bonds with the surface cluster atoms. Yet the elastic contrast is sufficient for nanosphere-like eigenmodes to build up and hybridize with the sound modes, leading to a Van Hove-like singularity in the corresponding density of vibrational states. In fact, a similar picture was proposed to explain the origin of the controversed low-frequency excitations observed in the Raman spectra of glasses ("boson peak") (Duval et al. 1990, 2007). According to this picture, the glass bears an elastic inhomogeneity at the nanometer scale where cohesive amorphous nanodomains feature nanoparticle-like localized dynamics (Tanguy et al. 2002; Rossi et al. 2005).

Most of the nanoparticles considered in both the present and previous sections can be classified as hard condensed matter nanoparticles. As rich as the spectroscopy of nanoparticles vibrations can be, several attempts have been made to extend it to soft matter and in particular to biological matter. P.-G de Gennes and M. Papoular were among the first authors to theoretically investigate the existence of nanoparticle-like vibrations in globular biological objects (de Gennes and Papoular 1969). Several Raman experiments studies of globular proteins have been reported, with somewhat limited success (Painter et al. 1982). One complicating aspect with globular proteins is that their collective dynamics is very much coupled to that of the hydration water (Diehl et al. 1997) and their composite domain structure is rather open. Such is not the case of other biological nanoparticles like viruses. Transmission electron microscopy or atomic force microscopy characterizations have long established that viruses are compact structures with a variety of sizes ranging between c.a. tens and a few hundreds of nanometers. In addition, they are known to have high size and shape monodispersities that should favor the detection of their particle modes. Both theoretical (Ford 2003; Fonoberov and Balandin 2004; Saviot et al. 2004a; Saviot et al. 2007) and numerical analyses (Tama and Brooks 2005) of virus vibrations modes have predicted their existence and how they might play a key role in the life cycles of viruses or in their use in bionanotechnology. In spite of beginning inconclusive experimental characterizations through inelastic light scattering (Stephanidis et al. 2007a), the detection of nanoparticle vibration modes in biological matter opens up challenging prospects in the field of nanobiophysics.

11.4 Summary and Future Perspective

Considerable progress has been achieved during the last 30 years regarding the observation of acoustic vibrations of nanoparticles. This is mainly due to the tremendous advances in the synthesis of nanoparticles. While the first observations of acoustic vibrations in nanoparticles were performed on rough metallic surfaces, recent ones could benefit from nanoparticles with unmatched size, shape, and crystallinity control. The growing ability of tailoring the synthesis of nanoparticles and the concomitant development of new materials where nanoparticles tend to replace atoms as constitutive building units, open up challenging perspectives for the investigation of nanoparticles' vibrations as a mean of characterization of collective phenomena. For instance, the study of nanoparticles' vibrations within 3D ordered arrays of metal nanocrystals (so-called supracrystals) has evidenced vibrational coherence effects that intrinsically depend on the supercrystalline quality (Courty et al. 2005; Duval et al. 2005a). The propagation of acoustic waves in such nanobased materials (Lisiecki et al. 2008), and particularly the existence of phononic band gaps in the meV regime (like their equivalent at sub-meV energies in submicrometer nanoparticle assemblies (Cheng et al. 2006) are exciting issues to the world of nanophysics.

Advances in the detection techniques are another important factor. Low-frequency Raman scattering experiments now routinely benefit from multichannel acquisition thanks to CCD detectors contrary to the first monochannel setups using photomultipliers. As a result, the acquisition time of such spectra has been shortened significantly. At the same time, the more frequent usage of Brillouin setups such as six-pass tandem Fabry–Pérot interferometers provides unmatched frequency resolution. The complementary femtosecond pump-probe measurements recently opened the door to single particle measurements. While inelastic light scattering measurements on unique nanoparticles much smaller than the wavelength of light has not been achieved yet, micro-Brillouin measurements have already been obtained for larger nanoparticles.

As a result of these advances in both the synthesis of nanoparticles and the detection of their acoustic vibrations, detailed features can now be experimentally observed which require improved models to take into account the detailed structure of the nanoparticles. Such models have been presented here and compared to the common homogeneous and isotropic spherical model. The changes resulting from simple nonspherical shapes (spheroids) but also from the presence of an embedding matrix and from the absence of crystal twinning have been detailed and models have been described to account for them. While all these effects have been properly taken into account with continuum mechanics only, atomistic calculations are starting to play an

increasing role especially for extremely small nanoparticles. Such approaches are required to quantify the effects of surface relaxation, adsorption of species at the surface of the nanoparticles, the presence of crystal twins, or to propose a more accurate description of the nanoparticle–matrix interface for example. The sustained advances in the preparation of nanoparticles and the detection of their acoustic vibrations will require such refinements to be taken into account sooner rather than later.

References

Bachelier, G. and A. Mlayah. 2004. Surface plasmon mediated Raman scattering in metal nanoparticles. *Phys. Rev. B* 69:205408.

Berger, A. 1993. Prolate silver particles in glass surfaces. *J. Non-Cryst. Solids* 163:185.

Bragas, A. V., C. Aku-Leh, and R. Merlin. 2006. Raman and ultrafast optical spectroscopy of acoustic phonons in CdTe$_{0.68}$Se$_{0.32}$ quantum dots. *Phys. Rev. B* 73:125305.

Chamberlain, M. P., C. Trallero-Giner, and M. Cardona. 1995. Theory of one-phonon Raman scattering in semiconductor microcrystallites. *Phys. Rev. B* 51:1680.

Cheng, W., S.-F. Ren, and P. Y. Yu. 2005a. Erratum: Microscopic theory of the low frequency Raman modes in germanium nanocrystals [*Phys. Rev. B* 71, 174305 (2005)]. *Phys. Rev. B* 72:059901(E).

Cheng, W., S.-F. Ren, and P. Y. Yu. 2005b. Microscopic theory of the low frequency Raman modes in germanium nanocrystals. *Phys. Rev. B* 71:174305.

Cheng, W., J. J. Wang, U. Jonas, W. Steffen, G. Fytas, R. S. Penciu, and E. N. Economou. 2005c. The spectrum of vibration modes in soft opals. *J. Chem. Phys.* 123:121104.

Cheng, W., J. Wang, U. Jonas, G. Fytas, and N. Stefanou. 2006. Observation and tuning of hypersonic bandgaps in colloidal crystals. *Nat. Mater.* 5:830.

Combe, N., J. R. Huntzinger, and A. Mlayah. 2007. Vibrations of quantum dots and light scattering properties: Atomistic versus continuous models. *Phys. Rev. B* 76:205425.

Cottancin, E., G. Celep, J. Lermé, M. Pellarin, J. R. Huntzinger, J. L. Vialle, and M. Broyer. 2006. Optical properties of noble metal clusters as a function of the size: Comparison between experiments and a semi-quantal theory. *Theor. Chem. Acc.* 116:514.

Courty, A., A. Mermet, P. A. Albouy, E. Duval, and M. P. Pileni. 2005. Vibrational coherence of self-organized silver nanocrystals in f.c.c. supra-crystals. *Nat. Mater.* 4:395.

de Gennes, P. G. and M. Papoular. 1969. *Polarisation Matière et Rayonnement*. Société française de physique ed. Paris, France: P.U.F., pp. 243–258.

Del Fatti, N., C. Voisin, F. Chevy, F. Vallée, and C. Flytzanis. 1999. Coherent acoustic mode oscillation and damping in silver nanoparticles. *J. Chem. Phys.* 110:11484.

Del Fatti, N., C. Voisin, D. Christofilos, F. Vallée, and C. Flytzanis. 2000. Acoustic vibration of metal films and nanoparticles. *J. Phys. Chem. A* 104:4321.

Diehl, M., W. Doster, W. Petry, and H. Schober. 1997. Water-coupled low-frequency modes of myoglobin and lysozyme observed by inelastic neutron scattering. *Biophys. J.* 73(5):2726–2732.

Dubrovskiy, V. A. and V. S. Morochnik. 1981. Natural vibrations of a spherical inhomogeneity in an elastic medium. *Izv. Earth Phys.* 17:494.

Duval, E. 1992. Far-infrared and Raman vibrational transitions of a solid sphere: Selection rules. *Phys. Rev. B* 46:5795–5797.

Duval, E., A. Boukenter, and B. Champagnon. 1986. Vibration eigenmodes and size of microcrystallites in glass: Observation by very low frequency Raman scattering. *Phys. Rev. Lett.* 56:2052.

Duval, E., A. Boukenter, and T. Achibat. 1990. Vibrational dynamics and the structure of glasses. *J. Phys.: Condens. Matter* 2:10227.

Duval, E., A. Mermet, A. Courty, P. A. Albouy, and M. P. Pileni. 2005a. Coherence effects on Raman scattering from self-organized Ag nanocrystals: Theory. *Phys. Rev. B* 72:085439.

Duval, E., L. Saviot, A. Mermet and D. B. Murray. 2005b. Continuum elastic sphere vibrations as a model of low lying optical modes in icosahedral quasicrystals. *J. Phys: Condens. Matter* 17:3559.

Duval, E., A. Mermet, and L. Saviot. 2007. Boson peak and hybridization of acoustic modes with vibrations of nanometric heterogeneities in glasses. *Phys. Rev. B* 75:024201.

Eringen, A. C. and E. S. Suhubi. 1975. *Elastodynamics*. Vol. II. Academic, New York, pp. 804–833.

Fonoberov, V. A. and A. A. Balandin. 2004. Low-frequency vibrational modes of viruses used for nanoelectronic self-assemblies. *Phys. Stat. Sol. (b)* 241(12):R67–R69.

Ford, L. H. 2003. Estimate of the vibrational frequencies of spherical virus particles. *Phys. Rev. E* 67:051924.

Goupalov, S. V., L. Saviot, and E. Duval. 2006. Comment on Infrared and Raman selection rules for elastic vibrations of spherical nanoparticles. *Phys. Rev. B* 74:197401.

Gupalov, S. V. and I. A. Merkulov. 1999a. Theory of Raman light scattering by nanocrystal acoustic vibrations. *Phys. Solid State* 41:1349.

Gupalov, S. V. and I. A. Merkulov. 1999b. Theory of Raman Scattering by nanocrystal acoustic vibrations. *Fiz. Tverd. Tela (St. -Peterburg)* 41:1473.

Hartland, G. V. 2007. Is perfect better? *Nat. Mater.* 6:716.

Iuga, M., G. Steinle-Neumann, and J. Meinhardt. 2007. Ab-initio simulation of elastic constants for some ceramic materials. *Eur. Phys. J. B* 58:127.

Ivanda, M., K. Babocs, C. Dem, M. Schmitt, M. Montagna, and W. Kiefer. 2003. Low-wave-number Raman scattering from CdS_xSe_{1-x} quantum dots embedded in a glass matrix. *Phys. Rev. B* 67:235329.

Kriebig, U. and M. Volmer. 1995. *Optical Properties of Metal Clusters*. Springer, Berlin, Germany.

Kuok, M. H., H. S. Lim, S. C. Ng, N. N. Liu, and Z. K. Wang. 2003. Brillouin study of the quantization of acoustic modes in nanospheres. *Phys. Rev. Lett.* 90:255502.

Lamb, H. 1882. On the vibrations of an elastic sphere. *Proc. London Math. Soc.* 13:189.

Li, Y., H. S. Lim, S. C. Ng, Z. K. Wang, M. H. Kuok, E. Vekris, V. Kitaev, F. C. Peiris, and G. A. Ozin. 2006. Micro-Brillouin scattering from a single isolated nanosphere. *Appl. Phys. Lett.* 88:023112.

Lindsay, S. M., M. W. Anderson, and J. R. Sandercock. 1981. Construction and alignment of a high performance multipass vernier tandem Fabry-Perot interferometer. *Rev. Sci. Instrum.* 52(10):1478–1486.

Lisiecki, I., V. Halté, C. Petit, M. P. Pileni, and J.-Y. Bigot. 2008. Vibration dynamics of supra-crystals of cobalt nanocrystals studied with femtosecond laser pulses. *Adv. Mater.* 20:1.

Liu, J., L. Ye, D. A. Weitz, and P. Sheng. 1990. Novel acoustic excitations in suspensions of hard sphere colloids. *Phys. Rev. Lett.* 65:2602.

Liu, T.-M., J.-Y. Lu, H.-P. Chen, C.-C. Kuo, M.-J. Yang, C.-W. Lai, P.-T. Chou et al. 2008. Resonance-enhanced dipolar interaction between terahertz photons and confined acoustic phonons in nanocrystals. *Appl. Phys. Lett.* 92:093122.

Margueritat, J., J. Gonzalo, C. N. Afonso, A. Mlayah, D. B. Murray, and L. Saviot. 2006. Surface plasmons and vibrations of self-assembled silver nanocolumns. *Nano Lett.* 6:2037–2042.

Margueritat, J., J. Gonzalo, C. N. Afonso, G. Bachelier, A. Mlayah, A. S. Laarakker, D. B. Murray, and L. Saviot. 2007. From silver nanolentils to nanocolumns: Surface plasmon-polaritons and confined acoustic vibrations. *Appl. Phys. A* 89:369.

Margueritat, J., J. Gonzalo, C. N. Afonso, U. Hrmann, G. Van Tendeloo, A. Mlayah, D. B. Murray, L. Saviot, Y. Zhou, M. H. Hong, and B. S. Luk'yanchuk. 2008. Surface enhanced Raman scattering of silver sensitized cobalt nanoparticles in metal dielectric nanocomposites. *Nanotechnology* 19:375701.

Mariotto, G., M. Montagna, G. Viliani, E. Duval, S. Lefrant, E. Rzepka, and C. Mai. 1988. Low-energy Raman scattering from silver particles alkali halides. *Europhys. Lett.* 6:239.

Martini, I., J. H. Hodak, and G. V. Hartland. 1998. Spectroscopy and dynamics of nanometer-sized noble metal particles. *J. Phys. Chem. B* 102:6958.

Montagna, M. 2008. Brillouin and Raman scattering from the acoustic vibrations of spherical particles with a size comparable to the wavelength of the light. *Phys. Rev. B* 77:045418.

Murray, D. B. and L. Saviot. 2004. Phonons in an inhomogeneous continuum: Vibrations of an embedded nanoparticle. *Phys. Rev. B* 69:094305.

Murray, D. B., C. Netting, L. Saviot, C. Pighini, N. Millot, D. Aymes, and H.-L. Liu. 2006. Far-infrared absorption by acoustic phonons in titanium dioxide Nanopowders. *J. Nanoelectron. Optoelectron.* 1:92–98.

Muskens, O., D. Christofilos, N. Del Fatti, and F. Vallée. 2006. Optical response of a single noble metal nanoparticle. *J. Opt. A: Appl. Opt.* 8:S264.

Norris, A. N. 2006. The isotropic material closest to a given anisotropic material. *J. Mech. Mater. Struct.* 1:223.

Painter, P. C., L. E. Mosher, and C. Rhoads. 1982. Low-frequency modes in the Raman spectra of proteins. *Biopolymers* 21(7):1469–1472.

Palinginis, P., S. Tavenner, M. Lonergan, and H. Wang. 2003. Spectral hole burning and zero phonon linewidth in semiconductor nanocrystals. *Phys. Rev. B* 67:201307.

Papant, B., H. Portalés, L. Saviot, J. Lermé, B. Prével, M. Pellarin, E. Duval, A. Perez, and M. Broyer. 1993. Quadrupolar vibrational mode of silver clusters from plasmon-assisted Raman scattering. *Phys. Rev. B* 60:17107.

Penciu, R. S., M. Kafesaki, G. Fytas, E. N. Economou, W. Steffen, A. Hollingsworth, and W. R. Russel. 2002. Phonons in colloidal crystals. *Europhys. Lett.* 58:699.

Penciu, R. S., H. Kriegs, G. Petekidis, G. Fytas, and E. N. Economou. 2003. Phonons in colloidal systems. *J. Chem. Phys.* 118:5224. et rfrences incluses.

Pighini, C., D. Aymes, N. Millot, and L. Saviot. 2007. Low-frequency Raman characterization of size-controlled anatase TiO_2 nanopowders grown by continuous hydrothermal synthesis. *J. Nanopart. Res.* 9:309.

Portalès, H. 2001. Etude par diffusion Raman de nanoparticules métalliques en matrice diélectrique amorphe. PhD thesis, Université Claude Bernard Lyon 1, Lyon, France. http://tel.archives-ouvertes.fr/tel-00002169.

Portalès, H., L. Saviot, E. Duval, M. Fujii, S. Hayashi, N. Del Fatti, and F. Vallée. 2001. Resonant Raman scattering by breathing modes of metal nanoparticles. *J. Chem. Phys.* 115:3444.

Portalès, H., L. Saviot, E. Duval, M. Gaudry, E. Cottancin, M. Pellarin, J. Lermé, and M. Broyer. 2002. Resonant Raman scattering by quadrupolar vibrations of Ni-Ag core-shell nanoparticles. *Phys. Rev. B* 65:165422–165426.

Portalès, H., N. Goubet, L. Saviot, S. Adichtchev, D. B. Murray, A. Mermet, E. Duval, and M.-P. Piléni. 2008. Probing atomic ordering and multiple twinning in metal nanocrystals through their vibrations. *Proc. Natl. Acad. Sci. U.S.A* 105:14784–14789.

Ramirez, F., P. R. Heyliger, A. K. Rappé, and R. G. Leisure. 2008. Vibrational modes of free nanoparticles: From atomic to continuum scales. *J. Acoust. Soc. Am.* 123:709.

Rossi, B., G. Viliani, E. Duval, L. Angelani, and W. Garber. 2005. Temperature-dependent vibrational heterogeneities in harmonic glasses. *Europhys. Lett.* 71:256.

Saviot, L. 1995. Etude par diffusion Raman du confinement des vibrations optiques et acoustiques dans des nanoparticules de CdSe et de silicium poreux. PhD thesis, Université Claude Bernard Lyon 1, Lyon, France.

Saviot, L. and D. B. Murray. 2004a. The connection between elastic scattering cross sections and acoustic vibrations of an embedded nanoparticle. *Phys. Stat. Sol. (c)* 1(11):2634.

Saviot, L. and D. B. Murray. 2004b. Long lived acoustic vibrational modes of an embedded nanoparticle. *Phys. Rev. Lett.* 93:055506.

Saviot, L., B. Champagnon, E. Duval, I. A. Kudriavtsev, and A. I. Ekimov. 1996. Size dependence of acoustic and optical vibrational modes of CdSe nanocrystals in glasses. *J. Non-Cryst. Solids* 197:238.

Saviot, L., B. Champagnon, E. Duval, and A. I. Ekimov. 1998. Size-selective resonant Raman scattering in CdS doped glasses. *Phys. Rev. B* 57:341.

Saviot, L., D. B. Murray, A. Mermet, and E. Duval. 2004a. Comment on Estimate of the vibrational frequencies of spherical virus particles. *Phys. Rev. E* 69(2):023901.

Saviot, L., D. B. Murray, and M. C. Marco de Lucas. 2004b. Vibrations of free and embedded anisotropic elastic spheres: Application to low-frequency Raman scattering of silicon nanoparticles in silica. *Phys. Rev. B* 69:113402.

Saviot, L., C. H. Netting, and D. B. Murray. 2007. Damping by bulk and shear viscosity of confined acoustic phonons for nanostructures in aqueous solution. *J. Phys. Chem. B* 111:7457.

Saviot, L., C. H. Netting, D. B. Murray, S. Rols, A. Mermet, A.-L. Papa, C. Pighini, D. Aymes, and N. Millot. 2008. Inelastic neutron scattering due to acoustic vibrations confined in nanoparticles: Theory and experiment. *Phys. Rev. B* 78:245426.

Stephanidis, B., S. Adichtchev, P. Gouet, A. McPherson, and A. Mermet. 2007a. Elastic properties of viruses. *Biophys. J.* 93(4):1354–1359.

Stephanidis, B., S. Adichtchev, S. Etienne, S. Migot, E. Duval, and A. Mermet. 2007b. Vibrations of nanoparticles: From nanospheres to fcc cuboctahedra. *Phys. Rev. B* 76:121404.

Tama, F. and C. L. Brooks III. 2005. Diversity and identity of mechanical properties of icosahedral viral capsids studied with elastic network normal mode analysis. *J. Mol. Biol.* 345(2):299–314.

Tang, Y. and M. Ouyang. 2007. Tailoring properties and functionalities of metal nanoparticles through crystallinity engineering. *Nat. Mater.* 6:754.

Tanguy, A., J. P. Wittmer, F. Leonforte, and J.-L. Barrat. 2002. Continuum limit of amorphous elastic bodies: A finite-size study of low-frequency harmonic vibrations. *Phys. Rev. B* 66:174205.

Visscher, W. M., A. Migliori, T. M. Bell, and R. A. Reinert. 1991. On the normal modes of free vibration of inhomogeneous and anisotropic elastic objects. *J. Acoust. Soc. Am.* 90:2154.

Weitz, D. A., T. J. Gramila, and A. Z. Genack. 1980a. Anomalous low-frequency Raman scattering from rough metal surfaces and the origin of surface-enhanced Raman scattering. *Phys. Rev. Lett.* 45:355.

Weitz, D. A., T. J. Gramila, A. Z. Genack, and J. I. Gersten. 1980b. Anomalous low-frequency Raman scattering from rough metal surfaces and the origin of surface-enhanced Raman scattering. *Phys. Rev. Lett.* 45:355–358.

Zhao, J. and Y. Masumoto. 1999. Size dependence of confined acoustic phonons in CuCl nanocrystals. *Phys. Rev. B* 60:4481.

12

Superheating in Nanoparticles

Shaun C. Hendy
Industrial Research Ltd.

and

*MacDiarmid Institute for Advanced
Materials and Nanotechnology*

Nicola Gaston
Industrial Research Ltd.

12.1 Introduction

While the supercooling of liquids below their freezing temperature is a frequently observed phenomenon, the superheating of solids is far less common. The reason for this is that most solids tend to premelt at their surfaces at temperatures slightly below their melting point (Tartaglinoa et al., 2005). This surface premelting means that when the melting point is reached, the nucleation of liquid is not required for the solid to melt, so the solid cannot maintain a metastable, superheated state. It is surprising then that superheating can be observed in atomic clusters (Breaux et al., 2003; Neal et al., 2007; Shvartsburg and Jarrold, 2000) and nanoparticles (Schebarchov and Hendy, 2006), which possess very large surface-to-volume ratios and typically have melting points that are substantially below those of the bulk (Buffat and Borel, 1976).

Bulk solids melt at the temperature at which their free energy equals that of the corresponding liquid state. In bulk materials, the melting temperature is strongly correlated with the energy of cohesion of the bulk. Indeed, the relationship between the melting temperature and the cohesive energy of elemental solids is approximately linear as shown in Figure 12.1.

What happens when we reduce the number of atoms from the infinite limit to consider finite clusters? The cohesive energy per atom will generally remain constant for all atoms that are sufficiently far from the surface for their environment will be unchanged from that of the bulk. However, the atoms at the surface are more weakly bound than bulk atoms, and therefore as the particle size (and number of atoms) decreases and the proportion of atoms at the surface increases, the average cohesive energy per atom will decrease. Given the above linear relationship between melting temperature and cohesive energy, it is relatively straightforward then to predict that the melting point will also decrease. This is the basic argument first put forward by Pawlow of melting point depression (Pawlow, 1909).

This argument can be framed more precisely in terms of surface energies. As the surface energy of a liquid–vapor interface (γ_{lv}) will be less in general than that of the solid (γ_{sv}), the energetic balance between the solid and the liquid states in particles with a high surface area-to-volume ratio will depend on particle size. Thus, as the particle size decreases, the free energy difference between the solid and the liquid particles will also decrease, lowering the temperature at which the liquid droplet becomes thermodynamically stable relative to the solid particle. In spherical particles, the surface area-to-volume ratio is inversely proportional to the particle radius, so one might expect that the decrease in the melting temperature will also be in inverse proportion to the particle radius. Such behavior has been observed experimentally, and via computer simulations, in many different nanoparticle systems for particle sizes as small as 2–3 nm (Ben David et al., 1995; Buffat and Borel, 1976). This behavior is expected to hold for free particles, as well as supported particles, provided interactions between the substrate and the particle do not significantly favor one phase over the other (Hendy, 2007).

In smaller particles (<300 atoms), often called atomic clusters, it may however not be appropriate to think of the particle

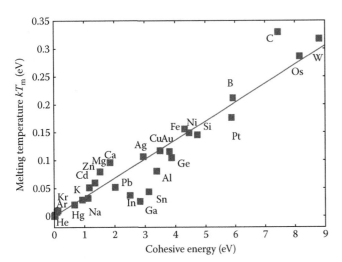

FIGURE 12.1 The correlation between melting point and cohesive energy is shown for a range of elemental solids. It is clear that the overall relationship is linear, and that gallium (Ga), tin (Sn), and aluminum (Al) are among the stronger outliers.

in terms of a surface and a volume. In such clusters, it is known that the atomic and electronic structure can change significantly as the cluster size decreases (Baletto and Ferrando, 2005). Both atomic and electronic shell closings can result in strong variations in cohesive energy with particle size. It also becomes difficult to define a melting point precisely in such systems. For instance, near a melting transition, small clusters are thought to fluctuate between the solid and the liquid states in a process known as dynamic coexistence (Berry et al., 1988). This has the effect of broadening the transition temperature as in an ensemble of small particles at temperatures in the transition region, a portion of them will typically be solid and a portion will be liquid. Nonetheless, a more precise transition temperature can be defined as the temperature at which the heat capacity reaches a maximum (Haberland, 2002); in bulk systems, the heat capacity diverges at the transition temperature.

The recent development of free cluster calorimetry (Breaux et al., 2003; Schmidt et al., 2001) has allowed these effects to be studied in detail, independent of any substrate. Such experiments have revealed that in small atomic clusters, consisting of tens to hundreds of atoms, the melting temperature can vary erratically and non-monotonically with particle size. In small sodium clusters, for example, large fluctuations in the melting point with cluster size have been observed (Schmidt et al., 1998). These fluctuations have been attributed to the occurrence of geometric shell closings at certain cluster sizes, which prevent the onset of diffusion and surface melting (Haberland et al., 2005). Remarkably, free cluster calorimetry has revealed that small clusters of Sn (Shvartsburg and Jarrold, 2000), Ga (Breaux et al., 2003), and Al (Neal et al., 2007) can remain solid at temperatures well above the bulk melting point.

Superheating has also been observed in simulations of larger metal nanoparticles terminated with particularly stable surface facets (Schebarchov and Hendy, 2006). Such surfaces, which

include both Al (111) (van der Gon et al., 1990) and (100) (Molenbroek and Frenken, 1994), are often called non-melting as they do not exhibit the surface melting (Carnevali et al., 1987) that precedes melting of the bulk in many materials. In macroscopic Pb crystals with non-melting (111) surface facets, superheating of up to 120 K above the melting point has been observed (Herman and Elsayed-Ali, 1992). Similar degrees of superheating have been predicted theoretically for the Al (111) surface (Di Tolla et al., 1995). While superheated macroscopic crystals are only metastable, simulations and theory suggest that superheated Al particles with diameters of several nanometers can be stable microcanonically (Schebarchov and Hendy, 2006) as the particles seek to avoid the energetic cost of phase coexistence.

In some circumstances, a substrate or an embedding matrix can also serve to increase the melting temperature of a nanoparticle. In general, one expects a substrate to reduce the surface energy of a nanoparticle, leading to an increase in cohesive energy and hence an increase in melting temperature (Qi and Wang, 2005). However, provided the particle remains discrete and does not react or mix with the substrate, the cohesive energy should remain bounded by the bulk cohesive energy and the melting temperature should remain below that of the bulk. Nonetheless, superheating has been observed in several instances for embedded nanoparticles (Zhong et al., 2001). This could occur, for example, if there were a sufficiently strong epitaxial relationship between the solid particle and the embedding matrix, so that γ_{sm}, the energy of the solid–matrix interface, was less than that of the liquid–matrix interface, γ_{lm}.

The purpose of this chapter is to review the evidence, both experimental and computational, for the superheating of free and supported nanoparticles, and to survey our current theoretical understanding of these phenomena. In this chapter, we focus almost entirely on metal nanoparticles, whose melting behavior has been studied extensively in the last decade. We also briefly discuss the superheating of nanoparticles embedded in a matrix. We begin by discussing some of the experimental techniques used in nanoparticle and cluster calorimetry. We then discuss superheating in bulk materials, which is closely related to the absence of surface melting in certain materials. This is followed by a discussion of melting point depression in small particles, and whether surface melting can occur in such particles. We then survey recent experimental findings of superheating in free atomic clusters, and computational findings of superheating in larger particles under microcanonical conditions. Finally, we briefly discuss superheating in embedded nanoparticles.

12.2 Techniques for Studying the Melting of Nanoparticles

In this section, we briefly describe some of the techniques that have been used to study the melting of clusters and nanoparticles. We will focus on techniques for studying free particles as these tend to be rather specialized. One way to determine the melting point of a solid is to measure its caloric curve in a process known as calorimetry. For a bulk material, a caloric curve can be determined by measuring the change of internal energy of a material as a function of its temperature. For instance, the material can be placed in a thermally insulated box, into which energy can be added in a controlled manner. One then measures the temperature T of the material as energy U is added, obtaining the caloric curve $U(T)$. At the melting point there will be a step in the caloric curve and a corresponding divergence in the heat capacity $c = \partial U/\partial T$. Below we discuss how caloric curves can be measured in nanoparticles and atomic clusters.

12.2.1 Free Particle Calorimetry

In nanoparticle calorimetry, there are of course a number of complications that arise when attempting to measure the melting point. Firstly, if the particle is supported by a substrate, then it is likely that the substrate will alter the caloric behavior of the particle either through contamination, or by interaction between the free surface of the particle and the surface of the substrate. Thus, there have been a number of free particle calorimetry techniques developed to study particles in the absence of a substrate. However as noted earlier, the melting transition in nanoparticles will be broadened due to the coexistence of solid and liquid clusters over a finite range of temperatures. This broadening includes the effects of dynamic coexistence (Berry et al., 1984) where individual clusters fluctuate between the solid and the liquid states over time.

The phenomenon of dynamic coexistence can be imagined as follows. For many elements, ranging from noble gases to metals, the most stable configuration of 13 atoms is the highly symmetric and compact icosahedron. We can imagine that this well ordered structure, existing in a deep well on the potential energy surface, is the nanoscale analogue of the solid state. If we imagine removing one of the surface atoms of the icosahedron and placing it on the outside of the structure, we can consider this to be an analogue of the liquid state where there will be many energetically similar configurations where the single atom on the outside of the cluster is relatively free to explore different surface configurations. There will be a small barrier to this atom repositioning itself in the icosahedron as the neighboring atoms will have relaxed around the vacant point on the surface. This situation is described in Figure 12.2. The result is, that in

an ensemble of small particles at temperatures in the transition region, a portion of them will typically be solid and a portion will be liquid. Nonetheless, a more precise transition temperature can be defined as the temperature at which the heat capacity reaches a maximum (Haberland, 2002) keeping in mind that it is over a range of temperatures about this melting temperature that particles can be found in a molten state.

With this caveat, a number of techniques have been developed to study melting in atomic clusters based on the construction of caloric curves. These techniques generally involve measuring the internal energy of a thermalized beam of clusters. Photofragmentation is a technique developed by Hellmut Haberland's group (Schmidt et al., 1997), where the energy of a thermalized population of mass-selected clusters is raised by a known amount using a laser until the clusters reach a temperature where evaporation of atoms from the clusters takes place. The temperature of the irradiated clusters can then be determined by increasing the cluster source temperature until the fragmentation behavior of both thermally and laser-heated clusters becomes the same. This gives the internal energy $U(T)$ of the clusters in the beam thermalized at a known temperature T. A caloric curve $U(T)$ and corresponding heat capacity $c(T)$ produced by the photofragmentation technique is shown in Figure 12.3 for a 192-atom Na cluster. A similar technique developed by Martin Jarrold's group (Breaux et al., 2003; Jarrold and Honea, 2007) uses collisions with a dense helium carrier gas to induce dissociation of their clusters. Again, they measure the internal energy $U(T)$ as a function of temperature to produce caloric curves and heat capacities.

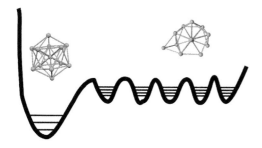

FIGURE 12.2 A schematic representation of dynamic coexistence of both solid and liquid clusters at the same temperature. (Adapted from Berry, R.S. et al., *Phys. Rev. A*, 30, 919, 1984.)

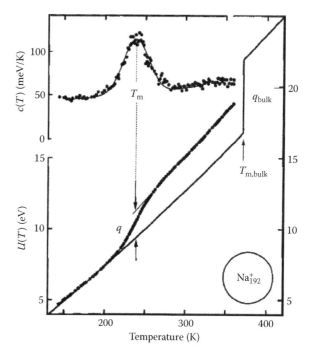

FIGURE 12.3 The caloric curve $U(T)$ and heat capacity $c(T)$ for a free 192-atom sodium cluster near the melting transition. (Reproduced from Schmidt, M. et al., *Nature*, 393, 238, 1998. With permission.)

As these methods require the use of mass spectrometry, they can only be used on charged clusters. Bachels et al. (2000) have developed another technique that can be used with neutral clusters. Changes in the internal energy of a cluster beam are measured using a sensitive pyroelectric foil with which a cluster beam collides, causing temperature increases which are detected by the corresponding jump in voltage across the foil.

These techniques can provide very precise measurements of the caloric curve, and in principle can be used to determine melting points. Even so, the broadening in temperature of the melting transition (see Figure 12.3) and the fact that multiple peaks in the heat capacity are observed in some circumstances (Breaux et al., 2005b; Neal et al., 2007) can make the identification of the melting point difficult both in principle (as it occurs over a range in temperatures due to dynamic coexistence) and in practice (as peaks associated with solid–solid transitions prior to melting in the clusters may overlap with the melting peak).

12.2.2 Other Experimental Techniques for Free Nanoparticles

Instead of focusing on the internal energy, there are numerous techniques that focus on other physical properties as a function of temperature. For example, Shvartsburg and Jarrold (2000) used the change in shape of tin clusters at the melting point to identify the melting transition. Small solid tin clusters adopt highly prolate shapes, which then revert to closely spherical shapes upon melting. As the cross-sectional area of a tin cluster will depend on its state, so will its mobility in a carrier gas. Thus by measuring the mobility of clusters thermalized at a known temperature the shape, and hence the state, of a cluster beam can be measured as a function of temperature.

12.2.3 Simulations of Nanoparticle Melting

Molecular dynamics simulation techniques have been used to study melting of clusters and nanoparticles for many years (Honeycutt and Andersen, 1987). Typically, these techniques rely on empirical potentials to describe the interactions between atoms, although more recently first principles approaches for calculating energies and forces such as density functional theory have also been used (for example see Reference Chacko et al., 2004). Caloric curves can be constructed either by gradual heating or cooling of a particle using an appropriate molecular dynamics thermostat (Nose, 1990), or by using a thermostat in conjunction with a technique such as parallel tempering (Neirotti et al., 2000), where replicas of a particle are equilibrated at a range of temperatures in parallel with Monte Carlo exchange of replicas to help ensure that ergodicity is achieved efficiently.

One of the drawbacks of molecular dynamics is that the timescales accessible are relatively short. While it can provide unrivaled atomistic detail of the dynamics and thermodynamics, simulations of longer than a fraction of a microsecond are not practical. Thus, one must be cautious when interpreting the results of molecular dynamics simulations: superheated solids that are metastable on the timescale of a simulation may be unstable on experimental timescales.

12.2.4 Supported and Embedded Nanoparticles

There are many more techniques that can be used to study the melting of supported and embedded nanoparticles, including electron microscopy (Lereah et al., 2001) and x-ray diffraction (Gråbaek et al., 1990). These methods identify the melting transition by allowing the observation of the breakdown of long-range order in the sample. However, it can be difficult to precisely measure the temperature of the sample especially if it is being heated by the electron beam for instance. It is also possible to use differential scanning calorimetry to study the melting of embedded nanoparticles in a bulk sample (Goswami and Chattopadhyay, 1995).

12.3 Superheating of Bulk Materials

In this section, we discuss superheating in bulk materials as a prelude to considering the possibility in nanoparticles. As noted in Section 12.1, the difficulty in producing superheated bulk solids is related to the phenomenon of surface melting. The theory of surface melting goes back at least to Pawlow's 1909 paper ("On the dependence of the melting point on the surface energy of a solid body") (Pawlow, 1909) and is the origin of the view (Cahn, 1986) that the superheating of solids is impossible, as stated by Frenkel (1946):

> It is well known that under ordinary circumstances an overheating (of a crystal), similar to the overheating of a liquid, is impossible. This peculiarity is connected with the fact that the melting of a crystal, which is kept at a homogeneous temperature, always begins on its free surface. The role of the latter must, accordingly, consist in lowering the activation energy, which is necessary for the formation of the liquid phase, i.e., of a thin liquid layer, down to zero.

Indeed, in bulk materials, melting is frequently initiated at a surface, and at temperatures below the melting point it has been observed that the bulk can coexist with a molten thin film. A well-known example of this is the Pb (110) surface (Frenken et al., 1986), which exhibits a thin molten film between the solid and the vapor interfaces at temperatures below the melting point. As the bulk melting temperature is approached, the thickness of this film (often called the quasi-liquid layer) diverges, leading to complete melting of the bulk at the melting temperature. This is known as surface melting.

Thus, the occurrence of surface melting will generally preclude the possibility of superheating in bulk solids as when the melting point is reached, nucleation of the liquid is not required to initiate melting. However, whether surface melting will actually occur depends on the relative magnitudes of the interfacial energies. If the sum of the surface energies of the solid–liquid and the liquid–vapor interfaces ($\gamma_{sl} + \gamma_{lv}$) is less than the surface energy of the solid–vapor interface γ_{sv}, then surface melting will lead to a decrease in total surface energy. Thus surface melting should occur at a temperature below the bulk melting temperature. Physically, the condition that $\gamma_{sv} > \gamma_{sl} + \gamma_{lv}$ corresponds to the situation where the melt fully wets the solid surface (i.e., the melt exhibits a contact angle of zero when in contact with the solid facet).

There are several techniques that have been employed for superheating a bulk solid, most of which control the role of the surface in melting. Early studies (Ainslie et al., 1961; Cornia et al., 1963; Tammann, 1910) used materials with particularly viscous melts, such that the propagation of the solid–liquid interface from the surface into the bulk was slowed considerably. In this way, materials such as quartz (Ainslie et al., 1961) were able to be superheated by up to 300 K. An alternative method was used in an early experiment on tin, which involved passing a current through a crystal of tin, while fanning the outside to prevent surface melting. In this way, the interior of the crystal was estimated to reach a temperature 2 K higher than the normal melting temperature (Khaikin and Bené, 1939).

However, when $\gamma_{sv} < \gamma_{sl} + \gamma_{lv}$, that is, when the melt only partially wets a surface facet, surface premelting is not expected to occur. In this case, surface melting would lead to an increase in surface energy, and as such it should not occur below the bulk melting temperature. Indeed, such surfaces, which include Pb (111) (Pluis et al., 1987), and both Al (111) (van der Gon et al., 1990) and (100) (Molenbroek and Frenken, 1994), are often called non-melting. Furthermore, in the absence of surface melting, superheating has been observed in a number of cases. For instance, the Pb (111) surface can be superheated by up to 120 K above the bulk melting temperature (Herman and Elsayed-Ali, 1992, 1994). In micron-sized Pb crystals, bounded only by non-melting (111) surface facets, superheating by ~3 K above the melting point has been observed (Heyraud and Métois, 1987). Similar degrees of superheating have been predicted theoretically for the Al (111) surface (Di Tolla et al., 1995) using molecular dynamics simulation, albeit on much shorter timescales.

What are the limits of metastable superheating that can occur via this mechanism? In the absence of surface melting or some heterogeneous nucleation site for the melt, some instability will eventually trigger nucleation at some temperature above the equilibrium melting temperature. Fecht and Johnson (1988) have suggested that in the bulk this point occurs when the entropy of the solid exceeds that of the liquid; Tallon has used MD simulations to suggest that it is a mechanical instability that destroys the superheated state (Tallon, 1989). In the presence of a free non-melting surface, Di Tolla et al. (1995) have argued that it is non-melting-induced faceting that eventually leads to surface melting and the melting of the bulk.

In summary, superheating can occur in bulk solids either through suitable chosen non-equilibrium conditions that prevent surface melting or in solids bounded by non-melting surfaces. In these cases, the superheated solid is at best metastable and melting will eventually occur either through the eventual nucleation of the liquid or through some instability in the solid itself.

12.4 Melting Point Depression in Nanoparticles and Atomic Clusters

In this section, we discuss the origin of the depression of the melting point in small particles. The dependence of the melting temperature on the particle size was demonstrated on small gold particles between 20 and 50 Å in size (Buffat and Borel, 1976). A significant depression of the melting temperature (15%–45%) was observed below 50 Å and the measured melting temperatures were shown to agree well with phenomenological models based on the experimentally known surface tension. For simplicity, if we assume that the density ρ of the solid and the liquid particle are identical, the difference between the Helmholtz free energy F_s of a spherically symmetric solid particle of radius R and the free energy F_l of the corresponding molten droplet is given by

$$\Delta F = F_s - F_l = \frac{4\pi R^3}{3}\left(f_s - f_l\right) + 4\pi R^2 \left(\gamma_{sv} - \gamma_{lv}\right), \quad (12.1)$$

where $f_{s(l)}$ is the free energy density of the solid (liquid). The particle will melt at the temperature T_m where $F_s = F_l$ which gives

$$T_m = T_c\left(1 - 3\frac{\gamma_{sv} - \gamma_{lv}}{\rho L R}\right). \quad (12.2)$$

if one expands $f_s - f_l$ as

$$f_s - f_l = -\rho L\left(\frac{1 - T}{T_c}\right), \quad (12.3)$$

where

    L is the latent heat of melting
    T_c is the bulk melting temperature

The result given here, that the reduction of melting temperature is in inverse proportion to particle size, has been observed in many systems and holds down to particle sizes of several nanometers or so (Buffat and Borel, 1976).

However, atomic clusters, consisting of less than several hundred atoms, exhibit irregular variations in many of their properties, including their melting temperature (Schmidt et al., 1998). For metal clusters, these are typically associated with shell closings in the jellium model which confer a particular stability (i.e., peaks in the cohesive energy at particular numbers of electrons, often referred to as "magic numbers"). Noble gas clusters, on the other hand, are prototypical examples of clusters that exhibit geometrical shell closings, that is, particular stability associated with a particular number of atoms being able to complete a well-packed cluster structure with a minimum number of surface atoms relative to the inner atoms. For smaller clusters, such geometric structures are usually icosahedral, although as the number of atoms increases the preference for icosahedral structures becomes weaker since the interaction between each additional shell of atoms and the core becomes weaker. This is the price paid for maintaining well-packed surfaces. For larger numbers of atoms, bulk-like packing becomes eventually preferable, such as face-centered cubic-type clusters. Therefore the variation of properties with cluster size can be due not only to "magic numbers" of a particular packing type, but additionally to competition between different structural motifs. In addition, it has been shown that depending on the property being studied, a combination of both electronic and geometrical shell closings may play a role (Schmidt et al., 1998). This seems to be particularly the case for properties which may involve both geometric contributions (e.g., surface structure) and electronic contributions (e.g., to the cohesive energy, since for the sodium clusters studied, the strongest determinant of cluster stability is the number of electrons). The occurrence of shell closings, whether electronic or geometric, at certain cluster sizes can prevent the onset of diffusion and surface melting (Haberland et al., 2005). Nonetheless, the manner in which this complex, small-scale melting behavior gives way to the more familiar phenomenology of the bulk melting transition remains poorly understood.

Thus, it is generally accepted that there are two size regimes when describing the melting behavior of small particles. In the first regime, particles are large enough that they can be regarded as bulk solids with high surface area-to-volume ratios, whereas in the second regime, particles are too small for this picture to hold. In the former, the melting temperature of a particle will be a smooth function of its surface area-to-volume ratio. In the latter case, where particles consist of fewer than several hundred atoms, it is expected that the melting temperature will depend sensitively and non-monotonically on particle size.

12.5 Surface Melting in Nanoparticles

It is clear that the absence or avoidance of surface melting is important for the successful superheating of bulk solids. However, the occurrence of surface melting in nanoparticles is not as well understood as it is in bulk materials. Indeed, in particles of radius $R \sim 3\gamma_{sl}/\rho L$, where γ_{sl} is the solid–liquid interfacial energy and ρL is the latent heat of melting per unit volume, the energy to melt a particle can be comparable to the cost of forming the solid–liquid interface. This radius is of the order of 1 nm for metals (Turnbull, 1950), if one uses bulk values for the material properties. Thus, in particles of this size, the gain in free energy from surface melting will be less than or equal to the energy needed to melt the solid surface layer.

Some insight can be gained by applying phenomenological models of surface melting for bulk materials to a spherically symmetric nanoparticle, as shown in Figure 12.4. In this approximation, the free energy of a nanoparticle of radius R covered by a liquid layer of melt with thickness $R - r$ can be written as

$$F = \frac{4\pi\left(R^3 - r^3\right)}{3}f_l + \frac{4\pi r^3}{3}f_s + 4\pi R^2\gamma_{lv} + 4\pi r^2\left(\gamma_{sl} + \Delta\gamma e^{-(R-r)/\xi}\right).$$

(12.4)

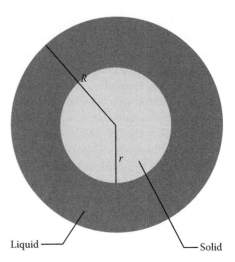

FIGURE 12.4 A surface-melted spherical particle of radius R. The thickness of the liquid layer is $R - r$.

where $\Delta\gamma = \gamma_{sv} - \gamma_{lv} - \gamma_{sl}$. The last term approximates the thickness dependence of the solid–liquid interfacial energy γ_{sl} (van der Veen et al., 1990) where ξ is a correlation length. The exponential form of the thickness dependence follows from a Ginzburg–Landau analysis of the premelting of flat surfaces (Pluis et al., 1990). We recall that if $\Delta\gamma > 0$, then the liquid will fully wet a planar solid surface. If $\Delta\gamma < 0$, then only partial wetting will occur.

Strictly speaking, the radial symmetry assumed in Equation 12.4 is only valid for $\Delta\gamma > 0$. For $\Delta\gamma < 0$, any molten surface layer will presumably not wet the solid cluster fully, leaving exposed facets (Schebarchov and Hendy, 2006). For simplicity, we will only deal with the case of $\Delta\gamma > 0$ where spherical symmetry can be assumed; it is possible to show that when $\Delta\gamma < 0$ surface melting should not occur.

Now we look for extrema in the free energy F for particular r. For $0 < r < R$, such extrema correspond to surface-melted configurations and if they are minima they correspond to stable or metastable surface-melted states. Using Equation 12.4, then we need to find solutions to

$$\frac{dF}{dr} = -4\pi r^2\left(f_l - f_s\right) + 8\pi r\gamma_{sl} + 4\pi r\Delta\gamma\left(2 + r/\xi\right)e^{-(R-r)/\xi} = 0$$

(12.5)

between 0 and R. Note that $r = 0$ is always an extremum and corresponds to the fully liquid cluster.

For surface melting to occur, a minimum in the free energy must appear at $r = R$. Setting $r = R$ in Equation 12.5, one can solve for the temperature, T_s, at which this can happen:

$$T_s = T_c\left(1 - \frac{\gamma_{sv} - \gamma_{lv}}{\rho L}\left(2 + \frac{R}{R_c}\right)\right)$$

(12.6)

where $R_c = \xi(\gamma_{sv} - \gamma_{lv})/\Delta\gamma$. Thus, F has a stationary point at $r = R$ and $T = T_s$. Computing the second derivative of F at $r = R$ and $T = T_s$, we find that

$$\frac{d^2F}{dr^2} = 4\pi \frac{\Delta\gamma}{\xi}\left(R^2 + 2\xi(R - R_c)\right). \qquad (12.7)$$

If $\Delta\gamma > 0$ and $R > R_c$, then the second derivative is positive so the extrema at $r = R$ and $T = T_s$ is a minimum, corresponding to the onset of a stable surface-melted state. Indeed, in the limit $R \to \infty$, we recover the criteria for surface melting on bulk solid surfaces, namely, that $\Delta\gamma > 0$ (Tartaglinoa et al., 2005). Indeed, if $\Delta\gamma < 0$, then $T_s > T_c$ and full melting precedes surface melting as for bulk surfaces (Di Tolla et al., 1996).

For a nanoparticle with finite radius R, the temperature T_s at which surface melting occurs differs from that of the equivalent bulk surface. Indeed, if $\Delta\gamma > 0$, then the surface melting temperature of a nanoparticle is less than that of the bulk surface. However, by comparing Equations 12.6 and 12.2 it is easily seen that $T_s > T_m$ for particles with radii $R < R_c$. Further, the free energy of the coexisting state is always greater than that of the liquid when $R < R_c$, so surface melting will not occur at all in clusters less than this critical size (Bachels et al., 2000).

This argument leads us to expect that surface melting will be less likely to occur in small particles, especially when $R < R_c$. However, it is not clear whether this argument, which relies on macroscopic concepts and quantities, will apply to small nanoparticles and clusters. Nanoparticles possess edges and vertices, in addition to facets, and in general it is not possible to assign facets to the surfaces of small clusters. As is discussed in Section 12.6, recent experiments on small Al clusters by Jarrold's group (Neal et al., 2007) have revealed that at certain cluster sizes (51, 52, 56, 61, and 83 atoms) the measured heat capacities cannot be fitted by a two-state (fully solid and fully liquid) model. This suggests that some intermediate state arises prior to complete melting and a strong candidate for this must be a partially melted state. Indeed, surface melting in particles of this size has been observed by molecular dynamics simulation (Cheng and Berry, 1992), although intermediate solid structures have also been observed in some simulations (Cleveland et al., 1998).

Although it would appear that metastable superheating stabilized by non-melting surface facets is possible in microsized Pb particles (Heyraud and Métois, 1987), such an effect is yet to be observed conclusively in nanoparticles. As we will see below, experiments on Al clusters by Jarrold's group (Neal et al., 2007) found degrees of superheating at several cluster sizes (in particular, the 48-atom Al cluster). However, we will see that these sizes are distinct from those where possible premelting behavior is observed. Thus, the evidence to date suggests that superheating and surface melting are indeed incompatible in small clusters. However, unlike in bulk systems, we cannot claim that it is the lack of surface melting that causes superheating in such instances; even in the absence of surface melting we would still expect that the melting temperature should be highly suppressed in such small clusters. Thus, in general, we expect that the absence of surface melting is a necessary condition for superheating in nanoparticles, but it is not sufficient.

12.6 Superheating of Atomic Clusters

In this section, we turn to the superheating of small clusters (<300 atoms), and we will see that further information about the nature of these nanoparticles is required in order to fully explain the phenomenon of superheating. As noted earlier, in clusters of this size we can no longer think of the cluster in terms of bulk and surface quantities, and the melting temperature is expected to depend sensitively on the number of atoms in a cluster.

12.6.1 Tin

Tin cluster cations with between 10 and 30 atoms were first shown to superheat in 2000 (Shvartsburg and Jarrold, 2000). The melting transition was observed by measuring the collision cross sections of the ions, which decrease upon melting due to a change from a prolate geometry to a spherical liquid droplet. It would also be possible for the change in cross section to occur due to a change in solid structure from prolate to oblate; however that possibility only means that the measured transition temperatures are a lower bound for the melting point, and as such remain a confirmation of superheating. While it was recognized by the authors of that work that the structures of the small clusters are very unlike that of the bulk material, it was also pointed out that the cohesive energies of these clusters are expected to be less than the bulk cohesive energy. Therefore, it does seem that at this point the correlation between melting temperature and cohesive energy that seems so clear for the bulk elements is no longer applicable. At what size exactly this begins to break down is an interesting question; for mesoscopic tin particles, the smooth melting point depression expected according to standard arguments is observed, and the regime change can only be said to occur between 30 and several thousand atoms. The combined experimental results showing melting point depression for larger tin particles and superheating for very small clusters are shown in Figure 12.5.

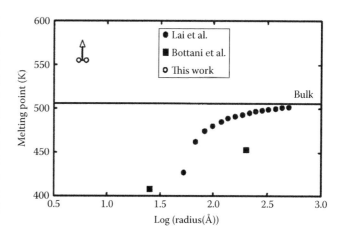

FIGURE 12.5 The melting points of tin clusters over a range of sizes. "This work" refers to the original work of Jarrold's group. (Reproduced from Shvartsburg, A.A. and Jarrold, M.F., *Phys. Rev. Lett.*, 85, 2530, 2000. With permission.)

It is worth noting that the authors suggested that similar effects should be at work for germanium and silicon clusters, which adopt similar structures: however the much higher melting points of bulk Ge and Si already are some indication that the bonding in the bulk materials is significantly different than in tin. For the clusters to superheat, a significant difference between the structures of the clusters and that of the bulk seems to be required.

12.6.2 Gallium

Indeed, it was not in germanium or silicon that the next case of superheating clusters was found, but in gallium, where cationic clusters with between 17 and 40 atoms were seen to remain solid well above the bulk melting point ($T_m = 303$ K) of gallium (Breaux et al., 2003). In this study, in contrast to that of tin, calorimetry measurements were performed on gallium clusters in order to produce curves of heat capacity vs. temperature, in which a peak provides a signature of melting. In actual fact, they track the value of the kinetic energy at which 50% of the clusters dissociate, which is a measure of the change in the internal energy with temperature and therefore a measure of the changing heat capacity. As for the Sn clusters, the origin of the superheating of these small gallium clusters is assumed to be structural.

Nonetheless, it is interesting to note that superheating has also been seen in macroscopic Ga samples. The 1937 study of Volmer and Schmidt (*Über den Schmelzvorgang-On melting*) (Peppiatt, 1975; Volmer and Schmidt, 1937) demonstrated a superheating of well-formed facets of gallium by 0.1 K. It was also reported that liquid gallium does not completely wet certain faces of the bulk solid, which suggests that certain facets of bulk Ga are nonmelting. However, in Figure 12.1 it is already clear that gallium is an outlier, requiring only half the amount of energy of a typical element (relative to the cohesive energy) before it melts.

The cohesive energy of the elemental solids is defined as the difference between the energy per atom of the solid and the energy of an individual atom in isolation. However gallium is rather unique among metals in that in the solid state it has a very unusual structure—an orthorhombic 8-atom unit cell composed of dimers which have a bond distance 15% shorter than the next nearest neighbor distance. As such, it has been called a "molecular metal," which really contains two distinct kinds of bonding: covalent bonds within the dimers which result in a pseudogap at the Fermi level, and metallic bonding formed by overlap of the electronic wavefunctions along the buckled planes perpendicular to the molecular bond of the dimer. As a consequence, conduction in gallium is highly anisotropic, with a much greater conductivity along the buckled planes than perpendicular to this direction (Gong et al., 1991; Heine, 1968).

If we estimate the "metallic" cohesive energy of the solid, by calculating the cohesive energy relative to the dimer, we immediately see that gallium moves much closer to the other elements, as shown in Figure 12.6. In fact, in this case we are still overestimating the cohesive energy relative to the melting temperature, since the dimers in the bulk will be distorted from the

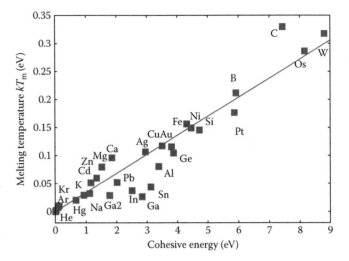

FIGURE 12.6 The correlation between melting point and cohesive energy is shown for a range of elemental solids. The cohesive energy of gallium calculated relative to the gallium dimer is included for comparison (labeled Ga2). It is clear that this results in a shift bringing gallium more in line with the other elements.

free dimer geometry. This argument certainly explains the low melting point of gallium relative to other metals, and due to the anisotropy of the lattice most likely relates to the surface superheating of gallium that was observed very early on (Volmer and Schmidt, 1937).

Density functional theory (DFT)–based molecular dynamical simulations have been performed on small gallium clusters with the same number of atoms as in the experiment, in order to probe the structures and bonding in these systems (Chacko et al., 2004). These simulations are able to reproduce melting temperatures in agreement with experiment, and describe the bonding between gallium atoms in these clusters as being strongly covalent, with only weak hybridization leading to a preference for 90° or 180° angles between bonds. One point that weakens their comparison with experiment is the utilization of neutral clusters for their simulations, which may cause some differences compared to the cationic clusters produced in experiment. In an additional paper (Breaux et al., 2005a), the authors of the original experimental paper (Breaux et al., 2003) extend their measurements to larger cationic clusters with between 29 and 55 atoms. Here, they point out that the structures of these small clusters may sometimes be amorphous and not well defined at all, as demonstrated by very flat curves of heat capacity relative to temperature, with no well-defined peak corresponding to a melting transition. They nonetheless conclude from a change in the cross sections that these amorphous clusters have indeed melted, and also at a higher temperature than the melting point of the bulk. They conclude that in these cases the melting transition is actually a second-order phase transition. This is an additional complication in the relationship between structure and melting point.

More recent theoretical investigations (Joshi et al., 2006), again DFT-based MD simulations performed on neutral clusters, conclude from a comparison of clusters with 30 and 31 atoms that the cluster with an odd number of atoms melts at a higher temperature due to a more ordered structure. It is significant that the number of atoms differs only by one, since the bulk crystallizes in a structure based on dimers. One may imagine that clusters with an even number of atoms might have a different underlying pattern of bonding than a cluster with an odd number of atoms. However, this interpretation is again limited by the comparison of neutral with cationic clusters.

Given that the first two observations of superheating in small clusters were for tin and gallium, it seems reasonable to extend such studies to neighboring elements in groups 13 and 14 of the periodic table.

12.6.3 Aluminum

Soon after the discovery of superheating in gallium clusters, similar experiments were performed on small aluminum clusters (Neal et al., 2007). The experiments show that superheating is possible in small clusters of aluminum but only for certain cluster sizes, with the superheating not as dramatic as in Ga but possibly as much as 100 K for one extreme case ($N = 48$), as shown in Figure 12.7. This is interesting since aluminum is quite different to gallium in the bulk, adopting a much more standard fcc metallic lattice, and consequently melts at a rather higher temperature ($T_m = 933$ K). It seems likely that further cases of superheating of clusters will be more like aluminum than gallium, if we accept the hypothesis that gallium is unique in the origin of the particularly low melting temperature of the bulk.

The phenomenon of dynamic coexistence can also be examined in the light of these experiments. In particular, the authors have employed a three-state model, which allows for the existence of intermediate states in the melting transition to describe their measured heat capacity curves. For some cluster sizes, the peaks in the heat capacity are highly symmetric and well described by a two-state model which assumes that the process is an equilibrium at the melting temperature

$$\text{Solid} \rightleftharpoons \text{Liquid} \qquad (12.8)$$

with the progress of the transition (with Gibbs free energy change ΔG_m) described by

$$K = \frac{[L]}{[S]} = \exp\left[-\frac{\Delta G_m}{k_B T}\right] \qquad (12.9)$$

which can be rewritten as

$$K(T) = \exp\left[-\frac{\Delta H_m}{k_B}\left(\frac{1}{T} - \frac{1}{T_m}\right)\right] \qquad (12.10)$$

given that at the midpoint of the transition, at T_m, $K(T_m) = 1$ and $\Delta G_m = 0$, so that the entropy change $\Delta S_m = \Delta H_m/T_m$, with ΔH_m being the enthalpy change for the melting transition. Then, the fraction of liquid clusters can be written as

$$\phi_L = \frac{K(T)}{1 + K(T)} \qquad (12.11)$$

Finally, the heat capacity C can be calculated as

$$C = (1 - \phi_L)S_S\frac{\Delta E_{MD}(T)}{\Delta T} + \phi_L S_S S_L\frac{\Delta E_{MD}(T)}{\Delta T} + \frac{\Delta(\phi_L \Delta H_m)}{\Delta T} \qquad (12.12)$$

Here S_S and S_L are scale factors that relate the heat capacities of the liquid and solid clusters to the internal energy estimated from the modified Debye model, E_{MD}. The application of such a model to the experimental heat capacities is shown in Figure 12.8 for a 39 atom cationic cluster. Also shown is the use of a three-state model:

$$\text{Solid} \rightleftharpoons \text{Intermediate} \rightleftharpoons \text{Liquid} \qquad (12.13)$$

which allows for the description of either premelting or postmelting phenomena. In the example shown, the shoulder on the high temperature side of the transition implies that the second

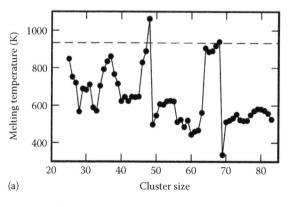
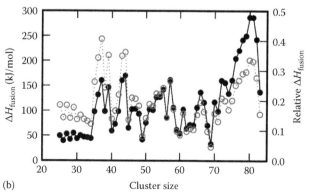

(a)   Cluster size   (b)   Cluster size

FIGURE 12.7 The melting points of aluminum clusters (a) compared to the bulk melting temperature (shown as a dashed line). Plot (b) is of the latent heat per cluster obtained from the two-state model (solid circles, left-hand scale) and also the relative latent heat (open circles, right-hand scale) which is the cluster latent heat divided by n times the bulk latent heat per atom. (Reproduced from Neal, C.M. et al., *Phys. Rev. B*, 76, 054113, 2007. With permission.)

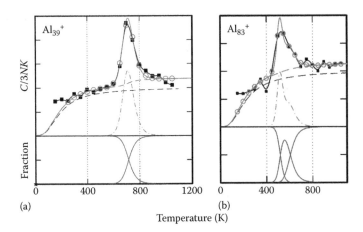

FIGURE 12.8 (a) The result of a two-state model fit to the experimental heat capacity vs. temperature curve. The upper part of the figure shows the measured heat capacities (solid points), the results of the fit to the two-state model (open circles), the heat capacity from the modified Debye model (dashed black curve), the calculated heat capacity from the two-state model (without the latent heat, dashed gray line), and the contribution from the latent heat is the peak below (dashed-dotted line). The lower part of the figure shows the proportion of solid and liquid particles as a function of temperature. (b) The three-state model reproduces the asymmetric peak seen for certain cluster sizes. (Reproduced from Neal, C.M. et al., *Phys. Rev. B*, 76, 054113, 2007. With permission.)

transition has a lower latent heat than for the first transition. In cases where the shoulder is on the low temperature side of the peak, it can be imagined that this is a sign of surface premelting, or a sign of a structural transition, where a different solid structure to the ground state may become entropically favored as the temperature is raised, and therefore exist briefly before melting. However, it is hard to imagine that there would be a larger latent heat for such a transition than for the melting transition, and thus the post-melting phenomena are more likely to be due to surface or partially melted intermediates. A similar phenomenon is indicated by the small dip in the heat capacity observed before the main peak in the heat capacity, which is assumed to be due to annealing processes, i.e., where the cluster is originally in a metastable state and finds the ground state only as the temperature is raised. These points have been removed from the fitting procedure for the two- or three-state models as this process is not included in the model.

12.6.4 Summary

Given that the number of systems of free clusters in which superheating has been observed remains to date very small, it is difficult to generalize about the origin of superheating and to what extent it may be observed in other systems. Based on the data in the works that we have mentioned, clusters of a certain size seem to melt at around the same temperature (for example, $T_m \sim 800\,K$ for Ga_{46}^+, $Ga_{45}Al^+$, Al_{46}^+, and also >550 K for Sn_{32}, 600 K for Al_{32}^+, 600 K for Ga_{32}^+), which suggests that the behavior of the small clusters may be quite similar across the different elements. To what extent this depends on the structure of these clusters is not yet completely clear. However, given that of the poor metals (the metals in groups 13–15 of the periodic table, Al, Ga, In, Tl, Sn, Pb, and Bi) aluminum has the highest melting temperature in the bulk, it is possible that In, Tl, Pb, and Bi clusters of similar sizes will also superheat.

12.7 Superheating in Larger Nanoparticles

In the specific case of an isolated nanoparticle, in only poor contact with a heat bath, the situation is somewhat more complicated. At this stage it is useful to distinguish between the canonical and the microcanonical caloric curves of nanoparticles. A microcanonical ensemble can be prepared by irradiating a dilute beam of cold particles with a laser, leaving the clusters in a narrow range of total energies E. If each isolated particle, prepared in this way, is in sufficiently poor contact with a heat bath, it will relax internally more rapidly than its temperature equilibrates with its environment. In such a situation the particle will follow a microcanonical caloric curve, $T(E)$ where T is the microcanonical temperature, as it slowly equilibrates with the heat bath.

In large particles, the canonical and microcanonical caloric curves will be identical. In small particles, however, the microcanonical and canonical caloric curves can differ. Perhaps, the most celebrated example of this is the discovery of negative heat capacities in the microcanonical calorimetry of small Na clusters. This was anticipated theoretically (Bixon and Jortner, 1989; Labastie and Whetten, 1990) and later confirmed experimentally (Schmidt et al., 2001). Negative heat capacities can arise due to S-bends in the microcanonical caloric curves, as shown in Figure 12.9. S-bends themselves arise as the particle tries to avoid the energetic cost of solid–liquid phase coexistence. Recall that in the microcanonical ensemble, the solid branch of the caloric curve is linked to the liquid branch by a solid–liquid coexistence line. As discussed earlier, the energetic cost of forming the solid–liquid interface can be substantial in a small particle, so the particle will avoid this by overheating on the solid branch and undercooling on the liquid branch, leading to an S-bend.

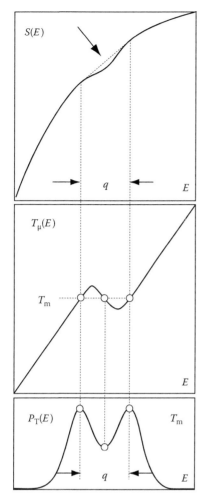

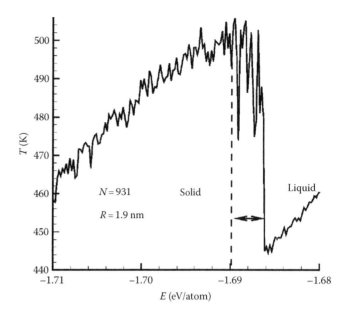

FIGURE 12.10 A microcanonical caloric curve for a 931-atom Pb nanoparticle near the melting point. The 931-atom particle does not exhibit a stable coexisting phase. The large fluctuations near the melting point signal the appearance of precritical liquid nuclei. (Reprinted from Hendy, S.C., *Phys. Rev. B*, 71, 115404, 2005. With permission.)

FIGURE 12.9 Three quantities are plotted as a function of the internal energy of a cluster, illustrating different manifestations of the same phenomenon (q is the latent heat, T_m the melting temperature). Top: The total entropy $S(E)$ having an inverted curvature dent (arrow), which is strongly exaggerated here. Such a structure is theoretically expected for a small particle. Middle: A back bending microcanonical caloric curve. The heat capacity becomes negative in the region with the negative slope. Bottom: The energy distribution $P_T(E)$ of a cluster ensemble close to its melting temperature. Because of the inverted curvature of the entropy the distribution becomes bimodal. (Reprinted from Schmidt, M. et al., *Phys. Rev. Lett.*, 86, 1191, 2001. With permission.)

While S-bends can occur in clusters and particles with surface melting and non-melting surfaces, they are likely to be more pronounced in particles with non-melting surfaces where the cost of forming the interface will be greater. Indeed, in sufficiently small particles, phase coexistence is thought to become completely unstable. For example, by using molecular dynamics simulations, Hendy (2005) found that in small Pb particles bounded by non-melting (111) facets phase coexistence was unstable in clusters with fewer than 1000 atoms. This is analogous to the disappearance of surface melting in small particles as discussed previously (Bachels et al., 2000). Avoidance of phase coexistence leads to overheating of the solid branch of the caloric curve and undercooling of the liquid branch, as shown in Figure 12.10.

Can this overheating result in superheating? There are two distinct ways in which it could do so. Firstly, if no phase coexistence occurs prior to melting then in principle the solid could remain metastable up to and above the bulk melting temperature. However, it is also possible that a superheated solid nanoparticle could be stable microcanonically, if it was unfavorable for it to undergo phase coexistence. Indeed, Schebarchov and Hendy (2006) have shown using a simple phenomenological model, that in principle, a nanoparticle that was bounded by non-melting surface facets could remain stable above the bulk melting temperature as it seeks to avoid phase coexistence. Further, using molecular dynamics, they found that a Al 4033-atom truncated octahedron structure, when described by the many-body glue potential for Al (Ercolessi and Adams, 1994), exhibited stable superheating prior to melting (see Figure 12.11) under microcanonical conditions. The truncated octahedron structure possesses both (111) and (100) surface facets, and with this potential $\Delta\gamma = \gamma_{sv} - \gamma_{sl} - \gamma_{lv}$ has been estimated as $\approx -2.3\,\text{meV}/\text{Å}^2$ for the (111) facets and $\approx -1.6\,\text{meV}/\text{Å}^2$ for the (100) facets (Di Tolla et al., 1995). With surface facets that satisfy $\Delta\gamma < 0$, as the 4033-atom particle seeks to avoid phase coexistence, they found that it can become superheated above the bulk melting temperature by between 5 and 26 K before melting begins. Finding that the superheated state could be obtained both by heating from the solid state, and by cooling from a solid–liquid state, they concluded that the superheated state was stable, rather than only metastable.

Furthermore, if the microcanonically superheated particle is then put in contact with a thermostat at the same temperature,

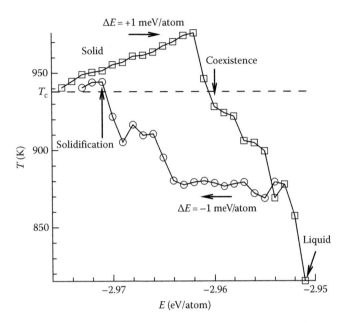

FIGURE 12.11 A microcanonical caloric curve for a 4033-atom Al cluster near the bulk melting point ($T_c = 939$ K using the potential). The curve was constructed both by heating from the solid state (squares) and cooling from a coexisting structure (circles) that emerges as the cluster approaches the liquid state. (Reprinted from Schebarchov, D. and Hendy, S.C., *Phys. Rev. Lett.*, 96, 256101, 2006. With permission.)

the particle is observed to melt after a short period of time. Again this suggests that the superheated particle is not merely trapped in a metastable superheated state, but rather that it is in a stable microcanonical state, which ceases to be stable when it can obtain the latent heat to melt from a heat bath.

This prediction has not been tested experimentally. It should be noted that the molecular dynamics simulations in Schebarchov and Hendy (2006) were applied to ideal truncated octahedral particles. It would be very difficult to prepare particles of such a size experimentally without introducing defects and stacking faults. These defects would lower the stability of the solid relative to the liquid and possibly suppress the superheating seen in the simulations here.

12.8 Superheating of Embedded Nanoparticles

As noted earlier, the melting temperature is suppressed in small particles because the difference in interfacial energies $\gamma_{sv} - \gamma_{lv}$ of the free solid and liquid particles is greater than zero. However, if the particle is embedded in a matrix, this is not necessarily the case. Provided the matrix itself has a higher melting temperature than that of the particle, the possibility of superheating exists if $\gamma_{sm} < \gamma_{lm}$, where γ_{sm} is the energy of the solid–matrix interface and γ_{lm} is that of the liquid–matrix interface. For example, this might be possible if there is a good epitaxial relationship between the solid particle and the matrix. Further, if this inequality held, the degree of superheating would be expected to *increase* linearly with inverse particle size.

There are a number of ways one can embed a nanoparticle in a matrix. These include ion implantation (Liu et al., 1998), precipitation during a quench (Goswami and Chattopadhyay, 1996), sol-gel techniques (Mikrajuddin et al., 2001), or ball milling (Rosner et al., 2003). Many of these techniques have been used to produce matrix–particle composites that can be heated above the bulk melting temperature of the particle material without melting (Mei and Lu, 2007). For instance, Zhang and Cantor (1991) fabricated In and Pb nanoparticles embedded in an Al matrix by precipitation and found superheating of up to 40 K above the bulk melting temperatures of In and Pb respectively. In both these cases, the lattice mismatch between the embedded particles (Pb and In) and the matrix (Al) is quite large (~25%) so they argued that it was less likely that $\gamma_{sm} < \gamma_{lm}$. However, Goswami and Chattopadhyay (1996) found that Pb particles in an Al matrix which exhibited cube-on-cube epitaxy could be superheated, whereas Pb particles of a similar size in a Ni matrix with little or no epitaxy could not be superheated. Furthermore, Sheng et al. (1998) have found the expected inverse relationship between degree of superheating and particle size in In, Sn, Bi, Cd, and Pb particles in an Al matrix.

Zhang and Cantor (1991) offered an alternative explanation for the superheating of embedded particles involving the lack of nucleation sites for the liquid at the ordered particle–matrix interface. This explanation for superheating is similar in a sense to that for superheating of macroscopic solid surfaces that are non-melting. In this case, one might argue that although $\gamma_{sm} > \gamma_{lm}$ so that the interfacial energy can be reduced by melting, in fact $\gamma_{sm} < \gamma_{sl} + \gamma_{lm}$ so that the nucleation of a liquid layer would not be favored, potentially leading to metastable superheating of the embedded particle.

Finally, we note that superheating has been observed in nanoparticles that have been coated with a higher melting point element. Daeges et al. (1986) coated Ag particles (120–160 μm in diameter) with Au, and a superheating of the Ag up to 25 K was observed for ~1 min. This effect has been seen in clusters and nanoparticles in molecular dynamics simulations (Broughton, 1991). Metallic nanoparticles encapsulated in graphitic shells have been superheated hundreds of degrees above their bulk melting temperatures (Banhart et al., 2003).

12.9 Conclusion

We have considered the evidence and theoretical understanding for superheating in metallic clusters and nanoparticles. Superheating has been observed in very small Ga, Sn, and Al clusters, although it is likely to be due to changes in the structure and chemical bonding in these small particles. Embedded nanoparticles can also be superheated above their bulk melting temperature if there is a good epitaxial relationship with the host material. However, in bulk solids, superheating is associated with non-melting surfaces, which allow the solid to remain metastable with respect to the liquid. This has yet to be observed in nanoparticles where the stability of the solid is reduced relative to the liquid by the finite contribution of the surface energy. There is some evidence from simulations that isolated

Al particles bounded by non-melting surfaces can remain stable microcanonically above the bulk melting temperature but this is yet to be confirmed experimentally.

Acknowledgment

The authors would like to thank the Royal Society of New Zealand's Marsden Fund (contract numbers IRL0602 and IRL0801).

References

Ainslie, N. G., J. D. MacKenzie, and D. Turnbull, 1961, *J. Phys. Chem.* **65**, 1718.

Bachels, T., H.-J. Güntherodt, and R. Schäfer, 2000, *Phys. Rev. Lett.* **85**, 1250.

Baletto, F. and R. Ferrando, 2005, *Rev. Mod. Phys.* **77**, 371.

Banhart, F., E. Hernández, and M. Terrones, 2003, *Phys. Rev. Lett.* **90**(18), 185502.

Ben David, T., Y. Lereah, G. Deutscher, R. Kofman, and P. Cheyssac, 1995, *Phil. Mag. A* **71**, 1135.

Berry, R. S., J. Jellinek, and G. Natanson, 1984, *Phys. Rev. A* **30**, 919.

Berry, R. S., T. L. Beck, H. L. Davis, and J. Jellinek, 1988, *Adv. Chem. Phys.* **70**, 75.

Bixon, M. and J. Jortner, 1989, *J. Chem. Phys.* **91**, 1631.

Breaux, G. A., R. C. Benirschke, T. Sugai, B. S. Kinnear, and M. F. Jarrold, 2003, *Phys. Rev. Lett.* **91**, 215508.

Breaux, G. A., B. Cao, and M. F. Jarrold, 2005a, *J. Phys. Chem. B Lett.* **109**, 16575.

Breaux, G. A., C. M. Neal, B. Cao, and M. F. Jarrold, 2005b, *Phys. Rev. Lett.* **94**, 173401.

Broughton, J., 1991, *Phys. Rev. Lett.* **67**, 2990.

Buffat, P. and J.-P. Borel, 1976, *Phys. Rev. A* **13**, 2287.

Cahn, R. W., 1986, *Nature* **323**, 668.

Carnevali, P., F. Ercolessi, and E. Tosatti, 1987, *Phys. Rev. B* **36**, 6701.

Chacko, S., K. Joshi, D. G. Kanhere, and S. A. Blundell, 2004, *Phys. Rev. Lett.* **92**, 135506.

Cheng, H.-P. and R. S. Berry, 1992, *Phys. Rev. A* **45**, 7969.

Cleveland, C. L., W. D. Luedtke, and U. Landman, 1998, *Phys. Rev. Lett.* **81**, 2036.

Cornia, R. L., J. D. MacKenzie, and D. Turnbull, 1963, *J. Appl. Phys.* **34**, 2245.

Daeges, J., H. Gleiter, and J. Perepezko, 1986, *Phys. Lett. A* **119**, 79.

Di Tolla, F. D., F. Ercolessi, and E. Tosatti, 1995, *Phys. Rev. Lett.* **74**, 3201.

Di Tolla, F. D., E. Tosatti, and F. Ercolessi, 1996, *Monte Carlo and Molecular Dynamics of Condensed Matter Systems*. SIF, Bologna, Italy, pp. 347–398.

Ercolessi, F. and J. B. Adams, 1994, *Europhys. Lett.* **26**, 583.

Fecht, H. J. and W. L. Johnson, 1988, *Nature* **334**, 50.

Frenkel, J., 1946, *Kinetic Theory of Liquids*. Clarendon, Oxford, U.K.

Frenken, J. W. M., P. M. J. Maree, and J. F. V. der Veen, 1986, *Phys. Rev. B* **34**, 7506.

van der Gon, A. W. D., R. J. Smith, J. M. Gay, D. J. O'Connor, and J. F. van der Veen, 1990, *Surf. Sci.* **227**, 143.

Gong, X. G., G. L. Chiarotti, M. Parinello, and E. Tosatti, 1991, *Phys. Rev. B* **43**, 14277.

Goswami, R. and K. Chattopadhyay, 1995, *Acta Metall.* **43**, 2837.

Goswami, R. and K. Chattopadhyay, 1996, *Appl. Phys. Lett.* **69**(7), 910.

Gråbaek, L., J. Bohr, E. Johnson, A. Johansen, L. Sarholt-Kristensen, and H. H. Andersen, 1990, *Phys. Rev. Lett.* **64**(8), 934.

Haberland, H., 2002, *Atomic Clusters and Nanoparticles: Les Houches Session LXXIII*. Springer, Berlin, Germany.

Haberland, H., T. Hippler, J. Donges, O. Kostko, M. Schmidt, and B. von Issendorff, 2005, *Phys. Rev. Lett.* **94**, 035701.

Heine, V., 1968, *J. Phys. C (Proc. Phys. Soc.)* **2**, 222.

Hendy, S. C., 2005, *Phys. Rev. B* **71**, 115404.

Hendy, S. C., 2007, *Nanotechnology* **18**, 175703.

Herman, J. W. and H. E. Elsayed-Ali, 1992, *Phys. Rev. Lett.* **69**, 1228.

Herman, J. W. and H. E. Elsayed-Ali, 1994, *Phys. Rev. B* **49**(7), 4886.

Heyraud, J. C. and J. J. Métois, 1987, *J. Cryst. Growth* **82**, 269.

Honeycutt, J. D. and H. C. Andersen, 1987, *J. Phys. Chem.* **91**, 4950.

Jarrold, M. F. and E. C. Honea, 2007, *J. Phys. Chem.* **95**, 9181.

Joshi, K., S. Krishnamurty, and D. G. Kanhere, 2006, *Phys. Rev. Lett.* **96**, 135703.

Khaikin, S. and N. Bené, 1939, *C. R. Acad. Sci. U.R.S.S.* **23**, 31.

Labastie, P. and R. L. Whetten, 1990, *Phys. Rev. Lett.* **65**, 1567.

Lereah, Y., R. Kofman, J. M. Penisson, G. Deutscher, P. Cheyssac, T. B. David, and A. Bourret, 2001, *Phil. Mag. B* **81**, 1801.

Liu, Z., H. Wang, H. Li, and X. Wang, 1998, *Appl. Phys. Lett.* **72**(15), 1823.

Mei, Q. S. and K. Lu, 2007, *Prog. Mater. Sci.* **52**, 1175.

Mikrajuddin, F. I., K. Okuyama, and F. G. Shi, 2001, *J. Appl. Phys.* **89**(11), 6431.

Molenbroek, A. M. and J. W. M. Frenken, 1994, *Phys. Rev. B* **50**, 11132.

Neal, C. M., A. K. Starace, and M. F. Jarrold, 2007, *Phys. Rev. B* **76**, 054113.

Neirotti, J. P., F. Calvo, D. L. Freeman, and J. D. Doll, 2000, *J. Chem. Phys.* **112**(23), 10340.

Nose, S., 1990, *J. Phys.: Condens. Matter* **2**, SA115.

Pawlow, P., 1909, *Z. Phys. Chem.* **65**, 1.

Peppiatt, S. J., 1975, *Proc. R. Soc. Lond. A* **345**, 401.

Pluis, B., A. W. D. van der Gon, J. W. M. Frenken, and J. F. van der Veen, 1987, *Phys. Rev. Lett.* **59**, 2678.

Pluis, B., D. Frenkel, and J. F. van der Veen, 1990, *Surf. Sci.* **239**, 282.

Qi, W. and M. Wang, 2005, *Mater. Lett.* **59**, 2262.

Rosner, H., P. Scheer, J. Weissmuller, and G. Wilde, 2003, *Phil. Mag. Lett.* **83**, 511.

Schebarchov, D. and S. C. Hendy, 2006, *Phys. Rev. Lett.* **96**, 256101.

Schmidt, M., R. K. W. Kronmuller, B. von Issendorff, and H. Haberland, 1997, *Phys. Rev. Lett.* **79**, 99.

Schmidt, M., R. Kusche, B. von Issendorff, and H. Haberland, 1998, *Nature* **393**, 238.

Schmidt, M., R. Kusche, T. Hippler, J. Donges, W. Kronmuller, B. von Issendorff, and H. Haberland, 2001, *Phys. Rev. Lett.* **86**, 1191.

Sheng, H. W., K. Lu, and E. Ma, 1998, *Acta Mater.* **46**(14), 5195.

Shvartsburg, A. A. and M. F. Jarrold, 2000, *Phys. Rev. Lett.* **85**, 2530.

Tallon, J. L., 1989, *Nature* **342**, 658.

Tammann, G., 1910, *Z. Phys. Chem.* **68**, 257.

Tartaglinoa, U., T. Zykova-Timana, F. Ercolessi, and E. Tosatti, 2005, *Phys. Rep.* **411**, 291.

Turnbull, D., 1950, *J. Chem. Phys.* **18**, 769.

van der Veen, J. F., B. Pluis, and A. D. van der Gon, 1990, *Kinetics of Ordering and Growth at Surfaces*. Plenum Press, New York, pp. 343–354.

Volmer, M. and O. Schmidt, 1937, *Z. Phys. Chem. B* **35**, 467.

Zhang, D. L. and B. Cantor, 1991, *Acta Metall. Mater.* **39**, 1595.

Zhong, J., L. H. Zhang, Z. H. Jin, M. L. Sui, and K. Lu, 2001, *Acta Mater.* **49**, 2897.

13

Spin Accumulation in Metallic Nanoparticles

Seiji Mitani
National Institute
for Materials Science

Kay Yakushiji
National Institute of Advanced
Industrial Science and Technology

Koki Takanashi
Tohoku University

13.1 Introduction

Spin accumulation is the deviation of spin population in nanostructures from its thermal equilibrium state. It is nowadays considered as one of the most important phenomena in the scientific and technological field of the so-called spintronics, which has emerged as a new branch of electronics since the 1990s and includes rich spin-related physics (Wolf et al. 2001, Awschalom et al. 2002, Zutic et al. 2004, Maekawa et al. 2006). The first example of primary functional properties in which spin accumulation plays an important role is current-perpendicular-to-plane giant magnetoresistance (CPP-GMR) investigated intensively in the early 1990s (e.g., Pratt 1991). Magnetoresistance (MR) effect is a change in resistance according to the magnetization configuration, and CPP-GMR occurs in metallic multilayers consisting of ferromagnetic layers separated by nonmagnetic spacers when an electrical current flows in the direction perpendicular to the layer planes. Physics of CPP-GMR is well described by the macroscopic model proposed by Valet and Fert (1993), considering spin accumulation around the interfaces in multilayers as well as spin-dependent bulk and interface scattering. Note that GMR has already been used for technical applications such as reading heads of high-density hard disc drives (HDD), and researches and development of GMR are continued for the demands of higher density HDD (e.g., Tsang et al. 1998, Takagishi et al. 2002).

Since the magnitude of CPP-GMR is given not only by spin accumulation but also by other factors such as spin-dependent interface and bulk scattering, the measurement of CPP-GMR is not a direct way to characterize spin accumulation in materials. Direct electrical detection of spin accumulation signals has been tried by using nonlocal spin injection techniques in metallic lateral devices. In the experiment by Jedema et al. (2001), clear and reasonable nonlocal electrical resistance was first observed at room temperature as an evidence of spin accumulation in metal, followed by a variety of similar and improved experimental results to date (e.g., Kimura and Otani 2007a,b). Other than CPP-GMR and nonlocal spin injection, spin Hall effect (SHE) is also of interest. SHE is a Hall effect in *nonmagnetic* materials which gives rise to spin accumulation, instead of a Hall voltage, in substances through their off-diagonal transport properties (Hirsch 1999, Takahashi et al. 2006, Takahashi and Maekawa 2008). SHE can be employed for conversion from electrical current to spin current and vice versa in spintronic devices. In 2004, the presence of SHE was first experimentally confirmed by optical detection of spin accumulation generated at the edges of a GaAs sample (Kato et al. 2004). More recently, spin accumulation due to SHE has also been examined in metallic systems in which electrical detection techniques enable us to characterize spin transport in detail (Valenzuela and Tinkham 2006, Seki et al. 2008), and thereby new topics of spin transport physics such as the proposal of resonant skew scattering mechanism for giant SHE (Guo et al. 2009) and the discovery of spin Zeebeck effect (Uchida et al. 2008) have appeared. Above mentioned are a few typical examples of phenomena closely related with spin accumulation while spin transport in any kind of nanostructure is influenced by the occurrence of spin accumulation to a certain degree.

Metallic nanoparticles are of particular interest with respect to spin accumulation since only a small amount of spin accumulation may be enough to bring about a large effect in their small volumes. Physical properties of metallic nanoparticles have been studied for a long time, many of which originate from discrete electronic energy levels formed in nanoparticles, known as Kubo effect (Kubo 1962). Concerning electron transport properties, the electrical charging effect is also crucial in nanoparticles, which arises from the addition of excess electrons into a nanoparticle with small capacitance, and it causes so-called single-electron tunneling (SET) behaviors represented by the Coulomb blockade and Coulomb staircase (Gravert and Devoret 1992). Since SET provides new principles for high-density memory devices and low power consumption transistors, the development of single-electron devices has been one of the central issues for future electronics (e.g., Yano et al. 1999). The importance of nanoparticles with a large charging energy against thermal fluctuation of the charged state is now growing for higher temperature operations of SET devices.

In this chapter, nonequilibrium spin accumulation in magnetic and nonmagnetic metal nanoparticles is described. Although creation and detection of spin accumulation can be made by electrical, optical, and other methods such as spin pumping by microwave excitation, here we focus on electrical methods that have actually been applied to metallic nanoparticles in recent studies. Before the main topics are discussed, fundamentals of spin accumulation (Section 13.2) and basic transport phenomena in nanoparticles (Section 13.3) are introduced separately. Then, what happens when spin accumulation occurs in nonmagnetic nanoparticles is described in Section 13.4. As a more complicated case, some remarkable phenomena on spin accumulation in ferromagnetic nanoparticles are presented in Section 13.5, followed by brief descriptions of related phenomena and potential applications (Section 13.6) and conclusions (Section 13.7).

13.2 Fundamentals of Spin Accumulation

Spin is an electron's degree of freedom different from charge, and thereby spatial distribution of spin can be given, being independent of charge distribution. Spin accumulation is defined as deviation from spin distribution at the thermal equilibrium state by injecting spin, and it modifies chemical potential of up- and down-spin electrons from their equilibrium states. An interface of ferromagnetic and nonmagnetic metals through which electrical current passes is one of the simplest systems for spin accumulation. Figure 13.1 depicts chemical potential of up- and down-spin electrons, spin density, and spin accumulation around the interface under an applied electrical field, where it is assumed that the interface resistance is negligible and electrical resistivity of the ferromagnetic metal is the same as that of the nonmetallic one for instantly understanding the concept of spin accumulation. In addition, although chemical potential of charged particles under electric field should be called electrochemical potential, here we call it chemical potential for simplicity. Electrical current in a ferromagnetic metal is spin polarized, i.e., the current densities of up- and down-spin electrons are not the same. When spin-polarized current comes to the interface from the ferromagnetic metal side in

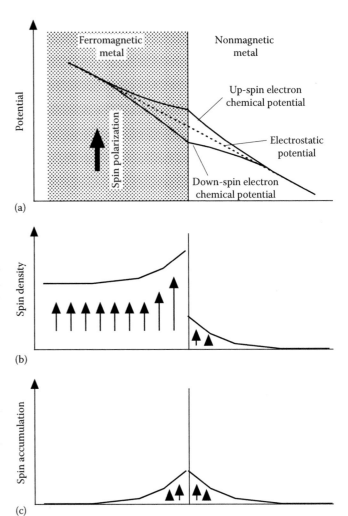

FIGURE 13.1 Schematic illustration of spin accumulation generated by electrical current passing through an interface of ferromagnetic and nonmagnetic metals; (a) chemical potential of up- and down-spin electrons, (b) nonequilibrium spin density, (c) nonequilibrium spin accumulation.

Figure 13.1, the ratio of the current densities of up- and down-spin electrons must become unity before reaching the interior of the nonmagnetic metal since that in the nonmagnetic metal is unpolarized. For majority (e.g., up) spin electrons, the current channel shrinks at the interface, so that the majority spin electrons are accumulated around the interface. For minority (e.g., down) spin electrons, on the other hand, the current channel is expanded. As a result, minority spin electrons around the interface are drawn out to the nonmagnetic side. Both the processes cause the increase of the difference between the majority and minority spin electrons' populations around the interface. This change in the spin population from the equilibrium state is spin accumulation in this case. At the same time, the spin accumulation modifies the chemical potential of up and down electrons, so that further accumulation of spin is suppressed. In the stationary state, a certain degree of spin accumulation and thereby chemical potential splitting are maintained by flowing current.

It is noted that the spin accumulation and the modification of chemical potential occur at the spatial area with the size of spin diffusion length, which is the characteristic distance for which an electron runs before its memory of spin state is lost. Another important feature is that the modification of the spatial densities for up- and down-spin electrons does not occur independently. They are related with one another under the charge neutrality condition in metals, i.e., no charge accumulation occurs associated with the modification. For further understanding, detailed description is given in review papers (e.g., Valet and Fert 1993, Johnson 2007, Takahashi and Maekawa 2008) and books (e.g., Takahashi et al. 2006, Zutic and Fabian 2006).

A tunnel junction is the counterpart of metallic interfaces which are almost transparent in charge transport. Figure 13.2a shows a schematic illustration of a double-barrier tunnel junction consisting of ferromagnetic metal right and left electrodes (F) and a thin center electrode of nonmagnetic metal (N), which is hereafter abbreviated as an F/N/F junction. While thin tunnel barriers allow electrons to move between the metallic electrodes owing to the quantum mechanical tunnel effect, electrical conductance through the barriers is much lower than those inside the metallic electrodes. As a result, when a bias voltage is applied between the right and left electrodes, chemical potential changes (=voltage drops) occur only at the barriers as shown in Figure 13.2a. Figure 13.2b shows spin-resolved density of states (DOS) of each electrode in case of antiparallel magnetization configuration, where DOS for majority(+) and minority(−) spin electrons in the right and left ferromagnetic electrodes and the nonmagnetic electrode are represented by D_F^+, D_F^-, and D_N, respectively ($D_F^+ > D_F^-$). Note that the majority spin of the left electrode corresponds to the spin-up state and, on the other hand, that of the right electrode corresponds to the spin-down state. If momentum and energy dependence of tunneling probability of an electron is neglected, the conductance between electrodes is simply proportional to the product of DOS of each electrode. Then, up- and down-spin electron currents coming into the center electrode are proportional to $D_F^+ D_N$ and $D_F^- D_N$, and those going out from the center electrode are proportional to $D_F^- D_N$ and $D_F^+ D_N$, respectively. Since $D_F^+ D_N > D_F^- D_N$, this means that spin-up polarized current comes into and spin-down polarized one goes out from the center electrode, resulting in the accumulation of up-spin electrons there (and also chemical potential splitting). For better

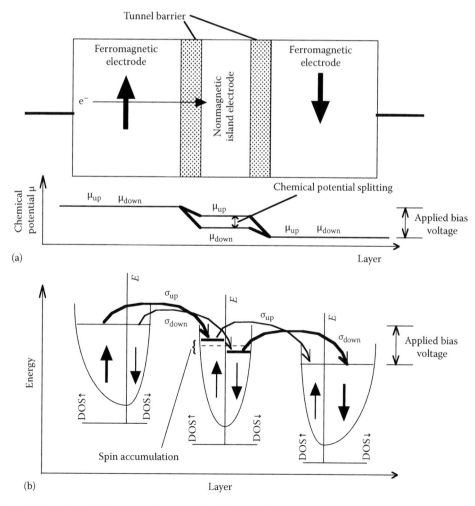

(a)

(b)

FIGURE 13.2 Schematic illustration of spin accumulation in a double-tunnel junction consisting of two ferromagnetic electrodes and a nonmagnetic island electrode; (a) junction structure and chemical potential diagram, (b) spin-resolved density of states and electron filling.

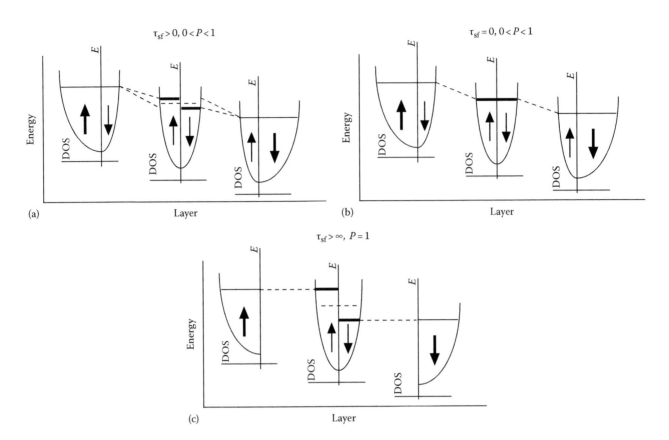

FIGURE 13.3 Spin accumulation in various F/N/F double-tunnel junctions; (a) finite spin relaxation and finite spin polarization, (b) limit of fast relaxation, (c) half metallic spin polarization and no spin relaxation.

understanding, Figure 13.3 shows some limiting cases of spin accumulation in terms of spin relaxation time τ_{sf} and spin polarization defined as $P = (D_F^+ - D_F^-)/(D_F^+ + D_F^-)$, where spin relaxation time is defined as the mean time between two successive events of spin flip. For a nonzero spin relaxation time, a certain magnitude of spin accumulation is induced in the center electrode (Figure 13.3a). For the limit of fast spin relaxation $\tau_{sf} \sim 0$, injected spins are immediately relaxed and lose their direction before injection, so that no spin accumulation occurs (Figure 13.3b). An interesting case is theoretically considered in the condition of no spin relaxation and half metallic spin polarization, i.e., $\tau_{sf} = \infty$, $P = 1$ (Figure 13.3c). In this case, the large spin accumulation occurs and the chemical potential of up- (down-) spin electrons corresponds to that of the left (right) electrode. An emphasis should be put on the fact that the chemical potential splitting *never* exceeds the voltage applied to the junction.

A remarkable consequence of spin accumulation in the F/N/F double-tunnel junction as shown in Figure 13.2 is the appearance of tunnel magnetoresistance (TMR), which is not identified with the conventional TMR described by the Julliere model (Julliere 1975). TMR ratio (defined as $(R_{antiparallel} - R_{parallel})/R_{parallel}$ where $R_{parallel}$ and $R_{antiparallel}$ are junction resistance in the parallel and antiparallel magnetization configurations of electrodes) in the Julliere model applicable to usual ferromagnetic tunnel junctions is expressed as

$$\text{TMR ratio} = \frac{2P_1 P_2}{1 - P_1 P_2}, \qquad (13.1)$$

where P_1 and P_2 are the spin polarization for the left and right ferromagnetic electrode (i.e., the junction indexes 1 and 2 represent left and right, respectively). This means that appearance of TMR requires nonzero spin polarization in electrode materials. In this sense, the F/N/F double-tunnel junction including a center electrode with zero spin polarization does not show any TMR since it can be regarded as a series circuit of F/N and N/F junctions. The distinct mechanism of TMR induced by spin accumulation will be discussed for the case of F/N-nanoparticle/F junctions in Section 13.4, and is also described in Maekawa et al. (2002).

13.3 Coulomb Blockade in Metallic Nanoparticles

Electron transport in a metallic nanoparticle electrically connected with leads through thin tunnel barriers is dramatically affected by the charging effect of the nanoparticle (Gravert and Devoret 1992). Figure 13.4a shows a schematic illustration of a double-tunnel junction consisting of right and left electrodes and a center island (nanoparticle) and its current–voltage characteristics (*I–V* curve). When even a single electron comes from the

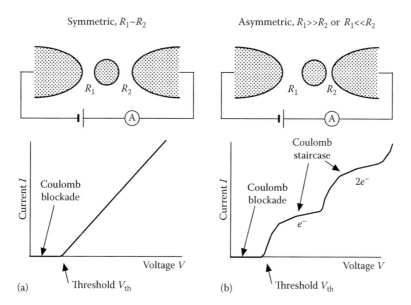

FIGURE 13.4 Current–voltage characteristics (*I–V* curves) exhibiting the Coulomb blockade and Coulomb staircases in double-tunnel junctions; (a) symmetric case $R_1 \sim R_2$, (b) asymmetric case $R_1 \gg R_2$ or $R_1 \ll R_2$.

left electrode into the nanometric island, the charge neutrality is significantly broken in the island and the electrostatic energy of island increases by the charging energy of a single electron. If the island (nanoparticle) is small enough to have a charging energy larger than both thermal fluctuation at the measurement temperature and the applied bias voltage, no current flows, i.e., the Coulomb blockade occurs. The charging energy E_c can be evaluated as

$$E_c = \frac{e^2}{2C}, \tag{13.2}$$

where

  e is the electron charge
  C is the capacitance of the double-tunnel junction (sum of the capacitance for each junction)

The capacitance includes the major contribution of self-capacitance of the nanoparticle (when typically ~1 nm or smaller in size). Therefore, the capacitance of an isolated metallic sphere gives a good approximation to roughly estimate the charging energy of nanoparticle-based double-tunnel junctions, and it is given as

$$C = \frac{1}{2\pi\varepsilon d}, \tag{13.3}$$

where

  ε is the dielectric constant of the surrounding material
  d is the diameter of the sphere

In case when d is 1 nm and ε is 10 times larger than the dielectric constant of vacuum (corresponding to ε for aluminum-oxide and magnesium-oxide), the temperature corresponding

to the charging energy $T = E_c/k_B$ is estimated to be ~1700 K by Equations 13.2 and 13.3, indicating that the Coulomb blockade can be observed at room temperature for nanoparticles with the sizes of 1 nm order. While a symmetric structure is considered in Figure 13.4a, Figure 13.4b describes an *asymmetric* case with significant differences in resistance and capacitance of the two barriers. In particular, since the magnitude of tunnel resistance has an exponential dependence on the thickness, it is likely that a large asymmetry in tunnel resistance appears in real samples. Then, the inverse of the tunneling rate, i.e., the interval of the tunneling events, is given by the product of tunnel resistance R and capacitance C, and the large difference in tunneling rate between the right and left junctions causes significant charge accumulation: for example, some excess electrons *dwell* in the island when the incoming tunneling rate is much larger than the outgoing one. Furthermore, the excess charge is quantized by the number of the excess electrons, and therefore the discrete charge accumulation occurs in the island as a function of bias voltage, resulting in staircase behavior of the *I–V* curve, called Coulomb staircase, as shown in Figure 13.4b. (The discreteness, in other words, reflects the formation of a bottleneck of current at the highly resistive barrier against the decay of a certain charge state in the island.) The Coulomb blockade and the Coulomb staircase are the two major effects of SET caused by the electrical charging effect of the island and the particle nature of electrons.

The orthodox theory of SET (Gravert and Devoret 1992) reproduces experimental observations of *I–V* curves with the Coulomb blockade and staircases. In this semi-classical theory for a double-tunnel junction, the electrostatic energy change due to a SET is first considered and is given as

$$E_i^{\pm} = \frac{e}{C}\left(\frac{e}{2} \pm ne \pm C_j V\right), \tag{13.4}$$

where

> i, j represent junction indexes, $(i, j) = (1, 2)$ or $(2, 1)$
>
> $+$ and $-$ represent forward and backward tunneling processes, respectively, along the direction of the applied electric field
>
> n is the number of excess charges
>
> C is the total capacitance of the island defined as $C = C_1 + C_2$ (C_1, C_2: capacitance of the left and right junctions)
>
> V is the bias voltage applied to the whole double junction

Statistical treatment of the energy change leads to the forward and backward tunneling rates $\Gamma_i^{\pm}$ as

$$\Gamma_i^{\pm} = \frac{G_i}{e^2} \frac{E_i^{\pm}(n, V)}{\exp\{E_i^{\pm}(n, V)/k_B T\} - 1}, \tag{13.5}$$

where

> G_i is the conductance of junction-i
>
> T is the absolute temperature

The probability $P(n, V)$ of the appearance of the charge state n at V is obtained by solving the master equation, i.e., the requirement that electronic charge is neither created nor annihilated for any tunneling process. Then, by using $P(n, V)$, current I is given as

$$I = -e \sum_{n=-\infty}^{\infty} p(n, V) \left[\Gamma_1^{+}(n, V) - \Gamma_1^{-}(n, V) \right]$$

$$= -e \sum_{n=-\infty}^{\infty} p(n, V) \left[\Gamma_2^{+}(n, V) - \Gamma_2^{-}(n, V) \right], \tag{13.6}$$

where $\sum_n P(n, V) = 1$.

At the end of this section, let us briefly discuss an important condition for the validity of the orthodox theory and for the experimental observation of Coulomb blockade and staircases. SET represented by the Coulomb blockade and staircases is a consequence of the particle nature of electrons. This particle nature disappears if electronic states in the island electrode (nanoparticle) and the right (or left) electrode are well coupled due to low tunnel resistance. In such a case, the wave function of an electron is not localized inside the island but spreads to the right (or left) electrode. Therefore, in order to observe SET, the tunnel resistance should be much larger than the resistance quantum R_K ($=h/e^2 \sim 25.8\,\text{k}\Omega$).

13.4 Spin Accumulation in Nonmagnetic Nanoparticles

The main topic of this section is interplay of spin accumulation and SET in nanoparticles. As mentioned in Section 13.1, it is a simple but important fact that nanoparticles have extremely small volumes so that even a single electron's spin brings about large spin accumulation per unit volume as well as a single-electron charge brings about a large electrostatic effect. In more details,

since the number of electronic states per unit energy in a substance is given by the product of DOS and its volume, nanoparticles have a small number of electronic states *per unit energy* so that injected electrons with a certain spin direction are likely to occupy considerably high energy states with the same spin direction over the Fermi level. Namely, a large chemical potential splitting can be realized by even small spin accumulation.

Although spin accumulation in nonmagnetic metal nanoparticles (induction of nonequilibrium magnetization) draws particular interest, only a few results have been reported on it to date. This is due to the difficulty in preparing samples including metallic nanoparticles connected to electrodes through thin tunnel barriers. Many efforts have been made for sample preparation, and spin accumulation in nonmagnetic metal nanoparticles has been examined through the observation of TMR induced by spin accumulation. In 2005, Zhang et al. reported TMR in Co/AlO/Al-nanoparticle/AlO/Co double-tunnel junctions (Zhang et al. 2005). Figure 13.5 shows their data of *I–V* and TMR curves. TMR of about 10% occurs at 4.2 K for the double-tunnel junctions including Al nanoparticles as a center island electrode although the Coulomb blockade is not so clear. They also tried to measure the Hanle effect to discuss the spin coherence time in Al nanoparticles. The Hanle effect is attributed to spin precession due to an external magnetic field perpendicular to the spin direction. Figure 13.6 shows the change in current for the Co/AlO/Al-nanoparticle/AlO/Co double-tunnel junctions, which is interpreted as the Hanle effect for electron spins injected into the Al nanoparticles. Bernand-Mantel et al. successfully observed TMR induced by spin accumulation in a *single* Au nanoparticle by using their unique experimental setup (Bernand-Mantel et al. 2006, Seneor et al. 2007). They used a conductive AFM tip to nanoindent a thin resist mask and obtained a nanometer-sized double-tunnel junction of Co/Al$_2$O$_3$/Au-nanoparticle/Al$_2$O$_3$/Co as shown in the inset of Figure 13.7a. Figure 13.7c shows *I–V* curves and differential conductance d*I*/d*V* at 4.2 K for one of the samples, in which the Coulomb blockade is clearly observed. The observation of the Coulomb blockade has an important meaning that no direct tunneling between two Co electrodes (not through the Au particles) contributes to the electron transport in the double-tunnel junction. Furthermore, the Coulomb staircase with a single oscillation period proves the realization of electrical transport and spin injection in a single Au nanoparticle. Figure 13.7b shows TMR for this sample at 20 meV and 4.2 K. They attributed the inverse TMR (negative sign in its definition in Section 13.2) of 7% to spin accumulation in a single Au nanoparticle. Another evidence for spin accumulation is observed in d*I*/d*V* for another sample. The peak shift between the parallel and antiparallel magnetization configurations of the Co electrodes, depicted in Figure 13.7a, can be interpreted as an effect of spin accumulation in an Au nanoparticle. Nogi et al. also observed TMR induced by spin accumulation in Au nanoparticles and a possible magnetoresistive effect caused by spin accumulation in Cr nanoparticles (Nogi et al. 2007). In a recent experiment, furthermore, correlation between *I–V* curves and TMR was examined (Mitani et al. 2008).

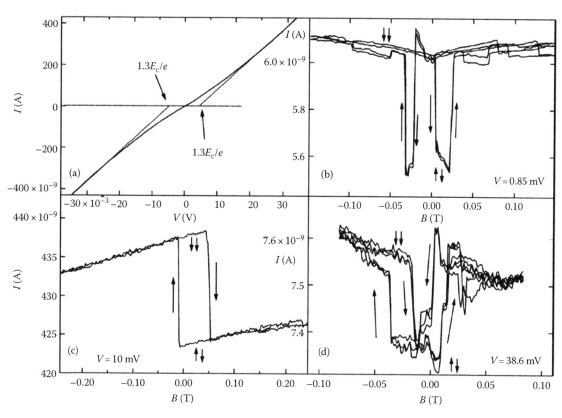

FIGURE 13.5 *I–V* curve and tunnel magnetoresistance (TMR) curves in a Co/AlO/Al-nanoparticle/AlO/Co double-tunnel junction. (Reprinted from Zhang, L.Y. et al., *Phys. Rev. B*, 72, 155445, 2005. With permission.)

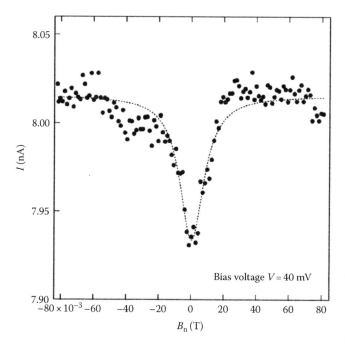

FIGURE 13.6 Change in current as a function of an applied magnetic field B_n normal to magnetization for a Co/AlO/Al-nanoparticle/AlO/Co double-tunnel junction. (Reprinted from Zhang, L.Y. et al., *Phys. Rev. B*, 72, 155445, 2005. With permission.)

Figure 13.8a illustrates the schematic sample structure of a Fe/MgO/Au-nanoparticle/MgO/Fe double-tunnel junction used in the experiments. A STM image for Au nanoparticles grown on a MgO tunnel barrier is shown in Figure 13.8b, in which the sizes of particle images are enlarged by a convolution effect with the STM tip. The mean diameter of the Au nanoparticles is evaluated from the deposition volume and the number density to be ~1 nm (Ernult et al. 2006). Figure 13.8c shows TMR curves at different bias voltages and 4.2 K. TMR induced by spin accumulation in Au nanoparticles reaches 12% at the bias voltage of 250 mV. Figure 13.8d shows bias voltage dependence of current and TMR at 4.2 K, indicating that TMR appears when current starts to flow above the threshold voltage of the Coulomb blockade. This behavior is quite reasonable because TMR is induced by spin accumulation generated by spin-polarized current with a certain magnitude, and it is not interpreted as TMR due to the direct tunneling between ferromagnetic electrodes. Therefore, this behavior is also considered as an evidence for TMR due to spin accumulation.

Physics of TMR induced by spin accumulation in an F/N-nanoparticle/F junction is briefly summarized in Figure 13.9. Figure 13.9a shows a schematic of the junction considered, which includes a center electrode with a small size. Figure 13.9b shows tunnel probabilities of electrons for each barrier and for each spin. The thicknesses of the arrows exhibiting tunneling events

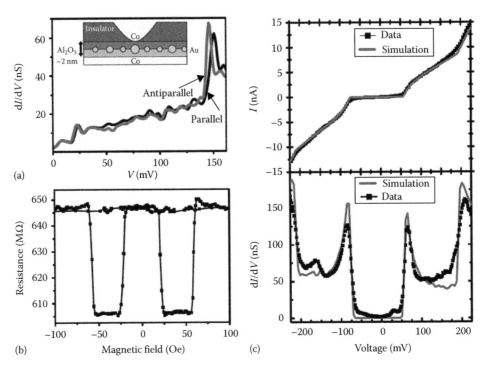

FIGURE 13.7 (a) Differential conductance (dI/dV) curves and sample structure for a Co/AlO/Au-nanoparticle/AlO/Co double-tunnel junction, (b) TMR curve for another sample at bias voltage of 20 mV, (c) I–V and dI/dV curves obtained for the sample for which TMR is shown in (b), in experiments and simulation. (From Seneor, P. et al., *J. Phys.: Condens. Matter*, 19, 165222, 2007. With permission.)

mean their magnitude. First, we assume that the electrical charging effect (Coulomb blockade) is negligible (Figure 13.9c). In this case, the origin of TMR can be understood straightforwardly from the difference in spin accumulation between parallel and antiparallel magnetization configurations: For the parallel magnetization configuration, no spin accumulation occurs since spin densities incoming to, and outgoing from, the island are balanced. For the antiparallel magnetization configuration, on the other hand, the incoming up-spin (down-spin) electron density is larger (smaller) than the outgoing one and consequently spin accumulation occurs as explained in Section 13.2. After spins are accumulated in the island, since the flow rate, i.e., tunneling rate, of each spin electron is proportional to the difference in chemical potential between the left and right sides of a tunnel barrier, the incoming up-spin (down-spin) electron current has been suppressed (enhanced) by the rise (descent) of the topmost occupied level for up-spin (down-spin) in the island, as shown in Figure 13.9c. Namely, the band filling and thereby the chemical potential for each spin are modified by accumulating spins, so that the incoming and outgoing flows of spins reaches a stationary (balanced) state and thus no further spin is accumulated. This modification of the chemical potential causes a difference in the total resistance of the double-tunnel junction between the parallel and antiparallel configurations, i.e., TMR. It is theoretically proved that the maximum TMR due to this mechanism is one half of TMR in the case when Julliere model is applied to the right and left ferromagnetic electrodes (e.g., Maekawa et al. 2002).

Next, let us consider the electrical charging effect on spin accumulation. When the electrical charging energy is large

enough to give rise to interplay of spin accumulation and SET, the phenomena become a little complicated as shown in Figure 13.9d: The charging energy E_c is added to the chemical potential in the island. For the tunneling from the left electrode to the island, the effective bias voltage (the drop in chemical potential) is reduced by E_c/e. Therefore, no tunneling occurs at the bias voltages below E_c/e, and tunneling current starts to flow when the bias voltages exceed E_c/e. For the tunneling from the island to the right electrode, on the other hand, the effective bias voltage is gained by E_c/e for both up- and down-spins. In this case, if the magnitude of spin accumulation is the same as the case without electrical charging effect (Figure 13.9c), the *ratio* of the chemical potential "splitting" between up- and down-spin electrons to the chemical potential "drop" between the island and the right electrodes is *smaller* than that in the case without electrical charging effect. This means that the effect of the chemical potential splitting, which is required to reach a stationary state for up- and down-spin currents, is smaller. Then, to compensate this effective reduction of chemical potential splitting, further spin accumulation occurs. Thus, the Coulomb blockade enhances spin accumulation in a nanoparticle.

Analytical calculations in the regime of the orthodox theory for symmetric double-tunnel junctions including a nonmagnetic island shows that spin accumulation is enhanced by a factor of 1–2 in the assistance of the Coulomb blockade and the maximum TMR corresponds to that in the Julliere model (Mitani et al. 2007). In more general, numerical results based on the orthodox theory are useful. The theoretical framework of spin-dependent SET with spin accumulation is basically the same as

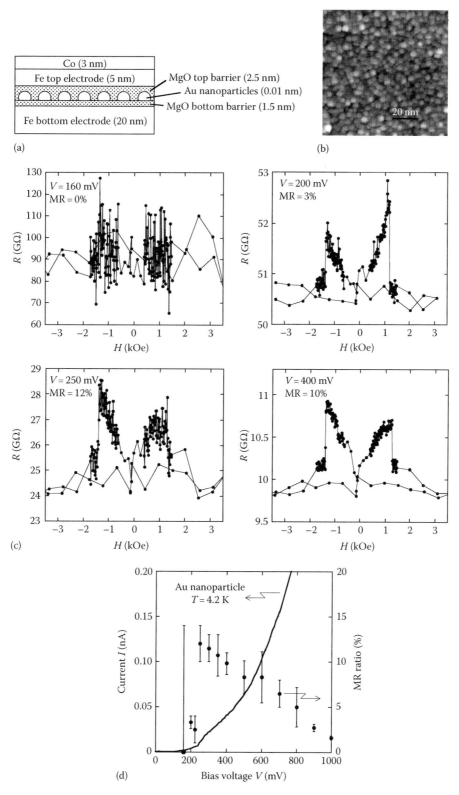

FIGURE 13.8 Structure and spin-dependent transport properties in a Fe/MgO/Au-nanoparticle/MgO/Fe double-tunnel junction; (a) schematic illustration of the sample structure, (b) STM image of Au nanoparticles grown on a MgO tunnel barrier (nominal thickness of Au: 0.04 nm), (From Ernult, F. et al., *Phase. Trans.*, 79, 717, 2006. With permission.) (c) TMR curves at different bias voltages and 4.2 K, (d) current *I* (solid line) and TMR (closed circle) as a function of bias voltage. (From Mitani, S. et al., *Appl. Phys. Lett.*, 92, 152509, 2008. With permission.)

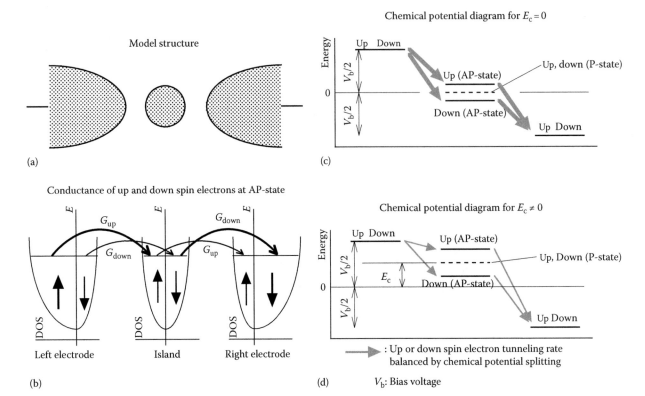

FIGURE 13.9 Schematic illustration for spin accumulation and chemical potentials shift in double-tunnel junctions; (a) model structure, (b) tunnel conductance for each junction and each spin channel at the state of antiparallel magnetization configuration (AP-state), (c) no Coulomb blockade case, (d) enhancement of spin accumulation due to the Coulomb blockade.

that without consideration of spin degree of freedom, which is described by Equations 13.4 through 13.6. However, $\Gamma_i^{\pm}$ and G_i are spin-dependent and the terms of chemical potential shift ΔE_F^{σ} (σ = up or down; chemical potential splitting = $\Delta E_F^{up} - \Delta E_F^{down}$) should be added to Equation 13.5. The tunneling rate revised for the case with spin accumulation is given as

$$\Gamma_{i\sigma}^{\pm} = \frac{G_{i\sigma}}{e^2} \frac{E_i^{\pm}(n,V) + \Delta E_F^{\sigma}}{\exp\left\{ (E_i^{\pm}(n,V) + \Delta E_F^{\sigma})/k_B T \right\} - 1}, \quad (13.7)$$

where $\Gamma_{i\sigma}^{\pm}$ and $G_{i\sigma}$ are the spin-dependent forward and backward tunneling rate and the spin-dependent conductance of junction-i, respectively. By solving the master equation, the probability $P(n,V)$ is obtained, and the currents for the parallel and antiparallel configurations are calculated. The detailed procedure and calculations were reported in literature (Brataas et al. 1999a,b, Maekawa et al. 2002, Weymann and Barnas 2003). Some examples of the calculated results by the orthodox theory including the effect of spin accumulation are shown in Figures 13.10 and 13.11 (Wang et al. 2007), where the limit of slow spin relaxation is assumed. Figure 13.10a shows chemical potential shift ΔE_F^{σ}, current I, and TMR, where TMR is defined as the so-called pessimistic definition ($= (R_{antiparallel} - R_{parallel})/R_{antiparallel} = (I_{parallel} - I_{antiparallel})/I_{parallel}$), calculated for an asymmetric F/N/F double-tunnel junction

with the Coulomb blockade. The spin polarization of right and left ferromagnetic electrodes is set to be 0.45. The capacitance and tunnel conductance ratios are chosen as $C_1/C_2 = 0.01$ and $G_1/G_2 = 10$ in Figure 13.10a. Therefore, the ratio of time constant of tunneling $(C_1/G_1)/(C_2/G_2)$ is 0.001. This large asymmetry leads to appearance of the discrete charge states, i.e., the Coulomb staircases, and the oscillation of ΔE_F^{σ} for antiparallel configuration and TMR. The magnitude of chemical potential splitting is 10–100 meV. For the rather symmetric case where $C_1/C_2 = 10$ and $G_1/G_2 = 100$ so that the ratio of $(C_1/G_1)/(C_2/G_2)$ is only 0.1, on the other hand, no oscillatory behavior is seen in ΔE_F^{σ}, I, and TMR, as shown in Figure 13.10b. The most interesting feature in Figure 13.10a and b is the TMR maximum. The TMR maximum (~33%) for the symmetric double junction is larger than that for the asymmetric one. Furthermore, it exceeds TMR for symmetric F/N/F junctions without the Coulomb blockade which is half of the value of the Julliere model, i.e., TMR = $P^2/(1 + P^2)$ = 16.8% in the pessimistic definition. Figure 13.11 shows a contour plot of the TMR maximum as a function of the ratios of C_1/C_2 and G_1/G_2. TMR of ~33% appears in the conditions of relatively symmetric junctions $(C_1/G_1)/(C_2/G_2)$ ~ the order of 1 and is almost the same as the value of the Julliere model (=33.7%). This is in good agreement with the analytical results mentioned above.

Finally, to discuss spin relaxation time (spin life time), let us revisit the observation that TMR in Au nanoparticles appears

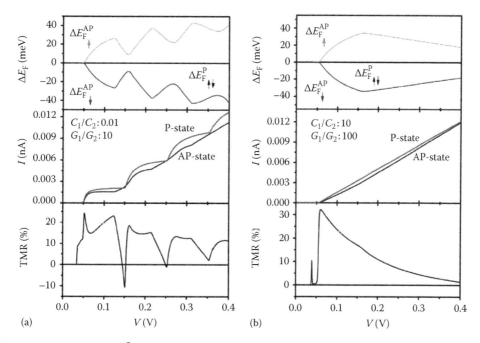

FIGURE 13.10 Chemical potential shift ΔE_F^σ, current I (for parallel (P-) and antiparallel (AP-) magnetization states) and TMR (=(I_{parallel} − $I_{\text{antiparallel}}$)/I_{parallel}) calculated for a F/N-nanoparticle/F junction; (a) asymmetric case, (b) rather symmetric case. (From Wang, H. et al., *Phys. Stat. Sol. (b)*, 244, 4443, 2007. With permission.)

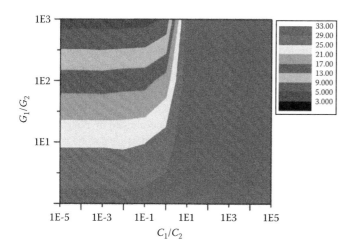

FIGURE 13.11 (See color insert following page 9-8.) Contour plot of the TMR maximum as a function of the ratios of C_1/C_2 and G_1/G_2 in F/N-nanoparticle/F junctions. (From Wang, H. et al., *Phys. Stat. Sol. (b)*, 244, 4443, 2007. With permission.)

above the Coulomb threshold (Mitani et al. 2008). As mentioned earlier (Sections 13.2 and 13.4), this is an evidence for spin accumulation in Au nanoparticles. Semi-quantitative analysis may give information of spin relaxation time in Au nanoparticles. The condition to retain stationary spin accumulation in a nanoparticle is that the time interval of incoming tunneling events of spin-polarized electrons is shorter than the spin relaxation time inside the nanoparticle. Namely, spin should be injected into the nanoparticle before the formerly

injected spin is relaxed. This idea gives semi-quantitative estimation of spin relaxation time as

$$\tau_{\text{sf}} \sim \frac{e}{I_c^{\text{TMR}}}, \qquad (13.8)$$

where I_c^{TMR} is the critical current for the observation of significant TMR. From the data in Figure 13.8d, $I_c^{\text{TMR}} \sim 0.1$ nA, so that $\tau_{\text{sf}} \sim 10$ ns. A similar result is obtained for the data in Figure 13.7b, where $I_c^{\text{TMR}} <$ the measurement current ~ 20 mV/600 MΩ and then $\tau_{\text{sf}} > 5$ ns. Therefore, the spin relaxation time obtained for Au nanoparticles seems to be surprisingly long since spin relaxation times for various bulky metals are estimated to be on the order of 0.1–10 ps. Although the mechanism of the large enhancement of spin relaxation time in Au nanoparticles is not clear, the discrete energy levels in nanoparticles might be related with the long spin relaxation time since similar enhancement was reported for semiconductor quantum dots in which electronic properties are governed by the definitely quantized energy levels (e.g., Fujisawa et al. 2001, 2002).

13.5 Spin Accumulation in Ferromagnetic Nanoparticles

Transport properties of ferromagnetic nanoparticles are also crucially influenced by spin accumulation since spin accumulation gives a chemical potential shift rather than a little additional spin polarization to the ferromagnetic densities of states.

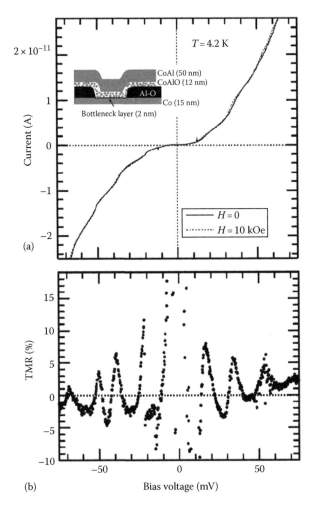

FIGURE 13.12 (a) *I–V* curves and (b) TMR measured for a Co/AlO/Co-nanoparticle/AlO/Co double-tunnel junction. (From Yakushiji, K. et al., *J. Appl. Phys.*, 91, 7038, 2002. With permission.)

Figure 13.12a and b shows *I–V* curves and TMR in an asymmetric Co/AlO/Co-nanoparticle/AlO/Co double-tunnel junction, and the inset is a schematic illustration of the sample structure (Yakushiji et al. 2002). *I–V* curves clearly show the Coulomb staircases, and TMR oscillates associated with the Coulomb staircases. A striking feature of TMR is the change in its sign, i.e., the appearance of inverse TMR, which has first been predicted as an effect of spin accumulation by Barnas and Fert (1998a). Figure 13.13a and b show experimental results of *I–V* curves and TMR for an asymmetric Al/AlO/Co-nanoparticle/AlO/Co double-tunnel junction (Yakushiji et al. 2005) in which an Al electrode is used and therefore the magnetization configuration may be simpler than that in the Co/AlO/Co-nanoparticle/AlO/Co junction. Similar to the observations in Figure 13.12, TMR oscillates associated with the Coulomb staircases, and inverse TMR appears. The chemical potential splitting and the TMR oscillation etc. in ferromagnetic nanoparticles can also be analyzed by the framework of the orthodox theory described in Section 13.4. Figure 13.14 shows chemical potential shifts, currents, and TMR calculated for a N/F-nanoparticle/F junction.

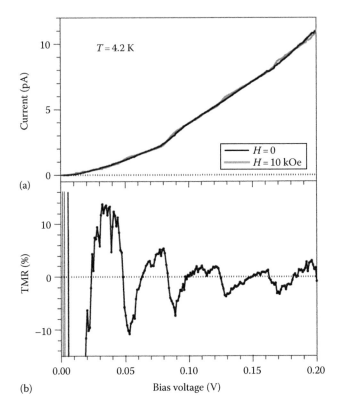

FIGURE 13.13 Experimental results of (a) *I–V* curves and (b) TMR for an asymmetric Al/AlO/Co-nanoparticle/AlO/Co double-tunnel junction. (Adapted from Yakushiji, K. et al., *Nat. Mater.*, 4, 57, 2005.)

I–V curves are significantly modified by the chemical potential shifts, resulting in the inverse TMR. Further discussion is given in Refs. (Yakushiji et al. 2005, 2007, Ernult et al. 2007).

While the spin relaxation time is evaluated by Equation 13.8, the effect of spin relaxation in a nanoparticle can be taken into account in the orthodox theory. In the slow relaxation limit, owing to the spin conservation law, incoming up-spin (down-spin) electron current should exactly be the same as outgoing up-spin (down-spin) electron current, i.e., $I_{1\sigma} = I_{2\sigma}$ (1,2: junction index, σ = up or down). For finite spin relaxation time, the incoming and outgoing currents with up- (or down-) spin satisfy the following relationship:

$$\frac{(I_{1\sigma} - I_{2\sigma})}{e} = \frac{D_\sigma \Omega}{\tau_{sf}} \Delta E_F^\sigma, \tag{13.9}$$

where

D_σ is DOS at Fermi level for spin σ in the ferromagnetic island (nanoparticle)

Ω is the volume of the island

Note that one should distinguish DOS for all electrons (applied to Equation 13.8) from that for tunneling electrons (applied to the evaluation of G and P) and that $I_{1up} + I_{1down} = I_{2up} + I_{2down}$ for the charge conservation although $I_{1up} \neq I_{2up}$. The currents and chemical potential shifts for up- and down-spin electrons are

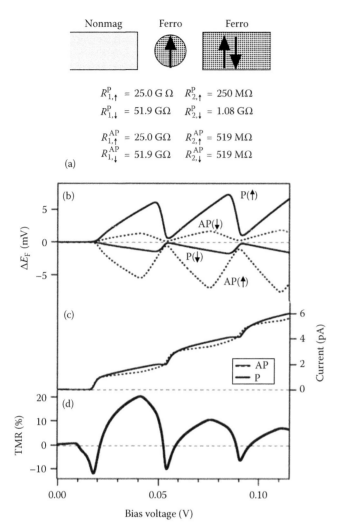

FIGURE 13.14 (a) Model and parameters for calculation, (b) chemical potential shift ΔE_F^σ, (c) current I and (d) TMR (= $(I_{parallel} - I_{antiparallel})/I_{parallel}$) calculated for a N/F-nanoparticle/F junction. The parameter set for the calculation corresponds to $R_1 = 17.0\,G\Omega$, $R_2 = 203\,G\Omega$ and tunnel spin polarization $P = 0.35$. (From Yakushiji, K. et al., *Phys. Rep.*, 451, 1, 2007. With permission.)

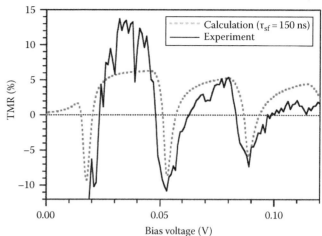

FIGURE 13.15 Bias-voltage dependence of TMR calculated for a N/F-nanoparticle/F junction with the assumption of the finite spin relaxation time $\tau_{sf} = 150\,ns$, in comparison with an experimental result (see Figure 13.13b). (Adapted from Yakushiji, K. et al., *Nat. Mater.*, 4, 57, 2005.)

that putting two or more nanoparticles in series between electrodes does not significantly change its physics (Imamura et al. 2000, Maekawa et al. 2002).

13.6 Related Phenomena and Potential Applications

The studies on current-induced magnetization reversal revealed that spin accumulation gives torque to a local magnetic moment (Zhang et al. 2002). Inoue and Brataas have theoretically studied magnetization reversal due to spin accumulation in ferromagnetic nanoparticles (Inoue and Brataas 2004). The theoretically evaluated spin accumulation is large enough to reverse the magnetization direction of typical ferromagnetic nanoparticles, showing that spin accumulation in nanoparticles is effective not only in TMR but also in current-induced magnetization reversal. Other novel phenomena are also expected to occur by using spin accumulation in nanoparticles. For example, the large effective field of spin accumulation may induce ferromagnetism or suppress superconductivity in nonmagnetic nanoparticles.

TMR induced by spin accumulation in nonmagnetic nanoparticles may provide a potential application based on the Coulomb blockade and nonvolatility of ferromagnetic magnetization. In the original idea for the combination of the Coulomb blockade and TMR, ferromagnetic nanoparticles are considered to be used because TMR signals are conventionally obtained for ferromagnetic materials. However, the sizes of nanoparticles which show the Coulomb blockade at room temperature are as small as ~1 nm, and therefore the problem of thermal fluctuation of magnetization arises even for the highest magnetic anisotropy materials such as FePt that is a candidate material for ultrahigh density hard disk media. To overcome this problem, the use of

determined self-consistently. Figure 13.15 shows TMR as a function of bias voltage, calculated for the Co/AlO/Co-nanoparticle/AlO/Al junction with the assumption of the finite spin relaxation time $\tau_{sf} = 150\,ns$, which reproduces the experimental result with the inverse TMR in a Co nanoparticle (Figure 13.13). The spin relaxation time of $\tau_{sf} = 150\,ns$ obtained in this analysis is significantly long, showing that Co nanoparticles has an enhanced spin relaxation time as well as Au nanoparticles. In original and review papers (Yakushiji et al. 2005, Ernult et al. 2007), more details were described and the numerical calculation proved that spin accumulation enhances TMR in ferromagnetic nanoparticles. Similar numerical calculations for spin-dependent transport in ferromagnetic nanoparticles have been made by other groups (e.g., Barnas and Fert 1998b, Majumdar and Hershfield 1998, Imamura et al. 1999, Martinek et al. 1999, 2002). It is noted nonmagnetic nanoparticles with TMR induced by spin accumulation may be effective. While the Coulomb blockade occurs at nonmagnetic nanoparticles, the stability of spin direction is given by ferromagnetic electrodes electrically coupled with the nanoparticles. A memory device based on this idea is proposed (Mitani and Takanashi 2008).

13.7 Conclusion

Spin accumulation is the deviation of the spatial distribution of spins from its thermal equilibrium state. In F/N-nanoparticle/F double-tunnel junctions, spin accumulation is induced in the nonmagnetic nanoparticle by flowing electrical current through the junction and gives rise to novel TMR effect in tunneling current from/to the nonmagnetic material. TMR of ~10% has been experimentally observed for Al and Au nanoparticles. In F/F-nanoparticle/F double-tunnel junctions, interplay of spin accumulation and spin-dependent SET causes enhancement, oscillation, and sign change of TMR. These phenomena can be reproduced by the numerical simulations using the orthodox theory of spin-dependent SET with spin accumulation. From a fundamental physics point of view, emphases should be placed on the enhancement of spin accumulation due to the Coulomb blockade in nanoparticles and the enhancement of spin relaxation time in nanoparticles. The former is theoretically important to explain the magnitude of TMR induced by spin accumulation in nonmagnetic nanoparticles, which is as large as TMR in conventional tunnel junctions. The latter plays a crucial role for the realization of significant spin accumulation in nanoparticles, and spin relaxation times in Au and Co nanoparticles have been experimentally estimated to be ~10 and 150 ns, respectively. Spin accumulation in nanoparticles could be an important phenomenon for future spintronics, including rich spin-related physics.

References

Awschalom, D. D., D. Loss, and N. Samarth. 2002. *Semiconductor Spintronics and Quantum Computation*. Springer-Verlag, Berlin, Germany.

Barnas, J. and A. Fert. 1998a. Effects of spin accumulation on single-electron tunneling in a double ferromagnetic microjunction. *Europhys. Lett.* 44: 85.

Barnas, J. and A. Fert. 1998b. Magnetoresistance oscillations due to charging effects in double ferromagnetic tunnel junctions. *Phys. Rev. Lett.* 80: 1058.

Bernand-Mantel, A., P. Seneor, N. Lidgi et al. 2006. Evidence for spin injection in a single metallic nanoparticle: A step towards nanospintronics. *Appl. Phys. Lett.* 89: 062502.

Brataas, A., Y. V. Nazarov, J. Inoue et al. 1999a. Spin accumulation in small ferromagnetic double-barrier junctions. *Phys. Rev. B* 59: 93.

Brataas, A., Y. V. Nazarov, J. Inoue et al. 1999b. Non-equilibrium spin accumulation in ferromagnetic single-electron transistors. *Europhys. J. B* 9: 421.

Ernult, F., S. Mitani, K. Takanashi et al. 2006. Self-assembled metallic nanoparticles for spin dependent single electron tunneling. *Phase Trans.* 79: 717.

Ernult, F., K. Yakushiji, S. Mitani et al. 2007. Spin accumulation in metallic nanoparticles. *J. Phys.: Condens. Matter* 19: 165214.

Fujisawa, T., Y. Tokura, and Y. Hirayama. 2001. Transient current spectroscopy of a quantum dot in the Coulomb blockade regime. *Phys. Rev. B* 63: 081304.

Fujisawa, T., D. G. Austing, Y. Tokura et al. 2002. Allowed and forbidden transitions in artificial hydrogen and helium atoms. *Nature* 419: 278.

Gravert, H. and M. H. Devoret. 1992. *Single Charge Tunneling*. Plenum Press, New York.

Guo, G. Y., S. Maekawa, and N. Nagaosa. 2009. Enhanced spin hall effect by resonant skew scattering in the orbital-dependent Kondo effect. *Phys. Rev. Lett.* 102: 036401.

Hirsch, J. E. 1999. Spin Hall effect. *Phys. Rev. Lett.* 83: 1834.

Imamura, H., S. Takahashi, and S. Maekawa. 1999. Spin-dependent Coulomb blockade in ferromagnet/normal-metal/ferromagnet double tunnel junctions. *Phys. Rev. B* 59: 6017.

Imamura, H., J. Chiba, S. Mitani et al. 2000. Coulomb staircase in STM current through granular films. *Phys. Rev. B* 61: 46.

Inoue, J. and A. Brataas. 2004. Magnetization reversal induced by spin accumulation in ferromagnetic transition-metal dots. *Phys. Rev. B* 70: 140406(R).

Jedema, F. J., A. T. Filip, and B. J. van Wees. 2001. Electrical spin injection and accumulation at room temperature in an all-metal mesoscopic spin valve. *Nature* 410: 345.

Johnson, M. 2007. Spin injection and accumulation in mesoscopic metal wires. *J. Phys.: Condens. Matter* 19: 165215.

Julliere, M. 1975. Tunneling between ferromagnetic films. *Phys. Lett. A* 54: 225.

Kato, Y. K., R. C. Myers, A. C. Gossard et al. 2004. Observation of the spin hall effect in semiconductors. *Science* 306: 1910.

Kimura, T. and Y. Otani. 2007a. Large spin accumulation in a permalloy-silver lateral spin valve. *Phys. Rev. Lett.* 99: 196604.

Kimura, T. and Y. Otani. 2007b. Spin transport in lateral ferromagnetic/nonmagnetic hybrid structures. *J. Phys.: Condens. Matter* 19: 165216.

Kubo, R. 1962. Electronic properties of metallic fine particles. *J. Phys. Soc. Jpn.* 17: 975.

Maekawa, S. 2006. *Concepts in Spin Electronics*. Oxford University Press, New York.

Maekawa, S., S. Takahashi, and H. Imamura. 2002. Theory of tunnel magnetoresistance. *Spin Dependent Transport in Magnetic Nanostructures*, eds. T. Shinjo and S. Maekawa, pp. 143–235. Taylor & Francis, Boca Raton, FL.

Majumdar, K. and S. Hershfield. 1998. Magnetoresistance of the double-tunnel-junction Coulomb blockade with magnetic metals. *Phys. Rev.* 57: 11521.

Martinek, J., J. Barnas, G. Michalek et al. 1999. Spin effects in single-electron tunneling in magnetic junctions. *J. Magn. Magn. Mater.* 207: L1.

Martinek, J., J. Barnas, S. Maekawa et al. 2002 Spin accumulation in ferromagnetic single-electron transistors in the cotunneling regime. *Phys. Rev B* 66: 014402.

Mitani, S. and K. Takanashi 2008. Tunnel magnetoresistance due to spin accumulation in nonmagnetic nanoparticles and its potential applications. *Trans. Mater. Res. Soc. Jpn.* 33: 295.

Mitani, S., H. Imamura, K. Takanashi et al. 2007. unpublished.

Mitani, S., Y. Nogi, H. Wang et al. 2008. Current-induced tunnel magnetoresistance due to spin accumulation in Au nanoparticles. *Appl. Phys. Lett.* 92: 152509.

Nogi, Y., H. Wang, F. Ernult et al. 2007. Preparation and magnetotransport properties of MgO-barrier-based magnetic double tunnel junctions including nonmagnetic nanoparticles *J. Phys. D: Appl. Phys.* 40: 1242.

Pratt, W. P., S. F. Lee, J. M. Slaughter et al. 1991. Perpendicular giant magnetoresistances of Ag/Co multilayers. *Phys. Rev. Lett.* 66: 3060.

Seki, T., Y. Hasegawa, S. Mitani et al. 2008. Giant spin Hall effect in perpendicularly spin-polarized FePt/Au devices. *Nat. Mater.* 7: 125.

Seneor, P., A. Bernand-Mantel, and F. Petroff. 2007. Nanospintronics: When spintronics meets single electron physics. *J. Phys.: Condens. Matter* 19: 165222.

Takagishi, M., K. Koi, M. Yoshikawa et al. 2002. The applicability of CPP-GMR heads for magnetic recording. *IEEE Trans. Magn.* 38: 2277.

Takahashi, S. and S. Maekawa. 2008. Spin current in metals and superconductors. *J. Phys. Soc. Jpn.* 77: 031009.

Takahashi, S., H. Imamura, and S. Maekawa. 2006. Spin injection and spin transport in hybrid nanostructures. *Concepts in Spin Electronics*, ed. S. Maekawa, pp. 343–370. Oxford University Press, New York.

Tsang, C. H., R. E. Fontana, T. Lin et al. 1998. Design, fabrication, and performance of spin-valve read heads for magnetic recording applications. *IBM J. Res. Dev.* 42: 103.

Uchida, K., S. Takahashi, K. Harii et al. 2008. Observation of the spin Seebeck effect. *Nature* 445: 778.

Valenzuela, S. O. and M. Tinkham. 2006. Direct electronic measurement of the spin Hall effect. *Nature* 442: 176.

Valet, T. and A. Fert. 1993. Theory of the perpendicular magnetoresistance in magnetic multilayers. *Phys. Rev. B* 48: 7099.

Wang, H., S. Mitani, K. Takanashi et al. 2007. Numerical simulation of spin accumulation and tunnel magnetoresistance in single electron tunnelling junctions with a nonmagnetic nanoparticle. *Phys. Stat. Sol. (b)* 244: 4443.

Weymann, I. and J. Barnas. 2003. Transport characteristics of ferromagnetic single-electron transistors. *Phys. Stat. Sol. (b)* 236: 651.

Wolf, S. A., D. D. Awschalom, R. A. Buhrman et al. 2001. Spintronics: A spin-based electronics vision for the future. *Science* 294: 1488.

Yakushiji, K., S. Mitani, K. Takanashi et al. 2002. Tunnel magnetoresistance oscillations in current perpendicular to plane geometry of CoAlO granular thin films. *J. Appl. Phys.* 91: 7038.

Yakushiji, K., F. Ernult, H. Imamura et al. 2005. Enhanced spin accumulation and novel magnetotransport in nanoparticles. *Nat. Mater.* 4: 57.

Yakushiji, K., S. Mitani, F. Ernult et al. 2007. Spin-dependent tunneling and Coulomb blockade in ferromagnetic nanoparticles. *Phys. Rep.* 451: 1.

Yano, K., T. Ishii, T. Sano et al. 1999. Single-electron memory for giga-to-tera bit storage. *Proc. IEEE* 87: 633.

Zhang, L. Y., C. Y. Wang, Y. G. Wei et al. 2005. Spin-polarized electron transport through nanometer-scale Al grains. *Phys. Rev. B* 72: 155445.

Zhang, S., P. M. Levy, and A. Fert. 2002. Mechanisms of spin-polarized current-driven magnetization switching. *Phys. Rev. Lett.* 88: 236601.

Zutic, I. and J. Fabian. 2006. Bipolar spintronics. *Concepts in Spin Electronics*, ed. S. Maekawa, pp. 43–92. Oxford University Press, New York.

Zutic, I., J. Fabian, and S. D. Sarma. 2004. Spintronics: Fundamentals and applications. *Rev. Mod. Phys.* 76: 323.

14

Photoinduced Magnetism in Nanoparticles

Vassilios Yannopapas
University of Patras

14.1 Introduction

One of the most fascinating fields of modern optics is that of optical metamaterials. They are man-made structures with electromagnetic (EM) properties which are not met in naturally occurring materials such as artificial magnetism and negative refractive index (Veselago, 1968; Pendry, 2000, 2004). The artificial magnetic response of the metamaterials can lead to strong paramagnetic (permeability $\mu > 1$) and diamagnetic response (permeability $\mu < 1$ or even $\mu < 0$) of the same metamaterial structure, in frequency regions where such a response is not met in naturally occurring materials, like the near-infrared and optical regions where ordinary materials with strong magnetic response are very rare. Magnetic activity in these regions of the EM spectrum is of great technological importance since it allows for the realization of devices such as compact cavities, adaptive selective lenses, tunable mirrors, isolators, converters, optical polarizers, filters, and phase shifters (Panina et al., 2002; Yen et al., 2004). Magnetic metamaterials realized at radio frequencies have already found application in magnetic resonance imaging (Wiltshire et al., 2001).

Most experimental realizations of magnetic metamaterials are based on metallic split-ring resonators and variations of such (Shelby et al., 2001; Linden et al., 2004; Yen et al., 2004). Due to its geometry, a split-ring resonator (see Figure 14.18) operates as an LC circuit and, under illumination, exhibits strong magnetic response around the LC resonance frequency. Experiments on the magnetic response of split-ring resonators have been mostly performed in the microwave and far-infrared regions. Their miniaturization in the micron or even nanoscale, so that they exhibit magnetic activity in the near-infrared and visible regimes, is still challenging since it requires advanced lithographic techniques.

However, magnetic activity in the infrared regime has been theoretically predicted for arrays consisting of less elaborate scatterers such as three-dimensional (3D) arrays of spherical particles (Wheeler et al., 2005; Yannopapas and Moroz, 2005; Jylhä et al., 2006). The emergence of magnetic response in these structures relies on a different mechanism than the excitation of a LC resonance as it is the case for the split-ring resonators. First of all, the materials which the particles are made from must exhibit a certain type of internal resonance around which the electric permittivity (dielectric function) of the material assumes very high values (O'Brien and Pendry, 2002; Huang et al., 2004). Such a resonance might be a phonon-polariton or an exciton-polariton resonance (see later), and the corresponding materials can be ionic or semiconductor materials. When an array of particles made from such a resonant material is illuminated by an EM wave of frequency around the internal resonance of the particles, strong polarization currents are generated within the particles resulting in a macroscopic magnetization of the array of particles. This phenomenon is described as photoexcitation-induced magnetism or photoinduced magnetism. The macroscopic magnetization of the array of particles is quantified by the strong resonant behavior of the effective (average) magnetic permeability μ_{eff} of the array. Around the resonance frequency of the particle, the magnetic permeability may become even negative; this effect is not expected in naturally occurring materials, and it is one of the salient features of the optical metamaterials.

The theoretical treatment of the structures discussed earlier is based on two different approaches. The first one is the effective medium treatment which provides directly the effective electric permittivity and magnetic permeability of a composite structure made, e.g., from spherical particles. It is an approximate theory since it essentially substitutes the actual inhomogeneous structure (metamaterial) with a homogeneous one whose EM response is similar to that of the actual (inhomogeneous) structure. As such, it is valid only in the limit of long wavelengths in which case the geometric details of the structure (particle radius, interparticle distance, and periodic or nonperiodic arrangement of the particles) are not "felt" by an EM wave of very long wavelength. The second approach is a rigorous EM theory which solves exactly the Maxwell's equations within a composite structure consisting of particles. Both theoretical methods are presented in Section 14.2 and applied to specific cases of photoinduced magnetic metamaterials in Section 14.3 of the present chapter. Section 14.4 reviews experiments on such materials and Section 14.5 concludes the chapter.

14.2 Theory

Both theoretical methods studied in this section (effective medium and layer-multiple scattering methods) rely on the scattering theory of EM radiation from a spherical particle (scatterer) which is known in literature as the Mie scattering theory. We will proceed, therefore, to a brief summary of Mie theory where the central quantity is the scattering T-matrix which essentially provides the EM field scattered from a particle in terms of the incident field. However, its usefulness goes beyond this, as it also provides the effective optical parameters ϵ_{eff}, μ_{eff} of a composite material containing the said scatterers. Before studying the core of Mie theory, we first introduce the EM spherical waves (multipole expansion of the EM field) as rigorous solutions of Maxwell's equations within a homogeneous medium.

14.2.1 Multipole Expansion of the EM Field

Let us consider a harmonic EM wave, of angular frequency ω, which is described by its electric-field component:

$$\mathbf{E}(\mathbf{r}, t) = \mathrm{Re}[\mathbf{E}(\mathbf{r})\exp(-i\omega t)]. \qquad (14.1)$$

In a homogeneous medium characterized by a dielectric function $\epsilon(\omega)\epsilon_0$ and a magnetic permeability $\mu(\omega)\mu_0$, where ϵ_0, μ_0 are the electric permittivity and magnetic permeability of vacuum, Maxwell equations imply that $\mathbf{E}(\mathbf{r})$ satisfies a vector Helmholtz equation, subject to the condition $\nabla \cdot \mathbf{E} = 0$, with a wave number $q = \omega/c$, where $c = 1/\sqrt{\mu\epsilon\mu_0\epsilon_0} = c_0/\sqrt{\mu\epsilon}$ is the velocity of light in the medium. The spherical-wave expansion of $\mathbf{E}(\mathbf{r})$ is given by (Jackson, 1975)

$$\mathbf{E}(\mathbf{r}) = \sum_{l=1}^{\infty}\sum_{m=-l}^{l}\left\{a_{lm}^{H}f_l(qr)\mathbf{X}_{lm}(\hat{\mathbf{r}}) + a_{lm}^{E}\frac{i}{q}\nabla\times\left[f_l(qr)\mathbf{X}_{lm}(\hat{\mathbf{r}})\right]\right\}, \qquad (14.2)$$

where a_{lm}^{P} ($P = E, H$) are coefficients to be determined. $\mathbf{X}_{lm}(\hat{\mathbf{r}})$ are the so-called vector spherical harmonics (Jackson, 1975) and f_l may be any linear combination of the spherical Bessel function, j_l, and the spherical Hankel function, h_l^{+}. The corresponding magnetic induction, $\mathbf{B}(\mathbf{r})$, can be readily obtained from $\mathbf{E}(\mathbf{r}, t)$ using Maxwell's equations:

$$\mathbf{B}(\mathbf{r}) = \frac{\sqrt{\epsilon\mu}}{c_0}\sum_{l=1}^{\infty}\sum_{m=-l}^{l}\left\{a_{lm}^{E}f_l(qr)\mathbf{X}_{lm}(\hat{\mathbf{r}}) - a_{lm}^{H}\frac{i}{q}\nabla\times\left[f_l(qr)\mathbf{X}_{lm}(\hat{\mathbf{r}})\right]\right\}, \qquad (14.3)$$

and it is not written down explicitly in the following discussion.

14.2.2 Scattering by a Single Scatterer

We are now in position to solve the problem EM scattering from a single sphere [Mie scattering theory (Jackson, 1975; Bohren and Huffman, 1984)], i.e., the determination of the expansion coefficients (like the a_{lm}^{P} of Equation 14.2) of the EM field scattered from the sphere when the latter is illuminated by a plane EM wave.

We consider a sphere of radius S, with its center at the origin of coordinates, and assume that its electric permittivity, ϵ_s, and/or magnetic permeability, μ_s, are different from those, ϵ_h, μ_h, of the surrounding homogeneous medium. An EM plane wave which is incident on this scatterer is described by Equation 14.2 with $f_l = j_l$ (since the plane wave is finite everywhere) and appropriate coefficients a_L^0, where L denotes collectively the indices Plm. That is,

$$\mathbf{E}^0(\mathbf{r}) = \sum_L a_L^0 \mathbf{J}_L(\mathbf{r}) \qquad (14.4)$$

where

$$\mathbf{J}_{Elm}(\mathbf{r}) = \frac{i}{q_h}\nabla\times j_l(q_h r)\mathbf{X}_{lm}(\hat{\mathbf{r}}), \quad \mathbf{J}_{Hlm}(\mathbf{r}) = j_l(q_h r)\mathbf{X}_{lm}(\hat{\mathbf{r}}) \qquad (14.5)$$

and $q_h = \sqrt{\epsilon_h\mu_h}\,\omega/c_0$. The coefficients a_L^0 depend on the amplitude, polarization, and propagation direction of the incident EM plane wave (Jackson, 1975).

Similarly, the wave that is scattered from the sphere is described by Equation 14.2 with $f_l = h_l^{+}$, which has the asymptotic form appropriate to an outgoing spherical wave: $h_l^{+} \approx (-i)^l\exp(iq_h r)/iq_h r$ as $r \to \infty$, and appropriate expansion coefficients a_L^{+}, namely,

$$\mathbf{E}^{+}(\mathbf{r}) = \sum_L a_L^{+}\mathbf{H}_L(\mathbf{r}) \qquad (14.6)$$

where

$$\mathbf{H}_{Elm}(\mathbf{r}) = \frac{i}{q_h}\nabla\times h_l^{+}(q_h r)\mathbf{X}_{lm}(\hat{\mathbf{r}}), \quad \mathbf{H}_{Hlm}(\mathbf{r}) = h_l^{+}(q_h r)\mathbf{X}_{lm}(\hat{\mathbf{r}}). \qquad (14.7)$$

The wavefield for $r > S$ is the sum of the incident and scattered waves, i.e., $\mathbf{E}^{out} = \mathbf{E}^0 + \mathbf{E}^+$. The spherical-wave expansion of the field $\mathbf{E}^I$ for $r < R$ (inside the sphere) is obtained in a similar manner by the requirement that it be finite at the origin ($\mathbf{r} = \mathbf{0}$), i.e.,

$$\mathbf{E}^I(\mathbf{r}) = \sum_L a_L^I \mathbf{J}_L^s(\mathbf{r}) \tag{14.8}$$

where $\mathbf{J}_L^s(\mathbf{r})$ are given from Equation 14.5 by replacing q_h with $q_s = \sqrt{\epsilon_s \mu_s} \, \omega / c_0$.

By applying the requirement that the tangential components of $\mathbf{E}$ and $\mathbf{H}$ be continuous at the surface of the scatterer, we obtain a relation between the expansion coefficients of the incident and the scattered field, as follows:

$$a_L^+ = \sum_{L'} T_{LL'} a_{L'}^0, \tag{14.9}$$

where $T_{LL'}$ are the elements of the so-called scattering transition T-matrix (Bohren and Huffman, 1984). Equation 14.9 is valid for any shape of scatterer; for spherically symmetric scatterers, each spherical wave scatters independently of all others, which leads to a transition T-matrix which does not depend on m and is diagonal in l, i.e., $T_{LL'} = T_L \delta_{LL'}$; it is given by

$$T_l^E(\omega) = \left[\frac{j_l(q_s r)\frac{\partial}{\partial r}(rj_l(q_h r))\epsilon_s - j_l(q_h r)\frac{\partial}{\partial r}(rj_l(q_s r))\epsilon_h}{h_l^+(q_h r)\frac{\partial}{\partial r}(rj_l(q_s r))\,\epsilon_h - j_l(q_s r)\frac{\partial}{\partial r}(rh_l^+(q_h r))\epsilon_s}\right]_{r=S} \tag{14.10}$$

$$T_l^H(\omega) = \left[\frac{j_l(q_s r)\frac{\partial}{\partial r}(rj_l(q_h r))\mu_s - j_l(q_h r)\frac{\partial}{\partial r}(rj_l(q_s r))\mu_h}{h_l^+(q_h r)\frac{\partial}{\partial r}(rj_l(q_s r))\mu_h - j_l(q_s r)\frac{\partial}{\partial r}(rh_l^+(q_h r))\mu_s}\right]_{r=S} \tag{14.11}$$

14.2.3 Effective-Medium Theory

A composite material of spherical scatterers, e.g., metallic or semiconductor nanoparticles in a polymer host or suspended in a liquid, can be described, in the subwavelength limit, as a homogeneous medium of effective relative permittivity ϵ_{eff} and effective relative permeability μ_{eff}. We assume that the scatterers possess a relative permittivity ϵ_s and relative permeability μ_s and are embedded in a host medium described by a relative permittivity ϵ_h and relative permeability μ_h. The volume filling fraction occupied by the scatterers (the percentage of space covered by the scatterers) is denoted by f. The effective parameters ϵ_{eff} and μ_{eff} can be calculated from the extended Maxwell–Garnett (EMG) theory (Doyle, 1989; Ruppin, 2000), which goes one step ahead the ordinary Maxwell–Garnett theory by incorporating characteristics of Mie scattering in the corresponding formulae of ϵ_{eff} and μ_{eff}, i.e.,

$$\epsilon_{eff} = \epsilon_h \frac{x^3 - 3if \, T_1^E}{x^3 + \frac{3}{2}if \, T_1^E}, \tag{14.12}$$

and

$$\mu_{eff} = \mu_h \frac{x^3 - 3if \, T_1^H}{x^3 + \frac{3}{2}if \, T_1^H}, \tag{14.13}$$

where $T_1^E(T_1^H)$ are the electric-dipole (magnetic-dipole) components of the scattering matrices of Equation 14.10 (Equation 14.11) for $l = 1$:

$$T_1^E(\omega) = \left[\frac{j_1(x_s)[xj_1(x)]'\epsilon_s - j_1(x)[x_s j_1(x_s)]'\epsilon_h}{h_1^+(x)[x_s j_1(x_s)]'\epsilon_h - j_1(x_s)[xh_1^+(x)]'\epsilon_s}\right], \tag{14.14}$$

$$T_1^H(\omega) = \left[\frac{j_1(x_s)[xj_1(x)]'\mu_s - j_1(x)[x_s j_1(x_s)]'\mu_h}{h_1^+(x)[x_s j_1(x_s)]'\mu_h - j_1(x_s)[xh_1^+(x)]'\mu_s}\right], \tag{14.15}$$

where

$j_1(h_1^+)$ is the spherical Bessel (Hankel) function of order one

$[xj_1(x)]' = d[zj_1(z)]/dz|_{z=x}$ etc.

x stands for the sphere size parameter $x \equiv \sqrt{\epsilon_h \mu_h} \, \omega S/c = 2\pi S/\lambda$

where

S denotes the sphere radius

λ is the wavelength in the host medium

Also, $x_s \equiv \sqrt{\epsilon_s \mu_s} \, \omega S/c = 2\pi S/\lambda_s$, where λ_s is the wavelength in the sphere medium. Equations 14.12 and 14.13 are valid in the quasi-static limit, i.e., provided that $x \ll 1$ but *not* necessarily $x_s \ll 1$, as it is the case for the ordinary Maxwell–Garnett formula (static limit) (Doyle, 1989; Ruppin, 2000).

The optical parameters ϵ_{eff} and μ_{eff}, given by Equations 14.12 and 14.13, can characterize the optical response of a composite material in the long-wavelength limit, i.e., in the limit where the wavelength of propagating EM radiation is much larger than the interparticle distance. When the wavelength is comparable to the interparticle distance (lattice constant for a periodic crystal of scatterers), then multiple scattering of light becomes the dominant process and the inhomogeneity of the composite material cannot be hidden "under the carpet" as it is done in the effective-medium theories (such as the EMG theory examined here), and a rigorous solution to Maxwell's equations is therefore needed. One of the most powerful techniques used for the solution of Maxwell's equations on inhomogeneous periodic or partially periodic crystals is the layer-multiple-scattering method considered next.

14.2.4 Layer-Multiple-Scattering Method

The layer-multiple-scattering (LMS) method is a very general and powerful method for solving Maxwell's equations and, thus, predicting the optical properties of composite structures such as photonic crystals and metamaterials. Since an analytic derivation and presentation of all the relevant equations and formulas is beyond the scope of this chapter, we will only provide a brief outline of the method. The interested reader should resort to more technical papers (Stefanou et al., 1998, 2000) where the method is thoroughly analyzed.

We consider a 3D crystal containing macroscopic scatterers arranged periodically in space. We can view the crystal as a succession of layers parallel to a given crystallographic plane of the crystal. The layers have the same two-dimensional (2D) periodicity (that of the chosen crystallographic plane) described by a 2D lattice:

$$\mathbf{R}_n = n_1\mathbf{a}_1 + n_2\mathbf{a}_2, \qquad (14.16)$$

where $\mathbf{a}_1$ and $\mathbf{a}_2$ are primitive vectors of the said plane (which is taken to be the xy plane), and $n_1, n_2 = 0, \pm1, \pm2, \ldots$. We may number the sequence of layers which constitute the infinite crystal, extending from $z = -\infty$ to $z = +\infty$, as follows: $\ldots -2, -1, 0, 1, 2, \ldots$. The $(N + 1)$th layer is obtained from the Nth layer by a primitive translation to be denoted by $\mathbf{a}_3$. Obviously, $\mathbf{a}_1$, $\mathbf{a}_2$, and $\mathbf{a}_3$ constitute a basis for the 3D space lattice of the infinite crystal.

We define the 2D reciprocal lattice corresponding to Equation 14.16:

$$\mathbf{g} = m_1\mathbf{b}_1 + m_2\mathbf{b}_2, \quad m_1, m_2 = 0, \pm1, \pm2, \ldots \quad (14.17)$$

where $\mathbf{b}_i \cdot \mathbf{a}_j = 2\pi\delta_{ij}$, and $i, j = 1, 2$. The reduced (k_x, k_y)-zone associated with the earlier discussion, which has the full symmetry of the given crystallographic plane, is known as the surface Brillouin zone (SBZ) [see, e.g., Modinos, 1984]. We define a corresponding 3D reduced $\mathbf{k}$-zone as follows:

$$\mathbf{k}_\parallel \equiv (k_x, k_y) \text{ within the SBZ}$$

$$-\frac{|\mathbf{b}_3|}{2} < k_z \le \frac{|\mathbf{b}_3|}{2}, \qquad (14.18)$$

where $\mathbf{b}_3 = 2\pi\mathbf{a}_1 \times \mathbf{a}_2/[\mathbf{a}_1 \cdot (\mathbf{a}_2 \times \mathbf{a}_3)]$ is normal to the chosen crystallographic plane. The reduced $\mathbf{k}$-zone defined by Equation 14.18 is of course completely equivalent to the commonly used, more symmetrical Brillouin zone (BZ), in the sense that a point in one of them also lies in the other or differs from such one by a vector of the 3D reciprocal lattice.

Let us now assume that we have a crystal consisting of non-overlapping spherical scatterers in a host medium of different dielectric function and let us look at the structure as a sequence of layers of spheres with the 2D periodicity of Equation 14.16. A Bloch wave solution, of given frequency ω and given $\mathbf{k}_\parallel$, of Maxwell's equations for the given system has the following form in the space between the Nth and the $(N + 1)$th layers (we write down only the electric-field component of the EM wave):

$$\mathbf{E}(\mathbf{r}) = \sum_{\mathbf{g}} \left\{ \mathbf{E}_{\mathbf{g}}^+(N)\exp\left[i\mathbf{K}_{\mathbf{g}}^+ \cdot (\mathbf{r} - \mathbf{A}_N)\right] \right.$$
$$\left. + \mathbf{E}_{\mathbf{g}}^-(N)\exp\left[i\mathbf{K}_{\mathbf{g}}^- \cdot (\mathbf{r} - \mathbf{A}_N)\right] \right\} \qquad (14.19)$$

with

$$\mathbf{K}_{\mathbf{g}}^\pm = \left(\mathbf{k}_\parallel + \mathbf{g}, \ \pm\left[q^2 - (\mathbf{k}_\parallel + \mathbf{g})^2\right]^{1/2}\right), \qquad (14.20)$$

where

q is the wavenumber
$\mathbf{A}_N$ is an appropriate origin of coordinates in the host region between the Nth and the $(N + 1)$th layers

A similar expression (with N replaced by $N + 1$) gives the electric field between the $(N + 1)$th and the $(N + 2)$th layers. Naturally the coefficients $\mathbf{E}_{\mathbf{g}}^\pm(N+1)$ are related to the $\mathbf{E}_{\mathbf{g}}^\pm(N)$ coefficients through the scattering matrices of the Nth layer of spheres. We have

$$E_{\mathbf{g}i}^-(N) = \sum_{\mathbf{g}'i'} Q_{\mathbf{g}i;\mathbf{g}'i'}^{\mathrm{IV}} E_{\mathbf{g}'i'}^-(N+1) + \sum_{\mathbf{g}'i'} Q_{\mathbf{g}i;\mathbf{g}'i'}^{\mathrm{III}} E_{\mathbf{g}'i'}^+(N)$$

$$\qquad (14.21)$$

$$E_{\mathbf{g}i}^+(N+1) = \sum_{\mathbf{g}'i'} Q_{\mathbf{g}i;\mathbf{g}'i'}^{\mathrm{I}} E_{\mathbf{g}'i'}^+(N) + \sum_{\mathbf{g}'i'} Q_{\mathbf{g}i;\mathbf{g}'i'}^{\mathrm{II}} E_{\mathbf{g}'i'}^-(N+1),$$

where $i = x, y, z$, and $\mathbf{Q}$ are appropriately constructed transmission/reflection matrices for the layer. For a detailed description of these matrices, which are functions of ω, $\mathbf{k}_\parallel$, the scattering properties of the individual scatterer (sphere), and the geometry of the layer, see Stefanou et al. (1998, 2000).

A generalized Bloch wave, by definition, has the property:

$$\mathbf{E}_{\mathbf{g}}^\pm(N+1) = \exp(i\mathbf{k} \cdot \mathbf{a}_3)\mathbf{E}_{\mathbf{g}}^\pm(N)$$
$$\mathbf{k} = \left(\mathbf{k}_\parallel, k_z(\omega, \mathbf{k}_\parallel)\right) \qquad (14.22)$$

where k_z may be real or complex. Substituting Equation 14.22 into Equation 14.21, we obtain

$$\begin{pmatrix} \mathbf{Q}^{\mathrm{I}} & \mathbf{Q}^{\mathrm{II}} \\ -[\mathbf{Q}^{\mathrm{IV}}]^{-1}\mathbf{Q}^{\mathrm{III}}\mathbf{Q}^{\mathrm{I}} & [\mathbf{Q}^{\mathrm{IV}}]^{-1}[\mathbf{I} - \mathbf{Q}^{\mathrm{III}}\mathbf{Q}^{\mathrm{II}}] \end{pmatrix} \begin{pmatrix} \mathbf{E}^+(N) \\ \mathbf{E}^-(N+1) \end{pmatrix}$$

$$= \exp(i\mathbf{k} \cdot \mathbf{a}_3) \begin{pmatrix} \mathbf{E}^+(N) \\ \mathbf{E}^-(N+1) \end{pmatrix} \qquad (14.23)$$

where $\mathbf{E}^\pm$ are column matrices with elements: $E_{\mathbf{g}_1 x}^\pm, E_{\mathbf{g}_1 y}^\pm$, $E_{\mathbf{g}_1 z}^\pm, E_{\mathbf{g}_2 x}^\pm, E_{\mathbf{g}_2 y}^\pm, E_{\mathbf{g}_2 z}^\pm, \ldots$. In practice we keep a finite number of $\mathbf{g}$-vectors (those with $|\mathbf{g}| < g_{\max}$, where $g_{\max}$ is a cutoff parameter) which leads to a solvable system of equations. The number of independent components of the electric field is in fact two-thirds of that discussed earlier, in view of the zero-divergence of the electric field, but we need not go into technical details here (Stefanou et al., 1998, 2000).

Equation 14.23 constitutes a typical eigenvalue problem; because the matrix on the left-hand side of Equation 14.23 is not Hermitian, its eigenvalues are in general complex numbers. We remember that ω and $\mathbf{k}_\parallel$ are given quantities and, therefore, the eigenvalues of the matrix on the left-hand side of Equation 14.23 determine k_z; depending on the number of $\mathbf{g}$-vectors we keep in the calculation, we obtain a corresponding number of k_z-eigenvalues for the given $\omega, \mathbf{k}_\parallel$. These eigenvalues of k_z looked upon as functions $k_z = k_z(\omega; \mathbf{k}_\parallel)$ of real ω, for given $\mathbf{k}_\parallel$, are known as the real-frequency lines in the complex k_z-space. We refer to them as the complex band structure of the crystal associated with the crystallographic surface defined by Equation 14.16. In all cases $[\mathbf{k}_\parallel, \mathrm{Re}(k_z)]$ lies in the reduced zone defined by Equation 14.18. To obtain a full picture of the ordinary frequency band structure, one needs to know all the real k_z-sections of the real-frequency lines for every $\mathbf{k}_\parallel$ in the irreducible part of the SBZ; in the remaining part of the SBZ they are determined by symmetry.

The on-shell method that has been described has a number of advantages over the traditional methods. The unit layer along the z-direction may consist not of one plane of spheres (as implied so far) but by a number of planes which may be different (the radii of the spheres and/or their dielectric functions may be different) as long as they have the same 2D periodicity. Moreover, we can easily calculate the transmission, reflection, and absorption coefficients of light incident at any angle on a slab of the crystal. For this purpose, we combine the **Q**-matrices of the different layers that make the slab into a final set of matrices that determines the scattering properties of the slab (Stefanou et al., 1998, 2000). The method applies equally well to nonabsorbing systems which have a well-defined frequency band structure, and to absorbing systems which do not have a well-defined frequency band structure (there can be no truly propagating waves in such systems). Finally one can deal with the slabs of crystals which contain impurity planes as long as the 2D periodicity parallel to the surface of the slab is retained.

14.3 Magnetic Activity in Crystals of Nonmagnetic Particles

Having established a rigorous theoretical armory in the previous section (Section 14.2), we are ready to confront with several cases of artificial composite (meta-) materials consisting of spherical particles. In particular, we are interested in inherently nonmagnetic particles which, however, exhibit strong magnetic activity, either paramagnetic or diamagnetic due to the excitation of internal material resonances by illumination with the EM radiation of proper frequency.

14.3.1 Phonon-Polaritonic Particles

We first examine the case of a 3D metamaterial of nonoverlapping polaritonic spheres, i.e., spheres made from a material supporting phonon-polariton excitations. Such materials are usually

optically active ionic crystals that support phonon-polaritons, i.e., simultaneously excited modes of the material polarization and the EM field (Ibach and Lüth, 2003). Usually, the phonon-polariton are excited when infrared radiation impinges on such a material inducing coupled oscillations of the polarization and the EM field. The dielectric function in such materials is provided by the Drude–Lorentz model (Ibach and Lüth, 2003):

$$\epsilon(\omega) = \epsilon_\infty \left(1 + \frac{\omega_L^2 - \omega_T^2}{\omega_T^2 - \omega^2 - i\omega\gamma_1} \right), \qquad (14.24)$$

where
  ϵ_∞ is the asymptotic value of the dielectric permittivity at high frequencies
  γ_1 is the loss factor

The respective ω_T and ω_L are the transverse and longitudinal optical phonon frequencies.

To begin with, losses are ignored in the polaritonic spheres by setting $\gamma_1 = 0$ in Equation 14.24. As can be easily inferred from Equations 14.12 and 14.13, the resonances of the T_1^H and T_1^E introduce resonances in μ_{eff} and ϵ_{eff}, respectively (although not at the same frequencies). Around these resonances, μ_{eff} and ϵ_{eff} may even assume negative values as a result of the strong electric and magnetic activity.

As a polaritonic material, the choice is LiTaO$_3$ whose permittivity is given by Equation 14.24 having $\omega_T = 26.7 \cdot 10^{12}$ rad/s, $\omega_L = 46.9 \cdot 10^{12}$ rad/s, and $\epsilon_\infty = 13.4$ (Schall et al., 1999). The radius of the spheres is $S = 2.8\,\mu\mathrm{m}$ and the volume filling fraction occupied by them is $f = 0.736$. Of course, such a high value of f is possible only in face-centered cubic (fcc) or hexagonal closest packing (hcp) arrangements. In Figure 14.1a, the real and imaginary parts

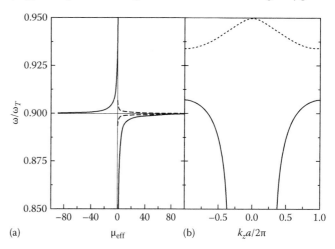

FIGURE 14.1 (a) The real (solid line) and imaginary (broken line) parts of μ_{eff} predicted by the extended Maxwell–Garnett theory and (b) the frequency band structure normal to the (001) surface of an fcc crystal of LiTaO$_3$ spheres ($S = 2.8\,\mu\mathrm{m}$, $\omega_T = 26.7 \cdot 10^{12}$ rad/s, $\omega_L = 46.9 \cdot 10^{12}$ rad/s, $\epsilon_\infty = 13.4$) in air, as calculated by the rigorous LMS method. The lattice constant is $a = 7.94\,\mu\mathrm{m}$. The solid curve in (b) refers to a degenerate band whilst the dotted one to nondegenerate band. (Reprinted from Yannopapas, V., *Appl. Phys. A*, 87, 259, 2007. With permission.)

of μ_{eff} are shown, as calculated from Equation 14.13, for the previous collection of LiTaO$_3$ spheres. A region where the real part of μ_{eff} is negative is identified, namely from $\omega/\omega_T = 0.900$ to $\omega/\omega_T = 0.948$. This region corresponds to the first resonance of T_1^H.

In order to ensure the validity of the EMG theory (Equations 14.12 through 14.15) against an ab initio method (LMS method), one can consider a periodic realization of the LiTaO$_3$ particle array examined previously. That is, it is assumed that the LiTaO$_3$ spheres occupy the sites of an fcc lattice. For the sphere parameters considered earlier, i.e., $S = 2.8\,\mu m$ and $f = 0.736$, the corresponding fcc lattice constant is $a = 7.94\,\mu m$. The crystal is viewed as a succession of planes parallel to the (001) surface of fcc.

The crystal is viewed as a stack of 2D crystal layers (planes of spheres) parallel to the (001) surface (which we assume parallel to the xy plane) of an fcc lattice. Note in passing that each crystal layer is a 2D square lattice with the lattice constant $a_0 = a/\sqrt{2}$. For any given $\mathbf{k}_\| = (k_x, k_y)$, one can calculate the real-frequency lines: $k_z = k_z(\omega, \mathbf{k}_\|)$. Figure 14.1b shows the frequency band structure, for $\mathbf{k}_\| = \mathbf{0}$ (normal to the (001) surface), for the above-mentioned polaritonic crystal. From Figure 14.1b, it is obvious that the LMS frequency band structure predicts a wide photonic band gap from $\omega/\omega_T = 0.907$ to $\omega/\omega_T = 0.934$. From $\omega/\omega_T = 0.934$ to $\omega/\omega_T = 0.949$, there exists a nondegenerate band, and as such, the transmission of normally incident light is negligibly small over the frequency region of this band (optically inactive band) (Stefanou et al., 1992). Within the region of the gap and of the optically inactive bands, one expects to observe a significantly attenuated transmission of light through finite slabs of the above-mentioned crystal (Yannopapas and Moroz, 2005). Therefore, one may assume that there exists an effective band gap extending in both regions, i.e., from $\omega/\omega_T = 0.907$ to $\omega/\omega_T = 0.949$. This effective gap almost coincides with the frequency region over which μ_{eff} is negative, i.e., from $\omega/\omega_T = 0.900$ to $\omega/\omega_T = 0.948$. The latter certifies the accuracy of the EMG theory.

14.3.2 Exciton-Polaritonic Particles

The magnetic response demonstrated in Figure 14.1 takes place around the transverse phonon frequency ω_T which lies within the near-infrared regime. Since all ionic materials similar to LiTaO$_3$, e.g., SiC, TlBr, TlCl, exhibit phonon-polariton excitations within this spectral regime (near infrared), if magnetic response in higher frequencies is needed, one has to choose a different class of polaritonic materials. Such a class of materials are the semiconductors exhibiting exciton-polariton excitations in visible frequencies and even below (Yannopapas and Vitanov, 2006). Excitons are bound states between an electron excited into the conduction band and the hole that remains in the valence band, with the Coulomb interaction being responsible for the binding energy (Toyozawa, 2003). An exciton can be described accurately by an hydrogenic atomic model with corresponding hydrogenic eigenstates and eigenspectrum. When an exciton is driven from its ground state to an excited state, we have a net polarization in the semiconducting material. If the polarization currents introduced are strong enough, a macroscopic magnetic

activity may result in when the semiconductor is illuminated by an externally incident light with frequency corresponding to a specific exciton transition.

Semiconductor materials with strong excitonic oscillator strength, leading to strong photoinduced polarizations currents are, e.g., Cu$_2$O possessing a 2P exciton line at 576.84 nm and CuCl possessing a Z$_3$ exciton line at 386.93 nm (Artoni et al., 2005). Around the exciton frequencies, the dielectric function of the above-mentioned semiconductors is given by

$$\epsilon_s(\omega) = \epsilon_\infty + A\frac{\gamma}{\omega_0 - \omega - i\gamma}. \tag{14.25}$$

The constant A of the Equation 14.25 is proportional to the exciton oscillation strength; for Cu$_2$O, $A = 0.02$ whilst for CuCl, $A = 632$. Since the exciton oscillation strength of the cuprite is rather weak, we will focus on the copper chloride. The rest of the parameters for CuCl that enter Equation 14.25 are (Artoni et al., 2005) $\epsilon_\infty = 5.59$, $\hbar\omega_0 = 3.363\,eV$, and $\hbar\gamma = 5\times10^{-5}\,eV$. The small value of the loss factor γ implies a very narrow exciton linewidth.

We consider a 3D array consisting of CuCl spheres in air. The radius of the spheres is $S = 20\,nm$ and the volume filling fraction occupied by the spheres $f = 0.74$ (close-packed arrangement). Of course, such a high value of f is possible only in fcc or hcp arrangements. The arrays of CuCl nanoparticles can be fabricated by colloidal crystallization (Orel et al., 2003) and ion implantation techniques (Fukumi et al., 1999).

Figure 14.2a shows the effective magnetic permeability μ_{eff} of the above collection of CuCl spheres, as calculated by Equations 14.13 and 14.15. It is evident that the composite material exhibits

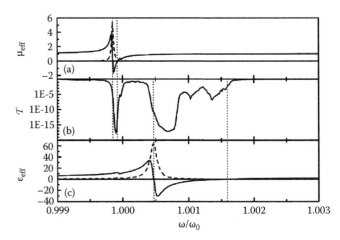

FIGURE 14.2 (a) Effective relative permeability μ_{eff} and (c) permittivity ϵ_{eff} of a collection of close-packed CuCl nanospheres with radius $S = 20\,nm$ in air, as obtained by the EMG theory. The real (imaginary) parts correspond to the solid (broken) lines. (b) Corresponding transmittance curve for light incident normally on a slab consisting of 16 (001) planes of an fcc crystal of the above-mentioned CuCl nanospheres, as calculated by the rigorous LMS method. The two pairs of vertical dotted lines define the negative $\Re\mu_{eff}$ and $\Re\epsilon_{eff}$ regions, respectively. (Reprinted from Yannopapas, V. and Vitanov, N.V., *Phys. Rev. B*, 74, 193304-1, 2006. With permission.)

paramagnetic ($\Re\mu_{eff} > 1$) as well as diamagnetic behavior ($\Re\mu_{eff} < 1$) around the exciton resonance ω_0 which corresponds to a wavelength of 386.93 nm, approximately 19 times larger than the sphere radius. Since the real part of μ_{eff} assumes values from −1.65 to 3.8, the structure under study is a true subwavelength magnetic metamaterial. The magnetic activity in the region of the exciton resonance stems from the enhancement of the displacement current inside each sphere which, in turn, gives rise to a macroscopic magnetization of the collection of CuCl spheres. Similar but less dramatic effects are expected for arrays of Cu_2O particles.

In order to test the validity of the EMG theory (Equations 14.12 through 14.15) by comparing it with the LMS method, we consider a periodic realization of the CuCl nanoparticle array examined previously. That is, we assume that the CuCl spheres occupy the sites of an fcc lattice. For the sphere parameters considered earlier, i.e., $S = 20$ nm and $f = 0.74$, the corresponding fcc lattice constant is $a = 56.58$ nm. The crystal is viewed as a succession of planes parallel to the (001) surface of fcc.

Figure 14.2b shows the transmittance of light incident normally on a finite slab of 16 (001) planes of CuCl nanospheres, as obtained by the LMS method. We identify two main gaps where transmittance is significantly suppressed. The first one extends roughly from $\omega/\omega_0 = 0.9997$ to 1.0 and the second from 1.0003 to 1.0016. The left pair of dotted vertical lines of Figure 14.2 corresponds to the area of negative $\Re\mu_{eff}$. It is evident that the negative $\Re\mu_{eff}$ region almost coincides with the first gap of the transmittance curve. This is what one should expect from a slab of a material which exhibits negative permeability in a certain frequency region. The second (wider) transmittance gap does not correspond to magnetic activity but to electric one. That is, in Figure 14.2c we show the effective permittivity ϵ_{eff}, as obtained from EMG theory (Equations 14.12 and 14.14). We observe that, again, the region where $\Re\epsilon_{eff}$ is negative (right pair of dotted vertical lines) correlates very well with the second gap of the transmittance curve. From this, we can infer that the EMG prediction of negative $\Re\mu_{eff}$ (and negative $\Re\epsilon_{eff}$) is verified by the rigorous LMS method.

Figure 14.3a and b shows the effective magnetic permeability μ_{eff} of a collection of CuCl spheres, as calculated by Equations 14.13 and 14.15, for different sphere radii (see caption of Figure 14.3). The volume filling fraction has been kept constant, i.e., $f = 0.74$ (close-packed structure), since it ensures the strongest magnetic response for a given sphere radius. Along with μ_{eff}, Figure 14.3c shows the corresponding transmittance curves for slabs of fcc crystals of CuCl spheres for the three different sphere radii. It is evident that as the radius of the CuCl nanoparticles increases, the magnetic resonance becomes more prominent. This is due to the fact that as the radius of the spheres increases, the sphere size parameter x increases accordingly, giving rise to a stronger resonance of T_1^H, and subsequently of μ_{eff}. One also observes that the magnetic resonance and the transmittance curves, accordingly, shift to lower frequencies as the radius increases. This is an expected result, since the poles ω_s of T_1^H (Equation 14.15) move away from ω_0 according to $\Delta\omega/\omega_0 \sim -(S/c)^2$, where $\Delta\omega = \omega_s - \omega_0$.

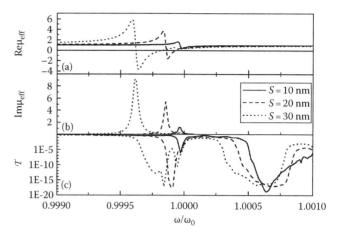

FIGURE 14.3 (a) Real and (b) imaginary parts of the effective relative permeability μ_{eff} of a collection of close-packed CuCl nanospheres in air, for different radii: $S = 10$ nm (solid line), 20 nm (dashed), and 30 nm (dotted). (c) Transmittance curves for light incident normally on a slab of 16 (001) fcc planes of the above-mentioned CuCl nanospheres, as calculated by the rigorous LMS method. (Reprinted from Yannopapas, V., *Appl. Phys. A*, 87, 259, 2007. With permission.)

As a further test of the validity of the EMG theory, Figure 14.4 compares the complex frequency band structure normal to the (001) of an infinite fcc crystal of close-packed CuCl spheres (the same spheres as those of Figure 14.2) with the dispersion lines resulting from ϵ_{eff} and μ_{eff} of Equations 14.12 and 14.13, i.e., $k = (\omega/c)\sqrt{\epsilon_{eff}(\omega)\mu_{eff}(\omega)}$. In Figure 14.4a where $\Re k_z$ is depicted, we observe that the agreement between the EMG theory and the LMS method is overall very good, with the exception of some frequency bands that lie from $\omega/\omega_0 = 1.0005$ to 1.0015, which EMG theory fails to reproduce. These bands are multipole, nondegenerate bands and, as such, they cannot be described by any dipole effective medium theory (Yannopapas et al., 1999). In Figure 14.4b,

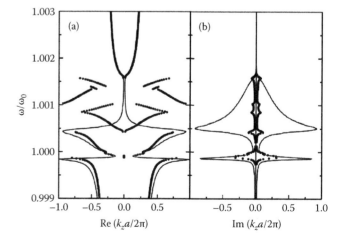

FIGURE 14.4 Full circles: Complex frequency band structure normal to the (001) surface of an fcc crystal of close-packed CuCl spheres ($S = 20$ nm) in air, as calculated by the rigorous LMS method. Solid lines: dispersion curves of the above-mentioned crystal, as obtained from the EMG theory. (Reprinted from Yannopapas, V. and Vitanov, N.V., *Phys. Rev. B*, 74, 193304-1, 2006. With permission.)

we see that the EMG theory does not reproduce the LMS results and predicts a much larger $\mathrm{Im}k_z$. At this point, the EMG theory is incorrect since such high values of $\mathrm{Im}k_z$ are not expected for a material (CuCl) which is almost lossless at this frequency region ($\gamma/\omega_0 \sim 10^{-5}$, see earlier). This is a known drawback of EMG theory (Yannopapas and Moroz, 2005) since the corresponding imaginary parts of ϵ_{eff} and μ_{eff} are associated with the extinction of the propagating beam due to scattering, and therefore cannot be used directly to evaluate the absorption or the heating rate (Ruppin, 2000).

Figure 14.5 examines the effect which has the arrangement of spheres on the magnetic response of the composite. That is, the transmittance curves for light incident normally on slabs containing the planes of spheres parallel to the (001) and (111) surfaces are shown. For the latter crystallographic direction, both fcc and hcp configurations have been considered. One observes that the magnetic response is practically independent of the positioning of the spheres since all curves almost coincide within the region of magnetic activity, except from a further deepening of the transmission dip which is evident for the (111) slabs. Furthermore, the almost perfect coincidence of the different transmittance curves of Figure 14.5, along with the very good agreement of EMG and LMS theory in Figure 14.4a, imply that the metamaterial under study exhibits an *isotropic* magnetic response.

Next, the effect of disorder in the magnetic activity of the metamaterial under study is examined. The most common type of disorder met in colloidal structures is stacking disorder (Vlasov et al., 2000). According to this, the self-assembled structures are neither pure fcc nor hcp structures, but they are actually random hexagonally close-packed (rhcp) structures, i.e., random mixtures of fcc and hcp configurations. The dash-dotted curve of Figure 14.5 corresponds to such an rhcp structure. It is clear that, within the region of magnetic activity, this curve coincides with the fcc and hcp ones, suggesting that stacking disorder does not affect the magnetic response of the structure.

Figure 14.6 examines the role of lattice vacancies (voids) in the magnetic response of the CuCl arrays. The transmittance curves for the slabs of the defected crystals (with voids) of Figure 14.6c are calculated by employing the average T-matrix approximation (ATA) within the LMS framework (Stefanou and Modinos, 1991a,b). As expected, the presence of vacancies restricts to some degree the magnetic activity of the composite as a result of the decrease in the volume filling fraction f. However, the activity is still prominent and experimentally measurable (the width of the transmittance dip is only slightly reduced), even for a void concentration of 20% which is an exaggeration of the actual amount of point defects in colloidal crystals (Vlasov et al., 2000).

In both Figures 14.5 and 14.6, one sees that, although the presence of disorder leaves almost unaffected the magnetic response, at the same time it alters considerably both the shape and the width of the transmission gap which results from the *electric* activity. As it is evident from the width of the resonances in μ_{eff} and ϵ_{eff} (see Figure 14.2), the magnetic resonance of single CuCl sphere is much more localized within the sphere than the electric one. If we view the interaction of the spheres under the prism of a tight-binding picture (Ibach and Lüth, 2003), the strong localization of the magnetic resonances about the spheres leads to a much weaker magnetic interaction between neighboring spheres. Therefore, the magnetic activity we observe is a mere addition of the magnetic resonances of individual CuCl spheres and is essentially not affected by the way the spheres are placed in space. On the other hand, the spatially broader electric resonances of each CuCl sphere overlap with those of neighboring spheres, giving rise to the bands of resonances which, of course, depend on the particular local arrangement of the spheres.

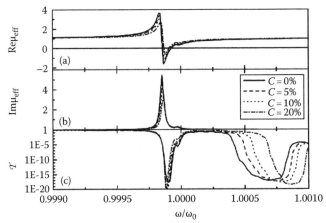

FIGURE 14.6 (a) Real and (b) imaginary parts of the effective relative permeability μ_{eff} of a collection of CuCl nanospheres ($S = 20$ nm) in air for different volume filling fractions: $f = 0.74$ (solid line), 0.70 (dashed), 0.66 (dotted), and 0.59 (dash-dotted). (c) Transmittance curves for light incident normally on different (001) slabs, 16 planes thick, of fcc crystals of close-packed CuCl spheres containing random voids of different concentrations: 0% (solid line), 5% (dashed), 10% (dotted), and 20% (dash-dotted), as obtained by the ATA-LMS method. These values of the void concentration correspond to the values of the volume filling fraction f of (a) and (b). (Reprinted from Yannopapas, V. and Vitanov, N.V., *Phys. Rev. B*, 74, 193304-1, 2006. With permission.)

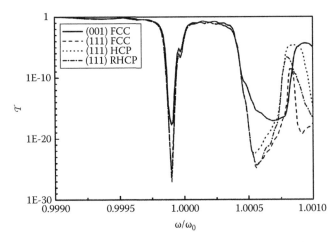

FIGURE 14.5 Transmittance curves for light incident normally on 16-planes-thick slabs of close-packed CuCl nanospheres ($S = 20$ nm) in air, with different stacking configurations, i.e., fcc (001) (solid line), fcc (111) (dashed), hcp (111) (dotted), and rhcp (111) (dash-dotted), as obtained by the LMS method. (Reprinted from Yannopapas, V. and Vitanov, N.V., *Phys. Rev. B*, 74, 193304-1, 2006. With permission.)

14.3.3 Plasmonic Meta-Atoms

As stated earlier, the basic prerequisite for the emergence of magnetism in arrays of nonmagnetic particles is the occurrence of huge values of the electric permittivity around a specific internal resonance of the particles. A different route for obtaining magnetic activity in the optical regime is based on a bottom-up approach (Rockstuhl et al., 2007) and it is outlined in Figure 14.7. A noble metal, e.g., silver, serves as the starting material. It provides a large negative real part of the permittivity (Figure 14.7a). If a small nanoparticle is made from silver, the latter supports surface-plasmons excitations for appropriate wavelengths (Figure 14.7b). If one considers a metamaterial made from a close-packed array of such silver nanoparticles, the corresponding effective relative electric permittivity possesses a strong Lorentz resonance around the collective surface-plasmon excitation (Figure 14.7c). The real part of the effective permittivity of this metamaterial exhibits large negative values for wavelengths less than the surface-plasmon wavelength but huge positive values for larger wavelengths. Therefore, a larger (macroscopic) sphere formed by this metamaterial exhibits strong Mie resonances (Figure 14.7d) within the spectral region of the huge permittivity, allowing for the observation of magnetic activity in a "meta-metamaterial" made from these macroscopic spheres (plasmonic "meta-atoms") (see Figure 14.7e). The observed magnetic activity will occur within the optical region since the surface-plasmon resonances of individual noble-metal nanoparticles also occur in this region.

The above arrays of meta-atoms made from the clusters of silver nanoparticles have been studied by both effective

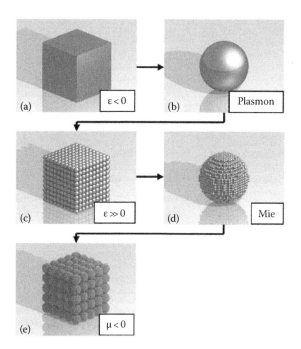

FIGURE 14.7 Conceptual sketch of the various levels used for constructing a meta-metamaterial. (Reprinted from Rockstuhl, C. et al., *Phys. Rev. Lett.*, 99, 017401-1, 2007. With permission.)

medium theory and the LMS method. In Figure 14.8, the effective magnetic permeability of a "meta-metamaterial" consisting of meta-atoms made from a cluster of silver nanoparticles of radius 2.9 nm (the silver nanospheres are assumed to occupy the sites of a cubic lattice with period 6 nm) is shown. The radius of

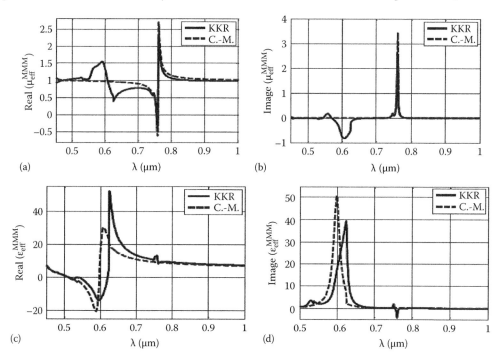

FIGURE 14.8 Effective material parameters of a meta-metamaterial that consists of spheres of radius 22.5 nm arranged in a cubic lattice of lattice constant 50 nm. The solid lines are obtained by the LMS method whilst the dotted ones by EMG theory. (Reprinted from Rockstuhl, C. et al., *Phys. Rev. Lett.*, 99, 017401-1, 2007. With permission.)

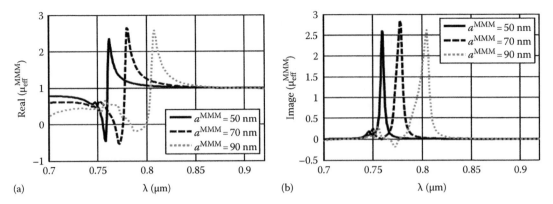

FIGURE 14.9 Effective permeability of a metamaterial as a function of the sphere (meta-atom) radius (see inset). The spheres are arranged in a cubic lattice. (Reprinted from Rockstuhl, C. et al., *Phys. Rev. Lett.*, 99, 017401-1, 2007. With permission.)

the meta-atoms is 22.5 nm and are arranged in a cubic lattice of period 50 nm. It is evident that the agreement of the effective-medium treatment with the rigorous one (LMS method) is overall very good. The strong magnetic activity is obvious whilst the real part of the permeability also assumes negative values as small as ≈ -0.5. From Figure 14.8c and d, it is also worth noting that the structure also shows remarkably high values of the effective permittivity.

The magnetic activity of the metamaterials of plasmonic meta-atoms can be tuned at around a prescribed frequency by varying the radius of the meta-atoms. Figure 14.9 shows the variation of the effective permeability (as obtained by the LMS method) with the meta-atoms' radius. One can see that the resonance wavelength and strength, the latter only slightly, increase with the radius. However, a further size increase of the radius of the meta-atoms causes a further shift at the expense of a lower strength because spheres are too large and the broadening of the resonance due to absorptive and radiation losses causes a degradation.

The magnetic metamaterial of plasmonic meta-atoms can be realized within a liquid-crystalline environment (Rockstuhl et al., 2007) by self-organization techniques [the functionalization of silver nanoparticles using thiol chemistry (Cseh and Mehl, 2006)] similar to those used in incorporating bucky balls (C_{60} clusters) (Dardel et al., 2001) in liquid crystals. By employing patterned light fields, the liquid crystals which contain the silver nanoparticles can form the desired plasmonic meta-atoms (Rockstuhl et al., 2007).

14.3.4 Metal–Dielectric–Metal Nanosandwiches

An alternative structure exhibiting photoinduced magnetism is a 2D periodic array of metal–dielectric–metal nanosandwiches, i.e., coaxial metallic nanodisks separated by a cylindrical dielectric spacer (see Figure 14.10). These structures exhibit a tunable plasmonic resonance in the optical regime (Su et al., 2006). The plasmon modes of the individual metallic nanodisks of the nanosandwich interact strongly with each other, giving rise to a symmetric and an antisymmetric mode

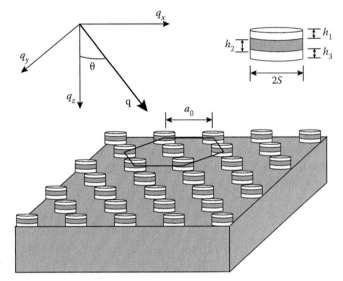

FIGURE 14.10 Schematic view of a hexagonal array of silver–silica–silver nanosandwiches on a quartz substrate. (Reprinted from Tserkezis, C. et al., *Phys. Rev. B*, 78, 165114-1, 2008. With permission.)

in the same fashion bonding, and antibonding orbitals emerge from the interaction of atomic orbitals in a diatomic molecule (Prodan et al., 2003). In the symmetric mode, the charge oscillations in each of the nanodisk are in phase whilst in the antisymmetric one, they are in opposite phase. Therefore, in the antisymmetric mode, an electric-current nanoloop is formed, giving rise to a small magnetic dipole which results in a macroscopic magnetization of the array of nanosandwiches (Tserkezis et al., 2008).

In order to verify the assumption of magnetic activity of the antisymmetric mode of the nanosandwich, an extension (Gantzounis et al., 2008) of the LMS method to nonspherical particles has been employed (Tserkezis et al., 2008). That is, due to the lack of a general effective-medium theory for nonspherical scatterers similar to the EMG theory presented here, one can follow a more indirect way: by inversion of the Fresnel formulas for the transmission and reflection coefficients t and r of a homogeneous slab of thickness D described by a refractive index n_2 and impedance z_2, placed between two semi-infinite media with n_1, z_1 and n_3, z_3, one can calculate (Smith et al., 2002)

$$z_2 = \pm \frac{\sqrt{(r-1)^2 - t^2 z_1 z_3}}{\sqrt{(r+1)^2 z_3^2 - t^2 z_1^2}} \quad (14.26)$$

and

$$\tan \frac{\beta}{2} = \pm i \sqrt{\frac{(r-1+t)[z_3(r+1)-tz_1]}{(r-1-t)[z_3(r+1)+tz_1]}} \quad (14.27)$$

where $\beta = (\omega/c)n_2 D$. The sign in Equation 14.26 is chosen so that the imaginary part of the impedance z_2 is always positive since we only deal with passive media. Also, the choice of the sign in Equation 14.27 is drawn on the basis of positive imaginary part of β which ensures an exponential decay of an inhomogeneous, outgoing wave. The effective parameters ϵ_{eff} and μ_{eff} of a composite metamaterial are deduced from an equivalent homogeneous slab described by $\epsilon_{eff} = n_2/z_2$ and $\mu_{eff} = n_2 z_2$, where z_2 and $n_2 = c\beta/\omega D$ are provided by Equations 14.26 and 14.27.

Figure 14.11a depicts ϵ_{eff} and μ_{eff} for an hexagonal array (of lattice constant 200 nm) of silver–silica–silver nanosandwiches with 50 nm radius (each silver nanodisk has 20 nm thickness and the silica spacer 40 nm), mounted on quartz substrate. From the graph of μ_{eff}, it is obvious that there is a strong magnetic activity around 2.1 eV, leading to a $\Re\mu_{eff}$ as small as −1.77 (Tserkezis et al., 2008). In Figure 14.11b, a map of $\Re\mu_{eff}$ as a function of the lattice constant of the array can be seen. It is obvious that when the nanosandwiches are very close (small values of the lattice constant), $\Re\mu_{eff}$ assumes more negative values in wider spectral

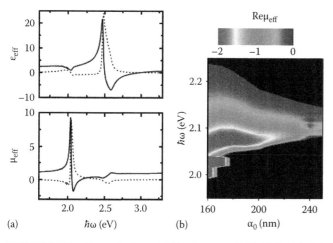

FIGURE 14.11 **(See color insert following page 9-8.)** (a) Real (solid lines) and imaginary (dashed lines) part of the effective permittivity, ϵ_{eff}, and permeability, μ_{eff}, of a hexagonal array, with lattice constant $a = 200$ nm, of silver–silica–silver nanosandwiches, with $S = 50$ nm, $h_1 = h_3 = 20$ nm, and $h_2 = 40$ nm (a, S, h_1, h_2, h_3 are defined in Figure 14.10), on a quartz substrate. (b) A map of the negative effective permeability of different hexagonal arrays of the nanoparticles described earlier. (Reprinted from Tserkezis, C. et al., *Phys. Rev. B*, 78, 165114-1, 2008. With permission.)

regions. In addition to the negative values of $\Re\mu_{eff}$, absorption losses are also dominant within the region of magnetic activity (Tserkezis et al., 2008).

14.4 Experimental Realization

The experiments on photoinduced magnetism in nanoparticles have been conducted so far mostly in the microwave and infrared regimes since the realization of the corresponding structures is far more easier in these regimes. However, these experiments are considered to be proof-of-principle experiments since they capture the underlying mechanism which generates the magnetic activity in artificial materials consisting of inherently nonmagnetic, naturally occurring materials. As stated earlier, the resonant behavior in the magnetic permeability is brought about by the induction of strong polarization currents within the particles which, in turn, give rise to macroscopic magnetization of the entire array of particles.

We note that the verification of the occurrence of a strong resonant behavior in the magnetic permeability corresponding to high positive and/or negative values of the latter is conducted indirectly. This is due to the fact that the direct magnetic measurements of the proposed metamaterials require sophisticated experimental setups. A more feasible procedure which is usually employed is to measure a quantity such as light transmittance/ reflectance from a finite slab of the metamaterial and compare directly with the numerical results of rigorous theoretical methods such as the LMS method examined earlier. If the agreement between theory and experiment is good and theory predicts strong resonances in the magnetic permeability, then one may safely conclude that the metamaterial under study indeed exhibits photoinduced magnetism. A strong resonant behavior of the magnetic permeability, which is accompanied by a region of negative values of its real part, is manifested as a transmittance minimum (gap) in the corresponding spectra, as shown theoretically earlier (negative $\Re\mu_{eff}$ means an imaginary wavevector and, thus, attenuation of an incoming light beam). However, experimentally, the presence of a minimum in the light-transmittance spectrum does not necessarily implies the existence of negative effective magnetic permeability in the same spectral region since the minimum might stem from negative values of the effective electric permittivity. In this case, experimentalists usually proceed to the fabrication of a metamaterial with negative refractive index based on the magnetic metamaterial under study. That is, within the existing lattice of particles exhibiting negative μ_{eff}, one incorporates a sublattice of other particles exhibiting negative ϵ_{eff}. When μ_{eff} and ϵ_{eff} become simultaneously negative, then the resulting effective refractive index n_{eff} becomes negative, too; in this case, the minimum of the transmittance curve of the magnetic lattice of particles becomes a maximum in the corresponding curve of the combined structure, indicating this way the presence of a negative μ_{eff}-band in this spectral region.

In recent experiments in the microwave regime (Zhao et al., 2008a,b), electromagnetically induced magnetism has been demonstrated in a 3D array of ceramic cubes of side length of 1 mm.

The cubes were made from $Ba_{0.5}Sr_{0.5}TiO_3$ (BST), a ceramic material which possesses a very high permittivity and low loss in the microwave regime. The interaction of the BST cubes with incident radiation induces strong polarization currents within their mass and, subsequently, remarkable magnetic activity in the whole structure, although BST itself is nonmagnetic. A simple cubic lattice of BST cubes was arrayed in a Teflon substrate with lattice constant 2.5 mm (see Figure 14.12). In the spectral region from 8 to 12 GHz, the array of BST particles exhibited strong magnetic

phenomena due to the huge value of the BST electric permittivity, i.e., 1600 + i4.8. Figure 14.13a shows the transmittance spectrum for EM microwave radiation incident on an array of BST cubes (dashed lines). It is evident that there exists a broad minimum around 8.25 GHz which corresponds to a region of negative effective magnetic permeability μ_{eff} according to theoretical predictions (Zhao et al., 2008a). In order to verify this, the experimentalists (Zhao et al., 2008a) have combined the array of BST particles with a network of metallic wires. It is well known that

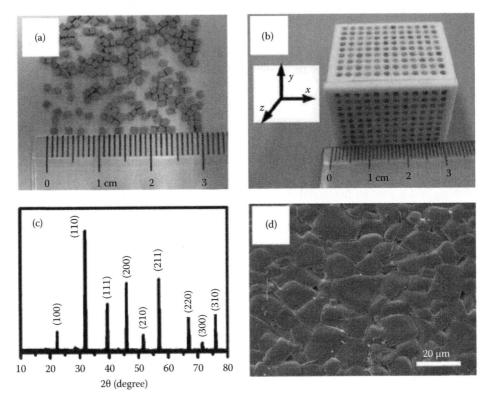

FIGURE 14.12 (a) Photograph of BST cubes with the side length of 1.0 mm. (b) Photograph of the simple cubic lattice of BST cubes arrayed in Teflon substrate. (c) XRD spectra of the BST cube. (d) SEM image. (Reprinted from Zhao, Q. et al., *Appl. Phys. Lett.*, 92, 051106-1, 2008b. With permission.)

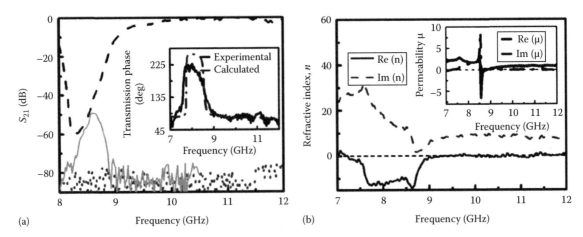

FIGURE 14.13 (a) Transmission for the BST cube array with side 0.75 mm only (dashed line), wire array only (dotted line), and the combination of BST cubes and wires (solid line). Inset of (a) shows the measured and calculated transmission phase of the sample with only one unit cell thickness along the propagation direction. (b) Retrieved refractive index and permeability (inset) based on the measured scattering parameters. (Reprinted from Zhao, Q. et al., *Phys. Rev. Lett.*, 101, 027402-1, 2008a. With permission.)

a network of metallic wires acts as a high-frequency pass filter as it allows the propagation of EM waves above a certain cutoff frequency. In the experiment of Zhao et al. (2008a), the network of metallic wires is designed so that it possesses a cutoff frequency well above the region of interest, i.e., above 12 GHz. For frequencies below the cutoff, the effective electric permittivity of the metallic network is negative. Therefore, if the effective magnetic permeability of the BST cube array is indeed negative, then the combined structure (BST cubes + metallic network) must exhibit a transmittance maximum within the same spectral region, more or less, where the BST cube array exhibits the transmittance minimum corresponding to negative μ_{eff}. This is indeed what Figure 14.13a shows: the metallic-wire network alone shows no transparency within the whole spectral region (dotted line). However, when it is combined with the BST cube array (solid line), a transmittance maximum emerges which is a clear demonstration of the negative magnetic permeability of the BST cube array. In the inset of Figure 14.13, one observes very good agreement between theory and experiment. Figure 14.13 shows the effective refractive index, as retrieved from the transmission spectra, whilst the inset shows the corresponding effective permeability that shows a very strong resonant behavior and assumes negative values within the region of the transmission gap.

The magnetic array of BST cubes possesses a certain amount of tunability of the resonance frequency of the effective permeability due to the variation of the electric permittivity of BST with temperature (Vendik and Zubko, 1997). That is, the BST permittivity decreases with increasing temperature, resulting in a blue shift of the corresponding resonance of the effective permeability, as predicted by the EMG effective-medium theory (Zhao et al., 2008b) and shown in Figure 14.14. At the same time, the strength of the resonance shows a maximum at about 10°C.

In the infrared regime, experiments on electromagnetically induced magnetism have been performed with SiC (silicon carbide) microparticles. Silicon carbide supports phonon-polariton excitations and possesses an electric permittivity given by the generic formula of Equation 14.24. The permittivity assumes high enough values close to the phonon-polariton excitation frequency so that a negative magnetic permeability is expected to occur. This is shown in Figure 14.15a where the real part of the permittivity possesses a maximum of about 200 around 800 cm^{-1}. Figure 14.15b shows the scanning electron micrograph of a SiC whisker (Schuller et al., 2007); the latter have typical dimensions of 1–2 μm in diameter and 40–100 μm in length.

The experimental identification of magnetic activity was also done indirectly in this case. Figure 14.16 shows the extinction spectra of radiation incident on the SiC whisker of 1 μm diameter for both polarization modes, as measured experimentally and calculated theoretically by Mie scattering theory (see Section 14.2.2). The agreement between theory and experiment is overall very good. The distinct peaks evident in both figures stem from the Mie resonances of a single SiC whisker (Schuller et al., 2007). Due to the high values of the SiC electric permittivity,

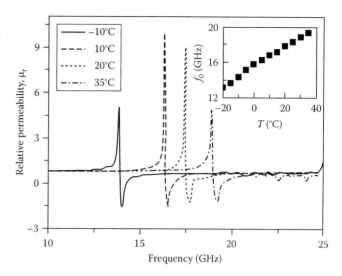

FIGURE 14.14 Calculated effective permeabilities of the dielectric composite under different temperatures. The inset shows the dependence of the first magnetic resonance frequency on the temperature. (Reprinted from Zhao, Q. et al., *Phys. Rev. Lett.*, 101, 027402-1, 2008a. With permission.)

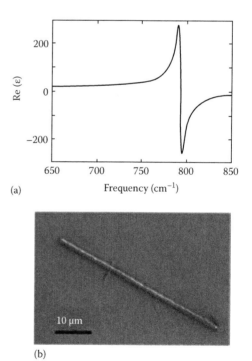

FIGURE 14.15 (a) Real part of the silicon carbide dielectric function. (b) Scanning electron micrograph of a SiC whisker. (Reprinted from Schuller, J.A. et al., *Phys. Rev. Lett.*, 99, 107401-1, 2007. With permission.)

an array of SiC whiskers is expected to possess strong resonant effective magnetic permeability around the Mie resonances. This is manifested in Figure 14.17 which shows the effective permeability for both the polarization modes of incident radiation (TE and TM modes) around Mie resonances. Therefore, since the agreement between theory and experiment is very good

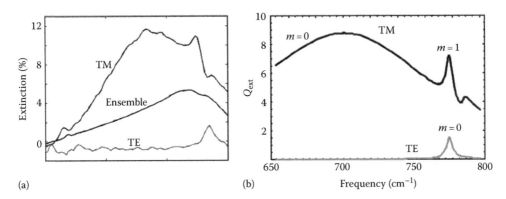

FIGURE 14.16 (a) Ensemble, TM (E field parallel to long axis) and TE (E field perpendicular to long axis) polarized extinction spectra of a 1 μm diameter whisker. (b) Calculated extinction efficiency of an infinitely long 1 μm diameter SiC cylinder for both TE and TM illumination. (Reprinted from Schuller, J.A. et al., *Phys. Rev. Lett.*, 99, 107401-1, 2007. With permission.)

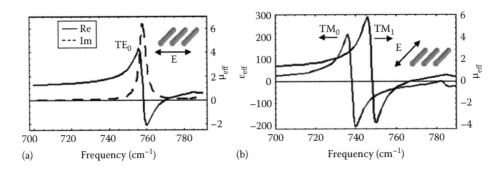

FIGURE 14.17 (a) Calculated effective permeability for a normal incidence TE illuminated array of infinitely long 1.5 μm diameter SiC rods. (b) Calculated permittivity (permeability) due to excitation of the zeroth (first) order TM mode. (Reprinted from Schuller, J.A. et al., *Phys. Rev. Lett.*, 99, 107401-1, 2007. With permission.)

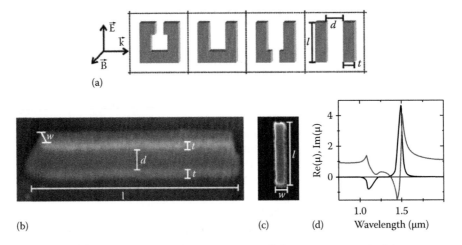

FIGURE 14.18 Schematic of the adiabatic transition from split-ring resonators (left) to cut-wire pairs (right) as magnetic atoms of optical metamaterials. (b) Electron micrograph (oblique-incidence view) of an actual cut-wire pair with $w = 150$ nm, $t = 20$ nm, $d = 60$ nm, and $l = 700$ nm; (c) corresponding top view. (d) Magnetic permeability for a single layer of cut-wire pairs, as obtained from Equations 14.26 and 14.27. (Reprinted from Dolling, G. et al., *Opt. Lett.*, 30, 3198, 2005. With permission.)

according to Figure 14.17, one expects that an array of SiC whiskers to become truly magnetic under radiation excitation.

Experiments on the photoinduced magnetization of particles in the near-infrared and optical regimes have been conducted by Dolling et al. (2005) and Shalaev et al. (2005).

In these experiments, the magnetic response does not rely on the induction of polarization currents within the particles but on the induction of displacement currents between a pair of metallic particles which acts as a miniaturized LC circuit. This mechanism has similarities with the one which is responsible for the magnetization of the metal–dielectric–metal nanosandwiches examined in Section 14.3.4. A pair of elongated metallic particles (metallic wires) is an attempt to miniaturize a split-ring resonator (top-down approach) which has been used as a magnetic atom in infrared and microwave experiments (see Figure 14.18). As it is evident from Figure 14.18d, the real part of the effective magnetic permeability assumes also negative values around the LC resonance frequency and reaches the value $\Re\mu_{eff} = -1.2$ at 1 μm (Dolling et al., 2005).

14.5 Conclusion

We have examined the phenomenon of photoinduced magnetism in arrays of nanoparticles which can be inherently nonmagnetic. This magnetic activity is manifested as a strong resonant behavior of the effective magnetic permeability of the array of nanoparticles which can lead to both paramagnetic or diamagnetic response with even negative permeability. The paramagnetic (diamagnetic) response occurs below (above) a specific frequency which corresponds to a specific resonance of the dielectric function of the material which the particles are made from. This resonance may stem from a specific material excitation (phonon-polariton or exciton-polariton) which introduces very high values of the corresponding dielectric function. The occurrence of strong magnetic response in the THz and infrared regimes is of paramount technological application since naturally occurring materials with magnetic response in these regions are particularly rare.

References

Artoni, M.; La Rocca, G.; Bassani, F. 2005. Resonantly absorbing one-dimensional photonic crystals. *Phys. Rev. E* 72: 046604-1–046604-11.

Bohren, C. F.; Huffman, D. R. 1984. *Absorption and Scattering of Light by Small Particles*. New York: Wiley.

Cseh, L.; Mehl, G. H. 2006. The design and investigation of room temperature thermotropic nematic gold nanoparticles. *J. Am. Chem. Soc.* 128: 13376–13377.

Dardel, B.; Guillon, D.; Heinrich, B.; Deschenaux, R. 2001. Fullerene-containing liquid-crystalline dendrimers. *J. Mater. Chem.* 11: 2814–2831.

Dolling, G.; Enkrich, C.; Wegener, M.; Zhou, J. F.; Soukoulis, C. M. 2005. Cut-wire pairs and plate pairs as magnetic atoms for optical metamaterials. *Opt. Lett.* 30: 3198–3200.

Doyle, W. T. 1989. Optical properties of a suspension of metal spheres. *Phys. Rev. B* 39: 9852–9858.

Fukumi, K.; Chayahara, A.; Kageyama, H.; Kadono, K.; Akai, T.; Kitamura, N.; Mizoguchi, H. et al. 1999. Formation process of CuCl nano-particles in silica glass by ion implantation. *J. Non-Cryst. Sol.* 259: 93–99.

Gantzounis, G.; Stefanou, N.; Papanikolaou, N. 2008. Optical properties of periodic structures of metallic nanodisks. *Phys. Rev. B* 77: 035101-1–035101-7.

Huang, K. C.; Povinelli, M. L.; Joannopoulos, J. D. 2004. Negative effective permeability in polaritonic photonic crystals. *Appl. Phys. Lett.* 85: 543–545.

Ibach, H.; Lüth, H. 2003. *Solid-State Physics*. Berlin, Germany: Springer.

Jackson, J. D. 1975. *Classical Electrodynamics*. New York: Wiley.

Jylhä, L.; Kolmakov, I.; Maslovski, S.; Tretyakov, S. 2006. Modeling of isotropic backward-wave materials composed of resonant spheres. *J. Appl. Phys.* 99: 043102-1–043102-7.

Linden, S.; Enrich, C.; Wegener, M.; Zhou, J. F.; Koschny, T.; Soukoulis, C. M. 2004. Magnetic response of metamaterials at 100 terahertz. *Science* 306: 1351–1353.

Modinos, A. 1984. *Field, Thermionic, and Secondary Electron Emission Spectroscopy*. New York: Plenum.

O'Brien, S.; Pendry, J. B. 2002. Photonic band-gap effects and magnetic activity in dielectric composites. *J. Phys.: Condens. Matter* 14: 4035–4044.

Orel, Z. C.; Matijevic, E.; Goia, D. V. 2003. Precipitation and recrystallization of uniform CuCl particles formed by aggregation of nanosize precursors. *Colloid Polym. Sci.* 281: 754–759.

Panina, L. V.; Grigorenko, A. N.; Makhnovskiy, D. P. 2002. Optomagnetic composite medium with conducting nanoelements. *Phys. Rev. B* 66: 155411-1–155411-17.

Pendry, J. B. 2000. Negative refraction makes a perfect lens. *Phys. Rev. Lett.* 85: 3966–3969.

Pendry, J. B. 2004. Negative refraction. *Contemp. Phys.* 45: 191–202.

Prodan, E.; Radloff, C.; Halas, N. J.; Nordlander, P. 2003. A hybridization model for the plasmon response of complex nanostructures. *Science* 302: 419–422.

Rockstuhl, C.; Lederer, F.; Etrich, C.; Pertsch, T.; Scharf, T. 2007. Design of an artificial three-dimensional composite metamaterial with magnetic resonances in the visible range of the electromagnetic spectrum. *Phys. Rev. Lett.* 99: 017401-1–017401-4.

Ruppin, R. 2000. Evaluation of extended Maxwell-Garnett theories. *Opt. Commun.* 182: 273–279.

Schall, M.; Helm, H.; Keiding, S. R. 1999. Far infrared properties of electrooptic crystals measured by THz time-domain spectroscopy. *Int. J. Infrared Millimet. Waves* 20: 595–604.

Schuller, J. A.; Zia, R.; Taubner, T.; Brongersma, M. L. 2007. Dielectric meta-materials based on electric and magnetic resonances of silicon carbide particles. *Phys. Rev. Lett.* 99: 107401-1–107401-4.

Shalaev, V. M.; Cai, W. S.; Chettiar, U. K.; Yuan, H. K.; Sarychev, A. K.; Drachev, V. P.; Kildishev, A. V. 2005. Negative index of refraction in optical metamaterials. *Opt. Lett.* 30: 3356–3358.

Shelby, R. A.; Smith, D. R.; Schultz, S. 2001. Experimental verification of a negative index of refraction. *Science* 292: 77–79.

Smith, D. R.; Schultz, S.; Markos, P.; Soukoulis, C. M. 2002. Determination of effective permittivity and permeability of metamaterials from reflection and transmission coefficients. *Phys. Rev. B* 65: 195104-1–195104-5.

Stefanou, N.; Modinos, A. 1991a. Scattering of light from a two-dimensional array of spherical particles on a substrate. *J. Phys.: Condens. Matter* 3: 8135–8148.

Stefanou, N.; Modinos, A. 1991b. Optical properties of thin discontinuous metal films. *J. Phys.: Condens. Matter* 3: 8149–8157.

Stefanou, N.; Karathanos, V.; Modinos, A. 1992. Scattering of electromagnetic waves by periodic structures. *J. Phys.: Condens. Matter* 4: 7389–7400.

Stefanou, N.; Yannopapas, V.; Modinos, A. 1998. Heterostructures of photonic crystals: Frequency bands and transmission coefficients. *Comput. Phys. Commun.* 113: 49–77.

Stefanou, N.; Yannopapas, V.; Modinos, A. 2000. MULTEM 2: A new version of the program for transmission and band-structure calculations of photonic crystals. *Comput. Phys. Commun.* 132: 189–196.

Su, K. H.; Wei, Q. H.; Zhang, X. 2006. Tunable and augmented plasmon resonances of $Au/SiO_2/Au$ nanodisks. *Appl. Phys. Lett.* 88: 063118-1–063118-3.

Toyozawa, Y. 2003. *Optical Processes in Solids*. Cambridge, U.K.: Cambridge.

Tserkezis, C.; Papanikolaou, N.; Gantzounis, G.; Stefanou, N. 2008. Understanding artificial optical magnetism of periodic metal-dielectric-metal layered structures. *Phys. Rev. B* 78: 165114-1–165114-7.

Vendik, G.; Zubko, S. P. 1997. Modeling the dielectric response of incipient ferroelectrics. *J. Appl. Phys.* 82: 4475–4483.

Veselago, V. G. 1968. The electrodynamics of substances with simultaneously negative values of ϵ and μ. *Sov. Phys.-Usp.* 10: 509–514.

Vlasov, Y. A.; Astratov, V. N.; Baryshev, A. V.; Kaplyanskii, A. A.; Karimov, O. Z.; Limonov, M. F. 2000. Manifestation of intrinsic defects in optical properties of self-organized opal photonic crystals. *Phys. Rev. E* 61: 5784–5793.

Wheeler, M. S.; Aitchison, J. S.; Mojahedi, M. 2005. Three-dimensional array of dielectric spheres with an isotropic negative permeability at infrared frequencies. *Phys. Rev. B* 72: 193103-1–193103-4.

Wiltshire, M. C. K.; Pendry, J. B.; Young, I. R.; Larkman, D. J.; Gilderdale, D. J.; Hajnal, J. V. 2001. Microstructured magnetic materials for RF flux guides in magnetic resonance imaging. *Science* 291: 849–851.

Yannopapas, V. 2007. Artificial magnetism and negative refractive index in three-dimensional metamaterials of spherical particles at near-infrared and visible frequencies. *Appl. Phys. A* 87: 259–264.

Yannopapas, V.; Moroz, A. 2005. Negative refractive index metamaterials from inherently non-magnetic materials for deep infrared to terahertz frequency ranges. *J. Phys.: Condens. Matter* 17: 3717–3734.

Yannopapas, V.; Vitanov, N. V. 2006. Photoexcitation-induced magnetism in arrays of semiconductor nanoparticles with a strong excitonic oscillator strength. *Phys. Rev. B* 74: 193304-1–193304-4.

Yannopapas, V.; Modinos, A.; Stefanou, N. 1999. Optical properties of metallodielectric photonic crystals. *Phys. Rev. B* 60: 5359–5365.

Yen, T. J.; Padilla, W. J.; Fang, N.; Vier, D. C.; Smith, D. R.; Pendry, J. B.; Basov, D. N.; Zhang, X. 2004. Terahertz magnetic response from artificial materials. *Science* 303: 1494–1496.

Zhao, Q.; Kang, L.; Du, B.; Zhao, H.; Xie, Q.; Huang, X.; Li, B.; Zhou, J.; Li, L. 2008a. Experimental demonstration of isotropic negative permeability in a three-dimensional dielectric composite. *Phys. Rev. Lett.* 101: 027402-1–027402-4.

Zhao, Q.; Du, B.; Kang, L.; Zhao, H.; Xie, Q.; Li, B.; Zhang, X.; Zhou, J.; Li, L.; Meng, Y. 2008b. Tunable negative permeability in an isotropic dielectric composite. *Appl. Phys. Lett.* 92: 051106-1–051106-3.

Optical Detection of a Single Nanoparticle

Taras Plakhotnik
The University of Queensland

15.1 Introduction

The ability to detect one by one and investigate small particles such as gold and latex spheres, semiconductor nanocrystals (also known as quantum dots), and single molecules opens new perspectives on applications and fundamental research. Changing physical and chemical environments affect the particle properties and enable nanosensing (Kong et al. 2000; Modi et al. 2003; Poncharal et al. 1999). Nanoparticles are considered as promising vehicles for controlled drug delivery (Allen and Cullis 2004; Huang et al. 2007) and for medical diagnostic applications (Sokolov et al. 2002). Tracing individual biomolecules and viruses provides unprecedented details for researchers working in the field of cellular biology and biochemistry (Seisenberger et al. 2001; Taton et al. 2000; Weiss 1999; Wieser and Schütz 2008; Wirth and Legg 2007; Xie and Trautman 1998). On the other hand, physicists and chemists are also interested in nanoparticles because their properties are different from those of bulk materials due to greatly enhanced quantum and surface effects (Weller 1993).

Although nanometer-sized particles can be observed with an electron microscope, special environmental conditions are required for this method to work. Electron microscopes operate only when samples are placed in vacuum. This puts severe limitations on biological and other applications. The main information obtained with electron microscopy is electron density in the sample and complementary methods are required for obtaining other characteristics. Some imaging techniques, such as nuclear magnetic resonance imaging (NMRI), are a long way from being capable to image even $1\,\mu m$ particles not to mention single molecules. The resolution of an atomic force microscope substantially degrades when imaging through soft tissues like cell membranes (de Jager and van Noort 2007). It turns out that optics and optical microscopy provide probably the most universal approach to the detection of nanoparticles under ambient conditions or embedded deeply into a transparent material. Moreover, it is quite common that optical properties of nanoparticles are of direct interest. In this chapter, particles of essentially sub-wavelength size will be of interest. Because the visible (optical) spectrum covers a range of wavelengths approximately between 400 and 750 nm, such particles should be smaller than 100 nm across. The attention will be mostly on the fundamental principles of optical detection.

The detection of a nanoparticle has two distinct steps, which are not always explicitly stated. One step includes design of instruments optimized to detect a change in an output of a photodetector caused by the presence of the particle. We will call this change a signal. The second step involves analysis to confirm that the detected signal is actually coming from the particle of our interest. It is quite obvious and expected that the signals sent by nanoparticles are very weak and can be easily obscured by *background* light, that is, light coming from any other objects and particles present in the sample/environments or directly from the light source used to illuminate the sample. The analytical step in the detection process is generally called *data analysis*. It would be little trouble if the background were a steady value, a not fluctuating offset additive to the signal in the detector output. The problem is that in practice the background is changing all the time and changing, strictly speaking, unpredictably. Such background adds *noise* to the photodetector output and, when the signal is too weak, no one can tell whether an increase/decrease in the output is caused by the signal or by the accidentally rising background. The signal itself also fluctuates due to the statistical character of light generation and detection thus making the distinction between the noise fluctuations and the signal even harder.

Interestingly, the possibility to reduce the steady background by simple arithmetical subtraction has been realized effectively only with the development of sensitive photoelectrical detectors. Photoelectrical detection of images obtained with ordinary bright-field illumination and monochromatic light allows detection of static 5 nm gold spheres within a few seconds of integration time and observation of 10 nm gold spheres moving in cells with video rate (Brabander et al. 1986). The trick is to apply appropriate settings to offset, contrast, and brightness. This idea is the essence of video enhancement microscopy, a technique which started its development more than half of a century ago and is still widely in use (Shotton 1988; Weiss et al. 1992).

The main goal of this chapter is to examine the limits set by nature on the sensitivity of the optical detection of nanoparticles. And it is meant to be a single nanoparticle in the analytical volume, not a large number of colloidal particles detected simultaneously (although the theoretical concepts discussed in this chapter are also applicable to ensemble measurements). The two steps in the detection process listed above are obviously interconnected. The more effective the optical scheme, the stronger the signal, and the easier its recognition and the data analysis. But how strong should the signal be? In this chapter, we describe the essential principles of both aspects of particle detection.

15.2 Propagation of Light Waves

Adequate understanding of optical detection requires basic knowledge of optics and, in particular, the understanding that light is an electromagnetic wave, which is characterized not only by its amplitude but also by its phase. In a *plane wave* propagating in a homogeneous medium along the *z*-axis of an orthogonal coordinate system, the electrical field vector **E** changes in time and space according to

$$\mathbf{E} = \mathbf{E}_1 \cos\left(\frac{2\pi n}{\lambda} z - \omega t + \varphi\right)$$

$$\equiv \mathbf{E}_1' \cos\left(\frac{2\pi n}{\lambda} z - \omega t\right) + \mathbf{E}_1'' \sin\left(\frac{2\pi n}{\lambda} z - \omega t\right) \quad (15.1)$$

where
> ω is its angular frequency
> λ is the wavelength in vacuum
> φ is the phase of the wave
> n is the refractive index of the medium

Trigonometry enables us to represent a wave with a phase as a sum of two waves where the phase is included in the amplitudes according to $E_1' = E_1 \cos(\varphi)$, $E_1'' = E_1 \sin(\varphi)$ rather than in the argument of the cosine function. Such representation of a wave will be convenient for the understanding of absorption and extinction of light by nanoparticles. It is also convenient to introduce, for brevity, a magnitude of a wavevector $k = 2\pi n/\lambda$. The direction of the wavevector **k** shows the direction of the phase velocity. In the following, we will use the convention where a vector is represented by a bold character and its length by the same character in *italics*. The magnitude E_1 is also called the amplitude of the wave.

Most photodetectors (including the human eye) respond to the square of the electrical field averaged over a time interval much longer than $2\pi/\omega$, the period of the field oscillations in the wave. This averaging will be represented symbolically by placing the expression which is averaged in angular brackets. For example, $\langle E(t)^2 \rangle$ is the average of $E(t)^2$. Such an averaged value multiplied by $\beta \equiv n\varepsilon_0 c \approx n/(377\,\Omega)$, where $\varepsilon_0 \approx 8.85 \times 10^{-12}\ \mathrm{C^2/N\ m^2}$ is the vacuum permittivity constant and c is the speed of light in vacuum, is called the *irradiance* or *power density* of light (expressed in units of W/m²) which we will denote as I. For the wave described by Equation 15.1, intensity equals $\beta E_1^2/2$ because $\cos^2(\omega t + \varphi)$ averaged over a long time interval equals 1/2. When two waves overlap on a photo sensitive area of the detector at $z = 0$, the resulting intensity reads

$$\beta \left\langle \left[E_1 \cos(\omega t + \varphi_1) + E_2 \cos(\omega t + \varphi_2) \right]^2 \right\rangle$$

$$= I_1 + I_2 + 2(I_1 I_2)^{1/2} \cos(\varphi_1 - \varphi_2) \quad (15.2)$$

where $I_{1,2} = \beta E_{1,2}^2/2$ and where we have taken into account that the time average of $2\cos(\omega t + \varphi_1)\cos(\omega t + \varphi_2)$ equals $\cos(\varphi_1 - \varphi_2)$ for time-independent phases. The last term in Equation 15.2 is called an interference term. The phase of a wave changes very rapidly (but slower than the field oscillation) when light is emitted by a thermal source, for example, by a hot filament in a conventional light bulb. This rapid change has a profound consequence for the detector response because the interference term averages out to zero unless the two phases change synchronously. They may do so if both waves are coming from the same source and the time delay between them is so short that the phase divergence is insignificant (much smaller than π). Note that the interference term also would be averaged out if the two waves had different frequencies. If $\cos(\varphi_1 - \varphi_2)$ is smaller than zero, the result of the interference will be destructive since the cross term proportional to $(I_1 I_2)^{1/2}$ will reduce the total detected intensity. If $\cos(\varphi_1 - \varphi_2)$ is larger than zero, the interference is constructive. In a more general case, the waves are partially coherent. This means that the interference term is somewhat less significant than $(I_1 I_2)^{1/2} \cos(\varphi_1 - \varphi_2)$ but is not negligible. When $\cos(\varphi_1 - \varphi_2) = 0$, the interference term disappears. This means that wave oscillating as $\cos(\omega t)$ does not interfere with a wave proportional to $\sin(\omega t) \equiv \cos(\omega t + \pi/2)$.

The plane wave considered so far is not a sufficiently good approximation in many practical cases because such a wave covers all space and carries infinite energy. All this certainly does not resemble a realistic picture. A better model for a wave is a so-called Gaussian beam. A Gaussian beam propagating in the z-direction is a beam where the electrical field depends on all three space coordinates and time according to (Mandel and Wolf 1995)

$$E = \frac{E_0}{\left(1 + \dfrac{z^2}{z_R^2}\right)^{1/2}} \exp\left(-\frac{x^2 + y^2}{w_0^2\left(1 + \dfrac{z^2}{z_R^2}\right)}\right)$$

$$\times \cos\left(kz + \frac{k(x^2 + y^2)}{2z\left(1 + \dfrac{z_R^2}{z^2}\right)} + \varphi(z) - \omega t\right) \quad (15.3)$$

where the so-called Rayleigh range $z_R = kw_0^2/2$. Thus, in every x, y plane (that is at a fixed value of z) and at any time, the field amplitude is not constant but decreases quite rapidly away from z-axis (see Figure 15.1).

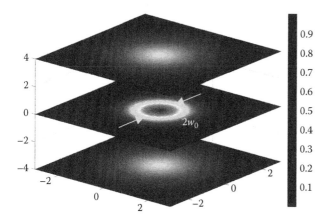

FIGURE 15.1 (See color insert following page 9-8.) Cross sections in x-y plane of a Gaussian beam propagating in vertical z-direction. The vertical bar shows coding of the amplitude of the electric field in the wave.

A good example of a Gaussian beam is a beam emitted by a laser. The energy flux through a cross section of such a beam is $P_0 = \pi w_0^2 \beta E_0^2/4$, where E_0 is measured in units of V m^{-1} and w_0 in units of m. The width of the Gaussian beam cross section is the narrowest at $z = 0$, the waist position, where the effective beam radius is w_0. The beam radius increases as the wave propagates. The phase shift $\varphi(z) = -\arccos\left[(1 + z^2/z_R^2)^{-1/2}\right]$ equals zero at $z = 0$ and approaches $-90°$ or $-\pi/2$ rad at very large values of z. This means that if the electrical field oscillates as $\cos(\omega t)$ at the waist position, it will oscillate as $\cos(kz - \omega t - \pi/2) \equiv -\sin(kz - \omega t)$ far from the waist. This fact is crucial for understanding light extinction, as explained in the following section.

A Gaussian beam remains Gaussian (with a different position and diameter of the waist) after passing through an optical system which has an optical axis (a single lens or a microscope objective typically satisfy this condition) and if a so-called paraxial approximation holds. This approximation does not work accurately for optical systems with extreme numerical apertures but still provides reasonable estimates. See Sheppard and Saghafi (1999) as an example for treatment of a non-paraxial case. A simple lens (of a diameter sufficient to intercept the whole beam and of focal length f) placed at distance s from the waist of the incoming beam transforms it into a beam with waist size w_0' located at distance s' such that $s' = \left[f^{-1} - (s - f)/(s^2 - sf + z_R^2)\right]^{-1}$ and $w_0' = w_0 f\left[(f - s)^2 + z_R^2\right]^{-1/2}$ (Self 1983). For example, if the input waist is at the front focal plane of the lens ($s = f$), the output waist is at the back focal plane ($s' = f$) and its size is $w_0' = f\lambda/(\pi w_0)$.

15.2.1 Natural Units of Power and Irradiance

According to quantum electrodynamics, the energy of a monochromatic field is always a multiple of photon energy quanta hc/λ, where $h \approx 6.6 \times 10^{-34}$ J/s is Planck's constant. Interaction between a wave and a particle may create or destroy a photon of a certain wavelength whose energy will be always hc/λ. This natural unit of energy converts the wave power and irradiance into photon rate and photon flux according to $R \equiv P\lambda/(hc)$ and $F \equiv I\lambda/(hc)$, respectively, and makes mathematical expressions more compact.

15.3 Interaction between Nanoparticles and Light

What happens when a particle is illuminated by a wave of light? If we neglect a possible feedback effect that the particle could have on the source of the wave (a very good approximation to the reality in most cases), the original (will be called *exciting* or *probing* in the following) wave will propagate as if nothing has changed but, in addition to the original wave, a new wave will be generated by the particle. This new wave is called a scattered wave (a general term for any kind of secondary emission). The scattered wave may be partially coherent and, therefore, may interfere with the exciting wave constructively or destructively causing substantial increase or decrease in the detector response. The magnitude of this effect depends on the degree of coherence and the relative intensities of the two waves. Complete cancellation (zero total intensity) can be achieved only at locations where the two intensities are equal, the waves are completely coherent, and their phases are shifted relative to each other by π.

When a particle is subject to an oscillating electrical field, electrons in the particle start to oscillate with the frequency of the field (at this point we neglect a possibility that an additional scattered wave may be of a different frequency and deal with this complication later). It is this oscillation which generates the scattered wave. If the particle is much smaller than the wavelength of the wave which it scatters, the emission of the particle can be modeled by emission of a point dipole representing an *induced dipole moment* whose time dependence is described by $\mathbf{p} = \mathbf{p}_0 \cos(\omega t)$. Electrical charge q with coordinates specified by a radius vector $\mathbf{r}_0 + \mathbf{d}/2$ and opposite charge $-q$ located at $\mathbf{r}_0 - \mathbf{d}/2$ represent a simple model of a point-dipole $\mathbf{p} = q\mathbf{d}$ if the length of vector $\mathbf{d}$ is very small compared to the distance from point $\mathbf{r}_0$ (the dipole location) to point $\mathbf{r}$, where the emission is detected. If this distance is also much longer than the wavelength (it is said in this case that the detection point is placed in a *far-field region*), the strength of the electrical field at point $\mathbf{r} \equiv (x, y, z)$ produced by a dipole located in a transparent media (refractive index n) at the origin of the coordinate system reads (Jackson 1998)

$$E = \frac{\pi p_0}{\varepsilon_0} \frac{\cos(kr - \omega t)}{r\lambda^2} \sin\theta \qquad (15.4)$$

The electrical field vector is perpendicular to the radius vector $\mathbf{r}$ and is in the same plane where the dipole moment vector and the radius vector are. The radiated power propagates with a speed of light c/n in all directions and the angular diagram of the energy flux (which is proportional to the square of the electrical field) is shown in Figure 15.2. Note that the point dipole does not emit in the direction parallel to the dipole moment. If the dipole moment is not rotating too quickly, this emission pattern can be used to determine its "instantaneous" orientation. For example,

a microscope objective near the dipole collects emission with efficiency dependent on the orientation of the dipole relative to the axis of the microscope objective.

The fraction of the power emitted in a solid angle Ω is $(3/8\pi)\int_\Omega \sin^2\theta\, d\Omega$ and if the angle between the dipole moment and the optical axis is Θ (see Figure 15.3), the collection efficiency equals (Plakhotnik et al. 1995)

$$C_d = \frac{1}{8}(4 - 3\cos\theta_{max} - \cos^3\theta_{max} + 3[\cos^3\theta_{max} - \cos\theta_{max}]\cos^2\Theta)$$

$$(15.5)$$

To collect 25% of the total emitted power, the collecting optics should intercept waves propagating in a solid angle defined by $\sin\theta_{max} = 0.8$ even for the most favorable orientation of the dipole at $\Theta = 90°$. The same objective collects only 10% of emission in the direction parallel to the dipole axis, that is, when $\Theta = 90°$. The dependence of the collection efficiency on the orientation of the dipole moment relative to the optical axis of the emission collecting optics is one of the reasons why the visible brightness of nanoparticles may be different even when they actually emit the same power.

A microscope objective is characterized conventionally by a numerical aperture defined as $NA = n\sin\theta_{max}$. The presence of the refractive index n enables the numerical aperture to be larger than 1 for immersion objectives (these are the objectives which are designed to work when their front lens is immersed in a liquid dielectric, usually oil or water). A higher numerical aperture improves resolution of the microscope (its ability to see two closely spaced small objects as two distinct spots) because the wavelength of light wave in a dielectric is n times shorter than the wavelength of the wave with the same frequency but propagating in vacuum.

But will an immersion objective improve the collection efficiency? Obviously, it will not be able to intercept light emitted downward, away from the objective. But as illustrated in Figure 15.3c, light emitted at angle θ will leave the dielectric slab propagating at angle θ' such that $\sin\theta' = n\sin\theta$. Therefore, the

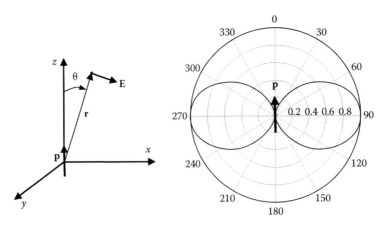

FIGURE 15.2 Angular dependence of the point-dipole emission in polar coordinates. The dipole oscillates vertically but the emission is proportional to $\sin^2\theta$. For example, the intensity of the beam propagating at 60° is 0.75 times the intensity of the beam propagating horizontally.

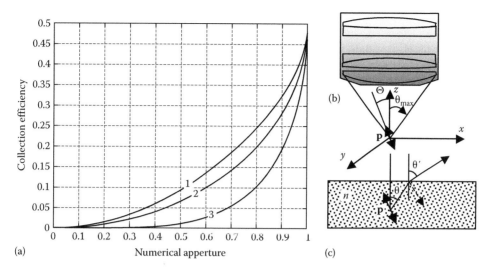

FIGURE 15.3 (a) Collection efficiency of a microscope objective for a point dipole emission and for an isotropic spherical wave. Curve 1 is calculated for collection in the direction perpendicular to the dipole axis ($\Theta = 90°$) and Curve 3 is calculated when the collection in the direction of the dipole axis ($\Theta = 0°$). The collection efficiency of a spherical wave is represented by Curve 2. Panel (b) illustrates the angles which are required to calculate the collection efficiency in Equation 15.5. Panel (c) shows that the emission is deflected from its original direction when the dipole is immersed in a slab with relative refractive index n. A part of the emitted energy will be reflected by the interface.

effective numerical aperture of the objective without immersion is $\sin \theta'_{max}/n$ not $\sin \theta'_{max}$. An immersion objective also eliminates the losses introduced by the reflection at the interface (a dashed line in Figure 15.3c). For example, the collection efficiency of a micro objective with NA = 0.95 placed in air will be only about 0.1 (corresponds to NA ≈ 0.6) for the averaged orientation of the dipole if the sample with a nanoparticle is covered with a glass slide ($n \approx 1.5$). If the dipole sits exactly on the surface separating the dielectric and air (or much closer than the wavelength), a more involved analysis shows that this will still have a significant effect on the collection efficiency due to the fact that such a dipole emits more energy into the medium with the higher refractive index (Lukosz 1979). Immersion makes the emission pattern identical to the one in vacuum and is shown in Figure 15.2.

The amplitude p_0 of the dipole oscillations depends on the frequency and the amplitude of the driving electrical field. Some particles scatter waves in a certain frequency range much more strongly than waves with frequencies outside that range. This phenomenon called a resonance significantly facilitates the detection because it enhances the signal relative to the background and makes it more specific. If the particles of interest have distinct resonances, they can be identified by recording and comparing data obtained with exciting waves of different frequencies (at resonance and away from the resonance). Also, the phase of the induced oscillations changes dramatically near the resonance frequency. This has a profound effect on interference between the scattered light and the oscillation-driving wave. In isotropic particles such as latex and metallic nanospheres, the direction of $\mathbf{p}_0$ depends on the direction of the electrical field. In a dye molecule at resonance with the driving field, the direction of $\mathbf{p}_0$ coincides with the direction of the *transition dipole moment* and therefore depends on the orientation of the molecule.

The electric field of the coherent dipole (it is assumed that the dipole is oriented along z-axis) emission driven by external field $E_0 \cos(\omega t)$ reads in a far-field region as

$$E_\theta = \left[a' \cos(kr - \omega t) + a'' \sin(kr - \omega t) \right] \frac{E_0 n}{2r\lambda} \sin \theta \quad (15.6)$$

where we have introduced two new parameters, a' and a'', to describe the response of the nanoparticle to the external field E_0. This expression is quite a general description of any elastic scattering such as Rayleigh scattering. Although resonance fluorescence always has a coherent part described by Equation 15.6, it also has an incoherent component which dominates the scattering signal at room temperatures. The values of $\varepsilon_0 n \lambda a' E_0/(2\pi)$ and $\varepsilon_0 n \lambda a'' E_0/(2\pi)$ are equal to $p_0 \cos \varphi$ and $p_0 \sin \varphi$ respectively, where φ is the phase shift between the driving field and the dipole oscillations (cf. Equation 15.4). The convenience of using a' and a'' is in their simple relation to the *extinction* and *scattering cross sections*.

The total power P_c (that is the power integrated over all directions) of the *coherent* emission by the induced dipole moment is

$$P_c = 2\pi \frac{a_c'^2 + a_c''^2}{3\lambda^2} I_0 \quad (15.7)$$

where

$\lambda \equiv \lambda/n$ is the wavelength of light in the media surrounding the particle

$I_0 = \beta E_0^2/2$ is the irradiance at the location of the particle

The concept of a cross section presents a scatter as an object which converts all the beam power crossing the cross section into something different. For example, a coherent scattering cross-section σ_c multiplied by the irradiance at the location of the scatter equals the power of the coherent scattered waves. Mathematically, this simply means $P_c \equiv \sigma_{cs}I_0$. Incoherent dipole emission can be represented by a classical *oscillating dipole* with a rapidly fluctuating phase and incoherent scattering cross section, σ_i determines the power of incoherent radiation $P_i \equiv \sigma_{is}I_0$. The total scattering cross section determines the total scattered power $P_s \equiv P_c + P_i \equiv \sigma_{ts}I_0$ etc. From Equation 15.7, it follows that

$$\sigma_{ts} = \frac{2\pi}{3\lambdabar^2}(a_c'^2 + a_c''^2) + \sigma_i \qquad (15.8)$$

If the particle is located in the waist region of a Gaussian beam as in Figure 15.4, the power detected by a photodetector placed, for example, on the axis of the Gaussian beam will be affected by the wave emitted by the particle. At a large distance from the beam waist, the field of the beam is proportional to $\sin(kr - \omega t)$ (see Section 15.2) and will interfere (constructively or destructively) only with the part of the dipole field proportional to $a_c'' \sin(kr - \omega t)$. The power measured by the detector will be $P = P_0 - a_c''I_0 \cos\Phi + C_d\sigma_{ts}I_0$, where C_d is the collection efficiency of the detector for the dipole emission (see Equation 15.5), Φ is the angle between the electrical field of the exciting wave and the dipole moment, and it is assumed that the detector intercepts the Gaussian beam entirely. If the presence of the particle reduces the detected power, then we should have $a_c'' > 0$.

The parameter a_c'' is also called the *extinction cross section*. This cross section can be experimentally determined according to $a_c'' = \varepsilon/N_A$, where $N_A \approx 6.02 \times 10^{23}$ mol^{-1} is the Avogadro's number, if the molar extinction coefficient ε is known for the particle. In units of cm^2, $a_c'' = 3.83 \times 10^{-21} \varepsilon$[cm^2], where ε is expressed in units of mol^{-1} L cm^{-1}.

Sometimes, extinction cross section is confused with *absorption cross section*. Only a part of the power taken from the probing beam will be transferred into other forms of energy (for example, heat). The rest will be scattered, that is, reemitted, in the form of optical waves. It is more logical to call only the part which is not reemitted as absorbed. The radiation quantum yield η is defined as

$$\eta \equiv \frac{P_c + P_i}{a_c''I_0} \geq a_c'' \frac{2\pi}{3\lambdabar^2} \qquad (15.9)$$

where P_i is the power of the incoherent dipole emission and the inequality is obtained by replacing the numerator with $(2\pi a_c''^2/3\lambdabar^2)I_0$ (see Equation 15.7). Therefore, the extinction cross section of any dipole-like nanoparticle is limited by the inequality

$$a_c'' \leq \eta\frac{3\lambdabar^2}{2\pi} \qquad (15.10)$$

The energy conservations principle insures that $\eta \leq 1$. The total scattering cross section can be expressed as $\sigma_{ts} = \eta a_c''$.

According to quantum theory, the particle can not reemit a fraction of the photon energy. The two options (heat production or scattering) are probabilistic in nature. The emission quantum yield η equals the probability that the reemission scenario will be realized.

The quantum yield may be affected by the environment to a large degree. It is known that some molecules are bright only under certain environmental conditions (a specific pH, viscosity of the solvent, etc.). The emission can be strongly quenched but if η were exactly zero, the particle would be undetectable because it would have zero extinction, absorption, and scattering cross sections (see Equations 15.9 and 15.10).

15.4 Optical Characteristics of Nanoparticles

First, we briefly describe the classification of different scattering waves resulting from the interaction of a nanoparticle with a probing wave.

15.4.1 Resonance Fluorescence

This is *resonance* scattering process resulting in *coherent* scattered wave at the frequency of the driving field. Resonance fluorescence can be envisaged as a result of interaction between a classical wave and a classical harmonic oscillator.

15.4.2 Rayleigh Scattering

This is non-resonance scattering process producing a *coherent* scattered wave at the frequency of the driving field. When the laser frequency approaches a resonance, Rayleigh scattering dramatically increases and gradually transforms into resonance fluorescence.

15.4.3 Photoluminescence

Luminescence can be understood as transition between quantized states in a particle. The energy of the emitted photon equals the energy difference between two states, an initial state (with higher energy) and a final state. This *incoherent* wave arises due to *spontaneous emission* of a photon either directly from the resonance state or after radiative or nonradiative relaxation to other excited states. Most of luminescence is emitted at a wavelength longer than the wavelength of the exciting wave. Such red or Stokes shifter emission can be spectrally separated from the exciting wave, Rayleigh scattering and resonance fluorescence.

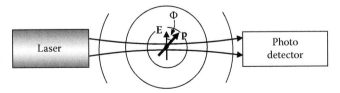

FIGURE 15.4 Design for classical extinction measurements employs a laser and a photo detector. The photo detector measures a change of the energy flux intercepted by its photo sensitive area.

15.4.4 Spontaneous Raman Scattering

This is a non-resonance scattering process resulting in *incoherent* scattered wave at a frequency different from the frequency of the driving field typically by a vibration frequency of the particle.

15.4.5 Contributions to Nanoparticle Scattering

For transparent dielectric nanoparticles, the strongest signal comes from *Rayleigh scattering*. The corresponding parameters in Equation 15.6 read

$$a'_c \approx \frac{\pi^2}{\lambda}\left(\frac{n_p^2 - n_m^2}{n_p^2 + 2n_m^2}\right)d^3 \quad \text{and} \quad a''_c \approx \frac{2\pi^5}{3\lambda^4}\left(\frac{n_p^2 - n_m^2}{n_p^2 + 2n_m^2}\right)^2 d^6 \quad (15.11)$$

respectively, where n_m is the refractive index of the surrounding material and n_p is the refractive index of the nanoparticle. Because these expressions are valid if $d \ll \lambda$, $a'_c \gg a''_c$ and usually a''_c can be set to be zero. But if it were exactly zero, the energy would not be conserved. No heat is generated by the particle ($\eta = 1$). The scattering cross section is much smaller than $3\lambda^2/(2\pi)$ and even smaller than the geometrical cross section of the particle. A diamond nanocrystal of 10 nm diameter in water has $\sigma_s \approx 10^{-8}\, 3\lambda^2/(2\pi)$ at $\lambda = 532$ nm. That is 10^8 times smaller than the upper limit of $3\lambda^2/(2\pi)$.

The scattering cross sections of gold or silver nanospheres may be enhanced significantly due to a so-called plasmon resonance effect. The frequency-dependent refractive index in metals can be expressed as $n_p = n_1(\omega) + in_2(\omega)$, where $i \equiv \sqrt{-1}$. Complex numbers allow treating the phase shift of the induced dipole oscillations in a mathematically elegant way. Factors a'_c and a''_c in Equation 15.6 then read

$$a''_c \approx \frac{6\pi^2 d^3}{\lambda}\frac{n_m^3 n_1 n_2}{(n_1^2 - n_2^2 + 2n_m^2)^2 + 4n_1^2 n_2^2}$$

and

$$\qquad\qquad (15.12)$$

$$a'_c \approx \frac{\pi^2 n_m d^3}{\lambda}\frac{(n_1^2 + n_2^2)^2 + (n_1^2 - n_2^2 - 2n_m^2)n_m^2}{(n_1^2 - n_2^2 + 2n_m^2)^2 + 4n_1^2 n_2^2}$$

Note if $n_2 = 0$ as in dielectric particles, then these expressions coincide with Equation 15.11. At a resonance $n_1^2(\omega) - n_2^2(\omega) + 2n_m^2(\omega) = 0$, the extinction cross section increases to $3\pi^2 d^3 n_m^2/(2\lambda n_1 n_2)$. For example, given bulk characteristics of gold at $\lambda \approx 520$ nm (Johnson and Christy 1972), a nanosphere ($d = 20$ nm, $n_1 \approx 1.2$, $n_2 \approx 2.6$) in water ($n_m \approx 1.33$) should have the extinction cross section $a''_c = 0.0013(3\lambda^2/2\pi)$.

Corrections are necessary to take into account the effect of the nanoparticle size on the refractive index (Hoevel et al. 1993). A 20 nm gold particle has quantum yield $\eta \approx 0.01$ (Jain et al. 2006) and its extinction cross section at resonance ($\lambda \approx 520$ nm) is of about 0.002 times the upper limit of $3\lambda^2/(2\pi)$. The emission of metallic particles has the same wavelength as the wavelength of the driving field and is completely coherent.

Metal nanoparticles consist of a very large number of "free" electrons moving in a potential field crated by ions. Such a system behaves rather as a classical harmonic oscillator. Therefore emission of metal nanoparticles can be classified as resonance fluorescence.

As the last example, consider a dye molecule (but the following is also applicable to quantum dots, color center). A molecule has a number of discrete energy levels (quantum states). The lowest possible energy is the ground state energy. The other states are excited states.

Resonance fluorescence usually results from excitation when the photon energy of the exciting wave matches the energy gap between the ground state and *one* of the excited states. The direction of the induced dipole moment coincides with the direction of the *transition dipole moment* (from the ground to the excited state) and its magnitude is proportional to cos Φ where Φ is the angle between the transition dipole and the direction of the electric field. Therefore, if the molecule does not rotate during the experiment, the extinction will depend as $\cos^2\Phi$ on the orientation of the molecule relative to the direction of the electrical field in the laser beam. If $\cos\Phi = 1$, the parameters in Equation 15.6 read

$$a''_c = \frac{3\bar{\lambda}^2}{2\pi}\frac{A_{eg}\Gamma}{2}\frac{1}{\Gamma^2 + (\omega - \omega_0)^2} \qquad (15.13)$$

and

$$a'_c = \frac{3\bar{\lambda}^2}{2\pi}\frac{A_{eg}}{2}\frac{\omega_0 - \omega}{\Gamma^2 + (\omega_0 - \omega)^2} \qquad (15.14)$$

where

  ω_0 is the resonance frequency of the molecule
  A_{eg} is the Einstein coefficient of spontaneous emission rate from the excited resonance state
  2Γ is the line width (full width at half maximum) of the absorption band

At exact resonance, $a''_c = 3\bar{\lambda}^2/(4\pi) \times A_{eg}/\Gamma$ and $a'_c = 0$. The ratio A_{eg}/Γ is always smaller than 2 to satisfy Equation 15.10. It can be on the order of 0.1 if the molecule is at liquid helium temperatures (Plakhotnik et al. 1997) but it is typically on the order of 10^{-7} for molecules in a liquid solvent at room temperatures. At room temperatures, coherent resonance scattering cross section is much smaller than the extinction cross section. The rest of the energy taken from the exciting wave may be released through a photoluminescence path or converted nonradiatively into heat.

The spontaneous emission time of photoluminescence is inversely proportional to the refractive index of the bulk material surrounding the nanoparticle (Nienhuis and Alkemade 1976). Only in extreme cases such as diamond where the refractive index is 2.5, this dependence is significant (Beveratos et al. 2001). But mostly the value of τ_{sp} depends on the particle nature. Dye molecules typically would have τ_{sp} within the range 1 ns < τ_{sp} < 10 ns (1 ns = 10^{-9} s). Radiative lifetime of ZnO nanocrystals is inversely proportional to their volume and lies in tens of picosecond (1 ps = 10^{-12} s) range (Fonoberov and Balandin 2004). Radiative lifetime of luminescent lanthanides is in the range of milliseconds (Selvin 2002).

In analogy to the spontaneous emission time, one can introduce the quenching time τ_q, the time it takes to return from the excited state to the ground state nonradiatively. The better the photon-emitting entity is protected from the influence of the environment, the longer the quenching time is. The emission quantum yield can be expressed in terms of τ_q and τ_{sp} as $\eta = \tau_q/(\tau_q + \tau_{sp})$. It is clear from this equation that a high quantum yield is more difficult to achieve if τ_{sp} is very long because this would require a very long quenching time as well. Note that $\eta\tau_{sp}$ is the relaxation time of the excited state. The average quenching time in melanin is as short as 0.17 ps (Nofsinger et al. 2001) but this time can be as long as 1 s in phosphorescent nanoparticles.

15.5 Saturation of the Signal

Can the brightness of a particle (that is the number of photons emitted by the particle per one second) be increased indefinitely? For example, by increasing the power of the laser used for illumination, one may expect to increase the number of detected photons. The problem is that in many cases the increase of the photon emission rate slows down and saturates when the excitation power gets higher. Saturation is a quantum effect caused by a delay (the spontaneous lifetime is a fundamental part of this delay) between disappearance of a photon from the exciting wave and return of the particle to its ground state.

The equations of Section 15.3 are based on the classical-induced dipole and break down when quantum effects become important. These equations can be fixed if the "cross sections" a_c' and a_c'' are multiplied by an intensity-dependent factor. Generally, this factor can be obtained only by solving quantum-mechanical equations describing behavior of the particle but in most cases, the result is a factor of $I_{sat}/(I + I_{sat})$, where I_{sat} is the *saturation intensity*. The total emitted power is then described by a simple equation

$$P_t = \eta a_c'' \frac{I_{sat}}{I + I_{sat}} I \equiv P_{max} \frac{I}{I + I_{sat}} \tag{15.15}$$

where the maximum emission power $P_{max} = \eta a_c'' I_{sat}$ (cf. Equation 15.10). The dependence described by Equation 15.15 is shown in Figure 15.5. The concept of the saturation intensity is very important for understanding the limitations on the achievable brightness.

The saturation intensity (and as a consequence the maximum emission rate) is strongly affected by any metastable state where a nanoparticle can be temporarily locked before returning to the ground state. Especially affected by this factor are molecules, but also quantum dots are not immune. A metastable state is a state with a very long lifetime τ_{ms}. A typical metastable state is a so-called triplet state in a molecule. As the excitation power increases, the molecule jumps faster to its excited state and then to the metastable state. The probability for a molecule that is

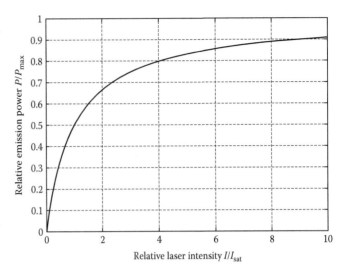

FIGURE 15.5 Dependence of the emission power on the intensity of the laser light used for excitation. The power increases proportionally to the excitation intensity as long as $I \ll I_{sat}$ but when $I \gg I_{sat}$ the dependence flattens. The emission power never exceeds P_{max}.

initially in the excited state to make transition to the metastable state within a very short time Δt is $v_{ms}\Delta t$. While the molecule is in the metastable state, it neither emits light nor absorbs light from the probing laser. Sooner or later the molecule returns back to its ground state. One can think about this process as reversible bleaching in contrast with true bleaching which permanently destroys the molecule.

The saturation photon flux and emission rates read

$$F_{sat} = \frac{1}{K_{ms} a_c'' \eta \tau_{sp}} \tag{15.16}$$

and

$$R_{max} = \frac{1}{K_{ms} \tau_{sp}} \tag{15.17}$$

where a dimensionless coefficient $K_{ms} \approx 1 + (1/2)\tau_{ms} v_{ms}$. The larger the probability and the longer the time spent in the metastable state are, the larger K_{ms} is. If the metastable state can be neglected, $K_{ms} = 1$. If there is more than one metastable state, then $K_{ms} \approx 1 + (1/2)\sum_i \tau_{ms}^{(i)} v_{ms}^{(i)}$, where the summation runs over all metastable states numbered with superscript i (Plakhotnik et al. 1997). For most organic molecules, a typical range of K_{ms} is between 1 and 100.

Note that if the molecule spends a significant time (compared to the excitation rate $a_c'' F$) in the resonance state, F_{sat} and R_{max} are two times smaller than predicted by Equations 15.16 and 15.17, which do not take into account stimulated emission.

The maximum emission rate does not depend on the emission yield η. This may look a bit counterintuitive but when the yield is low, the particle returns to its ground state quickly and can be also excited quickly provided that the exciting laser power is large enough to support the high excitation rate.

How many photons can a molecule emit at most? Given the range of K_{ms} and τ_{sp}, a typical maximum emission rate for a dye molecule is in the range of $10^8 - 10^5$ photons s^{-1}. It is obviously much harder to detect a particle with radiative lifetime like in lanthanides.

Metal and transparent dielectric nanoparticles are adequately modeled by classical harmonic oscillators whose amplitudes can increase indefinitely (of course, in extreme cases, the particles will be evaporated due to the excess heat or destroyed by electric breakdown caused by a too strong electrical field etc.) and therefore do not show any saturation.

15.6 General Description of Noise

Noise level of the detection system is as important as the magnitude of the signals. The noise depends on the average photon rate R hitting a photosensitive area of the detector and the time response or the integration time τ_d of the detector. Quite generally, σ^2 (the reader hopefully will not be confused here with a cross section which is also denoted by a Greek sigma), the variance of the noise in the output of an optical detector with quantum yield of ϕ can be split in three terms as follows:

$$\sigma^2 = v_0^2 \tau_d + \alpha \phi \tau_d R + v_p^2 \tau_d R^2 \tag{15.18}$$

This expression represents a Taylor's series expansion of the dependence of the noise variance on the detected power. The first term is called "dark" noise because this noise is present even in the absence of the signal. In modern photon counting detectors, this term is negligible in most practical cases. A good photodiode in the current measuring regime has an equivalent noise of $v_0^2 \approx 10^4$ s^{-1}. We skip the second term to return to it later and look first at the third term which is usually related to the intensity (power) fluctuations of the light source and also is called power noise. It is impossible to design a laser with absolutely stable output power. There will be always some technical instability in the apparatus. The integration-time dependence of the technical noise is quite complicated because the spectrum of technical noise may deviate significantly from a so-called *white noise* for which the spectral density of the noise is frequency independent (in Equation 15.18, $v_p R$ is the power density of the white noise). It is usually possible to identify a frequency where the technical noise is at a minimum. It is then advisable to modulate the intensity of the light source at such frequency and use synchronous detection with a narrow band centered at the selected frequency to reduce the noise.

In the simplest case, the noise is a *normally distributed* random variable. The main property of such a random variable x is that it is unpredictable but the probability P of a particular measurement to result in a value of x being somewhere between $x_0 - dx/2$ and $x_0 + dx/2$ obeys a normal probability distribution function (PDF) given by

$$P\left(|x - x_0| \leq \frac{dx}{2}\right) = \frac{1}{\sqrt{2\pi}\sigma} \exp\left(-\frac{(x_0 - \mu)^2}{2\sigma^2}\right) dx \tag{15.19}$$

In this expression, σ is called the standard deviation of the PDF and tells how quickly the probability decreases when the value of x deviates from μ which is the average value of x and also the most probable value of x. In the expression above, it is assumed that the interval dx is very small in comparison with σ. The notations on the left-hand side of the above expression will be repeatedly used throughout this chapter. Sometimes, this will be also written as $P(S|B)$, where S is a logical statement conditional on logical statement B. That is, $P(S|B)$ is the probability of S being true given that B is true.

The shape of the normal distribution is such that $P(|x - \mu| \leq \sigma) \approx 0.68$, $P(|x - \mu| \leq 2\sigma) \approx 0.95$, and $P(|x - \mu| \leq 3\sigma) \approx 0.997$. In other words, *approximately* 68% of the measured values will deviate from the most probable value by less than σ and *approximately* 95% of the measured values will deviate from μ by less than 2σ etc. The above numbers do not imply that by making N measurements, the number of results deviating from μ by more than 2σ will be exactly $0.05N$. If the probability to get a certain result in one measurement is p (in the example above $p = 0.05$), then the probability of getting M such results in N *statistically independent* measurements is described by a binominal distribution

$$P(M \mid N, p) = \frac{N!}{M!(N-M)!} p^N (1-p)^{N-M} \tag{15.20}$$

The mean value of M in this distribution equals pN and the standard deviation is $[Np(1-p)]^{1/2}$.

The second term in Equation 15.18 is the most fundamental of its origin. It is related to the quantum nature of light and is also called shot noise. The energy of electromagnetic radiation is carried by photons. In the case of a coherent wave, the number of photons detected within a certain period of time fluctuates even when all technical instabilities are eliminated. If the wave power is kept constant and the photons are counted for a time interval of τ_d, the result will be unpredictable. On average, the number of counted photons will be $R\tau_d$, the rate R being the average photon rate. The variance of the counted numbers will be simply equal to $R\tau_d$ the number of photons counted in average. A real photodetector differs from the idealized photon counting device by a smaller than 1 quantum yield ϕ, and a larger than 1 excess noise factor α. A smaller than 1 quantum yield reduces the average number of counted photons by factor ϕ as compared to the ideal photon counter. The value of α can be larger than 1 if the detector adds excess noise. This happens in photoelectrical detectors where primary photoelectrons are multiplied in an avalanche process. Example of such detectors are avalanche photodiodes, electron multiplying CCDs, and CCDs equipped with image intensifiers where $\alpha \approx 2$. For the following discussion, we will assume that $\alpha = 1$. The statistics of the detected photocounts are described by Poisson distribution $P(M|\mu) = e^{-\mu}\mu^M/M!$, which is very close to a Gaussian distribution when the average number of detected photons μ is large (in practice, already when this number is larger than 10 the Gaussian statistics can be used instead of Poisson distribution).

There are so-called nonclassical states of light where the quantum fluctuations of the photon numbers are much smaller than in a coherent beam but such nonclassical beams are very difficult to generate and to handle. These states will be briefly discussed later.

Statistics of photons scattered by a single molecule or a quantum dot can be sub-Poissonian (Mandel and Wolf 1995) but only when the integration time is much shorter or comparable to the lifetime of the excited state (this is not usually the case in practice). Otherwise, it is Poissonian or Gaussian for most practical purposes.

Except for the last term in Equation 15.18, the relative fluctuations of the detector response (if measured by the standard deviations of the response which is a square root of the variance) get smaller as the power of light increases. Thus, detection is greatly simplified for brighter particles, that is particles which emit more photons per given time.

An important characteristic of the signal quality is the so-called signal-to-noise ratio (SNR). This is the ratio of the mean amplitude of the signal to the standard deviation of the noise. If, for example, the SNR is 2, then *in a single measurement* with 95% confidence, we will be able to identify the presence of the signal even if it is superimposed on the noisy background with a normal PDF. However, this conclusion depends strongly on the assumptions about the PDF of the background noise. There is a reason (called the central limit theorem) to believe that the noise will indeed obey the normal distribution. But one should be very cautious when applying a normal PDF in practice. There is no warranty that the noise actually will be distributed normally especially far away from its most probable value.

It should be mentioned that all photodetectors have a limited dynamic range (the range from the maximum output to the noise level of the detector). A high value of the background (not just the related noise) may affect the detection because it effectively reduces the dynamic range.

15.7 Benchmarks for Extinction and Scattering Measurements

If photoemission yield η of a particle is smaller than 50%, the power absorbed by the particle is larger than the radiated power. Therefore, it seems that in this case detection of extinction of the laser beam should be easier than the detection of the wave scattered by the particle. Interestingly, historically, the first detection and spectroscopy of single molecules (Moerner and Kador 1989) has been achieved using absorption technique. However, followed up publications on this subject (Ambrose et al. 1991; Orrit and Bernard 1990; Shera et al. 1990) have exploited resonance-enhanced scattering (fluorescence) instead. This is because the figure of merit is not the magnitude of the signal but the SNR.

The two types of measurements are shown schematically in Figure 15.6a and b, respectively. A parameter $M \equiv 3\eta\lambda^2/(2\pi a_c'')$, which is always larger than 1 will be called the merit of scattering for particle detection. A reason for the name will become clear soon. The strong inequality $M \gg 1$ holds, for example, for molecules unless their emission is extremely quenched (a dye molecule at room temperature has $a_c'' \approx 10^{-17}$ cm^2 and hence its $M \approx 10^7 \eta$).

In an "ideal" extinction measurement, the noise is dominated by the shot noise of the laser beam $(\tau_d \phi R_0)^{1/2}$ while the signal equals $a_c'' \phi F_0 \tau_d$. Therefore, a basic expression for the SNR in this case reads

$$\mathrm{SNR}_{xt} = \frac{a_c''}{S}(\phi R_0 \tau_d)^{1/2} = \left(\frac{\phi a_c''}{S}\right)^{1/2} q_{xt}^{1/2} \qquad (15.21)$$

where

$S = \pi w_0^2$ is the probing beam cross section at the location of the particle (so that $F_0 = R_0/S$)

$q_{xt} \equiv \tau_d F_0 a_c''$ is the number of photons extinct from the probing beam during the measurement

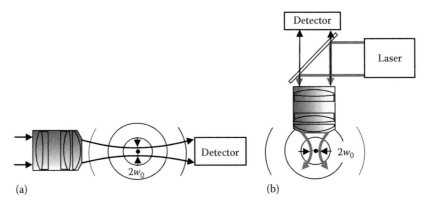

(a)   (b)

FIGURE 15.6 An arrangement of optics for scattering and absorption/extinction (this is shown with dotted lines) measurements. Gaussian beams are focused by the micro objectives. The best signal-to-noise ratio is achieved when the particle is placed at the center of the beam waist. (a) Absorption measurements. In the absence of the particle, the photon rate on the right detector is $R_0 = \pi w_0^2 F_0$, where F_0 is the photon flux in the center of the beam waist. (b) Scattering measurements. The objective is used both for illumination of the particle and for collection of its emission. Such a scheme is called epi-illumination. A beam splitter reflects the laser beam toward the particle and transmits the collected scattering toward the photo detector.

An "ideal" scattering measurement is background free and the noise is dominated by Poissonian statistics of the signal fluctuations. The signal is the scattered light (both coherent and incoherent) collected by the optics. The basic noise defined by the standard deviation (STD) of the signal equals the square root of the detected number of photons and correspondingly

$$SNR_s = (\eta C_d \phi a_c'' F_0 \tau_d)^{1/2} = (\eta \phi C_d)^{1/2} q_{xt}^{1/2} \quad (15.22)$$

It is easy to see that for any small particle $SNR_s/SNR_{xt} = (\eta C_d S/a'')^{1/2}$. Because $a_c'' = 3\eta \lambdabar^2/(2\pi M)$, one concludes that $SNR_s/SNR_{xt} = 2\pi C_d MS/(3\lambdabar^2)^{1/2}$. At the minimum beam cross section set by the wave nature of light, $S \approx \lambdabar^2/2$ and one gets $SNR_s/SNR_{xt} \approx (C_d M)^{1/2}$. If the collection efficiency of optics $C_d \approx 1$, absorption would just match scattering in terms of the SNR under most favorable $M = 1$. Hence, $SNR_s \gg SNR_{xt}$ for molecules, quantum dots, and transparent dielectric nanoparticles.

Why is it so much easier to detect absorption of light by objects much larger than the wavelength of the probing beam than scattering (as everyday experience suggests)? Such objects are made of a large number of small scatterers and the individual contributions from these scatterers will be added coherently and constructively in the forward direction. This will increase the extinction proportionally to the number of the small particles. Coherent scattering in directions other than forward will be relatively weak because the path length from the source of the exciting wave to the particle and then to the detector will depend on the position of the particle. On the detector, the phase of the wave scattered by a particle will be uncorrelated with that of other particles. This results in summation of intensities (not the amplitudes) of the fields created by every small particle constituting the large particle. The fundamental reason for the fact that it may be easier to measure extinction caused by a large object than the related diffused scattering is that the condition $SNR_s/SNR_{xt} = (S\eta/(Na''))^{1/2} \ll 1$ is easily achievable if N, the number of nanoparticles is a large number.

How close can a real experiment be to the above theoretical estimate? Figure 15.7 shows an image obtained from diamond nanocrystals spread on a surface of an ordinary quality prism where crystals as small as 37 nm in diameter are still visible. The detection limit may be significantly improved if a much higher quality surface is prepared. The scattering from the surface can be reduced by high-quality polishing or by using an atomically smooth surface of cleaved crystals. A substrate with atomic roughness used by Kukura et al. (2009) dramatically reduced the surface scattering so that the coherent part of scattering by a quantum dot at a room temperature could be detected.

The noise can be drastically reduced by using a beam with not fluctuating photon numbers (Mondal et al. 2008). Such a beam can be obtained with a single-photon source based, for example, on spontaneous parametric down-conversion (Kwiat et al. 1995). A single-photon source produces q_0 photons (quanta of energy) on demand, an exact number which does not fluctuate. If all the photons make their way to the detector but the efficiency of the photon detector is ϕ, then the detector will detect $q_0\phi$ photons in average. The probability for detecting q photons

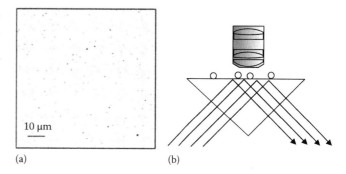

FIGURE 15.7 Scattering of diamond nanocrystals spread over a glass substrate. (a) Gray scale image shows the scattering of naonodiamonds. The smallest detectable particles have the scattering cross section as small as $\approx 10^{-14}$ cm². This limit is determined by the roughness of the glass surface scattering from which is the main source of the non vanishing background. (b) Diamonds were illuminated with the evanescent wave (on the side of air) produced by total internal reflection from the glass–air interface. (Reprinted from Colpin, Y. et al., *Opt. Lett.*, 31, 625, 2006. With permission.)

obeys a binominal distribution whose variance equals $\phi(1 - \phi)q_0$. Therefore, the SNR of the absorption measurements with an ideal single-photon source but a not ideal photon detector is (Mondal et al. 2008; Plakhotnik 2007)

$$SNR_{SPS} = \frac{a''}{S} \frac{\phi q_0}{\left[\left(1 - \phi + \frac{a''}{S}\phi\right)\phi\left(1 - \frac{a''}{S}\right)q_0\right]^{1/2}} \quad (15.23)$$

The SNR equals $\approx q_{xt}^{1/2}$ (that is the square root of the total number of extinct photons) in the case of an ideal photon detector $\phi = 1$. This surpasses the SNR of all previously described schemes. If the overall photon detection efficiency is 0.90, then $SNR_{SPS} \approx 3 \times SNR_{xt}$. This is only a threefold gain in comparison to the SNR achievable with a coherent beam. Note that the overall photon detection efficiency takes into account all the losses of the primary photons (reflection from any optical element, etc.) except for those caused by the presence of the nanoparticle, therefore 0.90 is an extremely good detection efficiency.

15.8 Interference of Scattered and Auxiliary Reference Beams

A generalization of the basic extinction measurement scheme is a scheme where the scattered wave of power R_s interferes with an auxiliary reference wave of a much larger power R_r. Essential elements of the experimental setup are shown in Figure 15.8. A version of this scheme has been first demonstrated for single molecules (Plakhotnik and Palm 2001) and later applied to other nanoparticles (Ignatovich and Novotny 2006; Lindfors et al. 2004). In its principles, the scheme is similar to differential interference contrast microscopy (Murphy 2001). Controversial claims about its superior SNR have been made due to an amplification effect appearing in the expression for the total detected power

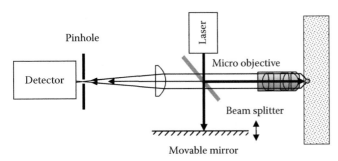

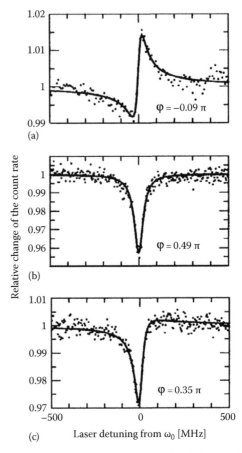

FIGURE 15.8 An interference experiment. A beam splitter directs laser light toward a sample with a nanoparticle embedded. A mirror is used to create a reference wave which interferes with the wave scattered by the particle. The phase difference between the reference wave and the scattered wave and the intensity of the reference wave can be tuned by changing the reflectivity and the position of the mirror. In practice, reflection from the front surface of the sample (although not always adjustable) can be used instead of the mirror.

$R_r + 2R_r^{1/2}R_s^{1/2}\cos^2(\varphi_b - \varphi_s) + R_s$, where the dependence on the signal is strongly enhanced by the middle term if R_r is large. Suppose however that the media (surrounding in Figure 15.8 the particle of interest) scatters backward a coherent wave whose intensity changes randomly in time and/or space. We may not know where the particle is and have to search for its location or have to determine its "arrival" time. When the signal is superimposed on such fluctuating background the cross-term in the middle does not make the detection much easier. Due to this term both the background fluctuations and the signal will be enhanced.

An example of an experimental result is shown in Figure 15.9. This experiment has been carried out with the sample immersed in liquid helium because the lower temperature increases the fraction of scattering, which is coherent with the laser wave by reducing the linewidth (see Equations 15.13 and 15.14). In the actual setup, no movable mirror has been used but a reflection from the front surface of the sample has been employed instead. There were many molecules in the sample and the phase shift between the reflected wave and the signal was different for different molecules because their positions relative to the reflecting surface were different.

For the noise analysis, we consider a reference wave as a *superposition of all coherent waves* hitting the detector (this includes reflections from all the interfaces, unwanted coherent scattering from the sample, etc.).

The power of the reference wave in units of photons per second is R_r. The STD of the noise in the detected signal is described by Equation 15.18. For simplicity, we assume that $\phi = 1$. The value of α may be larger than 1 especially if coherent scattering from the particles surroundings has strongly fluctuating intensity and interferes with the primary reference wave. The detected signal generally depends on the phase difference between the reference wave and the coherent wave scattered by the particle but at the most the signal equals $2\tau_d(C_d\eta_c R_r a_c''F_0)^{1/2} + C_d a_c''F_0\tau_d$. Therefore, the SNR in the scheme with an auxiliary beam reads

FIGURE 15.9 Interference pattern detected for different values of the phase shift (see panels a, b, and c) between the reference beam and the coherent emission of a single molecule and the detuning of the laser frequency from exact resonance at ω_0. The coherent part of the emission makes almost 100% of the total emission because the molecule is cooled to very low temperatures (1.6 K) to reduce the dephasing effect caused by the fluctuating environments. (Reprinted from Plakhotnik, T. and Palm, V., *Phys. Rev. Lett.*, 87, 183602, 2001. With permission.)

$$\text{SNR}_{\text{aux}} = \frac{2(C_d\eta_c R_r a_c''F_0)^{1/2} + C_d a_c''F_0}{(\alpha R_r + v_p^2 R_r^2 + v_0^2)^{1/2}}\tau_d^{1/2} \qquad (15.24)$$

with $v_0^2 \equiv \sigma_b F_0 + v_b^2\sigma_b^2 F_0^2 + v_e^2$ including all noisy contributions independent on R_r: a shot noise $\sigma_b F_0$ and a power noise $v_b^2\sigma_b^2 F_0^2$ of *incoherent* background caused by the laser illumination (σ_b denotes the corresponding cross section), and an electrical noise of the detector v_e^2. If the term $C_d a_c''F_0$ is neglected (this can be done if the reference beam is much stronger than the scattering from the particle), the expression for SNR has a maximum

$$\text{SNR}_{\text{aux}} = \frac{2(C_d\eta_c a_c''F_0\tau_d)^{1/2}}{(\alpha + 2v_p v_0)^{1/2}} \qquad (15.25)$$

when the reference beam power is $R_r = R_{ro} = v_0/v_p$. Note that $a_c''F_0\tau_d$ is simply the number of photons extinct from the probing beam by the nanoparticle and that the best SNR of this scheme is *at most* only two times better than the SNR in an ideal scattering measurement (see Equation 15.22). Moreover, if the electrical noise of the detector and the incoherent backgrounds are negligible in comparison to power fluctuations of the reference beam ($v_0 \rightarrow 0$), then $R_{ro} \rightarrow 0$ which means that no reference beam is the best option. Note, when $R_{ro} \rightarrow 0$, $C_d a_c'' F_0$ cannot be neglected.

The scheme allows reduction (but not elimination) of the negative effect on the particle delectability caused by a *noisy photodetector*. If v_e is the dominating contribution to v_0, then $R_{ro} = v_e/v_p$ is the right choice for the reference beam photon rate.

Laser illumination frequently makes noise caused by fluctuations of the parasitic coherent scattered/reflected wave much stronger than any other contribution to the noise. As has been explained above, no gain of the SNR can be achieved in such a case.

15.9 Cavity Enhancement

Placing an absorbing particle into a high-quality Fabry–Perot cavity (FPC) may substantially improve the SNR if compared to the direct absorption measurements and efforts that have been made toward detection of single-molecule absorption in a cavity (Horak et al. 2003; McGarvey et al. 2006). The method relies on the strong dependence of the cavity transmission on intracavity losses A_{inc}. There are two ways to measure the effect of the molecule on a cavity. One can measure either transmission of the cavity at a fixed wavelength, for example at a resonance frequency of the empty cavity, or a shift of the resonance frequency caused by the presence of a molecule. A Fabry–Perot cavity (FPC) is shown in Figure 15.10. In practice, a cavity may have very different design and have a shape of a microsphere or a microtoroid but that does not change the basic principles. Elementary consideration of multiple reflections from the mirrors proves the following results (Born and Wolf 1980; Ye and Lynn 2003).

The maximum transmission through the cavity is achieved at resonance conditions when the change of the wave phase during a round trip inside the cavity is a multiple of 2π, that is, when $m = 2\ell n/\lambda_m$, $m = 1, 2, \ldots$. The angular frequencies of two adjacent resonances are separated by a *free spectral range*, $\Delta\omega_{FSR} = \pi c/(\ell n)$. The width of the resonance is $\Delta\omega_W = \Delta\omega_{FSR}(1 - \Re + A_{inc})/\pi$, where A_{inc} is the relative loss of power in the cavity (other then transmission

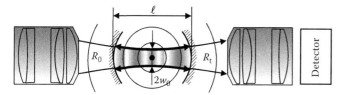

FIGURE 15.10 Cavity enhanced detection of absorption. Highly reflecting mirrors of curvature radius r form an optical cavity of length ℓ where light travels back and forth many times each time being absorbed by the particle inside the cavity. The laser field inside the cavity is much stronger than outside. The material inside the cavity introduces losses A_{inc} which include also losses due to the presence of the particle. The nodes are in the middle between two adjacent antinodes (darker regions). The minimal area of the cavity field cross section is $S = \pi w_0^2$.

through the mirrors) during the round trip and $\Re$ is the reflectivity of each of the cavity mirrors (we assume that mirrors are identical). The *cavity quality factor*, the ratio of the frequency to the width of the resonance equals $Q = \omega/\Delta\omega_W = 2\pi n\ell\lambda^{-1}(1 - \Re + A_{inc})^{-1}$. If a phase shift $\delta\varphi$ is caused by the presence of a particle in the cavity, the position of the resonance shifts by

$$\delta\lambda = \frac{\delta\varphi\lambda^2}{(4\pi\ell n)} \tag{15.26}$$

If R_0 is the photon rate sent toward the FPC and R_t is the rate transmitted at resonance conditions through the FPC, then

$$R_t = R_0 \frac{(1 - \Re)^2}{(1 - \Re + A_{inc})^2} \tag{15.27}$$

Strictly speaking, Equation 15.27 is applicable only if $A_{inc} \ll 1$ and $1 - \Re \ll 1$. We will assume that these conditions are valid.

The electromagnetic field inside the cavity is a spatially inhomogeneous standing wave. One can construct it by considering interference of two running waves and assuming that power of each is $R_t(1 - \Re)$. Due to the interference, nodes (where the electrical field is zero) and antinodes (where the field amplitude is at maximum) are separated by a quarter of the wavelength along the cavity axis. The intracavity buildup electrical field in the antinode equals the field in a free propagating beam with a photon rate of

$$R_{inc} = R_0 \frac{4(1 - \Re)}{(1 - \Re + A_{inc})^2} \tag{15.28}$$

In the following, the analysis is done for transmission measurements but similar ideas hold for the shift of resonances if the presence of the particle effects (directly or indirectly) the phase.

A small change in the cavity loss factor δA_{inc} causes a change in the output power

$$\delta R_t \approx \frac{-2R_0(1 - \Re)^2}{(1 - \Re + A_{inc})^3} \delta A_{inc} \tag{15.29}$$

The sensitivity of the cavity to the presence of a nanoparticle is at maximum when the particle is placed in antinodes where the extra loss per round trip added by the particle is $\delta A_{inc} = 2a_c''/S$. Thus the change in the transmitted power equals

$$\delta R_t \approx -\frac{4R_0(1 - \Re)^2}{(1 - \Re + A_{inc})^3} \frac{a_c''}{S} \tag{15.30}$$

The shot noise registered by the detector equals $(\tau_d R_t)^{1/2}$ while the signal is $\tau_d \delta R_t$. Therefore, the SNR reads

$$\text{SNR} \approx \frac{4(1 - \Re)(\tau_d R_0)^{1/2}}{(1 - \Re + A_{inc})^2} \frac{a_c''}{S} \approx \frac{4(\tau_d R_t)^{1/2}}{1 - \Re + A_{inc}} \cdot \frac{a_c''}{S}$$

$$\approx \frac{2(1 - \Re)^{1/2}(\tau_d R_{inc})^{1/2}}{1 - \Re + A_{inc}} \cdot \frac{a_c''}{S} \tag{15.31}$$

The variety of the expressions for SNR leads to apparently contradicting conclusions made in papers dealing with cavity effects. For example, the dependence of the SNR from $(1 - \Re + A_{inc})^{-1}$ may be considered linear or quadratic depending on the choice of the expression. Also, the optimum reflectivity depends on whether the optimization is sought keeping R_0, R_t, or R_{inc} as a constant. Assuming that R_{inc} does not change (which is a convenient choice if the saturation effects are to be considered), the best SNR is achieved when $\Re = 1 - A_{inc}$. This SNR can be expressed as follows:

$$SNR_{cav} = q_{xt}^{1/2}\left(\frac{a_c''}{SA_{inc}}\right)^{1/2} = \frac{SNR_{xt}}{A_{inc}^{1/2}} \qquad (15.32)$$

where $q_{xt} = \tau_d R_{inc} a_c''/S$ is the number of photons extinct from an ordinary probing beam of power R_{inc} if the measurements are done as shown in Figure 15.6a and SNR_{xt} is the SNR in that measurement. Note that the cavity forms an integration circuit with a characteristic response time of $\tau_{cav}/(1 - \Re + A_{cav})$, where τ_{cav} is the time which light takes to travel the cavity length. The above equations hold in a steady state regime, which is a good approximation if the measuring time is much longer than the response time.

The cavity with a molecule is equivalent to an object made of many virtual molecules (they are created by multiple reflections from the cavity mirrors). Therefore, the advantage of the cavity (compared to the direct extinction scheme) is proportional to the square root of the number of these molecules (see the discussion of extinction by a body much larger than the wavelength).

Now we compare SNR_{cav} and SNR_s. As an example, we take a symmetric confocal cavity of length ℓ for which $S = \lambda\ell/4$ (a general approximate expression $S = \lambda\ell/2 \times [r/(2\ell) - 1/4]^{1/2}$ can be used for a symmetric cavity (Siegman 1986) which is not too close to a concentric case $\ell = 2r$) and get

$$\frac{SNR_{cav}}{SNR_s} \approx \left(\frac{2\lambda}{\ell A_{inc}^{1/2} M}\right)^{1/2} \qquad (15.33)$$

Even in the limit set by diffraction, $S \approx \lambda^2/2$ (that is, $\ell = 2\lambda$ in the above expression), an extraordinary small value of $A_{cav} < 10^{-7}$ is required to achieve $SNR_{cav} > SNR_s$ for a molecule at room temperature with typical $M \approx 10^7$. Up to now, $A_{inc} \approx 10^{-6}$ has been achieved (Vahala 2003) only for mirrors in vacuum. The method should be much more suitable for particles like gold nanospheres where the M-factor is much larger.

15.10 Photothermal Detection

The method is based on the detection of a thermal lens created around an absorbing particle (Bialkowski 1996; Boyer et al. 2002; Tokeshi et al. 2001), and is a sensitive method for detecting weak absorption (see Figure 15.11). The reason for getting a better SNR in this case is that the scattering cross section of the lens can be much larger than the scattering cross section of the particle.

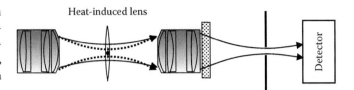

FIGURE 15.11 A generic photothermal detection scheme. A powerful laser beam is used to heat the nanoparticle. The increased temperature creates a thermal lens in the surrounding medium. This lens is then detected with a second probing beam. In this example, the spot size of the probe beam in the plane of a small aperture depends on the strength of the thermal lens and therefore the intensity of the probe beam transmitted through the aperture depends on the power of the laser heating the particle. A color filter in front of the pinhole blocks the heating laser light. The intensity of the heating laser can be modulated with a high frequency. This modulation can be detected using a narrow band circuit (synchronous detector) to reduce the noise.

It seems that there are no theoretical limits to the magnitude of the effect caused by the thermal lens but the practical limits are quite severe.

To estimate the effect which the thermal lens has on the probing beam, we calculate an average phase shift caused by the presence of the lens. The temperature increase around the absorbing particle (relative to the temperature at a very large distance from the particle) is $\delta T = \dfrac{a_c''I_0(1-\eta)}{2\pi r\kappa}$, where κ is the thermal conductivity of the surrounding media expressed in W m^{-1} K^{-1}, I_0 is the irradiance of the beam (expressed in W m^{-2}) illuminating the particle to produce heat, and r is the distance from the particle. The related change of the refractive index $\delta n = \delta T \cdot dn/dT$, where dn/dT is the thermal coefficient of the refractive index. The change of the refractive index Δn averaged over a sphere of radius r_0 equals

$$\overline{\delta n} = \frac{3}{r_0^3}\int_0^{r_0}\delta n \cdot r^2 dr = \frac{3}{2}\frac{a_c''I_0(1-\eta)}{2\pi\kappa r_0}\frac{dn}{dT}$$

and the resulting average phase shift is

$$\delta\varphi \approx 2\pi\frac{l\overline{\delta n}}{\lambda}$$

where

$$l = \frac{2}{r_0^2}\int_0^{r_0}(r_0^2 - r^2)^{1/2}r\,dr = \frac{2}{3}r_0$$

is an averaged path length of the light beam through the sphere. The final result for the phase shift reads

$$\delta\varphi \approx \frac{a_c''I_0(1-\eta)}{\kappa\lambda}\frac{dn}{dT}$$

The optical system detecting the thermal lens converts this phase shift into a change of power of the probing beam $\delta P_{LD} = P_{LD} \sin(\delta\varphi)$, where P_{LD} is the power of the beam used for lens detection. When the phase shift is much smaller than 1

$$\delta P_{LD} = \frac{a_c'' I_0}{\lambda\kappa}\frac{dn}{dT}(1-\eta)P_{LD} \qquad (15.34)$$

An estimate for the power change can also be derived by modifying (to account for absorption of only one particle) a known expression for the effect of the thermal lens on a probing beam in a bulk solution (Bialkowski 1996). Both approaches perfectly agree with each other.

Consider as an example a particle with $\eta \approx 0$ immersed in water where $\kappa - 0.6\,W\,m^{-1}\,K^{-1}$ and $dn/dT \approx -0.9 \times 10^{-4}\,K^{-1}$ (Tokeshi et al. 2001). Imagine also that a 1 W of radiation power is focused down to a diffraction limited spot of 0.5 μm across to make $I_0 \approx 300\,MW$ cm^{-2} and that the relative fluctuations of the probing beam power are 10^{-4}. Under these conditions, the minimum detectable cross section is $a_c'' \approx 6 \times 10^{-16}\,cm^2$. Detection of 2.5 nm gold spheres in polyvinyl alcohol (absorption cross section of about $4 \times 10^{-15}\,cm^2$) has been achieved in an experiment by Boyer et al. (2002) using a beam intensity $I_0 \approx 20\,MW\,cm^{-2}$ and 10 ms integration time. The detection limit of this method can be improved by increasing the pump beam power. However, the saturation effect sets a limit if one attempts to detect molecules with this method. In a molecule, the maximum effective value of $a_c'' I_0$ is limited by the saturation effects. Because the excited state lifetime is typically longer than 1 ps, $a'' I_0 \lesssim 10^{12}\,s^{-1}\,hc/\lambda$. For a wavelength of 500 nm, a molecule in water would cause a change of power $\delta P_{LD} \lesssim 10^{-4}\,P_{LD}$. The relative fluctuations of the probe power should be less than 10^{-4} to achieve an SNR of 1 for a molecule with 1 ps relaxation time. Single-molecule detection (SNR ≈ 3) has been demonstrated experimentally (Tokeshi et al. 2001) using benzene ($dn/dT \approx -6.5 \times 10^{-4}\,K^{-1}$) instead of water to improve the sensitivity. When the absorbing particles are imbedded in a scattering medium, the sensitivity of this method exclusively to absorption becomes a very important additional benefit.

Recently, a combination of the photothermal enhancement and a high-quality cavity (Armani et al. 2007) has been demonstrated in application to single molecule detection. The relative shift of the resonance can be estimated using Equations 15.26 and 15.28, and the expression for $\delta\varphi$ derived above. The final result reads

$$\frac{\delta\lambda}{\lambda} \approx \frac{a_c''(1-\eta)}{\pi n \ell S\kappa}\frac{dn}{dT}\frac{(1-\Re)P_0}{(1-\Re+A_{inc})^2} \qquad (15.35)$$

Similar to the discussion of Equation 15.31, the dependence on $1 - \Re + A_{inc}$ depends on the power (input, transmitted, or buildup) used in the expression.

15.11 Advanced Data Analysis

The fundamental detection limits we are aiming to establish in this chapter are the limits achieved when the data analysis extracts all the information hidden in the data stream. The amount of information that can be recovered from the data depends not only on the data itself but also on the theoretical models and background knowledge available. To appreciate the complexity and the challenge of the data analysis, look at the two panels in Figure 15.12.

The two panels represent images of 256×256 pixels in size. Most of the pixels contain only normally distributed noise with $\sigma = 1$. In panel a, 1 has been added to 100 pixels forming a vertical line in the middle of the image. Thus, for each of these pixels, the SNR is 1. In panel b, 3 has been added to 100 randomly distributed pixels. The SNR (per pixel) in the second image is three times larger that in the first one. However, a vertical line in panel A can be easily identified (for a better visibility, you can tilt the page and look at it at a small angle), but it is impossible to spot the location of the "elevated" pixels in panel b. There are about 200 pixels out of 256×256 pixels in the image where the noise deviates by more than 3σ from its mean value simply due to the statistics. Of course, simple statistical analysis will reveal that in the right image there are about 300 pixels where the value

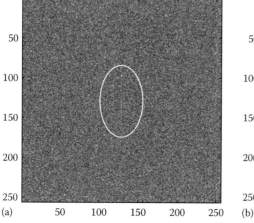
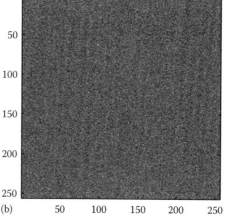

FIGURE 15.12 On panel (b), the signal is three times stronger than on panel (a). However, a vertical line (encircled by the ellipse) in the middle of panel (a) is clearly visible while exact identification of 100 bright dots in panel (b) is virtually impossible.

deviates by more than 3σ but it will be impossible to tell which 100 of these 300 pixels contain the signal. On the other hand, panel A has about 10,000 pixels where the signal is more than σ above the average. And yet the stripe is visible. It is important to recognize that the apparent visibility of the vertical line in Figure 15.12a is a consequence of the data analysis unconsciously done by the brain and based on prior "experience" in pattern recognition. On an abstract level, a straight line is not better or more logical than any other "random" pattern. Smart data analysis should obviously formally include such prior information. Data analysis is one of the directions where significant progress is expected in the nearest future. The following example will only outline how such an analysis can be done.

In a practical application, nanoparticle detection can be used to measure low concentrations of luminescent molecules in a solvent pumped though a narrow channel. Imagine that an optical system has been set up to detect luminescence from a small analytical region of the channel. Instead of measuring an average signal (which will be very noisy because most of the time there will be no molecules present in the analytical region), counting molecules seems to be an advantage. A simple approach to counting is to set a threshold for the detector output and to attribute every burst above the threshold to a molecule. This strategy has been exploited routinely (de Mello 2003). However, setting the threshold is not a trivial exercise. If the threshold is lifted, the probability increases for not recording detection when the molecule is actually present (false negative). If the threshold is lowered, the probability of detecting a false positive event increases.

The most logical approach to the data analysis is based on Bayes's theorem. For a good account on Bayesian statistics, see Gelman et al. (2004), Jaynes (2003), and Sivia (1996). Bayesian methods are getting more and more popular in relation to analysis of signals detected from individual nanoparticles (Donley and Plakhotnik 2001; Kou et al. 2005; McHale et al. 2004; Plakhotnik 2008; Witkoskie and Cao 2004; Yang 2008; Zhang and Yang 2005). As an example, we consider counting of nanoparticles described above. The following analysis is applicable to counting of any obscure events, for example, counting cells in a cell culture where a specific particle is present. A signal from the particle and noise are present in the output of an optical detector simultaneously. We collect data D by measuring N cells some of which may have one particle of our interest and our goal is to find the best possible estimate for the number of cells M where the particle is actually present. Bayes's theorem states that

$$P(M|D,B) = \frac{P(D|M,B)P(M|B)}{P(D|B)} \qquad (15.36)$$

relates conditional probability $P(M|D, B)$ to probability $P(D|M, B)$, which is easier to calculate. $P(M|B)$ equals an *a prior* probability that there were M such cells given relevant prior information B. If no background information is available (which we consider to be the case), we may set $P(M|B)$ as uniform, that is independent of M. The denominator $P(D|B)$ is also independent of M and, therefore, $P(M|D, B) \propto P(D|M, B)$, where

the omitted normalization constant can be obtained from the condition $\sum_{M=0}^{\infty} P(M|D,B) = 1$. To do any meaningful calculations, we need to know a bit about the particles and the noise. First, let us assume that the PDF of the signal and the noise are known functions. The probability density of the detector output x is $\rho_{\text{sig}}(x)$, if only signal is present, and $\rho_0(x)$ if only background noise is present. Assuming that the noise and the signal are two statistically independent variables, the PDF of the output if both background and the signal are present simultaneously reads $\rho_1(x) = \int_{-\infty}^{\infty} \rho_0(u)\rho_{\text{sig}}(x-u)\mathrm{d}u$. If in exactly M measurements out of N the particle is present, the distribution function of output x in a single measurement is $\rho = M/N \cdot \rho_1 + (N-M)/N \cdot \rho_0$. We then sort the measured values of x into K bins to find m_k, the number of measurements satisfying the condition $x_k < x \leq x_{k+1}$ for each bin ($k = 1, \ldots, K$).

The probability that a single measurement produces a result which fits in kth bin is

$$p_k(M) = \int_{x_k}^{x_{k+1}} \rho(x)\,\mathrm{d}x = \frac{M}{N}\cdot P\big(x_k < x \leq x_{k+1}\big|1\big)$$

$$+ \frac{N-M}{N}\cdot P\big(x_k < x \leq x_{k+1}\big|0\big) \qquad (15.37)$$

where 1 and 0 to the right from the vertical line indicate presence and absence of the particle, respectively. The probability that the whole series of measurements results in m_k counts in the kth bin (k runs between 1 and K) is a product of the above probabilities, $\prod_{k=1}^{K} p_k^{m_k}$ (neglecting the correlation between the probabilities is not strictly valid but is a good approximation) and, therefore,

$$P(M|D) \propto \prod_{k=1}^{K} p_k^{m_k}, \qquad (15.38)$$

where $\sum_{k=1}^{K} m_k = N$. Consider first a very simple counting strategy based on a threshold. In this case, the histogram consists of only two bins: bin 1 contains $x < x_{\text{th}}$ and bin 2 contains all values of $x \geq x_{\text{th}}$. The corresponding probability therefore reads

$$P(M|D) \propto \big[1 - p_2\big]^{N-m_2} p_2^{m_2},$$

where $p_2(M) = (M/N)\,P\,(x \geq x_{\text{th}}|1) + ((N-M)/N)\,P\,(x \geq x_{\text{th}}|0)$.

To find the most probable value of M, the derivative of the right-hand side of the above expression should be set to zero. Straightforward calculations reveal that the most probable value is close to

$$M_{mp} \approx \frac{m_2 - N\cdot P(x \geq x_{\text{th}}|0)}{P(x \geq x_{\text{th}}|1) - P(x \geq x_{\text{th}}|0)}. \qquad (15.39)$$

Note that the right-hand side of Equation 15.39 is not always an integer and can be even smaller than zero. Obviously, according to its physical meaning, M_{mp} should be an integer satisfying the condition $M_{mp} \geq 0$. Actually, $P(M|D)$ should be calculated for all physically possible values of M, and the most probable value of M will be the one for which $P(M|D)$ is the largest. If we repeat a number of times an experiment where the true number of the cells with the particle is M_t, the statistics of M_{mp} will be determined by the statistics of m_2. Because the mean value of m_2 obviously equals $M_t \cdot P(x \geq x_{th}|1) + (N - M_t) \cdot P(x \geq x_{th}|0)$, the mean value of M_{mp}, as defined by Equation 15.39, will be M_t. The standard deviation of M_{mp} obtained in such experiments equals the standard deviation of m_2 divided by $P(x \geq x_{th}|1) - P(x \geq x_{th}|0)$. This can be reduced to

$$STD(M_{mp})$$

$$\approx \frac{\sqrt{M_t \cdot P(x \geq x_{th}|1)P(x < x_{th}|1) + (N - M_t)P(x \geq x_{th}|0)P(x < x_{th}|0)}}{P(x \geq x_{th}|1) - P(x \geq x_{th}|0)}.$$

$$(15.40)$$

The optimum threshold value should minimize the relative error. Unfortunately, this value depends on the values of M_t, which would be unknown in a real experiment. If the value of M_t is not known, and we actually want to determine it from the data, then the optimal threshold cannot be found from Figure 15.13.

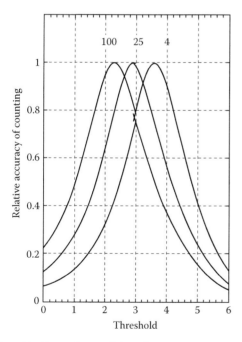

FIGURE 15.13 Dependence of the relative counting accuracy (M divided by the STD of the probability distribution for the estimated value of M) on the threshold value for different actual number of particles M. For all the curves, the STDs of the background noise and the signal are $\sigma_b = 1$ and $\sigma_s = 0.5$ respectively, the number of measurements $N = 10^5$, the mean amplitude of the signal $\mu_s = 3$. The actual number of particles is $M_t = 6500$, $M_t = 950$, and $M_t = 75$ for the curves from left to right. The curves are normalized to their peak values for easier comparison. The numbers above each curve are the maximum values of the relative accuracy.

As a detailed example, we consider a simple case when the noise and the signal are subject to uncorrelated Gaussian noise. The STD of the background noise σ_b is set to 1. Figure 15.14 shows how the counting accuracy depends on the choice of the threshold value. For example, if the threshold is taken at the maximum of the curve representing the case $M = 6500$ (that is at 2.3) but the actual number of particles was much smaller (around 75), then the counting error will be about 2.5 times larger than the one which can be achieved if the threshold is

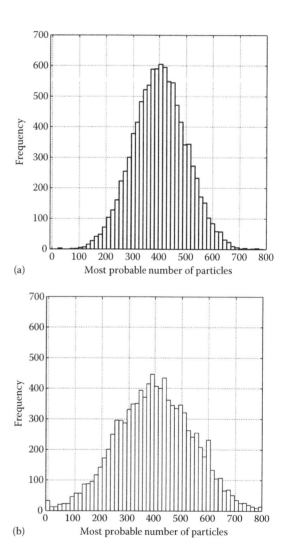

FIGURE 15.14 Results of computer simulations where normally distributed signal (mean amplitude 1.5, standard deviation 0.5) has been added to normally distributed noise (standard deviation 1). The simulated date represent a set of 10^5 measurements where signal has been added to the noise in 400 measurements selected at random from the whole set. (a) The data has been analyzed by sorting the data points into a 130-bin histogram and then searching the most probable value of M as given by Equation 15.38. (b) The data has been analyzed using an optimized threshold. This thought experiment has been repeated 10,000 times. The standard deviations of the most probable number of particles from the "exact" number of particles are 100 and 144 for the histograms shown in panels (a) and (b), respectively.

optimized for the lower number of particles (that is when the threshold is set at 3.65).

A better approach is to abandon the idea of a threshold and use a large number of bins instead. When the number of bins is large, the importance of their actual sizes and locations vanishes. Moreover, with a large number of bins the available information about the data statistics is used more effectively. This improves the accuracy of the estimate for M_t. An example is shown in Figure 15.14 where the improvement is of about 1.5 times even if compared to the optimized threshold.

This method can be easily generalized to a case when, for example, the mean value of the signal is unknown accurately. In this case, the analysis can be done by assuming a distribution of possible values of μ_s (this will require some prior knowledge about the detected particles). When $P(\mu_s|B)$ is assigned, the probability of any value of M can be calculated according to

$$P(M|D, B) \propto \int_0^{\mu_{max}} P(D|M, \mu_s, B)P(\mu_s|B)d\mu_s \quad (15.41)$$

The dependence of $P(M|D, B)$ on M tells all we can possibly know about the number of particles, given the data and the prior information.

15.12 Conclusion

The theory and the examples presented in this chapter do not cover the whole topic of nanoparticle detection by means of optical instruments. The methods used for such detection are developing and improving as the reader reads this conclusion. However, the basic principles covered here do not change so quickly and the chapter will stay up to date for quite a while. The attention was focused on single particles. The effect of an ensemble of particles and the light scattered by the ensemble can be described in the first approximation as a superposition of individual contributions. Therefore the presented theory is a basis for ensemble measurements. It is expected that the future progress will improve sensitivity of optical methods and will move single particle detection out of advanced laboratories to a more wide practice.

Acknowledgment

The author acknowledges Prof. Norman Heckenberg whose assistance and advice have helped to improve clarity of presentation and consistency in the terminology.

References

Allen TM, Cullis PR. 2004. Drug delivery systems: Entering the mainstream. *Science* 303:1818–1822.

Ambrose WP, Basche T, Moerner WE. 1991. Detection and spectroscopy of single pentacene molecules in a para-terphenyl crystal by means of fluorescence excitation. *J. Chem. Phys.* 95:7150–7163.

Armani AM, Kulkarni RP, Fraser SE, Flagan RC, Vahala KJ. 2007. Label-free, single-molecule detection with optical microcavities unlabeled target molecules. *Science* 317:783–787.

Beveratos A, Brouri R, Gacoin T, Poizat JP, Grangier P. 2001. Nonclassical radiation from diamond nanocrystals. *Phys. Rev. A* 64:061802.

Bialkowski SE. 1996. *Photothermal Spectroscopy Methods for Chemical Analysis.* New York: John Wiley & Sons.

Born M, Wolf E. 1980. *Principles of Optics.* Cambridge, U.K.: Cambridge University Press.

Boyer D, Tamarat P, Maali A, Lounis B, Orrit M. 2002. Photothermal imaging of nanometer-sized metal particles among scatterer. *Science* 297:1160–1163.

Brabander MD, Nuydens R, Geuens G, Moeremans M, Mey JD. 1986. The use of submicroscopic gold particles combined with video contrast enhancement as a simple molecular probe for the living cell. *Cell Motil. Cytoskeleton* 6:105–113.

Colpin Y, Swan A, Zvyagin AV, Plakhotnik T. 2006. Imaging and sizing of diamond nanoparticles. *Opt. Lett.* 31:625–627.

de Jager M, van Noort J. 2007. Atomic force microscopy. In *Encyclopedia of Life Sciences.* New York: John Wiley & Sons.

de Mello AJ. 2003. Seeing single molecules. *Lab Chip* 3:29N–34N.

Donley EA, Plakhotnik T. 2001. Statistics for single-molecule data. *Single Molecules* 2:23–30.

Fonoberov VA, Balandin AA. 2004. Radiative lifetime of excitons in ZnO nanocrystals: The dead-layer effect. *Phys. Rev. B* 70:195410.

Gelman A, Carlin JB, Stern HS, Rubin DB. 2004. *Bayesian Data Analysis.* Boca Raton, FL: Chapman & Hall/CRC.

Hoevel H, Frotz S, Hilger A, Kreibig U. 1993. Width of cluster plasmon resonances: Bulk dielectric function and chemical interface damping. *Phys. Rev. B* 48:18178–18188.

Horak P, Klappauf BG, Haase A, Folman R, Schmiedmayer J et al. 2003. Possibility of single-atom detection on a chip. *Phys. Rev. A* 67:043806.

Huang H, Pierstorff E, Osawa E, Ho D. 2007. Active nanodiamond hydrogels for chemotherapeutic delivery. *Nano Lett.* 7:3305–3314.

Ignatovich FV, Novotny L. 2006. Real-time and background-free detection of nanoscale particles. *Phys. Rev. Lett.* 96:013901.

Jackson JD. 1998. *Classical Electrodynamics.* New York: John Wiley & Sons.

Jain PK, Lee KS, El-Sayed IH, El-Sayed MA. 2006. Calculated absorption and scattering properties of gold nanoparticles of different size, shape, and composition: Applications in biological imaging and biomedicine. *J. Phys. Chem. B* 110:7238–7248.

Jaynes ET. 2003. *Probability Theory: The Logic of Science.* Cambridge, U.K.: Cambridge University Press.

Johnson PB, Christy RW. 1972. Optical constants of the noble metals. *Phys. Rev. B* 6:4370–4379.

Kong J, Franklin NR, Zhou C, Chapline MG, Peng S et al. 2000. Nanotube molecular wires as chemical sensors. *Science* 287:622–625.

Kou SC, Xie XS, Liu JS. 2005. Bayesian analysis of single-molecule experimental data. *Appl. Statist.* 54:469–506.

Kukura P, Celebrano M, Renn A, Sandoghdar V. 2009. Imaging a single quantum dot when it is dark. *Nano Lett.* 9(3):926–929.

Kwiat PG, Mattle L, Weinfurter H, Zeilinger Z, Sergienko AV, Shih Y. 1995. New high-intensity source of polarization-entangled photon pairs. *Phys. Rev. Lett.* 75:4337–4340.

Lindfors K, Kalkbrenner T, Stoller P, Sandoghdar V. 2004. Detection and spectroscopy of gold nanoparticles using supercontinuum white light confocal microscopy. *Phys. Rev. Lett.* 93:037401.

Lukosz W. 1979. Light-emission by magnetic and electric dipoles close to a plane dielectric interface. III. Radiation-patterns of dipoles with arbitrary orientation. *J. Opt. Soc. Am.* 69:1495–1503.

Mandel L, Wolf E. 1995. *Optical Coherence and Quantum Optics.* Cambridge, U.K.: Cambridge University Press.

McGarvey T, Conjusteau A, Mabuchi H. 2006. Finesse and sensitivity gain in cavity-enhanced absorption spectroscopy of biomolecules in solution. *Opt. Express* 14:10441–10451.

McHale K, Berglund AJ, Mabuchi H. 2004. Bayesian estimation for species identification in single-molecule fluorescence microscopy. *Biophys. J.* 86:3404–3422.

Modi A, Koratkar N, Lass E, Wei B, Ajayan PM. 2003. Miniaturized gas ionization sensors using carbon nanotubes. *Nature* 424:171–174.

Moerner WE, Kador L. 1989. Optical detection of single-molecules in solids. *Phys. Rev. Lett.* 62:2535–2538.

Mondal PP, Gilbert RJ, So PTC. 2008. Plasmon enhanced fluorescence microscopy below quantum noise limit with reduced photobleaching effect. *Appl. Phys. Lett.* 93:093901.

Murphy D. 2001. *Fundamentals of Light Microscopy and Digital Imaging*, pp. 153–168. New York: Wiley-Liss.

Nienhuis G, Alkemade CTJ. 1976. Atomic radiative transition probabilities in a continuous medium. *Physica B & C* 81:181–188.

Nofsinger JB, Ye T, Simon JD. 2001. Ultrafast nonradiative relaxation dynamics of eumelanin. *J. Phys. Chem. B* 105:2864–2866.

Orrit M, Bernard J. 1990. Single pentacene molecules detected by fluorescence excitation in para-terphenil crystals. *Phys. Rev. Lett.* 65:2716–2719.

Plakhotnik T. 2007. Seeing small. *J. Lumin.* 127:204–208.

Plakhotnik T. 2008. Testing hypothesis with single-molecules: Bayesian approach. In *Theory and Evaluation of Single-Molecule Signals*, eds. E Barkai, F Brown, M Orrit, H Yang. Hackensack, NJ: World Scientific.

Plakhotnik T, Palm V. 2001. Interferometric signatures of single molecules. *Phys. Rev. Lett.* 87:183602.

Plakhotnik T, Moerner WE, Palm V, Wild UP. 1995. Single-molecule spectroscopy: Maximum emission rate and saturation intensity. *Opt. Commun.* 114:83–88.

Plakhotnik T, Donley EA, Wild UP. 1997. Single-molecule spectroscopy. *Annu. Rev. Phys. Chem.* 48:181–212.

Poncharal P, Wang ZL, Ugarte D, de Heer WA. 1999. Electrostatic deflections and electromechanical resonances of carbon nanotubes. *Science* 283:1513–1516.

Seisenberger G, Ried MU, Endre T, Buning H, Hallek M, Brauchle C. 2001. Real-time single-molecule imaging of the infection pathway of an adeno-associated virus. *Science* 294:1929–1932.

Self SA. 1983. Focusing of spherical Gaussian beams. *Appl. Opt.* 22:658.

Selvin PR. 2002. Principles and biophysical applications of lanthandine-based probes *Annu. Rev. Biophys. Biomol. Struct.* 31:275–302.

Sheppard CJR, Saghafi S. 1999. Electromagnetic Gaussian beams beyond the paraxial approximation. *J. Opt. Soc. Am. A* 16:1381–1386.

Shera EB, Seitzinger NK, Davis LM, Keller RA, Soper SA. 1990. Detection of single fluorescent molecules. *Chem. Phys. Lett.* 174:553–557.

Shotton DM. 1988. Video-enhanced light microscopy and its applications in cell biology. *J. Cell Sci.* 89:129–150.

Siegman AE. 1986. *Lasers.* Oxford, U.K.: Oxford University Press.

Sivia DS. 1996. *Data Analysis: A Bayesian Tutorial.* New York: Oxford University Press.

Sokolov K, Follen M, Richards-Kortum R. 2002. Optical spectroscopy for detection of neoplasia. *Curr. Opin. Chem. Bio.* 6:651–658.

Taton TA, Mirkin CA, Letsinger RL. 2000. Scanometric DNA array detection with nanoparticle probes. *Science* 289:1757–1760.

Tokeshi M, Uchida M, Hibara A, Sawada T, Kitamori T. 2001. Determination of subyoctomole amounts of nonfluorescent molecules using a thermal lens microscope: Subsingle-molecule determination. *Anal. Chem.* 73:2112–2116.

Vahala KJ. 2003. Optical microcavities. *Nature* 424:839–846.

Weiss S. 1999. Fluorescence spectroscopy of single biomolecules. *Science* 283:1676–1683.

Weiss DG, Maile W, Wick R. 1992. Video microscopy. In *Light Microscopy in Biology*, ed. SJ Lacey, pp. 221–278. Oxford, U.K.: IRL Press.

Weller H. 1993. Colloidal semiconductor Q-particles: Chemistry in the transition region between solid state and molecules. *Angew. Chem. Int. Edit.* 32:41–53.

Wieser S, Schütz GJ. 2008. Tracking single molecules in the live cell plasma membrane—Do's and don't's. *Methods* 46:131–140.

Wirth MJ, Legg MA. 2007. Single-molecule probing of adsorption and diffusion on silica surfaces. *Annu. Rev. Phys. Chem.* 58:489–510.

Witkoskie JB, Cao J. 2004. Single molecule kinetics II. Numerical Bayesian approach. *J. Chem. Phys.* 121:6373.

Xie XS, Trautman JK. 1998. Optical studies of single molecules at room temperature. *Annu. Rev. Phys. Chem.* 49:441–480.

Yang H. 2008. Model-free statistical reduction of single-molecule time series. In *Theory and Evaluation of Single-Molecule Signals*, eds. E Barkai, F Brown, M Orrit, H Yang. Hackensack, NJ: World Scientific.

Ye J, Lynn TW. 2003. Application of optical cavities in modern atomic, molecular, and optical physics. *Adv. Atom. Mol. Opt. Phys.* 49:1–83.

Zhang K, Yang H. 2005. Photon-by-photon determination of emission bursts from diffusing single chromophores. *J. Phys. Chem. B* 109:21930–21937.

Second-Order Ferromagnetic Resonance in Nanoparticles

Derek Walton
McMaster University

16.1 Introduction

Magnetic materials can be described as an assembly of magnetic moments produced by electron spins on the magnetic ions in the material. At low temperatures, the spins are all parallel to one another in a ferromagnet, and antiparallel in an antiferromagnet. If the spins are canted, or the number of parallel spins is different from that of anti-parallel spins, the result is a ferrimagnet. Except for the anti-ferromagnet, the material can have a net magnetic moment in the absence of an external field.

The spins can tilt away from their equilibrium positions, and the exchange energy coupling them leads to spin waves. Tilting the spins reduces S_z, their component in the z direction, thereby reducing the magnetization. The spin-wave density increases as the temperature is raised, and, as a consequence, the magnetization, which is the sum of S_z, decreases with the rising temperature. Spin waves obey Bose–Einstein statistics at low densities, but they are quantized in the sense that the sum of the changes in S_z is equal to the change in the moment produced by flipping one spin. Since no more spins can be flipped once they are all reversed, the spin-wave density cannot be increased indefinitely. The spin-wave approximation becomes increasingly questionable as the temperature is raised. Nevertheless, it provides a useful basis for discussion and will be used here, even at temperatures close to the magnetic transformation temperature.

The electromagnetic field accompanying microwaves is strongly coupled, via its magnetic component to these small spin displacements. The microwave field is quantized, meaning that its energy can only be changed in quanta which equal the product of Planck's constant h, and the microwave frequency f. Similarly, the spin waves are quantized in the same way, and the quanta equal $hf_{\text{spin wave}}$; so if one microwave quantum creates one spin-wave quantum, the frequencies of the microwaves and spin waves must be equal. But more than one spin-wave quantum can be generated upon absorption of a microwave quantum, as long as the sum of the spin-wave quanta equal the microwave quantum, thus if two spin waves are created, $f = 2f_{\text{spin wave}}$, if three are created $f = 3f_{\text{spin wave}}$, and so on.

The lowest frequency spin wave is the uniform mode where all the moments rotate in unison, hence the wavelength is infinite. Higher frequency spin waves can only have wavelengths, λ, less than $\lambda = 2L/n$, where L is the dimension of the particle in the direction of propagation, and n is an integer. In the bulk, the wavelengths of all important spin waves are small compared to the sample dimensions, and there is an inverse quadratic relation between the spin-wave frequency, ω, and the square of the wavelength, λ, $\hbar\omega = 2JS(ak)^2$ (Kittel, 1968), where $k = 2\pi/\lambda$ is the wave vector, J is the magnetization density, S is the spin, and a the lattice constant. For very small particles, however, the longest possible wavelength, after the uniform mode (whose wavelength is infinite), equals twice the dimension of the particle, and can have a high frequency. The spin-wave frequencies of the different levels are $\hbar\omega_n = 8\pi^2 JS(a/\lambda)^2 \approx 8\pi^2 JS(an/L)^2$, and, in small particles, large separations in energy arise between spin-wave energy levels.

The excitation of the uniform mode in a sample whose spins are precessing about a magnetic field in the z direction takes place in a microwave field whose magnetic vector is at right angles to the z direction. If the frequency of the microwaves equals that of the uniform mode, the torque on the magnetic moment will tend to open the angle between it and the z axis. This decreases S_z, increasing the density of uniform-mode spin waves, which can then decay into other spin waves. The details of this process can be found in Sparks (1964).

The ferromagnetic resonance of small particles is a subject of considerable interest at present, however, all results are for the first-order absorption of microwave energy by the uniform mode. The absorption of a single microwave photon by a single spin wave is usually referred to as FMR; so 2FMR will be used here to describe the creation of pairs of spin waves with equal and opposite wave vectors. Our purpose is to discuss some important effects of particle size on the second-order process. Conservation conditions lead to completely different physics for the two processes. Since the first-order process creates a single magnon, the magnon wave vector must be zero, which means that only the uniform precession magnon at the center of the zone can be excited. The second-order process on the other hand excites pairs of magnons, and while the whole zone is available for macroscopic samples, for small grains magnons can only exist when the wavelength is a multiple of twice the scale of the grain. Thus, the dispersion relations are not continuous at low frequencies, and consist of the separated levels referred to above. The energy of the photon must be twice that of the magnons involved; so the minimum microwave frequency can be significant, e.g., for 30 nm magnetite grains, it is about 24 GHz, and the next level would be at 96 GHz; so it is not difficult to excite one level without affecting any others.

Hendriksen et al. (1993) have partially solved the problem of spin waves in small particles. Their theory shows that the lower levels are degenerate, corresponding to a range of wave vectors. The reason for this is that, because of the boundaries, the low frequency spin waves are not the plane waves that are characterized by a single value of the wavelength. The following is a brief summary of Hendriksen et al.'s treatment of the problem:

The Heisenberg Hamiltonian is $H = -\frac{1}{2} \sum_{i,j} J S_i \cdot S_j$, where J is the exchange energy, and S_i, S_j are the spins on sites i and j which, for simplicity, are assumed to be the nearest neighbors. The eigenstates of this Hamiltonian in the ground state, where all the spins are parallel, are found in terms of the raising and lowering operators (the z direction is parallel to the spins), $S^+ = S_x + iS_y$, $S^- = S_x - iS_y$ leading to the quantum mechanical equations of motion describing the propagation of a spin deviation $i\hbar \left(dS_i^+(t)/dt \right) = \left[S_i^+(t), H \right]$. These can, in a site-dependent random phase approximation be written as

$$\hbar \omega S_i^+ = -J \sum_j \frac{M_i + M_j}{2} S_j^+, \qquad (16.1)$$

where

M is the thermally averaged value of the spin at the site
ω is the angular frequency of the spin wave

The right hand side of Equation 16.1 is the same everywhere except on the boundaries. In a large sample, the boundaries can be ignored and a solution in terms of plane waves obtained (Kittel, 1968). For small particles, Hendriksen et al. diagonalize Equation 16.1 numerically, with the result that the eigenvalues consist of a series of energy levels. In general, because of the restricted size, the eigenvalues are complex, and the eigenstates are damped plane waves. Hendriksen et al., however, avoid this complication by symmetrizing the equations of motion that

results in purely real eigenvalues. An important result of the numerical analysis is that each level is degenerate, consisting of a series of wave vectors.

The levels are very widely spaced for small particles, for instance, the frequency separation between the uniform mode and the first level is approximately 16 GHz in a 40 nm particle of magnetite, and four times that for the second level. It is the exploitation of the resulting strong frequency dependence of the microwave absorption by such widely separated levels that will be explored in what follows.

As outlined above, the absorption process must conserve energy and wave vector: the microwave wave vector, $k = 2\pi/\lambda$, is essentially zero; so in the first-order process only spin waves with zero wave vector can interact with the microwaves, which restricts the absorption to the uniform mode, whose wavelength is infinite. In the second-order process, a single microwave photon is absorbed by two spin waves of equal and opposite wave vector, and all spin waves can absorb provided the microwave frequency (which must be twice the spin-wave frequency for energy conservation), is correct. This means that a given microwave frequency can only be absorbed by particles of a given size.

The second-order process is considerably more efficient than the first order one: an FMR was used in the very first experiments on microwave heating described in Walton et al. (1993): the microwave absorption by the uniform mode for the sample, a piece of archaeological ceramic, was found to extend from 2.45 to 2.85 GHz; so a magnetron taken from a microwave oven was used as the power source, supplying a cylindrical cavity resonant at the microwave frequency, providing an average microwave magnetic field of about 200 G (1 G = 10^3 A/m). This field resulted in an absorption by the sample of 6 W. In contrast, as will be discussed below, with 2FMR, a field of 13 G produced an absorption of 4 W; so 6 W would require about 20 G, and 2FMR is about an order of magnitude more efficient.

However, as the sample temperature increases, J decreases, and so does the frequency of the spin wave, and the uniform-mode resonance is lost. FMR was unable to increase the sample temperature more than ~100°C.

It would be expected that resonance would be lost and the heating would cease in 2FMR as well. But this argument does not take into account the broadening of the spin-wave levels, consisting of the intrinsic structural broadening plus that resulting from the decay of the spin waves into phonons and other spin waves (this is discussed in detail in Sparks, 1964). Most of these processes are temperature dependent, and the width of the level increases rapidly with temperature.

The matrix only interacts with the microwave electric field which is much less effective than the interaction of the magnetic field with the magnetic grains (this is clearly revealed by the small temperature rise of those samples that do not contain any magnetic grains able to be excited). The magnetic grains absorb the microwave energy that is then transmitted to the matrix. There is an energy barrier between the grains and the matrix that leads to a temperature difference between them while microwaves are being absorbed. How much hotter the grains get depends on the power absorbed. It is very hard to estimate the thermal conductivity of the barrier from first principles; so it is difficult to calculate the temperature difference even approximately. However, a qualitative feeling for its magnitude is revealed by the fact that in most cases a lava or a ceramic sample's magnetic moment is removed, indicating that the magnetite grains have reached their Neel temperature (~580°C), while the sample temperature remains at ~200°C or less.

The efficiency with which the magnetic grains absorb is illustrated by the fact that two samples, one of pure magnetite and the other a sample of lava rock containing a ~0.1% single-domain magnetite, in resonance with the microwaves, absorbed microwaves at the same rate (Walton, 2004). The experiment involved placing identically sized samples, 5 mm diameter by ~5 mm high cylinders glued to a quartz rod, in the center of a cylindrical resonant cavity tuned to the microwave frequency. The sample temperature was monitored with an infra-red pyrometer through a small hole in the side of the cavity. Microwaves were supplied by a solid state amplifier operating between 14 and 17 GHz fed by an HP signal generator. Coupling from the waveguide to the microwave cavity was through a 10 × 1 mm slot in the cavity. A moving vane was used to adjust the length of the slot in order to obtain an optimum coupling to the cavity. With the slot, it was possible to reduce the reflected power to negligible levels.

The results are shown in Figure 16.1. In fact, the identity between the heating curves obviously indicates that even the low concentration of magnetic particles in the lava was able to absorb all the incident radiation, if so, the heating should be independent of particle concentration above some minimum level. An attempt was made to determine this level with a series of samples of decreasing concentration: no change in the heating in the microwave cavity at a power level of 1 W was found between a 2% concentration of 50 nm magnetite, and surprisingly, 0.001%.

There is considerable difficulty in ensuring a uniform distribution of material at such low concentrations; so the experiment was not continued to lower concentrations.

If the lava is absorbing microwaves at the same rate as the solid magnetite, the power density in the magnetic grains in the lava must be ~1000 times the average delivered to the sample; so the initial rate of temperature increase of the grains would be ~10^3 times higher. The initial rate of sample temperature increase with 1 W supplied to the cavity was ~4 C/s; so, if isolated, the magnetic grains would heat up at ~4000 C/s. However, with a power of only 1 W supplied to the cavity, heat flow to the matrix would quickly limit their temperature increase. On the other hand if 66 W is supplied (the maximum our amplifier was capable of) the rate of increase will be ~264,000 C/s, and at that rate the grains would have reached their Neel temperature in, very roughly, ~0.002 s. The heat flow to the matrix would be negligible over such a short time, and it would be expected that those grains would be demagnetized immediately on reaching their blocking temperature. In that time only ~0.13 J would be delivered, of which ~40% would be absorbed by the magnetite particles. The volume of the sample is 0.1 cm³, the specific heat is ~1 J/°C cm³, so 0.05 J would only raise the sample temperature ~0.5°C.

The maximum sample temperature as a function of the time the microwaves are on, at 66 W power delivered to the cavity, is shown below for a sample of lava rock. There is a rapid initial increase as the magnetic grains absorb microwaves. The rapid increase stops when the grains reach their Neel temperature because spin waves do not exist above it, thus the grains will be held at the Neel temperature, and they will then heat the matrix at a slower rate. The rate must be slower because the grains can no longer absorb all the microwaves. It appears that, although the shortest time programmed was 0.01 s, there was no change in the sample temperature (outside experimental error) until 0.05 s, suggesting that the amplifier was incapable of responding faster than that (Figure 16.2).

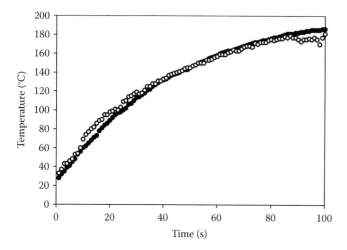

FIGURE 16.1 The increase in temperature with time for two samples, both weighing 0.1 g, with 1 W supplied to the cavity. One sample was a solid piece of magnetite, the other was lava rock containing ~0.1% magnetite. The similarity of the profiles is remarkable, indicating that although the concentration of magnetic material in the lava was three orders of magnitude less it was just as efficient in absorbing microwaves as the solid magnetite. (From Walton, D., *Appl. Phys. Lett.*, 85, 5367, 2004. With permission.)

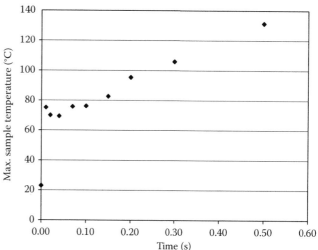

FIGURE 16.2 Temperature of sample immediately after turning off the microwave power, after leaving the power on for the time shown. The lack of change initially suggests that the amplifier response is limited to ~0.05–0.1 s.

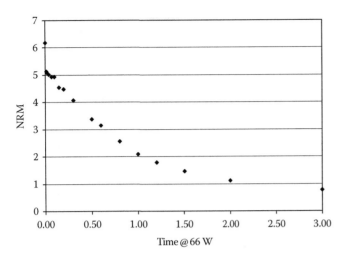

FIGURE 16.3 The decrease in magnetization after each of the heating steps of Figure 16.2.

Measurement of the sample's magnetic moment reveals that ~20% is demagnetized immediately, indicating that this fraction of the magnetic material is absorbing microwaves (Figure 16.3).

Some applications of the quantization of spin-wave levels, and the associated increased microwave absorption from the increased wave-vector range associated with the lowest levels will now be considered.

16.2 Hyperthermia Using 2FMR with Magnetic Nanoparticles

Presently available therapies for the treatment of cancer, such as chemotherapy and radiation, in attempting to attack the tumors also stress healthy tissue to a large degree (De Nardo and De Nardo, 2008). Therefore, it is desirable to find a way of targeting only the tumors. Magnetic nanoparticles provide a means for doing this because they can be delivered to the tumors themselves. Once there, they can be excited by an electromagnetic field that raises their temperature sufficiently for tumor necrosis (Pankhurst et al., 2003).

Magnetic fluid hyperthermia (MFH) cancer treatment, whereby a fluid containing magnetic nanoparticles is injected directly into tumors, has been investigated for some time. In these efforts, an alternating magnetic field with frequencies of hundreds of kilohertz was used to heat the nanoparticles and destroy the tumors. This minimally invasive procedure, unlike laser, microwave, and ultrasound hyperthermia, minimizes unnecessary heating in healthy tissues because only the magnetic nanoparticles absorb energy from the electromagnetic field. Ferrofluids consisting of magnetic nanoparticles dispersed in water or a hydrocarbon fluid can be injected into the tumors. For medical applications, the biocompatibility of both the fluid and the nanoparticles must be considered. Obviously, the magnetic material should not be toxic, a material that satisfies these requirements is magnetite (Fe_3O_4).

Another way to deliver magnetic material to the tumor is by attaching targeting antibodies to the magnetic nanoparticles. Instead of circulating throughout the body, the antibodies and nanoparticles would concentrate on the cancerous cells with receptors for the antibody.

Magnetic fluid hyperthermia cancer treatments must maintain therapeutic temperatures in diseased tissues for approximately 30 min (Pankhurst et al., 2003). Excessive heating, however, can cause unwanted charring. One way to ensure that this does not occur is to use nanoparticles with low-Curie temperatures (defined as the temperature at which the material loses its magnetic moment): heating takes place until they reach the Curie temperature and then stops. Nanoparticles with Curie temperatures near the therapeutic range can efficiently maintain temperatures between 42°C and 45°C and therefore enhance magnetic fluid hyperthermia. These self-regulating nanoparticles ensure diseased tissues reach the necessary temperatures while preventing excessive heating and damage to the surrounding healthy tissues.

Many studies have confirmed the feasibility of magnetic fluid hyperthermia cancer treatment. Hergt and others performed *in vitro* experiments to evaluate the heating effects of commercially available magnetite (Fe_3O_4) particles (Hergt et al., 1998). Study samples consisted of a sphere of compressed magnetite particles embedded in a large volume of KCl/Currageenan-gel. The samples were exposed to an alternating magnetic field ($H = 18$ G, frequency = 300 kHz) and the surface temperatures of the particle spheres were recorded. The heat generated depended heavily on the particle size and the microstructure. The results proved that even small amounts of particles with suitable properties can generate the heat required for magnetic fluid hyperthermia. Hilger and others used magnetic fluid hyperthermia to treat breast cancer in mice (Hilger et al., 2002). Fluid containing iron oxide particles was injected into tumors grown from human breast adenocarcinoma cells. The tumor and healthy tissue temperatures were monitored throughout the 4 min application of the alternating magnetic field ($H = 6.5$ G, frequency = 400 kHz). Therapeutic temperatures were obtained, but regions of insufficient heating were also observed. Cool spots were consistent with lower concentrations of magnetic fluid, emphasizing the importance of fitting the magnetic fluid distribution to the tumor's shape. Jordan and others also explored the effects of magnetic fluid hyperthermia on mammary carcinoma in mice (Jordan et al., 1997). Magnetic fluid containing magnetite particles was delivered by intratumoral injection (0.015 mg magnetite/mm³). A post-injection examination of the cancerous tissues showed deep fluid penetration. An alternating magnetic field ($H = 6–12.5$ G, frequency = 520 kHz) was applied for 20–30 min. Widespread death of cancerous cells was observed after MFH treatment. This study also tracked tumor growth in the 50 days following treatment. MFH treatment halted growth in some tumors, but did not impede growth in others. These differences were attributed to the inconsistent distribution of the magnetic fluid in the tumors.

Interstitial hyperthermia with ferrofluid injected in the vicinity of tumors has been used by Johansen et al. to treat prostate cancer (Johansen et al., 2005). The frequency used was 100 MHz, and the strength of the microwave field was 5 G. 1.5 g of magnetic particles 15 nm in size were injected into the prostate at 24 locations and the 5 G field achieved a specific absorption rate (SAR) of 0.3 W/cc. The prostate volume was 35 mL so the concentration of the magnetic material was very approximately 4%. After the field reached 5 G, it took 6 min for the tumor's temperature to rise to 6°C, at an average rate of 0.017°C/s, to 44°C.

At these low frequencies, energy absorption by the particles is due to the loss resulting from hysteresis, and is much less efficient than resonant microwave absorption by spin waves: in the experiment with the ceramic referred to above, a field of ~13 G was able to raise the temperature of a ceramic sample of volume 0.1 cc containing 0.1% (~0.07 g) magnetite at a rate of 3.5°C/s. To make a *very* rough estimate, make the crude assumption that there is not a great deal of difference between the thermal characteristics of the ceramic and the prostate, in which case a rate of 0.017°C/s would only require a field 0.062 G, a factor of 80 times less than that used by Johansen et al. The resultant decrease in the intensity of the microwave field would obviously be beneficial.

Because of the high frequencies involved, e.g., 6 GHz for 100 nm particles, interstitial application is probably the only practical method. On the other hand, the increasing attenuation in tissue as the microwave frequency increases could be an advantage—by selecting an appropriate frequency (and corresponding particle size) the microwave field could be largely confined to the tumor, and damage to healthy tissue surrounding the tumor could be minimized.

It is important to recognize that an interstitial application would take place by the insertion of a suitable antenna. At the frequencies of interest ~10 GHz, with a wavelength of $\lambda = 3$ cm, the near field of the antenna must be taken into account. The near field consists of the reactive-field region that extends to $1/2\pi$, followed by the radiative region that extends to $D^2/2\lambda$, where D is the dimension of the antenna. The far-field region follows where the attenuation is very strong so, for practical purposes, the scale of the volume affected by the microwaves can be taken as the size of the radiative region. If the dimension of the antenna is on the order of λ, this size is $\sim\lambda/2$ so, the effective distance from an antenna whose dimension is λ, is about $\lambda/2$. It is very difficult to calculate the intensity of the field in this region, and the resolution of that question is best done experimentally. Oscillations can occur in the radiative near field, and this is illustrated in results published by Tell (2000) that show a peak in the specific absorption rate for 1950 MHz microwaves in the head at a distance of 16–17 cm, and little attenuation before it. The dimensions of the antenna were not given, but the wavelength of the microwaves is 15.4 cm so, if the D is about one wavelength, it would appear that to a very rough approximation, the microwave energy density is constant in the near-field region. The magnetic transformation temperature can be lowered by alloying, so the possibility exists that the composition and the size of the particles can be designed for the absorption to be confined to the tumor, and to stop when a desired temperature is reached.

16.3 Geophysical Applications

Microscopic magnetite and titanomagnetite particles in rocks become magnetized when the rock cools in the earth's magnetic field, thereby creating a record of the direction and the intensity of that field when the rock was formed. Subsequent movement by geological or other processes will expose the rock to a field with a different orientation. Beginning with the smallest, and continuing with increasingly larger particles, the magnetic particles will begin to slowly reorient their magnetization to that of the new field, and the rock is partially remagnetized, or overprinted by the new field. The magnetization is now the vector sum of the remaining original moment (NRM) and the overprint. The magnitude of the overprint depends on the size and the composition of the rock, and can be considerable. In order to access the NRM, the overprint must be removed.

The smaller particles (<~1 μm) are single-domain particles, which relax to the new field direction with a relaxation rate that decreases exponentially with particle volume, and the smallest particles relax first followed by larger particles. The largest particles to relax are those whose relaxation time is on the order of the time the new field was present. The relaxation time is also inversely proportional to the temperature, so heating in the zero field to an elevated temperature can remove the very old overprints carried by the single-domain particles in a matter of minutes.

Larger particles consist of two or more magnetic domains, and they relax by domain boundary motion which is also thermally activated. However, it is possible for domain boundaries to become blocked, and during heating in the zero field will not return to their original position, so the overprint is not removed completely. The only solution is to remove the contribution from the multi-domains entirely. Doing this thermally, using an oven, requires heating in the zero field to above the Neel temperature which is clearly impractical: it would also remove the NRM. It is possible to demagnetize by applying an alternating magnetic field that is slowly reduced to zero (AF demag). In order to avoid demagnetizing the single-domain grains that are carrying the NRM, the maximum field must be low. A common misconception is that this can be done. However, data in the literature show that the task is not an easy one: while a maximum field of 600 G removes 85% of the moment in grains 1 μm in size, it also removes 83% of the moment carried by single-domain grains 0.037 μm in size (Dunlop and Ozdemir, 1997); so, a lower field must be used. A common practice is to demagnetize with a maximum field of 200 Oe, but this only removes 50% of the 1 μm moment (Dunlop and Ozdemir, 1997).

Multi-domain grains can also be partly demagnetized (although, in practice this is rarely complete) by cooling below their Verwey transition. Dunlop et al. report that this only increases the percentage removed by a subsequent AF demag with a maximum field of 200 G, to 65% in 1 μm grains (Dunlop and Ozdemir, 1997).

In spite of the above well-known difficulties, the vast majority of paleodirections (orientations of the earth's magnetic field when the rock first cooled, that provide the key information on plate tectonics) have been determined using AF demag to remove the unwanted overprints (McElhinny, 2000). AF demagnetization of multi-domain overprints has been used for many decades, and forms the basis for an understanding of many geological phenomena. However, the results described in Dunlop and Ozdemir (1997) strongly suggest that the overprints have not been completely removed, which raises questions about the validity of paleomagnetic directional data, particularly in very old rocks.

2FMR provides a simple and straightforward way to remove the magnetization of multi-domain grains without affecting the smaller single-domains: quantization of the spin-wave energy levels makes it possible to demagnetize the multi-domains using frequencies too low to be absorbed by the smaller single-domains. Using a frequency below that, which is absorbed by grains large enough for the quantization gaps to disappear, occurs when magnetite grains >~300 nm in size. These grains absorb at 8.8 GHz; so, these, and all larger grains would be demagnetized at that frequency, but the single-domains carrying the NRM would not absorb, and the NRM of the sample would not be affected provided the microwave power level is high enough to minimize sample heating, thereby preventing the single-domains from being thermally demagnetized.

16.4 Dating Archaeological Ceramics

The second application of 2FMR is, in a sense, the mirror image of the first, and uses the overprint carried by the single-domain grains for viscous dating (the gradual production of the overprint with time has been called "magnetic viscosity," and the resultant moment is referred to as a viscous remanent magnetization or VRM).

Archaeomagnetic materials acquire an NRM on cooling in a kiln. Subsequent use and eventual storage produces a viscous VRM proportional to, and in the orientation of, the local magnetic field during the time between the initial cooling, and their retrieval that differs in magnitude and orientation from the natural remanent magnetization (NRM) produced when the material first cooled. To a good approximation, all magnetic grains with relaxation times less than the age of the sample are reoriented by the environmental field. If the VRM can be measured, it can be used to date the material.

The magnetization is carried by small magnetic particles embedded in a nonmagnetic matrix. The major contribution, particularly in ceramics, comes from nanometer-sized single-domain grains. These grains are usually ferrimagnetic iron compounds, of which magnetite is the most important. They carry a ferromagnetic moment that lies along a preferred direction established by an anisotropy field in the grain, hence the magnetic ground state of single-domain grains is twofold degenerate. The magnetic vector can reverse its direction with a relaxation rate given by (Neel, 1955)

$$\omega = \nu \exp\left(-\frac{KV}{kT}\right), \tag{16.2}$$

where

ν is an attempt frequency on the order of 10^8 Hz
K is the anisotropy energy per unit volume
V is the volume of the particle
k is Boltzmann's constant
T is the temperature

Let n_V be the fractional alignment of the grains. By this we mean that if we multiply n_V by the number of grains and by the average moment of each grain, we will obtain the total moment. The equilibrium value in a small field **H**, and temperature T is $n_V = (JV/3kT)$ **H** (Kittel, 1968), where J is the saturation magnetization at T. During cooling in a geomagnetic field $\mathbf{H_G}$ this becomes $n_V(T_b) = (JV/3kT_b)\mathbf{H_G}$ where T_b is the "blocking temperature." This is the temperature at which the reversal rate given in Equation 16.1 can no longer follow the cooling, and the fractional alignment ceases to increase.

The resulting magnetic moment of the material is $\int_{V_{bl}}^{V_{max}} n_V JV N(V)\,dV$ (Walton, 1980), where $N(V)$ is the distribution of grain sizes, and V_{bl} a volume below which the grains are superparamagnetic and cannot contribute to a permanent sample moment.

As discussed above, after cooling, the material may find itself in a different field, $\mathbf{H_v}$, and the grains will start to relax to their new equilibrium distributions. However, the relaxation rate is an exponential function of the volume, and while small grains may reach equilibrium, large grains may never do so. The fractional alignment after a time t is given by (Williams and Walton 1988)

$$n_V(t) = n_V(T_b)e^{-\omega t} + n_V(T)(1 - e^{-\omega t})$$

$$= \frac{JVH_G}{6kT_b}\exp\left(\nu t e^{-KV/kT}\right) + \frac{JVH_v}{6kT}\left[1 - \exp\left(\nu t e^{-KV/kT}\right)\right]. \tag{16.3}$$

Because of the double exponentials, the two contributions are extremely strong functions of grain size, and the second term that represents the overprint is confined to the smaller grains. Thus, exclusive demagnetization of these grains can yield information as to the time the material has been exposed to $\mathbf{H_v}$. This is conventionally done thermally (Walton, 1983), and the material is demagnetized at successively higher temperatures until the contribution of $\mathbf{H_v}$ to the magnetic vector disappears. The time at elevated temperature is then converted to a time at ambient temperature using time–temperature relations deduced from Equations 16.2 and 16.3.

Alternating field demagnetization (AFD) was used in a first attempt to measure the VRM by Heller and Markert (1973). AFD entails subjecting the sample to alternating magnetic fields of increasing magnitude that are then allowed to decay. Thus, it demagnetizes the grains with the lowest coercivities first, and those with the largest last. However, the largest coercivities are those of the smaller single-domain grains, so a substantial fraction of the VRM would be expected to be demagnetized last. AFD is not widely used at present to reliably identify a VRM, and thermal methods are used exclusively in recent studies of viscous dating (Borradaile, 2003).

Attempts to use progressive thermal demagnetization are plagued by the problem that the VRM is small compared to the NRM, and when the sample is demagnetized thermally, the VRM is removed first. The small changes in the resultant of the two vectors are difficult to measure making it very difficult to determine how much, and when the VRM has been removed. The microwave procedure described below operates on quite a different principle: removing the NRM *first*, leaving the VRM isolated, and easily measured.

Viscous overprints are carried by single-domain grains and multi-domain grains with mobile domain boundaries. As the exposure time to $\mathbf{H}_v$, the local field where the ceramic is stored or buried, increases, single-domain grains with increasing relaxation times, i.e., larger grains, become magnetized. If, after the ceramic is retrieved, demagnetization of the sample is started at low frequencies, the grains carrying the NRM (and the multi-domains) will be demagnetized, leaving the VRM carried by the single-domains unaffected. As the frequency is increased, progressively smaller grains will be demagnetized. As long as these are carrying the NRM, changes in the sample's magnetic vector will be small. Eventually, a frequency will be reached that demagnetizes the smallest grains that carry the NRM. At that point, since the only magnetized grains left in the sample are the VRM carriers, the direction of magnetization will change. The frequency at which this occurs (VF) is proportional to the size of the largest grains carrying a VRM, which in turn is determined by the age of the sample. It is then a simple matter of calibration to relate VF to the age of the material.

The microwave frequency can be very precise, so the major uncertainty lies in determining VF. Some idea of the precision with which this can be accomplished was obtained by numerically demagnetizing a hypothetical sample with a VRM emplaced at right angles to the NRM. Figure 16.4 shows the change in the angle of the magnetization as the microwave frequency is increased for two cases– one in which the VRM is 10% of the NRM, and another in which it is 1%.

Interestingly the sharpness of the transition increases as the magnitude of the VRM decreases. This compensates to a degree for the greater uncertainty in measuring the angle resulting from the decreased signal.

It is worth observing that it would be well nigh impossible to use a VRM, that is 1% of the NRM for viscous dating, if conventional thermal methods are used. Indeed the VRM in ancient pottery seems to be of that order of magnitude because significant (outside measurement error) changes in the magnetic vector are very rarely seen. On the other hand, if the NRM is erased, a small VRM of that order is easily measured.

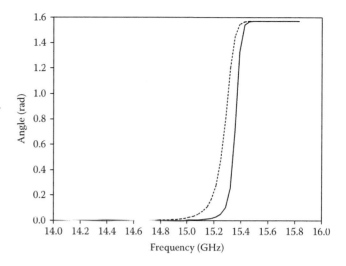

FIGURE 16.4 Results of a numerical calculation of the change in angle with frequency as a hypothetical sample is demagnetized that carries a VRM emplaced at right angles to the NRM that is 10% of the NRM (dashed line) or 1% (solid line).

To illustrate the preceding discussion, Figure 16.5 shows the dependence of VF on age for a series of hypothetical samples containing cubic magnetite grains.

The 2FMR dating method is expected to be most useful for ceramics; so, rather than trying to deduce the age from first principles, it would be more practical to simply determine the relationship between the frequency at which the NRM first disappears and the age of the sample with well-dated samples from the regions of interest. Scientific dating is usually inferior to dating by an archaeologist, but it is often useful in situations where the information is insufficient for an archaeologist to be able to date an object accurately, for instance, for many pre-Columbian ceramics.

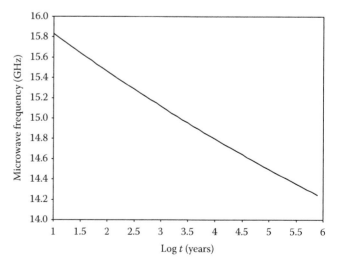

FIGURE 16.5 Minimum frequency required to start removing the overprint as a function of age, in the hypothetical sample of Figure 16.4.

16.5 Summary and Conclusions

The quantization of spin-wave levels reveals interesting new physics: first, the large separation between the lower levels in small particles, and second their large extension in the wave-vector space, that can explain the efficient heating of the nanoparticles by 2FMR. These properties lead to at least three applications for 2FMR that promise significant improvements:

1. Hyperthermia with 2FMR appears to be almost two orders of magnitude more efficient than the low-frequency methods used so far; it has the advantage of a much smaller exposure time, and the ability to restrict the microwave radiation to the tumor itself. A drawback is that it requires interstitial application.

2. The ability to remove the irreversible contributions of multi-domain grains would remove an important source of error in determinations of the orientation of the ancient geomagnetic field.

3. The ability to isolate and accurately measure the viscous magnetic moment produced during storage promises a scientific dating technique that is related to an intrinsic property of the material, and does not require the presence of contemporary organic material as does carbon dating, or the burial of a radiation monitor as does thermoluminescent dating.

The rapid heating produced by 2FMR has been used to provide an improved way to obtain estimates of the strength of the ancient geomagnetic field (Gratton et al., 2007). This is an improvement on the classical double-heating method used in paleomagnetism to access the ancient field: the sample is heated twice, once in a zero field, and then to the same temperature in a laboratory field. By comparing the NRM removed in the zero-field step with the moment produced by the laboratory field–step, an estimate of the ancient field can be made. The microwave application just substitutes the microwave amplifier for the furnace that, unfortunately, preserves some of the drawbacks of the double-heating method. This is discussed in Walton and Boehnel (2008).

Probably, the greatest weakness in the discussion presented here is the lack of experimental evidence. The conventional method for studying spin waves is neutron inelastic scattering, but this is difficult to perform on small particles, and while it should be doable on arrays of narrow wires, it is difficult to make a sample of sufficient volume. The three applications of 2FMR discussed above await funding for the experimental work to progress.

References

Borradaile, G. J., 2003, Viscous magnetization, archaeology and Bayesian statistics of small samples from Israel and England, *Geophys. Res. Lett.* 30, 1528 and references therein.

De Nardo, G. L. and S. J. De Nardo, 2008, Turning the heat on cancer, *Cancer Biother. Pharm.* 23(6).

Dunlop, D. J. and O. Ozdemir, 1997, *Rock Magnetism*, Cambridge University Press, Cambridge, U.K.

Gratton, M. N., J. Shaw., and L. L. Brown, 2007, Absolute palaeointensity variation during a precursor to the Matuyama-Brunhes transition recorded in Chilean lavas, *Phys. Earth Plan. Int.* 162, 61.

Heller, F. and H. Markert, 1973, The age of viscous remanent magnetization of Hadrian's wall *J. Roy. Astron. Soc.* 31, 395

Hendriksen, P. V., Linderoth, S., and P. A. Lindgard, 1993, Finite size modifications of the magnetic properties of clusters, *Phys. Rev. B* 48, 7259.

Hergt, R. et al., 1998, Physical limits of hyperthermia using magnetic fine particles. *IEEE Trans. Magnet.* 34(5), 3745–3754.

Hilger, I. et al., 2002, Thermal ablation of tumors using magnetic nanoparticles: An in vivo feasibility study. *Invest. Radiol.* 37(10), 580–586.

Jordan A, et al., 1997, Inductive heating of ferrimagnetic particles and magnetic fluids: Physical evaluation of their potential for hyperthermia. *Int. J. Hypert.* 9, 51–68.

Johannsen, M. et al., 2005, Clinical hyperthermia of prostate cancer using magnetic nanoparticles: Presentation of a new interstitial technique, *Int. J. Hypert.* 21(7), 637–647.

Kittel, C., 1968, *Introduction to Solid State Physics*, Wiley, New York.

McElhinny, M. W., 2000, *Palaeomagnetism: Continents and Oceans*, Academic Press, San Diego, CA.

Neel, L., 1955, Some theoretical aspects of rock magnetism, *Adv. Phys.* 4, 191.

Pankhurst, Q. A. et al., 2003, Applications of magnetic nanoparticles in biomedicine. *J. Phys. D: Appl. Phys.* 36, R167–R181.

Sparks, M., 1964, *Ferromagnetic Relaxation Theory*, McGraw Hill, New York.

Tell, R. A., 2000, in *Radio Frequency Radiation Dosimetry*, B. J. Klauenberg and D. Miklavcic (eds.), Kluwer Academic Publishers, Dordrecht, the Netherlands.

Walton, D., 1980, Time-temperature relationships in the magnetization of assemblies of single-domain grains, *Nature* 286, 245.

Walton, D., 1983, Viscous magnetization, *Nature* 305, 616.

Walton, D., 2004, Avoiding mineral alteration during microwave magnetization, *Geophys. Res. Lett.* 31, LO3606, doi:10.1029/2003.

Walton, D. and H. Boehnel, 2008, The microwave frequency method, *Phys. Earth Plan. Int.* 167, 145.

Walton, D., Share, J., Rolph, T. C., and Shaw, J., 1993, Microwave magnetization, *Geophys. Res. Lett.* 20, 109–111.

Walton, D., H. Boehnel., and D. J. Dunlop, 2004, Response of magnetic particles to microwaves, *Appl. Phys. Lett.* 85, 5367.

Williams, W. and D. Walton, 1988, Thermal cleaning of viscous magnetic moments, *Geophys. Res. Lett.* 15, 1089.

17

Catalytically Active Gold Particles

Ming-Shu Chen
Xiamen University

17.1 Introduction

Owing to its inert properties, gold is a well-known precious metal and popular in jewelry and decoration. Though it is being found to be active as a catalyst only in a recent few decades, the early examples existed in 1970s, though significantly less active than transition metals [1]. The significant effect of gold as a catalyst was studied by Haruta et al. in 1987 who found that gold supported on TiO_2 being very active for CO oxidation at low temperature [2] and oxidation of propene to propene oxide [3] (see Figure 17.1) and also by Hutchings et al. through the hydrochlorination of ethyne to vinyl chloride [4]; both reactions discussed were heterogeneous in nature.

In homogeneous catalysis, transition metals are traditional catalysts for organic synthesis [5,6]. In recent years, gold, often a cationic gold, has been found to be active for many organic reactions with much higher selectivities [7–9]. This has been reviewed by Hashmi et al. [10,11]. The most fundamental reactivity pattern in gold-catalyzed organic reactions is the activation of a C–C multiple bond, in most cases an alkyne, for the attack of a nucleophile. In 1998, Teles et al. found kinetically highly active cationic gold(I) catalysts for hydration of alkynes to substitute the traditional method using mercury(II) ions in aqueous sulfuric acid [12]. Gold catalyst in organic synthesis shows a high functional group tolerance, for example, alkynyl iodides or vinyl iodides as well as thioethers are tolerated, and high synthetic efficiency. Cationic gold was unique in catalytic asymmetric aldol reactions [13] and the addition of nucleophiles to alkynes [12,14].

The positional-selective oxidation of different alcohols and even carbohydrates with molecular oxygen is another important effect of gold catalysis [15]. Recently, gold in homogeneous catalysis was extended from the nucleophilic additions from alkynes to olefins for both the intramolecular additions of alcohols and the intermolecular addition of arenes [16,17].

The unique catalytic properties of supported nanosized-Au particles depend strongly on the support, particle size and shape, as well as other factors [2,4,18–20]. Numerous studies have addressed how the shape, structure, and electronic properties of Au particles correlate with the catalytic activity. For example, the size of Au nanoparticles and the properties of the support have been shown to be critical in giving rise to the unique catalytic activity. Furthermore, active sites have been proposed to reside at the interface between the Au nanoparticles and the oxide support. However, the size distribution and shapes of nanoparticles in oxide-supported catalysts often vary widely, and the particle structure, particularly the structure at the metal–oxide interface, is generally not well-defined. Moreover, Au nanoparticles typically sinter rapidly under reaction conditions [19]. Because of these complexities, the structure/active sites of supported Au catalysts remain unclear. The recent synthesis of highly ordered Au mono- and bilayer structures or "nanofilms" supported on a highly reduced and ordered titania surface and the exceptionally high catalytic activity for CO oxidation observed for the Au bilayer structure have aided significantly in defining the nature of the active site [21]. The ordered Au nanofilms are stable in ultrahigh vacuum (UHV) to 900 K, much higher than

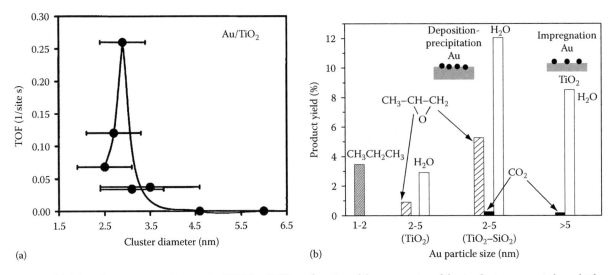

FIGURE 17.1 (a) CO oxidation turnover frequencies (TOFs) at 300 K as a function of the average size of the Au clusters supported on a high surface area TiO_2 support. The Au/TiO_2 catalysts were prepared by deposition-precipitation method, and the average cluster diameters were measured by TEM. (From Valden, M. et al., *Science*, 281, 1647, 1998. With permission.) (b) Influence of Au particle size on product yield during propene epoxidation using O_2 and H_2. (From Sinha, A.K. et al., *Topic Catal.*, 29(3–4), 95, 2004. With permission.)

the temperature (i.e., 700–800 K) at which supported Au nanoparticles rapidly sinter. These studies illustrate the importance of defects on the oxide support in stabilizing Au nanoparticles and the need for designing optimum oxide supports for Au and other noble metal catalysts.

17.2 Applications of Supported Gold Nanoparticles as Catalysts

17.2.1 Carbon Monoxide Oxidation

In 1987, Haruta et al. recognized that supported gold nanoparticles were highly active for the oxidation of CO at a temperature even significantly below 273 K [2], and this characteristic was crucially affected by support materials and preparation methods. Such a unique characteristic has not been observed in other metals. Detailed electron microscopy studies revealed that the very active catalysts comprised small gold nanoparticles ~2–4 nm in diameter. Goodman et al. using scanning tunneling microscopy (STM) studied the growth of gold nanoparticles on a TiO_2(110) surface [22]. Different size distribution of gold nanoparticles can be prepared by controlling the deposition amount of gold and postannealing. CO catalytic oxidation on such model surfaces of Au/TiO_2(110) also revealed the similar unique properties with strong size dependence, in which the highest CO_2 formation rate was observed on gold particles in a size of around 2.5–3 nm (Figure 17.2) [19]. Such model studies also demonstrated quantum size effect of gold nanoparticles. Because of the remarkable catalytic activity in the low-temperature CO oxidation, studies related to the supported gold nanoparticles have attracted numerous researchers in recent two decades.

Considerable emphasis has been placed on catalyst preparation, in particular, the nature of the support. The high activity

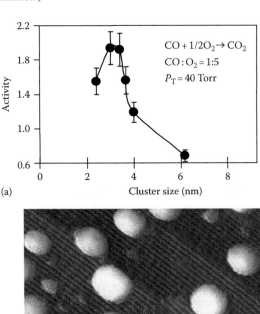

FIGURE 17.2 (a) CO oxidation turnover frequencies (TOFs) at 350 K as a function of mean particle size of the Au clusters on a TiO_2(110) surface. (From Valden, M. et al., *Science*, 281, 1647, 1998.) (b) An STM image of gold clusters on a TiO_2(110) surface with Au coverage of 0.25 ML. (From Valden, M. et al., *Science*, 281, 1647, 1998. With permission.)

was observed on α-Fe_2O_3- [23–27], TiO_2- [2,7,19–21,28–31], and CeO_2 [32–34]-supported gold catalysts. Among these, TiO_2 has been the most detailed studies [2,7,19–21,31,34]. The method of preparation is crucial in controlling the activity of the catalyst. The regular impregnation (a method in which an active component is added to support by stirring it with a metal salt solution) of titania resulted large gold particles (>10 nm) that are inactive for CO oxidation. Haruta et al. devised a deposition precipitation process, in which a support was stirred in a solution of a gold compound with the pH value adjusting by the addition of a base, thus leading to deposit of small gold nanoparticles onto the support surfaces [2]. A large batch of Au/TiO_2 catalyst was synthesized by DP process, and which became one of the World Gold Council's standard gold catalysts. Because of the less stability of the supported gold catalysts mainly due to sintering, research interest continues to be on the design of supported gold catalysts for the oxidation of CO at low temperature. Yan et al. develop a specific form of TiO_2 as a support [35]. Mesoporous structures, like MCM and SBA [36–38], as well as nanoparticles of SiO_2-Al_2O_3 [39] have been found to enhance the stability of gold catalysts.

The capability of CO catalytical oxidation at an ambient temperature leads supported gold catalysts in potentially removing the trace amounts of CO from toxic environments. Another particular application of gold catalysts is in fuel cells, especially in polymer electrolyte fuel cells (PEFCs) for electric vehicles operating at about 353–373 K [40–42]. Currently, hydrogen resource for fuel cell is mainly produced from coal, natural gas, methanol, and so on, by steam reforming and water gas shift (WGS) reactions. Trace amounts of residual CO can poison the Pt anode in PEFCs operating at low temperature. Hence, CO has to be removed from H_2 in the presence of water to ensure long cell lifetimes. Conceptually, CO oxidation is a simple catalytic method to remove CO. However, under excess moist H_2, oxidation of CO without oxidizing the hydrogen is particularly a difficult challenge in catalysis. The commonly used supported Pt, Ru, Pd, and Rh catalysts for selective oxidation require a temperature in the range of 423–473 K, which is much higher than the operation temperature for PEFCs. Furthermore, the selectivity for CO oxidation in H_2 is poor. Catalysts based on mixed oxides of copper are active at round 353 K, but supported gold catalysts display superior activities and selectivities [25,26,43–45]. These studies were not carried out at a mimic condition as in fuel cell, for example, with CO_2 and H_2O.

17.2.2 Hydrogenation Reactions

Supported gold catalysts are active for hydrogenation of alkenes as well as alkynes [1,4,46–48]. The first study was done by Bond et al. in 1973, in which alkenes were hydrogenated over an Au/SiO_2 with gold loading amount less than 0.01 wt% [46]. Low concentrations of Au supported on SiO_2 were also found to be active for the hydrogenation of 1-pentene, with the maximum activity observed at 0.04 wt% Au [47]. In contrast, Au γ-Al_2O_3 was inactive, indicating that hydrogenation reactions could be sensitive to both Au particle size and the nature of the

support, as has been observed in CO oxidation reaction. On a 5% Au/SiO_2, hydrogenation of ethene is of first order in hydrogen [47]. Hydrogenation of propene on Au/SiO_2 using deuterium (D_2) is much slower than using hydrogen (H_2), suggesting that breaking H–H bond is the rate-determining step [48].

The typical catalytic properties of supported gold nanoparticles in hydrogenation reaction would be the position-selective hydrogenation, like selective hydrogenation of α,β-unsaturated aldehydes [49]. A good example is the hydrogenation of acrolein over SiO_2-, ZrO_2-, TiO_2-, and ZnO-supported gold catalysts [49,50]. Bailie and Hutchings [51] found that Au/ZrO_2 and Au/ZnO catalysts are highly selective for the formation of crotyl alcohol through the hydrogenation of crotonaldehyde with selectivities up to 81% at conversions of 5%–10%, in which supported gold catalysts preferentially hydrogenate the C=O bond rather than the C=C bond. The studies by Claus and coworkers on the hydrogenation of acrolein were comprehensive and concentrated on designing catalysts comprising gold nanoparticles (ca. 5 nm diameter) [49,50].

$$H_2C = CH - CH = O + H_2 \rightarrow H_2C = CH - CH_2 - OH$$

17.2.3 Water Gas Shift Reaction

The WGS reaction is a catalytic reaction in which carbon monoxide reacts with water to form carbon dioxide and hydrogen over catalysts:

$$CO + H_2O \rightarrow CO_2 + H_2$$

This is an important industrial reaction to produce high-purity hydrogen for use in ammonia synthesis and many other applications. The WGS reaction was discovered by Italian physicist Felice Fontana in 1780. The reaction is slightly exothermic. The WGS reaction is sensitive to the reaction temperature, with the tendency to shift toward reactants as temperature increasing due to Le Chatelier's principle. The standard industrial process includes two stages, stage one being a high-temperature shift at 623 K and stage two being a low-temperature shift at 463–483 K, using iron oxide promoted with chromium oxide and copper on a mixed support composed of zinc oxide and aluminum oxide as catalysts, respectively [52].

Attempts to lower the WGS reaction temperature are very important to lower the residue CO, especially for hydrogen in fuel cell. Regarding the low-temperature activities of supported Au catalysts for CO oxidation, they may also be active for WGS reaction. The WGS activity of Au catalysts at low temperature was first reported by Andreeva and coworkers for Au/Fe_2O_3 catalysts [53], which were found to be enhanced by the addition of Ru [54] and metal oxides [55]. TiO_2-supported gold catalysts are also active for WGS reaction [56,57]. Then most of the interesting researches focused on Au/CeO_2 after Fu et al. reporting that it is more active than Au/TiO_2 [58]. Although Au/CeO_2 is significantly more active than commercial Cu catalysts, they are prone to deactivation caused by carbonates and/or formates blocking the active sites [59].

17.2.4 Selective Oxidation

Supported gold catalysts have been found to be particularly effective for epoxidation of alkenes and oxidation of alcohols. One of the potential applications is catalyzed propene epoxidation to propene oxide [3]. In contrast to the commercial process of epoxidation of ethene with dioxygen using a supported Ag catalyst [60], the epoxidation of propene is much difficult due to low selectivities over most catalysts investigated. Haruta and coworkers found that epoxidation of propene with dioxygen over supported gold catalysts can be enhanced by the addition of H_2 [3]. Initial selectivities were low but promising, and improvements were made by using different titanium-containing supports such as TS-1, Ti-zeolite β, Ti-MCM-41, and Ti-MCM-48 [61–64]. The active species-supported gold nanoparticles for epoxidation reaction are considered to be metallic gold with diameters much smaller than 2 nm [3]. Currently, these catalysts showed relatively short lifetimes but could be reactivated.

Compared with well-established platinum and palladium nanoparticles for selective oxidation of polyols, supported gold catalysts were found to be more effective [15,65]. The graphite-supported gold catalysts showed 100% selectivity for the oxidation of glycerol to glyceric acid using dioxygen as the oxidant under relatively mild conditions, with yields approaching 60% [64]. Under comparable conditions, supported Pd/C and Pt/C always produce C_3, C_2, and C_1 products in addition to glyceric acid. Other significant results were observed with Au/CeO_2 that catalyzes the selective oxidation of alcohols to aldehydes and ketones and of aldehydes to acids under relatively mild conditions using O_2 as the oxidant [66].

17.2.5 C–H Bond Activation

The activation of C–H bonds in alkanes and selective conversion is tough challenge in catalysis and of immense commercial significance, especially for light alkanes. The use of supported gold catalysts for C–H bond activation was initiated by Zhao et al. [67] for the oxidation of cyclohexane to cyclohexanol and cyclohexanone at 423 K. Selectivity of about 90% was achieved on an Au/ZSM-5 catalyst and more than 90% on an Au/MCM-41 catalyst. The oxidation of cyclohexane using O_2 is central to produce nylon-6 and nylon-6,6, the worldwide production of which exceeds 106 ton per year. Recently, Xu et al. reported that Au/C catalysts are as active as Pt and Pd catalysts for the selective oxidation of cyclohexane [68]. These studies demonstrated that supported gold catalysts are indeed active for the activation of C–H bonds with high selectivities.

17.2.6 Synthesis of Hydrogen Peroxide

Hydrogen peroxide, H_2O_2, is a mild oxidant used in many large-scale processes such as bleaching, fine-chemical industry, and as a disinfectant. Such widespread applications account for large demand of H_2O_2. Hydrogen peroxide is industrially produced by the sequential hydrogenation and oxidation of an alkyl anthraquinone [69]. Hence, the development of a new, highly efficient, and smaller-scale manufacturing process for H_2O_2 is of significant commercial interest.

Recent research showed that H_2O_2 can be synthesized directly from H_2 and O_2 over a supported Pd catalyst [70,71]. Au/Al_2O_3 and Au/SiO_2 catalysts were found to be effective for the direct synthesis, and are significantly enhanced by using supported Au/Pd alloys [63,70–72].

17.3 Interaction of Au with Oxide Supports

17.3.1 Importance of Surface Defects

Defects on oxide surfaces play a key role in the nucleation and growth of metal nanoparticles as well as in defining their electronic and chemical properties. Therefore, considerable work has focused on the characterization of surface defects and their interaction with metal atoms and particles [20]. The adsorption of probe molecules in conjunction with STM allows detailed characterization of surface defects. STM is especially useful in that it provides atomic-level information [73,74]. Figure 17.3a shows a high-resolution STM image of a single crystal rutile surface of titania, specifically the TiO_2(110) surface [31]. The bright spots along the rows correspond to the five-coordinate Ti atom sites, that is, the coordinately unsaturated Ti cations, as shown in the schematic of Figure 17.3b. The additional bright spots between the ordered rows are surface oxygen vacancies, so-called defects, as indicated by the red circles in Figure 17.3b. Defects of this kind can be created by sputtering with Ar^+ or by annealing in UHV [74]. Such defects have been found to markedly affect the adsorption energy, the particle shape, and the electronic structure of deposited Au nanoparticles and to influence their unique catalytic properties [19–21,31,73,75,76]. Theoretical calculations have demonstrated that Au particles bind more strongly to a defect-rich surface compared with a defect-deficient surface and that charge transfer may occur from the titania support to the Au particles [73]. High-resolution STM combined with density functional theoretical (DFT) calculations have confirmed that the bridging oxygen vacancies are the active nucleation sites for Au particles on titania and that each vacancy site can bind approximately three Au atoms on average [73]. The adsorption energy of a single Au atom on an oxygen vacancy site is more stable by 0.45 eV compared with the stoichiometric surface. Low-energy ion scattering (LEIS) spectroscopy has been used to examine the growth of Au on TiO_2(110) [75,76]. Two-dimensional (2-D) Au islands are initially formed on the titania surface up to a critical coverage that depends on the defect density, after which 3-D particles nucleate. The critical Au coverage at which transformation of Au particles from 2-D to 3-D occurs has been shown to be markedly dependent on the defect density, that is, the maximum coverage of 2-D domains correlates closely with the surface defect density. Au nanoparticles on defect-rich titania are found to be much more chemically active than those on defect-deficient titania [77].

Using ultraviolet photoemission spectroscopy (UPS), surface oxygen vacancies on a highly reduced titania surface have been

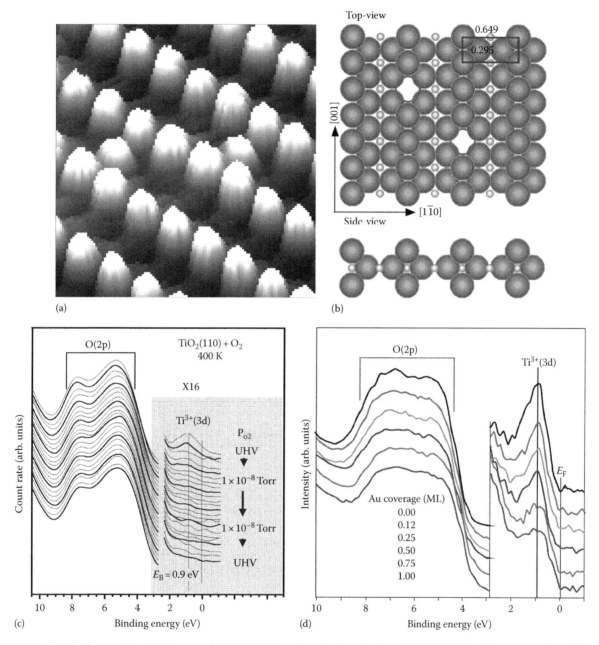

FIGURE 17.3 (a) A high-resolution STM image of $TiO_2(110)$, (b) a schematic structural model of $TiO_2(110)$ with empty circles indicating the surface oxygen vacancies, and (c and d) UPS spectra for a reduced $TiO_2(110)$ surface followed by exposing to oxygen at 400 K and deposition of various amounts of Au at room temperature. (From Chen, M.S. and Goodman, D.W., *Acc. Chem. Res.*, 39, 739, 2006. With permission.)

titrated by exposure to oxygen or Au deposition in our laboratory [31]. Emission from a Ti^{3+} electronic level at a binding energy of 0.9 eV is evident in the UPS data of Figure 17.3c and d. Note that on an oxygen vacancy site, two neighboring Ti atoms are reduced from Ti^{4+} to Ti^{3+}. These Ti^{3+} surface defects can be completely reoxidized by exposure to oxygen [31,74], as shown in Figure 17.3c. Following Au deposition onto the reduced titania surface, a decrease in the intensity of the Ti^{3+} state is apparent, showing that Au initially nucleates at the defect sites [31]. This suggests that Au atoms bond to the surface at oxygen vacancy sites, that is, bonding between Au and Ti^{3+}. The existence of Au–Ti bonds

is also evident in well-ordered Au mono- and bilayer films on a TiO_x/Mo(112) (see Section 17.3.3) [21] and in TiO_2-supported Au nanoparticles studied with extended x-ray absorption fine structure (EXAFS) [78]. In the latter, EXAFS shows a normal Au–Ti bond distance, whereas the Au–O bond distance is well in excess of the normal Au–O bond length, suggesting that Au binds with Ti rather than O at the Au–oxide interface.

Figure 17.4a shows STM images of 0.25 ML (1 ML: one monolayer corresponds to one Au atom per surface five-coordinate Ti^{4+}) of Au deposited on a $TiO_2(110)$ single crystalline surface at 300 K, followed by an anneal at 850 K for 2 min [19,31].

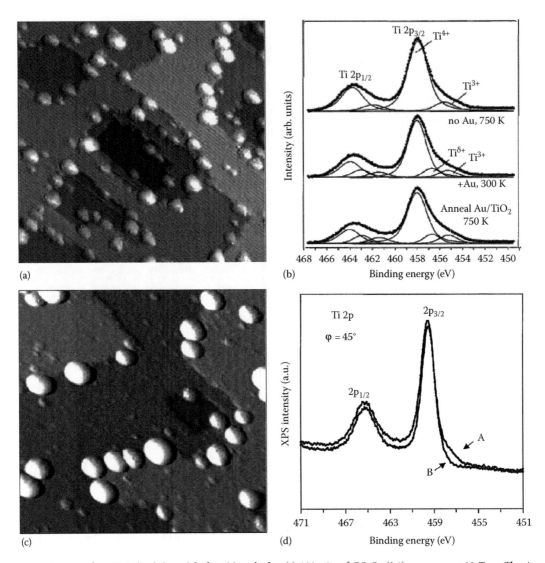

FIGURE 17.4 STM images of Au/TiO$_2$(110)-(1 × 1) before (a) and after (c) 120 min of CO:O$_2$ (2:1) exposure at 10 Torr. The Au coverage was 0.25 ML, and the sample was annealed at 850 K for 2 min before the exposures. All of the exposures are given at 300 K. The size of the images is 50 × 50 nm. (b) Ti 2p spectra taken before and after dosing 0.5 ML of gold to a TiO$_2$(110) surface. (d) XPS Ti 2p spectra for (a) and (c), respectively. (From Valden, M. et al., *Science*, 281, 1647, 1998; Rodriguez, J.A. et al., *J. Am. Chem. Soc.*, 124, 5242, 2002. With permission.)

The atomically resolved TiO$_2$(110) surface consists of flat terraces with atom rows separated by ~0.645 nm. 1-D and 2-D Au nanoparticles are apparent and nucleate at defect sites at very low Au coverages. At Au coverage of 0.25 ML, Au nanoparticles with specific morphologies are imaged as bright protrusions, indicating a relatively narrow particle size distribution. These pictures show that Au nanoparticles initially nucleate at defect sites, growing first from 1-D to 2-D, then finally to 3-D structures. Figure 17.4b displays Ti 2p spectra collected before and after dosing Au to a TiO$_2$(110) surface [77]. By using a photon energy of 625 eV, only the composition of the first two to three layers of the surface was measured [79]. The surface without gold (top) was annealed in UHV at 750 K for 2 min to induce significant amount of O vacancies, which were distributed from the surface to the bulk. The Ti 2p spectrum for this system is well fitted by a set of two doublets with p3/2 components at 458.03 (Ti^{4+}) and 455.96 eV (Ti^{3+}). After

deposition of Au on such surface at 300 K, the features between 454 and 456 eV gain relative intensity with respect to the main feature at ~458 eV. The resulting Ti 2p spectrum needs three doublets for a good fit (center of Figure 17.4b), 458.06, 456.93, and 455.41 eV corresponding to Ti^{4+} cations, the Ti^{3+} ions weakly oxidized (Ti$^{\delta+}$) by interaction with Au and Ti^{3+} species, respectively. Final annealing of the Au/TiO$_2$(110) surface at 750 K produces a clear increase in the signal covering the 454–456 eV region due to a rise in the intensity of the Ti^{3+} and Ti$^{\delta+}$ peaks. This phenomenon can be explained as the migration of O vacancies from the bulk to the surface of the oxide, since Au adatoms can enhance the relative stability of surface vacancies and modify the rate of vacancy exchange between the bulk and the surface of the oxide. Because of the presence of O vacancies in the interface, the electronic properties of the gold nanoparticles were perturbed, making it more chemically active.

17.3.2 Sintering of Au Nanoparticles

Because of the intrinsic properties of Au, its interaction with most metal oxides is relatively weak, in most cases weaker than the Au–Au bond. This is demonstrated with temperature-programmed desorption (TPD) of Au in conjunction with theoretical calculations, where the binding energy of Au to an oxide support is shown to be much smaller than the Au–Au bond. These relative energy differences lead to facile sintering of Au nanoparticles as a function of reaction time, that is, small, highly dispersed particles eventually convert to thermodynamically preferred larger particles [19]. This drawback has restricted commercialization of supported Au nanoparticles as catalysts. Accordingly, the thermal stability of oxide-supported Au nanoparticles has been a subject of extensive studies. It is generally agreed that the sintering of titania-supported Au nanoparticles occurs mainly via the Ostwald ripening mechanism [75], first explained by Wilhelm Ostwald in 1896. This mechanism describes the energetically preferred mechanism by which large particles grow larger, drawing material from smaller particles, which shrink. This thermodynamically driven spontaneous process occurs because larger particles are more energetically favored due to their greater volume to surface area ratio. As the system minimizes it overall energy, molecules/atoms on the surface of the smaller (energetically less favorable) particles diffuse and add to the larger particles. Therefore, the smaller particles continue to shrink, whereas the larger particles continue to grow. Note that particle diffusion/coalescence, where two or more particles merge to form a larger particle, may occur and even dominate under certain reaction conditions. Campbell et al. have developed an improved kinetic model, based on the pioneering model of Wynblatt and Gjostein, for sintering of supported metal nanoparticles that can be used to more accurately predict the particle distribution as a function of reaction time [75].

Sintering of supported Au nanoparticles was found to be significantly different under reaction conditions compared with a vacuum environment. Using STM, the sintering in UHV for Au particles with a size of 2–5 nm was shown to occur above 600 K. CO exposure has no apparent effect on the morphology of the Au/TiO$_2$(110), whereas significant changes occur after exposure to O$_2$ or CO:O$_2$ at or near room temperature, as shown in Figure 17.4A and C [19]. With exposures of CO:O$_2$, the Au particle density was greatly reduced as a result of sintering. X-ray photoemission spectra (XPS) show no changes in the chemical composition of the Au/TiO$_2$(110) surface, especially the feature corresponding to reduced titanium sites, before and after CO exposure; however, the partially reduced TiO$_2$(110) surface was oxidized after CO:O$_2$ (and O$_2$ alone, not shown) exposure (see Figure 17.4D). A small shoulder at the low binding energy side of the XPS Ti $2p_{3/2}$ peak, owing to the presence of Ti^{3+} species, was completely absent after 120 min CO:O$_2$ exposure at 300 K. Since the structural and surface chemical changes on exposure to O$_2$ and CO:O$_2$ were identical and the fact that there were no detectable changes after exposure to CO, it was concluded that the Au/TiO$_2$(110) surface exhibits an exceptionally high reactivity

toward O$_2$ at 300 K that promotes the sintering of the Au nanoparticles. In reaction studies, the Au nanoparticles exhibited a very high activity toward CO oxidation; however, the surface was effectively deactivated after reaction for 120 min. This deactivation is believed to be caused by O$_2$-induced agglomeration of the Au nanoparticles as seen in Figure 17.4.

Recently, it was also shown that an oxidized TiO$_2$ surface can bind small Au nanoparticles (less than 20 atoms) stronger than on a reduced TiO$_2$ surface, owing to the stabilization of covalent bonds as well as ionic bonding at troughs [80]. This specific case is different from that of catalytic active Au nanoparticles with an optimum particle size of 2–3 nm (a few hundred atoms) shown to sinter more rapidly under oxidizing condition as discussed above [19].

17.3.3 Design and Synthesis of Sinter-Resist Oxide Supports

Because supported Au catalysts typically deactivate by sintering, considerable effort has focused on the design and synthesis of sinter-resist oxide supports, for example, dispersion of Au nanoparticles on nanosized oxide supports or restricted movement in oxide nanopores. An example of such a model system is a Ti-doped SiO$_2$ film prepared on an Mo single crystal surface [81,82]. First, a well-ordered monolayer SiO$_2$ film with a well-defined c(2 × 2) low-energy electron diffraction (LEED) pattern was prepared on a single crystal Mo surface. Ti was then deposited on the SiO$_2$ film at room temperature followed by oxidation at ~850 K and an anneal at ~1050 K. An 8% Ti-doped surface is very flat and essentially free of 3-D Ti or TiO$_x$ clusters, with isolated bright spots at the step edges and on the terraces as shown in Figure 17.5a. Ti was found to incorporate into the surface forming Ti–O–Si linkages as evidenced by high-resolution electron energy loss spectroscopy (HREELS) [81] and STM [82]. The average number density of isolated Ti atoms on the SiO$_2$ terrace from the STM images is estimated to be ~3.0 × 10^{13}/cm^2. With an increase in the Ti coverage to 17%, the surface remains flat with the formation of reduced, i.e., TiO$_x$, 3-D islands on the SiO$_2$ terraces (Figure 17.5b). Au was found to nucleate primarily at the Ti defects when deposited on the 8% Ti-doped SiO$_2$ surface (Figure 17.5a). In contrast, Au primarily decorates the extremities of the TiO$_x$ islands when deposited on the 17%

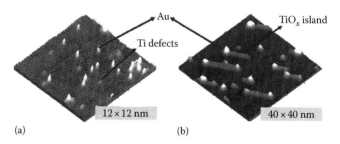

FIGURE 17.5 3-D STM images of (a) Au(0.04 ML)/TiO$_x$(8%)–SiO$_2$ and (b) Au(0.08 ML)/TiO$_x$(17%)–SiO$_2$ showing that both Ti defects and TiO$_x$ islands play a role as nucleation sites for Au nanoclusters. (From Min, B.K. et al., *J. Phys. Chem. B*, 108, 14609, 2004. With permission.)

Ti-doped SiO_2 surface (Figure 17.5b). Thermal sintering of Au particles on either of the Ti-doped surfaces was significantly inhibited compared with the corresponding sintering characteristics of Au on SiO_2. Adhesion between the Au particles and the support is clearly governed by the density of defects at the interface between the particle and the support. Indeed, recent experiments and theoretical calculations show that Au on titania actually stabilizes the presence of oxygen vacancies at the particle–support interfaces [77].

Recently, a highly ordered, reduced titania surface has been synthesized on an Mo(112) substrate [21]. The film was prepared by deposition of ~1 ML Ti onto a monolayer thickness SiO_2 film that grown on Mo(112) under UHV, following subsequent oxidation in $\sim 5 \times 10^{-8}$ Torr O_2 at 850 K, annealing at 1200 K and decomposition at 1400 K. An atomically resolved STM image, a LEED pattern, and a structural model for this reduced titania surface, designated as (8×2)-TiO_x, are shown in Figure 17.6. In this model, seven Ti atoms decorate every eight Mo atoms along the Mo(112) trough, binding to the surface via Ti–O–Mo bonds and to each other via Ti–O–Ti linkages. This structure is consistent with the atomically resolved STM image showing a pair of protruding Ti rows for each pair of troughs in the Mo(112) surface. Accordingly, the distance between the rows of each pair is somewhat smaller than the distance between each pair of rows. These unusual features are likely resulting from the fact that the row distance of the Mo substrate along the [−110] direction is 0.445 nm, much too long for Ti–Ti bonding via a Ti–O–Ti linkage. Thus, two rows of Ti displace laterally toward each other to form an effective Ti–O–Ti bond. The Ti atom density is therefore estimated to be 7/8 of that of the top layer of Mo atoms in Mo(112). This well-ordered titania film exhibits a single phonon feature at

84 meV in HREELS and is assigned to the Ti–O stretching mode, as shown in Figure 17.7a. Considering the energy loss at 95 meV for the Ti–O stretching mode for bulk TiO_2, the phonon feature at 84 meV is consistent with a reduced titania film as evident from the XPS data (Figure 17.7b), in which the Ti $2p_{3/2}$ binding energy is peaked at 455.5 eV, much lower than 458.1 eV which is observed for a TiO_2 thin film on Mo(112) [83]. Therefore, the oxidation state of Ti for this well-ordered thin film on Mo(112) was determined to be 3+.

Early studies using LEIS and STM for Au on TiO_2(110) have revealed that Au bonding on oxygen vacancy sites, that is, reduced TiO_2, is stronger than the corresponding bonding to fivefold coordinated Ti sites and bridging oxygen sites, that is, a stoichiometric surface [20,73]. On the rutile TiO_2(110) surface (see Figure 17.1b), the two Ti atoms nearest to an oxygen vacancy are reduced to Ti^{3+}, whereas for the (8×2)-TiO_x thin film, there is a full monolayer of reduced Ti^{3+} sites. Accordingly, strong binding between deposited Au and the TiO_x surface is anticipated. Indeed, on deposition of Au onto this (8×2)-TiO_x surface followed by an anneal at 900 K, Au completely wets the surface as indicated by the Au/Mo AES ratio, the corresponding ν_{CO} intensity, and as evidenced by the STM images shown in Figure 17.6 [21,84]. The optimal annealing temperature of 900 K was obtained by monitoring the Au/Ti, Au/Mo, and Ti/Mo AES ratio, and the related intensity of CO adsorption at 90 K as a function of annealing temperature for an ~1 ML Au deposited onto the TiO_x/Mo(112) surface at room temperature. 2-D Au islands were formed initially and increase in size with an increase in Au coverage. At 1 ML, large smooth terraces were imaged without the formation of 3-D particles, and a sharp (1×1) LEED pattern was apparent (inset of Figure 17.8). The two ordered structures,

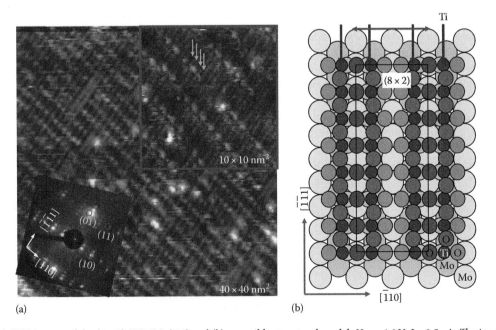

(a)   (b)

FIGURE 17.6   (a) STM images of the (8×2)-TiO_x/Mo(112) and (b) a possible structural model. U_S = +1.0 V, I = 0.5 nA. The insets show a (8×2) LEED pattern and a zoom in high-resolution STM image. Two pairs of arrows were shown to indicate the Ti rows in the proposed structural model and the protruding line seen in the STM image. (From Chen, M.S. et al., *Surf. Sci.*, 601, 632, 2007. With permission.)

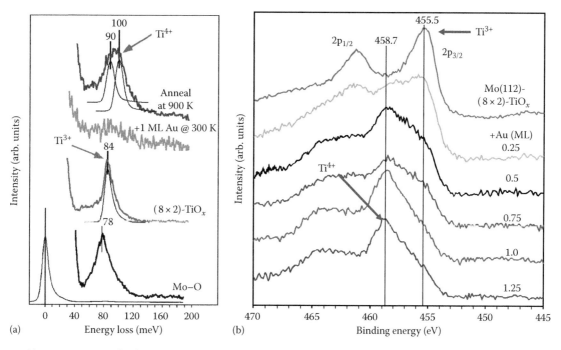

FIGURE 17.7 (a) HREELS spectra for the O/Mo(112), TiO$_x$/Mo(112), and Au/TiO$_x$/Mo(112) as indicated. (b) XPS Ti 2p spectra of Au on the TiO$_x$/Mo(112) with various coverages following an anneal at 900 K. (From Chen, M.S. et al., *Surf. Sci.*, 601, 632, 2007. With permission.)

FIGURE 17.8 STM images of various amounts of Au on (8 × 2)-TiO$_x$/Mo(112) after anneal at 900 K. U_S = +1.0 V, I = 0.5 nA. (From Chen, M.S. et al., *Surf. Sci.*, 601, 632, 2007. With permission.)

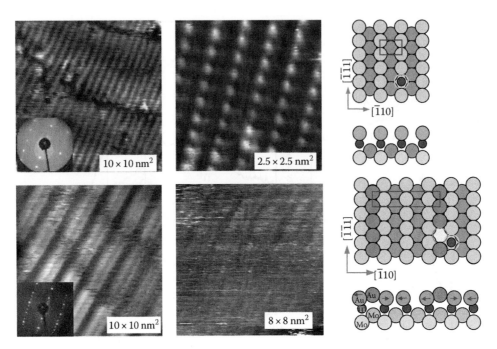

FIGURE 17.9 Atomic resolved STM images of Mo(112)-(1 × 1)-(Au, TiO$_x$) and Mo(112)-(1 × 3)-(Au, TiO$_x$) and corresponding structural models. (From Chen, M.S. et al., *Surf. Sci.*, 601, 632, 2007. With permission.)

(1 × 1) mono- and (1 × 3) bilayer, were formed at Au coverages of 1 and 1.3 ML, respectively. Figure 17.9 shows atomically resolved images of the (1 × 1) and (1 × 3) Au films. Protruding rows with spacings of ~4.5 and ~13.5 Å, corresponding to one and three times the Mo(112) surface spacing along the [110] direction, are consistent with the observed (1 × 1) and (1 × 3) LEED patterns. More detailed STM images show atomically resolved arrangements of the surface atoms, consistent with the proposed structural models of Figure 17.9.

The wetting of Au on this TiO$_x$ film is also evidenced by LEIS spectroscopy, in which the Au peak intensity increases linearly up to one monolayer, as compared with that of Au on Mo(112)

(Figure 17.10a) [84]. 2-D growth of Au up to 1 ML on the TiO$_x$/Mo(112) contrasts with the growth of Au on TiO$_2$(110) whereas 3-D clustering occurs at a coverage >0.2 ML, demonstrating that reduced titania is indeed important in binding of Au. TPD of Au from TiO$_x$ is compared with that from Mo(112) in Figure 17.10b [84]. At submonolayer Au coverages, only one desorption feature of Au from TiO$_x$ is evident with a peak desorption temperature between 1260 and 1310 K. This peak temperature is lower than the desorption maximum of ~1450 K for Au from Mo(112), thereby excluding the possibility of direct bonding between Au and substrate Mo atoms in the ordered Au/TiO$_x$/Mo(112) films. On increasing the Au coverage to greater than

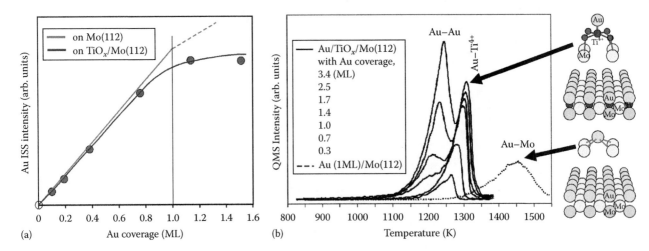

FIGURE 17.10 (a) LEIS intensity of Au as a function of Au coverage on the TiO$_x$/Mo(112) and Mo(112). (b) TPD spectra of Au on Mo(112) with coverage of ~1 ML and on TiO$_x$/Mo(112) with various Au coverages. The binding geometries of Au on the Mo(112) and TiO$_x$/Mo(112) were schematically shown for comparison. (From Chen, M.S. et al., *Surf. Sci.*, 601, 632, 2007. With permission.)

1 ML, a shoulder at ~1200 K appears, and is assigned to desorption from 3-D Au nanoparticles. These results confirm that the Au–TiO$_x$ binding energy is greater than the Au–Au binding energy, consistent with the unusual stability of these ordered Au nanofilms. The high binding energy between Au and the TiO$_x$/Mo(112) surface is the driving force for Au wetting this oxide support.

In Figure 17.7b, the XPS Ti 2p peaks are shown for Au on the TiO$_x$ film as a function of Au coverage [84]. The Ti 2p feature for the TiO$_x$ film is apparent with a Ti 2p$_{3/2}$ binding energy at ~455.7 eV. This peak position is ~2.4 eV lower than 458.1 eV observed for monolayer TiO$_2$ on Mo(112) and Mo(110), and thus is assigned to Ti^{3+} states, consistent with a single phonon feature observed at 84 meV by HREELS (Figure 17.7a). On deposition of Au, the Ti 2p$_{2/3}$ peak shifts markedly to a higher binding energy with a peak maximum at 458.8 eV, corresponding to a titanium oxidation state of 4+. Furthermore, the Ti–O phonon feature shifts from 84 to 95–100 meV, consistent with the presence of Ti^{4+} subsequent to formation of the (1 × 1) monolayer structure. An Au-induced Ti oxidation state from Ti^{3+} in (8 × 2)-TiO$_x$/Mo(112) to Ti^{4+} in (1 × 1)-Au/TiO$_x$/Mo(112) is attributed primarily to the restructuring of the TiO$_x$ film, leading to changes in the coordination geometries. This observation demonstrates the strong interaction between Au and the reduced titania surface.

17.4 Active Sites/Structure for CO Oxidation

17.4.1 Structure of the Active Sites

Catalytic activities of supported Au nanoparticles for CO oxidation are reported to be strongly dependent on the particle size, the particle shape, and the nature of the support [2,19–21,31]. From a variety of experimental observations, the corner and/or edge sites at the perimeter/contact area of the interface between the Au nanoparticles and the support are purported to serve as a unique site for reactant activation. These results imply that higher rates should result with a decrease in particle size. However, in fact, the catalytic rate decreases as the particle size decreases below 3 nm (with a thickness of two atomic layers) for supported Au nanoparticles (see Figure 17.11a) [19,31]. Schematics of the 1-D, 2-D, and 3-D Au structures with one, two, and three atomic layers in thickness are shown in the insets of Figure 17.11. Moreover, a lower catalytic rate for CO oxidation is found for the (1 × 1) monolayer compared with the rate for the (1 × 3) bilayer of Au/TiO$_x$/Mo(112) (see Figure 17.11b) [21]. These data suggest that a synergism between the first and the second layer is essential for the unique catalytic activity for supported Au nanoparticles. Note that the particle size required to achieve the best catalytic performance may be different in various systems due to different particle shapes, given that the more important factor is the requisite bilayer feature. Even with a similar number of Au atoms, the apparent size of a particle can vary substantially with particle morphology. For example, an Au$_{10}$ particle may be active if present as a 3-D structure, but inactive if present as planar 1-D or 2-D structure.

The rates computed on a per surface Au atom for CO oxidation obtained on the (1 × 3) bilayer Au nanofilm is around 45 times higher than rates reported for high-surface-area TiO$_2$-supported Au nanoparticles [21]. The rates for ordered bilayers, model nanoparticles two-atomic-layers in thickness, and the very best high-surface-area supported catalysts are compared in Figure 17.12 [31]. The blue bars of the histogram are the computed rates based on total Au. The rates obtained for the ordered bilayers are approximately one order of magnitude higher than the rates for the high-surface-area supported catalysts. As discussed above, assuming that the active site consists of a combination of the first and second layer Au atoms (see the insets of Figure 17.12), the rates, computed on a per active site basis from the corresponding particle structure, become comparable

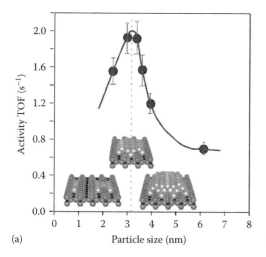

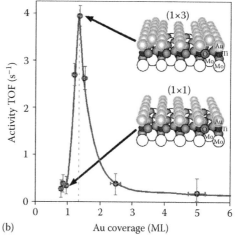

FIGURE 17.11 Catalytic activity for CO oxidation as a function of (a) particle size on the TiO$_2$(110), with CO:O$_2$ = 1:5 and a total pressure of 40 Torr, at 353 K. Schematic structural models for 1-D, 2-D, and 3-D structures with two-atomic and three-atomic layers thick Au particles on the TiO$_2$(110). (b) Au coverage on the Mo(112)-(8 × 2)-TiO$_x$, with CO:O$_2$ = 2:1 and a total pressure of 5 Torr, at room temperature. (From Chen, M.S. and Goodman, D.W., *Acc. Chem. Res.*, 39, 739, 2006. With permission.)

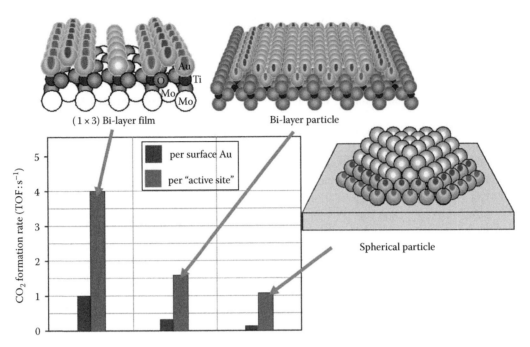

FIGURE 17.12 **(See color insert following page 9-8.)** Comparison of catalytic activities for CO oxidation on the Mo(112)–(1 × 3)–(Au, TiO$_x$), Au/TiO$_2$(110), and Au supported on high-surface-area TiO$_2$ with a mean particle size of ~3 nm. The corresponding structural models were shown with red and blue marks to indicate the active sites. (From Valden, M. et al., *Science*, 281, 1647, 1998; Chen, M.S. and Goodman, D.W., *Science*, 306, 252, 2004. With permission.)

as shown by the red bars of the Figure 17.10 histogram. It is noteworthy that on a per Au atom basis, the rates for the supported particles are decidedly lower than for the Au bilayer. This may arise due to: (1) the fact that various sizes and shapes of the particles coexist on the surface; (2) an electronic effect caused by particle contact area, particle shape, and so on; and/or (3) a steric effect in that the reactants have multidirectional access to the bilayer structure but only unidirectional access to the Au nanoparticles. The importance of the interface may well be reflected from an inverse catalyst system of TiO$_2$ nanoparticles on bulk Au(111) surface by Rodriguez et al. [85], in which metallic Au sites located nearby the oxide nanoparticles were shown to be the catalytically active sites. In a recent study by Kung's group [86], using Br ion as a probe, they also confirmed that not all CO adsorption sites on a gold nanoparticles are catalytic active sites, but that the perimeter Au atoms at/near the particle–support interface (perimeter) are active sites.

17.4.2 Nature of the Active Sites

The nature of the active sites particularly with respect to whether metallic Au or positive Au, that is, Au$^+$ and/or Au^{3+} is the active species remains controversial. Most studies, including those using model supported catalysts, have found that metallic or slightly negative Au can be attributed to their unique catalytic activities [2,19–21,31,73,75–77]. The electronic state of Au nanoparticles can be probed using CO adsorption in combination with infrared spectroscopy [31,87]. It has been shown that the v_{CO} mode shifts to lower frequency on electron-rich Au clusters

and to higher frequency on electron-deficient clusters relative to bulk Au, and that the extent of the shift can be correlated with the electronic charge on Au, as shown in Figure 17.13 [31]. The v_{CO} frequencies for CO adsorption on ordered Au monolayer and bilayer structures are compared with monolayer and multilayer Au on molybdenum single crystals, together with pertinent literature data, in Figure 17.13. For monolayer Au on reduced titania, a single v_{CO} mode corresponding to CO adsorbed on an atop Au atom is observed at ~2107 cm^{-1} for low CO exposures. For bilayer Au on reduced titania, both the first and second layer Au atoms are accessible to CO, thus a broad v_{CO} feature at 2109 cm^{-1} is observed, a feature that can be decomposed into two bands at 2107 and 2112 cm^{-1}. These two features correspond to atop adsorbed CO on first layer Au and to CO adsorbed on second layer Au. These results suggest that CO binds more tightly to first layer than to the second layer Au atoms. For CO adsorbed on multilayer Au, for example, eight monolayers of Au on single crystal molybdenum, the v_{CO} mode is found at 2124 cm^{-1}, a frequency identical to those found for CO on bulk crystal Au surfaces. On monolayer Au on molybdenum, where the Au has been shown to be negatively charged, the v_{CO} mode is at 2095 cm^{-1}. The extent of the electron transfer from the substrate molybdenum to Au was estimated to be ~0.08 electrons based on the charge transfer reported for Au/Mo(110). As displayed in Figure 17.13, v_{CO} frequencies at ~2124, 2112, 2107, and 2095 cm^{-1} are evident for low CO exposures on multilayer Au on molybdenum, for the second layer of Au in an Au bilayer on titania, for monolayer Au on titania, and for monolayer Au on Mo(112), respectively. These observed v_{CO} frequencies demonstrate that the Au films

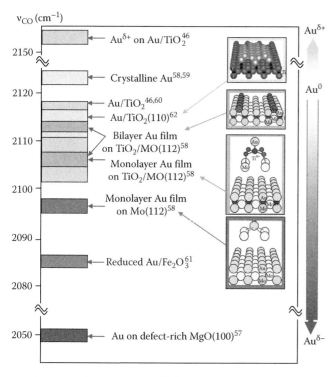

FIGURE 17.13 (See color insert following page 9-8.) Comparison of the stretching frequencies for CO adsorption on various supported Au catalysts. The indicated reference number in the figure was originated in Ref. [31]. (From Chen, M.S. and Goodman, D.W., *Acc. Chem. Res.*, 39, 739, 2006. With permission.)

on reduced titania are electron-rich, for example, $Au^{\delta-}$, and that the extent of electron transfer from the titania film to the Au is less than that for monolayer Au on Mo. Note that the surface arrangements of the atoms in monolayer Au on a single crystal Mo surface and monolayer Au on a reduced titania surface are similar, with the exception that in the former, Au binds directly to the substrate Mo, whereas in the latter, Au binds to the substrate via Ti. This sequence of ν_{CO} frequencies, or the extent of the electron-rich state of Au, is consistent with the order of the heats of adsorption for CO, that is, monolayer Au on molybdenum > monolayer Au on titania > multilayer or bulk-like Au. The electron-rich nature of Au nanoparticles is supported by theoretical calculations and ancillary experimental data. As shown in Figure 17.13, ν_{CO} frequencies are reported at 2120–2100 cm^{-1} for CO on Au particles supported on TiO_2, ~2088 cm^{-1} for Au on Fe_2O_3 (reduced FeO), and 2050 cm^{-1} on very small Au nanoparticles supported on defect-rich MgO. In contrast, CO adsorbed on positive Au exhibits a frequency ranging from 2148 to 2210 cm^{-1}.

Electron-rich Au nanoparticles are reported to adsorb O_2 more strongly and to activate the O–O bond via charge transfer from Au by forming a superoxo-like species, and also to facilitate activation of CO [20,31,75,88–90]. Furthermore molecularly chemisorbed oxygen on Au/TiO_2 at the surface was found to react directly with CO to form CO_2 without O_2 dissociation. This is consistent with electron-rich Au playing a critical role

in O–O bond activation, and the reaction pathway for CO oxidation on small Au particles proceeding via dioxygen species rather than atomic oxygen. A direct correlation has been found between the activity of Au particles for the catalytic oxidation of CO and the concentration of F-centers (defects) at the surface of an MgO support [21], implying a critical role of surface F-centers in the activation of Au in Au/MgO catalysts, see Figure 17.14 [91,92]. Moreover, an interesting evidence that negative charged Au is more active can be seen from that Au_6^- is capable of oxidizing CO at a rate 100 times greater than previously reported for model or commercial gold cluster-based catalysts [93]. Note that the gold cluster anions in here have undergone only a single reaction cycle at most, whereas a real catalyst can run over thousands of reaction cycles.

Positive Au, for example, Au^+ and Au^{3+}, have been reported to be active for CO oxidation by several groups [94,95]. Guzman and Gates using X-ray absorption near-edge structure (XANES) investigated the oxidation state of MgO-supported Au catalysts with particle sizes of ~3 nm during CO oxidation [94]. With various CO:O_2 reactant ratios, these authors found the CO_2 formation rate to increase with the concentration of Au^+. From these data, the authors proposed a critical role of Au^+ in supported Au catalysts for CO oxidation. It is noteworthy, however,

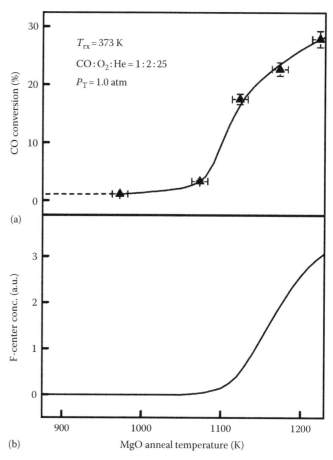

FIGURE 17.14 Catalytic activities (a) for CO oxidation over Au/MgO catalysts as a function of the density of F-center (b). (From Yan, Z. et al., *J. Am. Chem. Soc.*, 127, 1604, 2005. With permission.)

that this conclusion was based on data in which $CO:O_2$ reactant ratio was altered significantly, a change that could dramatically affect the reaction rate. For example, for CO catalytic oxidation on platinum-group metals, CO_2 formation is positive first order with respect to the O_2 partial pressure, with extremely high rates realized for oxygen-rich reaction conditions [96]. Furthermore, the observed rate of $\sim 2 \times 10^{-3}$ s^{-1} over MgO-supported Au particles by Gates et al. [94] is far lower than rates found for active metallic Au nanoparticles [31]. Hutchings et al. investigated the oxidation states of Au in α-Fe_2O_3-supported Au catalysts for CO oxidation using X-ray photoelectron, X-ray absorption, and Mössbauer effect spectroscopies [95]. These authors found that 5wt% Au/Fe_2O_3 prepared by coprecipitation of the respective hydroxides and drying at 393 K exhibits a high rate for CO oxidation, whereas a catalyst calcined at 673 K in air exhibits a relatively low rate. The former catalyst was determined to contain mainly cationic Au with a mean particle size from 3.8 to 7nm, whereas the latter catalyst contained primarily metallic Au with a mean particle size of 8.2nm. Hutchings et al. concluded that cationic Au may assist metallic Au in catalytic CO oxidation, but these authors could not conclude that metallic Au is inactive or that cationic Au is present in active Au catalysts. It has been demonstrated that the activity of supported Au catalysts strongly depends on the particle size with best performance achieved at $\sim$3 nm. The activity decreases dramatically with an increase in particle size as shown by Haruta et al. and Goodman et al. [2,19–21,31]. A metallic Au particle of 8.2nm intrinsically has a very low catalytic activity. It is noteworthy that in the Hutchings's experiments [95], on a 50 mg 5%wt Au/Fe_2O_3, a CO flow rate of 0.5 mL/min, 100% conversion corresponds to a TOF of $\sim 2 \times 10^{-3}$ (per Au site per second), a rate far lower than 0.1–4 observed for metallic Au catalysts [2,19–21,31].

Very recently, using similar techniques, Gates and Corma [78] investigated FeOx-doped Au/TiO_2 (Degussa P25) for CO oxidation using XANES and EXAFS. The transient XANES data recorded in flowing CO + O_2 show that cationic Au was rapidly reduced completely to zero-valent Au under CO oxidation conditions. These results provide strong evidence for the presence of zero-valent Au rather than cationic Au in the working catalysts. Note a specific rate of $\sim$0.1 CO_2 molecule per Au site per second was achieved in this system, a rate comparable to the rates reported for other oxide-supported Au catalysts [2,20,31]. In fact, Friend's group [97] has shown that CO can react with atomic oxygen on an Au surface at 70 K, consistent with Au$^+$ and Au^{3+} being strong oxidizers that could potentially oxidize CO, that is, be reduced by CO, at very low temperatures. Schwartz et al. [98] also showed that a higher catalytic rate correlates with fully reduced Au, and that after reduction, no reoxidation was observed under CO oxidation conditions, even in air at temperatures as high as 300°C. Using X-ray adsorption spectroscopy combined with XPS and FTIR, Kung et al. [99] concluded that for an Au/TiO_2 catalyst, metallic Au is necessary for high CO catalytic oxidation rates. With XANES and ^{13}C isotopic transient analysis, Davis et al. [100] also concluded that active Au/TiO_2 and Au/Al_2O_3 catalysts contain predominately metallic Au.

Altogether, this body of data shows that although cationic Au may be active for CO catalytic oxidation, its activity is decidedly lower than that of metallic Au nanoparticles.

17.4.3 Reaction Pathway

The activation and reaction of CO and O_2 molecules on supported Au nanoparticles is highly important in understanding the unique catalytic properties for gold catalysts. The key question is whether the support is involved directly in activating reactant molecules. It is generally agreed that the support plays an important role in stabilizing and defining the morphologies and electronic properties of Au nanoparticles [20,31]. The perimeter/contact area of the interface between the Au particles and the support has been proposed to serve as a unique reaction site where reactants are activated. From theoretical calculations, the support has been shown to play an active role in the bonding and activation of adsorbates bound to Au [88]. Molina and Hammer have proposed the active site to be low-coordinated Au atoms in combination with surface cations interacting simultaneously with an adsorbate [101].

The Au–TiO_2 interface and Au itself have been proposed for the activation of the O_2 molecule. Model catalytic studies for Au on $TiO_2(110)$ and $TiO_x/Mo(112)$ demonstrated that a bilayer structure is critical to the unique catalytic properties of Au for CO oxidation [19,21,31]. Under UHV conditions, the active Au nanoparticles and/or nanofilms were found to be electron-rich, that is, negatively charged. Furthermore, in the ordered Au monolayer and bilayer structures described above, the $Ti^{\delta+}$ of the support titania is not assessable to the reactants, since each surface Ti site binds directly to Au atoms located at the topmost surface [21]. The exceptionally high catalytic activities for CO oxidation observed on ordered bilayer Au thus strongly suggest an Au-only CO oxidation pathway, that is, that the oxide support itself may not need to directly involve the activation/reaction of reactant molecules. This reaction pathway is confirmed by DFT calculations that show the reaction sequence for CO oxidation for Au-only surface sites on a TiO_2-supported Au nanoparticle to have a similar activation energy (0.36–0.40 eV) as that involving the support [102]. Moreover, molecularly chemisorbed oxygen on Au/TiO_2 is found to be stable at the surface [89] and to react directly with CO to form CO_2 without requiring the dissociation of O_2 [90].

Using *in situ* XANES, van Bokhoven et al. [103] found clear evidence for charge transfer from small Au particles to oxygen along with partial depletion of the Au d band on exposing to oxygen. This leads to partially oxidized Au particles that can be reduced quickly by CO to CO_2. Under CO oxidation reaction conditions, CO is dominant on the surface, consistent with O_2 being activated on Au particles rather than the support. The results also imply that partially oxidized Au is present as a short-lived species under catalytic conditions. This also suggests that the rate-limiting step for CO oxidation on supported Au catalysts is the activation of O_2. The first layer Au on $TiO_2(110)$ surface was found to bind oxygen 40% more strongly than do Au particles using TPD [75].

17.5 Origins of the Unique Activities for Gold Nanoparticles

The unique catalytic properties of supported Au nanoparticles have motivated extensive experimental and theoretical studies with the aim of elucidating the origin of the special activity. Unfortunately, the data in the literature and the discussion vary widely; therefore, the nature of the active Au species/structure/ site remains unclear. In general, the origins of the catalytic activity of Au have been proposed to originate from one or more of three contributions: (1) presence of low-coordinated Au sites, (2) charge transfer between the support and Au, and/or (3) quantum size effects.

17.5.1 Low-Coordinated Au Sites

The catalytically active gold nanoparticles are typically less than 8 nm in diameter. Because the density of corner, edge, and/or surface Au atoms related to the number of total Au atoms in a particle increase with decreasing particle size, coordinatively unsaturated Au atoms at the corner, edge, and at the particle surface have been proposed to constitute the active sites [104–107]. Indeed, the binding of CO and O_2 are demonstratively higher at the corner or edge sites than on the terrace or on the smooth surface [104–106]. Moreover, DFT calculations show that oxygen and carbon monoxide can be only adsorbed on gold atoms with a coordination number less than 8 at certain condition [107]. This was confirmed by CO adsorption on Au/TiO_2, Au/ZrO_2, and Au/FeO using infrared spectroscopy [108,109]. The CO oxidation involves adsorption of CO and O_2; the presence of low-coordinated Au atoms was, therefore, regarded to be a key factor for the catalytic activity of Au nanoparticles [104]. However, a lower catalytic activity is found for monolayer Au (coordination number of 3–4) compared with bilayer Au [coordination number of ~6 (see Figure 17.11)] [21]. Furthermore, for supported Au nanoparticles, the catalytic rate decreases as the particle size decreases below 3 nm, although the relative densities of the corner and/or edge sites increase continuously with decreasing particle size (Figure 17.11) [19]. Therefore, low coordinated corner and/or edge sites are essential, but alone are not the single factor on which catalytic activity is dependent.

17.5.2 Charge Transfer between the Support and Au Nanoparticles

The interaction between Au and the support alters the electronic structure of Au nanoparticles and promotes their catalytic activities for low-temperature CO oxidation. In particular, defects on the oxide support are thought to play a key role in anchoring the Au particles and in transferring electronic charge to Au, with these two effects in combination contributing to the special catalytic activity [19–21,73,76,77,104,110–113]. The electronic state of Au nanoparticles can be probed using CO adsorption in combination with infrared spectroscopy, as discussed in Section 17.4.2. The ν_{CO} mode shifts to lower frequency on electron-rich Au clusters and to higher frequency on electron-deficient clusters relative to bulk Au, and that the extent of the shift can be correlated with the electronic charge on Au [87,114]. The ν_{CO} frequencies for CO adsorption on ordered Au monolayer and bilayer structures, together with pertinent literature data, are shown in Figure 17.13. For monolayer Au on reduced titania, a single ν_{CO} mode corresponding to CO adsorbed on an atop Au atom is observed at ~2107 cm^{-1} for low CO exposures, which is higher than that of 2096 cm^{-1} for Au/Mo(112). Regarding the similar arrangements of surface gold atoms in both $Au/TiO_x/Mo(112)$ and Au/Mo(112), Au atoms on TiO_x film are less negatively charged than 0.08 e for Au on Mo surface [115]. On the bilayer Au on reduced titania, two features corresponding to atop adsorbed CO on first layer Au and to CO adsorbed on second layer Au were observed. These results suggest that CO binds more tightly to first layer than to the second layer Au atoms. For CO adsorbed on multilayer Au, for example, eight monolayers of Au on single crystal Mo, the ν_{CO} mode is found at 2124 cm^{-1}, a frequency identical to those found for CO on single crystal Au surfaces [116]. On a negatively charged Au film [115], that is, monolayer Au on Mo, the ν_{CO} mode is found at 2095 cm^{-1}. As displayed in Figure 17.13, the νCO frequencies occur at ~2124, 2112, 2107, and 2095 cm^{-1} for low CO exposures on multilayer Au on Mo, the second layer of Au in an Au bilayer on titania, monolayer Au on titania, and monolayer Au on Mo, respectively. These observed ν_{CO} frequencies demonstrate that the Au films on reduced titania are electron-rich, for example, Au$^{\delta-}$, and that the extent of electron transfer from the substrate to the Au is less than that for monolayer Au on Mo. This sequence of ν_{CO} frequencies, or the extent of the electron-rich charge state of Au, is consistent with the order of the heats of adsorption for CO being: monolayer Au on Mo > monolayer Au on titania > multilayer or bulk-like Au [87].

The electron-rich nature of Au nanoparticles is supported by theoretical calculations [105,106] and ancillary experimental data [19,73,97,117]. Electron transfer to Au nanoparticles has been probed by laser excitation of TiO_2 nanoparticles coated with Au nanoparticles [118], by photoemission spectroscopy and STM [119]. As shown in Figure 17.13, ν_{CO} frequencies are reported at 2120–2100 cm^{-1} for CO on Au particles supported on TiO_2 [45,120], ~2088 cm^{-1} for Au on Fe_2O_3 (reduced FeO) [121], and 2050 cm^{-1} for very small Au particles supported on defect-rich MgO [114]. Electron-rich Au nanoparticles are reported to adsorb O_2 more strongly, to activate the O–O bond via charge transfer from Au by forming a superoxo-like species [122], and to facilitate activation of CO [104–106]. Furthermore, molecularly chemisorbed oxygen on Au/TiO_2 is found to be stable at the surface [89,122] and to react directly with CO to form CO_2 without O_2 dissociation [90]. This is consistent with electron-rich Au playing a critical role in O–O bond activation, and the reaction pathway for CO oxidation on small Au particles proceeding via a dioxygen species rather than atomic oxygen. A direct correlation has been found between the activity of Au particles for the catalytic oxidation of CO and the concentration of F-centers (defects) at the surface of an MgO support, implying a critical role of surface F-centers in the activation of Au in Au/MgO catalysts [97].

17.5.3 Quantum Size Effects

The size of Au nanoparticles has been shown to affect the electronic properties decisively [19,123,124]. Pronounced energy gap has been revealed by photoemission spectroscopy and DFT for a 20-atom Au cluster, which possesses a tetrahedral structure with atomic packing similar to that of bulk gold but with very different properties [125]. A metal-to-insulator transition is observed as the size of Au nanoparticles decreases below 3 nm by measuring the tunneling current as a function of the bias voltage (I–V) [19] and by measuring the local barrier height [124]. This behavior has also been observed for $Pd/TiO_2(110)$ [126], for Ag particles grown in nanopits on a graphite surface [127], and for Ag particles supported on $Al_2O_3/NiAl(110)$ [128]. The valence band structures of (1×1) monolayer and (1×3) bilayer Au films on reduced titania are significantly different from those of bulk Au [87], for example, that the electronic properties of the ordered Au films being quite different compared with bulk gold. These size dependences of the electronic and catalytic properties suggest that the pronounced structure sensitivity of CO oxidation on Au/TiO_2 relates to limited size or quantum size effects.

Quantum size effects have been invoked to account for the unique properties of nanometer-scale metallic particles relative to the bulk [129–132]. Ligand-stabilized metal nanoparticles in the size range of 1–2 nm were found to exhibit well-pronounced quantum size behavior [133]. Ag nanoparticles have been shown theoretically to exhibit three novel, size-dependent vibrational features compared with the bulk [134], and experimentally to reveal a series of equidistant resonances near the Fermi level with a decreasing energy separation with increasing cluster size [128]. Au nanoparticles less than 4 nm were found to exhibit significant quantum size effects with respect to the electronic configuration and the vibrational modes using Au-197 Mossbauer spectroscopy [135].

Quantum size effects have also been found to influence the thermodynamic properties of metallic nanoparticles [136], properties of superconductors [137], and chemisorptive properties of nanosized materials [104–106,115,138–141]. Charge transfer for CO adsorption depends critically on the size of an Na quantum dot [140]. Particle size effects were found for CO adsorption on Au deposits supported on FeO(111) grown on Pt(111) [139], and for CO and oxygen [139] on Au/TiO_2. Pronounced thickness-dependent variations in the oxidation rate induced by quantum-well states were observed for ordered films up to 15 atomic layers in thickness [142]. All these studies indicate significant quantum size effects on the electronic, chemical, and catalytic properties of metal nanoparticles.

17.6 Conclusions

Supported gold nanoparticles have been found to have many potential applications including CO oxidation, selective oxidation, hydrogenation, and WGS reaction. Position-selective oxidation or hydrogenation over gold nanoparticles exhibits remarkably higher selectivity compared with other transition metals. The strong dependence of the particle sizes for supported Au nanoparticles, particularly the lose of catalytic activity with an increase in particle size above 6–10 nm, requires extreme stability under reaction conditions for commercialization. Well-ordered Au bilayer nanofilms on a reduced titania thin film exhibit catalytic activity comparable with the most active Au nanoparticles, that is, the thickness of the particle rather than the particle diameter is the critical structural feature with respect to catalytic activity. These studies have shown that a bilayer Au structure is a critical feature for catalytically active Au nanoparticles, along with low-coordinated Au sites, support-to-particle charge transfer effects, and quantum size effects. The strong binding between Au and a reduced titania surface results in complete wetting by Au nanofilms of an oxide surface. The so-formed Au bilayer nanofilm (two atomic layers in thickness) exhibits unusual high activity for CO oxidation and increases the active site density by one to three orders of magnitude compared with typical high-surface-area supported Au catalysts. Cationic Au may be an active catalytic species; however, the activities of these sites are lower than that of metallic Au nanoparticles. Future studies should focus on the details of the sintering mechanism and the design/synthesis of functional oxide supports to enhance the gold-support interaction and to retard particle sintering.

Acknowledgment

This work is supported by National Natural Science Foundation of China (20873109), Major Project of Chinese Ministry of Education (309019) and Natural Science Foundation of Fujian Province, China (2008 J0168).

References

1. G. C. Bond, *Gold Bull.* 1972, 5, 11–13.
2. M. Haruta, T. Kobayashi, H. Sano, and N. Yamada, *Chem. Lett.* 1987, 16, 405–408.
3. T. Hayashi, K. Tanaka, and M. Haruta, *J. Catal.* 1998, 178, 566–575.
4. G. J. Hutchings, *J. Catal.* 1985, 96, 292–295.
5. K. C. Nicolaou, P. G. Bulger, and D. Sarlah, *Angew. Chem. Int. Ed.* 2005, 44, 4442.
6. K. C. Nicolaou, P. G. Bulger, and D. Sarlah, *Angew. Chem. Int. Ed.* 2005, 44, 4490.
7. A. S. K. Hashmi and G. J. Hutchings, *Angew. Chem. Int. Ed.* 2006, 45, 7896.
8. D. J. Gorin and F. D. Toste, *Nature* 2007, 446, 395.
9. E. Jiménez-Núñez and A. M. Echavarren, *Chem. Commun.* 2007, 333.
10. A. S. K. Hashmi, *Chem. Rev.* 2007, 107, 3180.
11. A. S. K. Hashmi and M. Rudolph, *Chem. Soc. Rev.* 2008, 37(9), 1766–1775.
12. J. H. Teles, S. Brode, and M. Chabanas, *Angew. Chem. Int. Ed.* 1998, 37, 1415.
13. Y. Ito, M. Sawamura, and T. Hayashi, *J. Am. Chem. Soc.* 1986, 108, 6405–6406.

14. Y. Fukuda and K. Utimoto, *J. Org. Chem.* 1991, 56, 3729–3731.

15. L. Prati and M. Rossi, *J. Catal.* 1998, 176, 552–560.

16. A. S. K. Hashmi, L. Schwarz, J.-H. Choi, and T. M. Frost, *Angew. Chem. Int. Ed.* 2000, 39, 2285–2288.

17. C.-G. Yang and C. He, *J. Am. Chem. Soc.* 2005, 127, 6966–6967.

18. A. K. Sinha, S. Seelan, S. Tsubota, and M. Haruta, *Top. Catal.* 2004, 29(3–4), 95–102.

19. M. Valden, X. Lai, and D. W. Goodman, *Science* 1998, 281, 1647.

20. M. S. Chen and D. W. Goodman, *Catal. Today* 2006, 111, 22.

21. M. S. Chen and D. W. Goodman, *Science* 2004, 306, 252.

22. X. Lai, T. P. St Clair, M. Valden, and D. W. Goodman, *Prog. Surf. Sci.* 1998, 59(1–4), 25–52.

23. S. Golunski, R. Rajaram, N. Hodge, G. J. Hutchings, and C. J. Kiely, *Catal. Today* 2002, 72, 107–113.

24. M. J. Kahlich, H. A. Gasteiger, and R. J. Behm, *J. Catal.* 1999, 182, 430–440.

25. B. Qiao and Y. Deng, *Chem. Commun.* 2003, 2192–2193.

26. G. Avgouropoulos, T. Ioannides, C. Papadopoulou, J. Batista, S. Hocevar, and H. K. Matralis, *Catal. Today* 2002, 75, 157–167.

27. P. Landon, J. Ferguson, B. E. Solsona, T. Garcia, S. Al-Sayari, A. F. Carley, A. A. Herzing et al., *J. Mater. Chem.* 2006, 16, 199–208.

28. F. Moreau, G. C. Bond, and A. O. Taylor, *J. Catal.* 2005, 231, 105–114.

29. R. Zanella, S. Giorgio, C. R. Henry, and C. Louis, *J. Phys. Chem. B* 2002, 106, 7634–7642.

30. W.-C. Li, M. Comotti, and F. Schueth, *J. Catal.* 2006, 237, 190–196.

31. M. S. Chen and D. W. Goodman, *Acc. Chem. Res.* 2006, 39, 739.

32. J. Guzman, S. Carrettin, and A. Corma, *J. Am. Chem. Soc.* 2005, 127, 3286–3287.

33. H. Sakurai, T. Akita, S. Tsubota, M. Kiuchi, and M. Haruta, *Appl. Catal. A* 2005, 291, 179–187.

34. M. S. Chen and D. W. Goodman, *Chem. Soc. Rev.* 2008, 37, 1860–1870.

35. W. Yan, B. Chen, S. M. Mahurin, V. Schwartz, D. R. Mullins, A. R. Lupini, S. J. Pennycook, S. Dai, and S. H. Overbury, *J. Phys. Chem. B* 2005, 109, 10676–10685.

36. Z. Konya, V. F. Puntes, I. Kiricsi, J. Zhu, J. W. Ager III, M. K. Ko, H. Frei, P. Alivisatos, and G. A. Somorjai, *Chem. Mater.* 2003, 15, 1242–1248.

37. K.-J. Chao, M.-H. Cheng, Y.-F. Ho, and P.-H. Liu, *Catal. Today* 2004, 97, 49–53.

38. C. Aprile, A. Abad, H. Garcia, and A. Corma, *J. Mater. Chem.* 2005, 15, 4408–4413.

39. M. Haruta, *J. New Mater. Electrochem. Syst.* 2004, 7, 163–172.

40. D. Cameron, R. Holliday, and D. Thompson, *J. Power Sources* 2003, 118, 298–303.

41. M. M. Maye, J. Luo, L. Han, N. N. Kariuki, and C.-J. Zhong, *Gold Bull.* 2003, 36, 75–83.

42. J. Zhang, Y. Wang, B. Chen, C. Li, D. Wu, and X. Wang, *Energy Convers. Manage.* 2003, 44, 1805–1815.

43. R. M. T. Sanchez, A. Ueda, K. Tanaka, and M. Haruta, *J. Catal.* 1997, 168, 125–127.

44. B. Schumacher, Y. Denkwitz, V. Plzak, M. Kinne, and R. J. Behm, *J. Catal.* 2004, 224, 449–462.

45. J. T. Calla and R. J. Davis, *Ind. Eng. Chem. Res.* 2005, 44, 5403–5410.

46. G. C. Bond, P. A. Sermon, G. Webb, D. A. Buchanan, and P. B. Wells, *J. Chem. Soc., Chem. Commun.* 1973, 444b.

47. P. A. Sermon, G. C. Bond, and P. B. Wells, *J. Chem. Soc. Faraday Trans.* 1979, 1, 75, 385–394.

48. A. Saito, and M. Tanimoto, *J. Chem. Soc. Faraday Trans.* 1988, 1, 84, 4115–4124.

49. P. Claus, *Appl. Catal. A* 2005, 291, 222–229.

50. P. Claus, A. Brueckner, C. Mohr, and H. Hofmeister, *J. Am. Chem. Soc.* 2000, 122, 11430–11439.

51. J. E. Bailie and G. J. Hutchings, *Chem. Commun.* 1999, 2151–2152.

52. N. Schumacher, A. Boisen, S. Dahl, A. A. Gokhale, S. Kandoi, L. C. Grabow, J. A. Dumesic, M. Mavrikakis, and I. Chorkendorff, *J. Catal.* 2005, 229, 265–275.

53. D. Andreeva, V. Idakiev, T. Tabakova, and A. Andreev, *J. Catal.* 1996, 158, 354–355.

54. A. Venugopal, J. Aluha, D. Mogano, and M. S. Scurrell, *Appl. Catal. A* 2003, 245, 149–158.

55. J. Hua, Q. Zheng, Y. Zheng, K. Wei, and X. Lin, *Catal. Lett.* 2005, 102, 99–108.

56. H. Sakurai, A. Ueda, T. Kobayashi, and M. Haruta, *Chem. Commun.* 1997, 271–272.

57. V. Idakiev, T. Tabakova, Z. Y. Yuan, and B. L. Su, *Appl. Catal. A* 2004, 270, 135–141.

58. Q. Fu, A. Weber, and M. Flytzani-Stephanopoulos, *Catal. Lett.* 2001, 77, 87–95.

59. C. H. Kim and L. T. Thompson, *J. Catal.* 2005, 230, 66–74.

60. G. Boxhoorn, Shell Internationale Research Maatschappij B.V., the Netherlands, EP 255975, 1988, p. 8.

61. T. A. Nijhuis, T. Visser, and B. M. Weckhuysen, *J. Phys. Chem. B* 2005, 109, 19309–19319.

62. N. Yap, R. P. Andres, and W. N. Delgass, *J. Catal.* 2004, 226, 156–170.

63. A. Zwijnenburg, M. Makkee, and J. A. Moulijn, *Appl. Catal. A* 2004, 270, 49–56.

64. M. D. Hughes, Y.-J. Xu, P. Jenkins, P. McMorn, P. Landon, D. I. Enache, A. F. Carley et al., *Nature* 2005, 437, 1132–1135.

65. S. Biella, G. L. Castiglioni, C. Fumagalli, L. Prati, and M. Rossi, *Catal. Today* 2002, 72, 43–49.

66. A. Abad, P. Concepción, A. Corma, and H. García, *Angew. Chem. Int. Ed.* 2005, 44, 4066–4069.

67. R. Zhao, D. Ji, G. Lu, G. Qian, L. Yan, X. Wang, and J. Suo, *Chem. Commun.* 2004, 904–905.

68. Y.-J. Xu, P. Landon, D. Enache, A. F. Carley, M. W. Roberts, and G. J. Hutchings, *Catal. Lett.* 2005, 101, 175–179.

69. H. T. Hess, I. Kroschwitz, and M. Howe-Grant, *Kirk-Othmer Encyclopedia of Chemical Engineering*, Vol. 13, Wiley, New York, 1995, p. 961.

70. P. Landon, P. J. Collier, A. F. Carley, D. Chadwick, A. J. Papworth, A. Burrows, C. J. Kiely, and G. J. Hutchings, *Phys. Chem. Chem. Phys.* 2003, 5, 1917–1923.

71. Y. F. Han and J. H. Lunsford, *J. Catal.* 2005, 230(2), 313–316.

72. T. Ishihara, Y. Ohura, S. Yoshida, Y. Hata, H. Nishiguchi, and Y. Takita, *Appl. Catal. A* 2005, 291, 215–221.

73. E. Wahlstrom, N. Lopez, R. Schaub, P. Thostrup, A. Ronnau, C. Africh, E. Laegsgaard, J. K. Norskov, and F. Besenbacher, *Phys. Rev. Lett.* 2003, 90, 026101.

74. M. A. Henderson, W. S. Epling, C. L. Perkins, C. H. F. Peden, and U. Diebold, *J. Phys. Chem. B* 1999, 103, 5328.

75. S. C. Parker and C. T. Campbell, *Top. Catal.* 2007, 44, 3.

76. F. Cosandey and T. E. Madey, *Surf. Rev. Lett.* 2001, 8, 73.

77. J. A. Rodriguez, G. Liu, T. Jirsak, J. Hrbek, Z. P. Chang, J. Dvorak, and A. Maiti, *J. Am. Chem. Soc.* 2002, 124, 5242.

78. S. Carrettin, Y. Hao, V. Aguilar-Guerrero, B. C. Gates, S. Trasobares, J. J. Calvino, and A. Corma, *Chem. Eur. J.* 2007, 13, 7771.

79. V. E. Henrich and P. A. Cox, *The Surface Science of Metal Oxides*, Cambridge University Press: Cambridge, U.K., 1994.

80. D. Matthey, J. G. Wang, S. Wendt, J. Matthiesen, R. Schaub, E. Laegsgaard, B. Hammer, and F. Besenbacher, *Science* 2007, 315, 1692.

81. M. S. Chen and D. W. Goodman, *Surf. Sci.* 2005, 574, 259–268.

82. B. K. Min, W. T. Wallace, and D. W. Goodman, *J. Phys. Chem. B* 2004, 108, 14609–14615.

83. D. Kumar, M. S. Chen, and D. W. Goodman, *Thin Solid Films* 2006, 515(4), 1475–1479.

84. M. S. Chen, K. Luo, D. Kumar, C. W. Yi, and D. W. Goodman, *Surf. Sci.* 2007, 601, 632.

85. J. A. Rodriguez, S. Ma, P. Liu, J. Hrbek, J. Evans, and M. Pérez, *Science* 2007, 318, 1757.

86. S. M. Oxford, J. D. Henao, J. H. Yang, M. C. Kung, and H. H. Kung, *Appl. Catal. A* 2008, 339, 180–186.

87. M. Chen, Y. Cai, Z. Yan, and D. W. Goodman, *J. Am. Chem. Soc.* 2006, 128, 6341.

88. X. Q. Gong, A. Selloni, O. Dulub, P. Jacobson, and U. Diebold, *J. Am. Chem. Soc.* 2008, 130(1), 370–381.

89. D. Stolcic, M. Fischer, G. Gantefor, Y. D. Kim, Q. Sun, and P. Jena, *J. Am. Chem. Soc.* 2003, 125, 2848.

90. J. D. Stiehl, T. S. Kim, S. M. McClure, and C. B. Mullins, *J. Am. Chem. Soc.* 2004, 126, 13574.

91. Z. Yan, S. Chinta, A. A. Mohamed, J. P. Fackler, Jr., and D. W. Goodman, *J. Am. Chem. Soc.* 2005, 127, 1604.

92. M. Sterrer, M. Yulikov, E. Fischbach, M. Heyde, H. P. Rust, G. Pacchioni, T. Risse, and H. J. Freund, *Angew. Chem. Int. Ed.* 2006, 45, 2630.

93. W. T. Wallace and R. L. Whetten, *J. Am. Chem. Soc.* 2002, 124, 7499.

94. J. Guzman and B. C. Gates, *J. Am. Chem. Soc.* 2004, 126, 2672.

95. G. J. Hutchings, M. S. Hall, A. F. Carley, P. London, B. E. Solsona, C. J. Kiely et al., *J. Catal.* 2006, 242, 71.

96. M. S. Chen, Y. Cai, Z. Yan, K. K. Gath, S. Axnanda, and D. W. Goodman, *Surf. Sci.* 2007, 601, 5326–5331.

97. B. K. Min, A. R. Alemozafar, D. Pinnaduwage, X. Deng, and C. M. Friend, *J. Phys. Chem. B* 2006, 110, 19833.

98. V. Schwartz, D. R. Mullins, W. Yan, B. Chen, S. Dai, and S. H. Overbury, *J. Phys. Chem. B* 2004, 108, 15782.

99. J. H. Yang, J. D. Henao, M. C. Raphulu, Y. Wang, T. Caputo, A. J. Groszek, M. C. Kung et al., *J. Phys. Chem. B* 2005, 109, 10319.

100. J. T. Calla, M. T. Bore, A. K. Datye, and R. J. Davis, *J. Catal.* 2006, 238, 458.

101. L. M. Molina and B. Hammer, *Phys. Rev. Lett.* 2003, 90, 206102.

102. I. N. Remediakis, N. Lopez, and J. K. Norskov, *Angew. Chem. Int. Ed.* 2005, 44, 1824.

103. J. A. van Bokhoven, C. Louis, J. Miller, M. Tromp, O. V. Safonova, and P. Glatzel, *Angew. Chem. Int. Ed.* 2006, 45, 4651.

104. N. Lopez, T. V. W. Janssens, B. S. Clausen, Y. Xu, M. Mavrikakis, T. Bligaard, and J. K. Norskov, *J. Catal.* 2004, 223, 232–235.

105. N. Lopez, J. K. Norskov, T. V. W. Janssens, A. Carlsson, A. Puig-Molina, B. S. Clausen, and J. D. Grunwaldt, *J. Catal.* 2004, 225, 86–94.

106. G. Mills, M. S. Gordon, and H. Metiu, *J. Chem. Phys.* 2003, 118, 4198–4205.

107. N. Lopez and J. K. Nørskov, *J. Am. Chem. Soc.* 2002, 124, 11262.

108. J. D. Grunwaldt, M. Maciejewski, O. S. Becker, P. Fabrizioli, and A. Baiker, *J. Catal.* 1999, 186, 458–469.

109. C. Lemire, R. Meyer, S. Shaikhutdinov, and H. J. Freund, *Angew. Chem. Int. Ed.* 2004, 43, 118–121.

110. S. C. Parker, A. W. Grant, V. A. Bondzie, and C. T. Campbell, *Surf. Sci.* 1999, 441, 10–20.

111. C. T. Campbell, S. C. Parker, and D. E. Starr, *Science* 2002, 298, 811–814.

112. Y. Wang and G. S. Hwang, *Surf. Sci.* 2003, 542, 72–80.

113. F. Cosandey, L. Zhang, and T. E. Madey, *Surf. Sci.* 2001, 474, 1–13.

114. B. Yoon, H. Häkkinen, U. Landman, A. S. Wörz, J. M. Antonietti, S. Abbet, K. Judai, and U. Heiz, *Science* 2005, 307, 403–407.

115. J. A. Rodriguez and M. Kuhn, *Surf. Sci.* 1995, 330, L657–L664.

116. D. C. Meier, V. Bukhtiyarov, and D. W. Goodman, *J. Phys. Chem. B* 2003, 126, 12668–12671.

117. H. J. Freund and G. Pacchioni, *Chem. Soc. Rev.* 2008, 37(10), 2224–2242.

118. V. Subramanian, E. E. Wolf, and P. V. Kamat, *J. Am. Chem. Soc.* 2004, 126, 4943–4950.

119. T. Minato, T. Susaki, S. Shiraki, H. S. Kato, M. Kawai, and K. I. Aika, *Surf. Sci.* 2004, 566, 1012–1017.

120. T. V. Choudhary, C. Sivadinarayana, C. C. Chusuei, A. K. Datye, J. P. Fackler Jr., and D. W. Goodman, *J. Catal.* 2002, 207, 247–255.

121. S. T. Daniells, A. R. Overweg, M. Makkee, and J. A. Moulijin, *J. Catal.* 2005, 230, 52–65.

122. B. Yoon, H. Hakkinen, and U. Landman, *J. Phys. Chem. A* 2003, 107, 4066–4071.

123. M. C. Daniel and D. Astruc, *Chem. Rev.* 2004, 104, 293–346.

124. Y. Maeda, M. Okumura, S. Tsubota, M. Kohyama, and M. Haruta, *Appl. Surf. Sci.* 2004, 222, 409–414.

125. J. Li, H. J. Zhai, and L. S. Wang, *Science* 2003, 299, 864–867.

126. C. Xu, X. Lai, G. W. Zajac, and D. W. Goodman, *Phys. Rev. B* 1997, 56, 13464–13482.

127. H. Hovel, B. Grimm, M. Bodecker, K. Fieger, and B. Reihl, *Surf. Sci.* 2000, 463, L603–L608.

128. N. Nilius, M. Kulawik, H. P. Rust, and H. J. Freund, *Surf. Sci.* 2004, 572, 347–354.

129. J. A. A. J. Perenboom, P. Wyder, and F. Meier, *Phys. Rep.* 1981, 78, 173–292.

130. W. P. Halperin, *Rev. Mod. Phys.* 1986, 58, 533–606.

131. F. M. Mulder, T. A. Stegink, R. C. Thiel, L. J. Dejongh, and G. Schmid, *Nature* 1994, 367, 716–718.

132. S. W. Chen, *J. Electroanal. Chem.* 2004, 574, 153–165.

133. G. Schmid, *Adv. Eng. Mater.* 2001, 3, 737–743.

134. A. Kara and T. S. Rahman, *Phys. Rev. Lett.* 1998, 81, 1453–1456.

135. P. M. Paulus, A. Goossens, R. C. Thiel, A. M. van der Kraan, G. Schmid, and L. J. de Jongh, *Phys. Rev. B* 2001, 64, 205418-1–205418-18.

136. Y. Volokitin, J. Sinzig, L. J. de Jongh, G. Schmid, M. N. Vargaftik, and I. I. Moiseev, *Nature* 1996, 384, 621–623.

137. Y. Guo, Y. F. Zhang, X. Y. Bao, T. Z. Han, Z. Tang, L. X. Zhang, W. G. Zhu et al., *Science* 2004, 306, 1915–1917.

138. V. A. Bondzie, S. C. Parker, and C. T. Campbell, *Catal. Lett.* 1999, 63, 143–151.

139. C. Lemire, R. Meyer, S. K. Shaikhutdinov, and H. J. Freund, *Surf. Sci.* 2004, 552, 27–34.

140. V. Lindberg and B. Hellsing, *J. Phys. C* 2005, 17, S1075–S1094.

141. W. T. Wallace and R. L. Whetten, *J. Phys. Chem. B* 2000, 104, 10964–10968.

142. L. Aballe, A. Barinov, A. Locatelli, S. Heun, and M. Kiskinova, *Phys. Rev. Lett.* 2004, 93, 196103.

18

Isoelectric Point of Nanoparticles

Rongjun Pan
Guangxi University of Technology

Kongyong Liew
*South-Central University
for Nationalities*

and

University Malaysia Pahang

18.1 Introduction

Nanomaterials have attracted worldwide attention due to their unique properties that differ significantly from bulk materials and their potential applications. It is likely that the development of nanoscience and nanotechnology would become a new area of growth for the twenty-first century's economy. Hence, investment in research and development in nanoscience and nanotechnology has been continually increased. It is reported that global investment for nanoscience and nanotechnology has come to 124 hundred billion dollars in 2006, with a growth rate of about 13% compared to that in 2005 (Minister of Science and Technology of China, 2008).

The purpose of nanoscience and nanotechnology, in the final analysis, is the utilization of nanomaterials. During the utilization, the formation of interface between the nanoparticles and other material(s) would be unavoidable (Figure 18.1). Therefore, factors such as dispersing condition, surface adsorption behavior, and interfacial properties between the particles and the dispersed phase will definitely determine the performance of the functionalized materials. To achieve homogeneous particle dispersion and desirable interface for nanoparticles is now an urgent challenge worldwide.

It is well known that the surface charge on the nanoparticles is a key parameter governing the electric behavior of particles in solution. This parameter determines the electrophoretic mobility, dispersing status, surface adsorption behavior, and interface properties of the particles within and without an external field, which acts to prevent or promote particle attraction and adhesion. The DLVO model describes the interaction between small colloidal particles in solution (Derjaguin and Laudau, 1941; Verwey et al., 1948). The total potential energy, which is a function of separation between two particles in a colloidal system, consists of a long-range electrostatic repulsion and a short-range, attractive van der Waals interaction, as illustrated in Figure 18.2. With higher surface charge, nanoparticles will be dispersed better in the matrix due to stronger static repulsion between particles, resulting in homogeneous suspension, hence better performance of nano-enhanced materials. Conversely, with lower surface charge, nanoparticles will tend to aggregate, hence the aggregated particles leading to deleterious influences.

The isoelectric point of nanoparticles is a very important parameter used to indicate the surface charge status. It is no exaggeration that all the performances of nanomaterials are in correlation with its isoelectric point. In this chapter, many aspects, including the basic conception of isoelectric point and point of zero charge, the origin of surface charge, the double electric layer around the particles when the interface forms, and the measurement and the prediction of isoelectric point of nanoparticles, are discussed.

18.2 Basic Concepts

There are two mechanisms in stabilizing the system of nanoparticles, viz., steric stabilization and electrostatic stabilization. Electrostatic stabilization is correlated to nanoparticles' surface charge significantly.

There are two concepts used to indicate nanoparticles' surface charge in colloidal system, namely, isoelectric point and point

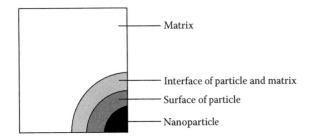

FIGURE 18.1 Schematic illustration of interface of nanoparticles dispersed in matrix.

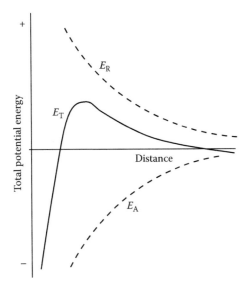

FIGURE 18.2 Curve of potential energy vs. distance between two particles. E_R is the repulsion energy, E_A is the attraction energy, and E_T is the total energy potential.

of zero charge. The *isoelectric point* is defined as the pH value at which the electrokinetic potential equals zero. The *points of zero charge* are defined as the pH values at which one of the categories of surface charge equals zero. Three operational categories of surface charge could be identified: (1) structural, denoted by σ_{st}; (2) adsorbed proton, σ_H; and (3) adsorbed ion, denoted by Δq. Thus, two points of zero charge can be defined: (1) point of zero net proton charge (PZNPC, $\sigma_H = 0$) and (2) point of zero net charge (PZNC, $\sigma_{st} + \sigma_H = 0$) (Li et al., 2004). For pure materials or those metal oxides without specific adsorption, the isoelectric point is equal to its point of zero charge (Kosmulski and Saneluta, 2004).

18.3 Origin of Nanoparticles' Surface Charge

It has been confirmed experimentally that nanoparticles' surface is positively or negatively charged. However, the origin of the surface charge is uncertain when the particles are dispersed in aqueous or in nonaqueous medium, even if in the same matrix.

18.3.1 Origin of Surface Charge in Aqueous Medium

Generally, when the nanoparticles are dispersed in aqueous medium, the origin of surface charge is considered to occur through the following routes.

18.3.1.1 Dissociation of Functional Group(s) of the Nanoparticles

Usually, some particles, such as those in polymer colloids, are charged because of the dissociation of functional group(s). For example, when nano-sized SiO_2 are dispersed in water with different pH value, equilibration with surface Si–OH will take place as follows:

$$\backslash\!\!\!-\!\!Si-OH_2^+ \underset{H^+}{\rightleftharpoons} \backslash\!\!\!-\!\!Si-OH \underset{OH^-}{\rightleftharpoons} \backslash\!\!\!-\!\!Si-O^- + H_2O$$

Therefore, the surface charge condition varies with the pH of the medium. Evidently, it is H^+ or OH^- that determines whether the surface is negatively or positively charged as well as the charge density of the nanoparticles when nano-sized SiO_2 are dispersed in aqueous medium. Hence, the ions of H^+ and OH^- are usually called *potential-determining ions* in such a situation.

18.3.1.2 Adsorption of Charged Ions

For those nanoparticles without ionizable functional groups, most of them are charged because of the adsorption of charged ions (Shen and Wang, 1997; Hunter, 2001). It has been confirmed experimentally that those ions that can produce insoluble compound(s) with any component of the nanoparticles will be preferentially adsorbed by the nanoparticles, which is generally named the selective adsorption of Fajans' rule (Fajans, 1923). Without the preferential ions, negative ions with weaker hydration activity will be firstly adsorbed while those with stronger hydration activity remain in the medium. This is the reason why most of the nanoparticles prepared by solution route are negatively charged. For instance, when nano-sized AgI particles are prepared by the reaction of $AgNO_3$ and KI solutions, Ag^+ or I^- will be preferentially adsorbed. With excessive $AgNO_3$, Ag^+ will be adsorbed and hence positively charged; conversely, I^- will be adsorbed and hence negatively charged.

18.3.1.3 Crystal Lattice Defect

For particle diameter reduced to nano sized, specific surface area increases sharply, resulting in considerable dangling bonds on the particle surface, causing a variation of isoelectric point (Sprycha et al., 1992). For nanocomposite, doping causes the replacement of certain ions and vacancy in the particles, changing the electron density and interactions, hence the charges.

18.3.2 Origin of Surface Charge in Nonaqueous Medium

For particles dispersed in nonaqueous medium, it is thought that the nanoparticles' surface charge originates from the friction between the particles and the matrix. Coehn demonstrated that dispersing medium with higher dielectric constant will charge the particles with lower dielectric constant positively (Coehn, 1898). This is the so-called Coehn rule. As the Coehn rule is not always obeyed, it is proposed that the origin of surface charge is caused by the selective adsorption of charged ions which originated from the slight ionization of the dispersing medium or the impurity of the matrix (Hunter, 2001).

18.4 Theories of Electric Double Layer

When nanoparticles are dispersed in matrix, an electric equilibrium will occur between the nanoparticles and the medium surrounding it because of the charged nanoparticles. In other words, an electric double layer will exist surrounding the particle, although the boundary and structure of each layer is uncertain (Tandon et al., 2008). For the electric properties to be understood adequately, it is necessary to know about the theory of electric double layer (Hiemenz and Rajagopalan, 1997; Shen and Wang, 1997; Zhao, 2008).

18.4.1 Helmholtz Model

Helmholtz demonstrated the electric double layer of colloids originally in 1879, in which he regarded the electric double layer as a parallel capacitor, which is schematically shown in Figure 18.3 (Helmholtz, 1979). As shown in Figure 18.3, on a positively charged surface, the negatively charged ions extend a very short distance (about 10^{-1} nm) away from the particle surface. Based on electrostatics, the following formula could be obtained:

$$\sigma = \frac{\varepsilon}{4\pi\delta}\varphi_0 \qquad (18.1)$$

where
 ε is the dielectric constant
 σ is the charge density of the surface
 δ is the distance between the two electrodes
 φ_0 is the surface potential of the particle

Therefore, the potential will decrease linearly as a function of the distance away from the positively charged surface. This model facilitated the understanding of the charged dispersing system initially. However, the Helmholtz model is not realistic and is unable to explain the difference between zeta (ζ) potential and surface potential.

18.4.2 Guoy–Chapman Model

Due to the weakness of Helmholtz model, Guoy and Chapman developed their model (Guoy, 1910; Chapman, 1913). They thought that the counterions are distributed dispersedly in matrix around the particles because of the equilibrium between electric attraction and thermal motion. The nearer the distance away from the particles is, the higher the concentration of the counterions will be due to the stronger electrostatic attraction.

When particles are dispersed in matrix, a thin layer of mixture containing counterions and matrix will cling to the particles tightly because of hydration. With the distance farther away from the charged surface, the interaction will be poorer. Hence, when electrophoresis occurs, the slipping surface between the particle and the matrix should locate in the electric double layer with certain distance (χ_ζ) away from the charged surface. The potential difference between χ_ζ and in the matrix (usually the potential is considered to be 0) is defined as *zeta potential* or *electrokinetic potential*. And the potential difference between the charged surface and in the matrix is defined as the *surface potential* (φ_0; see Figure 18.4). It is evident that the zeta potential is correlated with surface potential as well as the concentration of counterions in the zone between the charged surface and the slipping surface. With given surface potential, the higher the concentration of counterions is, the lower the zeta potential will be.

By using this model, not only zeta potential and surface potential can be easily differentiated, but also how electrolyte influences the zeta potential can be explained rationally. The drawback is that it cannot offer a reasonable explanation why zeta potential can change from positive or negative to the opposite, and why zeta potential can become higher than surface potential.

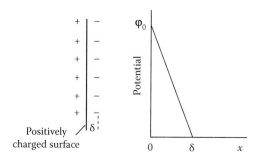

FIGURE 18.3 Schematic illustration of Helmholtz model and its variation of potential vs. the distance away from charged surface.

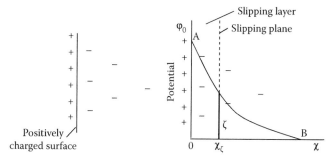

FIGURE 18.4 Schematic illustration of Guoy–Chapman model and its variation of potential vs. the distance away from the charged surface.

18.4.3 Stern Model

Stern furthered the electric double layer in 1924 by dividing the slipping layer into Stern layer and a layer similar to the slipping layer in Guoy–Chapman model (Stern, 1924). Stern layer, which is similar to that in Helmholtz model, clings firmly to the charged surface, and its thickness (denoted as δ) depends on the size of the adsorbed ions. It is thought that the slipping plane locates farther away than the Stern plane from the charged surface. Figure 18.5 shows the electric double layer of Stern model schematically and its variation of potential as a function of distance away from the charged surface. The intersection of potential curve and Stern plane indicates the Stern potential (φ_S).

When the surface is positively charged, Stern potential will be a little higher than zeta potential. Adversely, when the surface is negatively charged, Stern potential will be a little lower than zeta potential. Evidently, if the concentration of counterion is dilute enough, the diffusion layer will be extended sufficiently to cause a slight distinction between Stern potential and zeta potential. However, with higher concentration of counterions in matrix, significant distinction between the two potentials will occur. Additionally, when the adsorption on the surface of nonionic surfactant or polymer occurs, which correlates with steric stabilization, slipping plane will be extended farther away from the charged surface, hence significant distinction between the potentials. Thus, in the range of action of van der Waals attraction, the potential is high enough to resist the attraction, resulting in the efficient stabilization of nanoparticles. Therefore, surface charge also contributes to the steric stabilization of nanomaterials.

On the other hand, when high-value counterions or surfactant ions exist in matrix, the selective adsorption or specific adsorption of those counterions will occur, which results in the opposite potentials (Figure 18.6a). If electrostatic repulsion between the particles and co-ions could be overcome, the adsorption of co-ions could also occur, resulting in higher Stern potential than surface potential (Figure 18.6b).

Anyhow, Stern model could deal with the electrokinetics of particles qualitatively. However, it is difficult to quantitate the potential and to detail the structure of the adsorption layer, the

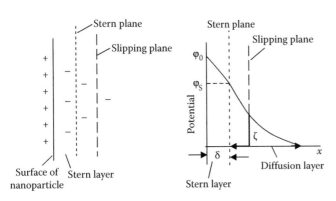

FIGURE 18.5 Schematic illustration of Stern model of electric double layers and its variation of potential as a function of distance away from the charged surface.

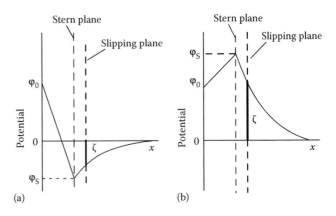

FIGURE 18.6 Potential variation of particles with adsorption of (a) high value of counterions and (b) co-ions.

variation of dielectric constant to ion concentration, and the inhomogeneous distribution of the surface charge.

18.4.4 Grahame Model

In 1947, Grahame perfected Stern model by dividing the layers into inner and outer electric layers (Grahame, 1953). The inner electric layer consists of inner Helmholtz layer and outer Helmholtz layer. Inner Helmholtz layer, which contains non-hydrated ions and matrix molecules, clings to the charged surface tightly; however, the outer Helmholtz layer, which contains hydrated ions, also clings to the charged surface but less tightly than the inner Helmholtz layer. The slipping plane locates between the outer Helmholtz layer and the matrix.

Obviously, the complex interplay between diffuse interfacial structures, interfacial chemistry, and slip phenomena results in electrokinetic behavior and is strongly dependent on both the medium and the particles considered. Given dispersing medium, the impact of isoelectric point of nanoparticles on the performance of functionalized materials can be crucial.

18.5 Determination of Isoelectric Point

When nanoparticles are dispersed in matrix, the properties of the suspension, including optical, electrical, viscous, and acoustic properties, will vary distinctly with the pH of medium close to or far away from the isoelectric point of nanoparticles. Based on these, the isoelectric point of nanoparticles could be determined.

18.5.1 Experimental Determination of Isoelectric Point

18.5.1.1 Electric Methods

Because of the close relationship between zeta potential and electrophoretic mobility, the measurements of zeta potential may be performed by imposing an electric field across a suspension of particles, measuring the resulting electrophoretic velocity of the particles, and then determining the isoelectric point, from which electrophoresis and electroosmosis were developed (Neale, 1946). Conversely, when the charged particles or the matrix move directionally because of the extrinsic influences, it will result in electric potential difference. Thus, sedimentation and streaming potential methods were developed.

18.5.1.1.1 Moving Boundary Electrophoresis

Moving boundary electrophoresis is usually used to determine the isoelectric point of colloids due to its convenience. Colloid particles will move at a constant velocity when equilibrium between frictional resistance and electrostatic force is reached (Figure 18.7). Thus, the following formulas could be given if the particles are spherical with a radius of r:

$$f = qE \qquad (18.2)$$

$$f' = 6\pi\eta rv \qquad (18.3)$$

$$f = f' = qE = 6\pi\eta rv \qquad (18.4)$$

where
   q is the electric quantity of the charged particles
   E is the electric intensity
   v is the electrophoresis velocity
   η is the viscosity of the suspension

Hence, it gives

$$\frac{v}{E} = \frac{q}{6\pi\eta r} \qquad (18.5)$$

where v/E is electrophoretic mobility. Conventionally, zeta potential is used to indicate the charged particles:

$$\zeta = \frac{q}{\varepsilon r} \qquad (18.6)$$

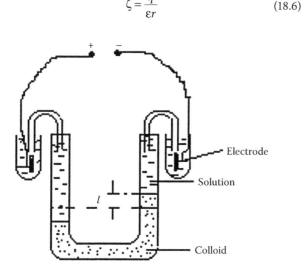

FIGURE 18.7 Schematic illustration of moving boundary electrophoresis.

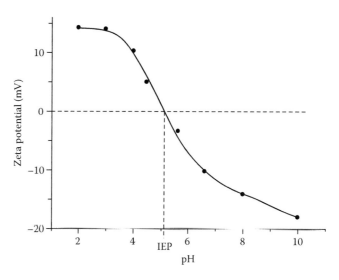

FIGURE 18.8 Zeta potential as a function of pH value.

where ε is the dielectric constant. When the above formula is put into Equation 18.5, it gives

$$\zeta = \frac{6\pi\eta v}{\varepsilon E} \qquad (18.7)$$

Therefore, if the electrophoresis velocity ($v = l/t$) can be measured, zeta potential can be calculated easily based on Equation 18.7. When zeta potentials in different pH values of dispersing matrix are obtained, the ζ–pH curve can be drawn, hence the achievement of isoelectric point (see Figure 18.8). For this method to be efficient, a clear boundary is needed.

For rod-like nanoparticles, a correction factor of 2/3 is needed. Thus, it gives

$$\zeta = \frac{4\pi\eta v}{\varepsilon E} \qquad (18.8)$$

Recently, capillary electrophoresis is used to determine the isoelectric point of nanoparticles. It is similar to the above method, although the capillary is used for electrophoresis.

18.5.1.1.2 Microelectrophoresis (or Particle Electrophoresis)

When charged particles are driven by electric field, the particles located at the same equipotential surface will move at the same velocity (Ayub et al., 1985; Doren et al., 1989). Being different from moving boundary electrophoresis, particle velocity is determined by the microscopic observation of single charged particle so as to obtain the electrophoretic mobility (see Figure 18.9). After electrophoretic mobility is measured, the zeta potential of the suspension could be achieved according to Equation 18.7 for spherical particles or Equation 18.8 for rod-like particles.

When series of suspensions with different pH values are obtained, the zeta potential corresponding to each suspension could be obtained. Based on the variation of zeta potential versus pH value, being similar to Figure 18.8, the isoelectric point of the charged particles could also be achieved. For the method

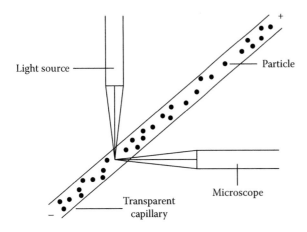

FIGURE 18.9 Schematic illustration of microelectrophoresis.

to be more practical, the applied electric field and time should vary with the electrophoretic mobility and conductivity of the suspension (Furusawa et al., 1990; Miller and Berg, 1991; Sprycha et al., 1992).

A great drawback of this method is that the observation object should be large enough (Smith and Narimatsu, 1993). However, with the development of science and technology, this method was developed by a combination of laser-scattering Doppler effect. By using this technique, microelectrophoresis can be promoted efficiently (Noordmans et al., 1993).

18.5.1.1.3 Isoelectric Focusing

When solution containing ampholyte and anti-convection matrix is placed in electric field, the pH gradient of solution will be stable due to the directional movement of ampholyte (Gelsema and Ligny, 1977; Bjellqvist et al., 1982; Thormann et al., 1986; Naydenov et al., 2006). With charged particles in this system, the particles will move to solution zone with certain pH value to become zero charged. The pH of the solution at this zone is the isoelectric point of the particles. Though thermal motion of the charged particles is unavoidable, once the particles move away from the pH zone, they will be forced back because they will be positively or negatively charged. Thus, the particles are "focused" on the isoelectric point zone (Figure 18.10). Hence, the isoelectric point of the particles could be determined by the pH value of the solution zone.

For the determination to be more successful, it has been developed by selecting chemical spacers (Sova, 1985; Tindall, 1986), by immobilizing pH gradient or by using capillary (Righetti et al., 1980; Righetti and Bossi, 1997a; Righetti, 2004).

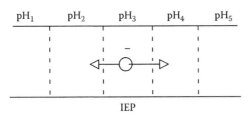

FIGURE 18.10 Schematic illustration of isoelectric focusing.

18.5.1.1.4 Electroosmosis

A schematic illustration of electroosmosis is shown in Figure 18.11. When electric field is applied to the suspension, relative motion will take place between the matrix and the charged particles (Smit and Stein, 1977; Srivastava and Lal, 1980; Kim et al., 1996). For spherical particles, Equation 18.7 holds. If the volume of the matrix flowed through the capillary during the given period of time (*t*) is *V* and the section area of the capillary is *A*, then the velocity, *v*, would be

$$v = \frac{V}{At} \tag{18.9}$$

Additionally, another formula could also be obtained based on Ohm's law:

$$\text{Potential} = IR = I\left(\frac{l}{KA}\right) \tag{18.10}$$

where
  K is the electric conductivity
  l is the distance
  I is the current

Putting Equations 18.9 and 18.10 into Equation 18.7, the following formula can be drawn:

$$\zeta = \frac{6\pi\eta v}{\varepsilon E} = \frac{6\pi\eta(V/At)}{\varepsilon(lI/KA)} = \frac{6\pi\eta KV}{\varepsilon Itl} \tag{18.11}$$

Thus, zeta potential could be measured after ε, V, K, l, and η are determined and hence the isoelectric point.

For this method to be more practical, particle concentration should be appropriate (Bowen and Jacobs, 1986), and the effect of the applied field should also be avoided (Ghowsi and Gale, 1991; Schatzel et al., 1991; Tandon and Kirby, 2008).

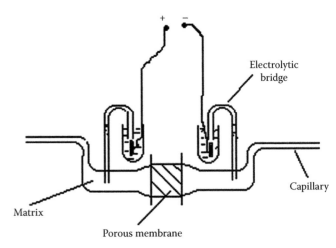

FIGURE 18.11 Schematic illustration of electroosmosis.

18.5.1.1.5 Streaming Potential and Sedimentation Potential Methods

Streaming potential and sedimentation potential methods are shown schematically in Figure 18.12.

As for streaming potential, which is the reverse of electroosmosis, liquid matrix will move directionally relative to the charged particles under the influence of external force, which was discovered by Quincke (1859). The zeta potential was calculated using a modified version of the Helmholtz–Smoluchowski equation (Fairbrother and Mastin, 1924; Schuch, 1989; Zhang et al., 2005):

$$\xi = \frac{dU}{dp} \frac{\eta}{\varepsilon \varepsilon_0} \sigma_B \tag{18.12}$$

where

dU/dp is the change in streaming potential versus flow pressure
η is electrolyte viscosity
$\varepsilon \varepsilon_0$ is the dielectric permittivity of the electrolyte solution
σ_B is the electrical conductivity of the bulk electrolyte solution

The bulk electrolyte conductivity may be substituted for the conductivity at the sample surface as long as the sample exhibits a negligible surface conductivity (Fairbrother and Mastin, 1924; Kitahara and Watanabe, 1984). This substitution can be made if the potential and resistance of the empty porous plug have been measured (Wang and Hubbe, 2001).

Being different from streaming potential, sedimentation potential, the reverse of electrophoresis, occurs due to dispersed particles' motion relative to the fluid under the influence of either gravity or centrifugation. The motion of charged particles in solution is very complex due to the deformation of the electric double layer resulting from the fluid motion, which is usually referred to as the *relaxation effect*, and gives rise to an induced electric potential difference. The sedimentation potential was first reported by Dorn in 1878 (Booth, 1954; Saville, 1982), and this is the reason why it is known by his name. For sedimentation method, a theory of sedimentation in a concentrated suspension of spherical colloidal particles was proposed by Levine (Levine et al., 1976). Hereafter, the theory was developed by Booth (Booth et al., 1954), Ohshima (Ohshima et al., 1984; Ohshima, 1998), Carrique (Carrique et al., 2001), and Keh (2002), so as to achieve the determination of zeta potential of identical spherical particles with arbitrary double-layer thickness. For spherical particles with a radius of a, the following formula could be obtained:

$$E_{\text{Sed}} = -\frac{\varphi(1-\varphi)}{(1+\varphi/2)} \frac{(\rho_P - \rho_0)}{K} \frac{\varepsilon_r \varepsilon_0 \xi}{\eta} g \tag{18.13}$$

where

φ is the particle volume fraction throughout the entire suspension
ρ_0 is the mass density of the liquid
ρ_P is the mass density of the particle
ε_r is the relative permittivity of the solution
ε_0 is the permittivity of a vacuum
η is the viscosity
K is the electric conductivity of the solution without particles
g is the gravitational acceleration

Particle volume fraction, φ, is given by

$$\varphi = \frac{(4/3)\pi a^3 N_P}{V} \tag{18.14}$$

where

a is the radius of sphere particles
V is the volume of the solution

Thus, when the sedimentation potential is measured, the zeta potential could be calculated, hence the isoelectric point of the particles from ξ–pH curve.

18.5.1.2 Optical Methods

Isoelectric point can also be determined by light scattering methods because when the charged particles are dispersed in matrix with different pH values, the optical properties of the suspension will vary with the aggregation condition. With a pH of the solution close to the isoelectric point of the particles, aggregation will occur greatly, hence a small particle concentration, which in turn results in a high transparency or a small scattering. Adversely, with a pH far away from the isoelectric point, the particles will be dispersed well in the matrix, resulting in a high particle concentration, hence poor transparency or high scattering. With a pH of the matrix right at the isoelectric point of the nanoparticles, the weakest scattering or the highest transparency will occur (Dougherty et al., 2008) (Figure 18.13a).

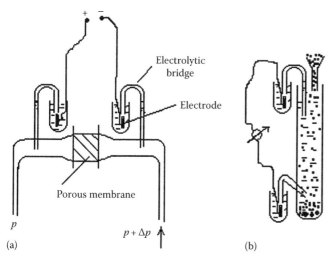

FIGURE 18.12 Schematic illustration of (a) streaming potential and (b) sedimentation potential.

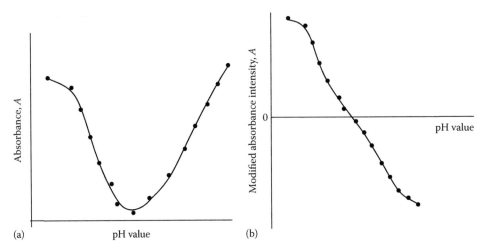

FIGURE 18.13 (a) Absorbance of the suspensions at a wavelength at which the prominent absorbance occurs as a function of pH value and (b) the modified curve after the absorbance was assigned negative value at higher pH than that for the lowest absorbance.

Because the weakest scattering of the highest transparency is difficult to be determined experimentally, Pan and co-workers (Pan et al., 2007) developed this method more practically by assuming that when the pH of the suspension is right at the isoelectric point of the nanoparticles, the absorbance will be extremely close to zero. Thus, another figure (Figure 18.13b) could be obtained by changing the sign for the absorbance to negative from the pH with the absorbance intensity much closer to the weakest. As showed in Figure 18.13b, a point of intersection at zero absorbance is obtained at a certain pH which indicates the isoelectric point of the nanoparticle. The isoelectric point obtained by using ultraviolet–visible (UV–vis) spectra well matched the result obtained by electrokinetic method (Pan et al., 2007).

18.5.1.3 Electroacoustic Method

In 1933, Debye predicted that a sound wave would generate an alternating electric field as it passed through an electrolyte, which was also found to occur in colloid suspension (Debye, 1933). Electroacoustic effect indicates the relationship between electrical and acoustic properties of charged particles dispersed in suspension, which is shown schematically in Figure 18.14. When an electric field is applied for the suspension, the charged particles will move back and forth, hence the origination of sonic wave. A relation between colloid vibration potential (CVP) and electrokinetic sonic amplitude (ESA) could be given by

$$\text{ESA} = \text{CVP} \cdot K^* = c\Delta\rho\varnothing f_g \mu_D \qquad (18.15)$$

where

K^* is the complex electric conductivity
c is the sonic velocity in matrix
$\Delta\rho$ is the density difference of both the particle and the matrix
$\varnothing$ is the volume fraction of the particle
f_g is the coupling factor of geometry and acoustics
μ_D is the motion velocity of the particles (Shen and Wang, 1997; Zhao, 2008)

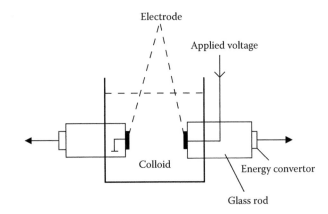

FIGURE 18.14 Schematic illustration of eletroacoustic apparatus.

Motion velocity μ_D is connected with zeta potential as follows:

$$\mu_D = \frac{2\varepsilon\xi}{3\eta}G(\alpha)(1+f) \qquad (18.16)$$

where

f is the complex factor
ε is the dielectric constant
η is the viscosity of the matrix
$G(\alpha)$ is an inertial factor which represents the effect of inertia forces on the dynamic mobility.

The factor of $(1+f)$ is proportional to the tangential electric field at the particle surface. For a given frequency, it depends on the permittivity of the particle and on α parameter.

$$G(\alpha) = \frac{1+(1+i)\sqrt{\dfrac{\alpha}{2}}}{1+(1+i)\sqrt{\dfrac{\alpha}{2}}+i\left(\dfrac{\alpha}{9}\right)\left(3+2\left(\dfrac{\Delta\rho}{\rho}\right)\right)} \qquad (18.17)$$

With low applied frequency, $G(\alpha) \approx 1$. Thus, the following formula named Smoluchowshi formula could be obtained. At this condition, μ_D correlates with the zeta potential only:

$$\mu_D = \frac{2\varepsilon\xi}{3\eta}(1+f) \tag{18.18}$$

For most small particles, f reduces to 0.5 because of the negligible effect of surface conductance (O'Brien et al., 1995). Thus, by using this technique, zeta potential can be obtained by the magnitude of μ_D. Hence, the isoelectric point of the particles even for particle suspensions with relative concentration that cannot be measured by conventional electrophoresis can be determined (Miller and Berg, 1991; Beattie and Djerdjev, 2000; Kosmulski et al., 2005), although it has been reported that the isoelectric point obtained by electroacoustic method may be lower than that obtained by electrophoresis method (Goetz and El-Aasser, 1992).

For the isoelectric point to be more credible, phase lag resulting from high frequency applied should be avoided.

18.5.1.4 Determination by Measuring the Contact Angle of Film

It has been reported recently that the surface isoelectric point of native air-formed oxide films on various metals can be determined by measuring contact angle at the hexadecane/aqueous solution interface as a function of the pH of the aqueous phase (McCafferty and Zettlemoyer, 1971). Hence, if the obtained nanoparticles are converted into film, the isoelectric point could also be measured.

For a solid surface S in contact with liquids 1 and 2, as shown in Figure 18.15, Young's equation gives

$$\pi + \gamma_{S2} = \gamma_{S1} + \gamma_{12}\cos\theta \tag{18.19}$$

where
π is the film pressure for liquid 2 on the solid surface S
γ is the interfacial tension between different phases

If liquid 2 is a hydrocarbon and liquid 1 is an aqueous phase of varying pH, the differentiation with respect to pH gives (McCafferty and Wightman, 1997).

$$0 = \frac{d\gamma_{S1}}{dpH} + \gamma_{12}\frac{d\cos\theta}{dpH} + \cos\theta\frac{d\gamma_{12}}{dpH} \tag{18.20}$$

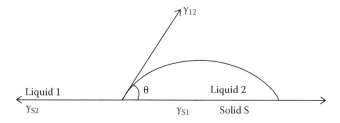

FIGURE 18.15 Schematic illustration of the two liquid-solid system. (From McCafferty, E. and Wightman, J.P., *J. Colloid Interface Sci.*, 194, 344, 1997. With permission.)

Because there is only a slight variation with pH in the interfacial tension of hexadecane versus aqueous solutions of various pH values, as an approximation, $d\gamma_{12}/dpH$ can be considered as zero (McCafferty and Wightman, 1997, 1998; McCafferty, 1999). Equation 18.20 becomes

$$0 = \frac{d\gamma_{S1}}{dpH} + \gamma_{12}\frac{d\cos\theta}{dpH} \tag{18.21}$$

or

$$-\frac{d\gamma_{S1}}{dpH} = \gamma_{12}\frac{d\cos\theta}{dpH} \tag{18.22}$$

The surface charge density σ at the oxide/solution interface is given by

$$\sigma = zF\left(\Gamma^{S1}_{H^+} - \Gamma^{S1}_{O^-}\right) \tag{18.23}$$

where $\Gamma^{S1}_{H^+}$ and $\Gamma^{S1}_{O^-}$ refer to the surface excesses of adsorbed protons MOH_2^+ and dissociated hydroxyl groups MO^- at the oxide/aqueous solution interface, and z equals 1 (but is carried along for completeness). Combining Equation 18.21, Young's equation, the Gibbs equation, and surface equilibria conditions pertaining to dissociation for hydroxylated oxide films lead (after some algebra) to the expression (McCafferty and Wightman, 1997)

$$\frac{d\cos\theta}{dpH} = \frac{1}{\gamma_{12}}\left[\frac{\sigma}{zF}\left(-2.303RT + zF\frac{d\psi}{dpH}\right) + \Gamma^{S1}_O zF\frac{d\psi}{dpH}\right] \tag{18.24}$$

where
θ is the contact angle at the hexadecane/aqueous solution interface
ψ is the potential at the oxide/solution interface

At the isoelectric point, the surface charge $\sigma = 0$, and the surface concentration of dissociated hydroxyl groups $\Gamma^{S1}_{O^-}$ is also zero so that Equation 18.24 gives

$$\left(\frac{d\cos\theta}{dpH}\right)_{\sigma=0} = 0 \tag{18.25}$$

Thus, the cosine of the contact angle will go through a minimum and the contact angle through a maximum as a function of the pH values of solution. The pH of the solution at which the minimum contact angle or the maximum cosine value occurs indicates the isoelectric point of the film, viz., the isoelectric point of the particles.

18.5.2 Theoretical Predictions

Theoretical predictions of the isoelectric point of nanoparticles are of importance due to the cost and the complication of experimental procedures. In previous works, the isoelectric points of many metal oxides were predicted (Parks, 1965; Carre et al., 1992; Kosmulski, 2001). It has been confirmed that the outermost surface of an oxide is covered with a layer of hydroxyl groups (McCafferty and Zettlemoyer, 1971; McCafferty and Wightman, 1997, 1998; Bolger, 1983). The Lewis acid–Lewis base properties of the hydroxyl groups on the oxide determine the surface charge when immersed in aqueous solutions (Tanabe, 1971). In aqueous solutions, the surface group of –MOH (M refers to the metal cation) will remain undissociated if the pH of the aqueous solution is the same as the isoelectric point of the oxide. If the pH is lower than the isoelectric point, the surface will acquire a positive charge:

$$-MOH + H^+ \rightleftharpoons -MOH_2^+$$

If the pH is higher than the isoelectric point, the surface will be negatively charged:

$$-MOH + OH^- \rightleftharpoons -MO^- + H_2O$$

or

$$-MOH \rightleftharpoons -MO^- + H^+$$

Thus, predictions for the isoelectric point of simple metal oxides could be made using an electrostatic model (Parks, 1965), which takes into account the surface charges (Parks, 1965; Hunter, 2001) originating from the amphoteric dissociation of surface –MOH groups and the adsorption of the hydrolysis products of $M^{z+}(OH)^{z-}$ (z is ionic charge):

$$IEP = B - 11.5\left[\frac{z}{R} + 0.0029(CFSE) + a\right] \quad (18.26)$$

$$R = 2r_o + r_+ \quad (18.27)$$

where
 r_o is the radius of oxygen ion (1.41 Å)
 r_+ is the radius of metal ion
 CFSE is the crystal field stabilization energy
 B and a are the parameters depending on the coordination number of metal ions

For instance, the isoelectric points of indium oxide, antimony oxide, and tin oxide could be predicted by this theory. Since Sn^{4+}, Sb^{5+}, and In^{3+} occupy octahedral interstices in SnO_2, Sb_2O_5, and In_2O_3, the coordination number for these metal ions is 6, and therefore B is equal to 18.6 and a is equal to zero (Parks, 1965). CFSEs were assumed to be zero in these calculations (Parks, 1965; Kosmulski, 2001; Szczuko et al., 2001; Sun et al., 2004). Thus, the isoelectric points of the three metal oxides are 9.37, 2.35, and 5.93, respectively.

For complex oxides like antimony tin oxide (ATO) and indium tin oxide (ITO), the isoelectric point can be described by (Carre et al., 1992)

$$IEP = \sum_i s_i IEP_i \quad (18.28)$$

where
 s_i is the mole fraction of ith component
 IEP_i is the isoelectric point of ith component

Take ITO nanoparticles with 10% of tin oxide as an example (Pan et al., 2007). Its isoelectric point will be

$$IEP = 0.61 \times 9.37 + 0.39 \times 5.93 = 8.03$$

Usually, the prediction value of isoelectric point may be a little lower or higher than that obtained experimentally. This difference is ascribed to several factors. First, some researchers (Dusastre and Williams, 1998; Szczuko et al., 2001) found an enrichment of certain component(s) on the surface of metal oxide(s). Another reason for differences between predicted and experimental values is the fact that the prediction neglects the contributions from the CFSE as well as the surface defects and nonstoichiometry (Logan, 1967; Parks, 1965). The predictions are considered approximate.

18.6 Summary

The isoelectric point of nanoparticles is an important parameter which determines the performance of materials functionalized by the addition of nanoparticles. Given interfacial charge and the attendant zeta potential, the impact of interfacial slip on micro-device performance can be crucial. Despite this, it has been shown that the complex interplay between diffuse interfacial structures, interfacial chemistry, and slip phenomena result in electrokinetic behavior that is difficult to predict, and is highly dependent on both the medium and the particles considered. The determination of the isoelectric point of nanoparticles can be achieved by various methods. The electrokinetic measurements of isoelectric point for nanoparticles may be especially difficult sometimes, since the electrokinetic potential is too low or particle dissolution occurs. Under these circumstances, the other experimental methods could serve as a substitute. Theoretical predictions that feature low cost and convenience can also offer qualitative results, giving their classical values.

References

Ayub, A. L.; Roberts, S. L.; Kwak, J. C. T. 1985. Adsorption of cationic surfactants on two size fractions of coal fines: Comparison of zeta potentials derived from microelectrophoresis and streaming potential measurements. *Colloids Surf.* 16:175–183.

Beattie, J. K.; Djerdjev, A. 2000. Rapid electroacoustic method for monitoring dispersion: Zeta potential titration of alumina with ammonium poly(methacrylate). *J. Am. Ceram. Soc.* 83: 2360–2364.

Bjellqvist, B.; Ek, K.; Righetti, P. G.; Gianazza, E.; Görg, A.; Westermeier, R.; Postel, W. 1982. Isoelectric focusing in immobilized pH gradients: Principle, methodology and some applications, *J. Biochem. Biophys. Methods* 6:317–339.

Bolger, J. C. 1983. Proceedings of the Second International Symposium on Adhesion Aspects of Polymeric Coatings. *Adhesion Aspects of Polymeric Coatings.* ed. K. L. Mittal, New York: Plenum Press, pp. 10–202.

Booth, F. 1954. Sedimentation potential and velocity of solid spherical particles. *J. Chem. Phys.* 22:1956–1968.

Bowen, W. R.; Jacobs, P. M. 1986. Electroosmosis and the determination of ζ potential: The effect of particle concentration. *J. Colloid Interface Sci.* 111:223–229.

Carre, A.; Roger, F.; Varinot, C. 1992. Study of acid/base properties of oxide, oxide glass, and glass–ceramic surfaces. *J. Colloid Interface Sci.* 154:174–183.

Carrique, F.; Arroyo, F. J.; Delgado, A. V. 2001. Sedimentation velocity and potential in a concentrated colloidal suspension: Effect of a dynamic Stern layer. *Colloids Surf. A: Physicochem. Eng. Asp.* 195:157–169.

Chapman, D. L. 1913. A contribution to the theory of electrocapillarity. *Philos. Mag.* 25(6):475–481.

Coehn, A. 1898. Ueber ein gesetcz der electricitätserregung. *Ann. Phys.* 64:217–228.

Debye, P. 1933. A method for the determination of the mass of electrolyte ions. *J. Chem. Phys.* 1:13–16.

Derjaguin, B. V.; Laudau, L. 1941. Theory of the stability of strongly charged lyophobic sols and of the adhesion of strongly charged particles in solutions of electrolytes. *Acta Physicochim. URSS* 14:633–662.

Doren, A.; Lemaitre, J.; Rouxhet, P. G. 1989. Determination of the zeta potential of macroscopic specimens using microelectrophoresis. *J. Colloid Interface Sci.* 130:146–156.

Dougherty, G. M.; Rose1, K. A.; Tok, J. B.-H. et al. 2008. The zeta potential of surface–functionalized metallic nanorod particles in aqueous solution. *Electrophoresis* 29:1131–1139.

Dusastre, V.; Williams, D. E. 1998. Sb(III) as a surface site for water adsorption on $Sn(Sb)O_2$, and its effect on catalytic activity and sensor behavior. *J. Phys. Chem. B* 102:6732–6738.

Fairbrother, F.; Mastin, H. 1924. CCCXII–studies in electro-endosmosis, part I. *J. Chem. Soc. Trans.* 125:2319–2330.

Fajans, V. K. 1923. Struktur und deformation der elektronenhüllen in ihrer bedeutung für die chemischen und optischen eigenschaften anorganischer verbindungen. *Die Naturwissenschaften* 10:165–172.

Furusawa, K.; Chen, Q.; Tobori, N. 1990. A new reference sample for microelectrophoresis. *J. Colloid Interface Sci.* 137:456–461.

Gelsema, W. J.; Ligny, C.L. 1977. Isoelectric focusing as a method for the characterization of ampholytes. *J. Chromatogr. A* 130:41–50.

Ghowsi, K.; Gale, R. J. 1991. Field effect electroosmosis. *J. Chromatogr. A* 559:95–101.

Goetz, R. J.; El–Aasser, M. S. 1992. Effects of dispersion concentration on the electroacoustic potentials of o/w miniemulsions. *J. Colloid Interface Sci.* 150:436–452.

Grahame, D. C. 1953. Diffuse double layer theory for electrolytes of unsymmetrical valence types. *J. Chem. Phys.* 21:1054–1061.

Guoy, G. 1910. Sur la constitution de la charge électrique à la surface dùn electrolyte. *J. Phys.* 9(4):457–466.

Helmholtz, H. 1979. Studien über elektrische Grenzschichten. *Ann. Phys.* 7(3):337–382.

Hiemenz, P. C.; Rajagopalan, R. 1997. *Principles of Colloid and Surface Chemistry*, 3rd edn. New York: Marcel Dekker Inc.

Hunter, R. J. 2001. *Foundations of Colloid Science.* Oxford, NY: Oxford University Press.

Keh, H. J. 2002. Sedimentation, electrophoresis, and electric conduction in suspensions of charged composite particles. *Bull. Coll. Eng.* 84:59–66.

Kim, K. J.; Fane, A. G.; Nystrom, M.; Pihlajamaki, A.; Bowen, W. R.; Mukhtar, H. 1996. Evaluation of electroosmosis and streaming potential for measurement of electric charges of polymeric membranes. *J. Membrane Sci.* 116:149–159.

Kitahara, A.; Watanabe, A. 1984. *Electrical Phenomena at Interfaces: Fundamentals, Measurements, and Applications.* New York: Marcel Dekker Inc.

Kosmulski, M. 2001. *Chemical Properties of Materials Surface.* New York: Marcel Dekker Inc.

Kosmulski, M.; Saneluta, C. 2004. Point of zero charge/isoelectric point of exotic oxides: Tl_2O_3. *J. Colloid Interface Sci.* 280:544–545.

Kosmulski, M.; Rosenholm, J. B.; Saneluta, C.; Boczkowska, K. M. 2005. Electroacoustics and electroosmosis in low temperature ionic liquids. *Colloids Surf. A: Physicochem. Eng. Asp.* 267:16–18.

Levine, S.; Neale, G.; Epstein, N. 1976. The prediction of electrokinetic phenomena within multiparticle systems: II. Sedimentation potential. *J. Colloid Interface Sci.* 57:424–437.

Li, D.; Hou, W.; Li, S.; Hao, M.; Zhang, G. 2004. The isoelectric point and the points of zero charge of Fe-Al-Mg hydrotalcite–like compounds. *Chin. Chem. Lett.* 15:224–227.

Logan, R. K. 1967. π-p charge exchange polarization and the possibility of a second ρ meson. *Phys. Rev. Lett.* 18:259–263.

McCafferty, E. A. 1999. A surface charge model of corrosion pit initiation and of protection by surface alloying. *J. Electrochem. Soc.* 146:2863–2869.

McCafferty, E.; Wightman, J. P. 1997. Determination of the surface isoelectric point of oxide films on metals by contact Angle titration. *J. Colloid. Interface Sci.* 194:344–355.

McCafferty, E.; Wightman, J. P. 1998. Determination of the concentration of surface hydroxyl groups on metal oxide films by a quantitative XPS method. *Surf. Interface Anal.* 26:549–564.

McCafferty, E.; Zettlemoyer, A. C. 1971. Adsorption of water vapour on α-Fe_2O_3. *Discuss. Faraday Soc.* 52:239–254.

Miller, N. P.; Berg, J. C. 1991. A comparison of electroacoustic and microelectrophoretic zeta potential data for titania in the absence and presence of a poly (vinyl alcohol) adlayer. *Colloids Surf.* 59:119–128.

Minister of Science and Technology of China. 2008. *The World Development Report of Advanced Technology—2007*. Beijing, China: Science Press.

Naydenov, C.; Kirazov, E. P.; Kirazov, L. P.; Genadiev, T. T. 2006. New approach to calculating and predicting the ionic strength generated during carrier ampholyte isoelectric focusing. *J. Chromatogr. A* 1121:129–139.

Neale, S. M. 1946. Electrical double layer, the electrokinetic potential and the streaming current. *Trans. Faraday Soc.* 42:473–478.

Noordmans, J.; Kempen, J.; Busscher, H. J. 1993. Automated image analysis to determine zeta potential distributions in particulate microelectrophoresis. *J. Colloid Interface Sci.* 156:394–399.

O'Brien, R. W.; Cannon, D. W.; Rowlands, W. N. 1995. Electroacoustic determination of particle size and zeta potential. *J. Colloid Interface Sci.* 173:406–418.

Ohshima, H. 1998. Sedimentation potential in a concentrated suspension of spherical colloidal particles. *J. Colloid Interface Sci.* 208:295–301.

Ohshima, H.; Healy, T. W.; White, L. R. 1984. Electrokinetic phenomena in a dilute suspension of charged mercury drops. *J. Chem. Soc., Faraday Trans.* 80:1643–1667.

Pan, R.; Liew, K.; Xu, L.; Gao, Y.; Zhou, J.; Zhou, H. 2007. A new approach for determination of iso-electric point of nanoparticles. *Colloids Surf. A: Physicochem. Eng. Asp.* 305:15–21.

Parks, G. A. 1965. The isoelectric points of solid oxides, solid hydroxides, and aqueous hydroxo complex systems. *Chem. Rev.* 65:177–198.

Quincke, G. 1859. Über eine neue art electrischer ströme. *Ann. Phys.* 107(2):1–47.

Righetti, P. G. 2004. Determination of the isoelectric point of proteins by capillary isoelectric focusing. *J. Chromatogr. A* 1037:491–499.

Righetti, P. G.; Bossi, A. 1997a. Isoelectric focusing in immobilized pH gradients: Recent analytical and preparative developments. *Anal. Biochem.* 247:1–10.

Righetti, P. G.; Bossi, A. 1997b. Isoelectric focusing in immobilized pH gradients: An update. *J. Chromatogr. B: Biomed. Sci. Appl.* 699:77–89.

Righetti, P. G.; Gianazza, E.; Ek, K. 1980. New developments in isoelectric focusing. *J. Chromatogr. A* 184:415–456.

Saville, D. A. 1982. The sedimentation potential in a dilute suspension. *Adv. Colloid Interface Sci.* 16:267–279.

Schatzel, K.; Weise, W.; Sobotta, A.; Drewel, M. 1991. Electroosmosis in an oscillating field: Avoiding distortions in measured electrophoretic mobilities. *J. Colloid Interface Sci.* 143:287–293.

Schuch, M. 1989. Streaming potential in nature. *Lecture Notes in Earth Science* 27:99–107.

Shen, Z.; Wang, G. 1997. *Colloid and Surface Chemistry*, 2nd edn. Beijing, China: Chemical industry Press.

Smit, W.; Stein, H. N. 1977. Electroosmotic zeta potential measurements on single crystals. *J. Colloid Interface Sci.* 60:299–307.

Smith, R. W.; Narimatsu, Y. 1993. Electrokinetic behavior of kaolinite in surfactant solutions as measured by both the microelectrophoresis and streaming potential methods. *Miner. Eng.* 6:753–763.

Sova, O. 1985. Autofocusing—A method for isoelectric focusing without carrier ampholytes. *J. Chromatogr. A* 320:15–22.

Sprycha, R.; Jablonski, J.; Matijevi, E. 1992. Zeta potential and surface charge of monodispersed colloidal yttrium(III) oxide and basic carbonate. *J. Colloid Interface Sci.* 149:561–568.

Srivastava, M. L.; Lal, S. N. 1980. Electrokinetic studies on a thorium oxide membrane: I. Electroosmosis and electrophoresis. *J. Membrane Sci.* 7:21–37.

Stern, O. 1924. Zur theorie der elektrischen doppelschicht. *Z. Elektrochem.* 30:508–516.

Sun, J.; Velamakanni, B. V.; Gerberich, W. W.; Francis, L. F. 2004. Aqueous latex/ceramic nanoparticle dispersions: colloidal stability and coating properties. *J. Colloid Interface Sci.* 280:387–399.

Szczuko, D.; Werner, J.; Behr, G.; Oswald, S.; Wetzig, K. 2001. Surface–related investigations to characterize different preparation techniques of Sb–doped SnO_2 powders. *Surf. Interface Anal.* 31:484–488.

Tanabe, K. 1971. *Solid Acids and Bases*. London, U. K.: Academic Press.

Tandon, V.; Kirby, B. J. 2008. Zeta potential and electroosmotic mobility in microfluidic devices fabricated from hydrophobic polymers: 2. Slip and interfacial water structure. *Electrophoresis* 29:1102–1114.

Tandon, V.; Bhagavatula, S. K.; Nelson, W. C.; Kirby, B. J. 2008. Zeta potential and electroosmotic mobility in microfluidic devices fabricated from hydrophobic polymers: 1. The origins of charge. *Electrophoresis* 29:1092–1101.

Thormann, W.; Mosher, R. A.; Bier, M. 1986. Experimental and theoretical dynamics of isoelectric focusing: Elucidation of a general separation mechanism. *J. Chromatogr. A* 351:17–29.

Tindall, S. H. 1986. Selection of chemical spacers to improve isoelectric focusing resolving power: Implications for use in two–dimensional electrophoresis. *Anal. Biochem.* 159:287–294.

Verwey, E. J. W.; Overbeek, J. Th. G. 1948. *Theory of Stability of Lyophobic Colloids*. Amsterdam, the Netherlands: Elsevier.

Wang, F.; Hubbe, M. A. 2001. Development and evaluation of an automated streaming potential measurement device. *Colloids Surf. A: Physicochem. Eng. Asp.* 194:221–232.

Xie, Y.; Liang, Y.; Zhang, Y.; Qin, Y.; Wu, W.; He, Y. 2006. The improvement on the method of eletrophoretic focusing. *Chin. Lett. Biotechnol.* 17:70–71.

Zhang, Y.; Xu, T.; Fu, R. 2005. Modeling of the streaming potential through porous bipolar membranes. *Desalination* 81:293–302.

Zhao, Z. 2008. *Applied Colloid and Interface Science*. Beijing, China: Chemical industry press.

19

Nanoparticles in Cosmic Environments

Ingrid Mann
Kinki University

and

*Belgian Institute for
Space Aeronomy*

19.1 Introduction

A large fraction of the heavy chemical elements in the different environments of the Galaxy (as well as in other galaxies) is contained in small solid dust particles. Cosmic dust particles (occasionally called "grains") typically cover large size intervals including particles of a size smaller than 100 nm. The particles of a size smaller than 100 nm are denoted as nanoparticles and are the subject of this chapter.

We begin by describing briefly the evolution of the cosmic dust (Section 19.2). Starting with astronomical observations, the chapter then discusses the optical properties of nanoparticles and the supposed detections through interstellar extinction observations (Section 19.3.1), observations of the thermal emission brightness (Section 19.3.2), and observations of their photoluminescence (Section 19.3.3). The subsequent sections address physical interactions of nanoparticles (Section 19.4.1), the detection of nano-dust by hypervelocity impacts (Section 19.4.2), the detection of nano-dust in the vicinity of comets (Section 19.4.3), and nano-dust detection in the Earth's atmosphere (Section 19.4.4).

We then describe studies of collected cosmic samples in the laboratory (Section 19.5.1) and laboratory studies of the formation of nanostructures within larger solids (Section 19.5.2). This chapter concludes with a summary and a critical discussion of present knowledge and perspectives of future research (Section 19.6).

19.2 Cosmic Dust Evolution and Properties

The evolution path of cosmic dust is summarized graphically in Figure 19.1.

Late stellar evolution: Cosmic dust particles initially form in the late stages of stellar evolution. The hydrogen burning that determines the properties of main sequence stars ends when the hydrogen in the interior of the stars is exhausted. Low and intermediate mass stars (with masses smaller than about eight solar masses) evolve to giant stars, and stars with larger masses explode as supernovae. These stages are governed by different nuclear fusion reactions that lead to the formation of heavy chemical elements (Burbidge et al. 1957). The presence of heavy elements allows for the formation of dust. Dust particles condense in the cooling gas that expands outward into the surrounding space between the stars (see, e.g., Hoyle and Wickramasinghe 1970).

Interstellar medium (ISM): The ISM is filled with a tenuous gas with number densities that are smaller than in ultrahigh vacuum generated in the laboratory. It basically contains three phases with different properties (McKee and Ostriker 1977): a hot and ionized phase with gas temperature $T_{gas} \approx 5 \times 10^5$ K and gas number density $n_{gas} \approx 0.003$ cm^{-3}, a cold and neutral phase with $T_{gas} \approx 80$ K and $n_{gas} \approx 0.25$ cm^{-3}, and a warm phase with $T_{gas} \approx 10^4$ K and $n_{gas} \approx 100$ cm^{-3} that can be partly neutral or ionized.

Cosmic dust evolution

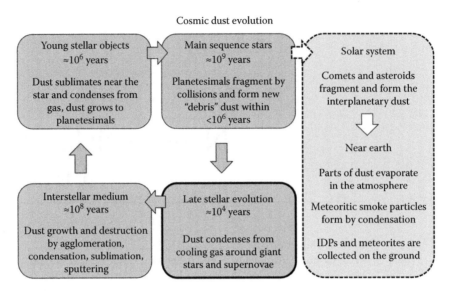

FIGURE 19.1 A sketch illustrating the major path of cosmic dust evolution, as discussed in the text, and the time that the dust remains in the respective stage. All the shown stages of dust evolution potentially contain nanoparticles.

Dust particles make up about 1% of the total mass in the ISM and are typically of sub-micrometer size. The observed properties of the dust particles imply a complex dust evolution in the ISM (Draine 2003). Interstellar dust particles experience different growth, destruction, and alteration processes that depend on the particular gas and radiation environment. The observed relative amount of the elements Fe, Al, Mg, and Si in the gas component varies for different regions of the ISM, indicating that these elements are condensed into dust particles in varying amounts. In regions where the relative velocities of dust particles are low, they may form larger agglomerate particles when they collide. In other regions, they fragment during mutual collisions with high relative velocities. The dust particles are also destroyed by sublimation and sputtering. The paths of surface reactions on the dust are largely unknown and different from reactions that are commonly known from laboratory experiments (see, e.g., Fraser et al. 2005). The estimated average lifetime of dust particles in the ISM is 10^8 years (Jones et al. 1994).

From molecular clouds to young stellar objects: The described three phases make up the majority of the observable regions of the ISM, but roughly half of the mass within the ISM is concentrated in molecular clouds. These are regions of low temperature ($T_{gas} \approx 80\,K$) and high density ($n_{gas} \approx 100\,cm^{-3}$), such that hydrogen is present as H_2 molecules. Volatile elements condense onto the dust and form icy mantles. Ultraviolet (UV) radiation induces chemical reactions in the icy compounds and complex organics form. Stars form by gravitational collapse and clusters of young stars are observed in molecular cloud regions. During star formation, the ISM dust is incorporated into the forming stars and the disks that surround them. The outer regions of the contracting cloud form a disk of gas and dust (protoplanetary disk, protoplanetary nebula) that partly accretes to the star. Out of the disk a planetary system may form (see, e.g., Takeuchi

2009). Dust evaporates in the inner zones of the protoplanetary disk and later recondenses. Dust particles grow by coagulation in the outer disk, and it is assumed that also a radial mixing transports dust particles from the inner regions outward. The aggregating dust forms planetesimals: objects of meters to thousands of kilometers in size. The evolution from molecular clouds to planetesimals involves complex chemical processes, partly traced by astronomical observations (see, e.g., van Dishoeck and Blake 1998). The vast majority of the gas component is either incorporated in planetesimals or removed in the later stage of the protoplanetary disks by thermal diffusion or by interactions with the stellar radiation and the stellar wind (see, for instance, Hollenbach et al. 2000).

Main-sequence stars and planetary systems: The protoplanetary disk stage ends with the disappearance of the gas component, and from there on, dust particles are produced by the fragmentation of planetesimals forming a circumstellar debris disk (see, e.g., Wyatt 2009). During this evolution of the disk from a protoplanetary disk to (possibly) a planetary system, the central star evolves to its main sequence stage. This is characterized by stable hydrogen burning. The stellar brightness and the stellar wind are less time-variable than in younger stars and the stellar wind has a low density. This is also the current stage of our solar system.

Dust in the solar system: The potential major sources of dust within our solar system are the small solar system bodies: trans-Neptunian objects, asteroids, and comets (see, e.g., Mann 2009). The dust from the two latter sources makes up the interplanetary dust cloud in the region of the terrestrial planets. Considering the interplanetary medium of the solar system at 1 AU (1 AU = the average distance between the Sun and the Earth), the ratio of dust mass to gas mass per unit volume is about unity, and the gas component consists mostly of the solar wind. The particle

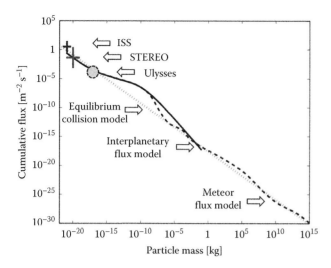

FIGURE 19.2 The cumulative dust flux as a function of mass near the Earth's orbit: The interplanetary dust flux model is primarily derived from impact measurements from spacecraft (Grün et al. 1985); the meteor flux model is primarily derived from meteor observations (Ceplecha et al. 1998). The equilibrium collision model describes an $m^{-5/6}$ power law distribution that evolves for a dust cloud in collision equilibrium. Fluxes of nanoparticles obtained onboard the International Space Station (ISS) (Carpenter et al. 2005), onboard STEREO (Meyer-Vernet et al. 2009), and onboard *Ulysses* (Wehry and Mann 1999) are shown, and these are further discussed in the text. (Adapted from Meyer-Vernet, N. et al., *Solar Phys.*, 256, 463, 2009.)

TABLE 19.1 Major Carbonaceous Materials in the ISM

Diamond	sp³—bonded carbon
Graphite	sp²—bonded carbon
Amorphous or glassy carbon	Mixed sp² and sp³—bonded carbon with short range order
HAC: hydrogenated amorphous carbon	Amorphous carbon with high hydrogen content
PAH: polycyclic aromatic hydrocarbon	sp²—bonded carbon with peripheral H
Aliphatic (chainlike) hydrocarbons	

Source: Adapted from Draine, B.T., *Annu. Rev. Astron. Astrophys.*, 41, 241, 2003.

flux of dust as a function of mass near the Earth's orbit is shown in Figure 19.2. Approximate number densities are 10^{-9} m^{-3} for 10 μm particles that make the majority of the optically observed dust and 10^{-5} m^{-3} for the smallest detected 10 nm particles, as is discussed later. The lifetime of the dust particles is less than about 10^{6} years. The lifetime is limited either by the Poynting–Robertson effect (deceleration of the dust by the tangential force of radiation pressure that arises in the frame of the moving particle) or by destruction through mutual collisions. Dust growth is negligible.

Dust flux near the Earth: Some objects crossing the Earth's orbit enter its atmosphere (Ceplecha et al. 1998). A large fraction of objects that enter the Earth's atmosphere evaporate as a result of entry heating. Sufficiently large objects lose only parts of their matter and the remnants reach the ground (*meteorites*). From the entering particles with a size less than 1 mm, the majority vaporize almost completely during entry. A fraction of the produced vapor recondenses as *meteoritic smoke particles*. Dust particles that are smaller than about 50 μm efficiently reradiate the entry heat and, therefore, experience only a moderate temperature increase. They slowly settle toward the ground (*interplanetary dust particles*, IDPs).

Dust observations: The different cosmic environments outlined above are studied by astronomical observations. Dust thermal emission observations of circumstellar and interstellar dust reveal dust composition, since, for small particles, emission features can be characteristic of the properties of the dust. The interstellar dust is also observed by its light scattering and by the polarization of the scattered light. Moreover, the interstellar extinction provides information about the size and the composition of dust (see discussion below). The dust in the solar system can also be studied with in situ experiments and with laboratory analyses of collected samples.

Dust properties: The major materials observed in cosmic dust are magnesium-rich silicates and silicon carbides, iron–nickel and iron–sulfur compounds, calcium and aluminum oxides, and carbonaceous species. Materials are denoted as carbonaceous when the majority of mass is contained in carbon atoms (cf. Draine 2003). The carbonaceous species are diamond, graphite, amorphous or glassy carbon, hydrogenated amorphous carbon (HAC), polycyclic aromatic hydrocarbon (PAH), and aliphatic hydrocarbons (see Table 19.1). The size of the carbonaceous particles is given in numbers of C atoms, which ranges from about 50 to 200, and an approximate size for a PAH containing 100 C atoms is 0.6 nm. In the ISM, about 20% of the carbon is contained in carbonaceous molecules, ices, and nanoparticles (as opposed to gas and larger dust), and the carbon in these larger molecules and nanoparticles is recognized only by certain typical carbon bonds (van Dishoeck 2008). Some features detected in distinct interstellar regions point to the existence of ice coatings on dust particles.

19.3 Scattering Properties of Nano-Dust and Astronomical Observations

The cross sections for scattering and for thermal emission of a solid particle drop steeply in the size range that is smaller than the considered wavelength (see Figure 19.3). The nanoparticles, being in a size range that is smaller than the wavelengths of most astronomical observations (the visible light, selected infrared spectral intervals, and beyond), evade most detection techniques when they are smaller than several 10 nm. While the absorption and emission of large particles occurs over a broad spectral range in the bulk solid, nanoparticles only absorb and emit photons in narrow spectral ranges that are determined by

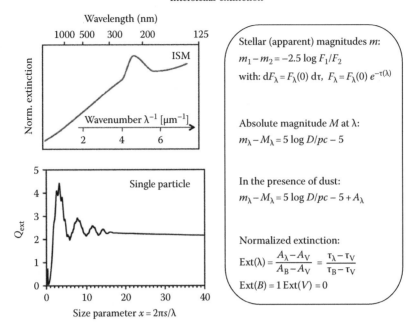

FIGURE 19.3 The interstellar extinction and its variation with wavelength is shown in the upper-left part of the figure. The flux density of a star observed at a distance r is reduced by traveling along the distance r with average optical thickness τ. Similar to the stellar brightness, the extinction is often given in magnitudes, as explained above on the right. The normalized extinction, A_λ, at a given wavelength is compared to the extinction A_B of the blue (mean wavelength 442 nm) and A_V visual (mean wavelength 540 nm) magnitudes. The lower-left part of the figure shows the extinction efficiency, Q_{ext}, calculated using the Mie theory for a $0.5\,\mu m$ radius silicate particle; $Q_{ext} = Q_{abs} + Q_{sca}$ decreases steeply for the size parameter $x < 1$, where $x = 2\pi s/\lambda$ with s being the radius of the particle and λ, the wavelength of light. (Mie calculations by courtesy of M. Koehler, University of Missouri, Columbia, MO.)

their atomic and molecular structures. Some of the methods to detect nanoparticles are (1) by scattered and absorbed light at a short wavelength, (2) by characteristic (thermal) emission, and (3) by luminescence, and these are discussed below.

19.3.1 Dust Light Scattering and Interstellar Extinction

The existence of dust in the ISM is evident from the interstellar extinction: an attenuation of starlight that is barely noticeable at $\lambda = 2\,\mu m$ increases at a shorter wavelength (Figure 19.3). Its final cutoff is not seen in observations at wavelengths as short as $\lambda = 200$ nm. The extinction is caused by scattering and absorption of interstellar dust between the star and the observer. Independent from the solid, out of which the particle consists the extinction efficiency, Q_{ext}, of a single particle of radius, s drops steeply for $s \ll \lambda$. Hence, dust particles smaller than ≈ 30 nm do not significantly contribute to the extinction.

For a single particle, the extinction efficiency, $Q_{ext}(s,\lambda)$, describes the amount of in-falling light that is deviated from its initial direction by absorption and scattering at the particle: $Q_{ext}(s,\lambda) = Q_{abs}(s,\lambda) + Q_{sca}(s,\lambda)$. The absorption efficiency is defined as $Q_{abs}(s,\lambda) = C_{abs}(s,\lambda)/\pi s^2$, where C_{abs} is the absorption cross section of the particle, and the scattering efficiency is defined as $Q_{sca}(s,\lambda) = C_{sca}(s,\lambda)/\pi s^2$, where C_{sca} is the scattering

cross section of the particle. For particles of a size of several 10 nm and larger, Q_{ext} is calculated applying light scattering theory and the optical properties (refractive indices) of the solid. The refractive indices are experimentally determined for the bulk solid material. The calculated $Q_{ext}(x)$ for a silicate particle is shown in the lower-left part of Figure 19.3, where Q_{ext} is given as a function of the size parameters $x = 2\pi s/\lambda$ and the steep decrease for $x \ll 1$ is obvious. The values are calculated using Mie theory, which provides the exact solution for the scattering of an electromagnetic wave at a sphere of given refractive index. The scattering of particles of a few nanometers in size, in contrast, is calculated numerically by considering the most important transitions in the system (see, e.g., Li and Draine 2001).

Many observed spectral extinction curves show an enhancement around $\lambda = 217.5$ nm. The wavelength of this "hump" seems fixed, while the width varies with the direction toward which the extinction is measured. Since carbonaceous particles have enhanced absorption cross sections around the range of the extinction hump (see Figure 19.4), carbonaceous nanoparticles and large carbonaceous molecules are usually assumed to cause the extinction hump (see Draine (2003) for a review). Aside from that, there are other models that explain the extinction around 217.5 nm: Bradley et al. (2005) compared the spectral variation of interstellar extinction to spectroscopic measurements of synthetic samples and of samples that were isolated from collected

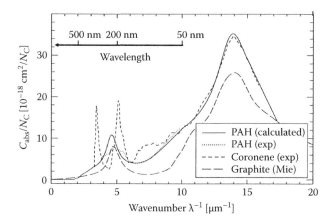

FIGURE 19.4 The absorption cross section for a suggested PAH particle model in the UV and Far UV: The cross sections are shown as a function of wavenumber and also shown are experimental data for coronene ($C_{24}H_{12}$) and PAH mixtures and an estimate based on Mie calculations for graphite particles. (From Li, A. and Draine, B.T., *Astrophys. J.*, 554, 778, 2001. With permission.)

IDPs. They found that organic compounds and amorphous silicates extracted from IDPs show similar profiles, and hence suggested that the extinction hump possibly arises from the size of the particles rather than their composition.

Note that in addition to the 217.5 nm feature, the extinction curve in the visible part of the spectrum includes a large number of weak features that are broader than atomic interstellar lines, yet not clearly attributed to the interstellar dust (Herbig 1995). Among others, small dust and large molecules are treated as carriers (Draine 2003).

19.3.2 Dust Temperature and Thermal Emission Brightness

In the absence of other energy sources, the temperature of an object in space is determined by the amount of absorbed (left side of Equation 19.1) and emitted light (right side of Equation 19.1):

$$\int Q_{abs}(s,\lambda)\pi s^2 J_\lambda \, d\lambda = \int Q_{abs}(s,\lambda) \, 4\pi s^2 \, B_\lambda(T_d) d\lambda \quad (19.1)$$

where

Q_{abs} is the efficiency for absorption, as explained above, and (applying Kirchhoff's rule) also for emission

s is the radius of the particle

λ is the wavelength of radiation

J_λ is the flux of the external radiation field

B_λ is the Planck function

T_d is the equilibrium temperature of dust

For large objects, $Q_{abs}(\lambda) \approx$ constant, and the object has a temperature close to that of a blackbody. The thermal emission as a function of the wavelength is also similar to that of a blackbody.

For small dust particles, Q_{abs} varies with λ, the variation being characteristic of the material and variations being stronger the smaller the particle. As a result, the equilibrium temperature, T_d, varies with the material and with the size of the dust; that is, the temperature exceeds that of a blackbody when $Q_{abs}(s,\lambda)$ is high at the wavelength where the particle absorbs radiation and low in the spectral range of its thermal emission. As a result of the higher temperature, the maximum of the thermal emission brightness shifts to a shorter wavelength.

Dust particles can reach even higher temperatures when they are stochastically heated. The equilibrium temperature of a dust particle evolves when the amount of absorbed and emitted radiation is equal per unit time. This applies when the flux of photons is large within the characteristic timescale of the optical transitions that follow the photon absorption. For dust with sizes less than 10 nm in the ISM, this is not the case (see, e.g., Draine 2004). Figure 19.5 shows the calculated temperature as a function of time for nano-dust particles in the interstellar radiation field. Particles with sizes below 20 nm are strongly heated by single photon absorption and that subsequently cool down again. The maximum temperature is far beyond the blackbody temperature and beyond the temperatures of the larger dust particles in the same radiation field.

Stochastic heating provided observational proof for the presence of nanoparticles in the ISM when the first scientific infrared satellite, IRAS, showed considerable emission brightness at 12 and 25 μm (Boulanger and Perault 1988). The thermal emission of stochastically heated nanoparticles explains the diffuse emission of ISM dust that is observed at wavelengths shorter than 60 μm. Figure 19.6 shows satellite observations of the diffuse ISM brightness in comparison to the calculated thermal emission for a model of ISM dust suggested by Li and Draine (2001) that consists of silicate and carbonaceous particles, the latter comprising graphite and PAHs. Large (≥25 nm) silicon and carbonaceous dust, denoted as Si-dust and C-dust in Figure 19.6, show a broad emission peak at wavelengths beyond approximately 60 μm. Particles <25 nm, denoted as Si-nano and C-nano, as a result of stochastic heating, show emission at shorter wavelengths. Carbonaceous nano-dust particles show strong emission bands and obscure the emission of nano-silicates. Figure 19.6 shows the case of the diffuse ISM at high galactic latitudes. The brightness within the galactic plane is further complicated by unresolved stellar brightness in the near infrared, but models for the dust emission are similar to the case shown here.

The emission from PAHs also explains the "ubiquitous infrared" (UIR) emission bands at wavelengths of 3.3, 6.2, 7.7, 8.6, and 11.3 μm. These bands are observed in the diffuse ISM, as shown in Figure 19.6, and also in other astronomical environments. Recent astronomical observations and dedicated laboratory experiments demonstrate the importance of PAHs for ISM evolution (Tielens 2008). Certain infrared emission bands were also assigned to interstellar and circumstellar diamonds based on comparison with laboratory measurements (Guillois et al. 1999).

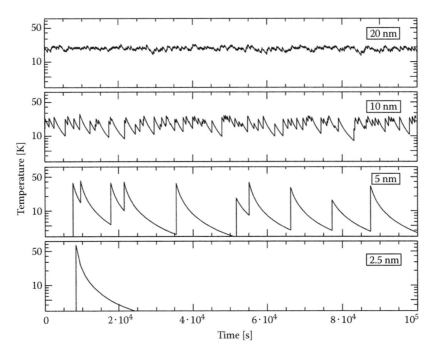

FIGURE 19.5 Temperature as a function of time of dust particles heated by the average interstellar radiation field in 1 day (1 day = 84,600 s). (From Draine, B.T., Astrophysics of dust in cold clouds, In *The Cold Universe, Saas-Fee Advanced Course*, Vol. 32, Lecture Notes 2002 of the Swiss Society for Astronomy and Astrophysics, Blain, A.W. et al. (eds.), Springer, Berlin, Germany, 2004, p. 213. With permission.)

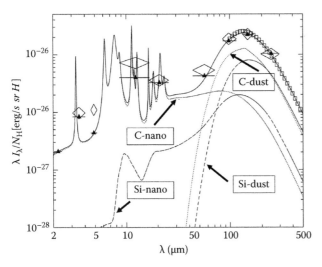

FIGURE 19.6 Observations and a suggested model of the diffuse ISM brightness at Galactic latitudes ≥25°: The squares show the submillimeter and microwave emission extrapolated from IRAS and COBE/DIRBE data (Finkbeiner et al. 1999), and the diamonds show the near-to-mid infrared data from COBE/DIRBE (Arendt et al. 1998). Shown in comparison is the ISM dust emission model suggested by Li and Draine (2001); the sum of the different emission components is shown as a solid line and the triangles show the conversion of this total brightness to the filter intervals of the COBE/DIRBE instrument. The contributions of the different dust components to the model are shown as dashed and dotted lines. Large (≥25 nm) silicate and carbonaceous dust, denoted as Si-dust and C-dust here, show a broad emission peak at wavelengths beyond approximately 60 μm. Particles <25 nm, denoted as Si-nano and C-nano, as a result of stochastic heating, show emission at shorter wavelengths. (From Li, A. and Draine, B.T., *Astrophys. J.*, 554, 778, 2001. With permission.)

19.3.3 Photoluminescence and the "Extended Red Emission"

A process that is particularly efficient for nanoparticles is photoluminescence, and this is often discussed to explain the extended red emission (ERE) that is observed in many astronomical objects. The ERE is an enhanced brightness in the red part of the visible spectrum. In most cases, the peak of the ERE is between 650 and 750 nm and the width ranges from 60 to 120 nm (numbers given by Witt and Vijh 2004). Photoluminescence caused by the interstellar radiation field can occur in many different environments, and therefore is a potential process to generate ERE.

The basic principle of photoluminescence is shown in Figure 19.7 for a semiconductor particle. Absorption of a UV photon

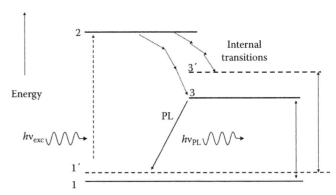

FIGURE 19.7 A sketch showing the basic principle of photoluminescence, a process that is highly efficient for some nanoparticles irradiated by UV photons, as further discussed in the text.

lifts an electron from the ground state near the upper level of the valence band, to a state in the conduction band. Internal transitions to an intermediate lower level in the conduction band follow. If no further internal transitions occur, then an optical transition to the ground state may follow, leading to photon emission. A similar process is also observed for large molecules or molecular ions. Photoluminescence does not occur when the bandgap can be crossed by internal transitions. This is, for instance, the case when the solid contains impurities. For nanoparticles, on the other hand, the probability of internal transitions is low, so that photoluminescence has a high efficiency. Moreover, as a result of quantum confinement, the bandgap increases for small particle sizes. Reducing the size of the particles increases the size of the bandgap, while it decreases the probability for internal transitions. For nano-sized particles, photoluminescence shifts to higher energies and its efficiency increases.

Ledoux et al. (2001) have studied the properties of nanoparticles of several carbon- and silicon-bearing materials with model calculations and laboratory measurements.

They found that the two considered types of silicon particles fall into the range of astronomical observations. Their derived bandgaps of different carbon- and silicon-bearing materials are shown in Figure 19.8. Bulk silicon has a bandgap of 1.17 eV, and the gap increases for decreasing particle size. The resulting

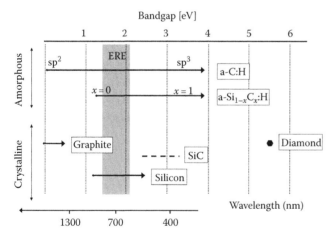

FIGURE 19.8 The bandgaps over which photoluminescence occurs and the corresponding wavelengths of emitted light for different carbon- and silicon-bearing materials that are discussed as cosmic dust components. The shaded range indicates the spectral interval of observed ERE. The two upper cases show the bandgap position of amorphous hydrogenated carbon (a-C:H) and its variation with the degree of sp^3 hybridization of the carbon atoms, and that for amorphous hydrogenated silicon carbide (a-Si$_{1-x}$C$_x$:H) and its variation with the concentration of C atoms. Below that, the bandgaps for crystalline material are shown. The energy range shown for SiC accounts for β-SiC with a bandgap of 2.4 eV and for α-SiC polytypes with bandgaps between 3.0 and 3.3 eV. Bulk silicon has a bandgap of 1.17 eV, and the arrow indicates the variation of bandgap with decreasing size of particles. For SiC, only the value for the bulk material is shown. The crystalline carbon energy gap also has an arrow indicating the small particle effect. (From Ledoux, G. et al., *Astron. Astrophys.*, 377, 707, 2001. Copyright ESO. With permission.)

wavelength range of photoluminescence for silicon nanoparticles agrees well with the observed range of ERE (shown as shaded area in Figure 19.8). While this seems to support the Si nanocrystals as carriers of the ERE, this does not agree with some other observational results (cf Witt et al. 2006). A potential option to explain observational results is also that nanoparticles form larger clusters or are attached to larger particles (Li and Draine 2002). Finally, we note that not all models to explain the ERE are based on nanoparticles.

19.4 Plasma Interactions of Nano-Dust and In Situ Measurements

The following section will address the interaction of nano-dust with the surrounding gas. In most cases, the surrounding gas is partially or fully ionized and is considered a plasma. This term will be used for the surrounding medium in the following discussion. As a result of the surface charge, the dynamics of dust is influenced by electromagnetic forces, and this becomes increasingly important for small particles. Surface charge and the resulting dust dynamics allow for the in situ detection of nanoparticles.

19.4.1 Dust Interactions and Dust Charging

Dust particles are charged as a result of impinging photons and plasma particles. Figure 19.9 shows the plasma temperatures that a dust particle encounters in moving from the local ISM toward the Sun. We use this to illustrate the surface charging in different plasma environments (see Kimura and Mann 1998). The solar system is embedded in a warm ISM phase ($T_{gas} = 10^4$ K). This interacts with the hot corona gas that extends from the Sun ($T_{gas} \approx 10^5$ K), and a region of high plasma temperature ($T_{gas} > 10^6$ K) builds up as a result of the interaction of these two plasma phases, as suggested, for instance, based on a gasdynamic

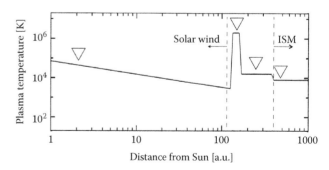

FIGURE 19.9 The plasma temperatures (model by Pauls and Zank 1996) at different regions around the Sun: the solar wind, the local ISM, and the transition region that forms as a result of the interaction between solar wind and ISM plasma. The triangles denote the locations for which the calculated dust surface charge is shown in Figure 19.10. The temperatures of electrons and alpha particles are assumed to be the same as those of protons shown here. For further discussion on dust charging, see Kimura and Mann (1998). (From Kimura, H. and Mann, I., *Astrophys. J.*, 499, 454, 1998. With permission.)

model by Pauls and Zank (1996). Note that the *Voyager* spacecraft are currently crossing this transition region, and experimental results will change our picture of the outer solar system plasma environment. The following considerations, yet, illustrate the influence of plasma temperature on dust surface charging.

The time variation of charge, Q, on a dust particle is

$$\frac{dQ}{dt} = \sum_k J_k \tag{19.2}$$

where J_k denote the different charging processes. The most common processes of dust charging in cosmic environments are photoelectron emission, impinging of plasma particles, secondary electron emission due to the impinging of plasma particles, and thermionic emission. Thermionic emission is only important in the case of high dust temperatures. The energies of impinging plasma particles are determined from the relative velocity between the plasma and the dust, or, in the case of high plasma temperatures and for particles at rest relative to the plasma, the thermal velocity determines the energy of impinging particles. The current of impinging plasma particles is further influenced by the attracting or repelling dust surface potential. The important plasma particles are electrons, protons, and α-particles, while ions of larger mass due to their low abundance generate negligible currents.

The equilibrium charge of a particle is found for $dQ/dt = 0$. The charged particle has an electrostatic potential, U with respect to the distant unperturbed plasma. This is shown in Figure 19.10 for interstellar dust that approaches the Sun. The dust motion arises from the Sun's velocity of about 25 km/s relative to the ISM. The differences between the cases of carbon and silicate arise from the material dependence of the yields for electron emission induced by impinging solar wind electrons and of the yields of

electron emission caused by absorbed photons, the yields being defined by the number of ejected electrons per absorbed photon or impinging plasma particle. Both materials have similar yields for emission of photoelectrons. As long as secondary electron emission is comparably unimportant, the dust surface potential is relatively constant over a wide range of sizes and is of the order of several volts. The total surface charge, Q, for a given surface potential, U, is

$$Q = 4\pi s\,\varepsilon_0 U \tag{19.3}$$

for a spherical dust particle of radius s, where ε_0 is the electric constant. For sizes of a few nanometers, the surface charge amounts to only few elementary charges.

A difference arises in the intermediate region between the solar wind and the ISM where the interaction between the two different plasmas streaming toward each other enhances the plasma temperature. As a result of the high plasma temperature, the secondary emission increases and dust particles attain a higher surface potential. Since secondary emission depends on the material, the surface charge is different for the shown silicate and graphite particles. The increase is prominent for dust sizes <10 nm. In this size range, the secondary electrons yield increase as a result of the small-size effect illustrated in Figure 19.11. The secondary electrons that are generated in the bulk material of a large dust particle are only ejected if they can reach the surface of the material before being stopped within the solid. The electrons that are produced in small particles can leave the particle in all directions, and this increases the rate of secondary electron emission.

Note that similar to the discussion of dust temperatures, the surface charge above was discussed for the equilibrium. This requires that the dust particles stay within the same plasma and radiation environment, and it requires that the dust charge flux

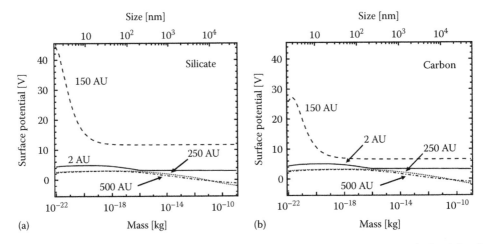

(a)   (b)

FIGURE 19.10 Surface potentials of interstellar dust grains in the vicinity of the Sun versus grain size calculated for plasma parameters at heliocentric distances 2 AU (solid line), 150 AU (dashed line), 250 AU (dotted line), and 500 AU (dash-dotted line) with the plasma temperatures shown in Figure 19.9. (a) Shows the case of silicate particles and (b) shows carbon particles, and calculations are based on assuming solid spherical particles. Secondary electron emission becomes important for high plasma temperatures (at 150 AU in this model). The enhanced surface potential for particle sizes smaller than 10 nm appears in this region of high plasma temperature and is caused by efficient secondary emission. (From Kimura, H. and Mann, I., *Astrophys. J.*, 499, 454, 1998. With permission.)

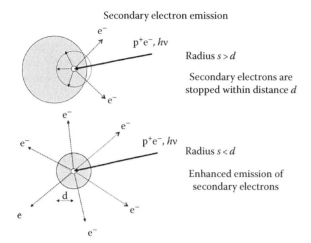

Secondary electron emission

FIGURE 19.11 The role of electron emission: The secondary electrons that are generated in small particles can leave the particle in all directions, and this increases the rate of secondary electron emission.

is sufficiently large per unit time. The surface charge of nanoparticles is often a few elementary charges, and the charge fluctuates stochastically. Moreover, the estimates for the different yields, which are needed to estimate the charge, are based on the measurements conducted on bulk solids and, therefore, may not apply for nano-sized scales. Indeed, laboratory studies show that particles smaller than a couple of nanometers experience different surface charging, but there are only few experimental results (see, e.g., Grimm et al. 2004) to prove this.

Conversely, the presence of dust particles influences the surrounding environment. This is particularly important for nanoparticles, since for a given small mass, they generate a large surface area to interact with the plasma. In the case of dust embedded in a cool gas, electrons that are ejected as a result of photon absorption have energies beyond the average energies of the plasma electrons, and therefore the photon flux onto the dust provides a mechanism to heat the gas. In hot plasma, on the other hand, the presence of dust particles leads to the deceleration of ions that pass through the dust and the charge balance is influenced since the ions charge-exchange when passing the dust (see Figure 19.12).

19.4.2 Nano-Dust Detection by Hypervelocity Impacts

There are no certain reports of the optical detection of nano-dust in the interplanetary medium of the solar system, and detection with in situ measurements is probably the more promising method (see Mann et al. 2007). Most dust measurements on spacecraft are based on the impact ionization of dust particles: The hypervelocity impact of dust particles onto a target produces a vapor that is partly ionized. The amount and time sequence of electrons and ions that are measured as a result of the dust impact are used to derive the mass and the speed of the dust particles. In order to generate a reliable signal, the dust impact should therefore produce a sufficient number of electrons and ions. Laboratory studies (McBride and McDonnell 1999) show that the produced charge, Q_{vapor}, is

$$Q_{vapor} \approx 0.7 m^{1.02} v^{3.48} \tag{19.4}$$

where m is the mass and v is the impact speed of the particle, and hence nano-dust is detected, if it is sufficiently fast to generate a detectable charge during impact.

The *Ulysses* space mission allowed for one of the longest dust measurements in the interplanetary medium. *Ulysses* was launched in 1988 and went from the Earth's orbit to Jupiter on the branch of an ellipsoid orbit. Through a flyby at Jupiter, the spacecraft was deflected into an orbit that is almost perpendicular to the ecliptic (that is the plane that goes through the center of the Sun toward which the orbits of the planets are concentrated) and has its perihelion (the closest distance to the Sun) at 1.3 AU and its aphelion (the farthest distance to the Sun) at 5.4 AU. During this Jupiter flyby, *Ulysses* firstly detected streams of dust particles being ejected from the vicinity of the planet (Grün et al. 1993).

The dust instrument on the *Ulysses* spacecraft measures particles in the mass range 10^{-19} kg < m < 10^{-10} kg. The detected stream particles are outside the range of calibrated data, and so are particles detected during later stages of the mission that

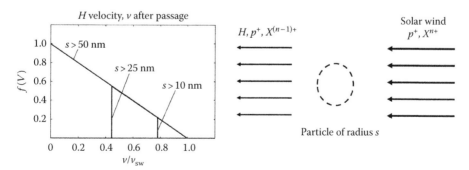

FIGURE 19.12 The impinging of solar wind particles (protons, p^+; and heavy ions, X^{n+}) on interplanetary dust: The protons are decelerated on a path length of 50 nm and implanted in the surface layers of large particles. The protons pass through nanoparticles, and the left side of the figure shows the velocity distribution, $f(v)$, of hydrogen atoms after passage through particles with radii $s = 10$ nm, $s = 25$ nm, and $s > 50$ nm. The incoming protons recombine during the passage. The calculations were made for SiO_2 particles and (proton) solar wind speed 450 km/s. Heavy solar wind ions also change their charge state during passage. (From Minato, T. et al., *Astron. Astrophys. Lett.*, 424, L13, 2004. Copyright ESO. With permission.)

possibly also originate from Jupiter (see Krüger et al. 2006 and references therein). It was suggested that small charged grains are deflected and accelerated outward from the Jupiter magnetosphere and reach *Ulysses* with speeds above 200 km/s. Considering the dust trajectories, Zook et al. (1996) inferred that the dust mass is 10^{-21} kg. Several mechanisms for generating the dust particles, for charging the dust in the magnetosphere, and for accelerating the dust particles in the magnetosphere and in the interplanetary magnetic field have been suggested since this observational discovery (see, for instance, Hamilton and Burns 1993). The analysis of further detections suggests that these particles most probably originate from the Jupiter's moon Io (Krüger et al. 2006). In a similar way, dust measurements on the *Cassini* spacecraft discovered streams of nano-dust ejected from Saturn; these dust particles are even smaller than the Jupiter stream particles (Kempf et al. 2005).

There is also evidence for the existence of high-velocity nano-dust within the solar system dust cloud (Meyer-Vernet et al. 2009). The plasma wave instrument STEREO/WAVES onboard the *Stereo A* spacecraft has recorded over 2 years a series of events that cannot be explained with normal plasma phenomena. The events appeared similar to the events detected during other space missions, which were caused by impacts and vaporization of dust particles. First attempts to explain the STEREO/WAVES data with impacts of micrometer-sized dust particles failed since the required flux to explain the data was far beyond the interplanetary dust flux (shown in Figure 19.2). The expected flux of nanometer-sized particles is of the order of magnitude that could explain the STEREO events, but when assuming typical impact velocities of 20 km/s (which results for dust particles in Keplerian orbits), the generated charge by impacts of nanometer-sized dust is too small to generate a detectable signal. Nano-dust impacting with significantly higher velocity could generate the observed signals. Equation 19.4 suggests that a nanoparticle impacting at 300 km/s generates the same charge

as a larger dust particle with four orders of magnitude greater mass. The 300 km/s velocity is far beyond the impact velocities for which Equation 19.4 was derived from laboratory measurements, but Meyer-Vernet et al. (2009) showed that the resulting charge production is plausible in terms of energy considerations and suggested that the measured events are due to the impact of nano-dust particles.

As far as the velocity of nanoparticles in the interplanetary medium is concerned, high velocities are likely to occur. The dust is subject to gravitational forces (mainly solar gravity), to solar radiation pressure force, and to electromagnetic forces. Dust of micrometer size and larger is predominantly influenced by gravity and moves in Keplerian orbits. Dust of several tenths of micrometer size is strongly influenced by the solar radiation pressure force, and particles are ejected from the inner solar system in hyperbolic orbits. Nanoparticles have a larger ratio, Q/m, of charge to mass and are more strongly influenced by electromagnetic forces than the larger dust. It is well established that small dust particles are deflected in the solar system magnetic field (Morfill and Grün 1979). Considering the dust production by collisions of larger objects and the thermal alteration of dust material in the inner solar system, Mann and Murad (2005) have shown that it is reasonable to assume that nano-dust be present in the interplanetary medium. Under certain conditions, the nanoparticles being released from larger dust particles that are in bound Keplerian orbits are accelerated by electromagnetic forces to speeds larger than 200 km/s (Figure 19.13). These accelerated nano-dust particles are suggested to explain the STEREO/WAVE events, and the derived flux is shown in Figure 19.2. The figure also shows the flux rate of particles that are ejected from the inner solar system by radiation pressure force. The velocities of these particles are only within a factor of 2 beyond the velocities of dust in bound Keplerian orbit, and they were identified within the *Ulysses* dust measurements by means of their direction of motion (Wehry and Mann 1999). The average flux of

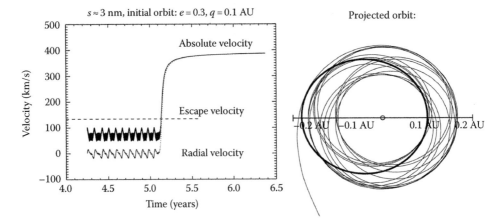

FIGURE 19.13 Trajectory of a 3 nm particle released from initial orbit with eccentricity 0.3 and perihelion 0.1 AU. The velocity as a function of time is shown on the left-hand side. The escape speed is shown with the dashed horizontal line. The assumed solar wind speed is 400 km/s. The projected trajectory is shown on the right-hand side, where the initial orbit is shown as a thick line and the evolving orbit shown with a thin line. The solar wind–induced electric field ejects the particle from the vicinity of the Sun. (Adapted from Mann, I. et al., *Planet. Space Sci.*, 55, 1000, 2007.)

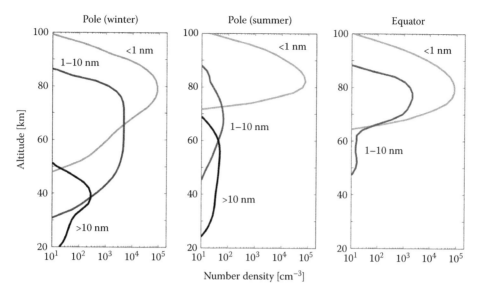

FIGURE 19.14 The calculated distribution of meteoritic smoke particles with radii $s < 1\,nm$, $1\,nm < s < 10\,nm$, and $s > 10\,nm$ for winter polar conditions (December 72°, north) and summer polar conditions (June 72°, north) and for the equator. (Courtesy of L. Megner, Department of Meteorology, Stockholm University, Stockholm, Sweden.)

nanometer-sized dust at 1 AU was recently also derived from the analysis of exposure foils that were located on the International Space Station (ISS) for about 2 years between 2002 and 2004 (Carpenter et al. 2005). These measurements provide no information about the velocity of particles.

19.4.3 Nano-Dust in the Vicinity of Comets

Several spacecraft encountered Halley's comet during its last visit to the inner solar system, and several dust measurements were carried out. The measurements on the *Vega* and *Giotto* spacecraft point to the presence of dust particles of 1 to 10 nm size (Utterback and Kissel 1990). A couple of years later, the survey of the x-ray satellite Röntgensatellit (ROSAT) discovered that comets are a regular source of x-ray emission (Dennerl et al. 1997). Though the majority of x-rays in comets is believed to originate from charge exchange between neutral gas and heavy solar wind ions, several studies suggest that dust-related processes involving nanoparticles can account for a fraction of several percent of the observed emission. Ip and Chow (1997) suggest that nano-dust particles due to their large surface charge are trapped by electromagnetic interaction in the cometary plasma and produce x-ray emission when they are destroyed by high-velocity collisions. Also electron-dust collisions (Northrop et al. 1997) and solar x-rays interacting with dust (Owens et al. 1998) are suggested as a source of x-rays from comets.

19.4.4 Nano-Dust in the Earth's Atmosphere

A large fraction of mass from objects that enter the Earth's atmosphere fragments and evaporates during entry. The meteoric vapor partly recondenses and forms small solid particles. These meteoric smoke particles of nanometer sizes and larger

are the predominant dust component in the 80–100 km altitude range (Hunten et al. 1980). Figure 19.14 shows the altitude profiles of nanoparticles suggested by model calculations taking into account condensation, transverse transport, and sedimentation near the Earth's pole for winter and for summer, and near the equator, where seasonal variation is small (Megner 2008).

Indeed several in situ measurements confirmed the presence of nano-dust at these altitudes after the first in situ measurements of dust from sounding rocket were reported by Havnes et al. (1996). The instruments carry entrance grids with applied electric potentials that shield the detectors from atmospheric electrons and positive ions, so that only charged dust particles and heavy ions enter the detector and these are measured by their electric current (see, e.g., Gelinas et al. 1998, Horanyi et al. 2000, Lynch et al. 2005, Rapp et al. 2005). The high dust fluxes compared to interplanetary medium conditions enable this method of detection. These in situ measurements are, however, limited to the detection of the accumulated electric current and do not distinguish single particle events, neither provide information about dust composition, size, or structure. The shock structure that forms in the atmosphere around the rocket deflects small particles from reaching the detector. This produces a cutoff at the lower-size end of detected particles.

19.5 Laboratory Measurements

A large number of laboratory methods are used to analyze cosmic materials: either meteorites or collected IDPs. The methods include imaging, chemical and mineralogical composition measurements, and analysis of isotope ratios. Some of the methods are destructive or partially destructive, and all of them require a sample mass of nanogram (10^{-12} kg); for some methods, even larger masses are required (Zolensky et al. 2000). For detailed analysis, hence dust in the micrometer-size range is required. Nonetheless, the laboratory studies provide some information on smaller structures. We will shortly address the laboratory studies of collected samples (Section 19.5.1), and then the formation of nanostructures within the larger bulk material (Section 19.5.2).

19.5.1 Laboratory Studies of Collected Samples

Both meteorites and collected IDPs are analyzed in great detail, and some information about nanostructures can be derived. A fraction of the collected IDPs appears as agglomerates of smaller "building stones" of average size close to 100 nm (see, e.g., Rietmeijer 2002). But the size range of substructures is broad, and FeS and FeNi compounds of sizes $\approx 5–10$ nm are commonly observed (Messenger, personal communication).

An especially interesting result is the separation of so-called presolar grains from meteorite material as well as from collected IDPs. Some of the dust particles that form during the later stages of stellar evolution mentioned in Section 19.2 are found in meteorites in our solar system and studied in the laboratory. These are highly refractory nanoparticles of several different materials. The measurements of specific relevant isotope ratios have shown that the presolar grains were formed before the formation of the solar system. These isotope ratios hugely differ from the ratios in the solar system material and can be associated with the isotope ratios that result from nuclear reactions in, for instance, certain classes of giant stars and supernovae (Zinner 1998). This requires that the particles survive within the ISM as well as in the protoplanetary disk during solar system formation. The size distributions of some of the identified presolar grains are shown in Figure 19.15 (for a discussion of the size distribution, the reader is referred to Ott 2003). The lack of particles with sizes smaller than $\cong 25$ nm (with diamonds being an exception) among the presolar dust grains that are found in meteorites possibly indicates that nanoparticles are more susceptible to destruction processes than larger dust particles. For the diamonds, because of their small size, isotope measurements are not carried out for the single grains, and it is sometimes discussed that they partly formed within the solar system (see Figure 19.15). Aside from that, crystals (e.g., TiC

and Fe_xNi_{1-x}) in the size range of 10 nm have been found inside presolar graphite grains (Bernatowicz et al. 1999).

19.5.2 Laboratory Measurements of Nanostructures

Laboratory measurements also show that nanostructures form within larger samples (see Figure 19.16). The required energy is, for instance, provided by moderate heating or the impact of energetic particles. The formation of diamonds within a meteoritic sample was, for example, observed after moderate heating of carbonaceous chondrites as well as in an ice mixture analogue of molecular cloud material (Kouchi et al. 2005). The heating transformation occurs on short timescales of minutes, and it may present a possible mechanism for producing nano-diamonds that have been identified in meteorites (mentioned above).

Dust material alteration may also occur within nano-dust at moderate temperatures.

Kamitsuji et al. (2004) reported the detection of Si nanocrystallites of about 10 nm diameter after heating a mixture of SiO_x particles with $x \approx 1$. Typical "stacking faults" of the Si nanocrystallites are observed in high-resolution images, and the electron diffraction images show the characteristic rings of the silicon cube structure. The Si nanocrystallites that are formed from SiO_x ($x \approx 1$) particles are still observed in the samples after heating to about 700°C and subsequent cooling to room temperature again. These Si nanocrystallites formed in SiO_2 particle samples survive heating to about 900°C.

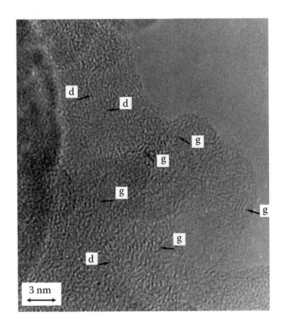

FIGURE 19.16 High-resolution transmission electron microscope (HRTEM) image of molecular cloud analogue material taken with 0.18 nm point resolution. The molecular cloud organic analog was formed by UV irradiation of an ice mixture at 12 K. The arrows d and g point to structures that are identified as diamond precursors and graphite nanocrystallites, respectively. (Courtesy of Y. Kimura, Tohoku University, Sendai, Japan.)

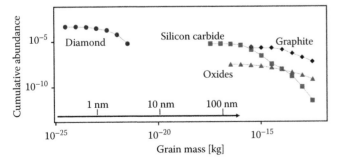

FIGURE 19.15 Cumulative size distributions of selected presolar grains and nano-diamonds identified in meteorite samples. (Adapted from Mann, I. and Kimura, H., *J. Geophys. Res.*, 105, 10317, 2000.)

The alteration of the internal composition within nanoparticles is also seen in laboratory measurements. In dust condensation experiments (Nuth et al. 2002), surface layers of oxide are observed to form instantaneously after nanoparticle formation; moreover, other surface reactions have been studied (Kimura et al. 2003). For nanoparticles in space, it was suggested that after the nucleation of SiO molecules from the gas phase, an inner core of Si that is surrounded by a mantle of SiO_2 would form (Witt et al. 1998).

19.6 Summary and Discussion

We summarize that all the stages of the evolution of cosmic dust from formation, modification, to destruction include the possibility of nanoparticles being present. The following observational findings point to the possible existence of nanoparticles:

- The ISM extinction in the UV should result from nanoparticles.
- The UIR emission bands observed in the diffuse ISM and also in other astronomical environments are explained with the characteristic emission of PAH particles.
- The diffuse emission of the ISM at wavelengths <60 μm is explained with stochastically heated dust of sizes <25 nm.
- Several of the proposed mechanisms to explain ERE that is observed in different astronomical objects are based on assuming nanoparticles.
- Nano-dust is detected with in situ measurements in the interplanetary medium, in near Earth space, and around other solar system planets.

Due to their large surface area for a small given mass, nanoparticles play an important role for the physical evolution in the ISM and in other cosmic environments. As far as the interactions of nanoparticles with atomic or molecular ions and neutral particles are concerned, most (but not all) of the theoretical studies start from assuming the material properties for the (macroscopic) bulk and then adapt to small-size effects. It is up to future research to further explore the domain where properties differ from that of the bulk solid material, and rather atomic, molecular, and intermolecular interactions are relevant to describe the particle. Further goals for future research on cosmic nanoparticles should be to accomplish the unambiguous detection of nanoparticles; to improve the understanding of the optical properties and, hence, the observational detection of nanoparticles; and to study dust growth, dust destruction, and dust internal evolution on nanoscales.

References

Arendt, R.G., Odegard, N., Weil, J.L., Sodroski, T.J., Hauser, M.G., Dwek, E., Kelsall, T. et al. 1998. The COBE diffuse infrared background experiment search for the cosmic infrared background. III. Separation of galactic emission from the infrared sky brightness. *The Astrophysical Journal* 508: 74–105.

Bernatowicz, T., Bradley, J., Amari, S., Messenger, S., Lewis, R. 1999. *New Kinds of Massive Star Condensates in a Presolar Graphite from Murchison*, 30th Annual Lunar and Planetary Science Conference, Houston, TX, March 15–29, 1999.

Boulanger, F. and Perault, M. 1988. Diffuse infrared emission from the galaxy. I—Solar neighborhood. *The Astrophysical Journal* 330: 964–985.

Bradley, J., Dai, Z.R., Erni, R., Browning, N., Graham, G., Weber, P., Smith, J. et al. 2005. An astronomical 2175 Å feature in interplanetary dust particles. *Science* 307: 244–247.

Burbidge, E.M., Burbidge, G.R., Fowler, W.A., and Hoyle, F. 1957. Synthesis of the elements in stars. *Reviews of Modern Physics* 29: 547–650.

Carpenter, J.D., Stevenson, T.J., Fraser, G.W., Lapington, J.S., and Brandt, D. 2005. Dust detection in the ISS environment using filmed microchannel plates. *Journal of Geophysical Research* 110: E05013.

Ceplecha, Z., Borovicka, J., Elford, W.G., Revelle, D.O., Hawkes, R.L., Porubcan, V., and Simek, M. 1998. Meteor phenomena and bodies. *Space Science Reviews* 84: 327–471.

Dennerl, K., Englhauser, J., and Trümper, J. 1997. X-ray emissions from comets detected in the Röntgen X-ray satellite all-sky survey. *Science* 277: 1625–1630.

Draine, B.T. 2003. Interstellar dust grains. *Annual Review of Astronomy and Astrophysics* 41: 241–289.

Draine, B.T. 2004. Astrophysics of dust in cold clouds. In *The Cold Universe, Saas-Fee Advanced Course*, Vol. 32, Lecture Notes 2002 of the Swiss Society for Astronomy and Astrophysics, eds. A.W. Blain, F. Combes, B.T. Draine, D. Pfenniger, and Y. Revaz, pp. 213–303. Berlin, Germany: Springer.

Finkbeiner, D.P., Davis, M., and Schlegel, D.J. 1999. Extrapolation of galactic dust emission at 100 microns to cosmic microwave background radiation frequencies using FIRAS. *The Astrophysical Journal* 524: 867–886.

Fraser, H.J., Bisschop, S.E., Pontoppidan, K.M., Tielens, A.G.G.M., and van Dishoeck, E.F. 2005. Probing the surfaces of interstellar dust grains: The adsorption of CO at bare grain surfaces. *Monthly Notices of the Royal Astronomical Society* 356: 1283–1292.

Gelinas, L.J., Lynch, K.A., Kelley, M.C., Collins, S., Baker, S., Zhou, Q., and Friedman, J.S. 1998. First observation of meteoritic charged dust in the tropical mesosphere. *Geophysical Research Letters* 25: 4047–4050.

Grimm, M., Langer, B., Schlemmer, S., Lischke, T., Widdra, W., Gerlich, D., Becker, U., and Rühl, E. 2004. New setup to study trapped nano-particles using synchrotron radiation, Eighth international conference on synchrotron radiation instrumentation. *AIP Conference Proceedings* 705: 1062–1065.

Grün, E., Zook, H.A., Baguhl, M., Balogh, A., Bame, S., Fechtig, H., Forsyth, R. et al. 1993. Discovery of Jovian dust streams and interstellar grains by the Ulysses spacecraft. *Nature* 362: 428–430.

Grün, E., Zook, H.A., Fechtig, H., and Giese, R.H. 1985. Collisional balance of the meteoritic complex. *Icarus* 62: 244–272.

Guillois, O., Ledoux, G., and Reynaud, C. 1999. Diamond infrared emission bands in circumstellar media. *The Astrophysical Journal Letters* 521: L133–L136.

Hamilton, D.P. and Burns, J.A. 1993. Ejection of dust from Jupiter's gossamer ring. *Nature* 364: 695–699.

Havnes, O., Trøim, J., Blix, T., Mortensen, W., Næsheim, L.I., Thrane, E., and Tønnesen, T. 1996. First detection of charged dust particles in the Earth's mesosphere. *Journal of Geophysical Research* 101: 10839–10848.

Herbig, G.H. 1995. The diffuse interstellar bands. *Annual Review of Astronomy and Astrophysics* 33: 19–74.

Hollenbach, D.J., Yorke, H.W., and Johnstone, D. 2000. Disk dispersal around young stars. In *Protostars and Planets IV*, eds. V. Mannings, A.P. Boss, and S.S. Russell, pp. 401–453. Tucson, AZ: University of Arizona Press.

Horanyi, M., Robertson, S., Smiley, B., Gumbel, J., Witt, G., and Walch, B. 2000. Rocket-borne mesospheric measurements of heavy (m > 10 amu) charge carriers. *Geophysical Research Letters* 27: 3825–3828.

Hoyle, F. and Wickramasinghe, N.C. 1970. Dust in supernova explosions. *Nature* 226: 62–63.

Hunten, D.M., Turco, R.P., and Toon, O.B. 1980. Smoke and dust particles of meteoric origin in the mesosphere and stratosphere. *Journal of the Atmospheric Sciences* 37: 1342–1357.

Ip, W.-H. and Chow, V.W. 1997. Note: On hypervelocity impact phenomena of microdust and nano X-ray flares in cometary comae. *Icarus* 130: 217–221.

Jones, A.P., Tielens, A.G.G.M., Hollenbach, D.J., and McKee, C.F. 1994. Grain destruction in shocks in the interstellar medium. *The Astrophysical Journal* 433: 797–810.

Kamitsuji, K., Ueno, S., Suzuki, H., Kimura, Y., Sato, T., Tanigaki, T., Kido, O., Kurumada, M., and Kaito, C. 2004. Direct observation of the metamorphism of silicon oxide grains. *Astronomy and Astrophysics* 422: 975–979.

Kempf, S., Srama, R., Horányi, M., Burton, M., Helfert, S., Moragas-Klostermeyer, G., Roy, M., and Grün, E. 2005. High-velocity streams of dust originating from Saturn. *Nature* 433: 289–291.

Kimura, H. and Mann, I. 1998. Electric charge of interstellar dust in the solar system. *The Astrophysical Journal* 499: 454–462.

Kimura, Y., Ueno, H., Suzuki, H., Tanigaki, T., Sato, T., Saito, Y., and Kaito, C. 2003. Dynamic behavior of Au cluster on the surface of silicon nanoparticles. *Physica E* 19: 298–302.

Kouchi, A., Nakano, H., Kimura, Y., and Kaito, C. 2005. Novel routes for diamond formation in interstellar ices and meteoritic parent bodies. *The Astrophysical Journal Letters* 626: L129–L132.

Krüger, H., Altobelli, N., Anweiler, B., Dermott, S.F., Dikarev, V., Graps, A., Grün, E. et al. 2006. Five years of Ulysses dust data: 2000–2004. *Planetary and Space Science* 54: 932–956.

Ledoux, G., Guillois, O., Huisken, F., Kohn, B., Porterat, D., and Reynaud, C. 2001. Crystalline silicon nanoparticles as carriers for the extended red emission. *Astronomy and Astrophysics* 377: 707–720.

Li, A. and Draine, B.T. 2001. Infrared emission form interstellar dust II, The diffuse interstellar medium. *The Astrophysical Journal* 554: 778–802.

Li, A. and Draine, B.T. 2002. Are silicon nanoparticles an interstellar dust component? *The Astrophysical Journal* 564: 803–812.

Lynch, K.A., Gelinas, L.J., Kelley, M.C., Collins, R.L., Widholm, M., Rau, D., MacDonald, E., Liu, Y., Ulwick, J., and Mace, P. 2005. Multiple sounding rocket observations of charged dust in the polar winter mesosphere. *Journal of Geophysical Research* 110: A03302.

Mann, I. 2009. Evolution of dust and small bodies: Physical processes. In *Small Bodies in Planetary Systems*, eds. I. Mann, A. Nakamura, and T. Mukai, *Lecture Notes in Physics*, Vol. 758, pp. 189–230. Berlin, Germany: Springer.

Mann, I. and Kimura, H. 2000. Interstellar dust properties derived from mass density, mass distribution, and flux rates in the heliosphere. *Journal of Geophysical Research* 105: 10317–10328.

Mann, I. and Murad, E. 2005. On the existence of silicon nanodust near the Sun. *The Astrophysical Journal Letters* 624: L125–L128.

Mann, I., Murad, E., and Czechowski, A. 2007. Nanoparticles in the inner solar system. *Planetary and Space Science* 55: 1000–1009.

McBride, N. and McDonnell, J.A.M. 1999. Meteoroid impacts on spacecraft: Sporadics, streams, and the 1999 Leonids. *Planetary and Space Science* 47: 1005–1013.

McKee, C.F. and Ostriker, J.P. 1977. A theory of the interstellar medium—Three components regulated by supernova explosions in an inhomogeneous substrate. *Astrophysical Journal* 218: 148–169.

Megner, L. 2008. Meteoritic aerosols in the middle atmosphere. Doctoral thesis. Stockholm, Sweden: Stockholm University.

Meyer-Vernet, N., Maksimovic, M., Czechowski, A., Mann, I., Zouganelis, I., Goetz, K., Kaiser, M.L., Bougeret, J.-L., and Bale, S.D. 2009. Dust detection by the wave instrument on stereo: Nanoparticles picked up by the solar wind? *Solar Physics* 256: 463–474.

Minato, T., Koehler, M., Kimura, H., Mann, I., and Yamamoto, T. 2004. Momentum transfer to interplanetary dust from the solar wind. *Astronomy and Astrophysics Letters* 424: L13–L16.

Morfill, G.E. and Grün, E. 1979. The motion of charged dust particles in interplanetary space: I—The zodiacal dust cloud; II—Interstellar grains. *Planetary and Space Science* 27: 1269–1292.

Northrop, T.G., Lisse, C.M., Mumma, M.J., and Desch, M.D. 1997. A possible source of the X-rays from comet Hyakutake. *Icarus* 127: 246–250.

Nuth, J.A., Rietmeijer, F.J.M., and Hill, H.G.M. 2002. Condensation processes in astrophysical environments: The composition and structure of cometary grains. *Meteoritics and Planetary Science* 37: 1579–1590.

Ott, U. 2003. The most primitive material in meteorites. In *Astromineralogy*, ed. T.K. Henning, *Lecture Notes in Physics*, Vol. 609, pp. 236–265. Berlin, Germany: Springer.

Owens, A., Parmar, A.N., Oosterbroek, T., Orr, A., Antonelli, L.A., Fiore, F., Schultz, R., Tozzi, G.P., Maccarone, M.C., and Piro, L. 1998. Evidence for dust-related X-ray emission from comet C/1995 O1 (Hale-Bopp). *Astrophysical Journal Letters* 493: L47–L50.

Pauls, H.L. and Zank, G.P. 1996. Interaction of a nonuniform solar wind with the local interstellar medium. *Journal of Geophysical Research* 101: 17081–17092.

Rapp, M., Hedin, J., Strelnikova, I., Friedrich, M., Gumbel, J., and Lübken, F.-J. 2005. Observations of positively charged nanoparticles in the nighttime polar mesosphere. *Geophysical Research Letters* 32: L23821.

Rietmeijer, F.J.M. 2002. Collected extraterrestrial materials: Interplanetary dust particles micrometeorites, meteorites, and meteoric dust. In *Meteors in the Earth's Atmosphere*, eds. E. Muradand and I.P. Williams, pp. 215–245. Cambridge, U.K.: Cambridge University Press.

Takeuchi, T. 2009. From protoplanetary disks to planetary disks: Gas dispersal and dust growth. In *Small Bodies in Planetary Systems*, eds. I. Mann, A. Nakamura, and T. Mukai, *Lecture Notes in Physics*, Vol. 758, pp. 1–35. Berlin, Germany: Springer.

Tielens, A.G.G.M. 2008. Interstellar polycyclic aromatic hydrocarbon molecules. *Annual Review of Astronomy and Astrophysics* 46: 289–337.

Utterback, N.G. and Kissel, J. 1990. Attogram dust cloud a million kilometers from comet Halley. *The Astronomical Journal* 100: 1315–1322.

van Dishoeck, E.F. 2008. Organic matter in space—An overview. In *Organics Matter in Space*, ed. S. Kwok, pp. 3–16. Proceedings International Union Symposium No. 251. Cambridge, U.K.: Cambridge University Press.

van Dishoeck, E.F. and Blake, G.A. 1998. Chemical evolution of star-forming regions. *Annual Review of Astronomy and Astrophysics* 36: 317–368.

Wehry, A. and Mann, I. 1999. Identification of ß-meteoroids from measurements of the dust detector onboard the Ulysses spacecraft. *Astronomy and Astrophysics* 341: 296–303.

Witt, A.N. and Vijh, U.P. 2004. Extended red emission: Photoluminescence by interstellar nanoparticles. In *Astrophysics of Dust*, ed. A.N. Witt, G.C. Clayton, and B.T. Draine, *ASP Conference Series*, Vol. 309, pp. 115–139. San Francisco, CA: Astronomical Society of the Pacific.

Witt, A.N., Gordon, K.D., and Furton, D.G. 1998. Silicon nanoparticles: Source of extended red emission? *The Astrophysical Journal Letters* 501: L111–L114.

Witt, A.N., Gordon, K.D., Vijh, U.P., Sell, P.H., Smith, T.L., and Xie, R.-H. 2006. The excitation of extended red emission: New constraints on its carrier from hubble space telescope observations of NGC 7023. *The Astrophysical Journal* 636: 303–315.

Wyatt, M.C. 2009. Dynamics of small bodies in planetary systems. In *Small Bodies in Planetary Systems*, eds. I. Mann, A. Nakamura, and T. Mukai, *Lecture Notes in Physics*, Vol. 758, pp. 37–70. Berlin, Germany: Springer.

Zinner, E. 1998. Stellar nucleosynthesis and the isotopic composition of presolar grains from primitive meteorites. *Annual Review of Earth and Planetary Sciences* 26: 147–188.

Zolensky, W.E., Pieters, C., Clark, B., and Papike, J.J. 2000. Small is beautiful: The analysis of nanogram-sized astromaterials. *Meteoritics and Planetary Science* 35: 9–29.

Zook, H.A., Grün, E., Baguhl, M., Hamilton, D.P., Linkert, G., Liou, J.-C., Forsyth, R., and Phillips, J. 1996. Solar wind magnetic field bending of Jovian dust trajectorie. *Science* 274: 1501–1503.

III

Nanoparticles in Contact

20

Ordered Nanoparticle Assemblies

Aaron E. Saunders
University of Colorado at Boulder

Brian A. Korgel
The University of Texas at Austin

20.1 Introduction

Nanoparticles can be made from many different kinds of materials, including metals and semiconductors, with diameters ranging from as small as about 1 nm in diameter up to micrometers (Kwon and Hyeon 2008). When nanoparticles are smaller than 10 nm or so, they exhibit physical properties that differ from the bulk material (Alivisatos 1996). These properties are size dependent and can be tuned to a significant extent by modifying the dimensions of the nanoparticles. Because their properties depend not only on their composition but also on their size, they represent a unique and interesting class of materials.

Synthetic methods for nanoparticles have been developed that can provide exquisite size control, yielding very narrow size distributions, and control of the shape from spheres to rods to disks (Lee et al. 2007). These techniques generally rely on nanoparticle surface modification with molecular coatings that prevent aggregation and stabilize their size and shape. The organic molecular coating enables their dispersion in solvents for convenient manipulation and deposition on substrates and is vital to their self-assembly into ordered superlattices. Dispersions of nanoparticles are optically clear and stable, without gravitational settling of nanoparticles over time.

When nanoparticles have a sufficiently narrow size distribution, they can be assembled like building blocks into ordered periodic arrays, or superlattices, like atoms crystallized into a solid. Because of their size, they diffuse rapidly when dispersed in a solvent; the superlattice formation does not require special external forces and can occur by the simple evaporation of a solvent from a dispersion of nanoparticles dropped onto a surface. Interest in nanoparticle assemblies arises in part from the fact that they are composite materials with macroscopic properties that evolve from those of the individual nanoparticles and their collective interactions, which can be adjusted to some extent by manipulating how they are assembled. With their new properties, nanoparticle assemblies are being explored for use in thin-film transistors (TFTs), sensors, floating gate memory elements, light emitting diodes (LEDs), thermoelectrics, and photovoltaics (PVs) (Markovich et al. 1999).

This chapter describes the fundamentals of nanoparticle assembly, beginning with a description of the nanoparticles themselves and their chemical makeup. Methods of characterizing nanoparticle assemblies are described, with a focus on small-angle x-ray scattering (SAXS), which is one of the most powerful techniques for determining the size of nanoparticles and the structure of nanoparticle assemblies. Finally, the underlying forces that drive self-assembly are discussed. Superlattice assembly occurs spontaneously, driven by the thermodynamics of the system. The interparticle interactions determine these thermodynamics and depend on their size and surface coating, which can be manipulated, thus providing a convenient model system to study the phase behavior of simple fluids (Gelbart and Ben-Shaul 1996, Gast and Russel 1998).

20.2 Background

20.2.1 Nanocrystal Superlattice

Figure 20.1 shows transmission and scanning electron microscopy images of ordered assemblies of gold nanoparticles, or superlattices. These superlattices were formed by dropping a small volume of solvent with dispersed nanoparticles onto a substrate and then letting the solvent evaporate. The superlattices form spontaneously. Because the nanoparticles dispersed

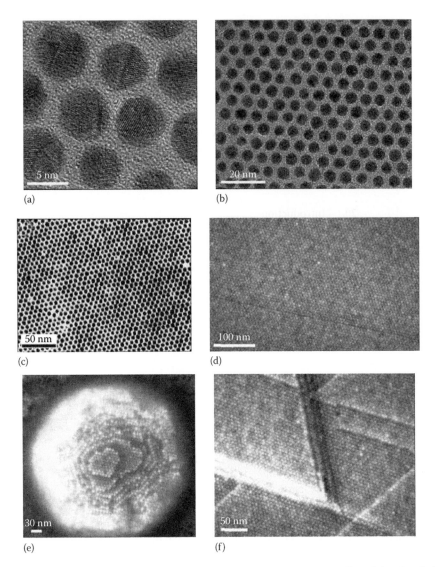

FIGURE 20.1 TEM and SEM images of gold nanocrystal superlattices. The nanocrystals are sterically stabilized with dodecanethiol. The dodecanethiol layer around each nanocrystal maintains the observed separation between the solid gold cores. Images (a–c) are TEM images and (d–f) are SEM images.

in solvents are called "colloids," these superlattices are sometimes referred to as "colloidal crystals." Colloidal crystals have been studied for nearly a century, but the focus of this research until the early 1990s was on relatively large, micrometer, and sub-micrometer-diameter colloidal particles.

20.2.2 Historical Background

20.2.2.1 Opals and Colloidal Crystals of Submicrometer Particles

Opals are naturally occurring superlattices consisting of uniform spherical silica particles that can range between 170 and 350 nm in diameter. These natural assemblies of sub-micrometer-diameter particles were studied in the 1930s by Levin and Ott using x-ray techniques (Levin and Ott, 1933), and in 1946, Copisarow and Copisarow described the opal as an array of colloidal silica particles

(Copisarow and Copisarow 1946). This description was confirmed by electron microscopy in 1964 by Sanders (Sanders, 1964). The primary interest of this research was to understand the characteristic color, or "fire," of opals from different parts of the world, and these structural studies confirmed a century-old hypothesis that the brilliant color produced by opals upon light exposure results from the diffraction of light by the 3-D colloidal array.

Throughout the 1980s, researchers studied colloid crystallization systematically, as synthetic routes to size-monodisperse colloids became available. These particles were much larger than the gold nanoparticles shown in Figure 20.1, with diameters of hundreds of nanometers. These colloids were also typically dispersed in water and stabilized with adsorbed ions, giving rise to repulsive, electrostatic double-layer forces. These long-range repulsive forces countered the shorter range van der Waals attractions between particles and prevented their aggregation.

By tuning the refractive index of the solvent, the van der Waals attractions could be minimized so that the particles interacted basically as hard spheres with a weak repulsion, and two types of interparticle potentials were considered in great detail to understand colloidal ordering in these studies: the hard-sphere potential and long-range electrostatic repulsion. The electrostatic repulsion between particles could be tuned by adjusting the ionic strength of the solvent and transitions between different ordered structures, like from a face-centered cubic (fcc) to a body-centered cubic (bcc) structure, were observed (Gast and Russel 1998). By changing the refractive index of the solvent, the interparticle potentials could be adjusted to mimic nearly perfectly hard spheres. Experimental measurements of crystallization kinetics and the general phase behavior were compared directly to model calculations (Gast and Russel 1998).

20.2.2.2 Nanoparticle Assemblies

Ordered assemblies of nanoparticles made with much smaller diameters, of about 10 nm or less, began to be studied in detail in the early 1990s (Murray et al. 2000). Sterically stabilized colloidal particles, of diameters of 10 nm or less, were found to spontaneously organize into close-packed colloidal crystals, or superlattices, by simply allowing the solvent to evaporate from a dispersion. The nanoparticles were too small to be affected by gravity and their very fast diffusion enabled the nanoparticles to arrange themselves into quasi-equilibrium structures within the time allowed by the evaporation of a thin film of solvent. Simply by drop casting of a dispersion on a TEM grid for imaging would reveal relatively extended and ordered close-packed monolayers when the nanoparticles were sufficiently monodisperse. Some of the earliest examples of this, include studies in 1989 by Schmid and Lehnert (1989), who observed Au nanocrystal superlattice formation and Bentzon et al. (1989) who observed ordered close-packed monolayers of iron oxide nanocrystals. Perhaps, the most striking early example of nanocrystal self-assembly into superlattices was from Bawendi's group in 1996, when they demonstrated ordered superlattice assemblies of CdSe nanocrystals, with confirmation of a relatively long-range order by small-angle x-ray scattering (SAXS) in addition to electron microscopy images (Murray et al. 1995). These studies reiterated that non-aggregating, size monodisperse colloidal particles organize into superlattices at high-volume fractions and this motivated many research groups to develop effective methods to produce monodisperse nanocrystals and to understand their assembly.

20.3 State of the Art

20.3.1 Organic Monolayer-Stabilized Nanoparticles

Synthetic advances now enable a wide variety of organic monolayer-coated nanoparticles, including metals and semiconductors (Klimov 2004). These nanoparticles are usually crystalline and are referred to as nanocrystals, with diameters less than about 20 nm or so. Their surfaces are coated with organic ligands

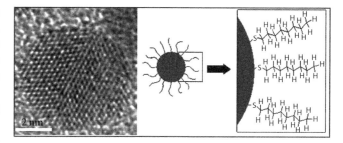

FIGURE 20.2 (Left) TEM image of a silver (Ag) nanocrystal. The crystalline nanoparticle is resting on an amorphous carbon support, appearing as the background behind the crystalline core. The nanocrystal is coated with dodecanethiol, which is a twelve carbon hydrocarbon chain with a thiol (–SH) functional group. The thiol chemisorbs to the Ag surface and the nanocrystals are coated with a monolayer of these hydrophobic ligands.

that maintain their size and prevent aggregation by steric stabilization. Figure 20.2 shows a TEM image of a dodecanethiol-stabilized silver nanocrystal. The crystalline core is visible in the TEM image. It rests on a carbon support and the hydrocarbon coating is not visible in the image because it blends in with the carbon support. Dodecanethiol is a 12-carbon long hydrocarbon with a thiol (–SH) functional group at its end. The thiol bonds to the silver surface and tethers the hydrocarbons onto the nanocrystal surface. In a good solvent, like toluene or chloroform, these hydrocarbons are very soluble and extend into the solvent. This extended ligand layer then provides a steric layer that pushes the inorganic cores apart and disperses them. Even after the solvent is removed, the organic ligands are present on the nanocrystal surfaces and keep them separated in the superlattice. Since the inorganic core diameters are relatively small, this brush layer can be quite thin and still provide dispersibility and prevent aggregation, of the order of nanometers.

Organic monolayer-stabilized nanocrystals are typically made by the process of arrested precipitation. Usually in a solvent, a material is precipitated by decomposing or converting molecular reactants to a crystalline solid. Capping ligands are added to the reactions—the organic molecules with functional groups that adsorb to the inorganic material—to passivate the nanocrystal surfaces and prevent their uncontrolled growth, thus stabilizing their size and preventing aggregation. Nanocrystals can be produced in aqueous or organic solvents by this process, and then dispersed in solvents to create nanocrystal "inks." These nanocrystals tend to be relatively stable in air, even for materials that oxidize readily in air, due to the stabilizing ligand layer bonded to the inorganic surface. Perhaps most importantly, arrested precipitation can yield significant quantities of nanoparticles with the very narrow size distributions required for ordered superlattice formation.

The use of capping ligands is also quite important for nanoparticle assembly, because they provide the ability to reversibly disperse and precipitate the nanoparticles without irreversible aggregation. This is generally not possible with charge-stabilized colloids, as the adsorbed molecular layer that provides the stabilization in solution is too thin to prevent the inorganic surfaces of the particles from touching when the solvent is removed. This is not necessarily a problem for much larger particles hundreds of nanometers in diameter, but nanoparticles in the 2–10 nm diameter range are much more susceptible to coalescence due to their small size, and a passivation layer is necessary to prevent irreversible aggregation of nanoparticles in this size range when the solvent is removed.

20.3.2 Nanocrystal Superlattice Characterization

When the size distribution is sufficiently narrow, sterically stabilized nanocrystals self-assemble into periodic arrays, or superlattices, when the solvent is removed (by evaporation for example). Figure 20.1 shows examples of electron microscopy images of gold nanocrystal superlattices. Transmission and scanning electron microscopy (TEM and SEM) provide straightforward techniques to characterize the order of the assemblies. These techniques, however, only provide a snapshot of the structure of the materials with a relatively narrow field of view. Small angle x-ray scattering (SAXS) provides a powerful technique for characterizing the average structure of the assemblies. Both techniques are complementary and when used in concert, provide an accurate measure of the structure of the assemblies.

20.3.2.1 Nanocrystal Characterization by TEM and SEM

Figure 20.1 shows high-resolution scanning electron microscopy (HRSEM) and transmission electron microscopy (HRTEM) images of nanocrystals ordered into arrays in two and three dimensions. In TEM imaging, the beam passes through the sample, and therefore must be relatively thin to obtain a meaningful image. TEM provides essentially a two-dimensional image of a three-dimensional structure. When nanocrystals are ordered into a monolayer, TEM is an excellent tool, providing a good resolution of the nanoparticle size, shape, and structural order. TEM also provides very high resolution images. The best TEM images, acquired with highly monochromatic field emission electron sources and aberration correction, can provide a resolution at the atomic scale.

When nanoparticles stack into thicker structures, like a colloidal crystal, TEM becomes less useful, as the beam must penetrate through the entire sample to obtain meaningful images. For thicker assemblies, SEM is a better tool. In SEM, electrons reflected from the sample surface provide images that have a three-dimensional appearance and reveal topographical details about the sample that TEM cannot provide. SEM has a lower resolution than TEM, but scanning electron microscopes can now provide images with a resolution of only a few nanometers, and individual nanocrystals that are as small as 4 nm or so can be resolved in superlattices. One challenge with SEM imaging, however, is the charging of the sample under the electron beam. In the past, samples required coating with metal. This could damage the sample. Now, lower energy high-resolution

SEMs exist for imaging uncoated samples and these are ideal tools for examining nanocrystal superlattices. Nonetheless, the nanocrystals are coated with organic ligands and the samples can charge under the microscope if the nanocrystals are made of insulating and semiconductor materials and it can be a challenge to obtain images that resolve individual nanocrystals in superlattices for these materials.

20.3.2.2 Small Angle X-Ray Scattering (SAXS)

TEM and SEM provide direct images of nanoparticle assemblies to provide a real-space snapshot of the order in the sample. SEM and TEM allow a direct visualization of nanoparticles on a substrate, but these techniques provide the visualization of only a very small fraction of the nanoparticles in the entire sample. In contrast to electron microscopy, SAXS probes a large sample volume and provides an ensemble average of the characteristic size of the nanoparticles and their order. TEM and SEM are nonetheless very powerful techniques because they also provide direct evidence of the size and shape of the nanoparticles; whereas, SAXS provide only averaged information about the nanoparticles. Ideally, it is best to study nanoparticle samples using both imaging techniques like TEM and SEM and scattering methods, like SAXS, to accurately characterize the sample.

Another useful aspect of SAXS is that it can be used to characterize both solid films of nanoparticles and nanoparticles dispersed in solvents. For imaging by TEM or SEM, the nanoparticles must be deposited onto a substrate, and removing the sample can lead to aggregation and coalescence, changing the size distribution significantly. SAXS measurements of dispersions provide direct information about not only the nanocrystal size distribution, but also how the nanocrystals might be interacting in the solvent (Korgel et al. 1998, Saunders and Korgel 2004.

Figure 20.3 shows examples of SAXS data acquired for a sample of sterically stabilized iron oxide nanocrystals. In this data set, SAXS measurements were acquired for a dispersion to determine the size distribution of the nanocrystals and then for a superlattice formed by evaporating the solvent to determine the structural order of the assembly. In an SAXS measurement, the scattered x-ray intensity is measured as a function of the scattering angle θ. Typically, the data is plotted as a function of the scattering vector $q = (4\pi/\lambda) \sin(\theta/2)$, which relates also to the x-ray wavelength, λ. q is also convenient since it relates directly to the characteristic d-spacing as $q = 2\pi/d$. The angular dependence of the scattered intensity $I(q)$, can be expressed by $I(q) \propto P(q)S(q)$ (Glatter and Kratky 1982).

The scattering of x-rays due to the difference in the electron density between the nanocrystal and the surrounding media (either a solvent or a vacuum) is accounted for by a shape factor, $P(q)$. The shape factor for single homogeneous spherical particles of radius R, for example, can be expressed as $P(q) = [3(\sin(qR) - qR \cos(qR))/(qR)^3]^2$. Kerker (1969) provides an extensive list of shape factors determined for particles of a variety of shapes. In dilute dispersions, the total scattered intensity $I(q)$, can be approximated as the sum of the scattering from *non-correlated*

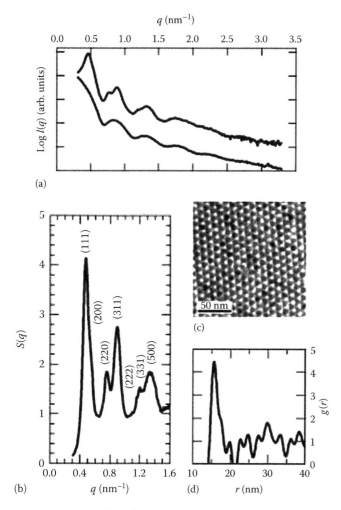

(a)

(b)

(c)

(d)

FIGURE 20.3 Small-angle x-ray scattering (SAXS) data for a dispersion of iron oxide nanocrystals (a, bottom curve), and the same particles dried into an ordered lattice (a, top curve). A TEM image of a superlattice made using the same particles as shown in (c). The structure factor $S(q)$, can be determined from the data in (a), and the peaks can be indexed to show that the nanocrystals order into a face-centered cubic (fcc) arrangement with a lattice constant $a \approx 23$ nm (b). (d) The Fourier transform of $S(q)$ provides the radial distribution function $g(r)$.

individual particles. The scattering intensity is a strong function of size, depending on the particle radius as $I(R) \propto R^6$. All nanocrystal samples are not perfectly monodisperse and exhibit a size distribution. To determine the average diameter and size distribution, the shape of the nanocrystals and the form of the size distribution must be assumed. This is one reason why complementary TEM and SEM images are also typically needed in concert with SAXS to provide an accurate picture of the nanocrystal sample. The scattering data for a sample with a size distribution (for independently scattering nanocrystals) then takes the form,

$$I(q) \propto \int P(q)N(R)R^6 dR, \qquad (20.1)$$

where $N(R)$ is the size distribution of the particle ensemble. For example, $N(R)$ can be assumed to be a Gaussian distribution,

$$N(R) = \frac{1}{\sqrt{2\pi\sigma^2}} \exp\left(-\frac{(R-\bar{R})^2}{2\sigma^2}\right), \qquad (20.2)$$

where

σ is the standard deviation
$\bar{R}$ is the average radius

Multiparticle effects, *diffraction*, arise when the arrangement of particles deviates from an entirely random distribution, i.e., spatial correlations exist between the locations of two or more particles. These correlations may be significant, as in the case of a superlattice, in which nanocrystals are densely packed in an ordered array, and dominate the scattering profile, or they could be subtle, as in a dispersion in which nanocrystals are slightly sticky, with weak aggregation. The effects of these correlations manifest themselves in the scattering as deviations from $P(q)$, and are contained in the structure factor $S(q)$.

As shown in Figure 20.3, SAXS provides a way to measure both the shape factor and the structure factor independently, provided that the nanocrystals can be dispersed well without significant aggregation. $P(q)$ is determined from the SAXS of the dispersion—the bottom scattering curve in Figure 20.3a. In this example, the iron oxide nanocrystals were dispersed in toluene, which is a good solvent for these hydrophobic nanocrystals. (It should be noted that some solvents, like Cl-containing solvents, such as chloroform, strongly absorb x-rays, and are not suitable for SAXS measurements). $S(q)$ can then be determined for the superlattice by dividing the total scattering (the upper curve in Figure 20.3a) by the measured $P(q)$. Figure 20.3b shows the diffraction profile obtained using this procedure. These diffraction peaks can then be indexed in the same way that an x-ray diffraction pattern of a crystalline solid would be indexed to determine the superlattice structure. The superlattice of iron oxide nanocrystals in Figure 20.3 indexes to a face-centered cubic (fcc) superlattice with a lattice constant of $a_{SL} = 23$ nm.

The structure factor can also be converted to a radial distribution function $g(r)$,

$$g(r) = 1 + \frac{1}{2\pi^2 nr}\int q[S(q)-1]\sin(qr)dq. \qquad (20.3)$$

$g(r)$ expresses the probability of finding a particle a distance r away from an arbitrary central particle in a system with particle number density n. Application of Equation 20.3 to the structure factor data results in the plot of $g(r)$ shown in Figure 20.3d.

Figure 20.3c shows a TEM image of a superlattice of the iron oxide nanocrystals examined by SAXS. A lattice constant for the superlattice can also be determined from the TEM image. Only one specific lattice plane, however, is imaged by TEM. In Figure 20.3c, this is a (111) plane of the fcc superlattice. For a cubic crystal, the interplanar d-spacing relates to the lattice constant a_{SL} as

$$a_{SL} = d_{hkl}\sqrt{h^2 + k^2 + l^2}. \qquad (20.4)$$

The superlattice lattice constant, a_{SL}, relates to the center-to-center nearest neighbor spacing, Δ, determined from the TEM, as $a_{SL} = \Delta\sqrt{2}$. SAXS and TEM provide complementary measures of the interparticle spacing in nanoparticle assemblies. There is usually a slight deviation between the two methods, due in part to the fact that TEM images are usually obtained for monolayers of nanocrystals, whereas SAXS exams a thicker, three-dimensional film in which the structure is not significantly affected by the underlying substrate.

20.3.3 Nanocrystals as Soft Spheres

The ligands are a key component of nanocrystal materials and occupy a significant amount of space in the superlattice. Consider a typical nanocrystal, say a 5 nm diameter Au nanocrystal that is coated with a monolayer of dodecanethiol. A fully extended C_{12} hydrocarbon chain is about 1.7 nm long. Therefore, if we consider both the volume of the inorganic core ($V_{Au} = (4/3)\pi R^3 = 65.5\,nm^3$) and estimate the volume occupied by the ligands ($V_{ligands} = [(4/3)\pi(R + \delta)^3] - [(4/3)\pi R^3] = 244.9\,nm^3$), it is obvious that the ligands occupy a significant volume of the nanoparticle—the volume occupied by the ligands is more than three times larger than the volume occupied by the inorganic core. In the superlattice, these ligands fill all the space around the nanocrystals.

SAXS measurements enable a determination of the volume excluded by the ligands. SAXS measurements of the nanocrystal dispersions are sensitive only to the inorganic core of the nanoparticle and the average size determined from the measurement is the inorganic nanocrystal core size, ignoring the ligands. The lattice constant measured by SAXS for the superlattice, on the other hand, contains information about the volume occupied by the ligands. From these two pieces of information, a detailed understanding of the ligands can be obtained.

In the example shown in Figure 20.3, the SAXS measurement of the dispersion of iron oxide nanocrystals gave an average diameter (d_p) of 14.9 nm and the center-to-center nearest neighbor spacing measured from the SAXS of the superlattice is $\Delta = a_{SL}/\sqrt{2} = 16.3$ nm. Therefore, the nearest neighbor edge-to-edge separation between particles is $\delta = \Delta - d_p = 16.3 - 14.9 = 1.4$ nm. This distance is shorter than the length of the fully extended chain. This is because there is a significant amount of volume in the interstitial spaces between nanocrystals that must be filled by the ligands. Consider that in an fcc lattice, the spherical particles only occupy 74% of the available volume. The unit cell has a volume of $a_{SL}^3 = 23^3 = 12,167\,nm^3$. Since the fcc unit cell has an equivalent of 4 nanocrystals within it, the volume occupied by the inorganic cores is only $4V_{inorganic} = 4 \times ((4/3)\pi R^3) = 6928\,nm^3$—only 57% of the superlattice is filled by the inorganic cores. The ligands fill the remaining space in the superlattice.

The ligands fill the intervening space between the inorganic cores in two different ways. First of all, there is a relatively rigid layer of ligands, essentially forming a shell between the nanocrystals; it is this ligand "shell" that intervenes directly between the neighboring inorganic cores and has a thickness of $\delta/2$, or

0.7 nm in this case. This ligand shell creates an *effective* hard-sphere radius of the nanocrystals of $R_{eff} = R + (\delta/2)$. These spheres occupy (four nanocrystals per unit cell) 9070 nm³ per unit cell volume, or ~74% of the unit cell. The remaining portion of the capping ligand layer is flexible and fills the free volume gaps in the fcc lattice that account for the remaining 26% of the unit cell. By accounting for this additional volume occupied by the ligands (3163 nm³), the ligand coverage per particle can be deduced. In this case, the ligands occupy 1310 nm³ per particle. The iron oxide nanocrystals are coated with oleic acid, which is a C_{18} hydrocarbon chain. The volume occupied per ligand can be estimated using the relation (Israelachivili 1992):

$$v = (27.4 + 26.9n)\times 10^{-3}\,nm^3. \tag{20.5}$$

Each oleic acid molecule occupies approximately 0.51 nm³. Therefore, there are approximately 2520 capping ligand molecules per particle. Since the total surface area of each inorganic core is $4\pi R^2 = 697.5\,nm^2$, each capping ligand occupies 0.28 nm². A typical circular footprint for a hydrocarbon molecule is ~0.17 nm² (Israelachivili 1992). Therefore, in this particular nanocrystal sample, the capping ligand layer is about 25%–30% more sparse than a close-packed monolayer.

These nanocrystals have been termed "soft spheres" in comparison to hard sphere particles (Korgel et al. 1998). Part of the capping ligand layer ends up being relatively rigid, while the outermost part of the ligand layer is flexible and can fill space as needed in the superlattice. The analogy to hard-sphere particles provides some insight into the forces that drive self-assembly.

20.3.4 Forces Directing Nanoparticle Assembly

20.3.4.1 Hard-Sphere Assembly and Quasi-Equilibrium Structure

The simplest conceptual model of sterically stabilized nanocrystals is to consider the particles as hard spheres, interacting with interparticle potential $\Phi(r)$,

$$\Phi = \begin{cases} \infty & \text{if } r \leq 2R \\ 0 & \text{if } r > 2R \end{cases}, \tag{20.6}$$

where

r is the center-to-center interparticle separation
R is the particle radius

Hard spheres are like billiard balls, they are particles that exclude a finite volume but exhibit an infinite repulsive force upon touching. The nanocrystals of course, are not truly hard spheres because the ligand shell provides a relatively soft boundary that can be compressed to a limited extent. There is also in reality a small attraction between the nanocrystals, but these attractive forces are relatively small; in fact, they must be less than kT since there is no aggregation in solution. These attractions can be considered to some extent as a perturbation on the hard-sphere potential. At any rate, as a starting point, it is use-ful to consider the hard-sphere potential for the nanocrystals, as it provides a useful conceptual framework for understanding nanocrystal self-assembly. The basic conclusion from this model is that the nanocrystals undergo a disorder–order phase transi-tion when the volume fraction in the dispersion becomes suf-ficiently dense.

The hard-sphere potential leads to a relatively straightforward estimation of the free energy of the solid (ordered) and fluid (dis-ordered) phases (Feynman 1972):

$$F - -kT \ln \left(\sum_{\text{configurations}} \exp \left(- \sum_{(i,j,\text{bonds})} \frac{\Phi_{i,j}}{kT} \right) \right), \quad (20.7)$$

where

k is Boltzmann's constant
T is temperature

$\Phi(r)$ is either 0 or ∞, depending on whether the nanocrystals overlap or not—i.e., the particles *cannot* overlap, and they do not recognize the presence of other particles when they are not touching. This leads to a useful expression,

$$\exp \left(- \sum_{(i,j,\text{bonds})} \frac{\Phi_{i,j}}{kT} \right) = \begin{cases} 0 & \text{if } r \le 2R \\ 1 & \text{if } r > 2R \end{cases} \quad (20.8)$$

and therefore, the free energy depends only on the possible con-figurations, or packing geometries in the fluid—it depends only on the *entropy* of the system. This is a very interesting result, revealing that entropy can drive the ordering of spherical par-ticles at a high-volume fraction. When the concentration in the dispersion is low, a disordered collection of particles can achieve many more configurations than any ordered phase and is ther-modynamically favored. However, at relatively high densities, the nanoparticles can actually achieve greater free volume by ordering than remaining in a disordered state. In 1957, Alder and Wainwright (1957) showed that a collection of hard spheres indeed undergoes a first order order–disorder phase transition (i.e., *freezing*) at a volume fraction of 0.49. This is much lower than the fcc lattice, which has a sphere volume fraction of 0.74. The particles order, despite the fact that interparticle forces are purely *repulsive*. The phase transition arises from the packing entropy.

The ordering of hard spheres is essentially a "disorder-avoiding" phase transition. By ordering into a lattice, the particles can actually sample more "states" and reach a higher packing den-sity than if they were in an amorphous, or glassy, state (Gelbart Ben-Shaul 1996). Consider that the maximum packing fraction of a disordered collection of hard spheres (~0.67) is significantly less than that of the packing fraction in an (fcc) lattice (0.74). In the ordered lattice, the spheres have less *configurational* entropy in the ordered lattice than in the disordered fluid, but exhibit a higher free-volume entropy (Gelbart Ben-Shaul 1996).

This is an important concept that applies to nanocrystal dispersions and explains in part the observation of superlat-tice formation when dispersions of monodisperse nanocrystals are dried on a substrate. As the solvent evaporates, the nano-crystal volume fraction increases and can eventually undergo a phase transition from the disordered fluid dispersion to a close-packed array. *In situ* SAXS measurements of evaporat-ing concentrated dispersions of C_{12}-coated silver nanocrystals have indicated that the particles do indeed spontaneously self-assemble during solvent evaporation, provided they are suffi-ciently size-monodisperse and well stabilized from aggregation (Connolly et al. 1998).

This analysis is important for another reason. It reveals that the nanoparticle assemblies can be treated—at least, to a first approximation—as thermodynamically favorable structures. The kinetics of the assembly process is not responsible for the ordering of the nanoparticles (although it can certainly disrupt it). This assumption of *quasi-equilibrium* can be checked by comparing the diffusion rates of the nanocrystals with the char-acteristic rate of solvent evaporation. Korgel et al. (1998) showed that the characteristic time for nanoparticle diffusion is faster than the characteristic rate for solvent evaporation and thus, the nanoparticles can diffuse rapidly enough to explore many possi-ble configurations, as needed to settle into the equilibrium struc-ture. In a concentrated nanocrystal dispersion, a 5 nm diameter particle can diffuse up to 10 nm during the time it takes for the solvent to evaporate 10 nm in thickness; the nanocrystals can diffuse and sample a significant available phase space and it can be argued that nanocrystal ordering can be treated as an equilib-rium (or quasi-equilibrium) problem. It is worth reiterating that these rapid diffusion rates are qualitatively different than large 0.1–10 μm colloids that require days to settle slowly from solu-tion to form colloidal crystals due to their slow diffusion times.

20.3.4.2 Soft Spheres: A Better Estimate of Attractive and Repulsive Forces

The steric repulsion due to the ligands is relatively short-ranged and a hard-sphere potential provides important insight about nanoparticle self-assembly; however, the interparticle potential in reality is softer. There is also a weak attraction between par-ticles due to van der Waals forces, which depends on the nano-crystal size (Korgel et al. 1998). The total interaction potential depends on the core size, the material, and the ligand length, as well as the ligand solvation by the solvent.

These attractive forces can influence assembly, and in some cases, prevent it. As an extreme case, consider nanoparticles that stick when they collide in solution. These nanoparticles assem-ble into fractal aggregates (Witten and Sander 1981). Although in a rigorous sense, there is order in the structure of a fractal, it is not periodic. In general, interparticle attractions that exceed kT lead to disorder and aggregation. Interparticle attractions that are less than kT do not necessarily disrupt order but can influ-ence it. For example, monolayers of polydisperse nanocrystals drop cast from a good solvent, exhibit size segregation, resulting in rafts of particles with the largest particles at the center and the smallest located around the periphery due to size-dependent interparticle attractions, from van der Waals and lateral capillary forces (Ohara et al. 1995, Rabideau et al. 2007).

The two-particle interaction potential Φ_{total}, can be estimated as the sum of a repulsive steric repulsion Φ_{steric} and an attractive van der Waals potential Φ_{vdW}:

$$\Phi_{total} = \Phi_{vdW} + \Phi_{steric}. \tag{20.9}$$

By convention, negative interaction potentials indicate an attraction between the nanoparticles. The van der Waals attraction is negative (attractive) at all interparticle separations and induces aggregation without the ligand layer coating the nanocrystals.

The van der Waals interaction results from a dipole-induced dipole attraction, which can be estimated for two equally sized spherical particles with radius R acting across a medium as

$$\Phi_{vdw} = -\frac{A_{121}}{6}\left[\frac{2R^2}{d^2-4R^2} + \frac{2R^2}{d^2} + \ln\left(\frac{d^2-4R^2}{d^2}\right)\right], \tag{20.10}$$

where d is the center-to-center separation. The Hamaker constant A_{121} characterizes the electronic interactions between particles of the same substance across a medium. A_{121} is directly related to the Hamaker constants of the nanoparticle material acting across a vacuum and particles of the supercritical medium acting across a vacuum, A_{11} and A_{22} respectively, by $A_{121} = \left(\sqrt{A_{11}} - \sqrt{A_{22}}\right)^2$. The Hamaker constants for typical nanocrystal core materials can be found in literature; values for A_{22}, are density dependent and can be calculated from Lifshitz theory (Israelachvili 1992):

$$A_{22} = \frac{3}{4}k_BT\left(\frac{\varepsilon_1-1}{\varepsilon_1+1}\right)^2 + \frac{3h\nu_e}{16\sqrt{2}}\frac{(n_1^2-1)^2}{(n_1^2+1)^{3/2}}, \tag{20.11}$$

where

ε_1 and n_1 are the dielectric constant and the refractive index of the medium, respectively

h is Planck's constant

ν_e is the maximum electronic absorbance frequency in the ultraviolet, typically taken to be 3×10^{15} s^{-1}, and $\varepsilon_{vacuum} = 1$ and $n_{vacuum} = 1$

The Hamaker constant depends on the polarizability of the nanoparticles and the surrounding medium. The Hamaker constant for metals interacting across a vacuum have been determined. For example, the tabulated value for bulk silver is $A_{11} = 2.185$ eV. For silver nanoparticles in hexane, $A_{131} = 0.91$ eV. Metal nanoparticles have relatively large Hamaker constants due to large polarizabilities; compare, for example, the Hamaker constants of polystyrene (in water) of 0.087 eV and mica (in water) of 0.125 eV.

The repulsive forces between nanocrystals arise due to the interactions of ligands with solvent molecules and through ligand–ligand interactions on interacting nanocrystals (*steric repulsion*) (Israelachvili 1992). The total steric repulsive energy

can be estimated as the sum of osmotic Φ_{osm}, and elastic Φ_{elas}, contributions: $\Phi_{steric} = \Phi_{osm} + \Phi_{elas}$ (Vincent et al. 1986). The osmotic term describes interactions between the nanocrystal tails and the solvent and is dependent upon the length of the capping ligand l:

$$\Phi_{osm} = \frac{4\pi Rk_BT}{v_{solv}}\phi^2\left(\frac{1}{2}-\chi\right)\left(l-\frac{d-2R}{2}\right)^2 \quad l<d-2R<2l, \tag{20.12}$$

$$\Phi_{osm} = \frac{4\pi Rk_BT}{v_{solv}}\phi^2\left(\frac{1}{2}-\chi\right)\left[l^2\left(\frac{d-2R}{2l}-\frac{1}{4}-\ln\left(\frac{d-2R}{l}\right)\right)\right]$$
$$d-2R<l, \tag{20.13}$$

where

d is the center-to-center interparticle separation

v_{solv} is the molecular volume of the solvent

ϕ is the volume fraction profile of the stabilizer extending from the particle surface

ϕ can be determined for good solvent conditions, in which the ligands are fully extended. ϕ decreases radially due to the curvature of the nanocrystal surface. ϕ can be calculated using the geometric equation for cylinders with a ligand cross-sectional area (SA_{thiol}) of 14.5 Å, extending radially from a curved surface with radius $R + z$:

$$\phi(z) = \frac{SA_{thiol}R}{\theta_{thiol}(R+z)}, \tag{20.14}$$

where

θ_{thiol} is the surface area per thiol head group, which represents the binding density

z is the radial distance from the metal surface (Shah et al. 2002)

The ligand solubility is considered in Φ_{osm} by the Flory–Huggins interaction parameter, χ. In the Flory–Huggins theory, $\chi = 1/2$ typically represents the boundary between a good solvent ($\chi < 1/2$) and a poor solvent ($\chi > 1/2$). When $\chi > 1/2$, Φ_{osm} is negative (attractive) due to poor ligand solubility, resulting in nanoparticles aggregation. The Flory–Huggins parameter can be estimated using solubility parameters or comparing cohesive energy densities. In a good solvent, the ligands are fully extended and fluctuate rapidly in the solvent. As ligands on colliding nanocrystals begin to overlap, there is a repulsive force that pushes the particles apart. This repulsion is of a slightly longer range than the fully extended ligands due to the osmotic pressure that builds between the nanocrystals as the effective ligand concentration increases between the nanocrystals. As nanocrystals approach very closely, an *elastic repulsion* occurs that is very strongly repulsive and results from large amounts of ligand overlap.

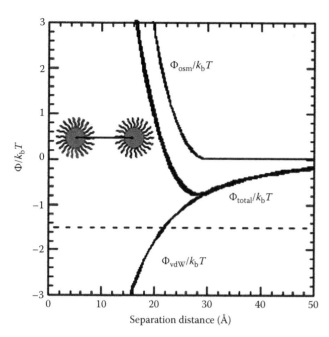

FIGURE 20.4 An example of the pair interaction potential calculated for two dodecanethiol-capped silver nanocrystals.

The elastic repulsive energy Φ_{elas}, originates from the entropy loss that occurs upon compression of the stabilizing ligands, which is only important at interparticle separations in the range, $d - 2R < l$:

$$\Phi_{elas} = \frac{2\pi R k_b T l^2 \phi \rho}{MW_2}\left\{x\ln\left[x\left(\frac{3-x}{2}\right)^2\right] - 6\ln\left(\frac{3-x}{2}\right) + 3(1-x)\right\}.$$

(20.15)

ρ and MW_2 represent the ligand density and molecular weight, and $x = (d - 2R)/l$. Since the elastic term represents the physical compression of the stabilizer, this term is repulsive at all the solvent conditions. The dispersion stability is essentially controlled by the osmotic term since it becomes effective as soon as the ligands start to overlap, $l < d - 2R < 2l$, and the elastic term does not contribute significantly to Φ_{total} until the ligands are forced to compress, i.e., when $d - 2R < l$.

Φ_{total} depends on the nanocrystal size, ligand composition, length and graft density, and solvent condition. Figure 20.4 shows an example of the pair-wise interparticle potential for a silver nanocrystal capped with dodecanethiol. Additionally, the van der Waals attraction depends on the nanocrystals size, with larger nanocrystals experiencing much stronger attractions at equivalent edge-to-edge separations than smaller nanocrystals. Provided that the attractive potential is not larger than kT, the nanocrystals can order.

20.3.4.3 Other Sources of Interactions

Other contributions to the nanocrystal interactions in addition to van der Waals attraction and steric repulsion might also be important in certain cases, and can significantly influence nanoparticle assembly in some cases. For example, recent measurements have shown that nanoparticles can have permanent charges associated with them, leading either to attraction or repulsion (Shevchenko et al. 2006a). Nanocrystals made from magnetic or semiconducting materials can also have permanent dipole moments, of either magnetic or electronic origin (Talapin et al. 2007). These dipole interactions can be strong enough to induce directional aggregation, as in the case of PbSe nanowire formation by an oriented attachment of nanocrystals (Cho et al. 2005). Dipole interactions are also believed to be a source of nanorod and nanowire bundling that is often observed in assemblies of these nanoparticles (Ghezelbash et al. 2006).

In polar solvents, charged groups on the terminal end of ligands give rise to electrostatic interactions between nanoparticles. Typically, acids or alcohols are used to impart a negative charge to the terminal functional group, while amines provide a positive-end group (Korgel and Monbouquette 1997). While it is obvious that like charges will strongly repel each and in fact typically give rise to excellent colloidal stability, the use of mixtures of nanocrystals with oppositely charged functional groups can be used for self-assembly. Kalsin et al. (2006) reported the assembly of binary mixtures of metal nanocrystals with oppositely charged ligands in aqueous media, resulting in the formation of large, well-ordered colloidal crystals. Recent work also suggests that the choice of capping ligand may also impart electrical charges on nanoparticles, which can give rise to a Coulombic stabilization of binary nanocrystal superlattices (Shevchenko et al. 2006a).

20.3.4.4 Influence of Shape

Nanoparticles with non-spherical shape form new types of assemblies with an orientational order between nanoparticles, in addition to their spatial order. Onsager originally showed that entropy can lead to new phases with an orientational order (Onsager 1949). Assemblies of nanodisks, nanorods, and nanowires can exhibit liquid crystal phases, including nematic (orientational order, but no positional order) and smectic (both orientational and positional ordering) phases for nanorods and nanowires, and columnar phases for nanodisks. Figures 20.5 and 20.6 show ordered assemblies of nanorods, nanowires, and nanodisks. Entropy is also largely responsible for these ordered assemblies.

Tiled patterns of self-assembled nanorods like those in Figure 20.5a are often observed when well-dispersed, monodisperse nanorods are deposited on a substrate by evaporating the solvent. These 35 nm long NiS nanorods have assembled into a close-packed monolayer with regions of orientational order. When the nanorods are longer, such as the 200–300 nm long CdS nanowires in Figure 20.5b and c, there is orientational ordering, but often the nanowires bundle together, forming oriented aggregates of nanowires. This bundling relates to the van der Waals attraction between the nanowires. In some cases, there may also be dipole–dipole interactions between neighboring nanorods resulting from a permanent dipole resulting from

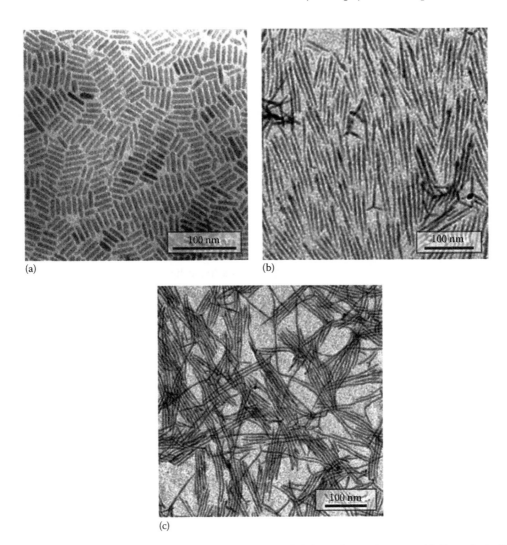

FIGURE 20.5 Examples of nanorod ordering. NiS nanorods orient into extended chains during deposition (a). The order in films of CdS nanorods depends strongly upon the aspect ratio: short nanorods can align into liquid crystal phases (b) while only short-range order is present for long nanorods (c).

the anisotropic crystal structure of the material, such as wurtzite CdS, which has polar crystallographic facets exposed at each end of the nanowire (Ghezelbash et al. 2006).

Nanodisks have also been shown to form assemblies with both positional and orientational orders (Sigman et al. 2003, Saunders et al. 2006). Figure 20.6 shows examples of copper sulfide (CuS and Cu_2S) nanodisks assembled into columnar arrays. The interparticle attraction between the faces of these nanocrystals can be quite strong, and relatively long chains of nanodisks can form. The cause of this chain formation is still under investigation; while the shape of the particle leads to asymmetric van der Waals attractions that would favor this type of face-to-face packing, the observed length of the chains suggests that additional forces may also be a factor. One possibility is the presence of a permanent electronic dipole moment in the nanodisk; the alignment of dipole moments on neighboring nanocrystals would provide an additional energetic stability for these structures (Sigman et al. 2003).

The columnar order of the copper sulfide nanodisks has been confirmed by SAXS. The columnar superlattice is a simple hexagonal unit cell. The disks stack into columns, which are arranged into a periodically ordered hexagonal close-packed array.

20.3.4.5 Binary Superlattice Formation

Binary superlattices (BSLs) can be formed with mixtures of nanoparticles of two different sizes. By mixing large and small nanocrystals with different radius ratios, a very diverse range of superlattices that are the structural analogs to NaCl, CuAu, AlB_2, $MgZn_2$, $MgNi_2$, Cu_3Au, Fe_4C, $CaCu_5$, CaB_6, $NaZn_{13}$, and cub-AB_{13} have been observed (Saunders and Korgel 2005, Shevchenko et al. 2006b). There has been significant computational and theoretical work on understanding the phase behavior of colloidal mixtures with bimodal size distributions, taking the particles as hard spheres (Cottin and Monson 1995). In some cases, these predictions agree well with the observed structures formed by nanocrystals and entropy provides one driving force

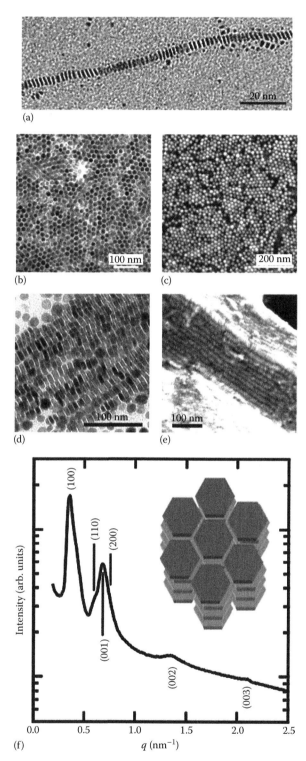

FIGURE 20.6 Nanodisks of copper sulfide self-assemble into ordered chains of disks stacked face-to-face when evaporated from dilute solutions (a). During evaporation of concentration dispersions, ordered columnar liquid crystal phases form (f, inset). TEM (b, d) and SEM (c, e) show disks stacking with the columns oriented perpendicular (b, c) and parallel (d, e) to the substrate. The SEM image in (e) shows a close-up of the top facet in the colloidal crystal in Figure 20.1h. SAXS data (f) from a film of disks shows peaks corresponding to a columnar phase; the hexagonal unit cell has dimensions $a = 20.9\,nm$ and $c = 9.4\,nm$.

for BSL formation. However, many of the observed structures deviate significantly from simple expectations based on the packing geometry (Shevchenko et al. 2006a,b). Additionally, kinetic factors are known to be much more important in BSL formation than in the case of the superlattice formation of monodisperse nanocrystals (Doty et al. 2002, Rabideau and Bonnecaze 2005). The details of how such a wide variety of structures forms requires much more study, but factors including electrostatic charge and depletion attraction between free-capping ligands have been suggested to play a role in aiding BSL assembly (Shevchenko et al. 2006b, Smith et al. 2009).

One particular BSL structure, however, that is now well understood and appears to be quite stable is the AlB_2 analog, or the AB_2 phase, where the diameter of A is larger than B (Smith et al. 2009). Figure 20.7 shows images of an AB_2 BSL constructed of 6.1 nm diameter Au nanocrystals mixed with 11.5 nm diameter iron oxide nanocrystals. The Au nanocrystals are coated with dodecanethiol and the iron oxide nanocrystals are coated with oleic acid. Most of the BSL structures observed to date have been studied by TEM, with relatively limited amounts of sample. The AB_2 BSL structure has been examined by SEM and SAXS as well, and these measurements have confirmed the long-range structural order of the assembly. The AB_2 BSL structure is readily explained based on the packing density of nanocrystals in the lattice (Smith et al. 2009). The large nanocrystals form a simple hexagonal sublattice, which is a relatively sparse assembly that would not typically be a stable structure for a monodisperse collection of spheres. However, the smaller nanoparticles infiltrate the simple hexagonal sublattice, filling the interstitial sites. In a perfect, simple, hexagonal AB_2 assembly, the nanoparticles fill 78% of the available space, which is more densely packed than an fcc lattice. The ideal size ratio for the AB_2 structure is 0.53. The radius ratio of the inorganic cores of the Au and iron oxide nanoparticles is 0.53. The ideal size ratio for NaCl and ZnS structures are 0.732 and 0.414. Although BSLs with NaCl structure have been observed (Saunders and Korgel 2005), the same long-range order has not yet been achieved. Therefore, the "design rules" for BSLs remain to be elucidated. Certainly, the packing entropy is one important consideration. Interparticle attractions, however, are what most likely contribute to stabilizing BSL structure. Furthermore, the approach to BSL assembly has been reported to be critical to the formation of BSLs with a long-range order, as opposed to the phase separation of small and large nanocrystals (Smith et al. 2009).

20.4 Critical Discussion

One aspect of nanoparticle assembly not discussed in any great detail in this chapter is the role of dynamics. Most of the presentation here focused on the quasi-equilibrium nature of the nanoparticle superlattices in order to understand their structure. Many of the observed structures can indeed be understood in terms of simple packing arguments and thermodynamic concepts. This is an important starting point for understanding how to control self-assembly. However, this is really only the

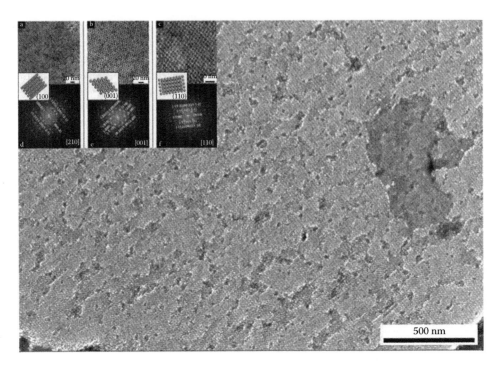

FIGURE 20.7 SEM image of a simple hexagonal AB$_2$ BSL of 6.1 nm diameter Au nanocrystals and 11.5 nm diameter iron oxide nanocrystals. (Inset) TEM images of similar BSLs formed on a TEM grid. (Courtesy of Danielle Smith.)

beginning of the story, as the kinetics of the assembly process can indeed influence the structure, and in fact, lead to a wide array of new types of structural order. For example, networked nanoparticles assembled in the shape of rings and honeycomb networks can form as a result of how solvent evaporates, by dewetting (Ohara and Gelbart 1998) or Marangoni flow (Stowell and Korgel 2001). In one striking example of how kinetic process can lead to order, evaporative cooling of the solvent interface of a drop cast concentrated dispersion of nanoparticles in humid air can lead to water droplet condensation that can lead to incredibly ordered arrays of micrometer scale holes in nanoparticle films (Saunders et al. 2004). The challenge with controlling and understanding these dynamic processes of nanoparticle assembly is that often very subtle differences in experimental conditions can lead to dramatically different structures. Nonetheless, a variety of robust kinetic effects exists that offer potentially new and unexplored methods of assembling nanoparticles with a unique structural order.

20.5 Summary

The underlying principles that determine the structure and the kinetics of the assembly process of these ordered nanoparticle assemblies are similar to those of the much larger submicrometer particles, but the smaller size of the nanoparticles leads to much faster crystallization kinetics and there is no gravitational settling of particles. Underlying the studies of ordered nanoparticle assemblies is the ability to make very monodisperse, sterically stabilized nanoparticles. There are at least two fundamental requirements for the nanoparticles to order: (1) good

steric stabilization in the solvent with interparticle interactions less than kT to prevent sticking and disordered aggregation; and (2) a narrow size distribution, of less than ~10% standard deviation about the mean diameter. Nanoparticles with these characteristics tend to assemble into ordered, periodic arrays.

20.6 Future Perspective

Following the demonstration of binary nanocrystal assemblies, which mimic bulk-crystal structures of binary alloys and compounds, one may envision the possibility of more complex assemblies corresponding to perovskites or other materials. Simulations, based on the optimization of the packing density of self-assembled structures, have predicted several such structures, consisting of the general stoichiometry AB_xC_y (Stucke and Crespi 2003). Although such assemblies will likely require the ability to further decrease nanocrystal polydispersity over a wide range of sizes, it is not unreasonable to believe that such structures will eventually be produced experimentally. As with the BSLs discussed above, however, the kinetics of the assembly process can easily lead to disorder and robust assembly *processes* require discovery to obtain these potentially thermodynamically stable structures.

The role of interparticle interactions in the self-assembly process is still relatively poorly understood. Additional work must be done to measure these interactions and to determine how the assembly of particles is affected. This additionally applies to new systems, such as platelet-sphere (Shevchenko et al. 2006a) or nanorod-sphere assemblies, an example of which is shown in Figure 20.8. The co-assembly of metal and semiconductor

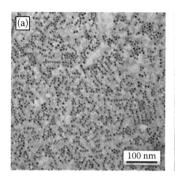
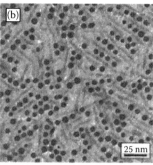

FIGURE 20.8 TEM images of the unusual phase behavior observed when mixing spherical Au nanocrystals and CdS nanorods. Contrary to the simulation results of hard-sphere mixtures (Adams et al. 1998), the particles here order into "peapod" structures with linear chains of spherical particles constrained between parallel nanorods. (Courtesy of Ali Ghezelbash.)

nanoparticles with different shapes leads to the formation of "peapod" structures—chains of spherical nanocrystal "peas" alternating between nanorods. Such assemblies differ significantly from structures predicted from simulations based on hard spheres and hard rods, which typically predict phase separation (Adams et al. 1998) and offer an intriguing example of the rich phase behavior of nanocrystal systems that has yet to be explored.

References

Adams, M., Dogic, Z., Keller, S. L., and Fraden, S. 1998. Entropically driven microphase transitions in mixtures of colloidal rods and spheres. *Nature* 393: 349–352.

Alder, B. J. and Wainwright, T. E. 1957. Phase transition for a hard sphere system. *J. Chem. Phys.* 27: 1208–1209.

Alivisatos, A. P. 1996. Semiconductor clusters, nanocrystals, and quantum dots. *Science* 271: 933–937.

Bentzon, M. D., Van Wonterghem, J., Morup, S., Tholen, A., and Koch, C. J. W. 1989. Ordered aggregates of ultrafine iron-oxide particles—Super crystals. *Phil. Mag. B* 60: 169–178.

Cho, K. S., Talapin, D. V., Gaschler, W., and Murray, C. B. 2005. Designing PbSe nanowires and nanorings through oriented attachment of nanoparticles. *J. Am. Chem. Soc.* 127: 7140–7147.

Connolly, S., Fullam, S., Korgel, B. A., and Fitzmaurice, D. 1998. Time-resolved small angle X-ray scattering studies of nanocrystal superlattice self-assembly. *J. Am. Chem. Soc.* 120: 2969–2970.

Copisarow, A.C. and Copisarow, M. 1946. Structure of hyalite and opal. *Nature* 157: 768–769.

Cottin, X. and Monson, P. A. 1995. Substitutionally ordered solid-solutions of hard-spheres. *J. Chem. Phys.* 102: 3354–3360.

Doty, R. C., Bonnecaze, R. T., and Korgel, B. A. 2002. Kinetic bottleneck to the self-organization of bidisperse hard disk monolayers formed by random sequential adsorption. *Phys. Rev. E* 65: 061503.

Feynman, R. P. 1972. *Statistical Mechanics; A Set of Lectures.* Reading, MA: W. A. Benjamin.

Gast, A. P. and Russel, W. B. 1998. Simple ordering in complex fluids—Colloidal particles suspended in solution provide intriguing models for studying phase transitions. *Phys. Today* 51: 24–30.

Gelbart, W. M. and Ben-Shaul, A. 1996. The "new" science of "complex fluids." *J. Phys. Chem.* 100: 13169–13189.

Ghezelbash, A., Koo, B., and Korgel, B. A. 2006. Self-assembled stripe patterns of CdS nanorods. *Nano Lett.* 6: 1832–1836.

Glatter, O. and Kratky, O. 1982. *Small Angle X-Ray Scattering.* New York: Academic Press.

Israelachvili, J. *Intermolecular & Surface Forces*, 2nd edn. 1992. New York: Academic Press.

Kalsin, A. M., Fialkowski, M., Paszewski, M., Smoukov, S. K., Bishop, K. J. M., and Grzybowski, B. A. 2006. Electrostatic self-assembly of binary nanoparticle crystals with a diamond-like lattice. *Science* 312: 420–424.

Kerker, M. 1969. *The Scattering of Light, and Other Electromagnetic Radiation.* New York: Academic Press.

Klimov, V. I. 2004. *Semiconductor and Metal Nanocrystals: Synthesis and Electronic and Optical Properties.* New York: Marcel Dekker, Inc.

Korgel, B. A. and Monbouquette, H. G. 1997. Quantum confinement effects enable photocatalyzed nitrate reduction and neutral pH using CdS nanocrystals. *J. Phys. Chem. B* 101: 5010–5017.

Korgel, B. A., Fullam, S., Connolly, S., and Fitzmaurice, D. 1998. Assembly and self-organization of silver nanocrystal superlattices: Ordered 'soft spheres.' *J. Phys. Chem. B* 102: 8379–8388.

Kwon, S. G. and Hyeon, T. 2008. Colloidal chemical synthesis and formation kinetics of uniformly sized nanocrystals of metals, oxides, and chalcogenides. *Acc. Chem. Res.* 41: 1696–1709.

Lee, D. C., Smith, D. K., Heitsch, A. T., and Korgel, B. A. 2007. Colloidal magnetic nanocrystals: Synthesis, properties and applications. *Annu. Rep. Prog. Chem., Sect. C: Phys. Chem.* 103: 351–402.

Levin, I. and Ott, E. 1933. X-ray studies of opals, glass and silica gel. *Zeitschrift fur Kristallographie* 85: 305–318.

Markovich, G., Collier, C. P., Henrichs, S. E., Remacle, F., Levine R. D., and Heath, J. R. 1999. Architectonic quantum dot solids. *Acc. Chem. Res.* 32: 415–423.

Murray, C. B., Kagan, C. R., and Bawendi, M. G. 1995. Self-organization of CdSe nanocrystallites into 3-dimensional quantum-dot superlattices. *Science* 270: 1335–1338.

Murray, C. B., Kagan, C. R., and Bawendi, M. G. 2000. Synthesis and characterization of monodisperse nanocrystals and close-packed nanocrystal assemblies. *Ann. Rev. Mater. Sci.* 30: 545–610.

Ohara, P. C. and Gelbart, W. M. 1998. Interplay between hole instability and nanoparticle array formation in ultrathin liquid films. *Langmuir* 14: 3418–3424.

Ohara, P. C., Leff, D. V., Heath, J. R., and Gelbart, W. M. 1995. Crystallization of opals from polydisperse nanoparticles. *Phys. Rev. Lett.* 75: 3466–3469.

Onsager, L. 1949. The effects of shape on the interaction of colloidal particles. *Ann. N. Y. Acad. Sci.* 51: 627–659.

Rabideau, B. D. and Bonnecaze, R. T. 2005. Computational predictions of stable 2D arrays of bidisperse particles. *Langmuir* 21: 10856–10861.

Rabideau, B. D., Pell, L. E., Bonnecaze, R. T., and Korgel, B. A. 2007. Observation of long range orientational order in monolayers of polydisperse colloids. *Langmuir* 23: 1270–1274.

Sanders, J. V. 1964. Colour of precious opal. *Nature* 204: 1151–1153.

Saunders, A. E. and Korgel, B. A. 2004. Second virial coefficient measurements of dilute gold nanocrystal dispersions using small angle X-ray scattering. *J. Phys. Chem. B* 108: 16732–16738.

Saunders, A. E. and Korgel, B. A. 2005. Observation of an AB phase in bidisperse nanocrystal superlattices. *Chem. Phys. Chem.* 6: 61–65.

Saunders, A. E., Shah, P. S., Sigman, M. B. et al. 2004. Inverse opal nanocrystal superlattice films. *Nano Lett.* 4: 1943–1948.

Saunders, A. E., Ghezelbash, A., Smilgies, D.-M., Sigman, M. B., and Korgel, B. A. 2006. Columnar self-assembly of colloidal nanodisks. *Nano Lett.* 6: 2959–2963.

Schmid, G. and Lehnert, A. 1989. The complexation of gold colloids. *Angew. Chem. Intl. Ed.* 28: 780–781.

Shah, P. S., Husain, S., Johnston, K. P., and Korgel, B. A. 2002. Role of steric stabilization on the arrested growth of silver nanocrystals in supercritical carbon dioxide. *J. Phys. Chem. B* 106: 12178–12185.

Shevchenko, E. V., Talapin, D. V., Kotov, N. A., O'Brien, S., and Murray, C. B. 2006a. Structural diversity in binary nanoparticle superlattices. *Nature* 439: 55–59.

Shevchenko, E. V., Talapin, D. V., Murray, C. B., and O'Brien, S. 2006b. Structural characterization of self-assembled multifunctional binary nanoparticle superlattices. *J. Am. Chem. Soc.* 128: 3620–3637.

Sigman, M. B., Ghezelbash, A., Hanrath, T., Saunders, A. E., Lee, F., and Korgel, B.A. 2003. Solventless synthesis of monodisperse Cu_2S nanorods, nanodisks, and nanoplatelets. *J. Am. Chem. Soc.* 125: 16050–16057.

Smith, D. K., Goodfellow, B., Smilgies, D. M., and Korgel, B. A. 2009. Self-assembled simple hexagonal AB_2 binary nanocrystal superlattices: SEM, GISAXS and defects. *J. Am. Chem. Soc.* 131: 3281–3290.

Stowell, C. and Korgel, B. A. 2001. Self-assembled honeycomb networks of gold nanocrystals. *Nano Lett.* 1: 595–600.

Stucke, D. P. and Crespi, V. H. 2003. Predictions of new crystalline states for assemblies of nanoparticles: Perovskite analogues and 3-D arrays of self-assembled nanowires. *Nano Lett.* 3: 1183–1186.

Talapin, D. V., Shevchenko, E. V., Murray, C. B., Titov, A. V., and Kral, P. 2007. Dipole-dipole interactions in nanoparticle superlattices. *Nano Lett.* 7: 1213–1219.

Vincent, B., Edwards, J., Emmett, S., and Jones, A. 1986. Depletion flocculation in dispersions of sterically-stabilized particles (soft spheres). *Coll. Surf.* 18: 261–281.

Witten, T. A. and Sander, L. M. 1981. Diffusion-limited aggregation, a kinetic critical phenomenon. *Phys. Rev. Lett.* 47: 1400–1403.

21

Biomolecule-Induced Nanoparticle Aggregation

Soumen Basu
University of Alabama

Tarasankar Pal
Indian Institute of Technology

21.1 Introduction

A burst of research activities has been seen in recent years for the synthesis and characterization of metal nanoparticles (NPs), which arise from their numerous possible applications in physics, chemistry, biology, materials science, and their different interdisciplinary fields (Bohren and Huffman 1983, Henglein 1993, Palato et al. 1994, Schmid 1994, Falkenhagen 1995, Kreibig and Vollmer 1995, Robert and Rao 1996, Jana and Pal 1999, Jana et al. 1999, 2001, Link and El-Sayed 1999, Gaponik et al. 2000, Pradhan et al. 2001). Because of their versatility in application, while evolution of the dispersion of small metallic particles with a tight size distribution is important, assembly of individual NPs into well-defined aggregates has recently become a widely pursued objective (Murray et al. 2000, Fendler 2001, Niemeyer 2001a,b, Katz and Willner 2004). Most recently, emphasis has been placed on organizing or assembling metal NPs into defined architectures, mainly for two reasons. First, such metal NP aggregates can display rich optical and electrical characteristics that are distinctly different from a simple collection of individual particles or the extended solid. Second, in relation to emerging electronic technologies, more sophisticated nanostructures are in demand [e.g., nanowires, nanotubes, and their two-dimensional (2D) and 3D NP assemblages].

The organization and patterning of inorganic NPs into 2D and 3D functional structures is a potential route to chemical, optical, magnetic, and electronic devices with useful properties. Many approaches have been described for the formation of 2D and 3D arrays of metal and semiconductor NPs. The aggregation of NPs induced by specific biological interactions attracts interest as a self-assembly process for the construction of complex nanostructures that exhibit new collective properties. Several reasons support the concept of utilizing biomolecules as building blocks of NP structures: (1) The diversity of biomolecules enables the selection of building units of predesigned size, shape, and functionality. (2) The availability of chemical and biological means modify and synthesize biomolecules. For example, the synthesis of nucleic acids of predesigned composition and shape, the elicitation of monoclonal antibodies, or the modification of proteins by genetic engineering to allow the construction of biomaterials for the directed assembly of NPs. (3) Enzymes may act as biocatalytic tools for the manipulation of the biomaterials. The hydrolysis of proteins, the scission or ligation of DNA, or the replication of nucleic acids may be employed as tools for the assembly of NP architectures through manipulation of the biomaterial. (4) Mother Nature has developed routes for the repair of biomolecules that may be applied to stabilize the biomolecule–NP structures. (5) NPs that are cross-linked with enzyme units may generate biocatalytic assemblies of predesigned functionality. These different features of the biomolecule cross-linking units provide a flexible means of generation of NP structures of tunable physical, chemical, and functional properties. For the generation of biomolecule cross-linked NPs, two types of biomolecule-functionalized NPs with complementary units should participate in the assembly process. Biomaterials utilized in the fabrication of such biomolecule–NP aggregates include biological protein host–guest pairs such as biotin–streptavidin (STV) (Cobbe et al. 2003), antigen–antibody (Mann et al. 2000), and complementary oligonucleotide pairs (Niemeyer 1999, Bashir 2001, Cobbe et al. 2003). A large variety of methods including optical methods such as differential light-scattering spectroscopy (Bogatyrev et al. 2002) have been used to study the biospecific assembly of NPs with proteins and oligonucleotides.

21.2 Protein-Based Aggregation of NPs

Protein-based recognition systems can be used to organize inorganic NPs into network aggregates. For instance, the interaction between D-biotin and the biotin-binding protein STV was utilized to induce the aggregation of NPs. Metal and semiconductor NPs can be functionalized with biotin derivatives by one of several synthetic routes. In the simplest method, thiol or disulfide derivatives of biotin are directly adsorbed onto metal NPs (e.g., Au, Ag). Alternatively, NPs can first be coated with an organic "shell" (e.g., by the polymerization of a trialkoxysilyl derivative or by polymer adsorption) and then covalently modified with biotin, for example, by carbodiimide coupling.

In recent years, there has been growing interest among chemists to incorporate complementary receptor–substrates sites into the molecules and to attach complementary receptor–substrates sites to the surface of metal NPs (Mann et al. 2000). The highly specific recognition properties of antibodies and antigens make them excellent candidate molecules for the programmed assembly of NPs in solution. Indeed, the versatility of using preformed NPs in association with antigen–antibody engineering should make it possible to assemble a wide range of NP-based structures with specific cross-linked structures, compositions, and macroscopic structures (Scheme 21.1).

STV–biotin binding is an ideal model for protein–substrate nanocrystal assembly because the complex has one of the largest free energies of association known for noncovalent binding of a protein and small ligand in aqueous solution ($K_a > 10^{14}$ M^{-1}). Moreover, there exists a range of readily accessible analogues with K_a values 10^0–10^{15} M^{-1} that are extremely stable over a wide range of temperature and pH. In principle, reversible cross-linking of biotinylated NPs should occur because the tetrameric structure of STV provides a connecting unit for 3D aggregation. The use of STV–biotin recognition motif in nanocrystal aggregation has been reported by Mann et al. (2000). Gold nanocrystals were functionalized by chemisorption of a disulfide biotin analogue (DSBA) and then cross-linked by multisite binding on subsequent addition of protein, STV (Scheme 21.1). The assembly of gold nanocrystals was monitored using dynamic light scattering, which yields an average hydrodynamic radius of all particles in solution. Addition of STV gave a rapid increase in the average hydrodynamic radius that obeyed the expected power law ($p < 0.0001$) for diffusion-limited aggregation. Aggregation was also monitored by a red to purple color change in the sol that was attributed to the distant-dependent optical properties of gold NPs. Transmission electron microscopy (TEM) images showed that the unmodified gold nanocrystals were present mostly as single particles, whereas the

biotin-coated particles underwent aggregation following addition of STV. Subsequently, the nanocrystals were invariably separated from each other by approximately 5 nm, consistent with the separation expected for STV cross-linking. The results indicate that aqueous dispersions of gold nanocrystals, possessing a narrow size distribution, can be readily modified by chemisorption of DSBA and is subsequently assembled by molecular recognition.

Letsinger and colleagues (Park et al. 2001) have reported a method to synthesize NPs–protein assembly utilizing 13 nm gold particles, STV, and biotinylated oligonucleotide to explore these hypotheses and some of the physical and chemical properties of the resulting new bioinorganic materials. The NP–protein assembly relies on three building blocks: STV complexes to four biotinylated oligonucleotides (1-STV), oligonucleotide-modified gold NPs (2-Au), and a linker oligonucleotide (3) that has half of its sequence complementary to 1 and the other half complementary to 2 (Scheme 21.2). Aggregates with similar properties could be formed by both methods, but premixing 3 and 1-STV facilitates aggregate formation. Since a 13 nm gold particle is substantially larger than STV ($4 \times 4 \times 5$ nm^3) (Mirkin et al. 1996, Niemeyer et al. 1998, Connolly and Fitzmaurice 1999, Li et al. 1999, Shenton et al. 1999), a 1:20 molar ratio of gold NP to STV was used to favor the formation of an extended polymeric structure rather than small aggregates composed of a few NPs or a structure consisting of a single gold NP functionalized with a hybridized layer of STV. It was seen that when 1-STV, 2-Au, and 3 were mixed at room temperature, no significant particle aggregation took place, even after 3 days, as evidenced by an unperturbed UV-vis spectrum of the solution. However, raising the temperature of the solution (53°C) to a few degrees below the melting temperature (T_m) of the DNA interconnects resulted in the growth of micrometer-sized aggregates and the characteristic redshift and dampening of the gold surface plasmon resonance associated with the particle assembly (Mirkin et al. 1996). It is now known theoretically and experimentally that when individual spherical gold particles come into close proximity with one another, electromagnetic coupling of clusters becomes effective and may lead to complicated extinction spectra depending on the size and shape of the formed cluster aggregate by splitting of the plasma resonance. The optical properties of the metallic NPs are mainly determined by two contributions: (1) the properties of the particles acting as well-isolated individuals and (2) the collective properties of the whole ensemble. Thus, in an ensemble of large number of particles, if the particles come close together, the oscillating electrons in one particle feel the electric field due to oscillation of the electrons in the surrounding particles and this leads to a collective plasmon oscillation of the aggregated systems. Under such situation, the isolated-particles approximation breaks down and the electromagnetic interactions between the particles play a determining role to offer a satisfactory description of the surface plasmon oscillations.

Scheme 21.2b reported a method for utilizing DNA and its synthetically programmable sequence recognition properties to assemble NPs functionalized with oligonucleotides into preconceived architectures. First, the 13 nm Au NPs are modified with different alkanethiol functionalized oligonucleotides (2 and 4),

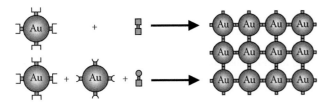

SCHEME 21.1 Schematic representation showing the formation of an idealized ordered structure using surface-attached antibodies and artificial antigens for cross-linking of gold NPs. (From Mucic, R.C. et al., *J. Am. Chem. Soc.*, 120, 12674, 1998; Srivastava, S. et al., *J. Am. Chem. Soc.*, 129, 11776, 2007; Perez, J.M. et al., *J. Am. Chem. Soc.*, 125, 10192, 2003. With permission.)

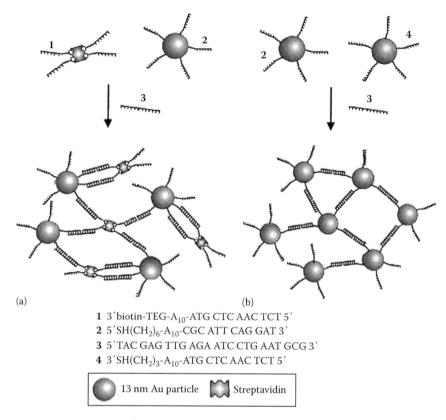

1 3′biotin-TEG-A$_{10}$-ATG CTC AAC TCT 5′
2 5′SH(CH$_2$)$_6$-A$_{10}$-CGC ATT CAG GAT 3′
3 5′TAC GAG TTG AGA ATC CTG AAT GCG 3′
4 3′SH(CH$_2$)$_3$-A$_{10}$-ATG CTC AAC TCT 5′

13 nm Au particle   Streptavidin

SCHEME 21.2 Schematic representation of DNA-directed assembly of Au NPs and STV. (a) Assembly of oligonucleotide-functionalized STV and Au NPs (Au–STV assembly). (b) Assembly of oligonucleotide-functionalized Au NPs (Au–Au assembly). (From Cobbe, S. et al., *J. Phys. Chem. B*, 107, 470, 2003. With permission.)

which are noncomplimentary. After that, the two oligonucleotide-modified gold NPs (**2**-Au and **4**-Au) are mixed together with a linker oligonucleotide (**3**) and the color of the solution immediate changed from red to purple which indicates the formation of close-packed assembly of Au NPs.

Connolly and Fitzmaurice (1999) have used the STV–biotin interaction to organize gold colloids that were functionalized by chemisorptive coupling to the DSBA 3. According to the generalized principle depicted in Scheme 21.3, the subsequent cross-linking was achieved by the addition of an STV as a linker. The immediate change of the sol from red to blue was indicative of the formation of oligomeric networks. This process was also monitored by dynamic light scattering, which showed that the average hydrodynamic radius of all particles in solution rapidly increased on STV-directed assembly. TEM images (Figure 21.1) revealed networks with an average of 20 interconnected particles that were separated by about 5 nm. Small angle x-ray scattering was employed to probe the solution structure of the networks. The above findings suggest that the STV–biotin system is versatile for developing novel strategies for assembling NPs in solution or on a substrate. The applicability of the STV–biotin system for generating supramolecular aggregates is enhanced by the availability of various biotin analogues (Sinha and Chignell 1979, Piran and Riordan 1990) and recombinant STV mutants (Sano and Cantor 1995, Sano et al. 1995, Reznik et al. 1996, Sano et al. 1996, 1998, Schmidt et al. 1996).

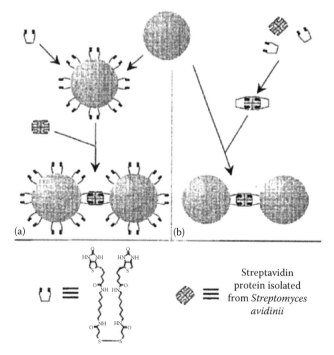

Streptavidin protein isolated from *Streptomyces avidinii*

SCHEME 21.3 Two routes for aggregation of gold nanocrystals (large spheres) using STV and a disulfide-biotin analogue. (a) Gold nanocrystals are modified by chemisorption of the DSBA and aggregation induced by addition of STV. (b) Alternatively, aggregation is induced by chemisorption of the DSBA bound to STV.

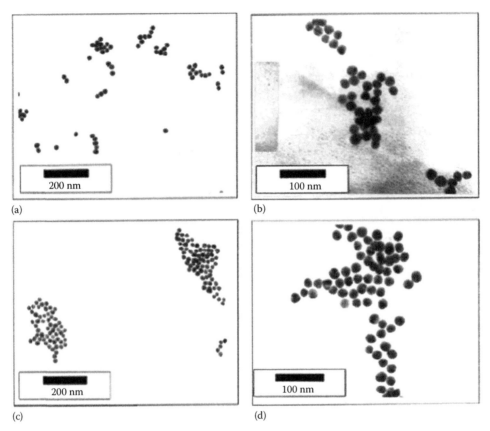

FIGURE 21.1 TEM images of (a) and (b) unmodified gold nanocrystals. It is observed that the nanocrystals are present as isolated particles or small aggregates, and that in the case of the small aggregates the constituent nanocrystals are touching; (c) and (d) biotin-modified gold nanocrystals after STV-induced aggregation. It is noted that the constituent nanocrystals are separated.

Mann and coworkers described a new biomolecular-derived route to the self-assembly of inorganic NPs using the recognition properties of surface-attached antibodies (Shenton et al. 1999). Their strategy involved the attachment of IgE or IgG antibodies with specificities to dinitrophenyl (DNP) and biotin, respectively, to individual Au NPs, followed by the addition of bivalent antigens with appropriate double-headed functionalities. Antigens with homo- (DNP–DNP) or hetero- (DNP–biotin) Janus structures connected by at least an eight-atom spacer were synthesized for this purpose. The formation of specific antibody–antigen cross-links between the particles results in the formation of metallic or bimetallic aggregates comprising covalently linked Au, or Au and Ag NPs (Scheme 21.4). In addition, they showed that higher-order structures in the form of macroscopic filaments of the self-assembled NPs are produced under certain conditions.

Belcher and coworkers used phage display techniques to select 12-mer peptides with a binding specificity for distinct semiconductor surfaces (Whaley et al. 2000). On the basis of a combinatorial library of about 109 random 12-mer peptides, phage clones were selected for their specific binding capabilities to one of five different single-crystal semiconductors: GaAs(100), GaAs(111), InP(100), and Si(100). These substrates were chosen to allow the systematic evaluation of peptide–substrate interactions (Figure 21.2). Specific peptide binding was found that is selective for the crystal composition (e.g., binding to GaAs, but not to Si) and crystal face [e.g., binding to GaAs(100), but not to GaAs(111)].

Rotello and coworkers (Srivastava et al. 2007) have demonstrated the fabrication of bionanocomposites based on ferritin and synthetic NPs (Scheme 21.5). Magnetic (FePt) and nonmagnetic (Au) NPs were used to assemble ferritin into near monodisperse bionanocomposites. This assembly process provided discrete essentially monodisperse aggregates that feature controlled interparticle spacing. In these materials, the magnetic dipoles of the synthetic and biological components interact, as manifested by changes in the blocking temperature (T_B), net magnetic moment, remanence, and coercivity of the resulting composites.

21.3 DNA-Directed Aggregation of NPs

DNA is particularly suitable to serve as a construction material in nanosciences (Seeman 1999, Niemeyer 2000). Despite its simplicity, the enormous specificity of the adenine–thymine (A–T) and guanine–cytosine (G–C) Watson-Crick hydrogen bonding allows the convenient programming of artificial DNA receptors. The power of DNA as a molecular tool is enhanced by the ability to synthesize virtually any DNA sequence by automated methods and to amplify any DNA sequence from microscopic to macroscopic quantities by means of polymerase chain reaction. Another

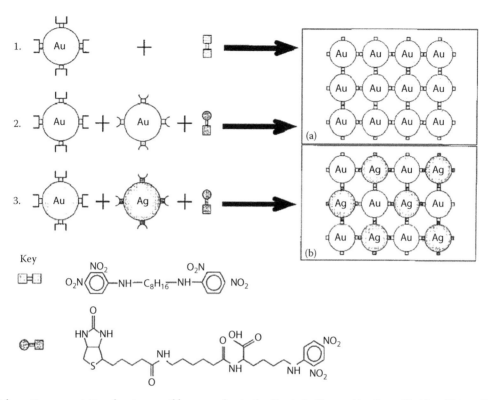

SCHEME 21.4 Schematic representation showing possible approaches to the directed self-assembly of metallic (a) and bimetallic (b) macroscopic materials using antibody–antigen cross-linking of inorganic NPs. The structures shown are idealized; in reality, the materials are highly disordered. (1) Au NPs with surface-attached anti-DNP IgE antibodies and homo-Janus DNP–DNP antigen connector. (2) Au NPs with either surface-attached anti-DNP IgE or antibiotin IgG antibodies and hetero-Janus DNP–biotin antigen. (3) 1:1 mixture of Au/anti-DNP IgE and Ag/antibiotin IgG NPs in association with DNP-biotin bivalent antigen.

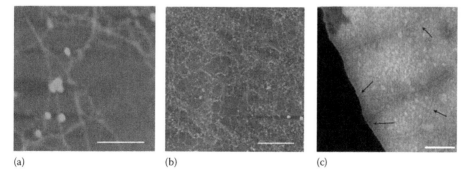

FIGURE 21.2 AFM and TEM analysis of peptide-semiconductor recognition. (a) and (b) AFM images of G1–3 phage bound to an InP(100) substrate. (a) Individual phage and their attached Au NPs. Scale bar, 250 nm. (b) Image showing the uniformity of phage coverage on the InP surface. Scale bar, 2.5 mm. (c) TEM image of G1–3 phage recognition of GaAs. Individual phage particles are indicated with arrows. Scale bar, 500 nm. (From Mucic, R.C. et al., *J. Am. Chem. Soc.*, 120, 12674, 1998. With permission.)

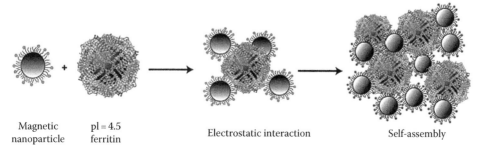

SCHEME 21.5 Magnetic NPs (FePt) assembled with ferritin via electrostatic interaction.

very attractive feature of DNA is the great mechanical rigidity of short double helices, so that they behave effectively as a rigid rod spacer between two tethered functional molecular components on both ends. Moreover, DNA displays a relatively high physicochemical stability. Finally, nature provides a complete toolbox of highly specific biomolecular reagents, such as endonucleases, ligases, and other DNA-modifying enzymes, which allow for the processing of DNA material with atomic precision and accuracy on the angstrom level. No other (polymeric) material offers these advantages, which are ideal for molecular constructions in the range of about 5 nm up to a few micrometers.

Letsinger and coworkers showed (Mucic et al. 1998) how DNA-directed assembly strategy can be used to prepare binary (two-component) network materials comprising two different-sized oligonucleotide-functionalized NPs (Scheme 21.6). Importantly, the reported proof of concept suggests that this strategy could be extended easily to a wide variety of multicomponent systems, in which NP building blocks that vary in chemical composition or size are arranged in space on the basis of their interactions with complementary linking DNA (Figure 21.3).

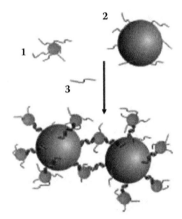

1 3′HS$(CH_2)_3$O(O′)OPO-ATG-CTC-AAC-TCT
2 3′TAG-GAC-TTA-CGC-OP(O)(O′)O$(CH_2)_6$SH
3 5′TAC-GAG-TTG-AGA-ATC-CTG-AAT-GCG

SCHEME 21.6 DNA-directed assembly, to prepare binary (two-component) network materials comprising two different-sized oligonucleotide-functionalized NPs.

Fitzmaurice and coworkers (Cobbe et al. 2003) reported a unique approach to the self-assembly of gold nanocrystals in solution using DNA oligomers (Scheme 21.7). By modifying gold nanocrystals with a biotin analogue, the nanocrystals are programmed to recognize and bind selectively STV–DNA conjugates. The addition of gold nanocrystals, similarly programmed to recognize and bind selectively the complementary STV–DNA conjugate, is expected to result in nanocrystal assembly. DNA duplex formation between the two complementary DNA oligomers bound to the individual nanocrystals drives the assembly process. The addition of a DNA linker or template molecule to initiate the assembly process is also unnecessary. This unique approach allows the kinetics of nanocrystal assembly to be controlled. Moreover, it is possible to terminate aggregation of a dispersion of nanocrystals at any time by the addition of single-stranded DNA oligomers. Consequently, it is possible to control the rate of formation and size of the resultant assemblies.

Mirkin et al. (1996) have used DNA hybridization to generate repetitive nanocluster materials. Two noncomplementary oligonucleotides were coupled in separate reactions with 13 nm gold particles by means of thiol adsorption (Scheme 21.8). A DNA duplex molecule that contains a double-stranded region and two cohesive single-stranded ends, which are complementary to the particle-bound DNA, was used as a linker. The addition of the linker duplex to a mixture of the two oligonucleotide-modified colloids led to the aggregation and slow precipitation of a macroscopic DNA-based colloidal material. The reversibility of this process was demonstrated by the temperature-dependent changes of the UV-vis spectroscopic properties. Since the colloids contained multiple DNA molecules, the oligomerized aggregates were well-ordered and 3D linked, as deduced from the TEM analysis. Images of 2D, single-layer aggregates revealed close-packed assemblies of the colloids with uniform particle separations of about 6 nm, which corresponds to the length of the DNA linker duplex.

Further studies of DNA-linked gold NP assemblies concerned the influence of the DNA spacer length on the optical (Storhoff et al. 2000) and electrical (Park et al. 2000) properties of the networks. The experiments provided evidence that the linker length kinetically controls the size of the aggregates, and that the optical properties of the NP assemblies are governed by the size of the aggregate

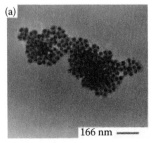

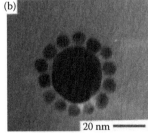

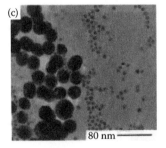

FIGURE 21.3 TEM images of the binary NP network materials supported on holey carbon grids: (a) an assembly generated from **1**-modified 8 nm particles, **2**-modified 31 nm particles, and linking oligonucleotide **3**; (b) a NP satellite structure obtained from the reaction involving 120:1 **1**-modified 8 nm particles/**2**-modified 31 nm particles and linking oligonucleotide **3**; and (c) **1**-modified 8 nm particles and **2**-modified 31 nm particles mixed together without linking oligonucleotide **3** (From Mandal, S. et al., *Langmuir*, 17, 6262, 2001. With permission.). (From Whaley, S.R. et al., *Nature*, 405, 665, 2000. With permission.)

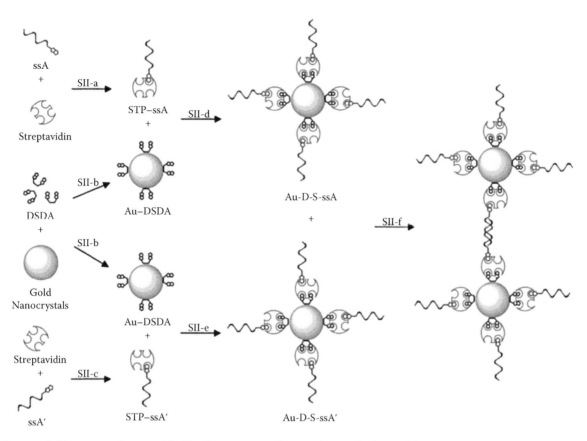

SCHEME 21.7 Gold nanocrystals are modified by chemisorption of DSDA. They are further modified by addition of complementary STV–DNA conjugates to separate sets of nanocrystals. Subsequent combination of both sets of nanocrystals results in duplex formation between the complementary DNA-modified nanocrystals and aggregation of the dispersion.

(Storhoff et al. 2000). These materials show semiconductor properties that are not influenced by the linker lengths (Park et al. 2000). Despite these advances, very little is known about the manipulation and tailoring of such NP networks, for instance, on ways to influence the structure and topography of the DNA hybrid materials subsequent to their formation by self-assembly. To control the stoichiometry and architecture of nanomaterials, Alivisatos and coworkers synthesized well-defined monoadducts from commercially available 1.4 nm gold clusters that contain a single reactive maleimido group and thiolated 18-mer oligonucleotides (Alivisatos et al. 1996, Loweth et al. 1999). Subsequent to purification, these conjugates allowed the rational construction of well-defined nanocrystalline molecules by means of DNA-directed assembly with a single-stranded template that contains the complementary sequence stretches. Depending on the template, the DNA-nanocluster conjugates were assembled to generate the head-to-head and head-to-tail homodimeric target molecules in approximately 70% purity.

More recently, multiple gold nanocrystal aggregates were generated by DNA-directed assembly (Loweth et al. 1999). The preparations, which contain up to three nanoclusters of different sizes organized in several ways, were then purified by electrophoresis. TEM characterization indicated that the nanocrystal molecules have a high flexibility. UV-vis absorbance measurements indicated changes in the spectral properties of the NPs as

a consequence of the supramolecular organization (Loweth et al. 1999). The concept of using DNA as a framework for the precise spatial arrangement of molecular components was carried out originally with the covalent conjugates of single-stranded DNA oligomers and STV protein (Niemeyer et al. 1994).

The STV–DNA conjugates were used as model systems for a variety of important fundamental studies on the DNA-directed assembly of macromolecules (Niemeyer 2001a,b). In addition to their model character, the covalent DNA–STV conjugates are also convenient as versatile molecular adapters in the nanoproduction of supramolecular assemblies. The covalently attached oligonucleotide moiety supplements the four native biotin-binding sites of STV with a specific recognition domain for a complementary nucleic acid sequence. This bispecificity allows the use of the DNA–STV conjugates as adapters for the assembly of basically any biotinylated compound along a nucleic acid template (Niemeyer 2001a,b). As an example, the strong biotin–STV interaction and the specific hybridization capabilities of the DNA–STV conjugates **3** were used to organize gold nanoclusters (Scheme 21.9) (Niemeyer et al. 1998). In this work, 1.4 nm gold clusters that contain a single amino substituent were derivatized with a biotin group, and the biotin moiety was used subsequently to organize the nanoclusters into the tetrahedral superstructure, defined by the geometry of the biotin-binding sites of the STV.

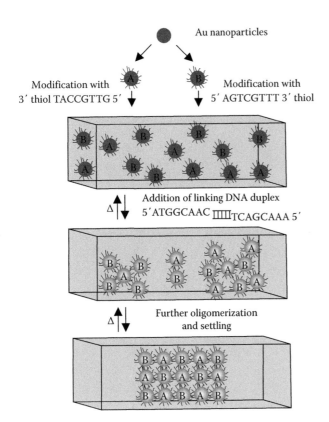

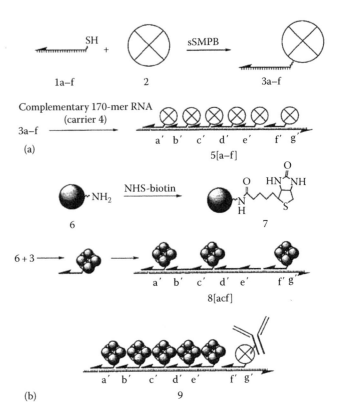

SCHEME 21.8 Scheme showing the DNA-based colloidal NP assembly strategy (the hybridized 12-base-pair portion of the linking duplex is abbreviated as IIII). If a duplex with a 12′-base-pair overlap but with "sticky ends" with four base mismatches is used in the second step, no reversible particle aggregation is observed. The scheme is not meant to imply the formation of a crystalline lattice but rather an aggregate structure that can be reversible annealed. Δ is the heating above the dissociation temperature of the duplex.

Subsequently, the nanocluster-loaded proteins self-assemble in the presence of a complementary single-stranded nucleic acid carrier molecule, thereby generating novel biometallic nanostructures such as **8** (Scheme 21.9). Since the DNA–STV conjugates can be used as a molecular construction kit, this approach even allows the combined assembly of inorganic and biological components to produce functional biometallic aggregates such as **9**, which contain an immunoglobulin molecule (Scheme 21.9). The functionality of the antibody in this aggregate allows the targeting of the biometallic nanostructures to specific tissues, substrates, or other surfaces. This approach impressively demonstrates the applicability of protein–ligand interaction and DNA hybridization for the nanoconstruction of novel inorganic/bioorganic hybrid systems (Niemeyer et al. 1998).

21.4 Other Biomolecule-Induced NPs Aggregation

Recently, we (Basu et al. 2007) have reported the aggregation of gold NPs (Figure 21.4) by the addition of biomolecule, glutathione (GSH), which can bind with gold NPs by its amine group

SCHEME 21.9 Schematic representation of the DNA–protein hybrids. (a) Generation of supramolecular aggregates from DNA–STV conjugates **3**, obtained by covalent coupling of 5′-thiol-modified oligonucleotides **1** and STV **2**. The 3′-end of the oligonucleotide is indicated by an arrowhead and the spacer chains between DNA and protein by wavy lines. The conjugates **3** with nucleotide sequences **a–f** self-assemble in the presence of RNA **4**, which contains complementary sequence sections, to form supramolecular aggregates **5**. (b) Fabrication of biometallic aggregates by means of DNA–STV adapters **3**. Monoamino-modified 1.4 nm gold clusters **6** are converted into biotin derivatives, and the biotinylated clusters **7** are coupled with DNA–STV adducts **3**. The resulting hybrids are assembled in the presence of helper oligonucleotides **1** and RNA carrier **4** to form supramolecular aggregates **8**. (The letters in brackets indicate the protein components bound to the carrier.) Similarly, an antibody-containing aggregate **9** was constructed from gold-labeled **3a–e** and a conjugate from **3f** and biotinylated IgG, previously coupled in separate reactions.

FIGURE 21.4 Typical TEM images for (a) 13 and (b) 20 nm Au aggregates by using GSH as a molecular linker.

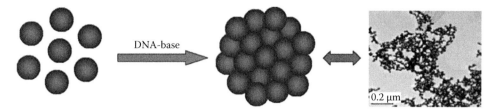

SCHEME 21.10 Schematic representation of the Ag NP aggregates by using DNA bases as molecular linker.

adjacent (α) to the carboxylic acid moiety (−COOH) and −SH group. We have shown the interaction between GSH and gold colloids at different pH, and it has been found that GSH can bind with gold NPs only at relatively low pH, but suppressed at intermediate and high pH. We have investigated the NP size effect on the nature of aggregation among the gold particles. This NP aggregation process occurs with a concomitant color change from red (dispersed gold NPs) to blue (aggregated networks), which can be monitored spectrophotometrically in solution.

We have also reported (Basu et al. 2008) a synthetic strategy for silver NP-DNA nucleobases inorganic–biological hybrid nano-assemblies (Scheme 21.10) and thereby to explore the strength of interaction of the nucleobases with metal surfaces. To the best of our knowledge, this is the first report of organizing silver NPs into periodic functional materials induced by DNA bases and to compare the strengths of interactions between the fundamental chemical components of DNA and silver NP surfaces from surface plasmon resonance spectroscopy and surface-enhanced Raman scattering signal intensity.

Sastry and coworkers (Mandal et al. 2001) have described the surface modification of aqueous silver colloidal particles with the amino acid cysteine and the cross-linking of the colloidal particles in solution (Figure 21.5). Capping of the silver particles with cysteine is accomplished by a thiolate bond between the amino

acid and the NP surface. The silver colloidal particles are stabilized electrostatically by ionizing the carboxylic acid groups of cysteine. Aging of the cysteine-capped colloidal solution leads to the aggregation of the particles via hydrogen bond formation between amino acid molecules located on neighboring silver particles. The aggregation is reversible upon heating the solution above 60°C. The rate of cross-linking of the silver particles via hydrogen bond formation may be accelerated by screening the repulsive electrostatic interactions between the particles using salt.

Gedanken and coworkers (Zhong et al. 2004) have described interactions between nanoscale Au colloids and two main types of organic functional groups, viz., alkanethiols and amino acids. The surface chemistry of particulate Au is dominated by electrodynamic factors related to its (negative) surface charge. In amino acids, the reactivity of the *R*-amine (adjacent to −COOH) is found to be pH-dependent. Linking via the *R*-amine is activated at low pH but suppressed at intermediate and high pH due to electrostatic repulsive forces between the Au surface and the charged carboxylate group or even the (formally neutral) polar carbonyl group in amides. However, dibasic amino acids can still be used to cross-link Au colloids at high pH. This offers a new way to organize Au NPs into extended architectures and functional materials over a wide range of pH (Figure 21.6).

Perez et al. (2003) have reported viral-induced nanoassembly of magnetic NPs due to the multivalent interactions between NP and virus for rapid, sensitive, and selective detection of a virus in solution. They show that these readily detectable magnetic changes can be used to directly detect viral particles at low concentrations (five viral particles in 10 μL) in biological samples. The developed magnetic viral nanosensors are composed of a

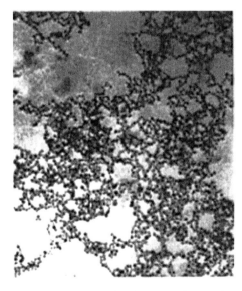

FIGURE 21.5 TEM micrographs of the silver hydrosol capped with cysteine as a function of time of aging *t* = 2 h. (From Mandal, S. et al., *Langmuir*, 17, 6262, 2001. With permission.)

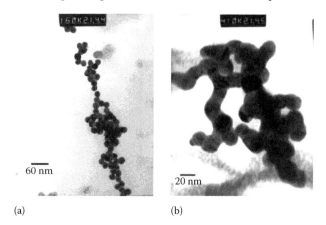

(a)　　　　　　　　　(b)

FIGURE 21.6 (a) Typical TEM image for Au NPs (20 nm) reacted with cysteine and (b) magnified detail of panel (a).

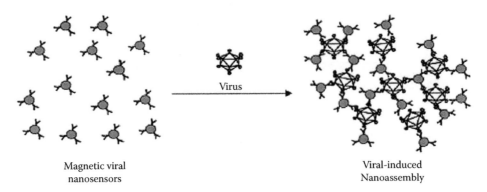

Magnetic viral
nanosensors

Virus

Viral-induced
Nanoassembly

SCHEME 21.11 Diagram of viral-induced nanoassembly of magnetic NPs [virus-surface-specific antibodies are immobilized on the magnetic NPs to create magnetic viral nanosensors. When exposed to viral particles in solution, clustering of the NPs occurs with a corresponding change in the MR signal (δT2)].

superparamagnetic iron oxide core caged with a dextran coating (Josephson et al. 1999) onto which virus-surface-specific antibodies were attached (Scheme 21.11). Attachment of anti-adenovirus 5 or anti-herpes simplex virus 1 antibodies was accomplished via protein G coupling, attached to the caged dextran via *N*-succinimidyl-3-(2-pyridyldithio)propionate.

21.5 Conclusion

This chapter has summarized recent advances in the rapidly developing area of biomolecule-induced NP aggregation. The fact that NPs and biomaterials such as enzymes, antibodies, DNA, or nucleic acids are of similar dimensions makes the hybrid systems attractive nanoelements or building blocks of nanostructures and devices. Since the NPs and biomolecules typically meet at the same nanometer length scale, this interdisciplinary approach will contribute to the establishment of a novel field, descriptively termed biomolecular nanotechnology or nanobiotechnology. Although biomolecules and inorganic materials can be chemically coupled by means of various methods, there is still a great demand for mild and selective coupling techniques that allow the preparation of thermodynamically stable, kinetically inert, and stoichiometrically well-defined bioconjugate hybrid NPs and their aggregates. Biomolecule-functionalized aggregate of NPs could be exploited for numerous applications in biomolecular electronics (Rawlett et al. 2003), biosensors (Fritzsche and Taton 2003, Muller et al. 2003), bioactuators (Patolsky et al. 2004, Weizmann et al. 2004), and medicine, namely, in photodynamic anticancer therapy (Samia et al. 2003), targeted delivery of radioisotopes (Lockman et al. 2003), drug delivery (Allen and Cullis 2004), electronic DNA sequencing, nanotechnology of gene-delivery systems (Cui and Mumper 2003, Salem et al. 2003, Luo et al. 2004), and gene therapy (Miller 2004). Combination of the unique properties of nanoobjects (such as NPs, nanorods, and their aggregates) and biomaterials provides unique opportunity for physicists, chemists, biologists, and material scientists to mold the new area of nanobiotechnology (Patolsky et al. 2004).

References

Alivisatos, A. P.; Johnsson, K. P.; Peng, X. et al. 1996. *Nature* 382: 609.

Allen, T. M.; Cullis, P. R. 2004. *Science* 303: 1818.

Bashir, R. 2001. *Superlattices Microstruct.* 29: 1.

Basu, S.; Ghosh, S. K.; Kundu, S. et al. 2007. *J. Colloid Interface Sci.* 313: 724.

Basu, S.; Jana, S.; Pande, S.; Pal, T. 2008. *J. Colloid Interface Sci.* 321: 288.

Bogatyrev, V. A.; Dykman, L. A.; Krasnov, Y. M.; Plotnikov, V. K.; Khlebtsov, N. G. 2002. *Colloid J.* 64: 671.

Bohren, C. F.; Huffman, D. R. 1983. *Absorption and Scattering of Light by Small Particles*; Wiley: New York.

Cobbe, S.; Connolly, S.; Ryan, D.; Nagle, L.; Eritja, R.; Fitzmaurice, D. 2003. *J. Phys. Chem. B* 107: 470.

Connolly, S.; Fitzmaurice, D. 1999. *Adv. Mater.* 11: 1202.

Cui, Z. R.; Mumper, R. J. 2003. *Crit. Rev. Ther. Drug Carrier Syst.* 20: 103

Falkenhagen, D. 1995. *Artif. Organs* 19: 792.

Fendler, J. H. 2001. *Chem. Mater.* 13: 3196.

Fritzsche, W.; Taton, T. A. 2003. *Nanotechnology* 14: R63.

Gaponik, N. P.; Talapin, D. V.; Rogach, A. L. 2000. *J. Mater. Chem.* 10: 2163.

Henglein, A. 1993. *Isr. J. Chem.* 33: 77.

Jana, N. R.; Pal, T. 1999. *Langmuir* 15: 3458.

Jana, N. R.; Sau, T. K.; Pal, T. 1999. *J. Phys. Chem. B* 103: 115.

Jana, N. R.; Gearhert, L.; Murphy, C. J. 2001. *Chem. Commun.* 617.

Josephson, L.; Tung, C. H.; Moore, A.; Weissleder, R. 1999. *Bioconjug. Chem.* 10: 186.

Katz, E.; Willner, I. 2004. *Angew. Chem. Int. Ed.* 43: 6042.

Kreibig, U.; Vollmer, M. 1995. *Optical Properties of Metal Clusters*; Springer: Berlin, Germany.

Li, M.; Wong, K. W.; Mann, S. 1999. *Chem. Mater.* 11: 23.

Link, S.; El-Sayed, M. A. 1999. *J. Phys. Chem. B* 103: 8410.

Lockman, P. R.; Oyewumi, M. O.; Koziara, J. M.; Roder, K. E.; Mumper, R. J.; Allen, D. D. 2003. *J. Control. Release* 93: 271.

Loweth, C. J.; Caldwell, W. B.; Peng, X.; Alivisatos, A. P.; Schultz, P. G. 1999. *Angew. Chem. Int. Ed.* 38: 1808.

Luo, D.; Han, E.; Belcheva, N.; Saltzman, W. M. 2004. *J. Control. Release* 95: 333.

Mandal, S.; Gole, A.; Lala, N.; Gonnade, R.; Ganvir, V.; Sastry, M. 2001. *Langmuir* 17: 6262.

Mann, S.; Shenton, W.; Li, M.; Connoly, S.; Fitzmaurice, D. 2000. *Adv. Mater.* 12: 147.

Miller, A. D. 2004. *Chem. Bio. Chem.* 5: 53.

Mirkin, C. A.; Letsinger, R. L.; Mucic, R. C.; Storhoff J. J. 1996. *Nature* 382: 607.

Mucic, R. C.; Storhoff, J. J.; Mirkin, C. A.; Letsinger, R. L. 1998. *J. Am. Chem. Soc.* 120: 12674.

Muller, R.; Csaki, A.; Fritzsche, W. 2003. *Tech. Mess.* 70: 582.

Murray, C. B.; Kagan, C. R.; Bawendi, M. G. 2000. *Annu. Rev. Mater. Sci.* 30: 545.

Niemeyer, C. M. 1999. *Appl. Phys. A* 68: 119.

Niemeyer, C. M. 2000. *Curr. Opin. Chem. Biol.* 4: 609.

Niemeyer, C. M. 2001a. *Angew. Chem. Int. Ed.* 40: 4128.

Niemeyer, C. M. 2001b. *Chem. Eur. J.* 7: 3188.

Niemeyer, C. M.; Sano, T.; Smith, C. L.; Cantor, C. R. 1994. *Nucleic Acids Res.* 22: 5530.

Niemeyer, C. M.; Burger, W.; Peplies, J. 1998. *Angew. Chem. Int. Ed.* 37: 2265.

Palato, L.; Benedetti, L. M.; Callegaro, L. 1994. *J. Drug Target.* 2: 53.

Park, S.-J.; Lazarides, A. A.; Mirkin, C. A.; Brazis, P. W.; Kannewurf, C. R.; Letsinger, R. L. 2000. *Angew. Chem. Int. Ed.* 39: 3845.

Park, S.-J.; Lazarides, A. A.; Mirkin, C. A.; Letsinger, R. L. 2001. *Angew. Chem. Int. Ed.* 40: 2909.

Patolsky, F.; Weizmann, Y.; Willner, I. 2004. *Nat. Mater.* 3: 692.

Perez, J. M.; Simeone, F. J.; Saeki, Y.; Josephson, L.; Weissleder, R. 2003. *J. Am. Chem. Soc.* 125: 10192.

Piran, U.; Riordan, W. J. 1990. *J. Immunol. Methods* 133: 141.

Pradhan, N.; Pal, A.; Pal, T. 2001. *Langmuir* 17: 1800.

Rawlett, A. M.; Hopson, T. J.; Amlani, I.; Zhang, R.; Tresek, J.; Nagahara, L. A.; Tsui, R. K.; Goronkin, H. 2003. *Nanotechnology* 14: 377.

Reznik, G. O.; Vajda, S.; Smith, C. L.; Cantor, C. R.; Sano, T. 1996. *Nat. Biotechnol.* 14: 1007.

Robert, H. D.; Rao, P. 1996. *J. Mater. Res.* 11: 2834.

Salem, A. K.; Searson, P. C.; Leong, K. W. 2003. *Nat. Mater.* 2: 668

Samia, A. C. S.; Chen, X.; Burda, C. 2003. *J. Am. Chem. Soc.* 125: 15736.

Sano, T.; Cantor, C. R. 1995. *Proc. Natl. Acad. Sci. USA* 92: 3180.

Sano, T.; Pandori, M. W.; Chen, X. M.; Smith, C. L.; Cantor, C. R. 1995. *J. Biol. Chem.* 270: 28 204.

Sano, T.; Reznik, G. O.; Szafranski, P. et al. 1996. *Proceedings of the 50th Anniversary Conference of the Korean Chemical Society*, Seoul, Korea, p. 359.

Sano, T.; Vajda, S.; Cantor, C. R. 1998. *J. Chromatogr. B* 715: 85.

Schmid, G. 1994. *Clusters and Colloids: From Theory to Applications*; VCH: Weinheim, Germany.

Schmidt, T. G. M.; Koepke, J.; Frank, R.; Skerra, A. 1996. *J. Mol. Biol.* 255: 753.

Seeman, N. C. 1999. *Trends Biotechnol.* 17: 437.

Shenton, W.; Davis, S. A.; Mann, S. 1999. *Adv. Mater.* 11: 449.

Sinha, B. K.; Chignell, C. F. 1979. *Methods Enzymol.* 62: 295.

Srivastava, S.; Samanta, B.; Jordan, B. J.; Hong, R.; Xiao, Q.; Tuominen, M. T.; Rotello, V. M. 2007. *J. Am. Chem. Soc.* 129: 11776.

Storhoff, J. J.; Lazarides, A. A.; Mucic, R. C.; Mirkin, C. A.; Letsinger, R. L.; Schatz, G. C. 2000. *J. Am. Chem. Soc.* 122: 4640.

Weizmann, Y.; Patolsky, F.; Lioubashevski, O.; Willner, I. 2004. *J. Am. Chem. Soc.* 126: 1073.

Whaley, S. R.; English, D. S.; Hu, E. L.; Barbara, P. F.; Belcher, A. M. 2000. *Nature* 405: 665.

Zhong, Z.; Patskovskyy, S.; Bouvrette, P.; Luong, J. H. T.; Gedanken, A. 2004. *J. Phys. Chem. B* 108: 4046.

Magnetic Nanoparticle Assemblies

Dimitris Kechrakos
School of Pedagogical and
Technological Education

22.1 Introduction

Magnetic nanoparticles (MNPs) are minute parts of magnetic materials with typical size well below 10^{-7} m. They are present in different materials found in nature such as rocks, living organisms, ceramics, and corrosion products, but they are also artificially made and used as the active component of ferrofluids, permanent magnets, soft magnetic materials, biomedical materials, and catalysts. Their diverse applications in geology, physics chemistry, biology, and medicine render the study of their properties of great importance to both science and technology.

In geology, the nature and origin of magnetic phenomena related to the presence of magnetite nanoparticles in rocks are of great interest to the palaeo-magnetist who searches for the geomagnetic record of rocks. The presence of magnetite particles associated with the trigeminal nerve in pigeons offers a reliable explanation to the Earth's magnetic field detection and the consequent navigation capability. In fine arts, the magnetic analysis of ancient paintings facilitates the reconstruction of the production techniques of ancient ceramics. In living organisms, the role of ferritin, an MNP per se, is important among the iron storage proteins. MNPs are also used as contrast agents in magnetic resonance imaging. Recent work has involved the development of bioconjugated MNPs, which facilitated specific targeting of these MRI probes to brain tumors. MNPs are also used as highly active catalysts, which has long been demonstrated by the use of finely divided metals in several reactions. Owing to their high surface-to-volume ratio, MNPs of iron are more efficient at waste remediation than bulk iron.

High-density magnetic data storage media provide a major technological driving force for further exploration of MNPs.

It is expected that if MNPs with diameter 5 nm can be used as individually addressed magnetic bits, magnetic data storage densities of 1 Tbit/in.^{2} would be achieved, namely, an order of magnitude higher than the present record (Moser et al. 2002). MNPs have also been demonstrated to be functional elements in magneto-optical switches, sensors based on giant magnetoresistance and magnetically controllable single electron transistor devices.

The most common preparation methods for MNPs produce assemblies with different structural and compositional characteristics that depend on the particular method adopted. Granular films, ferrofluids, and cluster-assembled films are characterized as assemblies with random order in MNP locations, while ordered arrays are found in patterned media (also known as magnetic dots) and self-assembled films. The MNP preparation methods are divided into top-down and bottom-up methods. In top-down methods, the NPs are formed from a larger system by appropriate physical processing, such as thermal treatment and etching. In bottom-up methods, the NPs are formed by an atomic nucleation process that takes place either in ultrahigh vacuum or in a liquid environment. The latter method relies on colloidal chemistry techniques and presently appears to be the most promising method for the production of nanoparticles with extremely narrow size distribution. Colloidal synthesis methods combined with self-assembly methods produce MNP samples with both size uniformity and long-range structural order. It is worth noticing that structural order in an MNP assembly is a decisive property for the production of ultrahigh-density storage media. Owing to their attractive features and their low cost, colloidal synthesis methods and self-assembly attract presently intense research activity in the field of MNP preparation (Petit et al. 1998, Murray et al. 2001, Willard et al. 2004, Darling and Bader 2005, Farrell et al. 2005).

The magnetic properties of MNPs and their assemblies provide a fascinating field for basic research, which is done on two different scales, the atomic and the mesoscopic. In the atomic scale, the properties of individual MNPs are examined, and they are revealed in samples with low particle concentration. In the mesoscopic scale, dense samples that exhibit collective magnetic behavior arising from interparticle interactions are examined. The study of the magnetic properties can be naturally divided into the investigation of the ground-state configuration (long-range order, disorder, etc.) and the excitations from it. Excitations can be either weak, for example at low temperature and weak external magnetic field, or strong, for example, close to a thermal phase transition or under a reversing magnetic field.

For individual MNPs, the ground-state configuration can differ remarkably from the parent bulk material in various ways. For example, owing to energy balance reasons, the abundance of magnetic domains that form in a bulk magnet can be replaced by a single domain (SD) in an MNP, which then becomes magnetically saturated even in the absence of an external magnetic field (Néel 1949). The application of an external field forces the atomic magnetic moments of an SD MNP to rotate coherently (Stoner and Wohlfarth 1948). Also, for temperature above a threshold, the direction of particle's magnetization fluctuates at random, making the particle behave as a molecule with a giant magnetic moment. The applications of this effect, known as superparamagnetism (SPM; Bean and Livingston 1959), are presently a lot, ranging from geology to medicine. Finally, we should remark that the above-described simplified picture of an SD MNP becomes invalid if one considers the crucial effect of the MNP surface. Reduced crystal symmetry and chemical disorder close to the surface can produce variations between the surface and interior magnetic structure and modify the overall response of the MNP to an applied field (Kodama 1999).

When MNPs form dense assemblies, interparticle interactions produce a collective behavior, by coupling the magnetic moments of individual MNPs. This fact renders, in most cases, even the determination of the ground-state configuration an intricate physical problem. The collective behavior of dense (interacting) assemblies is also reflected on the modified magnetic response of the assembly, compared to isolated MNPs. The most complex behavior occurs in samples with random morphology and long-range magnetostatic interactions. Various experimental measurements have been proposed to reveal the nature of the interparticle interactions, and various measuring protocols probe different aspects of the collective behavior. On the other hand, analytical models have difficulties in predicting or explaining the magnetic behavior of these interacting MNP assemblies, and most of the current research relies on numerical simulations.

In this chapter, we provide an introduction to the fundamental ideas and concepts pertaining to the magnetic properties of MNP assemblies. Emphasis is given to the response of MNP assemblies to an applied magnetic field and the related issue of magnetization reversal. This chapter is organized as follows: In Section 22.2, we discuss the magnetic properties of individual (isolated) MNPs. First, the condition under which an SD MNP is formed is derived, and then the magnetic response under an applied field is examined. The presentation is based on a simple theoretical model (Néel 1949, Stoner and Wohlfarth 1948). In Section 22.3, we give a brief overview of the most common magnetic characterization techniques and explain the information extracted from each one. In Section 22.4, we discuss the response of a dense MNP assembly to a magnetic field, when the interparticle interactions are important and lead to a collective behavior of the MNPs. Mean-field models and an introduction to modern numerical techniques (Monte Carlo, magnetization dynamics [MD]) to tackle this problem are presented. This chapter is summarized in Section 22.5, and the perspectives in this field are presented in Section 22.6.

22.2 Isolated Magnetic Nanoparticles

In this section, we derive the criterion for the formation of SD MNPs and examine the magnetization process at zero temperature by the coherent rotation of magnetization (Stoner-Wohlfarth [SW] model). The behavior of an MNP assembly at finite temperature is discussed, and the related concepts of SPM and blocking temperature are introduced. The effects of an applied dc magnetic field are examined within the simplest model assuming uniaxial anisotropy and bistability of particle moments (Néel model).

22.2.1 Single-Domain Particles

The ground-state magnetic structure of a ferromagnetic (FM) material is the outcome of the balance between three different types of energies, namely, the exchange (U_{ex}), the magnetostatic (U_m), and the anisotropy energy (U_a). The exchange interaction has its origin in the Pauli exclusion principle for electrons. Let the FM material be divided into small cubic elements, each one carrying a magnetic moment $\vec{\mu}_i$. The exchange interaction between the cubic elements favors the parallel alignment of neighboring magnetic moments, and it is written in the usual Heisenberg form as $U_{ex} = -(A/a^2)\sum_{ij}\cos\theta_{ij}$, where A is the stiffness constant, a is the lattice constant, and θ_{ij} is the angle between moments at sites i and j. The stiffness constant is related to the microscopic exchange energy J through the relation $A = zJS^2/a$, where S is the atomic spin and $z = 1, 2, 4$ for sc, fcc, and bcc lattices, respectively. The magnetostatic energy is the sum of Coulomb energies between the magnetic moments comprising the FM material. It can be expressed as $U_m = -\mu_0 \vec{H_d} \cdot \vec{M}/2$, where H_d is the *demagnetizing* field and M is the sample magnetization. The anisotropy energy is the energy required to orient the magnetization at an angle (θ) relative to certain fixed axes of the system, known as the *easy* axes. The microscopic mechanisms leading to anisotropy can be quite diverse, and the most common types of anisotropy found in FMs are as follows:

1. *Crystal* anisotropy. It arises from the combined effects of spin–orbit coupling and quenching of the orbital momentum that produce a preferred orientation of the magnetization along a symmetry axes of the underlying crystal. For a uniaxial material (e.g., hexagonal Co), it has the form $U_a = K_1 \sin^2 \theta + K_2 \sin^4 \theta + \cdots$, where K_1, K_2, … are the anisotropy constants, and θ is the angle between the magnetization direction and the easy axis. Typical values for cobalt are $K_1 = 4.5 \times 10^6 \, \text{J/m}^3$ and $K_2 = 1.5 \times 10^5 \, \text{J/m}^3$. For cubic crystals (e.g., fcc Fe, Ni), it reads $U_a = K_1 \left(a_1^2 a_2^2 + a_2^2 a_3^2 + a_3^2 a_1^2 \right) + K_2 a_1^2 a_2^2 a_3^2 + \cdots$, where a_1, a_2, a_3 are the direction cosines of the magnetization direction. Typical values for Fe are $K_1 = 4.8 \times 10^4 \, \text{J/m}^3$ and $K_2 = \pm 0.5 \times 10^4 \, \text{J/m}^3$.

2. *Stress* anisotropy. It is produced by the presence of stress in the sample, and it has a uniaxial character $U_a = K_\sigma \sin^2 \theta$, where $K_\sigma = (3/2) \lambda_i \sigma$, with λ_i the magnetically induced isotropic strain and σ the stress.

3. *Surface* anisotropy. This is caused by the presence of sample free boundaries, where the reduced symmetry and the presence of defects can induce additional anisotropy. It is important in MNPs because of the substantial surface-to-volume ratio.

4. *Shape* anisotropy. This occurs because, on the one hand, the demagnetizing field depends on the shape of the magnetized body and takes the lowest value along the longest axis of the sample, and on the other hand, U_m is minimized when M is parallel to H_d. As an example, consider a specimen in the shape of prolate spheroid with major axis c and minor axis a, magnetized at an angle θ with respect to c-axis. Then, $U_m = \mu_0/2[N_c(M\cos\theta)^2 + N_a(M\sin\theta)^2] = 1/2(N_c - N_a)M^2\sin^2\theta$, where N_c and N_a are the *demagnetizing* factors along the corresponding axes. This expression for U_m has the form of uniaxial anisotropy with $K_s = 1/2(N_c - N_a)M^2$. Typical cases are a spherical specimen with $K_s = 0$, an infinitely thin planar specimen with $N_\parallel = 0$ (in-plane) and $N_\perp = 1$, and an infinitely long (needle-shaped) specimen with $N_\parallel = 1/2$ (along the axis) and $N_\perp = 0$.

In studies of the magnetic properties of MNPs, it is a common practice, to describe, within the simplest approximation, the overall effect of the various anisotropy types by an *effective* uniaxial anisotropy term $U_a = K_{\text{eff}} \sin^2 \theta$. The constant K_{eff} accounts for the total effect of crystalline, surface, and shape anisotropy.

A bulk FM material is composed of many uniformly magnetized regions (*domains*). The direction of magnetization in different domains varies, and in a bulk sample it is randomly distributed leading to a nonmagnetized sample even at temperatures far below the Curie point. The formation of magnetic domains in FM materials results from the competition between the exchange and the magnetostatic energy. The former favors perfect alignment of neighboring moments and the latter is reduced by

breaking a uniformly magnetized body into as many as possible regions with opposite magnetization directions. The outcome of this competition is the formation of a certain number of domains in a sample with a particular orientation of the magnetization directions. A typical domain size in a bulk ferromagnet is $1 \, \mu\text{m}$.

Neighboring magnetic domains are separated by a region where the local magnetization gradually changes direction between the two opposite sides, known as *domain wall* (DW). DWs have finite width (δ_w) determined by the balance between the exchange and anisotropy energies. As an example, consider a one-dimensional model of a DW in a uniaxial material, where a 180° rotation of magnetization is distributed over N sites, as shown in Figure 22.1. The total energy per unit area reads

$$\sigma(N) = \sigma_{\text{ex}} + \sigma_a = JS^2 \left(\frac{\pi}{N}\right)^2 \left(\frac{N}{a^2}\right) + NaK_1. \qquad (22.1)$$

Minimization with respect to N leads to

$$\delta_w = Na = \pi \left(\frac{A}{K_1}\right)^{1/2}. \qquad (22.2)$$

For a typical exchange stiffness value $A \approx 10^{-11} \, \text{J/m}$), Equation 22.2 predicts $\delta_w \approx 0.4 \, \mu\text{m}$ for iron and $\delta_w \approx 60 \, \text{nm}$ for a magnetically harder material like cobalt. Substituting the result of Equation 22.2 into Equation 22.1 provides the areal energy density of the DW:

$$\sigma_w = 2\pi(AK_1)^{1/2}. \qquad (22.3)$$

Consider a finite sample of an FM material, with size d. As the size of the sample is reduced, the number of DWs it contains decreases, because fewer regions with opposite directions of magnetization are required to reduce the magnetostatic energy. Below a critical value of the system size, the sample does not contain any DW, and it is in an SD state exhibiting saturation magnetization (M_s). For a spherical particle, the critical diameter (d_c) can be estimated as follows: the SD state is stable when the energy needed to create a DW that spans the whole particle, $U_w = \sigma_w \pi r_2$, is greater than the magnetostatic energy gain from the reduction to a multidomain state, which is approximately

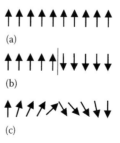

(a)

(b)

(c)

FIGURE 22.1 One-dimensional model of an FM. (a) Long-range order. (b) An infinitely thin DW (dashed line). The increase of exchange energy at the wall is higher than the decrease of the magnetostatic energy. (c) A 180° DW spread over $N = 10$ sites. The gradual rotation of atomic moments produces a state with lower total energy compared to (b).

equal to the magnetostatic energy stored in a uniformly magnetized sphere, $U_m = (1/3)\mu_0 M_s^2 V$, with M_s the saturation magnetization and $V = (4\pi/3)r^3$. The condition $U_w = U_m$ provides

$$r_c = 9\frac{(AK_1)^{1/2}}{\mu_0 M_s^2}. \tag{22.4}$$

For Fe, this approximation gives $r_c \approx 3\,nm$, which is by far too small. The reason is that the DW is assumed to have the same one-dimensional structure as in the bulk material. An improved calculation that considers a three-dimensional confinement of the DW provides for the critical radius:

$$r_c = \sqrt{\frac{9A}{\mu_0 M_s^2}\left[\ln\left(\frac{2r_c}{a}\right)-1\right]}. \tag{22.5}$$

In the case of Fe, numerical solution of Equation 22.5 gives $r_c \approx 25\,nm$, which is very close to more accurate micromagnetic calculations and the experimentally obtained value (Cullity 1972).

22.2.2 Magnetization by Coherent Rotation

The magnetization (M) of a bulk FM crystal that contains many magnetic domains changes under the application of an external magnetic field (H), a process known as *technical magnetization*. However, the value of M is not a unique function of H, and the state of the sample *prior* to the application of the field is important. This is the phenomenon of magnetic *hysteresis*, which is commonly depicted by drawing the $M - H$ dependence under a cyclic variation of the field from a positive to a negative and back to a positive saturation value (*hysteresis loop*). Two important characteristic values of a hysteresis loop are the *remanence* (M_r), namely, the magnetization after the removal of the saturating field, and the *coercivity* (H_c), namely, the field required for the magnetization to vanish. In a bulk FM crystal, the magnetization proceeds by two basic mechanisms, namely, DW motion (weak fields) and the rotation of magnetization (strong fields).

In MNPs, the change of magnetization under an applied field proceeds only by rotation, because the formation of DWs is energetically unfavorable. During the magnetization rotation, the atomic moments of the MNP remain parallel to each other, and the MNP behaves as a giant molecule carrying a magnetic moment of a few thousand Bohr magnetons ($\mu \sim 10^4 \mu_B$ for a 5 nm diameter Fe MNP). This process of magnetization is known as *coherent* rotation or SW model, after the authors who introduced and solved it (Stoner and Wohlfarth 1948). We discuss it briefly next. Consider an MNP with uniaxial (effective) anisotropy K_1 along an easy axis taken to be the z-axis (Figure 22.2). For an applied field that makes an angle θ_0 with the easy axis, we wish to determine the equilibrium position of the magnetic moment $\mu = M_s V$. Let $\vec{\mu}$ make an angle θ with the easy axis, then the total energy density reads

$$u = -K_1 \cos^2(\theta - \theta_0) - \mu_0 HM_s \cos\theta. \tag{22.6}$$

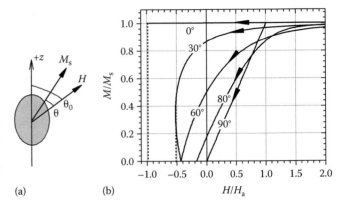

(a)   (b)

FIGURE 22.2   (a) Sketch of a MNP with uniaxial anisotropy along the z-axis and an applied field at an angle (θ_0) with respect to the easy axis. (b) Magnetization curves within the SW model for various field directions. The initial direction of the magnetization is taken along the field.

The equilibrium condition (zero torque) is

$$\frac{du}{d\theta} = 0 \Rightarrow 2K_1 \sin(\theta - \theta_0)\cos(\theta - \theta_0) + \mu_0 HM_s \sin\theta = 0 \tag{22.7}$$

and introducing the dimensionless quantity $h = H/H_a$ with the *anisotropy* field $H_a = 2K_1/\mu_0 M_s$, Equation 22.7 becomes

$$\sin(2(\theta - \theta_0)) + 2h\sin\theta = 0. \tag{22.8}$$

We define the reduced magnetization along the field $m = \mu\cos\theta/M_s V = \cos\theta$, and the solution of Equation 22.8 is written as

$$2m(1-m^2)^{1/2}\cos 2\theta_0 + \sin 2\theta_0(1-2m^2) + 2h(1-m^2)^{1/2} = 0. \tag{22.9}$$

The remanence ($h = 0$) and coercivity ($m = 0$) are readily obtained from Equation 22.9 as

$$m_r = \cos\theta_0 \quad \text{and} \quad h_c = \sin\theta_0\cos\theta_0. \tag{22.10}$$

For nonzero field values, Equation 22.9 is solved for h as a function of m and the data are shown in Figure 22.2. Consider the two extreme cases, namely, for $\theta_0 = 90°$ (hard-axis magnetization) and $\theta_0 = 0°$ (easy-axis magnetization). In the former case, the magnetization shows zero coercivity and a linear field dependence. In the latter case, the magnetization remains constant until the reversing field becomes equal to the anisotropy field, and then an *irreversible* jump of the reduced magnetization from $m = +1$ to $m = -1$ is seen. These extreme cases demonstrate the distinct mechanism of switching by rotation that can occur in an assembly. More generally, at an arbitrary field angle, an irreversible jump of the magnetization occurs at the so-called *switching* field (H_s) defined as the field value satisfying $dm/dh \rightarrow \infty$. At $H = H_s$, the local minimum of the total energy, corresponding to the higher energy state (magnetization opposite to the applied field) disappears, and the system jumps to the remaining minimum that corresponds to a magnetization direction along the field (see Figure 22.3). In other words, H_s is an instability point

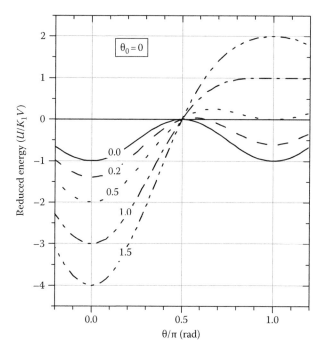

FIGURE 22.3 Dependence of total energy on the direction of the particle's moment (see Equation 22.6), for various strengths of the applied field ($h = H/H_a$). The energy minimum at $\theta = \pi$ becomes unstable at the switching field $h_s = 1$.

of the total energy, and it satisfies $du/d\theta = 0$ and $d^2u/d\theta^2 = 0$. In the SW model, the stability condition reads

$$\frac{d^2u}{d\theta^2} = 0 \Rightarrow \cos 2(\theta - \theta_0) \pm h \sin \theta = 0. \quad (22.11)$$

From Equations 22.8 and 22.11, we obtain for the switching field $h_s = H_s/H_a$

$$h_s = (\cos^{2/3} \theta_0 + \sin^{2/3} \theta_0)^{-3/2}. \quad (22.12)$$

By comparison of Equations 22.10 and 22.12, one finds that $h_c < h_s$ for $45° < \theta_0 < 90°$, namely, switching happens after the magnetization changes sign, while for field angles close to the easy axis, $0° < \theta_0 < 45°$, the magnetization changes sign by an irreversible jump ($h_c = h_s$). The physical distinction between h_s and h_c can be understood by the following example. Consider an SW particle under the application of a reversing field $h = h_c$, which brings the particle's moment $\bar{\mu}$ in a direction perpendicular to the field, so that $m = 0$. Then the field is switched off adiabatically. If $h_c < h_s$ (i.e., $45° < \theta_0 < 90°$), $\bar{\mu}$ will return back to the positive remanence value ($m = +1$), while if $h_c = h_s$ (i.e., $0° < \theta_0 < 45°$), $\bar{\mu}$ will jump to the negative remanence state ($m = -1$). The switching field of a hard (i.e., large anisotropy) magnetic material is a physical quantity with great technological interest in magnetic recording applications. In these, the information bit is stored in the direction of magnetization, and the switching field is the field required to write or erase this information.

Stoner and Wohlfarth (1948) also studied an assembly of isolated MNPs with easy axes directions distributed uniformly on a sphere (*random anisotropy model*, RIM). The reported values for the remanence and coercivity are

$$m_r = 0.5 \quad \text{and} \quad h_c = 0.48. \quad (22.13)$$

This result is particularly useful as random easy axis distribution is found in most MNP-based materials (granular films, cluster-assembled films, self-assembled arrays, etc.)

As a final remark, we remind that in the SW model thermal effects are ignored ($T = 0$), thus energy minimization with respect to the magnetic moment direction is a sufficient condition to determine the field-dependent magnetization at equilibrium. The magnetic behavior of SD particles at finite temperature is discussed in the following section.

22.2.3 Magnetic Behavior at Finite Temperature

How do thermal fluctuations affect the average magnetization direction of an isolated MNP? How does the presence of an applied field modify the magnetic response at finite temperature? Is the assembly magnetization stable in time, when the MNP moment is subject to thermal fluctuations? These points are briefly discussed next, along the lines of a model first studied by Néel (1949).

22.2.3.1 Superparamagnetism and Blocking Temperature

Consider an assembly of identical SD particles with uniaxial anisotropy. The energy (per particle) is $U = -K_1 V \cos^2 \theta$, where θ is the angle between the single particle magnetic moment $\bar{\mu}$ and the easy axis. The energy barrier that must be overcome for an MNP to rotate its magnetization is $E_b = K_1 V$. As first pointed out by Néel (1949), thermal fluctuations could provide the required energy to overcome the anisotropy barrier and spontaneously (i.e., without externally applied field) reverse the magnetization of an MNP from one easy direction to the other. This phenomenon can be thought of as a Brownian motion of a particle's magnetic moment. The assembly shows paramagnetic behavior; however, it is the giant moments of the MNPs that fluctuate rather than the atomic moments of a classical bulk paramagnetic material. This magnetic behavior of the MNPs is called *SPM* (Bean and Livingston 1959). At high enough temperature, $k_B T \gg K_1 V$, the anisotropy energy can be neglected and the assembly magnetization can be described by the well-known Langevin function $M = nM_s L(x)$, where n is the particle number density and $x = \mu_0 \mu H/k_B T$. Thus, the features serving as a signature of SPM are the scaling of magnetization curves with H/T, as dictated by the Langevin function, and the lack of hysteresis, that is, vanishing remanence and coercivity. Moreover, the major difference between classical paramagnetism of bulk materials and SPM is the weak fields ($H \sim 0.1\ T$) required to achieve the saturation of an MNP assembly magnetization M. This occurs because of the large particle moment ($\mu \sim 10^4\mu_B$) compared to the atomic moments ($\mu_{at} \sim \mu_B$).

The measurement of magnetization curves at sufficiently high temperatures can, in principle, be used to extract the particle moment μ. In practice, two complications arise. First, the presence of different particle sizes in any sample produces a convolution of the Langevin function with the volume distribution function. Second, interparticle interactions modify the reversal mechanism and the SW model needs extensions, which are discussed in Section 22.4.

At low temperatures, $k_B T \ll K_1 V$, the anisotropy barriers are very rarely overcome (weak thermal fluctuations), the assembly shows hysteresis, and this is called the *blocked* state.

One might now ask whether there exists a temperature value that draws the border between the blocked and the SPM state. Following Néel's arguments, we assume that thermal activation over the anisotropy barrier can be described within the *relaxation time* approximation (or *Arrhenius law*) as

$$\tau = \tau_0 \exp\left(\frac{K_1 V}{k_B T}\right), \tag{22.14}$$

where $1/2\tau$ is the probability per unit time for a reversal of $\vec{\mu}$. The intrinsic time τ_0 depends on the material parameters (magnetostriction constant, Young modulus, anisotropy constant, and saturation magnetization). Typical values are $\tau_0 \sim 10^{-10}$ to 10^{-9} s as obtained by Néel. To detect the superparamagnetic behavior experimentally, the MNP must be probed for a long-enough period of time to perform many switching events that would produce a vanishing small time-average magnetic moment. If τ_m is the measuring time window, the condition for SPM behavior is $\tau_m \gg \tau$. The strong (exponential) dependence of τ on temperature (see Equation 22.14) permits us to define a temperature value (or more precisely, a very narrow temperature range) above which the relaxation time is so small that SPM behavior is observed. This is called the *blocking temperature* (T_b) of the assembly, and is given by

$$T_b = \frac{K_1 V}{k_B} \ln\left(\frac{\tau_m}{\tau}\right). \tag{22.15}$$

For $T < T_b$, the particle moments fluctuate without switching direction (on average) and the assembly is in the blocked state exhibiting hysteresis. For $T > T_b$, the assembly is in the SPM state, hysteresis disappears, and thermal equilibrium is established. It is remarkable that the value of T_b depends on τ_m, which is a characteristic of the experimental technique adopted. For example, in dc susceptibility measurements $\tau_m \approx 100$ s, in ac susceptibility $\tau_m \approx 10^{-8}$ to 10^4 s, in Mössbauer spectroscopy $\tau_m \approx 10^{-9}$ to 10^{-7} s and in neutron spectroscopy $\tau_m \approx 10^{-12}$ to 10^{-8} s. Therefore, if T_b is of interest for a particular application, the measurement technique implemented must imitate the real conditions. For example, to study the reliability of magnetic storage media, dc magnetic measurements over a wide time window ($\tau_m \sim 10^2$–10^4 s) should be used, while to study magnetic recording speed, ac measurements are appropriate.

Brown (1963) extended the treatment of thermal activation over the anisotropy barrier, allowing also for fluctuations of μ transverse to the easy axis, which Néel has neglected, and obtained a different expression for τ_0. However, the common feature of both studies is the temperature and volume dependence of τ, so the final result, Equation 22.14, is referred to as the *Néel-Brown* model.

In a *polydisperse* assembly, the distribution of particle volumes $f(V)$ produces a corresponding distribution of blocking temperatures $f(T_b)$. Then, at a certain temperature T the assembly contains a mixture of blocked and SPM particles. The MNPs with volumes above a critical value, V_c, fulfill the requirement of strong thermal energy with respect to their anisotropy barrier, and are SPM, while those with $V \le V_c$ are blocked. From Equation 22.14, the critical volume reads $V_c = k_B T \ln(\tau_m/\tau_0)/K_1$. As explained above for T_b, also for V_c the experimental determination depends on the technique adopted. Most preparation techniques result in polydisperse samples and the problem of extracting the size distribution function from magnetic measurements, pioneered by Bean and Jacobs more than 50 years ago (Bean and Jacobs 1956) remains a difficult task mainly due to the complications introduced by interparticle interactions. Knobel and colleagues have recently reviewed this subject (Knobel et al. 2008).

22.2.3.2 Thermal Relaxation under an Applied Field

Consider an assembly of N identical MNPs with uniaxial anisotropy along the z axis and let their moments point initially along the $+z$ axis. Assume that a magnetic field H, weaker than the switching field, which is equal to H_a, is applied along the $-z$ axis. Then, the total energy per particle reads $U = -K_1 V \cos^2(\theta - \theta_0) + \mu_0 H M_s V \cos\theta$. It exhibits two nonequivalent local minima at $\theta = 0, \pi$ with values $U_\pm = -K_1 V \pm M_s V H$ and a maximum at $\theta = \pi/2$ with $U_{max} = K_1 V(H/H_a)^2$, as shown in Figure 22.4. The energy barriers and the corresponding relaxation times for the forward (+) and the backward (−) rotations are

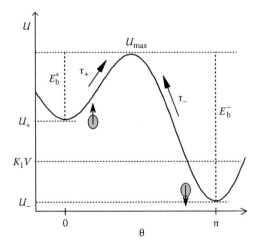

FIGURE 22.4 Total energy of an isolated particle with uniaxial anisotropy subject to a negative field parallel to the easy axis with value less than the switching field ($0 < H < H_s$). Energy barriers (E_b) and relaxation times (τ) for the forward (+) and the backward (−) process are not equal.

$$E_b^{\pm}(H) = K_1 V \left(\frac{1 \mp H}{H_a} \right)^2 \quad \text{and} \quad \tau_{\pm} = \tau_0 \exp\left(\frac{E_b^{\pm}}{k_B T} \right). \quad (22.16)$$

The change of τ_0 due to the field is much weaker than the change of the exponential factor and as such it is neglected in the above equation.

The blocking temperature, as measured within a time window τ_m, is reached when the observation time equals the *forward* relaxation time τ_+, because the latter corresponds to a moment flip from the initial state along $+z$ to the opposite direction, namely, a process that reduces the initial magnetization. From Equation 22.16 one obtains

$$T_b(H) = \frac{K_1 V (1 - H/H_a)^2}{k_B \ln(\tau_m/\tau)} \equiv T_b(0) \left(\frac{1-H}{H_a} \right)^2 \quad (22.17)$$

which indicates that the blocking temperature is reduced by the presence of a reverse field. By completely symmetric arguments one could show that T_b increases in the presence of a field with the same direction as the initial magnetization.

Since thermal fluctuations act in synergy to a reverse field in switching the moment of an MNP, it is expected that the coercivity of an assembly will decay with temperature. As discussed above, for a particle with its moment along the $+z$ axis, a reverse field ($0 < H < H_a$) reduces the barrier for reversal to the value E_b^+ given in Equation 22.16. If the field is strong enough, it will reduce the barrier to the value appropriate for superparamagnetic relaxation, namely, $k_B T \ln(\tau_m/\tau_0)$, and the (time average) magnetization will vanish. On the other hand, the reverse field that makes the magnetization vanish is by definition the coercive field. Therefore, the following relation holds

$$K_1 V \left(\frac{1 - H_c}{H_a} \right)^2 = k_B T \ln\left(\frac{\tau_m}{\tau_0} \right), \quad (22.18)$$

which, using Equation 22.15, provides the temperature dependent coercivity

$$H_c(T) = H_a \left[1 - \left(\frac{T}{T_b} \right)^{1/2} \right]. \quad (22.19)$$

The microscopic mechanism of thermal activation of the MNP moment over the anisotropy barrier produces a macroscopically measured time decay of the magnetization. We derive this dependence assuming that when a moment switches direction it continues to remain along the easy axis (Néel 1949). Then, at time t, N_+ particles occupy the lower minimum at $\theta = 0$, and the rest $N_- = N - N_+$ particles occupy the higher minimum at $\theta = \pi$. The time evolution of N_+ is governed by the *rate equation*

$$\frac{dN_+}{dt} = -\frac{N_+}{\tau_+} + \frac{N_-}{\tau_-}. \quad (22.20)$$

The magnetization per particle is given as $M(t) \equiv (2N_+(t)/N - 1)M_s$, and solution of Equation 22.20 provides

$$M(t) = M_\infty + (M_0 - M_\infty)\exp(-t/\tau) \quad (22.21)$$

with $1/\tau = 1/\tau_+ + 1/\tau_-$ being the reduced relaxation time and

$$M_\infty = \frac{\tau_+ - \tau}{\tau_+ + \tau_-} M_s \quad \text{and} \quad M_0 = \left(\frac{2N_+(0)}{N_-(0)} - 1 \right) M_s \quad (22.22)$$

the time asymptote and initial values of the particle magnetization, respectively. Equation 22.21 indicates that the magnetization decays exponentially toward the equilibrium value M_∞, reached as $t \to \infty$. In other words, equilibrium is reached when the population of the energy minima is proportional to the corresponding relaxation times ($N_+/N_- = \tau_+/\tau_-$), as dictated by Equation 22.20. When the applied field is strong enough ($H > H_s$) to produce only one minimum, thermal equilibrium is always reached. Obviously, in the absence of an external field, thermal equilibrium is reached when the two equivalent minima are equally populated ($N_+ = N_-$), resulting in a vanishing magnetization.

Notice that in Equation 22.20 we assumed bistability of the moment direction, which is a valid approximation provided the anisotropy barrier is high ($K_1 V \gg k_B T$). For lower anisotropy barriers or elevated temperatures ($K_1 V \approx k_B T$), the transverse fluctuations of $\vec{\mu}$, or, in other words, intra-valley motion around the energy minimum should be taken into account. A general treatment of thermal relaxation of SD MNPs was pioneered by Brown (1963) and extended to the case of an applied external field (Aharoni 1965, Coffey et al. 1998, Garanin et al. 1999).

If an assembly is polydisperse, characterized by a volume distribution $f(V)$, a distribution of blocking temperatures $f(T_b)$ exists. However, it still remains unclear if the mean value $\langle T_b \rangle$ is the appropriate blocking temperature of the assembly, which should be substituted, for example, in Equation 22.19. This point is discussed further in the literature (Nunes et al. 2004).

In a polydisperse assembly, a distribution of relaxation times $f(\tau)$ exists, with $f(\tau)d(\ln \tau)$ the probability of an MNP to have $\ln \tau$ in the range $(\ln \tau, \ln \tau + d(\ln \tau))$ and the normalization condition $\int_0^\infty f(\tau)d(\ln\tau) = 1$. In this case, the magnetization can be obtained by a superposition of the single-particle magnetization properly weighted, as follows:

$$M(t) = M_s \int_0^\infty \left[1 - \exp\left(\frac{-t}{\tau} \right) \right] \frac{f(\tau)}{\tau} d\tau, \quad (22.23)$$

where the term in brackets is the probability per unit time for a particle not to flip its moment. For a broad enough distribution, the observation time t will satisfy $\tau_1 \ll t \ll \tau_2$, where τ_1 and τ_2 are the minimum and maximum relaxation times of the assembly, respectively. Assuming a uniform distribution $f(\tau)$, it can be shown that the magnetization exhibits a logarithmic relaxation

$$M(H,t) = M(H,0) - S(H,T)\ln\frac{t}{\tau_0} \quad (22.24)$$

with S the *magnetic viscosity* of the system. Thus, polydispersity produces a much slower decay of magnetization with time.

The discussion so far refers to a field applied parallel to the easy axis. However, random anisotropy is most commonly found in MNP assemblies and the necessity to study the effect of a tilted field with respect to the easy axis arises. In this case, the calculation of the energy barriers and the relaxation time is a much more complicated task and no analytical solution exists. Numerical studies (Pfeiffer 1990) showed that the energy barrier for an applied field at an angle θ_0 to the easy axis can be approximately written as

$$E_{\rm b}(\theta_0) = K_1 V \left(\frac{1-H}{H_{\rm a}} \right)^{0.86+1.14 h_{\rm s}}, \qquad (22.25)$$

where $h_{\rm s}$ is given by Equation 22.12. In the limit of $\theta_0 = 0$, Equation 22.25 reduces to Equation 22.16.

The temperature dependence of the coercivity for a monodisperse assembly with random anisotropy has also been obtained numerically (Pfeiffer 1990) as

$$H_{\rm c}(T) = 0.48 H_{\rm a} \left[1 - \left(\frac{T}{T_{\rm b}} \right)^{0.77} \right], \qquad (22.26)$$

which at $T = 0$ reduces to the SW result of Equation 22.13. A detailed theoretical study of the relaxation time for a nonuniaxial applied field can be found in the review by Coffey and colleagues (Coffey et al. 1993).

As a concluding remark, the presence of polydispersity and random anisotropy makes the description of the magnetic behavior of an assembly intractable to exact analytical treatment. Instead, numerical approximations and simulation methods provide the alternative theoretical tools to study these systems. Numerical simulation approaches are introduced in Section 22.4.

22.3 Magnetic Measurements

Thermal relaxation has a dynamic character; therefore, the relation between the various relaxation times of the assembly and the measurement time is a decisive parameter for the outcome of a measurement. Additionally, if the assembly is not at equilibrium during the measurement, or if it changes its equilibrium state (e.g., by adiabatic changes of the applied field) the result of the measurements depends on the measurement protocol followed. In what follows we discuss two very common types of static measurements that reveal the temperature and field dependence of the magnetization and provide evidence for superparamagnetic relaxation. Dynamic measurements are not discussed in this article. The interested reader can find more on the physical principles behind the most common magnetic measurement techniques in the review of Dormann and colleagues (Dormann et al. 1997).

22.3.1 Field-Cooled (FC) and Zero-Field-Cooled (ZFC) Magnetization

This is a measurement protocol adopted for the investigation of the temperature-dependent magnetization of an assembly, and it reveals superparamagnetic behavior. It is performed in three stages. In the first, the sample is initially at a high enough temperature ($T_{\rm max}$) to ensure an SPM state, and it is cooled to low temperature ($T_{\rm min}$) to approach its ground state. In the second stage, a weak field is applied ($H \ll H_{\rm sat}$), the sample is heated up to $T_{\rm max}$, and the magnetization is measured as a function of temperature. This is the ZFC curve. In the third stage, the system is cooled down to $T_{\rm min}$, without removing the field, while the magnetization is recorded again, producing the FC curve. During cooling and heating, the temperature changes at the same constant rate. A typical ZFC-FC curve is shown in Figure 22.5. As the temperature rises, the blocked magnetic moments align easier along the applied field leading to an initial increase of the ZFC curve. However, as soon as thermal fluctuations push the moments over the anisotropy barrier, the thermal randomization of the moments produces a drop of the curve. Therefore, the peak of ZFC curve corresponds to the blocking temperature of the assembly. Notice that above $T_{\rm b}$ the ZFC and FC curves coincide, because the system is in thermal equilibrium and the heating (cooling) process is reversible. On cooling below $T_{\rm b}$ the moments remain partially aligned along the field, and the magnetization tends to a nonzero value. The magnetization vanishes at the ground state ($T_{\rm min}$) if the measuring field is very weak, a random distribution of the easy axes exists, and the assembly is noninteracting (dilute). Deviations from any of the above conditions produce a nonzero value for $M_{\rm ZFC}(T = 0)$. For isolated MNPs, the ZFC–FC curves are only weakly sensitive to the value of the applied field, provided that it is weak ($H \ll H_{\rm s}$).

22.3.2 Remanent Magnetization and Coercive Field

Remanent magnetization at a certain field, $M_{\rm r}(H)$, is measured after switching off the previously applied field H. In an assembly of MNPs, remanence arises because the moments of some particles that have rotated under an applied field, and to do so they have overcome an energy barrier, cannot rotate back to their original direction after the removal of the field. In a polydisperse assembly, at finite temperature T, only the blocked MNPs, namely, those with $T_{\rm b} < T$ contribute to the remanence. Therefore, $M_{\rm r}/M_{\rm s} = \int_{E_{\rm b,c}}^{\infty} f(E_{\rm b}) {\rm d} E_{\rm b}$ where $E_{\rm b,c} = K_1 V_{\rm c}$ is the critical barrier for SPM relaxation at temperature T. Taking into account that $T_{\rm b} \sim V$ (see Equation 22.15), we deduce that

$$\frac{{\rm d} M_{\rm r}(T)}{{\rm d} T} = f(T_{\rm b}), \qquad (22.27)$$

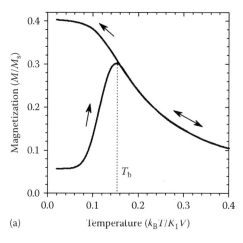
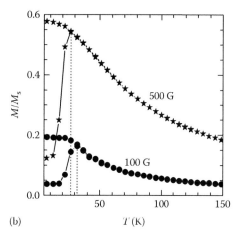

(a)　　　　　　　　　　　　　　　(b)

FIGURE 22.5 (a) Typical FC–ZFC magnetization curves. The curves join at the peak of the ZFC that corresponds to T_b. Arrows indicate the direction in which the measurements are taken. For $T > T_b$ the system is in thermal equilibrium and the heating process is reversible. (b) FC–ZFC curves for a dilute (noninteracting) assembly of Fe nanoparticles ($D = 3.0$ nm, $M_s = 1720$ emu/cm^3 and $K_1 = 2.4 \times 10^5$ erg/cm^3). The blocking temperature (dotted line) decreases weakly with increasing measuring field. The nonzero values of $M_{ZFC}(T = 0)$ are due to the finite value of the measuring field. Data produced by Monte Carlo simulations (Section 22.4.3). (Modified from Binns, C. et al., *Ag. Phys. Rev.*, B66, 84413, Figure 5, 2002.)

namely, the slope of $M_r(T)$ provides the barrier (or blocking temperature) distribution function of the assembly. There are three different measurement protocols for the remanent magnetization, as first suggested by Wohlfarth (1958):

1. *Thermoremanence TRM* (H, T), measured at the end of an FC process with field H from T_{max} down to the measuring temperature T.
2. *Isothermal remanence IRM* (H, T), measured at the end of ZFC process from T_{max} down to the measuring temperature T, at which a field H is applied and then removed.
3. *DC demagnetization remanence DcD* (H, T). First, a ZFC process from T_{max} down to the measuring temperature T is performed. Second, the sample is brought to *saturation remanence IRM* (∞, T). Third, a *reverse* field H is applied and then removed to leave the sample at the *DcD* (H, T) remanence.

Wohlfarth pointed out that for isolated MNPs the different remanent magnetizations are related as DcD(H) = IRM(∞) − 2 IRM(H). More interestingly, the deviations from this equality, defined as

$$\Delta M(H) = \text{DcD}(H) - [\text{IRM}(\infty) - 2 \cdot \text{IRM}(H)] \quad (22.28)$$

quantify the character and strength of interparticle interactions and are obtained experimentally (O'Grady et al. 1993). Positive ΔM values imply interactions with magnetizing character, and negative values indicate demagnetizing interactions. We should say that this is only a phenomenological characterization of the interactions, because Equation 22.28 does not provide any information about their microscopic origin. However, Equation 22.28 has been proved a standard tool for the quantification of interparticle interactions in complex MNP assemblies such as those used in modern industry of magnetic recording media

(granular films, particulate media). Interparticle interactions are discussed in Section 22.4.

22.4 Interacting Nanoparticle Assemblies

22.4.1 Introduction

The magnetic interactions that are present in bulk magnetic materials pertain to MNP assemblies, and they preserve their physical origin and characteristics. In particular, (direct) exchange between atomic moments separated by a few lattice constants can couple ferromagnetically or antiferromagnetically two MNPs via their surface atoms. Indirect exchange or Ruderman–Kittel–Kasuya–Yosida (RKKY) interaction exists between MNPs hosted in a metallic matrix, which provides free electrons required to mediate the interaction between the atomic moments of the MNPs. Finally, magnetostatic interactions, which are of minor importance in bulk magnets due to their weakness, become the dominant interactions in MNP assemblies with *well-separated* MNPs. This situation occurs for two reasons. First, the exchange interactions have a very short range (up to ~5 Å) and the RKKY interactions have an oscillating FM/AFM character with a period of a few angstroms, which renders to zero their average effect on the MNP volume, so both have a weak effect in interparticle coupling. On the other hand, magnetostatic interactions, in the lowest approximation, namely, the dipolar contribution, are proportional to the magnitude of the coupled magnetic moments, which for SD particles has an enormously large value compared to the atomic moments ($\mu_{MNP} \sim 10^3 \mu_B \sim 10^3 \mu_{atom}$).

Further on we discuss the effects of magnetostatic (dipolar) interactions on the magnetic properties of MNP assemblies and their interplay with single-particle anisotropy. The complexity of this problem arises from the *long-range* (~$1/d^3$, d being the interparticle distance) and *anisotropic* character of the dipolar interactions, namely, the dependence of interaction energy on the orientation of the moments relative to the bond joining the particle centers (Figure 22.6).

Understanding and controlling the effects of dipole–dipole interactions (DDI) in MNP assemblies is of paramount importance to modern technology of magnetic recording media for two opposite reasons. First, DDI couples the MNPs of an assembly. The ultimate goal in magnetic recording applications is to address each MNP individually and treat it as a magnetic bit. In this case, DDI have a parasitic role, and one wishes to estimate and reduce their impact in the magnetic properties of an assembly. On the contrary, magnetic logic devices have been proposed and built that exploit the magnetostatic coupling between ordered MNP arrays (linear or planar) to transfer a magnetic bit (usually a flipped moment) between two distant points in the array (Cowburn 2006). In this case, DDI are of central importance, and the goal is to enhance and tailor their effects.

Over the last two decades, many research groups have prepared and measured MNP assemblies in various forms (granular films, ferrofluids, cluster assembled films, self-assembled nanoparticles, lithographic arrays of magnetic dots) and studied the intrinsic factors (host and particle material, particle size, particle density) and the extrinsic factors (temperature, field, measurement protocol) that control the magnetic behavior. In many of these studies, the presence of magnetostatic interactions has been confirmed. Among the above-mentioned systems, the self-assembled MNPs prepared by a synthetic route offer the advantage of containing well-separated MNPs with a very narrow size distribution (σ_V ~ 5%–10%), so they are ideal systems to study DDI effects. Experimental observations on self-assembled MNPs that have been attributed to DDI include the reduction of the remanence

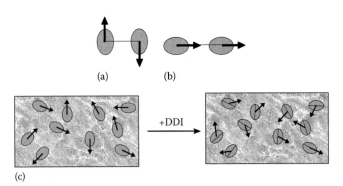

(a)   (b)

(c)

FIGURE 22.6 Ground-state configuration of MNPs with elliptic shape coupled by magnetostatic (dipolar) forces. The easy axis coincides with the long axis of the ellipse (shape anisotropy). Dipolar coupling induces anti-FM ordering when the moments are forced (by anisotropy) to remain normal to the bond, as in (a). FM ordering (nose to tail) is favored when the easy axes are parallel to the bond, as in (b). In an assembly with random anisotropy, as in (c), the moments are aligned along the local easy axes. Dipolar interactions have a complex effect, leading to the misalignment of the moments with respect to the local easy axis.

at low temperature (Held et al. 2001), the increase of the blocking temperature (Murray et al. 2001), the increase of the barrier distribution width (Woods et al. 2001), deviations of the ZFC magnetization curves from the Curie behavior (Puntes et al. 2001), and the difference between the in-plane and out-of-plane remanences (Russier et al. 2000). Long-range FM order in linear chains (Russier et al. 2003), and hexagonal arrays (Puntes et al. 2004; Yamamoto et al. 2008) of dipolar coupled SD MNPs has been demonstrated, supporting the existence of a *dipolar superferromagnetic* ground state, characterized by FM long-range order of the particle moments.

Investigations of the static and dynamic magnetic properties of dipolar interacting nanoparticle assemblies brought up fundamental issues related to the existence of a ground state that shares common features with *spin glasses*, such as slow relaxation, memory, and ageing effects (Sasaki et al. 2005). The latter are magnetic systems characterized by disorder and competing interactions that produce an energy landscape with many local minima, considered responsible for the occurrence of these effects. Dipolar interparticle interactions in dense and random nanoparticle assemblies are believed to cause a *spin-glass-like* behavior (Dormann et al. 1997).

Theoretical models have been developed in an effort to explain these observations and related previous ones in assemblies with randomly located MNPs (granular films, cluster-assembled films). On a microscopic level, the presence of DDI between MNPs modifies the magnetization switching mechanism, which for an isolated MNP obeys the Néel–Arrhenius model. When anisotropic MNPs are dipolar coupled, the reversal mechanism is determined by the interplay between the single-particle anisotropy energy ($E_a \sim K_1 V$) and the dipolar interaction energy ($E_d \sim \mu_i\mu_j/r_{ij}^3$). For weak interactions ($E_d \ll E_a$), the moments reverse *independently* by thermal activation over energy barriers, which are however modified due to DDI. This limiting case is treated within a mean-field approximation and is discussed in Section 22.4.2. For strong interactions ($E_d \gg E_a$), the single-particle reversal is no longer valid. The reversal of one particle can excite the reversal of others, and the assembly behaves in a collective manner. Many-body energy barriers exist in the system, with values that depend on the configuration of all moments. Their evaluation becomes a formidable task and numerical simulations offer in this case an indispensable tool. Numerical methods are briefly discussed in Section 22.4.3.

For a detailed review on the magnetic properties of dipolar interacting MNP assemblies, the reader is referred to the relevant literature (Dormann et al. 1997, Farrell et al. 2005, Kechrakos and Trohidou 2008, Knobel et al. 2008). The role of magnetostatic interactions in patterned magnetic media has been reviewed by Martín et al. (2003).

22.4.2 Mean Field Models

In an early attempt to include the effect of interparticle interactions in the thermal relaxation of MNPs, Shtrikman and Wohlfarth (1981) assumed that the single-particle anisotropy barrier of an MNP is increased by the Zeeman energy due to the *interaction field* H_{int} produced by the moments of neighboring particles. In this model, the Néel relaxation time is obtained from Equation 22.16 with the applied field H replaced by the interaction field H_{int}. The mean-field approximations consists in replacing H_{int} by its thermal average value, which in Néel's model is

$$\overline{H_{int}} = H_{int} \tanh\left(\frac{\mu_0 \mu H_{int}}{k_B T}\right) \approx \frac{\mu_0 \mu \overline{H_{int}^2}}{k_B T} \qquad (22.29)$$

the latter approximation being valid for weak interaction fields. The substitution of Equation 22.29 into Equation 22.16 gives

$$\tau = \tau_0 \exp\left[\frac{K_1 V + \mu_0^2 \mu^2 \overline{H_{int}^2}/k_B T}{k_B T}\right]. \qquad (22.30)$$

Using the approximation $1 + x \approx 1/(1 - x)$, we write Equation 22.30 in the following form:

$$\tau \approx \tau_0 \exp\left[\frac{K_1 V}{k_B (T - T_0)}\right] \qquad (22.31)$$

with $T_0 = \mu_0^2 \mu^2 \overline{H_{int}^2}/k_B K_1 V$. Equation 22.31, also known as the *Vogel–Fulcher law*, indicates that the relaxation time of an assembly of interacting MNPs is the same as that of the isolated MNPs at a *lower* temperature.

In the SW model, the temperature T_0, or equivalently the thermal average $\overline{H_{int}^2}$, is not related to the microscopic parameters of the assembly, that is, particle location, and is treated as a phenomenological parameter, fitted to experimental data. Dormann and colleagues (Dormann et al. 1988, 1997) developed a statistical model for the average barrier in a dipolar interacting assembly that quantifies the interaction field and provides for the single-particle energy barrier:

$$E_b = K_1 V + n_1 a_1 M_s^2 V \pounds\left(\frac{a_1 \mu^2}{V k_B T}\right) \qquad (22.32)$$

with

- n_1 the number of nearest neighbors of a particle
- $a_1 = x_v/\sqrt{2}$, x_v the volume concentration of the particles
- $\pounds(\cdot)$ the Langevin function

Equation 22.32 indicates that the anisotropy barrier is increased due to DDI, thus the model of Dormann et al. predicts an increase of the blocking temperature due to DDI. This model behavior has been observed in almost all types of MNP assemblies, with a few exceptions (Hansen and Morup 1998).

More recently, Allia et al. (2001) also used a phenomenological approach to describe a superparamagnetic assembly with weak DDI, namely, an assembly in a regime that the remanence and coercivity vanish, but the field-dependent magnetization varies with the concentration of MNPs indicating the presence of DDI. The authors (Allia et al. 2001) suggested that the dipolar field

changes at a high rate and in random direction and therefore acts similar to the thermal field. The effect is accounted for by an apparent increase of the system temperature. The magnetization at temperature T is given by $M = M_s \pounds[\mu H/k_B(T + T^*)]$, with T^* related to the average dipolar energy via $k_B T^* = n_1 \mu^2/d^3$ and obtained by a fitting procedure. This model interpreted successfully the magnetization behavior of Co nanoparticles in Cu matrix and established the existence of the interacting superparamagnet regime (Allia et al. 2001).

22.4.3 Numerical Techniques

The mean-field models have the advantage of providing analytical expressions suitable for extracting system parameters from the experimental data by a fitting process. However, they are not applicable to strongly dipolar systems, and they do not account for collective effects. Numerical techniques, on the other hand, have the major advantage that they treat rigorously the local and temporal statistical fluctuations of the macroscopic quantities characterizing the MNP assembly and provide an efficient interpolation scheme between the weak and the strong interaction regimes. We discuss briefly two most common numerical approaches, the Monte Carlo (MC) method and the MD method.

22.4.3.1 The Monte Carlo Method

Different algorithms that mimic thermal fluctuations of the degrees of freedom of a physical system by means of (pseudo) random numbers go under the umbrella of Monte Carlo techniques. In the case of MNPs, two widely used algorithms are the Metropolis Monte Carlo (MMC) and the Kinetic Monte Carlo (KMC). The former is appropriate for a description of the equilibrium behavior of an assembly, while the latter also accounts, within a certain timescale, for the transition to equilibrium. Both algorithms provide thermal averages of macroscopic quantities of interest in the canonical ensemble, that is, at constant temperature. To do so, a sampling of the phase space is performed; however, the sampling procedures differ, as outlined below.

The MMC algorithm samples the phase space, visiting preferentially states close to the equilibrium states (*importance sampling*). This is achieved when subsequently visited states form a Markov chain, meaning that the probability of visiting the next state depends only on the last visited one. To do so, one chooses the transition from state s to s' to occur with certainty, if it reduces the total energy ($E_{s'} \leq E_s$) and with a finite probability $p(s \rightarrow s') = \exp[-(E_{s'} - E_s)/kT]$, if it increases the total energy ($E_{s'} > E_s$). Thus the system is allowed to climb up energy barriers and slide down toward energy minima until it reaches eventually the global minimum.

In KMC, the system jumps from a state s at a local minimum to a new state s' being also a local minimum by overcoming a barrier E_b. The jump is performed within a predefined time step Δt with probability $p(\Delta t) = 1 - \exp(-\Delta t/\tau)$ where τ is the corresponding relaxation time with Arrhenius behavior, $\tau = \tau_0 \exp(E_b/kT)$.

In both algorithms, interparticle interactions are included by replacing the applied field H with the *total* field $H_i = H + \sum_{j(\neq i)} H_{int,ij}$, which includes the contribution from the *interaction* field $H_{int,ij}$. In contrast to mean-field theories, in MC and MD (see Section 22.4.3.2) techniques the interaction field is treated exactly, meaning that its value depends on the configuration of all the moments of the assembly, and it changes at each time-step.

An important distinction between the MC algorithms is that KMC simulates the relaxation of the system in physical time, while the time quantification of the MMC time step is possible only in the absence of interparticle interactions (Nowak et al. 2000, Chubykalo et al. 2003). However, a serious difficulty in KMC arises from the calculation of the local energy barrier required to obtain the transition probability. In an interacting system, the barrier depends on *all* degrees of freedom, and its calculation is a formidable task (Chubykalo-Fesenko and Chantrell 2004, Jensen 2006), usually performed in an approximate manner (Pfeiffer 1990, Chantrell et al. 2001). Furthermore, the KMC assumes that the system evolves through thermally activated jumps over energy barriers, an approximation that becomes invalid at elevated temperatures ($kT \sim E_b$), or when collective behavior becomes important, for example, in strongly interacting MNPs. Collective effects are better described within the MMC algorithm.

For a detailed description and technical implementation of MC algorithms, the interested reader could refer to the book by Landau and Binder (2000).

22.4.3.2 The Magnetization Dynamics Method

In this method, the equations of motion for the magnetic moments are integrated in time and time averages of the macroscopic magnetization are recorded. At zero temperature, the time evolution of a magnetic moment μ_i under a total field H_i is described by the Landau–Lifshitz–Gilbert (LLG) equation:

$$\frac{d\vec{\mu}_i}{dt} = -A(\vec{\mu} \times \vec{H}_i) - B_i \vec{\mu}_i \times (\vec{\mu}_i \times \vec{H}_i) \qquad (22.33)$$

with

$A \equiv \gamma/(1 + \alpha^2)$, $B_i \equiv \alpha\gamma/(1 + \alpha^2)\mu_i$, γ the gyromagnetic ratio
α a dimensionless damping parameter

The first term on the r.h.s. of Equation 22.33 is the torque term leading to precession around the field axis and the second one is a phenomenological damping torque that tends to align the precessing moment with the field H_i.

The dynamics at finite temperature are described by the introduction of an additional field $(H_{f,i})$ term in Equation 22.33 with stochastic character. H_f is assumed to have zero time average (*white noise*) and its values at different sites i, j or different instants t, t' are uncorrelated. The LLG equation augmented by the thermal field term is commonly referred to as the *Langevin* or *stochastic LLG* equation.

22.4.3.3 Timescale of Numerical Methods

The MC and MD techniques are complementary since they describe the thermal relaxation of magnetic properties in different timescales. In the MD method, the characteristic time is a fraction ($\sim10^{-2}$) of the precessional (Larmor) period ($\sim10^{-10}$ s), which implies that simulation times up to ~1 ns are presently attainable. Thus, MD is the appropriate scheme to investigate fast-relaxation phenomena, for example, the reversal path of magnetization under an applied short field pulse (Berkov 2002, Suess et al. 2002).

In KMC, the characteristic time is the single-particle relaxation time (see Equation 22.14), which is much larger than $\tau_0 \sim 10^{-10}$ s in the temperature range of interest ($kT \ll E_b$), a fact that makes the method suitable to treat slow-relaxation problems, for example, the thermal decay of magnetization in permanent magnets, a phenomenon that evolves within days or years (Van de Veerdonk et al. 2002).

Finally, when static magnetic properties are concerned, the system is at a stable (or metastable) state, and the MMC is a powerful and sufficient scheme to describe, for example, long-range order at the ground state or collective behavior at finite temperature (Kechrakos and Trohidou 1998, Jensen and Pastor 2003).

22.4.3.4 MMC Study of Dipolar Interacting Assemblies: A Case Study

In this section, we show typical results from MMC simulations of the magnetic properties of dipolar interacting MNP assemblies (Kechrakos and Trohidou 1998, 2002). Our system contains N identical SD NPs with diameter D and uniaxial anisotropy in a random direction. The MNPs are located randomly in space or on the vertices of a hexagonal lattice. The former is an appropriate model for granular samples, and the latter for self-assembled MNPs. The total energy of the system is

$$E = g\sum_{ij} \frac{\hat{S}_i \cdot \hat{S}_j - 3(\hat{S}_i \cdot \vec{\hat{R}}_{ij})(\hat{S}_i \cdot \vec{\hat{R}}_{ij})}{R_{ij}^3} - k\sum_i (\hat{S}_i \cdot \hat{e}_i)^2 - h\sum_i (\hat{S}_i \cdot \hat{H}), \qquad (22.34)$$

where
$\hat{S}_i$ is the magnetic moment direction (spin) of the ith particle
$\hat{e}_i$ is the easy-axis direction
R_{ij} is the center-to-center distance between particles i and j

Hats indicate unit vectors. The energy parameters entering Equation 22.34 are (1) the dipolar energy $g \equiv \mu_0^2\mu^2/4\pi d^3$, with $\mu = M_s V$ the particle moment and d the minimum interparticle distance, (2) the anisotropy energy $k \equiv K_1 V$, and (3) the Zeeman energy $h \equiv \mu_0\mu H$ due to the applied dc field H. The energy parameters (g, k, h) entering Equation 22.34, the thermal energy $k_B T$, and the treatment history of the sample determine the micromagnetic configuration at a certain temperature and field. The freedom to choose an arbitrary energy scale makes the numerical results applicable to a class of materials with the same

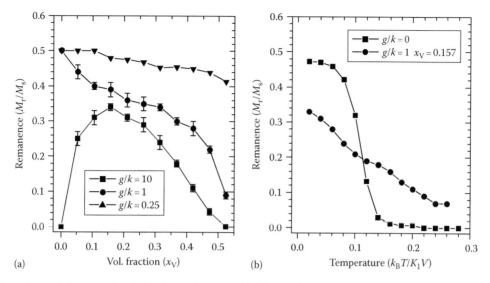

FIGURE 22.7 Dependence of the saturation IRM of a random assembly (a) on volume fraction of MNPs, at very low temperature ($t/k = 0.001$) (Modified from Kechrakos, D. and Trohidou, K. N., *Phys. Rev. B.*, 58, 12169, Figure 1, 1998.) and (b) on temperature, for fixed volume fraction (Modified from Kechrakos, D. and Trohidou, K. N., *Phys. Rev. B.*, 58, 12169, Figure 6, 1998.). The particles have random anisotropy. The data are obtained by MMC simulations.

parameter ratios rather than to a specific material. The crucial parameter that determines the transition from single-particle to collective behavior is the ratio of the dipolar to the anisotropy energy (g/k).

We show in Figure 22.7 the concentration and temperature dependence of the remanence magnetization of a random assembly. Notice in Figure 22.7a that weak DDI produce an increase of the remanence with concentration, while strong DDI have the opposite effect. Remarkably, the presence of free sample boundaries can reverse the increasing trend of the remanence, due to the presence of a demagnetizing field. When DDI are much stronger than single-particle anisotropy ($g/k \sim 10$), the remanence value is sensitive to the morphology of the assembly, as the peak around the percolation threshold indicates. This behavior is explained by the anisotropic character of DDI (see Figure 22.6). In Figure

22.7b, DDI interactions are shown to produce a much slower temperature decay, producing finite remanence values above the blocking temperature of the isolated MNPs. This result supports the predictions of the mean-field theory about the increase of the measured blocking temperature in dipolar interacting systems (Dormann et al. 1988).

In chemically prepared, self-assembled MNPs the possibility to control the interparticle separation by the variation of the surfactant (Willard et al. 2004) offers the possibility to study the dependence of T_b on interparticle spacing while preserving the geometrical arrangement of the assembly (hexagonal). In Figure 22.8, we show results for the ZFC magnetization (M_{ZFC}) and the blocking temperature as obtained from the peak of the $M_{ZFC}(T)$ curve for a hexagonal array of dipolar interacting MNPs with random anisotropy. Parameters corresponding to

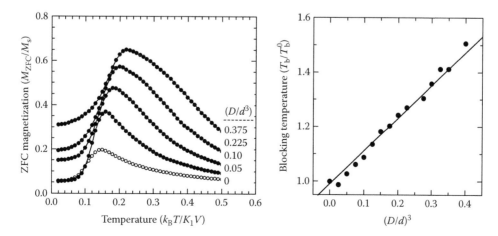

FIGURE 22.8 ZFC curves and blocking temperature for an ordered (hexagonal) assembly of identical MNPs with diameter D and center-to-center distance d. (a) Evolution of ZFC curves with decreasing d values. (b) Linear scaling of T_b with inverse cube of d. Data obtained by MMC simulations. (Modified from Kechrakos, D. and Trohidou, K. N., *Appl. Phys. Lett.*, 81, 4574, Figure 2, 2002.)

Co nanoparticles are used (Kechrakos and Trohidou 2002). The characteristic dependence of T_b on the inverse cube of interparticle spacing can be used as a proof of the dominant character of DDI in an assembly. Notice also that $M_{ZFC}(T \approx 0)$ assumes a positive value that increases with d values. This feature arises from the gradual formation of a long-range FM ground state, due to DDI.

More examples of MC or MD simulations and comparison to experiments on MNP assemblies can be found in the relevant literature (Vedmedenko 2007; Kechrakos and Trohidou 2008).

22.5 Summary

We have discussed the main theoretical concepts that pertain to the magnetization properties of isolated (noninteracting) nanoparticles and their assemblies. We estimated the critical radius for the formation of SD particles and studied the zero-temperature magnetization reversal mechanism of coherent rotation (SW model), the thermally activated reversal (Néel-Brown model), and the related phenomenon of SPM occurring above the blocking temperature. Complications arising from size polydispersity, distribution of easy axes directions, and applied field on the relaxation time for magnetization reversal were discussed. Two standard experimental techniques for (static) magnetic measurements, namely, the field and temperature dependence of magnetization were outlined. Finally, the subject of interparticle dipolar interactions was introduced along with the most common theoretical techniques used to analyze interacting systems. Examples from Monte Carlo studies of MNP assemblies were given.

22.6 Future Perspectives

The dynamic behavior of MNPs in the presence of interparticle interactions is expected to remain a topic of intense scientific and technological research in the coming years. The research effort is expected to focus on both the atomic scale properties of individual MNPs and on the mesoscopic properties of nanoparticle assemblies.

On the *atomic* scale, future goals will include (1) reduction of the magnetic particle size without violating thermal stability (*superparamagnetic limit*) at room temperature. The technological benefit from progress in this direction will be the development of magnetic data-storage media with higher areal density. Given that the SPM effect is not observed below a certain size of an MNP, due to disorder effects on the particle surface, the search for new high-anisotropy materials is required. Composite nanoparticles with a core-shell morphology (Skumryev et al. 2003) constitute an interesting perspective.

(2) Understanding and control of surface effects. With the reduction of particle size the contribution from surface moments becomes of increasing importance. The chemical structure of the surface (disorder, defects) controls the

magnitude and type of the surface anisotropy, which is usually much (up to ~10 times) larger than the core anisotropy. Synthetic methods can offer indispensable routes to surface structure modification. *Ab initio* electronic structure calculations are a valuable tool to predict the surface anisotropy values, and modeling of MNPs as multispin system will reveal complex magnetization reversal mechanisms beyond the SW model (Kachkachi et al. 2000). Experiments on individual nanoparticles (Wernsdorfer et al. 2000) offer a unique test of the above theories.

The future task on the *mesoscopic* scale will be to understand and control collective magnetic behavior in ordered nanostructures (self-assembled MNPs and magnetic patterned media). Ordered nanostructures include chemically prepared self-assembled MNPs and lithographically prepared magnetic patterned media. The chemical synthesis of MNPs and self-assembly (bottom-up approach) is a very promising and cost-effective method to produce ordered MNP arrays (Willard et al. 2004). However, deeper understanding and improvement of the self-assembly process is required in order to achieve larger (beyond $1 \, mm^2$) sample area with structural coherence. There is still a remaining problem as nanoparticles self-assemble into hexagonal arrays that are incompatible with the square arrangements required in industrial applications. A resolution to this problem could be the recently demonstrated templated assembly (Cheng et al. 2004). Lithographic patterning (top-down approach) offers better control over the geometrical aspects of the assembly but cannot yet produce nanostructures with size below ~100 nm (Martin et al. 2003). The increase of lithographic resolution is demanded in order to achieve patterned media with smaller (below 0.1 μm) characteristic size. On the measurements side, the improvement of existing techniques to probe mesoscopic magnetic order and excitations is demanded. Recent examples are the observation of mesoscopic sale magnetic order in self-assembled Co nanoparticles by an indirect method (small-angle neutron scattering) (Sachan et al. 2008) and by direct methods such as magnetic force microscopy (Puntes et al. 2004) and electron holography (Yamamoto et al. 2008). From the point of view of basic physics, ordered nanostructures constitute model systems to study collective magnetic behavior driven by magnetostatic interactions, because the size, the shape, and the spatial arrangement of the magnetic nanostructures are well controlled. Known phenomena are to be demonstrated on the mesoscopic scale and new ones possibly to be discovered. As a recent example, we refer to the observation of magnetic frustration in magnetostatically coupled magnetic microrods (Wang et al. 2006), a phenomenon previously met in bulk magnetic random alloys (spin glasses).

Finally, progress in numerical modeling will provide methods for bridging the atomic-scale and the mesoscopic-scale simulations. Such multiscale simulations point to the future of theoretical investigations in the field of relaxation in MNPs and have only recently started to appear (Yanes et al. 2007, Kazantseva et al. 2008).

Acknowledgment

The MC data presented in this chapter were obtained in collaboration with Dr. K.N. Trohidou at the Institute of Materials Science of NCSR 'Demokritos', in work supported by the Greek program for the Support of Researchers PENED (contract number 497) and the EU program Growth (contract number G5RD-CT-2001-00478, AMMARE).

References

Aharoni, A. 1965. Effect of a magnetic field on the superparamagnetic relaxation time. *Phys. Rev.* 177: 793.

Allia, P., Coisson, M., Tiberto, P. et al. 2001. Granular Cu–Co alloys as interacting superparamagnets. *Phys. Rev. B* 64: Art. No. 144420.

Bean, C. P. and Jacobs, I. S. 1956. Magnetic granulometry and super-paramagnetism. *J. Appl. Phys.* 27: 1448–1452.

Bean, C. P. and Livingston, J. D. 1959. Superparamagnetism. *J. Appl. Phys.* 30: 120S–129S.

Berkov, D. V. 2002. Fast switching of magnetic nanoparticles: Simulation of thermal noise effects using the Langevin dynamics. *IEEE Trans. Magn.* 38(5): 2489–2495.

Binns, C., Maher, M. J., Pankhurst, Q. A. et al. 2002. Magnetic behavior of nanostructured films assembled from preformed Fe clusters embedded. *Ag. Phys. Rev. B*66: 84413, Figure 5.

Brown, W. F. Jr. 1963. Thermal fluctuations of single-domain particle. *Phys. Rev.* 130: 1677–1686.

Chantrell, R. W., Walmsley, N., Gore J. et al. 2001. Calculations of the susceptibility of interacting superparamagnetic particles. *Phys. Rev. B* 63(2): Art. No. 024410.

Cheng, J. Y., Mayes, A. M., and Ross, C. A. 2004. Nanostructure engineering by templated self-assembly of block copolymers. *Nat. Mater.* 3: 823–828.

Chubykalo, O., Nowak, U., Smirnov-Rueda, R. et al. 2003. Monte Carlo technique with a quantified time step: Application to the motion of magnetic moments. *Phys. Rev. B* 67: Art. No. 064422.

Chubykalo-Fesenko, O. A. and Chantrell, R. W. 2004. Numerical evaluation of energy barriers and magnetic relaxation in interacting nanostructured magnetic systems. *J. Magn. Magn. Mater.* 343: 189–194.

Coffey, W. T., Cregg, P. J., and Kalmykov, Yu. P. 1993. On the theory of Debye and Néel relaxation of single domain ferromagnetic particles. *Adv. Chem. Phys.* 69: 263–315.

Coffey, W. T., Crothers, D. S. F., Dormann, J. L. et al. 1998. Effect of an oblique magnetic field on the superparamagnetic relaxation time: Influence of the gyromagnetic term. *Phys. Rev. B* 58: 3249–3266.

Cowburn, R. P. 2006. Where have all the transistors gone? *Science* 311: 183–184.

Cullity, B. D. 1972. *Introduction to Magnetic Materials.* Addison-Wesley Publishing Company, Reading, MA.

Darling, S. B. and Bader, S. D. 2005. A materials chemistry perspective on nanomagnetism. *J. Mater. Chem.* 15: 4189–4195.

Dormann, J. L., Bessais, L., and Fiorani, D. 1988. A dynamic study of small interacting particles: Superparamagnetic model and spin-glass laws. *J. Phys. C: Solid State Phys.* 21: 2015–2034.

Dormann, J. L., Fiorani, D., and Tronc, E. 1997. Magnetic relaxation in fine-particle systems. *Adv. Chem. Phys.* 98: 283–494.

Farrell, D., Cheng, Y., McCallum, R. W. et al. 2005. Magnetic interactions of iron nanoparticles in arrays and dilute dispersions. *J. Phys. Chem. B* 109: 13409–13419.

Garanin, A., Kennedy, E. C., Crothers, D. S. F. et al. 1999. Thermally activated escape rates of uniaxial spin systems with transverse field: Uniaxial crossovers. *Phys. Rev. E* 60: 6499.

Hansen, M. F. and Morup, S. 1998. Models for the dynamics of interacting magnetic nanoparticles. *J. Magn. Magn. Mater.* 184: 262–274.

Held, G. A., Grinstein, G., Doyle, H. et al. 2001. Competing interactions in dispersions of superparamagnetic nanoparticles. *Phys. Rev. B* 64: Art. No. 012408.

Jensen, P. J. 2006. Average energy barriers in disordered interacting magnetic nanoparticles. *Comp. Mater. Sci.* 35: 288–291.

Jensen, P. J. and Pastor, G. M. 2003. Low-energy properties of two-dimensional magnetic nanostructures: Interparticle interactions and disorder effects. *New J. Phys.* 5: Art. No. 68.

Kachkachi, H., Ezzir, A., Nogues, M. et al. 2000. Surface effects in nanoparticles: Application to maghemite γ-Fe_2O_3. *Eur. J. Phys. B* 14(4): 681–689.

Kazantseva, N., Hinzke, D., Nowak, U. et al. 2008. Towards multiscale modeling of magnetic materials: Simulations of FePt. *Phys. Rev. B* 77: Art. No. 184428.

Kechrakos, D. and Trohidou, K. N. 1998. Magnetic properties of dipolar interacting single-domain particles. *Phys. Rev. B* 58: 12169–12177.

Kechrakos, D. and Trohidou, K. N. 2002. Magnetic properties of self-assembled interacting nanoparticles. *Appl. Phys. Lett.* 81: 4574–4576.

Kechrakos, D. and Trohidou, K. N. 2008. Dipolar interaction effects in the magnetic and magnetotransport properties of ordered nanoparticle arrays. *J. Nanoscience Nanotech.* 8: 2929–2943.

Knobel, M., Nunes, W. C., and Sokolovsky, L. M. et al. 2008. Superparamagnetism and other magnetic features in granular materials. *J. Nanosci. Nanotechnol.* 8: 2836–2857.

Kodama, R. H. 1999. Magnetic nanoparticles. *J. Magn. Magn. Mater.* 200: 359–372.

Landau, D. P. and Binder, K. 2000. *A Guide to Monte Carlo Simulations in Statistical Physics.* Cambridge University Press, Cambridge, U.K.

Martín, J. I., Nogués, J., Liu, K. et al. 2003. Ordered magnetic nanostructures: Fabrication and properties. *J. Magn. Magn. Mater.* 256: 449–501.

Moser, A., Takano, K., Margulies, D. T. et al. 2002. Magnetic recording: Advancing into the future. *J. Phys. D: Appl. Phys.* 35: R157–R167.

Murray, C. B., Sun, S., Doyle, H. et al. 2001. Monodisperse 3d transition metal (Co, Ni, Fe) nanoparticles and their assembly into nanoparticle superlattices. *MRS Bull.* 26: 985–991.

Néel, L. 1949. Influence de fluctuations thermiques sur l'aimantation de grains ferromagnétiques trés fins. *Compt. Rend.* 228: 664–666.

Nowak, U., Chantrell, R. W., and Kennedy, E. C. 2000. Monte Carlo simulation with time-step quantification in terms of Langevin dynamics. *Phys. Rev. Lett.* 84: 163–166.

Nunes, W. C., Folly, W. S. D., Sinnecker, J. P., and Novak, M. A. 2004. Temperature dependence of the coercive field in single-domain particle systmes. *Phys. Rev. B* 70: Art. No. 014419.

O'Grady, K., EI-Hilo, M., and Chantrell, R. W. 1993. The characterization of interaction effects in fine-particle systems. *IEEE Trans. Magn.* 29: 2608–2613.

Petit, V., Taleb, A., and Pileni, M.P. 1998. Self-organization of magnetic nanosized cobalt particles. *Adv. Mater.* 10: 259–261.

Pfeiffer, H. 1990. Determination of anisotropy field distribution in particle assemblies taking into account thermal fluctuations. *Phys. Stat. Sol. (a)* 118: 295–306.

Puntes, V. F., Krishnam, K. M., and Alivisatos, A. P. 2001. Colloidal nanocrystal shape and size control: The case of cobalt. *Science* 291: 2115–2117.

Puntes, V. F., Gorostiza, P., Aruguete, D. M. et al. 2004. Collective behavior in two-dimensional cobalt nanoparticle assemblies observed by magnetic force microscopy. *Nat. Mater.* 3: 263–268.

Russier, V., Petit, C., Legrand, J. et al. 2000. Collective magnetic properties of cobalt nanocrystals self-assembled in a hexagonal network: Theoretical model supported by experiments. *Phys. Rev. B* 62: 3910–3916.

Russier, V., Petit, V., and Pileni, M. P. 2003. Hysteresis curve of magnetic nanocrystals monolayers: Influence of the structure. *J. Appl. Phys.* 93: 10001–10010.

Sachan, M., Bonnoit, C., Majetich, S. A. et al. 2008. Field evolution of magnetic correlation lengths in ε-Co nanoparticle assemblies. *Appl. Phys. Lett.* 92(15): Art. No. 152503.

Sasaki, M., Jönsson, P. E., Takayama, H. et al. 2005. Aging and memory effects in superparamagnets and superspin glasses. *Phys. Rev. B* 71(10): Art. No. 104405.

Shtrikman, S. and Wohlfarth, E. P. 1981. The theory of Vogel-Fulcher law of spin glasses. *Phys. Lett. A* 85(8,9): 467–470.

Skumryev, V., Stoyanov, S., Zhang, Y. et al. 2003. Beating the superparamagnetic limit with exchange bias. *Nature* 423: 850–853.

Stoner, E. C. and Wohlfarth, E. P. 1948. A mechanism of magnetic hysteresis in heterogeneous alloys. *Proc. R. Soc. Lond. A* 240: 599–642.

Suess, D., Schrefl, T., Scholz, W. et al. 2002 Fast switching of small magnetic particles *J. Magn. Magn. Mater.* 242: 426–429.

Van de Veerdonk, R. J. M., Wu, X. W., Chantrell, R. W. et al. 2002. Slow dynamics in perpendicular media. *IEEE Trans. Magn.* 38(4): 1676–1681.

Vedmedenko, E. Y. 2007. *Competing Interactions and Patterns in Nanoworld*. Wiley-VCH Verlag, Weinheim, Germany.

Wang, R. F., Nisoli, C., Freitas, R. S. et al. 2006. Artificial "spin ice" in a geometrically frustrated lattice of nanoscale ferromagnetic islands. *Nature* 439(7074): 303–306.

Wernsdorfer, W., Mailly, D., and Benoit, A. 2000. Single nanoparticle measurement techniques. *J. Appl. Phys.* 87(9): 5094–5096.

Willard, M. A., Kurihara, L. K., Carpenter, E. E. et al. 2004. Chemically prepared magnetic nanoparticles. *Int. Mater. Rev.* 49: 125–170.

Wohlfarth, E. P. 1958. Relations between different modes of acquisition of the remanent magnetization of ferromagnetic particles. *J. Appl. Phys.* 29: 595–596.

Woods, S. I., Kirtley, J. R., Sun, S. et al. 2001. Direct investigation of superparamagnetism in Co nanoparticle films. *Phys. Rev. Lett.* 87: 137205–137208.

Yamamoto, K., Majetich, S. A., McCartney, M. R. et al. 2008. Direct visualization of dipolar ferromagnetic domain structures in Co nanoparticle monolayers by electron holography. *Appl. Phys. Lett.* 93(8): Art. No. 082502.

Yanes, R., Chubykalo-Fesenko, O., Kachkachi, H. et al. 2007. Effective anisotropies and energy barriers of magnetic nanoparticles with Néel surface anisotropy. *Phys. Rev. B* 76(6): Art. No. 064416.

23

Embedded Nanoparticles

Leandro L. Araujo
The Australian National University

Mark C. Ridgway
The Australian National University

23.1 Introduction

"To embed" means *"to firmly fix in a surrounding mass"* or *"to enclose in supporting material."* The term *embedded nanoparticles* thus describes particles of nanometer dimensions (from 1 to ~100 nm) contained within a solid matrix. Such material systems will be the subject of this chapter, in which we present an overview of the growth methods, the fundamental properties, and the technological applications. Occasionally, the *embedded nanoparticle* terminology has been erroneously extended to nanoparticles (NPs) contained within a liquid, which we believe are better described as *solution-dispersed* or *colloidal nanoparticles.* These systems are not described here. Examples of embedded and colloidal Ge NPs, imaged with transmission electron microscopy (TEM), are shown in Figure 23.1a and b, respectively. This comparison highlights one advantage of an encapsulating matrix; specifically, NP contact interactions are absent in the former (embedded), while NP agglomeration due to strong van der Waals attractive forces is readily apparent in the latter (colloidal). In the historical and scientific discussion that follows, we will concentrate on issues of importance to embedded NPs in particular and thus not necessarily of relevance to other forms of NPs including, for example, free standing and colloidal.

The production of embedded NPs predates man given that they form in many minerals by natural processes without human intervention (Banfield and Navrotsky 2001; Hochella Jr. 2002). For example, Figure 23.1c is a high-resolution TEM image of magnetite (Fe_3O_4) NPs formed in olivine ($(Mg,Fe)_2SiO_4$) by natural oxidation (Banfield et al. 1990). Note the Moiré fringes resulting from the overlap of crystalline NPs or crystalline NPs and the crystalline matrix. For the specific case where NPs are single crystalline, it is customary to use the term *nanocrystal* (NC). All NCs can also be called NPs, but the inverse is not true

since NPs may also have polycrystalline, highly distorted or even amorphous structures.

Embedded NP production by man has a history much longer than many might expect. The famous Lycurgus Cup, a Roman relic from the fourth century AD now housed in the British Museum, changes in color from opaque green to translucent red as the illumination source changes from reflected to transmitted, respectively. This intriguing behavior, shown in the photographs of Figure 23.2a and b, stems from the absorption and the scattering of light by embedded Au and Ag NPs present in the glass (Maier and Atwater 2005; British Museum 2008). Specifically, the shorter wavelengths (green light) are readily scattered while the longer wavelengths (red light) are transmitted (Atwater 2007). Another early example includes Mesopotamian artisans of the ninth century AD producing embedded NPs in glassy films. By mixing Cu and Ag salts and oxides together with vinegar, ochre and clay on the surface of previously glazed pottery, glittering golden and copper colors were achieved upon annealing to ~600°C in a reducing atmosphere to form Cu and Ag NPs (Borgia et al. 2002; Pradell et al. 2008).

The intentional production of embedded NPs for technological applications, on the other hand, has a much more recent history. Though studies on very small particles date from the beginning of the twentieth century (for example, Zsigmondy 1914), it is generally accepted that physicist Richard Feynman of Caltech was responsible for triggering the boom in nanoscale research with his seminal presentation entitled *"There is Plenty of Room at the Bottom"* (Feynman 1960) presented at the 1959 Annual Meeting of the American Physical Society. With seemingly amazing foresight, Feynman discussed *"the problem of manipulating things on a small scale"* and described *"a field in which little has been done, but in which an enormous amount can be done... and that would have an enormous number of technical*

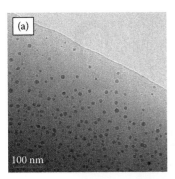
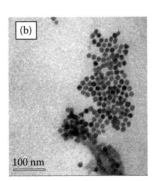

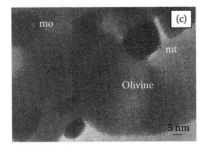

FIGURE 23.1 TEM images obtained for: (a) embedded Ge NPs in an SiO_2 matrix produced by ion implantation and thermal annealing; (b) colloidal Ge NPs prepared by high-temperature decomposition of tetraethylgermane (TEG) followed by dispersion in nonanoic acid, then deposited on a TEM grid. (Reprinted from Gerion, D. et al., *Nano Lett.*, 4, 597, 2004. With permission.); (c) naturally formed ~20 nm wide magnetite (mt) NCs in olivine (mo indicates Moiré fringes). (From Banfield, J.F. et al., *Contrib. Mineral. Petrol.*, 106, 110, 1990. With permission.)

FIGURE 23.2 (See color insert following page 9-8.) The Lycurgus cup seen in reflected light, green (a) and transmitted light, red (b). (Courtesy of the Trustees of the British Museum. With permission.)

applications." This field, of course, now encompasses nanoscience and nanotechnology. Beginning in the early seventies, theoretical and experimental studies of NPs (initially called *"superfine,"* *"ultrafine,"* *"ultramicroscopic"* or *"very small"* particles) proliferated around the globe and have grown exponentially in number

ever since. This extraordinary progress resulted, in part, from significant advances in both theory and experiment including the modeling of particles and the enhanced sensitivity and resolution attainable with analytical techniques suited to probe nanoscale phenomena, like electron microscopies, x-ray and electron diffraction, x-ray absorption spectroscopy (XAS), and small-angle x-ray scattering (SAXS) to name but a few.

Our interest in NPs, embedded and otherwise, is driven by both science and technology. As we describe in Section 23.3, nanometric-sized objects can exhibit properties not attainable with their bulk counterparts as a result of surface and/or quantum effects. For example, the photoluminescence emission from CdSe NPs spans a rainbow of colors, the latter being governed by NP size. When the NP is small relative to the Bohr exciton radius, the electron and the hole wavefunctions are *"quantum confined"* wherein the quantization of the bulk electronic bands yields discrete electronic transitions that move to a lower energy with an increasing NP size (Norris and Bawendi 1996). Exploiting this intriguing *nanoscience* for the betterment of society is the driving force behind the global *nanotechnology* boom. The potential uses of NPs are vast and ever increasing as our understanding of the science that governs their unique properties grows. In Section 23.4, we describe select state-of-the-art examples of embedded NP applications. For the CdSe NPs discussed above, assorted applications include sensors, solar cells, diodes, and lasers. Recently, their relevance and usefulness in biological and medical studies have been realized and exploited as, for example, cellular probes both *in vitro* and *in vivo* (Alivisatos et al. 2005). Their application to cancer diagnostics and treatment also shows much promise (Nie et al. 2007). Though the fundamental science that governs many metal NP applications (including the fourth and ninth century AD examples cited previously) remains the same, the technology has progressed significantly. The surface plasmon resonance (or the collective oscillation of conduction electrons in response to an external electromagnetic field), responsible for the deeply colored glasses and glazes popular in Roman times, now finds usage in the emerging field of plasmonics including, for example, electromagnetic energy transport with a sub-wavelength-scale mode confinement via plasmon waveguiding (Atwater et al. 2005).

As noted above, we restrict ourselves here to embedded NPs or those surrounded by a solid matrix. We have seen how such a matrix can inhibit NP agglomeration. Further advantages include superior mechanical, thermal, and chemical stability. On the other hand, disadvantages include the possibility of matrix–NP interactions. Ideally, the latter can be negated or at least minimized through a judicious choice of matrix material. For example, amorphous SiO_2 (a-SiO_2) or silica has widespread usage as a matrix material as a result of a stable and chemically inert nature in addition to a compatibility with Si-based electronics. Nonetheless, an inert matrix can still influence NP properties. For example, Ge NPs produced by ion implantation and thermal annealing are generally spherical when embedded in a-SiO_2 yet faceted when embedded in crystalline Al_2O_3, as shown in the TEM images of Figure 23.3a

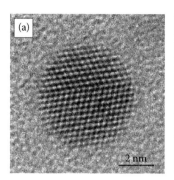
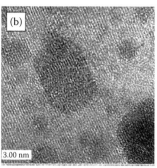

FIGURE 23.3 Illustrative high-resolution TEM images obtained for embedded Ge NPs produced by ion implantation and thermal annealing in: (a) an amorphous SiO_2 matrix, resulting in spherical NPs (Adapted from Xu, Q. et al., *J. Phys.: Conf. Ser.*, 61, 1042, 2007. With permission.) and (b) a crystalline Al_2O_3 matrix, resulting in faceted NPs. (Adapted from Xu, Q. et al., *Mater. Res. Soc. Symp. Proc.*, 880E, 22.1, 2005.)

and b, respectively. The NP/matrix interfacial energies clearly influence the shape of embedded NPs.

Embedded NPs can be readily formed from metals, semiconductors, and insulators, previously cited examples of which include Au, Ag, and CdSe NPs. In general, a scaling of properties (structural, optical, and many more) as a function of NP size is common to all. Differences in energy band gap necessarily yield differences in both properties and applications. As we have seen, the surface plasmon resonance is characteristic of particular metals while photoluminescence emission is characteristic of particular semiconductors (and insulators). To avoid both an excessively superficial approach and an overlap with the chapter entitled "Metallic Nanoparticles" within this very book, we shall further restrict ourselves to *embedded semiconductor NPs* and now describe their formation, properties, and applications.

23.2 Growth Methods

Here, we briefly present the most commonly used techniques for the growth of embedded NPs. The possibility of growing NPs into an existing solid matrix or growing the NPs and the matrix concomitantly and/or in immediately alternating steps makes the methods presented here more suited for the growth of embedded NPs than the most chemical methods, which in their turn are the techniques of choice for growing colloidal and solution-dispersed NPs (Brus 1986; Bruchez Jr. et al. 1998).

23.2.1 Ion Implantation

A common and effective means of fabricating nanoparticles in an embedding matrix is ion implantation. This technique has a long history of usage in the semiconductor industry for the introduction of both impurities and disorder as a means of altering the electrical properties of the implanted layer. The ability to control the implanted ion concentration and the depth of penetration is advantageous in an industry that necessitates such stringent reproducibility. The electrical properties of crystalline

semiconducting layers are easily varied with implanted ion concentrations of 0.1 at.% or less. To inhibit ion *channeling* during implantation, where ions are guided between atomic rows via a series of glancing elastic collisions with substrate atoms, crystalline substrates are typically aligned in a *non-channeling* orientation such that the incident ion beam direction is offset from the low-index crystallographic directions of the substrate. Channeling is most pronounced at low implantation energies and can result in a significantly enhanced ion penetration. With the substrate normal and ion-beam direction appropriately oriented to inhibit this effect, the implanted ion-depth distribution is equivalent to that achieved with an amorphous substrate. An example of an implanted ion-depth distribution, typically Gaussian-like, is shown in Figure 23.4.

Post-implantation thermal annealing is commonly required to increase the fraction of the implanted impurities on electrically-active substitutional lattice sites and minimize implantation-induced disorder within the implanted layer. Such disorder can be of many forms, varying from simple point defects—vacancies and interstitials—to the complete amorphization of the implanted layer. In semiconducting substrates, the structural disorder results primarily from elastic collisions between implanted ions and substrate atoms. When the latter acquire energy greater than that binding them in the lattice, these displaced atoms can, in turn, displace additional substrate atoms via a collision cascade. Implanted ions can also interact with substrate atoms through inelastic collisions resulting in the excitation or ionization of substrate atoms. For a given ion/substrate combination, the relative fractions of elastic and inelastic ion energy loss are governed by the ion velocity. Figure 23.4 also includes the depth distributions of these two energy loss processes. Inelastic interactions typically dominate at higher ion velocities (or equivalently at smaller depths of penetration).

An ion implanter generally consists of discrete sections for ion production, acceleration, mass-to-charge filtering, rastering,

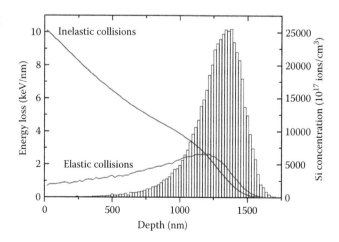

FIGURE 23.4 Calculations derived from the SRIM code (Zeigler et al. 1985) for Si ions of energy 1 MeV and fluence 1×10^{17} cm^{-2} incident on amorphous SiO_2 showing the depth distributions of elastic and inelastic energy loss processes in addition to the implanted Si ions.

and implantation of the target substrate. Negative ion sources are typically associated with tandem-type accelerators where ions are subsequently stripped of some or all electrons to become positive for a second stage of acceleration. Nonetheless, the direct implantation of negative ions can be beneficial in reducing charging effects in insulating substrates. Ions are accelerated though a potential difference that governs their energy and hence the depth of penetration in the target substrate. An electromagnet, with the field oriented perpendicular to the ion direction, is used to distinguish between both the desired element and the contaminants and also isotopes of the specific element. To insure lateral uniformity over the extent of the implanted layer, the ion beam is usually scanned rapidly in both the vertical and the horizontal planes across the target substrate. The latter is commonly housed in a dedicated chamber where the substrate orientation is optimized and the integrated ion charge is measured to accurately monitor the number of ions implanted in the substrate.

To achieve the desired change in *compositional* properties for the production of embedded nanoparticles in a matrix, higher implanted ion doses (typically 1–10 at.%) are necessary compared to those used to alter the *electrical* properties of a bulk semiconductor. For amorphous matrices such as silica (amorphous SiO_2), ion channeling is no longer of concern. With the appropriate ion/substrate combination, these high doses yield implanted ion concentrations well in excess of the solid solubility limits. Post-implantation annealing is again used to reduce disorder in the implanted layer, even in amorphous substrates where implantation-induced optical-active defects are common (Min et al. 1996). Annealing also results in implanted ion precipitation and diffusion, inducing nanoparticle nucleation and growth via an Ostwald ripening process. The nanoparticle number density plus depth and size distributions required for a given application are optimized through appropriate combinations of implantation and annealing parameters. Examples of NPs produced by ion implantation and thermal annealing are shown in Figures 23.1a and 23.3.

23.2.2 Sputter Deposition

The process of sputter deposition involves several steps to induce ejection (or *sputtering*) of atoms from a selected target and subsequent deposition of such atoms on a sample. Initially, high voltage is applied across a low-pressure gas (usually Ar for non-reactive sputtering and O_2 or N_2 for reactive sputtering) to create a plasma. DC (direct current) voltages are used when sputtering-conducting materials like metals (*dc-sputtering*), but it is necessary to apply AC (alternating current) voltages or RF (radio-frequency) signals when sputtering insulators to avoid charging at the target's surface (*rf-sputtering*). The ionized gas atoms are then accelerated against a target containing the material to be deposited and then ejected from the target by ballistic collision. The ejected atoms finally drift towards the sample and bond to it. A strong magnet is usually located below the target (and hence the process is sometimes called *magnetron sputtering*) in order to generate a magnetic field that confines the

plasma close to the target area, substantially increasing the ionization probability and the deposition rate. Unlike evaporation methods, sputtering does not require heating of the target and thus can be used for depositing a wider range of materials (Wasa 1992). Sputter deposition can produce uniform layers of a few nanometers thickness with very good quality and has been used to produce embedded NPs since the mid-1970s (Stewart 1977).

There are a couple of different ways to produce semiconductor NPs by sputter deposition, but in most cases the deposition must be followed by thermal annealing of the sample under appropriate atmosphere, temperature, and time conditions (or the sample must be heated during deposition). By changing the deposition and annealing parameters, the size and the crystallinity of the NPs can be controlled.

The method used in early studies (and still the most widely used in the present) consists of simultaneously sputtering the matrix and the NP elements onto the sample (thus called *co-sputtering*), as in Fujii et al. 1990, for example. A reasonably homogeneous film is initially formed and the ratio of the amount of sputtered matrix/NP elements influences the sequential formation of NPs under thermal annealing. A high-volume density of NPs (sample volume occupied by NPs compared to the total sample volume) can be achieved by co-sputtering, but their spatial distribution cannot be controlled, as illustrated in Figure 23.5a.

Another method has appeared more recently that consists of sputtering alternate thin layers of matrix and NP elements, as shown in Figure 23.5b for a 5 nm Ge layer deposited between two SiO_2 layers (Schmidt et al. 2007). Upon proper annealing conditions, NPs are formed as in Figure 23.5c. This approach produces a relatively small volume density of NPs, but yields very good control over the NP spatial distribution. In principle, the volume density of NPs in a sample can be increased by depositing several layers with the NP element stacked with matrix layers dividing them, but the diffusion effects and the interaction between the layers of the same element, as observed in Foss et al. 2007, must be carefully considered when applying such an approach.

23.2.3 Chemical Vapor Deposition

Chemical Vapor Deposition (CVD) encompasses several techniques based on the deposition of atoms by chemical reactions between precursor gases containing the elements of interest and the heated substrates. Precursor gases usually consist of molecules where organic compounds are bound to the element of interest; for example, *germane* (GeH_4) is used to deposit Ge layers and nanostructures. Upon contact with the heated substrate, germane decomposes by pyrolysis; while the Ge atoms bind to the substrate, H forms by-products that are carried away by flowing gas (N_2 or Ar, for example). Among the CVD methods, metal-organic CVD (MOCVD) and plasma-enhanced CVD (PECVD) are the most widely used for the growth of embedded NPs. Analogous to the production of embedded NPs by sputtering, CVD techniques can be used to co-deposit the species of matrix and the NPs or to deposit them separately in thin layers.

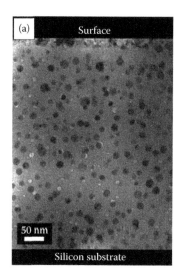

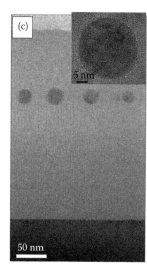

FIGURE 23.5 (a) TEM image of Ge NPs grown in SiO$_2$ by co-sputtering and thermal annealing (From Jensen, J.S. et al., *Appl. Phys. A*, 83, 41, 2006. With permission.); (b) and (c) as-sputtered 5 nm thick Ge layer encapsulated between SiO$_2$ layers and corresponding Ge NPs produced by the subsequent RTA of the sample. The insets show high-resolution images of the as-deposited layer and the produced NPs, respectively. (Reprinted from Schmidt, B. et al., *Nucl. Instrum. Meth. Phys. Res. B*, 257, 30, 2007. With permission.)

The mean size of the produced NPs also can be controlled by changing deposition parameters or post-annealing parameters in the case of PECVD.

MOCVD is so termed because it uses metal-organic compounds (molecules with organic elements binding to metallic ions) as precursors. It has been particularly used for the growth of III-V compound NPs (such as InAs and GaAs), as illustrated in Figure 23.6 (Tan et al. 2006). PECVD, in its turn, utilizes a plasma to enhance reaction rates of the precursors, allowing the use of lower substrate temperatures than in MOCVD. PECVD has been particularly used for the growth of group IV semiconductor NPs, like Ge and Si. Since the deposition occurs at lower temperatures, it usually must be followed by a thermal annealing of the sample to cause NP growth. Figure 23.7 shows a high-resolution TEM image of a Ge NP embedded in silica produced by PECVD followed by thermal annealing at 1010°C for 1 h under a N$_2$ flow (Agan et al. 2006).

CVD reactors must be operated under strict safety conditions because they usually involve very dangerous precursor gases like *silane* (SiH$_4$), which is extremely flammable and potentially pyrophoric (undergoes spontaneous combustion when exposed to air), *trimethylgallium* (Ga(CH$_3$)$_3$) which is pyrophoric and reacts violently with water, and *tertiarybutylarsine* (C$_4$H$_{11}$As), which is toxic by inhalation and pyrophoric.

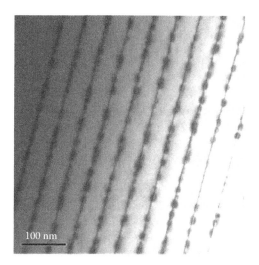

FIGURE 23.6 Cross-sectional TEM image of an MOCVD-produced QDIP (quantum-dot infrared photodetector) structure composed of 15 layers of In$_{0.5}$Ga$_{0.5}$As NPs (only 10 are shown) spaced by ~50 nm thick GaAs layers. (Adapted from Tan, H.H. et al., *IEEE J. Sel. Top. Quant. Electron.*, 12, 1242, 2006. With permission.)

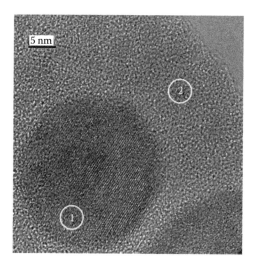

FIGURE 23.7 High-resolution TEM image of a Ge NC (1) embedded in amorphous SiO$_2$ (2). Crystalline planes are evident in the Ge nanocrystal. (From Ağan, S. et al., *Appl. Phys. A*, 83, 107, 2006. With permission.)

23.2.4 Other Methods

One of the early methods used for the production of Si NPs is *electrochemical dissolution* or *anodization of bulk crystalline wafers*. It consists in exposing the wafer to a combination of electrical current and chemical solutions (usually HF-based) that produces a high density of small holes running orthogonal to the surface. The resulting product was termed *porous Si* (Canham 1990). Similar results can be achieved using a light source in conjunction with or instead of an electrical current, the latter being simply called *wet etching* (Kartopu et al. 2008). But rather than producing well-defined particles in a homogeneous matrix (like in Figures 23.1 and 23.3, for example), anodization/wet etching of bulk semiconductors produces a porous structure with a mass of amorphous/highly damaged regions enclosing randomly shaped nanocrystallites, as illustrated in Figure 23.8.

Many other methods are available for the production of *free-standing* or *substrate-supported* NPs, which potentially could be turned into embedded NPs by a subsequent deposition of an adequate encapsulating matrix. For example:

- *Gas evaporation*, where a piece of bulk material is heated past the melting point until it evaporates and the resulting gas condensates as small particles on the surface of a substrate (Bostedt et al. 2003).
- *Cluster-beam evaporation*, where a beam composed of clusters of atoms ejected by supersonic expansion through the nozzle of a high-temperature crucible is directed to a substrate for the deposition (Sato et al. 1995).
- *Hypersonic plasma particle deposition* (*HPPD*), where vapor-phase precursors are injected into a thermal plasma generated by a DC arc torch and the resulting reacting mixture is quenched as it expands through a nozzle, after which particles are focused onto a substrate using an aerodynamic lens system (Perrey and Carter 2006).
- *Pulsed laser ablation* (*PLA*), where a bulk target is ablated by repeating laser pulses in a specific frequency, forming a plasma which is carried and cooled by an inert gas and condensed as NPs on a substrate surface. This technique can be coupled to a simultaneous CVD of a matrix to produce embedded rather than supported NPs, working like the co-deposition method described in the sputter deposition session (Ngiam et al. 1994).
- *Pulsed spark ablation*, which works like PLA but instead of laser pulses, the bulk target is vaporized by electric sparks of energy ~ 50–150 mJ (Saunders et al. 1993).
- *Microwave plasma deposition*, where a precursor gas containing the species of interest is decomposed into a plasma by microwave radiation and is blown through a convergent nozzle, by flowing inert gas, into a substrate (Takagi et al. 1990).

It is important to keep in mind that almost all the techniques applied to produce free-standing NPs are subject to the drawbacks of particle agglomeration and chemical impurity contamination. They require either very good vacuum (10^{-5} Pa or better) or an inert atmosphere (like pure Ar) in order to avoid oxidation of the produced NPs and can be seriously affected by uncontrollable particle agglomeration if more than one monolayer is deposited, which may result in the loss of the desired nanoscale properties due to particle-particle interactions.

23.3 Fundamental Properties

As mentioned in the introduction, intensive research efforts have been directed to the study of NPs (embedded and otherwise) in the last few decades, driven by the fact that NPs present several properties that are absent in their bulk counterparts. For example, bulk Si is an indirect band gap semiconductor and thus a very poor light emitter. On the other hand, Si NCs have proven not only to be efficient light emitters but also allow a selection of the emitted wavelength, becoming promising candidates for novel optoelectronic devices (Linnros 2005). Another example is the carrier multiplication (CM) phenomenon, where incoming photons create high-energy electron-hole pairs via impact ionization which, in their turn, transfer their excess energy to the creation of additional electron-hole pairs instead of dissipating it as lattice vibrations or heat (Luo et al. 2008). While CM is very poor in bulk semiconductors like GaAs and InP, it is much more efficient in GaAs and InP NCs, making them suited for the production of enhanced solar cells (Kamat 2008; Luo et al. 2008).

The origin of these and other properties of NPs can be summarized in a word: *size*. It is the limited dimensions and the limited number of atoms in NPs that allow for their interesting properties. In a broad sense, size effects can be assigned either to the higher number of surface atoms in a NP, generating a high *surface-to-volume ratio*, or to *quantum phenomena* originating from the discreteness of electronic states and the band gap variation with NC size. While the former are expected to scale with D^{-1} or $N^{-1/3}$ (D and N are the NP diameter and the total number of atoms in a NP, respectively), the latter is predicted to scale with D^{-2}. In the following, we present an overview of such effects

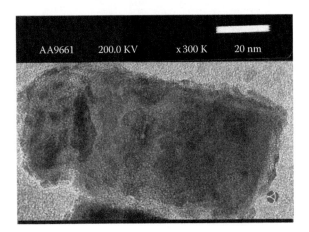

AA9661   200.0 KV   x 300 K   20 nm

FIGURE 23.8 TEM image of a Ge nanocrystallite produced by wet etching of bulk Ge. (Reprinted from Kartopu, G. et al., *J. Appl. Phys.*, 103, 113518/1, 2008. With permission.)

and comment on their impact on the structural, thermal, electronic, and optical properties of embedded NPs.

23.3.1 Surface Effects

In the macroscopic scale, the number of atoms located at the surface of a given material is so much smaller than the total number of atoms that surface-related effects play a minor role in determining the average properties of the material. Even for particles at the micron scale, as small as a sphere of $D \sim 1\,\mu m$, not more than 1% of the atoms reside at the surface. This scenario changes at the nanoscale, where the ratio F of the number of atoms at the surface N_S divided by the total number of atoms N, called *surface area-to-volume ratio* (or *surface-to-volume ratio* for short) $F = N_S/N$ becomes significant. A quick estimate of the increase in surface-to-volume ratio with a decreasing NP diameter is shown in Figure 23.9 for Ge NPs (assuming a perfect diamond lattice), where it can be seen that F evolves with D^{-1} and increases more drastically for D values of $\sim 7\,nm$ or less. Figure 23.9 also indicates that for NPs with $D \sim 15\,nm$ (or bigger) surface-to-volume ratio effects become much less pronounced. The three clusters with diamond-lattice ordering displayed in Figure 23.9 and the numbers on their left side illustrate the approximate number of atoms N for Ge NPs of the corresponding diameters. They do not represent real particles since the surface reconstruction is not taken into account and serve only as a rough visual guide. Although here Ge NPs are used as an example, the presented trends for F are valid for all NPs in general. Specific equations to estimate F can be derived for NPs of different shapes like cubes (Roduner 2006) and spheres (Jortner 1992).

A direct consequence of the increase in F is the decrease in the *coordination number* (CN) of an NP, defined as the average

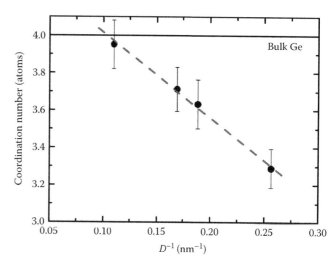

FIGURE 23.10 Decrease in the mean coordination number with the decrease in the mean size for an ensemble of Ge NPs embedded in silica, as determined from XAS measurements. The dashed line is a linear fit to the data, showing that the coordination number scales linearly with D^{-1}. The horizontal line indicates the coordination number for bulk crystalline Ge.

number of nearest neighbors around an atom or, equivalently, the mean number of atomic bonds per atom. Figure 23.10 shows the decrease in CN obtained from XAS measurements for Ge NP distributions with different mean sizes as produced by ion implantation (Araujo et al. 2008). The CN decreases linearly with D^{-1}, as expected from the increase in F.

Atoms located at the surface of an NP have fewer neighbors than atoms in the core, resulting in unsatisfied bonds. They become less stable for experiencing two different environments: from one side, they have the influence of the atoms in the NP core and from the other side, influences from the environment. In the case of colloidal or solution-dispersed NPs, surfactants readily bind to the NP surface (Modrow 2004). For embedded semiconductor NPs, the absence of chemical interaction with ligands or surfactants results more commonly in the formation of dangling bonds and other defects or less commonly in binding with matrix atoms, if the binding energies are favorable. In all cases, some degree of surface reconstruction occurs (Hamad et al. 1999; Pizzagalli and Galli 2002; Araujo et al. 2008; Djurabekova and Nordlund 2008), as illustrated in the molecular dynamics (MD) simulation of a free-standing Ge NP with 190 atoms ($D \sim 2\,nm$) shown in Figure 23.11a. The crystalline core still maintains the diamond structure but the gray-shaded superficial area shows significant atomic disorder. Unlike metals, semiconductors can easily assume an amorphous structure, both in the bulk and the NP forms (Ridgway et al. 2006). In fact, recent experimental (Araujo et al. 2008) and theoretical (Djurabekova and Nordlund 2008) results reinforce the suggestion from previous studies that surface reconstruction in embedded semiconductor NPs involves the formation of amorphous-like layers. Figure 23.11b and c show MD simulations for Si NPs embedded in SiO_2 with $D = 2.4$ and $1.3\,nm$, respectively, illustrating the increased disorder due to surface reconstruction

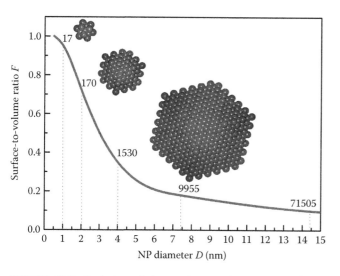

FIGURE 23.9 Evolution of the surface-to-volume ratio with NP diameter (curve) as estimated for Ge NPs embedded in SiO_2. The numbers right above the curve indicate the approximate number of atoms in the Ge NPs with the diameters indicated by the vertical dotted lines. The clusters give a rough visual idea of how the particles would look if they had a perfectly crystalline diamond structure.

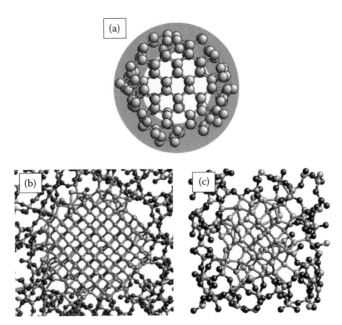

FIGURE 23.11 MD simulations showing cross sections of semiconductor NPs to illustrate the atomic disorder at the outer NP shells due to surface reconstruction. (a) Free-standing Ge NP with 190 atoms ($D \sim 2$ nm) showing a diamond-like crystalline structure at the core and a disordered surface shell highlighted in gray. (Reprinted from Pizzagalli, L. and Galli, G., *Mater. Sci. Eng. B*, 96, 86, 2002. With permission.) (b) and (c) Si NPs embedded in SiO_2 with diameters of 2.4 and 1.3 nm, respectively. Gray and black circles are silicon and oxygen atoms, respectively. (From Djurabekova, F. and Nordlund, K., *Phys. Rev. B*, 77, 115325/1, 2008.)

with the decrease in the NP size. The results in Figure 23.11c actually suggest that very small embedded Si NPs ($D \leq 2.0$ nm) cannot sustain a crystalline structure, which is in good agreement with experimental results for Ge NPs embedded in SiO_2 (Araujo et al. 2008). This surface reconstruction effect must be taken into account together with the increase in F as the NP size decreases, and renders models that consider NPs as small clusters with a perfectly bulk-like crystalline structure unsuited for the description of embedded semiconductor NPs with $D \sim 10$ nm or less. As it would be expected, other structural properties of embedded NPs like the crystallographic phase and the mean value and the variance of interatomic distance distributions are significantly affected by the increase in F and in surface reconstruction.

Surface effects also influence the thermal properties of NPs. While the majority of free-standing NPs (semiconductor and metallic) melts at increasingly *lower* temperatures than bulk as their size decreases (Baletto and Ferrando 2005), several embedded semiconductor NPs behave in an opposite way and melt at increasingly *higher* temperatures than bulk as their size decreases (Xu et al. 2006). Although no specific rule can be established thus far and exceptions exist for both cases, most models confirm that the melting of an NP starts at the surface (Kellermann and Craievich 2002; Xu et al. 2006). As a general trend, in freestanding NPs, the less-tightly bound surface atoms are not externally constricted and can have higher amplitudes of vibration, facilitating the melting process. Embedded NPs, on the other

hand, are closely surrounded by a matrix that (allied to surface reconstruction if any) can suppress their vibrational motion and limit their thermal expansion, as verified experimentally (Araujo et al. 2008). This can explain the *superheating* or melting at temperatures higher than the bulk-melting temperature observed for Ge NPs embedded in SiO_2 (Xu et al. 2006). When discussing the thermodynamic properties of NPs, it should be kept in mind that phase transitions in nanoscale systems are gradual, not sharp (Hill 1964); there are bands of temperature and pressure within which two or more cluster structures may coexist dynamically, similarly to the coexistence of chemical isomers (Baletto and Ferrando 2005; Roduner 2006).

Finally, the electronic and optical properties of embedded semiconductor NPs are also affected by surface effects, in particular the surface reconstruction mentioned earlier and the presence of a surrounding matrix (Daldosso et al. 2003). Surface reconstruction leads to the creation of energy levels within the energetically forbidden gap of the bulk counterpart, which can alter the electronic and optical properties of the NP by trapping electrons or holes (Alivisatos 1996). Furthermore, defects at the NP-matrix interface can act as radiative and nonradiative recombination centers and thus play a key role in the optoelectronic properties of embedded Si and Ge NPs (Pavesi et al. 2000; Singha et al. 2006). For example, nonbridging oxygen centers (defects where an oxygen atom binds to only one silicon atom and has an unpaired electron in its nonbonding $2p$ orbital) are know to significantly influence the observed photoluminescence from NPs embedded in SiO_2.

23.3.2 Quantum Confinement

The so-called *quantum confinement* effect refers to electrons and holes being spatially restricted in a material whose dimensions are on the order of or smaller than the *exciton Bohr radius* of that material. *Exciton* refers to the electron-hole pair created when an electron leaves the valence band and enters the conduction band, and the bulk *exciton Bohr radius* is the natural physical separation in a crystal between the electron in the conduction band and the hole it leaves behind in the valence band. Under quantum confinement conditions, energy levels may be treated as discrete and the *band gap* becomes size dependent, increasing as the NC size decreases, as illustrated in Figure 23.12. By definition, the band gap of a semiconductor is a region where electronic levels are forbidden, and it is given as the energy difference between the top of the valence band and the bottom of the conduction band. Confinement in one dimension is observed for the so-called *quantum wells* - films with thickness of a few tenths of nm but with macroscopic width and length, thus called "2-D structures" (Figure 23.13b). Confinement in two dimensions is verified for the so-called *quantum wires*, cylinder-like structures (such as nanorods) with the base dimensions measuring a few tenths of nanometers and the height measuring up to several microns, thus called "1-D structures" (Figure 23.13c). Finally, the confinement in all three dimensions happens for NCs (pyramidal, polyhedral, spherical, etc.) since all of their dimensions

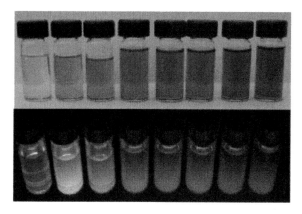

FIGURE 23.12 Schematic representation of the electronic levels in bulk (left) and nanocrystalline (right) semiconductors, illustrating the increase of band gap (vertical dashed arrow) and discreteness of levels (horizontal lines) characteristic of the quantum confinement effect in semiconductor NCs.

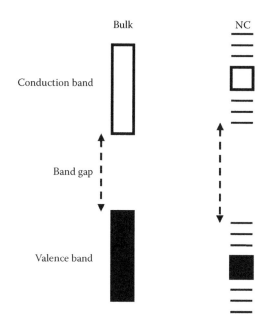

FIGURE 23.14 (See color insert following page 9-8.) Series of colloidal CdSe NC solutions illuminated with room light (top) and UV light (bottom). The NC size increases from left ($D \sim 1.5\,$nm, blue PL) to right ($D \sim 10\,$nm, red PL). (Reprinted from Rosenthal, S.J. et al., *Surf. Sci. Rep.*, 62, 111, 2007. With permission.)

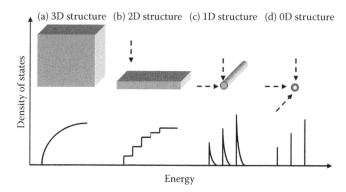

FIGURE 23.13 Schematic illustration of the confinement effect on a given band of a semiconductor as a function of the number of confinement dimensions. (a) Bulk material, no confinement; (b) thin film, confinement in one direction; (c) nanorod, confinement in two directions; (d) nanocrystal, confinement in all three directions. Full curves below each structure represent the density of the states and the dashed arrows around the structures indicate the confinement directions.

and fluorescence wavelengths of semiconductor NCs become size dependent. This is neatly illustrated in Figure 23.14, where the photoluminescence (PL) emitted by colloidal CdSe NCs excited by ultraviolet (UV) light is shown to shift from blue (~450 nm) to red (~650 nm) as the mean NC size increases from ~1.5 to 10 nm (Rosenthal et al. 2007). Quantum confinement effects were also observed in the near-infrared region for Ge NCs embedded in silica (Takeoka et al. 1998). As the NC mean size decreased from ~5 to 1 nm, the PL peak energy shifted monotonously from ~0.9 to 1.5 eV and the intensity increased by approximately two orders of magnitude. The correlated variation in PL and NC size indicates that quantum confinement effects are responsible for the observed PL, but this is not always the case for embedded NCs. Although enclosing the NCs in a dielectric matrix is advantageous to enhance the confinement of excitons, since the larger band gap of the matrix introduces an abrupt jump in the chemical potential at the interface, defects in the matrix and interface-related effects may hamper the luminescence from the NCs or act as luminescent centers at different wavelengths, making the task of assigning specific emission peaks to quantum confinement effects or matrix/interface states particularly tricky (Min et al. 1996).

23.4 Applications

In general, nanoparticles have a multitude of applications spanning all walks of life, from catalysis in chemical engineering to drug delivery in medical science. As noted before, such applications are enabled and driven by the change or perturbation of materials' properties that only become apparent when particles are scaled to nanometric dimensions. In this section, we continue to focus on embedded semiconductor nanoparticles. One very common but potentially little known application is the use of ZnO and TiO_2 nanoparticles in poly-ethylene terephthalate (PET) beverage containers. These two materials already have widespread use in sun creams due to their ability

are within few tenths of nanometers (Figure 23.13d). Because of that, semiconductor NCs are sometimes called "*0-D structures*" or "*quantum dots*" (QDs). A detailed description of quantum confinement effects and their corresponding theoretical models can be found in reviews available in the literature (Yoffe 2002).

The increase in the band gap with decreasing semiconductor NC size predicted by quantum confinement models has been verified experimentally for CdS (Vossmeyer et al. 1994) and Si (van Buuren et al. 1998) NCs, to quote two examples. A direct consequence of such variation in band gap is that the absorption to absorb harmful UV radiation. Their inclusion as UV blockers in the plastic from which bottles are manufactured can prolong the shelf life of the product contained therein. Below, we describe three additional technological applications of embedded semiconductor nanoparticles, with an emphasis on those of the most interest to the authors—electronic and photonic devices.

23.4.1 Semiconductor Nanoparticle-Based Nonvolatile Memory

Memory devices fabricated from bulk semiconductors are intrinsic to all modern computer technologies and can be of volatile or nonvolatile type. Volatile memory, such as Random Access Memory (RAM), loses stored information in the absence of power. Nonvolatile memory, for example, flash memory, retains data upon the loss of power but is typically inferior in terms of cost and performance. A means of improving these two parameters is thus of considerable economic and technological significance.

Flash memory is commonly based on a floating gate Metal Oxide Semiconductor (MOS) transistor structure, as shown schematically in Figure 23.15a. The floating gate, comprised of polycrystalline Si, is separated and isolated from the channel by the thin SiO_2 tunnel oxide while the control oxide separates the control gate from the floating gate. Charge travels through the tunnel oxide, in and out of the floating gate, via the Fowler–Nordheim tunneling process to achieve the programming and erasing functions. During a programming operation, the application of sufficient positive voltage on the control gate enables electron tunneling from the channel to the floating gate. Injected electrons remain trapped on the floating gate as a result of a gate-to-oxide energy barrier. Charge on the floating gate results in a change in threshold voltage and a read operation is achieved by measuring the *I-V* characteristics. To remove the charge from the floating gate for an erase operation, a negative voltage on the control gate induces electron tunneling from the floating gate to the channel.

Though flash is the most common form of nonvolatile memory, continued scaling of such devices is limited by charge retention considerations as governed by the tunnel oxide thickness. An alternative approach, first proposed by Tiwari et al. (1996), is the use of semiconductor nanoparticles in place of the continuous floating gate as shown in Figure 23.15b. As above, the injected charge retained in the NP ensemble alters the threshold voltage. For example, an NP ensemble with a density of $3-10 \times 10^{11}$ cm^{-2} and one stored electron per NP yields a detectable threshold voltage shift of 0.3–0.5 V (Hanafi et al. 1996). The use of a thinner tunnel oxide enables continued device scaling and is viable given the leakage through a defective pathway in the oxide that results in the discharge of only those NPs in close proximity to the pathway (while the rest of the NPs retain charge given that they are electrically isolated from each another). The reduction in tunnel oxide thickness yields a reduction in the programming voltage (and

hence power consumption) plus superior read/write times and a reliability relative to the conventional continuous floating gate nonvolatile memories (Ostraat et al. 2001).

The fabrication of the NP ensemble within a MOS structure can be achieved with the methods described previously, including the deposition and ion implantation techniques. An advantage of the latter is the ability to form nanoparticles directly within the tunnel oxide in close proximity to the channel and thus negate the need for deposition of the control oxide. Figure 23.16 shows a cross-sectional transmission electron microscopy image of an MOS transistor structure with Si NPs as the floating gate formed by ion implantation (Porti et al. 2007). A stringent control of the NP size, density, and depth distributions is of absolute necessity to minimize variation in the memory-operating characteristics including retention times and threshold voltages (Ostraat et al. 2001). Nanoparticles of diameter ~5 nm are considered optimal given that the effects of the Coulomb blockade on charge retention are negligible (Rao et al. 2004).

Si NPs have been the most intensively studied for use in nonvolatile memories given their compatibility with established CMOS fabrication processes. Nonetheless, the higher

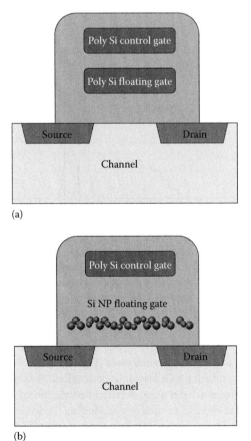

(a)

(b)

FIGURE 23.15 Schematic diagrams of an MOS transistor structure for use as nonvolatile memory with (a) a conventional continuous polycrystalline Si floating gate and (b) a semiconductor nanoparticle floating gate.

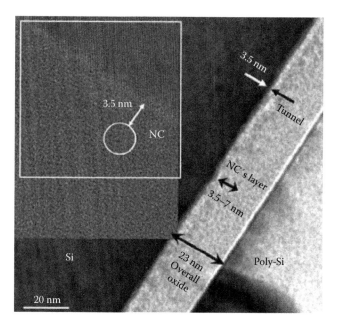

FIGURE 23.16 Cross-sectional transmission electron microscopy image of an MOS transistor structure with Si nanoparticles formed by the ion implantation as the floating gate. (Reprinted from Porti, M. et al., *J. Appl. Phys.*, 101, 064509/1, 2007. With permission.)

electron affinity of Ge may provide an advantageous alternative. Specifically, the lower conduction-band-minimum of Ge NPs relative to that of the Si channel yields superior charge retention for a given programming voltage as the tunneling probability from the NP to the channel is lowered (Tsoukalas et al. 2005). Another alternative under recent investigation is the use of InAs NPs formed directly on the tunnel oxide by molecular beam epitaxy (MBE) and subsequently embedded in SiO_2 by the PECVD of the control oxide (Hocevar et al. 2007). The retention time for electrons in such NPs was considerably higher (by four orders of magnitude) than that achieved with Si NPs as a result of the lower band gap relative to Si and the higher conduction band offset with SiO_2.

23.4.2 Semiconductor Quantum-Dot Lasers

The advantageous properties of Si and SiO_2 are such that the vast majority of electronic devices are fabricated from these materials. However, the indirect band gap of Si is not readily conducive

to photonic device fabrication, and as a consequence, the direct band gap compound semiconductors, such as the binary, ternary, and quaternary III-V alloys, dominate this market. The ability to tailor both the band gap and the lattice parameter by varying the alloy composition provides considerable application-specific flexibility. An example of a well-commercialized photonic device is the small, efficient, and reliable semiconductor diode laser. These devices can be rapidly modulated and readily tuned, generating widespread applications in the telecommunications field. Photon emission results from the radiative recombination of electrons and holes that cross the forward-biased junction. Facets cleaved perpendicular to the direction of the photon propagation reflect light back and forth through the active layer. Further electron-hole recombination yields additional photons and the stimulated-emission process.

Semiconductor lasers such as that shown schematically in Figure 23.17a (Henini and Bugajski 2005) are typically of a double heterostructure (DH) form where, for example, a GaAs layer is sandwiched between two $Al_xGa_{1-x}As$ layers. The smaller band gap and the larger refractive index of GaAs relative to $Al_xGa_{1-x}As$ yields, respectively, an electrical and an optical confinement in the active layer. In quantum well (QW) lasers, *quantum* confinement is achieved by reducing the thickness of the GaAs layer to a value comparable to the de Broglie wavelength of the charge carriers, restricting the motion of the latter from three to two dimensions. Confinement in all three dimensions can be achieved by replacing the continuous active layer with semiconductor nanoparticles (or *quantum dots* (QDs)) as shown in Figure 23.17b. As a consequence of confinement, the density of states (DOS) is modified dramatically, changing from parabolic-like continuous bands (DH) to step-like (QW) to quantized discrete levels (QD) as mentioned in Section 23.3 and illustrated in Figure 23.13.

The formation of embedded semiconductor NPs for use in a QD laser structure can be achieved with the deposition methods described previously, including MOCVD and MBE, with the aim of maximizing the number density while minimizing the defect density and the variation in size. When the lattice parameter of the NP and the matrix materials differ significantly, a rapid strain-induced change from a two-dimensional layer-by-layer growth to a three-dimensional island growth, termed the Stranski–Krastanov growth, is observed. With this self-assembly method, InAs QDs grown on a GaAs substrate are epitaxial and dislocation-free

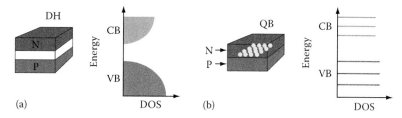

FIGURE 23.17 Schematic diagrams of semiconductor lasers with (a) a conventional continuous active layer of the double heterostructure (DH) type and (b) an active layer comprised of quantum dots (QD). The DOS with conduction and valence bands are shown on the right. (Reprinted from Henini, M. and Bugajski, M., *Microelectron. J.*, 36, 950, 2005. With permission.)

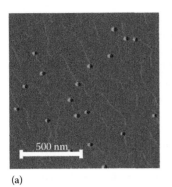
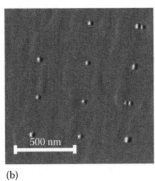

(a)   (b)

FIGURE 23.18 InAs nanoparticles on a GaAs substrate (a) without and (b) with lithographic seeding. (Reprinted from Shields, A.J., *Nat. Photon.*, 1, 215, 2007. With permission.)

with a very narrow size distribution (Shields 2007). The width of the latter necessarily influences the emission linewidth given the increase in band gap of semiconductor NPs with a decrease in NP size, as noted earlier. Figure 23.18 (Shields 2007) shows two atomic force microscopy images of InAs NP ensembles on a GaAs substrate. Those in Figure 23.18a appear preferentially located on surface terrace edges while the spatial positions of the NPs of Figure 23.18b have been defined by lithographic methods. Nanoscale pits formed on the GaAs substrate prior to the deposition subsequently seed NPs during the growth process.

Relative to QW lasers, those formed from QDs can yield a decrease in the emission linewidth, an increase in the modulation bandwidth and a decrease in the threshold current density (Bukowski and Simmons 2002). Furthermore, the temperature dependence of the threshold current density may also be suppressed. At low temperatures, the threshold current density is typically low but increases rapidly as the temperature rises above ~120 K due to leakage currents and the recombination in the cladding layer (Bukowski and Simmons 2002). The onset of this rise has been observed at higher temperatures in QD lasers relative to their QW laser counterparts (Fafard 2000).

23.4.3 Semiconductor Nanoparticle-Based Solar Cells

Photovoltaic (PV) solar cells have much potential as a means of reducing our reliance on fossil fuels and are thus of considerable environmental and economic significance. First-generation PV solar cells based on bulk single-crystal Si are efficient yet expensive. Second-generation cells formed from thin-film polycrystalline Si are more cost-effective yet less efficient. Considerable effort is now focused on third-generation cells with the aim of further reducing the cost per peak Watt of thin-film cells through an increase in their efficiency. Embedded semiconductor nanoparticles are one technology under investigation to achieve this goal.

The efficiency for converting solar photons to electronic carriers, measured as the ratio of the maximum power output to that input (the *ideal conversion efficiency*) is limited to ~< 30% for a given semiconductor material. In PV solar cells, photons of

below the band gap energy are not absorbed while the absorption of those exceeding this threshold can generate charge carriers. Nonetheless, for photons of energy greater than the band gap, their contribution to the cell potential is limited to the band gap energy with the remainder lost as heat through phonon generation (Bukowski and Simmons 2002). Increasing the number of band gaps is clearly a potential means of absorbing a greater fraction of the solar spectrum. Tandem cells combine cells of different thresholds in series, each absorbing different components of the solar spectrum and approaches based on both Si and InAs materials have been studied.

A means of forming Si-based structures with band gaps different from that of the bulk material would be advantageous from a processing compatibility perspective. Forming Si quantum dots in a silica matrix with an inter-dot separation sufficiently small to yield the overlap of carrier wavefunctions in adjacent dots can induce the confined electronic states to smear and form a miniband (Conibeer et al. 2006). A broad mini-band can yield a band gap that exceeds that of bulk Si and a tandem cell can thus be fabricated solely from engineered Si materials. The formation of wide band gap Si QDs via the sputter deposition of alternating layers of Si-rich SiO_2 and stoichiometric SiO_2 followed by thermal annealing has demonstrated much promise toward the eventual fabrication of Si tandem PV cells (Conibeer et al. 2006).

As an alternative, a self-assembled InAs/GaAs QD array within the intrinsic region of a *p-i-n* PV cell may have potential benefits for photon absorption in the low-energy part of the solar spectrum (Aroutiounian et al. 2001). The size and shape of the QDs can be tailored to maximize the absorption spectrum. Using multiple layers of QDs yields the vertical alignment of dots from one layer to the next, and as a consequence of such coupling, the electronic states are wire-like (Darhuber 1997). The electrons and holes photo-generated in the QDs can then be efficiently injected into the *p* and *n* regions of the cell and the conversion efficiency significantly increased (Aroutiounian et al. 2001).

References

Ağan, S., A. Celik-Aktaş, J. M. Zuo, A. Dana, and A. Aydinli. 2006. Synthesis and size differentiation of Ge nanocrystals in amorphous SiO_2. *Applied Physics A* 83: 107–110.

Alivisatos, A. P. 1996. Semiconductor clusters, nanocrystals and quantum dots. *Science* 271: 933–937.

Alivisatos, A. P., G. Weiwei, and C. Larabell. 2005. Quantum dots as cellular probes. *Annual Review of Biomedical Engineering* 7: 55–76.

Araujo, L. L., R. Giulian, D. J. Sprouster et al. 2008. Size-dependent characterization of embedded Ge nanocrystals: Structural and thermal properties. *Physical Review B* 78: 094112/1–094112/15.

Aroutiounian, V., S. Petrosyan, A. Khachatryan et al. 2001. Quantum dot solar cells. *Journal of Applied Physics* 89: 2268–2271.

Atwater, H. A. 2007. The promise of plasmonics. *Scientific American* 296: 56–63.

Atwater, H. A., S. A. Maier, A. Polman, J. A. Dionne, and L. Sweatlock. 2005. The new "p-n junction": Plasmonics enables photonic access to the nanoworld. *Materials Research Society Bulletin* 30: 385–389.

Baletto, F. and R. Ferrando. 2005. Structural properties of nanoclusters: Energetic, thermodynamic and kinetic effects. *Reviews of Modern Physics* 77: 371–423.

Banfield, J. F. and A. Navrotsky. 2001. Nanoparticles and the environment. *Reviews in Mineralogy and Geochemistry* 44: 1–349.

Banfield, J. F., D. R. Veblen, and B. F. Jones. 1990. Transmission electron microscopy of subsolidus oxidation and weathering of olivine. *Contributions to Mineralogy and Petrology* 106: 110–123.

Borgia, I., B. Brunetti, I. Mariani et al. 2002. Heterogeneous distribution of metal nanocrystals in glazes of historical pottery. *Applied Surface Science* 185: 206–216.

Bostedt, C., T. van Buuren, J. M. Plitzko, T. Möller, and L. J. Terminello. 2003. Evidence for cubic phase in deposited germanium nanocrystals. *Journal of Physics: Condensed Matter* 15: 1017–1028.

British museum link. 2008. Available at http://www.britishmuseum.org/explore/highlights/highlight_objects/pe_mla/t/the_lycurgus_cup.aspx

Bruchez Jr., M., M. Moronne, P. Gin, S. Weiss, and A. P. Alivisatos. 1998. Semiconductor nanocrystals as fluorescent biological labels. *Science* 281: 2013–2016.

Brus, L. 1986. Electronic wave functions in semiconductor clusters: Experiment and theory. *Journal of Physical Chemistry* 90: 2555–2560.

Bukowski, T. J. and J. H. Simmons. 2002. Quantum dot research: Current state and future prospects. *Critical Reviews in Solid State and Materials Sciences* 27: 119–142.

Canham, L. T. 1990. Silicon quantum wire array fabrication by electrochemical and chemical dissolution of wafers. *Applied Physics Letters* 57: 1046–1048.

Conibeer, G., M. L. Green, R. Corkish et al. 2006. Silicon nanostructures for third generation photovoltaic solar cells. *Thin Solid Films* 511–512: 654–662.

Daldosso, N., M. Luppi, S. Ossicini et al. 2003. Role of the interface region on the optoelectronic properties of silicon nanocrystals embedded in SiO_2. *Physical Review B* 68: 085327/1–085327/8.

Darhuber, A. A. 1997. Lateral and vertical ordering in multilayered self-organized InGaAs quantum dots studied by high resolution x-ray diffraction. *Applied Physics Letters* 70: 955–958.

Djurabekova, F. and K. Nordlund. 2008. Atomistic simulation of the interface structure of Si nanocrystals embedded in amorphous silica. *Physical Review B* 77: 115325/1–7.

Fafard, S. 2000. Quantum dot structures and devices with sharp adjustable electronic shells. *Physica E* 8: 107–116.

Feynman, R. P. 1960. There's plenty of room at the bottom. *Engineering and Science* 23: 22–36. Available online at http://www.zyvex.com/nanotech/feynman.html

Foss, S., T. G. Finstad, A. Dana, and A. Aydinli. 2007. Growth of Ge nanoparticles on SiO_2/Si interfaces during annealing of plasma enhanced chemical vapor deposited thin films. *Thin Solid Films* 515: 6381–6384.

Fujii, M., S. Hayashi, and K. Yamamoto. 1990. Raman-scattering from quantum dots of Ge embedded in SiO_2 thin films. *Applied Physics Letters* 57: 2692–2694.

Gerion, D., N. Zaitseva, C. Saw et al. 2004. Solution synthesis of germanium nanocrystals: Success and open challenges. *Nano Letters* 4: 597–602.

Hamad, K. S., R. Roth, J. Rockenberger, T. van Buuren, and A. P. Alivisatos. 1999. Structural disorder in colloidal InAs and CdSe nanocrystals observed by x-ray absorption near-edge spectroscopy. *Physical Review Letters* 83: 3474–3477.

Hanafi, H. I., S. Tiwari, and I. Khan. 1996. Fast and long retention-time nano-crystal memory. *IEEE Transactions on Electron Devices* 43: 1553–1558.

Henini, M. and M. Bugajski. 2005. Advances in self-assembled semiconductor quantum dot lasers. *Microelectronics Journal* 36: 950–956.

Hill, T. L. 1964. *Thermodynamics of Small Systems, Parts I and II.* Benjamin: Amsterdam, the Netherlands.

Hocevar, M., P. Regreny, A. Descamps et al. 2007. InAs nanocrystals on SiO_2/Si by molecular beam epitaxy for memory applications. *Applied Physics Letters* 91: 133114/1–3.

Hochella, M. F. Jr. 2002. Nanoscience and technology: The next revolution in the Earth sciences. *Earth and Planetary Science Letters* 203: 593–605.

Jensen, J. S., T. P. L. Pedersen, R. Pereira et al. 2006. Ge nanocrystals in magnetron sputtered SiO_2. *Applied Physics A* 83: 41–48.

Jortner, J. 1992. Cluster size effects. *Zeitschrift für Physik D-Atoms, Molecules and Clusters* 24: 247–275.

Kamat, P. V. 2008. Quantum dot solar cells: Semiconductor nanocrystals as light harvesters. *The Journal of Physical Chemistry C* 112: 18737–18753.

Kartopu, G., A. V. Sapelkin, V. A. Karavanskii et al. 2008. Structural and optical properties of porous nanocrystalline Ge. *Journal of Applied Physics* 103: 113518/1–7.

Kellermann, G. and A. F. Craievich. 2002. Structure and melting of Bi nanocrystals embedded in a B_2O_3-Na_2O glass. *Physical Review B* 65: 134204/1–6.

Linnros, J. 2005. Optoelectronics: Nanocrystals brighten transistors. *Nature Materials* 4: 117–119.

Luo, J.-W., A. Franceschetti, and A. Zunger. 2008. Carrier multiplication in semiconductor nanocrystals: Theoretical screening of candidate materials based on band-structure effects. *Nano Letters* 8: 3174–3181.

Maier, S. A. and H. A. Atwater. 2005. Plasmonics: Localization and guiding of electromagnetic energy in metal/dielectric structures. *Journal of Applied Physics* 98: 011101/1–10.

Min, K. S., K. V. Shcheglov, C. M. Yang et al. 1996. The role of quantum-confined excitons vs defects in the visible luminescence of SiO_2 films containing Ge nanocrystals. *Applied Physics Letters* 68: 2511–2513.

Modrow, H. 2004. Tuning nanoparticle properties: The x-ray absorption spectroscopic point of view. *Applied Spectroscopy Reviews* 39: 183–290.

Ngiam, S.-T., K. F. Jensen, and K. D. Kolenbrander. 1994. Synthesis of Ge nanocrystals embedded in a Si host matrix. *Journal of Applied Physics* 76: 8201–8203.

Nie, S., Y. Xing, G. J. Kim, and J. W. Simons. 2007. Nanotechnology applications in cancer. *Annual Review of Biomedical Engineering* 9: 257–288.

Norris, D. J. and M. G. Bawendi. 1996. Measurement and assignment of the size-dependent optical spectrum in CdSe quantum dots. *Physical Review B* 53: 16338–16346.

Ostraat, M. L., J. W. De Blauwe, M. L. Green et al. 2001. Ultraclean two-stage aerosol reactor for production of oxide-passivated silicon nanoparticles for novel memory devices. *Journal of the Electrochemical Society* 148: G265–G270.

Pavesi, L., L. D. Negro, C. Mazzoleni, G. Franzo, and F. Priolo. 2000. Optical gain in silicon nanocrystals. *Nature (London)* 408: 440–444.

Perrey, C. R. and C. B. Carter. 2006. Insights into nanoparticle formation mechanisms. *Journal of Materials Science* 41: 2711–2722.

Pizzagalli, L. and G. Galli. 2002. Surface reconstruction effects on atomic properties of semiconducting nanoparticles. *Materials Science and Engineering B* 96: 86–89.

Porti, M., M. Avidano, M. Nafría et al. 2007. Nanoscale electrical characterization of Si-nc based memory metal-oxide-semiconductor devices. *Journal of Applied Physics* 101: 064509/1–064509/9.

Pradell, T., J. Molera, A. D. Smith et al. 2008. The invention of lustre: Iraq 9th and 10th centuries AD. *Journal of Archaeological Science* 35: 1201–1215.

Rao, R. A., R. F. Steimle, M. Sadd et al. 2004. Silicon nanocrystal based memory devices for NVM and DRAM applications. *Solid State Electronics* 48: 1463–1473.

Ridgway, M. C., G. de M. Azevedo, R. G. Elliman et al. 2006. Preferential amorphisation of Ge nanocrystals in a silica matrix. *Nuclear Instruments and Methods in Physics Research B* 242: 121–124.

Roduner, E. 2006. Size matters: Why nanomaterials are different. *Chemical Society Reviews* 35: 583–592.

Rosenthal, S. J., J. McBride, S. J. Pennycook, and L. C. Feldman. 2007. Synthesis, surface studies, composition and structural characterization of CdSe, core/shell and biologically active nanocrystals. *Surface Science Reports* 62: 111–157.

Sato, S., S. Nozaki, H. Morisaki, and M. Iwase. 1995. Tetragonal germanium films deposited by the cluster-beam evaporation technique. *Applied Physics Letters* 66: 3176–3178.

Saunders, W. A., P. C. Sercel, R. B. Lee et al. 1993. Synthesis of luminescent silicon clusters by spark ablation. *Applied Physics Letters* 63: 1549–1551.

Schmidt, B., A. Mücklich, L. Röntzsch, and K.-H. Heinig. 2007. How do high energy heavy ions shape Ge nanoparticles embedded in SiO_2? *Nuclear Instruments and Methods in Physics Research B* 257: 30–32.

Shields, A. J. 2007. Semiconductor quantum light sources. *Nature Photonics* 1: 215–223.

Singha, A., A. Roy, D. Kabiraj et al. 2006. A hybrid model for the origin of photoluminescence from Ge nanocrystals in a SiO_2 matrix. *Semiconductor Science and Technology* 21: 1691–1698.

Stewart, G. R. 1977. Size effects in the electronic heat capacity of small Platinum particles embedded in silica. *Physical Review B* 15: 1143–1150.

Takagi, H., H. Ogawa, Y. Yamazaki, A. Ishizaki, and T. Nakagiri. 1990. Quantum size effects on photoluminescence in ultra-fine Si particles. *Applied Physics Letters* 56: 2379–2380.

Takeoka, S., M. Fujii, S. Hayashi, and K. Yamamoto. 1998. Size-dependent near-infrared photoluminescence from Ge nanocrystals embedded in SiO_2 matrices. *Physical Review B* 58: 7921–7925.

Tan, H. H., K. Sears, S. Mokkapati et al. 2006. Quantum dots and nanowires grown by metal–organic chemical vapor deposition for optoelectronic device applications. *IEEE Journal of Selected Topics in Quantum Electronics* 12: 1242–1254.

Tiwari, S., F. Rana, H. Hanafi et al. 1996. A silicon nanocrystals based memory. *Applied Physics Letters* 68: 1377–1340.

Tsoukalas, D., P. Dimitrakis, S. Kolliopoulou, and P. Normand. 2005. Recent advances in nanoparticle memories. *Materials Science and Engineering B* 124–125: 93–101.

van Buuren, T., L. N. Dinh, L. L. Chase, W. J. Siekhaus, and L. J. Terminello. 1998. Changes in the electronic properties of Si nanocrystals as a function of particle size. *Physical Review Letters* 80: 3803–3806.

Vossmeyer, T., L. Katsikas, M. Giersig et al. 1994. CdS Nanoclusters: Synthesis, characterization, size dependent oscillator strength, temperature shift of the excitonic transition energy, and reversible absorbance shift. *The Journal of Physical Chemistry* 98: 7665–7673.

Wasa, K. 1992. *Handbook of Sputter Deposition Technology: Principles, Technology, and Applications*. Norwich, NY: William Andrew Inc.

Xu, Q., I. D. Sharp, C. Y. Liao et al. 2005. Germanium nanocrystals embedded in sapphire. *Materials Research Society Symposium Proceedings* 880E: 22.1–22.6.

Xu, Q., I. D. Sharp, C. W. Yuan et al. 2006. Large melting-point hysteresis of Ge nanocrystals embedded in SiO_2. *Physical Review Letters* 97: 155701/1–4.

Xu, Q., I. D. Sharp, C. W. Yuan et al. 2007. Superheating and supercooling of Ge nanocrystals embedded in SiO_2. *Journal of Physics: Conference Series* 61: 1042–1046.

Yoffe, A. D. 2002. Low-dimensional systems: Quantum size effects and electronic properties of semiconductor microcrystallites (zero-dimensional systems) and some quasi-two-dimensional systems. *Advances in Physics* 51: 799–890.

Zeigler, J. F., J. P. Biersack, and U. Littmark. 1985. *The Stopping and Range of Ions in Matter*. New York: Pergamon Press.

Zsigmondy, R. 1914. *Colloids and the Ultramicroscope*. New York: John Wiley & Sons.

24

Coupling in Metallic Nanoparticles: Approaches to Optical Nanoantennas

Javier Aizpurua

Spanish Council for Scientific Research

and

Donostia International Physics Center

Garnett W. Bryant

National Institute of Standards and Technology

24.1 Introduction

Metal nanoparticles are now being exploited extensively in materials science, optics, chemistry, biology, and medicine because they interact strongly with visible and infrared light. They have proven to be excellent optical nanoantennas, acting as robust radiation receivers, localizers, and emitters that can connect light to the nanoworld. The optical properties of metallic nanoparticles are governed mainly by the excitation of electromagnetic surface modes, so-called surface plasmons. These modes are maintained by resonant oscillations of the surface charge density at the boundaries of the metal nanoparticle (see Figure 24.1). For simple metal nanoparticles, such as nanospheres or nanorods, the basic plasmon mode is a dipole charge oscillation, as shown in Figure 24.1, just as it would be in a traditional radio wave antenna.

The study and application of metal nanoparticle plasmonics have exploded in the last decade. The nanoparticle plasmon response can be tuned by the choice of the metal, the particle size, and the particle shape. This tunability makes metal nanoparticle plasmonics attractive for many applications. Metal nanoparticle plasmons can respond strongly to light because the metal conduction electrons near the surface contribute collectively to the response. Most importantly, the electromagnetic fields that are generated when plasmons on a metal nanoparticle couple to light can be greatly enhanced near the nanoparticle. Moreover, these fields can be squeezed and localized down to nanometer dimensions on the scale of the nanoparticle geometry, as illustrated in Figure 24.1a. The nanoscale focusing and localization

of light make metal nanoparticle plasmonics an important optics tool for use in probing and sensing the nanoscale objects. Such nanoscale optical probing is not possible with traditional optics, which is constrained to wavelength scale resolution by the diffraction limit of light.

Complex metallic structures made from coupled nanoparticles support more complex plasmon modes that result from mixing and distorting the intraparticle plasmon modes (see Figure 24.1b). This coupling can be used to further tune the emission, absorption, and directionality of light, especially on the nanoscale. As a result of the distortion of the intraparticle modes by the coupling, charge oscillations can be strongly localized and amplified where particles are tightly coupled, leading to even greater field enhancement and field localization for use in probing the nanoworld.

Ritchie first introduced the concept of surface plasmons when he predicted the existence of these modes in metallic films excited by fast electron beams [1]. However, the strong optical response and antenna properties provided by plasmons in metal nanoparticles have actually been exploited for many centuries, long before any knowledge of their existence. In the fourth century, Romans would mix gold with molten glass when making ornamental cups. Strong absorption and scattering of light around visible wavelengths of 520 nm by gold nanoparticles, which were acting as color centers in the glass, produced the ruby color of the glass. The stunning reds in stained glass found in cathedrals of the Middle Ages came from mixing gold with the glass. At that time, artisans would tune the optical response by empirically modifying the colloidal gold–glass mixtures, without knowing that they were controlling nanoparticle size and shape.

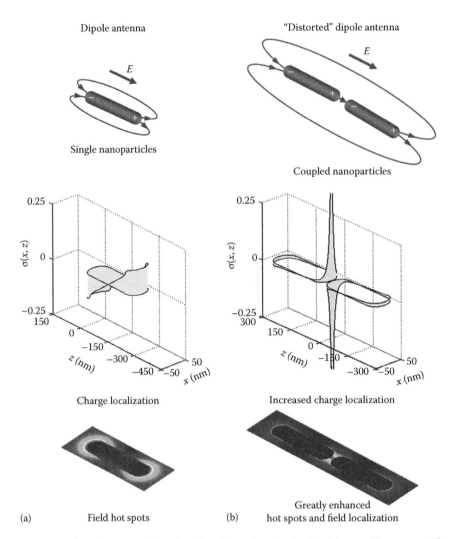

FIGURE 24.1 The fundamental surface plasmon modes and surface charge density $\sigma(x, z)$ in (a) a metallic nanoparticle with rod-like shape and (b) a pair of metallic particles with rod-like shape aligned along their axis, where the localization and enhancement of the surface charge density results from the coupling between the two rods. A schematic of the charge oscillation for the basic dipolar mode and the actual surface charge density are shown. Images of the corresponding near fields are displayed. Fields are greatly enhanced near the ends of the nanoparticles and in gaps between nanoparticles.

In the mid-nineteenth century, Faraday pointed out that the colloidal gold solutions most likely contained tiny gold particles responsible for their optical properties. In the early twentieth century, Mie developed the first description of the surface modes of spherical nanoparticles to describe optical scattering by small particles [2]. At that time, studies were focused on the absorption and scattering measurements of colloidal dispersions and cosmic dust, with little control of the structure and morphology of the nanoparticles involved. From the 1960s to the 1990s, many studies of surface plasmon excitations in nanoparticles involved nanoparticles interacting with charged beams (fast electrons mainly) [3]. Such experiments were meant to better understand the plasmonic response of small particles.

During the last two decades, the development of the capability to create metallic nanoparticles on demand has turned metal nanoparticle plasmonics into a powerful nanooptics tool for materials science, communications, biochemistry, and medicine.

Because of the development of additional sophisticated chemical [4–7] and lithographic [8–10] nanoparticle synthesis methods, nanoparticle plasmonics has advanced beyond the exploitation of single nanoparticles toward the use of complex structures, with greatly enhanced response due to interparticle coupling [11].

To highlight this rapid development of metal nanoparticle plasmonics, we point to some of the key applications. For example, the use of metal nanostructures with a "tuned" optical response and high local fields that can be employed to excite strongly nearby molecules has revolutionized field-enhanced spectroscopies such as surface-enhanced Raman spectroscopy [12] and surface-enhanced infrared absorption [13]. In both of these spectroscopies, coupled metallic nanoparticles act as very sensitive optical nanoantennas, pushing detection down to the single-molecule limit while maintaining specificity.

Chains of coupled metal nanoparticles, as shown in Figure 24.2, can be used to transport optical excitations on the nanoscale,

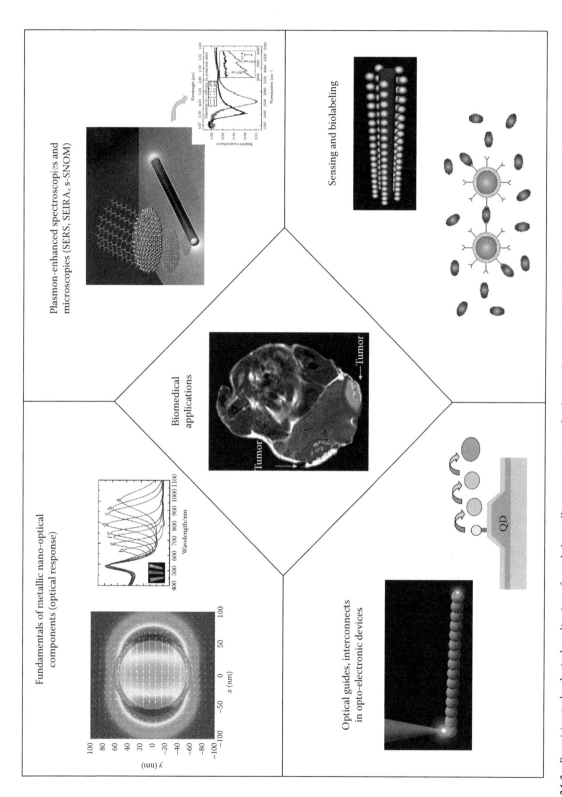

FIGURE 24.2 Promising technological application of coupled metallic nanostructures: fundamental nanooptics, field-enhanced spectroscopies and microscopies, communications, sensing and biolabeling, energy transfer, and biomedical applications. (Reproduced from (image in the center) Nanospectra Biosciences Inc., Houston, TX, http://www.nanospectra.com/images/tumor.jpg. With permission.)

breaking the diffraction limit for light propagation by 100-fold. Plasmonic guides, made from nanoparticles or combined with them, will provide controlled transmission for manipulating and switching optical signals on the nanoscale [14]. Nanoparticle plasmonics could further provide an effective interface connecting the rapid transfer of information via optics to the processing power of semiconductor electronics.

In addition, advanced materials composed of strongly coupled metal nanostructures are providing the first wave of metamaterials with extraordinary optical properties, such as negative refraction [16] and superlensing. However, the potential impact of nanoparticle plasmonics may be greatest in biological and medical applications. Biomolecules labeled with a metallic nanoparticle can be easily traced optically during biophysical processes, providing detailed information about the operation of molecular engines in organelles transport within the cell or the flow of energy during fluorescence resonance energy transfer [17]. In medical applications, functionalized nanoparticles with a conveniently tuned optical response are being used to signal the presence of cancer cells. Diagnosis is possible even in the early stages of a cancer because the nanoparticles provide a strong optical signal. At the same time, strong optical absorption by the nanoparticle makes it ideal as a tool for cancer therapy [18]. Optical heating of the functionalized nanoparticles provides a highly efficient way to kill attached cancer cells.

With this intense interest in nanoparticle plasmonics in mind, the goal of this chapter is to review and elucidate the principles that govern the plasmonics of metal nanoparticles. We will highlight the optical response that makes metal nanoparticles a compelling choice for the optical antennas of the nanoworld. We will especially stress the complex structures that physicists, chemists, and biologists are now beginning to investigate, where interparticle coupling plays a key role and allows the extreme control and localization of electromagnetic fields needed for nanoscale optics and sensing.

24.2 Background

Coupling light to plasmon excitations in matter is only possible for special geometries of the material. Light cannot couple directly to plasmon excitations of a flat, semiinfinite metal surface because energy and momentum cannot be conserved simultaneously. This is shown schematically by the solid and dashed dispersion lines for light and the surface plasmon, respectively, in Figure 24.3. There are several ways to provide the additional momentum needed for momentum conservation, so that the surface plasmon can couple to incident light. One way is to modify the planar metal surface by means of indentations or gratings that can provide "lattice" momentum to ensure momentum conservation. This is the approach followed when arrays of metal particles or holes in a surface are used to define novel metamaterials.

Momentum conservation is also possible when light is coupled to the plasmon excitations of a small metal particle. Here, the finite geometry provides the extra momentum needed for

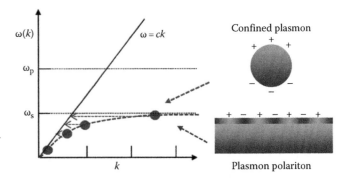

FIGURE 24.3 Energy ($\hbar\omega$) versus momentum (k) dispersion line of a surface plasmon propagating on a semi-infinite metal surface (dashed line). Closing the boundaries of a surface provides momentum (schematically shown through solid dots and dashed arrows) so that the localized plasmon can couple to the light line ($\omega = ck$). The bulk plasma frequency ω_p and the surface plasmon energy $\omega_s = \omega_p/\sqrt{2}$ are marked as dotted horizontal lines.

momentum conservation. Discrete excitations, known as metal nanoparticle plasmons, that can couple effectively to light are possible, as schematically shown by the dashed arrows in Figure 24.3. The finite geometry of the metal nanoparticles is essential for their enhanced optical response. The finite geometry allows coupling to external light; it provides a means to tune the energies of the plasmon excitations; and it leads to strong localization and enhancement of fields near sharp corners and points of the particles and in narrow gaps between particles. These are the key features of metal nanoparticles that are now being extensively exploited in nanooptics.

This section briefly reviews the basis for the optical response of simple nanoparticle systems that are the building blocks of more sophisticated plasmonic antenna structures. This also discusses how the optical response of the simple plasmonic antennas can be tuned based on (1) particle size, (2) particle shape, (3) coupling between particles, and (4) coupling to the environment.

The standard approach for describing the optical response of nanoparticles is the classical dielectric formalism [19,20], where the material is described, within linear optical response, by a bulk local dielectric function $\varepsilon(\omega)$. Boundary conditions for the potentials and fields at the particle surfaces establish the criteria that must be satisfied by the surface modes. The spatial distribution of the field associated with the excitation of a surface plasmon mode governs the characteristics of the near field and far field.

Two models will be used for $\varepsilon(\omega)$. In some of the examples discussed, the metal dielectric function will be characterized by a Drude-like function, which describes the response of the free-electron gas of the conduction band in metals:

$$\varepsilon = 1 - \frac{\omega_p^2}{\omega(\omega + i\gamma)} \qquad (24.1)$$

for frequency ω (i.e., energy $\hbar\omega$ and wavelength $\lambda = 2\pi c/\omega$ with c the speed of light). The bulk plasmon frequency of the metal ω_p is determined by the electronic density n of the material according to $\omega_p = \sqrt{4\pi n e^2/m_e}$, with e and m_e the charge and mass of the electron, respectively. γ is the plasmon damping in the metal. The Drude-like dielectric function resembles very well the response of a nearly free electron gas, as is the case for aluminum. For noble metals commonly used in plasmonics, such as gold and silver, the response has to be modified by the use of an additional contribution to the dielectric function that reproduces the effect of interband transitions. For other examples in this chapter, we will use empirical dielectric constants $\varepsilon(\omega)$ obtained from optical measurements to describe more accurately the full bulk dielectric response of the metal [21,22].

24.2.1 Effects of Particle Size: A Single Sphere

The most basic metal structure that can be used as an optical antenna is a simple, single-metal nanoparticle (often referred to as a dipolar particle). As pointed out in Section 24.1, the study of the optical response of spherical particles dates back to the beginning of the twentieth century. The electromagnetic fields for scattering by a sphere were obtained exactly by Mie in 1908 [2]. To gain insight, we first discuss here the simple case of a very small spherical particle of radius a, in which the electrostatic limit holds. In this limit, the particle is so much smaller than the wavelength of the incident light that the field is essentially constant in the particle, and electrostatics applies. We divide space into two regions, inside (in) and outside (out) the sphere, and expand the electrostatic potential Ψ for the field in each region in spherical coordinates (r, θ):

$$\Psi^{in}(r,\theta) = \sum_{l=0}^{\infty} \left(\frac{r}{a}\right)^l A_l P_l(\cos(\theta)) \quad \text{for } r < a, \qquad (24.2)$$

$$\Psi^{out}(r,\theta) = \sum_{l=0}^{\infty} \left(\frac{a}{r}\right)^{l+1} B_l P_l(\cos(\theta)) \quad \text{for } r > a. \qquad (24.3)$$

where A_l and B_l are the coefficients of the expansion and the $P_l(\cos(\theta))$ is the Legendre polynomials of the order l. When we apply the boundary conditions for the continuity of the potentials and the displacement field at the surface $r = a$, we obtain the following set of equations for each l term:

$$A_l = B_l \qquad (24.4)$$

$$\varepsilon_{sph} l A_l = -\varepsilon_{med}(l+1) B_l \qquad (24.5)$$

where ε_{med} and ε_{sph} are the dielectric constants of the surrounding medium and the sphere, respectively. By substituting Equation 24.4 into Equation 24.5, we obtain the dispersion relation for the electrostatic modes of a sphere: $\varepsilon_{sph} l + \varepsilon_{med}(l + 1) = 0$. If the dielectric response of the sphere ε_{sph} is described by a metallic Drude dielectric function (Equation 24.1), we obtain the l-polar

frequencies ω_l of the nonretarded surface modes of a metallic sphere with bulk plasma frequency ω_p:

$$\omega_l = \sqrt{\frac{l}{\varepsilon_{med} l + l + 1}} \omega_p \qquad (24.6)$$

This simple electrostatic approach is valid for very small particles ($a < 10$ nm). As the nanoparticle size increases, retardation effects due to the finite speed of light must be accounted for. When retardation is included, the modes are obtained from Maxwell's equations instead of the Laplace equation that describes electrostatics. The mode energy is reduced by retardation effects as the particles become larger. The exact solution for the electromagnetic modes, showing size-dependent plasmon energies, is obtained by solving the dispersion relations that include retardation effects:

$$k_{sph}^2 j_l(k_{sph}a)[k_{med}a \cdot h_l^{(1)}(k_{med}a)]' - k_{med}^2 h_l^{(1)}(k_{med}a)[k_{sph}a \cdot j_l(k_{sph}a)]' = 0 \qquad (24.7)$$

where $k_{med} = \omega/c\sqrt{\varepsilon_{med}}$ and $k_{sph} = \omega/c\sqrt{\varepsilon_{sph}}$. We plot in Figure 24.4b the evolution of these modes with the size of a metal sphere (radius a) characterized by a Drude-like response function (dependent on the plasma frequency ω_p). The first five multipolar modes are displayed in the figure.

For particles larger than a few nanometers, the mode energy is shifted noticeably from the electrostatic limit, reinforcing the need for a complete, fully retarded formulation of scattering by nanoantennas to obtain the correct energies (wavelengths) for the resonant modes.

The optical response of a metal nanosphere to a linearly polarized, incident, plane wave is displayed in Figure 24.4c. The shift of the energy of the peak response with size follows the trend of the plasmon modes. The balance between scattering and absorption processed is modified as the particle becomes larger, with larger scattering cross sections for larger particles. Essentially, absorption scales as the volume of the nanoparticle. Since scattering involves absorption followed by reemission, it scales as the square of the particle volume. Thus, scattering dominates the absorption for particles larger than approximately 5–10 nm.

Peaks in the optical response are related to the excitation of optically active plasmon modes (i.e., those modes with dipole moments along the polarization of the incident field). Extinction, scattering, and absorption peaks appear at lower energy (longer wavelength) for larger particles. For small particles ($a = 5$ nm), the only distinguishable peak is the dipolar peak at $\omega_{l=1} = \omega_p/\sqrt{3}$. As the particles increase in size, higher order peaks become apparent in the extinction spectrum. These multipolar peaks are clearly distinguishable here because of the ideal Drude response and the small damping used ($\gamma = \omega_p/10$). These higher order peaks are not so obvious in extinction spectra for gold and silver spherical particles. For gold and silver nanoparticles, the multipolar peaks are damped due to the more realistic, lossy response of the metal.

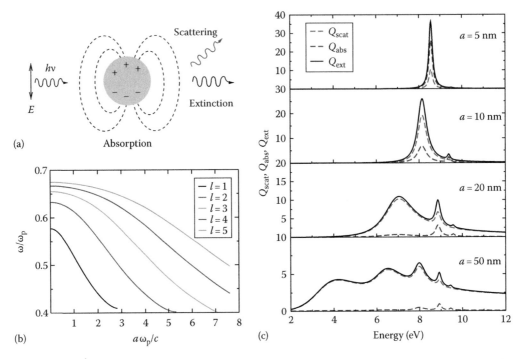

FIGURE 24.4 (a) Schematics of the scattering, absorption, and extinction processes in a metal nanosphere. (b) Surface plasmon modes of a metal sphere as a function of particle radius a. (c) Scattering, absorption, and extinction coefficients Q_{scat}, Q_{abs}, and Q_{ext} of metal particles of different sizes (radii a displayed in the insets). The bulk plasmon frequency is $\omega_p = 15.1$ eV, and the metal damping $\gamma = 0.15$ eV. Excitation of the dipolar mode ($l = 1$) is clear for small particles. Higher order modes ($l = 2, 3, 4,...$) are excited in larger particles. Scattering is much larger than the absorption in large particles.

24.2.2 Effects of Particle Shape

Surface plasmons are strongly affected by the particle shape. Spheroidal geometry is the simplest extension from spherical geometry. In the nonretarded limit for the optical response, Gans theory expresses the polarization of spheroids in terms of three depolarization factors P_x, P_y, and P_z for the spheroid axes x, y, and z, respectively (see inset in Figure 24.5). The parameter governing the optical response of a spheroid with long semiaxis L_z and short semiaxes $L_x = L_y$ in the electrostatic approach is the aspect ratio of the spheroid, L_x/L_z. If we express the ellipticity e of the spheroid in terms of this aspect ratio as $e = \sqrt{1 - (L_x/L_z)^2}$, the depolarization factors can be expressed as [23]

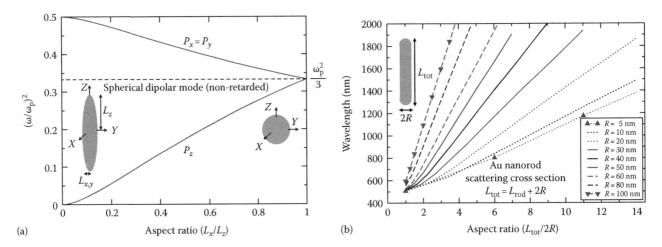

FIGURE 24.5 (a) Electrostatic dipolar modes of a spheroid as a function of aspect ratio between the short and the long semiaxis (L_x/L_z) for a Drude dielectric response of the spheroid. The limit of the spherical surface plasmon is recovered for $L_x/L_z = 1$. Longitudinal (P_z) and transverse ($P_x = P_y$) modes tend to zero and $\omega_p = \omega_s/\sqrt{2}$, respectively, as the aspect ratio tends to zero. (b) Full electrodynamical calculation of the longitudinal dipolar resonance position for a gold nanorod with hemispherical ends. The length of the rods is L_{total} and the width is $2R$. The aspect ratio for the rod is $L_{tot}/2R$.

$$P_z = \frac{1-e^2}{e^2}\left[\frac{1}{2e}\ln\left(\frac{1+e}{1-e}\right)-1\right] \qquad (24.8)$$

$$P_x = P_y = \frac{1-P_z}{2} \qquad (24.9)$$

The dipolar modes along each axis direction *x, y,* and *z* can be obtained for a lossless material from the following equation:

$$\frac{\varepsilon_{\text{spheroid}}}{\varepsilon_{\text{med}}} = \frac{P_j - 1}{P_j} \qquad (24.10)$$

where

j is a particular polarization direction *x, y,* or *z* of the incident field

$\varepsilon_{\text{spheroid}}$ and ε_{med} are the dielectric functions of the spheroid and the surrounding medium, respectively

For a metal spheroid described by a Drude-like function, the solutions for the dipolar-like surface modes for polarization along the *x, y,* and *z* directions ω_x, ω_y, and ω_z are directly related to the depolarization factors by $(\omega_i/\omega_p)^2 = P_i$. The curve in Figure 24.5a shows the modification of the modes as a function of the spheroid aspect ratio (L_x/L_z) in the nonretarded limit. For a sphere ($L_x/L_z = 1$), $P_x = P_y = P_z = 1/3$ and the electrostatic limit for a sphere, $\omega_x = \omega_y = \omega_z = \omega = \omega_p/\sqrt{3}$, is recovered.

In the electrostatic limit, the longitudinal mode shows a linear dependence of the resonance wavelength on the particle aspect ratio, independent of the actual particle size. This simple relation with the mode energy depending only on the particle shape holds only in the electrostatic limit. The spectral position and weight of a mode excited by a plane wave actually depends on both the particle geometry and the particle size, that is, on both the aspect ratio and the particle size.

In Figure 24.5b, we show the spectral resonances for gold nanorods with hemispherical ends from a fully retarded calculation. The dipolar resonance depends nearly linearly on aspect ratio but it also depends strongly on the actual particle size [24]. This complicated dependence on both size and shape must be accounted for to correctly map out the modes of even simple metal nanoparticles. More complicated shapes of nanoparticles can lead to further redshifts of the optical response. Sharper edges and tips in the nanoparticles, as in nanorice [25], nanocusps [26], nanocrescents [27,28], and nanostars [29] provide significant redshifting. In all these nanoparticles, the surface charge density is localized at the particle edges. In this case, the optical response is a combination of the plasmonic response of the material and the lightning rod effect due to the sharp features in the particle geometry.

24.2.3 Coupling between Particles

The field enhancement possible near a single metal nanoparticle is limited, even when the particle shape is optimized to enhance the local field. Typically, near fields close to a single metal particle are enhanced by factors of 10. A very effective way to further

tune the response and significantly increase the near fields is to couple several metal nanoparticles across small interparticle gaps, as pointed out in Section 24.1 (Figure 24.1). The coupling between the surface charge densities of different particles via Coulomb interaction produces large field enhancement of 100 or more when the particles are close. The Coulomb coupling also provides additional spectral tuning and control over nanoscale localization of the near fields. This coupling provides for effective engineering of the near fields, which is of great use in field-enhanced spectroscopies, sensing, and microscopy.

We illustrate the main effect of coupling between metal particles on the optical response (energy of the resonances) by following Schmeits and Dambly [30]. For the electrostatic limit, we present the equation of modes that governs the energy of the surface modes of two coupled nanospheres as a function of separation distance between the particles *d*. Following the schematic in Figure 24.6a, we describe the potentials $\Psi^{(i)}(r, \theta, \varphi)$ at each region *i* of space in the presence of two spherical particles of radius *a* and *b*, respectively, according to

$$\Psi^{(1)}(r,\theta,\varphi) = \sum_{l=0}^{\infty}\sum_{m=-l}^{m=+l}\left(\frac{r_1}{a}\right)^l A_{lm}Y_{lm}(\theta_1,\varphi)e^{im\varphi} \quad \text{for } r_1 < a, \quad (24.11)$$

$$\Psi^{(2)}(r,\theta,\varphi) = \sum_{l=0}^{\infty}\sum_{m=-l}^{m=+l}\left(\frac{r_2}{b}\right)^l D_{lm}Y_{lm}(\theta_2,\varphi)e^{im\varphi} \quad \text{for } r_2 < b, \quad (24.12)$$

$$\Psi^{(3)}(r,\theta,\varphi) = \sum_{l=0}^{\infty}\sum_{m=-l}^{m=+l}\left[\left(\frac{a}{r_1}\right)^{l+1}B_{lm}Y_{lm}(\theta_1,\varphi)+\left(\frac{b}{r_2}\right)^{l+1}C_{lm}Y_{lm}(\theta_2,\varphi)\right]e^{im\varphi}$$

$$(24.13)$$

where 1 and 2 refer to the regions inside the two particles and 3 refers to the region outside the particles. Y_{lm} are the spherical harmonics, and the subindexes of the radial and spherical angle coordinates $r_{1,2}$ and $\theta_{1,2}$, respectively, refer to particles 1 and 2. *m* is the azimuthal number describing the rotational symmetry.

The continuity conditions for the potential and the displacement field at the first sphere of radius *a* lead to the following matrix equations for each *m*

$$(A_l^m) = (D_l^m) + \sum_{k=m}^{l_{\max}}(P_{kl}^m)(C_k^m) \qquad (24.14)$$

$$(C_l^m) = \sum_{k=m}^{l_{\max}}(M_{kl}^m)(B_k^m) \qquad (24.15)$$

where A_l^m, B_l^m, C_l^m, and D_l^m are the $(l - m)$th row of the column matrices *A, B, C,* and *D* with $(l_{\max} - m)$ rows, being $l_{\max}$ the cutoff for the number of the multipolar terms included, and

$$P_{kl}^m = \frac{\alpha_{lm}}{\alpha_{km}}\frac{(l+k)!}{(l-m)!(k+m)!}\frac{b^{l+1}a^k}{(a+b+d)^{l+k+1}}, \qquad (24.16)$$

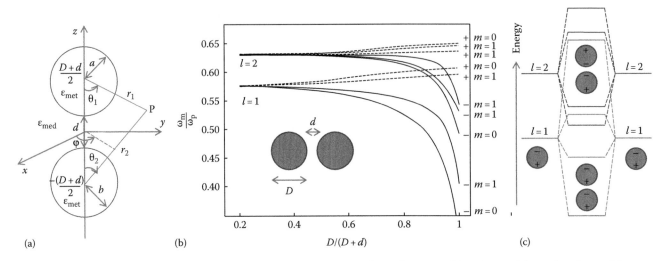

FIGURE 24.6 (a) Schematic of a dimer with the coordinates and parameters defining the system. (b) Evolution of the modes of a metal dimer ω_m, the metal characterized by a plasma frequency ω_p, as a function of the ratio between the particles size (D) and the separation distance (d). (c) Schematic of the hybridization of modes between two metallic nanoparticles. The polarization of the bonding (–) and antibonding (+) modes for coupling of the $l = 1$, $m = 0$ modes are drawn. Higher order modes coming from the hybridization of $l = 2$ modes are shown with darker lines.

$$M_{kl}^m = \frac{\alpha_{lm}}{\alpha_{km}} \frac{(\varepsilon_{med} - \varepsilon_{met})k}{k\varepsilon_{met} + (k+1)\varepsilon_{med}} \frac{(l+k)!}{(l-m)!(k+m)!} \frac{a^{l+1}b^k}{(a+b+d)^{l-k+1}} \tag{24.17}$$

are the kl elements of the matrices P and M connecting the terms with different multipolar order k and l ($l, k > m$). $(a + b + d)$ is the center-to-center particle separation, and

$$\alpha_{lm} = \left(\frac{2l+1}{4\pi} \frac{(l-m)!}{(l+m)!} \right)^{1/2} \tag{24.18}$$

The continuity conditions at the second sphere of radius b lead to a similar set of equations:

$$(D_l^m) = (C_l^m) + \sum_{k=m}^{l_{max}} (Q_{kl}^m)(B_k^m) \tag{24.19}$$

$$(B_l^m) = \sum_{k=m}^{l_{max}} (N_{kl}^m)(C_k^m) \tag{24.20}$$

where Q and N are matrices with similar expressions as P and M, but with the radius a and b permuted. From Equations 24.14, 24.15, 24.19, and 24.20, we obtain the matrix equation of the full system $(MN - 1)C = 0$. The condition for the existence of a nontrivial solution for the surface modes is given by

$$\det(MN - 1) = 0 \tag{24.21}$$

For the coupling of the dipolar modes ($l = 1$, $m = 0$) of two metallic particles with the identical diameter $D = 2a$, and with a separation distance d of the gap between the surfaces, characterized by a Drude-like dielectric function with plasma frequency ω_p, we obtain the frequency of the nonretarded mode as a function of the dimensionless parameter ($D/D + d$) that relates the center-to-center separation of the two particles $D + d$, with the particle diameter:

$$\left(\frac{\omega_{m=0;l=1}}{\omega_p} \right)_\pm^2 = \frac{1}{3} \left(1 \pm 2 \left(\frac{D/2}{D+d} \right)^3 \right) \tag{24.22}$$

In Figure 24.6b, we show the behavior of the dipolar mode ($l = 1$; $m = 0$) together with the evolution of other higher modes that are solutions of Equation 24.21 for different values of m and l. For each l and m, both symmetric (–, low energy) and antisymmetric (+, high energy) branches for coupled modes are obtained (see Figure 24.6b). The polarization scheme of these symmetries is displayed in Figure 24.6c.

The energy shifts produced by the coupling have a simple interpretation for small particles. This coupling of modes in metal structures has been recently described in the nonretarded approximation in the framework of what has been called the plasmon hybridization model [31]. By means of a Lagrangian description of the mechanical oscillations of the coupled electron plasma, the equation of surface modes becomes formally similar to the Schrödinger equation. In this description, the coupled plasmon modes result from the hybridization of the plasmon modes of the individual particles in the structure. These coupled modes can be described as bonding and antibonding modes, and the formalism is often referred to as plasmon chemistry because of the analogy to molecular orbital theory in chemistry [32].

In Figure 24.6c, we display the polarization of the first symmetric (–, bonding) and antisymmetric (+, antibonding) modes, which come from the hybridization of the $l = 1$, $m = 0$ modes of the original spherical nanoparticles. The symmetric mode, with charge of opposite sign piling up at the cavity, is the lowest energy mode. The antisymmetric mode, with charge of the same sign piling up at the cavity, is pushed up in energy, and is poorly excited by light, because the antisymmetric mode has zero dipole moment.

We have mentioned that the near fields in interparticle gaps can be strongly localized and squeezed down to nanoscale dimensions. This is one of the key features that makes coupled nanoparticles such useful nanoantennas. This enhancement can be understood in a coupled system easily in the electrostatic limit using a simple geometrical argument. In the electrostatic limit, we relate the incident field E_{inc} to the drop in potential. We consider two coupled particles separated by $D + d$ from center to center, where D is the diameter of each particle and d is the interparticle gap. In the absence of the particles, the potential drop ΔV across the distance $D + d$ is related to the incident field by $\Delta V = E_{inc} \cdot (D + d)$. When a pair of particles is present, the same drop in potential occurs just across the gap between the particles d; therefore, the local field E_{loc} between the particles is enhanced. Therefore, we can relate the incident field to the local field:

$$\Delta V \approx |E_{loc}|\,d \approx |E_{inc}|\,(D + d) \qquad (24.23)$$

Thus,

$$\left|\frac{E_{loc}}{E_{inc}}\right| \approx \frac{D + d}{d} \qquad (24.24)$$

With this simple approach, we can estimate the field enhancement in the gap as a function of D/d.

In Figure 24.7, we plot this simple geometrical expression together with full electromagnetic calculations of the field enhancement in the middle of a gap [33]. The agreement is outstanding for a wide range of separation distances.

24.2.4 Environment

When metal particles are embedded in a dielectric medium, the restoring force of the plasmon oscillations is reduced by the screening provided by the environment. As a result, the energy of the plasmon is lower (longer wavelength). This wavelength shift can be used for sensing because the relative shift of the plasmon is a measure of the dielectric function of the material surrounding the metal structures. In this way, it is possible to identify the presence of certain biomolecules by measuring with extinction spectroscopy the wavelength shifts of the localized surface plasmons [34]. When more sophisticated particles, such as nanorings, or coupled particles are used, the sensing capabilities of the nanoparticles can be improved [35]. In these sensor applications, the sensitivity of the measurement is determined by the wavelength shift of the plasmon per unit of refractive index change of the surrounding medium. In coupled systems with highly localized hot sites, that is, in regions of high near field that govern the response of the coupled system, environmental screening in the hot spot suffices to redshift the plasmon response of the system as if the whole structure was covered by the dielectric. This effect is shown in Figure 24.8. We present the far-field response of a system of two coupled metal nanoshells with a hot spot between the particles [36].

Covering the nanoshells with a molecular layer modeled as a dielectric with value $\varepsilon = 2$ redshifts the response of the coupled system from $\lambda = 830$ nm (black curve) to 950 nm (dark gray curve). Screening only a small portion of the area connecting both particles is enough to produce about 90% of the redshift ($\lambda = 935$ nm, light gray curve). This strong sensitivity stresses the importance of the local environment if the screening occurs in a region where the field is localized. This effect allows for the selective sensing of just a few molecules.

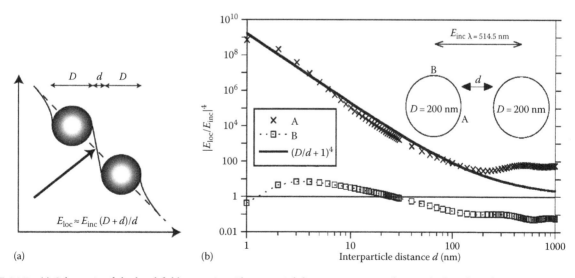

(a)   (b)

FIGURE 24.7 (a) Schematic of the local-field squeezing. The potential drop occurs across the gap d when the spheres are present. (b) Field enhancement to the fourth power produced by an incident plane wave linearly polarized along the silver dimer axis at a wavelength of 514 nm derived from the simple geometrical argument in (a) and compared with the values of the full electrodynamical calculation at 0.5 nm from the surface at the center of the gap (A). Values outside the cavity (B) are also shown. This simple rule provides a useful estimate for the field enhancement in coupled metal dimers. (Adapted from Xu, H.X. et al., *Proc. SPIE*, 4258, 35, 2001. With permission.)

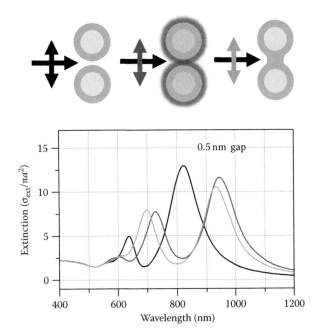

FIGURE 24.8 Extinction cross section of a gold nanoshell dimer separated by 0.5 nm (black curve). The cases where the nanoshells are totally covered by a screening layer of thickness 2 nm (dark gray curve) and only the hot spot at the cavity is covered by the same material (light gray curve) are also displayed. All the spectra are normalized to the geometrical area of one of the nanoshells with outer radius *a*. (Reprinted from Lassiter, J.B. et al., *Nano Lett.*, 8, 1212, 2008. With permission.)

24.3 Coupling in Realistic Nanoantennas

Coupling is a key approach for tuning the optical response of nanoantennas and engineering the field enhancement around them. We analyze now several antenna concepts in which different types of coupling are used to determine tuning, nanoscale field localization, and local field enhancement. Most of the calculations presented in this chapter are done using the boundary element method [37] to solve Maxwell's equations.

24.3.1 Intraparticle Coupling

The coupling between plasmons localized on different surfaces of the same metal particle has become a powerful tool to tune the spectral response of single scatterers. Two important examples of this intraparticle coupling are gold nanoshells and gold nanorings. In these structures, the coupling between plasmons located on the inner and outer walls of the structure produces an energy shift of the modes, which depends mainly on the separation between inner and outer walls. Just as discussed previously for single-metal spheres and dimers, a simple approximation for the energy of the surface modes in nanoshells can be obtained in the electrostatic limit. For nanoshells, the Laplace equation is solved by imposing the boundary conditions to the potentials and the displacement fields at both the outer radius r_{out} and the inner radius r_{int} of the nanoshell. By doing so, we obtain the

following dispersion relation for a nanoshell particle made with a metal shell, dielectric function ε_{met}, covering a core, dielectric function ε_{core}, with the particle surrounded by a medium with dielectric function ε_{med}:

$$l(l+1)\left(\frac{r_{int}}{r_{out}}\right)^{2l+1} = \frac{[l\varepsilon_{met} + (l+1)\varepsilon_{core}] \cdot [(l+1)\varepsilon_{met} + l\varepsilon_{med}]}{[\varepsilon_{met} - \varepsilon_{med}] \cdot [\varepsilon_{met} - \varepsilon_{core}]} \quad (24.25)$$

The solutions to this equation, for the metal shell characterized by a lossless Drude-like dielectric function $\varepsilon_{met} = 1 - \omega^2/\omega_p^2$ and with $\varepsilon_{med} = \varepsilon_{core} = 1$, are the surface plasmon modes of order *l* of a hollow shell:

$$\omega_{l\pm}^2 = \frac{\omega_p^2}{2}\left[1 \pm \frac{1}{2l+1}\sqrt{1 + 4l(l+1)\left(\frac{r_{int}}{r_{out}}\right)^{2l+1}}\right] \quad (24.26)$$

For every *l* mode, we obtain a symmetric (−) and an antisymmetric (+) solutions. The symmetric (antisymmetric) solution involves bonding (antibonding) surface charge density oscillations with charge of the same (opposite) sign piling up on the adjacent sides of the inner and outer walls [31]. We plot the energy of these modes as a function of r_{int}/r_{out} in Figure 24.9a. The symmetric (−) solution of the dipolar mode (*l* = 1) is the mode that is most strongly excited by incident light, generating a dipolar-like excitation with strong coupling between the plasmons on adjacent walls. We observe in Figure 24.9a the evolution of higher order modes ω_l as a function of the nanoshell thickness. These mode energies converge to the surface plasmon energy $\omega_p/\sqrt{2}$ as *l* increases. For a gold nanoshell, the energy of the modes varies over a wide range of the spectrum, spanning from the visible to the near-infrared, when the shell geometry is varied. This controlled coupling of inner and outer walls provides a powerful tool for tuning the response of these nanoparticles. The scattering coefficient for the shells of different thickness (different aspect ratio) is shown in Figure 24.9b. An outer radius of 50 nm has been used for this example. The inner radius has been varied from 10 up to 48 nm. As the inner radius approaches the outer radius, the energy of the modes decreases, that is, the plasmon wavelength is redshifted. A large redshift occurs for nanoshell thickness smaller than 10 nm ($r_{int} = 40$ nm and larger). This is the range of shell thicknesses where the intraparticle coupling becomes most effective as a tool to control the optical response.

To exploit this tool, precise fabrication techniques are needed to control the growth of the nanoshells. Control over such small interaction distances (<10 nm) can be obtained by chemical deposition methods, in which the growth of the shell can be monitored almost layer by layer [6].

Other structures with intraparticle coupling that controls the spectral response include metal thin films and nanorings [38]. In thin films, coupling between the top and bottom surfaces of the film generates bonding and antibonding solutions. For a nanoring, the coupling is between plasmons on the

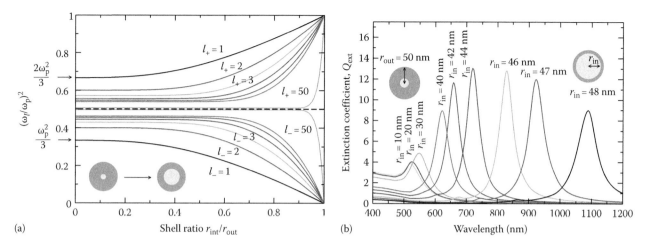

FIGURE 24.9 (a) Energy of the modes of a metal nanoshell as a function of the inner and outer radius of the shell obtained in the electrostatic approach. A redshift of the symmetric modes (−) is observed as the shell becomes thinner ($r_{int}/r_{out} \rightarrow 1$). (b) Full electrodynamical calculation of the extinction factor ($Q_{ext} = \sigma_{ext}/\sigma_{geom}$) of a gold nanoshell with an outer radius $r_{out} = 50$ nm and varying inner radius from $r_{int} = 10$ nm to $r_{int} = 48$ nm.

inner and outer walls of the ring. The couplings in nanorings and nanoshells give similar energy shifts. The main difference between nanorings and nanoshells is the cylindrical symmetry of the charge densities in the nanorings. Strong field enhancement in the vicinity of the inner and outer walls of a nanoring or nanoshell occurs near the plasmon resonance. We show top and side cross-sectional views of the field enhancement in the vicinity of a gold nanoring at resonance in Figure 24.10. The field enhancement is localized close to the walls of the ring and can be a factor of 15. A nanoring could serve as a nanocontainer that would enhance the sensitivity of molecular spectroscopy done on molecules deposited in it.

The tunable, enhanced response of these nanostructures has found extensive use in field-enhanced spectroscopies and biomedical applications, such as cancer therapy. Cancer therapy using nanoshells relies on the resonant absorption of energy by the nanoshell near-infrared plasmons (tuned by means of thin nanoshells). Local heating of tumor cells, attached to the

nanoshells due to the special functionalization of the nanoshell surface, leads to the selective destruction of tumor cells without collateral damage of nearby healthy tissue.

24.3.2 Interparticle Coupling

As explained in Section 24.3.1, field enhancement around single-metal nanoparticles can be engineered via intraparticle coupling. However, the field enhancements produced around these single structures are typically limited to factors of 10 or so (see Figure 24.10). An alternate approach that can more strongly increase the response of optical antennas is to couple plasmons from different particles located in proximity. This interparticle coupling can generate extremely large field enhancement in interparticle gaps, referred to as *hot spots* (high intensity), and electromagnetic cavities. In Section 24.2.3, we presented a simple rule to estimate the field enhancement in coupled metal dimers, relating the field enhancement to the ratio between the particle size and the

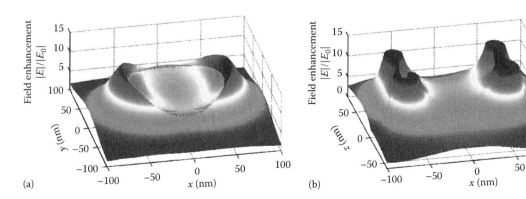

FIGURE 24.10 (See color insert following page 9-8.) Field enhancement in a gold nanoring for two cross sections through the center of the ring: top view (a) and side view (b).

separation distance between particles. Strongly enhanced fields in gaps are expected when gaps are small. Following this simple recipe, it is clear that complex structures of closely spaced, coupled particles should reveal dramatic effect from interparticle coupling. Coupled structures, such as disk dimers [39] and bowtie antennas [40], are often synthesized by lithographic methods, with the metal structures deposited on a substrate. With current lithographic techniques, the smallest interparticle gaps that can be fabricated controllably are about 10 nm. This currently limits the enhancement that can be achieved controllably. However, the potential of these structures to provide enhanced response is still remarkable especially if smaller gaps are achieved, as we will show now. In Figure 24.11, we compare the near-field amplitude at a resonance of three different gold nanorod structures. In the first case (Figure 24.11a), the local field of a 280 nm long, 80 nm wide gold nanorod is shown at resonance ($\lambda = 940$ nm). The field enhancement in this case is of the order of 15 times the incident field near the end of the rod, thereby giving intensity enhancements of around 200.

For comparison, a longer nanorod with approximately twice the length (570 nm) and the same width (Figure 24.11b) gives an enhancement of 30 at the resonance ($\lambda = 1695$ nm). The enhancement is about a factor of two larger because the effective dipole of the longer rod is doubled and due to a stronger lightning rod effect in this case. The resonance occurs at longer wavelengths for the longer rod, as previously described. A completely different situation occurs when two nanorods are coupled together. In Figure 24.11c, we show the field enhancement for a dimer made from the short rods with a 10 nm gap. These structures are often called optical gap antennas, due to the presence of the

gap in the center [41,42]. The possibility of loading the antenna gap with dielectrics or metals, similarly to concepts in radio-antenna theory, can be used to tune the antenna response in the visible [43,44]. Even if the total length of the gap antenna is the same as in the single rod antenna (Figure 24.11b), the optical response is very different, with different spectral peaks and localization of local fields. For the gap antenna (Figure 24.11c), a new mode appears, localized at the gap between rods with the expected redshift in the response ($\lambda = 1165$ nm), together with a large enhancement of the local field, as expected from the basic estimation in Equation 24.24. We observe in Figure 24.11c the enhancement factors of 75 in amplitude even for relatively large separation distances of 10 nm. The spectral evolution of the lowest order resonances in these two types of antennas is traced in Figure 24.12a, in which the position of the resonances wavelength is plotted as a function of the antenna gap. The single rod antenna shows a constant position of the dipolar peak, as a point of reference, but the response of the gap antenna redshifts as the gap decreases, distorting the dipole more strongly and piling up a large surface charge at the gap, as depicted in Figure 24.1. For very small gaps (<10 nm), the distorted dipolar resonance can be redshifted to the mid-infrared ($\lambda \approx 3000$ nm). As a result of the charge pile up, larger enhancements of up to 200–300 in amplitude can be obtained for the shortest realistic gaps [43,45], as observed in Figure 24.12b. Compared with single antennas, coupling in gap antennas provides a quantitative jump in the engineering of field enhancement.

Another approach for synthesizing strongly coupled structures with small gaps relies on chemical assembly. In colloidal chemistry, molecules can act as linkers between metal particles,

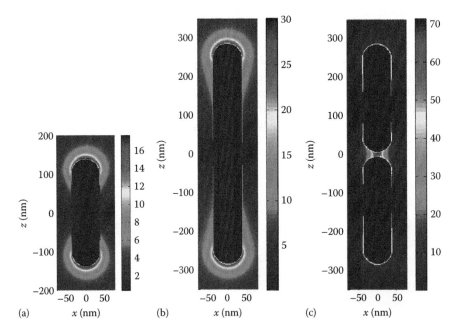

FIGURE 24.11 **(See color insert following page 9-8.)** Near-field distribution around three different gold nanorods at their respective resonance wavelengths. (a) A single nanorod of total length 280 and width 80 nm at a wavelength of $\lambda = 940$ nm. (b) A single gold nanorod with a total length of 570 and width 80 nm at a wavelength of $\lambda = 1695$ nm. (c) A pair of nanorods each of total length of 280 nm, longitudinally aligned with a 10 nm gap. The incident field is polarized along the rod axes.

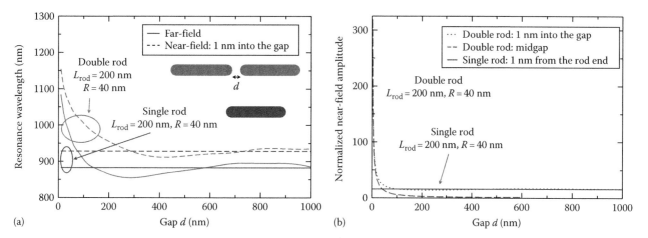

FIGURE 24.12 (a) Comparison of the resonance shifts for a single gold nanorod antenna and a gold gap antenna composed of the same nanorod segments of length $L_{rod} = 200$ nm and radius $R = 40$ nm as a function of the gap between both segments. (b) Near-field amplitude near the single rod antenna and in the gap of the gap antenna.

with the gaps determined by the molecule length. Dimer structures with gaps down to 1 or 2 nm are possible. This is the case for the silver dimer system presented in Figure 24.13. Such a structure can be used, for example, in field-enhanced spectroscopy to achieve single-molecule sensitivity [33].

In Figure 24.13b, we trace the evolution of the plasmon peaks as the separation between two silver particles is varied from the limit of widely separated dimers with weak coupling (gap $d = 40$ nm) to the limit of closely spaced dimers with strong coupling ($d = 0.5$ nm). When the particles are widely separated (black line), the optical response is very similar to the response of the spherical plasmon of an isolated silver particle ($\lambda = 390$ nm).

As the interparticle gap decreases, charge piles up at the gap and the scattering peak is redshifted. For these gap sizes, the main scattering peak corresponds to the excitation made by coupling and distorting the intraparticle dipolar modes, as illustrated in the schematic of Figure 24.13a. Higher order modes appear as the gap is reduced. These new additional modes come from coupling and distorting the higher order intraparticle plasmons leading to rapid, large charge oscillations near the gap [45]. Similar interparticle coupling is observed for many metal structures, such as in pairs of gold nanoshells, where similar redshifts as a function of separation have been reported [36]. In these coupled systems, tuning the gap provides control of the spectral response even for large interparticle separation [46].

24.3.3 Nanoparticles Coupled to a Continuum

Propagating surface plasmons on flat metal surfaces, thin metal films, and metal wires have a continuum of modes, which are a function of the propagation wavevector k, as we discussed briefly in Section 24.2. In contrast, localized surface plasmons on a metal nanoparticle have a discrete set of modes. An interesting situation arises when a discrete mode is coupled to a continuum of modes, giving rise to near-field localization of the propagating modes and Fano-like interference in the far-field response as a consequence of the coupling between the continuum of modes and the localized discrete mode. This effect is nicely illustrated when a tip, modeled as a dipolar particle, is coupled to an extended metal (Au here) surface. The tip (i.e., the particle) can be used to localize the response of the metal surface. The localization of the near field is a consequence of the interaction produced between the tip and its mirror image given by the reflection from the surface (see the schematic in Figure 24.14a). If we model the microscope tip by a virtual sphere of radius a_{tip}, with dielectric function ε_{tip}, surrounded by vacuum, the polarizability of this spherical tip α_{tip} is given by

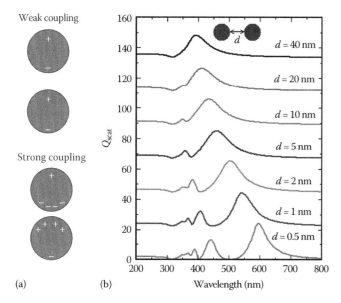

FIGURE 24.13 (a) Schematic of the weak (large gap) and the strong (small gap) coupling regimes in particle dimers. (b) Scattering coefficient (scattering cross section normalized to the geometrical area) of a set of silver dimers of radius 40 nm separated by a distance d. The incident light is polarized along the axis of the dimer.

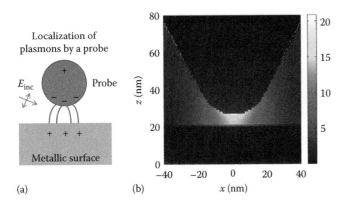

(a)   (b)

FIGURE 24.14 (a) Schematic of the localization of plasmons induced by the coupling between a metal surface and a dipolar particle. (b) Cross section of the near-field distribution of a silicon tip located on top of a gold surface in a typical scattering-type scanning near-field optical microscopy configuration. P-polarized light is incident from the left at a 45° angle. The dipole-like tip localizes the plasmon field at the gold surface producing high contrast of the gold surface with respect to a dielectric surface. A 20-fold enhancement of the localized field amplitude can be obtained even with a dielectric tip.

$$\alpha_{\text{tip}} = 4\pi a_{\text{tip}}^3 \frac{\varepsilon_{\text{tip}} - 1}{\varepsilon_{\text{tip}} + 2} \qquad (24.27)$$

When such a dipole probe is located a distance d from the surface, the interaction of the tip with its mirror image modifies the effective polarizability of the system $\alpha_{\text{tip-surf}}^{\text{eff}}$. For polarization perpendicular to the surface, the effective polarizability of the tip–surface system becomes

$$\alpha_{\text{tip-surf}}^{\text{eff}} = \frac{\alpha_{\text{tip}}(1 + \beta)}{1 - (\alpha_{\text{tip}}\beta / 16\pi(a_{\text{tip}} + d)^3)} \qquad (24.28)$$

with $\beta = (\varepsilon_s - 1)/(\varepsilon_s + 1)$, the electrostatic planar surface response of a substrate characterized by ε_s. A realistic example of the realization of such a system is the interaction of a probe tip in scattering-type scanning near-field optical microscopy (s-SNOM).

In Figure 24.14b, we observe the near-field distribution at the junction between a realistic Si tip, such as commonly used in s-SNOM, and a gold surface. The mirror image of the Si tip produces the localization of the near field to the nanometric scale, thus providing a mechanism to obtain material contrast with nanoscale resolution [47]. This is the basic mechanism of ultra-resolution in s-SNOM [48,49].

A similar approach based on the interaction of images can also be used to obtain coupled hybrid modes in a dielectric wire close to a metallic surface [50] and in metallic particles coupled to structures, such as metal nanowires, that support propagating modes [51]. In addition to the use in materials contrast microscopy, this localization of the optical signal could be important for the applications in optical switches and interconnects where a plasmon guide must transfer, absorb, or

emit an optical signal at specific points. In addition to the near-field localization, the interaction of a broadband plasmon (continuum) with the narrow band of a discrete state (e.g., localized plasmon, electron-hole state in a quantum dot, or a molecular dipole) can give rise to Fano-like profiles in far-field spectroscopy [52]. This is a common feature in extinction spectroscopy that can be clearly identified by the presence of symmetric and asymmetric dips in the extinction peaks profiles.

24.3.4 Metal Nanoparticles Coupled to Dipole Emitters

Metal nanoparticle optical antennas have been used extensively to couple light to and from a dipole emitter, for example, a molecule, an atom, or a quantum dot. Depending on the distance of the dipole emitter to the metallic antenna, the antenna can amplify the emission rate provided the emission is near resonance with the antenna or quench the emission [53]. When the emitter is close to the metal surface, metal absorption is dominant and light emission can be quenched by this absorption, as reported for emitters near metal surfaces [54] and metal nanoparticles. For larger separations, the emission can be enhanced and the dipole radiative decay rate is increased by the field enhancement near the metal surface [55]. For these larger separations (more than a few nanometers), the dipole emitters only weakly affect the emission properties of the antenna, mainly changing slightly the directionality of the emission. This has been observed, for example, in enhanced molecular fluorescence in the proximity of metal nanorods acting as $\lambda/4$ antennas (antennas of length $L = \lambda/4$) [56]. These effects are also exploited when metal antennas are used to control the emission properties (quantum yield and polarization) of bioemitters [57–60].

Enhancement of dipole emission can be observed for a variety of optical antenna configurations. We illustrate this effect for a gold nanoring acting as an antenna for dipole emission in Figure 24.15.

The dipole emitter is placed near the nanoring, and the modified spontaneous emission rate is compared with the emission rate for an isolated emitter. The spontaneous emission is strongly enhanced near the resonant wavelength of the gold nanorings (two nanorings of different heights h are considered).

As discussed so far, when a weak light field excites a metal nanoparticle that is coupled to a dipole emitter, the response of the pair is shifted by the coupling. Moreover, the emission from the dipole emitter can be enhanced or quenched depending on the interparticle separation. This weak field limit is the condition normally employed for optical spectroscopy. However, the response of the pair can be very different for a strong incident light field, which can drive the emitter through coherent Rabi oscillations before emission or quenching occurs [62,63]. In this strong field limit, the quantum character of the emitter becomes paramount. In this limit, optical response of the pair depends on both the electromagnetic response of the coupled pair and the quantum evolution of the dipole emitter. Strong coupling between the *quantum* dipole emitter and the metal nanoparticle

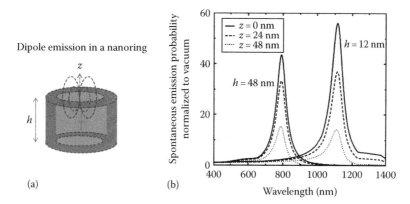

FIGURE 24.15 (a) Schematic of a dipole emitter in the proximity of an optical antenna, such as a gold nanoring. (b) Spontaneous emission probability of a point dipole located at three different heights ($z = 0$, 24, and 48 nm) from the center of a gold nanoring for ring heights $h = 48$ and 12 nm. The nanoring thickness is 16 nm in both cases. The emission is normalized to the emission of an isolated dipole. (Adapted from Aizpurua, J. et al., *J. Quant. Spectrosc. Rad. Transf.*, 89, 11, 2004. With permission.)

complicates the response. Fano interference can induce transparency in the metal nanoparticle response to the incident field, despite being in the strong field limit. The response of the quantum emitter can become nonlinear due to the self-coupling of the quantum emitter back onto itself via the coupling of the emitter to the metal nanoparticle. These issues are just now being investigated and they are opening a new realm for metal nanoparticle plasmon nanooptics where quantum optics meets nanooptics. This will be increasingly important as the revolution of quantum information processing and measurement takes hold and will drastically augment the range of antenna applications provided by metal nanoparticle antennas.

24.3.5 Chains of Nanoparticles

Coupling between plasmons on two metal nanoparticles leads to substantial shifts and modification of the modes with strong localization of the fields in the interparticle gaps. Coupling even more particles together leads to additional significant modifications of the modes, providing new flexibility to tailor the electromagnetic response and field localization [14,64–66]. Coupling long chains of particles opens up the possibility to transport plasmonic excitations along the chains. This excitation transport provides a means to move optical excitations along nanoscale structures, well below the sizes set by the diffraction limit for light (i.e., on a much smaller scale than usual photon transport in microscale waveguides). Thus, these chains are being studied extensively as potential optical switches and interconnects for optical signaling on the nanoscale [67]. Experiments have also shown that the field localization on a chain with a finite number of particles can be controlled by the plasmon wavelength with the field being localized to just a few particles at the beginning or end of a chain depending on the wavelength [68]. This opens additional ways to control and exploit the near-field localization.

For an infinite chain of metal particles, a broad band of modes, characterized by wavevector k, is generated, giving rise to a propagating plasmon along the chain. These propagating plasmons depend both on the nature of the intraparticle plasmon modes and on the separation between the particles d that governs the interparticle coupling. For an infinite chain of small metal nanoparticles of radius a that can be individually characterized by their dipolar response (the dipolar resonance of the nanoparticles appearing at $\omega = \omega_p/\sqrt{3}$), the dispersion relation for the transverse mode (polarization of the particles perpendicular to the chain axis) in the electrostatic limit is [69]

$$\omega^2 = \frac{\omega_p^2}{3}\left(1 + \left(\frac{a}{d}\right)^3 \sum_{j=1}^{\infty} \frac{2\cos(jkd)}{j^3}\right) \qquad (24.29)$$

(thin line labeled with T in Figure 24.16). This dispersion relation becomes much more complicated when retardation is included and the full electrodynamical solution is obtained. When retardation is included, the phase of the intraparticle response is important and intraparticle excitations can interfere constructively or destructively, giving rise not only to a shift of the dispersion lines (both longitudinal and transverse modes) but also to an anticrossing of the transversal modes, as seen in Figure 24.16 [69].

The dispersion of the longitudinal mode (L) obtained from full electrodynamical solutions is blue- or redshifted with respect to the electrostatic limit, depending on the location of the mode with respect to the light line. For the transverse mode (T), an anticrossing of the dispersion is obtained, modifying the simple crossing obtained in the electrostatic limit of Equation 24.29. These results stress the need for a full electromagnetic description when dealing with propagating solutions that involve complex interference of the phases of the induced excitations.

The decay length of these propagating modes is a critical parameter for any application that requires signal propagation over long distances without significant radiation loss. Both transverse and longitudinal modes can have propagation lengths that exceed 1 μm, for modes below the light line. Near and above the light line where the modes can couple more effectively to light, the radiative damping is enhanced and the decay

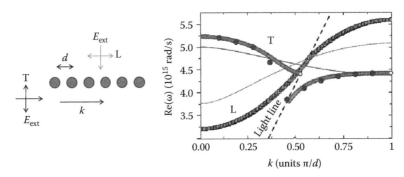

FIGURE 24.16 Dispersion of an infinite chain of metal particles. Transverse (labeled T) and longitudinal (labeled L) modes are displayed. Full numerical calculations are shown as thick lines. The electrostatic limit is shown as thin lines. A full calculation for 10 particles from Ref. [64] is displayed with circles. (Adapted from Koenderink, A.F. and Polman, A., *Phys. Rev. B*, 74, 033402, 2006. With permission.)

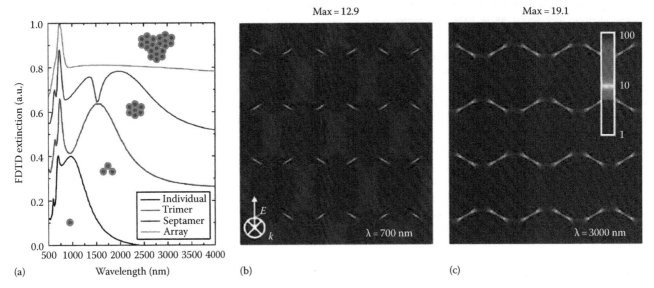

FIGURE 24.17 (**See color insert following page 9-8.**) (a) Extinction cross section of an individual gold nanoshell (black), a trimer (red), a septamer (blue), and an infinite hexagonal array (green curve) of gold nanoshells. (b) Near-field distribution for the infinite array at a wavelength of $\lambda = 700\,nm$ and (c) at a wavelength of $\lambda = 3000\,nm$. The gold nanoshells have an inner radius of $150\,nm$ and an outer radius of $172\,nm$. The separation between nanoshells is $8\,nm$. Similar local field enhancements are achieved between particles both for visible and infrared wavelengths. (Adapted from Le, F. et al., *ACS Nano*, 2, 707, 2008. With permission.)

length is shortened. Long decay lengths for the modes below the light line make particle chains a viable nanoscale paradigm for optical guiding.

Throughout this chapter, we have stressed the near-field localization and enhancement provided by particles, pairs of particles, and chains of particles. As shown in Figure 24.17, similar enhancement can be observed in the response of a more complicated distribution of coupled particles, such as in an infinite two-dimensional hexagonal array of gold nanoparticles [70]. Here, the enhancement can be realized over a broad spectral range. The response of the system gives considerable enhancement not only at visible frequencies (Figure 24.17b) but also at infrared frequencies (Figure 24.17c) because the dipolar interparticle coupling redshifts and broadens the response as the number of particles increases (see Figure 24.17a). The localization of the near field at infrared wavelengths (Figure 24.17c)

is a consequence of the distorted and redshifted plasmons localized at the interparticle gaps, increased by a pronounced lightning rod effect at mid-infrared wavelengths (due to the more conductive response of the metal). In the visible spectroscopy, the enhancement is a consequence of the coupling of higher order modes at the cavities (see Figure 24.17a and b). Because of this effective response over a broad range of the spectrum, these arrays can be used simultaneously for visible and infrared spectroscopy [71].

24.4 Summary and Future Perspectives

In this chapter, we have reviewed the basic understanding of metal nanoparticle plasmonics needed for exploiting these nanoparticles as building blocks for optical nanoantennas. These structures provide a robust, versatile paradigm for tailoring optical response and for concentrating and enhancing fields on the nanoscale. Both aspects are extremely useful for sensing, field-enhanced spectroscopy, near-field microscopy, optical signal propagation, and biomedical applications.

The plasmon response of metal nanoparticles is governed by particle size, particle shape, the environment of the particles, and any coupling between plasmons localized to different parts of the same particle (intraparticle coupling) or between plasmons on different, closely spaced particles (interparticle coupling). A small, spherical metal nanoparticle behaves like a simple dipole with isotropic response. For typical metals, this response is resonant at visible wavelengths. As the nanoparticle increases in size, this dipolar response redshifts, pushing the dipolar resonance toward the infrared, and additional, higher order modes contribute to the response. This dependence on size provides a way to tune the optical response to the spectral range of interest. When the particle shape is elongated in one direction, the response becomes anisotropic, with strong response along the particle axis and weaker transverse response. As the particle becomes more elongated, the longitudinal plasmonic response is further redshifted. Most importantly, the near fields at the sharper ends of the nanoparticle can be enhanced by an order of magnitude. This near-field enhancement opens up a wide range of antenna applications, in which a strong response is needed to amplify the optical signal of a weak emitter. Screening by the environment of the metal nanoparticle can further redshift the plasmon dipolar response. This can be a large shift, even for weak screening, provided that the perturbation by the environment is probed by the enhanced local fields of the nanoparticle. This sensitivity adds greatly to the utility of metal nanoparticles as optical nanoantennas for sensing, especially for approaching the single-molecule limit for detection. The plasmonic response of very small particles is well described by the electrostatic limit. However, as the particle size increases, retardation effects become important and a full electrodynamical description is needed to accurately describe the response.

In this chapter, we have stressed intra- and interparticle coupling as a way to control nanoparticle plasmonics because the optical response can be engineered over a wide spectral range and the localization of the field can be tailored precisely. In metallic nanoparticles, such as nanoshells and nanorings, intraparticle coupling between plasmons localized on different surfaces of the particle mixes and hybridizes the plasmon modes, similar to the hybridization that drives the formation of chemical bonds. This hybridization depends strongly on the spatial separation between the localized plasmons and can be controlled by tailoring the particle geometry. When plasmons from different particles are coupled across interparticle gaps, strong mixing, hybridization, and distortion of the modes also occur. A large pileup of the surface charge of the plasmon oscillations can occur at these gaps, resulting in even greater local field enhancements in the gaps between particles than at the ends of particles. This extreme field enhancement is now being extensively exploited in spectroscopies, such as Raman spectroscopy, which require large amplification of very weak signals.

The basic understanding and characterization of metal nanoparticle plasmonics is now being fully established. The ongoing challenge is to exploit successfully the capabilities of nanoparticle plasmonics in the wide range of applications that require nanoscale optical antennas. New areas that need to be tackled and developed include, for example, energy transfer in molecular processes near metal nanoparticles [72]; energy storage, as in solar cells in which efficiency might be greatly improved by the enhanced absorption of optical antennas [73]; and the combination of plasmonic structures with semiconductors, molecules, and photonic structures, in so-called hybrid systems, which explicitly utilize quantum emitters coupled with their own nanoantennas [62].

Different experimental groups are now making more extensive use of field localization to perform microscopy of plasmonic nanostructures [74–76]. Such techniques will be extended increasingly to image and probe spectral information about more complicated systems, especially biological species and processes, such as virus [77] and protein activity [78].

New phenomena associated with the quantum nature of the collective oscillations of the electrons that give the plasmon response are interesting because new issues are raised about nonlocality in the electronics of strongly coupled systems and the ultrafast response of plasmons [79]. The nonlinear optical response of plasmonic systems is just being developed and characterized and is already being exploited in, for example, second-harmonic generation [80] and four-wave mixing [81].

In addition to the clear importance of metal nanoparticle plasmonics for nanooptics, nanoscale plasmonics is important for a variety of other probe techniques. These include electrical detection of plasmons [82], light emission in scanning tunneling configurations [83], cathodoluminescence [84], and photoemission [85]. These are techniques in which the role of nanoscale plasmons in radiative decay can also strongly influence the physical processes.

The exploitation of metal nanoparticle plasmonics has a long history, dating back for many centuries, even though the basic understanding of nanoparticle plasmonics has only been developed recently. Applications of metal nanoparticles for nanoscale sensing, for spectroscopy at the single molecule limit, and for optical signaling at the nanoscale have taken off in recent years because the needed nanoparticle nanoantennas can be engineered and fabricated. An even broader range of applications should be expected as the use of metal nanoparticle plasmonics extends from classical nanoantennas toward quantum nanoantennas, from the linear regime into the nonlinear regime, and down to ultrafast time scales, to complement nanoscale spatial resolution.

References

1. Ritchie, R. 1957, *Phys. Rev.* 106, 874.
2. Mie, G. 1908, *Ann. Phys.* 342, 881.
3. Batson, P. E. 1982, *Phys. Rev. Lett.* 49, 936–940.
4. Liz-Marzán, L. M. 2004, *Mater. Today* 7, 26–31.
5. Murphy, C. J. et al. 2005, *J. Phys. Chem. B* 109, 13857–13870.

6. Oldenburg, S. J. et al. 1998, *Chem. Phys. Lett.* 288, 243–247.
7. Banholzer, M. J. et al. 2008, *Chem. Soc. Rev.* 37, 885–897.
8. Haynes, C. L. and Van Duyne, R. P. 2001, *J. Phys. Chem. B* 105, 5599.
9. Rechberger, W. et al. 2003, *Opt. Commun.* 220, 137.
10. Fromm, D. P. et al. 2004, *Nano Lett.* 4, 957.
11. Pelton, M., Aizpurua, J., and Bryant, G. W. 2008, *Laser Photon. Rev.* 2, 136–159.
12. Xu, H. et al. 1999, *Phys. Rev. Lett.* 83, 4357.
13. Neubrech, F. et al. 2008, *Phys. Rev. Lett.* 101, 157403.
14. Maier, S. A. et al. 2001, *Adv. Mater.* 13, 1501.
15. Nanospectra Biosciences Inc., Houston, TX. http://www.nanospectra.com/images/tumor.jpg [Online].
16. Yao, J. et al. 2008, *Science*, 321, 930.
17. Lakowicz, J. R. 2005, *Anal. Biochem.* 337, 171.
18. Hirsch, L. R. et al. 2003, *Proc. Natl. Acad. Sci. U S A* 100, 13549.
19. Kelly, K. L. et al. 2003, *J. Phys. Chem. B* 107, 668–677.
20. Myroshnychenko, V. et al. 2008, *Chem. Soc. Rev.* 37, 1792–1805.
21. Johnson, P. B. and Christy, R. W. 1972, *Phys. Rev. B* 6, 4370.
22. Palik, E. D. *Handbook of Optical Constants of Solids.* New York: Academic, 1985.
23. Link, S., Mohamed, M. B., and El-Sayed, M. A. 1999, *J. Phys. Chem. B* 103, 3073.
24. Bryant, G. W., García de Abajo, F. J., and Aizpurua, J. 2008, *Nano Lett.* 8, 631–636.
25. Wang, H. et al. 2006, *Nano Lett.* 6, 827–832.
26. Onuta, T.-D. et al. 2007, *Nano Lett.* 7, 557–564.
27. Lu, Y. et al. 2005, *Nano Lett.* 5, 119–124.
28. Bukasov, R. and Shumaker-Parry, J. S. 2007, *Nano Lett.* 7, 1113–1118.
29. Hao, F. et al. 2007, *Nano Lett.* 7, 729–732.
30. Schmeits, M. and Dambly, L. 1991, *Phys. Rev. B* 44, 12706.
31. Prodan, E. et al. 2003, *Science* 302, 419.
32. Nordlander, P. et al. 2004, *Nano Lett.* 4, 899.
33. Xu, H. X. et al. 2001, *Proc. SPIE* 4258, 35.
34. Willets, K. A. and Van Duyne, R. P. 2007, *Annu. Rev. Phys. Chem.* 58, 267–297.
35. Larsson, E. M. et al. 2007, *Nano Lett.* 7, 1256–1263.
36. Lassiter, J. B. et al. 2008, *Nano Lett.* 8, 1212.
37. García de Abajo, F. J. and Howie, A. 1998, *Phys. Rev. Lett.* 80, 5180.
38. Aizpurua, J. et al. 2003, *Phys. Rev. Lett.* 90, 057401.
39. Atay, T., Song, J. H., and Nurmikko, A. V. 2004, *Nano Lett.* 4, 1627–1631.
40. Schuck, P. J. et al. 2005, *Phys. Rev. Lett.* 94, 017402.
41. Mühlschlegel, P. et al. 2005, *Science* 308, 1607–1609.
42. Alu, A. and Engheta, N. 2008, *Nat. Photon.* 2, 307.
43. Aizpurua, J. et al. 2005, *Phys. Rev. B* 71, 235420.
44. Alu, A. and Engheta, N. 2008, *Phys. Rev. Lett.* 101, 043901.
45. Romero, I. et al. 2006, *Opt. Express* 14, 9988–9999.
46. Olk, P. et al. 2007, *Nano Lett.* 7, 1736–1740.
47. Hillenbrand, R. and Keilmann, F. 2000, *Phys. Rev. Lett.* 85, 3029.
48. Cvitkovic, A., Ocelic, N., and Hillenbrand, R. 2007, *Nano Lett.* 7, 3177–3181.
49. Kim, Z. H., Liu, S.-H. A., and Leone, S. R. 2007, *Nano Lett.* 7, 2258–2262.
50. Oulton, R. F. et al. 2008, *Nat. Photon.* 2, 496–500.
51. Knight, M. W. et al. 2007, *Nano Lett.* 7, 2346–2350.
52. Sturm, K., Schülke, W., and Schmitz, J. R. 1992, *Phys. Rev. Lett.* 68, 228–231.
53. Anger, P., Bharadwaj, P., and Novotny, L. 2006, *Phys. Rev. Lett.* 96, 113002.
54. Ford, G. W. and Weber, W. H. 1984, *Phys. Rep.* 113, 195–287.
55. Muskens, O. L. et al. 2007, *Nano Lett.* 7, 2871–2875.
56. Taminiau, T. H. et al. 2007, *Nano Lett.* 7, 28–33.
57. Seelig, J. et al. 2007, *Nano Lett.* 7, 685–689.
58. Bek, A. et al. 2008, *Nano Lett.* 8, 485–490.
59. Mackowski, S. et al. 2008, *Nano Lett.* 8, 558–564.
60. Moerland, R. J. et al. 2008, *Nano Lett.* 8, 606–610.
61. Aizpurua, J. et al. 2004, *J. Quant. Spectrosc. Radiat. Transf.* 89, 11–16.
62. Zhang, W., Govorov, A. O., and Bryant, G. W. 2006, *Phys. Rev. Lett.* 97, 146804.
63. Artuso, R. D. and Bryant, G. W. 2008, *Nano Lett.* 8, 2106–2111.
64. Weber, W. H. and Ford, G. W. 2004, *Phys. Rev. B* 70, 125429.
65. Li, K., Stockman, M. I., and Bergman, D. J. 2003, *Phys. Rev. Lett.* 91, 227402.
66. García de Abajo, F. J. 2007, *Rev. Mod. Phys.* 79, 1267.
67. Brongersma, M. L., Hartman, J. W., and Atwater, H. A. 2000, *Phys. Rev. B* 75, 16356–16359.
68. de Waele, R., Koenderink, F., and Polman, A. 2007, *Nano Lett.* 7, 2004–2008.
69. Koenderink, A. F. and Polman, A. 2006, *Phys. Rev. B* 74, 033402.
70. Le, F. et al. 2008, *ACS Nano* 2, 707–718.
71. Wang, H., Kundu, J., and Halas, N. J. 2007, *Angew. Chem., Int. Ed.* 46, 9040–9044.
72. Govorov, A. O. and Carmeli, I. 2007, *Nano Lett.* 7, 620–625.
73. Ferry, V. E. et al. 2008, *Nano Lett.*, 8, 4391–4397.
74. Hillenbrand, R. et al. 2003, *App. Phys. Lett.* 83, 368–370.
75. Esteban, R. et al. 2008, *Nano Lett.* 8, 3155–3159.
76. Rang, M. et al. 2008, *Nano Lett.* 8, 3357–3363.
77. Brehm, M. et al. 2006, *Nano Lett.* 6, 1307–1310.
78. Höppener, C. and Novotny, L. 2008, *Nano Lett.* 8, 642–646.
79. Kubo, A. et al. 2005, *Nano Lett.* 5, 1123.
80. Canfield, B. K. et al. 2007, *Nano Lett.* 7, 1251–1255.
81. Danckwerts, M. and Novotny, L. 2007, *Phys. Rev. Lett.* 98, 026104.
82. Vlaminck, I. D. et al. 2007, *Nano Lett.* 7, 703–706.
83. Schull, G., Becker, M., and Berndt, R. 2008, *Phys. Rev. Lett.* 101, 136801.
84. Nelayah, J. et al. 2007, *Nat. Phys.* 3, 348–353.
85. Douillard, L. et al. 2008, *Nano Lett.* 8, 935–940.

Metal–Insulator Transition in Molecularly Linked Nanoparticle Films

Amir Zabet-Khosousi
University of Toronto

Al-Amin Dhirani
University of Toronto

25.1 Introduction

Films comprising metallic grains embedded in insulating matrices represent an important class of electronic materials. Conventional granular films prepared by thermal evaporation, sputtering, chemical vapor deposition, electroless deposition, etc. have been extensively studied (Abeles et al. 1975). They are typically disordered and exhibit electronic properties that vary both widely and controllably, depending on the architecture and composition of their components. Three distinct regimes have been typically observed (Abeles et al. 1975):

1. A *thin film regime*, where the volume fraction of grains is small and the grains form isolated islands.
2. A *thick film regime*, where the concentration of grains is large and isolated islands merge, forming continuous pathways through the film.
3. A *transition regime*, where a transition from insulating to bulk regimes occurs. At a critical concentration of grains (percolation threshold), at least one sample spanning pathway is established throughout the film.

In conventional granular films, the film thickness influences the grain size and the inter-grain spacing, both of which strongly impact on film properties. For example, varying the spacing between metallic grains can induce a metal–insulator transition. The varying grain size in the thin-film regime influences single-electron charging energies and, in the thick-film regime, influences the length scale of the elastic scattering at grain surfaces. The latter can dominate film resistance and can lead to electron self-interference (Bergmann 1984). An independent control of the grain size and spacing in conventional granular materials can be achieved to a degree by co-depositing insulating and non-insulating material, but control over the uniformity of the grain size and spacing at the sub-nanometer scale is not generally possible.

Molecularly linked nanoparticle (ML-NP) films, on the other hand, represent a new, exquisitely controlled platform for systematically exploring structure–electronic property relationships in granular films (Zabet-Khosousi and Dhirani 2008). In this approach, an increased control is possible since nanoparticles (NPs) and molecules with desired properties can be separately synthesized by wet chemical methods and subsequently organized into assemblies. The separation of synthesis and assembly steps affords an independent control over the critical material parameters such as coupling and the arrangements of individual building blocks. The electronic and chemical properties of individual NPs depend on their size, shape, and composition and, therefore, are tunable using recent advances in synthetic methods. Single-electron charging energies and quantum mechanical energy levels, for example, are both controllable through the NP size. Coupling between NPs can be controlled by varying the molecular linkers. The thickness of films can be controlled by alternately self-assembling NPs and linker molecules on a surface. The arrangement of NPs is controllable to a degree through the assembly process. Also, the spatial disorder can be controllably introduced in otherwise-ordered systems by varying NP size distribution.

The flexibility to independently control the grain size, grain spacing, and film thickness combined with a broad range of choices for the NPs and molecules and the manner in which they can be arranged into materials has provided new opportunities in materials science. It has also enabled applications ranging from conductance switching (transistors) (Suganuma and Dhirani 2005) and information storage (Suganuma et al. 2005) to proof-of-principle demonstrations of chemical/biological sensing (Katz and Willner 2004), surface-enhanced Raman scattering (Freeman et al. 1995), conductive coatings (Goebbert et al. 1999), and catalysis (Schmid and Corain 2003).

In this chapter, we show that ML-NP films can serve as a new platform for studying two important phenomena in granular films, namely, Mott–Hubbard and percolation metal–insulator transitions (MITs). Materials are classified as metals or nonmetals depending on the behavior of their conductance as temperature tends to absolute zero. For nonmetals, conductance tends to zero; for metals, it tends to a finite value. This can be rationalized given their electronic band structures. In nonmetals, the energy bands are either completely filled or completely empty, and in metals, at least one band is partially filled. The latter can participate in conduction. Partially filled bands in nonmetals can be generated by exciting electrons over the energy gap, E_g, between the highest occupied (valence) band and the lowest unoccupied (conduction) band. At nonzero temperatures, there is a finite probability that some electrons will overcome the energy gap through thermal activation. Therefore, the conductivity of nonmetals tends to increase with temperature. The conductivity of a metal, on the other hand, tends to decline with the increasing temperature due to phonon scattering.

ML-NP films comprising insulating linker molecules and metallic NPs can exhibit bulk conductance with two extreme limiting behaviors: insulating or metallic. In the insulating limit, conductance through an ML-NP film is governed by tunneling through linker molecules and the single-electron charging of the NPs (Zabet-Khosousi and Dhirani 2008). The NPs can be considered as being connected by local conductances, g,

$$g \propto e^{-\beta s} e^{-E_c/k_B T}, \tag{25.1}$$

where

βis the tunneling decay constant
s is the inter-NP surface-to-surface separation
E_c is the single-electron charging energy

$$E_c = \frac{e^2}{2C}, \tag{25.2}$$

and C is the capacitance (Zabet-Khosousi and Dhirani 2008). The capacitance of an NP in an ML-NP film can be obtained by approximating the neighboring NPs as a conducting continuum separated from the central NP by an insulating shell, as shown in Figure 25.1. In this approximation, C is given by

$$C = 4\pi\varepsilon_0\varepsilon_r \left(\frac{1}{R} - \frac{1}{R+s}\right)^{-1} = 4\pi\varepsilon_0\varepsilon_r \frac{R(R+s)}{s}, \tag{25.3}$$

where

R is the NP radius
ε_0 is the permittivity of the vacuum
ε_r is the dielectric constant of the NP surrounding

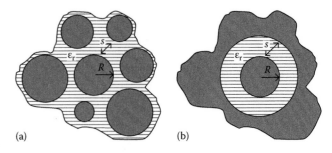

FIGURE 25.1 (a) Conducting NP of radius R separated from neighboring NPs by an average distance s. The NPs are embedded in an insulating medium with a dielectric constant ε_r. (b) NP in (a) is modeled as a sphere separated from a conducting NP, surrounded by an insulating shell with thickness s and dielectric constant ε_r. (Reprinted from Zabet-Khosousi, A. and Dhirani, A.-A., *Chem. Rev.*, 108, 4072, 2008. With permission.)

E_c, therefore, can be expressed as

$$E_c = \frac{e^2}{8\pi\varepsilon_0\varepsilon_r} \frac{s}{R(R+s)}. \tag{25.4}$$

We note that Equation 25.1 is only applicable when g is smaller than the quantum of conductance, g_Q:

$$g_Q = \frac{2e^2}{h} \approx (13 \text{ k}\Omega)^{-1}, \tag{25.5}$$

where h is Planck's constant. This requirement ensures that charges can flow by tunneling and yet are sufficiently localized so that the charge states of the NPs are well-defined. This can be understood in terms of the energy–time uncertainty principle:

$$\Delta E \Delta t \geq \frac{h}{2}. \tag{25.6}$$

Taking ΔE as the energy change due to (dis)charging, $\Delta E \approx E_c$, and Δt as the (dis)charging time due to tunneling, $\Delta t \approx RC$, we get

$$\frac{e^2}{2C}(RC) \geq \frac{h}{2} \Rightarrow R \geq \frac{h}{e^2}. \tag{25.7}$$

The factor of 2 in Equation 25.5 arises from the electron's spin degeneracy. For $g > g_Q$, Δt is very short such that ΔE becomes sufficiently large to overcome E_c. Single-electron charging effects are then suppressed as quantum couplings become strong, charges become delocalized, and the film exhibits metallic behavior (Devoret and Grabert 1992).

In order for a film of strongly coupled metal NPs to exhibit global metallicity, the film must cross a percolation threshold. At the threshold, strongly coupled NPs form at least one continuous sample-spanning metallic pathway. Although at or just above the threshold, the sample may be dominated by a nonmetallic conduction, as $T \to 0$, nonmetallic pathways shut down, and at absolute zero, conductance remains finite. Below the threshold, however, no sample-spanning metallic pathway exists, and overall, the assembly is nonmetallic.

We have studied MITs using multilayer films of Au NPs crosslinked with α,ω-alkanedithiols, $HS(CH_2)_nSH$, by independently varying inter-NP coupling (via n) (Zabet-Khosousi et al. 2006) and film thickness (Trudeau et al. 2002). In the thick-film limit, the length of the linker molecules is varied from ~0.5 nm ($n = 2$) to ~1.6 nm ($n = 10$) in increments of ~0.1 nm (one CH_2 unit). Films comprising short linkers ($n < 5$) exhibit metallic behavior, i.e. their conductances remain finite as $T \to 0$. Films comprising longer linkers ($n > 5$) exhibit thermally activated conductances and are nonmetals. Some films with $n = 5$ linkers exhibit metallic behavior and others nonmetallic behavior. The transition at $n = 5$ is explained in the context of a Mott–Hubbard model for MITs. The model is based on a lattice of atoms with electrons that interact via on-site repulsion and intersite coupling. As intersite coupling increases, initially localized electrons can become itinerant, resulting in an MIT. For ML-NP films comprising short linkers (e.g., $n = 4$), film thickness is varied through the number of NP/linker deposition cycles. Thin films (prepared with less than ~5 cycles) exhibit thermally activated conductances and are nonmetallic, whereas thick films (prepared with more than ~5 cycles) exhibit metallic behavior. The transition is modeled using an effective-medium approximation.

The ability to tune properties of ML-NP films through MITs has a remarkable implication that films with properties interpolating between metals and insulators may be viewed as being analogous to semiconductors. To explore this analogy, we briefly describe experiments in which ML-NP films near the MIT have been used as critical elements in novel field-effect transistors.

25.2 Film Preparation and Characterization

The NP films are prepared via stepwise self-assembly on nanometer-spaced Au electrodes. The small electrode spacing facilitates the crossing of the percolation threshold. Au NPs, 5.0 ± 0.8 nm in diameter and stabilized by tetraoctylammonium bromide (TOAB), are synthesized in toluene using an established procedure (Brust et al. 1998). Since TOAB is a weak-stabilizing ligand, Au NPs can be readily tethered to gold surfaces using alkanedithiols. Gold electrodes with a nanometer spacing can be fabricated using electromigration (see below) and can be functionalized with alkanedithiols by immersion in 0.5 mM ethanolic solutions of alkanedithiols for 1 h and rinsing. Multilayer ML-NP films on the electrodes are then prepared by alternate immersion in toluene solutions of Au NPs for ~30–60 min and 0.5 mM of alkanedithiols for ~10 min with intervening rinse steps. As film thickness increases, eventually the film bridges the gap between electrodes.

In order to monitor the growth of ML-NP films, films are also prepared on transparent glass substrates and conducting silicon substrates. The former are characterized using optical spectroscopy and the latter using scanning electron microscopy (SEM) and scanning tunneling microscopy (STM). To self-assemble NPs on the surface, the substrates are functionalized with a monolayer of 3-aminopropylmethyl-diethoxysilane, $NH_2(CH_2)_3Si(CH_3)(OCH_2CH_3)_2$. Subsequent layers of NPs and alkanedithiols are self-assembled as described above.

Figure 25.2 shows a UV/vis spectra of films of ~5 nm Au NPs linked with butanedithiol. The UV/vis spectra exhibit peak

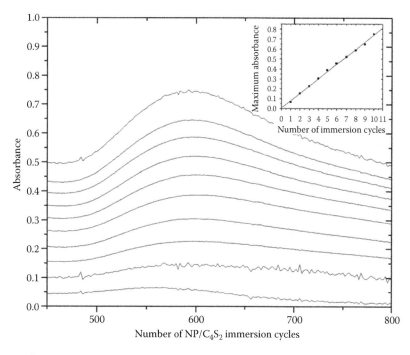

FIGURE 25.2 UV/vis spectra of butanedithiol-linked ~5 nm Au NP films. Films were prepared on glass with 1–10 immersion cycles. Inset: maximum absorbance versus number of immersion cycles.

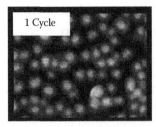
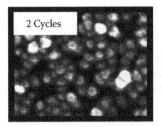
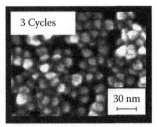

FIGURE 25.3 SEM images of butanedithiol-linked ~15 nm Au NP films. Films were prepared on silicon after 1, 2, and 3 immersion cycles. (Reprinted from Trudeau, P.E. et al., *J. Chem. Phys.*, 117, 3978, 2002. With permission.)

red shifts indicating a decrease in average inter-NP separation as the number of immersion cycles increases (Brust et al. 1998). Figure 25.2(inset) shows a linear increase in the maximum absorbance versus the number of cycles indicating that each immersion adds approximately the same amount of NPs. Ellipsometry measurements confirm that average film thickness increases approximately linearly with the number of cycles (Brust et al. 1998).

Figure 25.3 shows SEM images of films of ~15 nm Au NPs linked with butanedithiol after 1, 2, and 3 immersion cycles. The images show that after the first immersion cycle, NPs are mostly isolated. As the number of cycles increases, newly deposited NPs attach either to previously deposited NPs, forming "superclusters" of ML-NPs, or directly to bare regions of the substrate, seeding new superclusters. Figure 25.4 shows a scanning tunneling microscope (STM) image of a film of ~5 nm Au NPs linked with butanedithiol, prepared with four immersion cycles. Both the SEM and STM images show that ML-NP films are highly disordered. Films made with different lengths of the linker molecules exhibit similar trends.

FIGURE 25.4 STM image of a butanedithiol-linked ~5 nm Au NP film. The film was prepared on a doped-silicon/silicon-oxide substrate with four immersion cycles. Tip bias and current set point are −1.2 V and 0.1 nA, respectively. (Reprinted from Suganuma, Y. and Dhirani A.-A., *J. Phys. Chem. B*, 109, 15391, 2005. With permission.)

Nanometer-spaced electrodes may be fabricated as follows (Zabet-Khosousi and Dhirani 2007): First, glass sides are cleaned by immersion in a hot Piranha solution (3:1 H_2SO_4:H_2O_2) for 30 min, rinsing thoroughly with deionized water and drying with N_2. Then, two electrodes 100 μm wide, 4 mm long, and separated by a 100 μm gap are created by depositing 3 nm Cr followed by 100 nm Au through a shadow mask. The electrodes are deposited by resistively heating metals in a vacuum chamber at an initial pressure of ~1 μTorr and a rate of ~0.03 nm s⁻¹ for Cr and ~0.3 nm s⁻¹ for Au. After each deposition, the chamber is backfilled with nitrogen to ambient pressure. The shadow mask consists of 100 μm wide slits machined in a 150 μm thick metal shim. The mask is held in position using solid wires and screws (see Figure 25.5). A magnet wire is tightly attached to the mask using screws, and is oriented perpendicular to the slits, shadowing part of the slit and resulting in a gap in the Cr/Au. After the Cr/Au deposition, the magnet wire is cut and an additional 15 nm of Au is deposited. This creates a thin wire bridging the electrodes, resulting in a structure that resembles a fuse. A nanogap in the thin wire is created by applying a voltage and passing a large current through the wire. At sufficiently high voltages, typically 3–5 V, current suddenly drops and a gap is created due to electromigration: The high current density induces heating and momentum transfer from conducting electrons to metal ions, which, in turn, give rise to a gap that can be as small as a few nanometers in width (Park et al. 1999).

Figure 25.6 shows changes in the conductance of ML-NP films versus the number of NP/alkanedithiol immersion cycles. The conductances generally increase after each cycle because of the increase of film thickness as discussed above. For $n = 2$ and 3, as the number of cycles increases, conductances change initially very slowly, then rapidly, and finally at a constant rate. These observations suggest that a percolation transition occurs (see Section 25.4), and eventually in the bulk limit, the number of current pathways increases in proportion to the average thickness of the film. As n increases, the region of rapid change seems to occur at a lower number of cycles, until for $n \geq 5$ this region is no longer observed. For large n, conductance varies nonuniformly with the number of cycles. Variation in the rates of increase may be due to changes in the orientation and conformation of alkanedithiols on NPs' surfaces (Snow et al. 2002).

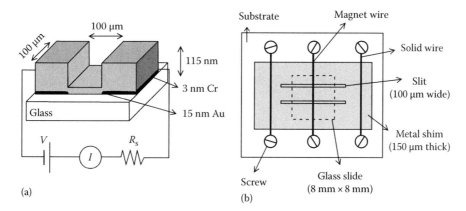

FIGURE 25.5 (a) Schematic of the junction and the electromigration circuit. (b) Schematic of the electrode deposition set-up. (Reprinted from Zabet-Khosousi, A. and Dhirani, A.-A., *Nanotechnology*, 18, 455305, 2007. With permission.)

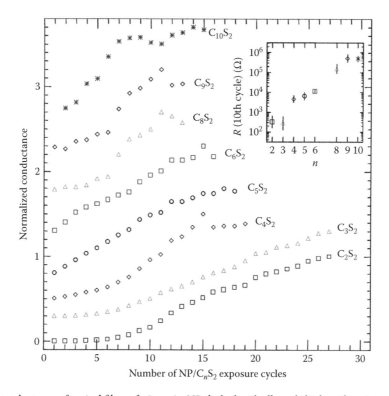

FIGURE 25.6 Normalized conductance of typical films of ~5 nm Au NPs linked with alkanedithiols with various n as a function of the number of immersion cycles. Conductances are normalized with respect to their maximum values. Offsets are added for clarity. Inset: film resistance, R, after the 10th cycle as a function of n. (Reprinted from Zabet-Khosousi, A. et al., *Phys. Rev. Lett.*, 96, 156403, 2006. With permission.)

25.3 Mott–Hubbard Metal–Insulator Transition

Figure 25.7 shows resistances, R, of multilayer films of ~5 nm Au NPs cross-linked with alkanedithiols with various n versus temperature. At low T, two distinct types of behaviors are observed: films with $n \leq 4$ exhibit finite R and are *metallic*; films with $n \geq 6$ exhibit rapidly increasing R and are *nonmetallic*. Both types of behaviors are observed among films with

$n = 5$. At intermediate T, the temperature coefficient of resistance, TCR, provides another means to compare the behavior of samples, where

$$\text{TCR} \equiv \frac{1}{R}\left(\frac{dR}{dT}\right). \tag{25.8}$$

Metals and nonmetals are generally known to exhibit positive and negative TCR, respectively. This trend is followed by films

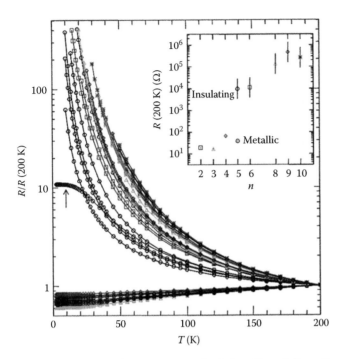

FIGURE 25.7 Normalized resistances of multilayer films of alkanedithiol-linked Au NPs versus temperature. Inset: Resistance of the films at 200 K as a function of n. (Reprinted from Zabet-Khosousi, A. et al., *Phys. Rev. Lett.*, 96, 156403, 2006. With permission.)

with $n \le 4$ and $n \ge 6$, respectively. For $n = 5$, samples with finite R at low T exhibit positive TCR (except one that is indicated by an arrow), and samples with rapidly increasing R at low T exhibit negative TCR. The inset shows R of the films at 200 K as a function of n. As n increases from 2 to 5, R changes by less than an order of magnitude for metallic samples. Going from nonmetallic to metallic samples with $n = 5$, R jumps by 2 orders. Thereafter, R changes by another ~2 orders for nonmetallic samples ($n \ge 5$). This change can be attributed to the exponential growth of tunneling resistances with distance

$$R \propto e^{\beta n}, \qquad (25.9)$$

where β is a constant. The observed β is ~0.9, in agreement with a reported value of ~1.0 ± 0.1 for alkanedithiols in single-molecule junctions (Tao 2006).

The above results can be explained in a context of a Mott–Hubbard MIT model, proposed originally for a lattice of hydrogen atoms (Mott 1990). Consider such a lattice at absolute zero and with variable lattice spacing, s. In the limit $s \to \infty$, an overlap between atomic wave functions is negligible, and electrons are localized on individual atoms. For conduction to occur, electrons have to transfer between neutral atoms, creating positively and negatively charged ions. This transfer requires an energy, U, arising from the difference between the ionization energy (IE) and the electron affinity (EA) of hydrogen atoms. U, known as the Hubbard energy, is the energy cost of transferring an electron from one atom to another and forming an electron-hole pair (Mott 1990):

$$U = \iint |\Psi(r_1)|^2 \frac{e^2}{4\pi\varepsilon r_{12}} |\Psi(r_2)|^2 \, d^3r_1 \, d^3r_2. \qquad (25.10)$$

$\Psi(r)$ is the wave function of the hydrogen atom. For a 1s state,

$$\Psi(r) \propto e^{-r/a_0}, \qquad (25.11)$$

where a_0 is the Bohr radius:

$$a_0 = \frac{4\pi\varepsilon_0 \hbar^2}{me^2}. \qquad (25.12)$$

U for hydrogen-like atoms has been evaluated and is given by

$$U = 0.625 \frac{e^2}{4\pi\varepsilon a_0}. \qquad (25.13)$$

Because of this energy cost, conduction at 0 K is suppressed, and the lattice is insulating.

For finite s, the overlap between atomic wave functions is nonzero and gives rise to energy bands as per the band theory of solids. The energy gap for conduction, E_g, then reduces to (see Figure 25.8)

$$E_g = U - \frac{\Delta_1 + \Delta_2}{2}. \qquad (25.14)$$

The widths, Δ_i, of energy bands, i, depend on the magnitude of overlap integrals, γ, between atomic wave functions:

$$\Delta_1 \approx \Delta_2 \approx \Delta \approx 2z\gamma, \qquad (25.15)$$

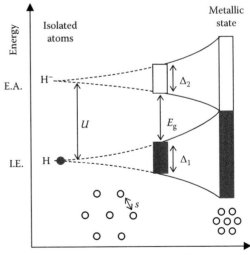

FIGURE 25.8 Evolution of energy levels in a lattice of hydrogen atoms during MIT. (Reprinted from Zabet-Khosousi, A. and Dhirani, A.-A., *Chem. Rev.*, 108, 4072, 2008. With permission.)

where z is the coordination number of atoms in the lattice. γ is given by

$$\gamma = \int \Psi_i^*(r) H \Psi_j(r) d^3 r, \qquad (25.16)$$

where

H is the Hamiltonian of the lattice
i and j represent nearest-neighbor sites at a distance s apart

For hydrogen atoms, the overlap energy integral is given by (Mott 1990)

$$\gamma \approx \frac{e^2}{4\pi\varepsilon a_0} \left(1 + \frac{s}{a_0}\right) e^{-s/a_0}. \qquad (25.17)$$

Note that the dependence of the pre-exponential term on s is negligible, compared with that of the exponential term. γ, therefore, increases exponentially with the decreasing s.

The bandwidths Δ_1 and Δ_2 also increase as s decreases, and at

$$U = \frac{\Delta_1 + \Delta_2}{2}, \qquad (25.18)$$

the energy gap for conduction disappears, and the lattice becomes metallic. Taking $\Delta_1 = \Delta_2$ and $z = 6$, the condition given by Equation 25.18 can be written as

$$U \approx 12\gamma. \qquad (25.19)$$

Using Equations 25.13 and 25.17 for U and γ, we get

$$s \approx 4.5 a_0, \qquad (25.20)$$

which is the Mott–Hubbard criterion for the onset of metallic behavior (Kittle 1986).

An analogous Mott–Hubbard MIT can be realized in ML-NP films where NPs serve as artificial atoms. Here, the energy gap arises from the NP charging energy (i.e. the Coulomb gap). The energy bands arise from the overlap between NP wave functions

$$\Psi_{NP}(r) \propto e^{-\kappa r}, \qquad (25.21)$$

where κ is the decay constant of the wave function outside of the NP. According to a step-potential model, κ is given by

$$\kappa \approx \frac{\sqrt{2m^*\phi}}{\hbar}, \qquad (25.22)$$

where

m^* is an effective mass of electrons
ϕ is the barrier height (i.e. the energy difference between the NP's Fermi level and the vacuum)
$\hbar$ is the reduced Planck's constant

κ provides a useful length scale in the Mott–Hubbard criterion for the MIT (see Equations 25.11 and 25.21 for wave functions):

$$a_0 \sim \frac{1}{\kappa}. \qquad (25.23)$$

Taking $\phi \approx 1.4$ eV and $m^* \approx 0.4 \times$ the mass of electrons reported for the gold–alkanedithiol–gold tunnel junctions (Wang et al. 2004), we obtain $\kappa \approx 4$ nm^{-1}. Applying the Mott–Hubbard criterion, we find a critical NP separation for MIT to be $\sim 4.5/(4 \text{ nm}^{-1}) = 1.1$ nm, which is consistent with the length of pentanedithiol linkers and the observation of the transition at $n = 5$.

To test whether the observed metallic behavior is a result of Mott–Hubbard MIT rather than of direct metal–metal contacts between NPs, the ML-NP films can be annealed under a nitrogen atmosphere. The annealed samples initially follow trends in R shown in Figure 25.7. Eventually, samples with $n \leq 5$ show sudden drops of 30%–50% in R at $100 \pm 20°C$ (see Figure 25.9). The mass spectroscopy and the electron microscopy of the annealed films have shown that annealing at sufficiently high temperatures releases dithiols, which in turn induces metal–metal contacts between NPs and drops in R (Fishelson et al. 2001). This suggests that before annealing, alkanedithiols indeed protect Au NPs from aggregation and point to an observed Mott mechanism of metallic behavior.

The observation of both metallic and insulating behaviors among $n = 5$ samples suggests that other parameters besides n can influence the MIT. They include distributions in NP sizes and inter-NP separations for a given n (due likely to presence of a solvent or TOAB, or an orientation of linkers), as well as fluctuations in electrostatic potentials due to trapped charges.

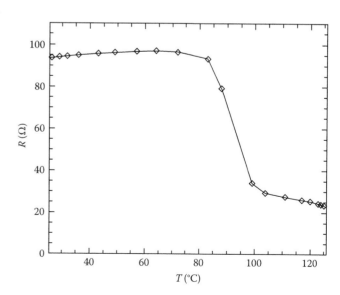

FIGURE 25.9 Resistance versus temperature during annealing a pentanedithiol-linked NP film.

Although bulk films with sufficiently short linker molecules behave as metals, they still exhibit electrical behavior that reflects a presence of their nanoscale components. Figure 25.10 shows normalized R versus T for metallic HS(CH$_2$)$_n$SH-linked Au NP films ($n \leq 5$) and thermally deposited 15 nm thick gold wires. Above 100 K, R varies linearly with T. TCR values, obtained by fitting straight lines to the data above 100 K, are shown in a lower inset. For comparison, a value of bulk Au is also shown. TCR of the NP films are in the range of ~0.001–0.002 K^{-1} and do not exhibit a systematic trend with n. The lack of a systematic trend likely arises due to a film disorder. However, TCR of the NP films are lower than the TCR of bulk Au (~0.0055 K^{-1}) by more than a factor of 2. Below 10 K, film resistances vary slowly (less than 1%). The lowered TCR of metallic ML-NP films arises because their conductivities are lower than that of bulk Au ($\sigma_{Au} = 4.5 \times 10^5 \, \Omega^{-1} \, cm^{-1}$). The highest conductivity reported for the ML-NP films is $\sigma \approx 2 \times 10^5 \, \Omega^{-1} \, cm^{-1} \approx 0.5\sigma_{Au}$, for films consisting of 15 layers of 4.8 nm Au NPs prepared using an ionic stepwise self-assembly method (Liu et al. 1998). The Au NPs were encapsulated with cationic polymer molecules, poly(diallyldimethylammonium chloride), and then attached to anionic polymer molecules, poly S-119, via electrostatic attractions. Studies of other ML-NP films have reported conductivities in the range of ~10^1–10^3 Ω^{-1} cm^{-1} (Musick et al. 2000, Wessels et al. 2004).

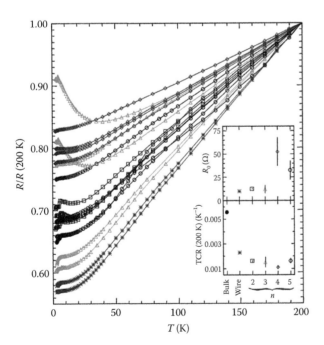

FIGURE 25.10 R of metallic films of HS(CH$_2$)$_n$SH-linked Au NPs ($n \leq 5$) and thin gold wires normalized to their values at 200 K versus T. Upper Inset: Residual resistance of metallic samples extrapolated as $T \to 0$ versus the type of samples. Lower Inset: TCR of the metallic samples at 200 K versus the type of samples. The same symbols are used to represent sample types in both insets (see abscissa) and in the main panel. (Reprinted from Zabet-Khosousi, A. et al., *Phys. Rev. Lett.*, 96, 156403, 2006. With permission.)

According to the Drude theory of metals (Kittle 1986), the conductivity, σ, of a metal is given by

$$\sigma = \frac{\eta e^2 \tau}{m}, \qquad (25.24)$$

where

η is the density of conducting electrons
τ is the mean free time

τ is determined by electron scattering, which can be categorized as elastic (e.g., impurity or defect scattering) and inelastic (e.g., electron–electron or electron–phonon scattering). The mean free time associated with these processes can be written as (Matthiessen's rule)

$$\frac{1}{\tau} = \frac{1}{\tau_{elastic}} + \frac{1}{\tau_{inelastic}}, \qquad (25.25)$$

and the conductivity of metals can be written as

$$\sigma^{-1} = \sigma_0^{-1} + \sigma(T)^{-1}. \qquad (25.26)$$

At very low temperatures, $\tau_{elastic}$ dominates since it is independent of temperature. At higher temperatures, $\tau_{inelastic}$ becomes significant and gives rise to a temperature-dependent conductivity (Kittle 1986). Assuming that the rate of inelastic scattering due to electron–phonon interactions increases as ~$k_B T$, where k_B is Boltzmann's constant and T is temperature, metallic conductivity decreases as ~$1/k_B T$ and resistivity increases linearly with T.

The observation of a lower TCR and σ in ML-NP films compared with bulk Au suggests that electron-scattering processes are strongly enhanced in ML-NP films (Dunford et al. 2006, Dunford and Dhirani 2008ab). Elastic electron-scattering decreases to zero-temperature conductivity (σ_0 in Equation 25.26), which in turn gives rise to smaller values of TCR. Temperature-independent elastic scattering dominates the conductivity of metallic ML-NP films since the sizes of NPs are typically much smaller than the mean free path, ℓ, of electrons in the bulk material. For example, in a film of 5 nm Au NPs, the time scale for elastic scattering can be estimated as $\tau_{elastic} \approx \ell/v_F \approx (5 \times 10^{-9} \, m)/(1.4 \times 10^6 \, m/s) \approx 3.6 \, fs$, where v_F is the Fermi velocity of electrons in gold. For bulk gold, $\ell \approx 41$ nm and $\tau_{elastic} \approx 29$ fs at 300 K (Crowell and Sze 1965). At 300 K, the time scale for inelastic scattering due to phonons is $\tau_{inelastic} \approx \hbar/k_B T \approx 25$ fs. The importance of elastic scattering in ML-NP films near the MIT has been shown by studies of the magnetoconductance of films using superconductor electrodes (Dunford et al. 2006).

25.4 Percolation-Driven Metal–Insulator Transition

ML-NP films comprising sufficiently short cross-linker molecules, such as 2-mercaptoethanol, 2-mercaptoethylamine, or 1,4-butanedithiol, to ensure strong inter-NP coupling, exhibit conductivities that depend strongly on the number of NP/linker

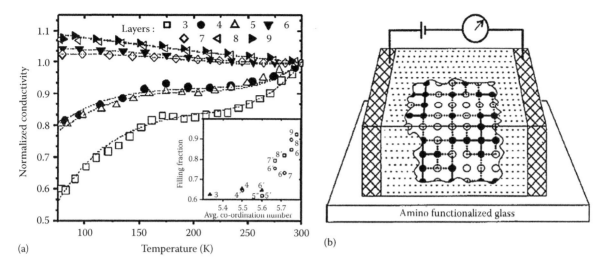

FIGURE 25.11 (a) Normalized conductivity versus temperature for butanedithiol-linked ~15 nm Au NP films prepared with 3–9 immersion cycles. Dashed lines are fits obtained using an effective-medium approximation model. (Inset) *p* versus *z*. Numerical labels correspond to the number of layers. Primes denote unshown conductivity data. (b) Lattice model for disordered arrays of NPs. Filled and open circles denote occupied sites and voids, respectively. Solid lines indicate metallic paths, and dashed lines indicate thermally activated paths. (Reprinted from Trudeau, P.-E. et al., *J. Chem. Phys.*, 117, 3978, 2002. With permission.)

immersion cycles. Figure 25.11a shows a normalized conductivity of films of butanedithiol-linked ~15 nm Au NPs prepared with 3–9 immersion cycles. Films prepared with ≤5 immersion cycles exhibit thermally assisted conductivities (i.e. $d\sigma/dT > 0$), whereas films prepared with ≥6 cycles exhibit metallic-like behavior (i.e. $d\sigma/dT < 0$). The latter were, strictly speaking, metallic since σ remains finite as $T \rightarrow 0$.

The percolation-driven insulator-to-metal transition can be modeled by treating the ML-NP film as a lattice of sites that are connected by random-valued conductances, g_{ij}, where i and j represent nearest-neighbor sites. In this approach, the disordered ML-NP film is first idealized as a lattice randomly filling the lattice sites. Figure 25.11b shows details of the model. NPs and voids are represented by filled and empty sites, respectively. If two adjacent sites i and j are both filled or both empty, then g_{ij} will be considered to be metallic (g_{m}) or insulating (g_{i}), respectively. If only one site is filled, then g_{ij} will be considered to be thermally activated (g_{t}). In a lattice with a fraction of filled sites p, the probability that two sites are connected by g_{m}, g_{t}, or g_{i} is given by p^{2}, $2p(1 − p)$, or $(1 − p)^{2}$, respectively. The distribution of local conductances is, therefore, given by

$$f(g_{ij}, p) = p^{2}\delta(g_{ij} - g_{m}) + 2p(1 - p)\,\delta(g_{ij} - g_{t})$$
$$+ (1 - p)^{2}\delta(g_{ij} - g_{i}), \qquad (25.27)$$

where δ represents Dirac's delta function. In the limit of $p \rightarrow 0$, all sites are empty and the lattice is an insulator. As $p \rightarrow 1$, all sites become occupied and the lattice becomes a metal. However, at $p = p_{c} < 1$, there are a sufficient number of filled sites that can

form a metallic pathway throughout the lattice. p_{c} is known as the percolation threshold and depends on the lattice geometry.

The effective conductance of the lattice, g_{eff}, can be determined using the effective-medium approximation (Kirkpatrick 1973). In this approximation, the average effect of the random g_{ij} is represented by an effective medium where all nearest-neighbor sites are connected by the equal conductances, g_{eff}. Assuming that the current between sites i and j remains the same, replacing g_{ij} with g_{eff} causes a local voltage, ΔV_{ij}, to be induced between the sites. g_{eff} is chosen such that ΔV_{ij} will average to zero. Using methods of network analysis, one can show that (Kirkpatrick 1973)

$$\Delta V_{ij} = V_{eff}\frac{g_{eff} - g_{ij}}{g_{ij} + (z/2 - 1)\,g_{eff}}, \qquad (25.28)$$

where

V_{eff} is the voltage drop between the adjacent sites in the effective medium

z is the coordination number of the lattice (e.g., $z = 4$ for square and $z = 6$ for cubic lattices in 2D and 3D, respectively)

Given the distribution $f(g_{ij}, p)$, the condition that the average of ΔV_{ij} should vanish yields

$$\sum_{\alpha = m, t, i}\frac{f(g_{\alpha}, p)(g_{eff} - g_{\alpha})}{g_{\alpha} + g_{eff}(z/2 - 1)} = 0. \qquad (25.29)$$

Taking $g_{i} = 0$, we obtain a quadratic equation for g_{eff} with roots

$$g_{eff} = \frac{-A \pm \sqrt{A^{2} + 4B}}{z - 2}, \qquad (25.30)$$

where

$$A = g_m\left(1 - p^2\frac{z}{2}\right) + g_t[1 - p(1-p)z],$$

(25.31)

$$B = g_m g_t\left[\frac{z}{2} - 1 - \frac{z}{2}(1-p)^2\right].$$

We take the positive root that gives the correct limiting result when $p \rightarrow 1$:

$$\lim_{p\rightarrow 1} g_{eff} = g_m.$$

(25.32)

The effective medium approximation with three types of conductances generates two thresholds (Pury and Cáceres 1997). The first threshold, p_t, arises from the requirement that $g_{eff} > 0$. Below this threshold, the current cannot flow. p_t can be obtained as follows:

$$g_{eff} > 0 \Rightarrow -A + \sqrt{A^2 + 4B} > 0 \Rightarrow B > 0 \Rightarrow p > 1 - \sqrt{1 - \frac{2}{z}}$$

$$\Rightarrow p_t = 1 - \sqrt{1 - \frac{2}{z}}.$$

(25.33)

Thus, p_t only depends on z: for $z = 4$, $p_t = 0.293$ and for $z = 6$, $p_t = 0.184$. The second threshold, p_m, corresponds to the onset of a metallic sample-spanning pathway. For $p_t < p < p_m$, the conductance is thermally assisted. Above p_m, both metallic and thermally activated pathways can be present. However, at sufficiently low temperatures, the thermally activated pathways shut down and the metallic pathways give rise to finite conductances as $T \rightarrow 0$. For $p \gg p_m$, the metallic pathways dominate and the lattice exhibits metallic behavior in a wide range of temperatures. To obtain p_m, we also set $g_t \rightarrow 0$. Equation 25.29 then becomes

$$g_{eff} = g_m\frac{p^2 z - 2}{z - 2}.$$

(25.34)

The requirement that $g_{eff} > 0$ now yields

$$p_m = \sqrt{\frac{2}{z}}.$$

(25.35)

p_m also only depends on z: For $z = 4$, $p_m = 0.707$, and for $z = 6$, $p_m = 0.577$.

Fits to the data, obtained using the effective-medium approximation, are shown as dashed lines in Figure 25.11a. The fits are obtained by taking

$$g_m = C_m\left[1 + \alpha(T - 300)\right]^{-1},$$

(25.36)

$$g_t = C_t \exp(-\beta s)\exp(E_\alpha/k_B T),$$

(25.37)

$$g_i = 0,$$

(25.38)

where

C_m and C_t are constants
α is the (TCR, see Equation 25.8)
β is the tunneling-decay constant ($\beta = 2\kappa$, where κ is defined by Equation 25.22)
s is the inter-NP separation
E_a is the activation energy
z is estimated by

$$z = 6 - \frac{2}{l},$$

(25.39)

where l is the number of layers and is approximated by the number of deposition cycles. Data for the nine-layer ML-NP film are used to determine α since this film is dominated by metallic conduction. Data for the five-layer film are used to determine C_m/C_t. Using physically-reasonable initial estimates ($\phi = 1\,eV$, $E_a = 28\,meV$, $s = 1\,nm$, $p = 0.6$), we determine C_m/C_t and interactively refine our estimates. The resulting parameters ($\phi = 0.21\,eV$, $E_a = 48\,meV$, $s = 1.66\,nm$) are used to fit the remaining data, using only p as a fit parameter.

The model satisfactorily describes the observed data over a wide range of temperatures. In samples with $p \approx 0.6$ (we were unable to observe the current below $p \approx 0.6$), the conductance is thermally assisted. As p increases, g_{eff} versus T exhibits signs of increased contributions from locally metallic transport. At $p = p_m \approx 0.65$–0.70, a sample-spanning metallic pathway is formed. Just beyond p_m, a combination of metallic and thermally activated transport is observed with the latter even dominating despite the films being fundamentally metallic (for example, see data for the seven-layer film in Figure 25.11a). As p increases further, the metallic transport dominates over a larger temperature range, and a limiting bulk behavior is observed (see Section 25.3). Although the model neglects a number of transport phenomena such as disorder-driven localization and variable-range hopping, it stresses that local fluctuations can generate highly variable local conductances, which in turn compete to determine overall film conductivity.

25.5 Applications

Observation of MIT in ML-NP films as a function of linker length or film thickness suggests a remarkable possibility of preparing materials with properties in between metals and insulators. Metals exhibit no energy gap (E_g) between the valence and the conduction bands, while insulators exhibit E_g that are very large compared to $k_B T$. Conventionally, materials with intermediate values of E_g (i.e. $E_g \sim k_B T$) are viewed as semiconductors and have found important electronic applications, notably in field-effect transistors (FETs). For semiconductors, charge carriers can be generated by thermal excitation across the energy gap; however, the density of the charge carriers is usually not so high as to cause a complete screening of the electric field inside the material. This property enables the control overflow of charge carriers via the application of a gate electric field in an FET configuration. Conductance can be switched between a maximum ("ON") and a minimum ("OFF") value as the gate electric field is varied.

From this perspective, ML-NP films with $E_g \sim k_B T$, too, can be viewed as (artificial) semiconductors and can be exploited as

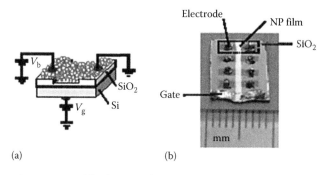

(a)

(b)

FIGURE 25.12 (a) Schematic of an ML-NP film device. (b) Photograph of a sample chip consisting of four devices. (Reprinted from Suganuma, Y. et al., *Nanotechnology*, 16, 1196, 2005. With permission.)

functional elements of FETs. E_g for molecularly linked NP films can be controlled via inter-NP coupling according to Equation 25.14 and via the size of superclusters of strongly coupled NPs according to Equation 25.4. In this section, two applications of "semiconducting" ML-NP films, namely, conductance switching and information storage, are briefly described. Films comprise butanedithiol-linked Au NPs, and as discussed earlier, exhibit an insulator to the metal transition as a function of the number of deposition cycles. Here, the films are prepared using four deposition cycles and are below the percolation threshold.

Figure 25.12a shows the schematic of the ML-NP film device. The ML-NP film is prepared on a silicon/silicon-oxide substrate. Two electrodes, source and drain, are attached to the film, and the silicon substrate is used as the gate electrode. The drain electrode is electrically grounded, and voltages are applied to the source and gate electrodes. Note that the devices are in macroscopic dimensions (Figure 25.12b).

Figure 25.13 shows differential conductance versus bias (V_b) and gate (V_g) voltages for a typical device at 77 K. At $V_g = 0$, a clear conductance suppression is observed as a dip at zero bias (Figure 25.13b). This is due to the single-electron charging of superclusters of cross-linked NPs in the film. A gate effect results in an approximately linear shift of the conductance

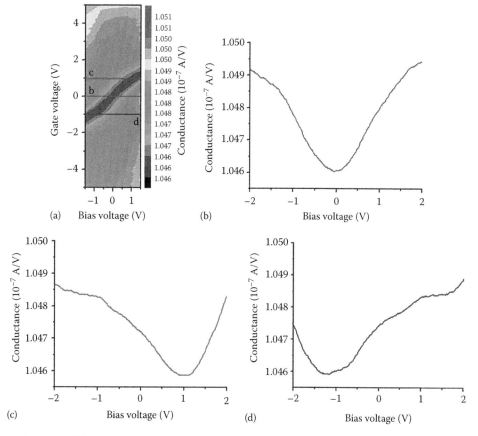

FIGURE 25.13 **(See color insert following page 9-8.)** (a) Differential conductance map as a function of V_b and V_g at 77 K. The map is obtained using a four-layer film of butanedithiol-linked Au NPs. (b–d) Differential conductance versus V_b at various V_gs, (b) $V_g = 0$ V, (c) $V_g = +1$ V, and (d) $V_g = -1$ V. (Reprinted from Suganuma, Y. et al., *Nanotechnology*, 16, 1196, 2005. With permission.)

dip away from zero bias. For example, at $V_g = +1$ and $-1\,V$, the conductance dip shifts to $V_b = +1$ and $-1\,V$, respectively (see Figure 25.13c and d). The variation of the conductance dip with V_g suggests that the gate voltage shifts the charging energy of the superclusters, in turn shifting the CB bias thresholds. At $V_b = 0$, conductance increases with increasing $|V_g|$. This feature illustrates the principle of conductance switching in FETs.

Another important application of semiconductors and FETs is information storage. The ML-NP film device can be used for this purpose as well. Figure 25.14a through c shows the effect of applying V_g of -5, 0 and $+5\,V$ as the device is cooled. Below a threshold temperature ($\sim$175 K), the CB gap remains shifted even if the gate voltage is subsequently turned off. This implies that the value of the applied V_g is effectively "recorded" in the conductance map of the device. The recorded value of V_g can then

be "read" through the shift in the CB gap. We also observe that upon warming the device, the CB gap becomes weaker and eventually at $\sim$175 K vanishes. Therefore, the stored information can be "erased."

Since the gate voltage can be varied continuously, the ML-NP film device can be used for analog memory storage. By applying multi-valued time-dependent V_g during cooling, it is possible to store multi-valued information. Figure 25.14d shows two examples. Applying cyclic time-dependent V_gs (see Figure 25.14d, center) during cooling can generate conductance maps at 77 K that resemble "$\bullet\bullet-$" and "$-$", that is, "U" and "T" in Morse code, respectively.

The ability to store values of V_g in the conductance maps can be attributed to the redistribution of the background charges. Above the threshold temperature, mobile background charges can redistribute in order to screen gate-induced electric fields. As the

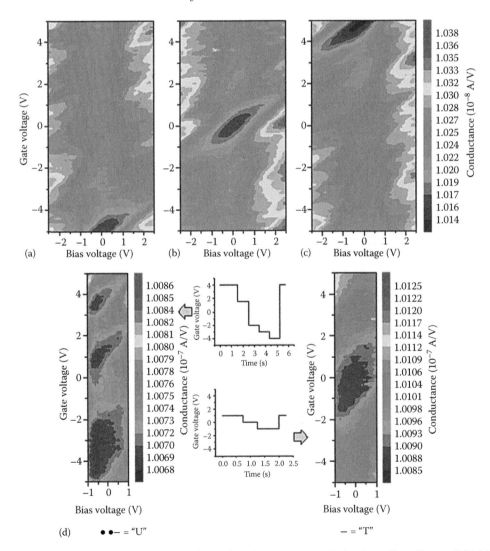

FIGURE 25.14 (See color insert following page 9-8.) Differential conductance maps of a four-layer film of butanedithiol-linked Au NPs as a function of V_b and V_g. The maps are obtained at 77 K after applying various gate voltages to the film as the film was slowly cooled. (a–c) Constant V_gs are applied to the film during cooling: (a) $V_g = -5\,V$, (b) $V_g = 0$, and (c) $V_g = +5\,V$. (d) Cyclic V_g (shown in the center) are applied to the film during cooling. The stored information in the conductance maps reading from $V_g = +5\,V$ toward $-5\,V$ resembles "$\bullet\bullet-$" (left) and "$-$" (right), which correspond to "U" and "T", respectively, in Morse code. (Reprinted from Suganuma, Y. et al., *Nanotechnology*, 16, 1196, 2005. With permission.)

temperature is lowered, eventually these charges can become trapped or "frozen", creating a charge-glass that generates gating fields even after the gate voltage is removed. There are several possible places where charges may be trapped, including on NPs themselves, linker molecules, SiO_2 substrate, and/or interfaces.

25.6 Conclusion

ML-NP films exhibit electronic properties that can be tuned from metallic to insulating by varying inter-NP coupling or film thickness. For thick ML-NP films, inter-NP coupling can be controlled by varying the length of the linker molecules. Films comprising short linkers ($n < 5$) are metallic, and films comprising longer linkers ($n > 5$) are not. The observed MIT as a function of n can be explained in the context of a Mott–Hubbard model and underscores the important role of linker molecules in influencing film properties. Thin films (prepared with less than ~5 cycles) are nonmetallic, whereas thicker films (prepared with more than ~5 cycles) are metallic. The transition as a function of film thickness can be explained using an effective-medium model. The model demonstrates the important role of percolation and competitive transport processes in determining film properties. ML-NP films near the MIT have intermediate properties, can be viewed as semiconductors and can be used to fabricate FETs. These results demonstrate the ability to control material properties over a wide range via nanoscale architecture and underscore the utility of ML-NP films as a platform for studying charge transport.

References

Abeles, B., Sheng, P., Coutts, M. D., and Arie, Y. 1975. Structural and electrical properties of granular metal films. *Advances in Physics*. 24: 407–461.

Bergmann, G. 1984. Weak localization in thin films. *Physics Reports*. 107: 1–58.

Brust, M., Bethell, D., Kiely, C. J., and Schiffrin, D. J. 1998. Self-assembled gold nanoparticle thin films with nonmetallic optical and electronic properties. *Langmuir*. 14: 5425–5429.

Crowell, C. R. and Sze, S. M. 1965. Ballistic mean free path measurements of hot electrons in Au films. *Physical Review Letters*. 15: 659–661.

Devoret, M. H. and Grabert, H. 1992. Introduction to single charge tunnelling. In *Single Charge Tunneling*, eds. H. Grabert and M. H. Devoret. New York: Plenum.

Dunford, J. L. and Dhirani, A.-A. 2008a. Conductance oscillations in molecularly linked Au nanoparticle film-superconductor systems. *Nanotechnology*. 19: 025202–025208.

Dunford, J. L. and Dhirani, A.-A. 2008b. Reflectionless tunneling at the interface between nanoparticles and superconductors. *Physical Review Letters*. 100: 147202–147205.

Dunford, J. L., Dhirani, A.-A., and Statt, B. 2006. Magnetoconductance of molecularly linked Au nanoparticle arrays near the metal-insulator transition. *Physical Review B*. 74: 115417–115422.

Fishelson, N., Shkrob, I., Lev, O., Gun, J., and Modestov, A. D. 2001. Studies on charge transport in self-assembled gold-dithiol films: Conductivity, photoconductivity, and photoelectrochemical measurements. *Langmuir*. 17: 403–412.

Freeman, R. G. et al. 1995. Self-assembled metal colloid monolayer: An approach to SERS substrates. *Science*. 267: 1629–1632.

Goebbert, C., Nonninger, R., Aegerter, M. A., and Schmidt, H. 1999. Wet chemical deposition of ATO and ITO coatings using crystalline nanoparticles redispersable in solutions. *Thin Solid Films*. 351: 79–84.

Katz, E. and Willner, I. 2004. Integrated nanoparticle-biomolecule hybrid systems: Synthesis, properties, and applications. *Angewandte Chemie International Edition*. 43: 6042–6108.

Kirkpatrick, S. 1973. Percolation and conduction. *Review of Modern Physics*. 45: 574–588.

Kittle, C. 1986. *Introduction to Solid State Physics*. New York: Wiley.

Liu, Y., Wang, Y., and Claus, R. O. 1998. Layer-by-layer ionic self-assembly of Au colloids into multilayer thin-films with bulk metal conductivity. *Chemical Physics Letters*. 298: 315–319.

Mott, N. F. 1990. *Metal-Insulator Transitions*. London, U.K.: Taylor & Francis.

Musick, M. D. et al. 2000. Metal films prepared by stepwise assembly. 2. Construction and characterization of colloidal Au and Ag multilayers. *Journal of Chemistry and Materials*. 12: 2869–2881.

Park, H., Lim, A. K. L., Alivisatos, A. P., Park, J., and McEuen, P. L. 1999. Fabrication of metallic electrodes with nanometer separation by electromigration. *Applied Physics Letters*. 75: 301–303.

Pury, P. A. and Cáceres, M. O. 1997. Tunneling percolation model for granular metal films. *Physical Review B*. 55: 3841–3848.

Schmid, G. and Corain, B. 2003. Nanoparticulated gold: Syntheses, structures, electronics, and reactivities. *European Journal of Inorganic Chemistry*. 3081–3098.

Snow, A. W. et al. 2002. Self-assembly of gold nanoclusters on micro- and nanoelectronic substrates. *Journal of Material Chemistry*. 12: 1222–1230.

Suganuma, Y. and Dhirani, A.-A. 2005. Gating of enhanced electron-charging thresholds in self-assembled nanoparticle films. *Journal of Physical Chemistry B*. 109: 15391–15396.

Suganuma, Y., Trudeau, P.-E., and Dhirani, A.-A. 2005. Multi-valued analogue information storage using self-assembled nanoparticle films. *Nanotechnology*. 16: 1196–1203.

Tao, N. J. 2006. Electron transport in molecular junctions. *Nature Nanotechnology*. 1: 173–181.

Trudeau, P.-E., Orozco, A., Kwan, E., and Dhirani, A.-A. 2002. Competitive transport and percolation in disordered arrays of molecularly-linked Au nanoparticles. *Journal of Chemical Physics*. 117: 3978–3981.

Wang, W., Lee, T., and Reed, M. A. 2004. Elastic and inelastic electron tunneling in alkane self-assembled monolayers. *Journal of Physical Chemistry B*. 108: 18398–18407.

Wessels, J. M. et al. 2004. Optical and electrical properties of three-dimensional interlinked gold nanoparticle assemblies. *Journal of American Chemical Society*. 126: 3349–3356.

Zabet-Khosousi, A. and Dhirani, A.-A. 2007. Shadow mask fabrication of micron-wide break-junctions and their application in single-nanoparticle devices. *Nanotechnology*. 18: 455305–455310.

Zabet-Khosousi, A. and Dhirani, A.-A. 2008. Charge transport in nanoparticle assemblies. *Chemical Reviews*. 108: 4072–4124.

Zabet-Khosousi, A. et al. 2006. Metal-insulator transition in films of molecularly linked gold nanoparticles. *Physical Review Letters*. 96: 156403–156406.

26

Tribology of Nanoparticles

Lucile Joly-Pottuz
University of Lyon

26.1 Introduction

Tribology (etymologically from the Greek "tribein," which means to rub, and "logos" a study) is the science of friction, wear, and lubrication. It is not a well-known science but it affects many areas of our daily life. The simple act of walking is governed by the friction of soles of shoes on the ground. Sports like skiing, skating, etc., are based on tribological phenomena. The brake of a car is provided by a friction between the two sides of the system (disk mounted on the wheel and brake pads), which must be raised to ensure effective braking, but if possible with low wear to save the pads.

However, most of the areas require a combination of low friction and low wear. To reduce friction, a third body (the lubricant) is needed between the two bodies in contact. For example, in the case of skating, a thin film of water formed between the ice and the skate acts as a lubricant and facilitates the sliding. But contrary to the previous example where the lubricant is provided inside the contact, in most cases it is necessary to supply the lubricant externally and to find the one that is best suited to a given application.

Today, improving the lubrication in the automotive field is a strong economic stake: the reduction of friction in the engines will lead to a reduction of gas consumption, the reduction of wear will increase their durability. The reduction in the gas consumption of engines is crucial at the moment. Indeed, the fuel reserves dwindle while the demand is constantly growing. But it is from an environmental point of view that improving the lubrication becomes very important. Indeed, since the 1979 Convention on air pollution and the Kyoto Protocol (1994), numerous research programs have been conducted to reduce transport pollution. The catalytic converters and particulate filters illustrate these advances. But much remains to be done to make our cars "clean."

Additives with a tribological action currently used in commercial lubricating oils are dithiocarbamate molybdenum (MoDTC) and zinc dithiophosphate (ZnDTP). These compounds are complex organic molecules containing sulfur and phosphorus. These two elements are known to be poisonous for catalytic converters because they hinder their proper functioning. Both additives also have other disadvantages. Their mechanism of action is based on the chemical reactions requiring high temperatures leading to the formation of compounds that will reduce the friction and limit the wear, but also to the formation of volatile harmful compounds. Furthermore, these compounds are only active at high temperatures. This means a critical period in cold start of the engines. Thus, it is necessary to find new additives that are environment-friendly and more efficient than those currently used.

Nanoparticles can be considered as modern lubricant additives. They present several major advantages compared to organic molecules currently used as lubricant additives:

- Their nanometer size allows them to enter easily the contact area, like molecules.
- They are immediately efficient even at ambient temperature. Thus, no induction period is necessary to obtain interesting tribological properties.

Several types of nanoparticles are now envisaged as lubricant additives. Nested nanoparticles (fullerenes) made on metal dichalcogenides or carbon were particularly studied because they are the nested structure of well-known lamellar compounds used in tribology (2H-MoS$_2$, graphite). Graphite has been studied for its tribological properties since 1950 (Savage 1948) and has been used as a solid lubricant for a long time. On a macroscopic scale, a friction coefficient of 0.1 was obtained with graphite, without any other lubricant and even at high temperature (Bowden and Tabor 1950). MoS$_2$ coatings are often used to lubricate surfaces in ultrahigh vacuum (space) conditions. But their efficiency depends on the oxygen content present in the coating composition (Fleischauer et al. 1999). Super-low friction coefficient (below 0.01) was obtained with an MoS$_2$ coating containing less than 1% of oxygen (Martin et al. 1993). The majority of MoS$_2$ films deposited in vacuum contain nonnegligible quantities of oxygen (10%–20% in atomic concentration) but are of interest for tribological properties. Basically, the good tribological properties of MoS$_2$ coatings are due to the formation of a transfer film on the counterface. MoS$_2$ coatings are preferentially used in ultrahigh vacuum since the environment has an influence on their tribological properties. Friction coefficient increases with moisture up to 65%, then decreases (Peterson et al. 1953). Indeed, in case of high moisture content, water molecules favored a cohesion of MoS$_2$ crystallites. Then, the transfer film, which controls the reduction of friction and the good properties of MoS$_2$ coatings, cannot be formed and the whole coating can be removed (Lancaster et al. 1990). From these results, we can think that the closed structure of MoS$_2$ will present advantageous tribological properties, even better than the corresponding lamellar MoS$_2$ structure. Indeed, fullerenes will present a very low quantity of oxygen (only the two first layers are slightly oxidized). Furthermore, the nanometer size of the sheets liberated during friction with fullerenes is a great advantage since these sheets can easily stick parallel to the surface. This will improve the movement by improving the shear between the two surfaces. On the contrary, MoS$_2$ sheets of the lamellar structure, with their micrometer size, can stick perpendicularly to the surface that is not favorable for an easy shear.

The nanoparticles tested in this chapter, carbon onions and inorganic fullerenes of MS$_2$ (IF-MS$_2$, M = Mo, W), have a spheroidal, nested structure. They have a multiwall structure (Figure 26.1) and their layers are made of curved carbon sheets or MS$_2$ sheets. Thus, they are composed of potentially lubricating sheets (of graphene or MS$_2$) without the disadvantage of containing dangling bounds on their edges, which are very reactive, as observed for lamellar MoS$_2$.

To evaluate the tribological properties of nanoparticles as lubricant additives, they were added at different concentrations (from 0.1 to 1 wt%) to a lubricating base oil (poly-alpha-olefin (PAO 4-PAO 6) base oil by ultrasonic bath. Dispersions of such nanoparticles in oil are not always stable for a long time without a dispersant additive. Two phenomena are responsible for the poor stability: aggregation and flocculation of the particles, sedimentation of particles and aggregates. van der Waals forces are responsible for aggregation of the nanoparticles. Gravitation forces make the particles fall down, but viscosity forces slow their motion and Archimede forces are opposed to gravitation forces. Basically, the sedimentation speed varies like the square of the radius of the particle. Thus, aggregates of nanoparticles will sediment faster than isolated ones. On the contrary, the Brownian motion can improve the stability of the nanoparticle dispersion. It corresponds to the movement of the particles due to the thermal motion caused by the collisions of the molecules of the liquid phase on the periphery of the particle. For small particles, the relation of Stokes–Einstein expresses the average distance x made by a particle during a time t. By considering all these phenomena, it is possible to evaluate the effect of the diameter of the nanoparticles on the stability of dispersion. The sedimentation time decreases when the mean diameter of the nanoparticles increases. This result is not surprising since van der Waals forces, responsible for nanoparticles aggregation, are more significant for nanoparticles with a large diameter. The mixing time of IF-WS$_2$ dispersed in paraffin oil can have an influence on the tribological properties of the dispersion (Moshkovith et al. 2007). The increase of the mixing time leads to a decrease of the size of nanoparticle aggregates and to a better reproducibility of friction experiments.

To evaluate the performances of a lubricant, tribological experiments are performed on a pin-on-flat tribometer consisting of a hemispherical pin sliding on a flat, and a few droplets of the lubricant are deposited on the flat before the experiment. Pin and flat are made of AISI 52100 steel (roughness: 25 nm). During friction, the pin is elastically deformed and the real contact surface of the pin is circular (see Figure 26.2). The diameter of this contact area corresponds to the Hertz calculated diameter. It depends on several parameters, material of both antagonist surfaces (Young's modulus E_1 and E_2 and Poisson's ratio v_1 and v_2 of surfaces), hemispherical pin diameter (R_1), and normal load (W), and can be calculated by using Equation 26.1:

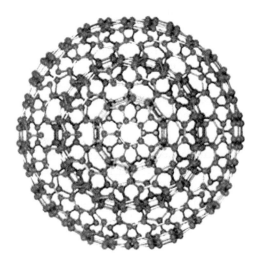

FIGURE 26.1 Schematic structure of inorganic fullerene of MS$_2$ or carbon onions.

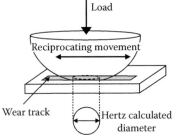

Load (N)	Contact pressure (GPa)	Calculated Hertz diameter (µm)
1	0.66	54
2	0.83	68
5	1.12	92
10	1.42	116

FIGURE 26.2 Pin-on-flat tribometer principle.

$$d = 2^* a = 2^* \left(\frac{3WR_1}{4E^*} \right)^{1/3} \qquad (26.1)$$

$$\frac{1}{E^*} = \frac{1 - v_1^2}{E_1} + \frac{1 - v_2^2}{E_2} \qquad (26.2)$$

The normal pressure generated on surface is maximum at the center of the contact zone and the lateral distribution is quadratic with a parabolic decrease toward edges. The maximum contact pressure can be determined from Equation 26.3:

$$P_{max} = \frac{3W}{2\pi \cdot a^2} = \left(\frac{6WE^{*2}}{\pi^3 R_1^2} \right)^{1/3} \qquad (26.3)$$

During friction process, the pin is elastically deformed and the real contact surface is circular. An observation by optical microscopy of the pin after friction gives information on the wear quantity. If wear is low, there are only some scratches on the pin. At the opposite, if the wear is important, the wear scar has a diameter larger than the Hertz calculated diameter and the pin is truncated and flattened. Friction coefficient is measured during friction test and corresponds to the ratio of the tangential force to the normal force:

$$\mu = \frac{F_t}{F_n} \qquad (26.4)$$

The lower is the friction coefficient, the easier the sliding between the two parts. Reducing the friction coefficient makes movement easier and energy is preserved. This is essential to reduce the consumption of the engine. Reduction of friction in engines is a real challenge. Friction coefficient of less than 0.02 can be achieved but cannot be transferred in an engine for practical reasons. With a friction coefficient of 0.02, a thrust of 20 g would be enough to move an object of 1 kg.

In this study, our goal was to reduce the friction coefficient below 0.1. Nanoparticles were studied as friction modifier and antiwear; this means that they were added to a base oil to reduce its friction coefficient and to reduce the wear observed during friction process. Wear is also an important problem since it causes the formation of debris that can be abrasive and increase wear. Presence of debris resulting from the wear of the parts in contact in an engine is the reason for oil change. So, reducing wear leads to space oil changes out, which will be good for the environment.

To evaluate the effect of the addition of nanoparticles in the base oil, two parameters were studied: the nanoparticle concentration in oil and the contact pressure inside the contact area. To study the lubrication mechanism of the nanoparticles, they were observed by transmission electron microscopy (TEM) first before friction, and after friction to study their transformation inside the contact area. Other analytical techniques like Raman spectroscopy and x-ray diffraction (XRD) are also very useful and gave important information on their structure.

26.2 Carbon Onions

26.2.1 Synthesis and Characterization

In 1992, Ugarte synthesized for the first time carbon onions by degradation of carbon soots under the electron beam irradiation in a TEM (Ugarte 1992). Several theories for the growth mechanism of carbon onions were considered (Ugarte 1995). Ugarte suggested that graphite sheets are formed starting from the surface of the nanoparticle, then proceeding into the core. Kuznetsov et al. studied the transformation of diamond nanoparticles into carbon onions and suggested that the formation of an outer graphite shell first occurs by the transformation of the (111) diamond planes into the (001) graphite planes (Kuznetsov et al. 1994). Thus the transformation starts from the surface and progressively proceeds to the center of the diamond nanoparticle. Other investigators have proposed another mechanism based on the formation of "spiroids" first, then a transformation into carbon onions (Qin et al. 1996, Ozawa et al. 2002). Ugarte's model was confirmed by several studies in the literature (Roddatis et al. 2002, Mykhaylyk et al. 2005). To explain the transformation of diamond into graphite, a "zipperlike" transformation mechanism was proposed (Kuznetsov et al. 1999). It is based on the opening of three cubic (111) diamond planes to form graphite sheets. Transformation of the diamond nanoparticle from the surface to its core suggests that the size of the carbon onion is directly related to the size of the initial diamond nanoparticle. However, diamond nanoparticles with a large diameter do not lead to the formation of carbon onions, but to the formation of graphite sheets (Hiraki et al. 2005).

Several synthesis methods for carbon onions have already been proposed. Several of them are based on the transformation of a specific carbon form into carbon onions, this transformation being often activated by electron beam irradiation. Ugarte synthesized the first carbon onions by irradiation of carbon soots. The conversion of carbon films under Al nanoparticles (Xu et al. 1998) or gold nanoparticles (Troiani et al. 2003) was reported. But diamond nanoparticles are most often used as a precursor. The transformation of diamond nanoparticles into carbon onions was studied by several investigators (Roddatis et al. 2002, Mykhaylyk et al. 2005). Annealing temperature of diamond has an influence on the structure of carbon onions obtained. Tomita showed that after annealing at 1700°C, nanoparticles are converted into carbon onions but a residual diamond core of the nanoparticles is conserved (Tomita et al. 2002). After annealing at 2000°C, the diamond core is effectively reduced in size but carbon onions become faceted.

Other synthesis methods were reported: arc discharge in water (Sano et al. 2001), and thermal reduction of glycerin in the presence of magnesium (Du et al. 2005). Recently, a synthesis method based on the decomposition of CH_4 on NiO/Al composite powder was reported (He et al. 2006a,b). This method leads to a massive production of carbon onions (1 g per hour) but the carbon onions contain a nickel core. Carbon onions of 15–40 nm in diameter containing a Fe core were synthesized by CVD (Wang et al. 2006).

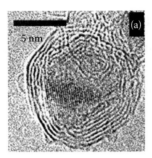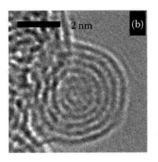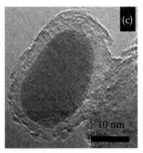

FIGURE 26.3　TEM image of carbon onions: (a) with diamond core, (b) without diamond core, (c) with a nickel core.

Carbon onions can be either fully graphitic-like, or contain a diamond or a metallic core. To fully study the tribological properties of carbon onions, three kinds of carbon onions were studied: fully graphitized carbon onions, carbon onions containing a diamond core, and carbon onions containing a nickel core. The first sample (named CO1) was obtained by annealing of diamond nanoparticles at 1600°C during 13 min and carbon onions containing a residual diamond core were preferentially formed. Typically, they have an average of 5–10 nm and a multilayer structure (Figure 26.3a). The presence of a residual diamond core was attributed to an incomplete graphitization. The second sample (named CO2) of carbon onions was synthesized by annealing diamond nanoparticles at 1700°C to obtain carbon onions without diamond core (Figure 26.3b) (Joly-Pottuz et al. 2008a). Carbon onions containing a nickel core were synthesized by decomposition of CH_4 on NiO/Al composite powder (Figure 26.3c).

The two first samples were fully characterized by electron energy-loss spectroscopy (EELS) and UV-Raman spectroscopy in order to confirm the presence of the diamond core in the first sample. EELS performed in a TEM is a useful technique to distinguish the different forms of carbon and has already been widely used to characterize such kind of nanoparticles but also single walled (Kuzuo et al. 1994) or multi walled carbon

nanotubes (Ajayan et al. 1993), C_{60} and C_{70} fullerenes (Sohmen et al. 1992). Several authors used this technique to follow the transformation of diamond into carbon onions (Tomita et al. 1999, Mykhaylyk et al. 2005), or the transformation of carbon onions into diamond under irradiation-induced compression (Redlich et al. 1998). Figure 26.4 presents results obtained by EELS on plasmon peak and carbon ionization edge. Basically, these two regions of the EELS spectrum are very useful to distinguish the different forms of carbon. Results on the low-loss plasmon peak clearly show a different shape between the different forms. Typically, this peak is located at 27 eV for graphite and at 33 eV for pure diamond (Egerton 1986). For graphite, a peak at 6 eV corresponding to the excitation of π electrons can also be observed. In our data, the spectrum of nanodiamond particles used as precursors for the carbon onion synthesis shows two peaks: the first one corresponds to amorphous carbon (22 eV) and a second one to diamond (33 eV). This is due to the presence of adventitious carbon at the periphery of nanodiamond particles and this has already been observed by Tomita et al. (1999). Plasmon peak of carbon onion with diamond core (CO1) also presents several contributions: one for the diamond structure and one for the graphite structure. A small peak can be seen at 6 eV, which is characteristic of graphitic carbon. These results confirm that the nanoparticle is composed of a diamond core

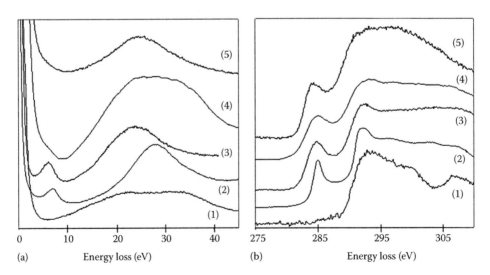

FIGURE 26.4　EELS spectrum of diamond nanoparticles (1), graphite (2), CO2 (3), CO1 (4), amorphous carbon (5). (a) Plasmon peak, (b) carbon K-edge. (From Joly-Pottuz, L. et al., *Tribol. Int.*, 41, 69, 2008a. With permission.)

surrounded by graphitic shells. The low-loss spectrum of carbon without diamond core (CO2) differs from the one of CO1 sample. The π plasmon peak is more intense and is centered at 6.1 eV, while this peak is at 7 eV for graphite. This difference has already been observed by Cabioc'h et al. (1997) and can be explained by the curvature of shells in the carbon onions: the coupling of electrons on the spherical shells being different from the coupling in the planar case (Yannouleas et al. 1996). No contribution of diamond is observed on the (σ + π) plasmon peak of CO2, showing that these carbon onions are completely graphitized.

At the carbon K-edge, a feature to distinguish easily graphite and diamond is the 1s/π^* transition (at 285 eV), which is present only for graphite (see Figure 26.4b). The EELS carbon K-edge observed for nanodiamond is very similar to the one of bulk diamond. The small peak visible at 285 eV confirms the presence of amorphous carbon around diamond nanoparticles. The 1s/π^* transition is observed in the spectrum of carbon onions. However, a broadening of this peak is observed compared to pure graphite and this can be explained by the absence of a long-range graphite-like order in the carbon structure of onions (Mykhaylyk et al. 2005).

UV-Raman spectroscopy was used to characterize the two samples since it is a very useful technique to detect the presence of diamond inside the particles. Indeed, UV–Raman spectroscopy is more sensitive to σ bonding present in all carbon structures (Gilkes

et al. 1998). By using a visible excitation wavelength, two broad peaks are obtained and it is not possible to distinguish between the two samples (Roy et al. 2003). Furthermore, the spectrum of nanodiamond contains a luminescence background that is difficult to remove (Sun et al. 2000). Figure 26.5 presents the spectrum obtained for the diamond nanoparticles used as a precursor and the two carbon onions samples, respectively. The results clearly give evidence for the presence of a diamond core, but only in the first sample CO1.

26.2.2 Tribological Properties

Few studies on the tribological properties of carbon onions have been reported so far in the literature. The addition of onions inside silver films does not have any effect on the friction coefficient, but an increase of lifetime of the coating by a factor of 15 was observed (Cabioc'h et al. 2002). The fact that carbon onions do not decrease friction coefficient of the silver films was explained by the fact that they are embedded inside the metal and consequently cannot roll. Used as solid lubricant between a steel ball and a silicon wafer, onions are able to give lower wear compared to graphite (Hirata et al. 2004). Low friction coefficients were also measured and carbon onions are efficient when their diameter becomes larger than surface roughness values. In the presence of moisture, carbon onions usually agglomerate, and the effect of their individual size is thus less significant. Carbon onions can form a layer between two surfaces providing low friction and low wear. Street et al. used carbon onions as lubricant additives in Krytox 143B, a perfluorinated polyether (PFPE) oil. Addition of carbon onions leads to a strong reduction of the friction coefficient (0.05 instead of 0.13) with a long lifetime of the oil, for tests run in air (Street et al. 2004).

The optimal concentration of carbon onions in the base oil was investigated (i.e., the concentration that leads to the smallest friction and wear). A concentration of 0.1 wt% leads to the same reduction of friction than a concentration of 1 wt% (friction coefficient equal to 0.09). The same tendency was observed for CO1 sample and CO2 sample. This point is very promising for future application of onions as lubricant additives since relatively low concentrations are sufficient to obtain interesting properties. Figure 26.6 compares friction coefficients obtained at several contact pressures for the two carbon onions samples:

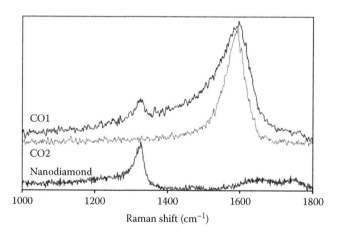

FIGURE 26.5 UV-Raman analysis of the two carbon onions samples. (From Joly-Pottuz, L. et al., *Tribol. Int.*, 41, 69, 2008a. With permission.)

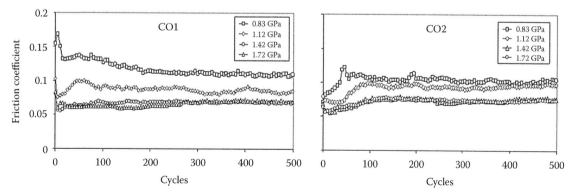

FIGURE 26.6 Comparison of the tribological performances of CO1 and CO2 at different contact pressures. (From Joly-Pottuz, L. et al., *Tribol. Int.*, 41, 69, 2008a. With permission.)

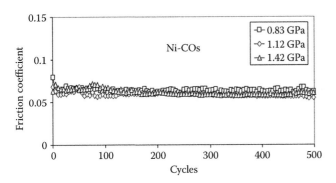

FIGURE 26.7 Friction coefficient obtained with a dispersion of Ni-COs at 0.1 wt% in PAO at several contact pressures. (From Joly-Pottuz, L. et al., *Tribol. Lett.*, 29, 213, 2008b. With permission.)

CO1 and CO2 (Joly-Pottuz et al. 2008a). Tribological performances of carbon onions CO1 and CO2 are similar, and friction coefficients of 0.06 are obtained with both samples at high contact pressures. A critical pressure of about 1.40 GPa is observed, to obtain friction coefficient lower than 0.06.

Tribological performances of Ni-COs were also studied at several contact pressures (Figure 26.7) (Joly-Pottuz et al. 2008b). Very low friction was obtained (below 0.08), even at low contact pressures. Ni-COs present better friction reducing properties than CO1 and CO2 on the whole pressure range.

Tribological properties of carbon onions were compared to those of other carbon forms: graphite, C_{60}, and nanodiamond particles used as a precursor for the synthesis of carbon onions. Table 26.1 summarized the friction coefficients obtained and the wear scar diameters measured after friction test. Friction coefficients obtained for the different carbon forms are quite similar except for Ni-COs which presents a lower friction coefficient. Concerning wear results, a comparison of the two carbon onions samples shows that the presence of a diamond core inside the carbon onions has a detrimental effect on the antiwear properties of carbon onions. Nanodiamond particles are abrasive and this can explain why the wear scar diameter is higher than with pure PAO base oil. Wear values measured in the presence of C_{60} are similar to those measured with carbon onions, but wear measured in the presence of graphite is much higher. The smallest wear value is obtained with Ni-COs. Tentatively, this result can be attributed to

the presence of the nickel core inside the carbon onions. To conclude on all these results, the presence of a diamond core inside the carbon onions has a detrimental effect on their tribological effect while a nickel core is found to have a beneficial effect.

To understand the lubrication mechanism of carbon onions, wear debris (particles collected inside the contact area after friction test) were characterized by TEM. Figure 26.8 presents a typical TEM image of one of these wear particles obtained after friction test. The selected area diffraction pattern (SAED) performed on the whole debris indicates the presence of many iron oxide nanoparticles (Joly-Pottuz et al. 2008c). These nanoparticles come from wear of steel counterparts. Inter-reticular distances measured for the new rings observed on the diffraction pattern of the wear debris fit well with distances reported for both magnetite iron oxide Fe_3O_4 (JCPDS data 19-0629) and/or maghemite iron oxide γ-Fe_2O_3 (JCPDS data 39-1346), which is an iron-deficient magnetite. Because of their quite similar distances in the diffraction pattern, it is not possible to distinguish between these two iron oxide species. However, no hematite structure (the most stable structure of iron oxide) is observed. In order to clarify if these nanoparticles have a maghemite and/or a magnetite structure, a quantitative EELS study was performed. A measured atomic ratio O/Fe of 1.54 ± 0.08 was obtained, that is more consistent with the maghemite composition (O/Fe = 1.5) than with the magnetite one (O/Fe = 1.33_3). Moreover, irradiation of the nanoparticles by electron beam was performed in order to better understand their structure. Before irradiation the diffractogram is consistent with a [1–10] projection of maghemite (or magnetite). After a few seconds of irradiation, a doubling of interplanar distance is clearly observed in the HRTEM image and is also visible on the diffractogram (Figure 26.9). This doubling is tentatively attributed to an ordering in maghemite crystal structure. This structure was already observed (Pecharroman et al. 1995) under the tetragonal form (space group: P 43 21 2), consisting in a *c* axis three times larger than the cubic parameter. Briefly, these experiments let us suppose that nanoparticles are preferentially composed of maghemite structure. This doubling of the distance is associated with a significant loss of oxygen during the irradiation process. The O/Fe ratio drops from 1.67 down to 0.2 within less than one minute. Since maghemite has an inverse spinel structure and contains iron vacancies, one can thus assume that vacancy ordering, promoted

TABLE 26.1 Friction Coefficients and Wear Scar Diameter (μm—in Italic) Measured with Carbon Forms Dispersed at 0.1 wt% in PAO at Several Contact Pressures

Contact Pressure (GPa)	PAO[a]	CO1	CO2	Ni-COs	Nanodiamond	Graphite	C_{60}	Hertz Diameter
0.83	0.27	0.11	0.10	0.06	0.11	0.11	0.11	68
	170	*120*	*90*	*75*	*175*	*130*	*100*	
1.12	0.1	0.09	0.09	0.06	0.1	0.09	0.09	92
	175	*140*	*115*	*92*	*185*	*145*	*125*	
1.42	0.08	0.07	0.07	0.06	0.09	0.09	0.09	116
	180	*150*	*135*	*116*	*195*	*155*	*145*	

[a] A dispersion of results was observed for pure PAO.

FIGURE 26.8 Wear particles observed after friction test with Ni-COs. (From Joly-Pottuz, L. et al., *Tribol. Lett.*, 30, 69, 2008c. With permission.)

by electron irradiation, takes place, leading to the period doubling. Furthermore, the presence of such vacancies may also help the departure of oxygen atoms under high electron flux irradiation. The distribution of the iron oxide nanoparticles and intact carbon onions inside the wear particles was studied by TEM imaging.

Dark-field images on the ring corresponding, respectively, to carbon onions and iron oxides were performed (Figure 26.10). The images were obtained on the part of the particles in Figure 26.8 that stands on a hole of the carbon film (to avoid contribution of the amorphous carbon film of the TEM grid). On Figure 26.10b, very small dots are observed in the wear particle material and they correspond to carbon onions. DF images of iron oxide contribution show the presence of well-distributed nanoparticles (10 nm) into the carbon onion network.

As suggested by Jin and Li, iron oxides (Fe_3O_4) act as lubricious oxides and FeOOH (goethite) can supply hydrogen on the counter surface (Jin and Li 2007). A similar mechanism is here proposed in the case of carbon onions with the formation of a carbon tribofilm containing lubricious iron oxides (maghemite) and possibly the presence of OH groups on their surface. Unfortunately, evidence for this mechanism only by TEM remains almost impossible.

To study the lubrication mechanism of Ni–COs, analyses were performed on the surface after friction test. Before friction, a metallic core composed of pure nickel is evidenced by high resolution TEM images and EELS analyses (Joly-Pottuz et al. 2008b). But it could be slightly oxidized at its surface as shown by XPS analysis. After friction, metallic nickel is also observed

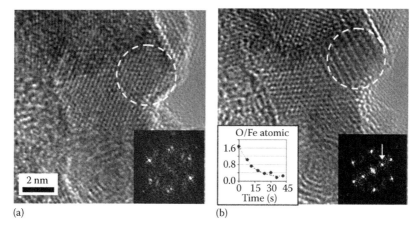

FIGURE 26.9 HRTEM image of an iron oxide nanoparticle (a) before irradiation and (b) after irradiation with the measured evolution of the atomic ratio O/Fe as a function of time. In both cases, the numerical diffractograms correspond to the circled area. Note the planar distance doubling in (b) associated with the additional spatial frequency arrowed in the diffractogram. An intact carbon onion can also be observed. (From Joly-Pottuz, L. et al., *Tribol. Lett.*, 30, 69, 2008c. With permission.)

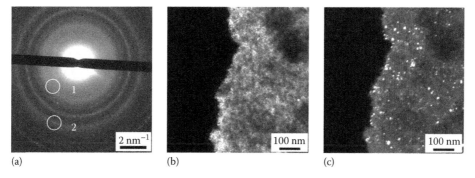

FIGURE 26.10 (a) The diffraction pattern of the wear particle and the positions of objective aperture used to perform the dark-field images, (b) and (c) dark-field images corresponding to positions 1 and 2, respectively.

in the wear track. These results suggest that carbon onions have been crushed inside the contact area and that consequently the Ni cores have been released inside the contact area. The XPS C 1s peak observed for pristine Ni–COs and on the tribofilm are dominated by a contribution at 284.8 eV corresponding to C–C/C–H bonding (sp² hybridized carbon). Only a slight increase of the peak width is observed and could correspond to the formation of some sp³ hybridized carbon. The tribofilm material can be compared to a coating formed of Ni-doped carbon material. This Ni-doped carbon material originated from crushed carbon onions and nickel nanoparticles and this kind of structure is thought to have interesting tribological properties. Indeed, good antiwear properties were already observed with metal-doped DLC film (Ti, Cr, Ni) (Zhou et al. 1997, Chang et al. 2002). The catalytic role of nascent nickel exposed is certainly crucial to decompose the PAO base oil into some other carbon species. Decomposition of PAO base oil was also suggested to be responsible for the good tribological properties observed for CO2 sample (Joly-Pottuz et al. 2008a). However, the lubricating mechanism clearly needs further investigation in the future.

To summarize, carbon onions present very interesting tribological properties when they are used as lubricant additives. Now they can be envisaged as promising additives for automotive lubrication. But other applications can also be envisaged. They present the advantage to be composed only of carbon and thus to be friendly for the environment. Lubrication mechanisms are not yet fully understood and more work is necessary. Furthermore, optimizing their synthesis methods can certainly be envisaged for gaining better tribological properties than those described in this chapter. For example, by improving the synthesis methods of sample CO1 to obtain sample CO2, carbon onions with better antiwear properties were obtained. Alternatively, carbon nanotubes presenting a cylindrical shape can also be envisaged as additives. Preliminary results obtained on their tribological performances are also promising (Joly-Pottuz and Ohmae 2008).

26.3 Inorganic Fullerenes of MS₂

26.3.1 Synthesis and Characterization

Inorganic fullerenes of metal disulfide MS_2 (M = Mo, W…) have a structure similar to carbon onions (several hollow spherical layers with different diameters fitting together to form an onion-like structure) but their sheets are basically made of MS_2. These structures are called "inorganic fullerene-like nanoparticles" or IF-MS_2 (Figure 26.11). They were first synthesized by Tenne (Tenne et al. 1998, Tenne 2003). These nanoparticles were first studied for their tribological and mechanical properties, but they also present interesting electrical properties (Azulay et al. 2006). In this chapter, inorganic fullerenes will be described by the term IF and the corresponding lamellar structure by the term 2H.

Feldman described first a model of growth of the IF-MoS_2 and IF-WS_2 (Feldman et al. 1996). Metallic oxides particles MO_3

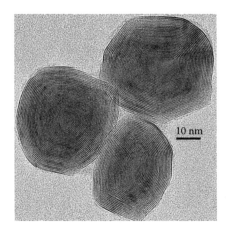

FIGURE 26.11 HRTEM micrograph of IF-MoS_2. (Courtesy of Tenne and Fleischer.)

(M = Mo, W) are reduced in a gas mixture (5% of H_2 and 95% of N_2) then react with H_2S. A layer of MS_2 is first formed at the periphery of the oxide particle (Figure 26.12a). This surface layer avoids aggregation and coalescence of the nanoparticles, which could involve the formation of macroscopic entities and 2H-MS_2. The fast diffusion of dihydrogen into the core of the particles allows a complete reduction of oxide to form MoO_2 or $W_{18}O_{49}$ (Figure 26.12b). This core is then transformed slowly and gradually into sulfide (Figure 26.12c). When the reaction is finished, metallic oxide originally present at the center of the IF structure has completely reacted. Synthesis methods of IF-WS_2 and IF-MoS_2 are different: MoO_3 particles are volatile at 700°C and the synthesis is a gas phase reaction; WO_3 particles are not sublimable below 1000°C, thus the synthesis is a solid–gas reaction. The kinetics of formation of these nanoparticles were studied (Feldman et al. 1998) and a new type of reactor was designed for the synthesis of IF-WS_2 in order to have a better contact between the solid particles and gases (Feldman et al. 2000). Other synthesis methods are possible: electric arc in water (Hu et al. 2004) and microwave-induced plasma (Brooks et al. 2006). IF-MoS_2 obtained by electric arc in water have a small diameter (5–30 nm) but contain a molybdenum-rich core. The presence of the molybdenum core may have an influence on the tribological properties of IF-MoS_2. Fullerenes obtained by microwave-induced plasma contain many defects.

IF-MoS_2 tested were synthesized at Weizmann Institute of Science (Rehovot, Israel), in Professor Tenne's laboratory. They have a mean diameter of 40 nm. IF-MoS_2 were elaborated by Nanomaterial, Ltd. company and have a mean diameter of 140 nm. Cizaire studied in detail the structure of these IF-MoS_2 (Cizaire 2003) and she described an assembly of MoS_2 nanocrystals of 10 nm length, contacting on all their edges to form irregular polyhedrons. X-ray diffraction (XRD) showed a dilatation of interlayer spacing by approximately 1% (distance between the 0002 reticular planes). This expansion, initially observed in the case of closed structures made of carbon (Saito et al. 1993), is generally attributed to the presence of residual stresses in the curved layers. Moreover, the number of atoms increases from one layer to another so that the layers are not commensurable.

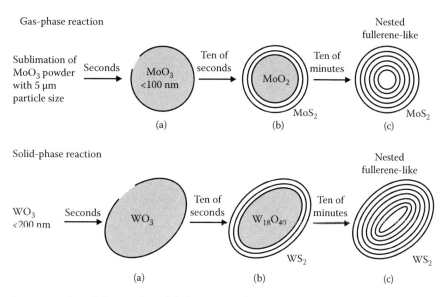

FIGURE 26.12   Schematic representation of the growth model of IF-MoS$_2$ and IF-WS$_2$. (From Feldman, Y. et al., *J. Am. Chem. Soc.*, 118. 5362, 1996. With permission.)

Thus, an expansion along the *c* axis is needed (Feldman et al. 1996). Observed by TEM, IF-WS$_2$ have a mean diameter of about 140 nm, larger than the one of IF-MoS$_2$. They are also more faceted, probably due to the higher number of walls. Srolovitz et al. calculated that beyond a certain number of layers, fullerenes cannot be spherical anymore (Srolovitz et al. 1995) and they become faceted. Srolovitz et al. showed that dislocations tend to gather to form grain boundaries (Srolovitz et al. 1994). Therefore, fullerenes can be described as an assembly of WS$_2$ nanocrystals. The presence of dislocations was also observed by TEM and dislocations lead to a reduction of mechanical strains inside the whole structure (Srolovitz et al. 1994).

Raman spectroscopy performed with an Argon laser (wavelength = 514.5 nm) allows to evidence the hexagonal structure of nanocrystals, with Mo atoms in a trigonal position (presence of the peak at 384 cm^{-1} not observed in the octahedral structure) (Cizaire 2003). However, the resolution in these spectra was not good enough to distinguish the IF structure from the 2H structure. Nevertheless, Raman spectroscopy is a very interesting technique to characterize metal dichalcogenide compounds. A Raman study of hexagonal MoS$_2$ shows the presence of three characteristic lines at 287, 383, and 409 cm^{-1} corresponding to the following modes: E_{1g}, E_{2g}^1, and A_{1g}, respectively. A fourth mode, called "rigid layer mode" is also possible (E_{2g}^2) but difficult to observe since it appears at a wave number lower than 20 cm^{-1}. Similar transitions are observed with WS$_2$, NbS$_2$, and TaS$_2$ (Table 26.2), but signals obtained for niobium and tantalum disulfides are too weak to be clearly observed (Mc Mullan et al. 1983, Hirata et al. 2001). Resonant Raman conditions can be obtained by using a wavelength near the absorption range of the structure. Resonant Raman conditions are of particular interest for IF-MoS$_2$ analysis as already noticed by Frey et al. (1998). Indeed the deconvolution of a new asymmetrical peak at 460 cm^{-1} allows

TABLE 26.2   Raman Active Modes for the MS$_2$ Lamellar Structures

Mode	Atoms	Direction of Vibration	MoS$_2$	WS$_2$	NbS$_2$	TaS$_2$
E_{1g}	S	Basal plane	287	327	260	243
E_{2g}^1	Mo + S	Basal plane	383	351	304	306
A_{1g}	S	*c* axis	409	420	379	381

to distinguish the fullerene from the lamellar structure. Thus, IF-MoS$_2$ were studied using a laser wavelength of 632.8 nm to be in resonant Raman conditions, while IF-WS$_2$ were studied at 514.5 nm wavelength (Figure 26.13).

Other analytical tools were used to characterize fullerenes before friction. XRD confirms an expansion of the interlayer distance and a shift of the (002) peak was observed for both fullerenes compared to the corresponding lamellar structure (0.2% for IF-WS$_2$). Such an expansion of the layer spacing in crystal associated with weak bonding between the layers is very common when the perfection of the layer stacking is lost. XANES analyses of IF-MoS$_2$ show the presence of sulfates at the top surface of IF (Joly-Pottuz and Dassenoy 2008). These results were confirmed by ToF-SIMS analyses of IF and 2H-WS$_2$. Nevertheless, this oxidation of sulfur species only concerns the very first layer, and fullerenes can be considered to be very pure and chemically inert.

26.3.2 Tribological Properties

Tribological properties of IF-MS$_2$ were focused on the case of IF-WS$_2$. A comparison between tribological properties of IF-WS$_2$, 2H-WS$_2$, and 2H-MoS$_2$ used as additives in oils, greases, or impregnated into porous matrix was made. Results definitely show that the IF structure presents better friction reducing properties than

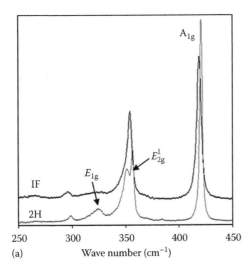

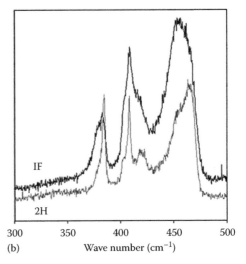

(a) Wave number (cm^{-1})   (b) Wave number (cm^{-1})

FIGURE 26.13 Raman spectra: (a) IF-WS$_2$ and 2H-WS$_2$ at 514.5 nm, (b) IF-MoS$_2$ and 2H-MoS$_2$ at 832.8 nm.

the lamellar one (Rapoport et al. 2003a–1999). These results are explained first by the fact that fullerenes avoid contact between metal asperities and second by the absence of dangling bonds making the fullerenes much more chemically stable. Indeed, the main problem with the lamellar structures like 2H-MoS$_2$ and 2H-WS$_2$ is the easy oxidation at the edges of crystals. Moreover, these structures easily adhere to surfaces, while fullerenes do not, as Rapoport showed by SFM (Rapoport et al. 1997). Furthermore, fullerenes can enter into the asperities, acting as reservoirs while lamellar structures are stuck perpendicular to the surface and a rapid deterioration during friction takes place (Rapoport et al. 2003b). The spherical structure of the IF avoids problems of orientation encountered with 2H platelets (Hu et al. 2005). Recently, Rapoport studied the effect of IF-WS$_2$ rubbed on alumina surface (Rapoport et al. 2005). He demonstrated that the IF-WS$_2$ are preserved in the pores of the sintered alumina and they are peeled-off, and supply the solid lubricant during a long-term test. Chhowalla studied the tribological properties of IF-MoS$_2$ coatings in the presence of moisture and compared the results with those obtained from 2H-MoS$_2$ coatings (Chhowalla et al. 2000). With IF coatings, lower friction coefficients were obtained because fullerenes do not present many dangling bonds.

The lubrication mechanism of IF-WS$_2$ is supposed to be based on the formation of a transfer film formed under the effect of shear but not under hydrostatic pressure, as observed in the case of two-mica surfaces (Drummond et al. 2001). Transfer films obtained from fullerenes are very thin and homogeneous while those obtained from lamellar structures are rough and disorganized. Greenberg compared the effects of addition of IF in oil under three lubrication regimes: hydrodynamic, mixed, and limit (Greenberg et al. 2004). Under mixed lubrication, IF were found to be the most efficient because a film is formed on the surface. In hydrodynamic regime, fullerenes do not have any interaction with surfaces. In the boundary lubrication regime, films formed on surfaces are quickly removed, due to contact severity.

Several applications are now envisaged for these nanoparticles. A company named Nanomaterials (headquarters in New York

(United States), research and development center in Rehovot (Israel)) launched the marketing of "Nanolub," a lubricant made up of a lubricating base (oil or grease) and the addition of IF-WS$_2$. This lubricant presents better tribological properties compared to grease or oil alone, or oils containing 2H-MoS$_2$ platelets (Fleischer et al. 2005). A mass production of 2000 tons/year is envisaged. IF-WS$_2$ nanoparticles can also lubricate orthodontic wires (Katz et al. 2006), for example Ni-P coatings containing IF-WS$_2$ are very useful to reduce the friction between the wires and the brackets glued on teeth. Zou confirms the improvement of Ni-P coatings with the addition of MS$_2$ fullerenes (Zou et al. 2006). Furthermore, toxicology tests of these nanoparticles have demonstrated that they are not toxic in oral administration or upon exposure to skin. Thus, these nanoparticles can be envisaged to lubricate many tribological systems in the future.

As for carbon onions, the effect of the concentration of IF on the tribological properties was studied for IF of MoS$_2$ and WS$_2$. An optimal concentration of 1 wt% was determined (Table 26.3). The effect of the contact pressure on their tribological properties was also studied (Table 26.4) and the pressure leads to a decrease of the friction coefficient. Better tribological results were obtained with IF-WS$_2$ as friction reducing and antiwear additives. Practically, no wear was observed in the case of IF-WS$_2$ since the wear scar diameter measured on the pin after friction process is equal to the Hertz calculated diameter (Joly-Pottuz et al. 2005).

The lubrication mechanism of IF was further studied by HRTEM and Raman spectroscopy, and TEM analyses of the wear particles reveal the formation of MS$_2$ sheets during friction. Very thin films composed of only a few WS$_2$ layers were

TABLE 26.3 Effect of the Concentration of Fullerenes in PAO on Their Tribological Properties ($P = 1.12$ GPa, $v = 2.5$ mm/s, $T = 20°C$)

	0	0.1 wt%	0.5 wt%	1 wt%	2 wt%
IF-MoS$_2$	0.16/0.09	0.06	0.06	0.04	0.04
IF-WS$_2$	0.16/0.09	0.08	0.07	0.06	—

TABLE 26.4 Values of Friction Coefficient (μ) and Wear Scar Diameters Obtained at Different Contact Pressures for Pure PAO, PAO + 1 wt% IF-WS$_2$, and PAO + 1 wt% IF-MoS$_2$

Pressure (GPa)	Ball Diameter (mm)	Hertz Calculated Diameter (μm)	PAO		IF-WS$_2$ 1% PAO		IF-MoS$_2$ 1% PAO	
			μ	Wear (μm)	μ	Wear (μm)	μ	Wear (μm)
0.33	12	54	0.26	85	0.06	65		
0.42	12	68	0.26	115	0.05	80		
0.52	12	86	0.16	150	0.05	95		
0.66	6	54	0.16	115	0.05	75	0.08	110
0.83	6	68	0.25	170	0.04	86	0.07	120
1.12	6	92	0.16/0.09	175	0.04	100	0.06	125
1.42	6	116	0.08	180	0.04	120	0.04	140
1.72	4.5	106	0.07	150	0.04	115		

observed similar to those observed after friction test with pure MoS$_2$ coatings (Martin et al. 1993). Incommensurate conditions can be reached with these thin films. For example, the wear particle shown in Figure 26.14 is composed of at least three superimposed WS$_2$ sheets with a rotational angle of 30° between two of them and an angle of 10° between the other two. With an angle of 30°, there is no atomic coincidence between adjacent sliding layers, so that shear is easier and friction can vanish, as predicted by Hirano et al. (1991). Incommensurate conditions observed for the wear particles after friction tests with IF-WS$_2$ can explain the very low friction coefficient measured with these fullerenes.

In situ Raman analyses were performed to better understand the mechanism of formation of sheets from the fullerenes structure. These experiments were carried out using an original device composed of a Raman spectrometer, a diamond anvil cell and a tribometer. The diamond anvil cell was used to submit the fullerenes

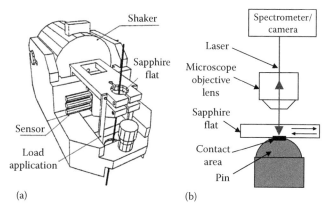

(a) (b)

FIGURE 26.15 (a) Schematic representation of the small alternative tribometer that can be installed under the Raman spectrometer, (b) principle of *in situ* Raman analysis. (From Joly-Pottuz, L. et al., *Appl. Phys. Lett.*, 91: 153107, 2007. With permission.)

to very high hydrostatic pressures (up to 35 GPa) in order to follow this structure under such high pressures. In the case of a tribological contact where the two contact materials are made of steel, Raman analyses are not possible. Thus, we developed an original pin-on-flat tribometer using a transparent flat made of sapphire. This tribometer is small enough to be installed directly under the optical microscope of a Raman spectrometer (Figure 26.15). Two kinds of Raman analyses were performed: analyses during friction tests to follow the formation of the lamellar structure and analyses on fullerenes only submitted to pressure in the contact (without shear) to follow their behavior when they are mainly submitted to an uniaxial pressure inside the contact. Results of the diamond anvil cell experiments with IF-WS$_2$ and 2H-WS$_2$ are presented in Figure 26.16. Spectra obtained after the release of the load are similar to those obtained for the pristine material. Nevertheless, a broadening of the peaks is observed and this is more marked in the case of IF indicating the creation of disorder in the structures. TEM observations of the IF collected after the experiment show the formation of some layers at the edge of IF. But this formation is not significant and can come from the peripheral zone where the pressure is not hydrostatic.

The behavior of fullerenes under pressure inside the contact area was studied by performing a loading–unloading experiment

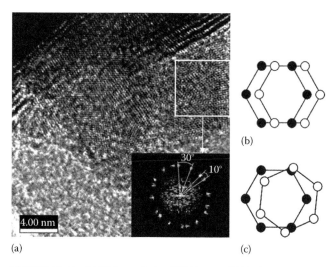

(a) (b)

(c)

FIGURE 26.14 (a) High-resolution TEM image of WS$_2$ sheets in incommensurate conditions. The diffractogram of the surrounded area shows the angle of 30° or 10° between the three sheets superimposed. (b and c) Diagrams of two layers of atoms stacked in a hexagonal network. In case (b), atoms coincide perfectly, friction is high. In the case (c), the layer of white atoms suffered a rotation. The atoms do not coincide and friction will be low. (From Joly-Pottuz, L. et al., *Tribol. Lett.*, 18, 477, 2005. With permission.)

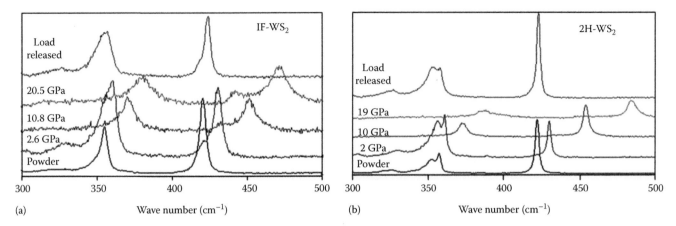

FIGURE 26.16   Raman spectra obtained during diamond anvil cell experiments with IF-WS$_2$ (a) and 2H-WS$_2$ (b).

(Joly-Pottuz et al. 2006). The process of loading consists first in recording the spectrum of the pristine powder deposited on the ball, then just after closing the contact, and finally after application of loads from 1 to 10 N. The unloading corresponds to the opposite process. In the simple contact configuration used here, the contact pressure can easily be estimated thanks to the theory of the contact mechanics between two solids (Johnson 1985) and by considering several points: sphere and plane initial geometry, purely static loading, purely elastic and isotropic behavior of the substrates, and infinitely smooth surfaces. Under normal loading, the solids are reversibly deformed with a circular contact zone. The normal pressure generated on surface is maximum at the center of the contact zone and the lateral distribution is quadratic with a parabolic decrease toward edges. The value of the maximum contact pressure was calculated from Equation 26.3.

Taking into account the mechanical characteristics of both the plane (sapphire) and the sphere (AISI 52100 steel), the contact geometry, and the normal load applied, we calculated the maximum contact pressure. We obtained P_{max} = 0.78, 0.98, 1.33, and 1.68 GPa, respectively, for W = 1.0, 2.0, 5.0, and 10.0 N. Figures 26.17 and 26.18 present the results obtained for the loading–unloading experiments with 2H and IF-WS$_2$. Spectra

after the experiments are different from those obtained with the pristine material. The differences are more pronounced in the case of IF and these results prove that the structures are changed during the loading–unloading experiments. In the case of 2H, an orientation of the sheets parallel to the surface occurs due to hydrostatic pressure and this orientation is very favorable to reduce friction because of the easy shear of one sheet on another. In the case of IF, the spectra after the experiment are very similar to those obtained after the experiment with 2H. We can thus conclude that WS$_2$ sheets are formed during the experiment and that these sheets are oriented. Fullerenes can undergo very high hydrostatic pressures (up to 35 GPa), but under uniaxial pressure, fullerenes are exfoliated to form sheets (Figure 26.19). This process can be assimilated to a "nut-cracker" process.

From the diamond anvil cell experiments (Figure 26.16) a relation between the Raman shift and the pressure can be determined for the E_{2g}^1 and A_{1g} peak, respectively: $1.5x + 356$ and $3.4x + 420.7$ (IF-WS$_2$); $1.6x + 356.6$; and $3.3x + 421.5$ (2H-WS$_2$). From these relations and the results obtained on the spectra, it becomes possible to determine the pressure inside the contact area. The contact pressure was calculated from the loading–unloading experiment with 2H and IF and compared

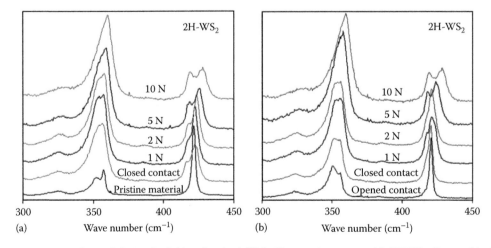

FIGURE 26.17   Raman spectra obtained during load (a) and unload (b) inside a static contact with 2H-WS$_2$ dispersed in PAO. A difference between spectra obtained before load and after unload can be observed.

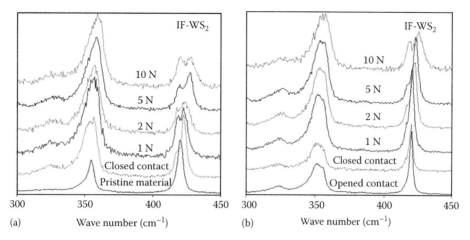

(a) Wave number (cm⁻¹)   (b) Wave number (cm⁻¹)

FIGURE 26.18 Raman spectra obtained during load (a) and unload (b) inside a static contact with IF-WS$_2$ dispersed in PAO. As in the case of 2H-WS$_2$, spectra before load and after unload are different.

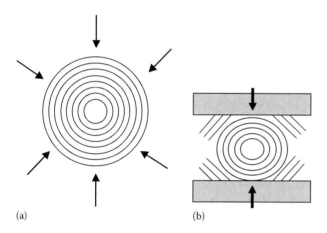

FIGURE 26.19 (a) Under high hydrostatic pressure, fullerenes are not structurally modified; (b) inside a static contact area, exfoliation of fullerenes to form sheets occurs. (From Joly-Pottuz, L. et al., *J. Appl. Phys.*, 99, 023524, 2006. With permission.)

to the pressures calculated from the Hertzian theory (Figure 26.20). Results calculated for 2H are very similar to the Hertz calculated pressure, while differences between the experimental and theoretical curves can be observed for IF. Several hypotheses can explain these differences. First, the analyzed location does not correspond to the center of the contact area. This would explain the low values obtained. The most probable hypothesis is that the pressure is not uniformly distributed in the contact but is higher on the fullerenes clusters as shown in Figure 26.21. The same experiments were also conducted with 2H-MoS$_2$ and IF-MoS$_2$ (Joly-Pottuz and Dassenoy 2008).

Raman analyses were also performed during friction tests to follow the structural evolution of fullerenes (Joly-Pottuz et al. 2007). We choose the same conditions than those used for a steel/steel contact: a frequency of 0.5 Hz and a temperature of 20°C. The tangential movement was stopped during the acquisition of the spectrum (approximatively 2 min), because friction reducing properties of IF-WS$_2$ are obtained from the very beginning of the test. Thus, we may assume that the active compound that leads to a reduction of friction (WS$_2$ sheets) is formed during the

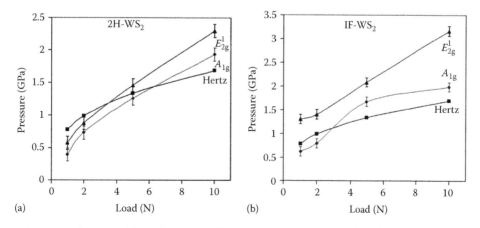

(a) Load (N)   (b) Load (N)

FIGURE 26.20 Contact pressures determined from the Raman shift observed on the spectra obtained during the loading–unloading experiment with (a) 2H-WS$_2$ and (b) IF-WS$_2$. These pressures are compared to those calculated from Hertzian theory. (From Joly-Pottuz, L. et al., *J. Appl. Phys.*, 99, 023524, 2006. With permission.)

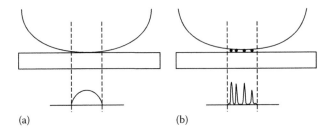

(a)   (b)

FIGURE 26.21 Distribution of the pressure inside (a) a theoretical contact and (b) a contact containing fullerenes.

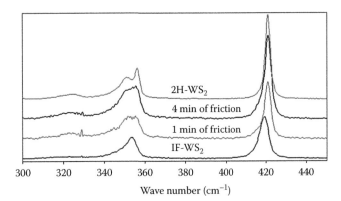

FIGURE 26.22 Raman spectra realized inside the contact area after 1 and 4 min of friction compared with the spectra of pristine IF-WS₂ and 2H-WS₂. (From Joly-Pottuz, L. et al., *Appl. Phys. Lett.*, 91, 153107, 2007. With permission.)

first minutes of the process. Figure 26.22 compares the spectra obtained inside the contact area after 1 and 4 min of friction with those of the pristine materials. Spectra obtained inside the contact area are different between 1 and 4 min of friction and the peak at 355 cm^{-1} becomes very similar to those observed on the spectrum of 2H-WS₂. Figure 26.23 presents a linear combination of 2H and IF spectra to simulate the shape of the peak after 1 and 4 min of friction. After 1 min, the proportion of 2H is about 38%.

After 4 min it increases to 50%. The presence of 2H inside the contact area after 1 min of friction explains the good tribological results obtained with IF-WS₂ from the beginning of the test. These results are also consistent with the loading–unloading experiments which show that, even without shear, IF are exfoliated into lamellar sheets. From these results and those obtained during the loading–unloading experiment, the evolution of the proportion of 2H inside the contact area can be represented (Figure 26.24).

TEM observations and Raman study clearly prove that the lubrication mechanism of IF is based on the formation of MS₂ single sheets inside the contact area. Experiments performed by Professor Stefan Scillag (Stockholm University) also confirm this lubrication mechanism. A TEM cross section of the tribofilm formed on the steel surface during friction process with IF-WS₂ was performed by focused ion beam (FIB). HRTEM observations of this sample show the lamellar structure of the tribofilm formed that is composed of a few WS₂ sheets oriented in the direction of sliding (Figure 26.25).

Basically, the lubrication mechanism of IF nanoparticles is similar to the one of molybdenum dithiocarbamate (MoDTC), an additive currently used in automotive lubrication as a friction modifier. After friction process with this molecule, MoS₂ were observed in a carbon matrix. These sheets come from the chemical decomposition of the molecule activated by high temperature and shear inside the contact area (Grossiord et al. 1998). The main advantage of fullerenes is their efficiency at ambient temperature and from the beginning of the test. There is no induction period whereas MoDTC needs to be thermally activated. Thus, MoDTC is not active at the cold start of the engines. This is a critical problem for hybrid systems composed of both electrical and thermal engines working alternatively, because the thermal engine is likely to run at low temperature very often. Therefore, it is essential to find additives that are active at low temperature.

The lubrication mechanism of fullerenes can be assimilated to a "drug delivery mechanism." Indeed, MS₂ single sheets are formed only inside the contact area where their presence is required, in the same way as a drug that delivers its active

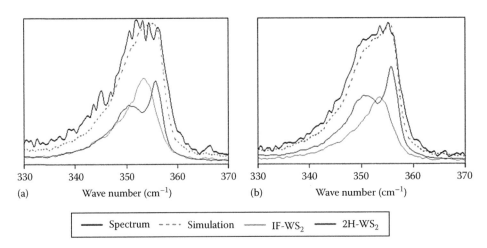

FIGURE 26.23 Raman spectra obtained during friction compared with simulations of a mixture of IF-WS₂ and 2H-WS₂: (a) after 1 min with 38% 2H-WS₂ and 62% IF-WS₂; (b) after 4 min with 50% 2H-WS₂ and 50% IF-WS₂.

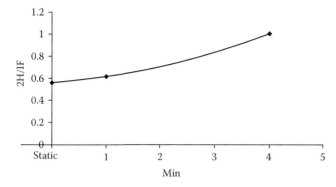

FIGURE 26.24 Evolution of the 2H ratio inside the contact area during a static experiment or during friction test with IF-WS₂.

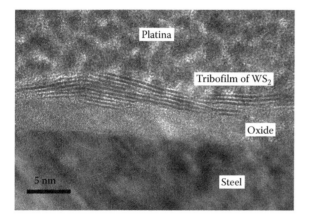

FIGURE 26.25 TEM image of a cross section performed by FIB on the tribofilm formed during friction process with IF-WS₂. The tribofilm is composed of WS₂ sheets. (Courtesy of Professor Stefan Scillag.)

FIGURE 26.26 Schematic model of the lubrication mechanism of fullerenes based on the formation of sheets inside the tribological contact.

substance at the right place in the human body. Figure 26.26 schematically describes the mechanism of the release of the single sheets inside the contact area.

Other mechanisms can also be responsible for the very low friction coefficient obtained with IF, such as bearing effect or chemical reactions. Observations of the wear scar on the pin after friction test with IF-WS₂ show the formation of a corona around the wear scar (Figure 26.27). This corona can have a bearing effect that can lead to a decrease of the pressure inside the contact area. In fact, due to the presence of this corona, the real contact area between the two surfaces is larger than the theoretical one and thus the maximum contact pressure decreases. This phenomenon was observed in the case of a ball of 12 mm in diameter. The wear scar presents a diameter of 120 μm, while the Hertz calculated diameter is equal to 136 μm. This means that the real contact area during friction test was smaller than the theoretical area. The change of the ball diameter leads to a change of the convergent angle (Figure 26.28) and fullerenes should enter more easily in the case of a high convergent angle (in the case of a small diameter ball). With a ball of 12 mm, fullerenes may not enter easily inside the contact area and may form a corona around the contact area. At 0.83 GPa, a friction coefficient of 0.03 was obtained with a 12 mm diameter ball while it was equal to 0.04 with a 6 mm diameter. Bearing effect may play an important role in the case of a 12 mm ball and plays also a role in the case of a 6 mm diameter ball. However, by avoiding the good penetration of fullerenes inside the contact convergent, bearing effect may have also a poor effect. More work is necessary to study this point. There is certainly a compromise between the bearing effect and the entrance of fullerenes inside the contact area to provide single sheets of MoS₂.

Chemical reactions can also be envisaged in the lubrication mechanism. WS₂ appears to be efficient under extreme pressure conditions. This behavior could be explained by the formation of tungsten carbide and iron sulfide. Iron sulfide is lubricious while tungsten carbide is hard and may prevent wear. In the case of IF-MoS₂, since Mo is not carbophilic the chemical reaction will not occur. More experiments are necessary to study this point in detail. Thus, IF-WS₂ would also be an interesting extreme-pressure additive.

IF can be considered as very promising lubricant additives. An important point now is to improve their dispersion in oil. Fullerenes tend to aggregate and aggregates easily settle in the oil. Moshkovitz studied the effect of the aggregate size on the tribological properties of IF-WS₂ dispersed in oil (Moshkovitz

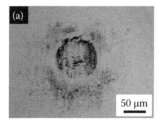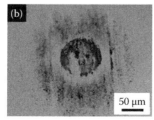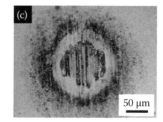

FIGURE 26.27 Wear scars observed on pins after friction tests at different number of cycles: (a) 20 cycles, (b) 50 cycles, and (c) 500 cycles. These results show the formation of a corona around the contact area.

FIGURE 26.28 Schematic representation of the convergent angle with pins of different diameters.

et al. 2007). An increase of the mixing time of the dispersion leads to a decrease of the aggregate size, and fullerenes enter the contact area more easily. Vibrations of the tribological systems have also an influence on the penetration of fullerenes inside the contact area, and an improvement of the tribological properties of fullerenes is also observed (Perfiliev et al. 2006). Vibrations lead to a variation of the gap between the rubbed surfaces and fullerenes, or fullerenes aggregates, easily enter the contact area. Industrials that commercialize lubricating oil are currently working on the improvement of dispersion of nanoparticles in oil. This demonstrates the increasing interest of nanoparticles as lubricant additives.

WS$_2$ and MoS$_2$ are the most efficient fullerenes. They present better tribological results than IF-NbS$_2$ and TaS$_2$ (Schuffenhauer et al. 2005). Other fullerenes made of metal dichalcogenides may also present interesting tribological properties and more investigations are necessary. WS$_2$ nanotubes also present interesting properties. Nanotubes made of metal dichalcogenides but also carbon nanotubes can be envisaged as lubricant additives (Joly-Pottuz and Dassenoy 2008, Joly-Pottuz and Ohmae 2008).

IF can also be used as coatings. First results under ultrahigh vacuum show the efficiency of such coatings, since superlubricity (friction coefficient below 0.01) was achieved at low temperature (Joly-Pottuz and Iwaki 2007).

26.4 Conclusion

Fullerenes present promising tribological properties. IF present peculiar friction reducing properties coupled with antiwear properties. Carbon onions are also very interesting. More studies are necessary to understand their lubrication mechanism. An interesting point is the formation of iron oxides that are lubricious. An important point for both nanoparticles is their efficiency at ambient temperature that is a main advantage compared to the additives currently used.

Even if the nanoparticles studied in that chapter are more expensive than the additives presently used, they can be considered as interesting additives for the future for two reasons. Automotive lubrication needs to change in the future to preserve the environment. Molecules currently used are dangerous and have to be replaced. Continual improvements of the synthesis processes of the nanoparticles will soon lead to mass production and drop in prices. Formulated oil and grease containing IF-WS$_2$ nanoparticles begins to be commercialized. This shows the growing interest of the additives manufacturers in this new way to lubricate engines.

26.5 Future Works

The formulated oils used for the lubrication of engines are complex formulations containing additives with other specific properties: detergency and antifoaming agent. The nanoparticles must be added to a formulated oil in order to check their efficiency in the presence of other additives. Or it is possible that a pure base oil containing only nanoparticles present the same properties than a fully formulated oil. Thus, a comparison of the efficiency of fully formulated oil and pure base oil containing only nanoparticles must be performed. This kind of experiments may be performed on a real engine in the research and development centre of an automotive manufacturer.

To fully understand the structural evolution of nanoparticles under pressure and shear, numerical simulations are of much interest. First results obtained in collaboration with Professor Susan B. Sinnott from the University of Florida (Joly-Pottuz et al. 2010) prove the utility of these experiments for further studies on the lubrication mechanism of nanoparticles.

Acknowledgments

The author would like to thank Dr. Niles Fleischer of Nanomaterials Ltd (Israel) and Professor Reshef Tenne of the Weizman Institute of Science (Israel) for providing the fullerenes sample and for their helpful discussions. Professor Nobuo Ohmae of Kobe University is also thanked for providing carbon onions and for the welcome in his laboratory.

The author would also like to thank Dr. Bruno Reynard, director of Laboratoire des Sciences de la Terre at Ecole Normale Supérieure de Lyon (UMR CNRS 5570), for his collaboration for the Raman study, and Gilles Montagnac for his support during the Raman measurements and his helpful discussions.

All members of the Laboratoire de Tribologie et Dynamique des Systèmes at Ecole Centrale de Lyon (UMR CNRS 5513) who participated in this work are deeply thanked, especially Professor Jean Michel Martin, Dr. Fabrice Dassenoy, Béatrice Vacher, Michel Belin, and Thierry Le Mogne.

The author would like to thank Professor Stefan Scillag of Stockholm University for its TEM image of the WS$_2$ tribofilm. This image is a good illustration of the lubrication mechanism of WS$_2$ fullerenes.

One part of this work was supported by the Japan Society for the Promotion of Science (JSPS).

References

Ajayan, P. M., Iijima, S., Ichihashi, T. 1993. Electron-energy-loss spectroscopy of carbon nanometer-size tubes, *Physical Review B* 47: 6859–6862.

Azulay, D., Kopnov, F., Tenne, R., Balberg, I., Millo, O. 2006. Observation of current reversal in the scanning tunneling spectra of fullerene-like WS$_2$ nanoparticles, *Nanoletters* 6: 760–764.

Bowden, F. P., D. Tabor. 1950. *The Friction and Lubrication of Solids*, Clarendon Press, Oxford, U.K.

Brooks, D. J., Douthwaite, R. E., Brydson, R., Calvert, C., Measures, M. G., Watson, A. 2006. Synthesis of inorganic fullerene (MS$_2$, M = Zr, Hf and W) phases using H$_2$S and N$_2$/H$_2$ microwave-induced plasmas, *Nanotechnology* 17: 1245–1250.

Cabioc'h, T., Girard, J. C., Jaouen, M., Denanot, M. F., Hug, G. 1997. Carbon onions thin film formation and characterization, *Europhysics Letters* 38: 471–475.

Cabioc'h, T., Thune, E., Riviere, J. P., Camelio, S., Girard, J. C., Guérin, P., Jaouen, M., Henrard, L., Lambin, P. 2002. Structure and properties of carbon onion layers onto various substrates, *Journal of Applied Physics* 91: 1560–1567.

Chang, Y. Y., Wang, D. Y., Wu, W. T. 2002. Catalysis effects of metal doping on wear properties of diamond-like carbon films deposited by a cathodic-arc activated deposition process, *Thin Solid Films* 420: 241–247.

Chhowalla, M., Amaratunga, G. A. J. 2000. Thin films of fullerene-like MoS$_2$ nanoparticles with ultra-low friction and wear, *Nature* 407: 164–167.

Cizaire, L. 2003. Lubrification par les nanoparticules (in French), PhD thesis, Génie des matériaux, Ecole Centrale de Lyon, No. 2003-26.

Drummond, C., Alcantar, N., Israelachvili, J., Tenne, R., Golan, Y. 2001. Microtribology and friction-induced material transfer in WS$_2$ nanoparticle additives, *Advanced Functional Materials* 11: 348–354.

Du, J., Liu, Z., Li, Z., Han, B., Sun, Z., Huang, Y. 2005. Carbon onions synthesized via thermal reduction of glycerin with magnesium, *Materials Chemistry and Physics* 93: 178–180.

Egerton, R. F. 1986. *Electron Energy-Loss Spectroscopy in the Electron Microscope*, Plenum Press, New York.

Feldman, Y., Frey, G. L., Homyonfer, M., Lyakhovitskaya, V., Margulis, L., Cohen, H., Hodes, G., Hutchison, J. L., Tenne, R. 1996. Bulk synthesis of inorganic fullerene-like MS$_2$ (M = Mo, W) from the respective trioxides and the reaction mechanism, *Journal of the American Chemical Society* 118: 5362–5367.

Feldman, Y., Lyakhovitskaya, V., Tenne, R. 1998. Kinetics of nested inorganic fullerene-like nanoparticle formation, *Journal of the American Chemical Society* 120: 4176–4183.

Feldman, Y., Zak, A., Popovitz-Biro, R., Tenne, R. 2000. New reactor for production of tungsten disulfide hollow onion-like (inorganic fullerene-like) nanoparticles, *Solid State Sciences* 2: 663–672.

Fleischauer, P. D., Lince, J. R. 1999. A comparison of oxidation and oxygen substitution in MoS$_2$ solid film lubricant, *Tribology International* 32: 627–636.

Fleischer, N., Genut, M., Zak, A., Rapoport, L., Tenne, R. 2005. Superior performance of inorganic fullerene-like nanospheres as EP/AW additives for oils and grease, Additives 2005, Dublin, Ireland, 05-07/04/2005.

Frey, G. L., Elani, S., Homyonfer, M., Feldman, Y., Tenne, R. 1998. Optical-absorption spectra of inorganic fullerene like MS$_2$ (M=Mo, W), *Physical Review B* 57: 6666–6671.

Gilkes, K. W. R., Sands, H. S., Batchelder, D. N., Milne, W. I., Robertson, J. 1998. Direct observation of sp^3 bonding in tetrahedral amorphous carbon UV Raman spectroscopy, *Journal of Non-Crystalline Solids* 227–230: 612–616.

Greenberg, R., Halperin, G., Etsion, I., Tenne, R. 2004. The effect of WS$_2$ nanoparticles on friction reduction in various lubrication regimes, *Tribology Letters* 17: 179–186.

Grossiord, C., Varlot, K., Martin, J. M., Le-Mogne, T., Esnouf, C., Inoue, K. 1998. MoS$_2$ single sheet lubrication by molybdenum dithiocarbamate, *Tribology International* 31: 737–743.

He, C. N., Zhao, N. Q., Du, X. W., Shi, C. S., Ding, J., Li, J. J., Li, Y. 2006a. Low-temperature synthesis of carbon onions by chemical vapor deposition using a nickel catalyst supported on aluminium, *Scripta Materialia* 54: 689–693.

He, C. N., Zhao, N. Q., Shi, C. S., Du, X., Li, J. J., Cui, L., He, F. 2006b. Carbon onion growth enhanced by nitrogen incorporation, *Scripta Materialia* 54: 1739–1743.

Hiraki, J., Mori, H., Taguchi, E., Yasuda, H., Kinoshita, H., Ohmae, N. 2005. Transformation of diamond nanoparticles into onion-like carbon by electron irradiation studied directly inside an ultra-high vacuum transmission electron microscope, *Applied Physics Letters* 86: 223101.

Hirano, M., Shinjo, K., Kaneko, R., Murata, Y. 1991. Anisotropy of frictional forces in muscovite mica, *Physical Review Letters* 67: 2642–2645.

Hirata, T., Ohuchi, F. S. 2001. Temperature dependance of the Raman spectra of 1T-TaS$_2$, *Solid State Communications* 117: 361–364.

Hirata, A., Igarashi, M., Kaito, T. 2004. Study of solid lubricant properties of carbon onions produced by heat treatment of diamond clusters or particles, *Tribology International* 37: 899–905.

Hu, J. J., Zabinski, J. S. 2005. Nanotribology and lubrication mechanisms of inorganic fullerene-like MoS$_2$ nanoparticles investigated using lateral force microscopy (LFM), *Tribology Letters* 18: 173–180.

Hu, J. J., Bultman, J. E., Zabinski, J. S. 2004. Inorganic fullerene-like nanoparticles produced by arc discharge in water with potential lubricating ability, *Tribology Letters* 17: 543–546.

Jin, Y., Li, S. 2007. Superlubricity of in-situ generated protective layer on worn metal surfaces in presence of Mg$_6$Si$_4$O$_{10}$(OH)$_8$. In *Superlubricity*, A. Erdemir, and J. M. Martin (eds.), pp. 447–471. Elsevier, Oxford, U.K.

Johnson, K. L. 1985. *Contact Mechanics*, Cambridge University Press, Cambridge, U.K.

Joly-Pottuz, L., Dassenoy, F. 2008. Nanoparticles made on metal dichalcogenides. In *Nanolubricants*, J. M. Martin and N. Ohmae (eds.), pp. 15–92. Wiley, New York.

Joly-Pottuz, L., Iwaki, M. 2007. Superlubricity of tungsten disulphide coatings in ultra high vacuum. In *Superlubricity*, A. Erdemir and J. M. Martin, (eds.), pp. 229–238. Elsevier, Oxford, U.K.

Joly-Pottuz, L., Ohmae, N. 2008. Carbon-based nanolubricants. In *Nanolubricants*, J. M. Martin and N. Ohmae (eds.), pp. 93–147. Wiley, New York.

Joly-Pottuz, L., Dassenoy, F., Belin, M., Vacher, B., Martin, J. M., Fleischer, N. 2005. Ultralow-friction and wear properties of IF-WS$_2$ under boundary lubrication. *Tribology Letters* 18: 477–485.

Joly-Pottuz, L., Martin, J. M., Dassenoy, F., Belin, M., Montagnac, G., Reynard, B., Fleischer, N. 2006. Pressure-induced exfoliation of inorganic fullerene-like WS$_2$ particles in a Hertzian contact, *Journal of Applied Physics* 99: 023524.

Joly-Pottuz, L., Martin, J. M., Belin, M., Dassenoy, F. 2007. Study of inorganic fullerenes and carbon nanotubes by in-situ Raman tribometry, *Applied Physics Letters* 91: 153107

Joly-Pottuz, L., Matsumoto, N., Kinoshita, H., Vacher, B., Belin, M., Montagnac, G., Martin, J. M., Ohmae, N. 2008a. Diamond-derived carbon onions as lubricant additives, *Tribology International* 41: 69–78.

Joly-Pottuz, L., Vacher, B., Le-Mogne, T., Martin, J. M., Mieno, T., He, C. N., Zhao, N. Q. 2008b. The role of nickel in Ni-containing nanotubes and onions as lubricant additives, *Tribology Letters* 29: 213–219.

Joly-Pottuz, L., Vacher, B., Ohmae, N., Martin, J. M., Epicier, T. 2008c. Anti-wear mechanism of carbon nano-onions as lubricant additives, *Tribology Letters* 30: 69–80.

Joly-Pottuz, L., Bucholz, E. W., Matsumoto, N., Phillpot, S. R., Sinnott, S. B., Ohmae, N., Martin, J. M. 2010. Friction properties of carbon nano-onions from experiment and computer simulations, *Tribology Letters* 37: 75–81.

Katz, A., Redlich, M., Rapoport, L., Wagner, H. D., Tenne, R. 2006. Self-lubricating coatings containing fullerene-like WS$_2$ nanoparticles for orthodontic wires and other possible medical applications, *Tribology Letters* 21: 135–139.

Kuznetsov, V. L., Chuvilin, A. L., Butenko, Y. V., Mal'kov, I. Y., Titov, V. M. 1994. Onion-like carbon from ultra-disperse diamond, *Chemical Physics Letters* 222: 343–348.

Kuznetsov, V. L., Zilberberg, I. L., Butenko, Y. V., Chuvilin, A. L., Segall, B. 1999. Theoretical study of the formation of closed curved graphite-like structures during annealing of diamond surface, *Journal of Applied Physics* 86: 863–870.

Kuzuo, R., Terauchi, M., Tanaka, M., Saito, Y. 1994. Electron energy loss spectra of single-shell carbon nanotubes, *Japan Journal of Applied Physics* 33: L1316–L1319.

Lancaster, J. K. 1990. A review of the influence of environmental humidity and water on friction, lubrication and wear, *Tribology International* 23: 371–389.

Martin, J. M., Donnet, C., Le-Mogne, T., Epicier, T. 1993. Superlubricity of molybdenum disulphide, *Physical review B* 48: 10583–10586.

McMullan, W. G., Irwin, J. C. 1983. Raman scattering from 2H and 3R-NbS$_2$, *Solid State Communications* 45: 557–560.

Moshkovith, A., Perfiliev, V., Verdyan, A., Lapsker, I., Popovitz-Biro, R., Tenne, R., Rapoport, L. 2007. Sedimentation of IF-WS$_2$ aggregates and a reproducibility of the tribological data, *Tribology International* 40: 117–124.

Mykhaylyk, O. O., Solonin, Y. M., Batchelder, D. N., Brydson, R. 2005. Transformation of nanodiamond into carbon onions: A comparative study by high-resolution transmission electron microscopy, electron energy-loss spectroscopy, x-ray diffraction, small-angle x-ray scattering, and ultraviolet Raman spectroscopy, *Journal of Applied Physics* 97: 074302.

Ozawa, M., Goto, H., Kusunoki, M., Osawa, E. 2002. Continuously growing spiral carbon nanoparticles as the intermediates in the formation of fullerenes and nanoonions, *Journal of Physical Chemistry B* 106: 7135–7138.

Pecharroman, C., Gonzalez-Carreno, T., Iglesias, J. E. 1995. The infrared dielectric properties of maghemite, γ-Fe2O3, from reflectance measurement on pressed powders, *Physics and Chemistry of Minerals* 22: 21–29.

Perfiliev, V., Moshkovith, A., Verdyan, A., Tenne, R., Rapoport, L. 2006. A new way to feed nanoparticles to friction interfaces, *Tribology Letters* 21: 89–93.

Peterson, M. B., Johnson, R. L. 1953. Friction and wear investigations of molybdenum disulfide I: Effect of moisture, *NACA TN 3055*.

Qin, L. C., Iijima, S. 1996. Onion-like graphitic particles produced from diamond, *Chemical Physics Letters* 262: 252–258.

Rapoport, L., Bilik, Y., Feldman, Y., Homyonfer, M., Cohen, S. R., Tenne, R. 1997. Hollow nanoparticles of WS$_2$ as potential solid-state lubricants, *Nature* 387: 791–793.

Rapoport, L., Feldman, Y., Homyonfer, M., Cohen, H., Sloan, J., Hutchison, J. L., Tenne, R. 1999. Inorganic fullerene-like material as additives to lubricants: Structure-function relationship, *Wear* 225–229: 975–982.

Rapoport, L., Leshchinsky, V., Volovik, Y., Lvovsky, M., Nepomnyashchy, O., Feldman, Y., Popovitz-Biro, R., Tenne, R. 2003a. Modification of contact surfaces by fullerene-like solid lubricant nanoparticles, *Surface and Coatings Technology* 163–164: 405–412.

Rapoport, L., Leschchinsky, V., Lapsker, I., Volovik, Y., Nepomnyashchy, O., Lvovsky, M., Popovitz-Biro, R., Tenne, R. 2003b. Tribological properties of WS$_2$ nanoparticles under mixed lubrication, *Wear* 255: 785–793.

Rapoport, L., Nepomnyashchy, O., Lapsker, I., Verdyan, A., Moshkovich, A., Feldman, Y., Tenne, R. 2005. Behavior of fullerene-like WS$_2$ nanoparticles under severe contact conditions, *Wear* 259: 703–707.

Redlich, P., Banhart, F., Lyutovich, Y., Ajayan, P. M. 1998. EELS study of the irradiation-induced compression of carbon onions and their transformation to diamond, *Carbon* 36: 561–563.

Roddatis, V. V., Kunetsov, V. L., Butenko, Y. V., Su, D. S., Schlögl, R. 2002. Transformation of diamond nanoparticles into carbon onions under electron irradiation, *Physical Chemistry Chemical Physics* 4: 1964–1967.

Roy, D., Chhowalla, M., Wang, H., Sano, N., Alexandrou, I., Clyne, T. W., Amarantunga, G. A. J. 2003. Characterisation of carbon nano-onions using Raman spectroscopy, *Chemical Physics Letters* 373: 52–56.

Saito, Y., Yoshikawa, T., Bandow, S., Tomita, M., Hayashi, T. 1993. Interlayer spacings in carbon nanotubes, *Physical Review B* 48: 1907–1909.

Sano, N., Wang, H., Chhowalla, M., Alexandrou, I., Amaratunga, G. A. J. 2001. Nanotechnology: synthesis of carbon "onions" in water, *Nature* 414: 506–507.

Savage, R. H. 1948. Graphite lubrication, *Journal of Applied Physics* 19: 1–10.

Schuffenhauer, C., Parkinson, B. A., Jin-Phillip, N. Y., Joly-Pottuz, L., Martin, J. M., Popovitz-Biro, R., Tenne R. 2005. Synthesis of fullerene-like tantalum disulfide nanoparticles by gas phase reaction and laser ablation, *Small* 1: 1100–1109.

Sohmen, E., Fink, J., Krätschmer, W. 1992. Electron energy-loss spectroscopy studies on C_{60} and C_{70} fullerite, *Zeitschrift für Physik B Condensed Matter* 86: 87–92.

Srolovitz, D. J., Safran, S. A., Tenne, R. 1994. Elastic equilibrium of curved thin films, *Physical Review E* 49: 5260–5270.

Srolovitz, D. J., Safran, S. A., Homyonfer, M., Tenne, R. 1995. Morphology of nested fullerenes, *Physical Review Letters* 74: 1779–1782.

Street, K. W., Marchetti, M., Wal, R. L. V., Tomasek, A. J. 2004. Evaluation of the tribological behavior of nano-onions in Krytox 143B, *Tribology Letters* 16: 143–149.

Sun, Z., Shi, J. R., Tay, B. K., Lau, S. P. 2000. UV Raman characteristics of nanocrystalline diamond films with different grain size, *Diamond and Related Materials* 9: 1979–1983.

Tenne, R. 2003. Advances in the synthesis of inorganic nanotubes and fullerene-like nanoparticles, *Angewandte Chemie International Edition* 42: 5124–5132.

Tenne, R., Homyonfer, M., Feldman, Y. 1998. Nanoparticles of layered compounds with hollow cage structures (inorganic fullerene-like structures), *Chemistry of Materials* 10: 3225–3238.

Tomita, S., Fujii, M., Hayashi, S., Yamamoto, K. 1999. Electron energy-loss spectroscopy of carbon onions, *Chemical Physics Letters* 305: 225–229.

Tomita, S., Burian, A., Dore, J. C., LeBolloch, D., Fujii, M., Hayashi, S. 2002. Diamond nanoparticles to carbon onions transformation: X-ray diffraction studies, *Carbon* 40: 1469–1474.

Troiani, H. E., Camacho-Bragado, A., Armendariz, V., Gardea-Torresday, J. L., Jose-Yacaman, M. 2003. Synthesis of carbon onions by gold nanoparticles and electron irradiation, *Chemistry of Materials* 15: 1029–1031.

Ugarte, D. 1992. Curling and closure of graphitic networks under electron-beam irradiation, *Nature* 359: 707–709.

Ugarte, D. 1995. Onion-like graphitic particles, *Carbon* 33: 989–993.

Wang, X., Xu, B., Liu, X., Guo, J., Ichinose, H. 2006. Synthesis of Fe-included onion-like Fullerenes by chemical vapor deposition, *Diamond and Related Materials* 15: 147–150.

Xu, B. S., Tanaka, S. I. 1998. Formation of giant onion-like fullerenes under Al nanoparticles by electron irradiation, *Acta Metallurgica* 46: 5249–5257.

Yannouleas, C., Bogachek, E. N., Landman, U. 1996. Collective excitations of multishell carbon microstructures: Multishell fullerenes and coaxial nanotubes, *Physical Review B* 53: 10225–10236.

Zhou, K., Cao, W., Zhang, Y., Liu, W. 1997. Structure and tribological properties of plasma-polymerizes nickel carbonyl films, *Thin Solid Films* 303: 89–93.

Zou, T. Z., Tu, J. P., Zhang, S. C., Chen, L. M., Wang, Q., Zhang, L. L., Ne, D. H. 2006. Friction and wear properties of electroless Ni-P-($IF-MoS_2$) composite coatings in humid air and vacuum, *Materials Science and Engineering A* 426: 162–168.

27

Plasmonic Nanoparticle Networks

Erik Dujardin
*Centre National de la
Recherche Scientifique*

Christian Girard
*Centre National de la
Recherche Scientifique*

27.1 Introduction

Self-assembled nanoplasmonics is a new interdisciplinary topic that aims at self-assembling, interconnecting, and characterizing resonant metallic nanostructures that are able to funnel, confine, and propagate light energy from a conventional laser source to a single molecular entity. Following this paradigm, several orders of magnitude in the miniaturization scale of optical devices, spanning from tens of micrometers down to the molecular scale can be expected. In this chapter, we describe some recent experimental and theoretical results on plasmonic structures made by self-assembly or surface deposition of colloidal metallic particles with the main objective of overcoming the current limitations of an exclusive top-down approach of plasmonics. More specifically, the interest of these objects for tailoring original near-field optical properties will be exposed (near-field optical confinement, local density of electromagnetic state squeezing, etc.). In particular, it is shown that a bottom-up approach is not only able to produce interesting nanoscale metallic building blocks and design their optical properties but is also able to easily produce complex superstructures that can bridge the submicron range gap between nanofabricated structures and single colloids, and which would be difficult to achieve by other means.

27.2 Background

Although known since ancient Rome as ingredients for stained glass, metallic colloids with typical sizes of a few tens of nanometers are traditionally traced back to the late 1850s' experiments by M. Faraday on the optical properties of gold films and suspensions [34]. Since then, the topic has expanded into a vast field of research at the crossroads of chemistry, physics, biology, and instrumentation. This chapter reflects this interdisciplinary approach by exploring, back and forth, theoretical concepts or tools to understand the interplay between metallic nanomaterials and tailored electromagnetic fields, on one hand and, on the other hand, by illustrating them with specific experimental approaches. Section 27.3 is dedicated to isolated and extremely simple clusters of metallic nano-objects while Section 27.4 exposes how self-assembly can bring these concepts to a higher level of complexity, and how theoretical description tools can be adapted to this change in length scale. Finally, Section 27.5 shows how information processing down to molecular scale with light can be envisioned beyond the diffraction limit using self-assembled plasmonic nanosystems.

Regarding the metallic nanoparticle synthesis, an exhaustive survey is far beyond the scope of this article. The reader willing to get a background overview on synthesis and general properties of metal colloids may advantageously read the book by Feldheim and Foss [36] and some more recent reviews among which are [21,31,33,51,83]. Similarly, the theory of plasmons localized in metallic nanoparticles can be traced back by the seminal work of Mie [91], but the reader will find it useful to refer to references [7,71,105] for background information.

27.3 Colloids and Localized Plasmons: Basic Concepts

It is well known that the physical properties of a macroscopic system differ considerably from the physical properties of the individual atoms that compose the materials. Consequently, it is quite natural to expect a change in the optical behavior when one performs a fragmentation of an object in smaller and smaller volumes. For example, in the case of noble metals or semiconductors, we can observe drastic changes in the electromagnetic responses of small particles compared to those of bulk materials. In the case of metals, these responses are generally controlled by the dynamical properties of free electrons confined within the particles [71]. To better apprehend the modern science of colloidal plasmonics, we dedicate this section to isolated nanoparticles and to the interaction between two colloidal particles. The first parts give a simple analytical description of plasmon physics while the last two parts give a practical overview of the synthetic means to prepare those objects and experimental approaches to characterize their optical behavior.

27.3.1 Spherical Metallic Particles

The theoretical study of the linear optical response properties of small metallic aggregates has been extensively investigated for the last 50 years. In the case of noble metal spherical particles, characterized by a permittivity $\epsilon_m(\omega_0)$, embedded in transparent materials of the dielectric constant ϵ_{env}, the interface between the two media introduces surface plasmon resonances, the frequency of which depends on the optical properties of the metal and the surroundings, as well as the topography of the particle surface. In many situations, single metal particles can be schematized by a sphere of radius a (Figure 27.1). Their optical response can then be described by a scalar frequency–dependent polarizability. In CGS units, it reads

$$\alpha(\omega_0) = a^3 \left(\frac{\epsilon_m(\omega_0) - \epsilon_{env}}{\epsilon_m(\omega_0) + 2\epsilon_{env}} \right) \qquad (27.1)$$

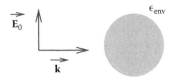

FIGURE 27.1 Schematic drawing of a single metallic nanoparticle of sub-wavelength size submitted to an external illumination field. The propagation direction is schematized by the wave vector **k**.

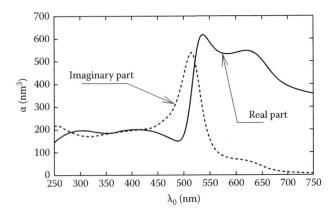

FIGURE 27.2 Example of numerical simulations performed with a single gold particle. Plot of the spectral variations of the dipolar polarizability (solid line: real part, dashed line: imaginary part.)

From relation (27.1), we can extract the spectra for both real and imaginary parts of $\alpha(\omega_0)$. From the experimental data by Palik [101], reliable representations of the dielectric constants $\epsilon_m(\omega_0)$ of noble metals can be achieved. In Figure 27.2, we have used this data to treat numerically the case of gold nanospheres. The two main peaks of Figure 27.2 indicate the position of the plasmon resonance. Owing to the spherical geometry assumed in the model, the spectral position of the resonance does not depend on the orientation of the electric field vector associated with the illumination light. In other words, each illumination configuration gives rise to the same plasmon oscillation that occurs when

$$\Re\left\{\epsilon_m(\omega_0) + 2\epsilon_{env}\right\} = 0. \qquad (27.2)$$

Consequently, the extinction spectra of a sample containing a large number of such non-interacting nanoparticles is given by the simple relation:

$$I_{ext}(\lambda_0) = \frac{8\pi^2}{\lambda_0} \Im\{\alpha(\omega_0)\}, \qquad (27.3)$$

where λ_0 represents the incident wavelength. In the particular case of isolated spherical metallic nanoparticles, we can see that the extinction spectra correspond to the spectral variation of the polarizability imaginary part (cf. dashed curve of Figure 27.2).

27.3.2 Elongated Metallic Particles

Unlike what happens with perfectly spherical particles, the extinction spectra of single nanorods exhibit two plasmon bands that correspond to electron oscillations along their length (longitudinal mode) and across their section (transverse mode). Once again, this shape effect can be described with a simple analytical model by using an anisotropic dynamical polarizability. In the absolute frame of Figure 27.3, the polarizability tensor is diagonal:

$$\alpha(\omega_0) = \begin{pmatrix} \alpha_\perp(\omega_0) & 0 & 0 \\ 0 & \alpha_\perp(\omega_0) & 0 \\ 0 & 0 & \alpha_\parallel(\omega_0) \end{pmatrix}, \qquad (27.4)$$

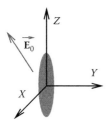

FIGURE 27.3 Schematic drawing of an ellipsoidal metallic particle illuminated by a light field characterized by the vector $\mathbf{E}_0$.

where the two independent components $\alpha_\perp(\omega_0)$ and $\alpha_\parallel(\omega_0)$ can be described by two Lorentzian functions:

$$\alpha_\perp(\omega_0) = \frac{\alpha_\perp^0 \omega_\perp^2}{\omega_\perp^2 - \omega_0(\omega_0 + i\Gamma_\perp)} \quad \text{and} \quad \alpha_\parallel(\omega_0) = \frac{\alpha_\parallel^0 \omega_\parallel^2}{\omega_\parallel^2 - \omega_0(\omega_0 + i\Gamma_\parallel)}. \tag{27.5}$$

where

- $\alpha_{\perp/\parallel}^0$ are the polarizability coefficients along the (OX) and (OZ) directions
- $\omega_{\perp/\parallel}$ labels both transverse and longitudinal plasmon resonances
- $\Gamma_{\perp/\parallel}$ represents the corresponding damping factors

By using these relations, we can obtain the extinction spectra of an ensemble of non-interacting and disordered nanorods:

$$I_{\text{ext}}(\lambda_0) = \frac{8\pi^2}{3\lambda_0} \Im\{2\alpha_\perp(\omega_0) + \alpha_\parallel(\omega_0)\}. \tag{27.6}$$

Finally, after substituting the two relations (Equation 27.5) in this expression, the extinction spectrum reveals two plasmonic bands centered around the plasmon resonances $\omega_\perp$ and $\omega_\parallel$:

$$I_{\text{ext}}(\lambda_0) = \frac{8\pi^2}{3\lambda_0} \left\{ \frac{2\alpha_\perp^0 \omega_\perp^2 \omega_0 \Gamma_\perp}{(\omega_\perp^2 - \omega_0^2)^2 + \omega_0^2 \Gamma_\perp^2} + \frac{\alpha_\parallel^0 \omega_\parallel^2 \omega_0 \Gamma_\parallel}{(\omega_\parallel^2 - \omega_0^2)^2 + \omega_0^2 \Gamma_\parallel^2} \right\}. \tag{27.7}$$

Remarkably, the amplitude and the position of these two bands strongly depend on the aspect ratio of the nanorod as demonstrated in Figure 27.4. In order to compute these spectra, we have expressed in Equation 27.6 both polarizability components $\alpha_\perp(\omega_0)$ and $\alpha_\parallel(\omega_0)$ from the bulk permittivity $\epsilon_{\text{m}}(\omega_0)$ of the metal. For a prolate spheroid, which has two short axes b and c of equal length and a long axis $a > b$, one obtains the two relations:

$$\alpha_\perp(\omega_0) = \frac{a^2 b}{3} \left(\frac{\epsilon_{\text{m}}(\omega_0) - \epsilon_{\text{env}}}{L_\perp \epsilon_{\text{m}}(\omega_0) + (1 - L_\perp)\epsilon_{\text{env}}} \right) \tag{27.8}$$

and

$$\alpha_\parallel(\omega_0) = \frac{a^2 b}{3} \left(\frac{\epsilon_{\text{m}}(\omega_0) - \epsilon_{\text{env}}}{L_\parallel \epsilon_{\text{m}}(\omega_0) + (1 - L_\parallel)\epsilon_{\text{env}}} \right), \tag{27.9}$$

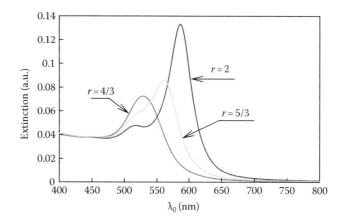

FIGURE 27.4 Sequence of three extinction spectra calculated from the quasi-static ellipsoidal model for increasing aspect ratio. The particle is a gold ellipsoid described with Johnson and Christy [64] permittivity.

where the two coefficients $L_\parallel$ and $L_\perp$ are defined by [7]

$$L_\parallel = \frac{1-e^2}{e^2}\left(\frac{1}{2e}\log\left(\frac{1+e}{1-1}\right)-1\right) \quad e^2 = 1 - \frac{b^2}{a^2} \quad L_\perp = 1 - 2L_\parallel. \tag{27.10}$$

27.3.3 Interaction between Localized Surface Plasmons

When a second metallic nanoparticle is placed in the immediate proximity of the first one (what is called the *near-field zone*), the strong optical coupling modifies the response of the pair (cf. Figure 27.5). The two local fields $\mathbf{E}(\mathbf{r}_a, \omega_0)$ and $\mathbf{E}(\mathbf{r}_b, \omega_0)$ at the center of the particles a and b, respectively, verify the two coupled equations:

$$\mathbf{E}(\mathbf{r}_a, \omega_0) = \mathbf{E}_0(\mathbf{r}_a, \omega_0) + \mathbf{T}(\mathbf{R}) \cdot \alpha(\omega_0) \cdot \mathbf{E}(\mathbf{r}_b, \omega_0)$$
$$\mathbf{E}(\mathbf{r}_b, \omega_0) = \mathbf{E}_0(\mathbf{r}_b, \omega_0) + \mathbf{T}(\mathbf{R}) \cdot \alpha(\omega_0) \cdot \mathbf{E}(\mathbf{r}_a, \omega_0)$$
$$\tag{27.11}$$

where $\mathbf{R} = \mathbf{r}_a - \mathbf{r}_b$ is the spacing vector between the two particles

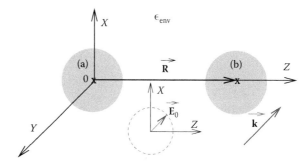

FIGURE 27.5 Geometry of two interacting nanoparticles excited by a plane wave propagating along the OY axis. The vector $\mathbf{E}_0$, parallel to the XOZ plane, represents the electric field part of the illumination wave.

$\alpha(\omega_0)$ represents their dynamical polarizabilities (assumed to be identical for the two particles), and defines $\mathbf{T}(\mathbf{R})$, the usual dipolar tensor:

$$\mathbf{T}(\mathbf{R}) = \frac{3\mathbf{R}\mathbf{R} - R^2\mathbf{I}}{R^5}. \qquad (27.12)$$

This equation can be simplified by introducing the axial symmetry of the system (cf. Figure 27.5). Indeed, if we choose the vectors $\mathbf{O}$ and $\mathbf{R}$ for $\mathbf{r}_a$ and $\mathbf{r}_b$, respectively, the tensor $\mathbf{T}(\mathbf{R})$ reduces to

$$\mathbf{T}(\mathbf{R}) = \frac{1}{R^3}\begin{pmatrix} -1 & 0 & 0 \\ 0 & -1 & 0 \\ 0 & 0 & 2 \end{pmatrix} \qquad (27.13)$$

The two vectorial equations (27.11) can then be rewritten as follows:

$$\mathcal{E}_0(\omega_0) = \mathcal{M}(\mathbf{R},\omega_0) \cdot \mathcal{E}(\omega_0) \qquad (27.14)$$

where $\mathcal{E}_0(\omega_0)$ is a super vector that contains the two incident fields at the particle locations:

$$\mathcal{E}_0(\omega_0) = \{\mathbf{E}_0(\mathbf{r}_a,\omega_0), \mathbf{E}_0(\mathbf{r}_b,\omega_0)\}, \qquad (27.15)$$

and $\mathcal{E}(\omega_0)$ contains the local field values:

$$\mathcal{E}(\omega_0) = \{\mathbf{E}(\mathbf{r}_a,\omega_0), \mathbf{E}(\mathbf{r}_b,\omega_0)\} \qquad (27.16)$$

For two identical particles, the matrix $\mathcal{M}(\mathbf{R},\omega)$ has a very simple form given by

$$\mathcal{M}(\mathbf{R},\omega_0) = \begin{pmatrix} \mathbf{I} & -\mathbf{A}(\mathbf{R},\omega_0) \\ -\mathbf{A}(\mathbf{R},\omega_0) & \mathbf{I} \end{pmatrix} \qquad (27.17)$$

with

$$\mathbf{A}(\mathbf{R},\omega_0) = \frac{\alpha(\omega_0)}{R^3}\begin{pmatrix} -1 & 0 & 0 \\ 0 & -1 & 0 \\ 0 & 0 & 2 \end{pmatrix} \text{ and } \mathbf{I} = \begin{pmatrix} 1 & 0 & 0 \\ 0 & 1 & 0 \\ 0 & 0 & 1 \end{pmatrix}$$
$$(27.18)$$

In order to obtain the local field at each particle center, the 6×6 matrix $\mathcal{M}(\mathbf{R},\omega_0)$ must be inverted. Due to the simple dimer configuration investigated in this section, this inversion can be performed analytically (Figures 27.6 and 27.7). That leads to

$$\mathcal{E}(\omega_0) = \mathcal{M}^{-1}(\mathbf{R},\omega_0) \cdot \mathcal{E}_0(\omega_0) \qquad (27.19)$$

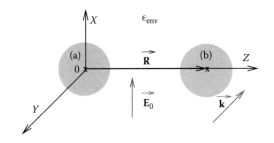

FIGURE 27.6   Illustration of the transverse illumination mode.

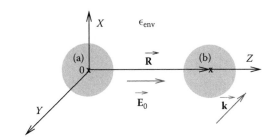

FIGURE 27.7   Illustration of the illumination configuration used to excite the longitudinal plasmon mode of the dimer.

with

$$\mathcal{M}^{-1}(\mathbf{R},\omega_0) =$$

$$\begin{pmatrix} \dfrac{R^6}{R^6 - \alpha^2(\omega_0)} & 0 & 0 & -\dfrac{\alpha(\omega_0)R^3}{R^6 - \alpha^2(\omega_0)} & 0 & 0 \\[2mm] 0 & \dfrac{R^6}{R^6 - \alpha^2(\omega_0)} & 0 & 0 & -\dfrac{\alpha(\omega_0)R^3}{R^6 - \alpha^2(\omega_0)} & 0 \\[2mm] 0 & 0 & \dfrac{R^6}{R^6 - 4\alpha^2(\omega_0)} & 0 & 0 & \dfrac{2\alpha(\omega_0)R^3}{R^6 - 4\alpha^2(\omega_0)} \\[2mm] -\dfrac{\alpha(\omega_0)R^3}{R^6 - \alpha^2(\omega_0)} & 0 & 0 & \dfrac{R^6}{R^6 - \alpha^2(\omega_0)} & 0 & 0 \\[2mm] 0 & -\dfrac{\alpha(\omega_0)R^3}{R^6 - \alpha^2(\omega_0)} & 0 & 0 & \dfrac{R^6}{R^6 - \alpha^2(\omega_0)} & 0 \\[2mm] 0 & 0 & \dfrac{2\alpha(\omega_0)R^3}{R^6 - 4\alpha^2(\omega_0)} & 0 & 0 & \dfrac{R^6}{R^6 - 4\alpha^2(\omega_0)} \end{pmatrix}$$

From this dynamical matrix, it is straightforward to find the plasmon resonances of the dimer. Two main illumination modes can be considered:

1. *Transverse mode*: This plasmon mode is excited by an electric field polarized perpendicularly to the dimer axis. In this configuration, the illumination field is defined by $\mathbf{E}_0(\mathbf{r}_a,\omega_0) = \mathbf{E}_0(\mathbf{r}_b,\omega_0) = (E_0(\omega_0), 0, 0)$ and the local field on each particle site is given by

$$E(\mathbf{r}_a,\omega) = E(\mathbf{r}_b,\omega) = \frac{R^3}{R^3 + \alpha(\omega_0)}E_0(\omega_0) \qquad (27.20)$$

The resonance frequency ω_T is reached when

$$\Re\{R^3 + \alpha(\omega_T)\} = 0. \qquad (27.21)$$

In order to supply a simple expression for ω_T we can use the analytical Lorentzian model for the dynamical polarizability:

$$\alpha(\omega_0) = \frac{\alpha_0 \Omega^2}{\Omega^2 - \omega_0^2} \qquad (27.22)$$

where Ω represents the plasmon frequency of the isolated particle. After applying the previous condition (27.21) and using Equation 27.22, one finds

$$\frac{\alpha_0 \Omega^2}{\Omega^2 - \omega_T^2} = -R^3, \qquad (27.23)$$

which gives the transverse angular frequency ω_T

$$\omega_T = \Omega\sqrt{1 + \frac{\alpha_0}{R^3}} \simeq \Omega\left(1 + \frac{\alpha_0}{2R^3}\right) \qquad (27.24)$$

In this configuration, the near-field interaction produces a blue shift of the plasmon resonance [87].

2. *Longitudinal mode*: This mode is excited by an electric field parallel to the dimer axis. By using a similar reasoning, one obtains the longitudinal plasmon frequency:

$$\omega_L = \Omega\sqrt{1 - \frac{2\alpha_0}{R^3}} \simeq \omega_0\left(1 - \frac{\alpha_0}{R^3}\right), \qquad (27.25)$$

that is red shifted with a R^{-3} dependence of the interparticle spacing. Within this simple interacting dipole model, we can observe that the peak splitting defined by $\Delta\omega = \omega_T - \omega_L$ saturates for $R = 2a$ (that corresponds to the physical contact between the particle surfaces) at the value $\Delta\omega = 3\Omega\alpha_0/16a^3$ and progressively vanishes when the two particles move away from each other.

27.3.4 Morphological Tailoring of Plasmons in Colloids

While the physical phenomena responsible for the bright and robust colors of colloidal suspensions of noble metals were rationalized a mere century ago, these properties have been empirically used for millennia. The convergence of the ancient chemical knowledge and the powerful physical description has fostered a booming experimental exploration of the ways to tailor the optical properties of metal colloids in the past 30 years. Numerous approaches can be used to produce finely divided metal particles, yet the most versatile strategy to produce particles with sizes ranging roughly between 200 and 2 nm doubtlessly consists in reducing precursor metal salts in a solvent in the presence of a stabilizing agent, in order to produce colloidal suspensions. In this section, we describe how the morphology of plasmonic metal (usually gold) nanoparticles can be modified and how it reflects on the spectral and spatial design of their optical properties. If the reduction of noble metal (copper, silver, gold, etc.) salts to metallic species is left uncontrolled, black powder or pale and shiny bulk materials are obtained. Intense colors will appear only if the reduction is done under control to yield nanometer-scale particles with a capped surface. Such particles are then spherical, in the first approximation, because of the highly isotropic symmetry group of these *fcc* metals.

The archetypal Türkevitch method consists in vigorously stirring a boiling aqueous solution of $HAuCl_4$, prior to fast addition of a citrate solution [123]. The citrate reduces the gold(III) ions and acts as a capping group that protects the nanoparticles. Since all nuclei are generated simultaneously and subsequently grow alongside until the complete depletion of the solution, the end population of nanoparticles is uniform in size. The size itself can be selected by choosing the molar ratio between gold salt and citrate. The lesser the citrate amount is introduced, the larger the average final particle size will be. A tutorial to obtain typical 13 ± 2 nm spherical gold nanoparticles is given in Figure 27.8. Citrate-capped nanoparticles are obtained by refluxing 100 mL of a 1 mM aqueous solution of $HAuCl_4$ to which 10 mL of a 38.8 mM trisodium citrate solution are added quickly while stirring vigorously and refluxed for an additional 15 min after addition. The typical ruby red tint of the obtained 13 nm gold colloids (Figure 27.8b) is due to a plasmon absorption band centered at a wavelength of 520 nm corresponding to the maximum of the imaginary part of the gold polarizability (Figure 27.2). A number of factors that can modify the features of this resonance band have been thoroughly studied. Although controlling the size would be the easiest way to tailor the nanoparticle optical properties, the plasmon band is only very marginally shifted toward lower energies by 0.5 nm when the particle size increases by 1 nm [81,82]. This can be understood from Equation 27.1 that shows that the resonance is determined by the cancellation of the denominator real part of the polarizability that does not depend strongly on the particle size. Alternatively, the absorption peak is red shifted when the refractive index of the surrounding medium (solvent, capping molecular layer or covering inorganic shell) is increased at an approximate rate of 100 nm per unit [82,93,110]. Since the surface plasmon band of 15 nm silver colloid is located at 440 nm, alloying gold and silver in the nanoparticles will proportionally displace the resulting surface plasmon between the extreme values [68,82]. Temperature has very little direct effect [71].

A decade ago, it was realized that one of the most effective approach to plasmon engineering in colloids was to break the confinement symmetry by modifying the nanoparticle aspect ratio from spherical to rod shape [81,84]. As described in Section 27.3.2, this symmetry breaking lifts the plasmon mode degeneracy and gives rise to two distinct transverse and longitudinal bands. Experimentally, the spherical gold nanoparticles are often lentil-shaped, {111} penta-tetrahedrally twinned nanocrystals with a small rim of {100} faces. The specific adsorption of the stabilizing surfactant on the {100} faces inhibits their growth in favor of the {111} subset, which result in penta-twinned nanorods. The first, template-free synthesis of gold nanorods (typically 10×50 nm) was reported by Wang et al. [12] who used a sacrificial gold anode to produce gold ions in the presence of surfactants (cetyltrimethylammonium bromide, $C_{16}TAB$, and tetraoctylammonium bromide, tC_8AB). The ions are subsequently reduced electrochemically but the higher affinity of $C_{16}TAB$ for the

---

11-nm Citrate stabilized Au nanoparticles

<u>Chemicals</u>

Hydrochloric acid (HCl) 97% . . . . 0.3 L
Nitric acid (HNO$_3$) 70% . . . . .0.1 L
Hydrogen tetrachloroaurate (III) trihydrate (HAuCl$_4$, 3H$_2$O) . . . . 40 mg
Tri-sodium citrate dihydrate (HOC(CO$_2$Na(CH$_2$CO$_2$Na)$_2$, 2H$_2$O). . . . 110 mg
Deionised (18 MΩ) water . . . . 1 L (for rinsing glassware) + 100 mL + 10 mL

<u>Preparation</u>

1. Mix the HCl and HNO$_3$ acids to get a fresh aqua regia solution (**CAUTION, highly corrosive**!). Clean all glassware to be used with this solution. Rinse copiously with deionized water. Dry in oven prior to use.
2. Dissolve HAuCl$_4$ into 100 mL of deionized water.
3. Dissolve the sodium citrate into 10 mL of deionized water.
4. Bring the gold solution to boiling temperature while stirring vigorously with a reflux unit fitted. Then, add the citrate solution quickly, at once (the solution changes from pale yellow to deep red).
5. Keep on boiling under reflux for 15 min with continous stirring.
6. Let it cool down to room temperature and then filter through a 0.2 μm membrane.

<u>Storage</u>

Store the resulting deep ruby red solution in a fridge at 4°C.

(a)

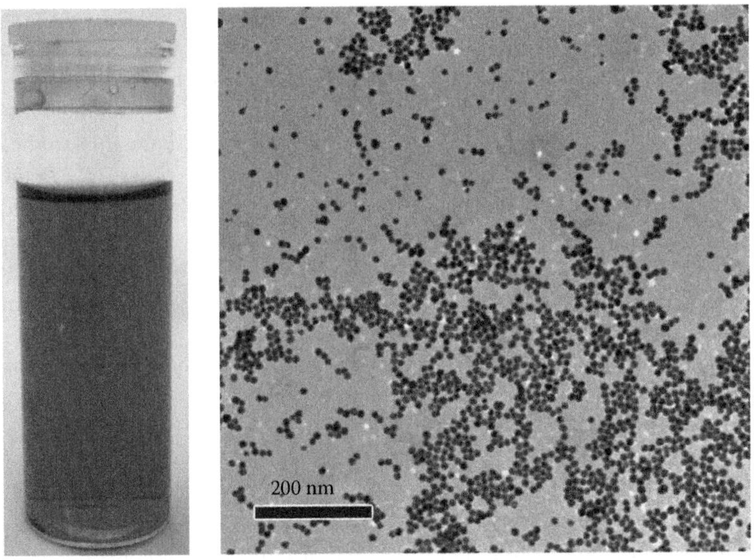

(b)   (c)

FIGURE 27.8 (a) Tutorial protocol for the synthesis of citrate-stabilized, 11 nm gold nanoparticles. (b) Typical gold colloid sol exhibiting the characteristic deep red color. (c) TEM image of 11 nm gold nanoparticles with a standard deviation of about 10%.

{100} crystal planes induce the anisotropic growth [136]. These principles have been further mastered by Murphy and coworkers who discovered that isotropic gold [12,60,61] or silver [61] seeds could grow into nanorods and nanowires with controlled aspect ratios by strictly decoupling the nucleation and growth stages. 4 nm gold seeds are produced by reducing gold(III) precursor with sodium borohydride. These seeds are then dispersed into a first growth solution containing C$_{16}$TAB as a crystal habit modifier and ascorbic acid as a mild reducing agent. Aliquots of growth solutions are successively mixed with fresh growth solutions. For gold, this protocol results in nanorods with an aspect ratio of 15–20, a typical width of 10–20 nm, and a width monodispersity within 10% [63,94]. Variations on this generic method have successfully overcome some of its initial limitations. Spherical byproducts are always present but could be limited by stabilizing

the seed with C$_{16}$TAB or adjusting the pH of the growth solution, and longer aspect ratios were obtained by mixing C$_{16}$TAB with either benzyldimethylhexadecylammonium chloride or silver ions [99]. Interestingly, the role of the iodide, a potential impurity in C$_{16}$TAB, was recently uncovered giving a new way to control the morphology. In the complete absence of iodide, spherical nanoparticles are produced, even by seeding into a C$_{16}$TAB-containing growth solution. For iodide concentration comprised between 2.5 and 10 μM, nanorods dominate (45% yield), while a majority (65% yield) of triangular prisms are produced for iodide concentrations exceeding 25 μM and up to 75 μM [92]. Alternatively, the mixture of shapes (spheres, platelets, and rods) obtained by the standard seed-mediated protocol can be sorted by a post-synthesis process combining centrifugation and redissolution in a Au(III)/C$_{16}$TAB solution. Indeed, gold nanoparticles are readily oxidized by

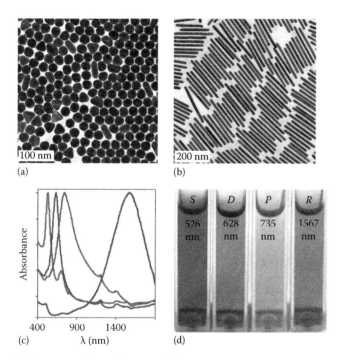

FIGURE 27.9 **(See color insert following page 9-8.)** Morphological control of gold colloid plasmonic properties. (a) Transmission electron microscopy (TEM) image of gold nanospheres. (b) TEM image of high aspect ratio gold nanorods. (c) Normalized extinction spectra of solutions of isolated spheres (S, red), nanodisks (d, blue), platelets (P, turquoise), and nanorods (R, brown). The photograph shows the corresponding solutions in deuterated water. (Adapted from Khanal, B.P. and Zubarev, E.R., *J. Am. Chem. Soc.*, 130, 12634, 2008. With permission.)

Au(III) in the presence of C_{16}TAB, [107] but the reactivity of platelets and spheres is twice higher than that of nanorods, so that fully dispersible by-products are obtained while nanorods still settle upon gravitational sedimentation. After separation, the supernatant contains 90% of nanodisks and nanoparticles (Figure 27.9a) while the nanorods are 99% pure (Figure 27.9b). Further, centrifugation allows to get separated solutions of spheres, flat discs, faceted platelets, and nanorods, the spectral properties of which are clearly tuned (Figure 27.9d), with maximal extinction shifted from 526 to 628, 735, and even to 1567 nm, respectively [69].

The tuning of the longitudinal plasmon mode energy toward longer wavelengths became achievable by extending the high aspect ratio nanorods synthesis toward long and thin nanowires. Silver nanowire with micrometer length and aspect ratios of up to 400 were produced by Murphy et al. by the same seed-mediated method as the gold nanorods but at a higher pH that modified the protonation of the mild reducing agent, ascorbic acid [61]. Mild reducing conditions were also a key element to obtain ultra-thin gold nanowires that were simultaneously produced by three groups by simply dissolving Au(III) salt into oleylamine (1-amino-9-octadecene) [55,86,126]. Single crystalline, 1.6–3 nm diameter wires grow along the {111} direction to reach lengths of up to 4 μm with or without the addition of oleic acid or silver ions.

While longitudinal growth results in a second plasmon mode that can be tuned between visible and infrared wavelengths and

thus provides an efficient approach to the spectral design of the near-field and far-field optical properties of metal colloids, exploring more complex morphology allows a spatial design of the plasmon modes. For example, silver and gold triangular prisms [100,102] show a similar spectra trend as the nanorods with essentially a single broad plasmon band shifting from 600 to near-infrared range as the side edge length increases (Figure 27.10a) [9]. As shown from the DDA calculations in Figure 27.10b, the effect of truncation from triangle prisms to hexagonal platelets results in a reverse trend, where the larger the truncated portion is, the farther the red shift of the main resonance peak [68]. However, the most appealing prospect of nanoplates might not be their spectral feature but rather the spatial distribution of their plasmon modes illustrated in Figure 27.10c and d. Indeed, when confined in a triangular shape, the surface electron gas resonates with the incident light for in-plane modes that will have maximal amplitude at the triangle tips [68,97]. This is clearly revealed by DDA calculations and readily observed in electron energy-loss spectroscopy amplitude maps that directly probe the spatial distribution of the plasmon modes. This aspect is further detailed in Section 27.5.1.

With the morphological control of colloidal metals, the extension of plasmonic design is also envisioned in the third dimension. Cubes, truncated cubes, and the resulting polyhedra were recently produced by adding, for example, silver nitrate [114] or phosphotungstic acid [137] for gold nanoparticles, poly(vinyl pyrrolidone) (PVP) [118,119] or polyethyleneglycol (PEG) [15,14] for silver colloids. Cubic colloids show two plasmon bands although not as well separated as nanorods or triangular prisms. Yet, similarly to the triangle to hexagon transition, the progressive truncation from cube to icosahedra and higher polyhedra is reflected in the extinction spectrum by the red shift of the low energy mode that eventually merges with the high-energy one into the symmetrical band characteristic of the spherical nanoparticles. While polyhedra may display spontaneous 2D and 3D self-assembly properties that could be exploited for building more complex plasmonic architectures, the spatial confinement of plasmon modes will not be as efficient as for the triangles owing to the absence of sharp angles. In this respect, the synthesis of multipods and branched nanoparticles could provide a family of nanoparticles with spectral tunability and high spatial confinement. Tri- and tetrapods were obtained by the slow reduction (L-ascorbic) of Au(I)–C_{16}TAB complex while multipods were produced by the reduction of Au(III) by tannic acid. In both cases, the presence of colloidal silver nanoplates was required [13,16]. While twinning probably plays an essential role in generating multiple branches growing from an initial nucleus, [32] no rationale has been proposed for the formation of these structures and a large number of experimental conditions for producing branched, star-shaped colloids has been reported in recent years [4,13,62,70,73,95,115,129,131,138]. The far-field optical absorption signature of these complex structures can be described as the composition of a symmetric band at 550–600 nm and a second resonance that can easily be displaced from a shoulder at 650 nm to a well-separated band at approximately 800 nm as seen on Figure 27.11. The sharpness of the protruding branches reflects directly on

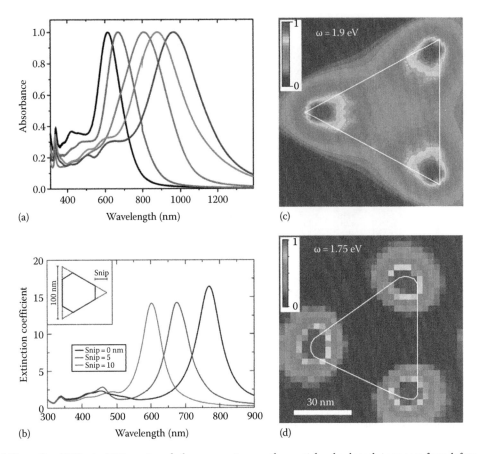

FIGURE 27.10 (a) Normalized UV–vis–NIR spectra of silver nanoprisms as the particle edge length increases from left to right. (Reproduced from Pastoriza-Santos, I. and Liz-Marzan, L.M., *J. Mater. Chem.*, 18, 1724, 2008. With permission.) (b) DDA simulations of the extinction efficiency of 16 nm-thick snipped triangular nanoprisms as the snip length increases from left to right (0, 10, and 20 nm). (Reproduced from Kelly, K.L. et al., *J. Phys. Chem. B*, 107, 668, 2003. With permission.) (c). Simulated electron energy loss spectroscopy (EELS) amplitude map of the 1.90 eV resonance of a nanoprism with 78 nm long sides. (Reproduced from Nelayah, J., et al., *Nat. Phys.*, 3, 348, 2007. With permission.) (d) Distribution of the mode centered at 1.75 eV, in the spectra of a triangular particle. The outer contour of the particle is shown as a white line. The amplitude of the mode is measured at its maximum after subtraction of the zero loss peak and subsequently fitted with a Gaussian distribution. The color scales in (c) and (d) are linear and in arbitrary units. (Reproduced from Nelayah, J. et al., *Nat. Phys.*, 3, 348, 2007. With permission.)

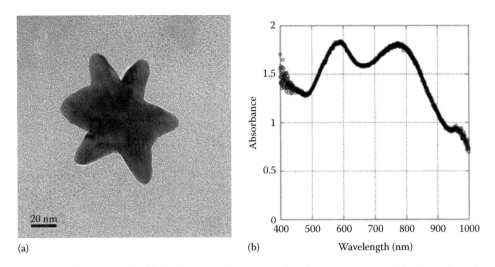

FIGURE 27.11 Star-shaped gold nanocrystals. (a) TEM images demonstrate that the nano-stars are defective and consist of multiple crystal domains. (b) The extinction spectrum of the nano-star solution exhibits broad visible and NIR peaks. (Reproduced from Nehl, C.L. et al., *Nano Lett.*, 6, 683, 2006. With permission.)

the local enhancement of the electromagnetic field making these complex and imperfect structures ideal candidates for Surface-Enhanced Raman Spectroscopy (SERS) studies and, consequently, for sensor applications.

Far from being exhaustive, this brief overview illustrates that nanoscale metallic objects are now readily available in the hands of material chemists with a high degree of control over size and shape. The far- and near-field optical properties are thus precisely tailored by a good morphological control that determines the spectral resonances and the spatial confinement. Nevertheless, the extremely small size of these structures impedes the study of their near-field optical properties and makes it difficult to interface them with light sources. The colloidal chemistry answer to this observation is to suggest the spontaneous organization of these nano-objects into well-defined architectures by exploiting the principles of self-assembly. Therefore it is important to examine how plasmons are affected when colloidal particles are brought into close vicinity as detailed in the next section (Section 27.3.5).

27.3.5 Plasmons in Interacting Colloids

In Section 27.3.3, we have seen that the interaction of two plasmonic nanoparticles results in the blue shift of the original resonance and the appearance of a second red-shifted band that emerges from the dipolar reinforcement of the longitudinal excitation along the pair axis. The dipolar coupling, which varies rapidly with the interparticle distance, has been explored on lithographically produced nanoparticle dimers [120] but the resolution limitations of lithography makes it challenging, if not impossible, to explore interparticle distances smaller than 10 nm. Conversely, the self-assembly of colloidal nanocrystals with a dielectric gap maintained in the 1–10 nm regime has been reported by many approaches. Here, the challenge is to restrict the assembly to small oligomers (di-, tri-, or tetramers) and not to propagate the self-assembly to the larger structures detailed in Section 27.4. If one replaces the small stabilizing molecules, such as citrate, by larger macromolecules, the kinetic interplay between particle growth and surface adsorption can be biased so that gold nanoparticles start to aggregate before full colloidal stabilization occurs. Two strategies can be proposed: either preformed polymer chains added to the growth medium interact with the nuclei that grow until the small cluster of nanoparticles is encapsulated or the initial stage of aggregation is quenched by the *in situ* templated polymerization. The first option has been observed with amine-rich polyelectrolyte [128], while the second one has been demonstrated with block-copolymers [130] and silica [78,127]. A more sophisticated approach consists in adding a ditopic molecular spacer to a nanoparticle suspension so that both ends of the linker are attached to one particle only. To ensure that extended aggregation does not occur, it is essential to allow only one linker per nanoparticle, which can be done by combining dilution and steric hindrance. Double stranded DNA oligonucleotides have been shown to produce di- or tritopic linkers that are able to link at most two (respectively three) gold nanoparticles [133].

In Figure 27.12a, nanoparticles where first functionalized with a single carboxylic anchoring group so that the subsequent addition of the diamine linker could at most join two particles. In agreement with the preceding theoretical description, the absorption spectra of the isolated and dimerized nanoparticles in Figure 27.12b clearly show the splitting of the initial plasmon band (540 nm) into two modes (540 and 690 nm). In contrast to the model in Section 27.3.3, the native 520 nm mode is not blue shifted primarily because gold is not a perfect metal described by the Drude model. While the polar functionalization of spherical nanoparticles is non-trivial (See Section 27.4), gold nanorods usually show a higher reactivity at their tips. Hence, dithiolated molecules reacting at the tips induce an end-to-end coupling and oligomerization into chains is prevented by using moderate concentrations of the linker. The dimer formation (Figure 27.12c) has a direct consequence on the plasmon coupling as seen in Figure 27.12d where the initial 2-band spectrum of the nanorods (transverse and longitudinal resonances) continuously evolves into the three-band spectrum of coupled nanorods, where the lowest energy band at 840 nm results from the coupling of the longitudinal plasmon modes [134]. Note that the 720 nm isosbestic point is characteristic of the equilibrium between the two monomeric and dimeric species. The intensity and large shift of the 840 nm peak are the direct consequence of the strong interparticle coupling made possible by the small but finite gap defined by the capping layer and the short linker molecules. Surface directed and capillary force mediated self-assembly methods have been shown to produce complex architectures and form submicrometer beads [135], yet its equivalent process with small metallic nanoparticles has only been explored recently but could yield an alternative approach [106].

These approaches will allow the study of interacting plasmons in a size domain yet unexplored but self-assembly can be pursued to form larger architectures that can bridge the gap with efficient nanofabricated plasmonic waveguides. The following section (Section 27.4) explores the prospects of using self-assembled superstructures made of crystalline plasmonic particles to confine light, enhance the local electromagnetic field, and potentially guide light toward molecular targets.

27.4 Light Confinement in Self-Assembled Plasmonic Architectures

The previous section (Section 27.3) clearly illustrates how colloidal synthesis can tailor the spectral characteristics of plasmonic nanoparticles and clusters of nanoparticles. However, the integration of these nanometric objects from liquid suspensions into devices poses a major addressability challenge. Yet, colloids can be organized into higher order architectures by following self-assembly principles that expands the embryonic examples of dimers and trimers to chains of nanoparticles and even further to chain networks, up to several micrometers. Therefore, by combining different self-assembling and self-organizing (bottom-up)

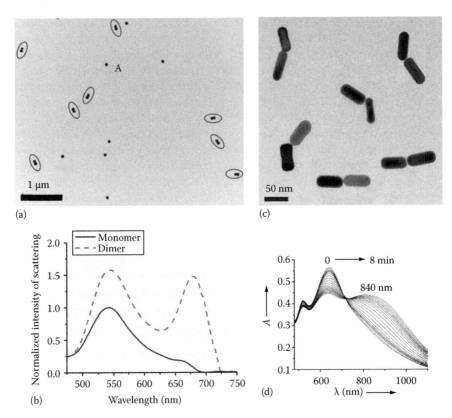

FIGURE 27.12 (a) TEM images of assembled Au nanoparticle dimers (circled, separation of the dimers: less than 2 nm). (b) Dark field scattering of a single Au nanoparticle and a single dimer placed on an indium tin oxide (ITO)-coated quartz plate. (Reproduced from Yim et al., *Nanotechnology*, 19, 435606, 2008. With permission.) (c) TEM images of Au nanorod dimers (8 min after the addition of PDT, 75 μL of the solution was drop-casted onto the TEM grid) and (d) the corresponding absorption spectral changes of Au nanorods (0.12 nM) with an average aspect ratio of 2.7 on addition of 0.8 μM 1,2-phenylenedimethanethiol (PDT). (Reproduced from Pramod, P. and Thomas, K.G., *Adv. Mater.*, 20, 4300, 2008. With permission.)

approaches with lithography, nano-inking and other top-down methods, it becomes conceivable to produce integrated multi-scale plasmonic architectures. The fabricated architectures can be optically characterized at the different scale levels by suitable far-field, near-field [24,46,56–58], and single molecule methods [38,40] or electron energy loss spectroscopy methods (EELS) [39,96]. From these new systems, specific key applications such as ultrasensitive sensors, ultra-small interconnects; enhanced spectroscopy and nanoscale microscopy could be tested.

The investigation of fundamental aspects and the optimiza-tion of the functionalities of bottom-up plasmonic superstruc-tures require an intensive support of both theoretical modeling and numerical simulations. For example, understanding how plasmons couple with each other and with nearby molecules [2,5,38] will be one of the major questions associated with future *bottom-up plasmonics* [3,6]. Among several challeng-ing theoretical objectives, we can mention (1) accurate evalu-ation of dissipation in the developed multiscale plasmonic structures; (2) study of the consequences of disorder on the efficiency of self-assembled plasmonic architectures; and (3) a realistic description of molecular coupling with the plasmonic surroundings. In this section, we will revisit some elementary mechanisms leading to light confinement in the vicinity of such systems.

For closely packed particle arrays, the coupling implies essentially the evanescent fields that tail-off each individual particle (evanescent coupling), while, for particle distances greater than a few tens of nanometers, the interaction relies on the interference of their scattered dipolar fields (dipole–dipole or far-field coupling). From these new systems, specific key applications such as ultrasensitive sensors, ultra-small intercon-nects, enhanced spectroscopy and nanoscale microscopy could be tested. In the following paragraphs, recent experimental methods to self-assemble nanoparticles or nanorods into linear chains are exposed before generalizing the previous theoretical approach on simulating the optical properties of dimers to lon-ger chains. Stepping up in complexity, we then examine the case of the extended nanoparticle networks that could be bridging superstructures between nanofabricated plasmonic waveguides and individual colloids.

27.4.1 Nanoparticle Chains

The aggregation of spherical nanoparticles usually yields the most symmetrical and compact structures in both planar and 3D environments. The production of nanoparticle chains implies a symmetry breaking where isometric nanoparticles are interacting in a vectorial way. A rather straightforward

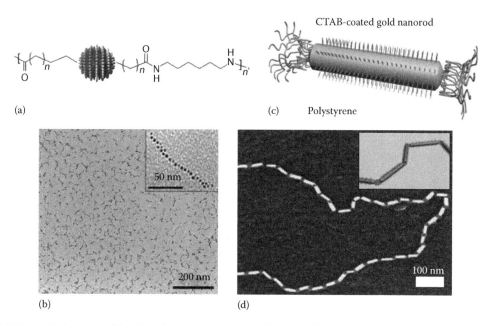

(a)

(c)   Polystyrene

CTAB-coated gold nanorod

(b)   (d)

FIGURE 27.13 (a) Schematic depiction of the chain formed of rippled particles linked through pole functionalization. (b) TEM images of chains that compose the precipitate obtained when 11-mercaptoundecanoic acid (MUA) pole-functionalized rippled NPs are reacted with 1,6-diaminohexane (DAH) in a two-phase reaction. (Reproduced from DeVries, G.A. et al., *Science*, 315, 358, 2007. With permission.) (c) Schematics of an amphiphilic gold nanorod carrying a double layer of CTAB along the longitudinal side (the 100 facets) and polystyrene molecules grafted to both ends. (d) SEM image of the self-assembled nanorod chains in the dimethyl formamide/water mixture at water contents of 20 wt%, respectively. The inset shows a sketch of the nanorod assembly. (Reproduced from Nie, Z.H. et al., *Nat. Mater.*, 6, 609, 2007. With permission.)

symmetry-breaking method consists in self-assembling nanoparticles on a linear template such as, for example, dendrimer-peptide [20] or polymer strands [113]. The limitation of this approach is the heterogeneity of the interparticle, spacing which weakens the surface plasmon coupling compared to the divalent nanoparticles reported before. A very elegant alternative approach consists in using the penta-twinned crystalline structure of most so-called "spherical" colloidal gold nanoparticles to specifically functionalize the nanoparticle poles with specific assembly-inducing moieties. Figure 27.13a and b illustrates this approach by the group of F. Stellacci [23,59].

Citrate-capped nanoparticles are functionalized with a mixed monolayer of 1-nonanethiol and 4-methylbenzenethiol that segregate into rings on the surface of the nanoparticle. To place-exchange at the polar defects, the particles are dissolved in a solution containing 40 molar equivalents (relative to the moles of particles) of 11-mercaptoundecanoic acid. The exposed carboxylic moieties can then be coupled using a ditopic 1,6-diaminohexane bridge, which results in the construction schematized in Figure 27.13a and illustrated by the TEM images of linear nanoparticle chains shown in Figure 27.13b. The typical dimensions are 10–20 nanoparticles per chains with a interparticle spacing in the order of 1–2 nm depending on the linker molecule [54].

This experimental example points to a need to further generalize the previous developments of Section 27.3.3 to the description of the optical properties of a finite chain formed with N metallic particles. This generalization can be easily achieved in the framework of the well-known Coupled Dipole Approximation (CDA) also called Discrete Dipole Approximation (DDA).

Historically, this formalism was introduced in astrophysics, more than 40 years ago, to analyze the optical properties of interstellar dust particles [103]. Many improvements of the original techniques were proposed and described in the recent literature [26,27,48,50,124]. Draine and Flatau [28] have made an interesting and complete review of this technique. In the beginning of the 1990s, this formalism has been applied to the theoretical investigation of the image formation in near-field optical microscopy [42,43,66,67,74]. The system under study is represented by the linear chain (depicted in Figure 27.14) composed of N nanoparticles of similar polarizabilities $\alpha(\omega_0)$. Each of these microscopic component acquires a fluctuating dipole moment induced by a local electric field $\mathbf{E}_i(\mathbf{r}_i, \omega_0)$:

$$\mathbf{m}(\mathbf{r}_i, \omega_0) = \alpha(\omega_0)\mathbf{E}(\mathbf{r}_i, \omega_0) \qquad (27.26)$$

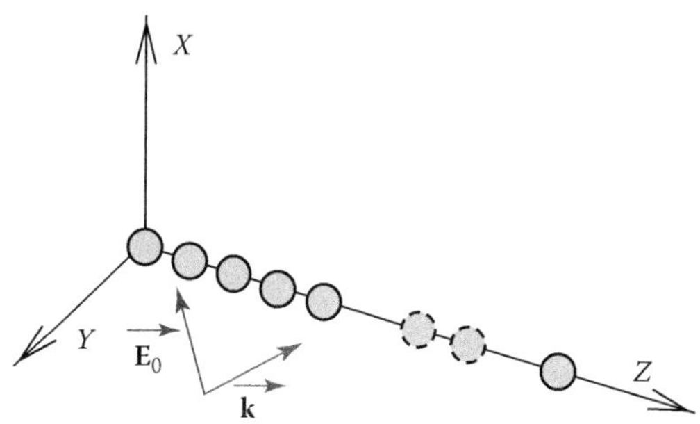

FIGURE 27.14 Schematic drawing of a metallic nanoparticle chain of finite length submitted to an external illumination field of arbitrary polarization. The propagation direction is schematized by the wave vector **k**.

that is self-consistently modified by the presence of the others. The many-body interactions between the particles can be introduced by writing N implicit linear equations [42,52,68,77]:

$$\mathbf{E}(\mathbf{r}_i,\omega_0)=\mathbf{E}_0(\mathbf{r}_i,\omega_0)+\sum_{j\neq i}\mathbf{S}_0(\mathbf{R}_{ij},\omega_0)\cdot\alpha(\omega_0)\cdot\mathbf{E}(\mathbf{r}_j,\omega_0), \quad (27.27)$$

in which $\mathbf{S}_0(\mathbf{R}_{ij},\omega_0)$ (where $\mathbf{R}_{ij}=\mathbf{r}_i-\mathbf{r}_j$) labels the retarded dipolar propagator in vacuum. In the Cartesian frame of Figure 27.14, its nine (α,β) components are given by the general relation

$$\mathbf{S}_{0,\alpha,\beta}(\mathbf{R}_{ij},\omega_0)=\exp(ikR_{ij})\times \quad (27.28)$$

$$\left\{-\frac{k^2}{R_{ij}^3}(R_{ij,\alpha}R_{ij,\beta}-R_{ij}^2\delta_{\alpha,\beta})+\left(\frac{1}{R_{ij}^5}-\frac{ik}{R_{ij}^4}\right)(3R_{ij,\alpha}R_{ij,\beta}-R_{ij}^2\delta_{\alpha,\beta})\right\}, \quad (27.29)$$

with $k=n\omega_0/c$ (where n represents the optical index of the surroundings). The N self-consistent equations (27.27) can be rewritten as 3×3 matrix equations:

$$\sum_{j=1}^{N}\mathbf{A}_{i,j}(\omega_0)\cdot\mathbf{E}(\mathbf{r}_j,\omega_0)=\mathbf{E}_0(\mathbf{r}_i,\omega_0), \quad (27.30)$$

with

$$\mathbf{A}_{i,j}(\omega_0)=\mathbf{I}\delta_{i,j}-\alpha_i(\omega_0)\cdot\mathbf{T}(\mathbf{R}_{i,j},\omega_0), \quad (27.31)$$

After inversion of the linear system (27.31), it is possible to compute both the local fields $\mathbf{E}(\mathbf{r}_i,\omega_0)$ and the electric polarization vector $\mathbf{P}(\mathbf{r}_i,\omega_0)=\alpha_i(\omega_0)\cdot\mathbf{E}(\mathbf{r}_i,\omega_0)$ inside the nanoparticle. Furthermore, for ensembles of N very small spherical nanoparticles, the CDA approach can be reduced by associating just one dipole per particle [43,77]. An example based on this simplification is provided in Figure 27.15 where the extinction coefficient:

$$C_{ext}(\lambda_0)=\frac{8n\pi^2}{\lambda_0|\mathbf{E}_0|^2}\sum_{j=1}^{N}\mathrm{Im}\left\{\mathbf{E}_0^*(\mathbf{r}_j,\lambda_0)\cdot\mathbf{P}(\mathbf{r}_j,\lambda_0)\right\} \quad (27.32)$$

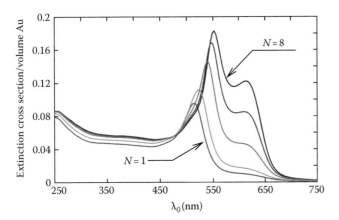

FIGURE 27.15 Examples of extinction spectra computed for gold particles chains from $N=1$ to $N=8$. Particle diameter $2a=13\,\text{nm}$; particle spacing 1 nm; optical index of the surroundings $n=1.33$.

is computed for gold nanoparticle chains of growing length (placed in an external medium of optical index $n=1.33$) versus the incident wavelength $\lambda_0=2\pi c/\omega_0$. The simulated extinction spectra of a dispersion of isolated 13 nm diameter Au nanoparticles (Figure 27.15); $N=1$ red curve) shows a single absorption peak at 520 nm. In other curves, simulations of one-dimensional superstructures based on the linear assembly of increasing numbers of similar particles produce spectra displaying a second plasmon peak revealing the plasmon longitudinal mode. In these simulations, where a perfect linear chain immersed in water ($n=1.33$) has been considered, both the intensity and the red shift of the longitudinal peak saturate rapidly when increasing the particle number. When this formulation is applied to more complex experimental situations (cf. Figure 27.18) where gold nanoparticles are coated by an effective medium of higher optical index, it is possible to reproduce the experimental features [79].

27.4.2 Nanorod Chains

More refined plasmonic design can arise from the combination of the morphological and self-assembly anisotropy described in Sections 27.3.4 and 27.4.1 by organizing metallic nanorods into chains as displayed in Figure 27.13 [35,98]. In this example, C_{16}TAB-stabilized single crystalline nanorods are functionalized at their apex, where the surfactant is more loosely adsorbed, by hydrophobic polystyrene-terminated thiols (Figure 27.13c). In THF/DMF/water ternary solvents (THF: tetrahydrofurane; DMF: dimethylformamide), the selective solvation of the C_{16}TAB-coated sides by DMF or water or the tips by THF or DMF results in the directional end-to-end assembly of nanorods for low-THF content mixtures Figure 27.13d. Similarly, K. G. Thomas et al. [122] showed that alkanethiols bearing a terminal carboxylic acid moiety (or, alternatively, cystein or α,ω-dithiols) could be grafted at the nanorod tips and chain formation could then be induced by varying the acetonitrile content in the aqueous solution [65,122]. The careful control of the self-assembly kinetics allows to monitor the change from the isolated nanorod absorption spectrum that displays two lateral and longitudinal plasmon modes, to the coupled nanorod spectrum that exhibits an extra band at lower energies, typically in the near-infrared region, resulting from the coupling of the longitudinal mode. The initial stage of this assembly process is the dimer formation detailed in Section 27.3.5 and Figure 27.12c and d but the coupled mode is moved further beyond 1000 nm upon chain growth as can be easily seen in the far-field absorption spectroscopy. In contrast, the near-field optical properties reported on single nanorods [25,56–58] have not yet been extended to nanorod chains, which would allow to determine the extent of the coupling between longitudinal modes. From electron microscopy analysis, it appears that the typical assembly contains a few aligned nanorods and the total chain length reaches at most one or two micrometer in length. This rather short length still hampers the easy integration of the chains into plasmonic devices but, so far, even more specific interactions such as streptavidin/biotin [10,49] or antigen/antibody [11] have failed to extend the chain length beyond a couple of micrometers, that is, about 10–30 nanorods. The next section introduces new extended colloidal networks that contain hundreds of nanoparticles while maintaining the 10 nm chain's width that confers to colloidal assemblies their asset compared to lithographically produced architectures.

27.4.3 Plasmonic Nanoparticle Networks (PNN)

We have recently reported the simple fabrication of complex and extended networks of interconnected chains of nanoparticles by a spontaneous self-assembly process [79] (cf. Figure 27.16). These networks, that we called Plasmonic Nanoparticles Networks

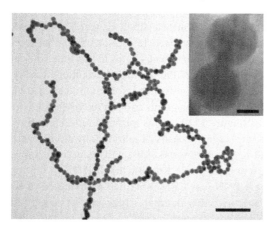

FIGURE 27.16 Example of TEM image showing the region of a self-assembled Au nanoparticle chain network (scale bar = 100 nm). Inset: High magnification TEM image showing organic material at the contact surface between two adjacent beads; scale bar = 5 nm. (From Girard, C. et al., *Phys. Rev. Lett.*, 97, 100801.1, 2006. With permission.)

(PNN), are multi-micrometer reticulated nets of one-nanoparticle-wide chains with a node-to-node distance of about 10–20 nanoparticles. High resolution TEM shows that almost all nanoparticles in the networks are penta-twinned single crystals separated by regular 1 nm gaps filled with organic matter. Although fragile, these structures are stable for weeks and can sustain elevated temperatures in solution. Moreover, their synthesis is straightforward as summarized in Figure 27.17. The standard 13 nm citrate capped gold nanoparticles described in Figure 27.8 are mixed with an aqueous solution of mercaptoethanol (HS–CH$_2$–CH$_2$–OH, MEA) at a typical molar ratio or one nanoparticle for 5000 MEA molecules. The solution is hand shaken and left still at room temperature for 24 h. After this delay, TEM images of the samples show the extended branched networks of nanoparticle chains (Figures 27.16 and 27.17c). Interestingly, upon addition of MEA, the light red colored solution turns purple. These optical properties of PNNs could be quantitatively monitored by recording the UV–visible absorbance spectrum as a function of time. This experiment is shown in Figure 27.18. The initial single plasmon band at 520 nm slowly decreases in intensity without energy shift at the same time as a shoulder emerges before shifting to low-energies and increasing in intensity. After 24 h, this new band stabilizes as a large peak centered on 700 nm. In first approximation, this new peak can be understood as the longitudinal mode emerging from the efficient coupling of nanoparticle surface plasmons by dipolar interactions between neighboring but non-touching nanoparticles, as described in the previous section. In a more sophisticated approach, the absorption spectra of realistic PNNs were precisely modeled by generalizing the Discrete Dipole Approximation (DDA) numerical scheme described in Section 27.4.1) [45,46]. Noticeably, the spectral evolution shows a very peculiar dependency on the Au: MEA molar ratio. Indeed, below 1500 MEA molecule per nanoparticle,

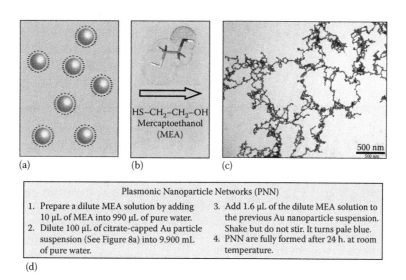

FIGURE 27.17 Self-assembly of plasmonic nanoparticle networks (PNN). (a) Citrate-stabilized gold nanoparticle (typical size 11 nm ± 0.2 nm) are mixed with (b) a 14.3 mM aqueous solution of mercaptoethanol (MEA) at a nanoparticle: MEA molar ratio exceeding 1: 5000. (c) Fully formed, single-particle branched chain networks are observed by TEM after 24 h at room temperature. (d) Detailed recipe to obtain PNN-form nanoparticles synthesized by the method in Figure 27.8a.

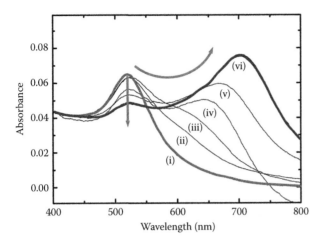

FIGURE 27.18 Time-dependent UV–vis spectra of an Au nanoparticle sol recorded at various times after addition of MEA at a nanoparticle: MEA molar ratio of 1: 5000. (i) 0 h, (ii) 3 h, (iii) 7 h, (iv) 24 h, (v) 48 h, and (vi) 72 h. Note the energy stability of the transverse plasmon mode at 520 nm. (Adapted from Lin, S. et al., *Adv. Mater.*, 17, 2553, 2005. With permission.)

no PNNs formation is observed. Beyond this onset, chains are formed, which rapidly expand into networks. The largest single particle chain networks are obtained for molar ratios comprised between 5,000 and 10,000 MEA molecule per nanoparticles [79]. When an extremely large excess of MEA is added, isotropic compact aggregates are formed and eventually precipitate. The current interpretation of this threshold in the self-assembly is directly linked to the symmetry breaking mechanism, which has to take place to form chains from isotropic nanoparticles and is illustrated in Figure 27.19. Basically, the citrate-stabilized nanoparticles are often composed of five single crystals (twins) assembled in a lentil-like particle the exposed faces of which are {111} planes linked with a small {100} rim exactly like the

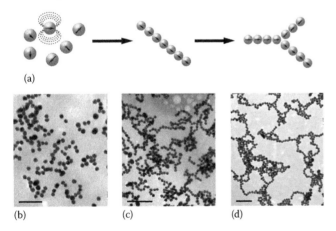

FIGURE 27.19 PNN formation mechanism within the dipolar fluid interpretation. (a) Schematics of the two main driving forces in the formation of branched chain networks from dipolar nanoparticle, namely dipole alignment in chains by enthalpic minimization and branching by entropic maximization. (b–d) TEM images of (b) isolated gold nanoparticles, (c) short linear chains, and (d) fully formed branched PNN. Scale bars are 100 nm. (From Lin, S. et al., *Adv. Mater.*, 17, 2553, 2005. With permission.)

seeds responsible for the formation of penta-twinned nanorods. It turns out that the MEA has a higher affinity for {111} faces where, besides the generic favorable interaction of the sulfur atom with the gold surface, the two-dimensional packing of the MEA molecules is perfectly suited for inter-molecular hydrogen bonding of the –OH moieties. In the case of MEA specifically, these interactions are dominant compared to the alkyl chains interactions that would prevail if the chains were longer than two carbon atoms or even more if the alkyl were replaced by an aryl group. The MEA hydrogen bonding acts as a locking mechanism that slowly stabilizes the self-assembled monolayer on {111}. At a molar ratio of 1:1500, there is just enough neutral MEA to cover slightly more than half the {111} faces so that the residual negatively charged citrate are confined into less than a half "hemi-sphere." This charge separation results into the formation of an induced surface dipole (Figure 27.19a). It is known, from magnetic dipolar particles, for example, that interacting dipoles tend to align and form chains to minimize the energy of the ensemble (Figure 27.19c). A further evolution of the system is the increase of its entropy which favors, at finite temperature, the formation of branching points as observed in the final branched networks (Figure 27.19d). This formation mechanism, which is intuitive for magnetic beads, is valid for any dipole-bearing colloid and has been observed for semiconducting nanoparticles bearing a permanent electrostatic dipole [117].

If these PNNs are easily obtained and studied in solution, they also constitute an interesting class of objects for self-assembled nanoplasmonic devices provided that they can be brought intact onto a substrate where the interfaced devices can be fabricated. The transfer of colloidal particle assemblies from a suspension onto a solid substrate has been a major challenge in recent years, if only for its importance in photonics, for example, in the fabrication of photonic bandgap devices by depositing a few layers of colloidal photonic crystal [125] or in using deposited colloids as masks for shadow evaporation ("nanosphere lithography") of plasmonic metal nanostructures [53]. However, the controlled deposition, spreading, and addressing of the colloidal superstructure, like PNNs remain challenging. Very recently, by systematically controlling the surface chemistry of the substrate, the deposition method and drying regime of the liquid colloidal solution, we have shown that it is possible to deposit intact and well-spread networks (see for example Ref. [8]). The experimental characterization of the near-field properties of the deposited PNN is ongoing and strongly motivated by the simulations performed on realistic PNN architectures deposited on a planar dielectric surface that can be investigated by applying the DDA method.

Let us consider the sample depicted in Figure 27.20 [139]. The system is a network of N identical sub-wavelength-sized nanoparticles of the same radius a and dipolar polarizabilities $\alpha(\omega_0)$. When the sample is illuminated by the electric field $\mathbf{E}_0(\mathbf{r}_i, \omega_0)$ associated with a surface evanescent wave generated by total internal reflection, each nanometric component acquires a fluctuating dipole moment induced by the local electric field $\mathbf{E}(\mathbf{r}_i, \omega_0)$:

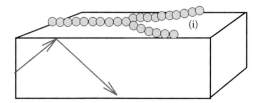

FIGURE 27.20 Schematic drawing of branched networks of single-nanoparticle chains deposited on a glass surface. (From Girard, C. et al., *New J. Phys.*, 10, 105016, 2008. With permission.)

$$\mathbf{P}(\mathbf{r}_i, \omega_0) = \alpha(\omega_0) \cdot \mathbf{E}(\mathbf{r}_i, \omega_0) \tag{27.33}$$

that is self-consistently modified by the presence of the others and of the surface. The many body interactions between the nanoparticles can be introduced by writing N implicit linear equations, similar to relation (27.27), by just replacing the propagator $\mathbf{S}_0(\mathbf{R}_{ij}, \omega_0)$, associated with the homogeneous medium, by

$$\mathbf{S}(\mathbf{R}_{ij}, \omega_0) = \mathbf{S}_0(\mathbf{R}_{ij}, \omega_0) + \mathbf{S}_{\text{surf}}(\mathbf{R}_{ij}, \omega_0) \tag{27.34}$$

where $\mathbf{S}_{\text{surf}}(\mathbf{R}_{ij}, \omega_0)$ accounts for the dipole image contributions generated by the surface [41]. Within the framework of the DDA description, the electric field outside the object is merely deduced from the self-consistent field inside the metallic particles, provided that we know the field propagator $\mathbf{S}(\mathbf{r}, \mathbf{r}', \omega)$ of the bare sample. One can write

$$\mathbf{E}(\mathbf{R}, \omega) = \mathbf{E}_0(\mathbf{R}, \omega) + \sum_j \mathbf{S}(\mathbf{R}, \mathbf{r}_j, \omega) \cdot \alpha(\omega) \cdot \mathbf{E}(\mathbf{r}_j, \omega). \tag{27.35}$$

This solution can be used to compute the normalized optical near-field intensity in the vicinity of the illuminated versus the position $\mathbf{R}$:

$$I(\mathbf{R}) = \frac{|\mathbf{E}(\mathbf{R}, \omega)|^2}{|\mathbf{E}_0(\mathbf{R}, \omega)|^2} \tag{27.36}$$

From this relation, we have simulated a sequence of three near-field optical maps of a PNN deposited on a transparent surface (cf. Figure 27.21). In the total internal illumination configuration of Figure 27.20, the incoming light is incident with an angle of 60° on the sample-to-air interface. The image sequence (Figure 27.21b, c, and d) is computed in the p-polarized mode for resonant wavelength $\lambda = 520\,\text{nm}$ (that can be identified in the extinction spectra). Three observation distances have been successively considered. At such short approach distances, we observe a marked confinement of light around all the branches of the network with brighter regions (orange in color figures) that reveals stronger lateral coupling between the plasmonic particles. Unlike what happens with larger lithographically designed chains of gold particles [72,109], where each gold pad generates individual bright spots in the vicinity of the particles, the optical near field spreads out over several nanoparticle sites without significant spatial modulation. The three near-field optical images of Figure 27.21 correspond to short range observation distances. Within this distance range, the field enhancement effect is still sufficient to reinforce the image contrast. These results clearly demonstrates that *chain networks self-assembled from crystalline noble metal nanoparticles* deposited on a transparent sample represent interesting objects for the sub-wavelength patterning of initially flat optical near fields. The optical patterning is optimized when working at incident wavelengths close to collective plasmon modes of the network.

27.5 Plasmonic Information Processing

Microelectronics overall supremacy over data storage and processing can, in part, be attributed to a successful pace in miniaturizing its elementary components while maintaining or even improving their performances for the past 60 years. However, power dissipation in current and future generations of transistors and interconnects has gradually appeared as a major concern for the electron-based information technology. While numerous alternatives have been proposed, none has yet achieved the

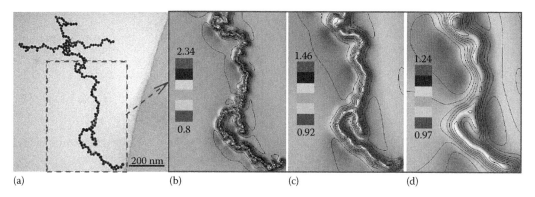

(a)   (b)   (c)   (d)

FIGURE 27.21 **(See color insert following page 9-8.)** (a) TEM image showing a self-assembled Au nanoparticle chain network deposited on a substrate. (b), (c), and (d) sequence of three optical near-field intensity maps computed in three consecutive planes parallel to the sample. The plane–sample distances are 20, 30, and 50 nm respectively. (From Girard, C. et al., *New J. Phys.*, 10, 105016, 2008. With permission.)

same level of maturity and consistency but photonics is now recognized as a performing partner to microelectronics. In this regards, conventional photonics lacks spatial resolution, but the conversion of photon energy to plasmonic excitation using noble metal surfaces opens up a new realm below the usual diffraction limit of light [45]. As discussed in the next paragraph, coupling elementary plasmon excitations increases the number of photonic states available in the close vicinity of the structures. Plasmonic information processing can be optimized by controlling this property because the knowledge of the Photonic Local Density of States (photonic LDOS) allows the design of new functionalities (plasmonic transmittance, routing, coupling with quantum system, etc.).

27.5.1 Photonic Local Density of States (Photonic LDOS)

The presence of material particles in the free space modifies the Local Density of States (LDOS) of electrons [121], photons [1], and other elementary excitations (plasmon–polaritons, excitons, etc.).

In optics, the role of the LDOS—also called *local mode density*—is well identified in the decay rate process of fluorescent molecules. This quantity also appears as a constant factor in Planck's law of blackbody radiation [76,85]. In a homogeneous medium, this factor consists in purely radiative eigenmodes, but while dealing with nanostructured samples non-radiative (evanescent) eigenmodes may also exist.

A recent series of papers describes LDOS optical measurements near plasmonics structures deposited on transparent planar surfaces [18,17,22,44,57,58,90]. Usually, the LDOS mapping is performed by scanning the excited dipolar tip of a Scanning Near-field Optical Microscope (SNOM) in the vicinity of the sample [22].

In order to define this scalar quantity, let us consider the physical system described in Figure 27.22. The electromagnetic modes sustained by the structure obey the eigenvalue equation:

$$-\nabla \wedge \nabla \wedge \mathcal{E}_n(\mathbf{r}, \omega_n) + k_n^2 \mathcal{E}_n(\mathbf{r}, \omega_n) = 0, \tag{27.37}$$

with $k_n = \omega_n/c$, and where $\mathcal{E}_n(\mathbf{r}, \omega_n)$ represents the amplitude of the normalized electric field associated with the nth electromagnetic mode:

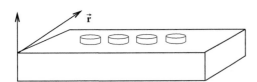

FIGURE 27.22 Schematic representation of four cylindrical metallic nanostructures supported by a plane solid surface. The vector **r** represents an observation point above the sample used for LDOS calculation.

$$\int |\mathcal{E}_n(\mathbf{r}, \omega_n)|^2 \, d\mathbf{r} = 1. \tag{27.38}$$

The number $N(\omega)$ of eigenvalues ω_n of magnitude lower than ω can be expressed as a sum of Heaviside distributions $\theta(\omega - \omega_n)$:

$$N(\omega) = \sum_n \theta(\omega - \omega_n). \tag{27.39}$$

From this definition, it is easy to deduce the density of modes $\rho(\omega)$ of the system:

$$\rho(\omega) = \frac{dN(\omega)}{d\omega} = \sum_n \delta(\omega - \omega_n). \tag{27.40}$$

Usually, it is useful to define a local density of modes $\rho(\mathbf{r}, \omega)$, the so-called LDOS, by multiplying each term of $\rho(\omega)$ by the squared modulus of the corresponding electric field mode:

$$\rho(\mathbf{r}, \omega) = \sum_n |\mathcal{E}_n(\mathbf{r}, \omega_n)|^2 \delta(\omega - \omega_n). \tag{27.41}$$

With this definition, the relation between $\rho(\omega)$ and $\rho(\mathbf{r}, \omega)$ simply reads

$$\rho(\omega) = \int \rho(\mathbf{r}, \omega) \, d\mathbf{r}. \tag{27.42}$$

In contrast to what happens with a perfect photon cavity that mainly sustains bound states, semi-infinite open systems like the one depicted in Figure 27.22, display continuous spectra of eigenmodes and eigenvalues. In this case, working with Green dyadic functions or field propagators, defined by Equation 27.34, greatly facilitates the effective computation of the optical LDOS of a complex system. The physical link between $\rho(\mathbf{r}, \omega)$ and the field propagator $S(\mathbf{r}, \mathbf{r}, \omega)$ associated to the entire system is established as follows. We introduce the following representation of the Dirac delta function ($\Im$ denotes the imaginary part):

$$\delta(x) = -\frac{1}{\pi} \lim_{\gamma \to 0} \Im \frac{1}{x + i\gamma}, \tag{27.43}$$

and the closure formula (**I** is the unit dyadic)

$$\sum_n \mathcal{E}_n(\mathbf{r}, \omega_n) \mathcal{E}_n^*(\mathbf{r}', \omega_n) = \mathbf{I}\delta(\mathbf{r} - \mathbf{r}'). \tag{27.44}$$

From the differential equation that defines the field-susceptibility of the whole system (with $k = \omega/c$)

$$-\nabla \wedge \nabla \wedge S(\mathbf{r}, \mathbf{r}', \omega) + k^2 S(\mathbf{r}, \mathbf{r}', \omega) = -4\pi k^2 \mathbf{I}\delta(\mathbf{r} - \mathbf{r}') \tag{27.45}$$

and the above closure relation (27.44), we can obtain the spectral representation of S as a function of the modal amplitudes $\mathcal{E}_n$:

$$S(\mathbf{r}, \mathbf{r}', \omega) = -4\pi k^2 \sum_n \left(\frac{\mathcal{E}_n(\mathbf{r}, \omega_n)\mathcal{E}_n^*(\mathbf{r}', \omega_n)}{k^2 - k_n^2} \right). \quad (27.46)$$

Comparing the relations (27.46) and (27.41), and using the identity $\delta(\omega - \omega_n) = 2\omega\delta(k^2 - k_n^2)/c^2$, we find

$$\rho(\mathbf{r}, \omega) = \frac{1}{2\pi^2\omega} \Im\{Tr S(\mathbf{r}, \mathbf{r}, \omega)\}. \quad (27.47)$$

When dealing with plasmonic structures that generally exhibit significant dissipative properties, Equation 27.47 does not represent *stricto sensu* the LDOS of the sample because non-radiative decay channels introduce additional contributions. Nevertheless, this relation remains valid to study the fluorescence decay rate of molecules located near the plasmonic components. As detailed in Ref. [3], the total decay rate Γ of a fluorescence level ω_0 of the transition dipole μ_0 is proportional to $\rho_\alpha(\mathbf{r}, \omega_0)$ at the location $\mathbf{r}$ of the molecule:

$$\Gamma(\mathbf{r}) = \frac{4\pi^2\omega_0\mu_0^2}{\hbar} \rho_\alpha(\mathbf{r}, \omega_0). \quad (27.48)$$

where $\rho_\alpha(\mathbf{r}, \omega_0)$ defines the partial LDOS along the α direction:

$$\rho_\alpha(\mathbf{r}, \omega) = \frac{1}{2\pi^2\omega} \Im\{S_{\alpha,\alpha}(\mathbf{r}, \mathbf{r}, \omega)\}. \quad (27.49)$$

In order to apply relation (27.47) to realistic plasmonic systems, the main difficulty lies in the computation of the $S(\mathbf{r}, \mathbf{r}', \omega)$ dyadic above the sample. When dealing with colloidal plasmonic nanostructures deposited on a surface, the current developments of computation methods in real space provides powerful tools to derive the electromagnetic response $S(\mathbf{r}, \mathbf{r}, \omega)$ (the procedure is detailed in Ref. [45]). From the numerical data describing

the spatial behavior of the imaginary part of the Green dyadic function $S(\mathbf{r}, \mathbf{r}, \omega)$, we can model maps or LDOS spectra and partial LDOS in the vicinity of self-assembled *bottom-up* plasmonic structures deposited on a plane surface. In Figure 27.23, the LDOS spectrum of a single gold nanoprism is computed at 80 nm from its center. In the visible frequency range, the spectrum appears with a series of plasmon resonances beginning with the standard transverse mode occurring at $\lambda_0 = 550$ nm. The other peaks reveal both lateral and multipolar modes associated with the structures. The sub-wavelength sizes of the gold particle is responsible for the high density of evanescent states localized around the nanoprism that is as well illustrated by the two LDOS images encapsulated in Figure 27.23. This concentration of modes explains the high contrast observed in these two maps where the three corners of the triangle display significant enhancements. More precisely, according to the spectrum that predicts a larger LDOS intensity around $\lambda_0 = 550$ nm than around $\lambda_0 = 850$ nm, we observe a better confinement around the structure for the wavelength $\lambda_0 = 550$ nm. Note that, the two LDOS maps of Figure 27.23 display features similar to those observed in Ref. [96] and reproduced in Figure 27.10 because high resolution electron energy loss spectroscopy, applied near plasmonic particles, delivered signals proportional to the photonic LDOS [39].

To generalize the concept of photonic LDOS previously introduced to PNN configurations, we merely have to rewrite Equation 27.47

$$\rho(\mathbf{r}, \omega) = \frac{1}{2\pi^2\omega} \Im\left[Tr S(\mathbf{r}, \mathbf{r}, \omega)\right], \quad (27.50)$$

together with a new form of field-susceptibility $S(\mathbf{r}, \mathbf{r}', \omega)$ expressed in terms of the individual polarizabilities $\alpha_j(\omega)$ ($j = 1, N$) associated with each nanoparticle that composes the network:

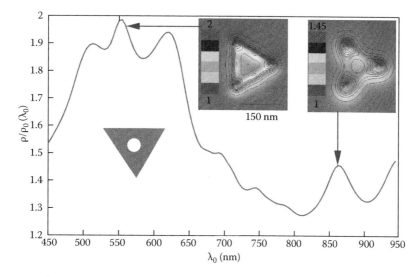

FIGURE 27.23 LDOS spectrum of a single gold nanoprism deposited on a glass planar surface. The spectrum is computed just above the triangle center (large white dot). The inset figures represent the corresponding LDOS maps computed in a plane located at 50 nm above the particle at the wavelengths $\lambda_0 = 560$ and 850 nm, respectively. The side of the nanoprism is 150 nm and its thickness 37.5 nm.

$$\mathcal{S}(\mathbf{r},\mathbf{r}',\omega) = \mathbf{S}(\mathbf{r},\mathbf{r}',\omega) + \sum_{j=1}^{N}\alpha_j(\omega)\mathbf{S}(\mathbf{r},\mathbf{r}_j,\omega)\cdot\mathcal{S}(\mathbf{r}_j,\mathbf{r}',\omega). \quad (27.51)$$

For large N numbers of plasmonic particles, this self-consistent equation must be solved numerically (for example by applying the Dyson sequence method described in Ref. [89] or by applying a simple matrix inversion method. However, more tractable expressions can be obtained when considering single or isolated particles adsorbed on planar surfaces [19]. From relations (27.50) and (27.51), four LDOS maps computed above a gold-PNN deposited on a glass planar surface are presented in Figure 27.24. In addition, in order to facilitate the graphic representation of this function $\rho(\mathbf{r},\omega)$, it has been normalized with respect to the spatially independent vacuum LDOS:

$$\rho_0(\omega) = \frac{\omega^2}{\pi^2 c^3}. \quad (27.52)$$

Consequently, the color scales in Figure (27.24a, b, c, and d) indicate the variation range of the dimensionless normalized function $\rho(\mathbf{r},\omega)/\rho_0(\omega)$. The sequence of images begins in the mesoscopic regime with two planar maps computed at 150 and

100 nm from the sample, respectively (cf. Figure 27.24a and b). In such a distance range, the LDOS variation is weak and the PNN is surrounded by pseudo–periodic ripples (ring like modes) that dominate the image pattern. To visualize the gradual transition between mesoscopic and nanoscopic regimes, the two next maps of Figure 27.24 present the same data computed when decreasing the observation distance at only 50 and 25 nm from the metal chains. In these two examples, we can observe a drastic decrease of the ripples together with a significant enhancement of the normalized LDOS in the immediate vicinity of the metal particles (see figure color bars). Indeed, entering the sub-wavelength range dramatically increases the access to the high density of the evanescent optical modes that tend to be localized around the individual spherical particles of the PNN. For example, at 25 nm, an enhancement factor of 10 can be expected above the structure.

27.5.2 Light Propagation in Nanoparticle and PNN

For about 10 years, numerous experimental studies based on plasmonic excitations exploit ensembles of metallic nanoparticles arranged as regular arrays microfabricated at the surface of

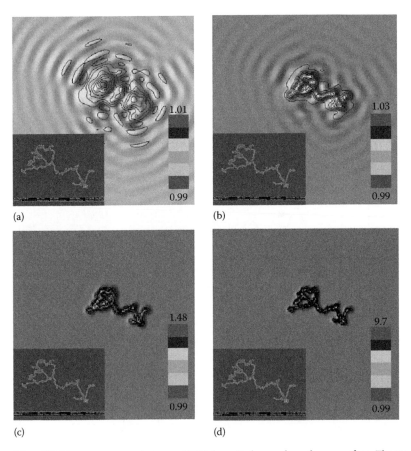

(a)   (b)

(c)   (d)

FIGURE 27.24 Sequence of four LDOS maps computed above a PNN deposited on a glass planar surface. The inset presents the SEM image of the structure. The maps have been computed in four consecutive planes (Z = constant) parallel to the sample: (a) Z = 150 nm, (c) Z = 100 nm, (c) Z = 50 nm, and finally (d) Z = 25 nm. The wavelength is 530 nm and the data are normalized with respect to the free space local density of state.

dielectric or semiconductor samples [37,72,75,80,108,111,116,132]. In fact, like with colloidal particles, each individual nanostructure rounds up the electron gas in the three dimensions and gives rise to localized surfaces. As already mentioned in Sections 27.3 and 27.4, the plasmon frequencies depend mainly on the particle geometry, the polarization direction of the exciting field with respect to the particle axis, and the dielectric functions of both the particle and the environment [7,26,30,77,122]. Similar to what happens with molecules [29], the radiation efficiency of metallic nanoparticles can be controlled by the local density of electromagnetic states available in the vicinity of the surroundings. In addition, their plasmon resonances can be significantly shifted when the particles are located very close to a metallic surface sustaining *surface plasmon polariton modes* (see previous Sections).

In spite of their sub-wavelength dimensions, nanodots of noble metals deposited on an optically transparent surface can be used to guide visible light over several micrometers [88]. More precisely, under permanent and localized illumination, a transmission band where the optical near field becomes commensurable with the metal arrangement can be open and used to guide light over the micrometric scale. In spite of significant losses (scattering and electron dissipations), such structures could be good candidates for the near-field optical addressing of single molecules in coplanar geometry [47].

Exploiting a similar phenomenon with colloidal nanowires, Dickson and Lyon were able to focus a laser spot by through-objective total internal reflection illumination on one end of 20 nm wide, 1–15 μm long nanowires made of pure silver or having one silver- and one gold sections [24]. By choosing the irradiation wavelength, the authors showed that light was emitted from the non-illuminated end, which can be interpreted either as surface plasmon propagation or as excitation of a surface plasmon mode that is strongly localized at the pure silver nanorod ends. However, when a silver–gold segmented nanowire is illuminated in a similar way, light is emitted from the non-illuminated tips only when the laser spot is focused on the gold end, a possible evidence of surface plasmon propagation though a transmissive Au → Ag boundary with far reaching propagation in the silver part. Since these first achievements, other nice plasmonic near-field guiding experiments have been reported in recent literature [25,104,112].

Finally, concerning self-assembled plasmonic particles, no guiding evidence has been reported with the PNN's deposited on dielectric surfaces up to now. Nevertheless, current experimental works have indicated that these new plasmonic objects could be excellent candidates to convey optical energy with an unprecedented lateral resolution [8].

27.6 Summary and Future Perspective

In this chapter, we have discussed the potential applications of self-assembled colloidal plasmonic architectures for realizing an unprecedented level of field confinement and enhanced light–matter interaction at the nanoscale. These objects open a new physics field for a mature chemistry domain. Following up along this concept, the simple fabrication of complex and extended networks of interconnected lattices of metallic nanoparticles could generate unique and unusual sub-wavelength patterning of the optical near field. Several impacts are associated with light confinement around plasmonic structures. The most direct consequence is the improvement of the quality of the near-field optical imaging and the increased local field enhancement, and finally the light energy storage in tiny volumes of matter. The optical physics related to the control of light confinement might have also important impact on the future solution for the miniaturization of both chemical and biological plasmonic sensors.

Finally, in terms of accumulated knowledge related to the mechanisms of the light confinement, current concepts mainly rely on a description of the local dielectric constant of the material (spectra, field maps, and photonic LDOS). Additional physical effects, such as non-local response, localized surface states, and effects due to the limited electron mean free path can play a significant role for structures with dimensions below 10 nm. Until now, most experiments on structures are not spatially resolved enough to quantify those effects. This is of fundamental importance for the understanding of the limits that can be achieved in terms of field confinement. *Bottom-up* structures will significantly contribute in this direction with realistic applications in near-field optical research.

Acknowledgments

The authors cordially thank A. Arbouet, G. Baffou, F. Bonell, G. Colas des Francs, R. Marty, R. Quidant, and A. Sanchot for useful scientific exchanges. They also would like to thank Prof. S. Mann, Dr. M. Li, and H. Guo, for stimulating discussions and a continued fruitful collaboration. In addition, we have benefited from the computing facilities provided by the massively parallel center CALMIP of Toulouse. E. D. acknowledge the financial support from the European Research Council (Grant contract ERC-2007-StG-203872).

References

1. G. S. Agarwal. Quantum electrodynamics in the presence of dielectrics and conductors. iii. Relations among one photon transition probabilities in stationary and nonstationary fields, density of states, the field-correlation functions, and surface-dependent response functions. *Phys. Rev. A*, 11:253–264, 1975.

2. J. Azoulay, A. Débarre, A. Richard, and P. Tchénio. Quenching and enhancement of single-molecule fluorescence under metallic and dielectric tips. *Europhys. Lett.*, 51:374–380, 2000.

3. G. Baffou, C. Girard, E. Dujardin, G. Colas des Francs, and O. J. F. Martin. Molecular quenching and relaxation in a plasmonic tunable system. *Phys. Rev. B*, 77:121101(R), 2008.

4. O. M. Bakr, B. H. Wunsch, and F. Stellacci. High-yield synthesis of multi-branched urchin-like gold nanoparticles. *Chem. Mater.*, 18:3297–3301, 2006.

5. W. L. Barnes. Fluorescence near interfaces: The role of photonic mode density. *J. Mod. Opt.*, 45:661–699, 1998.

6. C. R. Bennett, J. B. Kirk, and M. Babiker. Green–function theory of spontaneous emission and cooperative effects in a superlattice of thin metal layers. *Phys. Rev. A*, 63:63812-1–63812-9, 2001.

7. C. F. Bohren and D. R. Huffman. *Absorption and Scattering of Light by Small Particles*. Wiley, New York, 1983.

8. F. Bonell, E. Dujardin, C. Girard, M. Li, S. Mann, and R. Péchou. Processing and near-field optical properties of self-assembled plasmonic nanoparticle networks. *J. Chem. Phys.*, 130:034702-6, 2009.

9. A. Brioude and M. P. Pileni. Silver nanodisks: Optical properties study using the discrete dipole approximation method. *J. Phys. Chem. B*, 109:23371–23377, 2005.

10. K. K. Caswell, J. N. Wilson, U. H. F. Bunz, and C. J. Murphy. Preferential end-to-end assembly of gold nanorods by biotin-streptavidin connectors. *J. Am. Chem. Soc.*, 125:13914–13915, 2003.

11. J. Y. Chang, H. M. Wu, H. Chen, Y. C. Ling, and W. H. Tan. Oriented assembly of Au nanorods using biorecognition system. *Chem. Commun.*, 8:1092–1094, 2005.

12. S. S. Chang, C. W. Shih, C. D. Chen, W. C. Lai, and C. R. C. Wang. The shape transition of gold nanorods. *Langmuir*, 15:701–709, 1999.

13. H. M. Chen, C. F. Hsin, R. S. Liu, J. F. Lee, and L. Y. Jang. Synthesis and characterization of multi-pod-shaped gold/silver nanostructures. *J. Phys. Chem. C*, 111:5909–5914, 2007.

14. J. Y. Chen, J. M. McLellan, A. Siekkinen, Y. J. Xiong, Z. Y. Li, and Y. N. Xia. Facile synthesis of gold-silver nanocages with controllable pores on the surface. *J. Am. Chem. Soc.*, 128:14776–14777, 2006.

15. J. Y. Chen, B. Wiley, Z. Y. Li, D. Campbell, F. Saeki, H. Cang, L. Au, J. Lee, X. D. Li, and Y. N. Xia. Gold nanocages: Engineering their structure for biomedical applications. *Adv. Mater.*, 17:2255–2261, 2005.

16. S. H. Chen, Z. L. Wang, J. Ballato, S. H. Foulger, and D. L. Carroll. Monopod, bipod, tripod, and tetrapod gold nanocrystals. *J. Am. Chem. Soc.*, 125:16186, 2003.

17. C. Chicanne, T. David, R. Quidant, J.-C. Weeber, Y. Lacroute, E. Bourillot, A. Dereux, G. Colas des Francs, and Ch. Girard. Imaging the local density of states of optical corrals. *Phys. Rev. Lett.*, 88:097402-4, 2002.

18. G. Colas des Francs, C. Girard, and A. Dereux. Theory of near-field optical imaging with a single molecule as light source. *J. Chem. Phys.*, 117:4659–4665, 2002.

19. G. Colas Des Francs, C. Girard, M. Juan, and A. Dereux. Energy transfer in near-field optics. *J. Chem. Phys.*, 123:174709-7, 2005.

20. J. Cornelissen, R. van Heerbeek, P. C. J. Kamer, J. N. H. Reek, N. Sommerdijk, and R. J. M. Nolte. Silver nanoarrays templated by block copolymers of carbosilane dendrimers and polyisocyanopeptides. *Adv. Mater.*, 14:489, 2002.

21. M. C. Daniel and D. Astruc. Gold nanoparticles: Assembly, supramolecular chemistry, quantum-size-related properties, and applications toward biology, catalysis, and nanotechnology. *Chem. Rev.*, 104:293, 2004.

22. A. Dereux, Ch. Girard, C. Chicanne, G. Colas des Francs, T. David, E. Bourillot, Y. Lacroute, and J.-C. Weeber. Sub-wavelength mapping of surface photonic states. *Nanotechnology*, 14:935–938, 2003.

23. G. A. DeVries, M. Brunnbauer, Y. Hu, A. M. Jackson, B. Long, B. T. Neltner, O. Uzun, B. H. Wunsch, and F. Stellacci. Divalent metal nanoparticles. *Science*, 315:358–361, 2007.

24. R. M. Dickson and L. A. Lyon. Unidirectional plasmon propagation in metallic nanowires. *J. Phys. Chem. B*, 104:6095–6098, 2000.

25. H. Ditlbacher, A. Hohenau, D. Wagner, U. Kreibig, M. Rogers, F. Hofer, F. R. Aussenegg, and J. R. Krenn. Silver nanowires as surface plasmon resonators. *Phys. Rev. Lett.*, 95:257403-4, 2005.

26. W. T. Doyle. Optical properties of a suspension of metal spheres. *Phys. Rev. B*, 39:9852–9858, 1989.

27. B. T. Draine. The discrete-dipole approximation and its applications to interstellar graphite grains. *Astrophys. J.*, 333:848–872, 1988.

28. B. T. Draine and P. J. Flatau. Discrete–dipole approximation for scattering calculations. *J. Opt. Soc. Am. A*, 11:1491–1499, 1994.

29. K. Drexhage. Interaction of light with nonmolecular dye layers. *Prog. Opt. XII*, 163–232, 1974.

30. E. Dujardin, L. B. Hsin, C. R. C. Wang, and S. Mann. DNA-driven self-assembly of gold nanorods. *Chem. Commun.*, 1264–1265, 2001.

31. M. A. El-Sayed. Some interesting properties of metals confined in time and nanometer space of different shapes. *Acc. Chem. Res.*, 34:257, 2001.

32. J. L. Elechiguerra, J. Reyes-Gasga, and M. J. Yacaman. The role of twinning in shape evolution of anisotropic noble metal nanostructures. *J. Mater. Chem.*, 16:3906–3919, 2006.

33. S. Eustis and M. A. El-Sayed. Why gold nanoparticles are more precious than pretty gold: Noble metal surface plasmon resonance and its enhancement of the radiative and nonradiative properties of nanocrystals of different shapes. *Chem. Soc. Rev.*, 35:209, 2006.

34. M. Faraday. Experimental relations of gold (and other metals) to light. *Philos. Trans. R. Soc. Lond.*, 147:145, 1857.

35. D. Fava, Z. Nie, M. A. Winnik, and E. Kumacheva. Evolution of self-assembled structures of polymer-terminated gold nanorods in selective solvents. *Adv. Mater.*, 20:4318–4322, 2008.

36. D. L. Feldheim and C. A. Foss Jr. *Metal Nanoparticles: Synthesis, Characterization, and Applications*. Marcel Dekker, New York, 2000.

37. N. Félidj, J. Aubard, G. Lévi, J. R. Krenn, M. Salerno, G. Schider, B. Lamprecht, A. Leitner, and F. R. Aussenegg. Controlling the optical response of regular arrays of gold particles for surface–enhanced Raman scattering. *Phys. Rev. B*, 65:075419-9, 2002.

38. H. G. Frey, S. Witt, K. Felderer, and R. Guckenberger. High–resolution imaging of single fluorescent molecules with the optical near-field of a metal tip. *Phys. Rev. Lett.*, 93:200801-4, 2004.

39. F. J. Garcia de Abajo and M. Kociak. Probing the photonic local density of states with electron energy loss spectroscopy. *Phys. Rev. Lett.*, 100:106804-4, 2008.

40. J. M. Gerton, L. A. Wade, G. A. Lessard, Z. Ma, and S. R. Quake. Tip-enhancement fluorescence microscopy at 10 nanometer resolution. *Phys. Rev. Lett.*, 93:180801-4, 2004.

41. C. Girard. Near field in nanostructures. *Rep. Prog. Phys.*, 68:1883–1933, 2005.

42. C. Girard and X. Bouju. Coupled electromagnetic modes between a corrugated surface and a thin probe tip. *J. Chem. Phys.*, 95:2056–2064, 1991.

43. C. Girard and D. Courjon. Model for scanning tunneling optical microscopy: A microscopic self-consistent approach. *Phys. Rev. B*, 42:9340–9349, 1990.

44. C. Girard, T. David, C. Chicanne, A. Mary, G. Colas des Francs, E. Bourillot, J. C. Weeber, and A. Dereux. Imaging surface photonic states with a circularly polarized tip. *Europhys. Lett.*, 68:797–803, 2004.

45. C. Girard and E. Dujardin. Near-field optical properties of *top-down* and *bottom-up* nanostructures. *J. Opt. A: Pure Appl. Opt.*, 8:S73–S86, 2006.

46. C. Girard, E. Dujardin, M. Li, and S. Mann. Theoretical near-field optical properties of branched plasmonic nanoparticle networks. *Phys. Rev. Lett.*, 97:100801.1–100801.4, 2006.

47. C. Girard and R. Quidant. Near-field optical transmittance of metal particle chain waveguides. *Opt. Express*, 12:6141–6146, 2004.

48. G. H. Goedecke and O'Brien. Scattering by irregular inhomogeneous particles via the digitalized Green's function algorithm. *Appl. Opt.*, 27:2431–2438, 1988.

49. A. Gole and C. J. Murphy. Biotin-streptavidin-induced aggregation of gold nanorods: Tuning rod-rod orientation. *Langmuir*, 21:10756–10762, 2005.

50. J. J. Goodman, B. T. Draine, and P. J. Flatau. Application of the fast Fourier–transformation techniques to the discrete–dipole approximation. *Opt. Lett.*, 16:1198–1200, 1991.

51. M. Grzelczak, J. Perez-Juste, P. Mulvaney, and L. M. Liz-Marzan. Shape control in gold nanoparticle synthesis. *Chem. Soc. Rev.*, 37:1783, 2008.

52. E. Hao, S. Li, R. C. Balley, S. Zou, G. C. Schatz, and J. T. Hupp. Optical properties of metal nanoshells. *J. Phys. Chem. B*, 108:1224–1229, 2004.

53. C. L. Haynes and R. P. Van Duyne. Nanosphere lithography: A versatile nanofabrication tool for studies of size-dependent nanoparticle optics. *J. Phys. Chem. B*, 105:5599–5611, 2001.

54. Y. Hu, O. Uzun, C. Dubois, and F. Stellacci. Effect of ligand shell structure on the interaction between monolayer-protected gold nanoparticles. *J. Phys. Chem. C*, 112:6279, 2008.

55. Z. Y. Huo, C. K. Tsung, W. Y. Huang, X. F. Zhang, and P. D. Yang. Sub-two nanometer single crystal Au nanowires. *Nano Lett.*, 8:2041–2044, 2008.

56. K. Imura, T. Nagahara, and H. Okamoto. Characteristic near-field spectra of single gold nanoparticles. *Chem. Phys. Lett.*, 400:500–505, 2004.

57. K. Imura, T. Nagahara, and H. Okamoto. Imaging of surface plasmon and ultrafast dynamics in gold nanorods by near-field microscopy. *J. Phys. Chem. B*, 108:16344–16347, 2004.

58. K. Imura, T. Nagahara, and H. Okamoto. Plasmon mode imaging of single gold nanorods. *J. Am. Chem. Soc.*, 126:12730–12731, 2004.

59. A. M. Jackson, J. W. Myerson, and F. Stellacci. Spontaneous assembly of subnanometre–ordered domains in the ligand shell of monolayer-protected nanoparticles. *Nat. Mater.*, 3:330–336, 2004.

60. N. R. Jana, L. Gearheart, and C. J. Murphy. Wet chemical synthesis of high aspect ratio cylindrical gold nanorods. *J. Phys. Chem. B*, 105:4065–4067, 2001.

61. N. R. Jana, L. Gearheart, and C. J. Murphy. Wet chemical synthesis of silver nanorods and nanowires of controllable aspect ratio. *Chem. Commun.*, 617–618, 2001.

62. G. H. Jeong, Y. W. Lee, M. Kim, and S. W. Han. High-yield synthesis of multi-branched gold nanoparticles and their surface-enhanced Raman scattering properties. *J. Colloid Interface Sci.*, 329:97–102, 2009.

63. C. J. Johnson, E. Dujardin, S. A. Davis, C. J. Murphy, and S. Mann. Growth and form of gold nanorods prepared by seed-mediated, surfactant-directed synthesis. *J. Mater. Chem.*, 12:1765–1770, 2002.

64. P. B. Johnson and R. W. Christy. Optical constants of noble metals. *Phys. Rev. B*, 12:4370–4379, 1972.

65. S. T. S. Joseph, B. I. Ipe, P. Pramod, and K. G. Thomas. Gold nanorods to nanochains: Mechanistic investigations on their longitudinal assembly using alpha,omega-alkanedithiols and interplasmon coupling. *J. Phys. Chem. B*, 110:150–157, 2006.

66. O. Keller, S. Bozhevolnyi, and M. Xiao. On the resolution limit of near-field optical microscopy. In D. W. Pohl and D. Courjon, editors, *Near-Field Optics*. Volume E 242 of *NATO ASI*, pp. 229–237, NATO, Kluwer, Dordrecht, the Netherlands, 1993.

67. O. Keller, M. Xiao, and S. Bozhevolnyi. Configurational resonances in optical near-field microscopy: A rigorous point-dipole approach. *Surf. Sci.*, 280:217–230, 1992.

68. K. L. Kelly, E. Coronado, L. L. Zhao, and G. C. Schatz. The optical properties of metal nanoparticles: The influence of size, shape, and dielectric environment. *J. Phys. Chem. B*, 107:668–677, 2003.

69. B. P. Khanal and E. R. Zubarev. Purification of high aspect ratio gold nanorods: Complete removal of platelets. *J. Am. Chem. Soc.*, 130:12634–12635, 2008.

70. C. G. Khoury and T. Vo–Dinh. Gold nanostars for surface–enhanced Raman scattering: Synthesis, characterization and optimization. *J. Phys. Chem. C*, 112:18849–18859, 2008.

71. U. Kreibig and M. Vollmer. *Optical Properties of Metal Clusters*. Springer, Berlin, Germany, 1995.

72. J. R. Krenn, A. Dereux, J. C. Weeber, E. Bourillot, Y. Lacroute, J. P. Goudonnet, G. Schider et al. Squeezing the optical near-field by plasmon coupling of metallic nanoparticles. *Phys. Rev. Lett.*, 82:2590–2593, 1999.

73. P. S. Kumar, I. Pastoriza-Santos, B. Rodriguez-Gonzalez, F. J. Garcia de Abajo, and L. M. Liz–Marzan. High-yield synthesis and optical response of gold nanostars. *Nanotechnology*, 19:015606, 2008.

74. B. Labani, C. Girard, D. Courjon, and D. Van Labeke. Optical interactions between a dielectric tip and a nanometric lattice: Implication to near field scanning microscopy. *J. Opt. Soc. Am. B*, 7:936–943, 1990.

75. B. Lambrecht, G. Schider, R. T. Lechner, H. Ditlbacher, J. R. Krenn, A. Leitner, and F. R. Aussenegg. Metal nanoparticle gratings: Influence of dipolar particle interaction on the plasmon resonance. *Phys. Rev. Lett.*, 84:4721–4724, 2000.

76. L. D. Landau and E. M. Lifshitz. *Statistical Physics*, 3rd edn. Pergamon Press, London, U.K., 1960.

77. A. A. Lazarides and G. C. Schatz. DNA-linked metal nanosphere materials: Structural basis for the optical properties. *J. Phys. Chem. B*, 104:460–467, 2000.

78. W. Y. Li, P. H. C. Camargo, X. M. Lu, and Y. N. Xia. Dimers of silver nanospheres: Facile synthesis and their use as hot spots for surface-enhanced Raman scattering. *Nano Lett.*, 9:485–490, 2009.

79. S. Lin, M. Li, E. Dujardin, C. Girard, and S. Mann. One-dimensional plasmon coupling by facile self-assembly of gold nanoparticles into branched chain networks. *Adv. Mater.*, 17:2553–2559, 2005.

80. S. Linden, J. Kuhl, and H. Giessen. Controlling the interaction between light and gold nanoparticles: Selective suppression of extinction. *Phys. Rev. Lett.*, 86:4688–4691, 2001.

81. S. Link and M. A. El-Sayed. Size and temperature dependence of the plasmon absorption of colloidal gold nanoparticles. *J. Phys. Chem. B*, 103:4212–4217, 1999.

82. S. Link and M. A. El-Sayed. Spectral properties and relaxation dynamics of surface plasmon electronic oscillations in gold and silver nanodots and nanorods. *J. Phys. Chem. B*, 103:8410–8426, 1999.

83. L. M. Liz-Marzan. Tailoring surface plasmons through the morphology and assembly of metal nanoparticles. *Langmuir*, 22:32, 2006.

84. C. Lofton and W. Sigmund. Mechanisms controlling crystal habits of gold and silver colloids. *Adv. Funct. Mater.*, 15:1197–1208, 2005.

85. R. Loudon. *The Quantum Theory of Light*, 3rd edn. Oxford University Press, London, U.K., 2000.

86. X. M. Lu, M. S. Yavuz, H. Y. Tuan, B. A. Korgel, and Y. N. Xia. Ultrathin gold nanowires can be obtained by reducing polymeric strands of oleylamine-AuCl complexes formed via aurophilic interaction. *J. Am. Chem. Soc.*, 130:8900, 2008.

87. S. A. Maier, P. G. Kik, and H. A. Atwater. Optical pulse propagation in metal nanoparticle chain waveguide. *Phys. Rev. B*, 67:205402-1–205402-5, 2003.

88. S. A. Maier, P. G. Kik, H. A. Atwater, S. Meltzer, B. E. Koel, and A. A. G. Requicha. Local detection of electromagnetic energy transport below the diffraction limit in metal nanoparticle plasmon waveguides. *Nat. Mater.*, 2:229–232, 2003.

89. O. J. F. Martin, C. Girard, and A. Dereux. Generalized field propagator for electromagnetic scattering and light confinement. *Phys. Rev. Lett.*, 74:526–529, 1995.

90. J. Michaelis, J. Mlynek, C. Hettisch, and V. Sandoghdar. Optical microscopy using a single-molecule light source. *Nature*, 405:325–328, 2000.

91. G. Mie. Contribution to the optics of turbid media, particularly of colloidal metal solutions. *Annalen der Physik*, 25:377–445, 1908.

92. J. E. Millstone, W. Wei, M. R. Jones, H. J. Yoo, and C. A. Mirkin. Iodide ions control seed-mediated growth of anisotropic gold nanoparticles. *Nano Lett.*, 8:2526–2529, 2008.

93. P. Mulvaney. Surface plasmon spectroscopy of nanosized metal particles. *Langmuir*, 12:788–800, 1999.

94. C. J. Murphy, T. K. San, A. M. Gole, C. J. Orendorff, J. X. Gao, L. Gou, S. E. Hunyadi, and T. Li. Anisotropic metal nanoparticles: Synthesis, assembly, and optical applications. *J. Phys. Chem. B*, 109:13857–13870, 2005.

95. C. L. Nehl, H. W. Liao, and J. H. Hafner. Optical properties of star-shaped gold nanoparticles. *Nano Lett.*, 6:683–688, 2006.

96. J. Nelayah, M. Kociak, O. Stephan, F. J. Garcia de Abajo, M. Tence, L. Henrard, D. Taverna, I. Pastoriza-Santos, L. M. Liz-Marzan, and C. Colliex. Mapping surface plasmons on a single metallic nanoparticle. *Nat. Phys.*, 3:348–353, 2007.

97. J. Nelayah, O. Stephan, M. Kociak, F. J. de Abajo, L. Henrard, I. Pastoriza-Santos, L. M. Liz-Marzan, and C. Colliex. Mapping surface plasmons on single metallic nanoparticles using sub-nm resolved eels spectrum-imaging. *Microsc. Microanal.*, 13:144–145, 2007.

98. Z. H. Nie, D. Fava, E. Kumacheva, S. Zou, G. C. Walker, and M. Rubinstein. Self-assembly of metal-polymer analogues of amphiphilic triblock copolymers. *Nat. Mater.*, 6:609–614, 2007.

99. B. Nikoobakht and M. A. El-Sayed. Preparation and growth mechanism of gold nanorods (nrs) using seed-mediated growth method. *Chem. Mater.*, 15:1957–1962, 2003.

100. N. Okada, Y. Hamanaka, A. Nakamura, I. Pastoriza-Santos, and L. M. Liz-Marzan. Linear and nonlinear optical response of silver nanoprisms: Local electric fields of dipole and quadrupole plasmon resonances. *J. Phys. Chem. B*, 108:8751–8755, 2004.

101. D. Palik. *Handbook of Optical Constants of Solids*. Academic Press, New York, 1985.

102. I. Pastoriza-Santos and L. M. Liz-Marzan. Colloidal silver nanoplates. State of the art and future challenges. *J. Mater. Chem.*, 18:1724–1737, 2008.

103. E. M. Purcell and C. R. Pennypacker. Scattering and adsorption of light by spherical grains. *Astrophys. J.*, 186:705–714, 1973.

104. A. L. Pyayt, B. Wiley, Y. Xia, A. Chen, and L. Dalton. Integration of photonic and silver nanowire plasmonic waveguides. *Nat. Nanotechnol.*, 3:660–665, 2008.

105. H. Raether. *Plasmons on Smooth and Rough Surfaces and on Gratings.* Springer, Berlin, Germany, 1988.

106. T. P. Rivera, O. Lecarme, J. Hartmann, E. Rossitto, K. Berton, and D. Peyrade. Assisted convective-capillary force assembly of gold colloids in a microfluidic cell: Plasmonic properties of deterministic nanostructures. *J. Vac. Sci. Technol. B*, 26:2513–2519, 2008.

107. J. Rodriguez-Fernandez, J. Perez-Juste, P. Mulvaney, and L. M. Liz-Marzan. Spatially-directed oxidation of gold nanoparticles by Au(III)-CTAB complexes. *J. Phys. Chem. B*, 109:14257–14261, 2005.

108. M. Salerno, N. Félidj, J. R. Krenn, A. Leitner, F. R. Aussenegg, and J. C. Weeber. Near-field optical response of a two-dimensional grating of gold nanoparticles. *Phys. Rev. B*, 63:165422-6, 2001.

109. M. Salerno, J. R. Krenn, A. Hohenau, H. Ditlbacher, G. Schider, A. Leitner, and F. R. Aussenegg. The optical near-field of gold nanoparticle chains. *Opt. Commun.*, 248:543–549, 2005.

110. V. Salgueirino-Maceira, F. Caruso, and L. M. Liz-Marzan. Coated colloids with tailored optical properties. *J. Phys. Chem. B*, 107:10990–10994, 2003.

111. L. Salomon, C. Charbonnier, F. de Fornel, P. M. Adam, P. Guérin, and F. Carcenac. Near-field study of mesoscopic au periodic samples: Effect of the polarization and comparison between different imaging modes. *Phys. Rev. B*, 62:17072–17083, 2000.

112. A. W. Sanders, D. A. Routenberg, B. J. Wiley, Y. Xia, E. R. Dufresne, and M. A. Reed. Observation of plasmon propagation, redirection, and fan-out in silver nanowires. *Nano Lett.*, 6:1822–1826, 2006.

113. R. Sardar and J. S. Shumaker-Parry. Asymmetrically functionalized gold nanoparticles organized in one-dimensional chains. *Nano Lett.*, 8:731–736, 2008.

114. D. Seo, J. C. Park, and H. Song. Polyhedral gold nanocrystals with O-h symmetry: From octahedra to cubes. *J. Am. Chem. Soc.*, 128:14863–14870, 2006.

115. J. Sharma, Y. Tai, and T. Imae. Synthesis of confeito-like gold nanostructures by a solution phase galvanic reaction. *J. Phys. Chem. C*, 112:17033–17037, 2008.

116. C. Sönnichsen, S. Geier, N. E. Hecker, G. von Plessen, J. Feldmann, H. Ditlbacher, B. Lambrecht et al. Spectroscopy of single metallic nanoparticles using total internal reflection microscopy. *Appl. Phys. Lett.*, 77:2949–2951, 2000.

117. Z. Y. Tang, N. A. Kotov, and M. Giersig. Spontaneous organization of single CdTe nanoparticles into luminescent nanowires. *Science*, 297:237–240, 2002.

118. A. Tao, P. Sinsermsuksakul, and P. D. Yang. Polyhedral silver nanocrystals with distinct scattering signatures. *Angew. Chem.-Int. Edit.*, 45:4597–4601, 2006.

119. A. R. Tao, S. Habas, and P. D. Yang. Shape control of colloidal metal nanocrystal. *Small*, 4:310–325, 2008.

120. D. ten Bloemendal, P. Ghenuche, R. Quidant, I. G. Cormack, P. Loza-Alvarez, and G. Badenes. Local field spectroscopy of metal dimers by TPL microscopy. *Plasmonics*, 1:41–44, 2006.

121. J. Tersoff and D. R. Hamann. Theory of the scanning tunneling microscope. *Phys. Rev. B*, 31:805–813, 1985.

122. K. G. Thomas, S. Barazzouk, B. I. Ipe, S. T. S. Joseph, and P. V. Kamat. Uniaxial plasmon coupling through longitudinal self-assembly of gold nanorods. *J. Phys. Chem. B*, 108:13066–13068, 2004.

123. J. Turkevich, P. C. Stevenson, and J. Hillier. Nucleation and growth process in the synthesis of colloidal gold. *Discuss. Faraday Soc.*, 11:55–75, 1951.

124. V. V. Varadan, A. Lakhtakia, and V. K. Varadan. Scattering by three-dimensional anisotropic scatterers. *IEEE Trans. Antennas Propagat.*, 37:800–802, 1989.

125. Y. A. Vlasov, X. Z. Bo, J. C. Sturm, and D. J. Norris. On-chip natural assembly of silicon photonic bandgap crystals. *Nature*, 414:289–293, 2001.

126. C. Wang, Y. J. Hu, C. M. Lieber, and S. H. Sun. Ultrathin Au nanowires and their transport properties. *J. Am. Chem. Soc.*, 130:8902, 2008.

127. H. L. Wang, K. Schaefer, and M. Moeller. In situ immobilization of gold nanoparticle dimers in silica nanoshell by microemulsion coalescence. *J. Phys. Chem. C*, 112:3175–3178, 2008.

128. S. T. Wang, J. C. Yan, and L. Chen. Formation of gold nanoparticles and self-assembly into dimer and trimer aggregates. *Mater. Lett.*, 59:1383–1386, 2005.

129. W. Wang, X. Yang, and H. Cui. Growth mechanism of flowerlike gold nanostructures: Surface plasmon resonance (SPR) and resonance Rayleigh scattering (RRS) approaches to growth monitoring. *J. Phys. Chem. C*, 112:16348–16353, 2008.

130. X. J. Wang, G. P. Li, T. Chen, M. X. Yang, Z. Zhang, T. Wu, and H. Y. Chen. Polymer-encapsulated gold-nanoparticle dimers: Facile preparation and catalytical application in guided growth of dimeric ZnO-nanowires. *Nano Lett.*, 8:2643–2647, 2008.

131. H. L. Wu, C. H. Chen, and M. H. Huang. Seed-mediated synthesis of branched gold nanocrystals derived from the side growth of pentagonal bipyramids and the formation of gold nanostars. *Chem. Mater.*, 21:110–114, 2009.

132. G. A. Wurtz, J. S. Im, S. K. Gray, and G. P. Wiederrecht. Optical scattering from isolated metal particles and arrays. *J. Phys. Chem. B*, 107:14191–14198, 2003.

133. H. Yao, C. Q. Yi, C. H. Tzang, J. J. Zhu, and M. S. Yang. DNA-directed self-assembly of gold nanoparticles into binary and ternary nanostructures. *Nanotechnology*, 18:7, 2007.

134. P. Pramod and K. G. Thomas. Plasmon coupling in dimers of Au nanorods. *Adv. Mater.*, 20:4300–4305, 2008.

135. Y. D. Yin, Y. Lu, B. Gates, and Y. N. Xia. Template-assisted self-assembly: A practical route to complex aggregates of monodispersed colloids with well-defined sizes, shapes, and structures. *J. Am. Chem. Soc.*, 123:8718–8729, 2001.

136. Y. Y. Yu, S. S. Chang, C. L. Lee, and C. R. C. Wang. Gold nanorods: Electrochemical synthesis and optical properties. *J. Phys. Chem. B*, 101:6661–6664, 1997.

137. J. H. Yuan, Y. X. Chen, D. X. Han, Y. J. Zhang, Y. F. Shen, Z. J. Wang, and L. Niu. Synthesis of highly faceted multiply twinned gold nanocrystals stabilized by polyoxometalates. *Nanotechnology*, 17:4689–4694, 2006.

138. X. Q. Zou, E. B. Ying, and S. J. Dong. Seed-mediated synthesis of branched gold nanoparticles with the assistance of citrate and their surface-enhanced Raman scattering properties. *Nanotechnology*, 17:4758–4764, 2006.

139. C. Girard, E. Dujardin, G. Baffou, and R. Quidant. Shaping and manipulation of light fields with bottom-up plasmonic structures. *New J. Phys.*, 10:105016, 2008.

IV

Nanofluids

28

Stability of Nanodispersions

Nikola Kallay
University of Zagreb

Tajana Preočanin
University of Zagreb

Davor Kovačević
University of Zagreb

28.1 Introduction

Nanodispersion is a special kind of colloid dispersion containing dispersed nanoparticles. Terms "colloid" and "colloidal" are not always exactly defined but commonly refer to small entities. A colloid entity has at least one colloidal dimension. Colloid particles, such as small spheres, have three colloidal dimensions. Colloidal thin films have two colloidal dimensions while thin fibers are one-dimensional colloids. Colloid dispersion is a system in which colloid particles are dispersed in a medium. The question is, what is colloidal dimension? A simple, but not always correct, answer is "below 1 µm and above the molecular size of, e.g., 1 nm." Another answer is more correct, but less clear: "Particles are colloidal if they exhibit colloidal behavior." This "definition" obviously requires clarification. Under "colloidal behavior" one understands that colloid particles exhibit not only the behavior of a (macro) phase but, at the same time, also the behavior of molecules. Macroscopic properties are, e.g., defined structure, surface plane, interfacial tension, and size large enough for molecular medium to behave as homogeneous phase of defined viscosity. Molecular properties are related to Brownian motion, mutual collisions of particles. It should be noted that colloid particles are often aggregates of nanoparticles.

This chapter is about small solid colloid particles (nanoparticles) dispersed in a homogeneous liquid medium. What is special about nanoparticles? It may be said that they do not exhibit macroscopic behavior. Their internal structure is not necessarily the same as that of macrocrystals, the surface plane can be hardly defined, etc. There is no general agreement about the size under which particles may be considered as nanoparticles. Some authors would even misuse the term nanoparticle, but one may say that nanoparticles should be smaller than 10 nm or—more strictly, smaller than 5 nm.

A dispersion of colloid and nanoparticles can be stable, which means that the particles do not undergo aggregation or sedimentation. The dispersion is thermodynamically stable if the aggregates are unstable. In most cases, the stability of dispersion is due to a very slow aggregation process. In such kinetically stable dispersions, pronounced mutual repulsion of particles takes place. However, once the repulsion barrier is overcome, attraction at the closest contact prevails. Dispersion stability could be achieved by adsorption of chains protecting the close contact of interacting particles, but the system could be stable also due to the electrostatic repulsion caused by the surface charge. Numerous studies have been published on the different aspects of colloid stability (e.g., see references Frens and Heuts 1988, Holthoff et al. 1996, Behrens et al. 1998, Hiemstra and Riemsdijk 1999, Vorkapic and Matsoukas 1999, Baldwin and Dempsey 2001, Kallay and Žalac 2001), but only few of them deal with systems of small nanoparticles (e.g., Kallay and Žalac 2002, Brant et al. 2005, Navarro et al. 2008). In this chapter, the electrostatic effects on the stability of nanodispersions are analyzed.

28.2 Background

28.2.1 Surface Charge

A surface exposed to a liquid medium gets charged due to the interactions of surface sites with ions from the bulk of the solution (Lyklema 1995). Ions causing the surface charge are called potential-determining ions (pdi). For sparingly soluble salts like silver chloride, the pdi are constituent ions, i.e., silver and chloride ions. Silver ions get bound to surface chloride sites while chloride ions bind to silver sites (Kallay et al. 2008). Depending on the extent of these two processes, the surface is positively or negatively charged. In the excess of silver ions, the surface is positively charged and in the excess of chloride ions, the surface is negatively charged. For metal oxides and hydroxides in an aqueous environment, hydronium and hydroxide ions are pdi. The hydrated metal oxide surface may bind or release H$^+$ ions, leading to either positive or negative surface charge, respectively. These ions are pdi for most organic surfaces. In basic solutions, surface ≡COOH groups release H$^+$ ions, while in acidic solutions surface ≡NH$_2$ groups bind them. Again, the surface is positively charged at low pH, while the surface is negative at high pH. Since the electrostatic charge plays a dominant role in dispersion stability, this phenomenon will be covered in more detail. Also, in the case of nanodispersions, the charge distribution among particles markedly reduces the stability.

28.2.2 Surface Complexation Model

Equilibrium at the solid–liquid interface is commonly interpreted by the surface complexation model (SCM) or the site binding model (Lützenkirchen 2006). This theoretical approach will be demonstrated on the example of metal oxide aqueous interface. For metal oxides, the pdi are H$^+$ and OH$^-$ ions. Within the SCM, the 2-pK mechanism (Yates et al. 1974, Davis et al. 1978) will be demonstrated. Accordingly, surface charge is a result of two-step protonation (reactions 1 and 2) of the surface groups produced by hydration of the metal oxide surface (Kallay et al. 2004):

$$\equiv MO^- + H^+ \rightarrow \equiv MOH; \quad K_1^\circ \tag{28.1}$$

$$\equiv MOH + H^+ \rightarrow \equiv MOH_2^+; \quad K_2^\circ \tag{28.2}$$

where

≡M denotes the metal atom incorporated in the solid surface
K_1° and K_2° are thermodynamic equilibrium constants of the corresponding surface reactions

Generally, the thermodynamic equilibrium constant K° depends only on temperature and pressure and is defined in terms of equilibrium activities (a) of species J which are involved in the chemical reaction (Mills et al. 1998):

$$K^\circ = \prod_J a_J^{v_J} \tag{28.3}$$

where v denotes the stoichiometric number being positive for products and negative for reactants. For interfacial species S, the activity may be defined (Kallay et al. 2004) in terms of surface concentration Γ (amount of surface species per surface area), so that

$$a_S = \gamma_S \frac{\Gamma_S}{\Gamma^\circ} = \gamma_S \{S\} \tag{28.4}$$

where γ is the activity coefficient, and curly brace denotes the relative value of surface concentration with respect to the chosen standard value. There is no recommended value for standard surface concentration, but $\Gamma^\circ = 1$ mol m^{-2} seems to be a suitable

choice (Kallay et al. 2004). Consequently, {S} becomes the numerical value of surface concentration of species S expressed in moles per square meter: $\{S\} = \Gamma_S$ (mol m^{-2}). The activity coefficient of surface species S (γ_S) is defined through the difference in chemical potentials of real (μ^{real}) and ideal (μ^{id}) states at the surface

$$RT \ln \gamma_S = \mu_S^{real} - \mu_S^{id} = z_S F \Psi \tag{28.5}$$

where

F denotes the Faraday constant
R is the gas constant
T is the thermodynamic temperature
z_S is the charge number of surface species S
Ψ is the electrostatic potential affecting their state at the interface

Ideal state corresponds to the zero value of the electrostatic potential Ψ, equal to that in the bulk of the solution. The above equation can be rewritten as

$$\gamma_S = \exp\left(\frac{z_S F \Psi}{RT}\right) \tag{28.6}$$

In the case of two-step protonation (2-pK mechanism) described by reactions (28.1) and (28.2), the activity coefficients of charged interfacial species ≡MO$^-$ and ≡MOH$_2^+$ involved in surface reactions are

$$\gamma(\equiv MO^-) = \exp\left(\frac{-F \Psi_0}{RT}\right) \tag{28.7}$$

$$\gamma(\equiv MOH_2^+) = \exp\left(\frac{F \Psi_0}{RT}\right) \tag{28.8}$$

where Ψ_0 is the (inner) surface potential affecting their state at the interface. The activity coefficient of neutral ≡MOH species is equal to 1. Accordingly, thermodynamic equilibrium constants for the first (1) and second (2) steps of protonation are defined on the basis of Equations 28.3 through 28.8 as

$$K_1^\circ = \exp(F\Psi_0/RT) \cdot \frac{\{\equiv MOH\}}{\{\equiv MO^-\} \cdot a_{H^+}} \tag{28.9}$$

$$K_2^\circ = \exp(F\Psi_0/RT) \cdot \frac{\{\equiv MOH_2^+\}}{\{\equiv MOH\} \cdot a_{H^+}} \tag{28.10}$$

Effective surface charge is reduced by association of anions A$^-$ and cations C$^+$ from the bulk of solutions (reactions A and C), respectively:

$$\equiv MOH_2^+ + A^- \rightarrow \equiv MOH_2^+ \cdot A^-; \quad K_A^\circ \tag{28.11}$$

$$\equiv MO^- + C^+ \rightarrow \equiv MO^- \cdot C^+; \quad K_C^\circ \tag{28.12}$$

By introducing the activity coefficients into expressions for thermodynamic equilibrium constants of counterion surface association reactions (28.11) and (28.12) and by taking into account that interfacial ion pairs behave as oriented dipoles with charged ends being exposed to different potentials, Ψ_0 and Ψ_β, one obtains

$$\gamma(\equiv\!MO^- \cdot C^+) = \exp\!\left(\frac{F(\Psi_\beta - \Psi_0)}{RT}\right) \qquad (28.13)$$

$$\gamma(\equiv\!MOH_2^+ \cdot A^-) = \exp\!\left(\frac{-F(\Psi_\beta - \Psi_0)}{RT}\right) \qquad (28.14)$$

Accordingly, the corresponding thermodynamic equilibrium constants are

$$K_C^\circ = \frac{\exp(F(\Psi_\beta - \Psi_0)/RT)\cdot\{\equiv\!MO^- \cdot C^+\}}{\exp(-F\Psi_0/RT)\cdot\{\equiv\!MO^-\}\cdot a_{C^+}}$$
$$= \exp\!\left(\frac{F\Psi_\beta}{RT}\right)\cdot\frac{\{\equiv\!MO^- \cdot C^+\}}{\{\equiv\!MO^-\}\cdot a_{C^+}} \qquad (28.15)$$

$$K_A^\circ = \frac{\exp(-F(\Psi_\beta - \Psi_0)/RT)\cdot\{\equiv\!MOH_2^+ \cdot A^-\}}{\exp(F\Psi_0/RT)\cdot\{\equiv\!MOH_2^+\}\cdot a_{A^-}}$$
$$= \exp\!\left(\frac{-F\Psi_\beta}{RT}\right)\cdot\frac{\{\equiv\!MOH_2^+ \cdot A^-\}}{\{\equiv\!MOH_2^+\}\cdot a_{A^-}} \qquad (28.16)$$

Charge densities in the inner 0-plane, σ_0, and outer β-plane, σ_β, are defined by

$$\sigma_0 = F(\Gamma(\equiv\!MOH_2^+) - \Gamma(\equiv\!MO^-) + \Gamma(\equiv\!MOH_2^+ \cdot A^-)$$
$$- \Gamma(\equiv\!MO^- \cdot C^+)) \qquad (28.17)$$

$$\sigma_\beta = F(\Gamma(\equiv\!MO^- \cdot C^+) - \Gamma(\equiv\!MOH_2^+ \cdot A^-)) \qquad (28.18)$$

while the net surface charge density, σ_s, i.e., the effective charge directly bound to the surface per surface area, being equal in magnitude but opposite in sign with respect to the so-called surface charge density of diffuse layer σ_d, is

$$\sigma_s = -\sigma_d = \sigma_0 + \sigma_\beta = F(\Gamma(\equiv\!MOH_2^+) - \Gamma(\equiv\!MO^-)) \qquad (28.19)$$

The zero charge condition at the surface is expressed by three quantities, i.e., by the point of zero charge (p.z.c., pH_{pzc}) corresponding to $\sigma_0 = 0$, by the isoelectric point (i. e. p., pH_{iep}) corresponding to the electrokinetic potential $\zeta = 0$ (and $\sigma_s = 0$), and by the point of zero potential (p.z.p., pH_{pzp}) corresponding to $\Psi_0 = 0$. In the absence of specific adsorption of ions, and in the case of negligible or symmetric association of counterions, all three points coincide and correspond to the state in which

all electrical properties diminish ($\sigma_0 = \sigma_s = 0$, $\Psi_0 = \zeta = 0$). This electroneutrality condition (pH_{eln}) is related to the protonation equilibrium constants (Kallay et al. 2007) by

$$pH_{eln} = \frac{1}{2}\log(K_1^\circ \cdot K_2^\circ) \qquad (28.20)$$

The electroneutrality condition can be experimentally achieved at low electrolyte concentration (i.e., low ionic strength), and in such a case,

$$pH_{eln} = pH_{pzc} = pH_{iep} = pH_{pzp}; \quad I_c \to 0 \qquad (28.21)$$

Electrostatically stabilized dispersions contain sufficiently charged particles. Representative quantities are the net surface charge density σ_s and the electrostatic potential at the onset of diffuse layer Ψ_d, which may be approximated by the outer surface potential Ψ_β. The potential at the onset of the diffuse layer cannot be experimentally obtained. However, it is possible to measure the electrokinetic potential ζ which is related to Ψ_d by the Gouy–Chapman equation based on the Poisson–Boltzmann distribution of ionic species at the charged interface (Holmberg et al. 2002, Butt 2003)

$$\Psi_d = \frac{2RT}{F}\ln\!\left(\frac{\exp(-\kappa l_e) + th(F\zeta/4RT)}{\exp(-\kappa l_e) - th(F\zeta/4RT)}\right) \qquad (28.22)$$

where l_e is the separation distance of the electrokinetic slip plane (e-plane) from the d-plane ($l_e \approx 1$ nm), and κ is the Debye–Hückel parameter given by

$$\kappa = \sqrt{\frac{2I_cF^2}{\varepsilon RT}} \qquad (28.23)$$

the value of which depends on the permittivity of the medium ε ($\varepsilon = \varepsilon_0\varepsilon_r$) and on the ionic strength I_c:

$$I_c = \sum c_i z_i^2 \qquad (28.24)$$

For completely dissociated 1:1 electrolytes (such as NaCl), the ionic strength is equal to their concentration.

The Gouy–Chapman theory (Holmberg et al. 2002, Butt 2003) provides also the relationship between the potential at the onset of the diffuse layer Ψ_d and the surface charge density of the diffuse layer σ_d, which is equal in magnitude to the net surface charge density σ_s:

$$\sigma_s = -\sigma_d = -\sqrt{8RT\varepsilon I_c}\sinh\!\left(\frac{-F\Psi_d}{RT}\right) \qquad (28.25)$$

The following equation holds for spherical particles of radius r:

$$\sigma_s = -\sigma_d = \frac{\Psi_d\varepsilon(1 + r\kappa)}{r} \qquad (28.26)$$

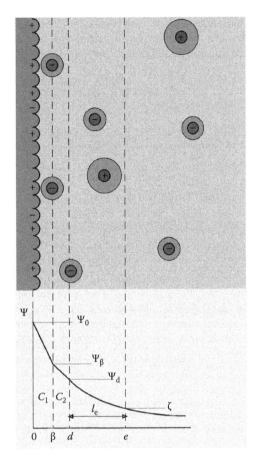

FIGURE 28.1 General model of the electrical interfacial layer. (Reproduced from Kovačević, D. et al., *Croat. Chem. Acta*, 80, 287, 2007. With permission.)

In addition to electrokinetic data providing Ψ_d and σ_s, it is possible to measure the surface charge density σ_0 by potentiometric titration, and also the surface potential Ψ_0 by means of a corresponding single-crystal electrode (Kallay et al. 2005, 2007) as well as to calculate potential Ψ_β, which is approximately equal to Ψ_d. For that purpose, one should apply the concept of the inner layer capacitor, the value of which is approximately 1 F m^{-1}. The structure of the electrical interfacial layer (EIL) at the solid–liquid interface is represented in Figure 28.1.

28.2.3 Charge Distribution

The charge of a particle is equal to the product of the surface charge density and surface area. Since the ionic interfacial equilibrium is related to the surface charge density, it is obvious that small particles bear a low charge. Even in a colloid dispersion containing identical particles with regard to composition, shape, and size, the particle charge will be distributed around an average value. In the case of relatively large colloid particles, this phenomenon is not expected to influence the stability of the system. However, in the case of small nanoparticles, the charge is low so that the charge distribution may substantially influence the stability of the system. As the equilibrium state at the interface

is a complex problem, the problem of charge distribution has not been solved. It is possible to consider a nanoparticle as a molecule of the valence equal to the number of active surface sites. If the valence is m, then the stepwise charging process for particles Pa due to the adsorption of cations (e.g., H$^+$) may be written as

$$\text{Pa} + \text{H}^+ \rightarrow (\text{PaH})^+; \quad K_1^\circ \qquad (28.27)$$

$$(\text{PaH})^+ + \text{H}^+ \rightarrow (\text{PaH}_2)^{2+}; \quad K_2^\circ \qquad (28.28)$$

$$(\text{PaH}_2)^{2+} + \text{H}^+ \rightarrow (\text{PaH}_3)^{3+}; \quad K_3^\circ \qquad (28.29)$$

$$.....$$

$$(\text{PaH}_{m-1})^{(m-1)+} + \text{H}^+ \rightarrow (\text{PaH}_m)^{m+}; \quad K_n^\circ \qquad (28.30)$$

In addition, counterions, anions A$^-$, associate with surface charge groups. Eighty to ninety percent of the original charge is usually compensated due to counterion association. For example, an average particle with n adsorbed cations may bind several counterions:

$$(\text{PaH}_n)^{n+} + \text{A}^- \rightarrow (\text{PaH}_n\text{A})^{(n-1)+}; \quad K_{n,1}^\circ \qquad (28.31)$$

$$(\text{PaH}_n\text{A})^{(n-1)+} + \text{A}^- \rightarrow (\text{PaH}_n\text{A}_2)^{(n-2)+}; \quad K_{n,2}^\circ \qquad (28.32)$$

$$(\text{PaH}_n\text{A}_{p-1})^{(n-p+1)+} + \text{A}^- \rightarrow (\text{PaH}_n\text{A}_p)^{(n-p)+}; \quad K_{n,p}^\circ \qquad (28.33)$$

The adsorption constants for different steps are mutually related. The binding energy (enthalpy) may be taken to be equal for all the steps. Conformation entropy, however, will depend on the fraction of occupied sites. Also, the electrostatic potential affecting the state of the adsorbed ion will depend on the total charge at the interface. Similar consideration will apply to the association of counterions. It may be concluded that the total charge of particles will differ and will be distributed around a certain average value.

28.3 Stability—Aggregation Rate

Dispersed systems are considered as stable if the aggregation of dispersed particles is so slow that no change in the dispersity could be detected, e.g., in one month. The stability could be achieved by adsorption of chain molecules protecting direct contact between particles. Also, charged particles exhibit mutual repulsion so that collisions do not always result in aggregation.

Nanodispersions are known to be less stable compared to ordinary colloid dispersions. The main reason is the relatively low electric charge of nanoparticles (de Gennes 1998). The effect of particle size (Higashitani et al. 1991, Kobayashi et al. 2005) and polydispersity (Semmler et al. 2000, Kallay et al. 2009) on aggregation of nanoparticles, as well as on aggregation of different particles (heteroaggregation), was examined (Hiemstra and Riemsdijk 1999, Lyklema and Duval 2005, Lin et al. 2006).

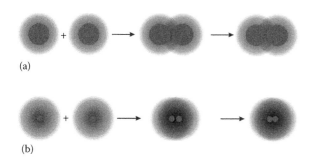

(a)

(b)

FIGURE 28.2 Overlap of electrical interfacial layers in the course of collision of two ordinary colloid particles (a) and of two nanoparticles (b). (Reproduced from Kallay, N. et al., *Croat. Chem. Acta.*, 82, 531, 2009. With permission.)

Modeling of aggregation on molecular level (Taboada-Serrano et al. 2005) and Monte Carlo simulations (Taboada-Serrano et al. 2006) of electrostatic surface interactions were also performed. In the course of collision of ordinary colloid particles, the overlap of the diffuse part of interfacial layers is partial, while it is almost complete in the case of nanoparticles (Figure 28.2). Therefore, the classical DLVO (Verwey and Overbeek 1948, Lyklema 1999, Behrens et al. 2000) theory is no more applicable for studying the kinetics of nanoparticle aggregation.

Two colliding nanoparticles form a common ionic cloud in a similar way as two interacting ions. For this reason, the Brønsted theory (Brønsted 1922, Christiansen 1922, Lyklema 1995), which was developed for the primary salt effect on the kinetics of ionic reactions, was applied (Kallay et al. 2002).

In his theory, Brønsted considered two reaction steps. In the first step, two ions form an ionic pair, while the second step yields the final products. Ion pairing and dissociation is fast so that the first step is equilibrated. The second step is slow, thus determining the overall rate being proportional to the concentration of ion pairs determined by the equilibrium constant. The activity coefficients are introduced by using the Debye–Hückel theory (Atkins and de Paula 2006). Application of the Brønsted theory to the kinetics of reaction of charged ionic species and nanoparticles is described as follows.

Reaction of two charged species A^{z_A} and B^{z_B} can be represented by the following two-step mechanism:

$$A^{z_A} + B^{z_B} \rightarrow AB^{z_{AB}} \rightarrow \text{product(s)} \qquad (28.34)$$

where z denotes the charge number. The charge number of the intermediate $AB^{z_{AB}}$ is equal to the sum of charge numbers if interacting species

$$z_{AB} = z_A + z_B \qquad (28.35)$$

The formation of the intermediate is fast and equilibrated, while the second step is slow and thus the rate-determining step. Accordingly, the rate of reaction v is given by

$$v = k'[AB^{z_{AB}}] \qquad (28.36)$$

where the square bracket denotes concentration and k' is the rate coefficient (constant) of the second step. The equilibrium constant of the first step is defined as

$$K^{\neq} = \frac{\gamma_{AB^{z_{AB}}}[AB^{z_{AB}}]}{\gamma_{A^{z_A}}[A^{z_A}]\gamma_{B^{z_B}}[B^{z_B}]} \qquad (28.37)$$

The concentration of the intermediate is thus

$$[AB^{z_{AB}}] = K^{\neq} \frac{\gamma_{A^{z_A}}\gamma_{B^{z_B}}}{\gamma_{AB^{z_{AB}}}}[A^{z_A}][B^{z_B}] \qquad (28.38)$$

Introduction of (28.38) into (28.36) yields

$$v = k'K^{\neq} \frac{\gamma_{A^{z_A}}\gamma_{B^{z_B}}}{\gamma_{AB^{z_{AB}}}}[A^{z_A}][B^{z_B}] \qquad (28.39)$$

The activity coefficients for charged species could be calculated by means of the Debye–Hückel theory (Atkins and de Paula 2006). For example, the activity coefficient of species B^{z_B} is given by

$$-\lg\gamma_{B^{z_B}} = \frac{z_B^2 A_{DH}I_c^{1/2}}{1 + ab_{DH}I_c^{1/2}} \qquad (28.40)$$

where a is the minimum center-to-center separation distance between interacting species and A_{DH} and b_{DH} are Debye–Hückel constants that depend on the temperature and the electric permittivity of the medium ε as

$$A_{DH} = \frac{2^{1/2}}{8\pi L \ln 10} \cdot \left(\frac{F^2}{\varepsilon RT}\right)^{3/2} \qquad (28.41)$$

$$b_{DH} = \left(\frac{2F^2}{\varepsilon RT}\right)^{1/2} \qquad (28.42)$$

For aqueous solutions at 25°C, $A_{DH} = 0.509\,\text{mol}^{-1/2}\,\text{dm}^{3/2}$ and $b_{DH} = 3.28\,\text{nm}^{-1}$.

According to (28.36) and (28.39), the rate of reaction v is proportional to the product of concentrations of interacting particles $[A^{z_A}][B^{z_B}]$

$$v = k \cdot [A^{z_A}] \cdot [B^{z_B}] \qquad (28.43)$$

where k is the rate constant (coefficient) of the process given by

$$k = k_0 \frac{\gamma_{A^{z_A}}\gamma_{B^{z_B}}}{\gamma_{AB^{z_{AB}}}} \qquad (28.44)$$

and

$$k_0 = k'K^{\neq} \qquad (28.45)$$

Application of the Brønsted concept to aggregation of nanoparticles requires the consideration of the electrostatic effect on the value of the $K^{\neq}$ since the charge of the particles depend on the composition of the medium. The chemical contribution $K_{ch}^{\neq}$ may be taken as a constant, while the electrostatic contribution $K_{el}^{\neq}$ may be approximated by the Coulomb equation

$$-RT \ln K_{el}^{\neq} = \Delta_r G_{el} = \frac{z_A e \cdot z_B e}{4\pi\varepsilon(r_A + r_B)} L \qquad (28.46)$$

where

   L is the Avogadro constant used to recalculate the energy to the "molar basis"

   $r_A + r_B$ is the minimum center-to-center separation distance between interacting particles

By introducing the Faraday constant, one obtains

$$\lg K^{\neq} = \lg K_{ch}^{\neq} - \frac{z_A z_B F^2}{4\pi\varepsilon LRT(r_A + r_B)\ln 10} \qquad (28.47)$$

According to Equations 28.34, 28.39, and 28.43 through 28.47, the following relationship, applicable to dispersed nanoparticles, is obtained

$$\lg k = \lg k_{diff} - \frac{z_A z_B F^2}{4\pi\varepsilon LRT(r_A + r_B)\ln 10} + \frac{2 z_A z_B A_{DH} I_c^{1/2}}{1 + a b_{DH} I_c^{1/2}} \qquad (28.48)$$

Distance a in Equation 28.48 denotes the minimum center-to-center separation of interacting charged species (particles and ions) that can be approximated by the radius of the nanoparticles and

$$k_{diff} = k' K_{ch}^{\neq} \qquad (28.49)$$

where k_{diff} denotes the rate coefficient (constant) of the rapid aggregation when each collision results in aggregation. Such a condition is achieved at high ionic strength

$$k_{diff} = \lim_{I_c \to \infty} k \qquad (28.50)$$

The rate coefficient of rapid diffusional aggregation for aggregation of particles A and B in the absence of repulsion, k_{diff}, can be obtained from the Smoluchowski theory (Smoluchowski 1916, Shaw 2000):

$$k_{diff} = \frac{k_B T}{3\eta}\left(\frac{1}{r_A} + \frac{1}{r_B}\right) \cdot (r_A + r_B) \qquad (28.51)$$

where k_B is the Boltzmann constant ($k_B = R/L$). From Equations 28.48 through 28.51 one obtains

$$\lg k = \lg k_{diff} - z_A z_B \cdot \left(\frac{\beta}{r_A + r_B} - 2 A_{DH} \frac{I_c^{1/2}}{1 + a b_{DH} I_c^{1/2}}\right) \qquad (28.52)$$

where

$$\beta = \frac{F^2 \ln 10}{4\pi\varepsilon LRT} \qquad (28.53)$$

Quantitative measure of dispersion stability is the stability coefficient W. It is defined as the ratio of the rate coefficient for rapid aggregation (each collision results in aggregation) and the rate coefficient of the examined system. It is directly related to the collision efficiency. For example, $W = 1000$ means that 1 of 1000 collisions results in aggregation. According to above equations, the stability coefficient for nanodispersions is given by

$$\lg W = \lg\left(\frac{k_{diff}}{k}\right) = z_A z_B \cdot \left(\frac{\beta}{r_A + r_B} - 2 A_{DH} \frac{I_c^{1/2}}{1 + a b_{DH} I_c^{1/2}}\right) \qquad (28.54)$$

The above equations have a simpler form for monodisperse systems ($r = r_A = r_B$).

28.4 Effect of Charge Distribution

One of the characteristics of nanodispersions is the charge distribution among the particles. It is clear that even in dispersions of particles identical in size and shape, the charge of the particles cannot be the same for all of them (Kallay 1976, 1977). This phenomenon is more pronounced in nanodispersions. For example, if the average charge number of nanoparticles is $z = 5$, then one may expect a high population of particles $z = 5$, but also significant portions of particles bearing the charge $z = 4$ and $z = 6$. In addition, some of them may exhibit charge more apart from the mean value of 5. Since electrostatic repulsion is responsible for the stability of dispersions, in the absence of surfactants, the charge distribution may significantly affect the stability of nanodispersions. The effect of charge distribution on the stability of nanodispersions or on the aggregation rate of nanoparticles was analyzed by numerical simulation on the basis of Equation 28.54. Nanodispersion of identical spherical particles with a certain charge distribution was assumed. Charge distribution was defined as follows: the concentration of nanoparticles B bearing an average charge number z is $[B^z]$. Concentration of particles bearing charge numbers $z - 1$ and $z + 1$ is

$$[B^{z+1}] = [B^{z-1}] = f \cdot [B^z] \qquad (28.55)$$

Concentrations of particles bearing charges even more apart from the average value are

$$[B^{z+2}] = [B^{z-2}] = f \cdot [B^{z+1}] = f^2 \cdot [B^z] \qquad (28.56)$$

$$[B^{z+3}] = [B^{z-3}] = f \cdot [B^{z+2}] = f^3 \cdot [B^z] \qquad (28.57)$$

Accordingly, the absence of charge distribution will be characterized by $f = 0$, while wider charge distributions will be characterized by higher f values, being always below 1.

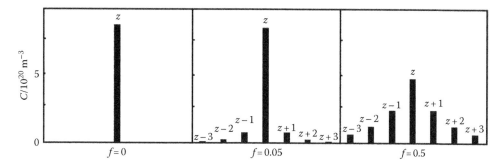

FIGURE 28.3 Different distributions of particles in nanodispersions for different f values. (Reproduced from Kallay, N. et al., *Croat. Chem. Acta.*, 82, 531, 2009. With permission.)

The concentration of particles of the average charge number z is related to the total concentration of particle [B] by

$$[B^z] = \frac{[B]}{1 + 2f + 2f^2 + 2f^3 + \ldots} \qquad (28.58)$$

The effect of parameter f on the distribution of particles is demonstrated in Figure 28.3.

According to Equations 28.43 and 28.58, the overall rate of aggregation of nanoparticles B exhibiting charge distribution is a sum of individual rates of differently charged particles

$$v = \sum k_{i,j} [B^{z_i}] \cdot [B^{z_j}] = k_{diff} \cdot \sum W_{i,j}^{-1} \cdot [B^{z_i}] \cdot [B^{z_j}] \qquad (28.59)$$

The diffusional rate corresponding to uncharged particles is given by

$$v_{diff} = k_{diff} \cdot [B]^2 \qquad (28.60)$$

Accordingly, the effective stability coefficient for nanosystems exhibiting charge distribution is

$$W_{eff} = \frac{v_{diff}}{v} = \frac{[B]^2}{\sum W_{i,j}^{-1} \cdot [B^{z_i}] \cdot [B^{z_j}]} \qquad (28.61)$$

The effect of pH and ionic strength was examined for model metal oxide nanoparticles of radius 1 nm. Average charge numbers of these model particles were estimated using the electrokinetic data for chromium hydroxide (Matijević 2002). The potential at the onset of diffuse layer Ψ_d was approximated by the electrokinetic potential ζ. The effective surface charge density, σ_s, was calculated by Equation 28.28 using the Gouy–Chapman theory for spherical geometry (Holmberg et al. 2002, Butt 2003). The average charge number of nanoparticles of radius r is equal to

$$z = \frac{4r^2 \pi \sigma_s}{F} \qquad (28.62)$$

The isoelectric point was taken to be at pH = 8.5. For nanoparticles of $r = 1$ nm at ionic strength of 10^{-3} mol dm^{-3}, the average charge number decreases from approximately 4 at pH = 3 to 0 at pH = 8.5, being negative in the basic region almost up

to −4 at pH = 11. For the same nanoparticles at pH = 3, the average charge number decreases with ionic strength from 4, at ionic strength of 10^{-3} mol dm^{-3}, to almost 0 at $c = 1$ mol dm^{-3}.

Figure 28.4 displays the calculation of the effective stability coefficients at low ionic strength of 10^{-3} mol dm^{-3} as a function of pH. Calculations were performed for different charge distributions. In the absence of charge distribution ($f = 0$), the system is stable in the pH region apart from the isoelectric point, i.e., at $7 < $ pH $ > 10$.

At the isoelectric point, the particles are uncharged and the system is unstable for any f value. The stability of the system significantly decreases as the charge distribution becomes wider. This effect is noticeable up to $f = 0.5$. Further increase of f above 0.5 has no significant effect on stability. The reason for such an effect lies in the fact that aggregation of particles of low charge approaches the fast regime of rapid aggregation and cannot be

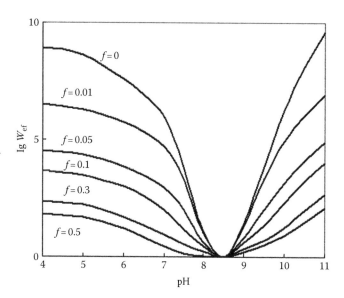

FIGURE 28.4 Effect of pH on the stability of a model metal oxide aqueous nanodispersion ($r = 1$ nm) at $T = 298$ K and ionic strength of 10^{-3} mol dm^{-3}. Calculations were performed for the absence of charge distribution ($f = 0$), as well as for wider distributions ($0 < f < 1$). No significant effect on the stability was observed above $f = 0.5$. (Reproduced from Kallay, N. et al., *Croat. Chem. Acta.*, 82, 531, 2009. With permission.)

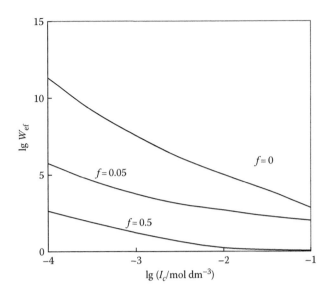

FIGURE 28.5 Effect of ionic strength on the stability of a model metal oxide aqueous nanodispersion ($r = 1\,nm$) at $T = 298\,K$ and pH = 3. Calculations were performed for the absence of charge distribution ($f = 0$), as well as for wider distributions ($0 < f < 1$). No significant effect on the stability was observed above $f = 0.5$. (Reproduced from Kallay, N. et al., *Croat. Chem. Acta.*, 82, 531, 2009. With permission.)

further accelerated. It is clear that charge distribution markedly reduces the stability of nanodispersions.

The effect of ionic strength on the stability of a nanodispersion was examined for the same system at pH = 3. In the absence of charge distribution ($f = 0$), the system is stable at low ionic strength, i.e., below 10^{-3} mol dm^{-3} (Figure 28.5).

The stability decreases with ionic strength and almost disappears at ionic strengths above 10^{-1} mol dm^{-3}. At high ionic strength, the electrostatic repulsion diminishes so that the system becomes completely unstable. In the region of low ionic strength, the stability of the system decreases again as the charge distribution becomes wider, which is noticeable up to $f = 0.5$. The same behavior was observed for larger nanoparticles. The only difference was that such systems were more stable.

The analysis performed in this study clearly shows that the charge distribution among nanoparticles markedly reduces the stability of nanodispersions. There is, however, an additional reason why nanodispersions do not exhibit high stability. For monodisperse systems of uniform particles bearing the same charge, the integration of Equation 28.59 and introduction of mass concentration of the dispersed phase, being equal to mass of solid phase divided by volume of dispersion (m/V), of density ρ result in the following expression for the time necessary to reduce the concentration of primary particles to half of the initial value ("half-time" of aggregation) $t_{1/2}$:

$$t_{1/2} = \frac{1}{[B]_0 k} = \frac{W}{[B]_0 k_{diff}} = \frac{4\pi\rho}{3\,(m/V)} \cdot \frac{W}{k_{diff}} \cdot r^3 \qquad (28.63)$$

It is clear that for a given stability coefficient W, the time needed to reduce particle concentration to half of the initial concentration

$[B]_0$ is markedly lower for smaller particles. Also, the particle number concentration in a nanodispersion might be so high that the second-order kinetic regime does not hold.

The above analysis allows the conclusion that nanodispersions are less stable compared to the ordinary colloid dispersions due to higher particle number concentration, but also due to the low charge and distribution of charges among dispersed nanoparticles.

28.5 Conclusion

Stability of nanodispersions, due to mutual electrostatic repulsion among particles, is as a rule lower with respect to ordinary colloid particles of size above 50 Å due to their low charge. The charge distribution among nanoparticles also reduces the stability. Stability is additionally reduced due to high particle number concentration. According to the described theoretical model, nanodispersions could be electrostatically stabilized at the condition of low ionic strength (low electrolyte concentration) if they are sufficiently charged. The stability coefficient may be predicted by the theoretical model based on the extended Brønsted concept.

Acknowledgment

The financial support from the Ministry of Science, Education and Sports of the Republic of Croatia (project No. 119-1191342-2961) is gratefully acknowledged.

References

Atkins, P. W. and de Paula, J., 2006, *Physical Chemistry*, 8th edn., Oxford University Press, Oxford, U.K.

Baldwin, J. L. and Dempsey, B. A., 2001, Effects of Brownian motion and structured water on aggregation of charged particles, *Colloids Surf. A* 177: 111–122.

Behrens, S. H., Borkovec, M., and Schurtenberger, P., 1998, Aggregation in charge-stabilized colloidal suspensions revisited, *Langmuir* 14: 1951–1954.

Behrens, S. H., Christl, D. I., Emmerzael, R., Schurtenberger, P., and Borkovec, M., 2000, Charging and aggregation properties of carboxyl latex particles: Experiments versus DLVO theory, *Langmuir* 16: 2566–2575.

Brant, J., Lecoanet, H., and Wiesner, M. R., 2005, Aggregation and deposition characteristics of fullerene nanoparticles in aqueous systems, *J. Nanoparticle Res.* 7: 545–553.

Brønsted, J. N., 1922, Theory of chemical reaction velocity, *Z. Physik. Chem.* 102: 169–207.

Butt, H.-J., 2003, *Physics and Chemistry of Interfaces*, Wiley-VCH, Berlin, Germany.

Christiansen, J. A., 1922, Velocity of bimolecular reactions in solution, *Z. Physik. Chem.* 113: 35–52.

Davis, J. A., James, R. O., and Leckie, J. O., 1978, Surface ionization and complexation at the oxide/water interface. I. computation of electrical double layer properties in simple electrolytes, *J. Colloid Interface Sci.* 63: 480–499.

Frens, G. and Heuts, J. J. F. G., 1988, The double layer potential Φ_δ as a rate determining factor in the coagulation of electrocratic colloids, *Colloids Surf.* 30: 295–305.

de Gennes, P.-G., 1998, Nanoparticles and dendrimers: Hopes and illusions, *Croat. Chem. Acta* 71: 833–836.

Hiemstra, T. and van Riemsdijk, W. H., 1999, Effect of different crystal faces on experimental interaction force and aggregation of hematite, *Langmuir* 15: 8045–8051.

Higashitani, K., Kondo, M., and Hatade S., 1991, Effect of particle size on coagulation rate of ultrafine colloidal particles, *J. Colloid Interface Sci.* 142: 204–213.

Holmberg, K., Shah, D. O., and Schwuger, M. J., 2002, *Handbook of Applied Surface and Colloid Chemistry*, John Wiley & Sons Ltd., West Sussex, U.K.

Holthoff, H., Egelhaaf, S. U., Borkovec, M., Schurtenberger, P., and Sticher, H., 1996, Coagulation rate measurements of colloidal particles by simultaneous static and dynamic light scattering, *Langmuir* 12: 5541–5549.

Kallay, N., 1976, Adsorption of ions on small spheres at low ionic strengths, *Croat. Chem. Acta* 48: 271–276.

Kallay, N., 1977, Adsorption of ions by colloids in electrolyte solutions, *Croat. Chem. Acta* 50: 209–217.

Kallay, N. and Žalac, S., 2001, Introduction of the surface complexation model into the theory of colloid stability (Authors' Review), *Croat. Chem. Acta* 74: 479–497.

Kallay, N. and Žalac, S., 2002, Stability of nanodispersions: A model for kinetics of aggregation of nanoparticles, *J. Colloid Inerface Sci.* 253: 70–76.

Kallay, N., Preočanin, T., and Žalac, S., 2004, Standard states and activity coefficients of interfacial species, *Langmuir* 20: 2986–2988.

Kallay, N., Dojnović, Z., and Čop, A., 2005, Surface potential at hematite-water interface, *J. Colloid Interface Sci.* 286: 610–614.

Kallay, N., Preočanin, T., and Ivšić, T., 2007, Determination of surface potential from the electrode potential of a single-crystal electrode, *J. Colloid Interface Sci.* 309: 21–27.

Kallay, N., Šupljika, F., and Preočanin, T., 2008, Measurement of the surface potential at silver chloride aqueous interface by means of the single crystal electrode, *J. Colloid Interface Sci.* 327: 384–387.

Kallay, N., Preočanin, T., and Kovačević, D., 2009, Effect of charge distribution on the stability of nanodispersions, *Croat. Chem. Acta.*, 82: 531–535.

Kobayashi, M., Juillerat, F., Galletto, P., Bowen, P., and Borkovec, M., 2005, Aggregation and charging of colloidal silica particles: Effect of particle size, *Langmuir* 21: 5761–5769.

Kovačević, D., Preočanin, T., Žalac, S., and Čop, A., 2007, Equilibria in the electrical interfacial layer revisited, *Croat. Chem. Acta* 80: 287–301.

Lin, W., Kobayashi, M., Skarba, M., Galletto, P., Mu, C., and Borkovec, M., 2006, Heteroaggregation in binary mixtures of oppositely charged colloidal particles, *Langmuir* 22: 1038–1047.

Lützenkirchen, J., 2006, *Surface Complexation Modelling*, in *Interface Science and Technology* series, Academic Press, London, U.K.

Lyklema, J., 1995, *Fundamentals of Interface and Colloid Science*, Volume II: *Solid-Liquid Interface*, Academic Press, London, U.K.

Lyklema, J. and Duval, J. F. L., 2005, Hetero-interaction between Gouy-Stern double layers: Charge and potential regulation, *Adv. Colloid Interface Sci.* 114–115: 27–45.

Lyklema, J., van Leeuwen, H. P., and Minor, M., 1999, DLVO-theory, a dynamic re-interpretation, *Adv. Colloid Interface Sci.* 83: 33–69.

Matijević, E., 2002, A critical review of electrokinetics of monodispersed colloids, in: A. Delgado (Ed.) *Interfacial Electrokinetics and Electrophoresis*, Marcel Dekker, Inc., New York.

Mills, I., Cvitaš, T., Homann, K., Kallay, N., and Kuchitsu, K., 1998, *Quantities, Units and Symbols in Physical Chemistry*, 2nd edn., Blackwell Scientific Publications, Oxford, U.K.

Navarro, E., Baun, A., Behra, R., Hartmann, N. B., Filser, J., Miao, A-J., Quigg, A., Santschi, P. H., and Sigg, L., 2008. Environmental behavior and ecotoxicity of engineered nanoparticles to algae, plants, and fungi, *Ecotoxicology* 17: 372–386.

Semmler, M., Rička, J., and Borkovec, M., 2000, Diffusional deposition of colloidal particles: Electrostatic interaction and size polydispersity effects, *Colloids Surf. A* 165: 79–93.

Shaw, D. J., 2000, *Introduction to Colloid and Surface Chemistry*, 4th edn., Elsevier Science, Oxford, U.K.

Smoluchowski, M., 1916, Versuch einer mathematischen Theorie der Koagulationskinetik kolloider Lösungen, *Z. Physik. Chem. (Leipzig)* 17: 129–135.

Taboada-Serrano, P., Chin, C.-J., Yiacoumi, S., and Tsouris, C., 2005, Modeling aggregation of colloidal particles, *Curr. Opin. Colloid Interface Sci.* 10: 123–132.

Taboada-Serrano, P., Yiacoumi, S., and Tsouris, C., 2006, Electrostatic surface interactions in mixtures of symmetric and asymmetric electrolytes: A monte carlo study, *J. Chem. Phys.* 125: 054716.

Verwey, E. J. W. and Overbeek, J. Th., 1948, *Theory of the Stability of Lyophobic Colloids*, Elsevier, Amsterdam, the Netherlands.

Vorkapic, D. and Matsoukas, T., 1999, Reversible agglomeration: A kinetic model for the peptization of titania nanocolloids, *J. Colloid Interface Sci.* 214: 283–291.

Yates, D. E., Levine, S., and Healy, T. W., 1974, Site-binding model of the electrical double layer at the oxide/water interface, *J. Chem. Soc. Faraday Trans. I* 70: 1807–1818.

29

Liquid Slip at the Molecular Scale

Tom B. Sisan
Northwestern University

Taeil Yi
Northwestern University

Alex Roxin
Columbia University

Seth Lichter
Northwestern University

29.1 Introduction

29.1.1 Observations of Slip

The no-slip boundary condition, which states that liquids and solids share an identical tangential velocity at a liquid–solid interface, is a mainstay of fluid mechanics. A century of macroscopic measurements on liquid flows has consistently confirmed the no-slip condition. However, Navier, who derived the equations for bulk fluid flow in 1823, proposed that fluids could slip relative to solid boundaries, Figure 29.1 (Navier, 1823). Navier's early study hinted that macroscopic measurements might conform to the no-slip prediction while admitting a small but undetectable amount of slip. Contrary to macroscopic flows, a small amount of slip can have serious consequences for nanoscale flows, on the design of nanoscale flow devices, and on our understanding of cellular-level biological flows. As experimental techniques improved, what was formerly undetectable became measurable. Molecular-scale slip has now been measured using optical techniques, inferred from force and flow measurements and confirmed using molecular dynamics simulations (see Table 29.1).

This chapter discusses slip and provides answers to the questions, "What are the molecular mechanisms of slip?" "What types of molecular motions result in slip?" "What are the parameters which control slip?" We start with the question, "Why should we care about slip?"

29.1.2 Applications

For flows through carbon nanotubes, slip can be significant (Holt et al., 2006; Majumder et al., 2005), with flow rates 10's or 100's of times that which would occur if the slip were absent. (See Thomas and McGaughey (2008) for a critique of the above two experimental works. See also Berezhkovskii and Hummer (2002) and Hummer et al. (2004) for molecular dynamics simulations of nanotube slip.) Flows through carbon nanotubes and other micro- and nanoscale channels offer possibilities for separating and sorting solute molecules by size and other properties (Brady-Estévez et al., 2008; Corry, 2008; Formasiero et al., 2008). However, the commercial use of small pores is limited by their extremely high flow resistance. For example, semi-permeable membranes have been used for many years as a means of desalination, but it takes a huge amount of energy to push a city's volume of water through microscopic desalinating pores. The presence of slip and concomitant reduced drag offer the prospect of dramatically reduced energy costs that may make a wide range of separation processes economically feasible (Eijkel, 2007; Eijkel and van den Berg, 2005; Gad-el-Hak, 1999; Sholl and Johnson, 2006; Urbakh et al., 2004).

There is recently renewed interest in the design of low-friction bearings and tribological surfaces that can lead to reductions in energy costs for rotating machinery and other devices in which there are lubricated surfaces moving against one another. Designing surfaces to facilitate slip may lead to advantages in lower friction, lower wear, and longer operating lifetimes (Choo et al., 2007). Additionally, mixing slip and no-slip surfaces can increase the bearings' load capacity (Wu et al., 2006).

Molecular-scale liquid slip comes into play in a wide range of other technologies such as nanoscale mixing and chemical reactors, lab-on-a-chip, integrated chips known as micro-electro-mechanics systems (MEMS), medical diagnostics, chemical and toxin sensing, water filtration, power generation including battery design, synthetic biochannels, drug delivery, nanopipettes, AFMs, and more (Prakash et al., 2008).

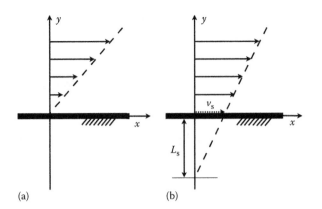

(a)   (b)

FIGURE 29.1 Boundary conditions. (a) For the no-slip boundary condition, the liquid velocity, shown by the arrows, goes to zero at the surface of the bounding solid. (b) A slip boundary. Molecular dynamics simulations, thermodynamic arguments based on mobility, and physical measurements indicate that slip occurs as a discontinuous jump in velocity between the solid surface and the first liquid layer, yielding a slip velocity v_s. The slip length L_s is defined as the distance below the solid surface at which the velocity profile extrapolates to zero.

29.1.3 Other Reviews

Several reviews have appeared that cover different aspects of slip. Potential applications of flow in nanoscale devices, both man-made and natural, can be found in Eijkel and van den Berg (2005). Experimental work is reviewed in Neto et al. (2005) with a compilation of earlier measurements of slip in Lauga and Stone (2003). A broad survey of experimental measurements plus theoretical approaches is available in Lauga et al. (2005), while Ellis and Thompson (2004) focuses on visco-elastic effects. Slip driven by gradients other than pressure or shear as well as the theory of slip near equilibrium via the fluctuation dissipatation theorem can be found in Bocquet and Barrat (2007). A review that focuses on slip at the moving contact line and superhydrophobicity can be found in Voronov et al. (2007). Finally, Braun and Naumovets (2006) review tribological friction, including solid-on-solid sliding and sliding due to adsorbed layers, with numerical simulation results and a phenomenological approach, based on the Frenkel–Kontorova equation, which is similar to the point of view taken here.

29.2 Background

29.2.1 Early Ideas about Slip

In his paper, Navier sets forth not only the equations of fluid mechanics but also its boundary conditions (Navier, 1823). At a solid surface, Navier finds that there is a discontinuity in the speed between the solid and the fluid, where the jump in speed at the liquid–solid interface is proportional to the shear stress, $v_s = \mu(\partial u/\partial y)/b_s$ where b_s is a material property of the solid, $\partial u/\partial y$ is the shear rate, and μ is the bulk viscosity. The ratio μ/b_s has dimensions of length and so can be replaced by another material constant, the *slip length* L_s, which depends on both the fluid and the solid:

$$v_s = L_s \dot{\gamma} \tag{29.1}$$

where we have written the shear rate $\partial u/\partial y$ in terms of its common designation $\dot{\gamma}$. Hence, the search was on to apprehend if (29.1) is the proper form for the boundary condition and, if so, to evaluate the slip length and understand the mechanisms by which slip occurs. The slip length according to (29.1) has a simple geometric interpretation, being the distance below the solid surface at which the velocity profile, if extrapolated, would reach zero, Figure 29.1.

Stokes, in his formulation of what would become known as the Navier–Stokes equations, took a continuum approach (Lamb, 1945). Navier's formulation, on the other hand, though predating atomic theory, was based on particle–particle interactions. Slip, from its initiation, was founded on the molecular nature of matter.

Though this chapter is concerned with liquids, we note that in 1879 Maxwell formulated a slip boundary condition for gaseous flow by considering that the angular distribution of molecules reflecting off a wall could be considered as arising from two populations (Maxwell, 1879; Sokhan et al., 2001). Some molecules are reflected from the wall with no preferred direction; hence, their average contribution to slip is zero. The second population of molecules preserves the stream-wise component of their incident trajectory, as they would in reflecting from a perfectly smooth surface; as the net-incident motion is in the downstream direction, these molecules make a positive contribution to the

TABLE 29.1 Slip Has Been Observed or Inferred from a Diverse Set of Experimental Observations and Numerical Simulations

Method	References
Fluorescent recovery after photo-bleaching (FRAP)	Pit et al. (2000)
Particle image velocimetry (PIV)	Tretheway and Meinhart (2002) and Lumma et al. (2003)
Surface forces apparatus (SFA)	Thompson and Robbins (1990b), Zhu and Granick (2002), Urbakh et al. (2004), Cottin-Bizonne et al. (2005), Zhu and Granick (2001), and Bonaccurso et al. (2003)
Atomic force microscopy (AFM)	Craig et al. (2001) and Vinogradova and Yakubov (2003)
Quartz crystal resonators (QCR)	Krim and Widom (1998) and Krim (2002)
Molecular dynamics simulations (MD)	Koplik et al. (1989), Thompson and Robbins (1990a), and Thompson and Troian (1997)

slip velocity. The net amount of slip depends on the relative amounts of these two populations.

Navier's boundary condition quickly fell into disuse, not because it was shown to be incorrect, but rather because even if it were correct, experimental measurement and practical experience overwhelmingly suggested that the slip length L_s was so small as to lead to the simpler no-slip condition $v_s = 0$ as a sensible approximation. It is easy to understand why this is so. For a pressure-driven flow in a circular tube of radius R, we find that the flow rate is

$$Q_{\text{Poiseuille}} = -\frac{\pi R^4}{8\mu} \frac{dp}{dx} \left(1 + \frac{4L_s}{R} \right) \qquad (29.2)$$

where dp/dx is the pressure gradient along the axis. So, unless the slip length is a significant fraction of the tube radius, its effect on the flow rate would not be measurable.

29.2.2 Renewed Interest in Slip

Interest in slip resurfaced with the demonstration that, when solved using the usual no-slip condition, the moving contact line (as at the edge of a droplet) possesses a non-integrable stress singularity (Huh and Scriven, 1971). A physical consequence of this mathematical singularity is that an infinite force would be required to move a contact line. Hence, raindrops would not slide down windowpanes and divers would be unable to penetrate through the surface of the pool into which they were headed. In order to reduce the stress singularity to a physically meaningful value, it was speculated that within a small patch of surface near the contact line, the no-slip condition needed to be replaced by a slip-boundary condition. Interest in slip was further promoted as new technologies were proposed based on small-scale flows in micro- and nanochannels (Eijkel, 2007) and as physical theories were sought for biological flow in cells (Hummer et al., 2004). It was not that these flows had unduly large slip lengths, rather the device size R had shrunk so that the ratio L_s/R in (29.2) was large enough to make slip-effects significant. Furthermore, new measurement techniques emerged which could measure small speeds close to walls (Neto et al., 2005).

29.2.3 Enhanced Mobility at the Liquid–Solid Interface

Mobility is a measure of how easily liquid molecules move under a given force. Since in many situations, liquid slip occurs at the liquid–solid interface, the mobility of the liquid molecules adjacent to the solid is an active area of study. Tolstoi postulated that enhanced mobility is due to the presence of vacancies, that is, molecular-sized holes (Blake, 1990; Ellis et al., 2003; Sokhan et al., 2001). The thermodynamic properties at a liquid–solid interface can facilitate the creation of these vacancies (deGennes et al., 2002). Hence, the mobility at the liquid–solid interface can be much greater than within the bulk liquid. As the mobility of

the liquid adjacent to the solid can be so much higher than the bulk's, the change in speed from the solid to the first liquid layer can be much greater than the change in speed from one layer to another within the bulk. For the flow of water through carbon nanotubes, the flow appears like a solid plug of water sliding through the nanotube. This large change in velocity across the liquid–solid interface appears as a discontinuous jump, what we call slip.

Hoffman developed a picture (Hoffman, 1983) similar to Tolstoi's, based on Eyring's theory of rate processes (Glasstone et al., 1941), in which liquid molecules hop along the substrate from one minima of the substrate potential to another. As his interest was in contact-line motion, the bias for forward diffusion in Hoffman's model is provided by the difference in the dynamic and the static contact angles. Other researchers have modified and refined these concepts of surface diffusion (de Ruijter et al., 1999; Hayes and Ralston, 1994; Petrov and Petrov, 1992; Ruckenstein and Dunn, 1977).

Mobility at a solid surface has been treated using the Green–Kubo theory, in which time correlations of the liquid density are used to determine the slip length in the limit of small applied shear rate (Barrat and Bocquet, 1999a). The theory predicts a nonzero mobility for the fluid atoms in contact with the wall. The dominant contribution to the slip length is found from liquid molecule correlations within the first liquid layer. This finding is similar to Tolstoi's and Hoffman's implicit assumption that the dynamics are dominantly tangent to the solid. The Green–Kubo predictions were validated by comparison with a molecular dynamics simulation (Priezjev and Troian, 2006). A critical commentary on the theory can be found in Petravik and Harrowell (2007). Underlying all these studies of slip is the unique behavior of the liquid molecules adjacent to the solid surface. Table 29.2 summarizes the models discussed above. We go further into our method of studying this enhanced mobility in Section 29.3.2.

29.2.4 Apparent Slip: Slip due to Gas Enrichment at the Solid

There has been a healthy skepticism about measurements of slip. After all, the no-slip condition has been and remains a stalwart of fluid mechanics. Questions have arisen, especially for hydrophobic surfaces, as to whether the liquid is in contact with the solid or is offset from the solid by an intervening gas. If such an intervening gas region is present, then slip can be ascribed to the usual no-shear boundary condition at a liquid–gas interface, and there is no need to postulate a new slip phenomenon at the liquid–solid interface. This is true even if the gas regions take up only a small portion of the solid surface. In fact, the question has arisen as to whether small pockets of trapped gas, called nanobubbles, may be responsible for the significant variation found between different experimental studies under apparently similar conditions. (Cottin-Bizonne et al., 2005; Huang et al., 2006). Nanobubbles have indeed been observed on hydrophobic surfaces (Tyrrell and Attard, 2001). It has also been shown

TABLE 29.2 Slip Length and Slip Velocity as Determined by the Molecular Models Discussed in Section 29.2.3

Slip length (Blake, 1990)

$$L_s = \delta \left[e^{\alpha S \sigma_{lv}(1-\cos\theta^0)/k_B T} - 1 \right],$$

where

    δ is the average distance between adjacent molecular layers
    S is the effective surface area of the hole
    α is the fraction of S composed of solid surface
    θ^0 is the equilibrium contact angle
    σ_{lv} is the liquid–vapor surface tension
    k_B is the Boltzmann factor
    T is the temperature of the liquid

Slip speed at the moving contact line for wetting liquids (Hoffman, 1983)

$$v_s = 2 l_{hop} \left(\frac{k_B T}{h} \right) \exp \left(\frac{-G}{k_B T} \right) \sin h \left\{ \frac{\sigma_{lv}(\cos\theta^0 - \cos\theta_{dynamic})}{N k_B T} \right\},$$

where

    h is the Planck's constant
    $\theta_{dynamic}$ is the dynamic contact angle
    G is the activation energy for a liquid molecule to hop the distance l_{hop} between adjacent adsorption sites on the solid surface (with no forcing applied)
    N is the number density per area of liquid–gas interface
    other variables are as in the entry above

Slip length via the fluctuation dissipation theorem (Barrat and Bocquet, 1999b)

$$L_s \sim \frac{D_{q\parallel}^*}{S_l(q) c_{LS}^2 \rho_c \sigma^2},$$

where

    $0 \leq 0.5 c_{LS} \leq 1$ measures the strength of interaction between the liquid and the solid
    ρ_c is the density of the first liquid layer
    σ is a measure of the size of liquid molecules
    $S_l(q)$ is the structure function of the first liquid layer
    $D_{q\parallel}^*$ is the in-plane diffusion coefficient

TABLE 29.3 Models of Slip Length in the Presence of a Gas Film between the Liquid and the Solid

Knudsen-regime gas layer (deGennes, 2002)

$$L_s = \left(\frac{2\pi m}{k_B T} \right)^{1/2} \left(\frac{\mu}{\rho} \right),$$

where

    ρ is the density of the gas of molecular mass m
    μ is the viscosity of the liquid
    k_B is the Boltzmann's constant
    T is the temperature

Reduced-viscosity model (Vinogradova, 1995)

$$L_s = h_g \left(\frac{\mu}{\mu_g} - 1 \right)$$

where

    h_g is the thickness of the gas film
    μ_g is the viscosity of the gas film
    μ is as in the entry above

It seems reasonable that otherwise carefully controlled experiments, purportedly with a liquid–solid interface, may have had intervening gaseous artifacts. It has been shown that for pressure-driven flow through tubes and channels, assuming that the walls are contaminated by nanobubbles reproduces the observation that slip length scales approximately with device size (Lauga and Stone, 2003). However, other observations seem immune from problems with nanobubbles: carbon nanotubes show huge slip lengths; experiments with degassed liquids show slip; molecular dynamics studies show true-liquid slip; thin mono- or bi-layers of physisorbed atoms show slip in quartz-crystal microbalance experiments; and, even strongly hydrophilic surfaces, for which nanobubbles are unlikely, require a means to remove the singularity at the moving contact line. It is to the description of genuine liquid–solid slip that we continue our exposition.

29.3 Dynamics of Slip

We set out a minimalist atomistic model of liquid slip that shows, albeit in a simple form, the mechanisms of slip, the presence of distinct regimes of slip, how and where transitions occur from one type of slip to another, and how parameters can be chosen to enhance or suppress slip. Note that though (29.1) already provides a simple expression for slip, it yields no means to evaluate the amount of slip. That is, (29.1) can not answer, "What is the value of L_s and how does it depend on the parameters?"

29.3.1 Molecular Structure near a Solid

To summarize the ideas presented in Section 29.2, it has been found that the mobility in the layer adjacent to the wall is much greater than that in the bulk. Hence, slip preferentially occurs between the wall and the liquid adjacent to it (and not in a diffuse liquid layer extending several molecular diameters into the liquid). This view is also supported by molecular dynamics simulations,

that dissolved gas will preferentially migrate to a solid surface to form, in the case of hydrophobic walls, a molecularly thin layer enriched over the bulk value by up to two orders of magnitude in concentration of the dissolved gas (Dammer and Lohse, 2006). The presence of a gaseous layer at the wall can considerably enhance the amount of slip, see Table 29.3.

The excess concentration of gas at the wall described above occurs over flat surfaces and has been investigated, in part, to isolate artifacts in experimental measurements of slip. On the other hand, the geometry of the wall may be purposefully designed to trap the intervening gas between the liquid flow and the wall. Micro-textured surfaces, often referred to as super-hydrophobic surfaces, with arrays of pillar-like structures at different length scales, forming a fractal-like surface, have been found to have dramatic effects on slip in nanochannels (Feuillebois et al., 2008), as well as on flows with moving contact lines (Patankar, 2004; Quéré, 2008; Rauscher and Dietrich, 2008).

in which it is frequently observed that the continuum equations of Newtonian fluid dynamics hold within the interior of a liquid, and that slip is entirely due to a discontinuity in speed adjacent to the solid (Koplik et al., 1989; Thompson and Troian, 1997).

Furthermore, liquid density is not constant near a molecularly smooth surface. Rather, the liquid tends to align itself in a layer-like structure, as has been observed in physical measurement (Mo et al., 2005; Yu et al., 2001) and in molecular dynamics simulations (Koplik et al., 1989) (see Figure 29.2). Thus, it is sensible to use the high density relative to the bulk density to identify a first liquid layer adjacent to the solid, and to single it out for further analysis.

Within the first liquid layer, the density is non-uniform along the solid surface (Figure 29.2). Liquid molecules spend most of their time at positions in between solid atoms, where the potential energy is minimum (as described in Section 29.3.2). These locations are energetically favored by the liquid molecules. If a vacancy forms at such a location, a neighboring liquid molecule, pushed by shear and by interactions with neighboring liquid molecules, will preferentially hop into this location.

Taking together the concepts and observations regarding slip, we find that slip is a phenomena (a) occurring within the first liquid layer, and (b) relying on the increased mobility within the first liquid layer. We now formulate these concepts into a

dynamical model of slip, which reveals that the increased mobility occurs via the creation of vacancies in the first liquid layer and the hopping of the liquid molecules along the solid substrate into newly created vacancies. Vacancies are observed, in molecular dynamics simulations, to be the most common initiator of slip. Much less frequent is an excess of molecules crowding into one site, leading to the hopping of the extra molecule along the surface, continuing until it diffuses into the bulk and restores a stable commensuration ξ (see Section 29.3.4.1). Vacancies and crowdings together are referred to as defects.

In this presentation, to further simplify the problem, we consider that the flow geometry is two-dimensional and the first liquid layer is reduced from two dimensions to simply a single-file chain of liquid molecules driven along a one-dimensional solid surface, yielding Frenkel–Kontorova dynamics.

29.3.2 Frenkel–Kontorova Dynamics

How do the forces on the N molecules in the first liquid layer determine the positions x_i of these $i = 1 \ldots N$ molecules as a function of time t, Figure 29.3? The one-dimensional first liquid layer is subject to the following forces:

1. Shear: Molecules in the bulk collide with molecules in the first layer. Momentum transfer from the bulk contributes a force $\eta_{LL}\,\dot{\gamma}$, where η_{LL} is the bulk viscosity times the interfacial area per liquid molecule, and the shear rate is

$$\dot{\gamma} = \frac{V - \dot{x}_i}{\delta} \tag{29.3}$$

where δ is the mean spacing between the first liquid layer and the layer above it moving at the mean velocity V.

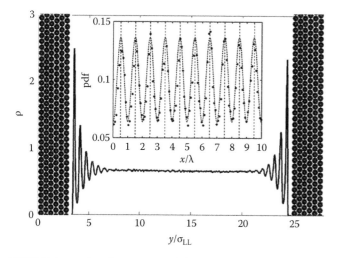

FIGURE 29.2 The inhomogeneity of the liquid density as determined by two-dimensional molecular dynamics simulation usings LAMMPS (Plimpton, 1995). The main figure shows the variation in liquid density ρ as a function of height y across a Couette channel. The height is non-dimensionalized with respect to the molecular size σ_{LL}. Bounding the flow are solid walls composed of four layers of atoms, as shown. Especially prominent are the large variations in density close to the walls. The liquid molecules closest to the wall—out to the y-distance of the first minimum in density—comprise what is called the first liquid layer. The inset shows the variations of density, in terms of the probability density function (pdf), within the first liquid layer as a function of distance x along the channel wall. The solid–sine curve is a fit to the data. The thin vertical lines show the positions midway between the locations of the solid atoms that compose the layer adjacent to the liquid. Liquid molecules spend most of their time in positions between solid atoms.

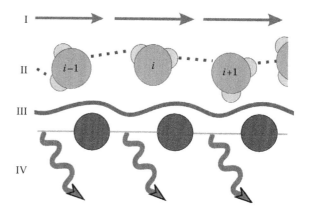

FIGURE 29.3 The formulation of the Frenkel–Kontorova model (29.4). Three liquid molecules, labeled $i - 1$, i, $i + 1$, of the N along the solid surface, are shown. The four forces acting on them are labeled I–IV, following the numbering used in Section 29.3.2. The chain of liquid molecules is forced by a transfer of momentum from the overlying liquid moving at speed V, as indicated by the arrows, I. The interactions between neighboring-liquid molecules is shown as the dotted lines, II. The horizontal sinusoidal line III is a schematic of the periodic variations in the substrate potential due to the solid atoms. The motion of the liquid over the solid transfers momentum to the solid, IV, which is dissipated as heat.

2. Nearest neighbors: Molecules in the first liquid layer collide with each other. This force could be modeled with a Lennard–Jones, dipole–dipole, or coulomb force as appropriate, as is done generally in molecular dynamics simulations. However, we use a simple spring force between nearest-neighbor molecules. A spring force affords the advantage of easier analytical approaches and is suitable to this introduction. So, the molecule at x_i is subject to a force $k(x_{i+1} - (x_i + a))$ from its neighbor to the right plus a force $k(x_{i-1} - (x_i - a))$ from its neighbor to the left where k is the magnitude of the spring constant and a is the mean spacing of the liquid molecules.

3. Substrate: As the liquid molecules move over the atoms which compose the solid surface they feel a washboard-like force. For a crystalline solid with constant lattice spacing λ, the x-component of the interaction force can be written as a Fourier series (Steele, 1973), which we truncate at the first term, yielding a force $(2\pi h/\lambda) \sin(2\pi x_i/\lambda)$, where h is the strength of the interaction.

4. Friction: There is a viscous damping of the first liquid layer stemming from its collisions with the solid, transferring its energy into randomized lattice vibrations and electronic excitations in the solid. This is modeled by a frictional force $\eta_{LS}\dot{x}_i$, proportional to the molecular speed with a friction factor η_{LS} (Pit et al., 2000).

These forces are illustrated graphically in Figure 29.3. Putting the four forces together into Newton's equation of motion yields,

$$m\ddot{x}_i = \eta_{ff}(V - \dot{x}_i) + h\sin\left(\frac{2\pi}{\lambda}x_i\right) - \eta_{fs}\dot{x}_i + k(x_{i+1} - 2x_i + x_{i-1})$$

(29.4)

This equation is known as the driven-damped Frenkel–Kontorova equation (Braun and Kivshar, 1998).

29.3.3 Simulating Systems Using the FK Model

Modeling any system computationally is an art which requires, among other things, choosing the number of atoms and molecules to include in the simulation, and how to handle the boundaries of the system—the interaction of the system with the rest of the world. In representing large-scale systems with vast numbers of molecules, end effects are negligible and so to reduce computations as much as possible, a smaller system with periodic boundary conditions is used. For small-scale systems, such as the example of flow in short carbon nanotubes to be discussed later, more care is needed in choosing the boundary conditions at the entrance and exit of the nanotube. In using the FK model, an additional challenge is in properly reducing the system to one dimension. First, the potential height h must be chosen so that the substrate force experienced by the molecules in one-dimension is analogous to those experienced in three-dimensions. Secondly, the motion of the molecules in the first liquid layer perpendicular to the direction of motion

must be accounted for (see Section 29.3.4). We now describe some of the important concepts for the one-dimensional FK model.

29.3.3.1 FK Ground State

Let $N(t)$ be the instantaneous number of liquid molecules in the first liquid layer and q be the number of substrate wavelengths. Then, $\xi = q/N$ is a measure of the frustration or *commensurability* of the liquid vis-a-vis the solid. When ξ = integer, there is one liquid molecule per integral number of substrate sites, and the liquid molecules tend to reside in the minima of the potential energy between the locations of the solid atoms. When ξ = non-integer, the liquid molecules are perturbed from the minima of the substrate potential. For any commensurability, there is a lowest-energy configuration, which is called the ground state.

29.3.4 vdFK Model: Adding Diffusive Flux to the FK Model

We now modify the FK model by allowing the number of molecules in the first liquid layer to fluctuate, which we term the variable-density Frenkel–Kontorova (vdFK) model.

29.3.4.1 vdFK Ground State

Though the liquid–solid interaction tends to fix the number of liquid molecules at a particular value, there are fluctuations about that value due to diffusion normal to the layer. The very definition of a liquid state demands that the number $N(t)$ of molecules in the first liquid layer vary with time. A liquid molecule, now at a particular location, will soon be elsewhere due to diffusion into and out of that location. Consider, for example, a portion of a solid with 100 sites covered by 100 liquid molecules. The liquid molecules will preferentially occupy the low-energy sites, Figure 29.2 (Lichter et al., 2004; Thompson and Robbins, 1990a). If diffusion into the bulk removes one liquid molecule from the first liquid layer, then ξ changes from 100/100 to 99/100, Figure 29.4. This slight mismatch is critical to slip. When the liquid is mismatched to the solid, slip can commence at low-shear rates, as we discuss in Section 29.3.6.4. The vdFK model then, has a most-likely lowest-energy value of ξ, but ξ changes during the course of a simulation—each value of ξ with its own ground state, and slip velocity (as we show later).

As described in the caption to Figure 29.4, defects can be created and destroyed by a particle flux into and out of the first liquid layer. The number of defects D, is given by $D = q - N(t)$. Particle flux can create more defects or destroy currently present defects.

The diffusive flux of molecules into and out from the first liquid layer can be made equivalent to a random walk. Consider that there are N_0 molecules in the first liquid layer's ground state. Now, consider the difference $Q(t) = N(t) - N_0$ from one time step to the next: changes of $(+1, -1, 0)$ atoms are analogous to (right, left, in-place) random steps, and so summing the

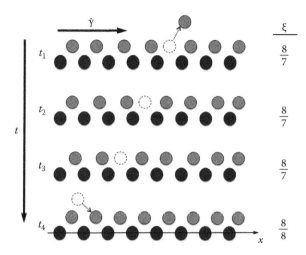

FIGURE 29.4 The flux of liquid into and out of the first liquid layer acts to initiate and terminate defect slip. The darker circles represent the array of atoms in the solid. Shown here is a schematic of the propagation of a defect due to the drift of a molecule from the first liquid layer into the bulk. Before time t_1, $\xi = 1 = 8/8$: there are eight liquid molecules, lighter circles, over a solid substrate of eight solid atoms, so each liquid can reside in the low-energy site between two solid atoms. At time t_1 a molecule drifts out of the first liquid layer leaving a vacancy, shown by the dashed circle, and $\xi = 8/7$. Under the action of the shear $\dot{\gamma}$ and the liquid–liquid interaction (not shown in this figure), the molecule upstream of the vacancy hops into it, creating a vacancy at its just-vacated site. The next molecule upstream can hop into this newly created vacancy. The sequential hopping results in the upstream propagation of a vacancy and a downstream propagation of a liquid molecule. This is the basic mechanism of defect slip, in which there is a net flux of mass downstream along the solid. The propagation of the defect terminates at time t_4 when, by chance, a molecule diffuses in from the bulk, annihilating the vacancy and restoring $\xi = 1$. In a similar process, an extra molecule can initially diffuse into the first liquid layer, leading to a crowding defect.

changes in $Q(t)$ is analogous to the distance traveled in a random walk. Random walks are often portrayed in terms of a drunk staggering to the right or left from a lamppost. We can give the drunk a ground state by tethering him to the lamppost with an elastic tether. If he strays too far from the lamppost, his further excursions are made less likely. Hence, we arrive at our model of diffusion normal to the solid, in which the probability of the first liquid layer losing or gaining a molecule at any time depends on how far the first liquid layer is already from the ground state:

$$P_+ = P_0 + \alpha Q \cdot H[-Q], \quad P_- = P_0 + \alpha Q \cdot H[+Q] \quad (29.5)$$

where

$P_+(P_-)$ is the probability of adding (removing) a molecule
Q is the current surplus of molecules (a negative Q indicates a deficit of molecules)
$H[n]$ is the heaviside step function that is unity for positive argument and zero otherwise

P_0 is the probability of adding or removing a molecule in the absence of the ground state, and α measures the strength of bias imposed by the ground state, that is, the strength of the tether.

Estimates for the parameters P_0 and α can be obtained from molecular dynamics data. There are many ways to model diffusion with a ground state, but (29.5) is one of the simplest (Martini et al., 2008b; Roxin, 2004).

It is computationally and analytically expedient to rearrange (29.4) into a non-dimensional form:

$$\ddot{x}_i = \tilde{f} + \sin(x_i) - \tilde{\eta}\dot{x}_i + \tilde{k}(x_{i+1} - 2x_i + x_{i-1}) \quad (29.6)$$

where $\tilde{f} = \tilde{\eta}_{LL}\tilde{V}$, $\tilde{\eta} = \tilde{\eta}_{LL} + \tilde{\eta}_{LS}$, and the tilde over bar indicates a non-dimensional parameter. Non-dimensional length is measured in units of $\lambda/(2\pi)$, and non-dimensional time is measured in units of $(\lambda/(2\pi))\sqrt{m/h}$.

Equations 29.5 and 29.6 make up the vdFK model. Results of the vdFK model are obtained through computer simulations. Properties associated with slip are obtained by averaging the molecule velocities over time.

As with molecular dynamics simulations, the vdFK model is a dynamic model of molecular motion, in which trajectories $x_i(t)$ of the liquid molecules are computed from the forces acting on each molecule. A significant difference between the vdFK model and molecular dynamics simulations is that the vdFK model reduces a three-dimensional computation to the one-dimensional problem of solving for the dynamics of the first liquid layer. Once the slip is determined from the one-dimensional problem, the remainder of the flow field can be computed using the continuum equations of fluid mechanics. The reduction in dimensionality yields a huge savings in computational effort. Additionally, analytical approximations and simple-scaling results of the vdFK model can be easily found, as are discussed in Section 29.3.6.

29.3.5 Converting between vdFK and MD Units

Results from the vdFK model can be compared with molecular dynamics simulation results and with physical experiments. To make this comparison possible, the parameters in the vdFK model need to be converted into the parameters used in numerical simulations or those available to experimentalists (Roxin, 2004). Molecular dynamics simulations typically use the energy scale ϵ_{LL} (ϵ_{LS}) between liquid (liquid and solid) molecules to characterize the strength of the attraction, and use $\sigma_{LL}(\sigma_{LS})$, the Lennard–Jones distance at which the liquid–liquid (liquid–solid) intermolecular potential is equal to zero, as the length scale.

The strength of the liquid–liquid interaction $\tilde{k}$, used in the Frenkel–Kontorova model, can be expressed in molecular dynamics parameters as

$$\tilde{k} = \frac{\lambda^2}{4\pi^2 h} k = \frac{\lambda^2}{4\pi^2 h} \frac{72}{2^{4/3}} \frac{\epsilon_{LL}}{\sigma_{LL}^2} \quad (29.7)$$

The amplitude of the wall potential, h, is determined by first computing the potential at a point (x, l) a distance l above the solid array of Lennard–Jones atoms spaced λ in x and s in y:

$$V_{LJ}(x) = 4\epsilon_{LS} \sum_{i=0}^{M_L} \sum_{j=-M}^{M} \left[\left(\frac{\sigma_{LS}}{r_{ji}} \right)^{12} - \left(\frac{\sigma_{LS}}{r_{ji}} \right)^{6} \right] \quad (29.8)$$

where $r_{ji} = (x - j\lambda)^2 + (l + is)^2$. The index j counts the $2M$ atoms along a particular layer, and i indexes the M_L solid layers that are separated by the distance s. The cosine transform of $V_{LJ}(x)$ is computed and the coefficient for the lowest mode is set equal to h.

The non-dimensional forcing parameter is expressed in terms of dimensional parameters as

$$\tilde{f} = \frac{\eta_{LL} \lambda}{2\pi} \frac{V}{h} \quad (29.9)$$

where the two-dimensional viscous coefficient is approximated as $\eta_{LL} = \mu\sigma$ and the wall potential h is determined as above.

The forcing $\tilde{f}$, which, as seen from (29.9), is proportional to the speed V of the adjacent bulk liquid, is not directly under experimental control. However, mathematically, $\tilde{f}$ is the bifurcation parameter governing the character of solutions to the Frenkel–Kontorova equation (Braun and Kivshar, 1998). So, we use $\tilde{f}$ in discussing results from the vdFK model. In typical macroscopic applications, the shear rate is used as a control parameter. For example, in a Couette flow apparatus, the shear rate can be set by choosing the particular speed at which the walls move. But, for flows with slip, specifying the wall speed does not, a priori, specify the shear rate. It can be seen from (29.3) that the shear rate depends on both V and the amount of slip v_s. Hence, in contrast to macroscopic flows, for small-scale flow, shear rate may not be a useful control parameter. When we present molecular dynamics results, we use the *effective* shear rate $\dot{\gamma}_{eff}$, which is a controllable parameter.

The damping rate η_{LS}, due to the solid, is difficult to relate to the properties of the numerical parameters. Nonetheless, it is important to correctly evaluate this parameter, as the amount of slip in the high-shear rate limit depends directly on its value [see (29.11)]. Presently, η_{LS} is often used as a fitting parameter, but methods have been presented to evaluate it from first principles (Persson, 1998) or from numerical simulation data (Sokhan et al., 2001, 2002).

It should be recalled that the model presented here is for a two-dimensional flow–where the transverse direction is just one molecule wide–which is achievable in a numerical experiment, but is not feasible for physical experiments. So, when comparing with physical experiments, the results here serve only as a guide.

29.3.6 Results from the vdFK Model

The flux from the bulk into the first liquid layer, as described by (29.5), can initiate and terminate slip (see Figure 29.4). However, once initiated, the dynamics of slip is independent of flux. That

is, we can regard N as fixed while solving (29.6), until (29.5) dictates that N be changed. With this simplification, we now take a look at (29.6) in several cases in which simple solutions can be found, or which illustrate critical concepts about slip (Braun and Kivshar, 2004).

29.3.6.1 Three Regimes of Slip

One important result from the vdFK model is the presence of transitions between the three different regimes of slip: no slip, defect slip, and global slip. In the no-slip regime, the liquid atoms are trapped by a strong liquid–solid interaction in their ground state. In the defect slip regime, only the defects can move, as shown in Figure 29.4. In the global slip regime, all molecules slide over the solid together, regardless of ξ. The three regimes are present in Figure 29.5.

29.3.6.2 ξ = Integer

At low values of the forcing parameter $\tilde{f}$, there are no propagating solutions to (29.6). Hence, there is no slip: liquid molecules remain trapped near the minima of the substrate potential. What happens as $\tilde{f}$ is increased depends on the commensurability ξ. When there is an integer number of substrate wavelengths λ per liquid molecule, $\xi = n$ = integer, all the molecules are evenly spaced. Hence, the force due to liquid–liquid interactions is zero. The maximum force on a molecule from the substrate is maximum$(-\partial U_{sub}/\partial x_i) = 1$. So, slip is zero until $\tilde{f}$ rises past the critical value $f_c = 1$. Then, all liquid molecules become dislodged and move together in global slip.

The slip velocity for global slip at forces just above the critical force, $\tilde{f} = 1 + \epsilon$, can be calculated as follows. Considering $\epsilon \ll 1$,

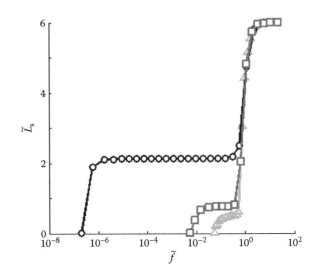

FIGURE 29.5 vdFK results showing slip length as a function of shear rate. At low enough forcing there is no-slip, $\tilde{L}_s = 0$. The first plateau on each curve is due to defect slip. The transition from no-slip to defect slip occurs over a range of forcings $\tilde{f}$ that covers six orders of magnitude, though $\tilde{k}$ varies only by a factor of ten: $\tilde{k} = 0.5 \,\triangle$, $1 \,\square$, $5 \,\bigcirc$. The second plateau is due to global slip. The transition to global slip is insensitive to the value of $\tilde{k}$ and occurs for all three curves at $\tilde{f} = 1$. At high forcings, the slip length asymptotes to the value given by (29.11).

the molecules move slowly along the substrate. The sinusoidal substrate potential yields a force that alternates in sign, resulting in the molecule moving faster when the substrate force and applied force, $\tilde{f}$, are in the same direction, and slower when the substrate force opposes $\tilde{f}$. Most of a molecule's time is spent near the point where the force from the substrate potential maximally opposes the applied force $\tilde{f}$. The total time taken for a molecule to traverse one wavelength of the substrate, is approximately the time taken to traverse this bottleneck portion of the trajectory. This gives the slip velocity for $\tilde{f}$ near 1 as

$$\tilde{v}_s = \frac{\sqrt{2(\tilde{f}-1)}}{\tilde{\eta}} \qquad (29.10)$$

29.3.6.3 Slip at High Shear Rates

The slip velocity computed from (29.10) is valid just after the onset of slip. As the forcing increases, those terms in (29.4) that are proportional to the slip velocity or to the forcing become increasingly larger than the other terms, and we are left with a limiting approximation for the high-shear rate slip velocity

$$v_s \equiv \langle \dot{x}_i \rangle = \frac{\eta_{LL} V}{\eta_{LL} + \eta_{LS}} \qquad (29.11)$$

From this slip velocity and the shear rate (29.3), the asymptotic value of the slip length at high shear is found to be

$$L_s(V \gg 1) = \frac{\eta_{LL}}{\eta_{LS}}\delta \qquad (29.12)$$

In contrast, some molecular dynamics studies of slip find that the slip length is unbounded at high shear. In those studies, the solid is composed of atoms fixed into place and so there is no capacity for the solid to absorb energy due to collisions with the liquid, resulting in an absence of damping. Then by (29.12), as $\eta_{LS} \rightarrow 0$, $L_s \rightarrow \infty$. On the other hand, real walls have a finite η_{LS} and hence a bounded slip as suggested by (29.12). The lesson here is that in order to accurately model high shear rate flows, the solid must faithfully reproduce the heat transfer from the liquid to the solid (Martini et al., 2008a).

29.3.6.4 $\xi \neq$ Integer

Even if we carefully choose the parameters in order to achieve ξ = integer, this value will not prevail. Recall that there is a continual flux of molecules from the bulk, and the number $N(t)$ of molecules in the first liquid layer is continually changing. So, the instantaneous value of ξ, defined in terms of the number of liquid molecules relative to the number of substrate sites, changes as new molecules arrive from the bulk or depart from the first liquid layer [see (29.5)].

When $\xi \neq$ integer, a second type of slip occurs at a much smaller value of the forcing $\tilde{f}$. Consider what happens when a molecule drifts into the bulk from the first liquid layer with $\xi = 1$. Now, the number of molecules becomes one less than the number of low-energy sites on the substrate. Near the location where the molecule is removed, the liquid–liquid interaction increases, pushing the molecules toward the vacant minima of the substrate potential. While, at a large distance from the location of the vacancy, the liquid–liquid interaction remains nearly unaffected, and the liquid molecules remain at the substrate potential minima, Figure 29.4.

An antikink (or kink) is the name given to the localized perturbations in displacement surrounding a vacancy (or crowding defect). The width of the kink or antikink is the spatial extent of the disturbance in molecular positions on inserting (or removing) the extra molecule. As you may intuitively guess, a larger value of the spring constant $\tilde{k}$ produces a wider defect. If $\tilde{k}$ is not too large, the defect is schematically as shown in Figure 29.4, where the two molecules on either side of the removed molecule have measurable displacements away from their perfect registry with the solid, while the other molecules remain nearly unperturbed by the presence of the defect.

So, in defect slip, most of the molecules remain stationary (subject to thermal motion) while only the few molecules within the small width of the defect translate. The change in inter-molecular spacing, and the consequent perturbation of the nearby molecules from the substrate potential minima, aids the translation of the defect. Consequently, the defect motion commences at a lower forcing than does the global slip (Figure 29.5). For systems in which the liquid–liquid interaction is large (i.e., in which $\tilde{k}$ is large, as is typically found for liquids), the critical forcing for defect slip is (Braun and Kivshar, 1998)

$$\tilde{f}_{PN} = 16\pi\tilde{k}e^{-\pi^2\sqrt{\tilde{k}}} \qquad (29.13)$$

The subscript in the equation above refers to the Peierls–Nabarro barrier (Braun and Kivshar, 1998). The Peierls–Nabarro energy is the minimum energy needed to adiabatically switch the stable-equilibrium configuration of liquid molecules to the higher-energy unstable configuration. Once $\tilde{f} > \tilde{f}_{PN}$, the work done over one substrate wavelength by the forcing becomes greater than the Peierls–Nabarro energy, and the defect can overcome the energetic barrier of the hills and troughs of the substrate potential and propagate through them, Figure 29.6 (Coopersmith and Fisher, 1983; Frank and van der Merwe, 1949).

29.3.6.5 Solitons

We have found that when a defect is present, it can hop from site to site under a relatively small forcing. As the defect moves forward one substrate wavelength, the configuration of molecules returns to its initial configuration, just shifted one substrate wavelength downstream. So, then, the process of hopping can recur sequentially across multiple sites, Figure 29.4.

Here, we consider a liquid with a density that is close to but not equal to $\xi = 1/$integer. Each hop of the defect along the solid adds to the amount of slip. So, the slip length will be proportional to the speed of the defects. To find this speed, it is convenient to take (29.4) to its continuum limit. This limit can be viewed as reducing the spacing between molecules, while

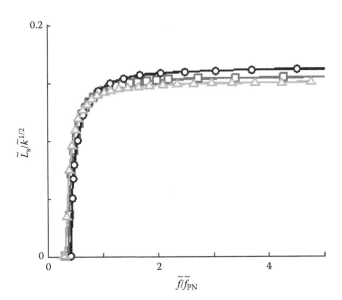

FIGURE 29.6 The vdFK results, from Figure 29.5, at the low forcings for which defect slip takes place, can be rescaled so that they lie nearly along a single curve. Slip length in the defect regime is proportional to $\tilde{k}^{1/2}$ as discussed in Section 29.3.6.5. The forcing $\tilde{f}$ scales with the Peierls–Nabarro force needed to propagate a defect, (29.13). For an explanation of the symbols see Figure 29.5.

increasing the inter-molecular interaction (described by k). In this limit, the discrete molecules merge into a continuum that can be described by the sine–Gordon partial differential equation (Braun and Kivshar, 1998)

$$\frac{\partial^2 U}{\partial t^2} - \frac{\partial^2 U}{\partial x^2} + \sin U = \tilde{f} - \tilde{\eta}\frac{\partial U}{\partial t} \qquad (29.14)$$

where the continuum displacement $U(x)$ has replaced the set of discrete displacements x_i, and has been non-dimensionalized. The propagating defects of the discrete Frenkel–Kontorova Equations 29.4 appear in the sine-Gordon equation as solitons, localized propagating waves. The soliton velocity is given by McLaughlin and Scott, 1978

$$\tilde{V}_{\text{defect}} = \frac{1}{\sqrt{1 + (4\tilde{\eta}/(\pi\tilde{f}))^2}} \qquad (29.15)$$

As long as the density of defects is low enough to keep the defects from interacting strongly, the slip velocity is proportional to the number of defects, that is, it is linear with $1/\xi$. This can be seen in Figure 29.7, in which the line stemming from the points (ξ = integer, $v_s = 0$) are linear. When $1/\xi$ is far from an integer value such that the kinks interact, the dynamics can be computed using an approach based on the hull function (Coopersmith and Fisher, 1988).

From (29.15), as $\tilde{f} \to 0$, $\tilde{V}_{\text{defect}} \sim \tilde{f}$. We rewrite (29.1) as $L_s = v_s/\dot{\gamma}$, and note that $\dot{\gamma}$ is a measure of the forcing $\tilde{f}$. Hence, for small forcing, $L_s \sim v_s/\dot{\gamma} \sim \tilde{V}_{\text{defect}}/\tilde{f} \sim \tilde{f}/\tilde{f} \sim 1$. So, we find that as both the soliton speed and the shear are proportional to the forcing, the

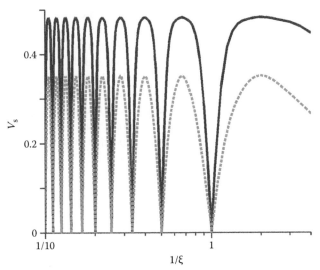

FIGURE 29.7 Slip velocity vs. $1/\xi$ for a forcing amplitude f that is less than f_c for global slip. When the commensurability ξ is integer, the curves touch down to a zero-slip velocity. However, when defects are present, defect slip occurs in proportion to the number of defects. So, on either side of a zero value, slip velocity increases linearly. As ξ becomes still farther from an integer value, defects no longer contribute additively and account needs to be taken for defect–defect interactions. There are an infinite number of points where slip is zero. Only a finite number are shown here. The dark solid curve is at a higher k than the light dashed curve.

slip length will be constant. This simple scaling agrees with the usual observation that over the range for which defect slip occurs, the slip length is constant, as can be seen from the vdFK results, Figure 29.5, as well as in molecular dynamics simulations, Figure 29.8, (Koplik et al., 1989; Lichter et al., 2004; Thompson and Troian, 1997).

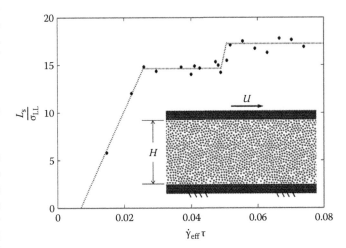

FIGURE 29.8 Slip length as a function of effective shear rate for the Couette flow shown in the inset. The effective shear rate, U/H, is non-dimensionalized with respect to the time scale τ of the molecular vibrations. After increasing from a zero slip length at low shear rate, there are two plateaus in the value of the slip length. The lower one is due to defect slip and the upper one is due to global slip (see Figure 29.9).

A back-of-the-envelope approximation can illustrate the importance of defects. We consider a 7 nm diameter nanotube. Liquid adjacent to the nanotube wall would experience periodic variations in the substrate potential of approximately 10^{-22} J with a corrugation wavelength of approximately 2 Å. For water, slip would commence when the parameter f reached approximately 1 fN. For a nanotube of 100 μm length, this would be achieved by a pressure drop of over 10^5 atm. However, (29.13) shows that the force required to move a single defect in a water-molecule chain (for which $\tilde{k} \sim 20$) is 16 orders of magnitude smaller.

Figures 29.5 and 29.6 can be used to summarize the characteristics of slip as revealed by the vdFK model. Figure 29.5 shows the slip length as a function of the forcing for three values of the non-dimensional $\tilde{k}$ which measures the strength of the liquid–liquid interaction k relative to the liquid–solid interaction h. As discussed in Section 29.3.6.2, the transition to the global slip depends only on the forcing and the strength of the liquid–solid interaction. As we see, the transition from defect slip to global slip, at $\tilde{f} \simeq 1$ is independent of $\tilde{k}$. However, the transition from no-slip to defect slip varies by approximately six orders of magnitude for the three values of $\tilde{k}$ shown. After the transition to defect slip, there is a plateau in the value of the slip length that is different for each curve. The vdFK model provides the scaling parameters to collapse these curves onto one master curve. The transition point to defect slip scales with $\tilde{f}_{PN}$ and, from (29.15) after restoring dimensions, the amount of slip $L_s \sim k^{1/2}$. Using these characteristic values to scale the results in Figure 29.5, we find that the curves fall nearly along a single curve, Figure 29.6.

29.3.7 Designing for Slip

The availability of a dynamical model for slip, simple scaling relationships, and closed-form results for limiting cases allows us to identify the parameters that can aid or abet slip. Experimentalists can use these results as rules of thumb for designing or improving nanoscale devices. We summarize these results in Table 29.4.

29.3.8 Molecular Dynamics Results

Molecular dynamics simulations confirm the findings from the vdFK model. Figure 29.8 shows the slip length versus the effective shear rate for the Couette flow geometry shown in the inset. The effective shear rate $\dot{\gamma}_{eff}$ is the speed of the top wall U divided by the height H of the channel. While the shear rate $\dot{\gamma}$ in the central part of the channel is often used as the independent variable in these types of plots, as we noted in Section 29.3.5, $\dot{\gamma}$ is not at the control of an experimentalist, while $\dot{\gamma}_{eff}$ is. Figure 29.9 shows that the molecular trajectories from molecular dynamics simulations match that given by vdFK simulations: low values of forcing slip occur by the propagation of defects, while at higher levels of forcing, global slip occurs. Taken together, we see that there is (1) a range of shear rates with a defect slip and a nearly constant slip length, (2) a transition to a region of global slip which (3) asymptotes to a higher value of slip length. These three characteristics from molecular dynamics simulations match the predictions from the vdFK model.

29.3.9 Other Aspects of Slip

The discussion here has been concerned with the flow of simple liquids over molecularly smooth surfaces. There are many other considerations that affect slip and may be used to aid in the design of slip devices. For example, surface texture may directly affect slip (Panzer et al., 1992; Priezjev and Troian, 2006). There are many types of driving forces that could have been used to generate slip at the solid surface (see Table 29.5). It is likely that the vdFK formalism presented here is a suitable model to describe these other forcing mechanisms. Also, there is evidence that the concepts developed here remain applicable to more complex liquids, while showing a range of additional phenomena (Ma et al., 1995; Martini et al., 2008b; Priezjev and Troian, 2004; Strunz and Elmer, 1998).

TABLE 29.4 What Can Be Done to Increase Slip?

Slip Length Increases As	Comments
$\eta_{LL} \uparrow$	Increasing the viscosity will increase the slip length, as can be seen in (29.11), which pertains to very high shear rates
$\eta_{LS} \downarrow$	Reducing the interaction between the liquid and the solid will increase the slip length. For instance, polar molecules may interact stronger on a metal surface because of electron drag and result in decreased slip. Likewise, on a "soft" solid more phonons may be excited, increasing damping. A rigid, non-conducting solid is best for slip
$k \uparrow$	Large k dramatically lowers the energy barrier to defect slip (Figure 29.5) due to a reduced Peierls–Nabarro barrier. Thus, the defects move at a smaller applied force. The defects also move faster at a given force
$h \downarrow$	If the liquid–solid attraction is small, the washboard force that opposes liquid molecule translation is smaller, increasing slip
ξ incommensurate	Commensurate registry between liquid and solid allows the liquid molecules to find low energy positions on the solid from which they are difficult to dislodge. When ξ is incommensurate, independent localized defects, which propagate easily, exist. The further ξ is from an integer value, the better

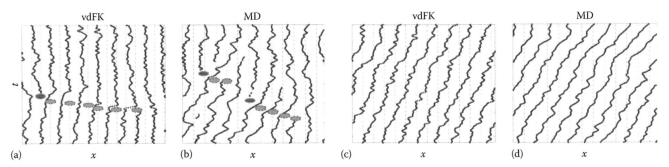

FIGURE 29.9 Molecular trajectories, position x versus time t, for the vdFK model compared with those of the first liquid layer in a molecular dynamics simulations of the Couette flow shown in the inset in Figure 29.8. Two frames are results from the vdFK model, while two are from molecular dynamics simulations. The thin vertical dashed lines show the positions midway between the solid molecules in the vdFK model (in a and c) and midway between the locations of the surface layer of solid atoms in the molecular dynamics simulation (in b and d). At low forcing, shown by the panels (a) and (b), slip is due to defect propagation. In the lower right, in both (a) and (b), the second-to-last molecule diffuses away, as shown by the termination of its trajectory. This creates a vacancy, highlighted by the oval, which propagates upstream, until a molecule drifts back into the first liquid layer, as shown by the appearance of a new trajectory. At high forcing, shown by the panels (c) and (d), global slip occurs, in which all molecules slip at approximately the same speed, as shown by their nearly parallel trajectories.

TABLE 29.5 Fields along the Solid Which Drive Slip

Thermal gradient on gaseous slip (Maxwell, 1879)

$$v_s = \frac{3}{4} \frac{\mu}{\rho T} \frac{\partial T}{\partial x},$$

where

  T is the temperature
  μ is the viscosity
  ρ is the density of the gas

Chemical potential (Ruckenstein and Rajora, 1983)

$$v_s = -\frac{D}{n_L k_B T} \frac{\partial \varphi}{\partial x},$$

where

  D is the surface diffusion coefficient
  n_L is the number density of the liquid per volume
  φ is the chemical potential
  k_B is the Boltzmann's constant
  T is the temperature

Electroosmotic slip (Joly et al., 2004; Eijkel, 2007)

$$v_s = \frac{\epsilon \zeta E}{\mu} \left(1 + \frac{L_s}{L_D}\right) \frac{V_0}{\zeta},$$

where

  ϵ is the permittivity
  ζ is the zeta potential
  E is the electric-field strength
  L_D is the Debye length
  μ is the viscosity
  V_0 is the surface potential

Note: The x-direction is tangent to the solid and the gradients are evaluated at the liquid–solid interface.

29.4 Summary

The notion of slip began with the advent of fluid mechanics. Since its inception, it has been based on the molecular nature of the solid and the liquid. We have developed these ideas into a dynamical theory of slip, in which two distinct types of slip are seen. In one, interactions between neighboring liquid molecules are critical in creating a localized defect that propagates along the periodic substrate potential. This defect involves the propagation of only a small number of molecules at a time through a background of non-propagating liquid. At higher levels of forcing, the entire first liquid layer slips, leading to larger values of slip length.

Liquids are typically portrayed as being completely formless and chaotic, while the stereotypical solid is regular and orderly. The first liquid layer is a chimera with aspects of both bulk liquid and solid. Its molecules are indistinguishable due to the inexorable interchange with the bulk, yet they are layered and regularly spaced when averaged over long times. It is the presence of an ordered solid substrate that fixes preferred locations for hopping, while the diffusive flux into and out of the layer weakens the order and can initiate slip. It is this special structure, fixed but friable, which gives the liquid–solid interface its interesting physics and its technological promise.

References

Barrat, J. L. and Bocquet, L. Influence of wetting properties on hydrodynamic boundary conditions at a fluid/solid interface. *Faraday Discuss.*, 112:119–127, 1999a.

Barrat, J. L. and Bocquet, L. Large slip effects at a nonwetting fluid-solid interface. *Phys. Rev. Lett.*, 82:4671–4674, 1999b.

Berezhkovskii, A. and Hummer, G. Single-file transport of water molecules through a carbon nanotube. *Phys. Rev. Lett.*, 89:064503, 2002.

Blake, T. D. Slip between a liquid and a solid: D. M. Tolstoi's (1952) theory reconsidered. *Colloid Surf.*, 47:135–145, 1990.

Bocquet, L. and Barrat, J. L. Flow boundary conditions from nano- to micro-scales. *Soft Matter*, 3:685–693, 2007.

Bonaccurso, E., Butt, H. J., and Craig, V. S. J. Surface roughness and hydrodynamic boundary slip of a Newtonian fluid in a completely wetting system. *Phys. Rev. Lett.*, 90:144501, 2003.

Brady-Estévez, A. S., Kang, S., and Elimelech, M. A single-walled-carbon-nanotube filter for removal of viral and bacterial pathogens. *Small*, 4:481–484, 2008.

Braun, O. M. and Kivshar, Y. S. Nonlinear dynamics of the Frenkel–Kontorova model. *Phys. Rep.*, 306:1–108, 1998.

Braun, O. M. and Kivshar, Y. S. *The Frenkel-Kontorova Model*. Berlin, Germany: Springer-Verlag, 2004.

Braun, O. M. and Naumovets, A. G. Nanotribology: Microsopic mechanisms of friction. *Surf. Sci. Rep.*, 60:79–158, 2006.

Choo, J. H. et al. A low friction bearing based on liquid slip at the wall. *J. Trib.*, 129:611–620, 2007.

Coopersmith, S. N. and Fisher, D. S. Pinning transition of the discrete sine-Gordon equation. *Phys. Rev. B*, 28:2566–2581, 1983.

Coopersmith, S. N. and Fisher, D. S. Threshold behavior of a driven incommensurate harmonic chain. *Phys. Rev. A*, 38:6338–6350, 1988.

Corry, B. Designing carbon nanotube membranes for efficient water desalination. *J. Phys. Chem. B*, 112:1427–1434, 2008.

Cottin-Bizonne, C., Cross, B., Steinberger, A., and Charlaix, E. Boundary slip on smooth hydrophobic surfaces: Intrinsic effects and possible artifacts. *Phys. Rev. Lett.*, 92:056102, 2005.

Craig, V. S. J., Neto, C., and Williams, D. R. M. Shear-dependent boundary slip in an aqueous Newtonian liquid. *Phys. Rev. Lett.*, 87:054504, 2001.

Dammer, S. F. and Lohse, D. J. Gas enrichment at liquid-wall interfaces. *Phys. Rev. Lett.*, 96:206101, 2006.

de Ruijter, M. J., Blake, T. D., and De Coninck, J. Dynamic wetting studied by molecular modeling simulations of droplet spreading. *Langmuir*, 15:7836–7847, 1999.

deGennes, P. G. On fluid/wall slippage. *Langmuir*, 18:3413–3414, 2002.

deGennes, P. G., Brochard-Wyart, F., and Quéré, D. *Capillarity and Wetting Phenomena: Drops, Bubbles, Pearls, Waves*. New York: Springer-Verlag, 2002.

Eijkel, J. C. T. Liquid slip in micro- and nanofluidics: Recent research and its possible implications. *Lab. Chip*, 7:299–301, 2007.

Eijkel, J. C. T. and van den Berg, A. Nanofluidics: What it is and what we can expect from it? *Microfluid. Nanofluid.*, 1:249–267, 2005.

Ellis, J. S. and Thompson, M. Slip and coupling phenomena at the liquid-solid interface. *Phys. Chem. Chem. Phys.*, 6:4928–2938, 2004.

Ellis, J. S., Mchale, G., Hayward, G. L., and Thompson, M. Contact angle-based predictive model for slip at the solid-liquid interface of a transverse-shear mode acoustic wave device. *J. Appl. Phys.*, 94:6201–6207, 2003.

Feuillebois, F., Bazant, M. Z., and Vinogradova, O. I. Effective slip over superhydrophobic surfaces in thin channels. *Phys. Rev. Lett.*, 102:026001, 2009.

Formasiero, F. et al. Ion exclusion by sub-2-nm carbon nanotube pores. *Proc. Natl. Acad. Sci. USA*, doi:10.1073:pnas.0710437105, 2008.

Frank, F. C. and van der Merwe, J. H. One-dimensional dislocations. I. Static theory. *Proc. R. Soc. Lond. A*, 198:205–216, 1949.

Gad-el-Hak, M. The fluid mechanics of microdevices—The Freeman scholar lecture. *J. Fluids Eng.*, 121:5–33, 1999.

Glasstone, S., Laider, K. H., and Eyring, H. *The Theory of Rate Processes*. New York: McGraw-Hill, 1941.

Hayes, R. A. and Ralston, J. The molecular-kinetic theory of wetting. *Langmuir*, 10:340–342, 1994.

Hoffman, R. L. A study of the advancing interface: II. Theoretical prediction of the dynamic contact angle in liquid-gas systems. *J. Colloid Interface Sci.*, 94(2):470–486, 1983.

Holt, J. K. et al. Fast mass transport through sub-2-nanometer carbon nanotubes. *Science*, 312:1034–1037, 2006.

Huang, P., Guasto, J. S., and Breuer, K. S. Direct measurement of slip velocities using three-dimensional total internal reflection velocimetry. *J. Fluid Mech.*, 566:447–464, 2006.

Huh, C. and Scriven, L. E. Hydrodynamic model of steady movement of a solid/liquid/fluid contact line. *J. Colloid Interface Sci.*, 35:85–101, 1971.

Hummer, G., Rasaiah, J. C., and Noworyta, J. P. Water conduction through the hydrophobic channel of a carbon nanotube. *Nature*, 414:188–189, 2004.

Joly, L., Ybert, C., Trizac, E., and Bocquet, L. Hydrodynamics within the electric double layer on slipping surfaces. *Phys. Rev. Lett.*, 93:257805, 2004.

Koplik, J., Banavar, J. R., and Willemsen, J. F. Molecular dynamics of fluid flow at solid surfaces. *Phys. Fluids A*, 1(5):781–794, 1989.

Krim, J. Resource letter: FMMLS-1: Friction at macroscopic and microscopic length scales. *Am. J. Phys.*, 70:890–897, 2002.

Krim, J. and Widom, A. Damping of a crystal oscillator by an adsorbed monolayer and its relation to interfacial viscosity. *Phys. Rev. B*, 38:12184–12189, 1998.

Lamb, H. *Hydrodynamics*. New York: Dover, 1945.

Lauga, E. and Stone, H. A. Effective slip in pressure-driven stokes flow. *J. Fluid Mech.*, 489:55–77, 2003.

Lauga, E., Brenner, M. P., and Stone, H. A. Microfluidics: The no-slip boundary condition. In J. Foss, C. Tropea, and A. Yarin, editors, *Handbook of Experimental Fluid Dynamics*, chapter 15. Berlin, Germany: Springer, 2005.

Lichter, S., Roxin, A., and Mandre, S. Mechanisms for liquid slip at solid surfaces. *Phys. Rev. Lett.*, 93:086001, 2004.

Lumma, D., Best, A., Gansen, A., Feuillebois, F., Rädler, J. O., and Vinogradova, O. I. Flow profile near a wall measured by double-focus fluorescence cross-correlation. *Phys. Rev. E*, 67:056313, May 2003. doi: 10.1103/PhysRevE.67.056313.

Ma, W. J., Iyer, L. K., Vishveshwara, S., Koplik, J., and Banavar, J. R. Molecular-dynamics studies of systems of confined dumbbell molecules. *Phys. Rev. E*, 51:441–453, 1995.

Majumder, M., Chopra, N., Andrews, R., and Hinds, B. J. Nanoscale hydrodynamics: Enhanced flow in carbon nanotubes. *Nature*, 438:44, 2005.

Martini, A., Hsu, H. Y., Patankar, N. A., and Lichter, S. Slip at high shear rates. *Phys. Rev. Lett.*, 100:206001, 2008a.

Martini, A., Roxin, A., Snurr, R. Q., Wang, Q., and Lichter, S. Molecular mechanisms of liquid slip. *J. Fluid Mech.*, 600:257–269, 2008b.

Maxwell, J. C. On stresses in rarefied gases arising from inequalities of temperature. *Philos. Trans. R. Soc. Lond.*, 170:231–256, 1879.

McLaughlin, D. W. and Scott, A. C. Perturbation analysis of fluxon dynamics. *Phys. Rev. A*, 18(4):1652–1680, 1978.

Mo, H., Evmenenko, G., and Dutta, P. Ordering of liquid squalane near a solid surface. *Chem. Phys. Lett.*, 415:106–109, 2005.

Navier, M. Mémoire sur les lois du mouvement des fluides. *Mém de l'Acad. des Sci.*, 6: 389–422, 1823.

Neto, C., Evans, D. R., Bonaccurso, E., Butt, H. J., and Craig, V. S. J. Boundary slip in Newtonian liquids: A review of experimental studies. *Rep. Prog. Phys.*, 68:2859–2897, 2005.

Panzer, P., Liu, M., and Einzel, D. The effects of boundary curvature on hydrodynamics fluid flow-calculation of slip lengths. *Int. J. Mod. Phys. B*, 6:3251–3278, 1992.

Patankar, N. A. Mimicking the lotus effect: Influence of double roughness structures and slender pillars. *Langmuir*, 20:8209–8213, 2004.

Persson, B. N. J. *Sliding Friction Physical Principles and Applications*. Berlin, Germany: Springer, 1998.

Petravik, J. and Harrowell, P. On the equilibrium calculation of the friction coefficient for liquid slip against a wall. *J. Chem. Phys.*, 127:174706, 2007.

Petrov, P. G. and Petrov, J. G. A combined molecular-hydrodynamic approach to wetting kinetics. *Langmuir*, 8:1762–1767, 1992.

Pit, R., Hervet, H., and Léger, L. Direct experimental evidence of slip in hexadecane: Solid interfaces. *Phys. Rev. Lett.*, 85:980–983, 2000.

Plimpton, S. J. Fast parallel algorithms for short-range molecular dynamics. *J. Comp. Phys.*, 117:1–19, 1995.

Prakash, S., Piruska, A., Gatimu, E. N., Bolin, P. W., Sweedler, J. V., and Shannon, M. A. Nanofluidics: Systems and application. *IEEE Sens. J.* 8(5):441–450, 2008.

Priezjev, N. V. and Troian, S. M. Molecular origin and dynamic behavior of slip in sheared polymer films. *Phys. Rev. Lett.*, 92(1):018302, 2004.

Priezjev, N. V. and Troian, S. M. Influence of periodic wall roughness on the slip behaviour at liquid/solid interfaces: Molecular-scale simulations versus continuum predictions. *J. Fluid Mech.*, 554:25–46, 2006.

Quéré, D. Wetting and roughness. *Annu. Rev. Mater. Res.*, 38:71–99, 2008.

Rauscher, M. and Dietrich, S. Wetting phenomena in nanofluidics. *Annu. Rev. Mater. Res.*, 38:143–172, 2008.

Roxin, A. Five projects in pattern formation, fluid dynamics, and computational neuro-science. PhD thesis, Evanston, IL: Northwestern University, 2004.

Ruckenstein, E. and Dunn, C. S. Slip velocity during wetting of solids. *J. Colloid Interface Sci.*, 59:135–138, 1977.

Ruckenstein, E. and Rajora, P. On the no-slip boundary condition of hydrodynamics. *J. Colloid Interface Sci.*, 96:488–491, 1983.

Sholl, D. S. and Johnson, J. K. Making high-flux membranes with carbon nanotubes. *Science*, 312:1003–1004, 2006.

Sokhan, V. P., Nicholson, D., and Quirke, N. Fluid flow in nanopores: An examination of hydrodynamic boundary conditions. *J. Chem. Phys.*, 115:3878–3887, 2001.

Sokhan, V. P., Nicholson, D., and Quirkea, N. Fluid flow in nanopores: Accurate boundary conditions for carbon nanotubes. *J. Chem. Phys.*, 117:8531–8539, 2002.

Steele, W. A. The physical interaction of gases with crystalline solids. *Surf. Sci.*, 36:317–352, 1973.

Strunz, T. and Elmer, F. J. Driven Frenkel-Kontorova model. I. Uniform sliding states and dynamical domains of different particle densities. *Phys. Rev. E*, 58(2):1601–1611, 1998.

Thomas, J. A. and McGaughey, A. J. H. Reassessing fast water transport through carbon nanotubes. *Nano Lett.*, 8:2788–2793, 2008.

Thompson, P. A. and Robbins, M. O. Shear flow near solids: Epitaxial order and flow boundary conditions. *Phys. Rev. A*, 41:6830–6837, 1990a.

Thompson, P. A. and Robbins, M. O. Origin of stick-slip motion in boundary lubrication. *Science*, 250:792–794, 1990b.

Thompson, P. A. and Troian, S. M. A general boundary condition for liquid flow at solid surfaces. *Nature*, 389:360–362, 1997.

Tretheway, D. C. and Meinhart, C. D. Apparent fluid slip at hydrophobic microchannel walls. *Phys. Fluids*, 14:L9–L12, 2002.

Tyrrell, J. W. G. and Attard, P. Images of nanobubbles on hydrophobic surfaces and their interactions. *Phys. Rev. Lett.*, 87:176104, 2001.

Urbakh, M., Klafter, J., Gourdon, D., and Israelachvili, J. The nonlinear nature of friction. *Nature*, 430:525–528, 2004.

Vinogradova, O. I. Drainage of a thin liquid flim confined between hydrophobic surfaces. *Langmuir*, 11:2213–2220, 1995.

Vinogradova, O. I. and Yakubov, G. E. Dynamic effects on force measurements. 2. Lubrication and the atomic force microscope. *Langmuir*, 19:1227–1234, 2003.

Voronov, R. S., Papavassiliou, D. V., and Lee, L. L. Slip length and contact angle over hydrophobic surfaces. *Chem. Phys. Lett.*, 441:273–276, 2007.

Wu, C. W. et al. Low friction and high load support capacity of slider bearing with a mixed slip surface. *Trans. ASME*, 128:904–907, 2006.

Yu, C. J., Evmenenko, G., Richter, A. G., Kmetko, J., and Dutta, P. Order in molecular liquids near solid-liquid interfaces. *Appl. Surf. Sci.*, 182:231–235, 2001.

Zhu, Y. and Granick, S. Rate-dependent slip of Newtonian liquid at smooth surfaces. *Phys. Rev. Lett.*, 87:096105, 2001.

Zhu, Y. and Granick, S. Limits of the hydrodynamics no-slip boundary condition. *Phys. Rev. Lett.*, 88(10):106102, 2002.

30

Newtonian Nanofluids in Convection

Stéphane Fohanno
*Université de Reims
Champagne-Ardenne*

Cong Tam Nguyen
Université de Moncton

Guillaume Polidori
*Université de Reims
Champagne-Ardenne*

30.1 Introduction

The current rapid development of nanosciences originates from the observation of specific properties of the matter at a nanometer scale. This rapid growth has been made possible by the development of modern technological processes allowing for a drastic reduction in the cost of nanoparticles. As a result, numerous applications of nanosciences begin to appear. Among them, nanofluids constitute an emerging field in heat transfer. Nanofluids are a new class of fluids consisting in a suspension of nanometer-sized solid particles dispersed in a base (pure) liquid. Investigations on nanofluids concern several aspects such as the preparation of the nanofluids, the characterization of their thermal conductivity and their rheology, as well as the applications of these nanofluids for convective heat transfer purposes (Wang and Mujumdar 2007). Initially, research works devoted to nanofluids were mainly focused on the analysis of their thermal conductivity. For very low volume fractions of nanoparticles, some of these suspensions proved to be very efficient in order to improve, under some conditions, the heat transfer performance. Several experimental investigations have shown a strong enhancement of the thermal conductivity by addition of nanoparticles. Improvements of several tens of percents were

observed for nanoparticle volume fractions less than 5% (Wang and Mujumdar 2007).

Because of the improvements observed in terms of their heat transfer properties, nanofluids became naturally a potential candidate to replace conventional fluids in some applications of heat exchangers. Presently, conventional heat transfer fluids such as water, ethylene glycol, or oils are still widely used for heat exchange purposes in the industry or in building applications. An increase in the thermal loads can generally be handled by an increase in the heat exchange surface. However, conventional fluids remain penalized by their limited thermal properties, among which is a low thermal conductivity. As a consequence, they are no longer suitable for some modern applications requiring a high level of performance while keeping a reduced size of the thermal system such as for the cooling of microprocessors, micro-electromechanical systems, as well as to obtain fast transient regimes in heating systems. Therefore, several research works focusing on convective heat transfer, with nanofluids as working fluids, have been carried out during the last decade in order to test their potential for applications related to industrial heat exchangers.

Furthermore, when flowing, the sole thermal conductivity of the nanofluid is no longer sufficient to evaluate the heat transfer efficiency (Polidori et al. 2007, Keblinski et al. 2008). In the case of convective heat transfer, it is also necessary to determine if the thermal performance of the nanofluid will remain better than that of the base fluid in spite of the increasing pressure drop. In particular, one of the key points to be solved concerns the determination of the role of the viscosity on the resulting flow dynamics and heat transfer characteristics (Prasher et al. 2006). Additional investigations are also required to analyze the relative importance of all the fluid thermophysical properties playing a role in the qualification of the convective heat exchange.

The purpose of this chapter is to make a review on nanofluids focusing on their potential for convective heat transfer applications, i.e., heat transfer in a flowing fluid. First, thermophysical properties influencing heat transfer in single-phase fluids are presented in Section 30.2. Then, Section 30.3 provides a description of nanofluids including their composition and the modeling of their thermophysical properties. As nanofluids are considered for convective heat transfer applications, the different types of convection are described in Section 30.4 and are illustrated by the academic boundary layer flow configuration. Basic relationships characterizing the dynamical, as well as the thermal, behaviors of convective flows are provided in terms of heat flux laws, conservation equations, and dimensionless parameters. Section 30.5 provides an overview of research activities on convective heat transfer in nanofluids. Finally, an application of nanofluids to natural convection is presented in Section 30.6 in order to illustrate the complexity of the modeling of convection in nanofluids, which is strongly dependent on an accurate modeling of all the thermophysical properties.

30.2 Thermophysical Properties of Single-Phase Fluids

Heat transfer characteristics of a fluid are strongly dependent on its thermophysical properties and on its flow dynamics. In the present section, only fluids composed of a single continuous phase (gas or liquid) are considered, i.e., fluids that do not contain any discrete phase like droplets, bubbles, or solid particles. Furthermore, it is assumed that the fluid evolves without changing phase (no condensation or boiling). Here is a brief overview of the main thermophysical properties necessary to characterize heat transfer in a flowing fluid.

30.2.1 Energy Storage

The ability of a fluid to store energy depends on its density (ρ in kg/m³) and its specific heat at constant pressure (C_p in J/kg K). The product of these two parameters (ρC_p in J/m³) defines the additional amount of energy that can be stored in a unit volume of a fluid when its temperature is increased by 1°C.

30.2.2 Conduction: Fourier's Law

If there are temperature gradients in a fluid, energy is transferred within the fluid under the form of heat in the direction of decreasing temperature (Figure 30.1). This mode of heat transfer is termed conduction and is described by Fourier's law of heat conduction:

$$q'' = -k\frac{\partial T}{\partial y} \tag{30.1}$$

where y is the coordinate in the direction of the heat flux. This law indicates that the heat flux (q'' in W/m²) is proportional to the temperature gradient, the constant of proportionality being the thermal conductivity (k, in W/m K). Fourier's law of heat conduction applies equally to fluids and solid bodies.

Examples of thermal conductivities for common fluids and solids are provided in Table 30.1. One may note that thermal conductivities of metals or metal oxides are several orders of magnitude higher than those of fluids.

30.2.3 Viscous Forces: Newton's Law of Viscosity

When a fluid is in motion, viscous forces oppose a resistance to the flow (Figure 30.2). This resistance can be expressed by Newton's law of viscosity:

$$\tau = \mu\frac{\partial u}{\partial y} \tag{30.2}$$

that relates the shear stress (τ in N/m² or Pa) between adjacent fluid layers to the normal velocity gradient, or shear rate, ($\partial u/\partial y$ in s⁻¹) by

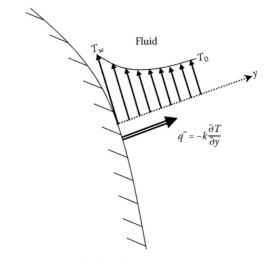

FIGURE 30.1 Fourier's law of heat conduction.

TABLE 30.1 Thermal Conductivities of Solids and Fluids at 20°C

Materials	Aluminum (Al)	Copper (Cu)	Alumina (Al₂O₃)	Water	Air	Ethylene Glycol (40%–60%)
k (W/m K)	237	401	36	0.60	0.0263	0.249

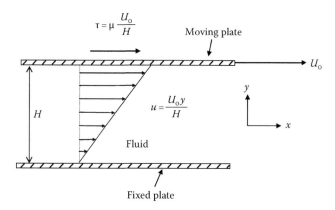

FIGURE 30.2 Newton's law of viscosity—Application to Couette flow.

means of a coefficient of proportionality (μ in kg/m s or Pa s) called the dynamic viscosity.

30.2.3.1 Newtonian Fluids

From this law, a Newtonian fluid is defined as a fluid whose relationship between the shear stress and the shear rate is linear (Figure 30.3). The dynamic viscosity of Newtonian fluids only depends on temperature and pressure, but not on the shear rate. Common fluids such as air, water, or oils have a Newtonian behavior.

30.2.3.2 Non-Newtonian Fluids

However, there exist a second category of fluids whose shear stress/shear rate relationship is no longer linear. For this second category of fluids, the viscosity is strongly dependent on the shear rate. These fluids are called non-Newtonian fluids. This second set of fluids is mainly composed of shear-thinning fluids (decreasing viscosity with increasing shear rate), shear-thickening fluids (increasing viscosity with increasing shear rate), and plastic fluids (yield stress to exceed in order to put the fluid in motion). The rheological behaviors of these fluids are sketched in Figure 30.3.

30.2.4 Prandtl Number

Fourier's law of heat conduction is sufficient to describe heat transfer in a fluid at rest. Now, when a fluid is flowing, heat conduction still exists. However, a second and important heat

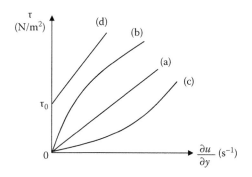

FIGURE 30.3 Common shear stress (τ)/shear rate ($\partial u/\partial y$) relationships of fluids. (a) Newtonian, (b) shear-thinning, (c) shear-thickening, (d) plastic (Bingham).

transfer mechanism appears when a fluid begins to flow. Indeed, a fluid flow corresponds to the displacement of a given mass of fluid. The displaced mass of fluid contains a certain amount of energy that is also transported (or convected) by the fluid flow. If heat is locally concentrated in a (high temperature) region of the fluid (e.g., close to a heated surface), the fluid motion will allow evacuating heat from this region faster than if there was only heat conduction. This additional heat transfer mechanism is termed convection and is further detailed in Section 30.4. The main parameter influencing the heat transfer characteristics of a fluid in motion is its Prandtl number (Pr). The Prandtl number is an important nondimensional parameter that compares the diffusivities of heat and momentum. Indeed, this parameter groups thermal and dynamical properties of the fluid and is given by $Pr = \mu C_p/k = \nu/a$, where $\nu = \mu/\rho$ is the kinematic viscosity (or momentum diffusivity) of the fluid (in m²/s) and $a = k/\rho C_p$ is the thermal diffusivity of the fluid (also in m²/s).

30.3 Nanofluids

30.3.1 Historical Background

As mentioned in Section 30.2.2, thermal conductivities of metallic particles are several orders of magnitudes higher than those of fluids. Therefore, the possibility of adding fine, but highly conductive, solid particles in a fluid in order to increase substantially the thermal conductivity, was already studied more than 100 years ago by Maxwell (1891). Assuming well-dispersed spherical particles, Maxwell proposed a model predicting the effective thermal conductivity of the liquid–solid suspension as a function of the base (pure) fluid and particle thermal conductivities and the particle volume fraction. Until the advent of nanotechnology in the early 1990s, which offered the possibility of manufacturing submicronic particles, or nanoparticles, only millimeter- or micron-sized particles were readily available at a reasonable cost. Nevertheless, the large size of the particles dispersed in the suspension presented several disadvantages such as rapid settling, clogging of the flow, and a strong increase in the pressure drop or erosion that hindered the use of such liquid–solid mixtures in heat exchangers. Since then, the rapid development of modern technological processes allowing for a drastic reduction in the cost of nanoparticles has permitted the development of a new class of liquid suspensions called "nanofluids" (Choi 1995).

30.3.2 What Is a Nanofluid?

A nanofluid is a liquid–solid mixture made up of nanometer-scale (<100 nm) solid particles dispersed into a base (pure) liquid such as water, ethylene glycol, or oils (Figure 30.4). Liquid–solid mixtures with nanoparticles present the advantage of being more stable and less subject to the above-mentioned drawbacks of larger particles.

A large variety of nanofluids combining various types of nanoparticles with different base fluids have experimentally been

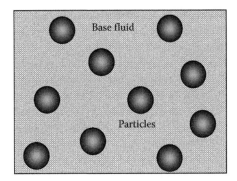

FIGURE 30.4 Schematic representation of a nanofluid.

studied (see, e.g., the reviews of Wang and Mujumdar (2007) and Murshed et al. (2008)). A nanofluid can be categorized according to its base fluid and its type of nanoparticles.

30.3.2.1 Base Fluids

Nanoparticles can be dispersed in most existing liquids. However, the base fluid is generally chosen among one of the widely used conventional heat transfer fluids such as water, ethylene glycol, or oils. The main thermophysical properties of these liquids are summarized in Table 30.2.

30.3.2.2 Characteristics of Nanoparticles

The size of nanoparticles is generally comprised between 1 and 100 nm. Several shapes of particles exist. Spherical and cylindrical shapes are frequent. To date, a large variety of solid nanoparticles have been tested for nanofluids applications. Metallic particles (copper, gold, silver ...) and oxide particles (alumina: Al_2O_3, copper oxide: CuO, SiO_2, titanium dioxide: TiO_2, etc.) are the most commonly used nanoparticles. One may also note the use of carbon nanotubes (CNT) whose tube shape differs significantly from the generally spherical shape of metallic or oxide particles.

30.3.3 Nanoparticle Concentration

One of the key parameter of a nanofluid is its concentration in nanoparticles. This concentration is generally given as the nanoparticle volume fraction (ϕ), which is defined by the ratio

$$\phi = \frac{\text{volume of nanoparticles}}{\text{volume of the suspension}} \tag{30.3}$$

30.3.4 Preparation of Nanofluids

There are mainly two types of methods to prepare nanofluids, namely, the one-step method and the two-step method.

One-step method:
This method consists in vaporizing a solid material under vacuum and then to directly condense its vapor into the liquid.

Two-step method:
In this method, dry primary nanoparticles are first produced and stored in the form of a nanopowder. Then, they are dispersed into the base liquid. The main drawback of this method is that dry nanoparticles agglomerate in the powder and it is difficult to break the agglomerates into primary particles. Therefore, the effective (or average) size of the particles becomes larger. Several methods are used in order to homogenize the suspension. Surfactants or dispersants may be used in order to prevent the formation of agglomerates and enhance the stability of the nanofluid. However, this may also affect the heat transfer properties of the nanofluid.

30.3.5 Thermal Conductivity of a Nanofluid

The thermal conductivity of a nanofluid (k_{nf}) depends on the particle volume fraction ϕ and the corresponding thermal conductivities of the base fluids (k_{bf}) and the solid nanoparticles (k_p). It is also a function of temperature like for any liquid and solid. Two approaches can be chosen in order to model the conductivity of a nanofluid: the microscopic approach or the macroscopic one. On the one hand, microscopic models consider the individual interactions of nanoparticles with the base liquid at a molecular level. On the other hand, macroscopic models consider nanofluids as a single phase with an effective thermal conductivity.

30.3.5.1 Macroscopic versus Microscopic Models

Several theories can be found in the literature in order to model the thermal conductivity of a liquid–solid mixture. One can distinguish macroscopic models from microscopic ones. Microscopic models are still in the early development stage, and the few data in the literature deduced from microscopic models applied to nanofluids such as alumina/water (see, e.g., Shukla and Dhir (2005)) indicate that for small temperature gradients, a simplified macroscopic model is largely sufficient to estimate the thermal conductivity. Furthermore, under some conditions, predicted thermal conductivity ratios obtained with macroscopic models have been found to be in quite good agreement with experimental results (see Lee et al. (1999)).

Macroscopic models are also more convenient for modeling convective heat transfer in nanofluids by using the single-phase approach. The single-phase approach considers the suspension as a single fluid with effective properties whereas the two-phase approach considers two separate phases: a dispersed phase (the nanoparticles) and a continuous phase (the base liquid).

TABLE 30.2 Thermophysical Properties of Water and Ethylene Glycol at 20°C

	ρ (kg/m³)	C_p (J/kg K)	k (W/m K)	β (K⁻¹)	μ (kg/m s)	Pr
Water	998.3	4182	0.60	2.06×10^{-4}	10^{-3}	6.96
Ethylene glycol	1116.6	2382	0.249	6.5×10^{-4}	2.14×10^{-2}	204

We will restrict the list of models presented below to macroscopic models.

30.3.5.2 Macroscopic Models

30.3.5.2.1 Maxwell Model

Assuming well-dispersed spherical particles, Maxwell (1891) proposed a model predicting the effective thermal conductivity of the liquid–solid suspension as a function of the base (pure) fluid and particle thermal conductivities and the particle volume fraction. This model is valid for small particle concentrations. The effective thermal conductivity (k_{eff}) model of Maxwell, also known as the Wasp model (Wasp 1977), is given by the following formula:

$$k_{eff} = \left(\frac{k_p + 2k_{bf} - 2\phi(k_{bf} - k_p)}{k_p + 2k_{bf} + \phi(k_{bf} - k_p)} \right) k_{bf} \qquad (30.4)$$

30.3.5.2.2 Hamilton and Crosser Model (1962)

The Hamilton and Crosser model is an extension of the Maxwell's theory accounting for the nonsphericity of particles through the use of a shape factor, n, defined as $n = 3/\psi$ with ψ being the particle sphericity (Hamilton and Crosser 1962). The sphericity ($\psi = A_e/A$) of a particle is defined as the ratio of the surface area (A_e) of the equivalent sphere having the same volume to the actual surface area (A) of the nonspherical particle.

The effective thermal conductivity of the Hamilton and Crosser model is given by

$$k_{eff} = \left(\frac{k_p + (n-1)k_{bf} - (n-1)(k_{bf} - k_p)\phi}{k_p + (n-1)k_{bf} + (k_{bf} - k_p)\phi} \right) k_{bf} \qquad (30.5)$$

Spherical particles have a sphericity $\psi = 1$ Corresponding to a shape factor $n = 3$. Using $n = 3$ in Equation 30.5 leads to the Maxwell equation (Equation 30.4). For nanofluids, the effective thermal conductivity (k_{eff}) corresponds to the nanofluid thermal conductivity (k_{nf}).

30.3.5.3 Limitations of the Models

Both the Maxwell and the Hamilton and Crosser models are limited by the fact that they only take into account the particle volume fraction, the particle shape, and the conductivities of the base fluid and the particles. They are not able to predict the anomalous increase in the thermal conductivity observed in some experimental investigations. Indeed, initial research works devoted to nanofluids mainly focused on the analysis of their thermal conductivity (Masuda et al. 1993, Choi 1995, Wang et al. 1999, Eastman et al. 2001). For very low volume fractions of nanoparticles, some of these suspensions proved to be very efficient in order to enhance, under some conditions, the heat transfer performance. An "anomalously" high increase of the effective thermal conductivity, beyond the predictions of the Maxwell's theory, has rapidly been noticed by several authors (Choi et al. 2001, Eastman et al. 2001). The Hamilton and Crosser

model was not sufficient either to explain this strong increase in the thermal conductivity. Therefore, several authors have attempted to provide a description of the underlying physical phenomena explaining such an increase in the thermal conductivity. Keblinski et al. (2002) explored four possible explanations: Brownian motion, liquid layering, ballistic heat transport, and clustering of nanoparticles. Keblinski et al. (2008) recently noticed that Maxwell's model for well-dispersed particles corresponds to the lower bound of the Hashin and Shtrikman (H–S) bounds for thermal conductivity (Hashin and Shtrikman 1962), the upper bound corresponding to clustered nanoparticles. They showed that most experimental data on the effective thermal conductivity of nanofluids fall within these limits. Nevertheless, despite the numerous experimental data gathered since more than a decade and the efforts in modeling, the development of a universal model for the effective thermal conductivity remains a challenging task. A more complete list of analytical models of the thermal conductivity is provided in Wang and Mujumdar (2007).

30.3.6 Viscosity of Nanofluids

The sole thermal conductivity is not sufficient to evaluate the convective heat transfer efficiency of nanofluids (Polidori et al. 2007, Keblinski et al. 2008). In particular, one of the key points to be solved concerns the determination of the role of the viscosity on the resulting flow dynamics and heat transfer characteristics (Prasher et al. 2006). As for single-phase fluids, the viscosity of a nanofluid will strongly depend on the rheological behavior of the nanofluids.

30.3.6.1 Rheological Behavior of Nanofluids

Nanofluids are suspensions of solid particles in a liquid. As a consequence, the shape and the concentration of particles will strongly affect the rheology of a nanofluid. According to the composition of the nanofluid, the rheological behavior may either be Newtonian (e.g., alumina-based or copper-based nanofluids for low volume fractions, Putra et al. 2003) or non-Newtonian with a shear-thinning character (e.g., CNT, Ding et al. 2006). Most nanofluids composed of spherical nanoparticles have a Newtonian behavior, provided the particle volume fraction is not too high. In the following, we will only consider Newtonian nanofluids. This will allow us to use some classical expressions for the viscosity of liquid–solid suspensions. However, due to the two-phase nature of the liquid–solid suspension and the small size of the nanoparticles, some effects such as the clustering of particles are not well taken into account. Like for the thermal conductivity, some discrepancies have been noticed between the models and experimental measurements.

30.3.6.2 Dynamic Viscosity Models

As for the thermal conductivity, the viscosity of a nanofluid does not only depend on the base-fluid viscosity and its temperature but also on the volume fraction and the shape of suspended nanoparticles. As a first approach, one may use classical relationships for liquid–solid mixtures that have been developed for dilute suspensions. The first model was developed by Einstein (1906). Brinkman (1952) proposed an improved formulation.

Einstein model:

This model was obtained under the assumption of a very dilute suspension of hard spheres with no interactions between the spheres. It is applicable for particle volume fractions less than 1%:

$$\frac{\mu_{nf}}{\mu_{bf}} = 1 + 2.5\,\phi \tag{30.6}$$

Brinkman model (1952):

Brinkman (1952) proposed a more elaborate model that is currently used in literature (Xuan and Roetzel 2000, Khanafer et al. 2003, Gosselin and da Silva 2004) for nanofluids:

$$\frac{\mu_{nf}}{\mu_{bf}} = \frac{1}{(1-\phi)^{5/2}} \tag{30.7}$$

Experimental models:

Due to the complexity of the rheological behavior of nanofluids, several models have directly been derived from experimental data in order to provide a better modeling of the viscosity of nanofluids. For example, Maïga et al. (2004, 2005) have recently proposed a viscosity model for water/alumina nanofluid:

$$\frac{\mu_{nf}}{\mu_{bf}} = 123\,\phi^2 + 7.3\,\phi + 1 \tag{30.8}$$

30.3.7 Other Thermophysical Properties

Introducing the particle volume fraction ϕ, the thermophysical properties of the nanofluid, namely, the density (ρ), volume coefficient of expansion (β), and specific heat (C_p), can be calculated from nanoparticle and base-fluid properties at the ambient temperature using the following classic formulas:

$$\rho_{nf} = (1-\phi)\rho_{bf} + \phi\rho_p \tag{30.9}$$

$$(\rho\beta)_{nf} = (1-\phi)(\rho\beta)_{bf} + \phi(\rho\beta)_p \tag{30.10}$$

$$(C_p)_{nf} = (1-\phi)(C_p)_{bf} + \phi(C_p)_p \tag{30.11}$$

where nf, bf, and p refer to the nanofluid, the base fluid, and the nanoparticle, respectively.

As mentioned in Buongiorno (2006), assuming that the nanoparticles and the base fluid are in thermal equilibrium, the nanofluid specific heat should be derived from

$$(\rho C_p)_{nf} = (1-\phi)(\rho C_p)_{bf} + \phi(\rho C_p)_p \tag{30.12}$$

However, several authors (Pak and Cho 1998, Maïga et al. 2004, 2005, Buongiorno 2006) prefer to use the simpler approach (Equation 30.11) to qualify the specific heat.

30.4 Convection

As nanofluids are considered for convective heat transfer applications, the different types of convection are described in this section and illustrated by the academic boundary layer flow configuration. First, a definition of convection is given in Section 30.4.1. Then, a description of the two main flow types (forced and buoyancy-induced) and of the flow regimes (laminar and turbulent) is provided in Sections 30.4.2 and 30.4.3, respectively. Next, convection modes are detailed in Section 30.4.4. Finally, heat flux laws and conservation equations applying to convective heat transfer are provided in Section 30.4.5.

30.4.1 Definition

Convection is a mode of heat transfer taking place within non-isothermal (presence of temperature gradients) fluids (liquids or gases) in motion. Generally, such a situation occurs when a fluid is flowing close to a wall whose surface temperature is different from that of the ambient fluid.

There are three main modes of convection: forced convection (FC), natural (or free) convection (NC), and mixed convection (MC). The convection mode will depend on the type of flow associated with the convective heat transfer: forced (FC), buoyancy-induced (NC) or a combination of both (MC).

30.4.2 Flow Types

30.4.2.1 Forced (or Imposed) Flows

This situation occurs when the fluid flow has a mechanical origin. The flow is imposed, or forced, by means of a pump, a fan, or a natural pressure gradient (e.g., wind). The flow is characterized by a reference or characteristic velocity. Examples of forced flows are external flows over solid surfaces (wings, external walls of buildings), flows past obstacles (bluff-bodies), and internal pipe flows (Figure 30.5).

30.4.2.2 Buoyancy-Induced Flows

The motion of the fluid has a thermal origin. The fluid flow is due to buoyancy forces induced by the spatial variations of the fluid density with temperature. If temperature gradients in the fluid are sufficiently high, density gradients in the fluid will create a vertical buoyancy force sufficient to overcome the viscous forces. Examples of buoyancy-induced flows are the boundary layer flow along a vertical heated plate, the flow in a differentially heated enclosure or the flow around a heated cylinder (Figure 30.6).

30.4.2.3 Forced and Buoyancy-Induced Laminar Boundary Layer Flows

Boundary layer flows are a category of external flows developing along a solid surface. This type of flows is encountered in many

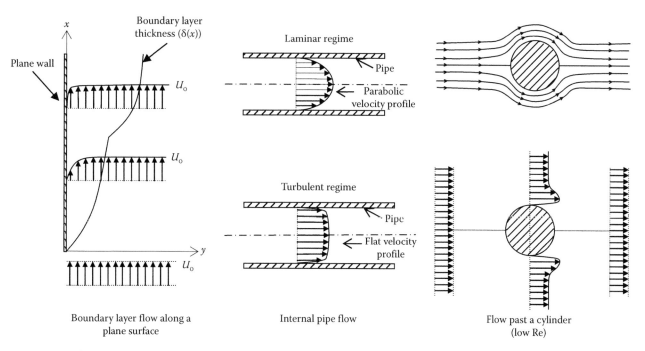

FIGURE 30.5 Schematic representations of forced flows.

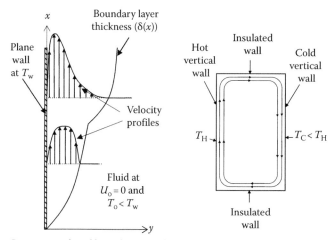

FIGURE 30.6 Examples of buoyancy-induced flows.

situations and may concern both forced and buoyancy-induced flows. One of the most simple boundary layer flow configurations is that of a fluid flowing along a plane vertical surface. The simplicity of the geometry presents the advantage of facilitating the description of the flow dynamics. The flow may be forced (Figure 30.7a) or induced by the temperature difference between the plane surface and the ambient fluid (Figure 30.7c). If the two flows occur simultaneously and are equally important, their combination may eventually lead to a mixed (or combined) flow configuration (Figure 30.7b).

The case of the buoyancy-induced boundary layer flow will serve as an example to describe the main characteristics of convection in nanofluids in Section 30.6.

30.4.2.3.1 Forced Boundary Layer Flow

A typical velocity profile of the forced boundary layer flow along a vertical plane surface is sketched in Figure 30.7a. The imposed velocity of the main forced flow is U_o. The region close to the wall where velocity gradients (in the direction normal to the wall) are important is called the dynamical boundary layer. In this region, the velocity goes from U_o (far from the wall) to 0 (at the wall) because of the action of the fluid viscosity (or friction). The region where viscous effects are significant is called the boundary layer and is comprised between the wall and a distance δ called the boundary layer thickness. δ is a function of the longitudinal coordinate (x).

30.4.2.3.2 Buoyancy-Induced Boundary Layer Flow

A typical velocity profile of the external boundary layer of a laminar buoyancy-induced flow along a vertical heated plane surface is sketched in Figure 30.7c. In this situation, the ambient fluid outside the boundary layer is at rest (no velocity). Contrary to the forced flow case, the velocity profile is strongly dependent on the temperature distribution close to the wall surface. A maximum of the velocity is observed within the boundary layer. The position of this maximum depends on the relative importance of viscous and buoyancy-induced forces. Like in forced convection, one can also define a dynamical boundary layer thickness (δ).

30.4.3 Flow Regimes

30.4.3.1 Definitions of Laminar and Turbulent Regimes

The fluid motion is mainly characterized by its flow regime, which can be either laminar or turbulent (Figure 30.8). In the laminar regime, fluid layers are moving one parallel to the other.

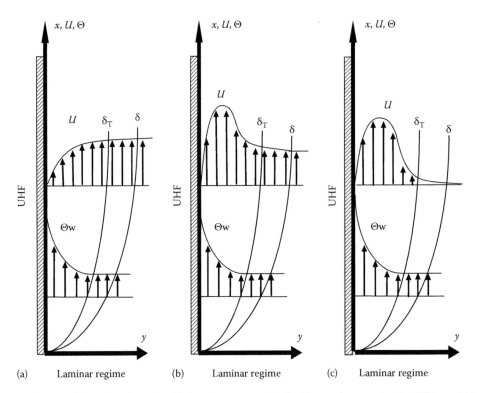

FIGURE 30.7 Aspects of external boundary layers for (a) forced convection (left), (b) mixed convection (middle), and (c) natural convection (right) − θ = *T* = Fluid temperature and θ$_w$ = *T*$_w$ = Wall Temperature.

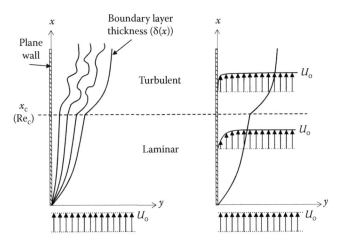

FIGURE 30.8 Flow regimes and corresponding velocity profiles in forced boundary layer flow.

For example, boundary layer flows sketched in Figure 30.7 are assumed laminar. In that case, there is essentially diffusion through the fluid layers so that mixing is limited.

However, there exists a second flow regime where the flow is no longer as well organized as in the laminar regime. Local and instantaneous velocity characteristics of the flow fluctuate a lot. This new regime is called turbulence and is found in most industrial flows. In the turbulent regime, mixing phenomena are intensified due to the action of turbulence. Because of this enhanced mixing, velocity profiles become flatter except very close to the wall where the velocity gradient is steeper (Figure 30.8). As a

consequence, wall friction as well as fluid-to-wall heat exchanges are strongly increased in the turbulent regime.

30.4.3.2 Characterization of the Flow Regime

For a given flow configuration, the flow regime depends on a characteristic nondimensional parameter: the Reynolds number (Re) for forced flows and the Grashof number (Gr) for buoyancy-induced flows.

30.4.3.2.1 Reynolds Number

In the case of forced flows, the Re is a measure of the ratio of inertial forces to viscous forces. The general expression of the Re is given by

$$\mathrm{Re} = \frac{U_0 L_0}{\nu} \tag{30.13}$$

where
  U_0 is a characteristic velocity of the imposed flow
  L_0 is a characteristic physical dimension (e.g., the plate length for an external boundary layer flow or the internal diameter for a pipe flow)
  $\nu = \mu/\rho$ is the fluid kinematic viscosity

30.4.3.2.2 Grashof Number

In the case of buoyancy-induced flows, the Gr is a measure of the ratio of buoyancy forces to viscous forces. The general expression of the Gr is given by

$$\mathrm{Gr} = \frac{g\beta(T_\mathrm{w} - T_\infty)L_0^3}{\nu^2} \qquad (30.14)$$

where

L_0 is a characteristic physical dimension (the plate height for an external boundary layer flow over vertical heated surface)

$(T_\mathrm{w} - T_\infty)$ is the temperature difference between the wall and the ambient fluid

g is the acceleration of gravity

β and ν are the volume coefficient of expansion and the kinematic viscosity of the fluid, respectively

30.4.3.2.3 Laminar-to-Turbulence Transition

The transition between the flow regimes occurs for a critical (transitional) number (Re_c or Gr_c) so that we obtain

- A laminar regime, if $\mathrm{Re} < \mathrm{Re}_c$ or $\mathrm{Gr} < \mathrm{Gr}_c$
- A turbulent regime, if $\mathrm{Re} > \mathrm{Re}_c$ or $\mathrm{Gr} > \mathrm{Gr}_c$

Values of Re_c or Gr_c differ according to the flow configuration. For boundary layer flows along a vertical plane surface, one will distinguish the global Re and Gr based on the total height (H) of the plate ($\mathrm{Re}_H = (U_0 H)/\nu$ and $\mathrm{Gr}_H = (g\beta(T_W - T_\infty)H^3)/\nu^2$) from the local values ($\mathrm{Re}_x = U_0 x/\nu$ and $\mathrm{Gr}_x = \dfrac{g\beta(T_w - T_\infty)x^3}{\nu^2}$ based on the abscissa x (i.e., the distance from the leading edge). Transition from laminar-to-turbulent flow will occur if the local Re, or Gr, exceeds the critical value Re_c or Gr_c.

In natural convection, the wall surface is sometimes heated or cooled with a constant heat flux (q_w''). In that case, a modified Gr, $\mathrm{Gr}^* = (g\beta q_w'' L_0^4)/k\nu^2$ based on q_w'', is defined.

30.4.4 Description of Convection

30.4.4.1 Convective Heat Transfer Modes

As mentioned in Section 30.4.1, there are three main modes of convection: forced convection, natural (or free) convection, and mixed convection.

Forced convection takes place when heat transfer results mainly from a flow imposed mechanically, whereas natural convection happens when heat transfer is mainly due to a buoyancy-induced flow. Mixed convection is a particular mode that occurs when mechanical and thermal causes coexist. Indeed, as soon as there is a temperature gradient in the fluid, one may expect buoyancy-induced flows to start. However, one will talk about natural convection only when buoyancy effects are prevailing. The limiting case when both forced convection and free convection are of equal importance is termed mixed (or combined) convection. This situation generally occurs when there are strong temperature gradients in the fluid and a small imposed velocity. The criterion defining the relative importance of natural and forced convections is the Richardson number, which is given by

$$\mathrm{Ri} = \frac{\mathrm{Gr}}{\mathrm{Re}^2} \qquad (30.15)$$

In summary, mixed convection occurs when $\mathrm{Ri} \cong 1$, natural convection prevails if $\mathrm{Ri} \gg 1$, and forced convection will be important if $\mathrm{Ri} \ll 1$.

A typical velocity profile of the laminar boundary layer mixed flow along a vertical heated plane surface is sketched in Figure 30.7b.

30.4.4.2 Convection in External Flows

When there is a temperature gradient between the solid surface and the main flow, one can also define a thermal boundary layer (Figure 30.7). The thermal boundary layer corresponds to the thin region close to the wall where temperature gradients are important and where heat transfer takes place. This layer is delimited by the wall surface on one side and the thermal boundary layer thickness (δ_T) on the other side. A thermal boundary layer develops when the wall surface is warmer or colder than the ambient fluid. Like with the dynamical boundary layer thickness (δ), a thermal boundary layer thickness (δ_T) can be defined both for forced convection and buoyancy-induced convection. δ_T is also a function of the longitudinal coordinate x.

In order to act on the temperature or the heat flux at the wall surface, the two most used thermal boundary conditions are the uniform wall temperature (UWT) condition and the uniform heat flux (UHF) condition. The UWT condition corresponds to a surface maintained at a constant temperature whereas the UHF condition corresponds to a surface heated or cooled with a constant heat flux.

30.4.4.3 Convection in Internal Flows

In a similar manner, velocity, and temperature gradients in the fluid layers close to the walls are found in internal flows (Figure 30.5). Convective heat transfer characteristics of the convection flow will strongly depend on these velocity and temperature profiles. Therefore, one must know not only the temperature difference ($T_\mathrm{wall} - T_\mathrm{fluid}$) but also the flow structure and temperature distribution in order to assess the convective heat exchange.

30.4.5 Modeling of Convective Heat Transfer

In order to model convective heat transfer, we need to know

- How heat is transferred between the wall surface and the fluid. This transfer will be modeled by heat flux laws (Fourier's law of conduction, Newton's law of cooling)
- How heat is transported in the fluid in motion. This will be obtained by application of the conservation principles resulting in the mass, momentum, and energy conservation equations (governing equations).

30.4.5.1 Heat Flux Laws

30.4.5.1.1 Fourier's Law

Here, we are interested in heat conduction from a solid surface to the fluid (in case of heating of the fluid) or from the fluid to the wall surface (cooling of the fluid).

Fourier's law of heat conduction for such a situation is given by

$$q'' = -k \frac{\partial T}{\partial n}\bigg|_{\text{wall}} \tag{30.16}$$

where

k is the thermal conductivity of the fluid
$\dfrac{\partial T}{\partial n}\bigg|_{\text{wall}}$ is the temperature gradient at the wall/fluid interface
   in the normal direction toward the fluid
q'' is the heat flux (W/m²)

30.4.5.1.2 Newton's Law of Cooling

In most cases, the temperature profile cannot be accessed easily. It is, therefore, easier to formulate the heat flux by a simple expression relating directly the heat flux to the temperature difference between the wall surface and the ambient fluid. Newton's law of cooling provides such a relation and is given by the following expression:

$$q'' = h(T_{\text{wall}} - T_{\text{fluid}}) \tag{30.17}$$

where h (in W/m² K) is the convective heat transfer coefficient that directly relates the heat flux q'' to the wall surface-to-fluid temperature difference ($T_{\text{wall}} - T_{\text{fluid}}$).

This formulation is convenient because of its simplicity. However, h depends on several parameters such as the flow-field, the physical dimensions, the temperature gradient, and the thermophysical properties of the fluid. These parameters can be grouped in dimensionless numbers, which are the Re or the Gr to account for the flow regime, and the Prandtl number, Pr = $\mu C_p/k$, for the thermophysical properties of the fluid.

30.4.5.1.3 Nüsselt Number

The convective heat transfer coefficient is often given under the form of a nondimensional parameter called the Nüsselt number (Nu). The definition of the Nüsselt number is

$$\text{Nu} = \frac{hL_0}{k} \tag{30.18a}$$

where L_0 is the same characteristic physical dimension as that defined for the Re or the Gr. The Nüsselt number is essentially a function of Re and Pr in forced convection, and of Gr and Pr in natural convection. The product Gr·Pr forms the Rayleigh number (Ra).

The Nüsselt number roughly compares the relative importance of convective heat transfer with respect to pure conduction (i.e., the case of a fluid at rest).

In forced convection, the convective heat transfer coefficient is also given by the nondimensional Stanton number (St):

$$\text{St} = \frac{h}{\rho C_p U_0} \tag{30.18b}$$

30.4.5.2 Conservation Equations (Governing Equations)

The fluid motion and the convective heat transfer are governed by the conservation of mass, momentum, and energy equations. For the sake of brevity, these equations are given below under the assumption of a two-dimensional, steady-state flow of an incompressible Newtonian fluid in the laminar regime and with negligible viscous dissipation. Furthermore, properties of the fluid are assumed constant. Thus, the equations, expressed in Cartesian coordinates, are as follows:

Conservation of mass equation or continuity equation:

$$\frac{\partial u}{\partial x} + \frac{\partial v}{\partial y} = 0 \tag{30.19}$$

Conservation of momentum equation:

$$x\text{-direction: } \rho\left(u\frac{\partial u}{\partial x} + v\frac{\partial u}{\partial y} \right) = -\frac{\partial P}{\partial x} + \mu\left(\frac{\partial^2 u}{\partial x^2} + \frac{\partial^2 u}{\partial y^2} \right) + \sum f_{bx} \tag{30.20a}$$

$$y\text{-direction: } \rho\left(u\frac{\partial v}{\partial x} + v\frac{\partial v}{\partial y} \right) = -\frac{\partial P}{\partial y} + \mu\left(\frac{\partial^2 v}{\partial x^2} + \frac{\partial^2 v}{\partial y^2} \right) + \sum f_{by} \tag{30.20b}$$

with $\sum f_{bi}$: sum of the body forces per unit volume in the *i*th-direction.

Conservation of energy equation:

$$\rho C_p\left(u\frac{\partial T}{\partial x} + v\frac{\partial T}{\partial y} \right) = k\left(\frac{\partial^2 T}{\partial x^2} + \frac{\partial^2 T}{\partial y^2} \right) \tag{30.21}$$

For an external boundary layer flow along a plane vertical heated wall (see Figure 30.7), the boundary layers are assumed to be very thin $\delta \ll x$ and $\delta_T \ll x$, so that gradients in the transverse (*y*) direction are stronger than in the main flow (*x*) direction (i.e., $\partial/\partial x \ll \partial/\partial y$). Furthermore, there is no pressure gradient across the boundary layer or along the flow, so that the set of equations reduces to

$$\frac{\partial u}{\partial x} + \frac{\partial v}{\partial y} = 0$$

$$u\frac{\partial u}{\partial x} + v\frac{\partial u}{\partial y} = \frac{\mu}{\rho}\left(\frac{\partial^2 u}{\partial y^2} \right) + g\beta(T - T_0)$$

$$\rho C_p\left(u\frac{\partial T}{\partial x} + v\frac{\partial T}{\partial y} \right) = k\left(\frac{\partial^2 T}{\partial y^2} \right)$$

where $g\beta(T - T_0)$ corresponds to the contribution of the buoyancy-induced force in the case of natural or mixed convection, and T_0 is the reference temperature of the ambient fluid. This set of equations will be used in the application of Section 30.6.

30.5 Convective Heat Transfer in Nanofluids

30.5.1 State of the Art

Conventional heat transfer fluids such as water, ethylene glycol, or oils are widely used for heat exchange purposes in the industry or in building applications. An increase in the thermal loads can generally be handled by an increase in the heat exchange surface. However, conventional fluids remain penalized by their limited thermal properties, among which is a low thermal conductivity. As a consequence, they are no longer suitable for some modern applications requiring a high level of performance while keeping a reduced size of the thermal system such as for the cooling of microprocessors or other electronic components (Nguyen et al. 2007a), micro-electromechanical systems, as well as to obtain fast transient regimes in heating systems. Therefore, several research works focusing on convective heat transfer, with nanofluids as working fluids, have been carried out during the last decade in order to test their potential for applications related to industrial heat exchangers.

During the past decade, research works focusing on convective heat transfer, with nanofluids as working fluids, have been carried out in order to test their potential for applications related to industrial heat exchangers. It is now well known that in forced convection (Pak and Cho 1998, Maïga et al. 2005, Buongiorno 2006, Ding et al. 2006, Heris et al. 2006, Nguyen et al. 2007a or b) using nanofluids can produce a considerable enhancement of the heat transfer coefficient that increases with the increasing nanoparticle volume fraction. As concerns natural convection or mixed convection, fewer results have been published in the literature until recently. Moreover, these investigations have been carried out on very diverse flow configurations: two-dimensional enclosures (Khanafer et al. 2003, Nnanna 2007, Ho et al. 2008), external boundary layers (Polidori et al. 2007), inside a horizontal cylinder (Putra et al. 2003), between horizontal parallel plates or discs (Kim et al. 2004, Wen and Ding 2005, 2006, Tzou 2008). Furthermore, the conclusions from the few published results in the literature seem to be controversial. For example, for a buoyancy-driven flow in a two-dimensional enclosure, Khanafer et al. (2003) have numerically found that the nanofluid heat transfer rate increases with the increase of nanoparticle volume fraction. On the other hand, the experimental study by Putra et al. (2003) for a natural convection case of copper and alumina–water nanofluids inside a horizontal differentially heated cylinder has shown an apparently paradoxical behavior of significant heat transfer deterioration. Wen and Ding (2005), using titanium dioxide nanoparticles, have also observed experimentally such deterioration in the natural convective heat transfer. Polidori et al. (2007) have tested two viscosity models. They have shown that external natural convection heat transfer is not solely characterized by the nanofluid effective thermal conductivity but that the viscosity also plays a key role. No definite conclusions could be drawn about the heat transfer enhancement because of the high sensitivity to the viscosity model. A few studies have also been carried out on mixed convection (Mirmasoumi and Behzadmehr 2008).

30.5.2 Influence of Nanofluid Viscosity and Other Thermophysical Properties

Although the use of nanofluids in natural, mixed, or forced convection seems extraordinarily promising, several aspects of nanofluid flow and heat transfer characteristics remain to be explored. Indeed, when flowing, the sole thermal conductivity of the nanofluid is no longer sufficient to evaluate the heat transfer efficiency (Polidori et al. 2007, Keblinski et al. 2008). In particular, one of the key points to be solved concerns the determination of the role of the viscosity on the resulting flow dynamics and heat transfer characteristics (Prasher et al. 2006). This comes from the fact that the rheology of nanofluids is quite singular in the sense that the viscosity of a nanofluid does not only depend on its temperature but also on the volume fraction and the shape of suspended nanoparticles. Keblinski et al. (2008) suggest that particle aggregation required to significantly enhance thermal conductivity also increases the fluid viscosity. Despite its importance in convective heat transfer, the particular rheology of nanofluids has not been given as much attention as the thermal conductivity. Most studies have been carried out only recently (Masuda et al. 1993, Wang et al. 1999, Das et al. 2003, Putra et al. 2003, Ding et al. 2006, Prasher et al. 2006, Avsec and Oblak 2007, Chen et al. 2007, Chevalier et al. 2007, Namburu et al. 2007, 2009, Nguyen et al. 2007b). Like for the thermal conductivity, no definite model has yet been proposed for the viscosity. According to the kind of nanofluids, the rheological behavior may either be Newtonian (e.g., alumina-based or copper-based nanofluids for low volume fractions, Putra et al. 2003, Heris et al. 2006) or non-Newtonian with a shear-thinning character (e.g., CNT, Ding et al. 2006). Basically, a strong increase in the viscosity is observed with an increase in the nanoparticle volume fraction. Thus, in the case of convective heat transfer, it is necessary to determine if the thermal performance of the nanofluid will remain better than that of the base fluid in spite of the increasing pressure drop. Furthermore, the complexity of the rheological behavior of nanofluids has presently led to the development of several viscosity models. These models still need to be tested and experimentally verified in real heat transfer situations. In addition, in the case of natural convection, the dependence of the physical properties with temperature will also have to be taken into account. From these observations, it clearly appears that additional investigations are required to analyze the relative importance of the thermophysical properties (volume expansion coefficient, dynamic viscosity, Prandtl number) in the qualification of the convective heat exchange. In addition, the influence of the specific properties of nanofluids both on transient flow regimes and on the transition between laminar and turbulent flow regimes is not yet well assessed.

30.6 Application

30.6.1 Introduction

The aim of this section is to enhance the discussion on the use of nanofluids for convective heat transfer applications. In order to reach this objective, an example of the application of nanofluids to natural convection is presented in order to illustrate the complexity of the modeling of convection in nanofluids and its dependence on the accurate modeling of all the thermophysical properties.

The following presentation will consist in a theoretical analysis of natural convection in external boundary layers subdivided into two main parts. The analysis is restricted to Newtonian nanofluids and is based on a macroscopic modeling under the assumption of constant thermophysical properties. The first part of the analysis presents the development of a theoretical model, derived from the integral formalism, and describing laminar natural convection along a uniformly heated vertical plate. Only one type of nanofluid (water–γ-Al$_2$O$_3$) is considered in this part. Two viscosity models proposed in the literature for alumina–water nanofluids are tested in order to compare their effects on the heat transfer efficiency. The second part of the analysis is dealing with the role of the flow regime (laminar or turbulent) in the use of Newtonian nanofluids in external free convection. Two Newtonian nanofluids, ethylene glycol–γ-Al$_2$O$_3$ and water–γ-Al$_2$O$_3$, are considered here. From the results of the first part, the same type of viscosity model is chosen for both nanofluids. The turbulent theory is established by extending laminar results to the turbulent regime. Then, the transition between the laminar and turbulent regimes is investigated by paying attention to the evolution of the critical Rayleigh number (or threshold) separating the two flow regimes as a function of the particle volume fraction of the nanofluids. Finally, concluding remarks are provided about the way nanofluids can enhance or not heat transfer in natural convection.

30.6.2 Position of the Problem

The physical system consists of a heated vertical plate under the thermal condition of uniform heat flux density (UHF) where laminar natural convection occurs (Figure 30.7c). The theoretical model assumes sufficiently small (realistic) temperature gradients across the boundary layer, so that thermophysical properties of the nanofluid are assumed to be constant except for the density variation in the buoyancy force, which is based on the incompressible fluid Boussinesq approximation. Since the solid particles have reduced dimension (~40 nm) and since they are easily fluidized, these particles can be considered to have a fluid-like behavior (Xuan and Roetzel 2000, Maïga et al. 2005). Furthermore, by assuming a local thermal equilibrium state and

negligible motion slip between the particles and the continuous phase, the nanofluid can be treated as a common pure fluid. Thus, one may expect that the classic theory for single-phase fluids can be extended to nanofluids.

In order to formulate the mathematical modeling foundation, the integral formalism has been derived based on the assumption that distinctive scaling lengths are considered for the dynamical and thermal boundary layer thicknesses, and that the ratio $\Delta = \delta_T/\delta$ of the thermal boundary layer thickness (δ_T) to that of the dynamical one (δ) is dependent only on the Prandtl number (Polidori et al. 2000) and not on the flow regime. The complete theoretical development associated with such an integral approach has been previously presented in detail for the UHF condition in Polidori et al. (2003) and Varga et al. (2004) for both laminar and turbulent free convection regimes. Therefore, for the sake of space, only the analytical solutions for the relative boundary layer thickness and the convective heat transfer coefficient are presented.

30.6.3 Laminar Free Convection Modeling

It is precised that the subscripts p, bf, nf, and r refer, respectively, to the particles, the base fluid, the nanofluid, and the ratio "nanofluid/base fluid" of the parameter considered. Usual integral forms of the boundary layer momentum and energy conservation equations can be directly extended to nanofluids:

$$\begin{cases} \dfrac{\partial}{\partial x} \displaystyle\int_0^{\delta_{nf}} U^2 \, dy = g\beta_{nf} \displaystyle\int_0^{(\Delta\delta)_{nf}} (T - T_\infty) \, dy - \nu_{nf} \left(\dfrac{\partial U}{\partial y} \right)_{y=0} \\ \dfrac{\partial}{\partial x} \displaystyle\int_0^{(\Delta\delta)_{nf}} (T - T_\infty) U \, dy = \dfrac{\nu_{nf}}{Pr_{nf}} \left(\dfrac{\partial T}{\partial y} \right)_{y=0} \end{cases} \quad (30.22)$$

Solving analytically system (30.22) with physically correct fourth-order polynomial profiles for flow velocity and temperature across their respective hydrodynamic and thermal boundary layers (Polidori et al. 2000, Varga et al. 2004) leads to a seventh-order polynomial in terms of $\Delta_{nf}(Pr_{nf})$:

$$\Delta_{nf}^7 - \frac{799}{126}\Delta_{nf}^6 + \frac{225}{14}\Delta_{nf}^5 - \frac{134}{7}\Delta_{nf}^4 + \frac{20}{3}\Delta_{nf}^3 + \frac{10}{9Pr_{nf}} = 0 \quad (30.23)$$

The values of the Δ_{nf} ratio for different Prandtl numbers are given in Table 30.3 and cover the physical situation, where $\delta_T(x) \le \delta(x)$ corresponding to usual fluids such that $Pr \ge 0.636$.

For conciseness, the details of the theoretical approach to access the heat transfer coefficient h are not repeated here. Using Fourier's law, the convective coefficient is expressed for the UHF surface conditions as follows:

TABLE 30.3 Effective Boundary Layer Thickness

Pr_{nf}	0.636	0.7	0.8	1	3	5	7	10	20	50	100
Δ_{nf}	1	0.977	0.947	0.901	0.733	0.680	0.653	0.630	0.597	0.574	0.565

$$h_{nf}(UHF) = \left[\frac{2g\beta_{nf}\varphi_w k_{nf}^4}{27(9\Delta_{nf}-5)\Delta_{nf}^4 v_{nf}^2 x}\right]^{\frac{1}{5}} \qquad (30.24)$$

In order to assess the influence of the particle volume concentration on a reference heat transfer, let us build the average Nüsselt number along the wall in terms of the base-fluid Gr:

$$\overline{Nu_{nf}} = \frac{\overline{h_{nf}}L}{k_{bf}} \qquad (30.25)$$

Thus, the average Nüsselt number calculation yields

$$\overline{Nu_{nf}^\star} = \frac{6}{5}\left[\frac{2\beta_r \, k_r^4}{27 \, v_r^2(9\Delta_{nf}-5)\Delta_{nf}^4}Gr_{bf}^\star\right]^{\frac{1}{5}} \qquad (30.26)$$

for the UHF surface condition, where $Gr_{bf}^\star$ is the modified Gr defined as follows:

$$Gr_{bf}^\star = \frac{g\beta_{nf}\varphi_w L^4}{k_{bf} v_{bf}^2} \qquad (30.27)$$

The convective heat transfer performance is called ε and defined as

$$\varepsilon(\%) = 100\left[\frac{X_{nf}}{X_{bf}}-1\right] \qquad (30.28)$$

where X is equally well the heat transfer parameter h or $\overline{Nu^\star}$.

Reporting either Equation 30.24 or Equation 30.26 in expression (30.25) and introducing the function

$$f(\Delta) = \frac{9\Delta_{bf}^5 - 5\Delta_{bf}^4}{9\Delta_{nf}^5 - 5\Delta_{nf}^4} \qquad (30.29)$$

yields the following characterization of the ε-function

$$\varepsilon(\%) = 100\left[\left(\frac{\beta_r k_r^4}{v_r^2}f(\Delta)\right)^\alpha - 1\right] \qquad (30.30)$$

where the subscript "r" refers to the nanofluid/base-fluid ratio and the α-constant depends on the surface condition and is 1/5 for the UHF conditions.

30.6.4 Thermophysical Properties of the γ-Al$_2$O$_3$/H$_2$O Nanofluid

Introducing the particle volume fraction ϕ, the thermophysical properties of the nanofluid, namely, the density, specific heat capacity, and volume coefficient of expansion have been

calculated from nanoparticle and base-fluid properties at the ambient temperature using Equations 30.28 through 30.30.

The Maxwell, or Wasp, model has been used for calculating the effective thermal conductivity of the nanofluid (Xuan and Li 2000, Xuan and Roetzel 2000):

$$k_r = \frac{k_{nf}}{k_{bf}} = \frac{k_p + 2k_{fb} - 2\phi(k_{bf}-k_p)}{k_p + 2k_{fb} + \phi(k_{bf}-k_p)} \qquad (30.31)$$

To illustrate the fact that the heat transfer performance of the nanofluids is not solely characterized by the effective thermal conductivity, we have considered the two following different models for the viscosity usually mentioned in the literature for oxide particle nanofluids. Model (I) is attributed to Brinkman (1952) and is currently used in literature (Xuan and Roetzel 2000, Khanafer et al. 2003, Gosselin and da Silva 2004) for nanofluids flowing under the natural convection regime as

$$\text{Model (I)} \quad \frac{\mu_{nf}}{\mu_{bf}} = \frac{1}{(1-\phi)^{5/2}} \qquad (30.7)$$

Model (II) is derived from experimental data and has recently been proposed by Maïga et al. (2004, 2005) to provide a better modeling of such nanofluids as

$$\text{Model (II)} \quad \frac{\mu_{nf}}{\mu_{bf}} = 123\phi^2 + 7.3\phi + 1 \qquad (30.32)$$

The same authors proposed the following formula for the effective Prandtl number of alumina–water nanofluid:

$$\frac{Pr_{nf}}{Pr_{bf}} = 82.1\phi^2 + 3.9\phi + 1 \qquad (30.33)$$

Variations of the thermal expansion coefficient, the conductivity, and the kinematic viscosity are shown in Table 30.4 and displayed in Figure 30.9 for the water-based nanofluid containing γ-Al$_2$O$_3$ with average diameter 42 nm initially at 20°C. Numerically deduced from Equation 30.23, the boundary layer parameter $f(\Delta)$ is also presented.

Several points can be noted from Figure 30.9, where the normalized parameters are plotted as a function of the nanoparticle fraction. Results are presented only for fractions up to 4% as no experimental data could be found in the literature concerning the rheological behavior of the nanofluid above this value. First, one can note that the effective thermal conductivity and volumetric expansion coefficient vary linearly with particle loading under the conditions of the present work. A very significant increase in thermal conductivity is observed, with a conductivity enhancement of ~12% for the particle volume fraction 4%. These results do agree with those of Eastman et al. (2001) for 35 nm average diameter Al$_2$O$_3$ particles. A *contrario*, a deterioration of the volumetric expansion coefficient of ~14% is achieved at the same 4% particle loading. Another major observation is that the behavior of the normalized dynamic viscosity is highly dependent on the model

TABLE 30.4   Physical Properties of the γ-Al$_2$O$_3$/H$_2$O Nanofluid at 20°C

| | Volume Fraction (%) of the γ-Al$_2$O$_3$/H$_2$O Nanofluid | | | | | |
	H$_2$O Only	Al$_2$O$_3$ Only	1	2	3	4
ρ (kg/m³)	998.3	3880	1027.12	1055.93	1084.75	1113.57
C_p (J/kg K)	4182	773	4147.91	4113.82	4079.73	4045.64
k (W/m K)	0.60	36	0.617	0.635	0.653	0.671
k_r	—	—	1.029	1.058	1.088	1.119
β (1/K)10⁻⁶	206	5	198.41	191.23	184.43	177.99
β_r	—	—	0.963	0.928	0.895	0.864
Model (I)						
v_I(m²/s) 10⁻⁶	1.00	—	0.997	0.994	0.993	0.993
v_r(I)	1.00	—	0.997	0.994	0.993	0.993
Pr(I)	6.96	—	6.88	6.80	6.73	6.66
Δ_{UHF}(I)	0.654	—	0.655	0.656	0.657	0.657
Model (II)						
v_{II}(m²/s)10⁻⁶	1.00	—	1.055	1.130	1.224	1.335
v_r(II)	1.00	—	1.055	1.130	1.224	1.335
Pr(II)	6.96	—	7.28	7.73	8.29	8.96
Δ_{UHF}(II)	0.654	—	0.651	0.647	0.642	0.637

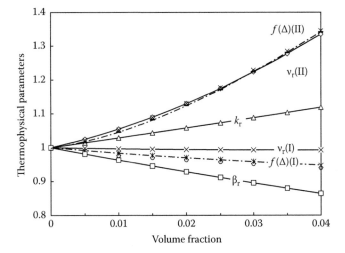

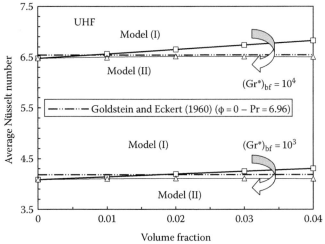

FIGURE 30.9   Thermophysical parameters of the γ-Al$_2$O$_3$/H$_2$O nanofluid.

FIGURE 30.10   The average Nüsselt number versus the particle loading.

used. Using model (I) yields a quasi-constant effective dynamic viscosity on the particle-loading range while model (II) gives a drastic augmentation of the dynamic viscosity of ~34% at a 4% particle loading. One will also note that the normalized dynamical parameter $f(\Delta)$ presents an evolution close to that of the dynamic viscosity, due to the fact that $f(\Delta)$ is Prandtl number dependent.

30.6.5 Heat Transfer Results with the γ-Al$_2$O$_3$/H$_2$O Nanofluid

The average Nüsselt number (Equation 30.26) based on the base-fluid thermal conductivity is plotted in Figure 30.10 for the UHF condition and two Gr's in the laminar regime. To validate the analytical modeling, the classic correlation for external

boundary layer natural convection attributed to Goldstein and Eckert (1960)

$$\overline{Nu}^*_{\phi=0} = 0.703 \, (Pr \, Gr^*)^{0.2} \tag{30.34}$$

is also reported, showing a close agreement for the reference case $\phi = 0$.

For the UHF thermal surface condition, the average Nüsselt number varies linearly with the volume fraction. This behavior is similar to that mentioned in Khanafer et al. (2003) and shows a significant heat transfer enhancement inherent to the use of Model (I). The results have also shown that the effective viscosity plays a key role in the natural convection heat transfer, as illustrated by the curves built with Model (II). Indeed, under the present work conditions, one can state that the use of

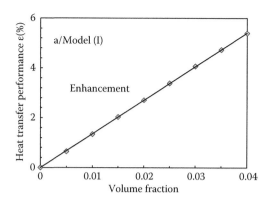
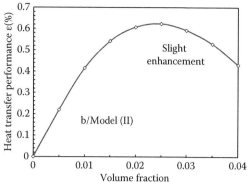

FIGURE 30.11 Heat transfer performance versus the particle loading for the two viscosity models (I) and (II).

nanoparticles for the purpose of heat transfer enhancement in natural convection with Model (II) seems illusory.

In order to get more details about these conclusions, Figure 30.11 presents the evolution of the ε-heat transfer performance versus the particle concentration. The first observation is that if Model (I) predicts an important enhancement of the nanofluid performance, a more contrasted conclusion can be drawn with Model (II). For the latter, a slight enhancement occurs in the particle-loading range with a maximum value of about 0.6% for a 2.5% volume fraction followed by a slight trend to a deterioration phenomenon. We recall that due to the lack of experimental data, the study is restricted to volume fractions up to 4%. One will notice that the extrapolation of the curves beyond the 4% volume fraction should yield to degradation in the heat transfer process as long as the nanofluid behavior remains Newtonian. As seen in Equation 30.30, what qualifies the heat transfer performance in external boundary layer natural convection is the ratio $(\beta_r k_r^4 / \nu_r^2) f(\Delta)$. Under the present adopted conditions, the following rough approximation $f(\Delta) \approx \nu_r$ can be established from Figure 30.9 independently of the thermal viscosity model. With such an approximation, a direct analysis of the new ratio $(\beta_r k_r^4 / \nu_r)$ seems sufficient to predict the heat transfer performance.

The above numerical results have eloquently shown that the use of Newtonian nanofluids for the purpose of heat transfer enhancement in laminar natural convection situations is not obvious, as such enhancement is dependent not only on nanofluids' effective thermal conductivities but also on their viscosities as well. The effect due to the kinematic viscosity seems to be dominant in the natural convective heat transfer in the way that such property will act more or less favorably on the enhancement process according to the Model used. Hence, caution should be taken in analyzing results, and further research efforts are, indeed, needed to develop suitable rheological models to predict the effective viscosity of such mixtures.

30.6.6 Turbulent Free Convection Modeling

Let us consider now the role the flow regime plays in the use of Newtonian nanofluids in external free convection. To make the analysis more complete, two Newtonian nanofluids, ethylene glycol–γ-Al$_2$O$_3$ and water–γ-Al$_2$O$_3$, are considered in this section.

For the modeling, results for the effective viscosity and Prandtl number are those proposed by Maïga et al. (2004, 2005) derived from model (II)

$$\frac{\mu_{nf}}{\mu_{bf}} = 306\phi^2 - 0.19\phi + 1 \tag{30.35}$$

and

$$\frac{Pr_{nf}}{Pr_{bf}} = 254.3\phi^2 - 3\phi + 1 \tag{30.36}$$

for ethylene glycol–γ-Al$_2$O$_3$ nanofluid.

Assuming that the turbulent boundary layer starts from the leading edge of the wall, the time-averaged integral forms of the boundary layer equations for the conservation of momentum and energy can be directly extended to nanofluids as

$$\begin{cases} \dfrac{\partial}{\partial x} \displaystyle\int_0^{\delta_{nf}} U^2 dy = g\beta_{nf} \displaystyle\int_0^{(\Delta\delta)_{nf}} (T - T_\infty) dy - (\nu + \nu_t)_{nf} \left(\dfrac{\partial U}{\partial y}\right)_{y=0} \\[4mm] \dfrac{\partial}{\partial x} \displaystyle\int_0^{(\Delta\delta)_{nf}} (T - T_\infty) U dy = -(a + a_t)_{nf} \left(\dfrac{\partial T}{\partial y}\right)_{y=0} \end{cases} \tag{30.37}$$

where ν_t and a_t denote, respectively, the eddy diffusivity of both momentum and heat.

Since there are no results in literature dealing with the growth of viscous and thermal layers, the present turbulent theory has been established under the assumption that the boundary layer laminar results could be easily extended to the turbulent regime, so that relation (30.23) remains still valid.

Deriving the Colburn analogy and using convenient wall shear stress under adequate velocity and temperature profiles (Kakaç and Yener, 1995, Varga et al., 2004) within the boundary layers give, after calculation, the turbulent Nüsselt number defined as

$$\left. Nu_{nf}^* \right|_{TURB} = 0.0631 \left(Ra_{bf}^*\right)^{\frac{2}{7}} \left[\frac{\sqrt{\Pi_\Delta Pr_{nf}}}{\Delta_{nf}} \left(1 + \frac{0.0823}{\Pi_\Delta Pr_{nf}^{10/15}}\right) \frac{k_r \nu_r^2}{\beta_r Pr_r} \right]^{-\frac{2}{7}} \tag{30.38}$$

where Π_Δ is a function of the Δ boundary layer ratio

$$\Pi_\Delta = \Delta_{nf}^{\frac{8}{7}}\left(\frac{7}{72} - \frac{7}{60}\Delta_{nf} + \frac{21}{253}\Delta_{nf}^2 - \frac{14}{435}\Delta_{nf}^3 + \frac{7}{1332}\Delta_{nf}^4\right) \quad (30.39)$$

In the laminar range, The Nüsselt number can also be written from Equation 30.26 as

$$\left.Nu_{nf}^*\right|_{LAM} = (Ra_{bf}^*)^{\frac{1}{5}}\left[\frac{2}{27(9\Delta_{nf} - 5)\Delta_{nf}^4 Pr_{nf}}\frac{\beta_r Pr_r}{k_r v_r^2}\right]^{\frac{1}{5}} \quad (30.40)$$

30.6.7 Laminar-to-Turbulent Transition Threshold

For lack of a physical criterion to define the transitional region, the present theoretical criterion used for transition to turbulence is similar to that chosen in Arpaci and Kao (2001) and Varga et al. (2004) corresponding to the location where laminar and fully developed turbulent flows interact together. Because transition is not a mathematical event, the most this analysis can do is to predict the order of magnitude of this transition.

We present in Figures 30.12 and 30.13 the evolutions of the local Nüsselt number with the base-fluid Rayleigh number in the particle-loading range, for the two nanofluids under study. The evolutions are built by the use of Equations 30.38 and 30.40, which materialize the two laminar and turbulent flow regimes. On the two graphs, a discontinuous line is used to materialize the laminar-to-turbulent threshold. We recall that this threshold is considered under a mathematical criterion corresponding to the point of intersection of two functions (30.387) and (30.40).

It can be seen that when the base-fluid Rayleigh number increases, the heat transfer parameter increases more, especially

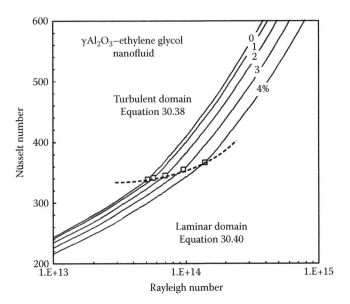

FIGURE 30.13 Nüsselt number for ethylene glycol–γ-Al$_2$O$_3$ versus the base-fluid Rayleigh number.

as the flow becomes turbulent ($Nu_{nf} \propto (Ra_{bf}^*)^{2/7}$ in the turbulent regime and $Nu_{nf} \propto (Ra_{bf}^*)^{1/5}$ in the laminar one).

About the particle volume fraction, what is significant is that, under the conditions of the study, the use of nanofluids to enhance the heat transfer seems illusory. The more the volume fraction is, the less the heat transfer is enhanced. This fact confirms a previous published work (Putra et al. 2003) in which the authors mentioned that unlike conduction or forced convection, a systematic and definite deterioration in natural convective heat transfer had been found to occur.

At a given particle loading and in the base-fluid Rayleigh number range wherein the two nanofluids present a laminar behavior in flowing, for $Ra_{bf}^* < 3.10^{11}$, comparing the two ethylene glycol–γ-Al$_2$O$_3$ and water–γ-Al$_2$O$_3$ nanofluids gives an enhancement about 6% in the heat transfer in favor of the ethylene glycol–Al$_2$O$_3$. On the other hand, what is singular is that the opposite phenomenon is observed for $Ra_{bf}^* > 5.10^{13}$ where the two nanofluids evolve in turbulent regime. Using this same ethylene glycol–γ-Al$_2$O$_3$ nanofluid induces important heat transfer degradation about 31% in the turbulent regime, at a given volume fraction.

To have a precise idea about the way the laminar-to-turbulent threshold evolves with the use of nanofluids, Figure 30.14 presents its evolution versus the particle loading. It can be seen that a similar trend is observed for the two nanofluids under study. Increasing the particle volume fraction induces an augmentation of the critical Rayleigh number, which results in a delay in appearance of turbulence. So, the concept of increasing the heat transfer with an earlier start of laminar-turbulent transition does not seem valid according to the results of the present study.

Thus, at the mathematical transition point, equaling Equations 30.38 and 30.40 gives the critical Rayleigh number, which is found to be only nanofluid Prandtl number–dependent.

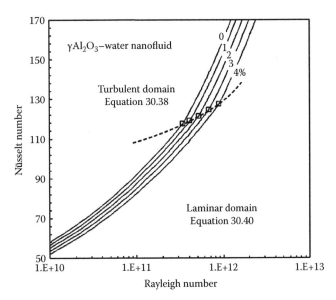

FIGURE 30.12 Nüsselt number for water–γ-Al$_2$O$_3$ versus the base-fluid Rayleigh number.

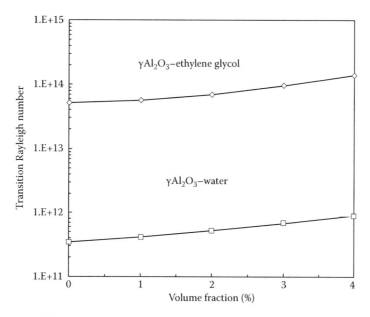

FIGURE 30.14 Critical (transition) Rayleigh number versus the nanofluid volume fraction.

Indeed, it is recalled that the boundary layer parameters Δ_{nf} and Π_Δ are also only Prandtl number dependent:

$$\left.Ra^*_{nf}\right|_c = 2.30\ 10^{11} \left[\frac{\Pi_\Delta^5}{(9\Delta_{nf} - 5)^7 \Delta_{nf}^{38}\ Pr_{nf}^2} \left(1 + \frac{0.0823}{\Pi_\Delta Pr_{nf}^{10/15}}\right)^{10} \right]^{\frac{1}{3}}$$

(30.41)

The evolution of the critical nanofluid Rayleigh number $\left.Ra^*_{nf}\right|_c$ is reported in Figure 30.15. The numerical results deduced from

Figures 30.12 and 30.13 are also reported by paying attention to transpose these results with the use of the relation

$$Ra^*_{nf} = Ra^*_{bf}\ \frac{\beta_r\ Pr_r}{k_r\ \nu_r^2}$$

(30.42)

It can be seen that the evolution versus the nanofluid Prandtl number is easily and suitably fitted by the correlation proposed by Varga et al. (2004) for single-phase fluid, namely,

$$\left.Ra^*_{nf}\right|_c = 1.32\ 10^{10} Pr_{nf}^{1.58}$$

(30.43)

This correlation gives a very interesting order of magnitude estimate of the transition threshold and should be extended to other kinds of Newtonian nanofluids.

30.6.8 Conclusion

The laminar and turbulent natural convection boundary layer along a vertical wall heated with a UHF has been theoretically investigated for ethylene glycol–γ-Al_2O_3 and water–γ-Al_2O_3 nanofluids in a wide range of Prandtl numbers (Pr = 6.96–262.42). The particle volume fraction is considered up to 4% to ensure a Newtonian behavior of the nanofluids. The modeling is based on the integral formalism taking into account the consideration of distinct thicknesses for both thermal and dynamical boundary layers by assuming that previous laminar effects can be easily extended to the turbulent region. The approach is based on both a macroscopic modeling and the use of an empirical viscosity model.

Despite the lack of data for comparison, deduced results have been shown to be in reasonable agreement with the few earlier

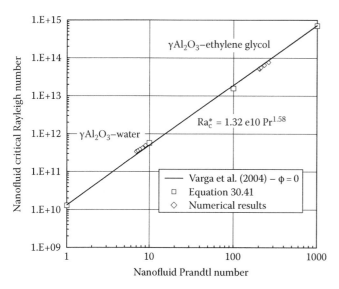

FIGURE 30.15 Critical (transition) Rayleigh number versus the nanofluid Prandtl number.

results in literature about the way the nanofluid can enhance or not heat transfer in natural convection.

Several conclusions can be drawn from this theoretical analysis:

1. Contrary to the general belief, it has been found that natural convection heat transfer in laminar regime is not solely characterized by the effective thermal conductivity. In fact, in external boundary layer natural convection, heat transfer is found to be characterized by the parameter group $(\beta_r k_r^4/v_r^2) f(\Delta)$, where $f(\Delta)$ is the boundary layer parameter function only of the mixture Prandtl number. Our results have shown that in natural convection, special care has to be exercised in drawing generalized conclusions about the heat transfer enhancement, because the sensitivity to the viscosity model used seems undeniable. The use of the Brinkman model for the dynamic viscosity yields a systematic and significant heat transfer enhancement, regardless of the particle concentration. On the other hand, the use of the experimental correlation for the viscosity leads only to a weak enhancement (less than 1%) with a trend to a deterioration phenomenon with increasing the particle concentration in the considered range.

2. Whatever the nanofluid (ethylene glycol–γ-Al$_2$O$_3$ or water–γ-Al$_2$O$_3$), increasing the nanoparticle volume fraction leads to a degradation in the external free convection heat transfer, compared to the base-fluid reference. This confirms previous conclusions about similar analyses and tends to prove that the use of nanofluids remains illusory in external free convection.

3. Flow regime is a parameter that seems to influence notably the heat transfer. Indeed, at a given volume fraction, comparing the two ethylene glycol–γ-Al$_2$O$_3$ or water–γ-Al$_2$O$_3$ nanofluids gives an enhancement about 6% in the heat transfer in the laminar regime in favor of the ethylene glycol–γ-Al$_2$O$_3$. *A contrario*, using this same nanofluid induces important heat transfer degradation about 31% in the fully turbulent regime.

4. Whatever the nanofluid (ethylene glycol–γ-Al$_2$O$_3$ or water–γ-Al$_2$O$_3$), increasing the particle volume fraction induces an augmentation of the critical nanofluid Rayleigh number, which results in a delay in appearance of turbulence.

5. Knowledge of the nanofluid Prandtl number allows one to use the correlation $Ra^\star_{nf}|_c = 1.32 \ 10^{10} Pr_{nf}^{1.58}$ to qualify the critical Rayleigh number.

References

Arpaci V.S. and Kao S.-H. 2001. Foundations of buoyancy driven heat transfer correlations. *Trans. ASME J. Heat Transfer* **123**: 1181–1184.

Avsec J. and Oblak M. 2007. The calculation of thermal conductivity, viscosity and thermodynamic properties for nanofluids on the basis of statistical nanomechanics. *Int. J. Heat Mass Transfer* **50**: 4331–4341.

Brinkman H.C. 1952. The viscosity of concentrated suspensions and solutions. *J. Chem. Phys.* **20**: 571–581.

Buongiorno J. 2006. Convective transport in nanofluids. *J. Heat Transfer* **128**: 240–250.

Chen H., Ding Y., and Tan C. 2007. Rheological behaviour of nanofluids. *New J. Phys.* **9** (367).

Chevalier J., Tillement O., and Ayela F. 2007. Rheological properties of nanofluids flowing through microchannels. *Appl. Phys. Lett.* **91**: n°233103.

Choi S.U.S. 1995. Enhancing thermal conductivity of fluids with nanoparticles. In *Developments and Applications of Non-Newtonian Flows*, D.A. Siginer and H.P. Wang (Eds.), FED-vol. 231/MD-vol. 66, ASME, New York, pp. 99–105.

Choi S.U.S., Zhang Z.G., Yu W., Lockwood F.E., and Grulke E.A. 2001. Anomalous thermal conductivity enhancement in nanotube suspensions. *Appl. Phys. Lett.* **79**: 2252–2254.

Das S.K., Putra N., and Roetzel W. 2003. Pool boiling characteristics of nano-fluids. *Int. J. Heat Mass Transfer* **46**: 851–862.

Ding Y., Alias H., Wen D., and Williams R.A. 2006. Heat transfer of aqueous suspensions of carbon nanotubes (CNT nanofluids). *Int. J. Heat Mass Transfer* **49**: 240–250.

Eastman J.A., Choi S.U.-S., Li S., Yu W., and Thompson L.J. 2001. Anomalously increased effective thermal conductivities of ethylene glycol-based nanofluids containing copper nanoparticles. *Appl. Phys. Lett.* **78**: 718–720.

Goldstein R.J. and Eckert E.R.G. 1960. The steady and transient free convection boundary layer on a uniformly heated vertical plate. *Int. J. Heat Mass Transfer* **1**: 208–218.

Gosselin L. and da Silva A.K. 2004. Combined heat transfer and power dissipation optimization of nanofluid flows. *Appl. Phys. Lett.* **85**: 4160–4162.

Hamilton R.L. and Crosser O.K. 1962. Thermal conductivity of heterogeneous two-component systems. *I&EC Fundam.* **1**: 182–191.

Hashin Z. and Shtrikman S. 1962. A variational approach to the theory of the effective magnetic permeability of multiphase materials. *J. Appl. Phys.* **33**: 3125.

Heris S.Z., Etemad S.Gh., and Esfahany M.N. 2006. Experimental investigation of oxide nanofluids laminar flow convective heat transfer. *Int. Commun. Heat Mass Transfer* **33**: 529–535.

Ho C.J., Chen M.W., and Li Z.W. 2008. Numerical simulation of natural convection of nanofluid in a square enclosure: Effects due to uncertainties of viscosity and thermal conductivity. *Int. J. Heat Mass Transfer* **51**: 4506–4516.

Kakaç S. and Yener Y. 1995. *Convective Heat Transfer*, 2nd edn., CRC Press, Boca Raton, FL.

Keblinski P., Philpot S.R., Choi S.U.S., and Eastman J.A. 2002. Mechanisms of heat flow in suspensions of nano-sized particles. *Int. J. Heat Mass Transfer* **45**: 855–863.

Keblinski P., Prasher R., and Eapen J. 2008. Thermal conductance of nanofluids: Is the controversy over? *J. Nanopart. Res.* **10**: 1089–1097.

Khanafer K., Vafai K., and Lightstone M. 2003. Buoyancy-driven heat transfer enhancement in a two-dimensional enclosure utilizing nanofluids. *Int. J. Heat Mass Transfer* **46**: 3639–3653.

Kim J., Kang Y.T., and Choi C.K. 2004. Analysis of convective instability and heat transfer characteristics of nanofluids. *Phys. Fluids* **16**: 2395–2401.

Lee S., Choi S.U.-S., Li S., and Eastman J.A. 1999. Measuring thermal conductivity of fluids containing oxide nanoparticles. *J. Heat Transfer* **121**: 280–289.

Maïga S.E.B., Nguyen C.T., Galanis N., and Roy G. 2004. Heat transfer behaviours of nanofluids in a uniformly heated tube. *Superlattices Microstruct.* **35**: 543–557.

Maïga S.E.B., Palm S.M., Nguyen C.T., Roy G., and Galanis N. 2005. Heat transfer enhancement by using nanofluids in forced convection flows. *Int. J. Heat Fluid Flow* **26**: 530–546.

Masuda H., Ebata A., Teramae K., and Hishinuma N. 1993. Alteration of thermal conductivity and viscosity of liquid by dispersing ultra-fine particles (dispersions of Al_2O_3, SiO_2, and TiO_2 ultra-fine particles). *Netsu Bussei (Japan)* **7**: 227–233.

Maxwell J.C. 1891. *A Treatise on Electricity and Magnetism*, 3rd edn., Clarendon Press, Oxford, U.K.

Mirmasoumi S. and Behzadmehr A. 2008. Numerical study of laminar mixed convection of a nanofluid in a horizontal tube using two-phase mixture model. *Appl. Thermal Eng.* **28**: 717–727.

Murshed S.M.S., Leong K.C., and Yang C. 2008. Thermophysical and electrokinetic properties of nanofluids: A critical review. *Appl. Thermal Eng.* **28**: 2109–2125.

Namburu P.K., Kulkarni D.P., Misra D., and Das D.K. 2007. Viscosity of copper oxide nanoparticles dispersed in ethylene glycol and water mixture. *Exp. Thermal Fluid Sci.* **32**: 397–402.

Namburu P.K., Das D.K., Tanguturi K.M., and Vajjha R.S. 2009. Numerical study of turbulent flow and heat transfer characteristics of nanofluids considering variable properties. *Int. J. Thermal Sci.* **48**: 290–302.

Nguyen C. T., Roy G., Gauthier C., and Galanis N. 2007a. Heat transfer enhancement using Al_2O_3–water nanofluid for an electronic liquid cooling system. *Appl. Thermal Eng.* **27**: 1501–1506.

Nguyen C.T., Desgranges F., Roy G. et al. 2007b. Temperature and particle-size dependent viscosity data for water-based nanofluids—Hysteresis phenomenon. *Int. J. Heat Fluid Flow* **28**: 1492–1506.

Nnanna A.G.A. 2007. Experimental model of temperature-driven nanofluid. *J. Heat Transfer* **129**: 697–704.

Pak B.C. and Cho Y. 1998. Hydrodynamic and heat transfer study of dispersed fluids with submicron metallic oxide particles. *Exp. Heat Transfer* **11**: 151–170.

Polidori G., Mladin E.C., and de Lorenzo T. 2000. Extension de la méthode de Karman-Pohlhausen aux régimes transitoires de convection libre, pour Pr > 0.6. *C.R. Acad. Sci. Paris – Série IIb.* **328**: 763–766.

Polidori G., Popa C., and Mai T.H. 2003. Transient flow rate behaviour in an external natural convection boundary layer. *Mech. Res. Commun.* **30**: 615–621.

Polidori G., Fohanno S., and Nguyen C.T. 2007. A note on heat transfer modelling of Newtonian nanofluids in laminar free convection. *Int. J. Thermal Sci.* **46**: 739–744.

Prasher R., Song D., Wang J., and Phelan P. 2006. Measurements of nanofluid viscosity and its implications for thermal applications. *Appl. Phys. Lett.* **89**: 133108.

Putra N., Roetzel W., and Das S.K. 2003. Natural convection of nanofluids. *Heat Mass Transfer* **39**: 775–784.

Shukla R.K. and Dhir V.K. 2005. Study of the effective thermal conductivity of nanofluids. *Proceedings of the ASME International Mechanical Engineering Congress and Exposition, IMECE05-80281*, Orlando, FL.

Tzou D.Y. 2008. Thermal instability of nanofluids in natural convection. *Int. J. Heat Mass Transfer* **51**: 2967–2979.

Varga C., Fohanno S., and Polidori G. 2004. Turbulent boundary-layer buoyant flow modeling over a wide Prandtl number range. *Acta Mech.* **172**: 65–73.

Wang X.-Q. and Mujumdar A.S. 2007. Heat transfer characteristics of nanofluids: A review. *Int. J. Thermal Sci.* **46**: 1–19.

Wang X., Xu X., and Choi S.U.S. 1999. Thermal conductivity of nanoparticle-fluid mixture. *J. Thermophys. Heat Transfer* **13**: 474–480.

Wasp F.J. 1977. *Solid-Liquid Slurry Pipeline Transportation*, Trans. Tech Publications, Berlin, Germany.

Wen D. and Ding Y. 2005. Formulation of nanofluids for natural convective heat transfer applications. *Int. J. Heat Fluid Flow* **26**: 855–864.

Wen D. and Ding Y. 2006. Natural convective heat transfer of suspensions of titanium dioxide nanoparticles (nanofluids). *IEEE Trans. Nanotechnol.* **5**: 220–227.

Xuan Y. and Li Q. 2000. Heat transfer enhancement of nanofluids. *Int. J. Heat Fluid Flow* **21**: 58–64.

Xuan Y. and Roetzel W. 2000. Conceptions for heat transfer correlation of nanofluids. *Int. J. Heat Mass Transfer* **43**: 3701–3707.

31

Theory of Thermal Conduction in Nanofluids

Jacob Eapen
North Carolina State University

31.1 Introduction

Nanofluids are engineered nanocolloidal suspensions for enhanced thermal transport. Colloids are ubiquitous. They are found in nature, such as in living cells, and are also commonly encountered in the chemical and biological industries. The modern science of colloids, which dates back to the early studies of Faraday in the mid-eighteenth century (Tweney, 2006), is multidisciplinary with overlapping domains in physics, chemistry, material science, and engineering (Hiemenz and Rajagopalan, 1997; Hunter, 1992). Traditionally, colloids are investigated for their diffusive and rheological properties. The diffusive properties are of paramount interest to the biological sciences while a fundamental understanding of both diffusive and shear behaviors is essential to the chemical industry.

The study of heat transport in solid dispersions is relatively recent. The thermophysical and transport properties of magnetic colloids (ferrofluids) with nanoparticles as small as 4 nm were reported in the 1970s and 1980s (see, for example, Fertman, 1987; Fertman et al., 1987; Popplewell et al., 1982). The viscosity and the thermal conductivity of ferrofluid suspensions were observed to increase with nanoparticle volume fraction and decrease with increasing temperature. Ahuja in 1975 (Ahuja, 1975a,b) showed that micrometer-sized polystyrene suspensions in aqueous glycerin and sodium chloride can increase the heat transfer by a factor of two under laminar flow conditions with marginal changes in the friction factor. In the 1990s, with the advent of nanotechnology, it has become possible to manufacture colloids with a variety of nanosized particles (such as oxides, carbides, and metals) that have the potential to exhibit superior thermophysical and transport properties. This raises the exciting possibility of employing nanofluids for enhancing the heat transfer in systems where a fluid is used as a medium to transfer energy.

The study of thermal conduction of nanofluids has a practical as well as an intrinsic value. On the practical front, enhanced thermal conductivities can potentially augment the heat transfer characteristics in engineering systems. On the theoretical front, the conduction mechanisms in colloidal systems, especially at nanoscales, have not been investigated in great detail thus far. This chapter elucidates the nonequilibrium principles of thermal conduction in colloidal systems followed by results from molecular dynamics simulations. The classical mean-field (or effective-medium) theories, which are discussed next, highlight *two limiting bounds*—an upper bound and a lower bound—of the classical Maxwell theory (Hashin and Shtrikman, 1962). The lower bound corresponds to a colloidal configuration of well-dispersed nanoparticles, and the upper bound represents a linear or a fractal nanoparticle arrangement (Eapen et al., 2010). Detailed comparisons to experimental data indicate that the nanofluid thermal conductivity is largely dependent on whether the nanoparticles stay dispersed in the base fluid, form a chain-like or fractal arrangement, or assume an intermediate configuration. Finally, the implications of thermal diffusion on aggregation and indirect effects on thermal conductivity are discussed (Gharagozloo et al., 2008; Hiemenz and Rajagopalan, 1997; Keblinski et al., 2008; Prasher et al., 2006b,c; Wensel et al., 2008).

31.1.1 Thermal Conduction Experiments and Model Predictions

Masuda and co-workers in Japan (Masuda et al., 1993) reported that the thermal conductivity of ultrafine suspensions of alumina, silica, and other oxides in water can increase by as much as 30% for low particle volume fraction (less than 5%). In the United States, Stephen Choi (Choi, 1995) in 1995 at the Argonne National Laboratory proposed to construct a new class of engineered fluids called nanofluids for superior heat transfer capabilities using dispersed nanoparticles. Since then, a series of experiments (a majority using hot-wire measurement systems) have been performed with nanofluids, and thermal conductivity enhancements and anomalous trends have been reported by several research groups.

The early experiments showed a fascinating increase (up to 40%) in the thermal conductivity with low nanoparticle volume fraction (<1%) (Eastman et al., 2001; Patel et al., 2003; Wang et al., 1999). These enhancements were observed to be larger than the predictions from the classical theory of Maxwell (see Figure 31.1) for well-dispersed nanoparticles. Larger enhancements were also reported in recent years (Chon et al., 2005; Chopkar et al., 2006, 2007; Hwang et al., 2007a; Hong et al., 2005; Kang et al., 2006; Li and Peterson, 2006, 2007a; Li and Xuan, 2006; Murshed et al., 2005, 2006; Penas et al., 2008; Philip et al., 2008; Schmidt et al., 2008b; Zhu et al., 2006, 2007). At the same time, several experiments showed modest conductivity enhancements (Ju et al., 2008; Keblinski et al., 2005; Putnam et al., 2006; Rusconi et al., 2006; Singh et al., 2009; Timofeeva et al., 2007; Venerus et al., 2006; Xie et al., 2002a; Zhang et al., 2006a,b), which were consistent with the Maxwell theory for well-dispersed nanoparticles.

In the Maxwell theory for well-dispersed nanoparticles (Maxwell, 1881), the ratio of effective thermal conductivity

of the nanofluid to that of the base fluid is given by Eapen et al., (2007c):

$$\frac{\kappa}{\kappa_f} = \frac{1 + 2\beta\phi}{1 - \beta\phi} \tag{31.1}$$

where

ϕ is the nanoparticle volume fraction
$\beta = [\kappa]/(\kappa_p + 2\kappa_f)$
$[\kappa] \equiv \kappa_p - \kappa_f$ is the difference between the thermal conductivities of the nanoparticle and the base fluid

If a finite temperature discontinuity exists at the nanoparticle–fluid interface, Maxwell model would still apply, provided the substitution is made $\kappa_f \to \kappa_f + \alpha\kappa_p$ (on the right-hand side), where $\alpha = 2R_b\kappa_f/d$, R_b is the interfacial thermal resistance, and d is the nanoparticle diameter (Benveniste, 1987; Nan et al., 1997). Subsequent mean-field theories have modified the original Maxwell model to take into account the effects of nanoparticle shape (Phelan et al., 2005). In the dilute limit, and for $\kappa_f \ll \kappa_p$, the enhancement in the thermal conductivity is a function of the volume fraction alone as shown by

$$\left(\frac{\kappa}{\kappa_f}\right)_{\beta\phi\to 0} = 1 + 3\beta\phi \approx 1 + 3\phi \tag{31.2}$$

In addition to larger-than-expected thermal conductivities, recent experiments also reveal other deviations from the Maxwell theory. The thermal conductivity is shown to exhibit an inverse dependence on the nanoparticle size (Chon et al., 2005; Hong et al., 2006; Kim et al., 2007; Li and Peterson, 2007a) and a quasi-linear dependence on the temperature (Das et al., 2003b; Li and Peterson, 2006; Li et al., 2008). Interestingly, there appears to be a fundamental difference between the thermal

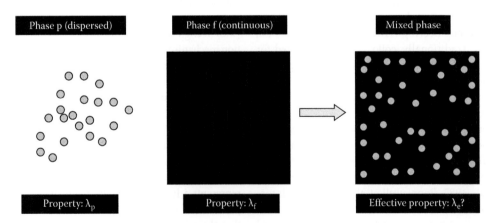

FIGURE 31.1 In the most recognized version of the classical Maxwell theory, the nanoparticles (dispersed phase p) are assumed to be well-dispersed in a base fluid (continuous phase f). The effective thermal conductivity (and other properties such as electrical conductivity and thermal diffusivity) in the colloidal state (mixed phase) are determined by combining all the nanoparticles into a single equivalent sphere. Since *continuous* conduction paths are formed through base fluid, the effective thermal conductivity of the above representation is biased toward that of the base fluid. In the dilute limit, the effective thermal conductivity is independent of the constituent thermal conductivities. The possibility of continuous conduction paths along nanoparticles (with linear or fractal-like agglomeration) is also allowed by the Maxwell theory. The details are discussed in Section 31.3.

conduction behavior of solid composites and nanofluids. In the former, smaller dispersed (or filler) particles, especially those in the nanometer size range, significantly reduce the matrix thermal conductivity. In some cases, the thermal conductivity is reduced well below that of the base material (Every et al., 1992) while in others the enhancement is mostly suppressed (Sofian et al., 2001). The solid composite behavior is easily explained through the interfacial thermal resistance, R_b, which has an inverse dependence on the particle diameter (Geiger et al., 1993; Hasselman and Donaldson, 2000; Nan et al., 1997). Thus, decreasing the filler particle size will reduce the effective thermal conductivity of the solid composites and *vice versa*. For the case of nanofluids, experimental data indicate that the thermal conductivity increases with decreasing nanoparticle size, a behavior that is clearly at odds with the Maxwell model (Chon et al., 2005; Hong et al., 2006; Kim et al., 2007; Li and Peterson, 2007a). A few experiments have also shown that the nanofluid thermal conductivity is not correlated in a simple manner to that of the nanoparticle, as predicted by the Maxwell model (Hong et al., 2005; Zhu et al., 2006). A limiting behavior at higher volume fractions is also observed in nanofluids, which is qualitatively different from that in solid composites. While the thermal conductivity displays a quadratic or power law behavior at higher volume fractions for solid composites (Pal, 2007; Zhang et al., 2007), it is known to rise rapidly at lower volume fractions for nanofluids and then saturate at higher volume fractions (Hong et al., 2005; Murshed et al., 2005; Zhu et al., 2006).

Several mechanisms have been recently proposed to account for the excess thermal conductivity beyond the Maxwell prediction for well-dispersed nanoparticles. These include Brownian motion of the nanoparticles (Bhattacharya et al., 2004; Kumar et al., 2004), fluid convection at the microscales (Jang and Choi, 2004, 2007; Kleinstreuer and Li, 2008; Koo and Kleinstreuer, 2004; Li and Peterson, 2007b; Patel et al., 2005; Prasher et al., 2005, 2006a; Shukla and Dhir, 2008; Yang, 2008), liquid layering at the particle–fluid interface (Avsec and Oblak, 2007; Tillman and Hill, 2007; Xie et al., 2005; Xue, 2003; Xue and Xu, 2005; Yu and Choi, 2003), nanoparticle shape (Gao and Zhou, 2006; Zhou and Gao, 2006), low-dimensional cluster aggregation (Eapen et al., 2007c, 2010; Gharagozloo et al., 2008; Keblinski et al., 2008; Prasher et al., 2006b,c; Timofeeva et al., 2007; Xu et al., 2006a), or a combination of these mechanisms (Feng et al., 2007; Ren et al., 2005; Wang et al., 2003; Xu et al., 2006b; Xuan et al., 2004). Most experimentally tested nanofluids have some level of aggregation that can range from chain-like configuration to large nanoparticle clumps. A particular focus of this chapter, therefore, is to investigate the effect of linear or fractal-like nanoparticle aggregation on the nanofluid thermal conductivity.

There are several other heat transfer properties/mechanisms such as the convective heat transfer coefficient (Bergman, 2009a; Buongiorno, 2006; Kolade et al., 2009; Wen and Ding, 2004b; Williams et al., 2008; Xuan and Li, 2003; Zeinali Heris et al., 2007); viscosity (Chevalier et al., 2007; Garg et al., 2008; Namburu et al., 2007; Prasher et al., 2006d; Schmidt et al., 2008a;

Tsai et al., 2008); specific heat (Bergman, 2009b; Zhou and Ni, 2008); thermal expansion (Nayak et al., 2009b); boiling, wetting, and critical heat flux (Bang et al., 2005; Chon et al., 2006; Das et al., 2003a; Kedzierski, 2009; Kim et al., 2008, 2009; Kumar and Milanova, 2009; Milanova and Kumar, 2005, 2008; Narayan et al., 2007; Pauliac-Vaujour et al., 2008; Sefiane, 2006; Vancea et al., 2008); and suppression of flow instabilities in flow loops (Nayak et al., 2008, 2009a) which are of interest to the thermal industry. These nanofluid properties, while critical to heat transfer applications (Bergman, 2009a), are not covered in this chapter. Several recent reviews present wide ranging discussions on the properties and the postulated transport mechanisms of nanofluids (Choi, 2009; Eastman et al., 2004; Murshed et al., 2008; Trisaksri and Wongwises, 2007; Wang and Mujumdar, 2007; Wen et al., 2009).

31.1.2 Colloidal Characteristics, Aggregation, and Thermal Conductivity

In a nanofluid complex physical and chemical adsorption mechanisms are expected at the nanoparticle surface. An uncharged surface can acquire charges either by preferential adsorption of an ion from the solution or through the ionization or dissociation of a surface group (Hiemenz and Rajagopalan, 1997). A charged colloidal particle is generally surrounded by ions that are of opposite sign. Even though there is an overall charge neutrality, the interface itself can affect several transport processes. The surrounding ions can move under the influence of thermal diffusion or an external field (Hunter, 1992). The distribution of charges on the particles and the presence of counterbalancing charges in the solution are known as the *electrical double layer*. Often, the double layer consists of two parts—an inner compact region (called Stern layer), which consists of ions that are bound tightly to the surface and an outer "diffuse" region where ions are free to move. The electric potential at the interface between the Stern and the diffusive layers is called the *zeta potential*. The zeta potential, which is a function of the surface charge of the colloidal particles and the adsorbed layer in the Stern layer, is an indicator of colloidal stability.

It is evident from the experiments thus far that the thermal conductivity enhancements are simply not a function of the volume fraction alone. It is shown that (Wensel et al., 2008; Xie et al., 2002b) the thermal conductivity in an aqueous media can be a function of pH (a measure of the hydrogen ions) of the colloidal system. The experiments by Lee et al. (2006) also show that the thermal conductivity is proportional to the product of the charges' site density and the ion density. These experiments highlight the importance of surface charges on the effective nanofluid thermal conductivity.

The diffusion of ions in the double layer will not have any significant influence on the nanofluid thermal conductivity (see Section 31.2.1). This is easily seen by comparing the time scales of diffusion and thermal conduction. The question that remains is: how does the surface chemistry affect the nanofluid thermal conduction process? It is well-known in the colloidal community that surface chemistry alters the aggregation structure of colloidal particles. A large number of investigations on colloids show that the dispersed phase exists in the form of aggregated structures, and not as primary particles (Hiemenz and Rajagopalan, 1997). By controlling the kinetics, several investigators have developed configurations with linear, chain-like or fractal structures that extent from several tens of nanometers to hundreds of nanometers (Carrillo et al., 2003; González, 1993; Huang and Yang, 1999; LaRosa and Cawley, 1992; Kellner and Kohler, 2005; Kovalchuk et al., 2008; Lee and Furst, 2006; Majolino et al., 1989; Martin et al., 1990; Meakin, 1992; Meakin and Deutch, 1987; Meakin et al., 1985; Odriozola et al., 2003; Starchev and Stoylov, 1993; Thurn-Albrecht et al., 1999; Weitz and Oliveria, 1984; Weitz et al., 1984, 1985; Wilcoxon et al., 1989; Wiltzius, 1987). These fractal structures are observed for a variety of colloidal dispersions (even at dilute volume fractions) and are dependent on the solution chemistry, surface charges, and the thermal (Brownian) motion of the nanoparticles. External fields such as gravity and temperature disrupt the quiescent conditions that are needed for fractal growth. Gravitational field, in general, supports sedimentation which leads to nonfractal or clumps of nanoparticles while external fields such as temperature and momentum can potentially break up the aggregates. Thus, for experimentally tested nanofluids, there is a competition between the growth of fractal-like structures, coalescence, sedimentation, and fragmentation (Meakin, 1992). The thermal conductivity (and other properties such as viscosity and diffusion) of a nanofluid system is sensitive to the geometrical configuration and connectivity of the aggregated structures (Eapen et al., 2010; Prasher et al., 2006b,c). The theory of thermal conduction with aggregated configurations is reviewed in Section 31.3 along with detailed comparisons to experimental data.

There exists an unusually large spread in the experimental data for nanofluids with the same constituents (see, for example, the spread in data for alumina and copper oxide data in Figure 31.6). In a recent International Nanofluid Properties Benchmark Exercise (INPBE), several experimental techniques, including hot-wire and optical methods, are compared with blind nanofluid samples (Buongiorno et al., 2009). The study concludes that all common thermal conductivity measurement systems are similar in accuracy and precision eventhough several outliers indicate the difficulties associated with accurate thermal conductivity measurements in complex colloidal systems.

31.2 Theory

31.2.1 Linear Phenomenological Theory

Transport relationships are typically determined between fluxes which develop in response to gradients in a system. A simple example is Fourier's law of thermal conduction, which states that the heat flux is proportional to the gradient in the temperature. The *proportionality constant* is known as the thermal

conductivity, which is a property of the medium. At the macroscopic or continuum level, thermal conductivity is simply regarded as an empirical constant, which, in principle, can be accessed by experiments.

Thermal conduction belongs to the more general family of transport phenomena which arise when a system initially in equilibrium is perturbed by an external force or disturbance. A system is said to be in equilibrium when there are no macroscopic gradients (such as flow of mass, momentum, and energy), and the intensive macroscopic state variables such as density and temperature are invariant in time. When such a system is disturbed by an external or internal disturbance, the system is considered to be in a nonequilibrium state. Local thermodynamic equilibrium (LTE) is often invoked for systems that are marginally out of equilibrium. LTE states that all the thermodynamic functions exist for each element of the system, and the thermodynamic quantities for a nonequilibrium system are the same functions of the local state variables as the corresponding equilibrium quantities (Fitts, 1962). LTE does not follow from the first and second laws of thermodynamics but is an additional assumption. If the gradients in the thermodynamics functions are small or if the time variation of these functions is slow compared to the slowest relaxation of the system, then LTE can be considered to be a good approximation. A *linear* phenomenological theory is usually employed to characterize the nonequilibrium transport processes. This theory, which is adapted from de Groot and Mazur (1984), Evans and Morriss (1990), Fitts (1962), and Hanley (1969), is described briefly in this chapter to highlight the coupled transport in nanofluids.

The well-known phenomenological relationships for mass and momentum transfer are given by Fick's law of diffusion and Newton's law of viscosity, respectively. As previously stated, Fourier's law is used to describe thermal conduction. When two or more gradients are present in the system, then the above constitutive relationships are not valid, in general. For example, if both temperature and concentration gradients are present, then energy can be transported by both conduction and diffusion of the atoms or molecules. As a generalization, the linear phenomenological theory postulates that the fluxes are linear homogeneous functions of the corresponding gradients. For an *n*-component system, this linearity is expressed as (de Groot and Mazur, 1984)

$$\mathbf{J} \equiv \mathbf{L}\hat{\mathbf{X}}; \quad \mathbf{J}_i = \sum_{k=1}^{n} L_{ik} \hat{X}_k \quad (i = 1, 2, ..., n) \quad (31.3)$$

where

    $\mathbf{J}$ and $\hat{\mathbf{X}}$ are the generalized flux and gradient vectors, respectively

    $\mathbf{L}$ is a matrix containing the phenomenological coefficients

The cross coefficients L_{ik} and L_{ki} are postulated to be equal in the linear phenomenological theory. This equality, also known as Onsager's reciprocity condition, can be derived from statistical mechanics on the basis of microscopic irreversibility. When $L_{ik} = 0$, the familiar constitutive relationships are recovered. The fluxes and gradients (or the thermodynamic "forces") are connected to the second law of thermodynamics through the entropy production (σ) which is given by

$$\sigma = \sum_{i=1}^{n} J_i \hat{X}_i \qquad (31.4)$$

It is possible by linear transformation to create new fluxes and gradients which satisfy Onsager's reciprocity conditions. For example, heat flux in a multicomponent system cannot be uniquely defined as it is impossible to separate the internal energy flux into purely diffusive and conductive terms. This can be readily seen from the microscopic picture, as described by Fitts (1962). The total energy of a system of a certain volume is the sum of its kinetic and potential energy. If a molecule leaves the system boundary, then the kinetic energy loss may be attributed to diffusion. If, however, a molecule inside the boundary transfers its kinetic energy to a molecule which is outside the boundary, then this loss can be accounted as the heat loss. So for the kinetic energy, there is no ambiguity in partitioning the total energy flux. It is, however, not so evident for the potential energy which is the sum of all the potential energies interactions. In other words, the kinetic energy of a molecule can be uniquely defined while the potential energy can only be defined for a system of molecules. Often, the potential energy interaction is divided equally between two molecules if they have a pair-wise interaction. There is no rigorous justification for this arbitrary procedure, even though it is a useful procedure in the microscopic theoretical development. Thus, changes in the potential energy when a molecule escapes the boundary or when it interacts with one another within the system cannot be uniquely partitioned into diffusive and conductive contributions.

A common definition for the heat flux follows from the second law of thermodynamics. For an n component system, the heat flux is given by (de Groot and Mazur, 1984)

$$\mathbf{J}_q = \hat{\mathbf{J}}_q - \sum_{k=1}^{n} h^k \mathbf{J}^k \qquad (31.5)$$

where
  $\hat{\mathbf{J}}_q$ is the heat flux which is usually measured in an experiment
  $\mathbf{J}_q$ is the reduced or conductive heat flux
  $\mathbf{J}^k$ and h^k are the mass flux and partial enthalpy of the kth component of the system, respectively

Note that the difference between $\hat{\mathbf{J}}_q$ and $\mathbf{J}_q$ represents heat transfer due to diffusion. This choice leads to (de Groot and Mazur, 1984; Hanley, 1969)

$$\mathbf{J}_q = L_{qq} \mathbf{X}_q + \sum_{k=1}^{n-1} L_{qk} \mathbf{X}^k \qquad (31.6)$$

$$\mathbf{J}^i = L_{iq} \mathbf{X}_q + \sum_{k=1}^{n-1} L_{ik} \mathbf{X}^k \qquad (31.7)$$

with

$$\mathbf{X}_q = \frac{-\nabla T}{T^2} \qquad (31.8)$$

$$\mathbf{X}^k = \frac{-\nabla(\mu^k - \mu^n)_T}{T} \qquad (31.9)$$

where
  $\mathbf{J}$ is the mass flux
  T is the temperature
  μ is the chemical potential
  the superscripts k and n stand for any two species

For a simple binary colloidal system with components (α, β), the phenomenological relationships reduces to

$$\mathbf{J}_q = -L_{qq} \frac{\nabla T}{T^2} - L_{q1} \frac{1}{T} \nabla(\mu^\alpha - \mu^\beta)_T \qquad (31.10)$$

$$\mathbf{J}^\alpha = -L_{1q} \frac{\nabla T}{T^2} - L_{11} \frac{1}{T} \nabla(\mu^\alpha - \mu^\beta)_T \qquad (31.11)$$

In experiments, diffusion is associated with a mass concentration gradient (x) rather than the chemical potential. By invoking Gibbs–Duhem identity (de Groot and Mazur, 1984)

$$\sum_{k=\alpha}^{\beta} x^k (\nabla \mu^k)_{T,p} = 0 \qquad (31.12)$$

and noting

$$(\nabla \mu^\alpha)_{T,p} = \left(\frac{\partial \mu^\alpha}{\partial x^\alpha} \right)_{T,p} \nabla x^\alpha \qquad (31.13)$$

the following expression in the matrix form, $\mathbf{J} = \mathbf{L}\hat{\mathbf{X}}$, is obtained:

$$\begin{pmatrix} \mathbf{J}_q \\ \mathbf{J}^\alpha \end{pmatrix} = - \begin{bmatrix} \dfrac{L_{qq}}{T^2} & L_{q1} \dfrac{1}{Tx^\beta} \left(\dfrac{\partial \mu^\alpha}{\partial x^\alpha} \right)_{T,p} \\ \dfrac{L_{1q}}{T^2} & L_{11} \dfrac{1}{Tx^\beta} \left(\dfrac{\partial \mu^\alpha}{\partial x^\alpha} \right)_{T,p} \end{bmatrix} \begin{pmatrix} \nabla T \\ \nabla x^\alpha \end{pmatrix} \qquad (31.14)$$

From the second law of thermodynamics, the necessary conditions for nonnegative entropy production are as follows: (1) $\mathbf{L}$ should be semipositive definite and (2) $\mathbf{L}$ is symmetric (Onsager reciprocity condition). These conditions result in (de Groot and Mazur, 1984)

$$L_{1q} = L_{q1}, \quad L_{11} \geq 0, \quad L_{qq} \geq 0 \tag{31.15}$$

$$L_{qq}L_{11} - \frac{1}{4}(L_{1q} + L_{q1})^2 = L_{qq}L_{11} - (L_{q1})^2 \geq 0 \tag{31.16}$$

The above relationships provide the limits placed by the second law of thermodynamics on the phenomenological transport coefficients. The diffusion flux **J** in the phenomenological equations is generally defined in the center-of-mass (CM) coordinate frame. It is defined as

$$\mathbf{J}^\alpha = \rho^\alpha (\mathbf{v}^\alpha - \mathbf{v}_{CM}) \tag{31.17}$$

where
- ρ is the total mass density
- $\mathbf{v}_{CM}$ is the barycentric or CM velocity

According to a theorem by Prigogine (de Groot and Mazur, 1984; Hanley, 1969), the diffusion flux can be defined in any frame of reference which leaves the entropy production invariant. For practical purposes, diffusion can, therefore, be measured in lab coordinates with respect to an arbitrary flow field $\mathbf{v}^\theta$.

The phenomenological coefficients (L) are related to experimental transport coefficients as (de Groot and Mazur, 1984)

$$\kappa = \frac{L_{qq}}{T^2} : \text{thermal conductivity} \tag{31.18}$$

$$D'' = \frac{L_{q1}}{\rho x^\alpha x^\beta T^2} : \text{Dufour coefficient} \tag{31.19}$$

$$D_T = \frac{L_{1q}}{\rho x^\alpha x^\beta T^2} : \text{thermal diffusion coefficient} \tag{31.20}$$

$$D_{\alpha\beta} = \frac{L_{11}}{\rho x^\beta T}\left(\frac{\partial \mu^\alpha}{\partial x^\alpha}\right)_{T,p} : \text{mutual diffusion coefficient} \tag{31.21}$$

In the above expressions, D_T, D'', and $D_{\alpha\beta}$ stand for the thermal diffusion coefficient, the Dufour coefficient, and the mutual (binary) diffusion coefficient, respectively. The density of the system is given by ρ. Thermal diffusion coefficient (D_T) accounts for the flow of matter with a temperature gradient while the Dufour coefficient (D'') is a measure of the inverse effect, which is the flow of heat due to the concentration gradient. The ratio of D_T to $D_{\alpha\beta}$ is known as the Soret coefficient (S_T). κ is the thermal conductivity of the colloidal system, and it is clear that it is a native property to the nanofluid without any contribution from the diffusion. For this reason, κ will be referred to as the intrinsic thermal conductivity of the nanofluid. As will be discussed subsequently, typical experiments do not measure the intrinsic thermal conductivity. Now, the set of flux relationships can be written in matrix form as

$$\begin{pmatrix} \mathbf{J}_q \\ \mathbf{J}^\alpha \end{pmatrix} = - \begin{bmatrix} \kappa & \rho_1 \left(\dfrac{\partial \mu^\alpha}{\partial x^\alpha}\right)_{T,p} TD'' \\ \rho x^\alpha x^\beta D_T & \rho D_{\alpha\beta} \end{bmatrix} \begin{pmatrix} \nabla T \\ \nabla x^\alpha \end{pmatrix} \tag{31.22}$$

With the law of entropy, the Onsager reciprocity relationship and the positivity of $(\partial \mu^\alpha / \partial x^\alpha)_{T,p}$, the following conditions on the transport coefficients can be derived (de Groot and Mazur, 1984):

$$\kappa \geq 0, \quad D_{\alpha\beta} \geq 0 \tag{31.23}$$

$$(T\rho(x^\alpha)^2 x^\beta)\left(\frac{\partial \mu^\alpha}{\partial x^\alpha}\right)_{T,p} \frac{D_T^2}{\kappa D_{\alpha\beta}} \leq 1 \tag{31.24}$$

The phenomenological theory, therefore, shows that the thermal conductivity and the mutual diffusion constant should be non-negative at all times while allowing for sign changes in the thermal diffusion and Dufour coefficients. If the thermal diffusion coefficient is positive, it means that the nanoparticles are diffusing from a higher temperature to a lower temperature and *vice versa*. The same analogy holds for the Dufour effect.

As seen before, there are multiple definitions for heat flux which lead to different effective thermal conductivities for the system. In an experiment, measurement can be made at the beginning ($\nabla x^\alpha = 0$) or at the end ($\mathbf{J}^\alpha = 0$). The four thermal conductivities for these limiting conditions can be written as (de Groot and Mazur, 1984)

$$\mathbf{J}_q = -\zeta \nabla T, \quad \hat{\mathbf{J}}_q = -\hat{\zeta} \nabla T \quad (\nabla x^\alpha = 0) \tag{31.25}$$

$$\mathbf{J}_q = -\lambda \nabla T, \quad \hat{\mathbf{J}}_q = -\hat{\lambda} \nabla T \quad (\mathbf{J}^\alpha = 0) \tag{31.26}$$

The reduced heat flux ($\mathbf{J}_q$) is associated with two thermal conductivities (ζ, λ) and the heat flux ($\hat{\mathbf{J}}_q$) with ($\hat{\zeta}$, $\hat{\lambda}$). The difference between the heat fluxes given in Equation 31.5 is solely due to the heat carried by the diffusing particles. It can be shown that (de Groot and Mazur, 1984)

$$\zeta = \kappa, \quad \hat{\zeta} = \kappa + D_T(h^\alpha - h^\beta)\rho x^\alpha x^\beta \tag{31.27}$$

$$\lambda = \kappa - \frac{(D_T)^2}{D_{\alpha\beta}}\left(\frac{\partial \mu^\alpha}{\partial x^\alpha}\right)\rho(x^\alpha)^2 x^\beta T,$$

$$\hat{\lambda} = \lambda = \kappa - \frac{(D_T)^2}{D_{\alpha\beta}}\left(\frac{\partial \mu^\alpha}{\partial x^\alpha}\right)\rho(x^\alpha)^2 x^\beta T \tag{31.28}$$

Note that $\hat{\mathbf{J}}_q$ is the heat flux which is usually measured in an experiment. Thus, at the end of the measurement when the mass fluxes are zero, the reduced thermal conductivity (λ) is *always* less than the intrinsic thermal conductivity (κ). It is interesting to note that the typical experiments do not measure the intrinsic thermal conductivity (κ). This difference in the thermal conductivities is usually small (few percents) for typical colloids, as shown by a simple estimation below for the condition, $\nabla x^\alpha = 0$:

$$\Delta\kappa \equiv \left|\hat{\zeta} - \kappa\right| = \left|D_T\right|\left|h^\alpha - h^\beta\right|\rho x^\alpha x^\beta$$

$$= O(10^{-10})\frac{m^2}{sK} \times O(10^4)\frac{J}{kg} \times O(10^3)\frac{kg}{m^3} \times 0.1 \times 0.9 = O(10^{-4})\frac{W}{mK}$$

$$\text{(31.29)}$$

In the above estimation, water is assumed as the base fluid with a nanoparticle mass fraction of 10%. A Soret coefficient of $O(0.1)$ is assumed (Gharagozloo et al., 2008; Putnam et al., 2006) along with the mutual diffusion coefficient $D_{\alpha\beta}$ conservatively bounded by the self-diffusivity of the water molecules $[O(10^{-9})\ m^2/s]$. The enthalpy of nanoparticles is neglected for estimating the upper bound on the change in thermal conductivity (NIST, 2007). A similar estimate for the second condition ($\mathbf{J}^\alpha = 0$) also results in a negligible change in the thermal conductivity (calculations not shown). Thus, it is clear that the diffusion of nanoparticles in a temperature gradient does not influence the thermal conductivity of a nanofluid significantly. The above conclusion is in agreement with several published reports (for example, refer Gharagozloo et al., 2008; Keblinski et al., 2002; Nie et al., 2008; Timofeeva et al., 2007).

The mutual diffusion coefficient ($D_{\alpha\beta}$), which arises from Fick's law, is different from the Brownian self-diffusion coefficients, D_α and D_β (which arise from the random self-walk of the constituent atoms or molecules). In the Hartley and Crank model, the mutual (or binary) diffusion coefficient of an ideal nanofluid is related to the self-diffusion coefficients as (Bertucci and Flygare, 1975; Hartley and Crank, 1949)

$$D_{\alpha\beta} = n_\alpha D_\beta + n_\beta D_\alpha \qquad \text{(31.30)}$$

where n_α and n_β are the mole fractions of the components α (nanoparticles) and β (base fluid), respectively. Generally, the mole fraction of nanoparticles is small, and hence $D_{\alpha\beta} \approx D_\alpha$. Note that ideality and dilute conditions are not satisfied in many experimentally tested nanofluid systems.

Thermodiffusion (D_T), mutual diffusion ($D_{\alpha\beta}$), and Brownian self-diffusion of the nanoparticles directly affects the aggregation processes, and thus, indirectly influences the thermal conductivity and viscosity of nanofluid systems. As discussed earlier, aggregation is a strong function of the diffusion processes (Gmachowski, 2002; Hess et al., 1986; Kim and Kramer, 2006; Weitz et al., 1984; Wiltzius, 1987). Thus, the mechanisms of slow diffusional processes (of the order of seconds to hours) are extremely important to understand the thermal and viscous transport mechanisms (Gharagozloo et al., 2008; Keblinski et al., 2008; Prasher et al., 2006c).

The correspondence between the experimentally observed conductivities and heat fluxes, and those that can be theoretically determined, is not always clear. The following sections will address this concern by following the appropriate developments in statistical mechanics and linear response theory.

31.2.2 Microscopic Origin of Heat Fluxes and Thermal Conductivity

The linear phenomenological theory described in the previous section lays the foundation for the thermal transport in a colloid. The multiplicity of heat flux and thermal conductivity definitions are unique to multicomponent systems. The fluxes and transport coefficients, which are described so far, are all macroscopic in nature, and they are expected to be derived through experiments. They, however, are ultimately dependent on the molecular mechanisms, and several frameworks exist where all the fluxes and transport coefficients are derivable from microscopic physics. The kinetic theory of Boltzmann (Chapman and Cowling, 1970) and the extension by Bogolubov (Hanley, 1969), Chapman, and Enskog and the time correlation formalism (Boon and Yip, 1991; Chandler, 1987; Evans and Morriss, 1990; McQuarrie, 2000; Zwanzig, 1965) developed by Green, Kubo, Callen, Zwanzig, Mori, and others are two such frameworks.

31.2.2.1 Time Correlation Functions

A theoretically apposite approach for characterizing the transport process in colloids is through the linear response theory (Chandler, 1987; Hanley, 1969), a powerful and a general framework suitable for analyzing all transport phenomena regardless of the physical state. In this theory, the transport coefficients can be computed from the time correlation functions of (instantaneous) microscopic fluctuations. In an equilibrium state, spontaneous fluctuations occur all the time, and time correlations are spatiotemporal functions that describe these fluctuations (Boon and Yip, 1991). For a given dynamical variable (A), the fluctuation (δA) can be defined as

$$\delta A(t) = A(t) - \left\langle A \right\rangle \qquad \text{(31.31)}$$

where $\left\langle A \right\rangle$ denotes the ensemble average of A. Under equilibrium conditions, the ensemble average of the fluctuations in A is zero. However, the ensemble average of the correlations between fluctuations is not zero, in general. In the linear response theory, it is shown that the correlations between such fluctuations under equilibrium conditions are related to nonequilibrium transport properties. Specifically, the thermal conductivity of a nanofluid is related to the autocorrelation of the heat flux fluctuations. A formal definition of a correlation function of two fluctuations δA and δB, which vary in time t, is given by Evans and Morriss (1990)

$$C_{AB}(t) = \int d\Gamma f_N^0 \delta A(t + \tau)\delta B(\tau) \equiv \left\langle \delta A(t + \tau)\delta B(\tau) \right\rangle \quad \text{(31.32)}$$

where f_N^0 is the equilibrium distribution function which corresponds to the steady unperturbed equations of motion with no explicit time dependence. Classical time correlations are not dependent on a particular choice of time, and hence, they are the functions of only the differences in time (called the stationary property). The correlation function now can be expressed as

$$C_{AB}(t) = \int d\Gamma f_N^0 \delta A(t) \delta B(0) \equiv \langle \delta A(t) \delta B(0) \rangle \qquad (31.33)$$

where the angular brackets denote the ensemble average. When the same variable is considered, the resulting expression becomes an autocorrelation function, which is given by

$$C_{AA}(t) = \int d\Gamma f_N^0 \delta A(t) \delta A(0) \equiv \langle \delta A(t) \delta A(0) \rangle \qquad (31.34)$$

The correlation functions described so far are based on the ensemble average but can be connected to time averages through the ergodic hypothesis. Originally developed by Boltzmann and Maxwell, the ergodic hypothesis imparts a dynamical basis to statistical mechanics. Simply put, it postulates an equivalence between the ensemble average and the time average of an observable. The ergodic hypothesis for time correlation functions can be expressed as

$$C_{AB}(t) = \int d\Gamma f_N^0 \delta A(t) \delta A(0) = \langle \delta A(t) \delta B(0) \rangle \equiv \lim_{t \to \infty} \frac{1}{t} \int_0^\infty \delta A(t) \delta B(0) \, dt$$

$$(31.35)$$

This is an important result from a practical point of view because the time correlation functions can be evaluated from the knowledge of the time history of the variables. For example, in molecular dynamics (MD) simulations, the ergodic hypothesis is exploited to compute ensemble averages. Time correlation functions have a similar status as that of the partition function in equilibrium statistical mechanics (McQuarrie, 2000). The partition function is unique for a particular ensemble, but separate time correlation functions exist which correspond to different nonequilibrium states. Unlike the partition function, time correlation functions (strictly speaking, their mathematical analogue in frequency space) are directly accessible to experiments such as neutron and x-ray scattering (Boon and Yip, 1991).

31.2.2.2 Onsager's Regression Hypothesis and Linear Response

The connection between the time correlation functions and transport properties comes from the celebrated Onsager's regression hypothesis which states that *the relaxation of macroscopic nonequilibrium disturbances is governed by the same laws as the regression of spontaneous microscopic fluctuations in an equilibrium system* (Chandler, 1987). This hypothesis essentially means that there is fundamentally no way of distinguishing between a nonequilibrium system, say a system with thermal gradient relaxing to its equilibrium state (which corresponds to zero gradients), and the correlations of the spontaneous fluctuations at equilibrium. An important result from the linear response theory is the fluctuation–dissipation (FD) theorem proved in 1951 by Callen and Welton (Chandler, 1987). Onsager's regression hypothesis is simply a consequence of the FD theorem. Later, the importance of FD theorem is brought out in the context of Brownian dynamics (BD) simulations of nanofluids.

In the linear response theory, the thermal conductivity is expressed as a time correlation of instantaneous heat flux fluctuations (McQuarrie, 2000). The following subsection will examine the statistical–mechanical basis of instantaneous heat flux (which also arises in the theory of direct molecular dynamics simulations (Evans and Morriss, 1990). The approach follows from the work of Irving and Kirkwood (1950).

31.2.2.3 Instantaneous Heat Fluxes

Consider a homogeneous system with N number of atoms or molecules (which can be in any physical state). At any instant of time, the microscopic mass density (ρ), momentum density ($\mathbf{j}$), and energy density (e) can be written as (Hanley, 1969)

$$\rho(\mathbf{r}, t) = \sum_{k=1}^{N} m_k \delta(\mathbf{r}_k - \mathbf{r}) \qquad (31.36)$$

$$\mathbf{j}(\mathbf{r}, t) = \sum_{k=1}^{N} \mathbf{p}_k \delta(\mathbf{r}_k - \mathbf{r}) \qquad (31.37)$$

$$e(\mathbf{r}, t) = \left(\sum_{k=1}^{N} \frac{\mathbf{p}_k \cdot \mathbf{p}_k}{2m_k} + \sum_{j=1}^{N} \sum_{k>j}^{N} \Phi(r_{jk}) \right) \delta(\mathbf{r}_k - \mathbf{r}) \qquad (31.38)$$

where
 m is the mass
 p is the momentum
 Φ is the potential energy

There is a total of N number of atoms each tagged by k. The three-dimensional delta functions place the atoms at a given macroscopic radius $\mathbf{r}$, and they have units of $1/V$ where V is the volume. By placing the delta function, the densities are expressed per unit volume. The volume V is such that it is small in comparison to macroscopic scale but sufficiently larger than the molecular dimensions. It is also tacitly implied in the energy density formulation that the potential energy interactions are pair-wise (jk) and it is shared equally among the atoms.

The Liouville equation gives the time derivative of any dynamical variable $A(\mathbf{r}^N, \mathbf{p}^N)$. It is expressed as (Hanley, 1969)

$$\dot{A} = \sum_{k=1}^{N} \frac{\mathbf{p}_k}{m_k} \cdot \nabla_{\mathbf{r}_k} A - \sum_{k=1}^{N} \sum_{k>j}^{N} \nabla_{\mathbf{r}_k} \Phi(r_{jk}) \cdot \nabla_{\mathbf{p}_k} A \qquad (31.39)$$

Using Equations 31.39 and 31.36, the time rate of change of density can be derived as (Hanley, 1969)

$$\dot{\rho}(\mathbf{r}, t) = -\nabla_{\mathbf{r}} \cdot \mathbf{j}(\mathbf{r}, t) \qquad (31.40)$$

where $\mathbf{j}$ is the momentum density. Equation 31.40 simply shows that mass, at any given point in time, is conserved. Now the time derivative of energy density is given by

$$\dot{e} = \sum_{k=1}^{N} \frac{\mathbf{p}_k \cdot \mathbf{p}_k}{2m_k} \frac{\mathbf{p}_k}{m_k} \cdot \nabla_{\mathbf{r}_k} \delta(\mathbf{r}_k - \mathbf{r})$$

$$+ \sum_{j=1}^{N} \sum_{k>j}^{N} \Phi(r_{jk}) \frac{\mathbf{p}_k}{m_k} \cdot \nabla_{\mathbf{r}_k} \delta(\mathbf{r}_k - \mathbf{r})$$

$$+ \sum_{j=1}^{N} \sum_{k>j}^{N} \Phi(r_{jk}) \left[\delta(\mathbf{r}_j - \mathbf{r}) - \delta(\mathbf{r}_k - \mathbf{r}) \frac{\mathbf{p}_k}{m_k} \cdot \nabla_{\mathbf{r}_k} \Phi(r_{jk}) \right] \quad (31.41)$$

Note the peculiar difference in the delta function for the third term. It can be further simplified by assuming that the delta functions are analytical. This allows a Taylor series expansion as

$$\delta(\mathbf{r}_k - \mathbf{r}) - \delta(\mathbf{r}_j - \mathbf{r}) = -\mathbf{r}_{jk} \cdot \nabla_{\mathbf{r}} \Theta_{kj} \delta(\mathbf{r}_k - \mathbf{r}) \quad (31.42)$$

where the operator Θ_{kj} is given by (Evans and Morriss, 1990)

$$\Theta_{kj} = 1 - \frac{1}{2!} \mathbf{r}_{kj} \cdot \nabla_{\mathbf{r}} + \cdots + \frac{1}{n!} \left[-\mathbf{r}_{kj} \cdot \nabla_{\mathbf{r}} \right]^{n-1} \quad (31.43)$$

After several steps, the energy density rate can be expressed as a microscopic (instantaneous) conservation law (Hanley, 1969) given by

$$\dot{e}(\mathbf{r}, t) = -\nabla_{\mathbf{r}} \cdot \hat{\mathbf{j}}_q(\mathbf{r}) \quad (31.44)$$

where the right side is expressed as a gradient of heat flux $\hat{\mathbf{j}}_q$ which is given by

$$\hat{\mathbf{j}}_q(\mathbf{r}, t) = \sum_{k=1}^{N} \frac{\mathbf{p}_k \cdot \mathbf{p}_k}{2m_k} \frac{\mathbf{p}_k}{m_k} \delta(\mathbf{r}_k - \mathbf{r})$$

$$+ \sum_{j=1}^{N} \sum_{k>j}^{N} \mathbf{I}\Phi(r_{jk}) \cdot \frac{\mathbf{p}_k}{m_k} \delta(\mathbf{r}_k - \mathbf{r})$$

$$- \sum_{j=1}^{N} \sum_{k>j}^{N} \left[\mathbf{r}_{jk} \otimes \nabla_{\mathbf{r}_{jk}} \Phi(r_{jk}) \right] \cdot \frac{\mathbf{p}_k}{m_k} \delta(\mathbf{r}_k - \mathbf{r}) \quad (31.45)$$

Equation 31.45 represents the instantaneous heat flux per unit volume, and Fourier's law for heat conduction can now be expressed as $\hat{\mathbf{j}}_q = -\kappa \nabla T$. On simplification, the heat flux reduces to

$$\hat{\mathbf{j}}_q(\mathbf{r}, t) = \overbrace{\sum_{k=1}^{N} \frac{p_k^2}{2m_k} \mathbf{v}_k \delta(\mathbf{r}_k - \mathbf{r})}^{K} + \overbrace{\sum_{j=1}^{N} \sum_{k>j}^{N} \mathbf{I}\Phi(r_{jk}) \cdot \mathbf{v}_k \delta(\mathbf{r}_k - \mathbf{r})}^{P}$$

$$+ \overbrace{\sum_{j=1}^{N} \sum_{k>j}^{N} \left[\mathbf{r}_{jk} \otimes \mathbf{F}(r_{jk}) \right] \cdot \mathbf{v}_k \delta(\mathbf{r}_k - \mathbf{r})}^{C} \quad (31.46)$$

where
  $\mathbf{v}$ is the velocity
  $\mathbf{F}$ is the force

It can be seen that the heat flux vector can be decomposed into three modes, the flux carried by the kinetic energy (K), flux carried by the potential energy (P), and the flux carried by the collisions or the work done by the stress tensor (C). Recall that in the linear response theory, the *fluctuation* of the heat flux is the quantity of interest. For a single component system, Equation 31.41 remains unchanged if the variables are changed to fluctuations. In this case, the fluctuation in the heat flux becomes

$$\delta\hat{\mathbf{j}}_q(\mathbf{r}, t) = \overbrace{\sum_{k=1}^{N} \left(\frac{p_k^2}{2m_k} - \left\langle \frac{p_k^2}{2m_k} \right\rangle \right) \mathbf{v}_k \delta(\mathbf{r}_k - \mathbf{r})}^{\delta K(\text{flux})}$$

$$+ \overbrace{\sum_{j=1}^{N} \sum_{k>j}^{N} \mathbf{I}\left(\Phi(r_{jk}) - \left\langle \Phi(r_{jk}) \right\rangle \right) \cdot \mathbf{v}_k \delta(\mathbf{r}_k - \mathbf{r})}^{\delta P(\text{flux})}$$

$$+ \overbrace{\sum_{j=1}^{N} \sum_{k>j}^{N} \left(\mathbf{r}_{jk} \otimes \mathbf{F}(r_{jk}) - \left\langle \mathbf{r}_{jk} \otimes \mathbf{F}(r_{jk}) \right\rangle \right) \cdot \mathbf{v}_k \delta(\mathbf{r}_k - \mathbf{r})}^{\delta C(\text{flux})}$$

$$(31.47)$$

Thus, the fluctuations in the heat flux can be decomposed as (Eapen et al., 2007b)

$$\delta\hat{\mathbf{j}}_q(\mathbf{r}, t) = \delta\hat{\mathbf{j}}_q^K(\mathbf{r}, t) + \delta\hat{\mathbf{j}}_q^P(\mathbf{r}, t) + \delta\hat{\mathbf{j}}_q^C(\mathbf{r}, t) \quad (31.48)$$

Note that $\hat{\mathbf{j}}_q(\mathbf{r}, t) = \delta\hat{\mathbf{j}}_q(\mathbf{r}, t)$ because of momentum conservation under equilibrium conditions. For example, the kinetic term can be expanded as

$$\delta\hat{\mathbf{j}}_q^K(\mathbf{r}, t) = \sum_{k=1}^{N} \left(\frac{p_k^2}{2m_k} \right) \mathbf{v}_k \delta(\mathbf{r}_k - \mathbf{r}) - \left\langle \frac{p_k^2}{2m_k} \right\rangle \sum_{k=1}^{N} \mathbf{v}_k \delta(\mathbf{r}_k - \mathbf{r}) \quad (31.49)$$

The last term is identically zero for a single component system. The same holds for potential and collision terms. For multicomponent systems, there exists a microscopic analogue for the diffusional heat flux given by Equation 31.5. The instantaneous multicomponent heat flux is given by (Hanley, 1969)

$$\mathbf{j}_q(\mathbf{r}, t) = \hat{\mathbf{j}}_q(\mathbf{r}, t) - \sum_s \left\langle h^s \right\rangle \mathbf{j}^s(\mathbf{r}, t) \quad (31.50)$$

where
  s is the number of components in the system
  $\mathbf{j}_q$ is the reduced (conductive) heat flux which takes the *instantaneous* diffusion into account

The term h stands for the specific partial enthalpy and the second term on the right denotes the partial enthalpy flux. Comparing Equation 31.50 with Equation 31.5, it can be seen that there is a one-to-one correspondence between the macroscopic and microscopic heat flux expressions; i.e., the ensemble average of Equation 31.50 leads to $\mathbf{J}_q = \hat{\mathbf{J}}_q - \sum_{k=1}^{n} h^k \mathbf{J}^k$. Usually in experiments, the diffusional (or enthalpic) heat flux is negligible. However, the instantaneous diffusional heat flux $\left(\sum_s \langle h^s \rangle \mathbf{j}^s(\mathbf{r},t) \right)$ is, in general, *not* negligible. The full expansion for a two-component system will result in (Hoheisel, 1987)

$$
\mathbf{j}_q(\mathbf{r},t) = \left(\frac{1}{2} \sum_{k=\alpha}^{\beta} \sum_{i=1}^{N_k} m_i^k (v_i^k)^2 \mathbf{v}_i^k \right.
$$
$$
+ \sum_{k=\alpha}^{\beta} \sum_{l=\alpha}^{\beta} \sum_{i=1}^{N_k} \sum_{j>i}^{N_l} \left[\mathbf{I}\Phi(r_{ij}^{kl}) + \mathbf{r}_{ij}^{kl} \otimes \mathbf{F}_{ij}^{kl} \right] \cdot \mathbf{v}_i^k
$$
$$
\left. - \sum_{k=\alpha}^{\beta} \langle h^k \rangle \sum_{i=1}^{N_k} \mathbf{v}_i^k \right) \delta(\mathbf{r}_i - \mathbf{r}) \qquad (31.51)
$$

where α and β are the two components in a colloidal system. Note that the partial enthalpy is expressed per molecule instead of mass. Also note that Equation 31.51 needs to be recast in the fluctuations to be applicable in linear response theory. This aspect is discussed in the next subsection.

31.2.2.4 Partial Enthalpy in a Binary Nanofluid

The evaluation of partial enthalpies in Equation 31.51 now requires simplifying assumptions. The mean enthalpy can be computed as the sum of the mean kinetic and potential energies, and the virial (Perronace et al., 2002; Vogelsang and Hoheisel, 1987). Strictly speaking, this is true only for ideal (equipotential) mixtures. However, its use in the linear response theory for nonideal mixtures (a metallic or oxide nanofluid is typically nonideal) can be justified by noting that Equation 31.51 can be reformulated in terms of the *fluctuations* of the heat flux constituents, K, P, and C (see Equation 31.47). The partial enthalpy of component α can be expressed as

$$
\langle h^{\alpha} \rangle = \overbrace{\left\langle \sum_{k=1}^{N_\alpha} \frac{1}{2} m_k v_k^2 \right\rangle}^{K} + \overbrace{\left\langle \sum_{j}^{N_\alpha} \sum_{k>j}^{N_\alpha} \Phi(r_{jk}) + \frac{1}{2} \sum_{j}^{N_{\alpha\beta}} \sum_{k>j}^{N_{\alpha\beta}} \Phi(r_{jk}) \right\rangle}^{P}
$$
$$
+ \overbrace{\left\langle \sum_{j}^{N_\alpha} \sum_{k>j}^{N_\alpha} r_{jk} F(r_{jk}) + \frac{1}{2} \sum_{j}^{N_{\alpha\beta}} \sum_{k>j}^{N_{\alpha\beta}} r_{jk} F(r_{jk}) \right\rangle}^{C} \qquad (31.52)
$$

Note that the potential energy (P) [and virial (C)] is summed among two groups of atoms—the atoms of component α (N_α) and α-atoms which interact with β-atoms ($N_{\alpha\beta}$). As noted before, it is implicitly assumed that the potential and virial interactions can be expressed pair-wise, and they are equally shared between atoms. A similar expression can be written for component β. Notice that the partial enthalpy in Equation 31.52 has three components (K, P, and C) which are congruent to the three modes in the heat flux for a single component system.

Vogelsang et al. (1989) has shown that the partial enthalpies based on the molecular quantities are significantly

different from those calculated with more accurate thermodynamic simulations. However, the *sum* of the partial enthalpies calculated by the molecular approximation is equal to the *sum* calculated by the thermodynamic approximation. In a nanofluid system, this observation offers a certain advantage. Calling the last term in Equation 31.51 as the enthalpy flux ($\mathbf{j}_h$), the following expression can be written for a two-component colloid:

$$
\mathbf{j}_h(\mathbf{r},t) = -\sum_{k=\alpha}^{\beta} h^k \sum_{i=1}^{N_k} \mathbf{v}_i^k \delta(\mathbf{r}_i - \mathbf{r})
$$
$$
= -h^\alpha \sum_{i=1}^{N_\alpha} \mathbf{v}_i^\alpha \delta(\mathbf{r}_i - \mathbf{r}) - h^\beta \sum_{i=1}^{N_\beta} \mathbf{v}_i^\beta \delta(\mathbf{r}_i - \mathbf{r}) \quad (31.53)
$$

where α and β denote the solid and the fluid atoms, respectively. Since the net momentum is zero (under equilibrium conditions), the enthalpy flux can be written (for equal masses) as

$$
\mathbf{j}_h(\mathbf{r},t) = -\sum_{i=1}^{N_\alpha} \mathbf{v}_i^\alpha (h^\alpha - h^\beta) \, \delta(\mathbf{r}_i - \mathbf{r}) \qquad (31.54)
$$

In a colloid, the potential energy and the virial of the solid atoms is generally much higher (in magnitude) than that of the fluid atoms, and the total enthalpy is dominated by the enthalpy of the solid atoms alone. Then, for equal masses, Equation 31.54 can be approximated as

$$
\mathbf{j}_h(\mathbf{r},t) = -(h^\alpha - h^\beta)\sum_{i=1}^{N_\alpha} \mathbf{v}_i^\alpha \delta(\mathbf{r}_i - \mathbf{r}) \approx -(h^\alpha)\sum_{i=1}^{N_\alpha} \mathbf{v}_i^\alpha \delta(\mathbf{r}_i - \mathbf{r})
$$
$$
\approx -(h^\alpha + h^\beta)\sum_{i=1}^{N_\alpha} \mathbf{v}_i^\alpha \delta(\mathbf{r}_i - \mathbf{r}) \qquad (31.55)
$$

or

$$
\mathbf{j}_h(\mathbf{r},t) \approx -(h^{\text{total}})\sum_{i=1}^{N_\alpha} \mathbf{v}_i^\alpha \delta(\mathbf{r}_i - \mathbf{r}) \qquad (31.56)
$$

Thus, within the stated assumptions, an accurate knowledge of the instantaneous partial enthalpies is not essential to get a good estimate of the instantaneous enthalpy heat flux if one of the enthalpies dominates the sum. Another simplifying assumption is the use of a single mean enthalpy for each species. Clearly, the mean enthalpy of the surface atoms of a nanoparticle is different from those of the interior. However, for nanoclusters comprising mostly surface atoms, the mean enthalpy of all cluster atoms will be similar to each other (Eapen et al., 2007b).

Under equilibrium conditions, the instantaneous enthalpy flux for a *single* component system is zero at all times. This is a consequence of having a net zero momentum for the system. For binary systems, the instantaneous enthalpy flux is *not* zero, in general. So there is a nontrivial contribution from the transport of mean enthalpy to the total microscopic heat flux. In typical nonequilibrium experiments for colloids, the ensemble-averaged enthalpy (or diffusional) flux is generally small compared to that from the other three contributions, *K*, *P*, and *C*. The enthalpy flux, thus, plays an important role in the instantaneous microscopic heat flux but not for the ensemble average.

Recall that the FD theorem provides a connection between the macroscopic relaxation and the fluctuations at equilibrium conditions. It can be shown that Equation 31.51 can be written in terms of fluctuating components, provided that the mean instantaneous enthalpy is considered to be a sum of the average kinetic energy, potential energy, and the virial, as shown in Equation 31.52. First consider the kinetic energy flux (for a two-component system), which is given by

$$\tilde{\mathbf{j}}_q^K = \frac{1}{2}\sum_{i=1}^{N_\alpha} m_i^\alpha (v_i^\alpha)^2 \mathbf{v}_i^\alpha - \left\langle \sum_{k=1}^{N_\alpha} \frac{1}{2} m_k v_k^2 \right\rangle \sum_{i=1}^{N_\alpha} \mathbf{v}_i^\alpha + \frac{1}{2}\sum_{i=1}^{N_\beta} m_i^\beta (v_i^\beta)^2 \mathbf{v}_i^\beta$$

$$- \left\langle \sum_{k=1}^{N_\beta} \frac{1}{2} m_k v_k^2 \right\rangle \sum_{i=1}^{N_\beta} \mathbf{v}_i^\beta \quad (31.57)$$

The tilde in Equation 31.57 and the absence of delta functions indicate that the heat flux is volume averaged over the whole system. Note the differences in the definition for heat fluxes which, in general, can be expressed as

$$\tilde{\mathbf{j}}_q = \int_V \hat{\mathbf{j}}_q(\mathbf{r})d\mathbf{r} \quad (31.58)$$

where *V* is the volume of the system. With the enthalpy as the sum of kinetic and potential energies and virial, the kinetic energy heat flux can now be written as

$$\tilde{\mathbf{j}}_q^K = \frac{1}{2}\sum_{i=1}^{N_\alpha} m_i^\alpha (v_i^\alpha)^2 \mathbf{v}_i^\alpha - \sum_{i=1}^{N_\alpha} \left\langle \sum_{k=1}^{N_\alpha} \frac{1}{2} m_k^\alpha (v_k^\alpha)^2 \right\rangle \mathbf{v}_i^\alpha + \frac{1}{2}\sum_{i=1}^{N_\beta} m_i^\beta (v_i^\beta)^2 \mathbf{v}_i^\beta$$

$$- \sum_{i=1}^{N_\beta} \left\langle \sum_{k=1}^{N_\beta} \frac{1}{2} m_k^\beta (v_k^\beta)^2 \right\rangle \mathbf{v}_i^\beta \quad (31.59)$$

that is

$$\tilde{\mathbf{j}}_q^K = \sum_{i=1}^{N_\alpha} \left\{ \frac{1}{2} m_i^\alpha (v_i^\alpha)^2 - \left\langle \sum_{k=1}^{N_\alpha} \frac{1}{2} m_k^\alpha (v_k^\alpha)^2 \right\rangle \right\} \mathbf{v}_i^\alpha$$

$$+ \sum_{i=1}^{N_\beta} \left\{ \frac{1}{2} m_i^\beta (v_i^\beta)^2 - \left\langle \sum_{k=1}^{N_\beta} \frac{1}{2} m_k^\beta (v_k^\beta)^2 \right\rangle \right\} \mathbf{v}_i^\beta \quad (31.60)$$

or

$$\tilde{\mathbf{j}}_q^K = \sum_{i=1}^{N_\alpha} (\delta E_i^{K,\alpha})\mathbf{v}_i^\alpha + \sum_{i=1}^{N_\beta} (\delta E_i^{K,\beta})\mathbf{v}_i^\beta = \sum_{j=\alpha}^\beta \sum_{i=1}^{N_j} \left(K_i^j - \left\langle K_i^j \right\rangle \right)\mathbf{v}_i^j$$

$$(31.61)$$

The kinetic energy flux (*K*) is, thus, expressed solely in terms of fluctuations in the kinetic energy on a per-atom basis. If potential and virial (collision) can be identified for single particles, then the total flux can be written as

$$\dot{\mathbf{j}}_q = \delta\tilde{\mathbf{j}}_q = \sum_{j=\alpha}^\beta \sum_{i=1}^{N_j} \left(\frac{1}{2} m_i(v_i^j)^2 - \left\langle \frac{1}{2} m_i(v_i^j)^2 \right\rangle \right)\mathbf{v}_i^j$$

$$+ \sum_{k=\alpha}^\beta \sum_{l=\alpha}^\beta \sum_{i=1}^{N_k} \sum_{j>i}^{N_l} \left[\mathbf{I}\left(\Phi\left(r_{ij}^{kl}\right) - \left\langle \Phi\left(r_{ij}^{kl}\right) \right\rangle \right) \right.$$

$$\left. + \left(\mathbf{r}_{ij}^{kl} \otimes \mathbf{F}_{ij}^{kl} - \left\langle \mathbf{r}_{ij}^{kl} \otimes \mathbf{F}_{ij}^{kl} \right\rangle \right) \right] \cdot \mathbf{v}_i^k \quad (31.62)$$

Thus, the instantaneous heat flux for a binary system can also be expressed as the *fluctuations* of three modes, *K*, *P*, and *C*. As discussed earlier, the partitioning of potential energy and virial into contributions from single atoms is not evident, especially if the system is nonisotopic (non-equipotential) or has many body interactions. It can be shown that many-body interactions can be cast into the form of pair-wise interactions (Li, 2000). The same procedure can be performed for the virial, in principle. Thus, the total heat flux can now be written explicitly in terms of fluctuations of each atom, as shown in Equation 31.62.

31.2.2.5 Thermal Conductivity of Colloids with Linear Response Theory

The derivation for thermal conductivity in multicomponent systems is slightly involved, and hence, only the final expression is given here. For cubic isotropic materials, the thermal conductivity of the system can be expressed as (McQuarrie, 2000)

$$\kappa = \frac{1}{3Vk_B T^2} \int_0^\infty \left\langle \delta\tilde{\mathbf{j}}_q(t)\delta\tilde{\mathbf{j}}_q(0) \right\rangle dt \quad (31.63)$$

where $\delta\hat{\mathbf{j}}_q$ is the volume-averaged fluctuations of the heat flux vector, which is given by

$$\delta\tilde{\mathbf{j}}_q = \tilde{\mathbf{j}}_q = \int_V \left(\frac{1}{2}\sum_{k=\alpha}^\beta \sum_{i=1}^{N_k} m_i^k (v_i^k)^2 \mathbf{v}_i^k \right.$$

$$+ \sum_{k=\alpha}^\beta \sum_{l=\alpha}^\beta \sum_{i=1}^{N_k} \sum_{j>i}^{N_l} \left[\mathbf{I}\Phi(r_{ij}^{kl}) + \mathbf{r}_{ij}^{kl} \otimes \mathbf{F}_{ij}^{kl} \right] \cdot \mathbf{v}_i^k$$

$$\left. - \sum_{k=\alpha}^\beta \left\langle h^k \right\rangle \sum_{i=1}^{N_k} \mathbf{v}_i^k \right) \delta(\mathbf{r}_i - \mathbf{r})d\mathbf{r} \quad (31.64)$$

A notable aspect of Equation 31.63 is that it satisfies the constraint of fluctuations in the heat flux. Similarly, the heat flux vector expression in Equation 31.64 satisfies the microscopic heat flux relationship given by $\mathbf{j}_q(\mathbf{r},t) = \hat{\mathbf{j}}_q(\mathbf{r},t) - \sum_s \langle h^s \rangle \mathbf{j}^s(\mathbf{r},t)$, where s is the number of components. The former is a necessity arising from the linear response theory while the latter is a constraint arising from instantaneous energy conservation.

The preceding time correlation approach is applicable to heterogeneous media such as colloids as long as there is a clear scale separation in the system, and an accurate estimate of the partial enthalpy can be estimated. As previously mentioned, the linear response theory is based on the idea that microscopic expressions for fluxes and transport properties can be made within a small volume such that it is smaller than the macroscopic dimensions but large enough to have local thermodynamic equilibrium. The local region, which depends on the specific system which is being considered, will have identifiable properties and thermodynamic functions. The only requirement on the local region is that it needs to have enough atoms for a representative structure and an equilibrium distribution for the microscopic variables.

31.2.2.6 Ensemble Averages of Microscopic Fluxes

Ensemble averaging makes the microscopic fluxes equivalent to those which are experimentally accessible. However, there are subtleties involved in their use in the theory and simulations. In the linear response theory, the *instantaneous* microscopic fluxes are theoretically appropriate while in nonequilibrium molecular dynamics simulations, an expression for the ensemble average (or time average) is essential. For a binary system (α, β), ensemble averaging on instantaneous heat flux vector results in

$$\langle \mathbf{j}_q(\mathbf{r},t) \rangle = \left\langle \left[\frac{1}{2} \sum_{k=\alpha}^{\beta} \sum_{i=1}^{N_k} m_i^k \left(v_i^k\right)^2 \mathbf{v}_i^k \right. \right.$$
$$+ \sum_{k=\alpha}^{\beta} \sum_{l=\alpha}^{\beta} \sum_{i=1}^{N_k} \sum_{j>i}^{N_l} \left[\mathbf{I}\Phi\left(r_{ij}^{kl}\right) + \mathbf{r}_{ij}^{kl} \otimes \mathbf{F}_{ij}^{kl} \right] \cdot \mathbf{v}_i^k$$
$$\left. \left. - \sum_{k=\alpha}^{\beta} \langle h^k \rangle \sum_{i=1}^{N_k} \mathbf{v}_i^k \right] \right\rangle \quad (31.65)$$

Note that the above expression is easily generalized for a multicomponent system. The issue of the operator Θ_{ij} (see Equation 31.43) arises in the ensemble averaging for stress tensor and heat flux vector. Using nonequilibrium molecular dynamics (NEMD) simulations, it is shown that the higher order terms in Θ_{ij} are insignificant for evaluating heat fluxes in binary colloidal systems (see Appendix).

31.2.2.7 Energy Conserving Brownian Dynamics

The linear response relationship Equation 31.63 is general enough to calculate the thermal conductivity of all states of matter, including colloids. The fluctuations that are required in the time correlation functions are accessible from equilibrium

molecular dynamics (EMD) simulations. In MD simulations, a system of atoms moves according to Newton's equations of motion (Allen and Tildesley, 1994). The appropriate correlations are then averaged over many time origins and initial conditions (Rapaport, 2004). The greatest limitation, however, is that MD simulations are restricted to small length and time scales (hundreds of nanometers and nanoseconds, respectively). As such, MD simulations cannot access the physical scales of real experiments with the current computing capabilities.

An alternative way to bypass the computational limitations of MD is to employ energy-conserving BD simulations. The classical BD is a mesoscopic simulation technique, akin to MD, in which coarse-grained particles mimic the large clusters of physical molecules. From a nanofluid perspective, BD involves modeling coarse-grained nanoparticles acted upon by three forces, conservative, dissipative, and random. The random force mimics the effect of base fluid molecules on the nanoparticles while the dissipative forces allow FD theorem to be satisfied. The nanoparticles then move according to Newton's second law of motion. BD takes advantage of two very different time scales—a fast time scale set by the relaxation of the lighter base fluid atoms, and a slow time scale arising from the relaxation of the heavier nanoparticle particles (Hansen and McDonald, 1986). By eliminating the explicit interactions of fluid and solid atoms, BD, in principle, can probe the slow hydrodynamic time scales associated with the nanofluids.

Classical BD simulations on thermal transport of nanofluids have been reported by several groups (Bhattacharya et al., 2004; Gupta and Kumar, 2007; Jain et al., 2009). The forces on the solute particles are evaluated from a two-body potential with empirical constants fitted from experimental data. However, there are several fundamental constraints that preclude the use of classical BD simulations for evaluating the thermal conductivity of nanofluids.

As the equation of motion, a classical BD simulation uses the Langevin equation, which is given by (Hansen and McDonald, 1986; Turq et al., 1977)

$$\frac{\mathbf{v}_i(t)}{dt} = -\gamma_i \mathbf{v}_i(t) + \mathbf{R}_i(t) + \mathbf{X}_i(t) \quad (31.66)$$

where

v is the velocity of the Brownian nanoparticle
R and **X** are the random and external forces, respectively

The constant γ stands for a dissipative friction coefficient. In the original derivation, Langevin assumed that the rate of change of momentum is proportional to the momentum itself but acting in an opposite direction. The random force has a zero mean and it is uncorrelated with velocity. In thermal equilibrium, the friction coefficient is related to the autocorrelation between the random forces. These are stated as

$$\langle \mathbf{R}(t) \rangle = 0, \quad \langle \mathbf{R}(t)\mathbf{v}(0) \rangle = 0, \quad \gamma = \frac{\beta}{3m} \int_0^\infty \langle \mathbf{R}(t)\mathbf{R}(0) \rangle \quad (31.67)$$

The connection between the friction coefficient (γ) and random force arises from the FD theorem.

The drawback of the classical BD simulations for evaluating the hydrodynamic interactions and thermal transport behavior comes from two conceptual constraints: momentum and energy conservation. It can be immediately noticed that Equation 31.66, when applied to a pair of colliding Brownian nanoparticles, does not conserve momentum. Thus, the dynamics of classical BD is diffusive and not hydrodynamic. Since the only conserved variable is mass, the only pertinent transport coefficient is diffusion (or properties that intimately depend on diffusivity such as ionic conductivity) (Español, 1995). BD simulations also lack Galilean invariance (Marsh et al., 1997b). Thus, viscosity and thermal conductivity are not *defined* in a classical BD simulation.

Momentum conservation is explicitly enforced in dissipative particle dynamics (DPD) (Avalos and Mackie, 1997; Español, 1995; Espanol and Warren, 1995). The momentum conservation is maintained for each collision by the following modification to the dissipative force term (Avalos and Mackie, 1999):

$$\mathbf{F}_{ij}^{D}(t) = \gamma_{ij}\hat{\mathbf{r}}_{ij} \otimes \hat{\mathbf{r}}_{ij}\cdot(\mathbf{v}_j - \mathbf{v}_i) \qquad (31.68)$$

where

$\hat{\mathbf{r}}_{ij}$ is the unit vector pointing in the direction of $(\mathbf{r}_j - \mathbf{r}_i)$

$\gamma_{ij}\hat{\mathbf{r}}_{ij} \otimes \hat{\mathbf{r}}_{ij}$ is a friction tensor which is a function of the separation distance between the particles (r_{ij})

Since the interaction is now pair wise (*ij*), momentum and angular momentum are conserved in each interaction or collision, and this simple modification makes BD appropriate for investigating hydrodynamic fields (such as flow fields). Thus, in a DPD simulation, both diffusion and viscosity are *bonafide* transport coefficients, but not thermal conductivity.

To simulate thermal conductivity, a mesoscale internal energy variable (u) needs to be defined (Avalos and Mackie, 1997, 1999; Español, 1997) as follows:

$$\dot{u}_i = \sum_j \frac{1}{2}(\mathbf{v}_j - \mathbf{v}_i)\cdot(\mathbf{F}_{ij}^{D} - \mathbf{F}_{ij}^{R}) + \dot{q}_{ij}^{D} - \dot{q}_{ij}^{R} \qquad (31.69)$$

where $\dot{q}_{ij}$ is defined as a mesoscopic, scalar heat flow between particles i and j. The internal energy of a Brownian nanoparticle can change in two ways: it can either come from the work done by the dissipative force (the conversion of mechanical energy into internal energy) or through heat transfer between the Brownian nanoparticles (arising from a temperature difference). Similar to the forces, the heat flow is also defined as dissipative and random; the latter is a strict requirement from the FD theorem.

Unlike in MD simulations, heat flow and internal energy need to be explicitly specified for BD thermal transport. In MD simulations, all forces are conservative, and the governing phase–space trajectories evolve according to time-symmetric Liouville equation (Evans and Morriss, 1990). On the other hand, BD trajectories pertain to those of the Fokker–Planck equation which

are discontinuous and stochastic (Marsh et al., 1997a). This has an important ramification in the use of correlation functions that are derived from the Liouville equation (McQuarrie, 2000). For example, it is shown recently (Ernst and Brito, 2006) that the linear response (also known as Green–Kubo) relationship for thermal conductivity (Equation 31.63) needs a correction in BD simulations. In addition, the linear response formalism for thermal conductivity requires the instantaneous heat flux expression. Unlike what have been assumed in a couple of prior investigations, the instantaneous enthalpy flux is not insignificant (even though the ensemble-averaged value can be negligible). As discussed in the appendix, nonequilibrium simulations (which are applicable for both MD and BD) can, however, compute the heat flux exactly without a prior knowledge of the enthalpy flux. Also note that the deviations from equilibrium conditions that arise from the thermal and hydrodynamic perturbations in nonequilibrium simulations are generally insignificant (Chantrenne and Barrat, 2004).

The first energy-conserving BD simulation on nanofluids was performed by He and Qiao (2008) to investigate the effect of Brownian motion on the effective nanofluid thermal conductivity. Both momentum and energy conservation laws were satisfied along with the FD relationships. When the equations of motion and stochastic constraints were rigorously satisfied, thermal conductivity of well-dispersed nanofluids was seen to be consistent with the Maxwell theory (Equation 31.1), and independent of the Brownian motion of the nanoparticles or nanoscale convection set by the motion of the nanoparticles (He and Qiao, 2008). This result highlights the importance of enforcing momentum and energy conservation, as well as FD relationships, in BD for thermal transport.

31.3 Thermal Conductivity Models

Theoretical models predate large-scale molecular and mesoscale simulation methods. As discussed in Section 31.1, the earliest theory for thermal conductivity (originally derived for electrical conductivity) was proposed by Maxwell in the late nineteenth century. Since then, several effective media or mean field models have been proposed. In this final section, a brief survey is made on the mean-field models. An analysis of Maxwell's original work indicates that his theory predicts two bounds, an *upper* and a *lower bound* (also derived by Hashin and Shtrikman (HS) using variational principles (Hashin and Shtrikman, 1962). The lower bound corresponds to a nanofluid configuration where all the nanoparticles are well-dispersed, while the upper bound represents a nanofluid with fractal-like nanoparticle arrangement (Eapen et al., 2010). While the lower bound (Equation 31.1) has been extensively quoted in the nanofluid literature, the upper bound has not received much attention. When the nanoparticles are well-dispersed, the conduction paths arise predominantly through the base fluid. In contrast, additional conduction paths emerge if the nanoparticles can form linear, fractal-like configuration. Thus, the Maxwell bounds correspond to maximally biased theoretical limits; the upper bound represents maximally biased conduction through the nanoparticles while the lower bound represents maximally biased conduction through the base fluid (Carson et al., 2005). When there is a symmetric bias, the effective thermal conductivity is given by the Bruggeman effective medium theory.

31.3.1 Series and Parallel Modes of Conduction

The simplest and, perhaps, the most intuitive models are the series and parallel modes of thermal conduction. In the former, the conducting paths, namely those through the base fluid and through the nanoparticles, are assumed to be in series, and in the latter, they are regarded to be in parallel (see Figure 31.2). The effective thermal conductivities for the two models are given by DeVera and Strieder (1977)

$$\frac{1}{\kappa^{=}} = \frac{1-\phi}{\kappa_f} + \frac{\phi}{\kappa_p} \tag{31.70}$$

$$\kappa^{\parallel} = (1-\phi)\kappa_f + \phi\kappa_p \tag{31.71}$$

where $\kappa^{=}$ and $\kappa^{\parallel}$ are the series and parallel mode thermal conductivities, respectively. Note that a typical nanofluid system is not homogeneous. In the dilute limit, the former is a function of the volume fraction alone, while the latter, as seen from the following relationships, is also a function of the constituent thermal conductivities:

$$\left(\frac{\kappa^{=}}{\kappa_f}\right)_{\phi \to 0} = 1 + \phi \tag{31.72}$$

$$\left(\frac{\kappa^{\parallel}}{\kappa_f}\right)_{\phi \to 0} = 1 + \phi\frac{\kappa_p}{\kappa_f} \tag{31.73}$$

From Equation 31.73, it is clear that the enhancement in the parallel mode can be much larger than that of series mode if $\kappa_p \gg \kappa_f$.

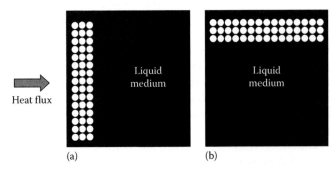

FIGURE 31.2 A two-dimensional representation of series (a) and parallel (b) modes of conduction paths for nanofluids. The parallel mode represents the most efficient way of heat conduction in a binary nanocolloidal (nanofluid) system.

31.3.2 Maxwell Upper and Lower Bounds

The series and parallel bounds are with configurations that are not homogeneous or isotropic. HS have derived a set of bounds which is most restrictive on the basis of volume fraction alone for a homogeneous and isotropic system. Any improvement on these bounds would require additional knowledge on the statistical variations of the dispersed medium. The bounds for nanofluid thermal conductivity are given by (Hashin and Shtrikman, 1962)

$$\kappa_f\left(1 + \frac{3\phi[\kappa]}{3\kappa_f + (1-\phi)[\kappa]}\right) \le \kappa \le \left(1 - \frac{3(1-\phi)[\kappa]}{3\kappa_p - \phi[\kappa]}\right)\kappa_p \tag{31.74}$$

It is assumed that $\kappa_p > \kappa_f$, or otherwise, the upper and lower bounds would simply reverse. Notice that Equation 31.1, which coincides with the lower HS bound when $\kappa_p > \kappa_f$ and with the upper bound in the opposite case, is rigorously exact to first order in ϕ, as evident from the dilute limit, $\kappa_p = \kappa_f(1 + 3\beta\phi)$. Physically, the lower limit corresponds to a set of well-dispersed nanoparticles in a fluid matrix while the upper limit corresponds to large pockets of fluid separated by linked or chain-like nanoparticles, as shown in Figure 31.3.

Note that the upper bound can also be derived from the assumptions placed in the original Maxwell theory. Instead of treating the fluid as the continuous phase, and the solid nanoparticles as the dispersed phase, the roles can be reversed. For very high nanoparticle volume fractions, this configuration is easily visualized. However, for dilute volume fractions, the necessity of having a percolating or fractal-like nanoparticle configuration becomes evident, as shown in Figure 31.4. For the lower bound, the nanoparticles are always well-dispersed, and therefore, the effective conductivity is biased toward the conduction paths in the surrounding fluid (Carson et al., 2005).

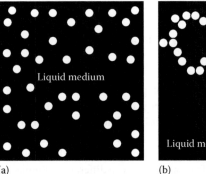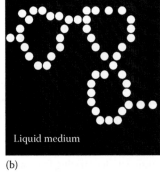

FIGURE 31.3 A two-dimensional representation of the nanocolloid configuration for (a) the lower bound and (b) the upper bound. Mathematically, both bounds are equivalent in the sense that one of the phases provides a continuous thermal conduction path. For the lower bound, the base fluid provides the continuous conduction path, while for the upper bound, fractal-like or chain-like agglomeration can provide the same. Note that the effective thermal conductivity in both configurations is maximally biased toward that of the continuous phase.

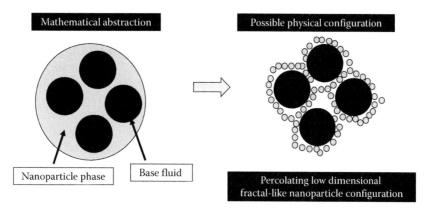

FIGURE 31.4 The mathematical abstraction in the Maxwell theory and possible physical configuration for the upper bound with the nanoparticle phase providing the continuous conduction path. For nanofluids with low volume fractions, such a configuration can exist only if the nanoparticles form percolating, fractal-like configuration embedding large volumes of base fluid. If the nanoparticle thermal conductivity is higher than the base fluid, the effective nanofluid thermal conductivity can be significantly enhanced by such linear, fractal-like configurations. Experimentally tested nanofluids are mostly in aggregated state, and thus, most of the enhancements beyond the Maxwell limit come from limited percolating effects (which also can manifest as enhancement from nonspherical composite particles). Note that large nanoparticle clumps will not provide additional thermal conduction paths. Thus, it is important to differentiate the difference between arbitrary clumping (that occurs from the settling of nanoparticles following ultrasonification, say) and engineered percolating nanoparticle configurations.

Likewise, the upper bound is biased toward the conduction paths along the percolating nanoparticles. The lower Maxwell bound (κ^{MX-}), thus, lies closer to the thermal conductivity of the series mode, while the upper bound (κ^{MX+}) approaches that of the parallel mode. If the configuration is neutral, i.e., neither favoring the series nor the parallel mode, then the effective thermal conductivity (κ^0) would lie in between lower and upper Maxwell bounds. This approach, attributed to Bruggeman and also sometimes known as the effective medium theory (EMT), predicts the thermal conductivity in the implicit form given by (Hashin and Shtrikman, 1962)

$$(1-\phi)\left(\frac{\kappa_f - \kappa}{\kappa_f + 2\kappa}\right) + \phi\left(\frac{\kappa_p - \kappa}{\kappa_p + 2\kappa}\right) = 0 \qquad (31.75)$$

In a nanofluid, the unbiased configuration would be a mix of well-dispersed nanoparticles and linear aggregation. All the mean-field models, thus, correspond to the different configurations of the dispersed medium. It can be shown for $\kappa_p > \kappa_f$ (DeVera and Strieder, 1977; Hashin and Shtrikman, 1962):

$$\kappa^{=} < \kappa^{MX-} < \kappa^0 < \kappa^{MX+} < \kappa^{\|} \qquad (31.76)$$

where κ^0 is the asymmetric (Bruggeman) thermal conductivity.

The recent model of Prasher et al. (2006b) assumes a linear, chain-like cluster configuration for the nanoparticles, and is very similar to the upper Maxwell configuration. Interfacial thermal resistance has not been taken into account in any of these models yet, and, if applicable, it is easily incorporated (Nan et al., 1997). The interfacial resistance always reduces the effective thermal conductivity, and hence, the bounds presented here are the highest for the appropriate configurations (Torquato and

Rintoul, 1995). The dilute limits for the upper Maxwell bound and Prasher et al. model are given by

$$\left(\frac{\kappa^{HS+}}{\kappa_f}\right)_{(\phi\kappa_p/\kappa_f)\to 0} = 1 + \frac{2\phi}{3}\left(\frac{\kappa_p}{\kappa_f}\right) \qquad (31.77)$$

$$\left(\frac{\kappa^{Pr}}{\kappa_f}\right)_{(\phi\kappa_p/\kappa_f)\to 0} = 1 + \frac{\phi}{3}\left(\frac{\kappa_p}{\kappa_f}\right) \qquad (31.78)$$

where κ^{Pr} is the thermal conductivity predicted by the Prasher et al. model. The above limits and parallel mode limit (Equation 31.73) are identical, except for the prefactor.

In most experimentally tested nanofluids, the nanoparticle configuration in the suspended state is generally unknown. Besides, it is not always possible to change the nanoparticle arrangement in a controlled manner. However, this is possible with magnetic nanofluids (or ferrofluids) where the alignment of nanoparticles can be controlled by intense magnetic fields. Such a study has been recently performed with Fe_3O_4 nanoparticles (Philip et al., 2007, 2008) and with Fe nanoparticles (Li et al., 2005). In these experiments, an external magnetic field was applied to the nanofluid system and thermal conductivity was measured as a function of applied field for different volume fractions. When the magnetic field was applied *parallel* to the temperature gradient, the thermal conductivity increased with increasing magnetic field. At very low magnetic fields, electron microscopy revealed a random arrangement of nanoparticles without a clearly identified structure (note that the electron microscopy evidence is indicative of the nanoparticle configuration in the liquid state but is not conclusive). The thermal conductivity of this nanofluid configuration was accurately predicted by the lower Maxwell bound. As the magnetic field was increased,

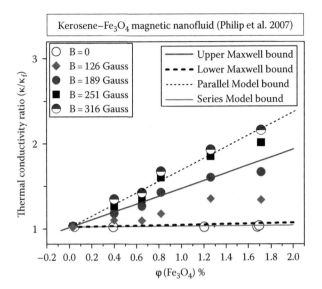

micrographs showed the emergence of linear, chain-forming clusters that eventually converged into long macroscopic chains at the highest magnetic fields. Expectedly, the thermal conductivity increased from the lower Maxwell bound to the parallel mode bound given by Equation 31.73. However, when the external magnetic field was applied *perpendicular* to the temperature gradient, interestingly, there was no dependence of the magnetic field on the thermal conductivity (Li et al., 2005). This observation is easily rationalized by noting that the perpendicular field generates nanoparticle chains in a direction perpendicular to the heat flow. Such a configuration is best described by the series mode which predicts a thermal conductivity very close to the base fluid itself. Thus, external magnetic fields induce a strongly anisotropic thermal conduction behavior in magnetic nanofluids.

The enhancement in the thermal conductivity of ferrofluids for different magnetic fields are delineated in Figure 31.5 (Philip et al., 2007). Also plotted are the four bounds discussed in this section. The correlation is striking. With increasing magnetic field, the enhancement progressively increases until it reaches the upper Maxwell bound and parallel bound. Along with the experiments performed by Li et al. (2005), this is, perhaps, the most conclusive and unambiguous experimental result that shows the effect of linear or fractal-like clustering on nanofluid thermal conductivity.

In Figure 31.6, the mean-field bounds for a large body of nanofluid data are depicted. The data set includes nanoparticles with relatively low κ (zirconia), moderate κ (alumina, copper oxide), and high κ (copper, carbon nanotubes). It also includes different base media, including water, ethylene glycol and oil, and nanoparticles with lower thermal conductivity relative to

FIGURE 31.5 Thermal conductivity of magnetic nanofluids (Fe_3O_4 nanoparticles dispersed in kerosene). At low magnetic fields (in the direction of temperature gradient), the nanoparticles are randomly dispersed and the effective thermal conductivity is well-described by the series or lower Maxwell bound. As the magnetic field increases, linear, chain-like nanoparticle structures evolve in the direction of the temperature field. Progressively, the thermal conductivity increases and approaches the parallel mode bound. External magnetic fields perpendicular to the temperature field do not change the nanofluid thermal conductivity significantly. The thermophysical data is given in Eapen et al. (2010).

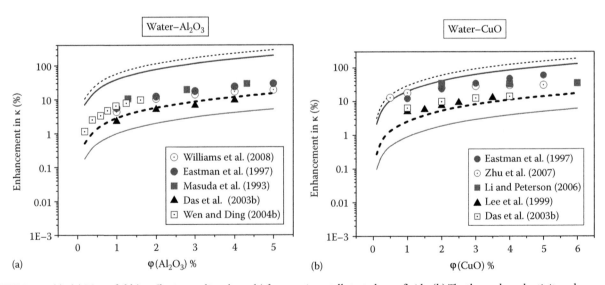

FIGURE 31.6 (a)–(g) Mean-field (or effective medium bounds) for experimentally tested nanofluids. (h) The thermal conductivity enhancements with molecular dynamics (MD) simulations of sub-nanometer solid particles. The thin-solid and the thin-dotted lines denote the enhancement in thermal conductivity with the series and parallel modes, respectively. The upper Maxwell bound is delineated by the thick-solid line while the lower Maxwell bound is given by thick-dashed line.

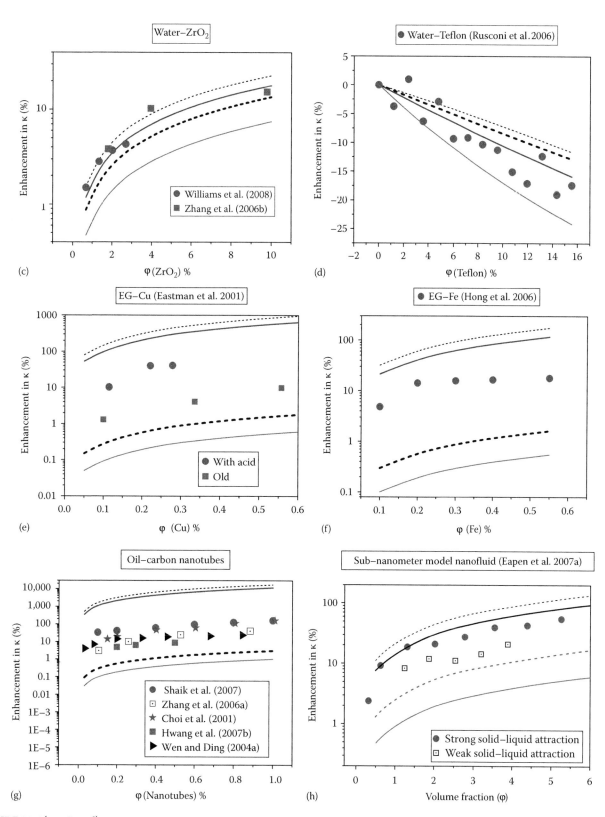

FIGURE 31.6 (continued)

the base media (Teflon or MFA in water). Several more comparisons are shown in (Eapen et al., 2010). Remarkably, most of the data lie between the Maxwell (HS) upper and lower bounds. In addition, MD and BD simulations also conform to the upper and lower Maxwell bounds with appropriate interfacial thermal resistance (Eapen et al., 2007a; Evans et al., 2006; He and Qiao, 2008; Vladkov and Barrat, 2006, 2008). Further, the nature of the thermal conduction in nanofluids is strikingly similar to that in liquid mixtures and solid nanocomposites (see Figure 31.7) (Eapen et al., 2010). Thus, there is overwhelming experimental and theoretical evidence to indicate that the nanofluid thermal conductivity is determined by the geometrical configuration of the nanoparticles. The thermophysical and transport properties used in the computations are taken from Eapen et al. (2010).

A notable feature is that only a small set of nanofluid data falls significantly below the lower Maxwell bound, even at very low volume fractions and with nanoparticles that are in the tens of nanometers. This behavior is very unlike that in solid composites where at low volume fractions and nanometer-sized filler particles, the effective thermal conductivity drops well below the series conduction bound. When the thermal conductivity of the dispersed medium becomes closer to that of the base media, the Maxwell bounds become narrower, as can be noted with zirconia and Teflon (MFA) nanofluids. For well-dispersed nanoparticles, the enhancement is consistent with the lower Maxwell bound. Since the Maxwell limit represents the maximum thermal conductivity that is possible with well-dispersed nanoparticles, it can be inferred that the interfacial thermal resistance

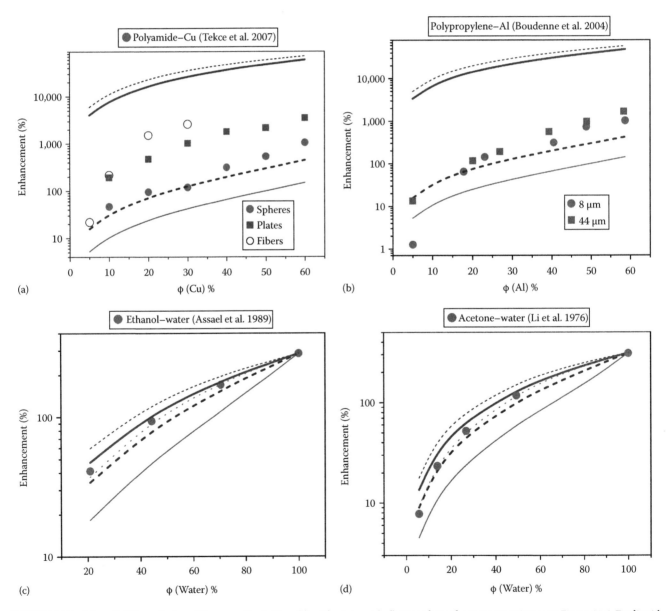

FIGURE 31.7 Mean-field bounds for solid composites (a,b) and liquid mixtures (c,d). Lines have the same meaning as in Figure 31.6. For liquid mixtures, not surprisingly, the symmetric or Bruggeman model (thin dotted line) best describes the mixture thermal conductivity.

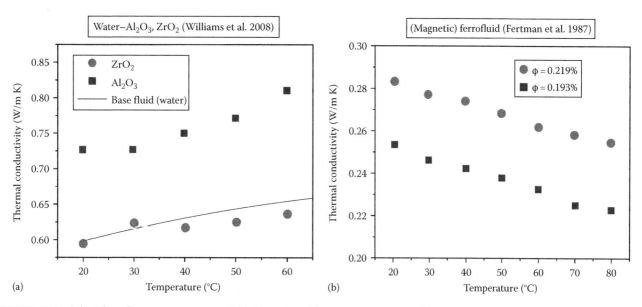

FIGURE 31.8 (a,b) Effect of temperature on nanofluid thermal conductivity. No strong correlation is observed between the thermal conductivity and the thermal motion of the nanoparticles. The base fluid thermal conductivity for ferrofluid is approximately 0.12 W/m K for the range of temperatures shown in (b).

for most reported nanofluids is negligible. A few exceptions, to a small degree, are noted (e.g., refer Timofeeva et al., 2007).

Effect of temperature: As discussed earlier, the nanofluid thermal conductivity increases with increasing temperature for water-based nanofluids (Das et al., 2003b; Williams et al., 2008). Ferrofluids, however, show an interesting behavior where the thermal conductivity *decreases* with increasing temperature (see Figure 31.8). In general, the nanofluid thermal conductivity has a loose correlation to that of the base fluid, as noted by Fertman et al. (1987), for ferrofluids, and more recently by Williams et al. (2008) for alumina and zirconia nanofluids.

31.3.3 Interfacial Thermal Resistance

The occurrence of an interfacial thermal (Kapitza) resistance at a liquid–solid interface has been experimentally evaluated by Cahill and coworkers who observed a bounding R_b of 0.67×10^{-8} and 2×10^{-8} K m² W⁻¹ for hydrophilic and hydrophobic interfaces, respectively (Ge et al., 2006). With nanofluids with carbon nanotubes, a large variation in R_b, ranging from a low 0.24×10^{-8} K m² W⁻¹ (Bryning et al., 2005) to a high 8.3×10^{-8} K m² W⁻¹ (Huxtable et al., 2003; Nan et al., 2004) that is comparable to R_b in a solid matrix [e.g., diamond–silicon composite having an R_b of 27×10^{-8} K m² W⁻¹ (Jagannadham and Wang, 2002)] is also reported. The large span in the R_b data and the near-zero R_b inferred from Figure 31.6 indicate an influence of the fluid interactions on the interfacial thermal resistance in nanofluids. Theoretical studies show that R_b attains relatively large values only when the liquid does not wet the solid surface (Barrat and Chiaruttini, 2003; Eapen et al., 2007a). For experimentally tested nanofluids, complete wetting may be a reasonable assumption for the dispersions of hydrophilic colloids (such as silica, and possibly for charged Teflon colloids), where particle solvation

is ensured by electrostatic forces (Eapen et al., 2010). The terms such as "hydrophobic" and "hydrophilic" are rather subtle, and the macroscopic concepts such as the contact angle may be a bit misleading. The rate of energy transfer would be indeed weaker if the liquid does not wet the solid, since in this case the liquid density in the interfacial layer would be depleted. Yet, from a microscopic point of view, what one may need to consider is the free energy of the insertion of the particle in the fluid. For a stable, nonaggregating colloidal dispersion, the latter is certainly negative (meaning, the particles are well-solvated). This means that even the particles made of a hydrophobic material such as Teflon can behave as "hydrophilic." The reason for this apparent paradox is related to the presence of the charged double layer, which leads to the formation of a solvation layer made of hydrated counterions, hindering solvent depletion in the interfacial layer (Eapen et al., 2010).

31.4 Conclusion

The data in Figures 31.5 and 31.6 strongly indicate that linear or fractal-like clustering effects are responsible for the thermal conductivity enhancements beyond the Maxwell prediction. Fundamentally, a nanofluid is a colloid and the predilection to aggregation is a prominent feature of all colloids. As shown by Weitz group in the early 1980s, aggregation (even for dilute colloids) is a function of time, temperature, surfactants (chemistry), and also the fractal dimension (Weitz and Oliveria, 1984; Weitz et al., 1984, 1985). Furthermore, the thermal diffusion of the nanoparticles has a perceptible influence on the nanofluid aggregation, and hence, on the thermal conductivity of the nanofluid itself (Gharagozloo et al., 2008). Future *in situ* experiments, which can probe both transport properties and the aggregation structure of the nanoparticles in the suspended state (such as with x-ray or neutron spectroscopy), can quantify the complex interactions in a nanofluid.

It is important, however, to note the difference between linear or fractal-like clustering, and arbitrary coalescence which leads to large nanoparticle clumps. The former can generate additional thermal conduction paths along the nanoparticles while the latter (which is inevitable, say, in the sedimentation processes aided by gravity) can only promote thermal conduction through the base fluid. In the latter case, the effective nanofluid thermal conductivity can be less than the Maxwell lower bound, even without a significant interfacial resistance. Hence, it is not surprising that clustering has been reported in several experiments that show enhancement and reduction in the nanofluid thermal conductivity (Hong et al., 2006; Zhu et al., 2006). It is, therefore, critical to identify and quantify the different types of nanofluid aggregation, coalescence, sedimentation, and fragmentation (Meakin, 1992) in the theoretical modeling of nanofluid transport properties.

Appendix

Non-equilibrium molecular dynamics (NEMD) simulations are performed to verify the analytical derivation for microscopic stress tensor and heat flux vector. NEMD simulations mimic an experimental procedure whereby a known heat flux (q'') is applied across two sections and the thermal conductivity is determined from Fourier's law as:

$$\kappa = -\frac{q''}{dT/dz} \tag{31.A.1}$$

where dT/dz is the temperature gradient. As with the linear response (or Green–Kubo) method, direct NEMD simulations are widely used for calculating the heat fluxes and thermal conductivities, and good conformity is generally observed between the methods (Schelling et al., 2002). Other synthetic NEMD methods (see, for example, Perronace et al., 2002; Sarman and Evans, 1992) provide alternate ways to compute the heat flux and thermal conductivity but are not considered in this chapter. A major limitation of NEMD stems from the extraordinarily large temperature gradients [$O(10^{10}$ K/m)] that are prescribed or generated in the system which are several orders higher than those observed in experiments. However, the deviation from equilibrium conditions, even with gigantic temperature gradients, is minimal.

A particularly simple way of direct simulation is through the imposed-flux method (Müller-Plathe, 1997) where a known heat flux is imposed on the system which generates a linear temperature profile at steady state. This method is compatible with periodic boundary conditions, conserves both energy and momentum, and experiences only limited perturbation effects. Reasonably accurate thermal conductivity estimates have been generated for atomic fluids, and with more complex fluids such as water and *n*-butane (Bedrov and Smith, 2000).

The simulation box is divided into many slabs perpendicular to one chosen direction, say z, with the edge slabs denoted as "cold" and the center slab as "hot." Periodic velocity exchanges are made between the atoms of these slabs such that the hottest atom in the cold slab is substituted with the coldest atom of the hot slab. This unphysical energy transfer generates a heat flux that flows from the middle to the edge slabs. At steady state, a linear temperature profile develops which is symmetric about the hot slab. The heat flux ($\hat{J}_q$) can be computed exactly with the known values of the velocities that are exchanged using the following expression:

$$\hat{J}_q = \frac{1}{2A_{xy}t} \sum_{\text{transfers}} \frac{m}{2}(v_{\text{h}}^2 - v_{\text{c}}^2) \tag{31.A.2}$$

where

A_{xy} is the cross-sectional area
t is the simulation time
m is the mass
v is the velocity
Subscripts h and c denote the hot and cold atoms.

The factor 2 in Equation 31.A.2 accounts for the heat flow in two directions about the center slab. In contrast, the time-averaged microscopic heat flux $\langle \tilde{j}_q \rangle$ for a binary system is given by

$$\langle \tilde{j}_q \rangle = \left\langle \left[\frac{1}{2} \sum_{k=\alpha}^{\beta} \sum_{i=1}^{N_k} m_i^k (v_i^k)^2 \mathbf{v}_i^k \right. \right.$$
$$\left. \left. + \sum_{k=\alpha}^{\beta} \sum_{l=\alpha}^{\beta} \sum_{i=1}^{N_k} \sum_{j>i}^{N_l} \left[\mathbf{I}\Phi(r_{ij}^{kl}) + \mathbf{r}_{ij}^{kl} \otimes \mathbf{F}_{ij}^{kl} \right] \cdot \mathbf{v}_i^k - \sum_{k=\alpha}^{\beta} \langle h^k \rangle \sum_{i=1}^{N_k} \mathbf{v}_i^k \right] \right\rangle$$

$$\tag{31.A.3}$$

As discussed in Section 31.2, the above expression assumes a spatial homogeneity which may not be strictly satisfied in a colloidal solution. Through NEMD simulations, the above microscopic expression will be compared to the exact heat flux given by Equation 31.A.2. Note that both heat fluxes are time-averaged quantities in NEMD simulations.

The model in this study consists of 5 solid clusters of 20 atoms each in a Lennard Jones (LJ) liquid (Allen and Tildesley, 1994) of 1948 atoms. Reduced units based on m, ε, and σ are used throughout in this section (referenced to base fluid). All the atoms have the same size (σ) and mass (m). The cluster atoms are held together by a finitely extendable nonlinear elastic (FENE) potential, which is given by (Evans et al., 2006; Grest and Kremer, 1986)

$$U_{\text{FENE}} = -A\varepsilon \ln \left[1 - \left(\frac{r}{B\sigma} \right)^2 \right] \tag{31.A.4}$$

where the constants A and B take the values 5.625 and 4.95, respectively. In addition to the above potential, the solid atoms also experience a standard LJ potential with parameters (ε, σ) (Allen and Tildesley, 1994; Eapen et al., 2007a). The simulations are carried out at a constant temperature of 1.0 and a volume corresponding to a pressure of 1.296. The density on an average is approximately 0.84. At this state point, the radial distribution function (*rdf*) indicates that the base fluid has a structure corresponding to that of a liquid. An exchange frequency of 1 in 60 time steps is found to be optimal to identify a statistically significant slope in the temperature profile. The equilibration is typically done for 100,000 iterations, and temperature in each slab is averaged for 150,000 iterations. Further averaging over 8–10 initial conditions are required to generate an acceptable linearity in the temperature profile (measured by the multiple correlation coefficient, R^2) (Eapen et al., 2007a).

Time-averaged heat flux: The objective of this exercise is to show that the microscopic binary heat flux expression in Equation 31.A.3 is suitable for use in colloids that are inherently inhomogeneous and nonideal. A strong (nonideal) solid–fluid (SF) cross-interaction strength of $\varepsilon_{SF}/\varepsilon = 7$ is prescribed for the interaction between the solid clusters and fluid atoms. Each half of the z-axis in the NEMD simulation cell is divided equally into 12–16 equi-sized slabs. In each slab, the microscopic heat flux is averaged over 150,000 iterations after an equilibration period of 100,000 iterations. In Figure 31.9, these slab heat flux estimates are compared to the exact heat flux given by Equation 31.A.2 developed across the hot and cold slabs.

The microscopic heat flux shows spatial oscillations which, after averaging, give a value of 0.1217. This is only 4.6% less than the exact value 0.1276 calculated from Equation 31.A.2. The standard deviation of the fluctuation is 0.0032, which is 2.6% of the average value. Similar results are obtained with different clusters, as shown in Table 31.1. Reasonable agreement is seen between the exact and the microscopic heat flux for the cases considered here, despite assuming spatial homogeneity. The result for model VI, which has one big nanoparticle of 100

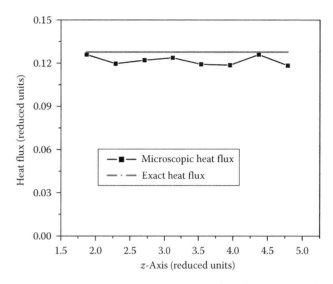

FIGURE 31.9 Comparison of microscopic heat flux computed by Equation 31.A.3 in different slabs with the exact value computed from Equation 31.A.2. For the sake of clarity, the exact value is shown as a continuous straight line.

atoms, is particularly noteworthy because the inhomogeneity has not resulted in a significant deviation in the average microscopic heat flux. This is not entirely surprising as the concept of homogeneity is always relative to the scales that are being considered. In monatomic fluids at molecular scales, there are large density fluctuations but by taking its Fourier transform, it is readily seen that these are transparent at the continuum scales (Boon and Yip, 1991). The delta functions in Equation 31.42 are slowly varying over the range where the potential is applicable. Thus, the operator Θ_{ij} in Equation 31.42, which represents the deviation due to inhomogeneity, represents a Taylor expansion of a small parameter r_{ij}/L, where r_{ij} is the atomic spacing and L is the largest wavelength of the heat flux vector. The ratio r_{ij}/L in the present simulation is $O(0.1)$, and thus, the operator Θ_{ij} does not have a significant effect on the computed time-averaged microscopic heat flux.

TABLE 31.1 Comparison of Microscopic Heat Flux (Equation 31.A.3) and Exact Estimate (Equation 31.A.2) Averaged over 150,000 Time Iterations and 10 Sets of Initial Conditions

Model	Cluster Arrangement	State (T, P)	$\dfrac{\varepsilon_{SF}}{\varepsilon}$	Exact $\hat{J}_q$	Microscopic $\langle \hat{j} \rangle_q$	%Δ
I	10 clusters with 10 atoms	1.0, 0.0	7	0.1198	0.1149	**−4.09**
II	10 clusters with 10 atoms	1.0, 1.3	7	0.1287	0.1247	**−3.11**
III	10 clusters with 10 atoms	1.0, 0.0	2	0.1114	0.0973	**−12.67**
IV	5 clusters with 20 atoms	1.0, 1.3	7	0.1276	0.1231	**−3.56**
V	5 clusters with 20 atoms	1.0, 1.3	2	0.1242	0.1222	**−1.61**
VI	1 cluster with 100 atoms	1.0, 1.3	7	0.2070	0.1972	**−4.7**

Note: All the cluster models have the same number fraction and use a constant energy algorithm.

Acknowledgments

JE wishes to acknowledge the interesting discussions with J. Buongiorno, W. Williams, S. Yip, R. Rusconi, R. Piazza, P. Keblinski, R. Prasher, J. Philip, J. Qiao, T. Bergman, D. Cahill, Ju Li and S. Choi. This work is partly funded by the US NRC Faculty Development Program.

References

Ahuja, A. S., 1975a, Augmentation of heat transport in laminar flow of polystyrene suspensions. I. Experiments and results, *J. Appl. Phys.* **46**:3408.

Ahuja, A. S., 1975b, Augmentation of heat transport in laminar flow of polystyrene suspensions. II. Analysis of the data, *J. Appl. Phys.* **46**:3417.

Allen, M. P. and Tildesley, D. J., 1994, *Computer Simulation of Liquids*, Clarendon Press, Oxford, U.K.

Assael, M. J., Charitidou, E., and Wakeham, W. A., 1989, Absolute measurements of the thermal conductivity of mixtures of alcohols with water, *Int. J. Thermophys.* **10**:793–803.

Avalos, J. B. and Mackie, A. D., 1997, Dissipative particle dynamics with energy conservation, *Europhys. Lett.* **40**:141.

Avalos, J. B. and Mackie, A. D., 1999, Dynamic and transport properties of dissipative particle dynamics with energy conservation, *J. Chem. Phys.* **111**(11):5267–5276.

Avsec, J. and Oblak, M., 2007, The calculation of thermal conductivity, viscosity and thermodynamic properties for nanofluids on the basis of statistical nanomechanics, *Int. J. Heat Mass Transf.* **50**:4331–4341.

Bang, I. C., Chang, S. H., and Baek, W.-P., 2005, Direct observation of a liquid film under a vapor environment in a pool boiling using a nanofluid, *Appl. Phys. Lett.* **86**:134107.

Barrat, J.-L. and Chiaruttini, F., 2003, Kapitza resistance at the liquid-solid interface, *Mol. Phys.* **101**:1605.

Bedrov, D. and Smith, G. D., 2000, Thermal conductivity of molecular fluids from molecular dynamics simulations: Application of a new imposed-flux method, *J. Chem. Phys.* **113**:8080.

Benveniste, Y., 1987, Effective thermal conductivity of composites with a thermal contact resistance between the constituents: Nondilute case, *J. Appl. Phys.* **61**:2840.

Bergman, T. L., 2009a, Analysis of heat transfer enhancement in minichannel heat sinks with turbulent flow using H_2O-Al_2O_3 nanofluids, *J. Electron. Packaging* **131**(2):021008-5.

Bergman, T. L., 2009b, Effect of reduced specific heats of nanofluids on single phase, laminar internal forced convection, *Int. J. Heat Mass Transf.* **52**(5–6):1240–1244.

Bertucci, S. J. and Flygare, W. H., 1975, Rough hard sphere treatment of mutual diffusion in binary liquid mixtures, *J. Chem. Phys.* **63**(1):1–9.

Bhattacharya, P., Saha, S. K., Yadav, A., Phelan, P. E., and Prasher, R. S., 2004, Brownian dynamics simulation to determine the effective thermal conductivity of nanofluids, *J. Appl. Phys.* **95**:6492.

Boon, J. P. and Yip, S., 1991, *Molecular Hydrodynamics*, Dover, New York.

Boudenne, A., Ibos, L., Fois, M., Gehin, E., and Majeste, J.-C., 2004, Thermophysical properties of polypropylene/aluminium composites, *J. Polymer Sci. Part B* **42**:722–732.

Bryning, M. B., Milkie, D. E., Islam, M. F., Kikkawa, J. M., and Yodh, A. G., 2005, Thermal conductivity and interfacial resistance in single-wall carbon nanotube epoxy composites, *Appl. Phys. Lett.* **87**:161909.

Buongiorno, J., 2006, Convective transport in nanofluids, *J. Heat Transf.* **128**:240.

Buongiorno, J., et al., 2009, A benchmark study on the thermal conductivity of nanofluids, *J. Appl. Phys.* **106**: 094312–14.

Carrillo, J. L., Donado, F., and Mendoza, M. E., 2003, Fractal patterns, cluster dynamics, and elastic properties of magnetorheological suspensions, *Phys. Rev. E* **68**(6):061509.

Carson, J. K., Lovatt, S. J., Tanner, D. J., and Cleland, A. C., 2005, Thermal conductivity bounds for isotropic porous materials, *Int. J. Heat Mass Transf.* **48**:2150–2158.

Chandler, D., 1987, *Introduction to Modern Statistical Mechanics*, Oxford University Press, New York.

Chantrenne, P. and Barrat, J.-L., 2004, Finite size effects in determination of thermal conductivities: Comparing molecular dynamics results with simple models, *J. Heat Transf.* **126**(4):577–585.

Chapman, S. and Cowling, T. G., 1970, *The Mathematical Theory of Non-Uniform Gases*, Cambridge University Press, London, U.K.

Chevalier, J., Tillement, O., and Ayela, F., 2007, Rheological properties of nanofluids flowing through microchannels, *Appl. Phys. Lett.* **91**(23):233103–3.

Choi, S. U. S., 1995, Enhancing thermal conductivity of fluids with nanoparticles, in: *Developments and Applications of Non-Newtonian Flows*, D. A. Siginer, and H. P. Wang (eds.), American Society of Mechanical Engineers, New York, pp. 99–105.

Choi, S. U. S., 2009, Nanofluids: From vision to reality through research, *J. Heat Transf.* **131**(3):033106-9.

Choi, S. U. S., Zhang, Z. G., Yu, W., Lockwood, F. E., and Grulke, E. A., 2001, Anomalous thermal conductivity enhancement in nanotube suspensions, *Appl. Phys. Lett.* **79**:2252–2254.

Chon, C. H., Kihm, K. D., Lee, S. P., and Choi, S. U. S., 2005, Empirical correlation finding the role of temperature and particle size for nanofluid (Al_2O_3) thermal conductivity enhancement, *Appl. Phys. Lett.* **87**:153107.

Chon, C. H., Paik, S. W., Tipton, J. B., and Kihm, K. D., 2006, Evaporation and dryout of nanofluid droplets on a microheater array, *J. Heat Transf.* **128**:735.

Chopkar, M., Das, P. K., and Manna, I., 2006, Synthesis and characterization of nanofluid for advanced heat transfer applications, *Scr. Mater.* **55**:549–552.

Chopkar, M., Kumar, S., Bhandari, D. R., Das, P. K., and Manna, I., 2007, Development and characterization of Al_2Cu and Ag_2Al nanoparticle dispersed water and ethylene glycol based nanofluid, *Mater. Sci. Eng. B* **139**:141–148.

Das, S. K., Putra, N., and Roetzel, W., 2003a, Pool boiling characteristics of nanofluids, *Int. J. Heat Mass Transf.* **46**:851–862.

Das, S. K., Putra, N., Thiesen, P., and Roetzel, W., 2003b, Temperature dependence of thermal conductivity enhancement for nanofluids *J. Heat Transf.* **125**:567–574.

de Groot, S. R. and Mazur, P., 1984, *Nonequilibrium Thermodynamics*, Dover Publications, Inc., New York.

DeVera, A. L. and Strieder, W., 1977, Upper and lower bounds on the thermal conductivity of a random, two-phase material, *J. Phys. Chem.* **81**:1783.

Eapen, J., Li, J., and Yip, S., 2007a, Beyond the Maxwell limit: Thermal conduction in nanofluids with percolating fluid structures, *Phys. Rev. E* **76**:062501.

Eapen, J., Li, J., and Yip, S., 2007b, Mechanism of thermal transport in dilute nanocolloids, *Phys. Rev. Lett.* **98**:028302.

Eapen, J., Williams, W. C., Buongiorno, J., Hu, L.-W., Yip, S., Rusconi, R., and Piazza, R., 2007c, Mean-field versus microconvection effects in nanofluid thermal conduction, *Phys. Rev. Lett.* **99**:095901.

Eapen, J., Rusconi, R., Piazza, R., and Yip, S., 2010, The classical nature of thermal conduction in nanofluids, *J. Heat Transf.* (in press).

Eastman, J. A., Choi, S. U. S., Li, S., Thompson, L. J., and Lee, S., 1997, Enhanced thermal conductivity through the development of nanofluids, *Proc. MRS Symp.* **457**:3–11.

Eastman, J. A., Choi, S. U. S., Li, S., Yu, W., and Thompson, L. J., 2001, Anomalously increased effective thermal conductivities of ethylene glycol-based nanofluids containing copper nanoparticles, *Appl. Phys. Lett.* **78**:718–720.

Eastman, J. A., Phillpot, S. R., Choi, S. U. S., and Keblinski, P., 2004, Thermal transport in nanofluids, *Annu. Rev. Mater. Res.* **34**(1):219–246.

Ernst, M. H. and Brito, R., 2006, New Green-Kubo formulas for transport coefficients in hard-sphere, Langevin fluids and the likes, *Europhys. Lett.* **73**:183–189.

Español, P., 1995, Hydrodynamics from dissipative particle dynamics, *Phys. Rev. E (Stat. Phys. Plasmas Fluids Relat. Interdisc. Top.)* **52**(2):1734–1742.

Español, P., 1997, Dissipative particle dynamics with energy conservation, *Europhys. Lett.* **40**:631–636.

Espanol, P. and Warren, P., 1995, Statistical mechanics of dissipative particle dynamics, *Europhys. Lett.* **30**:191–196.

Evans, D. J. and Morriss, G. P., 1990, *Statistical Mechanics of Non-Equilibrium Liquids*, Academic Press, London, U.K.

Evans, W., Fish, J., and Keblinski, P., 2006, Role of Brownian motion hydrodynamics on nanofluid thermal conductivity, *Appl. Phys. Lett.* **88**:093116.

Every, A. G., Tzou, Y., Hasselman, D. P. H., and Raj, R., 1992, The effect of particle size on the thermal conductivity of ZnS/diamond composites, *Acta Metall. Mater.* **40**(1):123–129.

Feng, Y., Yu, B., Xu, P., and Zou, M., 2007, The effective thermal conductivity of nanofluids based on the nanolayer and the aggregation of nanoparticles, *J. Phys. D: Appl. Phys.* **40**:3164–3171.

Fertman, V. E., 1987, Thermal and physical properties of magnetic fluids, *J. Eng. Phys. Thermophys.* **53**(3):1097–1105.

Fertman, V. E., Golovicher, L. E., and Matusevich, N. P., 1987, Thermal conductivity of magnetite magnetic fluids, *J. Magn. Magn. Mater.* **65**(2–3):211–214.

Fitts, D. D., 1962, *Nonequilibrium Thermodynamics*, McGraw-Hill Book Company, Inc., New York.

Gao, L. and Zhou, X. F., 2006, Differential effective medium theory for thermal conductivity in nanofluids, *Phys. Lett. A* **348**:355–360.

Garg, J., Poudel, B., Chiesa, M., Gordon, J. B., Ma, J. J., Wang, J. B., Ren, Z. F., et al., 2008, Enhanced thermal conductivity and viscosity of copper nanoparticles in ethylene glycol nanofluid, *J. Appl. Phys.* **103**(7):074301-6.

Ge, Z., Cahill, D. G., and Braun, P. V., 2006, Thermal conductance of hydrophilic and hydrophobic interfaces, *Phys. Rev. Lett.* **96**:186101.

Geiger, A. L., Hasselman, D. P. H., and Donaldson, K. Y., 1993, Effect of reinforcement particle size on the thermal conductivity of a particulate silicon-carbide reinforced aluminium-matrix composite, *J. Mater. Sci. Lett.* **12**:420–423.

Gharagozloo, P. E., Eaton, J. K., and Goodson, K. E., 2008, Diffusion, aggregation, and the thermal conductivity of nanofluids, *Appl. Phys. Lett.* **93**(10):103110-3.

Gmachowski, L., 2002, Aggregate restructuring and its effect on the aggregate size distribution, *Colloids Surf. A: Physicochem. Eng. Aspects* **207**(1–3):271–277.

González, A. E., 1993, Universality of colloid aggregation in the reaction limit: The computer simulations, *Phys. Rev. Lett.* **71**(14):2248.

Grest, G. S. and Kremer, K., 1986, Molecular dynamics simulation for polymers in the presence of a heat bath, *Phys. Rev. A* **33**:3628.

Gupta, A. and Kumar, R., 2007, Role of Brownian motion on the thermal conductivity enhancement of nanofluids, *Appl. Phys. Lett.* **91**(22):223102-3.

Hanley, H. J. M., 1969, *Transport Phenomena in Fluids*, H. J. M. Hanley (ed.), Marcel Dekker, New York.

Hansen, J.-P. and McDonald, I. R., 1986, *Theory of Simple Fluids*, Elsevier, London, U.K.

Hartley, G. S. and Crank, J., 1949, Some fundamental definitions and concepts in diffusion processes, *Trans. Faraday Soc.* **45**:801–818.

Hashin, Z. and Shtrikman, S., 1962, A variational approach to the theory of the effective magnetic permeability of multiphase materials, *J. Appl. Phys.* **33**:3125.

Hasselman, D. P. H. and Donaldson, K. Y., 2000, Role of size in the effective thermal conductivity of composites with an interfacial thermal barrier, *J. Wide Bandgap Mater.* **7**(4):306–318.

He, P. and Qiao, R., 2008, Self-consistent fluctuating hydrodynamics simulations of thermal transport in nanoparticle suspensions, *J. Appl. Phys.* **103**(9):094305-6.

Hess, W., Frisch, H. L., and Klein, R., 1986, On the hydrodynamic behavior of colloidal aggregates *Z. Phys. B: Condens. Matter* **64**:65–67.

Hiemenz, P. C. and Rajagopalan, R., 1997, *Principles of Colloid and Surface Chemistry*, Marcel Dekker, Inc., New York.

Hoheisel, C., 1987, *Theoretical Treatment of Liquids and Mixtures*, Elsevier, Amsterdam, the Netherlands.

Hong, K. S., Hong, T.-K., and Yang, H.-S., 2006, Thermal conductivity of Fe nanofluids depending on the cluster size of nanoparticles, *Appl. Phys. Lett.* **88**:031901.

Hong, T. K., Yang, H. S., and Choi, C. J., 2005, Study of the enhanced thermal conductivity of Fe nanofluids, *J. Appl. Phys.* **97**:064311.

Huang, C. and Yang, C. Z., 1999, Fractal aggregation and optical absorption of copper nanoparticles prepared by in situ chemical reduction within a Cu^{2+}-polymer complex, *Appl. Phys. Lett.* **74**(12):1692–1694.

Hunter, R. J., 1992, *Introduction to Modern Colloidal Science*, Oxford Science Publications, Oxford, U.K.

Huxtable, S. T., Cahill, D. G., Shenogin, S., Xue, L., Ozisik, R., Barone, P., Usrey, M., et al., 2003, Interfacial heat flow in carbon nanotube suspension, *Nat. Mater.* **2**:731–734.

Hwang, D., Hong, K. S., and Yang, H.-S., 2007a, Study of thermal conductivity of nanofluids for the application of heat transfer fluids, *Thermochim. Acta* **455**:66–69.

Hwang, Y., Lee, J. K., Lee, C. H., Jung, Y. M., Cheong, S. I., Lee, C. G., Ku, B. C., and Jang, S. P., 2007b, Stability and thermal conductivity characteristics of nanofluids, *Thermochim. Acta* **455**:70–74.

Irving, J. H. and Kirkwood, J. G., 1950, The statistical mechanical theory of transport processes. IV. The equations of hydrodynamics, *J. Chem. Phys.* **18**(6):817–829.

Jagannadham, K. and Wang, H., 2002, Thermal resistance of interfaces in AlN–diamond thin film composites, *J. Appl. Phys.* **91**(3):1224–1235.

Jain, S., Patel, H., and Das, S., 2009, Brownian dynamic simulation for the prediction of effective thermal conductivity of nanofluid, *J. Nanopart. Res.* **11**(4):767–773.

Jang, S. P. and Choi, S. U. S., 2004, Role of Brownian motion in the enhanced thermal conductivity of nanofluids, *Appl. Phys. Lett.* **84**:4316–4318.

Jang, S. P. and Choi, S. U. S., 2007, Effects of various parameters on nanofluid thermal conductivity, *J. Heat Transf.* **129**(5):617–623.

Ju, Y. S., Kim, J., and Hung, M.-T., 2008, Experimental study of heat conduction in aqueous suspensions of aluminum oxide nanoparticles, *J. Heat Transf.* **130**(9):092403-6.

Kang, H. U., Kim, S. H., and Oh, J. M., 2006, Estimation of thermal conductivity of nanofluid using experimental effective particle volume, *Exp. Heat Transf.* **19**:181–191.

Kedzierski, M. A., 2009, Effect of CuO nanoparticle concentration on R134a/lubricant pool-boiling heat transfer, *J. Heat Transf.* **131**(4):043205-7.

Keblinski, P., Phillpot, S. R., Choi, S. U. S., and Eastman, J. A., 2002, Mechanisms of heat flow in suspensions of nano-sized particles (nanofluids), *Int. J. Heat Mass Transf.* **45**(4):855–863.

Keblinski, P., Eastman, J. A., and Cahill, D. G., 2005, Nanofluids for thermal transport, *Mater. Today* **8**:36–44.

Keblinski, P., Prasher, R., and Eapen, J., 2008, Thermal conductance of nanofluids: Is the controversy over? *J. Nanopart. Res.* **10**(7):1089–1097.

Kellner, R. R. and Kohler, W., 2005, Short-time aggregation dynamics of reversible light-induced cluster formation in ferrofluids, *J. Appl. Phys.* **97**(3):034910-6.

Kim, J. and Kramer, T. A., 2006, Improved orthokinetic coagulation model for fractal colloids: Aggregation and breakup, *Chem. Eng. Sci.* **61**(1):45–53.

Kim, S. H., Choi, S. R., and Kim, D., 2007, Thermal conductivity of metal-oxide nanofluids: Particle size dependence and effect of laser irradiation, *J. Heat Transf.* **129**:298–307.

Kim, S. J., McKrell, T., Buongiorno, J., and Hu, L.-W., 2008, Alumina nanoparticles enhance the flow boiling critical heat flux of water at low pressure, *J. Heat Transf.* **130**(4):044501-3.

Kim, S. J., McKrell, T., Buongiorno, J., and Hu, L.-W., 2009, Experimental study of flow critical heat flux in alumina-water, zinc-oxide-water, and diamond-water nanofluids, *J. Heat Transf.* **131**(4):043204-7.

Kleinstreuer, C. and Li, J., 2008, Discussion: Effects of various parameters on nanofluid thermal conductivity. (Jang, S. P. and Choi, S. D. S., 2007, *ASME J. Heat Transf.* **129**, 617–623), *J. Heat Transf.* **130**(2):025501-3.

Kolade, B., Goodson, K. E., and Eaton, J. K., 2009, Convective performance of nanofluids in a laminar thermally developing tube flow, *J. Heat Transf.* **131**(5):052402-8.

Koo, J. and Kleinstreuer, C., 2004, A new thermal conductivity model for nanofluids *J. Nanopart. Res.* **6**:577–588.

Kovalchuk, N., Starov, V., Langston, P., Hilal, N., and Zhdanov, V., 2008, Colloidal dynamics: Influence of diffusion, inertia and colloidal forces on cluster formation, *J. Colloid Interface Sci.* **325**(2):377–385.

Kumar, R. and Milanova, D., 2009, Effect of surface tension on nanotube nanofluids, *Appl. Phys. Lett.* **94**(7):073107–3.

Kumar, D. H., Patel, H. E., Kumar, V. R. R., Sundararajan, T., Pradeep, T., and Das, S. K., 2004, Model for heat conduction in nanofluids, *Phys. Rev. Lett.* **93**:144301.

LaRosa, J. L. and Cawley, J. D., 1992, Fractal dimension of alumina aggregates grown in two dimensions, *J. Am. Ceram. Soc.* **75**(7):1981–1984.

Lee, M. H. and Furst, E. M., 2006, Formation and evolution of sediment layers in an aggregating colloidal suspension, *Phys. Rev. E (Stat. Nonlinear Soft Matter Phys.)* **74**(3):031401–11.

Lee, S., Choi, S. U. S., Li, S., and Eastman, J. A., 1999, Measuring thermal conductivity of fluids containing oxide nanoparticles, *J. Heat Transf.* **121**:280–289.

Lee, D., Kim, J. W., and Kim, B. G., 2006, A new parameter to control heat transport in nanofluids: Surface charge state of the particle in suspension, *J. Phys. Chem.* **110**:4323.

Li, C. C., 1976, Thermal conductivity of liquid mixtures, *AIChE J.* **22**:927–930.

Li, J., 2000, Modeling microstructural effects on deformation resistance and thermal conductivity, PhD thesis, Department of Nuclear Engineering, Massachusetts Institute of Technology, Cambridge, U.K.

Li, C. H. and Peterson, G. P., 2006, Experimental investigation of temperature and volume fraction variations on the effective thermal conductivity of nanoparticle suspensions (nanofluids), *J. Appl. Phys.* **99**:084314.

Li, C. H. and Peterson, G. P., 2007a, The effect of particle size on the effective thermal conductivity of Al_2O_3-water nanofluids, *J. Appl. Phys.* **101**:044312.

Li, C. H. and Peterson, G. P., 2007b, Mixing effect on the enhancement of the effective thermal conductivity of nanoparticle suspensions (nanofluids), *Int. J. Heat Mass Transf.* **50**:4668–4677.

Li, Q., Xuan, Y., and Wang, J., 2005, Experimental investigations on transport properties of magnetic fluids, *Exp. Therm. Fluid Sci.* **30**(2):109–116.

Li, Q. and Xuan, Y., 2006, Enhanced heat transfer behaviors of new heat carrier for spacecraft thermal management, *J. Spacecraft Rockets* **43**(3):687–690.

Li, C. H., Williams, W., Buongiorno, J., Hu, L.-W., and Peterson, G. P., 2008, Transient and steady-state experimental comparison study of effective thermal conductivity of Al_2O_3/water nanofluids, *J. Heat Transf.* **130**(4):042407-7.

Majolino, D., Mallamace, F., Migliardo, P., Micali, N., and Vasi, C., 1989, Elastic and quasielastic light-scattering studies of the aggregation phenomena in water solutions of polystyrene particles, *Phys. Rev. A* **40**(8):4665.

Marsh, C., Backx, G., and Ernst, M., 1997a, Fokker-Planck-Boltzmann equation for dissipative particle dynamics, *Europhys. Lett.* **38**:411.

Marsh, C. A., Backx, G., and Ernst, M. H., 1997b, Static and dynamic properties of dissipative particle dynamics, *Phys. Rev. E* **56**(2):1676.

Martin, J. E., Wilcoxon, J. P., Schaefer, D., and Odinek, J., 1990, Fast aggregation of colloidal silica, *Phys. Rev. A* **41**(8):4379.

Masuda, H., Ebata, A., Teramae, K., and Hishinuma, N., 1993, Alteration of thermal conductivity and viscosity of liquid by dispersing ultra-fine particles (Dispersion of γ-Al_2O_3, SiO_2, and TiO_2 ultra-fine particles), *Netsu Bussei (Japan)* **7**:227–233.

Maxwell, J. C., 1881, *A Treatise on Electricity and Magnetism*, 2nd edn., Claredon, Oxford, U.K.

McQuarrie, D. A., 2000, *Statistical Mechanics*, University Science Books, Sausalito, CA.

Meakin, P., 1992, Aggregation kinetics, *Phys. Scr.* **46**:295–311.

Meakin, P. and Deutch, J. M., 1987, Properties of the fractal measure describing the hydrodynamic force distributions for fractal aggregates moving in a quiescent fluid, *J. Chem. Phys.* **86**(8):4648–4656.

Meakin, P., Chen, Z.-Y., and Deutch, J. M., 1985, The translational friction coefficient and time dependent cluster size distribution of three dimensional cluster–cluster aggregation [sup a),b)], *J. Chem. Phys.* **82**(8):3786–3789.

Milanova, D. and Kumar, R., 2005, Role of ions in pool boiling heat transfer of pure and silica nanofluids, *Appl. Phys. Lett.* **87**:233107.

Milanova, D. and Kumar, R., 2008, Heat transfer behavior of silica nanoparticles in pool boiling experiment, *J. Heat Transf.* **130**(4):042401-6.

Müller-Plathe, F., 1997, A simple nonequilibrium molecular dynamics method for calculating the thermal conductivity, *J. Chem. Phys.* **106**:6082.

Murshed, S. M. S., Leong, K. C., and Yang, C., 2005, Enhanced thermal conductivity of TiO_2-water based nanofluids, *Int. J. Therm. Sci.* **44**:367–373.

Murshed, S. M. S., Leong, K. C., and Yang, C., 2006, Determination of the effective thermal diffusivity of nanofluids by the double hot-wire technique, *J. Phys. D: Appl. Phys.* **39**:5316–5322.

Murshed, S. M. S., Leong, K. C., and Yang, C., 2008, Thermophysical and electrokinetic properties of nanofluids-A critical review, *Appl. Therm. Eng.* **28**(17–18):2109–2125.

Namburu, P. K., Kulkarni, D. P., Dandekar, A., and Das, D. K., 2007, Experimental investigation of viscosity and specific heat of silicon dioxide nanofluids, *Micro Nano Lett.* **2**(3):67–71.

Nan, C.-W., Birringer, R., Clarke, D. R., and Gleiter, H., 1997, Effective thermal conductivity of particulate composites with interfacial thermal resistance, *J. Appl. Phys.* **81**:6692–6699.

Nan, C.-W., Liu, G., Lin, Y., and Li, M., 2004, Interface effect on thermal conductivity of carbon nanotube composites, *Appl. Phys. Lett.* **85**:3549–3551.

Narayan, G. P., Anoop, K. B., and Das, S. K., 2007, Mechanism of enhancement/deterioration of boiling heat transfer using stable nanoparticle suspensions over vertical tubes, *J. Appl. Phys.* **102**(7):074317-7.

Nayak, A. K., Gartia, M. R., and Vijayan, P. K., 2008, An experimental investigation of single-phase natural circulation behavior in a rectangular loop with Al_2O_3 nanofluids, *Exp. Therm. Fluid Sci.* **33**(1):184–189.

Nayak, A. K., Gartia, M. R., and Vijayan, P. K., 2009a, Nanofluids: A novel promising flow stabilizer in natural circulation systems, *AIChE J.* **55**(1):268–274.

Nayak, A. K., Singh, R. K., and Kulkarni, P. P., 2009b, Thermal expansion characteristics of Al_2O_3 nanofluids: More to understand than understood, *Appl. Phys. Lett.* **94**(9):094102-3.

Nie, C., Marlow, W. H., and Hassan, Y. A., 2008, Discussion of proposed mechanisms of thermal conductivity enhancement in nanofluids, *Int. J. Heat Mass Transf.* **51**(5–6):1342–1348.

NIST, 2007, National Institute of Standards and Technology (NIST), Fluid properties, http://webbook.nist.gov/chemistry/fluid/

Odriozola, G., Leone, R., Schmitt, A., Moncho-Jordá, A., and Hidalgo-Álvarez, R., 2003, Coupled aggregation and sedimentation processes: The sticking probability effect, *Phys. Rev. E* **67**(3):031401.

Pal, R., 2007, New models for thermal conductivity of particulate composites, *J. Reinf. Plast. Compos.* **26**(7):643–651.

Patel, H. E., Das, S. K., Sundararajan, T., Nair, A. S., George, B., and Pradeep, T., 2003, Thermal conductivities of naked and monolayer protected metal nanoparticle based nanofluids: Manifestation of anomalous enhancement and chemical effects, *Appl. Phys. Lett.* **83**:2931–2933.

Patel, H. E., Sundararajan, T., Pradeep, T., Dasgupta, A., Dasgupta, N., and Das, S. K., 2005, A micro-convection model for thermal conductivity of nanofluids, *Pramana - J. Phys.* **65**:863.

Pauliac-Vaujour, E., Stannard, A., Martin, C. P., Blunt, M. O., Notingher, I., Moriarty, P. J., Vancea, I., and Thiele, U., 2008, Fingering instabilities in dewetting nanofluids, *Phys. Rev. Lett.* **100**(17):176102-4.

Penas, J. R. V., de Zarate, J. M. O., and Khayet, M., 2008, Measurement of the thermal conductivity of nanofluids by the multicurrent hot-wire method, *J. Appl. Phys.* **104**(4):044314-8.

Perronace, A., Ciccotti, G., Leroy, F., Fuchs, A. H., and Rousseau, B., 2002, Soret coefficient for liquid argon-krypton mixtures via equilibrium and nonequilibrium molecular dynamics: A comparison with experiments, *Phys. Rev. E* **66**:031201.

Phelan, P. E., Bhattacharya, P., and Prasher, R. S., 2005, Nanaofluids for heat transfer applications, *Annu. Rev. Heat Transf.* **14**:255.

Philip, J., Shima, P. D., and Raj, B., 2007, Enhancement of thermal conductivity in magnetite based nanofluid due to chainlike structures, *Appl. Phys. Lett.* **91**(20):203108-3.

Philip, J., Shima, P. D., and Raj, B., 2008, Nanofluid with tunable thermal properties, *Appl. Phys. Lett.* **92**(4):043108-3.

Popplewell, J., Al-Qenaie, A., Charles, S. W., Moskowitz, R., and Raj, K., 1982, Thermal conductivity measurements on ferrofluids, *Colloid Polym. Sci.* **260**(3):333–338.

Prasher, R., Bhattacharya, P., and Phelan, P. E., 2005, Thermal conductivity of nanoscale colloidal solutions (nanofluids), *Phys. Rev. Lett.* **94**:025901.

Prasher, R., Bhattacharya, P., and Phelan, P. E., 2006a, Brownian-motion-based convective-conductive model for the effective thermal conductivity of nanofluids *J. Heat Transf.* **128**:588.

Prasher, R., Evans, W., Meakin, P., Fish, J., Phelan, P., and Keblinski, P., 2006b, Effect of aggregation on thermal conduction in colloidal nanofluids. *Appl. Phys. Lett.* **89**:143119.

Prasher, R., Phelan, P. E., and Bhattacharya, P., 2006c, Effect of aggregation kinetics on the thermal conductivity of nanoscale colloidal solutions (nanofluid), *Nano Lett.* **6**(7):1529–1534.

Prasher, R., Song, D., Wang, J., and Phelan, P., 2006d, Measurements of nanofluid viscosity and its implications for thermal applications, *Appl. Phys. Lett.* **89**(13):133108-3.

Putnam, S. A., Cahill, D. G., Braun, P. V., Ge, Z., and Shimmin, R. G., 2006, Thermal conductivity of nanoparticle suspensions, *J. Appl. Phys.* **99**:084308.

Rapaport, D. C., 2004, *The Art of Molecular Dynamics Simulation*, 2nd edn., Cambridge University Press, Cambridge, U.K.

Ren, Y., Xie, H., and Cai, A., 2005, Effective thermal conductivity of nanofluids containing spherical nanoparticles, *J. Phys. D: Appl. Phys.* **38**:3958–3961.

Rusconi, R., Rodari, E., and Piazza, R., 2006, Optical measurements of the thermal properties of nanofluids, *Appl. Phys. Lett.* **89**:261916.

Sarman, S. and Evans, D. J., 1992, Heat flow and mass diffusion in binary Lennard-Jones mixtures. II, *Phys. Rev. A* **46**:1960.

Schelling, P. K., Phillpot, S. R., and Keblinski, P., 2002, Comparison of atomic-level simulation methods for computing thermal conductivity, *Phys. Rev. B* **65**:144306.

Schmidt, A. J., Chiesa, M., Torchinsky, D. H., Johnson, J. A., Boustani, A., McKinley, G. H., Nelson, K. A., and Chen, G., 2008a, Experimental investigation of nanofluid shear and longitudinal viscosities, *Appl. Phys. Lett.* **92**(24):244107-3.

Schmidt, A. J., Chiesa, M., Torchinsky, D. H., Johnson, J. A., Nelson, K. A., and Chen, G., 2008b, Thermal conductivity of nanoparticle suspensions in insulating media measured with a transient optical grating and a hotwire, *J. Appl. Phys.* **103**(8):083529-5.

Sefiane, K., 2006, On the role of structural disjoining pressure and contact line pinning in critical heat flux enhancement during boiling of nanofluids, *Appl. Phys. Lett.* **89**:044106.

Shaikh, S., Lafdi, K., and Ponnappan, R., 2007, Thermal conductivity improvement in carbon nanoparticle doped PAO oil: An experimental study, *J. Appl. Phys.* **101**:064302.

Shukla, R. K. and Dhir, V. K., 2008, Effect of Brownian motion on thermal conductivity of nanofluids, *J. Heat Transf.* **130**(4):042406-13.

Singh, D., Timofeeva, E., Yu, W., Routbort, J., France, D., Smith, D., and Lopez-Cepero, J. M., 2009, An investigation of silicon carbide-water nanofluid for heat transfer applications, *J. Appl. Phys.* **105**(6):064306-6.

Sofian, N. M., Rusu, M., Neagu, R., and Neagu, E., 2001, Metal powder-filled polyethylene composites. V. Thermal properties, *J. Thermoplast. Compos. Mater.* **14**:20–33.

Starchev, K. and Stoylov, S., 1993, Structure of silica determined by use of a light-scattering method, *Phys. Rev. B* **47**(18):11725.

Tekce, H. S., Kumlutas, D., and Tavman, I. H., 2007, Effect of particle shape on thermal conductivity of copper reinforced polymer composites, *J. Reinf. Plast. Compos.* **26**:113–121.

Thurn-Albrecht, T., Meier, G., Müller-Buschbaum, P., Patkowski, A., Steffen, W., Grübel, G., Abernathy, D. L. et al., 1999, Structure and dynamics of surfactant-stabilized aggregates of palladium nanoparticles under dilute and semidilute conditions: Static and dynamic x-ray scattering, *Phys. Rev. E* **59**(1):642.

Tillman, P. and Hill, J. M., 2007, Determination of nanolayer thickness for a nanofluid, *Int. Commun. Heat Mass Transf.* **34**:399–407.

Timofeeva, E. V., Gavrilov, A. N., McCloskey, J. M., Tolmachev, Y. V., Sprunt, S., Lopatina, L. M., and Selinger, J. V., 2007, Thermal conductivity and particle agglomeration in alumina nanofluids: Experiment and theory, *Phys. Rev. E (Stat. Nonlinear Soft Matter Phys.)* **76**(6):061203-16.

Torquato, S. and Rintoul, M. D., 1995, Effect of interface on the properties of composite media, *Phys. Rev. Lett.* **75**(22):4067–4070.

Trisaksri, V. and Wongwises, S., 2007, Critical review of heat transfer characteristics of nanofluids, *Renewable Sust. Energ. Rev.* **11**(3):512–523.

Tsai, T.-H., Kuo, L.-S., Chen, P.-H., and Yang, C.-T., 2008, Effect of viscosity of base fluid on thermal conductivity of nanofluids, *Appl. Phys. Lett.* **93**(23):233121-3.

Turq, P., Lantelme, F., and Friedman, H. L., 1977, Brownian dynamics: Its application to ionic solutions, *J. Chem. Phys.* **66**(7):3039–3044.

Tweney, R. D., 2006, Discovering discovery: How Faraday found the first metallic colloid, *Perspect. Sci.* **14**:97.

Vancea, I., Thiele, U., Pauliac-Vaujour, E., Stannard, A., Martin, C. P., Blunt, M. O., and Moriarty, P. J., 2008, Front instabilities in evaporatively dewetting nanofluids, *Phys. Rev. E (Stat. Nonlinear Soft Matter Phys.)* **78**(4):041601-15.

Venerus, D. C., Kabadi, M. S., Lee, S., and Perez-Luna, V., 2006, Study of thermal transport in nanoparticle suspensions using forced Rayleigh scattering, *J. Appl. Phys.* **100**:094310.

Vladkov, M. and Barrat, J.-L., 2006, Modeling transient absorption and thermal conductivity in a simple nanofluid, *Nano Lett.* **6**:1224–1228.

Vladkov, M. and Barrat, J.-L., 2008, Modeling thermal conductivity and collective effects in a simple nanofluid, *J. Comput. Theor. Nanosci.* **5**:187–193.

Vogelsang, R. and Hoheisel, C., 1987, Thermal conductivity of a binary-liquid mixture studied by molecular dynamics with use of Lennard-Jones potentials, *Phys. Rev. A* **35**:3487.

Vogelsang, R., Hoheisel, C., Sindzingre, P., Ciccotti, G., and Frenkel, D., 1989, Computation of partial enthalpies of various Lennard-Jones model mixtures by NPT molecular dynamics, *J. Phys. Condens. Matter* **1**:957.

Wang, X.-Q. and Mujumdar, A. S., 2007, Heat transfer characteristics of nanofluids: A review, *Int. J. Therm. Sci.* **46**:1–19.

Wang, X., Xu, X., and Choi, S. U. S., 1999, Thermal conductivity of nanoparticle-fluid mixture, *J. Thermophys. Heat Transf.* **13**:474–480.

Wang, B.-X., Zhou, L.-P., and Peng, Z.-F., 2003, A fractal model for predicting the effective thermal conductivity of liquid with suspension of nanoparticles, *Int. J. Heat Mass Transf.* **46**:2665–2672.

Weitz, D. A. and Oliveria, M., 1984, Fractal structures formed by kinetic aggregation of aqueous gold colloids, *Phys. Rev. Lett.* **52**(16):1433.

Weitz, D. A., Huang, J. S., Lin, M. Y., and Sung, J., 1984, Dynamics of diffusion-limited kinetic aggregation, *Phys. Rev. Lett.* **53**(17):1657.

Weitz, D. A., Huang, J. S., Lin, M. Y., and Sung, J., 1985, Limits of the fractal dimension for irreversible kinetic aggregation of gold colloids, *Phys. Rev. Lett.* **54**(13):1416–1419.

Wen, D. and Ding, Y., 2004a, Effective thermal conductivity of aqueous suspensions of carbon nanotubes (carbon nanotube nanofluids), *J. Thermophys. Heat Transf.* **18**(4):481–485.

Wen, D. and Ding, Y., 2004b, Experimental investigation into convective heat transfer of nanofluids at the entrance region under laminar flow conditions, *Int. J. Heat Mass Transf.* **47**(24):5181–5188.

Wen, D. and Ding, Y., 2006, Natural convective heat transfer of suspensions of titanium dioxide nanoparticles (nanofluids), *IEEE Trans. Nanotechnol.* **5**(3):220–227.

Wen, D., Lin, G., Vafaei, S., and Zhang, K., 2009, Review of nanofluids for heat transfer applications, *Particuology* **7**(2):141–150.

Wensel, J., Wright, B., Thomas, D., Douglas, W., Mannhalter, B., Cross, W., Hong, H., Kellar, J., Smith, P., and Roy, W., 2008, Enhanced thermal conductivity by aggregation in heat transfer nanofluids containing metal oxide nanoparticles and carbon nanotubes, *Appl. Phys. Lett.* **92**(2):023110-3.

Wilcoxon, J. P., Martin, J. E., and Schaefer, D. W., 1989, Aggregation in colloidal gold, *Phys. Rev. A* **39**(5):2675.

Williams, W., Buongiorno, J., and Hu, L.-W., 2008, Experimental investigation of turbulent convective heat transfer and pressure loss of alumina/water and zirconia/water nanoparticle colloids (nanofluids) in horizontal tubes, *J. Heat Transf.* **130**(4):042412-7.

Wiltzius, P., 1987, Hydrodynamic behavior of fractal aggregates, *Phys. Rev. Lett.* **58**(7):710.

Xie, H., Wang, J., Xi, T., and Liu, Y., 2002a, Thermal conductivity of suspensions containing nanosized SiC particles, *Int. J. Thermophys.* **23**(2):571–580.

Xie, H., Wang, J., Xi, T., Liu, Y., Ai, F., and Wu, Q., 2002b, Thermal conductivity enhancement of suspensions containing nanosized alumina particles, *J. Appl. Phys.* **91**:4568.

Xie, H., Fujii, M., and Zhang, X., 2005, Effect of interfacial nanolayer on the effective thermal conductivity of nanoparticle-fluid mixture, *Int. J. Heat Mass Transf.* **48**:2926–2932.

Xu, J., Yu, B.-M., and Yun, M.-J., 2006a, Effect of clusters on thermal conductivity in nanofluids, *Chin. Phys. Lett.* **23**:2819–2822.

Xu, J., Yu, B., Zou, M., and Xu, P., 2006b, A new model for heat conduction of nanofluids based on fractal distributions of nanoparticles, *J. Phys. D: Appl. Phys.* **39**:4486–4490.

Xuan, Y. and Li, Q., 2003, Investigation on convective heat transfer and flow features of nanofluids, *J. Heat Transf.* **125**:151.

Xuan, Y., Li, Q., and Hu, W., 2004, Aggregation structure and thermal conductivity of nanofluids, *AIChE J.* **49**(4):1038–1043.

Xue, Q.-Z., 2003, Model for effective thermal conductivity of nanofluids, *Phys. Lett. A* **307**:313–317.

Xue, Q. and Xu, W.-M., 2005, A model of thermal conductivity of nanofluids with interfacial shells, *Mater. Chem. Phys.* **90**:298–301.

Yang, B., 2008, Thermal conductivity equations based on Brownian motion in suspensions of nanoparticles (nanofluids), *J. Heat Transf.* **130**(4):042408-5.

Yu, W. and Choi, S. U. S., 2003, The role of interfacial layers in the enhanced thermal conductivity of nanofluids: A renovated Maxwell model, *J. Nanopart. Res.* **5**:167–171.

Zeinali Heris, S., Nasr Esfahany, M., and Etemad, S. G., 2007, Experimental investigation of convective heat transfer of Al_2O_3/water nanofluid in circular tube, *Int. J. Heat Fluid Flow* **28**(2):203–210.

Zhang, X., Gu, H., and Fujii, M., 2006a, Effective thermal conductivity and thermal diffusivity of nanofluids containing spherical and cylindrical nanoparticles, *J. Appl. Phys.* **100**:044325.

Zhang, X., Gu, H., and Fujii, M., 2006b, Experimental study on the effective thermal conductivity and thermal diffusivity of nanofluids, *Int. J. Thermophys.* **27**:569–580.

Zhang, H., Ge, X., and Ye, H., 2007, Effectiveness of the heat conduction reinforcement of particle filled composites, *Model. Simul. Mater. Sci. Eng.* **13**:401–412.

Zhou, X. F. and Gao, L., 2006, Effective thermal conductivity in nanofluids of nonspherical particles with interfacial thermal resistance: Differential effective medium theory, *J. Appl. Phys.* **1000**:024913.

Zhou, S.-Q. and Ni, R., 2008, Measurement of the specific heat capacity of water-based Al_2O_3 nanofluid, *Appl. Phys. Lett.* **92**(9):093123–3.

Zhu, H., Zhang, C., Liu, S., Tang, Y., and Yin, Y., 2006, Effects of nanoparticle clustering and alignment on thermal conductivities of Fe_3O_4 aqueous nanofluids, *Appl. Phys. Lett.* **89**:023123.

Zhu, H. T., Zhang, C. Y., Tang, Y. M., and Wang, J. X., 2007, Novel synthesis and thermal conductivity of CuO nanofluid, *J. Phys. Chem. C* **111**:1646–1650.

Zwanzig, R. W., 1965, Time-correlation functions and transport coefficients in statistical mechanics, *Ann. Rev. Phys. Chem.* **16**:67.

32

Thermophysical Properties of Nanofluids

S. M. Sohel Murshed
University of Central Florida

Kai Choong Leong
Nanyang Technological University

Chun Yang
Nanyang Technological University

32.1 Introduction

Nanofluids belong to a new class of heat transfer fluids, which are engineered by dispersing nanometer-sized (typically less than 100 nm) solid particles, rods, or tubes in conventional heat transfer fluids such as water, ethylene glycol (EG), and engine oil (EO). In recent years, nanofluids have evoked immense interest from researchers of various disciplines because of their superior thermal properties and potential applications in diverse fields such as microelectronics, microfluidics, transportation, and biomedical. Nanofluids are found to possess higher thermal properties such as effective thermal conductivity and thermal diffusivity compared to their base fluids, and the magnitudes of these properties increase remarkably with increasing nanoparticle volume fraction. Particle size and shape as well as fluid temperature also have influence on the enhancement of the effective thermal conductivity of nanofluids. However, there are inconsistencies in reported experimental results and controversies in the proposed mechanisms for the enhanced thermal conductivity of nanofluids. The aim of this chapter is to present and discuss the thermophysical properties that include thermal conductivity, thermal diffusivity, specific heat, and viscosity of nanofluids under the influences of various factors such as concentration, size and shape of nanoparticles, and fluid temperature. The potential applications, synthesis, thermal conductivity mechanisms, and measurement techniques of nanofluids, together with a brief review of representative results from the literature on these properties, are also presented.

32.1.1 Background and Applications of Nanofluids

With ever-increasing thermal loads due to smaller features of microelectronic devices and larger power outputs, thermal management of microelectronic devices to maintain their desired performance and durability is one of the most important technical issues in many high-tech industries such as microelectronics, transportation, and manufacturing. The conventional method of increasing the cooling rate is to increase the area for exchanging heat with a heat transfer fluid. However, this approach requires an undesirable increase in the size of the thermal management system. In addition, the inherently poor thermophysical properties of traditional heat transfer fluids such as water, EG, or EO greatly limits the cooling performance. Thus, conventional methods for increasing heat dissipation are not suitable to meet the demand of these high-tech industries. There is, therefore, a need to develop advanced cooling techniques and innovative heat transfer fluids with better heat transfer performance than those presently available.

At room temperature, metals possess at least an order-of-magnitude higher thermal conductivity than fluids. For example, the thermal conductivity of copper at room temperature is about 700 times greater than that of water and about 3000 times greater than that of EO. Therefore, the thermal conductivities of fluids that contain suspended metallic or nonmetallic (oxide) particles would be expected to be significantly higher than those of conventional heat transfer fluids. As thermal conductivity of a fluid plays a vital role in the development of energy-efficient heat transfer equipment, numerous theoretical and experimental studies on increasing the thermal conductivity of liquids by suspending small particles have been conducted since the treatise by Maxwell more than a century ago (Maxwell, 1891). However, these studies on the thermal conductivity of suspensions have been confined to millimeter- or micrometer-sized particles. The major problems of such suspensions are the rapid settling of these particles, clogging of flow channels, and increased pressure drop in the fluid. If the fluid is kept circulating rapidly enough to prevent much settling, the microparticles would damage the walls of the heat transfer devices (e.g., pipes and channels) and wear them thin. In contrast, nanoparticles remain in suspension, and thereby reduce erosion and clogging.

Over the last several decades, scientists and engineers have attempted to develop fluids that offer better cooling or heating performance. However, the novel concept of a "nanofluid," which was coined at Argonne National Laboratory of USA by Choi and his coworkers in 1995 (Choi, 1995), is thought to meet the cooling challenges facing many high-tech industries. It should also be acknowledged that a Japanese research group (Masuda et al., 1993) reported the effective thermal conductivity and viscosity of several types of nanoparticles suspensions (i.e., nanofluids) as a function of particle volume fraction and temperature, even before the term "nanofluid" was coined at Argonne National Laboratory. From past investigations (Eastman et al., 2004; Wang and Mujumdar, 2007; Murshed et al., 2008a), nanofluids were found to exhibit significantly higher thermal properties, particularly thermal conductivity than those of base fluids. Thus, it is of great interest to utilize nanofluids for thermal system applications.

The impact of nanofluid technology is expected to be great, considering that heat transfer performance of heat exchangers or cooling devices is vital in numerous industries. For example, the transport industry has a strong incentive to reduce the size and weight of vehicle thermal management systems, and nanofluids can increase thermal transport of coolants and lubricants. When the nanoparticles are properly dispersed, nanofluids offer numerous benefits besides their substantially high effective thermal conductivity. These benefits include

1. Improved heat transfer and stability
2. Microchannel cooling without clogging
3. Miniaturized systems

The better stability of nanofluids will prevent rapid settling and will reduce clogging in the walls of heat transfer devices. The high thermal conductivity of nanofluids translates into higher energy efficiency, better performance, and lower operating costs. They can also reduce energy consumption for pumping heat transfer fluids. Thermal systems can be smaller and lighter. In vehicles, smaller components result in better gasoline mileage, fuel savings, lower emissions, and a cleaner environment. With the aforementioned highly desired thermal properties and potential benefits, nanofluids are thought to have a wide range of applications in numerous important fields such as microelectronics,

microfluidics, transportation, manufacturing, and medical. The details of the applications of nanofluids can be found in a paper by the authors (Murshed et al., 2008a).

32.1.2 Synthesis of Nanofluids

Nanofluids are mainly synthesized by two techniques, which are the two-step process and the direct evaporation technique or single-step process.

In the two-step process, dry nanoparticles are first produced by an inert gas condensation method, and they are then dispersed into a fluid. An advantage of the two-step process in terms of the eventual commercialization of nanofluids is that the inert gas condensation technique can produce large quantities of nanopowders. Small amounts of nanoparticles can also be synthesized by other techniques such as the sol-gel process, electrolysis metal deposition, and microdroplet drying. Proper dispersion techniques and a small volume fraction of nanoparticles are important to produce stable nanofluids. The morphology of nanoparticles such as mean particle size, particle shape, and size distribution also depend on the synthesis techniques.

The direct evaporation technique synthesizes nanoparticles and disperses them into a fluid in a single step. As with the inert gas condensation technique, this technique involves the vaporization of a source material under vacuum conditions. An advantage of this process is that nanoparticle agglomeration is minimized. The disadvantages are that the liquid must have a very low vapor pressure and that this technique can produce very limited amounts of nanofluids. Most researchers used the two-step process to produce nanofluids by dispersing commercial or self-produced nanoparticles in a liquid. Some efforts are also made to synthesis small quantities of sample nanofluids by other techniques in different laboratories. For example, Hong et al. (2005) produced Fe nanoparticles by a chemical vapor condensation process using iron carbonyl as a precursor under flowing helium atmosphere. By using the coprecipitation method, Zhu et al. (2006) prepared Fe_3O_4 (10 nm)/water-based nanofluids to investigate the effects of nanoparticle clustering and alignment on thermal conductivity. Nonetheless, regardless of the synthesis techniques used, nanoparticles in suspensions are prone to agglomerate and settle down, leading to a large size distribution and varying shapes of particles. Thus, surfactant and ultrasonication are commonly employed to ensure stable suspension with less agglomeration of nanoparticles.

32.2 Thermal Conductivity of Nanofluids

32.2.1 Models and Heat Transport Mechanisms

Since the treatise by Maxwell (1891), several models have been developed to predict the effective thermal conductivity of composites such as solid particle suspensions. These classical models, such as the Maxwell (1891) and Hamilton and Crosser (1962) models which were developed from the effective medium theory, have been verified by experimental data for mixtures with low concentrations of milli- or micrometer-sized particles. The Maxwell model was developed to determine the effective electrical or thermal conductivity of statistically homogeneous liquid–solid suspensions with low volume fraction, randomly dispersed, and uniformly sized spherical particles. By applying a shape factor, Hamilton and Crosser modified Maxwell's model for nonspherical particles. As a representative, the Maxwell model is given as

$$\frac{k_{eff}}{k_f} = \frac{k_p + 2k_f + 2\phi_p(k_p - k_f)}{k_p + 2k_f - \phi_p(k_p - k_f)}, \tag{32.1}$$

where

- k_{eff} is the effective thermal conductivity of the particle suspension
- ϕ_p is the volume fraction of particles
- k_f and k_p are the thermal conductivities of the base fluid and the particle, respectively

Except for some recent results (Putnam et al., 2006; Venerus et al., 2006), most experiments have shown that nanofluids exhibit anomalously high thermal conductivity which cannot be predicted accurately by these classical models. Therefore, many theoretical studies have recently been carried out to predict the anomalously increased thermal conductivity of nanofluids. Several models have been proposed by considering various mechanisms such as interfacial layering and the effect of particle movement with Maxwell model. However, these recently developed models have not been universally accepted and validated with a wide range of experimental results. A summary of most of the classical and recently developed models for the prediction of effective thermal conductivity of nanofluids is provided in a recent review article by the authors (Murshed et al., 2008a).

In order to explain the enhanced thermal conductivity of nanofluids, Wang et al. (1999) and Keblinski et al. (2002) proposed several mechanisms which were not considered by classical models. Besides particle surface properties, the microscopic motions of the nanoparticles due to the stochastic force (causing Brownian motion) and the interparticle potential force are also significant for the enhanced thermal performance of nanofluids (Wang et al., 1999). Four possible mechanisms for the anomalous increase in the thermal conductivity of nanofluids were elucidated by Keblinski et al. (2002). These are (1) Brownian motion of the nanoparticles, (2) liquid layering at the liquid/particle interface, (3) nature of the heat transport in the nanoparticles, and (4) the effect of nanoparticle clustering.

Due to Brownian motion, particles randomly move through the liquid and collide with other particles, thereby enabling strong transport of heat due to direct solid-solid interactions, which can increase the effective thermal conductivity. Brownian motion is characterized by the diffusion coefficient (D_b) of the particle (radius, r_p) suspended in an infinite liquid medium of viscosity (η), which is the well-known Stokes–Einstein equation given by

$$D_b = \frac{K_B T}{6\pi\eta r_p}, \tag{32.2}$$

where K_B is the Boltzmann's constant. However, a simple comparison of the time scales of Brownian and thermal diffusion will show that thermal diffusion is much faster than Brownian diffusion, even within the limits of extremely small particles. Thus, thermal diffusion is a more efficient mechanism than Brownian motion. In addition, by applying the kinetic theory of heat flow, it can be shown that the enhancement of thermal conductivity of nanofluids due to Brownian motion is not significant.

When the size of the nanoparticles in a fluid becomes less than the phonon mean free path, phonons no longer diffuse across the nanoparticle but move ballistically without any scattering. However, it is difficult to envision how ballistic phonon transport could be more effective than a very-fast diffusion phonon transport, particularly to the extent of explaining anomalously the high thermal conductivity of nanofluids. No work has been reported on the ballistic heat transport of nanofluids.

The basic idea of liquid layering around a nanoparticle, i.e., nanolayer, is that liquid molecules can form a layer around the solid particles, and thereby enhance the local ordering of the atomic structure at the interfacial region between the solid and liquid phases. Hence, the atomic structure of such a liquid layer is significantly more ordered than that of the bulk liquid. Given that solids, which have much ordered atomic structures, exhibit much higher thermal conductivity than liquids, the liquid layer at the interface would reasonably have a higher thermal conductivity than the bulk liquid. The nanolayer works as thermal bridge between the nanoparticle and its base fluid. Thus, the nanolayer is considered as an important factor that may enhance the thermal conductivity of nanofluids.

The effective volume of a cluster is considered much larger than the volume of the particles due to the lower packing fraction of the cluster, which is defined as the ratio of the volume of the solid particles in the cluster to the total volume of the cluster. Since heat can be transferred rapidly within such clusters, the volume fraction of the highly conductive phase (cluster) is larger than the volume of solid, thus increasing its thermal conductivity. In general, clustering may also exert a negative effect on heat transfer enhancement, particularly at a low volume fraction, by settling small particles out of the liquid and creating a large region of "particle-free" liquid with a high thermal resistance. Interestingly, the aggregation of nanoparticles in base fluid has recently been suggested as a key mechanism for the enhanced thermal conductivity of nanofluids (Prasher et al., 2006a).

Besides these mechanisms, the effects of particles interaction and surface chemistry for nanometer-sized particles could be significant in enhancing the thermal conductivity of nanofluids.

32.2.2 Measurement Techniques

Using modern electronic instrumentation and corrections to theoretical basis, the transient hot-wire (THW) method has evolved into an accurate method of determining the thermal conductivity of fluids. This method is well established and documented in the literature (Horrocks and McLaughlin, 1963; Haarman, 1971; Healy et al., 1976; Nagasaka and Nagashima, 1981). The THW method is based on the calculation of the transient temperature field around a thin wire (called the hot wire) due to the supply of a constant current through the wire. The wire that can be treated as a line source is surrounded by a sample medium whose thermal conductivity or thermal diffusivity is to be measured. The wire serves as both the heat source and the temperature sensor.

The heat transfer process of the hot-wire technique can be modeled as conduction of heat from an infinitely long (compared to its diameter), continuous line source and is governed by the transient heat conduction equation. The governing equation for radial transient heat conduction in a homogeneous and infinite medium is given by

$$\frac{\partial^2 \Delta T}{\partial r^2} + \frac{1}{r}\frac{\partial \Delta T}{\partial r} = \frac{1}{\alpha}\frac{\partial \Delta T}{\partial t}, \qquad (32.3)$$

where

$\Delta T = T - T_0$ is the temperature rise in the medium

T_0 is the initial temperature

T is the temperature in surrounding medium at time t and radial position r

α is the thermal diffusivity of the surrounding medium

Imposing the initial and boundary conditions by considering the physics of the problem and the domain geometry, this governing equation is solved for the temperature rise of hot-wire as follows (Murshed et al., 2005):

$$\Delta T = \frac{q}{4\pi k}\left[\ln t + \ln\frac{4\alpha}{a^2 C}\right], \qquad (32.4)$$

where

q is the heat generation rate per unit length of the wire

a is the wire radius

$C = \exp(\gamma)$ and $\gamma = 0.5772$ is Euler's constant

From the temperature rise of the wire given by Equation 32.4, the thermal conductivity of the medium (k) can be determined from

$$k = \frac{q/4\pi}{d\ln t/d\Delta T}. \qquad (32.5)$$

The advantage of the THW method lies in its elimination of natural convection effects. Besides its simple conceptual design, the THW method is fast compared to other techniques. In addition, this method can also be used to measure the thermal conductivity of electrically conducting media by applying a thin coating on the wire. Thus, most researchers, including the authors, used the THW method to measure the thermal conductivity of nanofluids. A schematic of the THW apparatus used in the authors' study is shown in Figure 32.1. The entire hot-wire experimental setup comprises several units, including power supply, Wheatstone bridge circuit, data acquisition and control system, and hot-wire cell which contains the test sample. The details of measurement procedure of a THW method can be found elsewhere (Murshed et al., 2005).

Although several studies also reported the use of other techniques such as the steady-state (Wang et al., 1999, 2003), temperature oscillation (Das et al., 2003), and the 3ω-wire (Yang and Han, 2006) methods to measure the effective thermal conductivity of nanofluids, these methods are not as accurate as the THW method. The temperature oscillation technique measures the thermal diffusivity, and calculates the thermal conductivity using the volumetric specific heat of the sample. Similar to the hot-wire method, the 3ω-wire method uses a metal wire suspended in a liquid. A sinusoidal current at frequency ω is passed through the metal wire and generates a heat wave at frequency 2ω, which is deduced by the voltage component at frequency 3ω. Even though the 3ω-wire method is not commonly used, it may be suitable to measure temperature-dependent thermal conductivity.

32.2.3 Properties of Commonly Used Base Fluids and Nanoparticles

In order to characterize the thermophysical properties such as thermal conductivity, thermal diffusivity, and viscosity of nanofluids, it is important to know these properties of both the nanoparticle and the base fluid. Therefore, the standard values of the thermophysical properties of these commonly used base

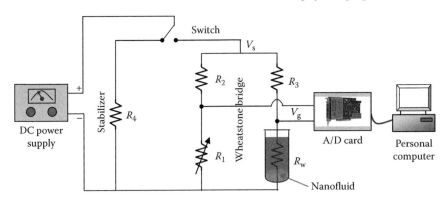

FIGURE 32.1 Schematic of THW experimental setup.

TABLE 32.1 Thermophysical Properties of Base Fluids at 300 K

Base Fluids	Thermal Conductivity, k (W/m K)	Thermal Diffusivity, α (m²/s)	Density, ρ (kg/m³)	Specific Heat, c_p (kJ/kg K)	Kinematic Viscosity, υ (m²/s)
Deionized water (DIW)	0.607	14.55×10^{-8}	998	4.2	9.2×10^{-7}
Ethylene glycol (EG)	0.255	9.385×10^{-8}	1111	2.4	18.1×10^{-6}
Engine oil (EO)	0.145	8.740×10^{-8}	884	1.9	9.44×10^{-4}

TABLE 32.2 Thermophysical Properties of Several Nanoparticles at 300 K

Nanoparticles	Thermal Conductivity, k (W/m K)	Thermal Diffusivity, α (m²/s)	Density, ρ (kg/m³)	Specific Heat, c_p (kJ/kg K)
TiO₂	8.04	2.9×10^{-6}	4000	0.711
CuO	17.65	5.17×10^{-6}	6500	0.525
Al₂O₃	39	11.9×10^{-6}	3970	0.775
Al	237	97.1×10^{-6}	2700	0.877
Cu	401	117×10^{-6}	8933	0.385

fluids and nanoparticles (Bolz and Tuve, 1973; Kaviany, 2002) are provided in Tables 32.1 and 32.2. Among the base fluids, water has the highest values of thermal properties while metals as usual have much higher thermal conductivity compared to their oxides.

32.2.4 Effect of Particle Volume Fraction, Particle Shape, and Base Fluids

In the last several years, many experimental investigations on the effective thermal conductivities of nanofluids containing different volume fractions, materials, and sizes of nanoparticles dispersed in different base fluids have been reported. Some key

results of the effective thermal conductivity of nanofluids from various research groups are presented in Figure 32.2. Several representative results on the thermal conductivity of nanofluids are elaborated, followed by results and the discussion of the work done by the authors.

As can be seen from Figure 32.2, Eastman et al. (1997) reported surprisingly about 44% increase in thermal conductivity of HE-200 oil by dispersing only 0.052 vol.% of Cu nanoparticles (35 nm) in it. They also showed that the thermal conductivity enhancement for 5 vol.% of Al₂O₃ (33 nm) and CuO (36 nm) nanoparticles in water were 29% and 60%, respectively. Wang et al. (2003) later showed 17% increase in the thermal conductivity for a loading of only 0.4 vol.% of same CuO nanoparticles

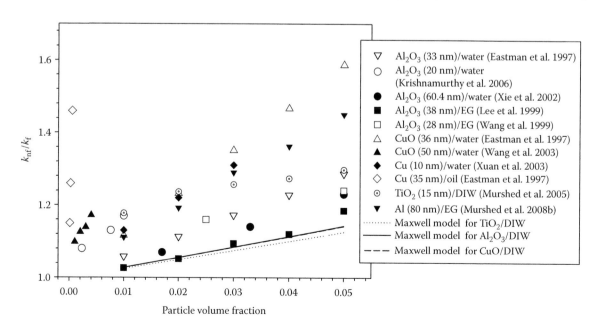

FIGURE 32.2 Comparison of enhanced thermal conductivity data of various nanofluids.

(50 nm) in water. For ethylene glycol-based CuO nanofluids and at 4% volumetric loading, a moderate enhancement of thermal conductivity (20%) was observed by Lee et al. (1999) and Eastman et al. (2001). By using the steady-state parallel plate method, the thermal conductivities of several types of nanofluids were measured by Wang et al. (1999). The Al_2O_3/EG-based nanofluids showed 18% increase in thermal conductivity at 4% particle volume fraction. In contrast, Xie et al. (2002) observed about 30% enhancement in the thermal conductivity for the same Al_2O_3/EG nanofluid at 5% volume fraction. Although the particles used by Xie et al. were twice as large, their results showed a much higher thermal conductivity than that of Wang et al. This discrepancy of results between Wang et al. (1999) and Xie et al. (2002) could be due to the different measurement methods and pH values of nanofluids used in both studies.

For the first time, Putnam et al. (2006) reported no anomalous enhancement of thermal conductivity of Au (4 nm)/ethanol-based nanofluids with very low particle volume fraction. Their observed maximum increase in thermal conductivity was 1.3% for 0.018% volumetric loading of such ultra-fine Au nanoparticles in ethanol. Their result is directly in conflict with the anomalous increase in thermal conductivity reported by Patel et al. (2003) for the same nanofluid and is also contrary to the substantial enhancement of the thermal conductivity of nanofluids reported by other researchers.

Except for Putnam et al. (2006), all other reported studies show that nanofluids exhibit much higher thermal conductivities than their base fluids, even when the volume fractions of suspended nanoparticles are very low, and they increase significantly with nanoparticle volume fraction. However, the increments of thermal conductivities are different for different types of nanofluids. Even for the same nanofluids, different research groups reported different enhancements. Besides particle volume fraction, the thermal conductivity of nanofluids also varies with the size, material of nanoparticles, as well as the base fluids. For instance, nanofluids with metallic nanoparticles were found to have a higher thermal conductivity than nanofluids with nonmetallic (oxide) nanoparticles. In contrast, some studies also reported that highly conductive nanoparticles are not always effective in enhancing the thermal conductivity of nanofluids (Hong et al., 2005, 2006).

The effective thermal conductivities of various types of nanofluids as a function of particle volume fraction, particle shape, and base fluids are measured by the authors. The THW method was employed and the overall measurement uncertainty was estimated to be within ±1.5%. The experimental apparatus was also calibrated by measuring the effective thermal conductivity of the base fluids. Due to the limited availability of different sizes of the same type of nanoparticles, TiO_2 nanoparticles of spherical (15 nm) and rod shape (10 × 40 nm) were used. A small amount (≈0.1 mM) of CTAB surfactant was added to all sample nanofluids for the better dispersion of nanoparticles. Results for TiO_2/deionized water (DIW)-based nanofluids (Figure 32.3) show a nonlinear relationship between thermal conductivity and particle volume fraction at lower volumetric loading (<2%) and a

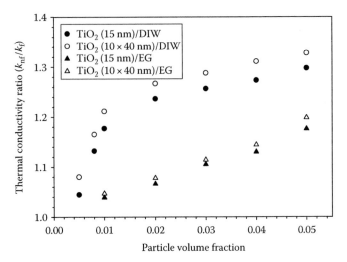

FIGURE 32.3 Effective thermal conductivity of TiO_2 nanofluids—effects of particle volume fraction, particle shape, and base fluid.

linear relationship at higher volumetric loading (≥2%). This non-linear behavior of nanofluids at lower volume fractions of nanoparticles may be due to the influence of the surfactant, long time sonication, and thus good dispersion of nanoparticles in the base fluid. Figure 32.3 illustrates that the nanofluid with rod-shaped TiO_2 nanoparticles exhibits larger thermal conductivity than that of the nanofluid with spherical-shaped nanoparticles. This is because of the larger shape factor and alignment of rod-shape nanoparticles compared to their spherical-shape counterparts.

The choice of base fluids also influences this enhancement. From Figure 32.3, it can be seen that the effective thermal conductivities of TiO_2 (15 nm)/DIW/EG-based nanofluids increase substantially with the increasing volumetric loading of nanoparticles. The thermal conductivity of TiO_2 (15 nm)/water nanofluids has a maximum enhancement of 29.7% for a particle volume fraction of 0.05 while the maximum increase of thermal conductivity of TiO_2 (15 nm)/EG-based nanofluids is 18% for the same particle volume fraction. This demonstrates that due to the higher value of the thermal conductivity of water than that of EG, water-based nanofluids exhibit much higher thermal conductivity compared to EG-based nanofluids.

Figure 32.4 illustrates that the effective thermal conductivities of Al_2O_3 (80 nm)/DIW/EG-based nanofluids also increase significantly with increasing volumetric loading nanoparticles. By suspending 0.05 volume fraction of Al_2O_3 (80 nm), nanoparticles in DIW and in EG, the maximum enhancements of thermal conductivities were 23% and 18%, respectively. This again indicates that water is a better base fluid compared to EG for these nanofluids. As shown in Figure 32.5, the thermal conductivity of Al (80 nm)/EG-based nanofluids has a maximum enhancement of 45% for a particle volume fraction of 0.05 while the maximum increase of thermal conductivity of Al/EO-based nanofluids is 31% for the same particle volumetric loading of 5%. Similar to the results presented in Figures 32.3 and 32.4, EG-based nanofluids also show much higher thermal conductivity than those of EO-based nanofluids.

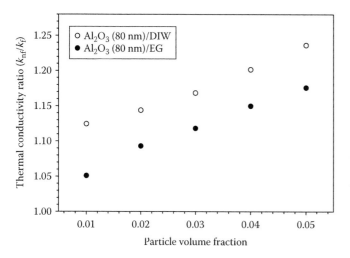

FIGURE 32.4 Enhancement of thermal conductivity of Al_2O_3 nanofluids with particle volume fraction.

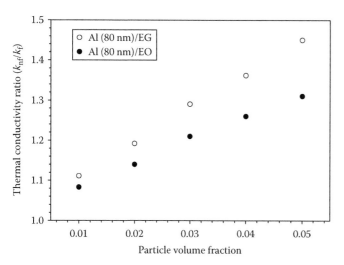

FIGURE 32.5 Enhancement of thermal conductivity of Al nanofluids with particle volume fraction.

The present results clearly demonstrate that suspending a small volume percentage of nanoparticles in base fluids significantly increases the effective thermal conductivities of nanofluids. However, for lower volume fractions (e.g., <2%) and very small-sized nanoparticles, the enhancement of thermal conductivity can be nonlinear. It was found that the higher the thermal conductivity of the base fluid, the larger the enhancement of the thermal conductivity of nanofluids. The reason lies in the nanolayering on the surface of nanoparticles and nanoparticle dispersion behavior as well as the effect of the dispersant (Murshed et al., 2005).

32.2.5 Effect of Particle Size

Particle size is an important parameter because shrinking particles down to the nanoscale not only increases the surface area relative to volume but also generates some nanoscale mechanisms in the suspensions. Both theoretical (Jang and Choi, 2004)

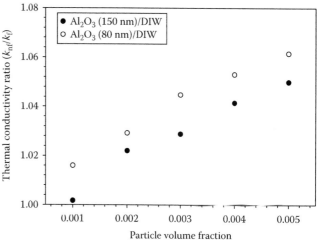

FIGURE 32.6 Effect of particle size on the enhancement of thermal conductivity of Al_2O_3 nanofluid.

and experimental evidences indicate that the effective thermal conductivity of nanofluids increases with decreasing particle size. Although particle size is significant, very few experimental studies have been reported on the effect of particle size on the thermal conductivity of nanofluids.

The thermal conductivities of nanofluids containing spherical shape Al_2O_3 nanoparticles of 80 nm and 150 nm are measured by the authors. Figure 32.6 illustrates that for the same base fluid, the smaller-sized (80 nm) Al_2O_3 nanoparticles show higher enhancement of thermal conductivity compared to the larger-sized (150 nm) particles. Chon and Kihm (2005) also found that nanofluids having smaller-sized Al_2O_3 particles have higher thermal conductivities than similar nanofluids with larger-sized nanoparticles. The reason is that the smaller-sized particles provide better dispersion and can also contribute to other nanoscale mechanisms.

32.2.6 Effect of Fluid Temperature

The fluid temperature can play an important role in enhancing the effective thermal conductivity of nanofluids. Although nanofluids are usually employed as coolants in a hotter environment, very few studies have been performed to investigate the effect of temperature on the effective thermal conductivity of nanofluids. Some key results are briefly presented here before discussing the present results.

In an early experiment, Das et al. (2003) reported a two- to fourfold increase in thermal conductivity over a temperature range of 21°C–51°C for nanofluids containing Al_2O_3 (38.4 nm) and CuO (28.6 nm) nanoparticles in water. The strong temperature dependence of thermal conductivity was attributed to the motion of nanoparticles. For Au (10–20 nm)/water-based nanofluids, the same group (Patel et al., 2003) also observed thermal conductivity enhancement of 5%–21% over a temperature range of 30°C–60°C at an extremely low volume concentration of 0.00026%. Later, Chon and Kihm (2005) showed that the thermal conductivity of Al_2O_3 (47 nm)/water-based nanofluids increase by only 6%–11% when the fluid temperature was increased from 31°C to 51°C. The Brownian motion of the nanoparticles was identified as a main mechanism for increasing the thermal conductivity of their nanofluids with fluid temperature. Whereas Li and Peterson (2006) reported that a small increase in fluid temperature from 27.5°C to 34.7°C results in a surprisingly three-fold increase in the enhanced thermal conductivity of the same Al_2O_3 (36 nm)/water-based nanofluids. Yang and Han (2006) measured the thermal conductivity of the suspensions of Bi_2Te_3 (20 × 170 nm) nanorods in FC72 and in oil (hexadecane) by using the 3ω-wire method. Interestingly, they observed a slight decrease in thermal conductivity with increasing fluid temperature, which is in contrast to the trend observed in nanofluids. The contrarian trend was claimed to be due mainly to the particle aspect ratio. Using the forced Rayleigh scattering technique, Venerus et al. (2006) measured the thermal conductivity of Au (22 nm)/water and Al_2O_3 (30 nm)/petroleum oil-based nanofluids. In contrast to all reported results, they found the level of thermal conductivity enhancement for these nanofluids to be independent of temperature.

Apart from the studies of Yang and Han (2006) and Venerus et al. (2006), most of the reported studies demonstrate that the thermal conductivity of nanofluids increases significantly with the fluid temperature. However, there is a lack of consistency among the reported results of different research groups. Hence, more experimental studies are needed to better understand the effect of temperature on the thermal conductivity of nanofluids.

The effect of the temperature on the effective thermal conductivity of several types of nanofluids is investigated by the authors. The THW system with a fluid heating facility was used to measure the thermal conductivity of nanofluids at different temperatures ranging from 20°C to 60°C (Murshed et al., 2008b). Figure 32.7 shows that at a temperature of 60°C, the effective thermal conductivity of Al_2O_3 (80 nm)/EG-based nanofluids increases by about 9% and 12% (compared to base fluid) for the nanoparticle

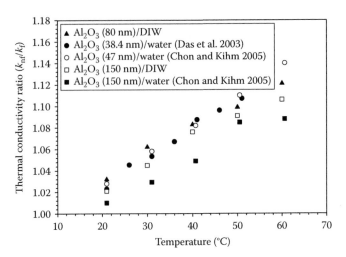

FIGURE 32.8 Comparison of temperature-dependent thermal conductivity of Al_2O_3 nanofluids at 1% particle volume concentration.

volumetric loadings of 0.5% and 1%, respectively. Under the same temperature, the enhancement of effective thermal conductivity of Al (80 nm)/EO nanofluids are 20% and 37% for volumetric loadings of 1% and 3% nanoparticles in the base fluid, respectively. Thus, the dependence of temperature on the thermal conductivity of nanofluids with metallic nanoparticles is more significant as compared to nanofluids with nonmetallic nanoparticles. Figure 32.8 compares the present results with results from the literature (Das et al., 2003; Chon and Kihm, 2005) for Al_2O_3 (different sizes)/water-based nanofluids at 1 vol.% of nanoparticles. It can be seen that except for Al_2O_3 (150 nm) of Chon and Kihm (2005), the results of all other sizes of this nanoparticle are reasonably consistent up to 50°C. For 150 nm nanoparticles, the observed differences between the authors' results and those of Chon and Kihm (2005) could be due to the different methods in preparing the nanofluids and the different measurement techniques used.

As shown in Figures 32.7 and 32.8, the experimental thermal conductivity values of different nanofluids increase significantly with fluid temperature. A linear increase in the effective thermal conductivity of nanofluids with temperature was also observed. This is because the higher fluid temperature intensifies the Brownian motion of nanoparticles and also decreases the viscosity of the base fluid. With an intensified Brownian motion, the contribution of micro-convection in heat transport increases, resulting in the higher enhancement in the thermal conductivity of nanofluids.

32.3 Thermal Diffusivity of Nanofluids

Despite the fact that thermal diffusivity is an important property especially in convective heat transfer applications, very little work has been performed on the determination of the effective thermal diffusivity of nanofluids. Wang et al. (2004) was the first to determine the thermal diffusivity of a nanofluid (i.e., CuO/water) by measuring its thermal conductivity and specific heat. Nonetheless, their calculated results showed

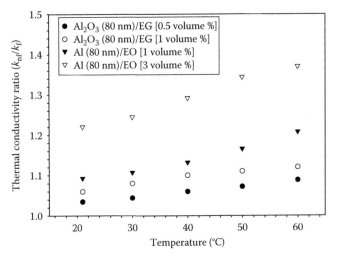

FIGURE 32.7 Temperature dependence of enhanced thermal conductivity of nanofluids. (Adapted from Murshed, S.M.S. et al., *Int. J. Therm. Sci.*, 47, 560, 2008b.)

severe fluctuation with particle volume fraction. Zhang et al. (2007) measured the thermal diffusivity of several nanofluids by the short THW method. The thermal diffusivity of their nanofluids increases with particle volume fraction and temperature. For example, by adding 6% mass fraction of Al_2O_3 nanoparticle in water, they observed about 23% increase in the thermal diffusivity of water at 30°C. Interestingly, at 5.3% volume fraction of CuO nanoparticle in water showed a constant 18% increase in thermal diffusivity at three different temperatures, i.e., 10°C, 20°C, and 30°C. However, their measurement uncertainty was as high as 5%.

The accurate measurement of thermal diffusivity is more complex for composite fluids such as nanofluids compared to solids or gases. Although there are several methods of measuring the thermal diffusivity of solids, gases, or liquids, no precise technique has been developed to measure the thermal diffusivity of nanofluids. This could be one of the reasons for very few studies on the thermal diffusivity of nanofluids. Apart from THW method, only a few techniques such as the temperature oscillation (Czarnetzki and Roetzel, 1995) and the thermal-wave cavity (Balderas-Lopez and Mandelis, 2001) techniques are often used to measure the thermal diffusivity of gases and liquids. Nevertheless, the applications of these methods are very limited due to complexities and difficulties to achieve accurate results, particularly for nanofluids. They are, therefore, not suitable for the convenient and accurate determination of the thermal diffusivity of nanofluids.

A double hot-wire (DHW) technique for the direct and precise measurement of the effective thermal diffusivity of several types of nanofluids was developed by the authors (Murshed et al., 2006). Figure 32.9 shows the schematic of the DHW experimental setup. The measuring principle of the DHW technique is based on the calculation of the transient temperature field by a sensor wire at some distance away from the source wire. The advantages of this technique are its simplicity and high measurement accuracy. The theoretical basis and details of the experimental procedure can be found elsewhere (Murshed et al., 2006).

Based on the calibration results of the base fluids, the measurement error was estimated to be within 1.2%, which indicates high accuracy of using this DHW technique. Figure 32.10 depicts the

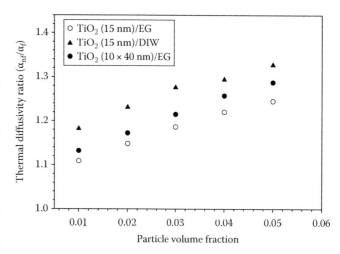

FIGURE 32.10 Effective thermal diffusivity of TiO_2 nanofluids with particle volume fraction (experimental data for TiO_2/EG are extracted from Murshed et al., 2006).

effective thermal diffusivity of EG- and DIW-based TiO_2 nanofluids as a function of particle volume fraction. For maximum 5% volume fraction of TiO_2 nanoparticles of 15 nm and 10×40 nm in EG, the maximum increases in effective thermal diffusivities were 25% and 29%, respectively. It can be seen from Figure 32.10 that TiO_2 (15 nm)/DIW-based nanofluids have higher thermal diffusivities compared to EG-based nanofluids. This is mainly due to the much higher thermal diffusivity of water compared to that of EG (Table 32.1). These results demonstrate that along with the particle volume fraction, particle shape and base fluid also have effects on the enhancement of the thermal diffusivity of the nanofluids. The particle shape and base fluid affect the effective thermal conductivity of nanofluids and the dispersion of particles, which in turn influence their effective thermal diffusivity.

Figure 32.11 presents the impact of particle types and base fluids on the thermal diffusivity of nanofluids. For nanofluids

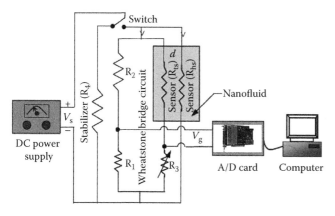

FIGURE 32.9 Schematic of DHW experimental setup. (Adapted from Murshed, S.M.S. et al., *J. Phys. D Appl. Phys.*, 39, 5316, 2006.)

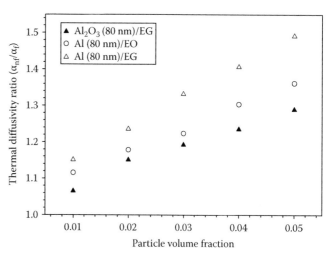

FIGURE 32.11 Effective thermal diffusivity of Al_2O_3 and Al nanofluids with particle volume fraction. (Adapted from Murshed, S.M.S. et al., *J. Phys. D Appl. Phys.*, 39, 5316, 2006.)

with Al (80 nm) nanoparticles of 5% volume fraction in EG and in EO, the maximum increases in thermal diffusivity were 49% and 36%, respectively. The EG-based nanofluid was found to have higher thermal diffusivity than that of the EO-based nanofluids due to the higher thermal diffusivity of EG as compared to the EO. Figure 32.11 also shows that the nanofluids with Al (80) nanoparticles exhibit significantly higher thermal diffusivity than that of Al_2O_3 (80) nanoparticles in the same base fluid (EG). This is due to the larger thermal diffusivity of Al compared to that of Al_2O_3 at room temperature (Table 32.2).

Figures 32.10 and 32.11 show that nanofluids exhibit significant increase in thermal diffusivity with the nanoparticle volume fraction. This is mainly because the effective thermal conductivity of nanofluids was found to be substantially higher than the value for the base fluids and increases with particle volume fraction. As mentioned previously, a small amount of CTAB surfactant was dissociated in the nanofluids. Surfactant may play a significant role in intensifying the enhancement of the thermal properties of nanofluids by ensuring better and stabilized dispersion of nanoparticles. Moreover, with the increased volumetric loading of nanoparticles, the specific heat of nanofluids decreases, resulting in an increase in the effective thermal diffusivity of nanofluids. The results also demonstrate that besides the particle volume fraction, base fluid and nanoparticle material influence the enhancement of the thermal diffusivity of nanofluids.

32.4 Specific Heat of Nanofluids

Although the specific heat of nanofluid is important in evaluating its thermal performance under flow conditions, very little effort has been made on determining this property. Only Zhou and Ni (2008) recently measured the specific heat of Al_2O_3 (45 nm) water-based nanofluid by using a differential scanning calorimeter. Their results showed that the specific heat of this nanofluid decreases significantly with particle volume fraction.

With known values of thermal conductivity (k), thermal diffusivity (α), and density (ρ) of nanofluids, their specific heat can then be determined from its definition as given by

$$c_{p-nf} = \frac{k_{nf}}{\alpha_{nf}\rho_{nf}} \tag{32.6}$$

where the density of the nanofluid can be obtained from

$$\rho_{nf} = \phi_p\rho_p + (1-\phi_p)\rho_f \tag{32.7}$$

and the subscripts nf, p, and f stand for nanofluid, particle, and base fluid, respectively.

The specific heats of several types of nanofluids were determined from Equation 32.6 by using the results of the effective thermal conductivity and thermal diffusivity of nanofluids, which were measured by the DHW method (Murshed et al., 2006). As can be seen from Figure 32.12, the specific heats of nanofluids decrease almost linearly with the increasing volume

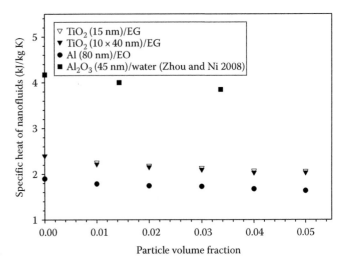

FIGURE 32.12 Specific heats of various nanofluids with particle volume fraction.

fraction of nanoparticles. Because of the much smaller specific heat of solid compared to that of liquid, the specific heat of a solid–liquid mixture decreases with an increase of the volumetric loading of solid particles. More work needs to be performed in order to better understand the mechanism of the specific heat of nanofluids.

32.5 Viscosity of Nanofluids

Viscosity is as critical as thermal conductivity in engineering systems that employ fluid flow. Pumping power is proportional to the pressure drop, which in turn is related to the viscosity.

32.5.1 Models for the Effective Viscosity

Other than very few experimental studies, no established model is available for predicting the effective viscosity of nanofluids. The Einstein model (Einstein, 1956) is commonly used to predict the effective viscosity of suspensions (η_{eff}) containing a low volume fraction of particles (usually <5 vol.%). The Einstein viscosity model is expressed as

$$\frac{\eta_{eff}}{\eta_f} = 1 + 2.5\phi_p \tag{32.8}$$

where

ϕ_p is the volume fraction of particle
η_f is the viscosity of the base fluid

Nielsen (1970) proposed a generalized power law-based model for the relative elastic moduli of composite materials, which is also widely used for relative viscosity. The simplified form of this model is given as

$$\frac{\eta_{eff}}{\eta_f} = (1 + 1.5\phi_p)e^{\phi_p/(1-\phi_m)} \tag{32.9}$$

where

ϕ_m is the maximum packing fraction

$[\eta]$ is the intrinsic viscosity ($[\eta] = 2.5$ for hard spheres)

For randomly dispersed spheres, the maximum close packing fraction is approximately 0.64. Both the Einstein (1956) and Nielsen (1970) models are used to predict the effective viscosity of nanofluids, which is shown in the following section.

32.5.2 Effect of Particle Volume Fraction

Before discussing the present results, a short review of some representative works on the viscosity of nanofluids follows. Among very few studies, Masuda et al. (1993) measured the viscosity of suspensions of different types of nanoparticles in water. They found that TiO_2 (27 nm) particles at a volumetric loading of 4.3% increased the viscosity of water by 60%. The viscosity of Al_2O_3 (80 nm)/DIW-based nanofluids was also found to increase by nearly 82% for the volumetric loading of 5%. For the same volumetric loading, a similar increment (86%) of the effective viscosity of Al_2O_3 (28 nm)/distilled water-based nanofluids was observed by Wang et al. (1999). They also found an increase of about 40% in the viscosity of EG at a volumetric loading of 3.5% of Al_2O_3 nanoparticle. Their results demonstrated that the viscosity of nanofluids depends on the dispersion method. In contrast, the viscosities of Al_2O_3 (13 nm)/water- and TiO_2 (27 nm)/water-based nanofluids at 10% particle volume fraction reported by Pak and Cho (1998) are about 200 and 3 times larger than that of the water, respectively. Such anomalous increase in the viscosity of their study was primarily attributed to the viscoelectric effect. The reasons for such large discrepancies are not clear. However, it could be due to difference in dispersion techniques and the sizes of nanoparticles used in each study. The viscosity results of these studies were significantly larger than the predictions from Einstein's model (Einstein, 1956).

In another study, Putra et al. (2003) measured the viscosity of Al_2O_3/water-based nanofluid as a function of shear rate and showed the Newtonian behavior of the nanofluids for volume percentages between 1% and 4%. They also found that the viscosity of nanofluid increases with the increase in particle volume fraction. Prasher et al. (2006b) reported the viscosities of Al_2O_3 nanofluids for various shear rates, temperature, and particle volume fraction. Their results demonstrate that viscosity is independent of shear rate, proving that the nanofluids are Newtonian in nature. It also shows that with increasing nanoparticle volume fraction, the viscosity increases. On the other hand, they found that viscosity is independent of temperature. This is contrary to the nature of liquid viscosity variation with the temperature.

Using a controlled rate rheometer (Contrases Low Shear 40 Model), the effective viscosity of nanofluids was measured by the authors (Murshed et al., 2008b). Results are compared with those predicted by the classical models as well as results from the literature. Figure 32.13 demonstrates that the viscosities of suspensions of TiO_2 and Al_2O_3 in water are independent of shear rates tested, which shows the Newtonian nature of these

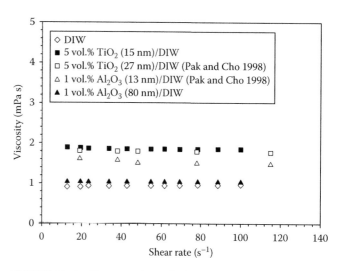

FIGURE 32.13 Viscosity of nanofluids as a function of shear rate.

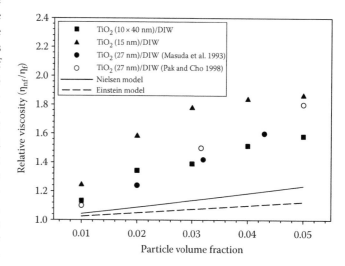

FIGURE 32.14 Comparison of relative viscosity of TiO_2 nanofluids with particle volume fraction.

nanofluids. The relative viscosity of TiO_2/water-based nanofluids from this study and those from the literature are shown in Figure 32.14. It can be seen that effective viscosity increases with particle volume fraction. For 15 nm TiO_2 nanoparticles in water, the maximum increase in viscosity was found to be 86% for a volume fraction of 0.05. For the same volume fraction of cylindrical shape (10 × 40 nm) TiO_2 nanoparticles, the viscosity of water is increased by 58% (Figure 32.14). This indicates that the shapes of the particles can influence the suspension viscosity. One reason for such an increase in viscosity is surface forces interaction among particles. It is through the surface charge that particle shape influences the electrostatic repulsive force. This repulsive force plays a significant role in altering the viscosity of nanofluids through the dispersion and sedimentation of nanoparticles. The particle shape can affect clustering and adsorption, which may also change the viscosity of nanofluids. In fact, TiO_2 (10 × 40 nm) nanoparticles in base fluids were found to be more stable than TiO_2 (15 nm) nanoparticles. The present results for these nanofluids are higher than those reported by Masuda et al. (1993) and Pak and Cho (1998). Although both of them used the same-sized TiO$_2$ nanoparticles of 27 nm and adjusted their suspensions to a high pH value (pH = 10), Pak and Cho observed higher viscosity values than those of Masuda et al. Figure 32.14 also demonstrates that the classical models are unable to predict the viscosity of these nanofluids.

Figure 32.15 presents the measured viscosities of Al$_2$O$_3$ (80 nm)/water-based nanofluids as a function of nanoparticle volume fraction and a comparison with the models of Einstein (1956) and Nielsen (1970). Al$_2$O$_3$ (80 nm) nanoparticles were found to increase the viscosity of water by nearly 82% for the maximum volumetric loading of 5%. A similar increment (86%) of the effective viscosity of Al$_2$O$_3$ (28 nm)/distilled water-based nanofluids was also observed by Wang et al. (1999) for the same volume fraction of 0.05. In spite of their smaller particle size (13 nm), Masuda et al.'s (1993) results showed a much larger increase in the viscosity of the same nanofluids. For example, their viscosity of Al$_2$O$_3$ (13 nm)/water-based nanofluids at 4.16 vol.% was about 1.68 times higher than the viscosity of water. Interestingly, the viscosity of the same nanofluid, i.e., Al$_2$O$_3$ (13 nm)/water measured by Pak and Cho (1998), was three times higher than that of water at 3 vol.% of nanoparticles. In this study, the maximum increase in viscosity was about 180% for 5% volumetric loading of Al$_2$O$_3$ (150 nm) nanoparticles. Figure 32.15 also demonstrates the effect of particle size on the viscosity of suspensions; the larger the particle size, the higher the viscosity. This is because of the particle size effect on its agglomeration in base fluid. Figure 32.15 also shows that the measured viscosities are severely underpredicted by these classical models.

Following the review and results presented earlier (Figures 32.14 and 32.15), it is confirmed that the viscosity of nanofluids increases significantly with particle volume fraction. However, there are discrepancies among the results from various research groups. The reasons for the discrepancies could be due to the difference in the size of the particle clusters, differences in the

dispersion techniques, and the use of surfactants. Indeed, the viscosity of nanofluids greatly depends on the methods used to disperse and stabilize the nanoparticle suspension. Like most of the reported studies, the classical models were also found to be unable to predict the effective viscosity of nanofluids. This is because these models considered only particle volume fraction, whereas the nanoparticles in fluids can easily form clusters and can experience surface adsorption due to surface forces. Clustering and adsorption increase the hydrodynamic diameter of nanoparticles, leading to the increase of relative viscosity. Besides the particle volume fraction and size, many other factors such as the nature of the particle surface, ionic strength of the base fluid, surfactants, pH value, and particle interaction forces may play significant roles to alter the viscosity of nanofluids. In general, the increment of viscosity can reduce the potential benefits of nanofluids. However, through thermal and hydraulic analysis, by comparing pressure drops at constant Nusselt number, Prasher et al. (2006b) showed that within four times increase in relative viscosity compared to the increase in the thermal conductivity, nanofluids are still superior heat transfer fluids than the base fluid.

32.5.3 Temperature Dependence of Viscosity

Figure 32.16 shows the effect of temperature on the viscosity of water and various nanofluids. The effective viscosities of sample fluids are found to decrease significantly and nonlinearly with fluid temperature. As indicated in Figure 32.16, nanofluids and DIW show almost the same trend of the decrease of viscosity with temperature, except for the large increase in the viscosity of Al$_2$O$_3$ nanofluid obtained by Pak and Cho (1998). However, these results of significantly decreasing the viscosity of nanofluids with temperature can make nanofluids even more attractive in

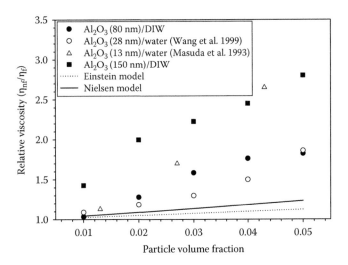

FIGURE 32.15 Comparison of relative viscosity of Al$_2$O$_3$ nanofluids with particle volume fraction.

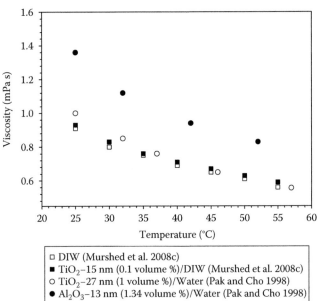

FIGURE 32.16 Effect of temperature on viscosity of nanofluids.

their applications in a heated environment. Thus, it is imperative to conduct more studies to understand the effect of temperature on the viscosity of nanofluids.

32.6 Summary and Perspective

The most focused thermophysical property of nanofluids so far is their effective thermal conductivity, which has been found to increase significantly with nanoparticle volume fraction. This effective thermal conductivity of nanofluids further increases with increasing fluid temperature. This makes nanofluids more attractive, particularly for its application as a coolant. Apart from the particle volume fraction and fluid temperature, the effects of other factors such as particle size and shapes, base fluid, and particle material on the effective thermal conductivity of nanofluids are also found to be significant.

Nanofluids showed much higher effective thermal diffusivity values than those of base fluids. Like thermal conductivity, the thermal diffusivity of nanofluids increases significantly with the increasing volume fraction of nanoparticles. The particle size and shape, together with particle material and base fluid, have effects on the thermal diffusivity of nanofluids. The specific heat of the nanofluids is found to decrease significantly with the increasing volume fraction of nanoparticles. Despite their importance, very little efforts have been made to determine the effective thermal diffusivity and specific heat of nanofluids.

The only inferior aspect of nanofluids is their viscosity which is also found to be substantially higher than their base fluids, and it increases with the volume fraction of nanoparticles. Such enhancement of viscosity may reduce the potential benefits of nanofluids. In order to keep the viscosity low, it is important to use low concentrations, small sizes, and well-dispersed nanoparticles in base fluids. Nevertheless, the viscosity of nanofluids is found to decrease nonlinearly with increasing fluid temperature, which is good news for the application of nanofluids in heated environments. The classical models are unable to predict the viscosity of nanofluids. Therefore, besides experimental investigations, theoretical studies on the viscosity of nanofluids are also important.

The field of nanofluids is just over a decade old and still in the research stage. Although results from various laboratories are scattered, it can be concluded that nanofluids exhibit enhanced thermophysical properties which increase with the increasing volumetric loading of nanoparticles and/or decreasing particle size. While these remarkable and promising thermophysical properties of nanofluids have been demonstrated, many challenges remain to be tackled. It is important to characterize nanoparticle size, size distribution, and dispersion in order to optimize the usefulness of the enhanced thermal performance of nanofluids. Besides keeping the focus on thermal conductivity, other properties such as thermal diffusivity and the viscosity of nanofluids should receive more attention. There is also a need to optimize both the thermal conductivity and viscosity of nanofluids for extracting their maximum benefits.

References

Balderas-Lopez, J. A. and Mandelis, A. 2001. A simple, accurate and precise measurement of thermal diffusivity in liquids using a thermal-wave cavity. *Review of Scientific Instruments* 72: 2649–2652.

Bolz, R. and Tuve, G. 1973. *Handbook of Tables for Applied Engineering Science*. Boca Raton, FL: The Chemical Rubber Co.

Choi, S. U. S. 1995. Enhancing thermal conductivity of fluids with nanoparticles. In *Developments and Applications of Non-Newtonian Flows*, eds. D. A. Siginer and H. P. Wang, pp. 99–105. New York: ASME Publishing.

Chon, C. H. and Kihm, K. D. 2005. Thermal conductivity enhancement of nanofluids by Brownian motion. *Journal of Heat Transfer* 127:810.

Czarnetzki, W. and Roetzel, W. 1995. Temperature oscillation techniques for simultaneous measurement of thermal diffusivity and conductivity. *International Journal of Thermophysics* 16: 413–422.

Das, S. K., Putra, N., Thiesen, P., and Roetzel, W. 2003. Temperature dependence of thermal conductivity enhancement for nanofluids. *Journal of Heat Transfer* 125: 567–574.

Eastman, J. A., Choi, S. U. S., Li, S., and Thompson, L. J. 1997. Enhanced thermal conductivity through the development of nanofluids. In *Proceedings of the Symposium on Nanophase and Nanocomposite Materials II*, eds. S. Komarneni, J.C. Parker, and H. Wollenberger, pp. 3–11. Boston, MA: Materials Research Society.

Eastman, J. A., Choi, S. U. S., Li, S., Yu, W., and Thompson, L. J. 2001. Anomalously increased effective thermal conductivities of ethylene glycol-based nanofluids containing copper nanoparticles. *Applied Physics Letters* 78: 718–720.

Eastman, J. A., Phillpot, S.R., Choi, S. U. S., and Keblinski, P. 2004. Thermal transport in nanofluids. *Annual Review of Materials Research* 34: 219–246.

Einstein, A. 1956. *Investigations on the Theory of the Brownian Movement*. New York: Dover Publications, Inc.

Haarman, J. W. 1971. A contribution to the theory of the transient hot-wire method. *Physica* 52: 605–619.

Hamilton, R. L. and Crosser, O. K. 1962. Thermal conductivity of heterogeneous two component systems. *Industrial and Engineering Chemistry Fundamentals* 1: 187–191.

Healy, J. J., de Groot, J. J., and Kestin, J. 1976. The theory of the transient hot-wire method for measuring thermal conductivity, *Physica* 82C: 392–408.

Hong, T., Yang, H., and Choi, C. J. 2005. Study of the enhanced thermal conductivity of Fe nanofluids. *Journal of Applied Physics* 97: 064311.

Hong, K. S., Hong, T., and Yang, H. 2006. Thermal conductivity of Fe nanofluids depending on the cluster size of nanoparticles. *Applied Physics Letters* 88: 031901.

Horrocks, J. K. and McLaughlin, E. 1963. Nonsteady state measurements of thermal conductivities of liquids polyphenyls. *Proceedings of the Royal Society of London* 273A: 259–274.

Jang, S. P. and Choi, S. U. S. 2004. Role of Brownian motion in the enhanced thermal conductivity of nanofluids. *Applied Physics Letters* 84: 4316–4318.

Kaviany, M. 2002. *Principles of Heat Transfer*. New York: John Wiley & Sons.

Keblinski, P., Phillpot, S. R., Choi, S. U. S., and Eastman, J. A. 2002. Mechanisms of heat flow in suspensions of nano-sized particles (nanofluids). *International Journal of Heat and Mass Transfer* 45: 855–863.

Krishnamurthy, S., Bhattacharya, P., Phelan, P. E., and Prasher, R. S. 2006. Enhanced mass transport in nanofluids. *Nano Letters* 6: 419–423.

Lee, S., Choi, S. U. S., Li, S., and Eastman, J. A. 1999. Measuring thermal conductivity of fluids containing oxide nanoparticles. *Journal of Heat Transfer* 12: 280–289.

Li, C. H. and Peterson, G. P. 2006. Experimental investigation of temperature and volume fraction variations on the effective thermal conductivity of nanoparticle suspensions (nanofluids). *Journal of Applied Physics* 99: 084314.

Masuda, H., Ebata, A., Teramae, K., and Hishinuma, N. 1993. Alteration of thermal conductivity and viscosity of liquid by dispersing ultra-fine particles (Dispersion of Al_2O_3, SiO_2 and TiO_2 ultra-fine particles). *Netsu Bussei* 4: 227–233.

Maxwell, J. C. 1891. *A Treatise on Electricity and Magnetism*. Oxford, U.K.: Clarendon Press.

Murshed, S. M. S., Leong, K. C., and Yang, C. 2005. Enhanced thermal conductivity of TiO_2-water based nanofluids. *International Journal of Thermal Sciences* 44: 367–373.

Murshed, S. M. S., Leong, K. C., and Yang, C. 2006. Determination of the effective thermal diffusivity of nanofluids by the double hot-wire technique. *Journal of Physics D: Applied Physics* 39: 5316–5322.

Murshed, S. M. S., Leong, K. C., and Yang, C. 2008a. Thermophysical and electrokinetic properties of nanofluids—A critical review. *Applied Thermal Engineering* 28: 2109–2125.

Murshed, S. M. S., Leong, K. C., and Yang, C. 2008b. Investigations of thermal conductivity and viscosity of nanofluids. *International Journal of Thermal Sciences* 47: 560–568.

Murshed, S. M. S., Tan S. H., and Nguyen, N. T. 2008c. Temperature dependence of interfacial properties and viscosity of nanofluids for droplet-based microfluidics. *Journal of Physics D: Applied Physics* 41: 085502.

Nagasaka, Y. and Nagashima, A. 1981. Absolute measurement of the thermal conductivity of electrically conducting liquids by the transient hot-wire method. *Journal of Physics E: Scientific Instruments* 14: 1435–1440.

Nielsen, L. E. 1970. Generalized equation for the elastic moduli of composite materials. *Journal of Applied Physics* 41: 4626–4627.

Pak, B. C. and Cho, Y. I. 1998. Hydrodynamic and heat transfer study of dispersed fluids with submicron metallic oxide particles. *Experimental Heat Transfer* 11: 151–170.

Patel, H. E., Das, S. K., Sundararajan, T., Nair, A. S., George, B., and Pradeep, T. 2003. Thermal conductivity of naked and monolayer protected metal nanoparticles based nanofluids: Manifestation of anomalous enhancement and chemical effects. *Applied Physics Letters* 83: 2931–2933.

Prasher, R., Evans, W., and Meakin, P. 2006a. Effect of aggregation on thermal conduction in colloidal nanofluids. *Applied Physics Letters* 89: 143119.

Prasher, R., Song, D., Wang, J., and Phelan, P. E. 2006b. Measurements of nanofluid viscosity and its implications for thermal applications. *Applied Physics Letters* 89: 133108.

Putnam, S. A., Cahill, D. G., Braun, P. V., Ge, Z., and Shimmin, R. G. 2006. Thermal conductivity of nanoparticle suspensions. *Journal of Applied Physics* 99: 084308.

Putra, N., Roetzel, W., and Das, S. K. 2003. Natural convection of nanofluids. *Heat and Mass Transfer* 39: 775–784.

Venerus, D. C., Kabadi, M. S., Lee, S., and Perez-Luna, V. 2006. Study of thermal transport in nanoparticle suspensions using forced Rayleigh scattering. *Journal of Applied Physics* 100: 094310.

Wang, X. Q. and Mujumdar, A. S. 2007. Heat transfer characteristics of nanofluids: A review. *International Journal of Thermal Sciences* 46: 1–19.

Wang, X., Xu, X., and Choi, S. U. S. 1999. Thermal conductivity of nanoparticle-fluid mixture. *Journal of Thermophysics and Heat Transfer* 13: 474–480.

Wang, B.-X., Zhou, L.-P., and Peng, X.-F. 2003. A fractal model for predicting the effective thermal conductivity of liquid with suspension of nanoparticles. *International Journal of Heat and Mass Transfer* 46: 2665–2672.

Wang, B.-X., Zhou, L.-P., and Peng, X.-F. 2004. Viscosity, thermal diffusivity and Prandtl number of nanoparticle suspensions. *Progress in Natural Science* 14: 922–926.

Xie, H., Wang, J., Xi, T., Liu, Y., Ai, F., and Wu, Q. 2002. Thermal conductivity enhancement of suspensions containing nanosized alumina particles. *Journal of Applied Physics* 91: 4568–4572.

Xuan, Y., Li, Q., and Hu, W. 2003. Aggregation structure and thermal conductivity of nanofluids. *AIChE Journal* 49: 1038–1043.

Yang, B. and Han, Z. H. 2006. Temperature-dependent thermal conductivity of nanorod-based nanofluids. *Applied Physics Letters* 89: 083111.

Zhang, X, Gu, H., and Fujii, M. 2007. Effective thermal conductivity and thermal diffusivity of nanofluids containing spherical and cylindrical nanoparticles. *Experimental Thermal and Fluid Science* 31: 593–599.

Zhou, S. Q. and Ni, R. 2008. Measurement of specific heat capacity of water-based Al_2O_3 nanofluid. *Applied Physics Letters* 92: 093123.

Zhu, H., Zhang, C., Liu, S., Tang, Y., and Yin, Y. 2006. Effects of nanoparticles clustering and alignment on thermal conductivities of Fe_3O_4 aqueous nanofluids. *Applied Physics Letters* 89: 023123.

33

Heat Conduction in Nanofluids

Liqiu Wang
The University of Hong Kong

Xiaohao Wei
The University of Hong Kong

33.1 Introduction

Heat conduction is the process of heat transfer from a high-temperature region to a low-temperature region through a body (gas, liquid, or solid) that is not in macroscopic relative motion. Heat conduction can also take place across the interface between two material bodies in physical contact, without macroscopic relative motion with each other, and at different temperatures.

The mechanism of heat conduction in gases and liquids has been well known as the transfer of kinetic energy of the molecular movement. Temperature is a macroscopic manifestation of the kinetic energy of molecules; thus, in a high-temperature region, the molecules have higher velocities than in some lower-temperature region. Molecules are in continuous random motion, colliding with one another and exchanging energy and momentum. Molecules have this random motion whether or not a temperature gradient exists. If a molecule moves from a high-temperature region to a region of lower temperature, it transports kinetic energy to the lower-temperature part of the body and gives up this energy through collisions with lower-energy molecules. On the other hand, heat conduction in solids comes from the *lattice vibrational waves* induced by the vibrational motions of the molecules positioned at relatively fixed positions in a periodic manner called a lattice and the energy transported via the *free flow of electrons* in the solid.

In the *phenomenological* approach for heat-conduction study, the *thermal conductivity* of a material is defined as the *rate of heat transfer through a unit thickness of the material per unit area per unit temperature difference*. The thermal conductivity thus reflects the ability of a material to conduct heat. A high value for thermal conductivity implies that the material is a good heat conductor, and a low value indicates that the material is a poor heat conductor. Typically, the value of liquid thermal conductivity is several orders lower than most of the metals or metal oxides

(Table 33.1). It would thus be logical to boost the conductivity of a fluid by using a suspension of particles of a highly conducting solid in it, the idea that was first suggested more than one century ago. Most of these early studies, however, used suspensions of millimeter- or micrometer-sized particles, which, while they showed some promise, suffered from problems such as poor suspension stability and channel clogging. Nanoparticles are the obvious substitute candidates; and thus studying the thermal conductivity of nanofluids, fluid suspensions of nanometer-sized particles, has grabbed the considerable attention of scientists and engineers in the last decade.

Recent experiments on nanofluids have shown substantial increases in thermal conductivity (Table 33.2; Choi et al. 2004, Das et al. 2008, Eastman et al. 2004, Peterson and Li 2006, Phelan et al. 2005, Sobhan and Peterson 2008, Wu et al. 2009). Suggested *microscopic* reasons for this significant enhancement include the nanoparticle Brownian motion effect (Bhattacharya et al. 2004, Jang and Choi 2004, Koo and Kleinstreuer 2004, Prasher et al. 2005, 2006a, Xuan et al. 2003), the liquid layering effect at the liquid–particle interface (Leong et al. 2006, Ren et al. 2005, Xie et al. 2005, Xue et al. 2004, Yu and Choi 2003, 2004), and the nanoparticle cluster/aggregate effect (Prasher et al. 2006b, Wang et al. 2003). As generally accepted (Choi et al. 2004, Das et al. 2008, Eapen et al. 2007, Eastman et al. 2004, Peterson and Li 2006, Phelan et al. 2005, Putnam et al. 2006, Rusconi et al. 2006, Sobhan and Peterson 2008, Wu et al. 2009), however, no conclusive explanation is available. Often, the explanation by one research group is confronted by others. There is also a lack of agreement between experimental results and between theoretical models. The fact that the conductivity enhancement comes from the presence of nanoparticles has directed research efforts nearly exclusively toward thermal transport at nanoscale. The classical heat-conduction equation has been *postulated* as the macroscale model but without adequate justification. Thermal conductivity is

TABLE 33.1   Thermal Conductivity Comparison

	Materials	Thermal Conductivity (W/m K)
Liquids	Carbon nanotube	200–6000
	Ethylene glycol	0.25
	Mineral oil	0.13
	Water	0.613
Metal oxide solids	Alumina	40
	Copper oxide	70
	Zinc oxide	60
Metallic solids	Aluminum	230
	Copper	398
	Silver	410

a *macroscale phenomenological* characterization of heat conduction and the conductivity measurements are not performed at the nanoscale, but rather at the macroscale. Therefore, interest should focus not only on what happens at the nanoscale but also on how the presence of nanoparticles affects the heat transport at macroscale.

In an attempt to isolate the mechanism responsible for the significant enhancement of thermal conductivity, a macroscale heat-conduction model in nanofluids has been recently developed from first principles by Wang and Wei (2008, 2009a,b). The model was obtained by scaling up a microscale model for heat conduction in nanoparticles and in base fluids. The approach for scaling-up is the volume averaging with help of multiscale theorems adequately discussed in Wang (2000a), Wang et al. (2008), and Whitaker (1999). The microscale model for the heat conduction in the nanoparticles and in the base fluids comes from the first law of thermodynamics and Fourier's law of heat conduction. The result shows that the presence of nanoparticles leads to a dual-phase-lagging heat conduction in nanofluids at macroscale with a potential of much higher thermal conductivity. Therefore, the presence of nanoparticles shifts the Fourier heat conduction in the base fluid into the dual-phase-lagging heat conduction in

nanofluids at the macroscale. The dual-phase-lagging heat conduction differs from the Fourier heat conduction mainly on its existence of thermal waves and possible resonance. Such waves and resonance come from the coupled conduction between the nanoparticles and the base fluids and are responsible for the extraordinary conductivity enhancement of nanofluids (Wang and Wei 2008, 2009a,b, Wang et al. 2008b).

In this chapter, we first present a brief review of some valid models for heat conduction. We, then, develop the theory of macroscale heat conduction in nanofluids. Finally, we generalize nanofluids into thermal-wave fluids in which heat conduction can support thermal waves.

33.2 Constitutive Relations of Heat Flux

By the second law of thermodynamics, there exists a physical quantity Q that, at a given time instant, is associated with each surface in a non-isothermal body. This quantity can be interpreted as the heat through the surface and has two fundamental properties: behaving additively on compatible material surfaces and satisfying the first law of thermodynamics. These two properties, when rendered precisely, imply the existence of a flux vector field q whose scalar product with the unit normal vector to the surface yields the surface density of the heat Q (Šilhavy 1985). q is therefore named the *heat flux density vector*, or the *heat flux* for short.

The relation between the heat flux q and the temperature gradient ∇T is called the *constitutive relation of heat flux*, or the *constitutive relation* for short. It is the most fundamental and important relation in heat conduction and is normally given by fundamental laws.

33.2.1 Fourier's Law

Fourier's law was the first constitutive relation of heat flux density and was proposed by the French mathematical physicist

TABLE 33.2   Reported Conductivity Ratio k/k_b

Base Fluids (Conductivity W/m K)	Nano Particles (Average Size; Concentration)	Maximum k/k_b	References
Water (0.613)	Al_2O_3 (<50 nm; up to 4.3 vol%)	1.08	Choi (1998)
	CuO (<50 nm; up to 3.4 vol%)	1.10	Choi (1998)
	Cu (75–100 nm; ethylene glycol	1.23	Liu et al. (2006)
	TiO_2 (15 nm; <5.0 vol%)	1.30	Murshed et al. (2005)
	C-MWNT (50 nm, 5 μm, 3 urn; 0.6 vol%)	1.38	Assael et al. (2004)
	Cu (18 nm; up to 5.0 vol%)	1.60	Xuan and Li (2000) and Eastman et al. (1998)
Ethylene glycol (0.252)	Fe (<10 nm; 6.0 vol%)	1.18	Hong et al. (2005)
	Al_2O_3 (<50 nm; up to 5.0 vol%)	1.18	Choi (1998)
	CuO (35 nm; up to 4.0 vol%)	1.21	Choi (1998)
	Cu (10 nm; up to 0.5 vol%)	1.41	Eastman et al. (1998)
Oil (0.145)	Cu (up to 100 nm; up to 7.6 vol%)	1.43	Xuan and Li (2000)

Note: k, nanofluid thermal conductivity; k_b, base-fluid thermal conductivity.

Joseph Fourier in 1807 based on experimentation and investigation (Wang 1994). For heat conduction in a homogeneous and isotropic medium, Fourier's law of heat conduction reads

$$q(r,t) = -k\nabla T(r,t), \tag{33.1}$$

where

r stands for the material point

t is the time

T is the temperature

∇ is the gradient operator

k is the thermal conductivity of the material, which is a thermodynamic property

By the state theorem of thermodynamics, k should be a function of two independent and intensive properties (normally pressure and temperature; Cengel and Boles 2006). The second law of thermodynamics requires that k is positive definite (Wang 1994, 1995, 2001). In engineering applications, we often take k as a material constant because variations in pressure and temperature are normally sufficiently small. The value of k is material dependent. If the material is not homogeneous or isotropic, k becomes a second-order tensor (Wang 1994, 1995, 1996, 2001). Along with the first law of thermodynamics, this equation gives the following classical *parabolic* heat-conduction equation:

$$\frac{1}{\alpha}\frac{\partial T}{\partial t} = \Delta T + \frac{1}{k}F. \tag{33.2}$$

Here

α is the thermal diffusivity of the material

F is the rate of internal energy generation per unit volume

Δ is the Laplacian

Fourier's law of heat conduction is an early empirical law. It assumes that q and ∇T appear at the same time instant t and consequently implies that thermal signals propagate with an infinite speed. If the material is subjected to a thermal disturbance, the effects of the disturbance will be felt instantaneously at distances infinitely far from its source. Although this result is physically unrealistic, it has been confirmed by many experiments that Fourier's law of heat conduction holds for many media in the usual range of heat flux q and temperature gradient ∇T (Wang 1994).

33.2.2 CV Constitutive Relation

With the development of science and technology such as the application of ultrafast pulse-laser heating on metal films, heat conduction appears in a range of high heat flux and high unsteadiness. The drawback of infinite heat propagation speed in Fourier's law becomes unacceptable. This has inspired the work of searching for new constitutive relations. Among many proposed relations (Wang 1994), the constitutive relation proposed by Cattaneo (1958) and Vernotte (1958, 1961)

$$q(r,t) + \tau_q \frac{\partial q(r,t)}{\partial t} = -k\nabla T(r,t) \tag{33.3}$$

is most widely accepted. This relation is named the CV constitutive relation after the names of the proposers. Here $\tau_q > 0$ is a material property and is called the *relaxation time*. The corresponding heat-conduction equation is thus

$$\frac{1}{\alpha}\frac{\partial T}{\partial t} + \frac{\tau_q}{\alpha}\frac{\partial^2 T}{\partial t^2} = \Delta T + \frac{1}{k}\left(F + \tau_q\frac{\partial F}{\partial t}\right). \tag{33.4}$$

Unlike its classical counterpart Equation 33.2, this equation is of *hyperbolic* type, characterizes the combined diffusion and wave-like behavior of heat conduction, and predicts a *finite speed*

$$V_{CV} = \sqrt{\frac{k}{\rho c \tau_q}} \tag{33.5}$$

for heat propagation (Wang and Zhou 2000, 2001, Wang et al. 2008a).

Note that the CV constitutive relation is actually a first-order approximation of a more general constitutive relation (single-phase-lagging model; Tzou 1992),

$$q(r, t + \tau_q) = -k\nabla T(r,t). \tag{33.6}$$

according to which the temperature gradient established at a point r at time t gives rise to a heat flux vector at r at a *later* time $t + \tau_q$. There is a finite built-up time τ_q for the onset of heat flux at r after a temperature gradient is imposed there. Thus, the τ_q represents the time lag needed to establish the heat flux (the result) when a temperature gradient (the cause) is suddenly imposed. The higher $\partial q/\partial t$ corresponds to a larger derivation of the CV constitutive relation from classical Fourier's law.

The value of τ_q is material dependent (Chandrasekharaiah 1986, 1998, Tzou 1997). For most solid materials, τ_q varies from 10^{-10} to 10^{-14} s. For gases, τ_q is normally in the range of $10^{-8} \sim 10^{-10}$ s. The value of τ_q for some biological materials and materials with nonhomogeneous inner structures can be up to 10^2 s (Beckert 2000, Kaminski 1990, Mitra et al. 1995, Peters 1999, Roetzel et al. 2003, Vedavarz et al. 1992). Therefore, the thermal relaxation effects can be of relevance even in common engineering applications where the time scales of interest are of the order of a fraction of a minute.

Three factors contribute to the significance of the second term in the hyperbolic heat-conduction equation (33.4): the τ_q value, the rate of change of temperature, and the time scale involved. The wave nature of thermal signals will be over the diffusive behavior through this term when (Tzou 1992)

$$\frac{\partial T}{\partial t} \gg \frac{T_r}{2\tau_q}\exp\left(\frac{t}{\tau_q}\right) \tag{33.7}$$

where T_r is a reference temperature. Therefore, the wave-like features will become significant when (1) τ_q is large, (2) $\partial T/\partial t$ is high, or (3) t is small. Some typical situations where hyperbolic heat conduction differs from classical parabolic heat conduction include those concerned with a localized moving heat source with a high intensity, a rapidly propagating crack tip, shock-wave propagation, thermal resonance, interfacial effects between dissimilar materials, laser material processing, and laser surgery (Chandrasekharaiah 1986, 1998, Joseph and Preziosi 1989, 1990, Tzou 1992, 1995a, 1997, Wang 1994, 2000b).

When $\tau_q \to \infty$ but $k_{\mathrm{eff}} = k/\tau_q$ is finite, the CV constitutive relation (33.3) and the hyperbolic heat-conduction equation (33.4) become (Joseph and Preziosi 1989)

$$\frac{\partial q(r,t)}{\partial t} = -k_{\mathrm{eff}}\nabla T(r,t), \qquad (33.8)$$

and

$$\frac{1}{\alpha_{\mathrm{eff}}}\frac{\partial^2 T}{\partial t^2} = \Delta T + \frac{1}{k_{\mathrm{eff}}}\frac{\partial F}{\partial t}, \qquad (33.9)$$

where

$\alpha_{\mathrm{eff}} = k_{\mathrm{eff}}/\rho c$

ρ and c are the density and the specific heat of the material, respectively

Therefore, when τ_q is very large, a temperature gradient established at a point of the material results in an *instantaneous heat flux rate* at that point, and vice versa. Equation 33.9 is a classical wave equation that predicts thermal-wave propagation with speed V_{CV}, like Equation 33.4. A major difference exists, however, between Equations 33.4 and 33.9: the former allows damping of thermal waves, the latter does not (Wang and Zhou 2000, 2001, Wang et al. 2008b).

33.2.3 Dual-Phase-Lagging Constitutive Relation

It has been confirmed by many experiments that the CV constitutive relation generates a more accurate prediction than classical Fourier's law. However, some of its predictions do not agree with experimental results either (Tzou 1995a, 1997, Wang 1994). A thorough study shows that the CV constitutive relation has only taken account of the fast-transient effects, but not the microstructural interactions. These two effects can be reasonably represented by the dual-phase lag between q and ∇T, a further modification of Equation 33.6 (Tzou 1995a, 1997):

$$q(r,t+\tau_q) = -k\nabla T(r,t+\tau_T). \qquad (33.10)$$

According to this relation, the temperature gradient at a point r of the material at time $t + \tau_T$ corresponds to the heat flux density vector at r at time $t + \tau_q$. The delay time τ_T is interpreted as being caused by the microstructural interactions (small-scale heat transport mechanisms occurring at the microscale, or small-scale effects of heat transport in space) such as phonon–electron

interaction or phonon scattering, and is called the *phase lag of the temperature gradient* (Tzou 1995a, 1997). The other delay time τ_q is interpreted as the relaxation time due to the fast-transient effects of thermal inertia (or small-scale effects of heat transport in time) and is called the *phase lag of the heat flux*. Both of the phase lags are treated as intrinsic thermal or structural properties of the material. The corresponding heat-conduction equation reads (Xu and Wang 2005)

$$\frac{1}{\alpha}\frac{\partial T(r,t')}{\partial t} = \Delta T(r,t'-\tau) + \frac{1}{k}F(r,t'), \quad t'=t+\tau_q, \quad \tau = \tau_q - \tau_T,$$

$$\text{for } \tau_q - \tau_T > 0 \quad \text{and} \quad t' > \tau_0, \qquad (33.11)$$

or

$$\frac{1}{\alpha}\frac{\partial T(r,t'-\tau)}{\partial t} = \Delta T(r,t') + \frac{1}{k}F(r,t'-\tau), \quad t'=t+\tau_T, \quad \tau = \tau_T - \tau_q,$$

$$\text{for } \tau_q - \tau_T < 0 \quad \text{and} \quad t' > \tau_T \qquad (33.12)$$

Unlike the relation (33.6) according to which the heat flux is the result of a temperature gradient in a transient process, the relation (33.10) allows either the temperature gradient or the heat flux to become the effect and the remaining one the cause. For materials with $\tau_q > \tau_T$, the heat flux density vector is the result of a temperature gradient. It is the other way around for materials with $\tau_T > \tau_q$. The relation (33.6) corresponds to the particular case where $\tau_q > 0$ and $\tau_T = 0$. If $\tau_q = \tau_T$ (not necessarily equal to zero), the response between the temperature gradient and the heat flux is instantaneous; in this case, the relation (33.10) is identical with classical Fourier's law (33.1). It may also be noted that while classical Fourier's law (Equation 33.1) is macroscopic in both space and time and the relation (33.6) is macroscopic in space but microscopic in time, the relation (33.10) is microscopic in both space and time. Also note that Equations 33.11 and 33.12 are of the delay and advance types, respectively. While the former has a wave-like solution and possibly resonance, the latter does not (Xu and Wang 2005). Both single-phase-lagging and dual-phase-lagging heat conduction has been shown to be admissible by the second law of extended irreversible thermodynamics (Tzou 1997) and by the Boltzmann transport equation (Cheng et al. 2008a, Xu and Wang 2005).

Expanding both sides of Equation 33.10 using the Taylor series and retaining only the first-order terms of τ_q and τ_T, we obtain the following constitutive relation that is valid at point r and time t:

$$q(r,t) + \tau_q \frac{\partial q(r,t)}{\partial t} = -k\left\{\nabla T(r,t) + \tau_T \frac{\partial}{\partial t}\left[\nabla T(r,t)\right]\right\}, \quad (33.13)$$

which is known as the Jeffreys-type constitutive equation of heat flux (Joseph and Preziosi 1989). In literature this relation is also called the *dual-phase-lagging constitutive relation*. When $\tau_q = \tau_T$, this relation reduces to classical Fourier's law (Equation 33.1), and it reduces to the CV constitutive relation (33.3) when $\tau_T = 0$.

Eliminating q from Equation 33.13 and the classical energy equation leads to the dual-phase-lagging heat-conduction equation that reads as follows, if all thermophysical material properties are assumed to be constant:

$$\frac{1}{\alpha}\frac{\partial T}{\partial t} + \frac{\tau_q}{\alpha}\frac{\partial^2 T}{\partial t^2} = \Delta T + \tau_T \frac{\partial}{\partial t}(\Delta T) + \frac{1}{k}\left(F + \tau_q \frac{\partial F}{\partial t}\right) \quad (33.14)$$

This equation is *parabolic* when $\tau_q < \tau_T$ (Wang and Zhou 2000, 2001, Wang et al. 2008b). Although a wave term $(\tau_q/\alpha)\partial^2 T/\partial t^2$ exists in the equation, the mixed derivative $\tau_T\partial(\Delta T)/\partial t$ completely destroys the wave structure. The equation, in this case, therefore predicts a nonwave-like heat conduction that differs from the usual diffusion predicted by the classical parabolic heat conduction (Equation 33.2). When $\tau_q > \tau_T$, however, Equation 33.14 can be approximated by Equation 33.4 and then predominantly predicts wave-like thermal signals.

The dual-phase-lagging heat-conduction equation (33.14) forms a generalized, unified equation that reduces to the classical parabolic heat-conduction equation when $\tau_T = \tau_q$, the hyperbolic heat-conduction equation when $\tau_T = 0$ and $\tau_q > 0$, the energy equation in the phonon scattering model (Guyer and Krumhansi 1966, Joseph and Preziosi 1989) when $\alpha = \tau_R c^2/3$, $\tau_T = (9/5)\tau_N$, and $\tau_q = \tau_R$, and the energy equation in the phonon–electron interaction model (Anisimòv et al. 1974, Kaganov et al. 1957, Qiu and Tien 1993) when $\alpha = k/(c_e + c_l)$, $\tau_T = c_l/G$, and $\tau_q = 1/G\left[(1/c_e) + (1/c_l)\right]^{-1}$. In the phonon scattering model, c is the average speed of phonons (sound speed), τ_R is the relaxation time for the umklapp process in which momentum is lost from the phonon system, and τ_N is the relaxation time for normal processes in which momentum is conserved in the phonon system. In the phonon–electron interaction model, k is the thermal conductivity of the electron gas, G is the phonon–electron coupling factor, and c_e and c_l are the heat capacity of the electron gas and the metal lattice, respectively. This, together with its success in describing and predicting phenomena such as ultrafast pulse-laser heating, propagation of temperature pulses in superfluid liquid helium, nonhomogeneous lagging response in porous media, thermal lagging in amorphous materials, and effects of material defects and thermomechanical coupling, heat conduction in nanofluids, bi-composite media and two-phase systems (Tzou 1997, Tzou and Zhang 1995, Vadasz 2005a,b,c, 2006a,b, Wang and Wei 2008, 2009a,b, Wang et al. 2008b) has given rise to the research effort on various aspects of dual-phase-lagging heat conduction (Tzou 1997, Wang and Zhou 2000, 2001, Wang et al. 2008b).

The dual-phase-lagging heat-conduction model that is based on Equation 33.14 has been shown to be well-posed in a finite region of n-dimensions ($n \geq 1$) under any linear boundary conditions including Dirichlet, Neumann, and Robin types (Wang and Xu 2002, Wang et al. 2001). Solutions of one-dimensional (1D) heat conduction has been obtained for some specific initial and boundary conditions by Antaki (1998), Dai and Nassar (1999), Lin et al. (1997), Tang and Araki (1999), Tzou (1995a,b, 1997), Tzou and Zhang (1995), Tzou and Chiu (2001). Wang and Zhou (2000, 2001) and Wang et al. (2008b) obtained analytical solutions for regular 1D, 2D, and 3D heat-conduction domains under essentially arbitrary initial and boundary conditions. The solution structure theorems were also developed for both mixed and Cauchy problems of dual-phase-lagging heat-conduction equations (Wang and Zhou 2000, Wang et al. 2001, 2008b) by extending those theorems for hyperbolic heat conduction (Wang 2000b). These theorems build relationships between the contributions (to the temperature field) by the initial temperature distribution, the source term, and the initial time-rate of the temperature change, uncovering the structure of the temperature field and considerably simplifying the development of solutions. Xu and Wang (2002) addressed thermal features of dual-phase-lagging heat conduction (particularly conditions and features of thermal oscillation and resonance and their contrast with those of classical and hyperbolic heat conduction). The issues associated with the Galilean principle of relativity have also been discussed by Cheng et al. (2008b) for both single- and dual-phase-lagging heat conduction models in moving media.

An experimental procedure for determining the value of τ_q has been proposed by Mengi and Turhan (1978). The general problem of measuring short-time thermal transport effects has been discussed by Chester (1966). Wang and Zhou (2000, 2001) and Wang et al. (2008b) developed three methods for measuring τ_q. Tzou (1997) and Vadasz (2005a,b, 2006a,b) developed an *approximate* equivalence between Fourier heat conduction in porous media and dual-phase-lagging heat conduction, and applied the latter to examine features of the former. Based on that equivalence, Vadasz (2005a,b,c, 2006a,b) showed that τ_T is always larger than τ_q in porous-media heat conduction so that thermal waves cannot occur according to the necessary condition for thermal waves in dual-phase-lagging heat conduction (Xu and Wang 2002). However, such waves are observed in casting sand experiments by two independent groups (Tzou 1997). In an attempt to resolve this difference and to build the intrinsic relationship between the two heat-conduction processes, Wang and Wei (2008) and Wang et al. (2008b) developed an *exact* equivalence between dual-phase-lagging heat conduction and Fourier heat conduction in two-phase systems subject to a lack of local thermal equilibrium. Based on this new equivalence, they also show the possibility of, and uncover, the mechanism responsible for the thermal oscillation in two-phase-system heat conduction.

Tzou (1995b, 1997) also generalized Equation 33.13, for $\tau_q \gg \tau_T$, by retaining terms up to the second order in τ_q but only the term of the first order in τ_T in the Taylor expansions of Equation 33.10 to obtain a τ_q-second-order dual-phase-lagging model:

$$q + \tau_q \frac{\partial q}{\partial t} + \frac{1}{2}\tau_q^2 \frac{\partial^2 q}{\partial t^2} = -k\left[\nabla T + \tau_T \frac{\partial}{\partial t}(\nabla T)\right]. \quad (33.15)$$

For this case the dual-phase-lagging heat-conduction equation (33.14) is generalized into

$$\frac{1}{\alpha}\frac{\partial T}{\partial t} + \frac{\tau_q}{\alpha}\frac{\partial^2 T}{\partial t^2} = \Delta T + \tau_T \frac{\partial}{\partial t}(\Delta T) + \frac{1}{k}\left(F + \tau_q \frac{\partial F}{\partial t}\right), \quad (33.16)$$

which is of hyperbolic type and thus predicts thermal wave propagation with a finite speed (Tzou 1995b, 1997)

$$V_T = \frac{1}{\tau_q}\sqrt{\frac{2k\tau_T}{\rho c}}. \qquad (33.17)$$

The thermal wave from Equation 33.4 is obviously different from that in Equation 33.16. While the former is caused only by the fast-transient effects of thermal inertia, the latter comes from these effects as well as the delayed response due to the micro-structural interaction. Tzou (1997) refers to the former wave as the CV wave and the latter wave as the T wave. By Equations 33.5 and 33.17, we have

$$V_T = \sqrt{\frac{2\tau_T}{\tau_q}}V_{CV}. \qquad (33.18)$$

Therefore, the T wave is always slower than the CV wave because Equations 33.15 and 33.16 are valid only for $\tau_q \gg \tau_T$. This has been shown by the heat propagation in superfluid helium at extremely low temperatures (Tzou 1997). It is interesting to note that Equation 33.15 is the simplest constitutive relation that accounts for the dual-phase-lagging effects and yields a heat-conduction equation of hyperbolic type. If the second-order term in τ_T is also retained, the resulting heat-conduction equation will no longer be hyperbolic (Tzou 1997). It is also of interest to note that Equation 33.16 closely resembles the energy equation describing the ballistic behavior of heat transport in an electron gas (Qiu and Tien 1993, Tzou 1997).

In this section, we have presented a brief review of some valid models for heat conduction from macro- to microscales: the *Fourier model* based on Equations 33.1 and 33.2, the *CV model* based on Equations 33.3 and 33.4, the *wave model* based on Equations 33.8 and 33.9, the *single-phase-lagging model* based on Equation 33.6, and the three *dual-phase-lagging models* [the dual-phase-lagging model based on Equations 33.10 through 33.12; the first-order dual-phase-lagging model based on Equations 33.13 and 33.14; the second-order dual-phase-lagging model based on Equations 33.15 and 33.16]. In literature, the *dual-phase-lagging model* usually refers to the first-order dual-phase-lagging model Equations 33.13 and 33.14, which is a generalized and unified model for heat conduction from macro- to microscales with the Fourier, wave, and CV models as its special cases.

33.3 Macroscale Heat Conduction in Nanofluids

The microscale model for heat conduction in nanofluids is well known. It consists of the field equation and the constitutive equation. The field equation comes from the first law of thermodynamics. The commonly-used constitutive equation is Fourier's law of heat conduction for the relation between the temperature gradient ∇T and the heat flux density vector $\mathbf{q}$ (Wang 1994).

For transport in nanofluids, the macroscale is a phenomenological scale that is much larger than the microscale and much smaller than the system length scale. Interest in the macroscale rather than the microscale comes from the fact that a prediction at the microscale is complicated due to the complex microscale structure of nanofluids, and also because we are usually more interested in large scales of transport for practical applications. Existence of such a macroscale description equivalent to the microscale behavior requires a good separation of length scales and has been well discussed by Auriault (1991) and Wang et al. (2008a).

To develop a macroscale model of heat conduction in nanofluids, the method of volume averaging starts with a microscale description (Wang 2000a, Whitaker 1999). Both conservation and constitutive equations are introduced at the microscale. The resulting microscale field equations are then averaged over a representative elementary volume (REV), the smallest differential volume resulting in statistically meaningful local averaging properties, to obtain the macroscale field equations. In the process of averaging, the *multiscale theorems* are used to convert integrals of gradient, divergence, curl, and partial time derivatives of a function into some combination of gradient, divergence, curl, and partial time derivatives of integrals of the function and integrals over the boundary of the REV (Wang 2000a, Wang et al. 2008b, Whitaker 1999). The readers are referred to Wang (2000a), Wang et al. (2008a), and Whitaker (1999) for the details of the method of volume averaging and to Wang (2000a) and Wang et al. (2008a) for a summary of the other methods of obtaining macroscale models.

Consider heat conduction in nanofluids with the base fluid and the nanoparticle denoted by β- and σ-phases, respectively. By the first law of thermodynamics and Fourier's law of heat conduction, we have the microscale model for heat conduction in nanofluids (Figure 33.1; Quintard and Whitaker 1993)

$$(\rho c)_\beta \frac{\partial T_\beta}{\partial t} = \nabla \cdot (k_\beta \nabla T_\beta), \quad \text{in the } \beta\text{-phase} \qquad (33.19)$$

$$(\rho c)_\sigma \frac{\partial T_\sigma}{\partial t} = \nabla \cdot (k_\sigma \nabla T_\sigma), \quad \text{in the } \sigma\text{-phase} \qquad (33.20)$$

$$T_\beta = T_\sigma, \quad \text{at the } \beta\text{–}\sigma \text{ interface } A_{\beta\sigma} \qquad (33.21)$$

$$\mathbf{n}_{\beta\sigma} \cdot k_\beta = \mathbf{n}_{\beta\sigma} \cdot k_\sigma \nabla T_\sigma, \quad \text{at the } \beta\text{–}\sigma \text{ interface } A_{\beta\sigma} \qquad (33.22)$$

Here T is the temperature. ρ, c, and k are the density, specific heat, and thermal conductivity, respectively. Subscripts β and σ refer to the β- and σ-phases, respectively. $A_{\beta\sigma}$ represents the area of the β–σ interface contained in the REV; $\mathbf{n}_{\beta\sigma}$ is the outward-directed surface normal from the β-phase toward the σ-phase, and $\mathbf{n}_{\beta\sigma} = -\mathbf{n}_{\sigma\beta}$ (Figure 33.1). To be thorough, Quintard and Whitaker (1993) have also specified the initial conditions and the boundary conditions at the entrances and exits of the REV; however, we need not do so for our discussion.

Next Quintard and Whitaker (1993) apply the superficial averaging process to Equations 33.19 and 33.20 to obtain

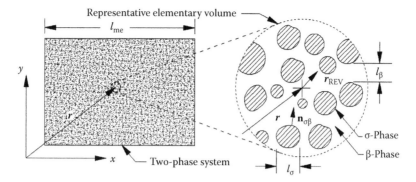

FIGURE 33.1 Nanofluids and representative elementary volume (REV).

$$\frac{1}{V_{REV}}\int\limits_{V_\beta}(\rho c)_\beta\frac{\partial T_\beta}{\partial t}dV = \frac{1}{V_{REV}}\int\limits_{V_\beta}\nabla\cdot(k_\beta\nabla T_\beta)\,dV, \qquad (33.23)$$

and

$$\frac{1}{V_{REV}}\int\limits_{V_\sigma}(\rho c)_\sigma\frac{\partial T_\sigma}{\partial t}dV = \frac{1}{V_{REV}}\int\limits_{V_\sigma}\nabla\cdot(k_\sigma\nabla T_\sigma)\,dV, \qquad (33.24)$$

where V_{REV}, V_β, and V_σ are the volumes of the REV, β-phase in REV and σ-phase in REV, respectively. We should note that the superficial temperature is evaluated at the centroid of the REV, whereas the phase temperature is evaluated throughout the REV. Neglecting variations of ρc within the REV and considering the system to be rigid so that V_β and V_σ are time independent, the volume-averaged form of Equations 33.19 and 33.20 are

$$(\rho c)_\beta\frac{\partial\langle T_\beta\rangle}{\partial t} = \langle\nabla\cdot(k_\beta\nabla T_\beta)\rangle, \qquad (33.25)$$

and

$$(\rho c)_\sigma\frac{\partial\langle T_\sigma\rangle}{\partial t} = \langle\nabla\cdot(k_\sigma\nabla T_\sigma)\rangle, \qquad (33.26)$$

where angle brackets indicate superficial quantities such as

$$\langle T_B\rangle = \frac{1}{V_{REV}}\int\limits_{V_\beta}T_\beta\,dV,$$

and

$$\langle T_\sigma\rangle = \frac{1}{V_{REV}}\int\limits_{V_\sigma}T_\sigma\,dV.$$

The superficial average, however, is an unsuitable variable because it can yield erroneous results. For example, if the temperature of the β-phase were constant, the superficial average

would differ from it (Quintard and Whitaker 1993). On the other hand, intrinsic phase averages do not have this shortcoming. These averages are defined by

$$\langle T_\beta\rangle^\beta = \frac{1}{V_\beta}\int\limits_{V_\beta}T_\beta\,dV, \qquad (33.27)$$

and

$$\langle T_\sigma\rangle^\sigma = \frac{1}{V_\sigma}\int\limits_{V_\sigma}T_\sigma\,dV. \qquad (33.28)$$

Also, intrinsic averages are related to superficial averages by

$$\langle T_\beta\rangle = \varepsilon_\beta\langle T_\beta\rangle^\beta, \qquad (33.29)$$

and

$$\langle T_\sigma\rangle = \varepsilon_\sigma\langle T_\sigma\rangle^\sigma, \qquad (33.30)$$

where ε_β and ε_σ are the volume fractions of the β- and σ-phases with $\varepsilon_\beta = 1 - \varphi$ and $\varepsilon_\sigma = \varphi$. φ is the volume fraction of the σ-phase defined by $\varphi = V_\sigma/V_{REV}$.

Quintard and Whitaker (1993) substitute Equations 33.29 and 33.30 into Equations 33.25 and 33.26 to obtain

$$\varepsilon_\beta(\rho c)_\beta\frac{\partial\langle T_\beta\rangle^\beta}{\partial t} = \langle\nabla\cdot(k_\beta\nabla T_\beta)\rangle, \qquad (33.31)$$

and

$$\varepsilon_\sigma(\rho c)_\sigma\frac{\partial\langle T_\sigma\rangle^\sigma}{\partial t} = \langle\nabla\cdot(k_\sigma\nabla T_\sigma)\rangle. \qquad (33.32)$$

Next Quintard and Whitaker (1993) apply the spatial averaging theorem (Theorem 40 in Wang et al. 2008a) to Equations 33.31

and 33.32 and neglect variations of physical properties within the REV. The result is

$$
\underbrace{\varepsilon_\beta(\rho c)_\beta \frac{\partial \langle T_\beta \rangle^\beta}{\partial t}}_{\text{accumulation}} = \underbrace{\nabla \cdot \left\{ k_\beta \left[\varepsilon_\beta \nabla \langle T_\beta \rangle^\beta + \langle T_\beta \rangle^\beta \nabla \varepsilon_\beta + \frac{1}{V_{\text{REV}}} \int_{A_{\beta\sigma}} \mathbf{n}_{\beta\sigma} T_\beta \, dA \right] \right\}}_{\text{conduction}}
$$

$$
+ \underbrace{\frac{1}{V_{\text{REV}}} \int_{A_{\beta\sigma}} n_{\beta\sigma} \cdot k_\beta \nabla T_\beta \, dA}_{\text{interfacial flux}}, \tag{33.33}
$$

and

$$
\underbrace{\varepsilon_\sigma(\rho c)_\sigma \frac{\partial \langle T_\sigma \rangle^\sigma}{\partial t}}_{\text{accumulation}}
$$

$$
= \underbrace{\nabla \cdot \left\{ k_\sigma \left[\varepsilon_\sigma \nabla \langle T_\sigma \rangle^\sigma + \langle T_\sigma \rangle^\sigma \nabla \varepsilon_\sigma + \frac{1}{V_{\text{REV}}} \int_{A_{\beta\sigma}} \mathbf{n}_{\beta\sigma} T_\sigma \, dA \right] \right\}}_{\text{conduction}}
$$

$$
+ \underbrace{\frac{1}{V_{\text{REV}}} \int_{A_{\beta\sigma}} n_{\beta\sigma} \cdot k_\sigma \nabla T_\sigma \, dA}_{\text{interfacial flux}}. \tag{33.34}
$$

By introducing the spatial decompositions, $T_\beta = \langle T_\beta \rangle^\beta + \tilde{T}_\beta$ and $T_\sigma = \langle T_\sigma \rangle^\sigma + \tilde{T}_\sigma$, and by applying scaling arguments and Theorem 40 in Wang et al. (2008a), Equations 33.33 and 33.34 are simplified into (Quintard and Whitaker 1993)

$$
\varepsilon_\beta(\rho c)_\beta \frac{\partial \langle T_\beta \rangle^\beta}{\partial t} = \nabla \cdot \left\{ k_\beta \left[\varepsilon_\beta \nabla \langle T_\beta \rangle^\beta + \frac{1}{V_{\text{REV}}} \int_{A_{\beta\sigma}} \mathbf{n}_{\beta\sigma} \tilde{T}_\beta \, dA \right] \right\}
$$

$$
+ \frac{1}{V_{\text{REV}}} \int_{A_{\beta\sigma}} n_{\beta\sigma} \cdot k_\beta \nabla \langle T_\beta \rangle^\beta \, dA
$$

$$
+ \frac{1}{V_{\text{REV}}} \int_{A_{\beta\sigma}} \mathbf{n}_{\beta\sigma} \cdot k_\beta \nabla \tilde{T}_\beta \, dA, \tag{33.35}
$$

and

$$
\varepsilon_\sigma(\rho c)_\sigma \frac{\partial \langle T_\sigma \rangle^\sigma}{\partial t} = \nabla \cdot \left\{ k_\sigma \left[\varepsilon_\sigma \nabla \langle T_\sigma \rangle^\sigma + \frac{1}{V_{\text{REV}}} \int_{A_{\beta\sigma}} \mathbf{n}_{\sigma\beta} \tilde{T}_\sigma \, dA \right] \right\}
$$

$$
+ \frac{1}{V_{\text{REV}}} \int_{A_{\beta\sigma}} \mathbf{n}_{\sigma\beta} \cdot k_\sigma \nabla \langle T_\sigma \rangle^\sigma \, dA
$$

$$
+ \frac{1}{V_{\text{REV}}} \int_{A_{\beta\sigma}} \mathbf{n}_{\sigma\beta} \cdot k_\sigma \nabla \tilde{T}_\sigma \, dA. \tag{33.36}
$$

After developing the closure for $\tilde{T}_\beta$ and $\tilde{T}_\sigma$, Quintard and Whitaker (1993) obtain a two-equation model:

$$
\varepsilon_\beta(\rho c)_\beta \frac{\partial \langle T_\beta \rangle^\beta}{\partial t} = \nabla \cdot \left\{ \mathbf{K}_{\beta\beta} \cdot \nabla \langle T_\beta \rangle^\beta + \mathbf{K}_{\beta\sigma} \cdot \nabla \langle T_\sigma \rangle^\sigma \right\}
$$

$$
+ ha_\upsilon \left(\langle T_\sigma \rangle^\sigma - \langle T_\beta \rangle^\beta \right), \tag{33.37}
$$

and

$$
\varepsilon_\sigma(\rho c)_\sigma \frac{\partial \langle T_\sigma \rangle^\sigma}{\partial t} = \nabla \cdot \left\{ \mathbf{K}_{\sigma\sigma} \cdot \nabla \langle T_\sigma \rangle^\sigma + \mathbf{K}_{\sigma\beta} \cdot \nabla \langle T_\beta \rangle^\beta \right\}
$$

$$
- ha_\upsilon \left(\langle T_\sigma \rangle^\sigma - \langle T_\beta \rangle^\beta \right), \tag{33.38}
$$

where h and a_υ come from modeling of the interfacial flux and are the film heat transfer coefficient and the interfacial area per unit volume, respectively. $\mathbf{K}_{\beta\beta}$, $\mathbf{K}_{\sigma\sigma}$, $\mathbf{K}_{\beta\sigma}$, and $\mathbf{K}_{\sigma\beta}$ are the effective thermal conductivity tensors, and the coupled thermal conductivity tensors are equal:

$$
\mathbf{K}_{\beta\sigma} = \mathbf{K}_{\sigma\beta}.
$$

When the system is isotropic and the physical properties of the two phases are constant, Equations 33.37 and 33.38 reduce to

$$
\gamma_\beta \frac{\partial \langle T_\beta \rangle^\beta}{\partial t} = k_{\beta\beta} \Delta \langle T_\beta \rangle^\beta + k_{\beta\sigma} \Delta \langle T_\sigma \rangle^\sigma + ha_\upsilon \left(\langle T_\sigma \rangle^\sigma - \langle T_\beta \rangle^\beta \right),
$$
$$
\tag{33.39}
$$

and

$$
\gamma_\sigma \frac{\partial \langle T_\sigma \rangle^\sigma}{\partial t} = k_{\sigma\sigma} \Delta \langle T_\sigma \rangle^\sigma + k_{\sigma\beta} \Delta \langle T_\beta \rangle^\beta + ha_\upsilon \left(\langle T_\sigma \rangle^\sigma - \langle T_\beta \rangle^\beta \right),
$$
$$
\tag{33.40}
$$

where

$\gamma_\beta = (1 - \varphi)(\rho c)_\beta$ and $\gamma_\sigma = \varphi(\rho c)_\sigma$ are the β-phase and σ-phase effective thermal capacities, respectively

$k_{\beta\beta}$ and $k_{\sigma\sigma}$ are the effective thermal conductivities of the β- and σ-phases, respectively

$k_{\beta\sigma} = k_{\sigma\beta}$ is the cross effective thermal conductivity of the two phases

The one-equation model is valid whenever the two temperatures $\langle T_\beta \rangle^\beta$ and $\langle T_\sigma \rangle^\sigma$ are sufficiently close to each other so that

$$
\langle T_\beta \rangle^\beta = \langle T_\sigma \rangle^\sigma = \langle T \rangle. \tag{33.41}
$$

This *local thermal equilibrium* is valid when any one of the following three conditions occurs (Quintard and Whitaker 1993, Whitaker 1999): (1) either ε_β or ε_σ tends to zero, (2) the difference in the β-phase and σ-phase physical properties tends to zero, (3) the square of the ratio of length scales $(l_{\beta\sigma}/L)^2$ tends to zero (e.g., steady, one-dimensional heat conduction). Here $l_{\beta\sigma}^2 = \left[\varepsilon_\beta \varepsilon_\sigma (\varepsilon_\beta k_\sigma + \varepsilon_\sigma k_\beta) \right] / (ha_\upsilon)$, and $L = L_\text{T} L_{\text{T1}}$ with L_T and L_{T1} as the characteristic lengths of $\nabla \langle T \rangle$ and $\nabla \nabla \langle T \rangle$, respectively, such that $\nabla \langle T \rangle = O\left(\Delta \langle T \rangle / L_T \right)$ and $\nabla \nabla \langle T \rangle = O\left(\Delta \langle T \rangle / L_{T1} L_T \right)$.

When the local thermal equilibrium is valid, Quintard and Whitaker (1993) add Equations 33.37 and 33.38 to obtain a one-equation model:

$$\langle \rho \rangle C \frac{\partial \langle T \rangle}{\partial t} = \nabla \cdot \left[\mathbf{K}_{\text{eff}} \cdot \nabla \langle T \rangle \right]. \tag{33.42}$$

Here $\langle \rho \rangle$ is the spatial average density defined by

$$\langle \rho \rangle = \varepsilon_\beta \rho_\beta + \varepsilon_\sigma \rho_\sigma, \tag{33.43}$$

and C is the mass-fraction-weighted thermal capacity given by

$$C = \frac{\varepsilon_\beta (\rho c)_\beta + \varepsilon_\sigma (\rho c) \sigma}{\varepsilon_\beta \rho_\beta + \varepsilon_\sigma \rho_\sigma}. \tag{33.44}$$

The effective thermal conductivity tenor is

$$\mathbf{K}_{\text{eff}} = \mathbf{K}_{\beta\beta} + 2\mathbf{K}_{\beta\sigma} + \mathbf{K}_{\sigma\sigma}. \tag{33.45}$$

The choice between the one-equation model and the two-equation model has been well discussed by Quintard and Whitaker (1993) and Whitaker (1999). They have also developed methods of determining the effective thermal conductivity tensor $\mathbf{K}_{\text{eff}}$ in the one-equation model and the four coefficients $\mathbf{K}_{\beta\beta}$, $\mathbf{K}_{\beta\sigma} = \mathbf{K}_{\sigma\beta}$, $\mathbf{K}_{\sigma\sigma}$, and ha_ν in the two-equation model. Their studies suggest that the coupling coefficients are on the order of the smaller of $\mathbf{K}_{\beta\beta}$ and $\mathbf{K}_{\sigma\sigma}$. Therefore, the coupled conductive terms should not be omitted in any detailed two-equation model of heat-conduction processes. When the principle of local thermal equilibrium is not valid, the commonly-used two-equation model in the literature is the one without the coupled conductive terms (Glatzmaier and Ramirez 1988):

$$\varepsilon_\beta (\rho c)_\beta \frac{\partial \langle T_\beta \rangle^\beta}{\partial t} = \nabla \cdot \left(\mathbf{K}_{\beta\beta} \cdot \nabla \langle T_\beta \rangle^\beta \right) + ha_\nu \left(\langle T_\sigma \rangle^\sigma - \langle T_\beta \rangle^\beta \right), \tag{33.46}$$

and

$$\varepsilon_\sigma (\rho c)_\sigma \frac{\partial \langle T_\sigma \rangle^\sigma}{\partial t} = \nabla \cdot \left(\mathbf{K}_{\sigma\sigma} \cdot \nabla \langle T_\sigma \rangle^\sigma \right) - ha_\nu \left(\langle T_\sigma \rangle^\sigma - \langle T_\beta \rangle^\beta \right). \tag{33.47}$$

On the basis of the above analysis, we now know that the coupled conductive terms $\mathbf{K}_{\beta\sigma} \cdot \nabla \langle T_\sigma \rangle^\sigma$ and $\mathbf{K}_{\sigma\beta} \cdot \nabla \langle T_\beta \rangle^\beta$ cannot be discarded in the exact representation of the two-equation model. However, we could argue that Equations 33.46 and 33.47 represent a reasonable approximation of Equations 33.37 and 33.38 for a heat-conduction process in which $\nabla \langle T_\beta \rangle^\beta$ and $\nabla \langle T_\sigma \rangle^\sigma$ are

sufficiently close to each other. Under these circumstances $\mathbf{K}_{\beta\beta}$ in Equation 33.46 would be given by $\mathbf{K}_{\beta\beta} + \mathbf{K}_{\beta\sigma}$ while $\mathbf{K}_{\sigma\sigma}$ in Equation 33.47 should be interpreted as $\mathbf{K}_{\sigma\beta} + \mathbf{k}_{\sigma\sigma}$. This limitation of Equations 33.46 and 33.47 is believed to be the reason behind the paradox of heat conduction in porous-media subject to lack of local thermal equilibrium analyzed by Vadasz et al. (2005). For an isotropic system with constant physical properties of the two phases, Equations 33.46 and 33.47 reduce to the traditional formulation of heat conduction in two-phase systems (Bejan 2004, Bejan et al. 2004, Nield and Bejan 2006, Vadasz 2005a):

$$\gamma_\beta \frac{\partial \langle T_\beta \rangle^\beta}{\partial t} = k_{e\beta} \Delta \langle T_\beta \rangle^\beta + ha_\nu \left(\langle T_\sigma \rangle^\sigma - \langle T_\beta \rangle^\beta \right), \tag{33.48}$$

and

$$\gamma_\sigma \frac{\partial \langle T_\sigma \rangle^\sigma}{\partial t} = k_{e\sigma} \Delta \langle T_\sigma \rangle^\beta - ha_\nu \left(\langle T_\sigma \rangle^\sigma - \langle T_\beta \rangle^\beta \right), \tag{33.49}$$

where we introduce the *equivalent* effective thermal conductivities $k_{e\beta} = k_{\beta\beta} + k_{\beta\sigma}$ and $k_{e\sigma} = k_{\sigma\sigma} + k_{\sigma\beta}$ for the β- and σ-phases, respectively, to take the above note into account. To describe the thermal energy exchange between solid and gas phases in casting sand, Tzou (1997) has also directly postulated Equations 33.48 and 33.49 (using k_β and k_σ rather than $k_{e\beta}$ and $k_{e\sigma}$) as a two-step model, parallel to the two-step equations in the microscopic phonon–electron interaction model (Anisimòv et al. 1974, Kaganov et al. 1957, Qiu and Tien 1993).

Rewrite Equations 33.39 and 33.40 in their operator form

$$\begin{bmatrix} \gamma_\beta \dfrac{\partial}{\partial t} - k_{\beta\beta} \Delta + h & -k_{\beta\sigma} \Delta - ha_\nu \\[2ex] -k_{\beta\sigma} \Delta - ha_\nu & \gamma_\sigma \dfrac{\partial}{\partial t} - k_{\sigma\sigma} \Delta + ha_\nu \end{bmatrix} \begin{bmatrix} \langle T_\beta \rangle^\beta \\[2ex] \langle T_\sigma \rangle^\sigma \end{bmatrix} = 0. \tag{33.50}$$

We then obtain an uncoupled form by evaluating the operator determinant such that

$$\left[\left(\gamma_\beta \frac{\partial}{\partial t} - k_{\beta\beta} \Delta + ha_\nu \right) \left(\gamma_\sigma \frac{\partial}{\partial t} - k_{\sigma\sigma} \Delta + ha_\nu \right) \right.$$
$$\left. - (k_{\beta\sigma} \Delta - ha_\nu)^2 \right] \langle T_i \rangle^i = 0, \tag{33.51}$$

where the index i can take β or σ. Its explicit form reads, after dividing by $ha_\nu (\gamma_\beta + \gamma_\sigma)$

$$\frac{\partial \langle T_i \rangle^i}{\partial t} + \tau_q \frac{\partial^2 \langle T_i \rangle^i}{\partial t^2} = \alpha \Delta \langle T_i \rangle^i + \alpha \tau_T \frac{\partial}{\partial t} \left(\Delta \langle T_i \rangle^i \right)$$
$$+ \frac{\alpha}{k} \left[F(\mathbf{r}, t) + \tau_q \frac{\partial F(\mathbf{r}, t)}{\partial t} \right], \tag{33.52}$$

where

$$\tau_q = \frac{\gamma_\beta \gamma_\sigma}{ha_\upsilon(\gamma_\beta + \gamma_\sigma)}, \quad \tau_T = \frac{\gamma_\beta k_{\sigma\sigma} + \gamma_a k_{\beta\beta}}{ha_\upsilon(k_{\beta\beta} + k_{\sigma\sigma} + 2k_{\beta\sigma})},$$

$$k = k_{\beta\beta} + k_{\sigma\sigma} + 2k_{\beta\sigma}, \quad \alpha = \frac{k_{\beta\beta} + k_{\sigma\sigma} + 2k_{\beta\sigma}}{\gamma_\beta + \gamma_\sigma}, \quad (33.53)$$

$$F(\mathbf{r},t) + \tau_q \frac{\partial F(\mathbf{r},t)}{\partial t} = \frac{k_{\beta\sigma}^2 - k_{\beta\beta}k_{\sigma\sigma}}{ha_\upsilon}\Delta^2 \langle T_i \rangle^i.$$

This is the dual-phase-lagging heat-conduction equation with τ_q and τ_T as the phase lags of the heat flux and the temperature gradient, respectively (Tzou 1997, Wang et al. 2008b). Here, $F(\mathbf{r},t)$ is the volumetric heat source. k, ρc, and α are the effective thermal conductivity, capacity, and diffusivity of nanofluids, respectively. Therefore, the presence of nanoparticles shifts the Fourier heat conduction in the base fluid into the dual-phase-lagging heat conduction in nanofluids at the macroscale. This is significant because all results regarding dual-phase-lagging heat conduction can, thus, be applied to study heat conduction in nanofluids.

The presence of nanoparticles gives rise to variations of thermal capacity, conductivity, and diffusivity, which are given by, in terms of ratios over those of the base fluid,

$$\frac{\rho c}{(\rho c)_\beta} = (1 - \varphi) + \varphi \frac{(\rho c)_\sigma}{(\rho c)_\beta}, \quad (33.54)$$

$$\frac{k}{k_\beta} = \frac{k_{\beta\beta} + k_{\sigma\sigma} + 2k_{\beta\sigma}}{k_\beta} \quad (33.55)$$

$$\frac{\alpha}{\alpha_\beta} = \frac{k}{k_\beta}\frac{(\rho c)_\beta}{\rho c}. \quad (33.56)$$

Therefore, $\rho c/(\rho c)_\beta$ depends *only* on the volume fraction of nanoparticles and the nanoparticle–fluid capacity ratio. However, both k/k_β and α/α_β are affected by the geometry, property, and dynamic process of nanoparticle–fluid interfaces. This dependency causes the most difficulty because it is the least precisely known feature of a nanofluid. The future research effort should thus focus on $(k_{\beta\beta} + k_{\sigma\sigma} + 2k_{\beta\sigma})/k_\beta$ to develop predicting models of thermophysical properties for nanofluids.

To show the possibility of conductivity enhancement, consider

$$\frac{\tau_T}{\tau_q} = 1 + \frac{\gamma_\beta^2 k_{\sigma\sigma} + \gamma_\sigma^2 k_{\beta\beta} - 2\gamma_\beta\gamma_\sigma k_{\beta\sigma}}{\gamma_\beta\gamma_\sigma(k_{\beta\beta} + k_{\sigma\sigma} + 2k_{\beta\sigma})}. \quad (33.57)$$

It can be large, equal, or smaller than 1 depending on the sign of $\gamma_\beta^2 k_{\sigma\sigma} + \gamma_\sigma^2 k_{\beta\beta} - 2\gamma_\beta\gamma_\sigma k_{\beta\sigma}$. Therefore, by the condition for the existence of thermal waves that requires $\tau_T/\tau_q < 1$ (Wang et al. 2008b, Xu and Wang 2002), we may have thermal waves in nanofluid heat conduction when

$$\gamma_\beta^2 k_{\sigma\sigma} + \gamma_\sigma^2 k_{\beta\beta} - 2\gamma_\beta\gamma_\sigma k_{\beta\sigma} < 0. \quad (33.58)$$

Note also that for heat conduction in nanofluids, there is a time-dependent source term

$$\frac{k_{\beta\sigma}^2 - k_{\beta\beta}k_{\sigma\sigma}}{ha_\upsilon}\Delta^2 \langle T_i \rangle^i$$

in the dual-phase-lagging heat conduction (Equation 33.52). Therefore, the resonance can also occur. These thermal waves and possibly resonance are believed to be the driving force for the conductivity enhancement. When $k_{\beta\sigma} = 0$ so that τ_T/τ_q is always larger than 1, thermal waves and resonance would not appear. The coupled conductive terms in Equations 33.39 and 33.40 are thus responsible for thermal waves and resonance in nanofluid heat conduction. It is also interesting to note that although each τ_q and τ_T is ha_υ-dependent, the ratio τ_T/τ_q is not. Therefore, the evaluation of τ_T/τ_q will be much simpler than τ_q or τ_T.

Lee et al. (1999) found that addition of 4% of Al_2O_3 particles increased thermal conductivity by a factor of 8%, while according to Eastman et al. (2004), CuO particles at the same volume fraction enhance the conductivity by about 12%. This is interesting because conductivity of CuO is less than that of Al_2O_3. The thermal wave theory can explain this since the conductivity enhancement k/k_β equals $(k_{\beta\beta} + k_{\sigma\sigma} + 2k_{\beta\sigma}/k_\beta)$ (Equation 33.55), which are strongly affected by nanofluids microstructures and interfacial properties/processes of nanoparticle–fluid interfaces.

33.4 Extension: Thermal-Wave Fluids

The present analysis of heat-conduction nature in the last section is not limited to nanofluids heat conduction, but valid for heat conduction in all two-phase systems. It can also be extended to heat conduction in a system involving more than two phases. Therefore, all multiphase fluids are candidates of *thermal-wave fluids in which heat conduction can support thermal waves and possible resonance*. The presence of such waves and resonance will enhance heat-conduction processes and, consequently, thermal conductivity significantly.

Thermal waves have been observed in casting sand experiments by two independent groups (Tzou 1997). Substantial increases in thermal conductivity have also been confirmed experimentally for the porous-media fluids (Aichlmayr and Kulacki 2006) and nanofluids (Das et al. 2008). However, the reported data of effective thermal conductivity are all in between those of two phases so that the k_β-enhancement appears only at $k_\sigma > k_\beta$. On the other hand, our theory shows that the k_β-enhancement can occur for all cases with $k_{\beta\beta} + k_{\sigma\sigma} + 2k_{\beta\sigma} > k_\beta$ (Equation 33.53). Therefore, it is possible to have some thermal-wave fluids that can support very strong thermal waves and resonance such that their conductivities are higher than those of two phases. We report here one of such thermal-wave fluids.

Our thermal-wave fluid is formed by emulsifying corn oil into distilled water with a small amount of Cetyl trimethyl Ammonium Bromide under ultrasonic disruption (Ultrasonic

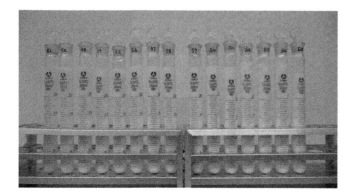

FIGURE 33.2 Oil/water emulsion (oil volume fraction from 0.5 to 14 vol%).

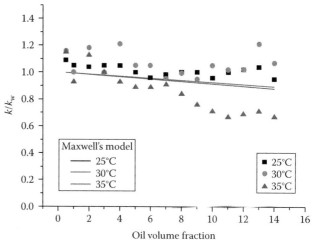

FIGURE 33.3 Variation of k/k_w with oil volume fraction and emulsion temperature (k: emulsion thermal conductivity; k_w: water thermal conductivity).

Cell Processor, Haishukesheng Ultrasonic Equipment Ltd). Loadings of corn oil droplets from 0.5 to 14 vol% are synthesized and tested. Figure 33.2 shows the picture of the synthesized thermal-wave fluid with 15 values of oil volume fractions 3 months after its preparation. The fluid is very stable, and no bulk phase separation has been observed. Note that microemulsions are generally thermodynamically stable; their free energy is even lower than that in the unmixed system (Hoar and Schulman 1943, Kumar and Mittal 1999). Furthermore, the microemulsions are also freeze/thaw recoverable. The average diameter of oil droplets are measured by dynamic light scattering system (Delsa Nano C, Beckman Coulter, United States) and listed in Table 33.3 for all 15 samples.

The conductivity ratio k/k_w measured by the standard transient hot-wire method (KD2, Therm Test Inc., Canada, see Wei et al. 2009 for the details of KD2 system) is shown in Figure 33.3 as a function of oil volume fraction and fluid temperature. Here k and k_w are the thermal conductivity of the thermal-wave

fluid and the water, respectively. The prediction by the Maxwell model is also plotted in Figure 33.3 for comparison (Das et al. 2008, Maxwell 1904). Remarkably, an extraordinary conductivity enhancement—up to a 21% increase at the oil volume fraction of 4% and 13% and the fluid temperature of 30°C—is obtained in the fluid after adding some oil with lower thermal conductivity. For most tested cases, an increase (rather than decrease) in thermal conductivity is achieved.

The oil/water emulsion conductivity predicted by using the Maxwell model shows a linear decrease with the increase of oil volume fraction and a negligible effect of emulsion temperature (Figure 33.3). The measured conductivity shows a strong sensitivity and a high nonlinearity to both the oil volume fraction and the temperature and is consistent with the theory of thermal waves and resonance (Tzou 1997, Wang et al. 2008b, Xu and Wang 2002).

33.5 Concluding Remarks

In an attempt to determine how the presence of nanoparticles affects the heat conduction at the macroscale and isolate the mechanism responsible for the significant enhancement of thermal conductivity, we have rigorously developed a macroscale heat-conduction model in nanofluids. The model was obtained by scaling up the microscale model for the heat conduction in the nanoparticles and in the base fluids. The approach for scaling-up is the volume averaging with help of multiscale theorems. The result shows that the presence of nanoparticles leads to a dual-phase-lagging heat conduction in nanofluids at the macroscale with a potential of much higher thermal conductivity. Therefore, the presence of nanoparticles shifts the Fourier heat conduction in the base fluid into the dual-phase-lagging heat conduction in nanofluids at the macroscale. This finding is significant because all results regarding dual-phase-lagging heat conduction in the literature can thus be applied to study heat conduction in

TABLE 33.3 Average Diameter of Oil Droplets Measured by Dynamic Light Scattering System

Oil Volume Fraction (%)	Average Diameter (nm)
0.5	189.0
1	180.3
2	183.3
3	176.9
4	196.0
5	173.2
6	188.3
7	146.0
8	140.2
9	146.1
10	179.3
11	173.9
12	177.1
13	169.3
14	193.7

nanofluids. This finding also raises the question of reliability for existing thermal conductivity data that were obtained based on the *hypothesized* Fourier heat conduction at the macroscale.

The dual-phase-lagging heat conduction differs from the Fourier heat conduction mainly on the existence of thermal waves and possible resonance. Such waves and resonance come from the coupled conduction of the nanoparticles and the base fluids and are responsible for the extraordinary conductivity enhancement. To confirm this experimentally, we have recently synthesized one novel kind of nanofluids by mixing the base liquid with some nanoelements with a *lower* conductivity than the base liquid: oil-in-water emulsion. This type of nanofluids can support much stronger thermal waves and resonance than all reported nanofluids, and consequently have an extraordinary water conductivity enhancement (up to 21%). Therefore, the nanofluids' conductivity enhancement in these nanofluids comes *mainly* from the thermal wave and resonance instead of the higher conductivity value of particles.

The dual-phase-lagging heat-conduction equation originates from the first law of thermodynamics and the dual-phase-lagging constitutive relation of heat flux density. It was developed in examining energy transport involving the high-rate heating in which the nonequilibrium thermodynamic transition and the microstructural effect become important associated with shortening of the response time. In addition to its application in the ultrafast pulse-laser heating, the dual-phase-lagging heat-conduction equation also arises in describing and predicting phenomena such as propagating of temperature pulses in superfluid liquid helium, nonhomogeneous lagging response in porous media, thermal lagging in amorphous materials, and effects of material defects and thermomechanical coupling. Furthermore, the dual-phase-lagging heat-conduction equation forms a generalized, unified equation with the classical parabolic heat-conduction equation, the hyperbolic heat-conduction equation, the energy equation in the phonon scattering model, and the energy equation in the phonon–electron interaction model as its special cases. This, with the rapid growth of microscale heat conduction of high-rate heat flux, has attracted the recent research effort on dual-phase-lagging heat conduction: its physical basis and experimental verification, well-posedness, solution structure, analytical and numerical solutions, methods of measuring thermal relaxation times, thermal oscillation and resonance, and equivalence with and application in two-phase-system heat conduction.

The dual-phase-lagging heat conduction has been shown to be admissible by the second law of the extended irreversible thermodynamics and by the Boltzmann transport equation. It is also proven to be well posed in a finite region of n-dimension ($n \geq 1$) under any linear boundary conditions including Dirichlet, Neumann, and Robin types. The solution structure theorems have been developed as well for both mixed and Cauchy problems of dual-phase-lagging heat-conduction equations. These theorems inter-relate contributions (to the temperature field) of the initial temperature distribution, the source term and the initial time-rate change of the temperature, uncover the structure of temperature field, and considerably simplify the development of solutions. The thermal oscillation and resonance in the dual-phase-lagging heat conduction have been examined in details. Conditions and features of underdamped, critically damped and overdamped oscillations have been obtained and compared with those in the classical parabolic heat conduction and the hyperbolic heat conduction. The condition for the thermal resonance is also available in the literature.

The macroscale theory of nanofluids heat conduction also generalizes nanofluids into thermal-wave fluids and leads to the experiment of extraordinary fluid conductivity enhancement by adding some fluid even with lower conductivity. Such new thermal-wave fluids also have long-term stability and can be produced in large quantities. Therefore, they can improve fluid conductivity and convective heat transfer more effectively than recently proposed nanofluids.

The future research effort should focus on (1) methods of determining the cross-effective thermal conductivity $k_{\beta\sigma}$, (2) $k_{\beta\sigma}$ correlation with nanofluids microstructures and interfacial properties/processes of nanoparticle–fluid interfaces, and (3) precise features of thermal waves and resonance in nanofluids and thermal-wave fluids.

Acknowledgment

The financial support from the Research Grants Council of Hong Kong (GRF718009 and GRF) is gratefully acknowledged.

References

Aichlmayr, H. T. and Kulacki, F. A. 2006. The effective thermal conductivity of saturated porous media. *Adv. Heat Transfer* 39: 377–460.

Anisimòv, S. I., Kapeliovich, B. L., and Perelman, T. L. 1974. Electron emission from metal surfaces exposed to ultrashort laser pulses. *Sov. Phys. JETP* 39: 375–377.

Antaki, P. J. 1998. Solution for non-Fourier dual phase lag heat conduction in a semi-infinite slab with surface heat flux. *Int. J. Heat Mass Transfer* 41: 2253–2258.

Assael, M. J., Chen, C. F., Metaxa, I., and Wakeham, W. A. 2004. Thermal conductivity of suspensions of carbon nanotubes in water. *Int. J. Thermophys.* 25: 971–985.

Auriault, J. L. 1991. Heterogeneous medium: Is an equivalent macroscopic description possible? *Int. J. Eng. Sci.* 29: 785–795.

Beckert, H. H. K. 2000. Experimental evidence about the controversy concerning Fourier or non-Fourier heat conduction in materials with a nonhomogeneous inner structure. *Heat Mass Transfer* 36: 387–392.

Bejan, A. 2004. *Convection Heat Transfer* (3rd edn.). New York: Wiley.

Bejan, A., Dincer, I., Lorente, A., Miguel, A. F., and Reis, A. H. 2004. *Porous and Complex Flow Structures in Modern Technologies*. New York: Springer.

Bhattacharya, P., Saha, S. K., Yadav, A., Phelan, P. E., and Prasher, R. S. 2004. Brownian dynamics simulation to determine the effect thermal conductivity of nanofluids. *J. Appl. Phys.* 95: 6492–6494.

Cattaneo, C. 1958. A form of heat conduction equation which eliminates the paradox of instantaneous propagation. *Comput. Rendus* 247: 431–433.

Cengel, Y. A. and Boles, M. A. 2006. *Thermodynamics: An Engineering Approach* (5th edn.). Boston, MA: McGraw-Hill.

Chandrasekaraiah, D. S. 1986. Thermoelasticity with second sound: A review. *Appl. Mech. Rev.* 39: 355–376.

Chandrasekharaiah, D. S. 1998. Hyperbolic thermoelasticity: A review of recent literature. *Appl. Mech. Rev.* 51: 705–729.

Cheng, L., Xu, M. T., and Wang, L. Q. 2008a. From Boltzmann transport equation to single-phase-lagging heat conduction. *Int. J. Heat Mass Transfer* 51: 6018–6023.

Cheng, L., Xu, M. T., and Wang, L. Q. 2008b. Single- and dual-phase-lagging heat conduction models in moving media. *J. Heat Transfer* 130: 121302/1–121302/6.

Chester, M. 1966. High frequency thermometry. *Phys. Rev.* 145: 76–80.

Choi, S. U. S. 1998. Nanofluid Technology: Current Status and Future Research. Korea-US Technical Conference on Strategic Technologies, Vienna, Austria, pp. 22–24.

Choi, S. U. S., Zhang, Z. G., and Keblinski, P. 2004. Nanofluids. In *Encyclopedia of Nanoscience and Nanotechnology*, ed. H. S. Nalwa, pp. 757–773. New York: American Scientific Publishers.

Dai, W. Z. and Nassar, R. 1999. A finite difference scheme for solving the heat transport equation at the microscale. *Numer. Methods Partial Diff. Eqs.* 15: 697–708.

Das, S. K., Choi, S. U. S., Yu, W. H., and Pradeep, T. 2008. *Nanofluids: Science and Technology*. Hoboken, NJ: John Wiley & Sons.

Eapen, J., Williams, W. C., Buongiorno, J., Hu, L. W., and Yip, S. 2007. Mean-field versus microconvection effects in nanofluid thermal conduction. *Phys. Rev. Lett.* 99: 095901.

Eastman, J. A., Choi, S. U. S., Li, S., Soyez, G., Thompson, L. J., and DiMelfi, R. J. 1998. Novel thermal properties of nanostructured materials. *International Symposium on Metastable Mechanically Alloyed and Nanocrystalling Materials*, Wollongong, Australia, pp. 7–12.

Eastman, J. A., Phillpot, S. R., Choi, S. U. S., and Keblinski, P. 2004. Thermal transport in nanofluids. *Annu. Rev. Mater. Res.* 34: 219–246.

Glatzmaier, G. C. and Ramirez, W. F. 1988. Use of volume averaging for the modeling of thermal properties of porous materials. *Chem. Eng. Sci.* 43: 3157–3169.

Guyer, R. A. and Krumhansi, J. A. 1966. Solution of the linearized Boltzmann equation. *Phys. Rev.* 148: 766–778.

Hoar, T. P. and Schulman, J. H. 1943. Transparent water-in-oil dispersions: The oleopathic hydro-micelle. *Nature* 152: 102–103.

Hong, T., Yang, H., and Choi, C. J. 2005. Study of enhanced thermal conductivity of Fe nanofluids. *J. Appl. Phys.* 97: 064311/1–064311/4.

Jang, S. P. and Choi, S. U. S. 2004. Role of Brownian motion in the enhanced thermal conductivity of nanofluids. *Appl. Phys. Lett.* 84: 4316–4318.

Joseph, D. D. and Preziosi, L. 1989. Heat waves. *Rev. Mod. Phys.* 61: 41–73.

Joseph, D. D. and Preziosi, L. 1990. Addendum to the paper heat waves. *Rev. Mod. Phys.* 62: 375–391.

Kaganov, M. I., Lifshitz, I. M., and Tanatarov, M. V. 1957. Relaxation between electrons and crystalline lattices. *Sov. Phys. JETP* 4: 173–178.

Kaminski, W. 1990. Hyperbolic heat conduction equation for materials with a nonhomogeneous inner structure. *J. Heat Transfer* 112: 555–560.

Koo, J. and Kleinstreuer, C. 2004. A new thermal conductivity model for nanofluids. *J. Nanopart. Res.* 6: 577–588.

Kumar, P. and Mittal, K. 1999. *Handbook of Microemulsion Science and Technology*. Boca Raton, FL: CRC Press.

Lee, S., Choi, S. U. S., Li, S., and Eastman, J. A. 1999. Measuring thermal conductivity of fluids containing oxide nanoparticles. *J. Heat Transfer* 121: 280–289.

Leong, K. C., Yang, C., and Murshed, S. M. S. 2006. A model for the thermal conductivity of nanofluids: The effect of interfacial layer. *J. Nanopart. Res.* 8: 245–254.

Lin, C. K., Hwang, C. C., and Chang, Y. P. 1997. The unsteady solutions of a unified heat conduction equation. *Int. J. Heat Mass Transfer* 40: 1716–1719.

Liu, M., Lin, M. C., Tsai, C. Y., and Wang, C. C. 2006. Enhancement of thermal conductivity with *Cu* for nanofluids using chemical reduction method. *Int. J. Heat Mass Transfer* 49: 3028–3033.

Maxwell, J. C. 1904. *A Treatise on Electricity and Magnetism*. Cambridge, U.K.: Oxford University Press.

Mengi, Y. and Turhan, D. 1978. The influence of retardation time of the heat flux on pulse propagation. *J. Appl. Mech.* 45: 433–435.

Mitra, K., Kumar, S., Vedavarz, A., and Moallemi, M. K. 1995. Experimental evidence of hyperbolic heat conduction in processed meat. *J. Heat Transfer* 117: 568–573.

Murshed, S. M. S., Leong, K. C., and Yang, C. 2005. Enhanced thermal conductivity of TiO_2-water based nanofluids. *Int. J. Therm. Sci.* 44: 367–373.

Nield, D. A. and Bejan, A. 2006. *Convection in Porous Media* (3rd edn.). New York: Springer.

Peters, A. G. F. 1999. Experimental investigation of heat conduction in wet sand. *Heat Mass Transfer* 35: 289–294.

Peterson, G. P. and Li, C. H. 2006. Heat and mass transfer in fluids with nanoparticle suspensions. *Adv. Heat Transfer* 39: 257–376.

Phelan, P. E., Bhattacharya, P., and Prasher, R. S. 2005. Nanofluids for heat transfer applications. *Annu. Rev. Heat Transfer* 14: 255–275.

Prasher, R., Bhattacharya, P., and Phelan, P. E. 2005. Thermal conductivity of nanoscale colloidal solutions (nanofluids). *Phys. Rev. Lett.* 94: 025901.

Prasher, R., Bhattacharya, P., and Phelan, P. E. 2006a. Brownian-motion-based convective-conductive model for the effective thermal conductivity of nanofluids. *J. Heat Transfer* 128: 588–595.

Prasher, R., Phelan, P. E., and Bhattacharya, P. 2006b. Effect of aggregation kinetics on the thermal conductivity of nanoscale colloidal solutions (nanofluid). *Nano Lett.* 6: 1529–1534.

Putnam, S. A., Cahill, D. G., Braun, P. V., Ge, Z. B., and Shimmin, R. G. 2006. Thermal conductivity of nanoparticle suspensions. *J. Appl. Phys.* 99: 084308.

Qiu, T. Q. and Tien, C. L. 1993. Heat transfer mechanisms during short-pulse laser heating of metals. *J. Heat Transfer* 115: 835–841.

Quintard, M. and Whitaker, S. 1993. One- and two-equation models for transient diffusion processes in two-phase systems. *Adv. Heat Transfer* 23: 369–464.

Ren, Y., Xie, H., and Cai, A. 2005. Effective thermal conductivity of nanofluids containing spherical nanoparticles. *J. Phys. D* 38: 3958–3961.

Roetzel, W., Putra, N., and Das, S. K. 2003. Experiment and analysis for non-Fourier conduction in materials with non-homogeneous inner structure. *Int. J. Thermal Sci.* 42: 541–552.

Rusconi, R., Rodari, E., and Piazza, R. 2006. Optical measurements of the thermal properties of nanofluids. *Appl. Phys. Lett.* 89: 261916.

Šilhavý, M. 1985. The existence of the flux vector and the divergence theorem for general Cauchy fluxes. *Arch. Rational Mech. Anal.* 90: 195–212.

Sobhan, C. B. and Peterson, G. P. 2008. *Microscale and Nanoscale Heat Transfer: Fundamentals and Engineering Applications*. Boca Raton, FL: CRC Press.

Tang, D. W. and Araki, N. 1999. Wavy, wavelike, diffusive thermal responses of finite rigid slabs to high-speed heating of laser-pulses. *Int. J. Heat Mass Transfer* 42: 855–860.

Tzou, D. Y. 1992. Thermal shock phenomena under high-rate response in solids. *Annu. Rev. Heat Transfer* 4: 111–185.

Tzou, D. Y. 1995a. A unified field approach for heat conduction from micro- to macro-scales. *J. Heat Transfer* 117: 8–16.

Tzou, D. Y. 1995b. The generalized lagging response in small-scale and high-rate heating. *Int. J. Heat Mass Transfer* 38: 3231–3240.

Tzou, D. Y. 1997. *Macro-to Microscale Heat Transfer: The Lagging Behavior*. Washington, DC: Taylor & Francis.

Tzou, D. Y. and Chiu, K. S. 2001. Temperature-dependent thermal lagging in ultrafast laser heating. *Int. J. Heat Mass Transfer* 44: 1725–1734.

Tzou, D. Y. and Zhang, Y. S. 1995. An analytical study on the fast-transient process in small scales. *Int. J. Eng. Sci.* 33: 1449–1463.

Vadasz, J. J., Govender, S., and Vadasz, P. 2005. Heat transfer enhancement in nano-fluids suspensions: Possible mechanisms and explanations. *Int. J. Heat Mass Transfer* 48: 2673–2683.

Vadasz, P. 2005a. Absence of oscillations and resonance in porous media dual-phase- lagging Fourier heat conduction. *J. Heat Transfer Trans. ASME* 127: 307–314.

Vadasz, P. 2005b. Explicit conditions for local thermal equilibrium in porous media heat conduction. *Transport Porous Media* 59: 341–355.

Vadasz, P. 2005c. Lack of oscillations in dual-phase-lagging heat conduction for a porous slab subject to imposed heat flux and temperature. *Int. J. Heat Mass Transfer* 48: 2822–2828.

Vadasz, P. 2006a. Exclusion of oscillations in heterogeneous and bi-composite media thermal conduction. *Int. J. Heat Mass Transfer* 49: 4886–4892.

Vadasz, P. 2006b. Heat conduction in nanofluid suspensions. *J. Heat Transfer* 128: 465–477.

Vedavarz, A., Mitra, K., Kumar, S., and Moallemi, M. K. 1992. Effect of hyperbolic heat conduction on temperature distribution in laser irradiated tissue with blood perfusion. *Adv. Bio. Heat Mass Transfer ASME HTD* 231: 7–16.

Vernotte, P. 1958. Les paradoxes de la théorie continue de l'equation de la chaleur. *Comput. Rendus* 246: 3154–3155.

Vernotte, P. 1961. Some possible complications in the phenomena of thermal conduction. *Comput. Rendus* 252: 2190–2191.

Wang, B. X., Zhou, L. P., and Peng, X. F. 2003. A fractal model for predicting the effective thermal conductivity of liquid with suspension of nanoparticles. *Int. J. Heat Mass Transfer* 46: 2665–2672.

Wang, L. Q. 1994. Generalized Fourier law. *Int. J. Heat Mass Transfer* 37: 2627–2634.

Wang, L. Q. 1995. Properties of heat flux functions and a linear theory of heat flux. *Int. J. Mod. Phys. B* 9: 1113–1122.

Wang, L. Q. 1996. A decomposition theorem of motion. *Int. J. Eng. Sci.* 34: 417–423.

Wang, L. Q. 2000a. Flows through porous media: A theoretical development at macroscale. *Transport Porous Media* 39: 1–24.

Wang, L. Q. 2000b. Solution structure of hyperbolic heat-conduction equation. *Int. J. Heat Mass Transfer* 43: 365–373.

Wang, L. Q. 2001. Further contributions on the generalized Fourier law. *Int. J. Transport Phenom.* 2: 299–305.

Wang, L. Q. and Wei, X. H. 2008. Equivalence between dual-phase-lagging and two-phase-system heat conduction processes. *Int. J. Heat Mass Transfer* 51: 1751–1756.

Wang, L. Q. and Wei, X. H. 2009a. Nanofluids: Synthesis, heat conduction and extension. *J. Heat Transfer* 131: 033102/1–033102/7.

Wang, L. Q. and Wei, X. H. 2009b. Heat conduction in nanofluids. *Chaos, Solitons Fractals* 39: 2211–2215.

Wang, L. Q. and Xu, M. T. 2002. Well-posedness of dual-phase-lagging heat conduction equation: higher dimensions. *Int. J. Heat Mass Transfer* 45: 1165–1171.

Wang, L. Q. and Zhou, X. S. 2000. *Dual-Phase-Lagging Heat-Conduction Equations*. Jinan, People's Republic of China: Shandong University Press.

Wang, L. Q. and Zhou, X. S. 2001. *Dual-Phase-Lagging Heat-Conduction Equations: Problems and Solutions*. Jinan, People's Republic of China: Shandong University Press.

Wang, L. Q., Xu, M. T., and Zhou, X. S. 2001. Well-posedness and solution structure of dual-phase- lagging heat conduction. *Int. J. Heat Mass Transfer* 44: 1659–1669.

Wang, L. Q., Xu, M. T., and Wei, X. H. 2008a. Multiscale theorems. *Adv. Chem. Eng.* 34: 175–468.

Wang, L. Q., Zhou, X. S., and Wei, X. H. 2008b. *Heat Conduction: Mathematical Models and Analytical Solutions*. Heidelberg, Germany: Springer-Verlag.

Wei, X. H., Zhu, H. T., and Wang, L. Q. 2009. CePO$_4$ nanofluids: Synthesis and thermal conductivity. *J. Thermophys. Heat Transfer* 23: 219–222.

Whitaker, S. 1999. *The Method of Volume Averaging*. Dordrecht, the Netherlands: Kluwer Academic.

Wu, D. X., Zhu, H. T., Wang, L. Q., and Liu, L. M. 2009. Critical issues in nanofluids preparation, characterization and thermal conductivity. *Curr. Nanosci.* 5: 103–112.

Xie, H., Fujii, M., and Zhang, X. 2005. Effect of interfacial nanolayer on the effective thermal conductivity of nanoparticle-fluid mixture. *Int. J. Heat Mass Transfer* 48: 2926–2932.

Xu, M. T. and Wang, L. Q. 2002. Thermal oscillation and resonance in dual-phase-lagging heat conduction. *Int. J. Heat Mass Transfer* 45: 1055–1061.

Xu, M. T. and Wang, L. Q. 2005. Dual-phase-lagging heat conduction based on Boltzmann transport equation. *Int. J. Heat Mass Transfer* 48: 5616–5624.

Xuan, Y. M. and Li, Q. 2000. Heat transfer enhancement of nanofluids. *Int. J. Heat Fluid Flow* 21: 58–64.

Xuan, Y. M., Li, Q., Zhang, X., and Hu, W. 2003. Aggregation structure and thermal conductivity of nanofluids. *AICHE J.* 49: 1038–1043.

Xue, L., Keblinski, P., Phillpot, S. R., Choi, S. U. S., and Eastman, J. A. 2004. Effect of liquid layering at the liquid-solid interface on thermal transport. *Int. J. Heat Mass Transfer* 47: 4277–4284.

Yu, W. H. and Choi, S. U. S. 2003. The role of interfacial layers in the enhanced thermal conductivity of nanofluids: A renovated Maxwell model. *J. Nanopart. Res.* 5: 167–171.

Yu, W. H. and Choi, S. U. S. 2004. The role of interfacial layers in the enhanced thermal conductivity of nanofluids: A renovated Hamilton-Crosser model. *J. Nanopart. Res.* 6: 355–361.

34

Nanofluids for Heat Transfer

Sanjeeva Witharana
University of Leeds

Haisheng Chen
University of Leeds

Yulong Ding
University of Leeds

34.1 Introduction to Nanofluids

Nanofluids are dilute liquid suspensions containing particles or particle assemblies that have at least one dimension smaller than 100 nm. Hence, a nanofluid consists of a base liquid and large numbers of tiny particles dispersed in the base liquid, as illustrated in Figure 34.1a. The base liquids can be water, ethylene glycol, mineral oil, refrigerant, or even mixtures of two or more liquids. The particles can be made of metal, metal oxide, carbon, carbide, and nitride. They can take spherical, rodlike, or tubular shapes, as shown in Figure 34.1b, that can be dispersed individually or in the form of aggregates (several individual particles stuck together) or in an entangled form (for long tubes or fibers), as illustrated in Figure 34.1c. Nanofluids can be transparent (Figure 34.2a), semitransparent (Figure 34.2b), or opaque depending on the properties and concentration of the dispersed particles. Nanofluids may contain a certain amount of surfactants or dispersants to enhance their stability.

The term "nanofluids" was first put forward by Dr. Stephen Choi (Choi 1995) although there was an earlier and independent report by Masuda et al. (1993) concerning nanoparticle suspensions. The initial stage of research on nanofluids was mainly conducted at the Argonne National Laboratories, United States, with a focus on thermal conductivity under macroscopically static conditions. The topic gained worldwide attention from the late 1990s and became a very hot topic from around 2002, as evidenced by the exponential growth in the number of publications. The popularity of the topic of nanofluids is associated

with some experimental observations of enhanced properties and behavior in heat transfer (Keblisnki et al. 2002), mass transfer (Krishnamurthy et al. 2006, Olle et al. 2006), wetting and spreading (Wasan and Nikolov 2003), and antimicrobial activities (Zhang et al. 2007a).

If their performance is established beyond doubt, nanofluids could have numerous potential applications. The enhanced thermal properties will attract small- to large-scale heating and cooling applications, from miniature electronics and automobiles to nuclear power plants (Wang and Mujumdar 2007, Ding et al. 2007a). Similarly, their antimicrobial behavior will ensure the controlling of harmful bacteria, making safer living environments (Zhang et al. 2007a).

Despite considerable research efforts and significant progress in the last few years, the fundamental understanding is still limited particularly for nanofluids under dynamic (flow) conditions as reflected by widespread scattering and disagreement in the published data and less-convincing arguments in the interpretation of data. The scope of this chapter is to provide an objective overview on (1) the transport properties of nanofluids, more specifically thermal conductivity and shear viscosity, and (2) the heat transfer of nanofluids under convective and boiling conditions. Sections 34.2 and 34.3 are devoted to the transport properties of nanofluids with Section 34.2 on thermal conductivity under macroscopically static conditions and Section 34.3 on shear viscosity. Sections 34.4 through 34.6 address the topics of forced convection, natural convection, and boiling heat transfer, respectively. Finally, Section 34.7 presents the concluding remarks.

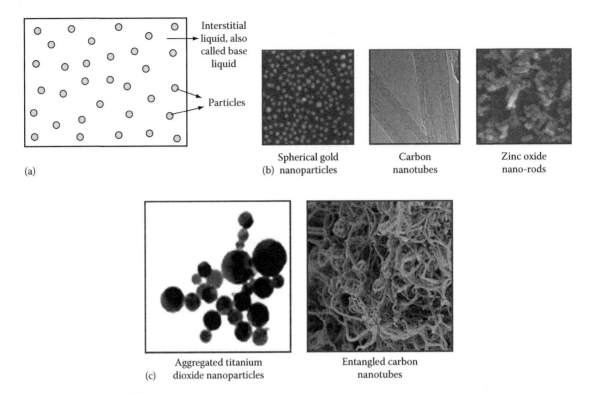

(a)

Interstitial liquid, also called base liquid

Particles

(b) Spherical gold nanoparticles

Carbon nanotubes

Zinc oxide nano-rods

(c) Aggregated titanium dioxide nanoparticles

Entangled carbon nanotubes

FIGURE 34.1 Definition of nanofluids.

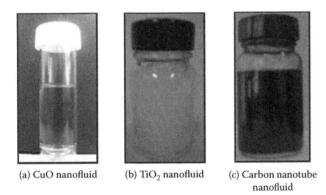

(a) CuO nanofluid   (b) TiO$_2$ nanofluid   (c) Carbon nanotube nanofluid

FIGURE 34.2 Photos of water-based nanofluid samples.

34.2 Transport Properties of Nanofluids: Thermal Conductivity

34.2.1 Concept of Thermal Conduction of Nanofluids and Measuring Techniques

34.2.1.1 Concept of Thermal Conduction of Nanofluids

Thermal conduction is one of three modes of transferring heat (thermal energy): conduction, convection, and radiation. The concept of convection is explained in Section 34.4. As the mode of radiation is insignificant in heat transfer of nanofluids at relatively low temperatures, it is not discussed. Thermal conduction takes place in solids and fluids, which are the two main constituents of nanofluids. Heat transfer by conduction is due

to three types of energy carriers: phonons, electrons, and molecules (Bird et al. 2002). A phonon is a quantized mode of vibration occurring in a rigid crystal lattice of a solid and is the main mechanism for thermal conduction in a nonmetallic material. In such a material, atoms are bound to each other by a series of bonds that behave like springs. In the presence of a temperature difference, the hot side of the material experiences more vigorous atomic movements, which are transmitted to the cooler side through the springs hence realizing the thermal energy transfer. For metals, there are free electrons. Movements and collisions of the electrons are the principal mechanism of thermal energy transfer as electrons in the hot side of the solids move faster than those on the cooler side. As electrons move much faster than phonons (the propagation of lattice vibration), conduction through electron collisions is more effective than that through lattice vibration. This is why metals generally are better heat conductors than ceramics. In fluids (liquids and gases), conduction occurs through collisions between freely moving molecules.

The effectiveness of thermal energy transfer through the conduction mode is quantified by thermal conductivity k, defined by the Fourier's law, $Q = k \cdot A \cdot (\Delta T/\Delta x)$, where $Q(W)$ is the rate of heat transferred across the cross-sectional area A (m^2) and ΔT(K) is the temperature difference between the hot and cold surfaces separated by a distance of Δx (m) as illustrated by Figure 34.3. Thermal conductivity of pure materials is regarded as a material property. Figure 34.4 illustrates the thermal conductivity of some materials at room temperature, commonly used in the formulation of nanofluids. Appreciate that the thermal conductivity of nanofluids is more accurately

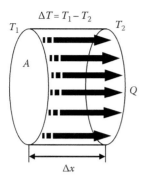

FIGURE 34.3 Definition of thermal conductivity based on Fourier's law.

called *effective thermal conductivity* as it is not a genuine material property. The reasons include the following: (1) Nanofluids are made of nanoscale solid particles and a base liquid, but both components lose their identities upon mixing and (2) Nanoparticles can have a dimension that is smaller than the

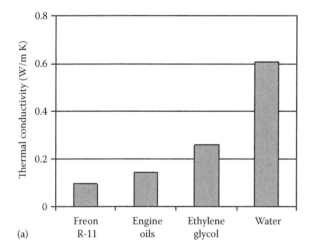

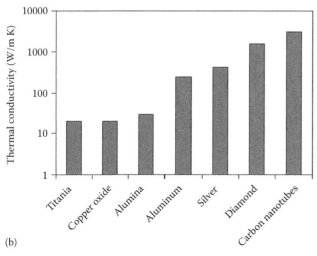

FIGURE 34.4 Thermal conductivities of some materials used in nanofluid formulations (data shown here are for 25°C; thermal conductivities of diamond and carbon nanotubes vary widely in the literature, so the values only represent order of magnitudes).

mean free path of the energy carriers, and hence the quantum effect may become important; see Section 34.2.3 for further discussion. For simplicity, however, we shall not differentiate the two terminologies in this chapter.

34.2.1.2 Measurement of Thermal Conductivity of Nanofluids

The measurement of the effective thermal conductivity is important to gain a more fundamental understanding of nanofluids and to find appropriate applications. This is reflected in the published work on the effective thermal conductivity, which, as mentioned earlier, accounts for the majority of papers in the nanofluids literature. An inspection of the literature shows that a number of techniques have been used to measure the effective thermal conductivity of nanofluids, including the conventional steady-state parallel-plate method, and transient-based hot-wire (Nagasaka and Nagashima 1981), oscillation (Czarnetzki and Roetzel 1995), and 3-ω (Cahill 1990, Yang and Han 2006) methods. In the following text, the steady-state parallel-plate and transient hot-wire methods are briefly discussed.

Steady-state parallel-plate method: This method is based on the Fourier's law explained in Section 34.2.1.1. The device for the measurement typically consists of two horizontally oriented parallel plates bounded by a well-insulated sidewall. The gap between the plates should be far smaller than the diameter of the plates. The gap is fully filled with the fluid to be measured. The upper plate is maintained at a higher temperature and the lower plate at a lower temperature. By measuring the rate of heat transfer from the upper plate to the lower plate and the temperature gradient across the fluid, one can obtain the thermal conductivity of the fluid according to Fourier's law. The reason for the use of horizontal plates and heating from the upper plate is to minimize natural convection effect, further explained in Section 34.5. The use of insulated side wall ensures one-dimensional temperature field. Measurements of thermal conductivity using the parallel-plate method are slow as one has to wait for the steady state to be established. This is likely to be the main reason why very few people have used such a method to measure the effective thermal conductivity of nanofluids.

Transient hot-wire methods: The hot-wire method is based on the measurement of temperature rise at a defined distance from a linear heat source (hot wire) embedded in a test sample (e.g., nanofluids). If the heat source is assumed to have a constant and uniform output along the length of the test sample, the thermal conductivity can be derived directly from the resulting change in the temperature over a known time interval. The hot-wire probe method utilizes the principle of the transient hot-wire method with the heating wire and the temperature sensor (thermocouple) encapsulated in a probe that electrically insulates the hot wire and the temperature sensor from the test sample. More details can be found from Nagasaka and Nagashima (1981). Measurements of thermal conductivity using the hot-wire method is very quick, which explains why the majority of the published studies on the effective thermal conductivity of nanofluids have chosen this method.

34.2.2 Thermal Conductivity of Nanofluids: Experimental Observations

As mentioned previously, the effective thermal conductivity of nanofluids has been dominating the literature in the past decade though this pattern began to change slightly over the last few years. A few reviews have been published over the period, for example, Keblinski et al. (2005), Das et al. (2006), Wang and Mujumdar (2007), Ding et al. (2007a), and Yu et al. (2008). The published data on thermal conductivity of nanofluids are mostly obtained at room temperature using either hot-wire or conventional parallel-plate methods. Figure 34.5 summarizes the room temperature data extracted from Lee et al. (1999), Eastman et al. (2001), Choi et al. (2001), Xie et al. (2002a,b), Biercuk (2002), Das et al. (2003a), Patel et al. (2003), Kumar et al. (2004), Assael et al. (2004), Zhang et al. (2007b), Wen and Ding (2004a,b, 2005a,b, 2006), Ding et al. (2006), and He et al. (2007). These data are a sample of the published experimental results. They represent aqueous as well as ethylene glycol and mineral oil–based base liquids and different types of nanoparticle materials. For comparison, a set of data for polymer-based carbon nanotube composite materials are also included. This type of materials can be argued to be similar to nanofluids from the viewpoint of fundamental physics though there is one important difference: nanoparticles in a composite do not enjoy the mobility that they would have in a nanofluid. As will be discussed below, the effect of nanoparticle mobility on the effective thermal conductivity of nanofluids is still a much debated area. A currently accepted view is that nanoparticle mobility plays a small role; see Section 34.2.3 for more discussion.

Figure 34.5 shows a significant degree of data scattering. In spite of the scattering, the presence of nanoparticles in fluids is seen to enhance the thermal conductivity, and the extent of enhancement depends on the nanoparticle material type and volume fraction. A higher volume fraction gives greater enhancement. A closer look at Figure 34.5 suggests that the data points can be approximately divided into two groups separated by two demarcation lines. The data points on the left hand side of the right line are for nanofluids made of metal nanoparticles and carbon nanotubes, whereas those on the right hand side of the left line are for nanofluids made of metal oxide and carbide nanoparticles. The region between the two demarcation lines represents overlapping between the two groups. Broadly speaking, the demarcation lines seem to indicate that the nanofluids made with high thermally conductive materials give a higher effective thermal conductivity. There are, however, deviations within each of the two regions. For example, the thermal conductivities of gold and copper are, respectively, 317 and 401 W/m K at room temperature, whereas the thermal conductivity of carbon nanotubes can be around 3000–6000 W/m K (Berber et al. 2000, Kim et al. 2001). The sequence of the three materials as shown in the left hand side of the band in Figure 34.5 is gold, carbon nanotubes, and copper. On the other hand, the thermal conductivities of CuO, alumina, and SiC at the room temperature are 20, 40, and 120 W/m K, respectively. However, the experimental data shown in Figure 34.5 indicate that copper oxide nanofluids give the highest enhancement, and little difference is seen between SiC and alumina nanofluids. Apart from possible measurement errors, particle size, shape, aggregation/entanglement, and interfacial resistance are believed to play a considerable role. Nevertheless, most of the publications only contain information of primary size of nanoparticles and/or shape obtained by electron microscopes. This does not represent the actual status of nanoparticles as they are prone to agglomerate and/or aggregate (and also entangling for nanotubes and nanofibers).

The experimental data have been compared with various macroscopic models developed for suspensions and composite materials on the basis of effective medium theory (Ding et al. 2007a). The results show that for spherical particles, all the original forms of the models give a predicted line that is slightly lower than the right demarcation line, and there is a very small difference between the original forms of the models within the range of particle concentration shown in Figure 34.5. This indicates that the original forms of the conventional models underpredict most nanofluids, particularly for Au, Cu, and CuO nanofluids. For carbon nanotube nanofluids, the models are found to provide an overprediction (Wen and Ding 2004a,b, Ding et al. 2007a,b). Current understanding of the underprediction is due to nanoparticle structuring, whereas the overprediction is due to the effect of interfacial resistance, both of which are not included in the conventional forms of the macroscopic models.

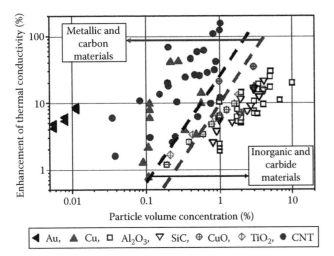

FIGURE 34.5 Effective thermal conductivity of nanofluids reported in the literature; representative data only extracted from Lee et al. (1999), Eastman et al. (2001), Choi et al. (2001), Xie et al. (2002a,b), Biercuk et al. (2002), Das et al. (2003a), Patel et al. (2003), Kumar et al. (2004), Assael et al. (2004), Zhang et al. (2007b), Wen and Ding (2004a,b, 2005a,b, 2006), Ding et al. (2006), and He et al. (2007).

34.2.3 Thermal Conductivity Enhancement of Nanofluids: Mechanisms

A number of mechanisms have been proposed for interpreting the experimentally observed thermal conduction enhancement (Keblinski et al. 2008). The most popular mechanisms include Brownian motion of nanoparticles (Patel et al. 2003, Kumar et al. 2004), interfacial ordering of liquid molecules at nanoparticle surfaces (Yu and Choi 2003), ballistic transport of energy carriers within individual nanoparticles (Keblinski et al. 2002), as well as nanoparticle structuring/networking (Keblinski et al. 2002, Nan et al. 2003, Wang et al. 2003, Prasher et al. 2006a,b).

Ballistic transport of energy carriers: Ballistic transport of energy carriers has been excluded as a possible mechanism for the enhanced thermal conductivity. This is because the thermal conductivity of nanoparticles decreases with decreasing particle size when the size becomes comparable to the mean free path of the energy carriers (Chen 1996).

Brownian motion: Brownian motion of nanoparticles could contribute to the thermal conduction enhancement in two ways: direct contribution due to motion of nanoparticles that transports heat and indirect contribution due to micro-convection of fluid surrounding individual nanoparticles. The direct contribution of Brownian motion has been shown theoretically to be insignificant as the timescale of the Brownian motion is about 2 orders of magnitude larger than that of the thermal diffusion of the base liquid (Keblinski et al. 2002). The indirect contribution

has also been shown to play a minute role (Evans et al. 2006). Furthermore, as nanoparticles are often in the form of agglomerates and/or aggregates, the Brownian motion is expected to play an even smaller role than expected. This conclusion is supported by experimental data shown in Figure 34.6, where the thermal conductivity enhancement is plotted as a function of temperature for nanofluids made of three types of metal-oxide nanoparticles. One can see that, except for the dataset of Das et al. (2003a) for CuO/H$_2$O nanofluids, the thermal conductivity enhancement is almost independent of temperature. Such weak temperature dependence suggests that the Brownian motion of nanoparticles is not a dominant mechanism for the thermal conductivity enhancement of nanofluids. The minor role of the Brownian motion is also supported by the lack of clear effect of the base liquid viscosity on the thermal conductivity enhancement of alumina-based nanofluids (Ding et al. 2007a,b).

Liquid molecular layering: At a solid–liquid interface, the liquid molecules could be significantly more ordered than they are in the bulk liquid. By analogy to the thermal behavior of crystalline solids, the ordered structure could be a mechanism of thermal conductivity enhancement (Keblinski et al. 2002). On such a basis, macroscopic models have been proposed to interpret the experimental data; such as by Yu and Choi (2003) and Wang et al. (2003). It is now clear that liquid–nanoparticle interface is one of the main factors that decrease (rather than increase) the effective thermal conductivity due to the so-called Kapitza interfacial resistance (Nan et al. 2003, Shenogin et al. 2004a,b, Gao et al. 2007). The effect of

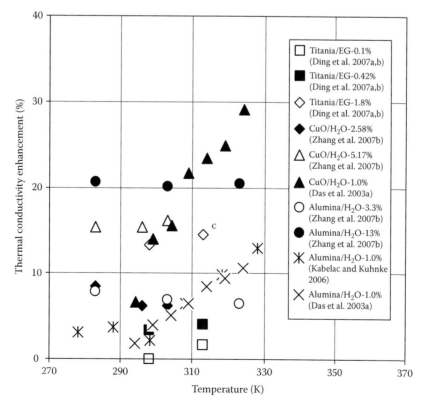

FIGURE 34.6 Effect of temperature on the thermal conductivity enhancement of nanofluids.

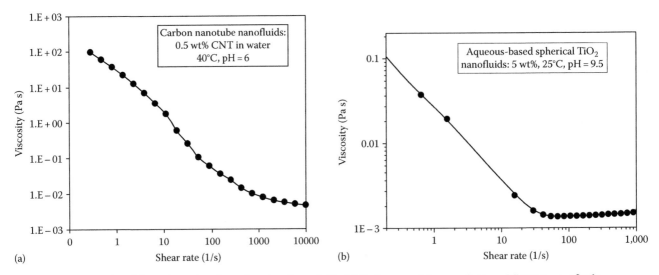

FIGURE 34.7   Examples of shear-dependent shear viscosity of nanofluids. (a) Carbon nanotube nanofluids and (b) TiO$_2$ nanofluids.

interfacial resistance on the overall effective thermal conductivity depends on the particle size (Keblinski et al. 2005, Prasher et al. 2005, Putnam et al. 2006, Gao et al. 2007). When particle size is relatively small in comparison with the characteristic length scale for the interfacial resistance, nanoparticles act as insulators. This may lead to instances where the thermal conductivity of a nanofluid becomes inferior to that of its base liquid (Putnam et al. 2006).

Nanoparticle structuring/aggregation: Recent studies have suggested that nanoparticle structuring/aggregation could be a dominant mechanism behind the experimentally observed thermal conductivity enhancement of nanofluids (Nan et al. 2003, Wang et al. 2003, Prasher et al. 2006a,b, Chen et al. 2007, Keblinski 2008). Using such a mechanism, well-dispersed nanoparticles in a fluid matrix gives the lowest thermal conductivity, whereas interconnected nanoparticles in the liquid enhances the thermal conduction. This can be understood from the viewpoint of circuit analysis; the well-dispersed situation is closer to conductors connected in a series mode, while the interconnected case is closer to those in a parallel mode (Keblinski et al. 2008). As a consequence, the key issue now is to obtain nanoparticle structural information, which can then be fed back to the conventional effective medium theories to predict the effective thermal conductivity of nanofluids. One of the efficient ways to describe the nanoparticle structural information is the fractal theory (Goodwin and Hughes 2000, Prasher 2006b, Chen et al. 2007), where one possible way for obtaining nanoparticle structural information is through rheological analysis to be discussed in Section 34.3.

34.3 Transport Properties of Nanofluids: Shear Viscosity

Shear viscosity plays an important role in determining the convective heat transfer coefficient and pressure drop of nanofluids. In spite of the importance, there are only a small number of published studies on the topic. The shear viscosity of nanofluids can be measured by using rheometers (Kwak and Kim 2005, Ding et al. 2006,

Prasher et al. 2006a, Chen et al. 2007) or viscometers (Praveen et al. 2007, Nguyen et al. 2008). A rheometer provides more information than a viscometer and is therefore preferred for nanofluids characterization. The reason for this is that nanofluids generally show non-Newtonian shear-thinning behavior, and hence their shear viscosities decrease with increasing shear rate (Chen et al. 2007).

Figure 34.7 shows two examples of shear dependence of the shear viscosity measured by the authors' group using a Bolin CVO rheometer. At a shear rate close to zero, a very high shear viscosity is observed, particularly for the carbon nanotube nanofluid, which can be attributed to the particle shape. Then the shear viscosity decreases rapidly with increasing shear rate and approaches a constant value at high shear rates. The high shear viscosity of nanofluids at low shear rates is due to nanoparticles structures. These structures are gradually destroyed with increasing shear rate, leading to a decreased shear viscosity at high shear rates. The high shear viscosity implies a high pressure drop and hence a high pumping power requirement for nanofluids. There is therefore a need to strike a balance between the benefit (enhanced heat transfer) and the penalty (increased pressure drop) when considering the use of nanofluids.

The degree of shear-dependence is also influenced by particle concentration, base liquid properties, particle size, and extent of particle structuring (Chen et al. 2007). As a consequence, the shear viscosity measured with rheometers can provide vital information about particle structuring, which, as mentioned in Section 34.2, can be fed into the classical effective medium-based theories to predict the effective thermal conductivity of nanofluids (Chen et al. 2009).

34.4 Heat Transfer Behavior: Forced Convective Heat Transfer of Nanofluids

Convective heat transfer refers to heat transfer between a fluid and a surface due to macroscopic motion of the fluid relative to the surface. This can be divided into two types: natural

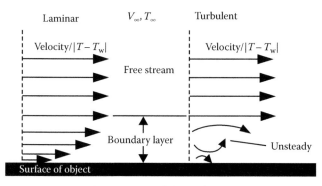

FIGURE 34.8 Nanofluids flowing across a solid surface with different temperatures showing the boundary layer.

convective heat transfer where fluid motion is induced by buoyancy, and forced convective heat transfer where fluid is forced through a confined region (e.g., pipes and channels) or across a confining wall. This section focuses on forced convection, leaving the natural convection to be discussed in Section 34.5. The forced convective heat transfer is quantified by the convective heat transfer coefficient, h, defined by Newton's law of cooling; $q = h \cdot (T_w - T_f)$, where q is the heat flux (in W/m²), and T_w and T_f are, respectively, the surface and bulk fluid temperatures (in K). Hence the heat transfer coefficient has units of watts per square meter kelvin. Figure 34.8 shows a nanofluid flowing across a solid surface at different temperatures. Due to viscous effect and temperature difference, hydrodynamic and thermal boundary layers form adjacent to the surface. A heat balance over a thin layer of fluid on the solid surface gives $h = k_f/\delta_t$ with k_f the thermal conductivity of the fluid and δ_t the boundary layer thickness. As the boundary layer thickness is affected by the flow and temperature fields of nanofluids, the heat transfer coefficient is obviously not a material property.

34.4.1 Experimental Results of Forced Convective Heat Transfer of Nanofluids

Out of the limited number of published studies on the forced convective heat transfer of nanofluids, the majority shows the enhancement of convective heat transfer (Xuan and Roetzel 2000, Li and Xuan 2002, Xuan and Li 2003, Jang and Choi 2006, Heris et al. 2007). A few studies show enhancement under certain conditions but little enhancement under other conditions (Pak and Cho 1998, Chein and Chuang 2007, Ding et al. 2007b, Lee and Mudawar 2007) while the others show little change or decrease in the convective heat transfer coefficient when nanoparticles are added to the base liquids (Yang et al. 2005, Ding et al. 2007b). These studies used pipe or channel flow and experiments were mostly done under the constant heat flux conditions. Review of these studies leads to the following observations:

a. The convective heat transfer coefficient of nanofluids has the highest value at the entrance but it decreases with increasing axial distance and tends to approach a constant value in the fully developed region. The entrance length depends on the properties and behavior of nanofluids. For a given nanofluids, the entrance length at low flow rates, for example, laminar flow for Newtonian fluids, is longer than that at high flow rates, for example, turbulent flow for Newtonian fluids.

b. Convective heat transfer coefficient can be enhanced or deteriorated depending on the nanofluids formulations and experimental conditions.

c. For cases where heat transfer enhancement is observed, nanofluids containing tubular or rodlike nanoparticles often give a higher enhancement of convective heat transfer coefficient in comparison with spherical or disc-like nanoparticles. Nanofluids made of lower viscous liquids (e.g., water) gives a higher heat transfer coefficient in comparison to that made of highly viscous liquids (e.g., mineral oil).

d. For cases where enhancement is observed, the convective heat transfer coefficient generally increases with increasing flow rate or increasing particle concentration, and the enhancement may exceed the extent of the thermal conductivity enhancement.

e. No clear trend has been found in the effect of particle size on the convective heat transfer coefficient of nanofluids.

f. For nanofluids made of particles with large aspect ratios, for example, carbon nanotubes, there seems to be a relationship between the rheological behavior and the convective heat transfer behavior.

In the following section, an attempt is made to explain the experimental results.

34.4.2 Why Enhancement in Some Cases but Deterioration in Other Cases?

A quantitative explanation to the experimental observations summarized in Section 34.4.1 is currently not possible. Therefore an attempt is made to give a qualitative interpretation. This will be done from both macroscopic and microscopic viewpoints. Consider a fluid flow with uniform velocity and temperature distributions enters a pipe. Due to friction between the fluid and the pipe wall, a hydrodynamic boundary layer will develop at the wall in which the flow velocity increases from zero at the wall to a maximum in a radial position depending on the axial position from the entrance. At a certain axial position from the entrance, the thickness of this boundary layer approaches a constant, and the flow is regarded as fully developed. Similarly, as the temperature of the fluid is different from the pipe wall, a thermal boundary layer will develop, though its thickness and the entry length may differ from the hydrodynamic boundary layer. Macroscopically, as mentioned previously, the forced convective heat transfer coefficient is given by $h = k_f/\delta_t$. This indicates that an increase in k_f or/and a decrease in δ_t can result in an increase of the convective heat transfer coefficient. Hence the entrance region, where boundary layer is thinnest, yields the highest convective heat transfer coefficient (Wen and Ding 2004b). As nanofluids have a higher thermal conductivity in comparison with their respective base liquids, the above expression also partly explains the enhanced convective heat transfer coefficient. However, it cannot adequately explain why the convective heat transfer coefficient enhancement is much higher than the thermal conduction enhancement in some cases, while there is no convective heat transfer enhancement in other cases despite considerable thermal conduction enhancement. In the following, an explanation is given from the microscopic point of view.

Microscopically, nanofluids are inhomogeneous species. There are at least two possible reasons for the inhomogeneity (Ding et al. 2006, 2007a). One is the presence of aggregates in nanofluids, which can be associated with either sintering during nanoparticle manufacturing or solution chemistry during nanofluids formulation. The former is often seen in processes involving elevated temperatures, for example, aerosol reactors. The resulting aggregates are too strong to be broken down to primary nanoparticles even with prolonged high energy processing. The latter is due to attraction between nanoparticles, for example, Van der Waal's attractive force and depletion phenomena. The aggregation can be controlled by solution chemistry and applying mechanical energy. The second reason is particle migration due to viscosity and velocity gradients. The experimental evidence of particle migration includes a longer entrance length for nanofluids (Merhi et al. 2005). There are also plenty of theoretical studies on particle migration; for example, Phillips et al. (1992) and Frank et al. (2003). If particles are very small, Brownian motion is strong and the effect of particle migration is negligible. If particles are large, for example, aggregates of hundreds of nanometers, the contribution of the Brownian motion is small, and a particle depletion region may exist at the wall region, which gives nonuniform distributions of particle concentration, viscosity, and thermal conductivity. The direct results of particle migration are lower particle concentration at the wall region and a thinner boundary thickness due to disturbance by the moving particles. This, according to $h = k_f/\delta_t$, can lead to three possible scenarios: (1) h is enhanced if the decrease in δ_t exceeds the decrease in k_f, (2) h does not change if the decrease in δ_t is equal to the decrease in k_f, and (3) h is reduced if the decrease in δ_t is lower than the decrease in k_f. This may qualitatively explain the experimental results.

34.5 Heat Transfer Behavior: Natural Convective Heat Transfer of Nanofluids

Natural convective heat transfer (also known as free convection) is caused by convection currents induced in a fluid surrounding a body without the application of external flow means. These convection currents are set up as a result of the temperature difference between the body and the fluid, which causes a change in the density of the fluid in the vicinity of the surface. Due to its inherent nature, the fluid mixing intensity in natural convection is far less than that of forced convection. As a consequence, the heat transfer coefficients are smaller in natural convection. In spite of this, natural convection of pure liquids has been extensively investigated because it implies no power consumption and a vast variety of applications.

Less than a handful of studies is found in the literature with regard to nanofluids heat transfer under natural convection conditions. By using a numerical technique, Khanafer et al. (2003) predicted that nanofluids have enhanced natural convective heat transfer. The enhancement was also observed experimentally by Nanna et al. (2005) for Cu/ethylene glycol nanofluids and by Nnanna and Routhu (2005) for alumina/water nanofluids. In contrary, Putra et al. (2003) found experimentally that the presence of nanoparticles in water systematically decreased the natural convective heat transfer coefficient. This is consistent with the experimental observations of Wen and Ding (2006) and Wen et al. (2006). The exact reasons for the divergence require further investigations.

34.6 Heat Transfer Behavior: Boiling Heat Transfer of Nanofluids

Boiling is a kind of phase-change heat transfer. As heat is being supplied to a liquid from a solid surface in contact with the liquid, the liquid gradually increases its temperature. First, it will be the natural convective currents that carry heat from the surface into the liquid. As the wall heat flux is increased, vapor bubbles begin to form on the surface, which is known as bubble nucleation. With further increase in heat flux, the bubbles detach and rise through the liquid. This state is known as nucleate boiling. The highest value of heat flux under which the nucleate boiling can persist is called the critical heat flux (CHF). Beyond this point, the vapor bubbles merge creating a vapor blanket and systematically drying out the liquid on the surface. Ultimately, this progresses into film boiling and the burnout of the heater surface. Hence CHF is an important turning point in nucleate boiling heat transfer. Since convection is the starting point, boiling has all the ingredients of convection, such as wall geometry, viscosity, density and thermal conductivity, expansion coefficient, and specific heat of the fluid. In addition, it is influenced by the heater surface characteristics, fluid's surface tension, latent heat of vaporization, pressure, and density. Boiling is preferred as a means of heat transfer due to its rapid heat transfer capability. In convective heat transfer, the heat flux is proportional to wall superheat. But in boiling heat transfer, it can be as large as three times the wall superheat (Rohsenow and Hartnett 1973).

A number of studies have been performed on boiling heat transfer of nanofluids (Das et al. 2003b,c, Tsai et al. 2003, You et al. 2003, Tu et al. 2004, Vassallo et al. 2004, Bang and Chang 2005, Wen and Ding 2005a, Wen et al. 2006, Kim et al. 2006a,b, 2007, Chopkar et al. 2008). It is widely agreed that the presence of nanoparticles in a liquid enhances the CHF. The mechanism of the CHF enhancement is attributed to the deposition and sintering of nanoparticles on the boiling surfaces, which increase the surface area, and the wettability.

There is a disagreement regarding boiling heat transfer of nanofluids in the nucleate regime. Wen and Ding (2005a) and Wen et al. (2006) observed the enhancement of boiling heat transfer under the nucleate regime for both titania and alumina nanofluids. This agrees with the observations by You et al. (2003) and Tu et al. (2004), but disagrees with those of Das et al. (2003b,c), Bang and Chang (2005), and Kim et al. (2006a) who reported the deterioration of boiling heat transfer in the nucleate regime. The possible reasons for the discrepancy could be the following:

a. As thermal conductivity and viscosity influence the heat transfer behavior of nanofluids in opposite ways, a combination of thermal conductivity enhancement and viscosity increase can cause either enhancement or deterioration of the heat transfer coefficient. However, there is little information in the published studies for making a conclusive assessment.

b. Stability of nanofluids and the presence of dispersant/surfactant affect the behavior of nanofluids, which are often not disclosed in published studies. For example, settling of nanoparticles in nanofluids with poor stability can change the properties of the boiling surface, and surfactant/dispersant may fail at elevated temperatures. This is supported by the recent work by Chopkar et al. (2008).

c. Boiling heat transfer consists of a number of subprocesses in parallel and/or series, including unsteady-state heat conduction, growth and departure of bubbles, and convection due to bubble motion and liquid refilling into cavities. These subprocesses are affected by parameters such as heater geometry, properties of the heater surface, the orientation of the heater, liquid subcooling, system pressure, and the mode in which the system is operated. Among these, the boiling surface properties are the key factors that influence the boiling heat transfer. The surface properties include surface roughness, surface wettability, and surface contamination as they all influence the number and distribution of active nucleation sites for bubble generation and their subsequent growth. In the published studies, however, surface roughness is the most often used parameter, and the interpretation of the effect of surface roughness on the boiling heat transfer has been based on the relative size of the suspended particles to the surface roughness. For example, Bang and Chang (2005) used a boiling surface of nanometer-scale roughness; hence, the sedimentation of particles was regarded to effectively increase the roughness of the surface, whereas a commercial cartridge heater with a micron-scale surface roughness was employed by Das et al. (2003b,c), the sedimentation of nanoparticles onto which was thought to decrease the effective surface roughness.

d. Different temperature measurement methods may lead to the different experimental results obtained by different investigators. For example, all thermocouples were welded on the outer surface of the cartridge heater by Das et al. (2003b,c). This may influence surface characteristics of the boiling surface as bubbles have a tendency to nucleate on the welded positions, and the measured temperature may not be representative of the boiling surface. Vassalao et al. (2004) used fine resistance wires for temperature measurements. Large uncertainties are expected for this sort of method as temperature is converted from the measured resistance of the heating wire against the standard temperature–resistance curve. Indeed, for boiling of pure water, more than 10°C deviation of superheat was observed under a fixed heat flux condition in different runs (Figure 34.1 in Vassallo et al. (2004)).

34.7 Concluding Remarks

This chapter gave a brief introduction to nanofluids with a specific focus on heat transfer applications. It covers transport properties of nanofluids, in particular thermal conductivity, shear viscosity, and heat transfer of nanofluids, under convective and boiling conditions. As far as the thermal conductivity is concerned, no new physics appears to be behind the experimentally observed thermal conductivity enhancement and viscosity increase. These can be interpreted by combining the structural information of nanoparticles with classical effective medium theories. However, at this point in time, there is no sufficient quantitative information to infer the dominant mechanisms that govern heat transfer enhancement under convective and boiling conditions.

References

Assael, M.J., Chen, C.F., Metaxa, I. et al. 2004. Thermal conductivity of suspensions of carbon nanotubes in water. *International Journal of Thermophysics* 25: 971–985.

Bang, I.C. and Chang, S.H. 2005. Boiling heat transfer performance and phenomena of Al$_2$O$_3$–water nano-fluids from a plain surface in a pool. *International Journal of Heat and Mass Transfer* 48: 2407–2419.

Biercuk, M.J. 2002. Carbon nanotube composites for thermal management. *Applied Physics Letters* 80: 2767–2769.

Bird, R.B., Stewart, W.E., and Lightfoot, E.N. 2002. *Transport Phenomena* (2nd edn.). New York: Wiley & Sons Inc.

Cahill, D.G. 1990. Thermal conductivity measurement from 30 to 750 K: The 3ω method. *Review of Scientific Instruments* 61: 802–808.

Chein, R. and Chuang, J. 2007. Experimental microchannel heat sink performance studies using nanofluids. *International Journal of Thermal Sciences* 46: 57–66.

Chen, G. 1996. Nonlocal and nonequilibrium heat conduction in the vicinity of nanoparticles. *ASME Journal of Heat Transfer* 118: 539–545.

Chen, H.S., Ding, Y.L., and Tan, C.Q. 2007. Rheological behaviour of nanofluids. *New Journal of Physics* 9(367): 1–25.

Chen, H.S., Witharana, S., Jin, Y. et al. 2009. Predicting thermal conductivity of liquid suspensions of nanoparticles (nanofluids) based on rheology. *Particuology,* 7(2): 151–157.

Choi, S.U.S. 1995. Enhancing thermal conductivity of fluids with nanoparticles, in: D.A. Siginer, H.P. Wang (Eds.), *Developments Applications of Non-Newtonian Flows.* FED-vol. 231/MD-vol. 66, New York: ASME, pp. 99–105.

Choi, S.U.S., Zhang, Z.G., Yu, W. et al. 2001. Anomalous thermal conductivity enhancement in nano-tube suspensions. *Applied Physics Letters* 79: 2252–2254.

Chopkar, M., Das, A.K., Manna, I. et al. 2008. Pool boiling heat transfer characteristics of ZrO_2-water nanofluids from a flat surface in a pool. *Heat and Mass Transfer* 44: 999–1004.

Czarnetzki, W. and Roetzel, W. 1995. Temperature oscillation techniques for simultaneous measurement of thermal-diffusivity and conductivity. *International Journal of Thermophysics* 16: 413–422.

Das, S.K., Putra, N., Thiesen, P. et al. 2003a. Temperature dependence of thermal conductivity enhancement for nanofluids. *Journal of Heat Transfer* 125: 567–574.

Das, S.K., Putra, N., and Roetzel, W. 2003b. Pool boiling characteristics of nano-fluids, *International Journal of Heat and Mass Transfer* 46: 851–862.

Das, S.K., Putra, N., and Roetzel, W. 2003c. Pool boiling of nanofluids on horizontal narrow tubes. *International Journal of Multiphase Flow* 29: 1237–1247.

Das, S.K., Choi, S.U.S., and Patel, H.E. 2006. Heat transfer in nanofluids—A review. *Heat Transfer Engineering* 27(10): 2–19.

Ding, Y.L., Alias, H., Wen, D.S. et al. 2006. Heat transfer of aqueous suspensions of carbon nanotubes (CNT nanofluids). *International Journal of Heat and Mass Transfer* 49: 240–250.

Ding, Y.L., Chen, H.S., Wang, L. et al. 2007a. Heat transfer intensification using nanofluids. *KONA Powder and Particle* 25: 23–38.

Ding, Y.L., Chen, H.S., He, Y.R. et al. 2007b. Forced convective heat transfer of nanofluids. *Advanced Powder Technology* 18: 813–824.

Eastman, J.A., Choi, S.U.S., Li, S. et al. 2001. Anomalously increased effective thermal conductivities of ethylene glycol-based nanofluids containing copper nanoparticles. *Applied Physical Letters* 78: 718–720.

Evans, W., Fish, J., and Keblinski, P. 2006. Role of Brownian motion hydrodynamcis on nanofluids thermal conductivity. *Applied Physical Letters* 88: 093116.

Frank, M., Anderson, D., Weeks, E.R. et al. 2003. Particle migration in pressure-driven flow of a Brownian suspension. *Journal of Fluid Mechanics* 493: 363–378.

Gao, L., Zhou, X., and Ding, Y.L. 2007. Effective thermal and electrical conductivity of carbon nanotube composites. *Chemical Physics Letters* 434: 297–300.

Goodwin J.W. and Hughes R.W. 2000 *Rheology for Chemists—An introduction.* Cambridge, U.K.: The Royal Society of Chemistry.

He, Y.R., Jin, Y., Chen, H.S. et al. 2007. Heat transfer and flow behaviour of aqueous suspensions of TiO_2 nanoparticles (nanofluids) flowing upward through a vertical pipe. *International Journal of Heat and Mass Transfer* 50: 2272–2281.

Heris, S.Z., Esfahany, M.N., and Etemad, S.G. 2007. Experimental investigation of convective heat transfer of Al_2O_3/water nanofluid in a circular tube. *International Journal of Heat and Fluid Flow* 28: 203–210.

Jang, S.P. and Choi, S.U.S. 2006. Cooling performance of a microchannel heat sink with nanofluids. *Applied Thermal Engineering* 26: 2457–2463.

Keblinski, P., Phillpot, S.R., Choi, S.U.S. et al. 2002. Mechanisms of heat flow in suspensions of nano-sized particles (nanofluids). *International Journal of Heat and Mass Transfer* 45: 855–863.

Keblinski, P., Eastman, J.A., and Cahill, D.G. 2005. Nanofluids for thermal transport. *Materials Today* 6: 36–44.

Keblinski, P., Prasher, R., and Eapen, J. 2008. Thermal conductance of nanofluids: Is the controversy over? *Journal of Nanoparticle Research* 10: 1089–1097.

Khanafer, K., Vafai, K., and Lightstone, M. 2003. Buoyancy-driven heat transfer enhancement in a two-dimensional enclosure utilizing nanofluids. *International Journal of Heat and Mass Transfer* 46: 3639–3653.

Kim, P., Shi, L., Majumdar, A. et al. 2001. Thermal transport measurements of individual multiwalled nanotubes. *Physical Review Letter* 87: 215502.

Kim, S.J., Bang, I.C., Buongiorno, J. et al. 2006a. Effects of nanoparticle deposition on surface wettability influencing boiling heat transfer in nanofluids. *Applied Physics Letters* 89: 153107-1–153107-3.

Kim, H., Kim, J. and Kim, M. 2006b. Experimental study on CHF characteristics of water-TiO_2 nanofluids. *Nuclear Engineering and Technology* 38: 61–68.

Kim, S.J., Bang, I.C., Buongiorno, J. et al. 2007. Surface wettability change during pool boiling of nanofluids and its effect on critical heat flux. *International Journal of Heat and Mass Transfer* 50: 4105–4116.

Krishnamurthy, S., Lhattacharya, P., Phelan, P.E. et al. 2006. Enhanced mass transport in nanofluids. *Nano Letters* 6(3): 419–423.

Kumar, D.H., Patel, H.E., Kumar, V.R.R. et al. 2004. Model for heat conduction in nanofluids. *Physical Review Letter* 93: 144301.

Kwak, K. and Kim, C. 2005. Viscosity and thermal conductivity of copper oxide nanofluid dispersed in ethylene glycol. *Korea-Australia Rheology Journal* 17: 35–40.

Lee, J. and Mudawar, I. 2007. Assessment of the effectiveness of nanofluids for single phase and two-phase heat transfer in micro-channels. *International Journal of Heat and Mass Transfer* 50: 452–463.

Lee, S., Choi, S., Li, S. et al. 1999. Measuring thermal conductivity of fluids containing oxide nanoparticles. *Journal of Heat Transfer* 121: 280–289.

Li, Q. and Xuan, Y.M. 2002. Convective heat transfer and flow characteristics of Cu-water nanofluids. *Science in China-Series E* 45: 408–416.

Masuda, H., Ebata, A., Teramae, K. et al. 1993. Alteration of thermal conductivity and viscosity of liquid by dispersed by ultra-fine particles(dispersion of γ-Al$_2$O$_3$, SiO$_2$, and TiO$_2$ ultra-fine particles). *Netsu Bussei (Japan)* 4: 227–233.

Merhi, D., Lemaire, E., and Bossis, G. 2005. Particle migration in a concentrated suspension flowing between rotating parallel plates: Investigation of diffusion flux coefficients. *Journal of Rheology* 49: 1429–1448.

Nagasaka, Y. and Nagashima, A. 1981. Absolute measurement of the thermal conductivity of electrically conducing liquids by the transient hot-wire method. *Journal of Physics E: Scientific Instruments* 14: 1435–1440.

Nan, C.W., Shi, Z., and Lin, Y. 2003. A simple model for thermal conductivity of carbon nanotube-based composites. *Chemical Physics Letters* 375: 666–669.

Nanna, A.G.A., Fistrovich, T., Malinski, K. et al. 2005. Thermal transport phenomena in buoyancy-driven nanofluids. In *Proceedings of 2005 ASME International Mechanical Engineering Congress and RD&D Exposition*, November 15–17, 2004, Anaheim, CA.

Nguyen, C.T., Desgranges, F., Galanis, N. et al. 2008. Viscosity data for Al$_2$O$_3$-water nanofluids—Hysteresis: Is heat transfer enhancement using nanofluids reliable? *International Journal of Thermal Sciences* 47: 103–111.

Nnanna, A.G.A. and Routhu, M. 2005. Transport phenomena in buoyancy-driven nanofluids—Part II. In *Proceedings of 2005 ASME Summer Heat Transfer Conference*, July 17–22, 2005, San Francisco, CA.

Olle, B., Bucak, S., Holmes, T.C. et al. 2006. Enhancement of oxygen mass transfer using functionalized magnetic nanoparticles. *Industrial & Engineering Chemistry Research* 45: 4355–4363.

Pak, B.C. and Cho, Y.I. 1998. Hydrodynamic and heat transfer study of dispersed fluids with submicron metallic oxide particles. *Experimental Heat Transfer* 11: 150–170.

Patel, H.E., Das, S.K., Sundararajan, T. et al. 2003. Thermal conductivities of naked and monolayer protected metal nanoparticle based nanofluids: Manifestation of anomalous enhancement and chemical effects. *Applied Physical Letters* 83: 2931–2933.

Phillips, R.J., Armstrong, R.C., Brown, R.A. et al. 1992. A constitutive equation for concentrated suspensions that accounts for shear-induced particle migration. *Physics of Fluids* 4: 30–40.

Prasher, R., Bhattacharya, P., and Phelan, P.E. 2005. Thermal conductivity of nanoscale colloidal solutions (nanofluids). *Physical Review Letters* 94: 025901.

Prasher, R., Song, D., and Wang, J. 2006a. Measurements of nanofluid viscosity and its implications for thermal applications. *Applied Physics Letter* 89: 133108.

Prasher, R., Evans, W., Meakin, P. et al. 2006b. Effect of aggregation on thermal conduction in colloidal nanofluids. *Applied Physics Letters* 89: 143119.

Praveen, K., Kulkarni, P., Misra, D. et al. 2007. Viscosity of copper oxide nanoparticles dispersed in ethylene glycol and water mixture. *Experimental Thermal and Fluid Science* 32: 397–402.

Putnam, P.A., Cahill, D.G., Braun, P.V. et al. 2006 Thermal conductivity of nanoparticle suspensions. *Journal of Applied Physics* 99: 084308.

Putra, N., Roetzel, W., and Das, S.K. 2003. Natural convection of nano-fluids. *Heat and Mass Transfer* 39: 775–784.

Rohsenow, W.M. and Hartnett J.P. 1973. *Handbook of Heat Transfer*. New York: McGraw Hill.

Shenogin, S., Bodapati, A., Xue, L. et al. 2004a. Effect of chemical functionalization on thermal transport of carbon nanotube composites. *Applied Physics Letters* 85: 2229–2231.

Shenogin, S., Xue, L.P., Ozisik, R. et al. 2004b. Role of thermal boundary resistance on the heat flow in carbon nanotube composites. *Journal of Applied Physics* 95: 8136–8144.

Tsai, C.Y., Chien, H.T., and Ding, P.P. 2003. Effect of structural character of gold nanoparticles in nanofluid on heat pipe thermal performance. *Materials Letters* 58: 1461–1465.

Tu, J.P., Dinh, N., and Theofanous, T. 2004 An experimental study of nanofluid boiling heat transfer. In *Proceedings of 6th International Symposium on Heat Transfer*, June 15–19, 2004, Beijing, China.

Vassallo, P., Kumar, R., and Damico, S. 2004. Pool boiling heat transfer experiments in silica-water nano-fluids. *International Journal of Heat and Mass Transfer* 47: 407–411.

Wang, X.Q. and Mujumdar, A.S. 2007. Heat transfer characteristics of nanofluids: A review. *International Journal of Thermal Sciences* 46: 1–19.

Wang, B.X., Zhou, L.P., and Peng, X.F. 2003. A fractal model for predicting the effective thermal conductivity of liquid with suspension of nanoparticles. *International Journal of Heat and Mass Transfer* 46: 2665–2672.

Wasan, D.T. and Nikolov, A.D. 2003 Spreading of nanofluids on solids. *Nature* 423: 156–159.

Wen, D.S. and Ding, Y.L. 2004a. Effective thermal conductivity of aqueous suspensions of carbon nanotubes (nanofluids). *Journal of Thermophysics and Heat Transfer* 18: 481–485.

Wen, D.S. and Ding, Y.L. 2004b. Experiment investigation into convective heat transfer of nanofluids at the entrance region under laminar flow conditions. *International Journal of Heat and Mass Transfer* 47: 5181–5188.

Wen, D.S. and Ding, Y.L. 2005a. Experimental investigation into the pool boiling heat transfer of aqueous based γ-alumina nanofluids. *Journal of Nanoparticle Research* 7: 265–274.

Wen, D.S. and Ding, Y.L. 2005b. Formulation of nanofluids for natural convective heat transfer applications. *International Journal of Heat and Fluid Flow* 26: 855–864.

Wen, D.S. and Ding, Y.L. 2006. Natural convective heat transfer of suspensions of TiO$_2$ nanoparticles (nanofluids). *Transactions of IEEE on Nanotechnology* 5: 220–227.

Wen, D.S., Ding, Y.L., and Williams, R.A. 2006. Pool boiling heat transfer of aqueous based TiO$_2$ nanofluids. *Journal of Enhanced Heat Transfer* 13: 231–244.

Xie, H.Q., Wang, J., Xi T. et al. 2002a. Thermal conductivity enhancement of suspensions containing nanosized alumina particles. *Journal of Applied Physics* 91: 4568–4572.

Xie, H.Q., Wang, J., Xi, T. et al. 2002b. Thermal conductivity of suspensions containing nanosized SiC particles. *International Journal of Thermophysics* 23: 571–580.

Xuan, Y.M. and Li, Q. 2003. Investigation on convective heat transfer and flow features of nanofluids. *Journal of Heat Transfer* 125: 151–155.

Xuan, Y.M. and Roetzel, W. 2000. Conceptions for heat transfer correlation of nanofluids. *International Journal of Heat and Mass Transfer* 43: 3701–3707.

Yang, B. and Han, Z.H. 2006. Temperature-dependent thermal conductivity of nanorod-based nanofluids. *Applied Physics Letters* 89: 083111.

Yang, Y., Zhong, Z.G., Grulke, E.A. et al. 2005. Heat transfer properties of nanoparticle-in-fluid dispersion (nanofluids) in laminar flow. *International Journal of Heat and Mass Transfer* 48: 1107–1116.

You, S.M., Kim, J.H., and Kim, K.H. 2003. Effect of nanoparticles on critical heat flux of water in pool boiling heat transfer. *Applied Physics Letters* 83: 3374–3376.

Yu, W. and Choi, S.U.S. 2003. The role of interfacial layers in the enhanced thermal conductivity of nanofluids: A renovated Maxwell model. *Journal of Nanoparticle Research* 5: 167–171.

Yu, W., France, D.W., Routbort, J.L. et al. 2008. Review and comparison of nanofluids thermal conductivity and heat transfer enhancements. *Heat Transfer Engineering* 29: 432–460.

Zhang, L.L., Jiang, Y., Ding, Y.L. et al. 2007a. Investigation into the antibacterial behaviour of suspensions of ZnO nanoparticles (ZnO nanofluids). *Journal of Nanoparticle Research* 9: 479–489.

Zhang, X., Gu, H., and Fujii, M. 2007b. Effective thermal conductivity and thermal diffusivity of nanofluids containing spherical and cylindrical nanoparticles. *Experimental Thermal and Fluid Science* 31: 593–599.

V

Quantum Dots

35

Core-Shell Quantum Dots

Gil de Aquino Farias
Universidade Federal do Ceará

Jeanlex Soares de Sousa
Universidade Federal do Ceará

35.1 Introduction

The physics of semiconductor heterostructures has attracted growing attention in the last few decades, both from the fundamental point of view as well as for their applications, which originally focused on the fabrication of optoelectronics devices. Modern growth techniques, such as molecular-beam epitaxy (MBE) and metal-organic chemical vapor deposition (MOCVD), have made possible the fabrication of layered materials with sharp, high-quality interfaces, and with dimensions comparable to the electron mean free path and the de Broglie wavelength. These artificial structures form an intriguing new class of materials, in which their macroscopic properties are the subject of design or control by varying the structural parameters or composition of the constituent layers. Due to the potential application of such systems, much work has been devoted to the understanding of their unique physical properties.

Due to new techniques available to grow semiconductors, the interest in complex heterostructures with geometries different from the planar emerged. With them, new and exciting applications in optoelectronics as well as in biomedicine also arose. These new heterostructures are constructed with coaxial semiconductor cylinders and concentric semiconductor spherical shells. They are also known as *core-shell structures*. Figure 35.1 shows a schematic of heterostructures with planar, cylindrical, and spherical geometries. Advanced optical measurements have shown that the optical and electronic properties of these systems present noticeable differences with respect to those with flat surfaces.

Planar heterostructures have been intensively investigated in the last few decades (for a review, see Refs. [1–4]), while core-shell heterostructures are reasonably new. To mention a few theoretical works, Kim et al. [5] have studied the differences between charge separation in planar and concentric coaxial structures. Tkach et al. [6] and Carvalho et al. [7] have studied a semiconductor superlattice with axial symmetry, i.e., a superwire, in which coaxial wires with alternating semiconductor layers form a periodic structure in the radial direction. In spherical geometry, Bryant et al. [8] have analyzed the problem of nanometer-size layered semiconductor quantum dots (QDs). Haus et al. [9] and Schooss et al. [10] analyzed a semiconductor superlattice with tridimensional radial symmetry, i.e., a superdot, constituted by alternating semiconductor layers in the radial direction. More recently, Ribeiro Filho et al. [11] investigated the electronic properties of quasi-periodic coaxial and radial superlattices.

On the experimental side, the technologies to fabricate planar heterostructures (like MBE, MOCVD) are so well developed that it is possible to grow from single to hundreds of heterostructures with a precise control of the individual thickness of the alternating layers. In fact, many applications based on planar heterostructures are commercially available (e.g., solid-state lasers, diodes, radiation detectors, and so on). On the other hand, the experimental fabrication of either coaxial or radial core-shell heterostructures is quite new, and the techniques are currently experiencing strong development efforts. In practice, only structures with a few shell layers are currently possible. However, the rich and intriguing physical phenomena arising from such structures are motivating enormous efforts to their

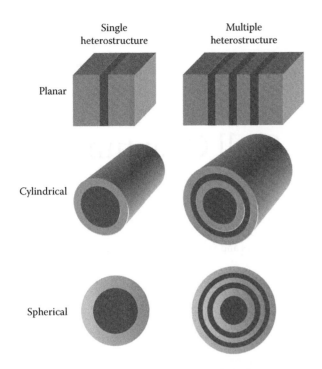

FIGURE 35.1 Schematics of single and multiple heterostructures with different geometries. Heterostructures with cylindrical and spherical geometries are also known as core-shell structures. Planar multiple heterostructures are also known as superlattices.

understanding and to their use in a broad spectrum of applications like conventional electronics, optoelectronics, biomedicine, and environmental and energy generation.

In this chapter, we present a comprehensive review of the basic properties of core-shell QDs. Basic physical models and mathematical tools that are able to capture the main physical characteristics experimentally observed are also presented. We attempted to make this text basic enough, yet profound, in order to make it easily accessible for researchers with different backgrounds. This chapter is organized as follows. Section 35.2 is dedicated to review recent developments on the fabrication and characterization of standalone and heterostructured colloidal QDs. In Section 35.3, a review of the basic concepts of quantum mechanics and solid-state theory is provided. In Section 35.4, we present a theoretical model to calculate the electronic structure and optical properties of core-shell QDs. The electronic structure of a few core-shell systems is presented in Section 35.5 and compared with experimental and theoretical data found in the literature. Finally, our conclusions will be drawn in Section 35.6.

35.2 State-of-Art in Core-Shell QDs

With recent advances in materials manipulation at the nanoscale, the degrees of freedom of charge carriers can be controlled to produce electron confinement in structures called *nanocrystals* (NCs) (quantum dots). These structures constitute a class of materials intermediate between molecular and bulk forms of matter.

Recent advances in the synthesis of highly monodisperse standalone and heterostructured colloidal nanocrystallites [12–19] have paved the way for numerous spectroscopic investigations that revealed the behavior of the electronic states of colloidal NCs with respect to their shape and size [14,20–23]. In fact, colloidal NCs are considered model systems to investigate the electronic structure of nanostructured materials. As the effective band gap increases and the NC size decreases, it is possible to tailor the electronic structure by means of shape and size control to produce desirable intra- and interband optical transitions. These features are useful for the development of novel optoelectronic devices with tunable emission or transmission properties and ultra-narrow spectral line widths. The unique optical properties of colloidal QDs make them promising building block for a number of applications in areas as different as polarized single-photon sources [14,15,24], biological detection and imaging [25,26], lasing [27–29], nonlinear optics [30,31], and photovoltaic (PV) cells [32–37].

Among the various materials, colloidal CdSe QDs are undoubtedly the most reported in the literature due to their tunable emission in the visible range and advantages in fabrication. Nowadays, CdSe colloidal NCs can be inexpensively grown with precise control over their shape and size [14]. However, the environment where NCs are embedded strongly affects their electronic structure and optical properties [38,39]. It was even pointed out that the QDs' electronic structure can be tuned through the manipulation of the surface ligands [40]. Chemically passivating the NC core with a thin shell of wide band gap semiconductor prevents chemical interaction with the inter-NC environment. Moreover, it allows substantial improvement of their optical stability and exhibits greater tolerance to processing conditions necessary for incorporation into solid-state structures.

Several wide band gap semiconductors (e.g., ZnS, CdS, ZnSe, and CdTe) have been epitaxially grown on the surface of CdSe colloidal NCs [19,41,42]. It is even possible to grow two different shell layers on the NC surface [18,43]. The use of such radial heterostructures (core-shell QDs) opens up new possibilities of further control of the QDs' electronic properties by means of electrons and holes wave function engineering. Depending on the combination of materials used in the core and shell regions, it is possible to control the relation position of electrons and holes. When electrons and holes are spatially separated between core and shell, it is said that the QD exhibits type II confinement; otherwise, it exhibits type I confinement.

35.3 Basic Concepts in Quantum Mechanics and Solid-State Theory

It is known that electrons moving solids are described by their total energy E and wave function $\Psi(\vec{r}, t)$, which is a complex function of the position and time that carries all dynamic information regarding electrons movement. It also has a probabilistic meaning: $|\Psi(\vec{r}, t)|^2$ represents the probability per unit volume to find the electron around position *r* at the instant *t*. This quantity is also known as *probability density*. Since the electron has to be somewhere, the probability density is normalized in such way that $\int |\Psi(\vec{r},t)|^2 d\vec{r} = 1$. The electron energy *E* and its wave function are obtained by solving the time-independent Schrodinger equation:

$$\left[-\frac{\hbar^2}{2m}\nabla^2 + V(\vec{r}) \right]\Psi_n(\vec{r}) = E_n\Psi_n(\vec{r}) \qquad (35.1)$$

Since the above equation does not include time dependence, it must be used only in problems where time is not an important parameter. In this equation, $\hbar = 1.05459 \times 10^{-34}$ J s is the reduced Planck constant, *m* is the electron mass, and $V(\vec{r})$ is the position-dependent potential energy landscape acting upon the electron. To put it simply, the Schrodinger equation provides all possible electron states, comprising an energy E_n, a wave function Ψ_n, and indexed by *n*, for a given potential $V(\vec{r})$. There are many other details concerning the quantum mechanical description of electrons. Although important, a complete revision of quantum mechanics is out of the scope of this chapter. For more detail on quantum mechanics, we recommend Refs. [44–47].

35.3.1 Energy Bands

It is known that electrons moving freely (without any potential acting on them) are described by the following wave function:

$$\Psi(\vec{r},t) = Ae^{i(\vec{k}\cdot\vec{r}-\omega t)} \qquad (35.2)$$

where
    k is $2\pi/\lambda$
    λ is the electron wavelength

A simple inspection of $|\Psi(\vec{r},t)|^2$ shows that the probability to find the electron is constant and position-independent, indicating that $\Psi(\vec{r},t)$ is essentially delocalized, and that it is equally probable to find the electron anywhere. It is important to note that the wave function in Equation 35.2 represents the particular case of an electron moving in a region with linear dimension *L* much larger than the electron wavelength λ. Since typical core-shell QD sizes are comparable to λ, one cannot use Equation 35.2 to describe the behavior of electrons in such systems. However, before stepping to the electron states of core-shell QDs, it is convenient to give a further look into the free electron problem ($L \gg \lambda$) to extract other useful information that is, in a first approximation, applicable in nanometer structures.

In contrast to the wave-like nature, electrons also have mass, which is a characteristic intrinsic to particles. The connection between the particle-like and wave-like behavior of electron, a phenomenon known as wave-particle duality [44–46], is provided by the de Broglie relationship:

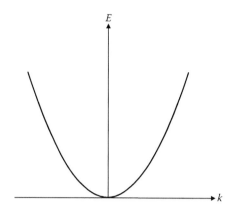

FIGURE 35.2 Energy dispersion relationship for a free particle.

$$\vec{p} = \hbar\vec{k} \qquad (35.3)$$

Since $\vec{p} = m\vec{v}$, and the kinetic energy of a particle with momentum $\vec{p}$ and mass *m* is given by $E = p^2/2m$, the electron energy whose wave function is given by Equation 35.1 is given by:

$$E = \frac{\hbar^2 k^2}{2m} \qquad (35.4)$$

The equation above is known as energy dispersion curves or band structure (see Figure 35.2). Such curves are very important to understand the physical properties of materials. Although Equation 35.4 is the energy dispersion of electrons moving freely in vacuum, all solids have their own band structure. It turns out that, for small values of *k*, the band structure of real materials is nearly parabolic, as in Equation 35.4, with some modifications.

Before discussing the modifications in Equation 35.4 that gives rise to the band structure of real materials, the reader should imagine himself as an electron moving in the interior of real materials. This movement may be likened to particles in a three-dimensional box with a very complicated interior. In real materials, at moderate temperatures there will be lattice defects (missing and/or impurity atoms, and so on), atoms vibrating around their lattice position, and other huge number electrons. Most of these electrons are bound to the host atoms and cannot move around. Others are released (due to temperature) from the valence shell of the host atoms and are free to move around the whole material volume. When electrons are released from host atoms, they leave behind an available electron state in the valence shell (hole state) that can be filled by an electron either moving or bound to neighbor atoms. Hole states behave like an electron with positive charge. Moving electrons occupy electronic states in the conduction band (CB), while the moving holes occupy electronic states in the valence band (VB). Thus, the band structure of real materials must be composed of the conduction and VB of energy. Finally, the reader should remember that electrons (moving or bound) are attracted to the positive nucleus of the host atoms, preventing them to move freely as they were in vacuum. The averaged interaction of electrons and holes with host atoms and other electrons and holes is felt as if they have

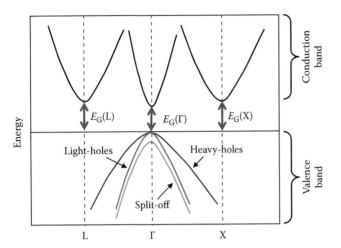

FIGURE 35.3 Main features of the energy versus momentum dispersion (band structure) of bulk solids for small values of the momentum.

an effective mass m^* which, for most materials, strongly differs from the electrons' rest mass in vacuum m_0.

Figure 35.3 depicts the main features of the band structure of real materials for small values of the wave vector k. In general, the curvature of the energy dispersion of electrons is different from the dispersion of holes. The minimum energy separation between the conduction and VB is known as energy band gap E_G. This energy indicates the amount of energy required to remove a bound electron to the host atom and put it to move in the material. This quantity also gives a primary indication of the metallic, semiconductor, or insulator character of materials. Large E_G (≥ 3 eV) indicates insulator materials, while small E_G (≈ 0 eV) indicates metallic materials. Intermediate values are the characteristic of semiconductor materials. The other interesting characteristic of real materials is that the k values of minimum of the CB, and of the maximum of the VB, may not coincide. If they coincide, the material is known as *direct band-gap material*. Otherwise, it is known as *indirect band-gap material*. Finally, due to coupling effects of orbital motion around the host atom and electron spin, there are two types of

holes: the light hole, with effective mass m_{lh}, and heavy hole, with effective mass m_{hh}.

35.3.2 Heterojunctions

Since core-shell QDs are composed of two different materials, each of them with their own band structures, it is necessary to understand how these band structures match at the interface. Figure 35.4 shows the band structure of two different materials before and after contact. In this figure, the energy gaps of the materials are different. Now imagine that both materials are the same. If one electron is moving in the CB of the left material toward right, it will cross the interface without actually feeling it. If the materials are different, the electron will see an abrupt step between the bottom of the CB in the left and right. If the bottom of CB in the right material is below the CB in the left, electrons will gain energy in the right. On the other hand, if the CB in the right is above CB in the left, electrons will see an energy barrier and will be reflected at the interface. The same analysis is also valid for holes.

In our core-shell QDs, the confinement potential will be determined by the relative alignment of the two materials. In a first approximation, the alignment of band structures are made with respect to the energy necessary to remove an electron in the CB of the compounding materials. This energy amount is called *electron affinity*. Thus, the energy barrier for electrons and holes between two different materials are obtained with

$$\Delta E_C = |\chi_1 - \chi_2| \tag{35.5}$$

$$\Delta E_V = |(E_{G1} + \chi_1) - (E_{G2} + \chi_2)| \tag{35.6}$$

The reader must be aware that this model is only valid when the lattice mismatch between the two materials is small. Otherwise, strain effects must be included. For a comprehensive view of strain effects, we suggest Refs. [48,49]. Figure 35.5 shows the schematics of the possible confinement profiles in core-shell QDs.

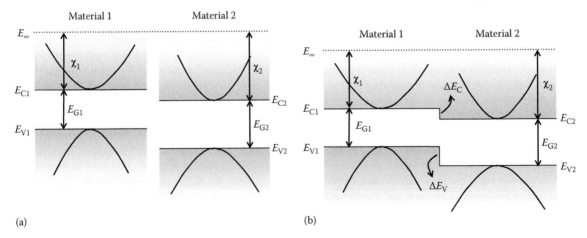

(a)   (b)

FIGURE 35.4 Band structure alignment near the interface between two different materials. (a) Before contact. (b) After contact.

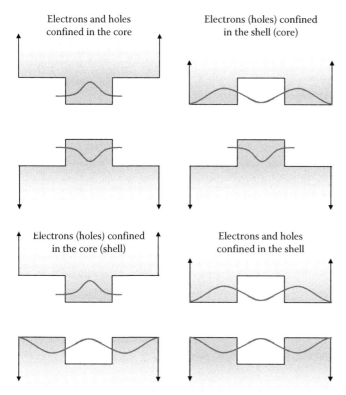

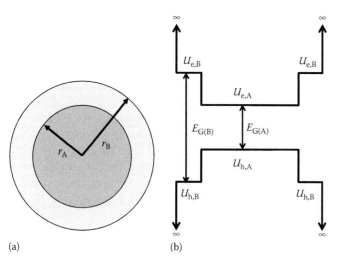

Electrons and holes confined in the core

Electrons (holes) confined in the shell (core)

Electrons (holes) confined in the core (shell)

Electrons and holes confined in the shell

FIGURE 35.5 Possible band structure alignment in core-shell QDs.

(a) (b)

FIGURE 35.6 (a) Schematics of a spherical core-shell QD. (b) Potential energy profile in a core-shell QD.

Depending on the band structure alignment, electrons and holes may be localized in different materials. If electrons and holes are localized in the same material, the structure exhibits type I confinement; otherwise, it exhibits type II confinement. We shall see later that the type of confinement in core-shell structures has profound consequences on their electronic and optical properties.

35.4 Modeling Core-Shell QDs

35.4.1 Electronic Structure

The simplest yet powerful method to calculate the electronic structure of core-shell QDs is based on the one-band effective mass model. In this model, the coupling between conduction and VB is disregarded. In this approximation, the following Schrodinger equation is solved separately for electrons and light and heavy holes:

$$\left[-\frac{\hbar^2}{2m_i}\nabla^2 + V_i(\vec{r}) \right]\Psi_{n,i}(\vec{r}) = E_{n,i}\Psi_{n,i}(\vec{r}) \qquad (35.7)$$

The subscript $i = e$, lh, hh indicates the type of carrier, m_i is the carrier effective mass, and $V_i(\vec{r})$ represents the confinement potential. $\Psi_{n,i}(\vec{r})$ and $E_{n,i}$ are the carrier's wave function and energy, respectively. Now suppose our core-shell QD is perfectly spherical, like the one depicted in Figure 35.6, where the inner and outer radii are given by r_A and r_B, respectively. Assuming that the potential is infinite in the exterior region, the confinement potential of a core-shell QD can be written as

$$V_i(\vec{r}) = \begin{cases} U_{i,A} & 0 \le r < r_A \\ U_{i,B} & r_A \le r < r_B \\ \infty & r \ge r_B \end{cases} \qquad (35.8)$$

In the equation above, $U_{i,A}$ and $U_{i,B}$ represent either the bottom of CB or the top of VB in the core and shell regions, respectively. Thus, their difference $|U_{i,A} - U_{i,B}|$ represents the confinement potential of the ith particle. Likewise, the effective mass also presents a similar expression:

$$m_i = \begin{cases} m_{i,A} & 0 \le r < r_A \\ m_{i,B} & r_A \le r < r_B \end{cases} \qquad (35.9)$$

Due to the spherical geometry, it is convenient to use spherical coordinates to solve Equation 35.7. In addition, one can assume that the wave function $\Psi(\vec{r}) = (r, \theta, \phi)$ can be written as the product of a radial function $R(r)$ and a spherical harmonic $Y_{l,m}(\theta, \phi)$, with orbital and magnetic quantum numbers l and m [50]. By substituting $\Psi(\vec{r}) = R(r)Y_{l,m}(\theta, \phi)$ in Equation 35.7 and using the following property of the spherical harmonics $Y_{l,m}(\theta, \phi)$:

$$\left[\frac{1}{\sin\theta}\frac{\partial}{\partial\theta}\left(\sin\theta\frac{\partial}{\partial\theta}\right) + \frac{1}{\sin^2\theta}\frac{\partial^2}{\partial\phi^2} \right]Y_{l,m} = -l(l+1)Y_{l,m} \qquad (35.10)$$

we obtain the following eigenvalue differential equation for the radial function $R(r)$:

$$\left[-\frac{\hbar^2}{2m_i(\vec{r})}\frac{d^2}{dr^2} + \left(V(r) + \frac{\hbar^2 l(l+1)}{2mr^2}\right) \right]F_{n,l}(r) = E_{n,l}F_{n,l}(r) \qquad (35.11)$$

where $F_{n,l}(r) = rR_{n,l}(r)$. For simplicity, the index i (used to identify the type of carrier) will be omitted from this point. The index l indicates that there will be different solutions for different values of the orbital quantum number l ($l = 0, 1, 2, 3$). Moreover, for a given l, each energy state is $2l + 1$ degenerated (same energy value, different wave functions), and identified by the magnetic quantum number m ($m = -l, -l + 1,\ldots, l$). States corresponding to the values $l = 0, 1, 2, 3$ are usually denoted by the symbols s, p, d, f, g.

Note that $E_{n,l}$ and $V(r)$ are constants, and the above equation has two different branches: (1) $E - V(r) > 0$ and (2) $E - V(r) < 0$, giving rise to the ordinary and modified spherical Bessel equations, respectively. The radial wave function in each QD region can be written in the following compact form:

$$R_{n,l}^{(A)}(r) = C_A \bar{j}_l(k_A r) + D_A \bar{n}_l(k_A r) \quad (0 \le r < r_A) \quad (35.12)$$

$$R_{n,l}^{(B)}(r) = C_B \bar{j}_l(k_B r) + D_B \bar{n}_l(k_B r) \quad (r_A < r r_B) \quad (35.13)$$

where C_A, D_A, C_B, D_B are constants to be determined. The wave number is defined as

$$k_{A<B} = \sqrt{\frac{2m(E_{n,l} - U_{A,B})}{\hbar^2}} \quad (35.14)$$

The functions $\bar{j}_l(u)$ and $\bar{n}_l(u)$ are given by

$$\bar{j}_l(u) = \begin{cases} j_l(u) & u \in \Re \\ i_l(|u|) & u \in \Im \end{cases} \quad (35.15)$$

$$\bar{n}_l(u) = \begin{cases} n_l(u) & u \in \Re \\ k_l(|u|) & u \in \Im \end{cases} \quad (35.16)$$

Here, $j_l(u)$ is an ordinary spherical Bessel function and $i_l(u)$ a modified spherical Bessel function of the first kind, $n_l(u)$ is a spherical Neumann function, and $k_l(u)$ a modified spherical Bessel function of the second kind [50].

In order to obtain the constants C_A, D_A, C_B, D_B and particle energies $E_{n,l}$, boundary and continuity equations for the wave function and probability current must be ensured. These conditions are

$$R_{n,l}^{(A)}(k_A r_A) = R_{n,l}^{(B)}(k_B r_A) \quad (35.17)$$

$$\frac{1}{m_A} \frac{dR_{n,l}^{(A)}}{dr}\bigg|_{r_A} = \frac{1}{m_B} \frac{dR_{n,l}^{(B)}}{dr}\bigg|_{r_A} \quad (35.18)$$

$$R_{n,l}^{(B)}(k_B r_B) = 0 \quad (35.19)$$

In addition, the wave function has to be finite for $r \to 0$ in order to fulfill the normalization condition. This requires that $D_B = 0$. After some cumbersome calculation, one obtains the following system of algebraic equations:

$$\begin{bmatrix} \bar{j}_l(k_A r_A) & -\bar{j}_l(k_B r_A) & -\bar{n}_l(k_A r_A) \\ \frac{1}{m_A}\frac{d\bar{j}_l(k_A r)}{dr}\big|_{r_A} & -\frac{1}{m_B}\frac{d\bar{j}_l(k_B r)}{dr}\big|_{r_A} & -\frac{1}{m_B}\frac{d\bar{n}_l(k_B r)}{dr}\big|_{r_A} \\ 0 & \bar{j}_l(k_B r_B) & \bar{n}_l(k_B r_B) \end{bmatrix} \begin{bmatrix} C_A \\ C_B \\ D_B \end{bmatrix} = 0$$

$$(35.20)$$

Solutions other than trivial ($C_A = C_B = D_B = 0$) are only possible if

$$\det \begin{pmatrix} \bar{j}_l(k_A r_A) & -\bar{j}_l(k_B r_A) & -\bar{n}_l(k_A r_A) \\ \frac{1}{m_A}\frac{d\bar{j}_l(k_A r)}{dr}\big|_{r_A} & -\frac{1}{m_B}\frac{d\bar{j}_l(k_B r)}{dr}\big|_{r_A} & -\frac{1}{m_B}\frac{d\bar{n}_l(k_B r)}{dr}\big|_{r_A} \\ 0 & \bar{j}_l(k_B r_B) & \bar{n}_l(k_B r_B) \end{pmatrix} = 0$$

$$(35.21)$$

The allowed carrier energies are the values for which the above equation is satisfied. Once the energies are known, one can solve Equation 35.20 combined with the following normalization condition:

$$\int_0^{r_A} dr r^2 \left| R_{n,l}^{(A)}(r) \right|^2 + \int_{r_A}^{r_B} dr r^2 \left| R_{n,l}^{(B)}(r) \right|^2 = 1 \quad (35.22)$$

to obtain the constants C_A, C_B, D_B, which fully determine the particle wave functions.

35.4.2 Optical Properties

The interaction of electromagnetic radiation with matter is one of the most intensively investigated fields in solid-state physics. This general denomination includes all processes where charged particles are under the action of an external electromagnetic field. From the corpuscular theory of light, it is known that any electromagnetic radiation is composed of small packages carrying energy named photons, which can be absorbed by electrons. Each photon has an energy of $E = \hbar\omega$, where ω is the frequency of oscillation of the electromagnetic field. The wavelength of the electromagnetic field is obtained with $c = \omega/k$ ($k = 2\pi/\lambda$), where c is the speed of light ($c = 3 \times 10^8$ m/s).

The most usual optical experiments are photoluminescence and optical absorption. Photoluminescence and optical absorption are closely related to light emission and absorption applications. Photoluminescence is the optical radiation emitted by a physical system (in excess of the thermal equilibrium black-body radiation) resulting from excitation to a nonequilibrium state by irradiation with light. If the incident photons have the energy of the same order of the material band gap, the incident radiation creates electron-hole (e–h) pairs, which will remain free until they are captured at an imperfection or recombine directly with a hole releasing the energy absorbed. Recombination can occur spontaneously or stimulated by the incident light beam. If the released energy is delivered as photons, the recombination process is named radiative, or nonradiative recombination otherwise. The spontaneous recombination has no connection with the incident photons, and can emit a photon in any direction with any polarization and frequency. However, the stimulated recombination is related to the absorption for obvious reasons. Since the incident light is absorbed in creating e–h pairs, most of the excitations occur near the surface, restricted to a region within a diffusion length (or absorption length) of the illuminated surface. Since the recombination radiation is subject to self-absorption, it will not propagate far from this region. It follows that most of the recombination radiation escapes through the nearby illuminated surface. Consequently, the vast majority of photoluminescence experiments are arranged to examine the light emitted from the irradiated side of the sample. This is often called *front-surface photoluminescence*. In thin samples with relatively low absorption of the recombination light, the back surface or transmission luminescence can also be examined. There are more sophisticated photoluminescence setups, like the photoluminescence of selective excitation technique, which is used to obtain the fine structure of the photoluminescence spectra.

Once an electron-hole pair is created by the absorption of a photon, they interact with each other by means of their opposite charges, forming a quasi-particle called *exciton*. The total energy of an exciton indicates the color (wavelength) of the light emitted by the quantum systems. Thus, it is an important quantity that must be addressed. In bulk materials, the exciton energy is given by

$$E_X = E_G - E_B \tag{35.23}$$

where

E_G is the band gap of the material

E_B is the exciton-binding energy that accounts for the attraction between the electron-hole pair

In bulk materials, the electron-hole pair resembles an hydrogen atom with different mass. In this case, the exciton-binding energy in bulk materials is calculated with the following expression:

$$E_B = -\frac{\mu_{eh}}{\kappa} \times 13.6 \text{ eV} \tag{35.24}$$

where $\mu_{eh}^{-1} = m_e^{-1} + m_h^{-1}$ and κ are the reduced mass and the dielectric constant of the material, respectively. For nanostructured materials, we have to account for the energy quantization. Thus, the total exciton energy in nanostructured materials is given by

$$E_X = E_e + E_h + E_G - E_B \tag{35.25}$$

where E_e and E_h are the energy states calculated for electrons and holes by solving Equation 35.19, and E_B is obtained with the following expression:

$$E_B = -\frac{e^2}{4\pi\varepsilon_0\kappa} \int\int \frac{|\Psi_e(\vec{r}_e)|^2 |\Psi_h(\vec{r}_h)|^2}{|\vec{r}_e - \vec{r}_h|} d\vec{r}_e d\vec{r}_h \tag{35.26}$$

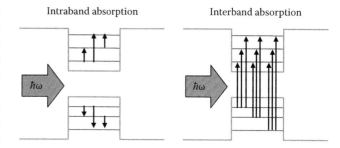

FIGURE 35.7 Different types of photons absorption in core-shell QDs, when illuminated by the light source with energy $E = \hbar\omega$.

Due to the discrete character of the energy levels E_e and E_h of nanostructures, the photon absorption may occur via different mechanisms regarding the energy range of the incident photons. In general, the electronic transitions caused by photons' absorption are divided into two types: intraband and interband transitions. A diagram with these processes is shown in Figure 35.7. Intraband transitions occur exclusively between states within either CB or VB. In this case, the energy of incident photons is very low (few meVs), raging in the infrared region (long wavelengths) of the electromagnetic spectrum. Interband transitions involve transitions between valence and CB. Here, the photons energy are, in general, larger than for intraband transitions. Interband transitions are only possible when the energy of the incident photons is larger than the band gap energy of the material investigated.

In the diagram shown in Figure 35.7, if an appropriate photon energy is supplied, electronic transitions can occur between two available electron states. Besides energy restrictions, there are other rules associated with the symmetry properties of the wave functions of the states involved in the transition. A compact form to indicate if a given transition is allowed is given by a quantity named oscillator strength:

$$f_{i,f} = \frac{2m_0}{E_f - E_i} \left| \langle \Psi_f | \hat{e} \cdot \vec{p} | \psi_i \rangle \right|^2 \tag{35.27}$$

where

E_i, Ψ_i and E_f, Ψ_f represent the energy and wave function of the initial and final states involved in the transition

$\hat{e}$ is the polarization vector of the incident light

$\vec{p}$ is the momentum operator given by $\vec{p} = -i\hbar\nabla$

The oscillator intensity is associated with the intensity of electronic transition when only energy and wave function symmetry considerations are taken into account. For instance, transitions for which $f_{i,f} = 0$ are prohibited, and no absorption/photoluminescence peak in the optical spectra will be observed for the radiation energy $E_f - E_i$. Therefore, absorption/photoluminescence peaks will be observed only if $f_{i,f} \neq 0$. Larger $f_{i,f}$ values indicate more intense absorption/photoluminescence peaks. The quantity represents the dipole moment associated to the $i \rightarrow f$ transition, and its evaluation depends on the type of approximation used to solve Schrodinger equation. In the effective mass approximation, all the atomic details are embedded in the values of the effective masses of the compound materials. The wave functions obtained are called *envelope wave functions*. Such wave functions do not include the periodic component related to the electronic behavior in the vicinity of the underlying atomic lattice, which are called Bloch wave functions. In order to precisely evaluate the dipole moment, the knowledge of this component is necessary [51]. However, it is still possible to evaluate qualitatively the oscillator strengths for intraband and interband transitions within the effective mass approximation with the following expressions:

$$f_{\text{intra}} = \frac{2m_0}{E_f - E_i} \left| \left\langle \Psi_f \left| \hat{e} \cdot \vec{p} \right| \psi_i \right\rangle \right|^2 \qquad (35.28)$$

$$f_{\text{inter}} \propto \frac{2m_0}{E_f - E_i} \left| \left\langle \Psi_f \left| \psi_i \right\rangle \right|^2 \qquad (35.29)$$

where the wave functions used in the expressions are the ones obtained within the effective mass approximation. Due to the normalization condition, the Bloch components do not appear in f_{intra}. On the other hand, the proportionality factor in f_{inter} involves the Bloch function. We suggest Refs. [51,52] for the readers interested in the detailed modeling of the interaction of quantum systems with light.

35.5 Results

Our discussion of the electronic structure will focus on CdSe-based core-shell QDs, which are widely reported in the literature. Among all materials used to cover CdSe NCs, we will discuss three different cases: CdS, ZnS, and CdTe. CdS and ZnS exhibit the lowest and highest type I confinement barriers with respect to the CdSe core, respectively. Thus, any other shell material is expected to present an intermediate behavior between CdS and ZnS. CdTe is our choice of compound for type II confinement. Besides that, many other compounds were successfully grown on the surface of standalone CdSe NCs [18,19]. The confinement profile of such core-shell structures is presented in Figure 35.8, and materials parameters used in this work are given in Table 35.1.

TABLE 35.1   Compilation of the Material Parameters Used in This Work

Parameters	CdSe	CdS	ZnS	CdTe
E_g (eV)	1.74 [53]	2.58 [54]	3.84 [55]	1.60 [57]
$V_e^{(0)}$ (eV)		0.32 [56]	1.44 [56]	0.40 [42]
$V_h^{(0)}$ (eV)		0.42 [56]	0.60 [56]	0.50 [42]
m_e/m_0	0.13	0.20	0.367	0.096 [57]
m_{hh}/m_0	0.45	0.80	0.569	0.400 [57]
ϵ/ϵ_0	9.1			7.1 [57]

35.5.1 CdSe/CdS and CdSe/ZnS QDs: Type I Confinement

Figure 35.9 displays the dependence of electron and hole energy states on the core size for a fixed shell thickness of 1 nm. For the sake of comparison, electron and hole states of an isolated CdSe QD with infinite confinement barriers are also shown (dashed lines). This figure shows the main qualitative features of the electronic structure of QDs. The first one, and most important, is that the energy of the confined states are inversely proportional to the QD size. For the isolated QD with infinite confinement barriers, they are given by

$$E_{n,l}^{(e,h)} = \frac{\hbar^2 x_{n,l}^2}{2m_{e,h}R^2} \qquad (35.30)$$

where $x_{n,l}$ represents the nth root of the spherical Bessel function $j_l(x)$. For noninfinite confinement barriers, the size dependence becomes weaker $E_{n,l}^{(e,h)} \propto R^{-\alpha}$, where $1 < \alpha < 2$. This feature is easily seen in Figure 35.9 for $l > 0$. Interestingly, the electronic states of isolated and core-shell QDs for $l = 0$ are nearly identical.

One notices that for the core-shell QDs, the figure only shows the states whose energies are smaller than the shell confinement barrier. The height of the confinement barriers strongly influences the number of confined states within the QD. For example, for small CdS/CdSe QDs ($R \leq 5\,\text{nm}$), which exhibits smaller confinement barriers in comparison to ZnS/CdSe, there are mostly three electron states below the confinement barrier ($n = 1$, $l = 0$, 1, 2). In the case of holes, although their confinement barrier is similar to the one for electrons, many more states are available (even for $n > 1$, not shown in Figure 35.9). This is a consequence of the larger effective mass of heavy holes, which lowers the total energy of the particle. As for ZnS/CdSe, which exhibits substantially larger confinement barriers for electrons, there are any more states available in the CB than in CdS/CdSe QDs.

Although Figure 35.9 does not show the energy states above the confinement barriers, it does not mean they do not exist. In our simple model, we applied the boundary condition $R_{n,l}(r_B) = 0$. This means that there is an infinite confinement barrier at the QD surface. So, there will be infinite energy states above the shell confinement barrier. However, in actual systems, these QDs are embedded within an arbitrary environment which can be vacuum, air, crystalline, or polymer matrix. Each of these

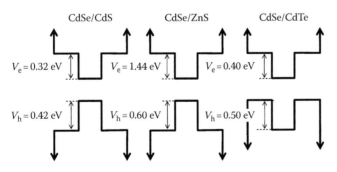

FIGURE 35.8   Confinement model of CdSe/CdS, CdSe/ZnS, and CdSe/CdTe core-shell QDs.

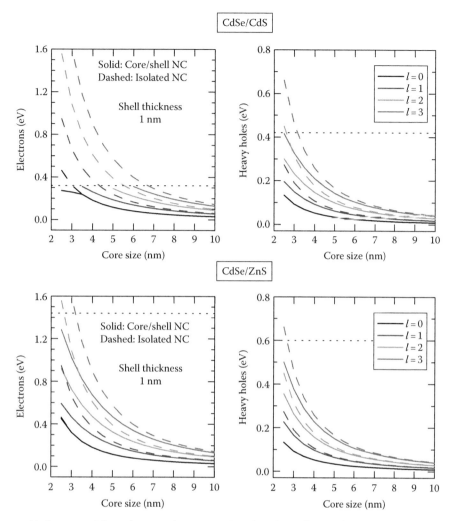

FIGURE 35.9 Electron and hole energies' dependence on the core size for CdS/CdSe and ZnS/CdSe core-shell QDs with a shell layer of 1 nm.

external environment have their own finite confinement barrier. Thus, once a particle occupies these states above the shell barrier, particles may escape from the QD. These states are called *quasi-bound states*, and they will not be addressed here. More information about the properties of quasi-bound states can be found in Refs. [58–60].

So far, we have discussed the dependence of the electronic structure on the core size for a fixed shell thickness. Now we investigate the role of the shell thickness for a fixed core size. These data are shown in Figure 35.10. One can see that ground-state energy ($l = 0$) does not depend on the shell thickness, regardless of the type of particle and shell material. This is not true for higher energy states. For low confinement barriers (CB of CdS/CdSe QD) and thin shell layers, the energy states for $l > 1$ is reasonably affected up to 1.5 nm of shell thickness. The reason behind this modification is the following. For a given energy, the wave function has an exponential decay behavior $\Psi \sim e^{kr}$ (where k is given by Equation 35.14) within the finite confinement barrier. The larger the value of k, the smaller the wavefunction penetration. If this penetration is smaller than the shell thickness, the confined state will not feel the external

infinite barrier at $r = r_B$ [where $R_{n,l}(r_B) = 0$], and the energy state will not be affected by the shell. If the wave function penetration is larger than the shell thickness, the state energy has to be increased to match the external boundary condition. The wave functions of electrons and heavy holes in CdSe/CdS and CdSe/ZnS QDs are depicted in Figure 35.11. Indeed, the lack of dependence on the shell width for $l = 0$ is caused by a negligible wave function penetration in the shell layer. One can also see that the wave function penetration ($l > 0$) is larger in CdSe/CdS due to its lower confinement barriers.

35.5.2 CdSe/CdTe QDs: Type II Confinement

Figure 35.12 shows the core radius and shell width dependence of electrons and heavy holes in CdSe/CdTe QDs. In this system, electrons are confined in the core, while holes are confined in the shell layer (see confinement model in Figure 35.8). One can see that the dependence of electron states with geometrical parameters is qualitatively similar to those in CdSe/CdS and CdSe/ZnS, in which electrons are also confined in the core. On the other hand, the dependence of hole states with geometrical parameters

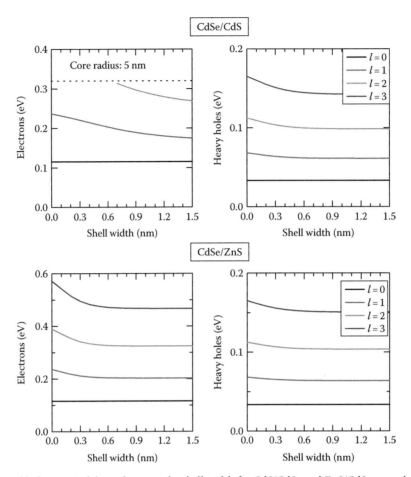

FIGURE 35.10 Electron and hole energies' dependence on the shell width for CdS/CdSe and ZnS/CdSe core-shell QDs with a core radius of 5 nm.

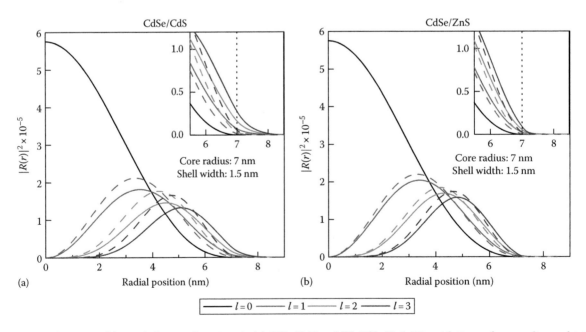

FIGURE 35.11 Electron and heavy-hole wave functions in (a) CdSe/CdS and (b) CdSe/ZnS QDs with 7 nm of core radius and 1.5 nm of shell width.

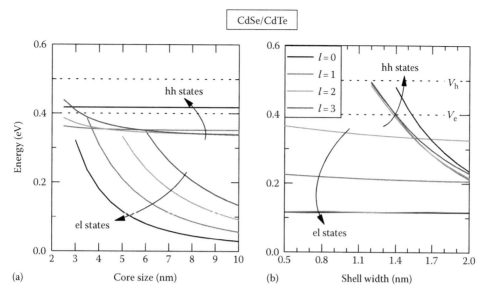

FIGURE 35.12 Electron and hole energies in CdSe/CdTe core-shell QDs: (a) core size dependence for a fixed shell width of 1.5 nm, (b) shell width dependence for a fixed core radius of 5 nm.

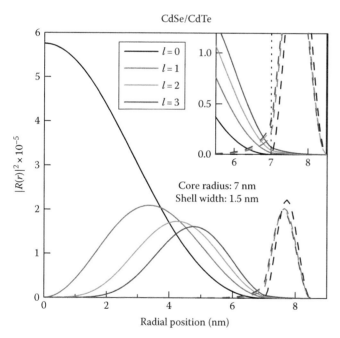

FIGURE 35.13 Electron and heavy-hole wave functions in CdSe/CdTe QD with 7 nm of core radius and 1.5 nm of shell width.

is rather different. Heavy-holes energies exhibit a weak dependence on the core radius. In fact, there is no dependence at all for $l = 0$, indicating that the heavy-hole wave function is totally confined in the shell layer with negligible penetration in the core. The weak dependence with core radius for $l > 0$ indicates that heavy-hole wave functions have small but nonnegligible penetration in the core. The heavy-hole wave functions are depicted in Figure 35.13. The dependence of heavy-hole states with shell thickness is appreciable. Changes of less than 1 nm cause an

energy variation of 0.3 eV. Interestingly, shell states with $l > 0$ have lower energies than with $l = 0$.

35.5.3 Optical Properties: Excitons

The total exciton energy in confined systems is given by Equation 35.25. In the particular case of our core-shell QDs, the ground-state exciton energy $E_X^{(0)}$ for type I (CdSe/CdS and CdSe/ZnS) and type II (CdSe/CdTe) systems are, respectively, given by

$$E_X^{(0)} = E_{G(CdSe)} + E_{1,0}^{(e)} + E_{1,0}^{(h)} + E_B \tag{35.31}$$

$$E_X^{(0)} = E_{G(CdSe)} - V_h + E_{1,0}^{(e)} + E_{1,0}^{(h)} + E_B \tag{35.32}$$

where the exciton-binding energy is obtained with

$$E_B = -\frac{e^2}{4\pi\varepsilon_0 \kappa_{CdSe}} \iint \frac{|\Psi_{1,0}^{(e)}(\vec{r}_e)|^2 |\Psi_{1,0}^{(h)}(\vec{r}_h)|^2}{|\vec{r}_e - \vec{r}_h|} d\vec{r}_e d\vec{r}_h \tag{35.33}$$

The calculation of the exciton-binding energy is cumbersome and can be performed by expanding the Coulomb potential $|\vec{r}_e - \vec{r}_h| - 1$ in a series of spherical harmonics:

$$\frac{1}{|\vec{r}_e - \vec{r}_h|} = \sum_{l=0}^{\infty} \frac{r_<^l}{r_>^{l+1}} \left(\frac{4\pi}{2l+1}\right) \sum_{m=-l}^{l} Y_{l,m}(\theta_e, \phi_e) Y_{l,m}^*(\theta_h, \phi_h) \tag{35.34}$$

where
$r_< = \min(r_e, r_h)$
$r_> = \max(r_e, r_h)$

By using the fact that $\Psi_{1,0}^{(i)}(\vec{r}) = R_{1,0}^{(i)}(r) Y_{0,0}^{(i)}(\theta, \phi)$, the ground-state exciton-binding energy integral can be reduced to

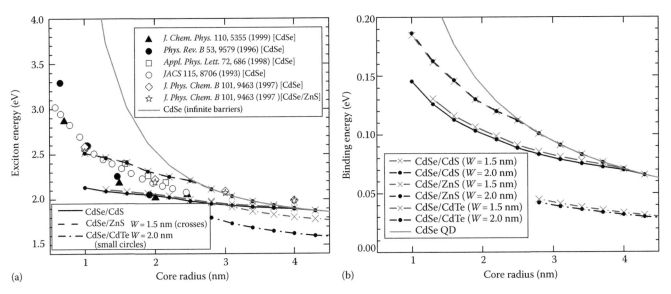

FIGURE 35.14 (a) Ground-state exciton energies for CdSe, CdSe/CdS, CdSe/ZnS, and CdSe/CdTe QDs. For the sake of comparison, experimental data from different authors in the literature are also shown [12,19,61–63]. (b) Size dependence of the exciton-binding energies.

$$E_B = -\frac{e^2}{4\pi\varepsilon_0\kappa_{CdSe}} \left(\int_0^\infty dr_e r_e \left| R_{1,0}^{(e)} \right|^2 \int_0^{r_e} dr_h r_h^2 \left| R_{1,0}^{(h)} \right|^2 \right.$$

$$\left. + \int_0^\infty dr_e r_e^2 \left| R_{1,0}^{(e)} \right|^2 \int_{r_e}^\infty dr_h r_h \left| R_{1,0}^{(h)} \right|^2 \right)$$

Once the electron and hole radial wave functions are known, the above integral can be evaluated with regular integration algorithms.

The size dependence of the ground-state exciton energy is depicted in Figure 35.14. The excitonic energies calculated with the infinite barriers model strongly differ from experimental data for small QDs, which is a well-known problem of the effective mass approximation. However, the effective mass model with finite confinement barriers provides reasonable agreement with experimental data for a wide range of QD sizes. Moreover, the agreement with the experimental CdSe/ZnS of Dabbousi et al. is remarkable [19]. Concerning the binding energy, which accounts for Coulomb attraction between electrons and holes, it is expected that the binding energies of standalone CdSe QDs are much larger than in core-shell QDs. The reason for this is simple. In core-shell QDs, the wave functions are spread over a larger volume, in comparison to bare CdSe QDs, due to the penetration in the shell layer. Binding energies in CdSe/ZnS are larger in comparison to CdSe/CdS because the lower confinement barriers in the latter favors larger wave functions' penetration in the shell. In the case of CdSe/CdTe, the binding energy is substantially smaller than in CdSe/CdS and CdSe/ZnS because the heavy-hole wave function is spatially separated from the electron wave function, increasing the average distance between electrons and holes. Interestingly, exciton energies as well as exciton-binding energies in type II systems are very sensitive to changes in the shell thickness.

35.6 Applications of Core-Shell QDs

Colloidal core-shell QDs are important for a number of applications in areas as diverse as biomedicine, optoelectronics, and power generation. This is due to their relatively simple and inexpensive methods of fabrication, which provide good control of shape and size, and their superior optical properties, exhibiting bright fluorescence and enhanced photostability up to several months, broad excitation spectra, and high absorption coefficients. In this section, some of the most important applications of core-shell QDs are reviewed and discussed.

35.6.1 Biomedical Applications

In the last decade, colloidal QDs have attracted enormous attention as a new class of fluorophores for diagnostic and sensoric applications. The unique optical properties of such nanoparticles lead to major advances in fluorescence detection and imaging in molecular and cell biology [64]. Colloidal QDs can be easily attached to biological molecules like peptides, proteins, and DNA [65–71]. QDs can act like nanoscale probes that can track the movement of cells and individual molecules as they move in their environment by means of conventional fluorescence measurements [72,73]. Such ability to observe and interact with complex systems in real time provides detailed information about fundamental mechanisms involved in the molecular and cellular changes associated with diseases. Due to the nanometer scale of QDs, they can readily interact with biomolecules on the cell surface and within the cell without altering its behavior and the biochemical properties of those molecules. Ref. [74] presents a detailed discussion of many biomedical applications of QDs and other nanostructured materials.

In order to fully appreciate the biomedical applications of colloidal QDs, the surface chemistry of the particle must be understood. QD's surfaces have to be protected and functionalized to provide biocompatibility, biostability, and solubility. In particular, attaching hydrophilic polymers to the surface greatly increases the solubility of the particles and can protect attached proteins from enzymatic degradation when used in vivo applications [75]. Hydrophilic polymers also increase the in vivo compatibility of nanoparticles. Uncoated nanoparticles are rapidly cleared from bloodstream by the reticuloendothelial system. On the other hand, nanoparticles coated with hydrophilic polymers exhibit longer half-lives in the bloodstream [76,77]. In fact, the ability to modify the QD's surfaces allows a large variety of molecular and biological entities to be bound to it. The different options of surface modification lead to QDs with different optical and chemical properties, and the possibility to target specific cell types or tissues. This is a necessary requirement for the multiplicity of applications that QDs can provide in diagnostics and sensorics.

Since water solubility is a necessary requirement for biomedical applications, two basic strategies were developed to achieve it. One approach completely replaces the surface ligands remaining from synthesis, and the other only caps the present ligands on the surface with hydrophilic polymers. Both strategies present advantages and disadvantages. The replacement of the original hydrophobic surface ligands leads to nanoparticles with small final diameter, only slightly larger than the bare QDs. However, it often results in poor quantum yields and modifications in the physicochemical properties and photostability of QDs. On the other hand, capping original surface ligands preserves the photoemission properties of QDs, but the final size of the particles are a few times larger than the original QD diameter. For a complete review of functionalization strategies of QDs, see Ref. [78].

35.6.2 Photovoltaic Applications

The field of photovoltaics gained new breath with the recent proposal and demonstration of NC-based PV devices [32,36]. Several characteristics make NCs an attractive option for the development of PV devices: (1) It is well known that the NCs band gap can be tailored to absorb light in the whole solar spectrum. (2) Enhanced impact ionization leads to multiple exciton generation (MEG) for each photon absorbed [33–35,79,80]. (3) Novel inexpensive chemical methods were developed to grow colloidal NCs that are highly compatible with solid-state device technology [38,39]. (4) Such NCs can be embedded in a semiconductor polymer film for the development of either flexible or nonflat devices, such that every surface might become a potential PV device. Finally, (5) contrary to other NC-based applications, NC size homogeneity is not mandatory. Actually, a size dispersion might be useful to enlarge the absorption window. Although MEG has been experimentally obtained many times with quantum yields (QYs) up to 700% [33–35,79,80], photocurrent QYs larger than 100% have not been reported so far [81].

The most probable configurations for the development of QD solar cells seem to be based on: (1) tridimensionally ordered array of QDs, in such way that the inter-QD distance is small enough to allow the formation of mini-bands through electronic coupling between neighboring QDs, or (2) QDs dispersed in organic semiconductor polymer matrices [36,82]. Nowadays, CdSe colloidal QDs can be inexpensively grown with precise control over their shape and size [14]. However, the environment where QDs are embedded strongly affects their electronic structure and optical properties [38,39]. Chemically passivating the QD core with a thin shell of wide band gap semiconductor prevents chemical interaction with the inter-QD environment. Moreover, it allows the substantial improvement of their optical stability and exhibits greater tolerance to processing conditions necessary for incorporation into solid-state structures.

Several wide band gap semiconductors (e.g., ZnS, CdS, and ZnSe) have been epitaxially grown on the surface of CdSe colloidal QDs [18,19]. Regardless of the shell material, incident photons are absorbed by the QDs, generating single or multiple e–h pairs. There are some possible relaxation channels for these excited QDs: (1) radiative recombination emitting photons with energies that inversely scale with the QD sizes, (2) nonradiative recombination through Auger processes or phonon emission [83], (3) quantum tunneling through the shell layer, and temperature-related effects like (4) thermionic emission and (5) thermal-assisted tunneling [84,85]. Depending on the relative efficiencies of such channels, different QD-based applications can be developed. For instance, faster radiative transitions in comparison to other processes lead to light-emitting applications, while efficient escape of e–h pairs may generate electrical currents, which are useful for the development of PV devices.

The photo-generated current in QD solar cells arises from the fraction of e–h pairs created by the absorption of photons that escape from the QDs through the shell layer before recombining either radiatively or not. The main escape processes are depicted in Figure 35.15. Due to the discreteness of the density of states, the energy difference between adjacent states in the conduction and VB is larger than the thermal activation energy k_BT even at room temperature, suppressing the occupation of excited states. This is particularly important in CdSe QDs because of the small carrier effective masses. In the case of small confinement barriers, the enhanced binding energy in QDs is several times larger than k_BT, so that temperature effects are not strong enough to dissociate confined e–h pairs. Thus, temperature-related processes can be ruled out and the main contribution to the photo-generated current can be considered as arising from the out-tunneling of ground-state e–h pairs. There are two obstacles competing to the out-tunneling of this e–h pair: shell barrier and Coulomb interaction. If these mechanisms are strong enough to hold the confined e–h pair during times of the same order of the recombination lifetime, the conversion efficiency is expected to be very low.

In order to identify the ideal conditions for which QD solar cells are able to work, it is necessary to compare the exciton tunneling times with their recombination lifetimes. It was recently shown that the exciton tunneling times are extremely sensitive to QD sizes, shell thickness, and confinement barrier heights. Depending on the combination of these quantities, the

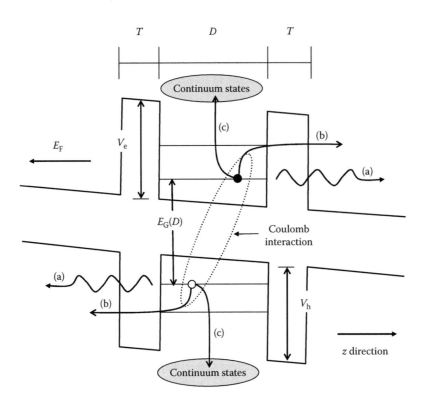

FIGURE 35.15 Schematics of the escape processes of an e–h pair in a single QD. T and D represent the shell thickness and core size of the QD, respectively. The CB and VB energy barriers are represented by V_e and V_h, respectively, and E_F is an external electric field. The figure shows the processes responsible for current generation observed in PV devices: (a) Ground-state tunneling, (b) thermal-assisted tunneling, and (c) thermionic emission. Processes (b) and (c) are highly dependent on temperature. (Adapted from Etteh, N.E.I. and Harrison, P., *Physica E*, 13, 381, 2002; Appenzeller, J. et al., *Phys. Rev. Lett.*, 92, 048301, 2004.)

out-tunneling times can be comparable to the recombination lifetimes, indicating that achieving efficient charge extraction depends on a careful selection of geometrical parameters [37].

35.7 Conclusions

Core-shell QDs are considered potential building blocks for a number of future technologies in the fields of biomedicine, optoelectronics, and PVs. Therefore, they represent an important subject for students, scientists, and engineers of different knowledge areas. In this chapter, we aimed to (1) review state-of-art research in core-shell QDs and (2) provide underlying principles and theoretical tools for nonspecialists to be able to study cores-shell QDs by themselves. Although simple, the theoretical tools presented here are powerful enough to reproduce many phenomena experimentally observed. We have shown that besides size control, covering QDs with another material offers additional degrees of tuning QD electronic properties, which are extremely sensitive to parameters like shell material and thickness. In fact, depending on the confinement barrier, the penetration length of the wave function in the shell is smaller than the shell thickness. Thus, the surface passivation has no influence on the electronic transitions that might occur in the core. This core isolation provided by the shell represents one of the main advantages of core-shell QDs.

References

1. M. G. Cottam, D. R. Tilley, *Introduction to Surface and Superlattices Excitations*, Cambridge University Press, Cambridge, U.K., 1989.
2. A. Mac Donald, in: C. R. Leavens, R. Taylor (Eds.), *Interfaces, Quantum Wells and Superlattices*, Plenum, New York, 1987.
3. E. L. Albuquerque, M. G. Cottam, *Polaritons in Periodic and Quasiperiodic Structures*, Elsevier, Amsterdam, the Netherlands, 2004.
4. G. H. Dohler, *Phys. Scr.* **24**, 430 (1981).
5. J. Kim, L. Wang, A. Zunger, *Phys. Rev. B* **56**, R15541 (1997).
6. N. V. Tkach, I. V. Pronishin, A. M. Makhanets, *Phys. Solid State* **40**, 514 (1998).
7. R. R. L. de Carvalho, J. Ribeiro Filho, G. A. Farias, V. N. Freire, *Superlattices Microstruct.* **25**, 221 (1998).
8. G. W. Bryant, P. S. Julienne, Y. B. Band, *Surf. Sci.* **361**, 801 (1996).
9. J. W. Haus, H. S. Zhou, I. Homma, H. Komiyama, *Phys. Rev. B* **47**, 1359 (1993).
10. D. Schooss, A. Mews, A. Eychmuller, H. Weller, *Phys. Rev. B* **49**, 17072 (1994).
11. J. Ribeiro Filho, R. R. L. de Carvalho, G. A. Farias, V. N. Freire, E. L. Albuquerque, *Physica B* **305**, 38 (2001).

12. C. B. Murray, D. J. Norris, M. G. Bawendi, *J. Am. Chem. Soc.* **115**, 8706 (1993).

13. O. I. Micic, J. R. Sprague, C. J. Curtis, K. M. Jones, J. L. Machol, A. J. Nozik, H. Giessen, B. Fluegel, G. Mohs, N. J. Peyghambarian, *Phys. Chem.* **99**, 7754 (1995).

14. X. Peng, L. Manna, W. Yang, J. Wickham, E. Scher, A. Kadavanich, A. P. Alivisatos, *Nature* **404**, 59 (2000).

15. X. Brokmann, E. Giacobino, M. Daham, J. P. Hermier, *Appl. Phys. Lett.* **85**, 712 (2004).

16. P. Reiss, J. Bleuse, A. Pron, *Nano Lett.* **2**, 781 (2002).

17. D. J. Milliron, S. M. Hughes, Y. Cui, L. Manna, J. Li, L.-W. Wang, A. P. Alivisatos, *Nature* **430**, 190 (2004).

18. D. V. Talapin, I. Mekis, S. Gotzinger, A. Kornowski, O. Benson, H. J. Weller, *J. Phys. Chem. B* **108**, 18826 (2004).

19. B. O. Dabbousi, J. Rodriguez-Viejo, F. V. Mikulec, J. R. Heine, H. Mattoussi, R. Ober, K. F. Jensen, M. G. Bawendi, *J. Phys. Chem. B* **101**, 9463 (1997).

20. M. Nirmal, C. B. Murray, M. G. Bawendi, *Phys. Rev. B* **50**, 2293 (1994).

21. M. Chamarro, C. Gourdon, P. Lavallard, A. I. Ekimov, *Jpn. J. Appl. Phys.* (Suppl. 34–1), 12 (1995).

22. A. L. Efros, M. Rosen, M. Kuno, M. Nirmal, D. J. Norris, M. G. Bawendi, *Phys. Rev. B* **54**, 4843 (1996).

23. U. Banin, O. Millo, *Annu. Rev. Phys. Chem.* **54**, 465 (2003).

24. X. Brokmann, G. Messin, P. Desbiolles, E. Giacobino, M. Dahan, J. P. Hermier, *New J. Phys.* **6**, 99 (2004).

25. P. Alivisatos, *Nat. Biotechnol.* **22**, 47 (2004).

26. J. K. Jaiswal, S. M. Simon, *Trends Cell Biol.* **14**, 497 (2004).

27. S. A. Ivanov, J. Nanda, A. Piryatinski, M. Achermann, L. Balet, I. V. Bezel, P. O. Anikeeva, S. Tretiak, V. I. Klimov, *J. Phys. Chem. B* **108**, 10625 (2004).

28. V. I. Klimov, A. A. Mikhailovsky, S. Xu, A. Malko, J. A. Hollingsworth, C. A. Leatherdale, H. J. Heisler, M. G. Bawendi, *Science* **290**, 314 (2000).

29. A. Piryatinski, S. A. Ivanov, S. Tretiak, V. I. Klimov, *Nano Lett.* **7**, 108 (2007).

30. B. Kraabel, A. Malko, J. Hollingsworth, V. I. Klimov, *Appl. Phys. Lett.* **78**, 1814 (2001).

31. M. A. Petruska, A. V. Malko, P. M. Volyes, V. I. Klimov, *Adv. Mater.* **15**, 610 (2003).

32. A. J. Nozik, *Physica E* **14**, 115 (2002).

33. R. D. Schaller, V. I. Klimov, *Phys. Rev. Lett.* **92**, 186601 (2004).

34. R. D. Schaller, M. Sykora, J. M. Pietryga, V. I. Klimov, *Nano Lett.* **6**, 424 (2006).

35. R. J. Ellingson, M. C. Beard, J. C. Johnson, P. Yu, O. I. Micic, A. J. Nozik, A. Shabaev, A. L. Efros, *Nano Lett.* **5**, 865 (2005).

36. W. U. Huynh, J. J. Dittmer, A. P. Alivisatos, *Science* **295**, 2425 (2002).

37. J. S. de Sousa, J. A. K. Freire, G. A. Farias, *Phys. Rev. B* **76**, 155317 (2007).

38. Z. Yu, L. Guo, T. Krauss, J. Silcox, *Nano Lett.* **5**, 565 (2005).

39. I. Mekis, D. V. Talapin, A. Kornowski, M. Haase, H. J. Weller, *J. Phys. Chem. B* **107**, 7454 (2003).

40. M. Soreni-Harari, N. Yaacobi-Gross, D. Steiner, A. Aharoni, U. Banin, O. Millo, N. Tessler, *Nano Lett.* **8**, 678 (2008).

41. M. A. Hines, P. Guyot-Sionnest, *J. Phys. Chem.* **100**, 468 (1996).

42. J. Li, L.-W. Wang, *Appl. Phys. Lett.* **84**, 3684 (2004).

43. L. Manna, E. C. Scher, L.-S. Li, A. P. Alivisatos, *J. Am. Chem. Soc.* **124**, 7136 (2002).

44. L. Pauling, E. B. Wilson Jr., *Introduction to Quantum Mechanics with Applications to Chemistry*, Dover, New York, 1963.

45. L. D. Landau, E. M. Lifshitz, *Quantum Mechanics (Non-Relativistic Theory)* 3rd edn., Butterworth-Heinemann, Newton, MA, 1977.

46. J. J. Sakurai, *Modern Quantum Mechanics*, revised edition, Addison-Wesley, New York, 1995.

47. P. Harrison, *Quantum Wells, Wires and Dots*, Wiley, Chichester, U.K., 2000.

48. C. Pryor, J. Kim, L. W. Wang, A. J. Williamson, A. Zunger, *J. Appl. Phys.* **83**, 2548 (1998).

49. V. A. Fonoberov, A. A. Balandin, *J. Appl. Phys.* **94**, 7178 (2003).

50. G. B. Arfken, H. J. Weber, *Mathematical Methods for Physicists*, Academic Press, San Diego, CA, 2005.

51. S. L. Chuang, *Physics of Optoelectronic Devices—Wiley Series in Pure and Applied Optics*, Wiley, New York, 1995.

52. M. Balkanski, R. F. Wallis, *Semiconductor Physics and Applications*, Oxford University Press, Oxford, U.K., 2000.

53. Y. D. Kim, M. V. Klein, S. F. Ren, Y. C. Chang, H. Luo, N. Samarth, J. K. Furdyna, *Phys. Rev. B* **49**, 7262 (1994).

54. Z. Yu, J. Li, D. B. O'connor, L.-W. Wang, P. F. Barbara, *J. Phys. Chem. B* **107**, 5670 (2003).

55. C. G. Van de Walle, *Phys. Rev. B* **39**, 1871 (1989).

56. B. S. Kim, M. A. Islam, L. E. Brus, I. P. Herman, *J. Appl. Phys.* **89**, 8127 (2001).

57. Y. Masumoto, K. Sonobe, *Phys. Rev. B* **56**, 9734 (1997).

58. A. Vasanelli, R. Ferreira, G. Bastard, *Phys. Rev. Lett.* **89**, 216804 (2002).

59. R. Oulton, J. J. Finley, A. I. Tartakovskii, D. J. Mowbray, M. S. Skolnick, M. Hopkinson, A. Vasanelli, R. Ferreira, G. Bastard, *Phys. Rev. B* **68**, 235301 (2003).

60. D. P. Nguyen, N. Regnault, R. Ferreira, G. Bastard, *Phys. Rev. B* **71**, 245329 (2005).

61. P. Guyot-Sionnest, M. A. Hines, *Appl. Phys. Lett.* **72**, 686 (1998).

62. L.-W. Wang, A. Zunger, *Phys. Rev. B* **53**, 9579 (1996).

63. E. Rabani, B. Hetényi, B. J. Berne, L. E. Brus, *J. Chem. Phys.* **110**, 5355 (1999).

64. C. M. Niemeyer, *Angew. Chem. Int. Ed. Engl.* **40**, 4128 (2001).

65. S. R. Whaley, D. S. English, E. L. Hu, P. F. Barbara, A. M. Belcher, *Nature* **405**, 665 (2000).

66. M. Bruchez Jr., M. Moronne, P. Gin, S. Weiss, A. P. Alivisatos, *Science* **281**, 2013 (1998).

67. W. C. W. Chan, S. M. Nie, *Science* **281**, 2016 (1998).

68. H. Mattoussi, J. M. Mauro, E. R. Goldman, G. P. Anderson, V. C. Sundar, F. V. Mikulec, M. G. Bawendi, *J. Am. Chem. Soc.* **122**, 12142 (2000).

69. G. P. Mitchell, C. A. Mirkin, R. L. Letsinger, *J. Am. Chem. Soc.* **121**, 8122 (1999).

70. S. Pathak, S. K. Choi, N. Arnheim, M. E. Thompson, *J. Am. Chem. Soc.* **123**, 4103 (2001).

71. M. Y. Han, X. H. Gao, J. Z. Su, S. Nie, *Nat. Biotechnol.* **19**, 631 (2001).

72. E. B. Voura, J. K. Jaiswal, H. Mattoussi, S. M. Simon, *Nat. Med.* **10**, 993 (2004).

73. X. Wu, M. P. Bruchez, *Methods Cell Biol.* **75**, 171 (2004).

74. S. E. McNeil, *J. Leukoc. Biol.* **78**, 585 (2005).

75. J. M. Harris, R. B. Chess, *Nat. Rev. Drug Discov.* **2**, 214 (2003).

76. I. Brigger, C. Dubernet, P. Couvreur, *Adv. Drug Deliv. Res.* **54**, 631 (2002).

77. S. M. Moghimi, J. Szebeni, *Prog. Lipid Res.* **42**, 463 (2003).

78. A. F. E. Hezinger, J. Teßmar, A. Göferich, *Eur. J. Pharm. Biopharm.* **68**, 138 (2008).

79. R. D. Schaller, V. M. Agranovich, V. I. Klimov, *Nat. Phys.* **1**, 189 (2005).

80. M. C. Beard, K. P. Knutsen, P. Yu, J. M. Luther, Q. Song, W. K. Metzger, R. J. Ellingson, A. J. Nozik, *Nano Lett.* **7**, 2506 (2007).

81. A. J. Nozik, *Inorg. Chem.* **44**, 6893 (2005).

82. M. Gratzel, *Nature* **414**, 338 (2001).

83. V. I. Klimov, A. A. Mikhailovsky, D. W. McBranch, A. C. Leatherdale, M. G. Bawendi, *Science* **287**, 1011 (2000).

84. N. E. I. Etteh, P. Harrison, *Physica E* **13**, 381 (2002).

85. J. Appenzeller, M. Radosavljevic, J. Knoch, Ph. Avouris, *Phys. Rev. Lett.* **92**, 048301 (2004).

36

Polymer-Coated Quantum Dots

Anna F. E. Hezinger
University of Regensburg

Achim M. Goepferich
University of Regensburg

Joerg K. Tessmar
University of Regensburg

36.1 Introduction

In the last decade, colloidal quantum dots (QDs) have drawn tremendous attention as a new class of fluorophores for a wide range of diagnostic and sensoric applications. Their unique optical properties have led to major advances in fluorescence detection and imaging in molecular and cell biology [1]. In developing QDs, it has become possible to link these inorganic semiconductor nanoparticles to biological molecules such as peptides [2], proteins [3–5], and DNA [6,7] for imaging purposes. They have also been adapted to perform as multicolor fluorescent labels for both in vitro and in vivo imaging [8,9]. QDs have also been successfully used as sensors for analytes ranging from small ions to complex molecules like sugars or even neurotransmitters [10–12].

The most commonly used QDs belong to the cadmium chalcogenide group, due to the ease of synthesis and handling. Their semiconductor nature gives rise to bright and stable fluorescence over broad excitation spectra with high absorption coefficients. These unique optical properties make QDs advantageous compared to common organic dyes and genetically engineered fluorescent proteins in many biological and biomedical applications. For example, QDs could be used for multiplexed imaging and long-term investigations, such as for cellular uptake studies or in vivo imaging, due to their tunable emission wavelength and a protracted photostability of up to several months [4,8]. Still, QD surfaces must be protected and functionalized to provide biocompatibility, biostability, and suitable surface properties for these applications.

The development of appropriate surface coatings represents a major step toward the applicability of QD systems. Coatings should provide three functions for the QDs: chemical and physical stabilization, the suppression of cellular toxicity, and the potential for further modifications via the attachment of certain surface groups. The continuous evolution of surface coatings began when the first water-soluble QDs were coated with mercaptopropionic acid, which was receptive to chemical functionalization using the free carboxylic group (Figure 36.1). These QDs were improved upon by the rapid development of a wide range of polymeric ligands and amphiphilic polymers that could be coordinated on top of the nanocrystal surface. These polymer and ligand coatings are aimed at a diversity of biological applications, and they even give rise to new fields of relevance such as lifetime imaging or single molecule detection. Because of their wide range of applicability, polymer and ligand coatings have been developed with variable properties. Moreover, two fundamentally different ways of creating surface coatings have emerged, each with its own advantages and disadvantages. This chapter will provide a summary and comparison of the different polymer-based coating strategies. It will also discuss the relevant polymers that are currently used for the modifications.

36.2 Biocompatible Quantum Dots

Present-day QD probes are photostable and water-soluble nanoparticulate systems that are capable of displaying strong luminescence and offer tunable size and emission wavelengths [13]. Progress in making better QDs can be ascribed to a series of technological developments that have provided new functionalities to candidate inorganic materials. Perhaps, the first such important development came with the highly crystalline and

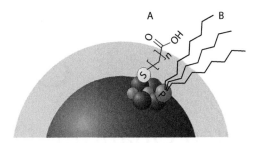

FIGURE 36.1 Schematic drawing of the QD surface with (A) a hydrophilic mercaptoalkane acid applied for water solubility and (B) a lipophilic TOPO ligand present from the synthesis.

monodisperse cadmium selenide nanocrystals that were synthesized in a hot coordinating solvent and introduced by Bawendi and coworkers in 1993 [14]. Another landmark followed when other different semiconductor materials were used as coatings to passivate the QD surface, thereby improving the photostability and brightness of these QDs [15]. In 1998, the first synthetic approaches to water-soluble semiconductor nanocrystals were published [3,4]. Today, QDs are composed of not only cadmium selenide (CdSe) but many other semiconducting materials derived from the II and VI elemental groups (e.g., CdTe, CdS, CdHg, ZnS) and III and V elemental groups (e.g., InAs, InP, GaAs) of the periodic table. With all of these different semiconductors in use, emissions of QDs can span the whole spectral range from ultraviolet to near-infrared [16–20] (Figure 36.2).

Water solubility, stability against oxidation and subsequent degradation, small diameters, and functionalizable groups are essential for the application of QDs in biological systems. Since unmodified nanocrystals possess extremely hydrophobic surface ligands like trioctylphosphine (TOPO) oxide or hexadecylamine resulting from the organometallic synthesis, they are not suited for biological applications. Due to this fact, the hydrophilization of the nanoparticle surface is an essential prerequisite for the application of QDs.

Since the first reports on water-soluble QDs were published, coating and capping strategies that provide a water-soluble shell have arisen, each having different effects on the overall properties of the modified particles. The strategies can be divided into

a distinct dichotomy of approaches. One approach completely replaces the surface bound ligands remaining from synthesis, while the other only caps the ligands that are present on the QDs with suitable amphiphilic polymers (Figure 36.3). Both approaches have advantages and disadvantages for the obtained water-soluble particles. Replacing the original hydrophobic surface ligands with amphiphilic ligands leads to particles with a significantly smaller final diameter. These composites are often only a few nanometers larger than the core QDs. Nevertheless, the exchange of the surface coating often results in poor quantum yields and strongly affects the physicochemical and photophysical stability of QDs in aqueous solutions. Surface capping chemistries, in contrast, retain the original surface ligands and therefore preserve the photophysical properties of the nanocrystals. However, this approach results in particles with a final size three or four times larger than the original nanocrystal diameter, which can be deteriorating for the later application.

The huge variety of different surface modifications results in QDs of very different optical and chemical properties. Indeed, this diversity is necessary for the multiplicity of applications that semiconductor nanocrystals are used for in diagnostics and sensorics. Properties like particle size and surface charge, as well as application-relevant parameters like chemical and photophysical stability, photoluminescence intensity, or cytotoxicity, have to be considered to choose the optimal system for each application.

The focus of the following chapter will be set on coating strategies with organic substances, mostly polymers or polymer derivatives. For completeness, it has to be mentioned that there are various other possibilities for inorganic coating of QDs with silica or titania. These coating strategies are also based on the same two principles of ligand exchange or ligand capping to anchor the inorganic coating on the nanocrystal surface. This is followed by the formation of another capping inorganic layer, shielding the QD and rendering it water soluble [21–32].

36.2.1 Effects of Surface Coating

Coatings can alter the overall QD properties. Alterations can be in the photophysical properties, the physicochemical properties, or the toxicological profile (Figure 36.4). Photophysical

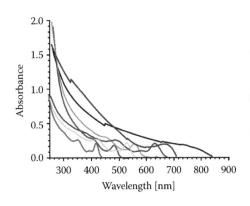

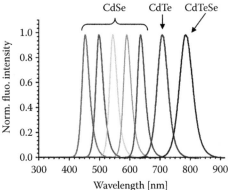

FIGURE 36.2 Absorbance and fluorescence spectra of CdSe/ZnS, CdTe, and CdTe/CdSe QDs of various sizes.

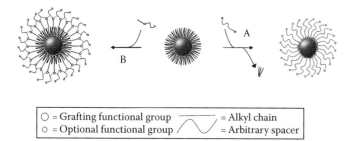

FIGURE 36.3 Scheme of the (A) ligand exchange and (B) the ligand capping strategy.

characteristics that may be affected include the emission wavelength, the quantum yield, and the photostability. The physicochemical aspects that may be varied influence the size, the charge, and the aggregation tendency of QDs in biological fluids and therefore mostly determine the stability of QD probes in different biological environments. The cytotoxicity of QDs is an essential point to consider especially in cell culture or for in vivo applications.

36.2.2 Photophysical Aspects

The natural QD capping, resulting from the synthesis, protects the surface against oxidation and can compensate for surface defects. Too many surface defects result in a decrease of quantum yield, because excitons can also emit their energy in a nonradiative manner. Additionally, the photostability is significantly and negatively influenced by photooxidation at the surface. Finally, the occurring surface oxidation is also responsible for an effect called "blueing" of the QDs, which is a shift of the emission wavelengths toward blue color [33–36]. In case the surface of a nanocrystal gets oxidized, the remaining emitting

semiconductor core gets smaller (Figure 36.5). When the core gets smaller, the emission wavelength shifts to higher energies and therefore smaller wavelengths [37,38]. Consequently, the exchange of the original capping causes an increased likelihood for the occurrence of damages due to an incomplete coverage and imperfect grafting of the newly added ligands. Additionally, thiol-containing ligands used in many approaches are themselves susceptible to the oxidation of the thiol group, leading to a subsequent detachment of the coating from the surface. Here again, the mere capping of the initial ligands on top of the QDs with amphiphilic polymers reduces the likelihood of surface defects and in most cases provides superior protection against oxidation due to the thicker shell on top of the particles.

36.2.3 Physicochemical Aspects

The physicochemical attributes of the nanocrystals that are affected by the chosen coating strategy include size, charge, and the aggregation stability of the particle suspension in biological systems. This makes the choice of the coating approach extremely important, as the physicochemical aspects have significant impact on the overall applicability of QDs. As previously mentioned, the ligand exchange method yields small diameter particles, but it also results in increased oxidation sensitivity of the thiol grafting ligands, which may result in an aggregation of the QDs, due to the loss of surface shielding. Therefore, in any application, there is a trade-off between advantageous and disadvantageous attributes that must be considered when determining a coating method. The beneficial small dimensions of QDs with exchanged ligands are most desirable in some applications, even with the accompanying low stability against aggregation. In other applications, capping the ligands, which produces comparatively large polymer-coated particles, is the chosen method

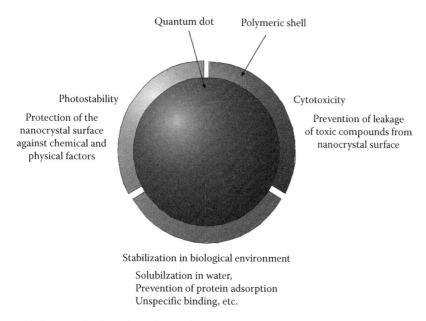

FIGURE 36.4 Functions and influences of polymeric coating on QDs.

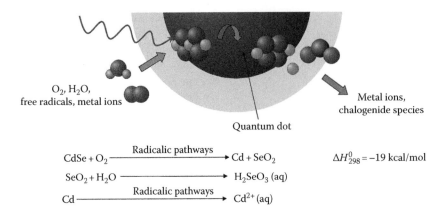

CdSe + O$_2$ $\xrightarrow{\text{Radicalic pathways}}$ Cd + SeO$_2$ $\Delta H^0_{298} = -19$ kcal/mol

SeO$_2$ + H$_2$O $\longrightarrow$ H$_2$SeO$_3$ (aq)

Cd $\xrightarrow{\text{Radicalic pathways}}$ Cd^{2+} (aq)

FIGURE 36.5 Schematic drawing and reaction scheme of the photooxidation on the nanocrystal surface of CdSe.

because it leads to good chemical stabilization of the surface and reliable protection against aggregation.

In both surface-coating methods, aqueous solubilization is achieved through added charged groups on the surface. The most commonly used chemical moieties for this are carboxyl and amino groups, both of which also hold the potential for further chemical functionalization with specific biomolecules. However, the use of these highly charged moieties also increases the likelihood of aggregation in biological environments caused by interactions with serum proteins or ions that are present in biological fluids. For example, anionic shell destabilization can occur due to high ionic strength of the aqueous solution, increased temperature, or complex salt mixtures, all of which reduce the repulsive forces of the ionic groups. To circumvent these issues, another frequently used technique for altering the physicochemical attributes of the particles is PEGylation of the existing polymer shell. PEGylation adds differently sized poly(ethylene glycol) (PEG) chains to the QDs, producing uncharged and mainly sterically stabilized colloids, while also reducing unspecific uptake in cells, preventing protein adsorption on the polymer shells and lowering the risk of agglomeration in biological fluids [39,40].

36.2.4 Toxicological Aspects

Because biological applications are at the forefront of the usage of QDs, their cytotoxicity is a tremendously important consideration. QD size, charge, and concentration, along with their outer shell bioactivity and oxidative and photolytic stress are all factors that, collectively and individually, can determine their cytotoxicity. For biological applications, the protection of the nanocrystal surface is not only important for probe stability, but it is also vital to prevent the leakage of cytotoxic semiconductor components from the inorganic core that may occur due to degradation or photooxidation. Also notable is that coating materials can also have toxic effects on cells, and this can be a significant issue with the use of amphiphilic substances, which can interfere, for example, with cellular membranes.

Surface oxidation takes place primarily through radical reactions of oxygen combined with UV irradiation. For cadmium-based QDs, this leads to the formation of chalcogenoxides

(e.g., SO$_2$, SO$_3$, SeO$_2$, TeO$_2$) and reduced cadmium. These chalcogenoxides can then desorb from the surface and dissolve in the aqueous surrounding (e.g., resulting in H$_2$SO$_3$, H$_2$SO$_4$, H$_2$SeO$_3$, TeO$_2$(aq)), leaving the residual reduced cadmium to be oxidized back to Cd^{2+} ions, and leading to the subsequent release of free cadmium ions [34–36] (Figure 36.5). These soluble Cd^{2+} ions are largely responsible for the toxicity that is ascribed to QDs. Consequently, as the impermeability to ions of the surrounding polymer shell increases, the overall cytotoxicity tends to decrease. However, at the low QD concentrations required for cellular experiments, most reports have not found adverse effects on cell viability, morphology, function, or development. While this does not mean that semiconductor nanocrystals are completely innocuous, a safe range for their use in biological applications does exist [41–43]. As progress with QD improvements continues, this safe range can be further extended through increasing quantum yields of the particles and decreasing detection limits and subsequently the required particle concentrations.

36.3 Ligand Exchange Strategies

Various molecules are suitable for ligand exchange, but all require a functional group to be grafted onto the nanocrystal surface. Thiol, amine, and phosphine groups are the main groups utilized here. Beyond the functional group, the rest of the molecule must solubilize the QD. As described herein, solubilization can be achieved with charged groups, hydrophilic spacers, or combinations of both.

36.3.1 Thiols

Thiol ligands, such as dithiols or thiol dendrimers, have been extensively studied, and are among the most prevalent strategies used for ligand exchange (Table 36.1, Figure 36.5). With thiol ligands, it is very easy to attain water solubility through the attachment of acidic ligands like thioglycolic acid, mercaptopropionic acid, and dihydrolipoic acid [44]. Here, a second functional group, specifically a carboxy group, is introduced, providing the possibility for further functionalization steps. Additionally, QDs that are modified in this manner can subsequently be covered

TABLE 36.1 Examples of Thiol Ligands and Polymers Used for Ligand Exchange with Their Mechanism of Interaction with the Semiconductor and the Intended Applications

Ligands	Mechanisms of Interaction	Applications	References
Mercaptoalkane acid	Monodentate thiol bond	Metal ion sensing	[91–93]
Dihydrolipoic acid	Bidentate thiol bond	Förster resonance energy transfer (FRET) experiments, ion sensing	[11,12]
PEGylated dihydrolipoic acid	Bidentate thiol bond	Cancer marker detection, live cell labeling, organelle tracking	[47,49,50]
Peptide or protein	Monodentate thiol group, leucine zipper, cystein domain, histidine tag	Tumor vascular imaging, intracellular targeting	[52,53]
Generation poly(amido amine) dendrite	Monodentate thiol bond	Transfection reagent	[57]
Poly(acrylic acid) based multidentate polymer	Multidentate thiol and/or amine bond		[58]

Note: Gray, coordination group toward QD; light gray, hydrophilic part; dark gray, can act as both.

using an oppositely charged polymer. For example, the attachment of a PEG group (PEGylation) becomes possible, as does coating with derivatives of poly(acryl amide) for further stabilization of the ligand shell [45].

Thiolated PEG polymers, obtained by the attachment of terminal thiol groups, are often synthesized for the coating of QDs [46–50]. The main advantages of the thiolated PEGs include easy synthesis, the ease of handling, and versatility for the applications. Because of these advantages, thiolated PEG ligands are widely used for solubilization. An extra benefit of PEG coatings is the reduced unspecific cellular uptake of the modified uncharged particles [43] that was mentioned earlier. Depending on the desired goal, varying polymer chain lengths and number of binding dentates can be used. The two most commonly used types are the monodentate [49,50] and bidentate [48] thiols. The latter ligand grafts more effectively on the nanocrystal surface, and it therefore provides enhanced stabilization of the nanocrystals

in aqueous solution. Still, these simple coating agents lead to reduced photoluminescence intensity of the nanoparticles and lack of long-term chemical stability of the thiol groups.

Just as synthetic polymers can attach onto the charged layer, as mentioned above, proteins can also be readily adsorbed by coatings [5]. While unspecific protein adsorption can be problematic, some groups have used the attachment of specific proteins onto the nanocrystal surface beneficially. The application of engineered peptides or proteins for functional coatings of QDs is a fast-growing frontier in nanocrystal modification. For these custom-designed proteins, biological targeting sequences are fused with the attachment domains for QDs. These attachment domains can include thiol-containing cysteine domains, cationic histidine tags [51], or leucine zipper peptides [5]. Beyond this, simple thiolated proteins are now utilized for direct attachment onto nanocrystal surfaces [52]. A further improvement involves the co-attachment of thiolated PEGs and engineered peptides on one particle surface [53]. This method provides specific binding, while also reducing the adsorption of other proteins on the QDs, thus enhancing their overall biocompatibility.

Another way to exchange existing surface ligands entails the application of grafting dendrons or dendrimers, which are three-dimensional, highly branched, and almost monodisperse macromolecules [54]. Dendrons and dendrimers themselves are core-shell nanostructures consisting of a core, which is the starting point for the stepwise polymerization, interior branching cells, and an exponentially increasing number of functional groups on the surface. The most commonly used dendrimers for nanocrystal coating are functionalized poly(amido amine) (PAMAM) polymers [55–57]. An important attribute of these cationic PAMAM polymers is their ability to effectively penetrate cell walls, which also makes them useful as commercial transfection agents. The PAMAM polymers possess a large number of primary and tertiary amine groups at the surface and in the interior branches of the molecule. These are known to allow for DNA complexation, and they can also be grafted to QD surfaces. When grafted to QDs, they improve the fluorescent properties of the modified semiconductor nanoparticles [58]. Despite this, they exhibit poor affinities for nanocrystal surfaces, and they do not provide stabilization against particle aggregation because of their charged groups. Due to this fact, PAMAM dendrimers must be further modified with additional thiol groups to improve their affinity for nanocrystal surfaces. Surprisingly, dendrimer-coated QDs seemed to transfect better than higher generation dendrons alone in first cell studies [57]. This might be explained by the altered particles sizes that are different from the free polymers, and it indicates that, the composite QD systems may also hold promise as an innovation for the transfection of cells.

New multidentate poly(acrylic acid) derivatives use thiol and amine groups for grafting on the nanocrystal surface. Multidentate ligands lead to a much denser polymer shell than can be achieved with monodentate ligands. This means that the resulting nanocrystals display sufficiently small hydrodynamic diameters. Additionally, grafted amine groups provide a further luminescence-enhancing effect [58].

36.3.2 Amines

Amine groups can also be used for the attachment of polymer coatings to nanocrystals (Table 36.2). As mentioned previously, amine groups only weakly bind to semiconductor surfaces; however, some researchers have successfully functionalized QDs with amine containing polymers. An example of this is poly(ethylene imine) (PEI) [59], which is also known for being an effective transfection agent. PEI-coated QDs show an effective lipophilic/hydrophilic phase transfer, while also displaying good solubility in other polar solvents. Unfortunately, PEI coatings also seem to enhance the photooxidation of the QDs, and they therefore increase the darkening of the nanocrystals. Another class of applied amine-containing polymers are poly(N,N-dimethylaminoethyl metharcylate)s, which exhibit ternary amine groups [60,61]. It was shown that these polymers not only effectively passivate the surface of the nanocrystals, but they also provide

TABLE 36.2 Examples of Amine Ligands and Polymers Used for Ligand Exchange with Their Mechanism of Interaction with the Semiconductor and the Intended Applications

Ligands	Mechanisms of Interaction	Applications	Reference
Branched poly(ethyleneimine)	Multidentate amine bond	Transfection agent	[59]
Poly(N,N-dimethylaminoethyl methacrylate)	Multidentate amine bond	—	[60]

Note: Gray, coordination group toward QD; light gray, hydrophilic part; dark gray, can act as both.

robust colloidal stabilization in various biological environments. Additionally, the polymer-coated particles exhibit an increase in quantum yields as compared to the uncoated ones. This can be ascribed here to the photoluminescence-enhancing effect of the amines.

36.3.3 Phosphines

To overcome the drawback of reduced photoluminescence intensity with thiol groups and the weak bonding of amine groups, phosphine-containing polymers have been developed. These polymers show more similarities to the original ligands used in synthesis (Table 36.3). In 2003, Bawendi et al. synthesized multidentate phosphine oxide polymers [62,63] composed of three sublayers enclosing the nanocrystal, an inner phosphine layer, a linking layer between the phosphine group and an outer functionalized layer. The attachment of phosphines here provides quantum yields up to 40%. Furthermore, the oligomeric outer layer was able to be modified with different functional moieties such as PEG chains [63]. This evidence shows that these multidentate ligands provide chemically stable and highly fluorescent QDs. A potential application for these particles might be lymph node mapping, due to their exceptionally small hydrodynamic radii of 15 to 20 nm, which would allow successful penetration through tissues.

Similar to the already mentioned dendrons and dendrimers, also poly(ether)s modified with aryl phosphine focal points have been developed [64]. Here, the incorporated phosphine group provides strong coordination to the surface without affecting the quantum yield of the particles. Moreover, the conic shape of the attached dendrons seems to be ideal for the adsorption onto the nanoparticles, because the formation of a closely packed polymer shell is possible. These obtained shells then effectively suppress subsequent diffusion of quenching substances like oxygen or other small ions from the surrounding solution to the nanocrystal surface.

36.4 Ligand Capping Strategies

A wide range of amphiphilic polymers for QD surface modification have been developed since the first publications describing water-solubilization using capping strategies (Figure 36.5). The common functionality among all of these polymers is the lipophilic part, which intercalates between the aliphatic chains of the present surface ligands and covers or encapsulates the whole QD with the original ligands from the synthesis still in place.

36.4.1 Amphiphilic Polymers

Amphipol triblock-copolymers of poly(acrylic acid) are an example of an amphiphilic polymer, which can be used for ligand capping (Table 36.4, A). This polymer is commercially applied to solubilize membrane proteins in aqueous solutions [65]. A few years ago, related diblock copolymers were developed for the preparation of biocompatible semiconductor nanocrystals on a large scale. For this system, the polymer shell is composed of octylamine-modified poly(acrylic acid) that is crosslinked with lysine (Table 36.4, B). These modified QDs can be further improved through PEGylation of the carboxylic groups to reduce unspecific binding to proteins or cells [66,67].

TABLE 36.3 Examples of Phosphine Ligands and Polymers Used for Ligand Exchange with Their Mechanism of Interaction with the Semiconductor and the Intended Applications

Ligands	Mechanisms of Interaction	Applications	Reference
Multidentate phosphine polymer	Multidentate phosphine bond	Lymph node mapping	[63]
Poly(ether) dendron	Monodentate phosphine bond	—	[64]

Note: Gray, coordination group toward QD; light gray, hydrophilic part.

TABLE 36.4   Examples of Amphiphilic Polymers Used for QD Capping with Their Applications

Capping Polymers	Applications	References
A	—	[65]

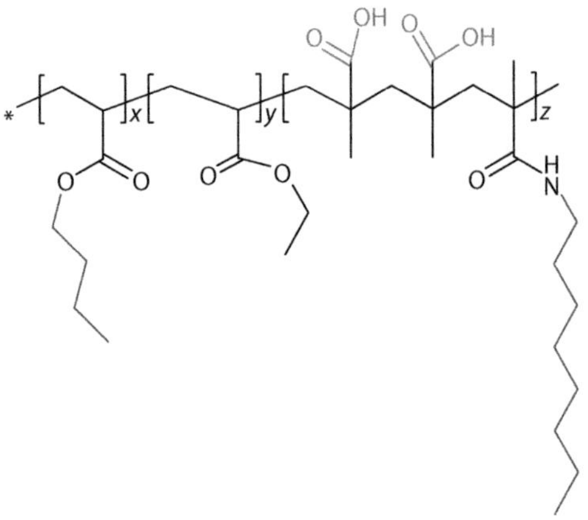

Triblock copolymer

| 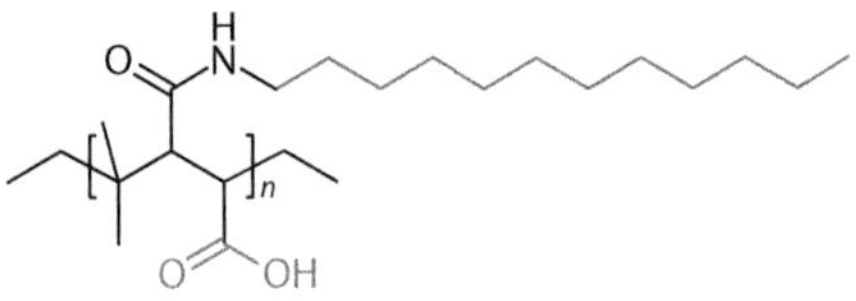 B | Multiphoton imaging in vivo, labeling of cancer markers and cellular targets, transfection experiments | [66,67] |

Poly(acrylic acid) derivate

| C | Transfection experiments | [68–70] |

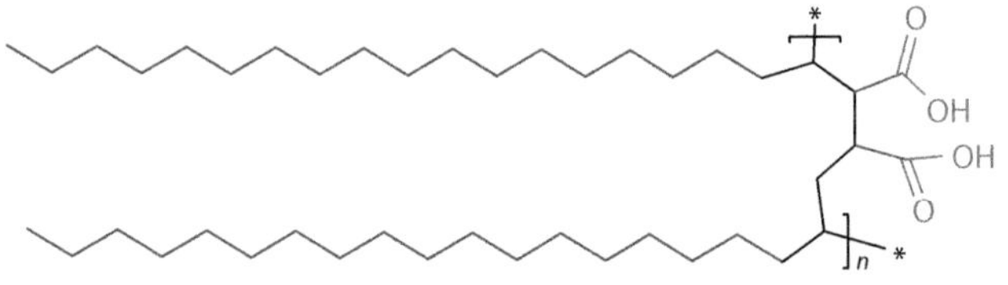

Poly(maleic acid-alt-1-olefin) derivative

| D | FRET experiments | [71] |

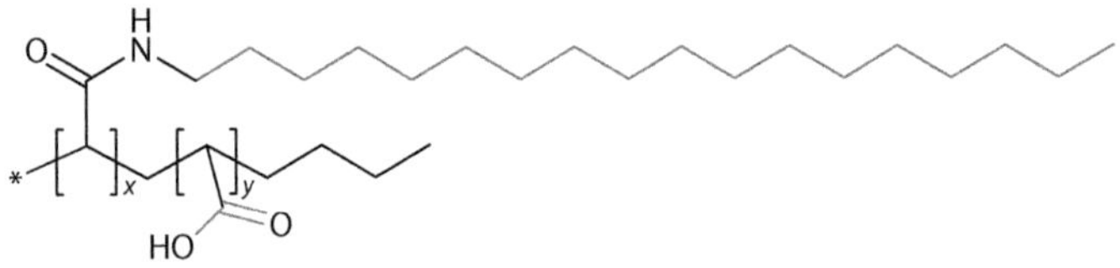

Poly(isobuthylene-alt-1-maleic acid) derivative

Note: Gray, coordination group toward QD; light gray, hydrophilic part.

Another amphipol type that is used for QD functionalization is the amphiphilic poly(maleic anhydride-alt-1-olefin)s with different alkyl chain lengths (Table 36.4, C) [68–70], which can be further crosslinked with a diamine to stabilize the polymer shell and can also be PEGylated to improve the particle stability. Poly(isobuthylene-alt-1-maleic acid) functionalized with dodecyl amine is also synthesized for the capping of QDs (Table 36.4, D) [71]. All of these amphipols have a hydrophilic backbone and hydrophobic side chains that interact with the aliphatic chains of the ligands present on the nanocrystal surface. This allows bridging between the lipophilic surface ligands and the hydrophilic solution. The solubilization of the nanoparticles in water is mainly promoted by the carboxylic groups of acrylic or maleic acid, which form the backbone of the amphipol shell. The shell architecture with the present functional group provides the possibility for further functionalization with antibodies or proteins, which are suitable for targeting cancer cells, through standard carbodiimide chemistry [66].

TABLE 36.5 Examples of Surfactant Used for QD Capping with Their Application

Capping Polymers	Applications	References
Phospholipid	Tracking of plasmid DNA, in vivo imaging, cell detection	[72,73,75]

Note: Gray, lipophilic part; light gray, hydrophilic part.

36.4.2 Micelles

A frequently used alternative to coating with amphiphilic polymers is the encapsulation of QDs in micelles, which can, for example, be composed of polymer conjugated surfactants (Table 36.5). The advantage of this method is the applicability of a wide variety of surfactants/lipids with different functionally terminated groups. PEG-derivatized phospholipids are used as micelle-building compounds due to their improved solubilization capability [72–74]. Other surfactants, such as lipids containing paramagnetic gadolinium complexes, can also be used for nanocrystal encapsulation [75,76]. These provide the capability for luminescence imaging as well as for MRI (magnetic resonance imaging). The QD-containing micelles preserve the optical properties of the encapsulated QDs and are also highly biocompatible. The drawback of using micelles is that only nanocrystals of predefined diameters and consequently emission wavelength can be encapsulated by certain micelle building surfactants or lipids. This limitation comes because the micellar size of a particular surfactant defines the inner free space available for the incorporation of the QDs [72].

36.4.3 Heterocyclic Amphiphiles

For sensoric applications of QDs, the use of different cyclodextrines can be advantageous. Here, interactions between the coating and the core are desirable and also necessary [77,78]. In the case of cyclodextrines, the hydrophobic pockets of the saccharide oligomers interact with the aliphatic chains of the TOPO present on the nanoparticle surface. Still, the immobilized cyclodextrines retain their capability of initiating molecular recognition. Due to this fact and the observation of fluorescence changes with analyte binding, this modification method appears to be very promising for sensing applications [78]. Another benefit of this capping strategy is the small diameter of the resulting QDs, which is achieved due to the small space requirements of cyclodextrines. A related approach to the creation of small water-soluble QDs without ligand exchange is the use of calixarenes, which are structurally similar organic polycyclic systems [79,80]. This coating also preserves the emission intensity of the QDs, while providing a small QD diameter. Calixarenes are cyclic oligomers based on a hydroxyalkylation product of a phenol and an aldehyde. It has been shown that calixarenes can be derivatisized with sugars or peptides to allow biological applications of these systems [79]. Derivatization with aliphatic and sulfonato groups was also achieved, and this allows for the optical detection of small molecular weight molecules such as acetylcholine [80] (Table 36.6).

36.5 Applications of Surface Coatings

The various coating methods, substances and the different characteristics of the resulting nanoparticles open a wide range of application areas, particularly in the fields of sensorics and diagnostics. For sensorical approaches, it is important to have a surface coating that allows interactions between analytes or reporter molecules with the QD. In contrast, diagnostic applications rely on biocompatibility, and special attention must be paid to cytotoxicity and the undesirable adsorption of proteins and possible subsequent particle aggregation.

36.5.1 Sensoric Applications

Sensoric applications of QDs are, in most cases, based on the interaction of an analyte molecule or ion with the nanocrystal surface. This interaction should lead to a change in the apparent fluorescent properties of the particle. As an example of this approach, QDs were coated with cysteine, thioglycolic acid, or related ligands for the detection of metal ions like Ag^+ [81], Cu^{2+} [82], Zn^{2+} [83], and also small toxic anions like cyanide [84]. Additionally, coated QDs were applied for optical temperature detection [85–87]. The conjugation of selective reagents or reporter molecules to the surface of luminescent nanocrystals is also utilized for QD probes. In particular, dihydrolipoic acid can be modified with functional moieties for selective K^+ [11] or glucose [12] sensing. Despite this promise, these approaches still seem to be restricted to a small number of analytes interacting with the surface coatings and the underlying QDs. Additionally, current QD systems possess low stability in biological systems and limited applicability in realistic sample arrangements due to the many possible interactions with similar ions present in solution.

The potential of QDs to be used in much more analyte-specific FRET (Förster resonance energy transfer)-based sensors can expand the applicability of semiconductor nanocrystals in sensorics. Here, the tunable wavelengths and the high quantum

TABLE 36.6   Examples of Heterocyclic Amphiphiles Used for QD Capping with Their Applications

Capping Polymers	Applications	Reference
β-Cyclodextrine	Molecular recognition	[78]
Calix[4]arene	Acetylcholine detection	[80]

Note: Gray, lipophilic part; light gray, hydrophilic part.

yields of the nanocrystals theoretically should enable efficient energy transfer with a wide number of conventional dyes. Indeed it has to be mentioned that FRET efficiencies obtained with QDs as donor species so far are still low compared to the efficiencies of common dyes. This may be attributed to the comparatively large size of even very thinly coated QDs, which makes it very complicated to bring the acceptor into close proximity of the donor for efficient FRET. Still, a variety of QD FRET applications have been developed and strategies aimed at improving the energy transfer were explored. Protein-binding sites have been studied via FRET investigations, whereas the acceptor dyes are bound to a protein binding site affording FRET when the assembly is adsorbed on the QD surface [88]. Therefore, different intracellular sensing applications can be achieved with QD-based FRET; pH, nuclear cleavage and protease activity can all be detected [89–92]. In addition, immunoassays for specific cancer marker detection have been developed [93].

Beyond conventional fluorescent dyes that are utilized as acceptors, several so-called dark quenchers, which are molecules or nanocrystals that do not re-emit light, were bound to QDs for FRET applications. This results in a detectable decrease of the fluorescence signal upon increasing FRET events. Examples of applications for these systems include the use of functionalized gold nanocrystals for DNA hybridization investigations [94,95], inhibition assays [96], glucose sensing [97] or, alternatively, the application of an organic dark quencher dye for pH sensing [98], and maltose binding assays [99].

36.5.2 Diagnostic Applications

Diagnostic approaches that utilize modified QDs are more dependent on impermeable polymer shells and efficient physical and biological shielding of the QD. It is essential that QDs be protected against unspecific adsorption of proteins and fast degradation leading to fluorescence loss. This can be achieved with densely packed polymer shells and subsequent PEGylation. These water-soluble and often-targeted QDs can be used for in vitro cellular imaging or for in vivo imaging of tumors. The applied in vivo imaging here is noninvasive, and it can detect deep tissue regions in mice and even larger species with high sensitivity and contrast without the use of radioactive radiation or larger instrumental setups, as is the case with computed tomography (CT). The intravenous injection of biocompatible QDs has been performed for blood vessel imaging [67], targeting of tissue-specific vascular markers [52], and lymph node mapping [100]. Diagnostic QD systems can also be used in the targeting of tumor cells in vivo using specific antibodies against Her2 markers [66]. Self-illuminating QDs represent another important innovation for in vivo imaging. These QDs need no external light for excitation, because they are modified with luciferase. In this system, the chemical energy of the substrate coelenterazine is converted into photon energy by the enzyme luciferase. This photon energy excites the QD through bioluminescence resonance energy transfer (BRET). With this excitation mechanism, autofluorescence is virtually eliminated, but the emitted photons are still absorbed or scattered in the surrounding tissue, making sensitive detection necessary [101–104].

In the area of cellular imaging, QD probes are used for the tagging of whole cells as well as for the investigation of single intracellular processes. Many cellular imaging studies are focused on membrane-specific markers, because these are easy to access and do not require the penetration of the membrane. Several attempts for the internalization of QDs in live cells have been made, however. Such approaches have made use of membrane translocation peptides [105], electroporation, or established transfection reagents [49]. The latter strategy for the internalization of QDs allows targeting of subcellular compartments like the mitochondria or nucleolus through the use of specific targeting peptides [52,106,107].

Recently, QD conjugates for combined cancer imaging and therapy have been developed. An aptamer, which can simultaneously target cancer cells and also binds an anticancer drug, was immobilized on the surface of the QD. The fluorescent properties of the drug quench the luminescence of the QD through a FRET mechanism. Upon the release of the drug, the luminescence of the nanocrystal is restored. Taken together, this system is a targeted QD imaging system that is capable of differential uptake, imaging, and cancer therapy [108].

36.6 Conclusions

A great variety of polymeric surface coatings for QDs are currently applied for a wide range of applications. All applications require distinct QD characteristics, and as discussed, these are adjustable through surface-coating polymers. Important characteristics like the size and photostability of water-soluble QDs strongly depends on the used capping strategy and the resulting particle architecture. Ligand exchange strategies can produce small particles, but these often lack long-term stability and photoluminescence

intensity. Their resistance against acids or bases and, in some cases, against chemical oxidation is very weak. Ligand-exchanged QDs have been used in cases where hydrodynamic diameter is the primary criterion for applicability such as in FRET experiments and other sensoric applications that depend on the accessibility of the nanocrystal surface, which can only be achieved by the attachment of small ligands. For ligand exchange procedures, the recent adaptation of phosphine groups has been very beneficial due to the improved stabilization of the nanocrystal surface and the additional surface passivation against oxidation. PEGylation can also provide further protection against unspecific protein absorption. For transfection experiments, a substitution with cationic polymers, for example, branched PEI or PAMAM dendrimers, can be very beneficial. In contrast to ligand exchange, ligand capping strategies effectively shield the nanocrystal surface, have subsequently low cytotoxicity and high stability in biological environments and are ideal for cellular and in vivo experiments. Ligand-capped QDs are best for studies that rely on sustained fluorescence in the presence of oxidizing agents and studies that demand low particle cytotoxicity. They are also useful for assays that are conducted in high salt concentrations. The protection of the QDs with an amphiphilic bilayer, such as phospholipids or amphiphilic polymers, is very useful. The amphiphilic capping can be easily modified with targeting sequences or proteins using carbodiimide chemistry. Despite the overall promise of QDs, no encapsulation method can be universally optimal for all biological and sensorical applications at once.

36.7 Future Outlook

The different polymeric surface coatings developed in the last decade combining biological materials with inorganic nanocrystals have been crucial for the successful use of QDs in cell and tissue imaging. Furthermore, they have created new systems in materials science for the controlled assembly of nanomaterials used in the biological environment. As research continues to produce different nanomaterials with novel unique properties, it will become possible to gather new multimodal imaging agents. Combining QDs for fluorescence imaging with MRI or CT contrast agents like Fe_2O_3 [109], FePt [110], or Gd complexes [75] will allow deep tissue imaging and fluorescence tracking of one system for sophisticated diagnostic applications. In the area of sensorics, QDs can also function as effective protein carrier or exciton donor for prototype self-assembled FRET nanosensors for the detection of many relevant signal molecules [65]. QDs could even drive more biosensors through a two-step FRET mechanism overcoming inherent donor–acceptor distance limitations [85]. Presently, intensity-based measurements with QDs have been employed in the fields of sensing and imaging. Indeed, lifetime-based methods with QDs will become more prevalent due to their superior resolution, independence from fluorescence intensity and consequently concentration at the detection point, and, finally, the possibility to out-gate the tissue autofluorescence present in all living biological systems. For these applications, QDs are an especially powerful tool due to their long excited state lifetimes compared to the common organic dyes and interfering tissue autofluorescence.

In the end, semiconductor nanocrystals will not overcome the use of conventional organic dyes in biological and sensorical applications, but they could complement dye deficiencies in particular approaches such as in vivo imaging. They also can be developed further for new applications like long-term imaging and lifetime measurements. Ultimately, adapting QD nanoparticles for biological use will teach us important lessons about creating future inorganic–organic hybrids for a myriad of other applications.

References

1. C.M. Niemeyer, Nanoparticles, proteins, and nucleic acids: Biotechnology meets materials science, *Angew. Chem. Int. Ed. Engl.* 40 (2001) 4128–4158.

2. S.R. Whaley, D.S. English, E.L. Hu, P.F. Barbara, A.M. Belcher, Selection of peptides with semiconductor binding specificity for directed nanocrystal assembly, *Nature* 405 (2000) 665–668.

3. M. Bruchez Jr., M. Moronne, P. Gin, S. Weiss, A.P. Alivisatos, Semiconductor nanocrystals as fluorescent biological labels, *Science* 281 (1998) 2013–2015.

4. W.C.W. Chan, S.M. Nie, Quantum dot bioconjugates for ultra-sensitive nonisotopic detection, *Science* 281 (1998) 2016–2018.

5. H. Mattoussi, J.M. Mauro, E.R. Goldman, G.P. Anderson, V.C. Sundar, F.V. Mikulec, M.G. Bawendi, Self-assembly of CdSe–ZnS quantum dots bioconjugates using an engineered recombinant protein, *J. Am. Chem. Soc.* 122 (2000) 12142–12150.

6. G.P. Mitchell, C.A. Mirkin, R.L. Letsinger, Programmed assembly of DNA functionalized quantum dots, *J. Am. Chem. Soc.* 121 (1999) 8122–8123.

7. S. Pathak, S.K. Choi, N. Arnheim, M.E. Thompson, Hydroxylated quantum dots as luminescent probes for in situ hybridization, *J. Am. Chem. Soc.* 123 (2001) 4103–4104.

8. M.Y. Han, X.H. Gao, J.Z. Su, S. Nie, Quantum-dot-tagged microbeads for multiplexed optical coding of biomolecules, *Nat. Biotechnol.* 19 (2001) 631–635.

9. X. Michalet, F.F. Pinaud, L.A. Bentolila, J.M. Tsay, J.J. Li, G. Sundaresan, A.M. Wu, S.S. Gambhir, S. Weiss, Quantum dots for live cells, in vivo imaging and diagnostics, *Science* 307 (2005) 538–544.

10. I.L. Mednitz, H.T. Uyeda, E.R. Goldman, H. Mattoussi, Quantum dot bioconjugates for imaging, labeling and sensing, *Nat. Mater.* 4 (2005) 435–446.

11. C.-Y. Chen, C.-T. Cheng, C.-W. Lai, P.-W. Wu, K.-C. Wu, P.-T. Chou, Y.-H. Chou, H.-T. Chiu, Potassium ion recognition by 15-crown-5 functionalized CdSe/ZnS quantum dots in H$_2$O, *Chem. Commun.* 3 (2006) 263–365.

12. D.B. Cordes, S. Gamsey, B. Singaram, Fluorescent quantum dots with boronic acid substituted viologens to sense glucose in aqueous solution, *Angew. Chem. Int. Ed.* 45 (2006) 3829–3832.

13. T.M. Jovin, Quantum dots finally come of age, *Nat. Biotechnol.* 12 (2003) 32–33.

14. C.B. Murray, D.J. Norris, M.G. Bawendi, Synthesis and characterization of nearly monodispers CdE (E = S, Se, Te) semiconductor nanocrystallites, *J. Am. Chem. Soc.* 115 (1993) 8706–8715.

15. M.A. Hines, P. Guynot-Sionest, Synthesis and characterization of strongly luminescing ZnS-Capped CdSe nanocrystals, *J. Phys. Chem.* 100 (1996) 468–471.

16. L.H. Qu, X.G. Peng, Control of photoluminescence properties of CdSe nanocrystals in growth, *J. Am. Chem. Soc.* 124 (2002) 2049–2055.

17. X.H. Zhong, Y.Y. Feng, W. Knoll, M.Y. Han, Alloyed Zn$_x$Cd$_{1-x}$S nanocrystals with highly narrow luminescence spectral width, *J. Am. Chem. Soc.* 125 (2003) 13559–13563.

18. R.E. Bailey, S.M. Nie, Alloyed semiconductor quantum dots: Tuning the optical properties without changing the particle size, *J. Am. Chem. Soc.* 125 (2003) 7100–7106.

19. S. Kim, B. Fisher, H.J. Eisler, M. Bawendi, Type-II quantum dots: CdTe/CdSe(core/shell) and CdSe/ZnTe(core/shell) heterostructures, *J. Am. Chem. Soc.* 125 (2003) 11466–11467.

20. B.L. Wehrenberg, C.J. Wang, P. Guyot-Sionnest, Interband and intraband optical studies of PbSe colloidal quantum dots, *J. Phys. Chem. B* 106 (2002) 10634–10640.

21. D. Gerion, F. Pinaud, S.C. Williams, W.J. PArak, D. Zanchet, S. Weiss, P.A. Alivisatos, Synthesis and properties of biocompatible water-soluble silica-coated CdSe/ZnS semiconductor quantum dots, *J. Phys. Chem. B* 105 (2001) 8861–8871.

22. A. Schroedter, H. Weller, Biofunctionalization of silica-coated CdTe and gold nanocrystals, *Nano Lett.* 2 (2002) 1363–1367.

23. T. Nann, P. Mulvaney, Single quantum dots in spherical silica particles, *Angew. Chem. Int. Ed.* 43 (2004) 5393–5396.

24. M.A. Petruska, A.P. Bartko, V.I. Klimov, An amphiphilic approach to nanocrystal quantum dot-titania nanocomposites, *J. Am. Chem. Soc.* 126 (2004) 714–715.

25. S.T. Selvan, T.T. Tan, J.Y. Ying, Robust, non-cytotoxic, silica-coated CdSe quantum dots with efficient photoluminescence, *Adv. Mater.* 17 (2005) 1620–1625.

26. Y. Yang, M. Gao, Preparation of fluorescent SiO$_2$ particles with single CdTe nanocrystal cores by the reverse microemulsion method, *Adv. Mater.* 17 (2005) 2354–2357.

27. M. Darbandi, R. Thomann, T. Nann, Single quantum dots in silica spheres by microemulsion synthesis, *Chem. Mater.* 17 (2005) 5720–5725.

28. R. Bakalova, Z. Zhelev, I. Aoki, H. Ohba, Y. Imai, I. Kanno, Silica-shelled single quantum dot micelles as imaging probes with dual or multimodality, *Anal. Chem.* 78 (2006) 5925–5932.

29. Z. Zehlev, H. Ohba, R. Bakalova, Single quantum dot-micelles with silica shell as potentially non-cytotoxic fluorescent cell tracers, *J. Am. Chem. Soc.* 128 (2006) 6234–6235.

30. A. Wolcott, D. Gerion, M. Visconte, J. Sun, A. Schwartzberg, S. Chen, J.Z. Zhang, Silica-coated CdTe quantum dots functionalized with thiols for bioconjugation to IgG proteins, *J. Phys. Chem. B* 110 (2006) 5779–5789.

31. T. Zhang, J.L. Stilwell, D. Gerion, L. Ding, O. Elboudwarej, P.A. Cook, J.W. Gray, A.P. Alivisatos, F.F. Chen, Cellular effect of high doses of silica-coated quantum dot profiled with high throughput gene expression analysis and high content cellomics measurements, *Nano Lett.* 6 (2006) 800–808.

32. S.T. Selvan, P.K. Prata, C.Y. Ang, J.Y. Ying, Synthesis of silica-coated semiconductor and magnetic quantum dots and their use in the imaging of live cells, *Angew. Chem. Int. Ed.* 46 (2007) 2448–2452.

33. W.G.J.H.M. van Sark, P.L.T.M. Frederix, A.A. Bol, H.C. Gerritsen, A. Meijerink, Bluening, bleaching, and blinking of single CdSe/ZnS quantum dots, *Chem. Phys. Chem.* 3 (2002) 871–879.

34. L. Sanhel, M. Haase, H. Weller, A. Henglein, Photochemistry of colloidal semiconductors. 20. surface modification and stability of strong luminescing CdS particles, *J. Am. Chem. Soc.* 109 (1987) 5649–5655.

35. J.E.B. Katari, V.L. Colvin, A.P. Alivisatos, X-ray spectroscopy of CdSe nanocrystals with applications to studies of the nanocrystal surface, *J. Phys. Chem.* 98 (1994) 4109–4117.

36. A.P. Alivisatos, Perspectives on the physical chemistry of semiconductor nanocrystals, *J. Phys. Chem.* 100 (1996) 13226–13239.

37. W.G.J.H.M. van Sark, P.L.T.M. Frederix, D.J. van den Heuvel, H.C. Gerritsen, Photooxidation and photobleaching of single CdSe/ZnS quantum dots probed by room-temperature time-resolved spectroscopy, *J. Phys. Chem. B* 105 (2001) 8281–8284.

38. W.G.J.H.M. van Sark, P.L.T.M. Frederix, D.J. van den Heuvel, A.A. Bol, J.N.J. van Lingen, C. de Mello Donea, H.C. Gerritsen, A. Meijerink, Time-resolved fluorescence spectroscopy study on the photophysical behavior of quantum dots, *J. Fluoresc.* 12 (2002) 69–75.

39. E.L. Bentzen, I.D. Tomlinson, J. Manson, P. Gresch, M.R. Warnement, D. Wright, E. Sanders-Bush, R. Blakely, S.J. Rosenthal, Surface modification to reduce nonspecific binding of quantum dots in live cell assays, *Bioconjug. Chem.* 16 (2005) 1488–1494.

40. A.M. Smith, H. Duan, M.N. Rhyner, G. Ruan, S. Nie, A systematic examination of surface coatings an the optical and chemical properties of semiconductor quantum dots, *Phys. Chem. Chem. Phys.* 8 (2006) 3895–3903.

41. H. Ron, A toxicological review of quantum dots: Toxicity depends on physicochemical and environmental factors, *Environ. Health Perspect.* 114 (2006) 165–172.

42. A.M. Derfus, W.C.W. Chan, S.N. Bhatia, Probing the cytotoxicity of semiconductor quantum dots, *Nano Lett.* 4 (2004) 11–18.

43. C. Kirchner, T. Liedl, S. Kudera, T. Pellegrino, A.M. Javier, H.E. Gaub, S. Stölzle, N. Fertig, W.J. Parak, Cytotoxicity of colloidal CdSe and CdSe/ZnS nanoparticles, *Nano Lett.* 5 (2005) 331–338.

44. J.K. Jaiswal, H. Mattoussi, J.M. Mauro, S.M. Simon, Long-term multiple color imaging of live cells using quantum dot bioconjugates, *Nat. Biotechnol.* 21 (2003) 47–51.

45. I. Potapova, R. Mruk, S. Prehl, R. Zentel, T. Basche, A. Mews, Semiconductor nanocrystals with multifunctional polymer ligands, *J. Am. Chem. Soc.* 125 (2003) 320–321.

46. H.T. Uyeda, I.L. Mednitz, J.K. Jaiswal, S.M. Simon, H. Mattoussi, Synthesis of compact multidentate ligands to prepare stable hydrophilic quantum dot fluorophores, *J. Am. Chem. Soc.* 127 (2005) 3870–3878.

47. E.-C. Kang, A. Ogura, K. Kataoka, Y. Nagasaki, Preparation of water-soluble PEGylated semiconductor nanocrystals, *Chem. Lett.* 33 (2004) 840–841.

48. S.K. Dixit, N.L. Goicochea, M.-C. Daniel, A. Murali, L. Bronstein, M. De, B. Stein, V.M. Rotello, C.C. Kao, B. Dragnea, Quantum dot encapsulation in viral capsids, *Nano Lett.* 6 (2006) 1993–1999.

49. A.M. Derfus, W.C.W. Chan, S.N. Bhatia, Intracellular delivery of quantum dots for live cell labeling and organelle tracking, *Adv. Mater.* 16 (2004) 961–966.

50. F. Hu, Y. Ran, Z. Zhou, M. Gao, Preparation of bioconjugates of CdTe nanocrystals for cancer marker detection, *Nanotechnology* 17 (2006) 2972–2977.

51. S.-Y. Ding, G. Rumbles, M. Lones, M.P. Tucker, J. Nedeljkovic, M.N. Simon, J.S. Wall, M.E. Himmel, Bioconjugation of (CdSe)ZnS quantum dots using a genetically engineered multiple polyhistidine tagged Cohesin/Dockerin protein polymer, *Macromol. Mater. Eng.* 289 (2004) 622–628.

52. M.E. Akerman, W.C.W. Chan, P. Laakkonen, S.N. Bhatia, E. Ruoslahti, Nanocrystal targeting in vivo, *PNAS* 99 (2002) 12617–12621.

53. W. Cai, D.-W. Shin, K. Chen, O. Gheysens, Q. Cao, S.X. Wang, S.S. Gambhir, X. Chen, Peptide-labeled near-infrared quantum dots for imaging tumor vasculature in living subjects, *Nano Lett.* 6 (2006) 669–676.

54. M.J. Cloninger, Biological applications of dendrimers, *Curr. Opin. Chem. Biol.* 6 (2002) 742–748.

55. B. Huang, D.A. Tomalia, Dendronisation of gold and CdSe/CdS (core-shell) quantum dots with tomalia type, thiol core, functionalized Poly(amido amine) (PAMAM) dendrons, *J. Luminesc.* 111 (2005) 215–223

56. B. Pan, F. Gao, R. He, D. Cui, Y. Zhang, Study on interaction between poly(amido amine) dendrimer and CdSe nanocrystal in chloroform, *J. Colloid Interface Sci.* 297 (2006) 151–156.

57. A.C. Wisher, I. Bronstein, V. Chechik, Thiolated PAMAM dendrimer-coated CdSe/ZnS nanoparticles as protein transfection agents, *Chem. Commun.* 15 (2006) 1637–1639.

58. A.M. Smith, S. Nie, Minimizing the hydrodynamic size of quantum dots with multifunctional multidentate polymer ligands, *J. Am. Chem. Soc.* 130 (2008) 11278–11279.

59. T. Nann, Phase-transfer of CdSe@ZnS quantum dots using amphiphilic hyperbranched polyethyleneimine, *Chem. Commun.* 13 (2005) 1735–1736.

60. M. Wang, J.K. Oh, T.E. Dykstra, X. Lou, G.D. Scholes, M.A. Winnik, Surface modification of CdSe and CdSe/ZnS semiconductor nanocrystals with Poly(N,N-dimethylaminoethyl methacrylate), *Macromolecules* 39 (2006) 3664–3672.

61. M. Wang, J.K. Oh, T.E. Dykstra, X. Lou, M.R. Salvador, G.D. Scholes, M.A. Winnik, Colloidal CdSe nanocrystals passivated by a dye-labeled multidentate polymer: Quantitative analysis by size-exclusion chromatography, *Angew. Chem. Int. Ed.* 45 (2006) 2221–2224.

62. S. Kim, M.G. Bawendi, Oligomeric ligands for luminescent and stable nanocrystal quantum dots, *J. Am. Chem. Soc.* 125 (2003) 14652–14653.

63. S.-W. Kim, S. Kim, J.B. Tracy, A. Jasanoff, M.G. Bawendi, Phosphine oxide polymer for water-soluble nanoparticles, *J. Am. Chem. Soc.* 127 (2005) 4556–4557.

64. B. Huang, D.A. Tomalia, Poly(ether) dendrons possessing phosphine focal points for stabilization and reduced quenching of luminescent quantum dots, *Inorg. Chim. Act.* 359 (2006) 1951–1966.

65. C. Luccardini, C. Tribet, F. Vial, V. Marchi-Artzner, M. Dahan, Size, charge, and interactions with giant lipid vesicles of quantum dots coated with an amphiphilic macromolecule, *Langmuir* 22 (2006) 2304–2310.

66. X. Wu, H. Liu, J. Liu, K.N. Haley, J.A. Treadway, J.P. Larson, N. Ge, F. Peale, M.P. Bruchez, Immunofluorescent labeling of cancer marker Her2 and other cellular targets with semiconductor quantum dots, *Nat. Biotechnol.* 21 (2003) 41–46.

67. D.R. Larson, W.R. Zipfel, R.M. Willams, S.W. Clark, M.P. Bruchez, F.W. Wise, W.W. Web, Water-soluble quantum dots for multiphoton fluorescence imaging in vivo, *Science* 300 (2003) 1434–1436.

68. T. Pellegrino, L. Manna, S. Kudera, T. Liedl, D. Koktysh, A.L. Rogach, S. Keller, J. Rädler, G. Natile, W.J. Parak, Hydrophobic nanocrystals coated with an amphiphilic polymer shell: A general route to water soluble nanocrystals, *Nano Lett.* 4 (2004) 703–707.

69. W.W. Yu, E. Chang, J.C. Falkner, J. Zhang, A.M. Al-Solmali, C.M. Sayes, J. Johns, R. drezek, V.L. Colvin, Forming biocompatible and non-aggregated nanocrystals in water using amphiphilic polymers, *J. Am. Chem. Soc.* 129 (2007) 2871–2879.

70. L. Qi, X. Gao, Quantum dot-amphipol nanocomplex for intracellular delivery and real-time imaging of siRNA, *ACS Nano* 2 (2008) 1403–1410.

71. M.T. Fernandez-Arguelles, A. Yakovlev, R.A. Sperling, C. Luccardini, S. Gaillard, A. Sanz Medel, J.-M. Mallet, J.-C. Brochon, A. Feltz, M. Oheim, W.J. Parak, Synthesis and characterization of polymer-coated quantum dots with integrated acceptor dyes as FRET-based nanoprobes, *Nano Lett.* 7 (2007) 2613–2617.

72. B. Dubertret, P. Skourides, D.J. Norris, V. Noireaux, A.H. Brivanlou, A. Libchaber, In vivo imaging of quantum dots encapsulated in phospholipid micelles, *Science* 298 (2002) 1759–1762.

73. C. Srinivasan, J. Lee, F. Papadimitrakopoulos, L.K. Silbart, M. Zhao, D.J. Burgess, Labeling and intracellular tracking of functionally active plasmid DNA with semiconductor quantum dots, *Mol. Ther.* 14 (2006) 192–200.

74. H. Fan, E.W. Leve, C. Scullin, J. Gabaldon, D. Tallant, S. Bunge, T. Boyle, M.C. Wilson, C.J. Brinker, Surfactant-assisted synthesis of water-soluble and biocompatible semiconductor quantum dot micelles, *Nano Lett.* 5 (2005) 645–648.

75. F. Boulmedais, P. Bauchat, M.J. Brienne, I. Arnal, F. Artzner, T. Gaocin, M. Dahan, V. Marchi-Artzner, Water-soluble PEGylated quantum dots: From a composite hexagonal phase to isolated micelles, *Langmuir* 22 (2006) 9797–9803.

76. G.A.F. van Tilborg, W.J.M. Mulder, P.T.K. Chin, G. Strom, C.P. Reutelingsperger, K. Nicolay, G.J. Strijkers, Annexin A5-conjugated quantum dots with a paramagnetic lipidic coating for the multimodal detection of apoptotic cells, *Bioconjug. Chem.* 17 (2006) 865–868.

77. J. Feng, S.-Y. Ding, M.P. Tucker, M.E. Himmel, Y.-H. Kim, S.B. Zhang, B.M. Keyes, G. Rumbles, Cyclodextrine driven hydrophobic/hydrophilic transformation of semiconductor nanoparticles, *Appl. Phys. Lett.* 86 (2005) 033108.

78. K. Palaniappan, S.A. Hackney, J. Liu, Supramolecular control of complexation-induced fluorescence change of water-soluble, β-cyclodextrine-modified CdS quantum dots, *Chem. Commun.* 23 (2004) 2704–2705.

79. T. Jin, F. Fujii, H. Sakata, M. Tamura, M. Kinjo, Calixarene-coated water-soluble CdSe-ZnS semiconductor quantum dots that are highly fluorescent and stable in aqueous solution, *Chem. Commun.* 22 (2005) 2829–2831.

80. T. Jin, F. Fujii, H. Sakata, M. Tamura, M. Kinjo, Amphiphilic p-sulfonatocalix[4]arene-coated quantum dots for the optical detection of the neurotransmitter acetylcholine, *Chem. Commun.* 34 (2005) 4300–4302.

81. J.-L. Chen, C.-Q. Zhu, Functionalized cadmium sulfide quantum dots as fluorescence probe for silver ion determination, *Anal. Chim. Act.* 546 (2005) 147–153.

82. M.T. Fernandez-Argülles, W.J. Jin, J.M. Costa-Fernandez, R. Pereiro, A. Sanz-Medel, Surface-modified CdSe quantum dots for the sensitive and selective determination of Cu(II) in aqueous solutions by luminescent measurements, *Anal. Chim. Act.* 549 (2005) 20–25.

83. Y. Chen, Z. Rosenzweig, Luminescent CdS quantum dots as selective ion probes, *Anal. Chem.* 74 (2002) 5132–5138.

84. W.J. Jin, M.T. Fernandez-Argueelles, J.M. Costa-Fernandez, R. Pereiro, A. Sanz-Medel, Photoactivated luminescent CdSe quantum dots as sensitive cyanide probes in aqueous solutions, *Chem. Commun.* 7 (2005) 883–885.

85. T.-C. Liu, Z.-L. Huang, H.-Q. Wang, J.-H. Wang, X.-Q. Li, Y.-D. Zhao, Q.-M. Luo, Temperature-dependent photoluminescence of water-soluble quantum dots for a bioprobe, *Anal. Chim. Act.* 559 (2006) 120–123.

86. S.F. Wuister, A. van Houselt, C. de Mello Donega, D. Vanmaekelbergh, A. Meijerink, Temperature antiquenching of the luminescence from capped CdSe quantum dots, *Angew. Chem. Int. Ed.* 43 (2004) 3029–3033.

87. P.A.S. Jorge, M. Mayeh, R. Benrashid, P. Cadas, J.L. Santos, F. Farahi, Quantum dots as self-referenced optical fiber temperature probes for luminescent chemical sensors, *Meas. Sci. Technol.* 17 (2006) 1032–1038.

88. A.R. Clapp, I.L. Medintz, J.M. Mauro, B.R. Fisher, M.G. Bawendi, H. Mattoussi, Fluorescence resonance energy transfer between quantum dot donors and dye-labeled protein acceptors, *J. Am. Chem. Soc.* 126 (2004) 301–110.

89. M. Susuki, Y. Yuzuru, H. Komatsu, K. Suzuki, K.T. Douglas, Quantum dot FRET biosensors that respond to pH, to proteolytic or nucleolytic cleavage, to DNA synthesis, or to a multiplexing combination, *J. Am. Chem. Soc.* 130 (2008) 5720–5725.

90. D. Zhou, L. Ying, X. Hong, E.A. Hall, C. Abell, D. Klenerman, A compact functional quantum dot-DNA conjugate: Preparation, hybridization, and specific label-free DNA detection, *Langmuir* 24 (2008) 1659–1664.

91. Y.S. Liu, Y. Sun, P.T. Vernier, C.-H. Liang, S.Y.C. Chong, M.A. Gundersen, pH-Sensitive photoluminescence of CdSe/ZnSe/ZnS quantum dots in human ovarian cancer cells, *J. Phys. Chem. C* 111 (2007) 2872–2878.

92. L. Shi, V. de Paoli, N. Rosenzweig, Z. Rosenzweig, Synthesis and application of quantum dots FRET-based protease sensors, *J. Am. Chem. Soc.* 128 (2006) 10378–10379.

93. K. Kerman, T. Endo, M. Tsukamoto, M. Chikae, Y. Takamura, E. Tamiya, Quantum dot-based immunosensor for the detection of prostate-specific antigen using fluorescence microscopy, *Talanta* 71 (2007) 1494–1499.

94. L. Dyadyusha, H. Yin, S. Jaiswal, T. Brown, J.J. Baumberg, F.P. Booy, T. Melvin, Quenching of CdSe quantum dot emission, a new approach for biosensing, *Chem. Commun.* 25 (2005) 3201–3203.

95. D. Zhou, J.D. Piper, C. Abell, D. Klenerman, D.-J. Kang, L. Ying, Fluorescence resonance energy transfer between a quantum dot donor and a dye acceptor attached to DNA, *Chem. Commun.* 38 (2005) 4807–4809.

96. E. Oh, M.-Y. Hong, D. Lee, S.-H. Man, H.C. Yoon, H.-S. Kim, Inhibition assay of biomolecules based on fluorescence resonance energy transfer (FRET) between quantum dots and gold nanoparticles, *J. Am. Chem. Soc.* 127 (2005) 3270–3271.

97. B. Tang, L. Cao, K. Xu, L. Zhuo, J. Ge, Q. Li, L. Yu, A new nanobiosensor for glucose with high sensitivity and selectivity in serum based on fluorescence resonance energy transfer (FRET) between CdTe quantum dots and Au nanoparticles, *Chem. Eur. J.* 14 (2008) 3637–3644.

98. M. Tomasulo, I. Yildiz, F.M. Raymo, pH-sensitive quantum dots, *J. Phys. Chem. B* 110 (2006) 3853–3855.

99. I.L. Medintz, A.R. Clapp, H. Mattoussi, E.R. Goldman, B. Fisher, J.M. Mauro, Self-assembled nanoscale biosensors based on quantum dot FRET donors, *Nat. Mater.* 2 (2003) 630–638.

100. S. Kim, Y.T. Lim, E.G. Soltesz, A.M. De Grand, J. Lee, A. Nakayama, Near-infrared fluorescent type II quantum dots for sentinel lymph node mapping, *Nat. Biotechnol.* 22 (2004) 93–97.

101. J.V. Frangioni, Self-illuminating quantum dots light the way, *Nat. Biotechnol.* 24 (2006) 326–328.

102. M.-K. So, C. Xu, A.M. Loening, S.S. Gambhir, J. Rao, Self-illuminating quantum dot conjugates for in vivo imaging, *Nat. Biotechnol.* 24 (2006) 339–343.

103. Y. Zhang, M.-K. So, A.M. Loening, H. Yao, S.S. Gambhir, J. Rao, HaloTag protein-mediated site-specific conjugation of bioluminescent proteins to quantum dots, *Angew. Chem. Int. Ed.* 45 (2006) 4936–4940.

104. Y. Gao, Y. Cui, R.M. Levenson, L.W.K. Chung, S. Nie, In vivo cancer targeting and imaging with semiconductor quantum dots, *Nat. Biotechnol.* 22 (2004) 969–976.

105. A. Hoshino, K. Fujioka, T. Oku, S. Nakamura, M. Suga, Y. Yamaguchi, Quantum dots targeted to the assigned organelle in living cells, *Microbiol. Immunol.* 48 (2004) 985–994.

106. F. Pinaud, X. Michalet, L.A. Bentonlila, J.M. Tsay, S. Doose, J.J. Li, G. Iyer, S. Weiss, Advances in fluorescence imaging with quantum dot bio-probes, *Biomaterials* 27 (2006) 1679–1687.

107. F.Q. Chen, D. Gerion, Fluorescent CdSe/ZnS nanocrystal-peptide conjugates for long-term, nontoxic imaging and nuclear targeting in living cells, *Nano Lett.* 4 (2004) 1827–1832.

108. V. Bagalkot, L. Zhang, E. Levy-Nissenbaum, S. Jon, P.W. Kantoff, R. Langer, O.C. Farokhzad, Quantum dot-aptamer conjugates for synchronous cancer imaging, therapy, and sensing of drug delivery based on bi-fluorescence resonance energy transfer, *Nano Lett.* 7 (2007) 3065–3070.

109. D.S. Wang, J.B. He, N. Rosenzweig, Z. Rosenzweig, Superparamagnetic Fe_2O_3 beads-CdSe/ZnS quantum dots core-shell nanocomposite particles for cell separation, *Nano Lett.* 4 (2004) 409–413.

110. H.W. Gu, R.K. Zheng, X.X. Zhang, B. Xu, Facile one-pot synthesis of bifunctional heterodimers of nanoparticles: A conjugate of quantum dot and magnetic nanoparticles, *J. Am. Chem. Soc.* 126 (2004) 5664–5665.

37

Kondo Effect in Quantum Dots

Silvano De Franceschi
Commissariat à l'Énergie Atomique

Wilfred G. van der Wiel
University of Twente

37.1 Introduction

The Kondo effect arises from the coherent, many-body interaction between a localized spin and a surrounding Fermi sea of electrons. The history of the Kondo effect started in 1934 with the experimental observation of an anomalous electrical resistivity minimum in gold samples by de Haas and collaborators (de Haas et al. 1934). During the following 30 years, the origin of this anomaly remained essentially unknown. It was the Japanese physicist Jun Kondo who provided the key to unravel the mystery (Kondo 1964). Kondo realized that the minimum in the electrical resistivity is associated with the presence of diluted magnetic impurities coupled to the conduction electrons in the nonmagnetic host metal. To model this coupling, Kondo assumed an antiferromagnetic exchange interaction J between a localized magnetic moment, $\mathbf{S}$, and the conduction electrons with spin, $\mathbf{s}$, as described by the so-called s-d model (Zener 1951):

$$H_{sd} = \sum_{\mathbf{k},\mathbf{k}'} J_{\mathbf{k},\mathbf{k}'}(S^+ c^+_{\mathbf{k},\downarrow} c_{\mathbf{k}',\uparrow} + S^- c^+_{\mathbf{k},\uparrow} c_{\mathbf{k}',\downarrow}) + S_z(c^+_{\mathbf{k},\uparrow} c_{\mathbf{k}',\uparrow} - c^+_{\mathbf{k},\downarrow} c_{\mathbf{k}',\downarrow}),$$

(37.1)

where S_z and $S^{\pm}$ (= $S_x \pm iS_y$) are the spin operators for the magnetic impurity with spin $\mathbf{S}$. Here we have only given the terms that describe spin-flip scattering processes, while the spin-conserving, potential scattering term (which can be formally eliminated) is omitted (Hewson 1993). The s-d model can be derived from the Anderson model in the appropriate parameter regime, as was demonstrated by Schrieffer and Wolff (1966).

Kondo's perturbative calculation of the electrical resistivity up to third order in J showed that the spin exchange interaction leads to singular scattering of the conduction electrons near the Fermi level. Perturbation theory breaks down due to the appearance of $\ln T$ terms that diverge at low temperature, T. Such

logarithmic divergence also appears in other physical quantities, such as the magnetic susceptibility, entropy, and specific heat (Hewson 1993). Combining the $\ln T$ divergence in the electrical resistivity (which increases at low temperature for an antiferromagnetic coupling) with the T^5 phonon contribution provided the explanation of the experimentally observed resistivity minimum at low temperature (Kondo 1964).

Kondo ascribed the divergence at low temperature to the presence of an internal quantum degree of freedom (the impurity spin), preventing the possibility to treat the scattering off a magnetic impurity as a single-particle problem, as in the case of scattering off static potential defects. Intuitively, we can understand this argument as follows. If at a certain instant of time the spin on the impurity is pointing upward, Pauli exclusion principle enables only spin-down electrons to hop on the impurity. The situation is reversed after an event of spin-flip scattering, that is, a scattering event in which the spin of the incoming electron differs from that of the outgoing one, and the impurity spin has to "flip" in order to conserve the total angular momentum. As a result, the Pauli principle establishes a correlation between scattering processes, implying that all the electrons from the Fermi sphere are collectively taking part in the interaction with the local impurity spin. In practice, this means that higher-order scattering processes need to be taken into account and added up together coherently in order to evaluate the scattering cross section. The logarithmic divergence in the $O(J^3)$ term found by Kondo arises from the coherent superposition of spin-flip scattering events, and it is ultimately determined by the energy sharpness ($\sim k_B T$) of the Fermi distribution function.

Kondo's pioneering work evoked a wave of theoretical activity throughout the following decades. In particular, one needed to find a more sophisticated theory to correctly explain the low-temperature behavior instead of the (unphysical) $\ln T$ divergence produced by perturbation theory, a quest referred to as the

"Kondo problem." Thus, the Kondo effect has become an important paradigm of condensed matter physics, a reference problem for testing many-body theories and techniques. Covering all the developments in the theoretical treatment of the Kondo problem goes well beyond the scope of this chapter. Yet we wish to briefly discuss a few important theoretical breakthroughs relevant to the discussion later on.

Abrikosov (1965) investigated whether the low-temperature divergence arising from the logarithmic terms could be removed by summing the leading-order logarithmically divergent terms in higher-order perturbation. However, the leading-order logarithmic sum fails, as it produces a divergence at finite temperature, T_K, referred to as the Kondo temperature. Although perturbation theory provided a good description for $T \gg T_K$, it clearly broke down for $T < T_K$.

The description of the physics below T_K builds on the idea of "scaling" introduced by Anderson in the late 1960s (Anderson 1967). In essence, higher-order excitations are eliminated perturbatively and taken into account in lower-order terms, to end up with an effective model valid on a lower energy scale. Also this model breaks down for $T \ll T_K$. However, based on analogies with other systems, one predicted that the coupling between the local moment and conduction electrons would become infinite at lower temperatures, implying a ground state in which the impurity is bound to a conduction electron in a spin singlet state.

It was Kenneth Wilson who overcame the shortcomings of perturbation theory with a method known as "numerical renormalization group theory" and confirmed the scaling hypothesis of the singlet ground state (Wilson 1974, 1975). His contribution was recognized in the award of the 1982 Nobel Prize in physics. The singlet is a many-body state, in the sense that many conduction electron spins contribute to the screening of the local moment. However, on average only one is bound to completely compensate the impurity spin.

Scaling and renormalization group theories have enabled to calculate the local density of states at a Kondo impurity. The singlet state was found to emerge as a peak in the local density of states at the Fermi level. The energy width of the peak is $\sim k_B T_K$ and hence directly related to the Kondo temperature. The zero-temperature limit is a Fermi liquid fixed point (Nozières 1974). The local spin degree of freedom is quenched and the scattering of electrons near the Fermi energy (those responsible for electrical conduction) can be treated in a single-particle picture. This behavior is associated with a contribution to the resistivity proportional to T^2. All the many-body correlations result in a renormalization of the density of states at the impurity.

In metals, the Kondo effect is due to the contribution of many independent magnetic impurities. Deviations from this picture start to occur for relatively high impurity concentrations due to the onset of interactions among impurities, mediated by the electron Fermi sea. In the opposite limit, the question arises whether the Kondo effect can also be studied at the level of a *single* impurity. And if so, what could we learn from such single-impurity experiments? To answer the above questions, experimentalists

have been exploring various strategies to locally probe the electronic properties at and around an individual impurity.

A first approach to this experimental challenge takes advantage of low-temperature scanning tunneling microscopy (STM). In 1998, two teams, one in the United States (Madhavan et al. 1998) and one in Europe (Li et al. 1998), succeeded almost simultaneously in measuring the electronic properties of a single Kondo impurity. In these experiments, the STM tip acts as a weakly coupled electrode, enabling to probe both the spatial and energy dependence of the local density of states (Figure 37.1). For a fixed lateral and vertical position of the tip, the derivative of the tunnel current, I, with respect to the tip voltage, V, is directly proportional to the local density of states at an energy eV from the Fermi level of the metal.

In a topographic STM image (obtained at constant tunnel current), magnetic impurities appear as small "bumps" on an atomically flat metal surface. The data shown in Figure 37.1b refer to Co atoms on Au. By cooling the sample down to 4–5 K, and positioning the STM tip right on top of a magnetic impurity, both teams found qualitatively similar dI/dV (V) characteristics, consisting of an asymmetric dip at zero applied bias (Figure 37.1c). This dip was interpreted as a direct evidence of the Kondo resonance. The occurrence of an asymmetric dip, as opposed to a Lorentzian peak, can be explained by a Fano-type interference between parallel tunneling paths: via the localized Kondo resonance and via

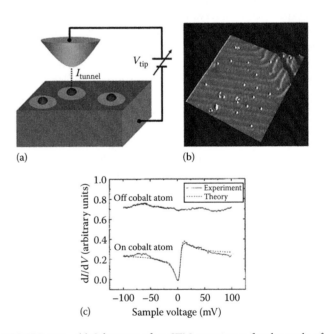

FIGURE 37.1 (a) Schematic of an STM experiment for the study of the Kondo effect in single magnetic impurities on a metal surface. (b) STM topography measurement of an atomically flat gold surface with dispersed cobalt impurities (bright spots). Measurement taken at constant tunnel current. (c) Tunneling spectroscopy measurement with the tip positioned on top of a single Co impurity. The Kondo effect produces an asymmetric dip in the differential conductance around zero bias voltage (lower trace). This dip gradually disappears by moving the tip away from the impurity (upper trace). (Reprinted from Madhavan, V. et al., *Science*, 280, 567, 1998. With permission.)

the continuum of surface states in the conduction band of the host metal (Fano 1961). The dip in dI/dV (V) becomes less pronounced when the tip is laterally shifted away from the impurity, and it vanishes completely at a distance of ~1 nm.

STM experiments rely on the fact that magnetic impurities (e.g., Co and Ce) giving rise to the Kondo effect in a bulk metal host, continue to do so when positioned *on top of* the metal surface. In fact, STM spectroscopy reveals that the Kondo temperature is significantly lower than that in the bulk case (T_K is estimated from the width of dI/dV dip). Due to the weak tunnel coupling, it is reasonable to assume that the STM tip does not perturb the local electronic properties and that a dI/dV measurement provides information on the Kondo impurity in equilibrium with the Fermi sea of the host metal.

By further approaching the STM tip, an antiferromagnetic coupling could in principle be established between the impurity spin and the conduction electrons of the tip itself. In this "invasive" configuration, the impurity spin would be screened by the simultaneous interaction with two spatially separated Fermi reservoirs. In this case, a finite bias voltage on the tip would drive the Kondo effect out of equilibrium, thereby creating an entirely new scenario, impossible to attain when the impurity is embedded in a metal host.

In practice, this strong coupling regime is difficult to achieve in an STM experiment. An alternative approach to realize such a type of system is to fabricate a device consisting of two metal "tips" separated by a sub-nanometer gap, and to position a magnetic impurity right in this gap. Four years after the first STM experiments, this idea was experimentally realized by two groups simultaneously, one at Cornell (Park et al. 2002) and the other at Harvard (Liang et al. 2002). These groups succeeded to fabricate molecular transistors by trapping individual paramagnetic molecules between two nanometer-scale gold electrodes. The molecules used in these experiments consisted of transition metal elements (providing the local spin degree of freedom) stabilized by organic ligands.

The realization of molecular transistors, however, remains very challenging. Due to their extremely small size (typically subnanometer), molecules require correspondingly narrow gaps between source and drain electrodes, which are very hard to realize in a reproducible fashion with conventional, top-down nanolithography techniques. Accurate positioning of individual molecules in these gaps, necessary for obtaining reliable transport properties, is still an open problem.

These difficulties can be circumvented by using nanostructures of much larger size, whose electronic properties can mimic those of magnetic impurities. Different ways exist to obtain such "artificial atoms" (Ashoori 1996). One of the most successful approaches relies on GaAs/AlGaAs heterostructures, containing a shallow (~100 nm deep) high-mobility two-dimensional electron gas (2DEG). The 2DEG electron density is typically ~10^{15} m^{-2}, corresponding to a Fermi wavelength of ~10 nm. Using electron-beam lithography followed by metal evaporation and lift off, narrow gate electrodes are patterned on the heterostructure surface as shown in Figure 37.2a. These electrodes are biased with negative gate voltages relative to the 2DEG. This causes local carrier depletion, leading to tunable lateral confinement within the 2DEG on the same scale as the Fermi wavelength (Figure 37.2b). A puddle of electrons can be defined with lateral dimensions of the order 100 nm and below. This is sufficiently small to create a discrete energy spectrum with typical level spacing of 0.1–1 meV. The discrete nature of the level spectrum becomes apparent at temperatures below ~1 K, which can be routinely obtained in ^{3}He-based cryogenic systems. The small puddle of electrons can then be regarded as an "artificial atom" or a "quantum dot" (QD). By properly adjusting the gate voltages on the surface electrodes, the QD can be electrically connected to extended regions of the 2DEG acting as source and drain contacts. The coupling is realized by quantum mechanical tunneling through the electrostatic potential barriers at the constrictions between the QD and the reservoirs.

What makes such QDs similar to magnetic impurities? The QD can contain a variable number of electrons, ranging typically from one to several tens. The Coulomb repulsion among the confined electrons is responsible for a charging energy, U, for adding (or removing) electrons to (from) the QD. As a result, the QD has usually one stable charge configuration, corresponding to a well-defined, integer number of confined electrons. This number can be tuned by using one or more gate voltages.

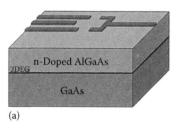 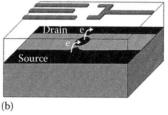 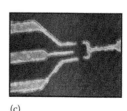

(a) (b) (c)

FIGURE 37.2 (a) Schematic of a lateral QD fabricated from a modulation-doped GaAs heterostructure. The heterostructure contains a GaAs/AlGaAs heterojunction with the upper AlGaAs layer doped with donor impurities. The electrons provided by the impurities are collected at the GaAs/AlGaAs interface forming a high-mobility 2DEG. Surface metal electrodes defined by electron-beam lithography can deplete the underlying 2DEG when biased negatively with respect to the 2DEG itself. (b) This local depletion is exploited to isolate a small puddle of electrons (the QD) between extended 2DEG regions (the source and drain contacts). (c) Scanning electron micrograph of the device used for the first observation of the Kondo effect in QD systems. (Reprinted from Goldhaber-Gordon, D. et al., *Nature*, 391, 156, 1998a. With permission.)

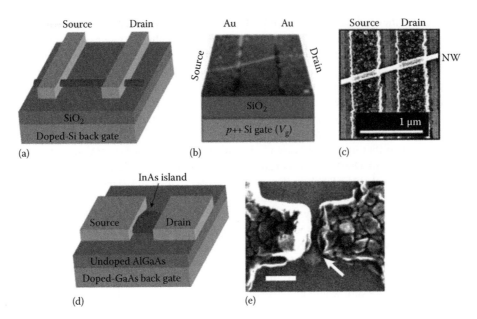

FIGURE 37.3 (a) Schematic of a QD device obtained by connecting an individual carbon nanotube (or, equivalently, a semiconductor nanowire) to a pair of metal electrodes (source and drain). The typical distance between these electrodes ranges between 0.1 and 1 μm. A heavily doped substrate is often used as a back-gate electrode to vary the electronic charge in the QD. (b) Atomic-force micrograph of a carbon nanotube QD device with a sketch of the underlying substrate. In this case, the distance between source and drain contacts is 0.3 μm. (Reprinted from Nygård, J. et al., *Nature*, 408, 342, 2000. With permission.) (c) Scanning electron micrograph of a similar device fabricated from an InAs semiconductor nanowire. (Reprinted from Sand-Jespersen, T. et al., *Phys. Rev. Lett.*, 99, 126603, 2007. With permission.) (d) Schematic of a QD device obtained by contacting a single self-assembled InAs island grown on a GaAs heterostructure. (e) Scanning electron micrograph of such devices. (Reprinted from Buizert, C. et al., *Phys. Rev. Lett.*, 99, 136806, 2007. With permission.)

A QD with a well-defined, odd number of electrons necessarily carries a half-integer spin (most likely a spin 1/2). As we will substantiate below, such a QD behaves as an artificial magnetic impurity that couples anti-ferromagnetically to the conduction electrons of the leads. Conceptually, this is identical to the case of a magnetic molecule between two nanoelectrodes. Characteristic energy scales such as the on-site charging energy U, and the mean level spacing are much smaller in GaAs-based QDs than in molecules. On the other hand, GaAs QDs enable a much higher degree of reproducibility and control over the relevant system parameters. Their unique versatility has been the key to a number of experimental achievements that have shone new light on Kondo physics.

The first breakthrough experiment in GaAs QDs was conducted at MIT in collaboration with researchers at the Weizmann institute (Goldhaber-Gordon et al. 1998a). The results were published only a few months before the STM experiments, and they were readily followed by consistent experimental achievements at Delft University of Technology (Cronenwett et al. 1998) and at the Max Planck Institute in Stuttgart (Schmid et al. 1998). Therefore, it is fair to say that 1998 has marked the beginning of a new era in the history of the Kondo effect. In a *Physics World* review article (Kouwenhoven and Glazman 2001), this new era is referred to as "*The revival of the Kondo effect.*"

GaAs-based heterostructures, however, are not the only possible approach to the realization of QDs. Following recent progress in nanoelectronics, new valuable alternatives have

emerged based on the use of various types of nanomaterials, such as carbon nanotubes and semiconductor self-assembled nanostructures. QDs are obtained by depositing or growing these nanomaterials on a suitable substrate, and contacting them individually with submicron scale metal electrodes defined by conventional nanolithography. Examples of these QD devices are shown in Figure 37.3.

In this chapter, we aim to provide an accessible overview of Kondo phenomena in QD systems. We start with a brief introduction of the theoretical framework. The following sections will cover different aspects of the Kondo effect, guiding the reader through the results of selected experiments. Given the large body of published work, we should like to stress that our selection, although quite broad, cannot provide a fully comprehensive view of this fascinating topic.

37.2 Basic Theory of Electron Transport in Quantum Dots

A schematic representation of a QD, highlighting its most important characteristics, is given in Figure 37.4a. The QD is depicted as a generic potential well with discrete energy levels connected to source and drain Fermi reservoirs via two tunnel barriers. We define the characteristic tunnel rates of the barriers as Γ_S/h and Γ_D/h, respectively, where h is Planck's constant. These rates determine the lifetime of the QD states, namely, the characteristic escape time, τ, of an electron placed on a QD level

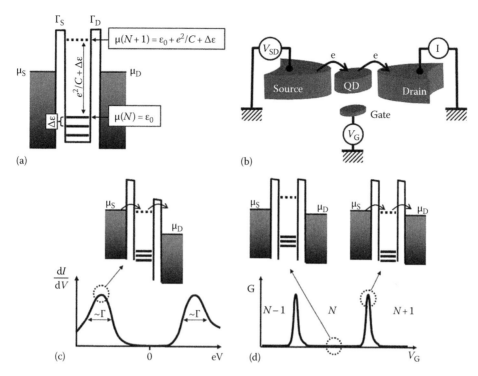

FIGURE 37.4   (a) Schematic of a generic QD device. (b) Energy diagram highlighting its most important parameters. (c) The qualitative behavior of the linear conductance, G, as a function of gate voltage, V_G, is shown in the lower panel. G is largely suppressed due to the Coulomb blockade effect (top left diagram), but it can be activated whenever the QD electrochemical potential is brought between the Fermi energies of the leads, so that two consecutive charge states become simultaneously accessible (top right panel). (d) Coulomb blockade can be overcome also by means of a sufficiently large source–drain voltage. The expected qualitative behavior is shown in the lower panel. A resonance in the differential conductance occurs as a result of the alignment of a QD level (e.g., the electrochemical potential for the successive charge state) with the Fermi energy of one of the leads (top panel). At low temperature, the width of this resonance is set by the lifetime broadening, Γ, of the QD state.

lying above the Fermi energy of the leads. Following Heisenberg's uncertainty principle, a finite lifetime translates into an energy broadening $\Gamma = h/\tau = \Gamma_S + \Gamma_D$ of the QD levels. We assume Γ to be smaller than the QD mean level spacing, $\Delta\varepsilon$. Due to the Coulomb repulsion among the electrons on the QD, there is an energy cost associated with the addition (or removal) of a single electron to (or from) the QD. This charging energy can be expressed in terms of the total capacitance of the QD, $C = C_S + C_D + C_G$, where C_S and C_D, are the QD-source and QD-drain capacitances, respectively, and C_G is the overall capacitance between the QD and surrounding gate electrodes. From classical electrostatics, we can write the ground state energy of a QD containing N electrons as

$$U(N) = \frac{\left(-|e|(N - N_0) + C_S V + C_G V_G\right)^2}{2C} + \sum_{n=1}^{N} E_n(B), \quad (37.2)$$

where

$-|e|$ is the electron charge

N_0 is the number of electrons in the dot at zero gate voltage

In systems such as GaAs QDs, the negative charge associated with N_0 electrons compensate the positive background charge originating from the ionized donors in the AlGaAs layer. We have defined the source–drain bias voltage as V, and the gate voltage as V_G. The terms $C_S V$ and $C_G V_G$ can change continuously and represent the charge on the dot that is induced by the bias voltage through the capacitance C_S and by the gate voltage through the capacitance C_G. The last term in Equation 37.2 is a sum over the occupied single-particle energy levels, ε_n, measured from the bottom of the conduction band. In the derivation of Equation 37.2, it has been implicitly assumed that the energy-level spectrum ε_n depends uniquely on the characteristics of the confinement potential and not on the number of electrons on the QD. All interaction effects among electrons are condensed into one single parameter, the electrostatic energy term e^2/C. This approximate model is commonly known as the constant interaction (CI) model.

Given the fact that the QD is in contact with two Fermi reservoirs, how many electrons will sit on it in the most energetically favorable configuration? To answer this question, we need to derive an expression for the electrochemical potential of the QD, $\mu_{dot}(N)$, which is defined as the minimum energy for adding the Nth electron. From Equation 37.2 we find

$$\mu_{dot}(N) \equiv U(N) - U(N-1)$$

$$= (N - N_0 - 1/2)\, e^2/C - |e|\, (C_S V + C_G V_G)/C + \varepsilon_N.$$

$$(37.3)$$

The first two terms describe the electrostatic contribution $-|e|\varphi_N$ with φ_N the electrostatic potential of the N-electron dot. The last term, ε_N, represents the chemical contribution $\mu_{ch}(N)$. Hence, $\mu_{dot}(N) = -|e|\varphi_N + \mu_{ch}(N)$. The most favorable number of electrons on the QD corresponds to largest N fulfilling the condition $\mu_{dot}(N) < \mu_S, \mu_D$. The addition of an extra electron at constant gate voltage lifts the electrochemical potential by an amount:

$$\Delta\mu_{dot}(N) = \mu_{dot}(N+1) - \mu_{dot}(N) = U(N+1) - 2U(N)$$

$$+ U(N-1) = e^2/C + \Delta\varepsilon. \tag{37.4}$$

Note that in the case of spin-degenerate levels, $\Delta\varepsilon$ vanishes every time N is increased from odd to even (the added electron fills up the same orbital state with opposite spin).

We define ε_0 (<0) as the energy position of the QD ground-state electrochemical potential with respect to the drain Fermi level (note that at equilibrium $\mu_S = \mu_D$). At low temperatures, such that $k_B T$ is much smaller than both $|\varepsilon_0|$ and ($e^2/C + \Delta\varepsilon + \varepsilon_0$), thermally induced charge fluctuations are suppressed. Then the number of confined electrons is a well-defined integer, that is, the integer minimizing the electrostatic energy. In this situation, a small bias voltage between source and drain cannot produce any current flow. This phenomenon is commonly known as Coulomb blockade. There are essentially two possibilities to obtain some current flow: to apply a sufficiently large bias voltage (Figure 37.4c) or to bring the electrochemical potential of the QD between the Fermi energies of the leads, which works also with a small bias voltage across the QD (Figure 37.4d). Any of these two options result in a sequential flow of electrons; an electron comes on the QD from the source and leaves to the drain before the following electron is allowed to enter the QD. This peculiar phenomenon is referred to as single-electron tunneling, and it is a direct consequence of the on-site Coulomb interaction.

Yet this is not the full story. In fact quantum mechanical charge fluctuations matter too, especially when the QD-lead tunnel coupling is strong leading to a relatively large lifetime broadening Γ of the QD states. To treat this problem, we introduce a simplified Hamiltonian for the QD system:

$$H_{QD} = \sum_{\mathbf{k},\sigma,\alpha} \xi_{\mathbf{k},\alpha} c^+_{\mathbf{k},\sigma,\alpha} c_{\mathbf{k},\sigma,\alpha} + \sum_\sigma \varepsilon_0 d^+_\sigma d_\sigma + \frac{e^2}{C} d^+_\uparrow d_\uparrow d^+_\downarrow d_\downarrow$$

$$+ \sum_{\mathbf{k},\sigma,\alpha} (V_{\mathbf{k},\sigma,\alpha} c^+_{\mathbf{k},\sigma,\alpha} d_\sigma + V^*_{\mathbf{k},\sigma,\alpha} d^+_\sigma c_{\mathbf{k},\sigma,\alpha}). \tag{37.5}$$

Here, $c^+_{\mathbf{k},\sigma,\alpha}$ creates an electron of energy $\xi_{\mathbf{k},\alpha}$, momentum $\mathbf{k}$, and spin σ ($=\downarrow,\uparrow$) in lead α ($=$ S, D); d^+_σ creates an electron of energy ε_0 and spin σ in the QD; $V_{\mathbf{k},\sigma,\alpha}$ is a tunneling matrix element. It is often reasonable to assume that $V_{\mathbf{k},\sigma,\alpha}$ does not depend on $\mathbf{k}$ and σ, that is, $V_{\mathbf{k},\sigma,\alpha} = \overline{V_\alpha}$. In this case, the tunnel rate $\Gamma_\alpha = 2\pi\rho_\alpha |\overline{V_\alpha}|^2$, where ρ_α is the electron density of states in lead α. Equation 37.5 is nothing but the Anderson Hamiltonian for a single-level impurity in a metal host (Anderson 1961) with the only difference that here the summations over delocalized states contain contributions from two distinct and spatially separated reservoirs. In this

model, N can only be 0, 1, or 2. This is clearly a simplification since N can easily exceed 2 in QDs with multiple electronic levels. Yet if energy levels are well separated from each other, retaining only the uppermost occupied level is a good approximation and Equation 37.5 can be used. In a seminal paper, Schrieffer and Wolff have shown that the Anderson model can be mapped onto the s-d model through a canonical transformation (Schrieffer and Wolff 1966). The resulting exchange coupling terms turn out to be antiferromagnetic: $J_{\mathbf{k},\mathbf{k}'} = V^*_\alpha V_{\alpha'}[1/(e^2/C + \varepsilon_0 - \varepsilon_{\mathbf{k}'}) + 1/(\varepsilon_\mathbf{k} - \varepsilon_0)]$ where $\alpha(\alpha') = $ L or R, $\mathbf{k} \in \alpha$, $\mathbf{k}' \in \alpha'$. This further clarifies the correspondence between the physics of a QD and that of a magnetic impurity in a metal host.

To examine the transport properties of a QD we shall distinguish between even and odd occupation. Let us begin with the $N =$ even case, for example, $N = 2$. As already anticipated, the first order in the tunnel coupling electron transport is blocked since $N = 2$ is the only energetically allowed state, and all the other charge states are inaccessible at low temperature ($k_B T \ll |\varepsilon_0|$). Charge fluctuations, however, can take place via higher-order tunneling processes. An example of such processes is shown in Figure 37.5a. Following the Heisenberg uncertainty relation, the QD can potentially lose an electron for a time $\sim h/|\varepsilon_0|$ before returning to its initial (stable) charge state. If the first electron escapes toward the

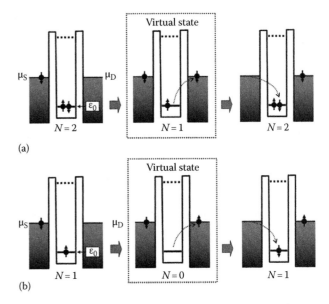

FIGURE 37.5 (a) Second-order elastic cotunneling process for a QD with two electrons and zero spin. This quantum coherent process takes an electron from source to drain via a one-electron intermediate state. The energy cost associated with this intermediate state corresponds to the electrochemical potential, ε_0, of the two-electron ground state. (b) Similar cotunneling process in the case of a one-electron, spin-1/2 ground state. In this case, the transfer of an electron can be accompanied by a spin flip. These types of cotunneling processes are responsible for the establishment of an antiferromagnetic exchange coupling between the QD spin and the spins of the conduction electrons in the leads. At low temperatures, this results in the complete screening of the local spin and the development of a Kondo resonance at the Fermi energy.

drain electrode and it is replaced by an electron from the source, the net result will be the transfer of an elementary charge from source to drain. Since the initial and final QD states have the same energy, this two-electron process is called elastic cotunneling. At finite bias, $eV = \mu_S - \mu_D > 0$, elastic cotunneling results in a small current. How large is this current? In the linear regime ($eV \sim k_B T$), only electrons within $\sim k_B T$ need to be taken into account. The transition rate for a second-order cotunneling process connecting an occupied state in the source to an empty state in the drain is proportional to $\left|\overline{V_L}\right|^2 \left|\overline{V_R}\right|^2 / \varepsilon_0^2$, that is, to $(\Gamma_L \Gamma_R / \varepsilon_0)^2$. Accordingly, elastic cotunneling will make a significant contribution to the QD conductance when $\Gamma \sim |\varepsilon_0|$. When rewritten as $\tau \sim h/|\varepsilon_0|$, this condition corresponds to the requirement that the Heisenberg uncertainty time is long enough to enable the two-electron tunneling process. What about higher-order contributions? If $\Gamma < |\varepsilon_0|$, these will be less and less important so that cotunneling can be treated as a small perturbation.

The situation is drastically different in the case of $N = $ odd. The difference arises from the contribution of cotunneling processes resulting in a flip of the QD spin. Figure 37.5b shows an example of second-order spin-flip process in the case of $N = 1$. Just as for a magnetic impurity in a nonmagnetic metal host, high-order spin-flip processes give rise to the Kondo effect. However, here it manifests itself as a $\ln T$ divergence of the conductance, G. Why not a diverging resistance as in the bulk metal case? The reason has to do with geometry. In a bulk metal, spin-flip processes mix electron waves with different wave vectors leading to an enhanced scattering cross section and hence a larger resistance. In a QD device, spin-flip cotunneling processes can mix electron waves on the two sides of the QD thereby favoring electron transport across the dot. Conductance, however, cannot grow indefinitely. If only one spin-degenerate dot level is involved in the cotunneling processes, G cannot exceed the quantum limit, $G_0 = 2e^2/h$, for a single perfectly conducting channel. Another way to visualize this phenomenon is by considering the effect on the local density of states. Spin-flip cotunneling processes lead to the appearance of the Kondo resonance opening up a conducting channel at the Fermi energy of the leads. In the zero-temperature limit, the conductance will approach the full transmission value G_0 only for symmetric tunnel barriers ($\Gamma_S = \Gamma_D$). This is referred to as the unitary limit. More in general, G will approach the limit $[4\Gamma_S \Gamma_D/(\Gamma_S + \Gamma_D)^2]G_0$.

With a strong asymmetry in the tunnel barriers (e.g., $\Gamma_S \gg \Gamma_D$), only one lead (the source) couples strongly to the QD giving rise to the Kondo effect. While the Kondo temperature, which depends on the sum $\Gamma_S + \Gamma_D$, can be relatively large, the QD conductance is necessarily very small. In this case, the second lead (the drain) can act as a weakly coupled electrode, similar to the tip of an STM. The density of states in the QD can then be directly probed by a dI/dV (V) measurement. A dI/dV peak is then expected at zero bias as a direct evidence of the Kondo resonance in the local DOS. Figure 37.6 summarizes some of the most significant fingerprints of the Kondo effect in QDs and a comparison with the main Kondo signatures in metals containing diluted magnetic impurities.

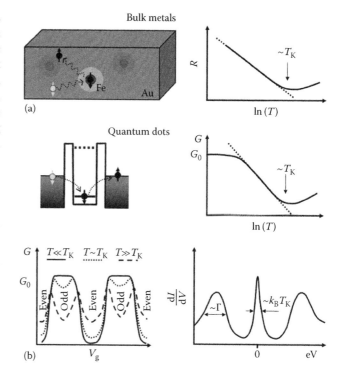

FIGURE 37.6 Main characteristic signatures of the Kondo effect in a QD as opposed to the Kondo effect in bulk metals containing diluted magnetic impurities. (a) In bulk metals, a Kondo resonance develops at each impurity site as a result of spin-flip scattering processes (left panel). The Kondo resonance increases the ability of each impurity to scatter electrons at the Fermi energy, resulting in a logarithmic divergence of the resistance at low temperatures (right panel). (b) In QDs, a Kondo resonance forms as a result of spin-flip cotunneling processes (top left panel). This opens up a tunneling path for electrons at the Fermi energy leading to an enhancement of the conductance (top right panel). In the low-temperature limit ($T \ll T_K$), the conductance saturates at a unitary value that depends on the ratio between the tunnel rates of the two barriers. For symmetric coupling, the unitary value equals the conductance quantum $G_0 = 2e^2/h$, that is, the conductance of a ballistic one-dimensional channel. In the simplest picture of a QD with well-separated energy levels, an ordinary spin-1/2 Kondo effect is expected for all odd-integer occupations of the QD. In the intermediate even-integer valleys, the QD ground state is a singlet and no Kondo effect is expected. Opposite temperature dependences of the linear conductance are consequently expected for the two parities resulting in a characteristic conductance pattern (bottom left panel). The formation of a Kondo resonance in the local density of states at the Fermi energy emerges as a zero-bias peak in the differential conductance whose width is set by the Kondo temperature (bottom right panel).

37.3 Kondo Effect in the Unitary Limit

In this section, we show how the Kondo features discussed above manifest themselves in an experiment performed by the authors of this chapter on a GaAs QD system. In 1998, the first experimental observations of the Kondo effect in semiconductor QDs were reported (Cronenwett et al. 1998, Goldhaber-Gordon et al. 1998a, Schmid et al. 1998). The experimental results followed reasonably well the expectations from the Anderson impurity model (Anderson 1961). However, in none of these studies the theoretically predicted unitary limit of conductance at $2e^2/h$ was reached. This unitary limit, corresponding to perfect transmission ($T = 1$) through the QD, was observed only two years later. First in a GaAs/AlGaAs-based QD structure (van der Wiel et al. 2000), and later also in a QD defined within a carbon nanotube (Nygård et al. 2000). In this section, we discuss the first experiment in which a strong Kondo effect is observed and the Coulomb blockade for electron tunneling is overcome completely by the Kondo effect.

The device in van der Wiel et al. (2000) was fabricated from an AlGaAs/GaAs heterostructure grown by molecular beam epitaxy on semi-insulating GaAs. A modulation-doped AlGaAs/GaAs heterojunction was used to create a 2DEG with an electron density of $3 \times 10^{15}\ m^{-2}$, 100 nm below the surface. The QD was formed inside a one-dimensional wire defined in the 2DEG by dry etching (top-left inset to Figure 37.7a). Quantum confinement was established by means of two transverse gates. We indicate by V_{gl} and V_{gr} the corresponding voltages. All measurements were performed in a dilution refrigerator with a base temperature $T = 15$ mK, using a standard lock-in technique with an AC voltage excitation between source and drain contacts of $3\ \mu V$.

Figure 37.7 summarizes the essential results from van der Wiel et al. (2000). In Figure 37.7a, the conductance, G, is plotted as a function of gate voltage V_{gl} for different temperatures. All traces were taken at a magnetic field $B = 0.4$ T perpendicular to the 2DEG. At base temperature (thick solid curve), the conductance in the valleys around $V_{gl} = -413$ mV and -372 mV reaches $2\ e^2/h$. In fact, the valleys are not observed anymore. For higher temperatures, two separate Coulomb peaks develop with increasing peak spacing and the valley conductance decreases. The adjacent Coulomb valleys show the opposite temperature dependence, that is, the conductance *increases* with increasing temperature. This even–odd parity suggests the pair-wise filling of the dot's discrete energy levels: that is, an unpaired spin (a *doublet*) in a valley with $N = $ odd, where we observe the Kondo effect, and a spin singlet for $N = $ even. In the lower right inset to Figure 37.7a, we show the differential conductance, dI/dV_{SD}, versus source–drain bias, V_{SD}, for different temperatures in the middle of the Kondo plateau at $V_{gl} = -413$ mV. The pronounced peak around $V_{SD} = 0$ reflects the Kondo resonance at the Fermi energy. We note that since the Zeeman splitting $g\mu_B B$ for $B = 0.4$ T is several times smaller than $k_B T_K$, the Kondo resonance is not split by the magnetic field. The peak height (i.e., the linear conductance) has

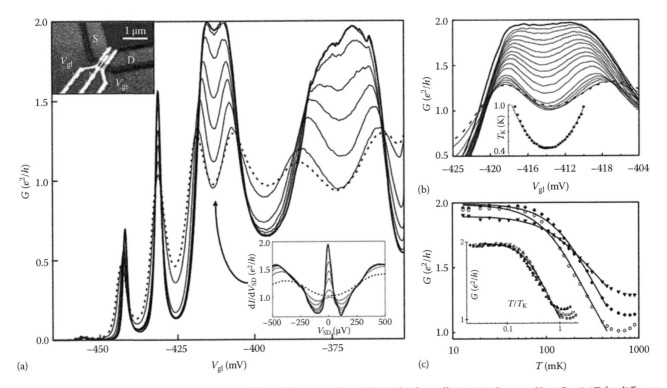

FIGURE 37.7 Experimental results for the Kondo effect in the unitary limit. (a) Coulomb oscillations in G versus V_{gl} at $B = 0.4$ T for different temperatures. $T = 15$ mK (thick solid trace) up to 800 mK (thick dashed trace). V_{gr} is fixed at −448 mV. Top left inset: atomic force microscope image of the device. Bottom right inset: differential conductance, dI/dV_{SD}, versus dc bias voltage, V_{SD}, for $T = 15$ mK (thick solid trace) up to 900 mK (thick dashed trace), at $V_{gl} = -413$ mV and $B = 0.4$ T. The Kondo resonance manifests itself as a peak in dI/dV_{SD} at zero bias. (b) $G(V_{gl})$ around the Kondo plateau in the case of optimized symmetric tunnel barriers. T ranges between 15 mK (thick solid trace) up to 800 mK (thick dashed trace). Inset: Kondo temperature, T_K, at the Kondo plateau as obtained from many fits as in (c); the solid line is a parabolic fit of $\log(T_K)$ versus V_{gl}. (c) $G(T)$ at fixed gate voltage for $V_{gl} = -411$ (solid diamonds), −414 (open circles) and −418 (solid triangles) mV. The solid curves are fits as explained in the text. Inset: G versus normalized temperature, T/T_K, for six different gate voltages. All traces are fitted to a single curve (solid line). (Reprinted from van der Wiel, W.G. et al., *Rev. Mod. Phys.*, 75, 1, 2003. With permission.)

a logarithmic temperature dependence with saturation at $2\,e^2/h$ for low temperature, corresponding to the unitary limit. The unitary limit implies that the transmission probability through the QD is equal to one. Although U is an order of magnitude larger than $k_B T_K$, the Kondo effect completely determines electron tunneling at low energies (i.e., $T \ll T_K$ and $eV_{SD} \ll k_B T_K$).

These measurements were taken after optimizing the two barrier gate voltages, V_{gl} and V_{gr}, in order to obtain nearly equal tunnel barriers. However, sweeping V_{gl}, as in Figure 37.7a, changes the left barrier much more effectively than the right one, and hence the barriers cannot be symmetric over the whole V_{gl}-range. For a quantitative comparison to theory, we chose V_{gr} such that, upon sweeping V_{gl}, we could obtain a flat plateau close to $2\,e^2/h$ (Figure 37.7b). The two discernable Coulomb oscillations at the highest temperatures have completely merged at low T. In Figure 37.7c, we fit G versus T for different gate voltages to the empirical function

$$G(T) = G_0 \left(\frac{T_K'^2}{T^2 + T_K'^2} \right)^s \tag{37.6}$$

This function is an analytical approximation to numerical renormalization-group (NRG) results for the Anderson impurity model (Costi et al. 1994, Goldhaber-Gordon et al. 1998b). $T_K' = T_K / \sqrt{2^{1/s} - 1}$ and s is a fit parameter that should be close to 0.2 for a spin-1/2 impurity. We find $s = 0.29$ and good agreement between experimental data and theoretical curves. The inset to Figure 37.7b shows T_K versus V_{gl} as obtained from many fits as in Figure 37.7c. $\log(T_K)$ follows a quadratic dependence on V_{gl} (i.e., ε_0) with a minimum in the middle of the conductance plateau. From the parabolic fit, we estimate $U = 0.5\,\text{meV}$ and $\Gamma = 0.23\,\text{meV}$ (see van der Wiel et al. 2000 for more details). All over the plateau, G is a universal function of the normalized temperature, T/T_K, regardless of the other energy scales, U, ε_0, and Γ, see inset to Figure 37.7c.

37.4 Kondo Effect Out of Equilibrium

In this section, we discuss in more depth the effect of a finite bias across a QD in the Kondo regime. As we anticipated in Section 37.2, when one of the contacts is weakly coupled (e.g., $\Gamma_S \gg \Gamma_R$), a measurement of the differential conductance versus source-drain voltage results in a zero-bias peak proportional to the Kondo resonance in the local DOS (see Figure 37.8a). (Note that the current, I, through the QD is proportional to the integrated local DOS and hence dI/dV is proportional to the DOS itself). Due to the small current, the applied bias does not perturb the Kondo effect, which is the result of the strong coupling between the QD and the source Fermi reservoir. In the case of symmetric coupling, however, both leads equally contribute to the Kondo effect (Figure 37.8b). Meir and coworkers predicted that at finite bias the Kondo resonance in the local DOS should split into two peaks aligned with the Fermi energies of the source and drain leads (Meir et al. 1993). Also in

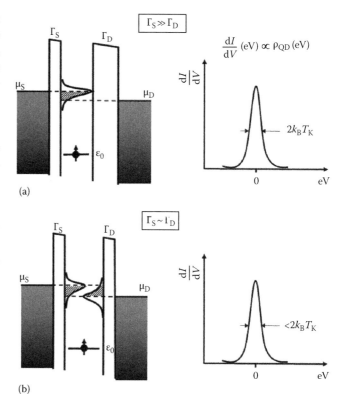

FIGURE 37.8 This figure illustrates the effect of finite bias voltage, V, across a QD in the Kondo regime. (a) For largely assymmetric tunnel barriers, the Kondo effect arises uniquely from the exchange correlations with the strongly coupled lead (source). A finite bias does not perturb significantly these correlations and the weakly coupled lead (drain) acts as a noninvasive probe of the local density of states. A zero-bias dI/dV peak proportional to the Kondo resonance in the local DOS is then expected. Calculations based on numerical renormalization group theory provide a value of $2k_B T_K$ of the full width at half maximum (FWHM) of the Kondo resonance. (b) This is no longer the case of comparable tunnel rates. Now both leads contribute to the Kondo effect and an applied bias induces a splitting of the Kondo resonance into two peaks aligned with the Fermi energies of the leads. Also in this case, a zero-bias dI/dV peak is found, but a simple proportionality relation with the Kondo resonance at equilibrium cannot be expected due to bias-induced decoherence. We speculate that the measured FWHM should be slightly smaller than $2k_B T_K$.

this case, the differential conductance exhibits a peak at zero bias, but this peak has no simple proportionality relationship with the Kondo resonance in the local DOS at zero bias. In fact, at finite bias, cotunneling processes can lead to the creation of electron–hole excitations in the leads, whose energy is provided by the potential difference across the dot. These excitations destroy the quantum coherence of the virtual tunneling processes necessary for the formation of a Kondo state. A progressive suppression of the Kondo split peaks is then to be expected at increasing bias voltage. It is natural to ask the following question: can the bias-induced splitting of the Kondo resonance be observed experimentally? Evidently, two-terminal measurements do not provide enough information. The first answer to this question came from a three-terminal experiment performed by the authors of this chapter and

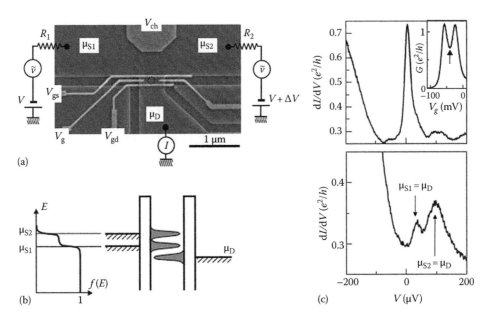

FIGURE 37.9 Experimental study of the Kondo effect out of equilibrium. (a) Scanning electron micrograph of the three-terminal mesoscopic device used for this experiment. The device consists of a lateral QD (dotted circle) coupled on the source side to the middle region a quasi-ballistic quantum wire. The measurement configuration is indicated. (b) A bias voltage, ΔV, across this wire (i.e., between source 1 and source 2) creates a double-step distribution function in the source reservoir. When the QD is in the Kondo regime, a split Kondo resonance is consequently expected. (c) This splitting is revealed by a measurement of the differential conductance, dI/dV, as a function of the (source 1)–drain voltage, V, at constant ΔV. Top panel: zero-bias resonance at $\Delta V = 0$. Bottom panel: split-resonance at $\Delta V = -150\,\mu V$. (Reprinted from De Franceschi, S. et al., *Phys. Rev. Lett.*, 89, 156801, 2002. With permission.)

coworkers. The device used for this experiment was fabricated from a 2DEG GaAs heterostructure of the type described above. It consists of a QD coupled on one side to the middle region of a quasi-ballistic quantum wire. A scanning-electron micrograph of the device is shown in Figure 37.9a. A finite bias voltage $\Delta V = (\mu_{S2} - \mu_{S1})/e$ applied across the quantum wire creates a double-step distribution function in the wire itself. This follows from the fact that electrons coming from the left contact do not interact with those coming from the right contact. Thus the QD effectively "sees" two electron reservoirs on the source side, with Fermi energies $\mu_{S1} = eV$ and $\mu_{S2} = e(V + \Delta V)$, plus an electron reservoir on the drain side, with Fermi energy $\mu_D = 0$. The ΔV-induced splitting of the Kondo resonance emerges from a measurement of the QD differential conductance dI/dV at constant ΔV. For $\Delta V = 0$, there is only one Fermi reservoir on the source side and a single zero-bias peak is observed (Figure 37.9c, upper panel) similar to the previously discussed two-terminal experiments. For a finite ΔV, two peaks emerge in the dI/dV (V) whose positions correspond to the alignment of μ_D with μ_{S1} and μ_{S2}, respectively. This is clearly seen in the lower panel of Figure 37.9c where $\Delta V = -150\,\mu V$. In these measurements, the drain contact is strongly coupled to the QD, resulting in a Kondo resonance as well as decoherence at finite bias across the dot. In other words, it does not act as a weakly coupled probe. More recently, Leturcq and coworkers at the ETH Zürich have succeeded in measuring the Kondo effect in a QD connected to three independent electrodes (Leturcq et al. 2005). By adjusting the corresponding tunnel barriers, they were able to realize a configuration characterized by two strongly coupled reservoirs,

giving rise to the Kondo effect, and a weakly coupled reservoir used as a probe for the local DOS. A split Kondo resonance was clearly observed as a result of a finite bias between the strongly coupled electrodes. These experiments show that the two DOS peaks arising from the bias-induced splitting of the Kondo resonance emerge at $e\Delta V \sim k_B T_K$ and remain visible up to bias voltages well above the Kondo energy scale. In Leturcq et al. (2005), split peaks were distinctively observed up to $e\Delta V \sim 10$–$20\,k_B T_K$.

Another way to split the Kondo resonance is to apply an external magnetic field, B. In the case of GaAs QDs, a magnetic field in the plane of the 2DEG couples almost exclusively to the spin degree of freedom, removing spin degeneracy. The Zeeman splitting of spin-1/2 states is $g\mu_B B$, where g is the electron g-factor and μ_B the Bohr magneton. Therefore, under an in-plane field a spin-flip cotunneling process implies a discrete change $\pm g\mu_B B$ in the QD energy. This energy change results into a splitting of the Kondo resonance into two peaks at $\pm g\mu_B B$ off the Fermi energy (see Figure 37.10a). Similar to the case of bias-induced splitting, the split peaks due to a finite Zeeman splitting are smaller than the single Kondo peak at $B = 0$. Again this is a consequence of decoherence. For $g\mu_B B > k_B T_K$, the zero-bias conductance is largely suppressed due to the removal of spectral weight at the Fermi energy. As opposed to the zero-field case, the dI/dV increases with the applied bias reaching a peak at $eV = g\mu_B B$ or $eV = -g\mu_B B$ (Figure 37.10b). This shows that bias is not always detrimental to the differential conductance. The data shown in Figure 37.10c illustrate the Zeeman splitting of the Kondo resonance in a QD defined in an InAs semiconductor nanowire. In an experimental

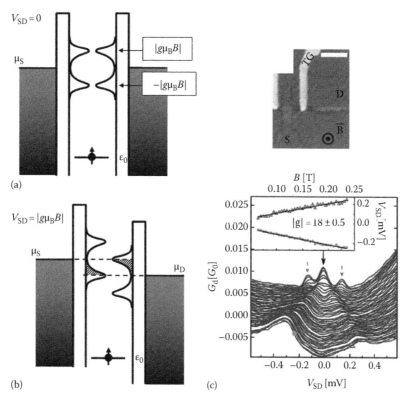

FIGURE 37.10 Zeeman splitting of the Kondo resonance. (a) Qualitative local DOS in a magnetic field. The removal of spin degeneracy results in a splitting of the Kondo resonance. For comparable tunnel rates, each lead contributes a pair of DOS peaks at $\pm g\mu_B B$ off the Fermi energy. In a measurement of the differential conductance as a function of source–drain bias, two peaks are found at $eV = \pm g\mu_B B$, that is, for the condition depicted in (b). An example of such a type of measurement is shown in (c) for a QD device fabricated from a single InAs nanowire (its scanning electron micrograph is shown in the upper inset). The dI/dV traces are taken for different magnetic fields, from $B = 0$ (upper trace) to $B = 0.24\,T$. The zero-bias peak at $B = 0$ is due to the Kondo resonance, while the two additional peaks indicated by small arrows originate from the superconducting nature of the source and drain contacts, and hence have nothing to do with the Kondo effect. At finite B, the Kondo peak splits due to the Zeeman effect while the side peaks quickly disappear due to the field-induced suppression of superconductivity. (Reprinted from Csonka, S. et al., *Nano Lett.*, 8, 3932, 2008. With permission.)

study carried out on GaAs QDs a peak splitting appreciably exceeding $|2g\mu_B B|$ was reported (Kogan et al. 2004).

So far, we have discussed three different ways to suppress the Kondo resonance: increasing temperature, magnetic field, or bias voltage. The underlying physical mechanisms are very different though. Temperature destroys the Kondo resonance by broadening the Fermi edges of the electron distribution function in the leads. An external magnetic field prevents elastic spin-flip cotunneling, due to the finite energy required to flip the local spin. An applied bias voltage affects the spin-flip cotunneling processes that bring electrons from source to drain. Such processes can be accompanied by the creation of electron–hole excitations at the expense of the potential energy drop between source and drain.

The Kondo effect out of equilibrium is a challenging theoretical problem. Different approaches have been proposed in order to account for the decoherence due to electron–hole excitations: from the so-called noncrossing approximation (Meir et al. 1993) to various renormalization group methods (Bulla et al. 2008). In a recent publication by Paaske and coworkers (Paaske et al. 2006), measurements of dI/dV vs. V performed on a carbon-nanotube

QD in a magnetic field were fitted to a perturbative renormalization group model resulting in excellent agreement.

37.5 Kondo Effect in a Multilevel Quantum Dot

It is known from experimental and theoretical studies on bulk metal systems that orbital degeneracy in magnetic impurities can have important consequences on the Kondo effect (Hewson 1993). For instance, it can be shown that the increased degeneracy, resulting from multiple orbital states, leads to an enhanced Kondo temperature since $T_K \sim e^{-1/n\nu J}$, where n is the number of degenerate states and ν the DOS in the electron sea at the Fermi energy (Hewson 1993). This is observed in the case of magnetic impurities with f-shell electrons. However, the enhancement of T_K is not the most interesting aspect. In this section, we will show how the versatility of QD systems has enabled the possibility to uncover new aspects of the Kondo effect associated with multi-orbital configurations.

In the previous sections, we considered the case of QDs with (single particle) energy levels well separated from each other. As a result, the QD has either a spin-1/2 or a spin-0 ground state, depending on whether the number of confined electrons is odd or even, respectively. Important deviations from this even–odd periodicity can occur when adjacent energy levels happen to be close to each other or even degenerate. In such a case, the QD can acquire a total spin moment larger than 1/2. Let us consider the simplest case of two nearly degenerate orbital levels. In the language of atomic physics, these two orbital levels form a four-fold degenerate shell (i.e., two orbital plus two spin states). The filling of this shell obeys the well-known Hund's first rule. For N = odd (i.e., one or three electrons on the shell), the ground state has spin 1/2, and it is fourfold degenerate. For N = even (i.e., two electrons on the shell), Hund's rule favors a spin-1 ground state with a threefold degeneracy. To be more precise, the QD ground state is a spin triplet whenever the exchange energy gain overcomes the single particle level spacing. By splitting the two orbital levels apart, the QD undergoes a transition from a spin-triplet to a spin-singlet ground state. Both electrons end up occupying the lowest lying orbital level, with opposite spin.

Singlet–triplet transitions have been experimentally observed in various types of QD systems, from GaAs QDs (van der Wiel et al. 1998) to carbon-nanotube and fullerene QDs. In fact, contrary to the case of "real" atoms, there are multiple experimental ways

to modify the energy-level spectrum of QDs and tune the energy spacing between quantized orbital levels. An effective approach consists of applying an external magnetic field that couples efficiently to the orbital degrees of freedom of the confined electrons. For instance, in the case of GaAs-based QDs a field of the order of ~0.1 T perpendicular to the QD plane can induce appreciable shifts in the orbital energies (note that for B = 0.4 T, a QD with a lateral size of $100 \times 100\,\text{nm}^2$ is threaded by a magnetic flux corresponding to one flux quantum, h/e). The tunable multilevel structure of QDs offers unprecedented opportunities for the study of the Kondo effect in unusual regimes. A rich scenario is found already for the simplest case of two closely spaced orbital levels. We discuss below the main experimental findings.

The first important discovery came in 2000 with the unexpected experimental observation of a strong Kondo resonance associated to a singlet–triplet degeneracy induced by an external magnetic field (Sasaki et al. 2000). This experimental result was obtained by the authors and their collaborators using a GaAs-based QD substantially different from those discussed above. The QD is defined in a double-barrier heterostructure sandwiched between heavily doped, n-type contact layers (the device is shown in Figure 37.11a and b). Electrons flow along the vertical

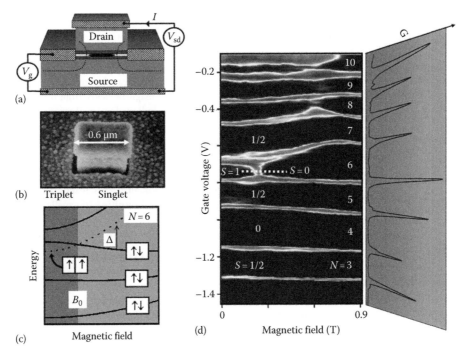

FIGURE 37.11 Singlet–triplet Kondo effect in a QD. (a) Schematic of the device. The QD is defined in a rectangular pillar etched out of a GaAs-based double-barrier heterostructure. The pillar has a lateral size of ~0.5 μm and a comparable height. A metal gate surrounding the base of the pillar controls lateral confinement and the number of electrons in the dot down to complete depletion. (b) Scanning electron micrograph of the device. (c) Qualitative single-particle energy-level spectrum of the QD in a magnetic field perpendicular to the dot plan. For six electrons on the dot the ground state is a spin triplet for $B < B_0$ and a spin singlet for $B > B_0$. The triplet configuration is favored when the exchange energy gain overcomes the energy cost for promoting an electron from the third to the fourth orbital level. The dashed line is obtained by subtracting the exchange energy from the fourth single-particle level. Therefore, B_0 is the field for which this dashed line crosses the third level. (d) Conductance as a function of perpendicular magnetic field and gate voltage. Coulomb peaks appear as almost horizontal lines (representative trace shown in the right panel). In the valley corresponding to six confined electrons, a triplet-to-singlet transition occurs at $B_0 \approx 0.22$ T. Due to the Kondo effect, this transition gives rise to a strong enhancement of the valley conductance. (Reprinted from Sasaki, S. et al., *Nature*, 405, 764, 2000. With permission.)

direction, that is, perpendicular to the QD plane. A gate electrode surrounding the double-barrier section is used to squeeze the 2DEG between in the double-barrier, thereby forming a small QD with few electrons. A triplet-to-singlet transition was observed for six electrons on the QD at a magnetic field, $B_c \approx 0.25\,\text{T}$. The qualitative electronic structure of the dot for $N = 6$ is shown in Figure 37.11c. The onset of a strong Kondo effect at the singlet-triplet degeneracy emerged as an enhanced conductance in the Coulomb valley for $N = 6$ (see Figure 37.11d). This phenomenon can be intuitively understood as a consequence of the increased possibilities for spin-flip cotunneling, following from the addition of a singlet state in resonance with the three triplet states (see Figure 37.12). We note that on the singlet side ($B > B_c$), no Kondo effect is expected while a very weak Kondo effect is detected on the triplet side ($B < B_c$). This Kondo effect follows from cotunneling processes flipping the spin between $S_z = 1$ to $S_z = -1$ via $S_z = 0$. The presence of a degenerate singlet state adds an alternative spin-flip path, leading to a more efficient exchange coupling and hence a much higher Kondo temperature.

A magnetic field perpendicular to the dot plane couples primarily to the orbital degrees of freedom. Yet a simultaneous Zeeman splitting of the triplet states has to occur. For moderate fields, such that $|g\mu_B B| < k_B T_K$, the effect of this splitting can be neglected. This was the case in (Sasaki et al. 2000), where $|g\mu_B B_c| \approx 5\,\mu\text{eV}$ and $k_B T_K \approx 15\,\mu\text{eV}$. At large fields, where triplet states are well separated from each other, another type of Kondo effect can be observed resulting from the degeneracy between the singlet and one of the triplet states (see Figure 37.13a and b). In view of

this twofold degeneracy, this integer-spin Kondo effect is analogous to a spin-1/2 Kondo effect, and it has a lower T_K as compared to the case of fourfold degeneracy. The first experimental observation of this singlet–triplet Kondo effect was reported for carbon-nanotube QDs. These types of nanodevices are obtained by contacting individual carbon nanotubes with a pair of metal electrodes acting as source and drain contacts. Each carbon nanotube consists of a graphene sheet rolled up along a certain axis defining the nanotube chirality. Depending on this chirality, the nanotube can be a one-dimensional metal or a one-dimensional semiconductor. When a nanotube is contacted by two metallic electrodes, its electronic spectrum becomes quantized also in the longitudinal direction, resulting in a discrete energy spectrum. Nygård and coworkers (Nygård et al. 2000) at the Niels Bohr Institute in Copenhagen reported the first evidence of the Kondo effect in such type of QD system (Figure 37.13c). They also showed that for an even number of electrons on the nanotube dot, the ground state is a singlet at zero magnetic field, and it turns into a triplet state at large magnetic fields orthogonal to the nanotube. In this orthogonal orientation, the field acts to remove spin degeneracy without affecting orbital energies. The singlet-to-triplet transition occurs when the lowest energy triplet state crosses the field-independent singlet state. At degeneracy, spin-flip cotunneling processes connecting the two spin states give rise to the Kondo effect.

Other more recent experiments have shown that a singlet–triplet transition can be enforced also at $B = 0$ using the local electric field produced by a gate voltage. The electric field modifies quantum confinement, causing the QD orbital levels to shift in energy. As shown in Figure 37.14, electrically induced singlet–triplet transitions have been reported for GaAs QDs (Kogan et al. 2003) as well as for fullerene QD devices (Roch et al. 2008). Interestingly, when the Kondo effect is present, the gate voltage drives the QD system through a quantum phase transition from a spin-0 state (on the singlet side) to a strongly correlated spin-1/2 state, resulting from a partial Kondo screening of the triplet spin moment (underscreened Kondo effect).

We have seen how the presence of two orbital states with a relatively small energy separation enables the observation of singlet–triplet Kondo effects for $N =$ even, namely, two electrons on the upper electronic shell. Orbital degeneracy can play an important role also for $N =$ odd, that is, for one or three electrons on the shell (note that three electrons is the same as one hole which is equivalent to one electron by virtue of particle–hole symmetry). Besides an enhancement of T_K, the most interesting physics emerges when the orbital quantum number is conserved during tunneling through the QD barriers. This conservation implies that the same (or at least a similar) orbital symmetry exists in the source and drain electrodes, and it is preserved by the tunneling Hamiltonian. In such a case, the orbital degree of freedom is entirely equivalent to the electron spin playing an active role in the establishment of Kondo correlations. The Kondo effect arises from cotunneling processes that flip the local spin, the orbital "pseudospin," or

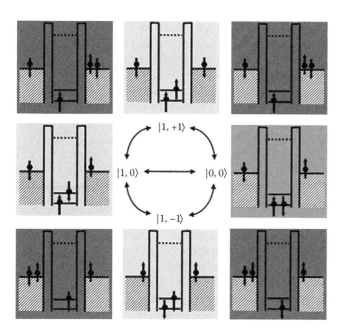

FIGURE 37.12 Second-order cotunneling processes in the presence of singlet–triplet degeneracy. The spin triplet states ($|1, -1\rangle$, $|1, 0\rangle$, and $|1, 1\rangle$) are represented in light grey panels, the spin singlet $|0, 0\rangle$ in dark grey. The four panels at the corners illustrate the intermediate, virtual states. (Reprinted from Sasaki, S. et al., *Nature*, 405, 764, 2000. With permission.)

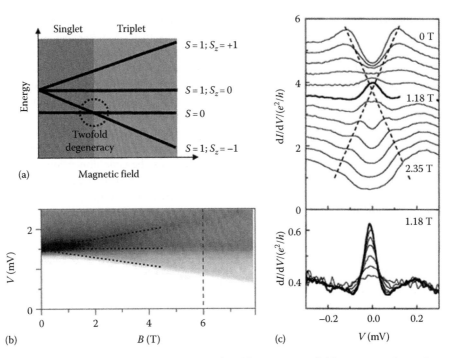

FIGURE 37.13 Twofold degeneracy at a singlet–triplet transition induced by a magnetic field acting mainly on the spin degree of freedom. (a) Qualitative energy diagram. While the energy of the singlet does not depend on the field, the triplet states do split. This splitting is clearly seen in the data just below. (b) dI/dV data for a carbon-nanotube QD with an even number of electrons. In this grey scale representation, dI/dV increases from white to black. (Reprinted from Paaske, J. et al., *Nat. Phys.*, 2, 460, 2006. With permission.) The lowest-energy triplet $|1, -1\rangle$ can in principle cross the singlet and become the ground state. A Kondo effect can manifest itself at the singlet–triplet crossing. (c) Experimental evidence of this singlet–triplet Kondo effect. This dI/dV data were obtained with the carbon nanotube device shown in Figure 37.3b and for an even number of confined electrons. A singlet–triplet transition of the type described in (a) is found at $B = 1.18\,\text{T}$ leading to a Kondo peak at zero bias. This peak splits when moving away from the degeneracy condition. (From Nygård, J. et al., *Nature*, 408, 342, 2000. With permission.)

both simultaneously. In the strong coupling limit ($T \ll T_K$), the Kondo Hamiltonian becomes invariant under transformations of the SU(4) unitary group in the four-dimensional Hilbert space defining the state of the QD. In this limit, orbital and spin degrees of freedom become fully entangled to each other forming a highly correlated state.

Let us now examine the consequences of SU(4) symmetry on measurable physical properties. An enhancement of the Kondo temperature is expected as compared to an ordinary spin-1/2 Kondo effect obeying SU(2) symmetry. This is merely a consequence of the higher degeneracy of the QD ground state, just like for the single-triplet Kondo effect discussed above. On the other hand, the expected unitary limit conductance for symmetric coupling is $2e^2/h$ as for a spin-1/2 Kondo effect. This corresponds to half of the expected unitary conductance for a single-triplet Kondo effect in the case of two symmetrically coupled conducting channels in each lead. In the latter case, electrons from both channels tunnel through the Kondo resonance without experiencing any backscattering. Why is this not possible in the SU(4) Kondo effect? To explain this, let us take a step back and analyze the meaning of unitary conductance. In the strong coupling limit, the local degeneracy originating from spin and/or orbital degrees of freedom is fully screened. Calculating the QD conductance in this limit reduces to the solution of a scattering problem, similar to the scattering of electrons off a static

impurity with no internal degrees of freedom. Only conduction electrons around the Fermi energy need to be taken into account. Using Landauer theory, the conductance takes the simple form: $G = \sum_\nu (e^2/h) \sin^2(\delta_\nu)$, where ν is an index identifying all spin (or pseudospin) quantum numbers, and δ_ν is the scattering phase shift for an electron of spin (or pseudospin) ν tunneling from source to drain. Using the Friedel Sum Rule, $\delta_\nu = \pi \langle N_\nu \rangle$, where $\langle N_\nu \rangle$ is the average occupation number of the ν state in the QD. In an ordinary SU(2) Kondo effect, there are two spin states occupied by one electron, that is, $\langle N_\nu \rangle = 1/2$ and $G = 2e^2/h$. In a singlet–triplet Kondo effect, there are four degenerate states occupied by two electrons, that is, $\langle N_\nu \rangle = 1/2$ and $G = 4e^2/h$. In an SU(4) Kondo effect, the four states resulting from spin and orbital degeneracy share only one electron, that is, $\langle N_\nu \rangle = 1/4$ and $G = 2e^2/h$. The same value is expected regardless of the amount of orbital mixing in the tunneling Hamiltonian (note that the establishment SU(4) symmetry requires an at least approximate conservation of the orbital quantum number during tunneling). Strong orbital mixing reduces the symmetry from SU(4) to SU(2). Therefore, a measurement of the linear conductance cannot discriminate between SU(4) and two-level SU(2). Choi and coworkers have theoretically shown that measurable differences can emerge in the nonlinear transport properties under an applied magnetic field (Choi et al. 2005). In general, the fourfold degeneracy is fully broken by a magnetic field coupling to both

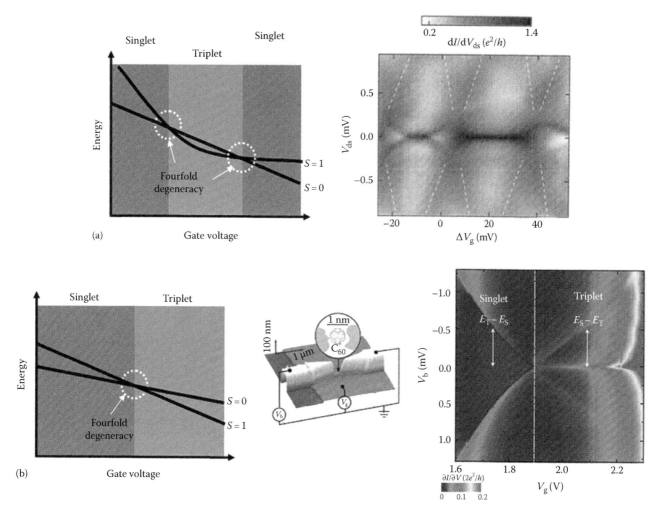

FIGURE 37.14 Singlet–triplet Kondo effect at zero magnetic field. In certain types of QD devices, the relative energies of orbital levels can be efficiently altered with a gate voltage. (a) An anti-crossing between two adjacent single-particle levels can result in a singlet–triplet–singlet transition (left panel). Experimental data obtained with the GaAs lateral QD shown in Figure 37.2c. The dI/dV is shown on grey scale as a function of gate voltage and source–drain bias. The edges of the Coulomb diamonds are highlighted by white, dashed lines. Kondo resonances are found inside two consecutive diamonds as dI/dV ridges (dark-grey lines) at zero bias voltage. The left diamond corresponds to an even number of electrons on the dot. Here the Kondo ridge is visible only in the central region of the diamond indicating a singlet–triplet–singlet transition of the type shown in the left panel. The Kondo effect originates from a fourfold singlet–triplet quasi-degeneracy. (From Kogan, A. et al., *Phys. Rev. B*, 67, 113309, 2003. With permission.) (b) A single gate-induced singlet–triplet transition (left panel) has been observed in a molecular single-electron transistor incorporating a C60 molecule. The device is shown in the central panel. The Kondo ridge on the triplet side (right panel) arises from true triplet degeneracy and not from a singlet–triplet degeneracy as in (a). (Reprinted from Roch, N. et al., *Nature*, 453, 633, 2008. With permission.)

orbital and spin magnetic moments (see Figure 37.15a). In the presence of SU(4) symmetry, calculations yield a multiple peak splitting of the Kondo resonance, as opposed to a twofold splitting in the case of SU(2). Such a fourfold splitting of the Kondo resonance has been reported for carbon-nanotube QDs where a twofold orbital degeneracy is inherently found (Jarillo-Herrero et al. 2005, Makarovski et al. 2007). Intuitively, this degeneracy can be associated with the two equivalent ways electrons can circle around the graphene cylinder, that is, clockwise and anti-clockwise. The resulting orbitals have opposite angular moments along the nanotube axis. As a result, their degeneracy can then be easily lifted by a longitudinal magnetic field. A conceptually similar situation occurs in vertical semiconductor QDs.

Experimental signatures of SU(4) symmetry were reported for this type of system (Sasaki et al. 2004). Yet while the orbital splitting of the Kondo resonance could be clearly identified, no Zeeman splitting was found probably due to the low characteristic temperature of the single-level spin-1/2 Kondo effect (Sasaki et al. 2004, Choi et al. 2005). Recently, signatures of SU(4) symmetry have been found also in the current noise properties of carbon nanotube QDs (Delattre et al. 2009).

We have insisted on the fact that SU(4) symmetry requires orbital degeneracy together with the conservation of the orbital quantum number during tunneling. This implies that an SU(4) Kondo effect should evolve into an orbital SU(2) Kondo effect when only spin degeneracy is lifted. This scenario has been

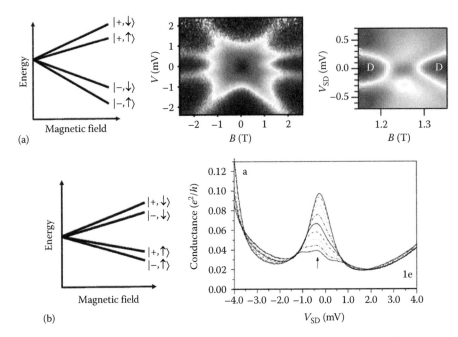

FIGURE 37.15 SU(4) and orbital Kondo effect. (a) In the central panel dI/dV is plotted (originally in blue/white/red color scale) as a function of magnetic field and source–drain bias. (Reprinted from Jarillo-Herrero, P. et al., *Nature*, 434, 484, 2005. With permission.) For a carbon nanotube QD with the magnetic field, B, approximately parallel to its axis. The zero-bias Kondo resonance splits into four peaks (light gray) at finite B, according to the simultaneous splitting of spin and orbital degeneracy as qualitatively shown in the left panel. The inner peaks arise from spin-flip cotunneling processes (from $(|-, \uparrow\rangle$ to $|-, \downarrow\rangle)$), while the outer peaks arise from cotunneling processes involving an orbital transition from $|-\rangle$ to $|+\rangle$. The presence of these outer peaks is a signature of SU(4) symmetry. A similar measurement is shown in the right panel for a GaAs vertical QD. (Reprinted from Sasaki, S. et al., *Phys. Rev. Lett.*, 93, 17205, 2004. With permission.) In this case, only the outer peaks are observed while spin-flip resonances appear to be too weak to be visible. (b) In carbon-nanotube QDs, a perpendicular magnetic field couples only to the spin degree of freedom. Therefore, a large Zeeman splitting can be obtained without affecting orbital degeneracy (left panel). In this way, it is possible to observe an SU(2) Kondo effect uniquely due to an orbital degeneracy. (Reprinted from Makarovski, A. et al., *Phys. Rev. B*, 75, 241407(R), 2007. With permission.) The Kondo resonance splits into three peaks: two outer peaks corresponding to spin-flip cotunneling, and one zero-bias peak due to orbital Kondo effect.

experimentally realized in carbon-nanotube QDs with the aid of large magnetic fields applied either parallel (Jarillo-Herrero et al. 2005) or perpendicular (Makarovski et al. 2007) to the nanotube axis. Data for the latter case are reported in Figure 37.15b. An elegant, fully "artificial" version of the orbital Kondo effect has been investigated by Weis and coworkers using electrostatically coupled QDs connected in parallel (Hübel et al. 2008 and references therein).

37.6 Concluding Remarks and Perspectives

The pioneering work of Goldhaber-Gordon and coworkers (Goldhaber-Gordon et al. 1998a) and the following exciting developments have shown that artificial nanostructures can serve as a powerful tool for the investigation of one of the most studied phenomena in the domain of strongly correlated electron systems. The Kondo effect has been found in a large variety of nanoscale devices, proving to be a very general phenomenon capable of producing substantial effects in the device transport properties.

In this chapter, we have chosen to focus on selected aspects of the Kondo effect with the goal to provide the less familiar reader with some essential background knowledge and an intuitive

understanding of the underlying physics. To conclude we wish to list some of the most interesting aspects of the Kondo effect that could not be addressed in this chapter.

1. *Phase shift in the resonant tunneling through a Kondo QD.* We have seen that in the Kondo regime the tunneling matrix element for each spin (or pseudospin) channel contains a phase shift that is related to the average occupation of the corresponding local state (Friedel Sum Rule). This is the phase shift an electron at the Fermi energy acquires by tunneling through dot. Measurements of this phase shift have been reported by us (van der Wiel et al. 2000) and Moty Heiblum's group (Ji et al. 2000, Zaffalon et al. 2008).

2. *Kondo effect under microwave irradiation.* It is known from early experiments with semiconductor QDs that an RF field can induce the formation of photon sidebands of QD energy levels (Oosterkamp et al. 1997, van der Wiel et al. 2002b). In the presence of the Kondo effect, analogous photon sidebands of the many-body Kondo resonance have been predicted (Hettler and Schoeller 1995, Platero and Aguado 2004). So far only one experimental group has reported the observation of this phenomenon (Kogan et al. 2004). The results of this group, however, contrast with those reported earlier by the authors of this chapter

and their collaborators, which indicate a dominant decoherence effect from the microwave field in combination with finite applied bias (Elzerman et al. 2000). The latter results are more in line with other theoretical predictions (Kaminski et al. 1999). We believe that more experiments are necessary to clarify this intriguing issue.

3. *Two-stage Kondo effect.* Under certain circumstances, the Kondo effect is not characterized by a single energy scale T_K, but, instead, by two energy scales T_{K1} and T_{K2} ($T_{K2} < T_{K1}$) resulting in a *two-stage* screening mechanism. This phenomenon requires that the QD has a spin state $S \geq 1$ coupled to single-channel leads. For $S = 1$, the first-stage screening process with characteristic energy scale T_{K1} is an underscreened Kondo effect, reducing the net spin from $S = 1$ to $S = 1/2$. The second stage, with a smaller energy scale T_{K2}, reduces the spin to $S = 0$, forming a spin singlet. For $T_{K2} \ll (T, V_{SD}) \ll T_{K1}$, the first-stage Kondo screening overcomes the Coulomb blockade, and G is expected to reach the unitary limit. For $(T, V_{SD}) < T_{K2}$, the second stage of screening interferes destructively with the first one, resulting in a suppression of G, ideally to zero. The signatures of a two-stage Kondo effect are a nonmonotonic temperature dependence of G and a sharp dip superimposed on the usual zero-bias resonance in dI/dV. There are different situations that can lead to a two-stage Kondo effect, which have been the subject of theoretical investigation (Pustilnik and Glazman 2001, Hofstetter and Schoeller 2002). Experimental evidence for a two-stage Kondo effect was reported for lateral GaAs QDs (van der Wiel et al. 2002a,b) and molecular single-electron transistors (Roch et al. 2008).

4. *Two-channel Kondo effect.* When a spin-1/2 magnetic impurity is coupled to two independent Fermi seas of electrons, an "overscreening" of its spin is theoretically predicted (Nozières and Blandin 1980, Zawadowski 1980). The low-temperature limit is an exotic state that cannot be described by Fermi liquid theory. The realization of this two-channel Kondo effect is challenging. A pioneering experiment was recently reported by the Goldhaber-Gordon group (Potok et al. 2006).

5. *Two-impurity Kondo effect.* Two localized spins can interact with each other either by direct exchange interaction or via an interceding electron sea. This second possibility applies also when the two spins are relatively far apart, and it is known as the Ruderman–Kittel–Kasuya–Yoshida (RKKY) interaction. Mesoscopic nanostructures, in particular double QDs (van der Wiel et al. 2003), can provide unprecedented opportunities for the investigation of the coupling between multiple artificial Kondo impurities under well-controlled and tuneable conditions. Only a few experiments on multiple Kondo dots have been reported so far (Jeong et al. 2001, Craig et al. 2004), and ample room is still left for further investigation.

6. *Kondo effect with competing electron–electron correlations.* The screening of a local spin by an electron reservoir can be severely affected when the reservoir is turned into a superconductor, a ferromagnetic metal, a Tomonaga–Luttinger liquid, or any other strongly correlated electron system. Such exotic competitions between different types of electron–electron correlations have been and remain the subject of fervid studies at both theoretical (Lee et al. 2008, Pereira et al. 2008, Rotter et al. 2008) and experimental (Buizert et al. 2007, Eichler et al. 2007, Hamaya et al. 2007, Sand-Jespersen et al. 2007, Hauptmann et al. 2008) level. The given references correspond to some of the most recent publications.

In spite of all the achievements in the past 10 years, many interesting problems remain to be addressed (Grobis et al. 2008). At the same time, the continuous progress in the science and technology of nanoscale systems is generating new opportunities for experiments, thereby fostering the development of this fascinating field.

Acknowledgments

We wish to thank L. Kouwenhoven for many stimulating discussions, G. Katsaros and P. Spathis for their help in the revision of this chapter. S.D.F. acknowledges financial support from the Agence Nationale de la Recherche through the "Chaire d'Excellence 2007" and the "Jeunes Chercheuses et Jeunes Chercheurs 2007" programmes.

References

Abrikosov, A.A. (1965). Renormalization group from diagrammatic calculations with Kondo spin represented by a pseudofermion. *Physics* 2: 5.

Anderson, P.W. (1961). Localized magnetic states in metals. *Phys. Rev.* 124: 41.

Anderson, P.W. (1967). Ground state of a magnetic impurity in a metal. *Phys. Rev.* 164: 352.

Ashoori, R.C. (1996). Electrons in artificial atoms. *Nature* 379: 413–419.

Buizert, C., Oiwa, A., Shibata, K. et al. (2007). Kondo universal scaling for a quantum dot coupled to superconducting leads. *Phys. Rev. Lett.* 99: 136806.

Choi, M-S., López, R., and Aguado, R. (2005). SU(4) Kondo effect in carbon nanotubes. *Phys. Rev. Lett.* 95: 067204.

Costi, T.A., Hewson, A.C., and Zlatic, V. (1994). Transport coefficients of the Anderson model via the numerical renormalization group. *J. Phys. Condens. Matter* 6: 2519.

Craig, N.J., Taylor, J.M., Lester, E.A. et al. (2004). Tunable nonlocal spin control in a coupled-quantum dot system. *Science* 304: 565–567.

Cronenwett, S.M., Oosterkamp, T.H., and Kouwenhoven, L.P. (1998). A tunable Kondo effect in quantum dots. *Science* 281: 540.

Csonka, S., Hofstetter, L., Freitag, F. et al. (2008). Giant fluctuations and gate control of the g-factor in InAs nanowire quantum dots. *Nano Lett.* 8: 3932–3935.

De Franceschi, S., Hanson, R., van der Wiél, W.G. et al. (2002). Out-of-equilibrium Kondo effect in a mesoscopic device. *Phys. Rev. Lett.*, 89: 156801–156804.

de Haas, W.J., de Boer J.H., and van den Berg, G.J. (1934). The electrical resistance of gold, copper and lead at low temperatures. *Physica* 1: 1115.

Delattre, T., Feuillet-Palma, C., Herrmann, L.G. et al. (2009). Noisy Kondo impurities. *Nat. Phys.* 5: 208–212.

Eichler, A., Weiss, M., Oberholzer, S. et al. (2007). Even-odd effect in Andreev transport through a carbon nanotube quantum dot. *Phys. Rev. Lett.* 99: 126602.

Elzerman, J.M., De Franceschi, S., and Goldhaber-Gordon, D. (2000). Suppression of the Kondo effect in a quantum dot by microwave radiation. *J. Low Temp. Phys.* 118: 375.

Fano, U. (1961). Effects of configuration interaction on intensities and phase shifts. *Phys. Rev.* 124: 1866.

Goldhaber-Gordon, D., Shtrikman, H., Mahalu, D. et al. (1998a). The Kondo effect in a single-electron transistor. *Nature* 391: 156–159.

Goldhaber-Gordon, D., Göres, J., Kastner, M.A. et al. (1998b). From the Kondo regime to the mixed-valence regime in a single-electron transistor. *Phys. Rev. Lett.* 81: 5225–5228.

Grobis, M., Rau, I.G., Potok, R.M. et al. (2008). Universal scaling in nonequilibrium transport through a single channel Kondo dot. *Phys. Rev. Lett.* 100: 246601–246604.

Hamaya, K., Kitabatake, M., Shibata, K. et al. (2007). Kondo effect in a semiconductor quantum dot coupled to ferromagnetic electrodes. *Appl. Phys. Lett.* 91: 232105.

Hauptmann, J.R., Paaske, J., and Lindelof P.E. (2008). Electric-field-controlled spin reversal in a quantum dot with ferromagnetic contacts. *Nat. Phys.* 4: 373–376.

Hettler, M.H. and Schoeller, H. (1995). Anderson model out of equilibrium: Time-dependent perturbations. *Phys. Rev. Lett.* 74: 4907–4910.

Hewson, A.C. (1993). *The Kondo Problem to Heavy Fermions.* Cambridge, MA: Cambridge University Press.

Hofstetter, W. and Schoeller, H. (2002). Quantum phase transition in a multilevel dot. *Phys. Rev. Lett.* 88: 016803.

Hübel, A., Held, K., Weis, J. et al. (2008). Correlated electron tunneling through two separate quantum dot systems with strong capacitive interdot coupling. *Phys. Rev. Lett.* 101: 186804.

Jarillo-Herrero, P., Kong, J., van der Zant, H.S.J. et al. (2005). Orbital Kondo effect in carbon nanotubes. *Nature* 434, 484.

Jeong, H., Chang, A.M., and Melloch, M.R. (2001). The Kondo effect in an artificial quantum dot molecule. *Science* 293: 2221–2223.

Ji, Y., Heiblum, M., Sprinzak, D. et al. (2000). Phase evolution in a Kondo-correlated system. *Science* 290: 779–783.

Kaminski, A., Nazarov, Yu.V., and Glazman, L.I. (1999). Suppression of the Kondo effect in a quantum dot by external irradiation. *Phys. Rev. Lett.* 83: 384–387.

Kogan, A., Granger, G., Kastner, M. A. et al. (2003). Singlet–triplet transition in a single-electron transistor at zero magnetic field. *Phys. Rev. B* 67: 113309.

Kogan, A., Amasha, S., and Goldhaber-Gordon, D. (2004). Measurements of Kondo and spin splitting in single-electron transistors. *Phys. Rev. Lett.* 93: 166602.

Kondo, J. (1964). Resistance minimum in dilute magnetic alloys. *Prog. Theor. Phys.* 32: 37.

Kouwenhoven, L.P. and Glazman, L. (2001). The revival of the Kondo effect. *Phys. World* 14: 33–38.

Lee, M., Jonckheere, T., and Martin, T. (2008). Josephson effect through an isotropic magnetic molecule. *Phys. Rev. Lett.* 101: 146804–146807.

Leturcq, R., Schmid, L., Ensslin, K. et al. (2005) Probing the Kondo density of states in a three-terminal quantum ring. *Phys. Rev. Lett.* 95: 126603–126606.

Li, J., Schneider, W.-D., Berndt R. et al. (1998). Kondo scattering observed at a single magnetic impurity. *Phys. Rev. Lett.* 80: 2893.

Liang, W., Shores M.P., Bockrath, M. et al. (2002). Kondo resonance in a single-molecule transistor. *Nature* 417: 725.

Madhavan, V., Chen, W., Jamneala, T. et al. (1998). Tunneling into a single magnetic atom: Spectroscopic evidence of the Kondo resonance. *Science* 280: 567.

Makarovski, A., Zhukov, A., Liu, J. et al. (2007). SU(4) and SU(2) Kondo effects in carbon nanotube quantum dots. *Phys. Rev. B* 75: 241407(R).

Meir, Y., Wingreen, N.S., and Lee, P.A. (1993). Low-temperature transport through a quantum dot: The Anderson model out of equilibrium. *Phys. Rev. Lett.* 70: 2601–2604.

Nozières, P. (1974). A "Fermi-liquid" description of the Kondo problem at low temperatures. *J. Low Temp. Phys.* 17: 31–42.

Nozières, P. and Blandin, A. (1980). Kondo effect in real metals. *J. Phys.* 41: 193.

Nygård, J., Cobden, D.H., and Lindelof, P.E. (2000). Kondo physics in carbon nanotubes. *Nature* 408: 342–346.

Oosterkamp, T.H., Kouwenhoven, L.P., Koolen, A.E.A. et al. (1997). Photon sidebands of the ground state and first excited state of a quantum dot. *Phys. Rev. Lett.* 78: 1536–1539.

Paaske, J., Rosch, A., Wolfle, P. et al. (2006). Non-equilibrium singlet–triplet Kondo effect in carbon nanotubes. *Nat. Phys.* 2: 460–464.

Park, J., Pasupathy, A.N., Goldsmith, J.I. et al. (2002). Coulomb blockade and the Kondo effect in single-atom transistors. *Nature* 417: 722.

Pereira, R.G., Laflorencie, N., Affleck, I. et al. (2008). Kondo screening cloud and charge staircase in one-dimensional mesoscopic devices. *Phys. Rev. B* 77: 125327–125341.

Platero, G. and Aguado, R. (2004). Photoassisted transport in nanostructures. *Phys. Rep.* 395: 1–157.

Potok, R.M., Rau, I.G., Shtrikman, H. (2006). Observation of the two-channel Kondo effect. *Nature* 446: 167–171.

Pustilnik, M. and Glazman, L.I. (2001). Kondo effect in real quantum dots. *Phys. Rev. Lett.* 87: 216601.

Roch, N., Florens, S., Bouchiat, V. et al. (2008). Quantum phase transition in a single-molecule quantum dot. *Nature* 453: 633–638.

Rotter, S., Türeci, H.E., Alhassid, Y. et al. (2008). Interacting quantum dot coupled to a Kondo spin: A universal Hamiltonian study. *Phys. Rev. Lett.* 100: 166601.

Sand-Jespersen, T., Paaske, J., Andersen, B.M. et al. (2007). Kondo-enhanced Andreev tunneling in InAs nanowire quantum dots. *Phys. Rev. Lett.* 99: 126603.

Sasaki, S., De Franceschi, S., Elzerman, J.M. et al. (2000). Kondo effect in an integer-spin quantum dot. *Nature* 405: 764.

Sasaki, S., Amaha, S., Asakawa, N. et al. (2004). Enhanced Kondo effect via tuned orbital degeneracy in a spin 1/2 artificial atom. *Phys. Rev. Lett.* 93: 17205.

Schmid, J., Weis, J., Eberl, K. et al. (1998). A quantum dot in the limit of strong coupling to reservoirs. *Physica B* 256–258: 182.

Schrieffer, J.R. and Wolff, P.A. (1966). Relation between the Anderson and Kondo Hamiltonians. *Phys. Rev.* 149: 491.

van der Wiel, W.G., Oosterkamp, T.H., Janssen, J.W. et al. (1998). Singlet-triplet transitions in a few-electron quantum dot. *Physica B* 256–258: 173–177.

van der Wiel, W.G., De Franceschi, S., Elzerman, J.M. et al. (2000). The Kondo effect in the unitary limit. *Science* 289: 2105.

van der Wiel, W.G., De Franceschi, S., Elzerman, J.M. (2002a). Two-stage Kondo effect in a quantum dot at a high magnetic field. *Phys. Rev. Lett.* 88: 126803.

van der Wiel, W.G., Oosterkamp, T.H., De Franceschi, S. et al. (2002b). Photon assisted tunneling in quantum dots. In *Strongly Correlated Fermions and Bosons in Low-Dimensional Disordered Systems*, eds. I.V. Lerner et al., pp. 43–68, Boston, MA/Dordrecht, the Netherlands/London, U.K.: Kluwer Academic Publishers, ISBN 1-4020-0748-5.

van der Wiel, W.G., De Franceschi, S., and Elzerman, J.M. (2003). Electron transport through double quantum dots. *Rev. Mod. Phys.* 75: 1.

Wilson, K.G. (1974). *Nobel Symposia* 24: 67. New York: Academic Press.

Wilson, K.G. (1975). The renormalization group: Critical phenomena and the Kondo problem. *Rev. Mod. Phys.* 47: 773.

Zaffalon, M., Bid, A., Heiblum, M. et al. (2008). Transmission phase of a singly occupied quantum dot in the Kondo regime. *Phys. Rev. Lett.* 100: 226601.

Zawadowski, A. (1980). Kondo-like state in a simple model for metallic glasses. *Phys. Rev. Lett.* 45: 211–214.

Zener, C. (1951). Interaction between the d shells in the transition metals. *Phys. Rev.* 81: 440.

38

Theory of Two-Electron Quantum Dots

Jan Petter Hansen
University of Bergen

Eva Lindroth
Stockholm University

38.1 Introduction

Two-electron lateral quantum dots is a quantum mechanical two-dimensional analogue of the three-dimensional helium atom. When the confining potential contains two wells, it compares correspondingly to a flat version of the H_2 molecule. Since electron–electron interaction plays an even more prominent role in two-dimensional systems, as compared to three-dimensional atoms and molecules, the study of quantal few-body structures in two dimensions is of fundamental interest in its own right. It may reveal new insights into the role of strongly correlated dynamics. Other confining potentials can also be constructed that have a less direct atomic or molecular analogue, for example, "quantum rings" (Viefers et al., 2003).

Two-electron quantum dots have also been suggested as carriers of quantum information (Loss and DiVincenzo, 1998). The basic quantum bit (qubit) is stored in either a spin state or a charge localized state of a double-dot system (Sælen et al., 2008), and several theoretical proposals exist (Hanson et al., 2007; Taylor et al., 2007; Waltersson et al., 2009). In a recent experimental demonstration (Koppens et al., 2006; Petta et al., 2004), a qubit was controlled and manipulated in the total spin state, that is, a singlet and triplet state of a quantum-dot molecule. Robust single-bit operations between these states were demonstrated on nanosecond timescales based on external voltage variations. A particular challenge in experiments is to minimize the interaction with sources leading to decoherence. These sources can be due to, for example, temperature-dependent fluctuations, spin–orbit interactions, and hyperfine interactions between the quantum-dot electrons and the "reservoir" of surrounding atoms and electrons. The interaction with such sources are state dependent, and a detailed knowledge of the spectrum of two-electron systems is therefore important also from a technological point of view.

In this chapter, we discuss two-electron quantum dots from a theoretical perspective based on accurate solutions of the Schrödinger equation. We first describe the model Hamiltonians and then discuss its spectra for various confining potentials, that is, quantum dots, quantum molecules, and quantum rings. Some examples of controlled dynamics between electronic states of the systems are also given. More complete reviews covering experimental methods and more overviews of quantum dots in general have, for example, been given by Hanson et al. (2007) and by Reimann and Mannineen (2002).

38.2 Two-Electron Model

The real physical quantum dot can be described as a spatial confinement of electrons in all three dimensions achieved through semiconductor processing techniques. Electrostatic gates can, for example, be used to laterally restrict a two-dimensional electron gas in a semiconductor heterostructure. The electron confined in the dot is thus embedded in the semiconductor material. This can be modeled through an effective electron mass, m^*, and a Coulomb interaction modified by the relative dielectric constant, ε_r, both specific for the host semiconductor. We consider here two-electron systems, that is, the Hamiltonian

$$H = h(x_1, y_1) + h(x_2, y_2) + \frac{e^2}{4\pi\varepsilon_r\varepsilon_0 r_{12}}. \tag{38.1}$$

The single-particle Hamiltonian, $h(x_i, y_i)$ describes the interaction of each electron with the confining potential, as well as with external electric and magnetic fields. We will here assume a constant magnetic field applied perpendicular to the dot. For such a field, the vector potential can be written as

$$\mathbf{A} = -\frac{1}{2}\mathbf{r} \times \mathbf{B} \qquad (38.2)$$

and thus

$$\frac{e}{m}\mathbf{A} \cdot \mathbf{p} + \frac{e^2}{2m}\mathbf{A}^2 = \frac{e}{2m^{\star}}\hat{\ell} \cdot \mathbf{B} + \frac{e^2}{8m}(x^2 + y^2)B^2 \qquad (38.3)$$

for a vector $\mathbf{r}$ in the xy-plane. With the electric field applied parallel to the x-axis we thus have

$$h(x, y) = -\frac{\hbar^2}{2m^{\star}}\nabla^2 + V(x, y) + \frac{e^2}{8m^{\star}}B^2(x^2 + y^2)$$

$$+ \frac{e}{2m^{\star}}B\hat{\ell}_z + g^{\star}\frac{e}{2m_e}B\hat{s}_z + eEx$$

$$= -\frac{\hbar^2}{2m^{\star}}\nabla^2 + V(x, y) + V_{\text{ext}}(x, y; \mathbf{B}, \mathbf{E}), \qquad (38.4)$$

where $g^{\star}$ is the effective gyromagnetic ratio of the semiconductor. The gate voltage applied to achieve lateral confinement results in a potential well, $V(x, y)$, but its precise form is not known. Self-consistent solutions of the combined Hartree and Poisson equations by Kumar et al. (1990) in the early 1990s indicated that for a dot with just a few electrons this confining potential is parabolic in shape at least to a first approximation. Since then a two-dimensional harmonic oscillator potential have been the standard choice for theoretical modeling:

$$V(x, y) = \frac{m^{\star}\omega^2}{2}(x^2 + y^2). \qquad (38.5)$$

The semiconductor parameters, $m^{\star}$ and ε_r, define appropriate length, time, and energy scales. The characteristic length scale is defined by

$$a_0^{\star} = \frac{4\pi\varepsilon_0\varepsilon_r\hbar^2}{m^{\star}e^2} = \frac{\varepsilon_r m_e}{m^{\star}}a_0, \qquad (38.6)$$

the energy scale by

$$1\,\text{Hartree}^{\star} = \frac{m^{\star}e^4}{(4\pi\varepsilon_0\varepsilon_r)^2\hbar^2} = \frac{m^{\star}}{\varepsilon_r^2 m_e}\text{Hartree}, \qquad (38.7)$$

and the timescale by

$$t_0^{\star} = \frac{(4\pi\varepsilon_0\varepsilon_r)^2\hbar^3}{m^{\star}e^4} = \frac{(\varepsilon_r)^2 m_e}{m^{\star}}t_0. \qquad (38.8)$$

For GaAs $m^{\star} \approx 0.067\,m_e$ and $\varepsilon_r \approx 12.4$ and thus $a_0^{\star} \approx 9.86\,\text{nm}$, 1 Hartree$^{\star} \approx 11.86\,\text{meV}$, and the typical time unit is $t_0^{\star} \approx 0.055\,\text{ps}$. The effective gyromagnetic ratio, corresponding to the bulk value in GaAs, is $g^{\star} = -0.44$. Strong dot confinement means then that $\hbar\omega \gg 1$ Hartree*, or equivalently that the characteristic length of the harmonic oscillator $\sqrt{\hbar/m^{\star}\omega} \ll a_0^{\star}$.

The problem at hand is to start with a *static* problem, consisting of the solution of the time-independent Schrödinger equation (for zero or time-independent external fields) to find its energy spectrum, E_k, and eigenstates, Ψ_k, defined by

$$H\Psi_k(\mathbf{r}_1, \mathbf{r}_2) = E_k\Psi_k(\mathbf{r}_1, \mathbf{r}_2). \qquad (38.9)$$

A variety of approximative methods can be used. For two-electron systems, the equation can be solved with very few approximations by diagonalization of the Hamiltonian in the space defined by a large basis set, cf. Section 38.4. When the external fields in the Hamiltonian depend on time, we need also to consider the *dynamic* problem, which amounts to solve the time-*dependent* Schrödinger equation:

$$i\hbar\frac{\partial}{\partial t}\Psi(\mathbf{r}_1, \mathbf{r}_2, t) = H\Psi(\mathbf{r}_1, \mathbf{r}_2, t). \qquad (38.10)$$

For example, for a time-dependent electric field, $V_{\text{ext}}(\mathbf{r}_1, \mathbf{r}_2, t) = \mathbf{E}(t) \cdot (\mathbf{r}_1 + \mathbf{r}_2)$. The wave function can be expanded in the eigenstates above (Equation 38.9):

$$\Psi(\mathbf{r}_1, \mathbf{r}_2, t) = \sum_k d_k(t)\Psi_k(\mathbf{r}_1, \mathbf{r}_2)\Sigma_k(S), \qquad (38.11)$$

where the total spin state $\Sigma S = 0, 1$ has been explicitly included. By projection, we obtain a first-order coupled set of equations for each probability amplitude, $d_k(t)$

$$i\hbar\dot{d}_k(t) = \sum_l d_l(t)V_{l,k}(t) + E_k d_k(t), \qquad (38.12)$$

where

$$V_{l,k}(t) = \delta_{S_l, S_k}\iint d^2\mathbf{r}_1 d^2\mathbf{r}_2 \Psi_l^{\star}(\mathbf{r}_1, \mathbf{r}_2)V_{\text{ext}}(\mathbf{r}_1, \mathbf{r}_2, t)\Psi_k(\mathbf{r}_1, \mathbf{r}_2). \qquad (38.13)$$

In the present example, the external coupling conserves the total spin, but this is not always the case (Nepstad et al., 2008). An alternative *adiabatic* method that originates from the description of slow atomic collisions (Bransden and Joachain, 2003) can be more convenient if the change in the time-dependent fields are relatively slow. In that case it can be more efficient and advantageous to expand the wave function in a basis of states, which at any instant of time are eigenfunctions of the dynamic Hamiltonian:

$$\left[\sum_{i=1,2}[h_{\text{int}}(\mathbf{r}_i) + V_{\text{ext}}(\mathbf{r}_i, t)] + \frac{e^2}{4\pi\varepsilon_r\varepsilon_0 r_{12}}\right]\theta_k((\mathbf{r}_1, \mathbf{r}_2, t) = \varepsilon_k(t)\theta_k(\mathbf{r}_1, \mathbf{r}_2, t).$$

$$(38.14)$$

When the time variation, for example, is in the electric field in the x direction, we have $\Sigma_{1,2}V_{\text{ext}}(\mathbf{r}_i, t) = \xi(t)(x_1 + x_2) = \xi(t)X$, the adiabatic expansion reads

$$\Psi(\mathbf{r}_1, \mathbf{r}_2, t) = \sum_k c_k(t)\theta_k(\mathbf{r}_1, \mathbf{r}_2; \xi(t)). \qquad (38.15)$$

Inserting Equation 38.15 into Equation 38.10 and using Equation 38.14, we find the governing equation for the coefficients:

$$\dot{c}_k(t) = \dot{\xi}(t)\sum_{j\neq k}\frac{\langle\theta_k|X|\theta_j\rangle}{\varepsilon_k - \epsilon_j}c_j(t) + \frac{i}{\hbar}\varepsilon_k(\xi(t))c_k(t). \qquad (38.16)$$

Note that the coupling, $\langle\theta_k|X|\theta_j\rangle$, results in avoided crossings of the energy curves, that is, $\varepsilon_k \neq \varepsilon_j$ for $j \neq k$. The adiabatic energies may however get close, and the formula above shows that this is essentially the regions where transitions take place (Nepstad et al., 2008).

38.3 Quantum Dots

For a parabolic confinement, as given in Equation 38.5, and in the absence of external fields, the two-electron Hamiltonian can be written in the center of mass, $\mathbf{R} = 1/2(\mathbf{r}_1 + \mathbf{r}_2)$, and relative motion, $\mathbf{r} = (\mathbf{r}_1 - \mathbf{r}_2)$, coordinates as (Taut, 1993; Zhu et al., 1997)

$$H = -\frac{\hbar^2}{4m^\star}\nabla_\mathbf{R}^2 + m^\star\omega^2\mathbf{R}^2 - \frac{\hbar^2}{m^\star}\nabla_\mathbf{r}^2 + \frac{1}{4}m^\star\omega^2\mathbf{r}^2 + \frac{e^2}{4\pi\epsilon_r\epsilon_0|\mathbf{r}|}. \qquad (38.17)$$

The total wave function then becomes separable as $\Psi(\mathbf{r}, \mathbf{R}) = \Psi_R^{N,M}(\mathbf{R})\Psi_r^{n,m}(\mathbf{r})$, where $\Psi_R^{N,M}(\mathbf{R})$ is an eigenfunction of the center of mass part of Equation 38.17 and $\Psi_r^{n,m}(\mathbf{r})$ is an eigenfunction to the relative motion part of Equation 38.17, and each are further separable in a radial and an angular part with quantum numbers $n(N)$ and $m(M)$ referring to the radial and angular degree of freedom, respectively. A state is thus characterized by the four quantum numbers (N, M, n, m) with $(n, N = 0, 1, \ldots)$ and $(m, M = 0, \pm 1, \ldots)$, and the total energy can be written as

$$E(N, M, n, m) = (2N + |M| + 1)\hbar\omega + (2n + |m| + 1)\hbar\omega + E_r(n, m), \qquad (38.18)$$

where the first term originates from the center of mass part of Equation 38.17, the second term originates from the harmonic oscillator part of the relative motion in Equation 38.17, and $E_r(n, m)$ accounts for the electron–electron interaction contribution to the energy. The spatial symmetry of the total wave function under exchange of particle one and particle two is given by the parity of $\Psi_r^{n,m}(\mathbf{r})$, and thus the spin singlets (triplets) will have even (odd) m.

The ground state is thus a singlet state with $m = 0$ followed by a doubly degenerate triplet state characterized with $m = \pm 1$

and so on. Energy spectrum classification and the relative role played by the electron–electron interaction in this system were first considered by Merkt et al. (1991), and later energy spectra were obtained by powerful nonperturbative methods (Drouvelis et al., 2003; Taut, 1993; Zhu et al., 1997). The role of the electron–electron interaction depends on the confinement strength. In Figure 38.1, we display the one-electron probability density

$$\rho(\mathbf{r}) = \int d^2\mathbf{r}_2 |\Psi(\mathbf{r}, \mathbf{r}_2)|^2 \qquad (38.19)$$

of the three lowest triplet states for two confinement strengths, $\hbar\omega = 1$ meV and $\hbar\omega = 0.125$ meV. We observe a clear "p-state" nature of the two lowest states but already for the third excited level the electron charge distributions differ. The widest well (right panel) favor a ring-like distribution compared to the centered distribution of the narrower one (left). The confinement strength can be characterized as the ratio between the typical harmonic oscillator size and the effective Bohr radius, the well in the right panel corresponds then to $\sqrt{\hbar/m^\star\omega}/a_0^\star \approx 10$ and is an

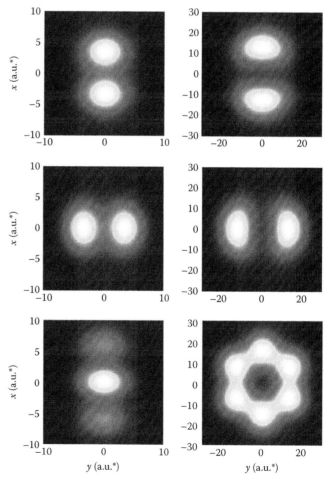

FIGURE 38.1 One-electron probability density of the three lowest triplet states of a two-electron quantum dot with harmonic confinement $\hbar\omega = 1.0$ meV (left) and $\hbar\omega = 0.125$ meV (right). (Reprinted from Sørngård, S., MSc thesis, University of Bergen, Bergen, Norway, 2009.)

example of very weak confinement. The system is then strongly dominated by the Coulomb interaction between the electrons. Wigner (1934) predicted that under such circumstances, that is, when the electron density is below a certain critical value, the electrons may crystallize and form a lattice. The lowest picture in the right panel shows precisely such an ordered spatial structure that we expect when a Wigner molecule has formed. References to several theoretical studies on Wigner crystallization can be found in the review by Reimann and Manninen (2002).

38.4 Quantum-Dot Molecules

A series of experimental studies has been performed based on quantum-dot double-well systems or so-called quantum-dot molecules (Koppens et al., 2006; Petta et al., 2004). Time-dependent external voltages can be used to load a precise number of electrons into one of the wells. The system is then transformed into a double well by time-dependent adiabatic switching of the gate voltages such that the initial character of the state is preserved. From a theoretical point of view, the double-well system is conveniently described by a two-center harmonic confining potential in one direction, in this case the $\hat{x}$ direction:

$$V(x, y) = \frac{1}{2} m^\star \omega^2 \min\left[x - \frac{d^2}{2} + y^2, \, x + \frac{d^2}{2} + y^2 \right], \quad (38.20)$$

which for $d = 0$ is identical to Equation 38.5. When $d \rightarrow \infty$ it becomes two separated identical wells, each again identical to Equation 38.5. At intermediate values of d, the potential has two minima leading to a two-center nature of the electronic states and a finite tunneling coupling between each well for electrons initially placed in one of the wells. This potential has been applied in a series of theoretical papers (Førre et al., 2006; Harju et al., 2002; Nepstad et al., 2008; Popsueva et al., 2007; Räsänen et al., 2007; Sælen et al., 2008; Wensauer et al., 2000).

In Figure 38.2, we display the 12 lowest energy curves as a function of d for two cases; with and without electron–electron interaction taken into account when diagonalizing the Hamiltonian. We show only the states with total $M_S = 0$ since the spectrum is nearly degenerate with respect to the direction of the spin. In the case without correlation, we obtain the energy levels of the two-dimensional harmonic oscillator both at $d = 0$ and at $d \rightarrow \infty$. In the latter case, the ground state is sixfold degenerate: With one electron in the ground state of *each* separate well, singlet as well as triplet states can form. In addition, both electrons can be in the same well, while the other is empty. In this case, the electrons must form a $S = 0$ state in order to occupy the lowest energy state and two degenerate states are formed from symmetrized combinations of singlet pairs in either of the wells.

When the electron–electron interaction is included, this degeneracy is lifted as seen in Figure 38.2b. We observe first that the ground state energy increases from 6 to 11.15 meV for $d = 0$ to 7.5 meV for $d = 80$ nm. As a consequence of the asymptotic behavior, we expect that at finite, but sufficiently large d, the energy spectrum will display a band structure separated by the confinement energy $\hbar\omega$. The ground state band will consist of one singlet and one triplet states correlating to the single electron ground state of each well. The excited bands will correspond to one, two, three, etc. excitons of each well, or covalent states corresponding to symmetrized products of the correlated the $d = 0$ states. In the figure, we observe the first excited band of single exciton states, four singlets, and four triplets formed by the four possible combinations of two separated two-dimensional harmonic oscillators. The bands are easily identified at the largest well separations displayed in the figure. On top of these two, singlet states follow, which are essentially two linear combinations of the ground state of separated wells with two electrons in each.

In more detail, we consider now the singlet states only: The ground state, $|g, +\rangle$, has positive parity and is essentially a linear combination of the ground states of two harmonic oscillators

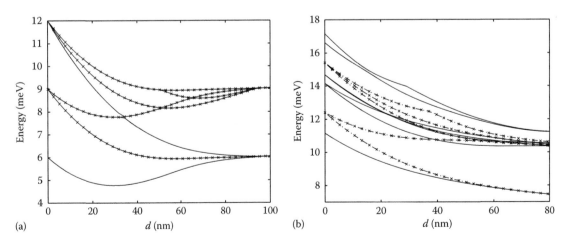

FIGURE 38.2 Quantum-dot molecule energy spectrum as a function of interdot distance, d, obtained by the diagonalization of the Hamiltonian (for total $M_S = 0$). The lowest energy states are shown for $\hbar\omega = 3$ meV without (a) and with (b) electron–electron interaction. Full lines: Singlet states. Broken lines with crosses: Triplet states. (Reprinted from Popsueva, V. et al., *Phys. Rev. B*, 76, 035303, 2007. With permission.)

centered at $\pm d/2$. The second band has four excited states that contain a single excited quanta $\left(n_x^L \text{ or } n_y^L \text{ or } n_x^R \text{ or } n_y^R = 1\right)$ and are of the form

$$\phi_{10}(\mathbf{r}_{1L})\phi_{00}(\mathbf{r}_{2R}) + \phi_{00}(\mathbf{r}_{1L})\phi_{10}(\mathbf{r}_{2R})$$

$$+ \phi_{10}(\mathbf{r}_{1R})\phi_{00}(\mathbf{r}_{2L}) + \phi_{00}(\mathbf{r}_{1R})\phi_{10}(\mathbf{r}_{2L}) \qquad (38.21)$$

with the L, R subscripts denoting left and right centered orbitals, respectively. The subscripts, $\phi_{n_x n_y}$, refer to the quantum numbers for the one-dimensional harmonic oscillator in the x- and y directions. The four singly excited states can be classified with respect to positive or negative parity as $\{|e_1, --\rangle, |e_2, -+\rangle, |e_3, +-\rangle, |e_4, ++\rangle\}$, where $\pm$ indicates the parity in the x and y coordinates, respectively, and e_i just stands for the ith excited singlet state. The third energy band contains two excited states of different x-parity, $\{|ion_5, ++\rangle, |ion_6, -+\rangle\}$, which are the ones that asymptotically correlate to the two-electron ground state of a single dot, again the indices just count the singlet states.

The single-particle electron densities of three states are shown in Figure 38.3. On the right panel, the conditional single-electron densities obtained by fixing all degrees of freedom except one x-component at a maximum of the wave function, and similarly for the other x-component (electron). The densities of the ground state $|g, +\rangle$ and excited state $|e, -\rangle$ (upper and middle panel) are seen to have the characteristics of two displaced eigenstates of the harmonic oscillator, and the conditional densities clearly show that "one" electron is centered in each dot. The state $|ion, -\rangle$ (lower panel), in contrast, shows a high probability that both electrons occupy the same dot, as discussed above.

We now discuss a scheme for the excitation of the ground state to a linear combination of the two upper "ionic" singlet states. The sum, or difference, of these two states corresponds to a localized two-electron state in one of the wells. Such a state can be created by applying three sequential laser pulses, that is, $V_{ext}(t) = Ei(t)(x_1 + x_2)$, $i = 1, 2, 3$. The first pulse transfers 50% of the ground state to the second excited state. The second pulse further transfers the population in state $|e, -\rangle$ to the lowest ionic state $|ion, +\rangle$. The last pulse transfers the part of the wave function remaining in the ground state directly to the $|ion, -\rangle$. At the end of the three pulses, we have produced the entangled state

$$\sim \frac{1}{\sqrt{2}}\left(|ion, -\rangle + e^{i\alpha}|ion, +\rangle\right) \qquad (38.22)$$

with α being a real number. Due to the energy separation between the two states, this describes a dynamic charge distribution of two electrons oscillating back and forth between the two wells. In Figure 38.4, we plot the population of the initial, final, and intermediate states in the excitation scheme. For more details on pulse parameters, we refer to Sælen et al. (2008). Note also that the excitation time can be reduced and the fidelity can be increased by optimizing the pulse parameters.

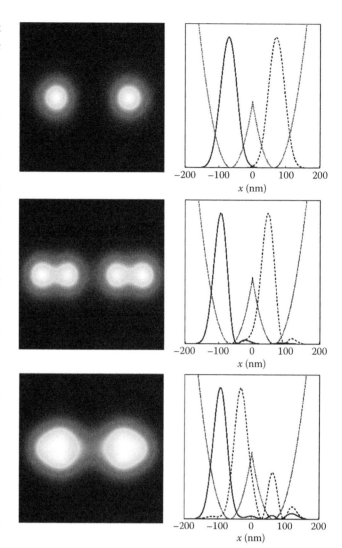

FIGURE 38.3 One-electron densities (left column) and conditional densities (right column) for a quantum-dot molecule. The conditional densities are evaluated at one maximum of the wave function, for each electron (solid and dashed line). Upper panel: ground state, middle panel: second excited state, lower panel: sixth excited state. The right column shows that the two upper states may be regarded as "covalent" states, whereas the lower state is an "ionic" state with both electrons in the same dot. The double dot potential is shown as dotted lines. The interdot distance is $d = 130$ nm. (Reprinted from Sælen, L. et al., *Phys. Rev. Lett.*, 100, 046805, 2008. With permission.)

38.5 Quantum Rings

A ring-formed confinement is obtained if a harmonic potential is displaced from the origin and then rotated around the z-axis:

$$V(x, y) = \frac{m^{\star}\omega^2}{2}(x - x_0)^2 + (y - y_0)^2, \qquad (38.23)$$

where $x_0 = y_0$ gives perfect polar symmetry. When the radius is large compared to the ring width, that is, $\sqrt{x_0^2 + y_0^2} \gg \sqrt{\hbar/m^{\star}\omega}$,

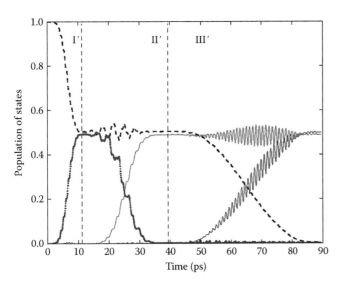

FIGURE 38.4 Population of quantum-dot molecule states during the sequence of pulses. The duration of the individual pulses is indicated by dashed vertical lines. The thin dashed line is the ground state population. The thick dotted line is the population of the excited state $|e, -\rangle$. The two lines that exhibit beats during pulse III are the population of the states $|ion, +\rangle$ (upper) and $|ion, -\rangle$ (lower). (Reprinted from Sælen, L. et al., *Phys. Rev. Lett.*, 100, 046805, 2008. With permission.)

the radial and angular motions start to decouple, and we have a so-called quasi one-dimensional ring. In the limit of complete decoupling, the one-electron equation can be solved analytically. A review on quasi one-dimensional rings, including the interesting many-electron case, has been given by Viefers et al. (2003). Experimental realizations of true few electron rings are, however, found in another regime. Lorke et al. (2000) have produced nanoscopic rings using self-assembly techniques. These rings have an inner radius of ≈ 20 nm or smaller, and a confining potential in the order of $\hbar\omega \approx 10$ meV. With these parameters, the ring radius is only slightly larger than the ring width. We expect the lowest energy states in such a potential to be ring-like, but for higher degrees of excitation, the eigenstates will approach those of a dot.

To visualize the two-particle states in a nanoscopic ring, we show (Waltersson et al., 2009) the *relative* probability density $\tilde{\rho}$ and the *relative* probability current $\tilde{j}$ in Figure 38.5. These are defined as

$$\tilde{\rho}(r_1, \phi_{rel}) \equiv \rho(r_1, \phi_1 - \phi_2) \tag{38.24}$$

$$\tilde{\mathbf{j}}(r_1, \phi_{rel}) \equiv \mathbf{j}(r_1, \phi_1 - \phi_2) \tag{38.25}$$

obtained from

$$\rho(\mathbf{r}_1) = \int d\mathbf{r}_2 \, |\Psi(\mathbf{r}_1, \mathbf{r}_2)|^2 \tag{38.26}$$

$$\mathbf{j}(\mathbf{r}_1) = \Re\left[\int d\mathbf{r}_2 \Psi^*(\mathbf{r}_1, \mathbf{r}_2)\left(-\frac{i\hbar}{m^*}\nabla_1\Psi(\mathbf{r}_1, \mathbf{r}_2)\right)\right] \tag{38.27}$$

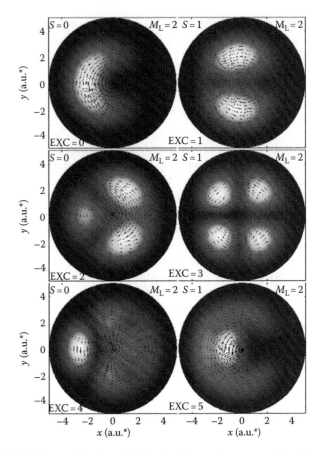

FIGURE 38.5 Relative probability densities and currents, see Equations 38.24 and 38.25, of the $M_L = 2$ quantum-ring system. The left column depicts singlets and the right column depicts triplets, starting with the lowest lying singlet (triplet) at the top left (right) corner, continuing with the first excited singlet (triplet) in the middle left (right) panel and so on. The label EXC = 0 is used for ground states, EXC = 1 for the first excited state, etc. The ring radius is $r_0 = 2$ a.u.* ≈ 19.6 nm and the confining potential is $\hbar\omega = 10$ meV. (Reprinted from Sælen, L. et al., *Phys. Rev. Lett.*, 100, 046805, 2008. With permission.)

through the coordinate transformation $\phi_1 \to \phi_{rel} = \phi_1 - \phi_2$. Since there is no preferred angle, ϕ, this is equivalent to freeze one electron at $\phi = 0$ and calculate the probability density (current) of the other one. It is the six lowest lying $M_L = 2$ states that are shown in Figure 38.5. The ground state has one relative current density peak, the first excited state has two peaks, etc., up to the third excited state. These vibrational excitations are expected in a quantum ring (Viefers et al., 2003). The fourth and fifth excited states, however, do not continue this quantum-ring pattern, indicating that these more energetic states are more dot-like. For the relative probability current, however, signs of deviation from ring behavior are seen earlier. While the radial component of the relative probability currents for a large (or quasi 1D) ring would approach (be) zero, the currents here show a rich structure. Already at the first excited state, and even more clearly in the higher lying states, we see complete departure from this circular shape. Even probability current vortices can be seen, that is, between the peaks in the third excited state. Hence, we are

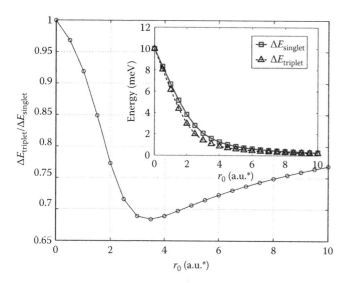

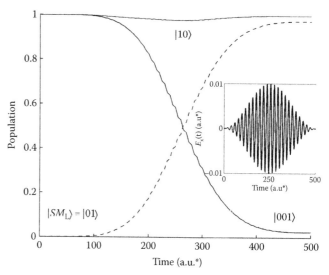

FIGURE 38.6 The quotient between the first excitation energy in the ring triplet system, $\Delta E_{triplet}$, and that in the singlet system, $\Delta E_{singlet}$, as a function of the ring radius r_0. The inset shows the absolute value of $\Delta E_{triplet}$ and $\Delta E_{singlet}$, also as a function of ring radius. Here, the effective Bohr radius = 1 a.u.* ≈ 9.8 nm. The confining potential is $\hbar\omega = 10$ meV. (Reprinted from Waltersson, E. et al., *Phys. Rev. B*, 79, 115318, 2009. With permission.)

FIGURE 38.7 The time development of the populations of the different states in a quantum ring when the central frequency, ω_L, of the pulse corresponds to the energy shift between the the two lowest states in the singlet system, $\Delta\varepsilon_{|SM_L\rangle} = \varepsilon_{|01\rangle} - \varepsilon_{|00\rangle} \approx 3.8$ meV implying a laser frequency of 0.9 THz. The population of the state $|SM_L\rangle = |00\rangle$ is almost completely transferred to $|01\rangle$, while the population of the $|10\rangle$ is seen to be nearly constant. The inset shows the x component of the electromagnetic pulse. (Reprinted from Waltersson, E. et al., *Phys. Rev. B*, 79, 115318, 2009. With permission.)

here in a region of strongly correlated electrons that still exhibit ring-like behavior.

The extra parameter in a ring compared with a dot provides an, at least in principle, easy way to tailor the energy spectrum of the quantum system. Figure 38.6 shows the ratio between the excitation energy (to the first excited state) in the triplet system relative to that in the singlet system as a function of ring radius. In the singlet system, the ground state has total angular momentum, $M_L = 0$, while the first excited state has $M_L = 1$. In the triplet system, it is vice versa. As discussed by Waltersson et al. (2009), different excitations energies in the triplet and the singlet systems can be used to construct a controlled NOT gate (CNOT). The idea is a setup where a change of orbital angular momentum takes place or not depending on if the total spin is zero or one, and where it is the excitation energy that distinguish the spin states. We would then have a qubit pair where one component is stored in the total electron spin and the other in the total angular momentum.

Figure 38.7 shows a simulation of the function of the CNOT. The quantum ring is exposed to a circularly polarized electromagnetic pulse $\mathbf{E}(t) = E(t)[\cos(\omega_L t)\hat{\mathbf{x}} \pm \sin(\omega_L t)\hat{\mathbf{y}}]$, where ω_L is the central frequency. The electric dipole interaction then couples neighboring M_L states ($\Delta M_L = 1$). The envelope $E(t)$ is taken as $E(t) = E_0 \sin^2(\pi t/T)$, which defines a pulse that lasts from $t = 0$ to $t = T$. Here, $T = 500$ a.u.* ≈28 ps and $E_0 \approx 0.01$ a.u.*, which corresponds to an intensity ~2.4×10^2 W/cm². To preferably induce transitions in the singlet system, the pulse central frequency, ω_L, is chosen to correspond to the energy shift between the two lowest states in this system $\Delta\varepsilon_{|SM_L\rangle} = \varepsilon_{|01\rangle} - \varepsilon_{|00\rangle} \approx 3.8$ meV. The driving laser frequency would then be $\omega_L/2\pi \approx 0.9$ THz. We

simulate the interaction with the electromagnetic pulse through the time-dependent Schrödinger equation, which is solved on the basis of fully correlated two-particle eigenstates to the ring Hamiltonian (see Section 38.6 below as well as the discussion around Equation 38.10). As can be seen in Figure 38.7, we observe a nearly complete transition to $|SM_L\rangle = |01\rangle$, from the initial state $|SM_L\rangle = |00\rangle$, with a small amount of unwanted population. Also shown is the time development of the population of an initial state $|SM_L\rangle = |10\rangle$, which is seen to be nearly constant. Thus, a CNOT is realized.

38.6 Computational Aspects

For the calculation of energy spectra, we apply direct diagonalization techniques. This is accurate and efficient for low lying states, and the basis functions allow also for the efficient calculation of the additional matrix elements needed in the time-dependent calculations. For the calculation of a large number of excited states, other techniques are more efficient (see Drouvelis et al., 2003; Taut, 1993; Zhu et al., 1997).

38.6.1 Analytical Basis Sets

In the case of two-electron quantum dots and quantum-dot molecules, we apply a basis of symmetrized spin singlet and triplet states as

$$\left|\Psi(\mathbf{r}_1, \mathbf{r}_2)\right\rangle = \sum_{j \geq i}^{n_{max}} c_{ij}\left|ij\right\rangle \otimes \left|S\right\rangle, \tag{38.28}$$

where

$$\langle \mathbf{r}_1, \mathbf{r}_2 \,|\, ij \rangle = \begin{cases} \dfrac{1}{\sqrt{2}}\Big[\phi_i(\mathbf{r}_1)\phi_j(\mathbf{r}_2) + (-1)^S \phi_j(\mathbf{r}_1)\phi_i(\mathbf{r}_2)\Big] & i \neq j \\[2mm] \phi_i(\mathbf{r}_1)\phi_j(\mathbf{r}_2) & i = j, \end{cases}$$

the c_{ij}'s are the expansion coefficients

$|S\rangle$ denotes the spin singlet or triplet state, that is, $|0\rangle$, $|1\rangle$

The functions defining each basis state are products of one-dimensional harmonic oscillator functions centered around the origin. With these basis functions, all matrix elements can be obtained analytically with explicit dependence on confinement strength and interwell distance d. The most complex calculations are those that involve the electron–electron interaction:

$$M_{K,L} = \left\langle \phi_{K_i}\phi_{K_j} \,\middle|\, \frac{1}{r_{12}} \,\middle|\, \phi_{L_i}\phi_{L_j} \right\rangle. \tag{38.29}$$

To solve this integral for arbitrary quantum numbers, we first express the electron–electron interaction as the Bethe integral:

$$\frac{1}{r_{12}} = \frac{1}{2\pi}\int \frac{d^2s}{s}\, e^{i\mathbf{s}\cdot\mathbf{r}_1}\, e^{-i\mathbf{s}\cdot\mathbf{r}_2}, \tag{38.30}$$

where $\mathbf{s} = (s_x, s_y)$. The integral of Equation 38.29 can thus be expressed as

$$M_{K,L} = \frac{1}{2\pi}\int \frac{d^2s}{s}\int d^2r_1 \phi_{K_i}(\mathbf{r}_1)\phi_{L_i}(\mathbf{r}_1)e^{i\mathbf{s}\cdot\mathbf{r}_1}\int d^2r_2 \phi_{K_j}(\mathbf{r}_2)\phi_{L_j}(\mathbf{r}_2)^{-i\mathbf{s}\cdot\mathbf{r}_2}, \tag{38.31}$$

which results in a combination of Laguerre polynomials and Gaussians $\exp(-s^2/2)$. The final integral over s can thus be integrated straightforwardly.

38.6.2 Systems with Circular Symmetry

When the confining potential has circular symmetry

$$V(r) = \frac{m^*\omega^2}{2}(r - r_0)^2, \tag{38.32}$$

where $r_0 = 0$ corresponds to a dot and $r_0 \neq 0$ to a circular ring, it is convenient to use polar coordinates. In the absence of any external electric field, the eigenfunctions to the single-particle Hamiltonian, Equation 38.4, then separate as

$$\left|\Psi_{nm_\ell m_s}\right\rangle = \left|u_{nm_\ell m_s}(r)\right\rangle\left|e^{im_\ell\phi}\right\rangle|m_s\rangle, \tag{38.33}$$

where $x = r\cos\phi$ and $y = r\sin\phi$. The radial part of the eigenfunction satisfies the equation

$$\left[-\frac{\hbar^2}{2m^*}\left(\frac{\partial^2}{\partial r^2} - \frac{m_\ell^2}{r^2}\right) + \frac{1}{2}m^*\omega_0^2(r - r_0)^2 + \frac{e^2}{8m^*}B^2r^2 \right.$$
$$\left. + \frac{e\hbar}{2m^*}Bm_\ell + g^*\frac{e\hbar}{2m_e}Bm_s - E \right] u_{nm_\ell}(r) = 0, \tag{38.34}$$

which can be solved with standard methods, although it is important to describe the region close to the potential minimum, $r = r_0$, well. Harmonic oscillator eigenstates, which are centered around $r = 0$, will, for example, not be very suitable to expand the solution when the ring grows large. A useful approach is instead that of B-splines. B-splines are piecewise polynomials of a chosen order k, defined on a so-called knot sequence, and they form a complete set in the space defined by the knot sequence and the polynomial order (deBoor, 1978). The important feature here is that the knot sequence can be adopted to the particular problem at hand. In the case of a ring, it can be dense around $r = r_0$, and then more sparse for smaller as well as larger r; see Waltersson and Lindroth (2007) and Waltersson et al. (2009) for more details. To solve the full two-particle equation, Equation 38.1, products of the eigenstates in Equation 38.33—antisymmetrized and coupled to a specific total spin—can be used as a basis. A form of the Coulomb interaction suitable for this purpose is obtained through expansion in cylindrical coordinates as suggested by Cohl et al. (2001):

$$\frac{1}{|\mathbf{r}_1 - \mathbf{r}_2|} = \frac{1}{\pi\sqrt{r_1 r_2}}\sum_{m=-\infty}^{\infty} Q_{m-1/2}(\chi)e^{im(\phi_1 - \phi_2)}, \tag{38.35}$$

where

$$\chi = \frac{r_1^2 + r_2^2 + (z_1 - z_2)^2}{2r_1 r_2}. \tag{38.36}$$

The two-dimensional form is obtained with $z_1 = z_2$ in (38.36). The $Q_{m-1/2}(\chi)$–functions are Legendre functions of the second kind and half-integer degree. Eigenstates to Equation 38.1 can now be found by diagonalization of the Hamiltonian matrix, that is, the matrix with elements:

$$H_{ij} = \left\langle \{ab\}_i \,\middle|\, h(1) + h(2) + \frac{e^2}{4\pi\epsilon_r\epsilon_0}\frac{1}{r_{12}} \,\middle|\, \{cd\}_j \right\rangle, \tag{38.37}$$

where Equation 38.35 can be inserted to get the last term in the following form:

$$\frac{e^2}{4\pi\epsilon_r\epsilon_0}\left\langle ab \,\middle|\, \frac{1}{r_{12}} \,\middle|\, cd \right\rangle = \frac{e^2}{4\pi\epsilon_r\epsilon_0}\left\langle u_a(r_i)u_b(r_j) \,\middle|\, \frac{Q_{m-1/2}(\chi)}{\pi\sqrt{r_i r_j}} \,\middle|\, u_c(r_i)u_d(r_j) \right\rangle$$

$$\times \left\langle e^{im_a\phi_i}e^{im_b\phi_j} \,\middle|\, \sum_{m=-\infty}^{\infty} e^{im(\phi_i - \phi_j)} \,\middle|\, e^{im_c\phi_i}e^{im_d\phi_j} \right\rangle$$

$$\times \left\langle m_s^a \,|\, m_s^c \right\rangle\left\langle m_s^b \,|\, m_s^d \right\rangle. \tag{38.38}$$

38.7 Concluding Remarks

In the present contribution, we have described the basic theoretical framework for two-electron quantum dots. We have discussed the spectrum of quantum dots, molecules, and rings. A precise understanding of the spectrum is a basic theoretical ingredient for successful progress in the experimental realization of quantum transport, metrology, and quantum information technology. This has been illustrated here by the demonstration of the possibility to fine-tune electronic excitations with the help of time-dependent electric fields that transfer an electron from one state to another with almost 100% probability.

Acknowledgments

We gratefully acknowledge support from the Norwegian Research Council (RCN) (J. P. Hansen), the Swedish Research Council(VR), and the Göran Gustafsson Foundation (E. Lindroth), as well as from the EU COST action CM0702; Chemistry with Ultrashort Pulses and Free-Electron Lasers: Looking for Control Strategies Through "Exact" Computations.

References

Bransden, B. H. and C. J. Joachain, 2003, *Physics of Atoms and Molecules*, 2nd edn. (Pearson Education Limited, Upper Saddle River, NJ).

Cohl, H. S., A. R. P. Rau, J. E. Tohline, D. A. Browne, J. E. Cazes, and E. I. Barnes, 2001, *Phys. Rev. A* **64**, 052509.

deBoor, C., 1978, *A Practical Guide to Splines* (Springer-Verlag, New York).

Drouvelis, P. S., P. Schmellcher, and F. F. Diakonos, 2003, *Eurphys. Lett.* **64**, 232.

Førre, M., J. P. Hansen, V. Popsueva, and A. Dubois, 2006, *Phys. Rev. B* **74**, 165304.

Hanson, R., L. P. Kouwenhoven, J. R. Petta, S. Tarucha, and L. M. K. Vandersypen, 2007, *Rev. Mod. Phys.* **79**, 1217.

Harju, A., S. Siljamäki, and R. M. Nieminen, 2002, *Phys. Rev. Lett.* **88**, 226804.

Koppens, F. H. L., C. Buizert, K. J. Tielrooij, I. T. Vink, K. C. Nowack, T. Meunier, L. P. Kouwenhoven, and L. M. K. Vandersypen, 2006, *Nature* **442**, 766.

Kumar, A., S. E. Laux, and F. Stern, 1990, *Phys. Rev. B* **42**, 5166.

Lorke, A., R. J. Luyken, A. O. Govorov, J. P. Kotthaus, J. M. Garcia, and P. M. Petroff, 2000, *Phys. Rev. Lett.* **84**, 2223.

Loss, D. and D. P. DiVincenzo, 1998, *Phys. Rev. A* **57**, 120.

Merkt, U., J. Huser, and M. Wagner, 1991, *Phys. Rev. B* **43**, 7320.

Nepstad, R., L. Sælen, and J. P. Hansen, 2008, *Phys. Rev. B* **77**, 125315.

Petta, J. R., A. C. Johnson, C. M. Marcus, M. P. Hanson, and A. C. Gossard, 2004, *Phys. Rev. Lett.* **93**, 186802.

Popsueva, V., R. Nepstad, T. Birkeland, M. Førre, J. P. Hansen, E. Lindroth, and E. Waltersson, 2007, *Phys. Rev. B* **76**, 035303.

Räsänen, E., A. Castro, J. Werschnik, A. Rubio, and E. K. U. Gross, 2007, *Phys. Rev. Lett.* **98**, 157404.

Reimann, S. M. and M. Manninen, 2002, *Rev. Mod. Phys.* **74**, 1283.

Sælen, L., R. Nepstad, I. Degani, and J. P. Hansen, 2008, *Phys. Rev. Lett.* **100**, 046805.

Sørngård, S., 2009, MSc thesis, University of Bergen, Bergen, Norway.

Taut, M., 1993, *Phys. Rev. A* **48**, 3561.

Taylor, J. M., J. R. Petta, A. C. Johnson, A. Yacoby, C. M. Marcus, and M. D. Lukin, 2007, *Phys. Rev. B* **76**, 035315.

Viefers, S., P. Koskinen, P. S. Deo, and M. Manninen, 2003, *Physica E* **21**, 1.

Waltersson, E. and E. Lindroth, 2007, *Phys. Rev. B* **76**(4), 045314.

Waltersson, E., E. Lindroth, I. Pilskog, and J. P. Hansen, 2009, *Phys. Rev. B* **79**, 115318.

Wensauer, A., O. Steffens, M. Suhrke, and U. Rössler, 2000, *Phys. Rev. B* **62**, 2605.

Wigner, E., 1934, *Phys. Rev.* **46**, 1002.

Zhu, J.-L., Z.-Q. Li, J.-Z. Yu, K. Ohno, and Y. Kawazoe, 1997, *Phys. Rev. B* **55**, 15819.

Thermodynamic Theory of Quantum Dots Self-Assembly

Xinlei L. Li
Zhongshan University

Guowei W. Yang
Zhongshan University

39.1 Introduction

Quantum dots (QDs), also called as nanocrystals, are a special class of semiconductor crystals that are composed of periodic groups of II–VI, III–V, or IV–IV materials, such as CdSe (Lee et al., 1998; Strassburg et al., 2000; Kratzert et al., 2001), InAs (Ebiko et al., 1998, 1999; Nakata et al., 2000; Yamaguchi et al., 2000; Joyce et al., 2001; Márquez et al., 2001; Krzyzewski et al., 2002; Migliorato et al., 2002; Bester and Zunger, 2003; Wasserman et al., 2003), InP (Seifert et al., 1996; Schmidbauer et al., 2002; Persson et al., 2003), and Ge QDs (Eaglesham and Cerullo, 1990; Kamins et al., 1997; Liu and Lagally, 1997; Ribeiro et al., 1998; Ross et al., 1998; Liu et al., 2000; Vailionis et al., 2000; Denker et al., 2001; Rastelli and Känel, 2002, 2003; Tersoff et al., 2002). Semiconductors derive their great importance from the fact that their electrical conductivity can be altered by adding an external stimulus, such as voltage. Therefore, semiconductors can be used to make critical parts of many different kinds of electrical circuits and optical applications. Semiconductor QDs, as a special structure, have their name because their small size causes quantum confinement and creates specific electronic states (Reithmaier et al., 2004; Badolato et al., 2005; Xu et al., 2007). Due to the quantum confinement and specific electronic states, QDs enable never before seen applications to science and technology.

Semiconductor QDs have already been used for optoelectronic applications, for example, exploiting the increased density of states and tunable energy levels due to quantum confinement (Chang et al., 1994). By controlling the material's composition and changing the size and shape of QDs, optoelectronic properties of QDs can be tuned. The electronic spectrum of the bulk is continuous, but that of QDs is not continuous. The special optoelectronic properties of QDs are different from those of the bulk system. Therefore, QD can be called a superatom, although it contains many atoms (about $10^5 - 10^6$ atoms; Shchukin and Bimberg, 1999). Because of the atomlike electronic spectrum of QDs, they have become an interesting and important object both for basic research and for device application. For example, QDs have been used as an active medium of semiconductor lasers to improve the laser performance (Kirstaedter et al., 1994; Bimberg et al., 1996; Ustinov et al., 1998; Grundmann, 2002; Ledentsov et al., 2002; Ustinov and Zhukov, 2002; Shchukin et al., 2003). QDs can also be allowed to construct new kinds of devices, e.g., cellular automata (Chen and Porod, 1995), single-electron transistors (Kastner, 1996), and lasers based on the resonant waveguiding effect (Ledentsov et al., 1996, 1997).

In the fields of biology, chemistry, and computer science, there are strong interests for researchers in nanotechnology. For example, Jokerst et al. developed the diagnostic instrumentation using the integration of semiconductor QDs into a portable microfluidic-based lymphocyte capture and detection device (Jokerst et al., 2008). This integrated system is capable of isolating and counting selected lymphocyte sub-populations from whole blood samples, which demonstrates the viability of incorporating QD detection schemes into a microfluidic analysis device for lymphocyte enumeration as a tool for monitoring HIV progression. On the other hand, in computer science, because the position of a single electron in a QD might attain several states, a QD could represent a byte of data. Alternatively, a QD might be used in more than one computational instruction at a time. Other applications of QDs include nanomachines, neural networks, and high-density memory or storage media.

39.2 Formation of QDs

39.2.1 Three Growth Modes in Epitaxy

Epitaxy is a method of depositing a monocrystalline film on a monocrystalline substrate, in which the deposited monocrystalline film can be called as epitaxial film or epitaxial layer. Epitaxial films may be grown from a gaseous source (gas phase epitaxy) or a liquid precursor (liquid phase epitaxy). The epitaxial film has an identical lattice structure and a similar orientation as those of the substrate because of the effect of the crystal substrate. It is different from other thin-films deposited on polycrystalline or amorphous films. Generally, if the epitaxial film has the same composition as that of the substrate, we name the process homoepitaxy, otherwise we name it heteroepitaxy.

In homoepitaxial growth, the different conglomerations of deposited atoms lead to various configurations. However, when the deposited temperature is so high that deposited atoms can diffuse easily on the substrate surface, various configurations would tend to a configuration that has the most stability. According to the thermodynamic view, the number of bonds between deposited atoms and substrate reaches a maximum in the case of layer-by-layer growth mode (two dimensional [2D]). Therefore, the 2D-growth mode has the most stability and is the most favorite in homoepitaxial growth.

In the case of heteroepitaxial growth, the growth mode becomes more complex than that in homoepitaxial growth. Growth mode in heteroepitaxial growth is determined mainly by the properties of the deposited material and substrate. Traditionally, there are three growth modes in the equilibrium theory of heteroepitaxial growth (Bauer, 1958), which are Frank–van der Merwe (FM; Frank and van der Merwe, 1949), Volmer–Weber (VM; Volmer and Weber, 1926), and Stranski–Krastanow (SK; Stranski and Krastanow, 1937) growth modes. These three growth modes may be also described visually as, respectively, layer-by-layer growth (2D), island growth (three dimensional [3D]), and layer-by-layer plus islands, as shown in Figure 39.1.

According to the well-known Young's equation (Figure 39.2), a contact angle should satisfy the condition $\gamma' = \gamma \cos \alpha + \gamma''$. Therefore, we can get the qualification for the layer-by-layer growth mode. The contact angle of the nuclei can be calculated by $\cos \alpha = (\gamma' - \gamma'')/\gamma$, here γ, γ', and γ'' are the surface energies of the nuclei and the substrate and the interface energy between the nuclei and the substrate per unit area (strictly speaking, they should be tensile forces). We find that the value of $\cos \alpha$ would be larger than 1 when $\gamma' \geq \gamma + \gamma''$, which means the contact angle is zero. i.e., a complete wet process, in the case of $\gamma' \geq \gamma + \gamma''$. Therefore, the layer-by-layer growth mode will happen when the surface energy of the substrate is higher than the sum of the surface energy of the deposited epitaxial film and the interface energy between them. We can further discuss other growth

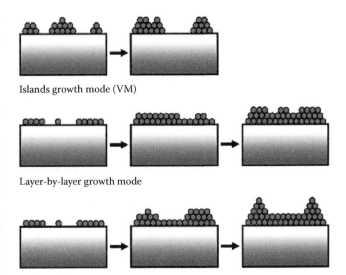

Islands growth mode (VM)

Layer-by-layer growth mode

Layer-by-layer plus islands growth mode

FIGURE 39.1 Three growth modes in heteroepitaxial growth.

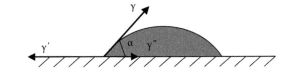

FIGURE 39.2 Schematic illustration of Young's equation.

modes based on Young's equation. When $\gamma' < \gamma + \gamma''$, $\cos \alpha$ is less than 1, which means that the nuclei have a definite contact angle with the substrate and can exist on the surface of the substrate. Furthermore, the deposited atoms should grow on the substrate surface in the island growth mode in the case that $\gamma' < \gamma + \gamma''$.

For the SK growth mode, a complete wet epitaxial layer first forms on the substrate where the case must be $\gamma' \geq \gamma + \gamma''$. However, in the case that $\gamma' \geq \gamma + \gamma''$, the epitaxial layer should grow up in the layer-by-layer mode. In principle, the layer-by-layer plus islands growth mode cannot occur if the epitaxial film and the substrate are a lattice match. Therefore, it is necessary to take the effect of strain into account in the investigation of the SK growth mode. In the next section, we will mainly introduce the SK growth mode and the formation of QDs in the growth mode.

39.2.2 Formation of QDs in the SK Growth Mode

It is improper to judge the growth mode only by comparing the surface energy and the interface energy of the epitaxial film and the substrate when the deposited film has a lattice mismatch with the substrate. Because of the lattice mismatch, the epitaxial film suffers from a compressive or tensile strain. In general, because the strain stored in epitaxial film would result in some defects (such as dislocation) in order to relieve the strain caused by mismatch at the very early stage, heteroepitaxy with a larger lattice mismatch is impossible. Therefore, the film with some defects cannot be named epitaxial film, and the growth cannot be called epitaxial growth. However, if the mismatch is sufficiently small, though the deposited film suffers from a compressive or tensile strain because of the lattice mismatch between the epitaxial material and substrate, defect-free growth can proceed in the initial deposited process. As epitaxial film grows, the strain stored in the film cannot be maintained and needs to be released by the formation of defects. Other than formation of defects in order to release strain caused by mismatch, if the deposited temperature is enough high, and growth rate is enough slow, i.e., the growth condition is close enough to equilibrium, morphological change is another pathway available for the release of strain. It has been proved by experiments and theories that the formation of 3D QDs can release effectively strain caused by lattice mismatch. The growth mode of QDs formation on epitaxial film is typical SK growth mode.

In SK growth mode, the complete wet epitaxial layer first appeared on the substrate surface is usually called wetting layer. When the wetting layer exceeds a critical thickness, QDs can form on the surface of wetting layer to release strain in which the reduction of the strain energy is usually called elastic relaxation energy or relaxation energy. An important question is that the formation of QDs can also lead to the increase of surface energy. If the reduction of strain energy caused by QDs is less than the increment of surface energy, it is also not favorable to form QDs. From thermodynamic point of view, the gain in elastic relaxation energy $\Delta E_{\text{relaxation}}$ caused by formation of a QD is proportional to the QD volume V, and the increment of surface energy, $\Delta E_{\text{surface}}$, is proportional to $V^{2/3}$. Therefore, the change of total energy caused by the formation of QD can be simply expressed as

$$\Delta E = A\gamma V^{2/3} - \kappa\varepsilon^2 A' V \tag{39.1}$$

where

  A and A' are coefficients that are determined by the shape of QD
  γ is the surface energy per unit area
  κ is the elastic modulus
  ε is the lattice mismatch

Obviously, as QD volume increases, the change of total energy first increases and then decreases, as shown in Figure 39.3.

When the volume of QD exceeds a critical value, the change of total energy will be less than 0. That is, the formation of QDs is more favorable thermodynamically.

Equation 39.1 shows simply the reasons for release of strain caused by the formation of QD in SK growth mode. However, Equation 39.1 cannot explain why QDs only appear when the wetting layer exceeds a critical thickness. In Section 39.2.3, we will discuss what determines the critical thickness of the wetting layer for QDs' formation and how to quantitatively calculate the critical volume of QDs.

39.2.3 Critical Condition of QDs Formation

The formation of QDs can effectively release strain caused by lattice mismatch between epitaxial film and substrate. When the reduction of strain energy caused by QDs is larger than the increment of surface energy, i.e., the change of total energy is less than 0, it is favorable thermodynamically to form QDs. The explanations only take the effect of formation of QDs into account; do not consider the change of the wetting layer during the growth process. Therefore, we cannot get the critical thickness of the wetting layer for QDs formation from Equation 39.1.

In order to explain why QDs only form when the wetting layer exceeds a critical thickness, we must consider the change of the wetting layer during the QD growth process. For an existing wetting layer with a certain thickness, there are only two possibilities for further growth. The first is to keep layer-by-layer growth, which results in the increase of thickness of wetting layer. The other one is QD formation on the wetting layer, as shown in Figure 39.4. Therefore, we can compare changes of energy caused by the two growth modes to identify which growth mode is more favorable. Here, we take Ge QD formation on Si (001) substrate as an example to illuminate the critical condition of QD formation in detail.

For bulk materials, the surface energy of Ge is lower than that of Si. When the Ge wetting layer is deposited on Si (001) substrate, surface energy of the Ge wetting layer reduces from that of Si (001) to its own as thickness increases, which has been proved in theory (Lu and Liu, 2005; Li and Yang, 2008). The thickness-dependent surface energy of the Ge wetting layer can be expressed approximately as the function of thickness θ ML, in which θ ML represents that the number of molecular layer is θ (Li and Yang, 2008)

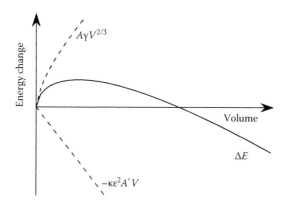

FIGURE 39.3 Change of total energy as a function of volume of QD.

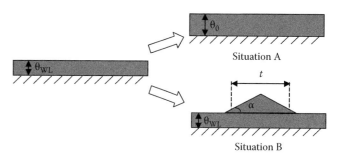

FIGURE 39.4 Two possibilities of further growth on a wetting layer.

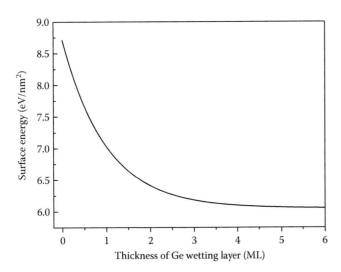

FIGURE 39.5 Surface-energy density of Ge wetting layer on Si (001) substrate as a function of thickness.

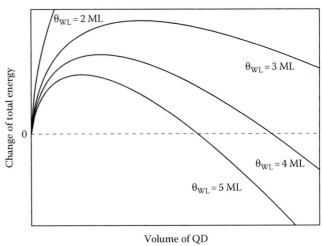

FIGURE 39.6 Change of total energy as a function of QD volume under a fixed QD density and θ_{WL} = 2 ML, 3 ML, 4 ML, and 5 ML.

$$\gamma(\theta) = \gamma_{Ge}^{\infty} + (\gamma_A^{\infty} - \gamma_B^{\infty})(1 - e^{-\theta}) \qquad (39.2)$$

The results calculated from Equation 39.2 are shown by Figure 39.5.

After knowing the relation between the surface energy and thickness of wetting layer, we can get the change of energy caused by the two growth situations, layer-by-layer growth (situation A) and QD formation (situation B) (Figure 39.4). For the situation A, the change of energy, ΔE_A, is mainly caused by the increase of thickness of wetting lay. Therefore, ΔE_A can be calculated by $\Delta E_A = (1/k)\left[\gamma(\theta_0) - \gamma(\theta_{WL})\right]$, where k is the density of QDs (the number of QDs per unit area). For the situation B, the change of energy, ΔE_B, is mainly determined by two factors, the increase of QDs' side face and elastic relaxation caused by QDs formation. So ΔE_B can be expressed as $\Delta E_B = t^2\gamma_s/\cos\alpha - t^2\gamma(\theta_{WL}) + \Delta E_{relaxation}$, where we consider that QDs have a pyramid shape. Therefore, the difference of ΔE_B and ΔE_A, $\Delta E = \Delta E_B - \Delta E_A$, can be written as

$$\Delta E = \frac{1}{k}\left[\gamma(\theta_{WL}) - \gamma(\theta_0)\right] + \frac{t^2\gamma_s}{\cos\alpha} - t^2\gamma(\theta_{WL}) + \Delta E_{relaxation} \quad (39.3)$$

In Equation 39.3, the surface energy per unit area of QDs' side face can be considered as a constant, and $\Delta E_{relaxation}$ is proportional to the QD volume V, which can be expressed as $\Delta E_{relaxation} = -\kappa\varepsilon^2 A'V$ (Tersoff and Tromp, 1993; Shchukin et al., 2004). According to conservation of mass, the volume of QD, V should be equal to $(\theta_0 - \theta_{WL})h_0/k$, where h_0 is the thickness of the Ge monolayer. So we can compare changes of energy of the two situations based on Equation 39.3. If ΔE is larger than 0, it is favorable to keep growth of layer-by-layer mode. Contrarily, it is favorable to form QDs on the wetting layer if ΔE is lower than 0.

Figure 39.6 shows the value of ΔE as a function of QD volume under different thicknesses of wetting layer and a fixed density of QDs. Definitely, when the thickness of wetting layer is too small, the value of ΔE is always larger than 0, which means that it is impossible to form QDs on wetting layer with a thickness of

less than a critical thickness, i.e., layer-by-layer growth mode, situation A, is favorable at the beginning stage of growth. As thickness of wetting layer increases, the value of ΔE can becomes less than 0 when volume of QD exceeds a critical volume, which means that the QDs can only form on the wetting layer when the wetting layer reaches a certain thickness. All the analytic results show that the growth process is the typical SK growth mode, in which the QDs can form only at a critical coverage.

The physical original that QDs only form on wetting layer with thickness of larger than a critical value is the balance between the thickness-dependent surface energy of wetting layer and the relaxation energy caused by QD formation. From Figure 39.5, we note that the rate of decrease of surface energy of wetting layer as thickness increases is higher in the case of small thickness than that in the case of large thickness, which means that layer-by-layer growth mode can effectively reduce the surface energy of wetting layer when the thickness of wetting layer is small. In this case, though the QD formation can also result in the decrease of total energy, the reduction caused by QD formation is lower than that caused by layer-by-layer growth. Therefore, at the initial growth stage, layer-by-layer growth mode is more favorable than QDs growth mode. When wetting layer exceeds a critical thickness, the rate of decrease of surface energy of wetting layer becomes very small, i.e., the effect of thickness-dependent surface energy of wetting layer becomes much weaker. At this rate, the relaxation energy of QDs plays a key role in the further growth process, and the difference of energies caused by the two growth modes, ΔE, can be less than 0 when QDs exceeds a certain volume (usually called critical volume for QD formation), which means that it is more favorable to QD formation at the later stage.

In this section, we introduce the formation of QDs. Once QDs form on the wetting layer, they can grow during further deposition process. In the process, the shape of QDs also varies with increasing volume of QDs. In Section 39.3, we will mainly introduce the shape transition of QDs during growth process.

39.3 Shape Transition of QDs

In typical semiconductor systems, such as Ge/Si system and InAs/GaAs system, QDs suffer from shape transition with increasing volume of QDs. Taking Ge QDs on Si substrate as an example, Ge QDs have two obvious shape transitions during the growth process, from pre-pyramid to pyramid (Kamins et al., 1997; Ribeiro et al., 1998; Vailionis et al., 2000) and from pyramid to dome (Ross et al., 1998, 1999; Liu et al., 2000; Denker et al., 2001; Rastelli and Känel, 2002), as shown in Figure 39.7. In this section, we will take the Ge/Si system as an example to introduce the shape transition of QDs.

39.3.1 Shape Transition from Pre-Pyramid to Pyramid

When Ge atoms are deposited on Si substrate and form QDs, the pre-pyramid QDs with a low contact angle usually appear first and then transform into pyramid shape QDs with a high contact angle with the QDs volume increasing. According to thermodynamic view, as mentioned above, the formation of QDs can effectively release elastic energy of QDs caused by the lattice mismatch between QDs and substrate, and is favorable when decrease of elastic energy of QDs is larger than the increase of surface energy caused by QD formation. Therefore, the origin of QD formation is essentially the balance of surface energy and relaxation energy of QDs. In fact, the shape transition from pre-pyramid to pyramid is also determined by the balance between surface energy and elastic relaxation energy of QDs. In detail, the size-dependent surface energy determines a low contact angle in the QDs with small volume at the initial stage of growth, while the elastic relaxation energy becomes more significant and induces a high contact angle in the QDs with large volume. In this section, we will mainly introduce quantitatively the shape transition from pre-pyramid to pyramid.

For a nominal coverage θ_0 where QDs with identical pyramidal shape and volume appear after the formation of the wetting layer whose thickness is θ, the total energy difference of a single QD between SK growth and imaginary layer-by-layer modes can be written as

$$\Delta E = \frac{1}{k}\left[\gamma(\theta) - \gamma(\theta_0)\right] + E_s - 4s^2\gamma(\theta) + \Delta E_r \qquad (39.4)$$

where

k is the QD density
E_s is the surface energy of the QD facets
s is half-base length
E_r is elastic relaxation energy of the QD

For a single pyramidal QD, the volume V should have the relation that $V = (4/3)s^3 \tan\alpha = (1/k)(\theta_0 - \theta)h_0$ according to the mass conservation for two types of the growth mode, here h_0 is the thickness of monolayer.

The first three terms in Equation 39.3 represent the surface energy difference caused by the QD formation. The surface energy densities of wetting and nominal layers, $\gamma(\theta)$ and $\gamma(\theta_0)$, can be easily calculated using Equation 39.2 in the previous section.

It is difficult to estimate the surface energy of the QD facets E_s because the surface energy density of the QD facets is changeful for the different contact angle. In order to quantitatively compute the surface energy E_s, we regard the QD facet as a step facet (Chen et al., 1997), as shown in Figure 39.8. Thus, the surface energy of the QD facets E_s can be divided into two parts: the surface energy of terraces E_{st} and the step edges creation energy E_{sc}, i.e., $E_s = E_{st} + E_{sc}$. The terraces have the same crystal face as that of the wetting layer surface (substrate surface), because the terraces grow in parallel with the substrate surface. Thus, the surface energy density of terrace has a similar relation with the distance between the terrace and the interface as between wetting layer and substrate as Equation 39.2. Accordingly, the total surface energy of the terraces E_{st} can be expressed as

$$E_{st} = \sum_{n=1}^{n_T}\left[\gamma(\theta_n)A_n\right] + \gamma(\theta_{n_T+1})A_{n_T+1} \qquad (39.5)$$

where

θ_n represents the distance between the surface of nth terrace and the interface
n_T is the total number of steps
A_n is the area of the nth terrace

Here θ_n, n_T, and A_n are given by $\theta_n = \theta + n - 1$, $n_T = [(s\tan\alpha/h_0]$, and $A_n = 4h_0\cot\alpha\left[2s - (2n-1)h_0\cot\alpha\right]$. The second term represents

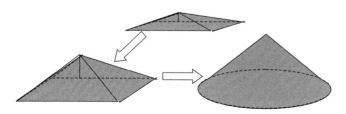

FIGURE 39.7 Shape transitions of QDs from pre-pyramid to pyramid to dome.

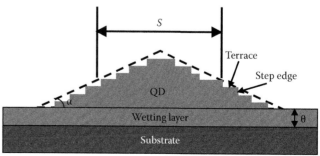

FIGURE 39.8 Schematic illustration of the shape of a coherent QD with a step facet on substrate surface.

the contribution from the surface energy of the top terrace, and θ_{n_T+1} and A_{n_T+1} can be written as $\theta_{n_T+1} = \theta + n_T$ and $A_{n_T+1} = 4(s - n_T h_0 \cot \alpha)^2$. The step edges creation energy E_{sc} contains two parts: The step creation energy and the repulsive interaction energy between steps (Poon et al., 1990; Chen et al., 1997) and can be written as

$$E_{sc} = 8 \sum_{n=1}^{n_T} \left[s - n h_0 \cot \alpha \right] \times \left[\lambda_0 + \lambda_d \left(\frac{a \tan \alpha}{h_0} \right)^2 \right] \quad (39.6)$$

where

λ_0 is the step creation energy per unit length
a is surface lattice constant
λ_d represents the energy of repulsive step–step interaction

After knowing the surface energy of QD facets, the quantitative relationship between size and shape of QDs under a fixed coverage can be presented by Equation 39.3. Thus, we can compare the total energy change of QDs with different contact angles and know which contact angle QDs is the steadiest under a certain volume.

Figure 39.9 shows the total energy change per unit volume as a function of volume of QDs for different contact angles. One can clearly find that QDs with low contact angle are preferred than QDs with high contact angle, when the volume of QDs is small. With increasing volume of QDs, QDs with high contact angle become more favorable than those with low contact angle. The results explain well the shape transition of QDs from pre-pyramid shape with a low contact angle to pyramid shape with a high contact angle.

As discussed in Section 39.5, the elastic relaxation energy at the top of QDs drives the QDs formation. However, the transition from a low contact angle to a high contact angle is determined by not only the elastic relaxation energy but also the size-dependent

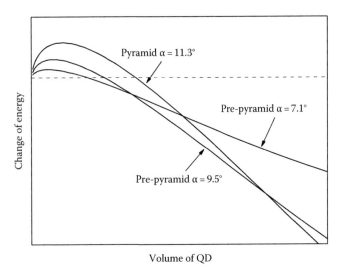

FIGURE 39.9 Total energy change as a function of volume of QDs for different contact angles.

surface energy. Firstly, the size-dependent surface energy has significant impact on the QDs shape. In the case of QDs with small volume, the total energy difference is mainly dominated by the size-dependent surface energy. Thus, the QD shape tends to minimize the surface energy, namely, to have a low contact angle. At the later stage of growth (QDs with large volume), the elastic relaxation becomes more significant and drives the QDs to have a high contact angle. Therefore, the theoretical analyses above show that the physical mechanisms of the shape transition from pre-pyramid shape with a low contact angle to pyramid shape with a high contact angle are actually the balance between surface energy and elastic relaxation energy.

39.3.2 Shape Transition from Pyramid to Dome

As QD volume increases further, the shape of QDs can transform from pyramid to dome. In detail, small QDs are square-based pyramids bounded by four {105} facts, whereas large QDs are multifaceted domes that have steeper facets. Based on thermodynamic view, we can adopt methods similar to those mentioned above, i.e., by comparing total energies of QDs with these two shapes.

Because both pyramid QDs and dome QDs form on an existent wetting layer, we can only compare the free energy of the formation of QD from a planar wetting layer, ΔE, which can be written as

$$\Delta E = E_s - E_w + \Delta E_r \quad (39.7)$$

where

E_s and E_w are the increase of surface energy of the QD facets and decrease of surface energy of wetting layer
ΔE_r is the elastic relaxation energy

When the QD is a pyramid, the change of the strain energy associated with the elastic relaxation is given (Tersoff and Tromp, 1993) $\Delta E_r = (-9/2) c V_p \tan \theta$, where V_p is the volume of a pyramid QD, $V_p = (1/6) t^3 \tan \theta$, and $c = (M_i \varepsilon)^2 (1 - \upsilon_s) / 2\pi G_s$, in which M_i and ε are the Young's modulus and the misfit strain of QD, and υ_s and G_s are the Poisson's ratio and shear modulus of substrate. Thus, the total energy change of a pyramid QD is

$$\Delta E_P = \left(\gamma_e \frac{1}{\cos \theta} - \gamma_s \right) \left(\frac{6 V_P}{\tan \theta} \right)^{2/3} - \frac{9}{2} c V_P \tan \theta \quad (39.8)$$

where γ_e and γ_s are the surface energy per unit area of the facets of QD and substrate.

When the QD is a dome, the surface energy of QD has the approximate relationship of the surface curvature: $\gamma_0 A + (1/2)\lambda \int (1/r) dA$, where A is the area of a defined surface, γ_0 is the intrinsic surface energy per unit area, λ is the curvature energy that is determined largely by the bulk density and nature of the constituent species, $(1/r)$ is the local curvature of surface area element dA. Accordingly, we attain $E_s = \gamma_0 A_e + \lambda \int (1/r) dA_e$

where A_e is the surface area of the QDs facets, and γ_0 is equal to γ_e of the pyramid QD with the same contact angle as that of the dome. The elastic relaxation energy of a dome, ΔE_r, can be written as $\Delta E_r = (-9/\sqrt{\pi})cV_D \tan\theta$ (Li et al., 2007), where V_D is the volume of a dome QD, $V_D = (1/3)\pi R^3 \tan\theta$. Thus,

$$\Delta E_D = \pi\left(\gamma_0 \frac{1}{\cos\theta} - \gamma_s\right)\left(\frac{3V_D}{\pi\tan\theta}\right)^{2/3} + \frac{2\pi\lambda}{\cos\theta}\left(\frac{3V_D}{\pi\tan\theta}\right)^{1/3}$$
$$- \frac{9}{\sqrt{\pi}}cV_D \tan\theta \tag{39.9}$$

We can compare the energies of QDs with two shapes, pyramid and dome, based on Equations 39.8 and 39.9. Figure 39.10 shows the theoretical results. Clearly, when volume of QD is less than a critical value, the energy change of pyramid QD is lower than that of dome QD, which means QD with pyramid shape is steadier than that with dome shape in this case. However, when volume of QDs is larger than the critical value, the energy change of dome QD becomes lower than that of pyramid QD, which means that QD with dome shape is steadier. Theoretical results show that the shape transition from pyramid to dome can occur with volume of QD increasing. The results agree well with the experimental observations (Ribeiro et al., 1998; Ross et al., 1998; Liu et al., 2000). The physical origin of the shape transition of strained QD from pyramid to dome is actually the balance between surface energy and relaxed energy of QD. The formation of strained QD is driven by the relaxation of strain energy at the stage of the surface energy expense. At the early stage of growth (QD with small volume), the elastic relaxation is not efficient for the shape of QD. The surface energy of dome is larger than that of pyramid. Thus, the equilibrium shape tends to pyramid. At the later stage of growth (QD with large volume), the elastic relaxation becomes more significant. The relaxation energy of dome is larger than that of pyramid with the same

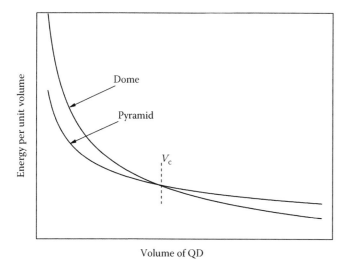

FIGURE 39.10 Energy per unit volume of pure Ge QDs with two typical shapes, pyramid and dome.

volume due to dome having high gradient. Hence, the energy change is lowered by the shape transition to gain the additional elastic relaxation. Although the surface energy of dome is larger than that of pyramid in this case, it is not significant due to large volume of QD. Thus, the QD tends to dome.

In this section, we introduce the shape transitions of QDs, from pre-pyramid to pyramid and pyramid to dome. When the total deposited amount is enough to form large QDs, the steady shape of QDs always is dome. The QDs with dome shape can grow steadily on the expense of the wetting layer. However, the QDs cannot grow without limit under a certain deposited amount. In the next section, we will introduce the final steady state of QDs.

39.4 Final Steady State of QDs

QDs can grow up on the expense of the wetting layer during deposited process. However, because the total deposited amount is finite, QDs cannot grow up without limit. During the growth process of QDs under a certain deposited amount, the thickness of wetting layer decreases with volume of QDs increasing, this leads to the increase of thickness-dependent surface energy of wetting layer. Meanwhile, as the volume of QDs increases, the interaction among QDs also becomes so strong that we cannot ignore the effect of interaction. Because the interaction energy among QDs and thickness-dependent surface energy of wetting layer restrict QDs growth, QDs cannot grow up without limit and have a final steady state. In this section, we mainly introduce the final steady state of QDs.

39.4.1 Interaction Energy among QDs

The elastic interaction energy between two QDs (i and j) with the same shape and volume can be written as (Shchukin et al., 2004)

$$E_{ij} = \frac{1+\upsilon}{1-\upsilon}\frac{1}{\pi}Y\varepsilon^2 \int_{V_D} dV \int_{V_D} dV' \frac{1}{r^3},$$

in which

υ and Y are the Poisson's ratio and the Young's modulus of QDs
r is the distance between dV and dV'

Thus, the elastic interaction energy among all the QDs in the area S is

$$E_{\text{interaction}} = \sum_j \sum_{i\neq j} E_{ij} = \frac{1}{2}k\sum_{j=2}^{k} E_{1j} \tag{39.10}$$

If the distance between the QDs, L, is much larger than the diameter of the QD ($L \gg 2R$), the interaction energy can be approximately written as

$$E_{ij} = \frac{1+\upsilon}{1-\upsilon}\frac{1}{\pi}Y\varepsilon^2 V_D^2 \frac{1}{L^3} \tag{39.11}$$

For the dome QDs, the elastic interaction energy between two QDs can be written as (Shchukin et al., 2004)

$$E_{ij} = \frac{1+\upsilon}{1-\upsilon}\frac{1}{\pi}Y\varepsilon^2 V_D^2 \frac{1}{L^3} F\left(\frac{\rho}{L}\right) \qquad (39.12)$$

where

$$\rho = (3\pi^{-1}\cot\theta V_D)^{1/3}$$

$F(\rho/L)$ is the correction factor which equals

$$F\left(\frac{\rho}{L}\right) = \sum_{s=0}^{\infty}\left\{\sum_{p=0}^{s}\left[\frac{(\rho/L)^s}{\Gamma(p+1)\Gamma(s-p+1)}\right]^2 \right.$$
$$\left. \times \frac{9}{4(p+1)(p+(3/2))(s-p+1)(s-p+(3/2))}\right\}\left[\frac{\Gamma((3/2)+s)}{\Gamma(3/2)}\right]^2 \qquad (39.13)$$

Figure 39.11 shows the calculated results according to Equations 39.10 and 39.11 where the volume of QD is 1000 nm³ and the contact angle is 11.3° in our calculation. Clearly, we can see that the two values of the interaction energy have an obvious difference when the two QDs are close each other. However, the interaction energy becomes almost equal to each other and close to zero with the increasing the distance. Therefore, we can neglect the interaction energy in the case of small QDs, because the distance L is much larger than the size of QD, and using Equation 39.13 for the large dome QDs, respectively. Note that, we neglect the deformation within the wetting layer due to the presence of QDs, because the deformation energy induced by QDs is too low for the case of small QDs, and too weaker than the interaction energy in the case of large QDs due to the small surface area and thinness of wetting layer.

39.4.2 Final Steady State of QDs

For QDs with large volume, the interaction among QDs cannot be neglected, especially when the distance between two QDs is small. Therefore, the change of energy caused by formation off a single QD should include the interaction energy with other surrounding QDs. In the previous section, we have shown that the difference of energies between two growth modes, layer-by-layer growth and QDs formation on wetting layer, can be written as $\Delta E = (1/k)[\gamma(\theta_{WL}) - \gamma(\theta_0)] + t^2\gamma_s/\cos\alpha - t^2\gamma(\theta_{WL}) + \Delta E_{relaxation}$ if the interaction energy among QDs is neglected at the initial growth stage. When we consider the effect of interaction among QDs in the case of large volume of QDs, the difference of energies between two growth modes becomes

$$\Delta E = \frac{1}{k}\left[\gamma(\theta_{WL}) - \gamma(\theta_0)\right] + \frac{t^2\gamma_s}{\cos\alpha} - t^2\gamma(\theta_{WL}) + \Delta E_{relaxation} + E_{interaction} \qquad (39.14)$$

Equation 39.14 has a similar means with Equation 39.3 when volume of QD is small. Through calculations using Equation 39.3, we can get the critical conditions of QDs formation. After considering the interaction among QDs, Equation 39.3 can be instead be replaced by Equation 39.14 to investigate the final steady state of QDs.

When the total deposited amount is a certain value, QDs can grow up on the expense of the wetting layer. The volume of QD has a relation with thickness of wetting layer as $\theta_{WL}h_0/k = \theta_0 h_0/k - V_D$. Obviously, when QD grows up, which lead to increase volume of QD, the wetting layer becomes more and more thin. Because of thickness-dependent surface energy of wetting layer, the increase of surface energy can restrict the QD growth. Simultaneously, interaction energy among QDs becomes more significant, which also restricts the QD growth. Figure 39.12 shows the calculated results based on Equation 39.14 about the final steady state, where the shape of QDs is considered to be dome. We can note that the total energy

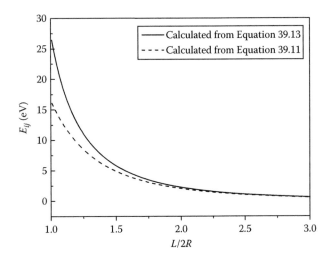

FIGURE 39.11 Interaction energy between two QDs as a function of distance.

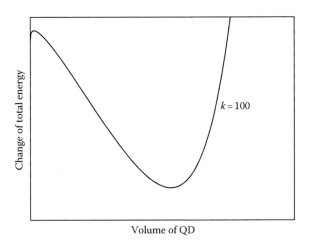

FIGURE 39.12 Total energy change as a function of QD size (domes) in the Ge/Si (001) system under a fixed deposited amount.

increase as the increasing of QD volume when QD volume exceeds a certain value, which means that it is difficult that the QD overruns the certain value, i.e., the final steady volume of QDs is the certain volume.

The physical origin of the limitation of QDs growth under a fixed deposited amount is that the interaction energy becomes significant because of the large volume of QDs and small distance between QDs and the thickness-dependent surface energy of wetting layer is effective when the QDs grow up to larger volume at the expense of the wetting layer. Therefore, the total energy will rise when the volume of QDs exceeds a certain value. There is a steady state with a minimum total energy.

39.5 Summary

Self-assembly of QDs in epitaxial growth is introduced from thermodynamic views. Semiconductor QDs, such as Ge and InAs QDs, can form on substrate surface after depositing a wetting layer. The growth mode which is typical SK growth mode can efficiently release strain caused by lattice mismatch between epitaxial material and substrate. Considering the contribution of thickness-dependent surface energy of wetting layer, at the initial stage, the layer-by-layer growth mode is the most favorable. The growth mode is determined by thickness-dependent surface energy of the wetting layer and can efficiently decrease the surface energy. When the thickness of the wetting layer exceeds a certain value, the rate of decrease of surface energy of the wetting layer becomes low. In this case, the relaxation caused by QDs formation plays a key role in the further growth process, and it is more favorable to QDs formation at the later stage.

During the growth process of QDs, they suffer from shape transition as increasing volume of QDs. The shape transitions are also from QD with low contact angle to QD with high contact angle. The physical origin of the shape transition from low contact angle to high contact angle is the balance between the surface energy and the relaxation energy caused by QDs formation. Surface energy has a significant influence when QDs have a small volume. Tendency to minimum of surface energy leads to the QDs with a low contact angle. However, when QDs have a large volume, the relaxation of QDs play key role in the shape, which require QDs to have a high contact angle.

QDs can grow at the expense of the wetting layer, but QDs cannot grow without limit under a certain total deposited amount. The interaction energy among QDs and the thickness-dependent surface energy of wetting layer restrict the growth of QDs. As QD volume increases, interaction becomes significant and the thickness-dependent surface energy of the wetting layer is effective when the QDs grow to a larger volume at the expense of the wetting layer. There is a steady state of QDs with a minimum total energy.

Acknowledgments

NSFC (50525206 and U0734004) and the Ministry of Education (106126) supported this work.

References

Alivisatos, A. P., 1996, Semiconductor clusters, nanocrystals, and quantum dots. *Science* **271**, 933.

Ando, T., Fowler, A. B., and Stern, F., 1982, Electronic properties of two-dimensional systems. *Rev. Mod. Phys.* **54**, 437.

Badolato, A., Hennessy, K., Atature, M. et al., 2005, Deterministic coupling of single quantum dots to single nanocavity modes. *Science* **308**, 1158.

Bauer, E., 1958, Phenomenological theory of crystal precipitation on surfaces. *Z. Kristallogr.* **110**, 372.

Bester, G. and Zunger, A., 2003, Compositional and size-dependent spectroscopic shifts in charged self-assembled $In_xGa_{1-x}As$/GaAs quantum dots. *Phys. Rev. B* **68**, 073309.

Bimberg, D., Ledentsov, N. N., Grundmann, M. et al., 1996, InAs-GaAs quantum dots: From growth to lasers. *Phys. Status Solidi B* **194**, 159.

Chang, H., Grundbacher, R., Jovanovic, D., Leburton, J.-P., and Adesida, I., 1994, A laterally tunable quantum dot transistor. *Appl. Phys.* **76**, 3209.

Chen, M. and Porod, W., 1995, Design of gate-confined quantum-dot structures in the few-electron regime. *J. Appl. Phys.* **78**, 1050.

Chen, K. M., Jesson, D. E., and Pennycook, S. J., 1997, Critical nuclei shapes in the stress-driven 2D-to-3D transition. *Phys. Rev. B* **56**, R1700.

Denker, U., Schmidt, O. G., Jin-Philipp, N.-Y., and Eberl, K., 2001, Trench formation around and between self-assembled Ge islands on Si. *Appl. Phys. Lett.* **78**, 3723.

Eaglesham, D. J. and Cerullo, M., 1990, Dislocation-free Stranski–Krastanow growth of Ge on Si(100). *Phys. Rev. Lett.* **64**, 1943.

Ebiko, Y., Muto, S., Suzuki, D. et al., 1998, Island size scaling in InAs/GaAs self-assembled quantum dots. *Phys. Rev. Lett.* **80**, 2650.

Ebiko, Y., Muto, S., Suzuki, D. et al., 1999, Scaling properties of InAs/GaAs self-assembled quantum dots. *Phys. Rev. B* **60**, 8234.

Frank, F. C. and van der Merwe, J. H., 1949, One-dimensional dislocations. I. Static theory. *Proc. R. Soc. Lond., Ser. A* **198**, 205.

Grundmann, M., Ed., 2002, *Nano-Optoelectronics: Concepts, Physics, and Devices* (Springer, Berlin, Germany).

Jokerst, J. V., Floriano, P. N., Christodoulides, N., Simmons, G. W., and McDevitt, J. T., 2008, Integration of semiconductor quantum dots into nano-bio-chip systems for enumeration of CD4 + T cell counts at the point-of-need. *Lab Chip*, **8**, 2079.

Joyce, P. B., Krzyzewski, T. J., Bell, G. R., and Jones, T. S., 2001, Surface morphology evolution during the overgrowth of large InAs–GaAs quantum dots. *Appl. Phys. Lett.* **79**, 3615.

Kamins, T. I., Carr, E. C., Williams, R. S., and Rosner, S. J., 1997, Deposition of three-dimensional Ge islands on Si(001) by chemical vapor deposition at atmospheric and reduced pressures. *J. Appl. Phys.* **81**, 211.

Kastner, M. A., 1996, In: *Proceedings of the 23rd International Conference on Physics of Semiconductors*, Berlin, Germany, edited by M. Scheffler and R. Zimmermann (World Scientific, Singapore), Vol. 1, p. 27.

Kirstaedter, N., Ledentsov, N. N., Grundmann, M. et al., 1994, Low threshold, large to injection laser emission from (InGa) As quantum dots. *Electron. Lett.* **30**, 1416.

Kratzert, P. R., Puls, J., Rabe, M., and Henneberger, F., 2001, Growth and magneto-optical properties of sub 10 nm (Cd, Mn)Se quantum dots. *Appl. Phys. Lett.* **79**, 2814.

Krzyzewski, T., Joyce, P., Bell, G., and Jones, T., 2002, Wetting layer evolution in InAs/GaAs(0 0 1) heteroepitaxy. Effects of surface reconstruction and strain. *Surf. Sci.* **517**, 8.

Ledentsov, N. N., Krestnikov, I. L., Maximov, M. V. et al., 1996, Ground state exciton lasing in CdSe submonolayers inserted in a ZnSe matrix. *Appl. Phys. Lett.* **69**, 1343.

Ledentsov, N. N., Krestnikov, I. L., Maximov, M. V. et al., 1997, Response to "Comment on Ground state exciton lasing in CdSe submonolayers inserted in a ZnSe matrix". *Appl. Phys. Lett.* **70**, 2776.

Ledentsov, N. N., Bimberg, D., Ustinov, V. M., Alferov, Z. I., and Lott, J. A., 2002, Quantum dots for VCSEL applications at $\lambda = 1.3\,\mu m$. *Physica E (Amsterdam)* **13**, 871.

Lee, S., Daruka, I., Kim, C. S., Barabasi, A.-L., Merz, J. L., and Furdyna, J. K., 1998, Dynamics of ripening of self-assembled II-VI semiconductor quantum dots. *Phys. Rev. Lett.* **81**, 3479.

Li, X. L. and Yang, G. W., 2008, Theoretical determination of contact angle in quantum dot self-assembly. *Appl. Phys. Lett.* **92**, 171902.

Li, X. L., Ouyang, G., and Yang, G. W., 2007, Thermodynamic theory of nucleation and shape transition of strained quantum dots. *Phys. Rev. B* **75**, 245428.

Liu, F. and Lagally, M. G., 1997, Self-organized nanoscale structures in Si/Ge films. *Surf. Sci.* **386**, 169.

Liu, C. P., Gibson, J. M., Cahill, D. G., Kamins, T. I., Basile, D. P., and Williams, R. S., 2000, Strain evolution in coherent Ge/Si islands. *Phys. Rev. Lett.* **84**, 1958.

Lu, G. H. and Liu, F., 2005, Towards quantitative understanding of formation and stability of Ge hut islands on Si(001). *Phys. Rev. Lett.* **94**, 176103.

Márquez, J., Geelhaar, L., and Jacobi, K., 2001, Atomically resolved structure of InAs quantum dots. *Appl. Phys. Lett.* **78**, 2309.

Migliorato, M. A., Cullis, A. G., Fearn, M., and Jefferson, J. H., 2002, Atomistic simulation of strain relaxation in $In_xGa_{1-x}As$/GaAs quantum dots with nonuniform composition. *Phys. Rev. B* **65**, 115316.

Nakata, Y., Mukai, K., Sugawara, M., Ohtsubo, K., Ishikawa, H., and Yokoyama, N., 2000, Molecular beam epitaxial growth of InAs self-assembled quantum dots with light-emission at 1.3 μm. *J. Cryst. Growth* **208**, 93.

Persson, J., Holm, M., and Pryor, C., 2003, Optical and theoretical investigations of small InP quantum dots in $Ga_xIn_{1-x}P$. *Phys. Rev. B* **67**, 035320.

Poon, T. W., Yip, S., Ho, P. S., and Abraham, F., 1990, Equilibrium structures of Si(100) stepped surfaces. *Phys. Rev. Lett.* **65**, 2161.

Rastelli, A. and Känel, H. von, 2002, Surface evolution of faceted islands. *Surf. Sci.* **515**, L493.

Rastelli, A. and Känel, H. von, 2003, Island formation and faceting in the SiGe/Si (001) system. *Surf. Sci.* **532–535**, 769.

Reithmaier, J. P., Sek, G., Löffler, A. et al., 2004, Strong coupling in a single quantum dot–semiconductor microcavity system. *Nature* **432**, 197.

Ribeiro, G. M., Bratkovski, A. M., Kamins, T. I., Ohlberg, D. A. A., and Williams, R. S., 1998, Shape transition of Germanium nanocrystals on a silicon (001) surface from pyramids to domes. *Science* **279**, 353.

Ross, F. M., Tersoff, J., and Tromp, R. M., 1998, Coarsening of self-assembled Ge quantum dots on Si(001). *Phys. Rev. Lett.* **80**, 984.

Ross, F. M., Tromp, R. M., and Reuter, M. C., 1999, Transition states between pyramids and domes during Ge/Si island growth. *Science* **286**, 1931.

Schmidbauer, M., Hatami, F., Hanke, M., Schaefer, P., Braune, K., Masselink, W. T., Köhler, R., and Ramsteiner, M., 2002, Shape-mediated anisotropic strain in self-assembled InP/ $In_{0.48}Ga_{0.52}P$ quantum dots. *Phys. Rev. B* **65**, 125320.

Seifert, W., Carlsson, N., Miller, M., Pistol, M. E., Samuelson, L., and Reine Wallenberg, L. R., 1996, In-situ growth of quantum dot structures by the Stranski–Krastanow growth mode. *Prog. Cryst. Growth Charact. Mater.* **33**, 423.

Shchukin, V. A. and Bimberg, D., 1999, Spontaneous ordering of nanostructures on crystal surfaces. *Rev. Mod. Phys.* **71**, 1125.

Shchukin, V., Ledentsov, N. N., and Bimberg, D., 2003, *Epitaxy of Nanostructures* (Springer, Berlin, Germany).

Shchukin, V. A., Bimberg, D., Munt, T. P., and Jesson Elastic, D. E., 2004, Interaction and self-relaxation energies of coherently strained conical islands. *Phys. Rev. B* **70**, 085416.

Stranski, I. N. and Krastanow, L., 1937, Zur Theorie der orientierten ausscheidung von ionenkristallen aufeinander. *Sitzungsber. Akad. Wiss. Wien, Math.-Naturwiss. Klasse* **146**, 797.

Strassburg, M., Deniozou, Th., Hoffmann, A. et al., 2000, Coexistence of planar and three-dimensional quantum dots in CdSe/ZnSe structures. *Appl. Phys. Lett.* **76**, 685.

Tersoff, J. and Tromp, R. M., 1993, Shape transition in growth of strained islands: Spontaneous formation of quantum wires. *Phys. Rev. Lett.* **70**, 2782.

Tersoff, J., Spencer, B. J., Rastelli, A., and Känel, H. von, 2002, Barrierless formation and faceting of SiGe islands on Si(001). *Phys. Rev. Lett.* **89**, 196104.

Ustinov, V. M. and Zhukov, A. E., 2002, GaAs-based long-wavelength lasers. *Semicond. Sci. Technol.* **15**, R41.

Ustinov, V. M., Weber, E. R., Ruvimov, S. et al., 1998, Effect of matrix on InAs self-organized quantum dots on InP substrate. *Appl. Phys. Lett.* **72**, 362.

Vailionis, A., Cho, B., Glass, G., Desjardins, P., Cahill D. G., and Greene, J. E., 2000, Pathway for the strain-driven two-dimensional to three-dimensional transition during growth of Ge on Si(001). *Phys. Rev. Lett.* **85**, 3672.

Volmer, M. and Weber, A., 1926, Nucleus formation in supersaturated systems. *Z. Phys. Chem. (Munich)* **119**, 277.

Wasserman, D., Lyon, S. A., Maciel, M. H. A., and Ryan, J. F., 2003, Formation of self-assembled InAs quantum dots on (110) GaAs substrates. *Appl. Phys. Lett.* **83**, 5050.

Xu, X. D., Sun, B., Berman, P. R. et al., 2007, Coherent optical spectroscopy of a strongly driven quantum dot. Science **317**, 929.

Yamaguchi, K., Yujobo, K., and Kaizu, T., 2000, Stranski–Krastanov growth of InAs quantum dots with narrow size distribution. *Jpn J. Appl. Phys.* **39**, L1245.

40

Quantum Teleportation in Quantum Dots System

Hefeng Wang
Purdue University

Sabre Kais
Purdue University

40.1 Introduction

The special quantum features such as superpositions, interference, and entanglement have revolutionized the field of quantum information and quantum computation. Quantum teleportation primarily relies on quantum entanglement, which essentially implies an intriguing property that two quantum correlated systems cannot be considered independent even if they are far apart. The dream of teleportation is to be able to travel by simply reappearing at some distant location. We have seen a familiar scene from science fiction movies: The heroes shimmer out of existence to reappear on the surface of a faraway planet. This is the dream of teleportation—the ability to travel from place to place without having to pass through the tedious intervening miles accompanied by a vehicle or an airplane. Although the teleportation of large objects still remains a fantasy, quantum teleportation has become a laboratory reality for photons, electrons, and atoms.[1–10]

By quantum teleportation an unknown quantum state is destroyed at a sending place while its perfect replica state appears at a remote place via dual quantum and classical channels. Quantum teleportation allows for the transmission of quantum information to a distant location despite the impossibility of measuring or broadcasting the information to be transmitted. The classical teleportation is like a fax in which one could scan an object and send the information so that the object can be reconstructed at the destination. In this conventional facsimile transmission, the original object is scanned to extract partial information about it. The scanned information is then sent to the receiving station, where it is used to produce an approximate copy of the original object. The original object remains intact after the scanning process. By contrast, in quantum teleportation, the uncertainty principle forbids any scanning process from

extracting all the information in a quantum state. The nonlocal property of quantum mechanics enables the striking phenomenon of quantum teleportation. Bennett and coworkers[28] showed that a quantum state can be teleported, provided one does not know that state, using a celebrated and paradoxical feature of quantum mechanics known as the Einstein–Podolsky–Rosen (EPR) effect.[11] They found a way to scan out part of the information from an object A, which one wishes to teleport, while causing the remaining part of the information to pass to an object B, via the EPR effect. In this process, two objects B and C form an entangled pair; object C is taken to the sending station, while object B is taken to the receiving station. At the sending station, object C is scanned together with the original object A, yielding some information and totally disrupting the state of A and C. The scanned information is sent to the receiving station, where it is used to select one of several treatments to be applied to object B, thereby putting B into an exact replica of the former state of A.

Quantum teleportation exploits some of the most basic and unique features of quantum mechanics: the teleportation of a quantum state encompasses the complete transfer of information from one particle to another. The complete specification of a quantum state of a system generally requires an infinite amount of information, even for simple two-level systems (qubits). Moreover, the principles of quantum mechanics dictate that any measurement on a system immediately alters its state, while yielding at most one bit of information. The transfer of a state from one system to another (by performing measurements on the first and operations on the second) might therefore appear impossible. However, it was shown that the property of entanglement in quantum mechanics, in combination with classical communication, can be used to teleport quantum states.

The application of quantum teleportation has been extended beyond the field of quantum communication. On the one hand, quantum teleportation can be implemented using a quantum circuit that is much simpler than that required for any nontrivial quantum computational task: the state of an arbitrary qubit can be teleported using as few as two quantum C-NOT gates. Thus, quantum teleportation is significantly easier to implement than even the simplest quantum computations if we are concerned only with the complexity of the required circuitry. On the other hand, quantum computing is meaningful even if it takes place very quickly and within a small region of space. The interest of quantum teleportation would be greatly reduced if the actual teleportation had to take place immediately after the required preparation. Quantum teleportation across significant time and space has been demonstrated with the technology that allows for the efficient long-term storage and purification of quantum information. The quantum teleportation of short distance will play a role in transporting quantum information inside quantum computers. People have shown that a variety of quantum gates can be created by teleporting qubits through special entangled states.[12,13] This allows the construction of a quantum computer based on just single qubit operations, Bell's measurement, and the GHZ states. A wide variety of fault-tolerant quantum gates have also been constructed. Gottesman and Chuang demonstrated a procedure that performs an inner measurement conditioned on an outer cat state.[12,13]

In quantum systems, interaction in general gives rise to entanglement. In this chapter, the entanglement in quantum dots system and its application for quantum teleportation will be discussed. We do not cover all the work that has been done in the field in this chapter. However, we chose a simple model to illustrate and introduce the subject. We present a model of quantum teleportation protocol based on one-dimensional quantum dots system. Three quantum dots with three electrons are used to perform teleportation: the unknown qubit is encoded using one electron spin on quantum dot A, the other two dots B and C are coupled to form a mixed space-spin entangled state. By choosing the Hamiltonian for the mixed space-spin entangled system, we can filter the space (spin) entanglement to obtain pure spin (space) entanglement, and after a Bell measurement, the unknown qubit is transferred to quantum dot B. Selecting an appropriate Hamiltonian for the quantum gate allows the spin-based information to be transformed into the charge-based information. The possibility of generalizing this model to the N-electron system is discussed. The Hamiltonian to construct the C-NOT gate will also be discussed in detail.

40.2 Entanglement

Ever since the appearance of the famous EPR experiment,[11] the phenomenon of entanglement,[14] which features the essential difference between classical and quantum physics,[15] has received wide theoretical and experimental attention.[15–22] Generally speaking, if two particles are in an entangled state then, even

if the particles are physically separated by a great distance, they behave in some respects as a single entity rather than as two separate entities. There is no doubt that entanglement has been lying in the heart of the foundation of quantum mechanics.[23]

Besides quantum computations, entanglement has also been the core of many other active research such as quantum teleportation,[6,24] dense coding,[25,26] quantum communication,[27] and quantum cryptography.[28] It is believed that the conceptual puzzles posed by entanglement—and discussed more than 50 years ago—have now become a physical source to brew completely novel ideas that might result in useful applications.

A big challenge faced by all the above-mentioned applications is to prepare the entangled states, which is much more subtle than classically correlated states. To prepare an entangled state of good quality is a preliminary condition for any successful experiment. In fact, this is not only a problem involved in experiments, but this also poses an obstacle to theories since the issue of how to quantify entanglement is still unsettled, which is now becoming one of the central topics in quantum information theory. Any function that quantifies entanglement is called an entanglement measure. It should tell us how much entanglement there is in a given multipartite state. Unfortunately there is currently no consensus as to the best method to define entanglement for all possible multipartite states. The theory of entanglement is only partially developed[23,29–32] and can only be applied in a limited number of scenarios, where there is unambiguous way to construct suitable measures. Two important scenarios are (1) the case of a pure state of a bipartite system, that is, a system consisting of only two components and (2) a mixed state of two spin-1/2 particles.

When a bipartite quantum system AB described by $H_A \otimes H_B$ is in a pure state, there is an essentially well-motivated and unique measure of entanglement between the subsystems A and B given by the von Neumann entropy S. If we denote the partial trace of ρ with $\rho_A, \in H_A \otimes H_B$ with respect to subsystem B, $\rho_A = Tr_B(\rho)$, the entropy of entanglement of the state ρ is defined as the von Neumann entropy of the reduced density operator ρ_A, $S(\rho) \equiv -Tr[\rho_A \log_2 \rho_A]$. It is possible to prove that for pure states, the quantity S does not change if we exchange A and B. So we have $S(\rho) \equiv -Tr[\rho_A \log_2 \rho_A] \equiv -Tr[\rho_B \log_2 \rho_B]$. For any bipartite pure state, if the entanglement $E(\rho)$ is said to be good, it is often required to have the following properties: (1) For separable states ρ_{sep}, $E(\rho_{sep}) = 0$. (2) Reversible operations performed on the two subsystems A and B alone do not change the entanglement of the total systems. (3) The most general local operations that one can apply are non-unitary. (4) The last property for a good measure of entanglement is that if we take two bipartite systems in the total state $\rho_t = \rho_1 \otimes \rho_2$, we should have $E(\rho_t) = E(\rho_1) + E(\rho_2)$. It is possible to show that the quantity S has all the above properties. Clearly, S is not the only mathematical object that meets the requirements (1)–(4), but, in fact, it is also accepted as the correct and unique measure of entanglement.

Generally, the strict definitions of the four most prominent entanglement measures can be summarized as follows[33]: (1) entanglement of distillation E_D; (2) entanglement of cost E_C;

(3) entanglement of formation E_F, and finally (4) relative entropy of entanglement E_R. The first two measures are also called operational measures while the second two measures do not admit a direct operational interpretation in terms of entanglement manipulations. It can be proved that if E is a measure defined on mixed states that satisfies the conditions for a good entanglement measure mentioned above, then for all states, $\rho \in (H^A \otimes H^B)$, $E_D(\rho) \leq E(\rho) \leq E_C(\rho)$, and both $E_D(\rho)$ and $E_C(\rho)$ coincide on pure states with the von Neumann reduced entropy as demonstrated above. For the fermion system, we chose to use Zanardi's measure,[34] which is given in Fock space as the von Neumann entropy.

40.3 Quantum Teleportation

Quantum teleportation is an entanglement-assisted teleportation. It is a technique used to transfer information on a quantum level, usually from one particle (or series of particles) to another particle (or series of particles) in another location via quantum entanglement. Its distinguishing feature is that it can transmit the information present in a quantum superposition, which is useful for quantum communication and computation.

More precisely, quantum teleportation is a quantum protocol by which the information on a qubit A (quantum bit, a two-level quantum system) is transmitted exactly (in principle) to another qubit B. This protocol requires a conventional communication channel capable of transmitting two classical bits, and an entangled pair (B, C) of qubits, with C at the origin location with A and B at the destination. The protocol has three steps: measure A and C jointly to yield two classical bits, transmit the two bits to the other end of the channel, and use the two bits to select one of four ways of recovering B.

The two parties are Alice (A) and Bob (B), and a qubit is, in general, a superposition of quantum state $|0\rangle$ and $|1\rangle$. Equivalently, a qubit is a unit vector in a two-dimensional Hilbert space. Suppose Alice has a qubit in some arbitrary quantum state $|\psi\rangle = \alpha|0\rangle + \beta|1\rangle$. Assume that this quantum state is not known to Alice, and she would like to send this state to Bob. A solution to this problem was discovered by Bennett et al.[28] The parts of a maximally entangled two-qubit state are distributed to Alice and Bob. The protocol then involves Alice and Bob interacting locally with the qubits in their possession and Alice sending two classical bits to Bob. In the end, the qubit in Bob's possession will be transformed into the desired state.

Alice and Bob share a pair of entangled qubits BC. That is, Alice has one half, C, and Bob has the other half, B. Let A denote the qubit Alice wishes to transmit to Bob. Alice applies a unitary operation on the qubits AC and measures the result to obtain two classical bits. In this process, the two qubits are destroyed. Bob's qubit, B, now contains information about C; however, the information is somewhat randomized. More specifically, Bob's qubit B is in one of four states uniformly chosen at random, and Bob cannot obtain any information about C from his qubit. Alice provides her two measured qubits that indicate which of the four

states Bob possesses. Bob applies a unitary transformation that depends on the qubits he obtains from Alice, transforming his qubit into an identical copy of the qubit C.

Suppose that the qubit A that Alice wants to teleport to Bob can be generally written as $|\psi\rangle_A = \alpha|0\rangle + \beta|1\rangle$. Alice and Bob share a maximally entangled state beforehand, for instance, one of the four Bell states:

$$
\begin{aligned}
\left|\Phi^+\right\rangle &= \frac{1}{\sqrt{2}}\left(|0\rangle_C \otimes |0\rangle_B + |1\rangle_C \otimes |1\rangle_B\right) \\[4pt]
\left|\Phi^-\right\rangle &= \frac{1}{\sqrt{2}}\left(|0\rangle_C \otimes |0\rangle_B - |1\rangle_C \otimes |1\rangle_B\right) \\[4pt]
\left|\Psi^+\right\rangle &= \frac{1}{\sqrt{2}}\left(|0\rangle_C \otimes |1\rangle_B + |1\rangle_C \otimes |0\rangle_B\right) \\[4pt]
\left|\Psi^-\right\rangle &= \frac{1}{\sqrt{2}}\left(|0\rangle_C \otimes |1\rangle_B - |1\rangle_C \otimes |0\rangle_B\right).
\end{aligned}
\tag{40.1}
$$

Alice takes one of the particles in the pair, and Bob keeps the other one. The subscripts C and B in the entangled state refer to Alice's or Bob's particle. We will assume that Alice and Bob share the entangled state $\left|\Phi^+\right\rangle = (1/\sqrt{2})\left(|0\rangle_C \otimes |0\rangle_B + |1\rangle_C \otimes |1\rangle_B\right)$. So, Alice has two particles (A, the one she wants to teleport, and C, one of the entangled pair), and Bob has one particle, B. In the total system, the state of these three particles is given by

$$
|\psi\rangle_A \frac{1}{\sqrt{2}}\left(|0\rangle_C |0\rangle_B + |1\rangle_C |1\rangle_B\right),
\tag{40.2}
$$

where subscripts A and C are used to denote Alice's system, and subscript B is used to denote Bob's system. This three-particle state can be rewritten in the Bell basis as

$$
\frac{1}{2}\left(\left|\Phi^+\right\rangle\left(\alpha|0\rangle + \beta|1\rangle\right) + \left|\Phi^-\right\rangle\left(\alpha|0\rangle - \beta|1\rangle\right) + \left|\Psi^+\right\rangle\left(\beta|0\rangle + \alpha|1\rangle\right) \right.
$$
$$
\left. + \left|\Psi^-\right\rangle\left(-\beta|0\rangle + \alpha|1\rangle\right)\right).
\tag{40.3}
$$

The teleportation starts when Alice measures her two qubits in the Bell basis. Given the above expression, the results of her measurement is that the three-particle state would collapse to one of the following four states (with equal probability of obtaining each)

$$
\begin{aligned}
&\left|\Phi^+\right\rangle\left(\alpha|0\rangle + \beta|1\rangle\right) \\[4pt]
&\left|\Phi^-\right\rangle\left(\alpha|0\rangle - \beta|1\rangle\right) \\[4pt]
&\left|\Psi^+\right\rangle\left(\beta|0\rangle + \alpha|1\rangle\right) \\[4pt]
&\left|\Psi^-\right\rangle\left(-\beta|0\rangle + \alpha|1\rangle\right).
\end{aligned}
\tag{40.4}
$$

Alice's two particles are now entangled to each other, in one of the four Bell states. The entanglement originally shared between Alice's and Bob's qubits is now broken. Bob's particle takes on one of the four superposition states shown above. Bob's qubit is now in a state that resembles the state to be teleported. The four possible states for Bob's qubit are unitary images of the state to be teleported.

The local measurement done by Alice on the Bell basis gives complete knowledge of the state of the three particles; the result of her Bell measurement tells her which of the four states the system is in. She simply has to send her results to Bob through a classical channel. Two classical bits can communicate which of the four results she obtained.

After Bob receives the message from Alice, he will know which of the four states his particle is in. Using this information, he can rotate the target qubit into the correct state $|\psi\rangle$ by applying the appropriate unitary transformation I, σ_Z, σ_X, or $i\sigma_Y$. Quantum teleportation using pairs of entangled photons[6,35–40] and atoms[8,9] have been demonstrated experimentally. There are also schemes suggesting the use of electrons to perform quantum teleportation.[4,7,41]

40.4 Entanglement in the One-Dimensional Hubbard Model

Quantum dots system is one of the proposals for building a quantum computer.[42,43] With dimensions ranging from a mere 1 nm to as much as 100 nm and consisting of anywhere between 10^3 and 10^6 atoms and electrons, semiconductor quantum dots are often regarded as artificial atoms. Charge carriers in semiconductor quantum dot are confined in all three dimensions, and the confinement can be achieved through electrical gating and/or etching techniques applied to a two-dimensional electron gas. To describe the quantum dots, a simple approximation is to regard each dot as having one valence orbital, the electron occupation could be $|0\rangle$, $|\uparrow\rangle$, $|\downarrow\rangle$ and $|\uparrow\downarrow\rangle$, with other electrons treated as core electrons.[44] The valence electron can tunnel from a given dot to its nearest neighbor obeying the Pauli principle and thereby two dots can be coupled together; this is the electron hopping effect. Another effect needs to be considered is the on-site electron–electron repulsion. A theoretical description of an array of quantum dots can be modeled by the one-dimensional Hubbard Hamiltonian:

$$H = -t \sum_{\langle ij \rangle, \sigma} c_{i\sigma}^\dagger c_{j\sigma} + U \sum_i n_{i\uparrow} n_{i\downarrow}, \qquad (40.5)$$

where

t stands for the electron hopping parameter
U is the Coulomb repulsion parameter for electrons on the same site
i and j are the neighboring site numbers
$c_{i\sigma}^\dagger$ and $c_{j\sigma}$ are the creation and annihilation operators

Entanglement using Zanardi's measure can be formulated as the von Neumann entropy given by

$$E_j = -Tr\left(\rho_j \log_2 \rho_j\right), \qquad (40.6)$$

where the reduced density matrix ρ_j is given by

$$\rho_j = Tr_j\left(|\Psi\rangle\langle\Psi|\right), \qquad (40.7)$$

where

Tr_j denotes the trace over all but the jth site
$|\Psi\rangle$ is the antisymmetric wave function of the fermion system

Hence, E_j actually describes the entanglement of the jth site with the remaining sites.

In the Hubbard model, the electron occupation of each site has four possibilities, there are four possible local states at each site, $|\nu\rangle_j = |0\rangle_j$, $|\uparrow\rangle_j$, $|\downarrow\rangle_j$, $|\uparrow\downarrow\rangle_j$. Since the Hamiltonian is invariant under translation, the local density matrix ρ_j of the jth site is site independent and is given by[45]

$$\rho_j = z|0\rangle\langle 0| + u^+|\uparrow\rangle\langle\uparrow| + u^-|\downarrow\rangle\langle\downarrow| + w|\uparrow\downarrow\rangle\langle\uparrow\downarrow| \quad (40.8)$$

with

$$w = \langle n_{j\uparrow} n_{j\downarrow}\rangle = Tr(n_{j\uparrow} n_{j\downarrow} \rho_j) \qquad (40.9)$$

$$u^+ = \langle n_{j\uparrow}\rangle - w, \quad u^- = \langle n_{j\downarrow}\rangle - w \qquad (40.10)$$

$$z = 1 - u^+ - u^- - w = 1 - \langle n_{j\uparrow}\rangle - \langle n_{j\downarrow}\rangle + w. \qquad (40.11)$$

The Hubbard Hamiltonian can be rescaled to have only one parameter U/t. The entanglement of the jth site with the other sites is given by[45]

$$E_j = -z \, Log_2 \, z - u^+ Log_2 \, u^+ - u^- Log_2 \, u^- - w \, Log_2 \, w. \quad (40.12)$$

For the one-dimensional Hubbard model with half-filled electrons, we have $\langle n_\uparrow\rangle = \langle n_\downarrow\rangle = 1/2$, $u^+ = u^- = 1/2 - w$, and the local entanglement is given by

$$E_j = -2w \log_2 w - 2\left(\frac{1}{2} - w\right)\log_2\left(\frac{1}{2} - w\right). \qquad (40.13)$$

For each site the entanglement is the same. Consider the particle–hole symmetry of the model, we can see that $w(-U) = \frac{1}{2} - w(U)$, so the local entanglement is an even function of U. As shown in Figure 40.1, the minimum of the entanglement is 1 as $U \to \pm\infty$. As $U \to +\infty$, all the sites are singly occupied the only difference is the spin of the electrons on each site, which can be referred as the spin entanglement. As $U \to -\infty$, all the sites are either doubly occupied or empty, which is referred as the space entanglement. The maximum entanglement is 2 at $U = 0$, which is the sum of

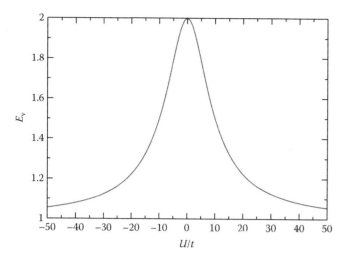

FIGURE 40.1 Local entanglement given by the von Neumann entropy E_v versus U/t for two sites two electrons.

the spin and space entanglement of the system. In Figure 40.1, we show the entanglement for two sites and two electrons, they qualitatively agree with that of the Bethe ansatz solution for an array of sites.[45]

40.5 Quantum Teleportation in Quantum Dots

Gittings and Fisher[46] showed that the entanglement in this system can be used in quantum teleportation. However, in their scheme both the charge and spin of the system are used to construct the unitary transformation. Here, we describe a different scheme to perform quantum teleportation. For two half-filled coupled quantum dots, under the conservation of the total number of electrons $N = 2$ and the total electron spin $S = 0$, a quantum entanglement of 2, two ebits (if each of the entangled particles is used to encode a qubit, the entangled joint states is called an ebit or entangled bit. Ebits are "shared resources" that require both particles) can be produced according to Zanardi's measure. Let us describe the teleportation scheme using three sites, A, B, and C. Suppose the qubit $|\Psi\rangle = \alpha|\uparrow\rangle + \beta|\downarrow\rangle$ will be teleported from site A (Alice), to site B (Bob), where the two sites B and C are in an entangled state:

$$|\Psi_{CB}\rangle = \frac{1}{\sqrt{2}}\left(c_{C\uparrow}^{\dagger} + c_{B\uparrow}^{\dagger}\right)\frac{1}{\sqrt{2}}\left(c_{C\downarrow}^{\dagger} + c_{B\downarrow}^{\dagger}\right)|0\rangle. \quad (40.14)$$

A spin-up electron and a spin-down electron are in a delocalized state on sites C and B. In the occupation number basis $|n_{C\uparrow} n_{C\downarrow} n_{B\uparrow} n_{B\downarrow}\rangle$, the state of the system can be written as

$$|\Psi_{CB}\rangle = \frac{1}{\sqrt{2}}\left(c_{C\uparrow}^{\dagger} + c_{B\uparrow}^{\dagger}\right)\frac{1}{\sqrt{2}}\left(c_{C\downarrow}^{\dagger} + c_{B\downarrow}^{\dagger}\right)|0\rangle\frac{1}{2}$$
$$\times\left(|0011\rangle + |1100\rangle + |1001\rangle + |0110\rangle\right). \quad (40.15)$$

From the state described by Equation 40.10, we can see that in the basis of $|n_{C\uparrow} n_{C\downarrow}\rangle$, there are four possible states: $|00\rangle$, $|11\rangle$, $|10\rangle$, $|01\rangle$. Corresponding to each of the states on site C, the states on site B are $|11\rangle$, $|00\rangle$, $|01\rangle$, $|10\rangle$ in the occupation number basis $|n_{B\uparrow} n_{B\downarrow}\rangle$. Under the restriction of the conservation of total number of electrons and total spin of the system, two ebits can be obtained, one is in the spatial degree of freedom, and the other is in the spin degree of freedom. In the basis of $|n_{C\uparrow} n_{C\downarrow} n_{B\uparrow} n_{B\downarrow}\rangle$, the two ebits are

$$\beta_0 = \frac{1}{\sqrt{2}}\left(|1100\rangle + |0011\rangle\right), \quad \beta_1 = \frac{1}{\sqrt{2}}\left(|1001\rangle + |0110\rangle\right). \quad (40.16)$$

These two ebits can be used in quantum teleportation. The C-NOT operation in the occupation number basis $|n_{A\uparrow} n_{A\downarrow} n_{C\uparrow} n_{C\downarrow}\rangle$ is given by

$$|1000\rangle \leftrightarrow |1011\rangle, |1010\rangle \leftrightarrow |1001\rangle, |01n_{C\uparrow} n_{C\downarrow}\rangle \leftrightarrow |01n_{C\uparrow} n_{C\downarrow}\rangle. \quad (40.17)$$

For the ebit β_0, in the quantum teleportation process, in basis $|n_{A\uparrow} n_{A\downarrow}\rangle |n_{C\uparrow} n_{C\downarrow} n_{B\uparrow} n_{B\downarrow}\rangle$, as shown in Figure 40.2, we have the initial state in the quantum dots:

$$|\Psi_0\rangle = \left(\alpha|10\rangle + \beta|01\rangle\right)\frac{1}{2}\left(|1100\rangle + |0011\rangle + |1001\rangle + |0110\rangle\right). \quad (40.18)$$

Alice performs the C-NOT operation on the two qubits she holds, using the source qubit as a control qubit and the half EPR qubit as target qubit:

$$|\Psi_1\rangle = \alpha|10\rangle\frac{1}{\sqrt{2}}\left(|0000\rangle + |1111\rangle\right) + \beta|01\rangle\frac{1}{\sqrt{2}}\left(|1100\rangle + |0011\rangle\right) \quad (40.19)$$

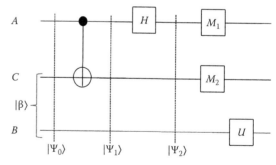

FIGURE 40.2 Quantum circuit for teleporting a qubit. The two top lines represent Alice's system, while the bottom line is Bob's system. β is an entangled pair of qubits Alice and Bob share. H represents a Hadamard transformation, M_1 and M_2 represent the measurement on the two top lines. U represents a unitary operation that Bob performs to rotate his qubit to the state Alice teleport. $|\Psi_0\rangle$ is the initial state for the whole system, $|\Psi_1\rangle$ is the state after Alice performs the C-NOT operation, and $|\Psi_2\rangle$ is the state after Alice performs the Hadamard operation on the initial qubit she holds. The outcome is the teleported state that Bob will get after performing a unitary operation according to the result of the measurement Alice made.

she performs the Hadamard operation on the initial qubit:

$$|\Psi_2\rangle = \alpha\left(|10\rangle + |01\rangle\right)\frac{1}{2}\left(|0000\rangle + |1111\rangle\right)$$

$$+ \beta\left(|10\rangle - |01\rangle\right)\frac{1}{2}\left(|1100\rangle + |0011\rangle\right). \qquad (40.20)$$

After these operations, Alice does the measurement M_1 and M_2 on the two qubits she holds, the following results will be obtained:

$$
\begin{array}{ll}
|M_1 M_2\rangle & |n_{B\uparrow} n_{B\downarrow}\rangle \\
|1011\rangle & \alpha|11\rangle + \beta|00\rangle \\
|1000\rangle & \alpha|00\rangle + \beta|11\rangle \qquad (40.21) \\
|0111\rangle & \alpha|11\rangle - \beta|00\rangle \\
|0100\rangle & \alpha|00\rangle - \beta|11\rangle.
\end{array}
$$

Then, after performing a unitary transformation using double-electron occupation and zero-electron occupation as basis, the source qubit can be obtained on site B. For this system, the Hamiltonian to perform the C-NOT operation is given by

$$H_{\text{C-NOT}} = |10\rangle_{AA}\langle 10|\left(|11\rangle_{CC}\langle 00| + |00\rangle_{CC}\langle 11|\right)$$

$$+ |01\rangle_{AA}\langle 01|\left(|11\rangle_{CC}\langle 11| + |00\rangle_{CC}\langle 00|\right)$$

$$= \frac{1}{2}\left(\sigma_Z^A + 1\right)\left(c_{C\uparrow}^\dagger c_{C\downarrow}^\dagger + c_{C\uparrow} c_{C\downarrow}\right)$$

$$+ \frac{1}{2}\left(1 - \sigma_Z^A\right)\left(c_{C\uparrow}^\dagger c_{C\downarrow}^\dagger c_{C\uparrow} c_{C\downarrow} + c_{C\uparrow} c_{C\downarrow} c_{C\uparrow}^\dagger c_{C\downarrow}^\dagger\right), \quad (40.22)$$

where σ_Z^A is the Pauli matrix. We can see that by using this Hamiltonian, the spin entanglement of the system is filtered, the space entanglement is used in the teleportation process. An important result is that the original state we want to teleport is in a superposition state of spin-up and spin-down electrons. However, after the teleportation process, the state we obtained on site B is a superposition state of double-electron occupation and zero-electron occupation. The information based on spin has been transformed to information based on charge, but the information content is not changed. It is well known that a difficult task in quantum information processing and spintronics is the measurement of a single electron spin;[47] in the scheme above, we changed the quantum information from spin based to charge based, thus makes the measurement fairly easier. This is also important in quantum computation based on electron spin since the readout can be easily measured.

The Hamiltonian for the C-NOT operation can be realized by constructing pulse sequences using the tools of geometric algebra. The tools of geometric algebra provide a useful means of constructing pulse sequences for quantum logic operations.[48]

This method is based on the use of primitive idempotents. The primitive idempotents, $E_\pm$, satisfy the following properties:

$$E_+ + E_- = 1, \quad (E_\pm)^2 = E_\pm, \quad E_+ E_- = 0. \qquad (40.23)$$

These idempotents can help simplify exponential operations as follows:

$$e^{A \cdot E_\pm} = e^A E_\pm + E_\mp, \quad (\text{if } [A, E_\pm] = 0). \qquad (40.24)$$

For spin-$\frac{1}{2}$ particles, the idempotents of interest are

$$E_\pm^i = \frac{1}{2}\left(1 \pm \sigma_Z^i\right), \quad E_\pm^{i,j} = \frac{1}{2}\left(1 \pm \sigma_Z^i \sigma_Z^j\right), \qquad (40.25)$$

where

   σ's are the Pauli matrices

   E_+^A is thus the density matrix for the A spin in the up state

   E_-^B is the density matrix for the B spin in the down state

Such operators have been useful in other NMR quantum computing experiments.[49]

$$E_+ = |0\rangle\langle 0|\begin{pmatrix} 1 & 0 \\ 0 & 0 \end{pmatrix} \qquad (40.26)$$

$$E_- = |1\rangle\langle 1|\begin{pmatrix} 0 & 0 \\ 0 & 1 \end{pmatrix} \qquad (40.27)$$

$$\sigma_x E_+ = |1\rangle\langle 0|\begin{pmatrix} 0 & 0 \\ 1 & 0 \end{pmatrix}, \quad \sigma_x E_- = |0\rangle\langle 1|\begin{pmatrix} 0 & 1 \\ 0 & 0 \end{pmatrix}. \qquad (40.28)$$

Using the definitions of E_+, E_-, and σ_x, the Hamiltonian for C-NOT gate can be rewritten in a simpler form. In this part, we transform the state representation from Fock space to the standard quantum computing representation: $|10\rangle = |\uparrow\rangle = |0\rangle$, $|01\rangle = |\downarrow\rangle = |1\rangle$. In the entangled pair, we define $|11\rangle_C = |\uparrow\downarrow\rangle = |0\rangle$ and $|00\rangle_C = |\emptyset\rangle = |1\rangle$. Then the Hamiltonian can be written as

$$H_{\text{C-NOT}} = E_+^A\left(\sigma_X^C E_-^C + \sigma_X^C E_+^C\right) + E_-^A\left(E_-^C + E_+^C\right) = E_+^A \sigma_X^C + E_-^A. \qquad (40.29)$$

The physical interpretation of the above equation is an instruction to perform the σ_X operation on site C if site A is spin up and to perform the identity operation if the state on site A is spin down.

The expression of the problem in terms of idempotents also makes the generation of the pulse sequence quite straightforward. The propagator for the C-NOT can be factorized into elements that can be physically applied. This is accomplished by first rewriting the propagator as

$$H_{\text{C-NOT}} = E_-^A + E_+^A \sigma_X^C = E_-^A + (i)(-i)E_+^A \sigma_X^C, \qquad (40.30)$$

which can be factorized into

$$H_{\text{C-NOT}} = \left(E_-^A - iE_+^A \sigma_X^C\right)\left(E_-^A + iE_+^A\right). \qquad (40.31)$$

Using the fact that the idempotents can be expressed as exponentials, the above expression becomes

$$H_{\text{C-NOT}} = e^{-iE_+^A \sigma_X^C \pi/2} \cdot e^{iE_+^A \pi/2} \tag{40.32}$$

This expression can be expressed as

$$H_{\text{C-NOT}} = e^{i\pi/4} \cdot e^{-i\sigma_X^C \pi/4} \cdot e^{-i\sigma_Z^A \pi/4} \cdot e^{i\sigma_Z^A \sigma_X^C \pi/4} \tag{40.33}$$

This is an exact expression for the propagator, and is also the pulse sequence for its implementation. Note here the basis for σ_X^C is different from the basis for σ_X^A, the basis for the former is double and zero occupation of site C, and the basis for the latter is spin-up and spin-down states, so in the operation σ_X^C will transform between state $|11\rangle$ and $|00\rangle$ in Fock space.

For another ebit β_1, in the quantum teleportation process, in basis $|n_{A\uparrow} n_{A\downarrow}\rangle |n_{C\uparrow} n_{C\downarrow} n_{B\uparrow} n_{B\downarrow}\rangle$, we have

$$|\Psi_0\rangle = \left(\alpha|10\rangle + \beta|01\rangle\right) \frac{1}{2}\left(|1100\rangle + |0011\rangle + |1001\rangle + |0110\rangle\right) \tag{40.34}$$

$$|\Psi_1\rangle = \alpha|10\rangle \frac{1}{\sqrt{2}}\left(|0101\rangle + |1010\rangle\right) + \beta|01\rangle \frac{1}{\sqrt{2}}\left(|1001\rangle + |0110\rangle\right) \tag{40.35}$$

$$|\Psi_2\rangle = \alpha\left(|10\rangle + |01\rangle\right) \frac{1}{2}\left(|0101\rangle + |1010\rangle\right)$$
$$+ \beta\left(|10\rangle - |01\rangle\right) \frac{1}{2}\left(|1001\rangle + |0110\rangle\right) \tag{40.36}$$

When Alice does the measurement M_1 and M_2, the following results will be obtained:

$$\begin{array}{cc} |M_1 M_2\rangle & |n_{B\uparrow} n_{B\downarrow}\rangle \\ |1001\rangle & \alpha|01\rangle + \beta|10\rangle \\ |1010\rangle & \alpha|10\rangle + \beta|01\rangle \\ |0101\rangle & \alpha|01\rangle - \beta|10\rangle \\ |0110\rangle & \alpha|10\rangle - \beta|01\rangle \end{array} \tag{40.37}$$

For this system, the Hamiltonian to perform the C-NOT operation is

$$H_{\text{C-NOT}} = |10\rangle_{AA}\langle 10|\left(|10\rangle_{CC}\langle 01| + |01\rangle_{CC}\langle 10|\right)$$
$$+ |01\rangle_{AA}\langle 01|\left(|01\rangle_{CC}\langle 01| + |10\rangle_{CC}\langle 10|\right)$$
$$= \frac{1}{2}\left(\sigma_Z^A + 1\right)\left(c_{C\uparrow}^\dagger c_{C\downarrow} + c_{C\downarrow}^\dagger c_{C\uparrow}\right)$$
$$+ \frac{1}{2}\left(1 - \sigma_Z^A\right)\left(c_{C\uparrow}^\dagger c_{C\uparrow} c_{C\downarrow}^\dagger c_{C\downarrow} + c_{C\downarrow}^\dagger c_{C\downarrow} c_{C\uparrow}^\dagger c_{C\uparrow}\right) \tag{40.38}$$

Then, after doing a unitary transformation using the electron spin up and spin down as basis, the source qubit can be recovered on site B. By using this Hamiltonian for the C-NOT operation, the space entanglement of the system is filtered, the spin entanglement is used in the process.

Using the geometric techniques of idempotents, the Hamiltonian for the C-NOT gate can be written in a simpler form. Here, we transform the representation of the qubit state from Fock space to standard quantum computing state: $|10\rangle = |\uparrow\rangle = |0\rangle$, $|01\rangle = |\downarrow\rangle = |1\rangle$. In the entangled pair, we define $|10\rangle_C = |\uparrow\rangle = |0\rangle$ and $|01\rangle_C = |\downarrow\rangle = |1\rangle$. Then the Hamiltonian can be rewritten as

$$H_{\text{C-NOT}} = E_+^A\left(\sigma_X^C E_-^C + \sigma_X^C E_+^C\right) + E_-^A\left(E_-^C + E_+^C\right) = E_-^A + E_+^A \sigma_X^C \tag{40.39}$$

The physical interpretation of the above equation is an instruction to perform the σ_X operation of site C if site A is spin up and to do the identity operation if the site A is spin down.

The propagator for the C-NOT operation can be constructed as follows, first rewriting the propagator as

$$H_{\text{C-NOT}} = E_-^A + E_+^A \sigma_X^C = E_-^A + (i)(-i)E_+^A \sigma_X^C \tag{40.40}$$

which can be factorized into

$$H_{\text{C-NOT}} = \left(E_-^A - iE_+^A \sigma_X^C\right)\left(E_-^A + iE_+^A\right) \tag{40.41}$$

Using the fact that the idempotents can be expressed as exponentials, the above expression becomes

$$H_{\text{C-NOT}} = e^{-iE_+^A \sigma_X^C \pi/2} \cdot e^{iE_+^A \pi/2} \tag{40.42}$$

This expression can be expressed as

$$H_{\text{C-NOT}} = e^{i\pi/4} \cdot e^{-i\sigma_X^C \pi/4} \cdot e^{-i\sigma_Z^A \pi/4} \cdot e^{i\sigma_Z^A \sigma_X^C \pi/4} \tag{40.43}$$

This is an exact expression for the propagator and is also a pulse sequence for its implementation. Note here the basis for σ_X^C is the same as for σ_X^A, the electron spin-up and spin-down states.

For $U \neq 0$, the state of the two-electron two-sites system can be described as follows:

$$|\Psi\rangle = a_1|1100\rangle + a_2|0011\rangle + b_1|1001\rangle + b_2|0110\rangle;$$
$$a_1^2 + a_2^2 + b_1^2 + b_2^2 = 1, \tag{40.44}$$

where $a_1 = a_2$, $b_1 = b_2$ because of the symmetry in the entangled pairs, such that the state can be written as

$$|\Psi\rangle = a\beta_0 + b\beta_1; \quad a^2 + b^2 = 1. \tag{40.45}$$

From the above analysis, we can see that in the case of using β_0 or β_1 as ebits, the unitary transformation is performed in the occupation number basis of $|n_{B\uparrow}\ n_{B\downarrow}\rangle$, using basis $|11\rangle$, $|00\rangle$ or $|10\rangle$, $|01\rangle$. We can select the basis separately, either charge or spin. We can also choose the Hamiltonian (one is related to the spin entanglement and the other is related to space entanglement) for the C-NOT operation, when the Hamiltonian for one ebit is chosen, the ebit corresponding to the other Hamiltonian is filtered.

If $U > 0$, the contribution of the spin entanglement to the total entanglement is greater than that of the space entanglement. The probability of getting the ebit $|\beta_1\rangle$ increases as U becomes larger. If $U < 0$, the contribution of the space entanglement to the total entanglement becomes greater than that of the spin entanglement, the probability of getting the ebit $|\beta_0\rangle$ increases as U becomes more negative. In the limit of U goes to $\pm\infty$, only spin entanglement or space entanglement will exist. This might be related to the spin charge separation in the Hubbard model.[50] In a previous study,[51] we showed that the maximum entanglement can be reached at $U > 0$ by introducing asymmetric electron hopping impurity to the system. This is very convenient in the quantum information processing. We can control the parameter U/t to increase the probability of getting either ebit.

40.6 Summary

We have proposed two schemes for the teleportation of a single qubit in quantum dots system modeled by the one-dimensional Hubbard Hamiltonian; two ebits are contained in the system and can be used in the teleportation process. Now we analyze the theoretical fidelity of these two teleportation schemes. The fidelity of teleportation is defined as the projection of the teleported state $|\psi'\rangle$ on site C to the initial state $|\psi\rangle = \alpha|0\rangle + \beta|1\rangle$ on site A, $|\langle\psi|\psi'\rangle|^2$. If Alice can distinguish all four possible measurement outcomes, the teleportation process can, in principle, be completed with a 100% success rate and is deterministic. If Alice, on the other hand, is only able to perform a partial measurement on her two particles, the success probability is less than 100% and the teleportation is probabilistic. In the first scheme, when the space entangled ebit is used, Alice does the measurement in charge basis. She can only distinguish on site C, whether it is doubly charged or has no charge. As a result, she can only distinguish two measurement results; thus, the fidelity of this scheme is 50%. In the second scheme, by using the spin entanglement, Alice does the measurement in spin basis, all four measurement results can be distinguished, thus the fidelity is 100%.

We discussed implementing quantum teleportation in three-electron system. For more electrons and in the limit of $U \to +\infty$, there is no double occupation, the system is reduced to the Heisenberg model, in the magnetic field. The neighboring spins will favor the antiparallel configuration for the ground state. If the spin at one end is flipped, then the spins on the whole chain will be flipped accordingly due to the spin–spin correlation, such that the spins at the two ends of the chain are entangled, a spin entanglement; this can be used for quantum teleportation and the information can be transferred through the chain. For $U \neq +\infty$, for the N-sites N-electron system with $S = 0$, the first

$N - 1$ sites entangled with the Nth site in the same way as that of the two-electron two-sites system: if the Nth site has 2 electrons, then the first $N - 1$ sites will have $N - 2$ electrons; if the Nth site has 0 electrons, then the first $N - 1$ sites will have N electrons; if the Nth site has 1 spin-up electron, then the total spin of the first $N - 1$ sites will be 1 spin down; if the Nth site has the 1 spin-down electron, then the total spin of the first $N - 1$ sites will be 1 spin up. So the same procedure discussed above can be used for quantum teleportation, but the new system with N-electrons is much more complicated than the previous three electron system. Moreover, Alice needs to control the first $N - 1$ sites and the source qubit. This situation is different from the spin chain. The correlation cannot be transferred from one end to the other.

We have studied the entanglement of an array of quantum dots modeled by the one-dimensional Hubbard Hamiltonian and its application in quantum teleportation. The entanglement in this system is a mixture of space and spin entanglement. The application of such an entanglement in quantum teleportation process has been discussed. By applying different Hamiltonian for the C-NOT operation, we can separate the ebit based on space entanglement or spin entanglement and apply it in quantum teleportation process. It turns out that if we use the ebit of the space entanglement, we can transform the spin-based quantum information to the charge-based quantum information, making the measurement fairly easy.

Efficient long-distance quantum teleportation is crucial for quantum communication and quantum networking schemes. Ursin[52] et al. have performed a high-fidelity teleportation of photons over a distance of 600 m across the River Danube in Vienna, with the optimal efficiency that can be achieved using linear optics. Another exciting experiment in quantum communication has also been performed by Ursin et al.[53,54] One photon is measured locally at the Canary Island of La Palma, whereas the other is sent over an optical free-space link to Tenerife, where the Optical Ground Station of the European Space Agency acts as the receiver. This exceeds previous free-space experiments by more than an order of magnitude in distance, and is an essential step toward future satellite-based quantum communication. Recently, decoy-state quantum cryptography over a distance of 144 km between two Canary Islands was demonstrated successfully. Such experiments also open up the possibility of quantum communication on a large scale using satellites.

The teleportation of single qubits is insufficient for a large-scale realization of quantum communication and quantum computation. Many scientists have developed and exploited teleportation of two-qubit composite system using a six-photon interferometer.[55] In this experiment, a six-photon interferometer has been exploited to teleport an arbitrary polarization state of two photons. The observed teleportation fidelities for different initial states are all well beyond the state estimation limit of 0.40. Not only does a six-photon interferometer provide an important step toward the teleportation of a complex system, but it will also enable future experimental investigations on a number of fundamental quantum communication and computation protocols.

References

1. R. L. de Visser and M. Blaauboer. Deterministic teleportation of electrons in a quantum dot nanostructure. *Phys. Rev. Lett.*, 96(24):246801, 2006.
2. J.-W. Pan, D. Bouwmeester, M. Daniell, H. Weinfurter, and A. Zeilinger. Experimental test of quantum nonlocality in three-photon Greenberger-Horne-Zeilinger entanglement. *Nature*, 403(3):515, 2000.
3. P. Chen, C. Piermarocchi, and L. J. Sham. Control of exciton dynamics in nanodots for quantum operations. *Phys. Rev. Lett.*, 87(6):067401, 2001.
4. F. de Pasquale, G. Giorgi, and S. Paganelli. Teleportation on a quantum dot array. *Phys. Rev. Lett.*, 93(12):12052, 2004.
5. J. H. Reina and N. F. Johnson. Quantum teleportation in a solid-state system. *Phys. Rev. A*, 63(1):012303, 2000.
6. D. Bouwmeester, J. Pan, K. Mattle, M. Eibl, H. Weinfurter, and A. Zeilinger. Experimental quantum teleportation. *Nature*, 390(11):575, 1997.
7. C. W. J. Beenakker and M. Kindermann. Quantum teleportation by particle-hole annihilation in the Fermi sea. *Phys. Rev. Lett.*, 92(5):056801, 2004.
8. M. D. Barrett, J. Chiaverinl, T. Schaetz, J. Britton, W. M. Itano, J. D. Jost, E. Knill et al. Deterministic quantum teleportation of atomic qubits. *Nature*, 429(17):737, 2004.
9. M. Riebe, H. Häffner, C. F. Roos, W. Hänsel, J. Benhelm, G. P. T. Lancaster, T. W. Körber et al. Deterministic quantum teleportation with atoms. *Nature*, 429(17):734, 2004.
10. H. Wang and S. Kais. Quantum teleportation in one-dimensional quantum dots system. *Chem. Phys. Lett.*, 421:338, 2006.
11. A. Einstein, B. Podolsky, and N. Rosen. Can quantum-mechanical description of physical reality be considered complete? *Phys. Rev.* 47:777, 1935.
12. G. Brassard, S. L. Braunstein, and R. Cleve. Teleportation as a quantum computation. *Phys. D*, 120(1):43, 1998.
13. D. Gottesman and I. Chuang. Demonstrating the viability of universal quantum computation using teleportation and single-qubit operations. *Nature*, 402:390–393, 1999.
14. E. Schrödinger. Discussion of probability relations between separated systems. *Proc. Camb. Philos. Soc.*, 31:555, 1935.
15. J. S. Bell. On the Einstein-Podolsky-Rosen paradox. *Physics*, 1:195–200, 1964.
16. C. H. Bennett, D. P. DiVincenzo, J. A. Smolin, and W. K. Wootters. Mixed-state entanglement and quantum error correction. *Phys. Rev. A*, 54:3824, 1996.
17. E. Hagley, X. Maytre, G. Nogues, C. Wunderlich, M. Brune, J. M. Raimond, and S. Haroche. Generation of Einstein-Podolsky-Rosen pairs of atoms. *Phys. Rev. Lett.*, 79:1, 1997.
18. Q. A. Turchette, C. S. Wood, B. E. King, C. J. Myatt, D. Leibfried, W. M. Itano, C. Monroe, and D. J. Wineland. Deterministic entanglement of two trapped ions. *Phys. Rev. Lett.*, 81:3631, 1998.
19. D. Bouwmeester, J.-W. Pan, M. Daniell, H. Weinfurter, and A. Zeilinger. Observation of three-photon Greenberger-Horne-Zeilinger entanglement. *Phys. Rev. Lett.*, 82:1345–1349, 1999.
20. C. Monroe, D. M. Meekhof, B. E. King, and D. J. Wineland. A Schrödinger cat superposition state of an atom. *Science*, 272:1131, 1996.
21. C. A. Sackett, D. Kielpinksi, B. E. King, C. Langer, V. Meyer, C. J. Myatt, M. Rowe et al. Experimental entanglement of four particles. *Nature*, 404:256–259, 2000.
22. A. Peres. *Quantum Theory: Concepts and Methods*. Kluwer Academic Publishers, Boston, MA, 1995.
23. G. Vidal, W. Dur, and J. I. Cirac. Entanglement cost of bipartite mixed states. *Phys. Rev. Lett.*, 89:027901, 2002.
24. D. Bouwmeester, K. Mattle, J.-W. Pan, H. Weinfurter, A. Zeilinger, and M. Zukowski. Experimental quantum teleportation of arbitrary quantum states. *Appl. Phys.*, 67:749, 1998.
25. C. H. Bennett and S. J. Wiesner. Communication via one- and two-particle operators on Einstein-Podolsky-Rosen states. *Phys. Rev. Lett.*, 69:2881, 1992.
26. K. Mattle, H. Weinfurter, P. G. Kwiat, and A. Zeilinger. *Phys. Rev. Lett.*, 76:4546, 1996.
27. B. Schumacher. Quantum coding. *Phys. Rev. A*, 51:2738, 1995.
28. C. H. Bennett, G. Brassard, C. Crepeau, R. Jozsa, A. Peres, and W. K. Wootters. Teleporting an unknown quantum state via dual classical and Einstein-Podolsky-Rosen channels. *Phys. Rev. Lett.*, 70:1895, 1993.
29. M. Horodecki, P. Horodecki, and R. Horodecki. Asymptotic manipulations of entanglement can exhibit genuine irreversibility. *Phys. Rev. Lett.*, 86:5844–5844, 2001.
30. M. Horodecki, P. Horodecki, and R. Horodecki. Separability of n-particle mixed states: Necessary and sufficient conditions in terms of linear maps. *Phys. Lett. A*, 283:1–7, 2001.
31. W. K. Wooters. Parallel transport in an entangled ring. *J. Math. Phys.*, 43:4307–4325, 2002.
32. M. A. Nielsen, C. M. Dawson, J. L. Dodd, A. Gilchrist, D. Mortimer, T. J. Osborne, M. J. Bremner, A. W. Harrow, and A. Hines. Quantum dynamics as a physical resource. *Phys. Rev. A*, 67:052301, 2003.
33. S. Kais. Reduced-Density-matrix mechanics with applications to many-electron atoms and molecules, *Advances in Chemical Physics*, Vol. 134, pp. 493, Edited by D. A. Mazziotti, Wiley, New York, 2007.
34. P. Zanardi. Quantum entanglement in fermionic lattices. *Phys. Rev. A*, 65:042101, 2002.
35. M. A. Nielson, E. Knill, and R. Laflamme. Complete quantum teleportation using nuclear magnetic resonance. *Nature*, 396:52, 1998.
36. D. Boschi, S. Branca, F. DeMartini, L. Hardy, and S. Popescu. Experimental realization of teleporting an unknown pure quantum state via dual classical and Einstein–Podolsky–Rosen channels. *Phys. Rev. Lett.*, 80:1121, 1998.

37. J. Pan, M. Daniell, S. Gasparoni, G. Weihs, and A. Zeilinger. Experimental demonstration of four-photon entanglement and high-fidelity teleportation. *Phys. Rev. Lett.*, 86:4435, 2001.

38. I. Marcikic, H. Riedmatten, W. Tittel, H. Zbinden, and N. Gisin. Long-distance teleportation of qubits at telecommunication wavelengths. *Nature*, 421:509, 2003.

39. D. Fattal, E. Diamanti, K. Inoue, and Y. Yamamoto. Quantum teleportation with a quantum dot single photon source. *Phys. Rev. Lett.*, 92:037904, 2004.

40. A. Furusawa and J. Sorensen. Unconditional quantum teleportation. *Science*, 282:706, 1998.

41. O. Sauret, D. Feinberg, and T. Martin. Electron teleportation with quantum dot arrays. *Eur. Phys. J. B*, 32:545, 2003.

42. D. Loss and D. P. DiVincenzo. Quantum computation with quantum dots. *Phys. Rev. A*, 57:120, 1998.

43. M. Friesen, M. P. Rugheimer, D. Savage, M. Lagally, D. van der Weide, and R. Joynt. Practical design and simulation of silicon-based quantum-dot qubits. *Phys. Rev. B*, 67:121301, 2003.

44. F. Remacle and R. D. Levine. From the cover: Architecture with designer atoms: Simple theoretical considerations. *PNAS*, 97:553, 2000.

45. S. Gu, S. Deng, Y. Li, and H. Lin. Entanglement and quantum phase transition in the extended Hubbard model. *Phys. Rev. Lett.*, 93:086402, 2004.

46. J. R. Gittings and A. J. Fisher. Describing mixed spin-space entanglement of pure states of indistinguishable particles using an occupation-number basis. *Phys. Rev. A*, 66:032305, 2002.

47. R. Ionicioiu and A. E. Popescu. Single spin measurement using spin-orbital entanglement. arXiv:quant-ph/0310047 v2, 2005.

48. M. D. Price, S. S. Somaroo, C. H. Tseng, J. C. Gore, A. F. Fahmy, T. F. Havel, and D. G. Cory. Construction and implementation of NMR quantum logic gates for two spin systems. *J. Magn. Res.*, 140:371–378, 1999.

49. Z. L. Madi, R. Bruschweiler, and R. R. Ernst. One- and two-dimensional ensemble quantum computing in spin Liouville space. *J. Chem. Phys.*, 109:10603–10611, 1998.

50. F. H. L. Essler, H. Frahm, F. Göhmann, A. Klümper, and V. E. Korepin. *The One-Dimensional Hubbard Model*. Cambridge University Press, Cambridge, U.K., 2005.

51. H. Wang and S. Kais. Entanglement and quantum phase transition in a one-dimensional system of quantum dots with disorders. *Int. J. Quantum Inf.*, 4(5):827, 2006.

52. R. Ursin, T. Jennewein, M. Aspelmeyer, R. Kaltenbaek, M. Lindenthal, P. Walther, and A. Zeilinger. Quantum teleportation across the Danube. *Nature*, 430:849, 2004.

53. R. Ursin, F. Tiefenbacher, T. Schmitt-Manderbach, H. Weier, T. Scheidl, M. Lindenthal, B. Blauensteiner et al. Entanglement-based quantum communication over 144 km. *Nat. Phys.*, 3(6):481, 2007.

54. M. Aspelmeyer, T. Jennewein, M. Pfennigbauer, W. R. Leeb, and A. Zeilinger. Long-distance quantum communication with entangled photons using satellites. *IEEE J. Select. Top. Quantum Electron.*, 9(6), 2003.

55. Q. Zhang, A. Goebel, C. Wagenknecht, Y. Chen, B. Zhao, T. Yang, A. Mair, J. Schmiedmayer, and J. Pan. Experimental quantum teleportation of a two-qubit composite system. *Nat. Phys.*, 2(10):678, 2006.

Index

Printed and bound by CPI Group (UK) Ltd, Croydon, CR0 4YY

24/10/2024

01778288-0017